ORTH DAKOTA
Bismarck
MINNESOTA
OUTH DAKOTA
apid City
Sioux Falls
Minneapolis
St. Paul
Mississippi River
IOWA
Des Moines
NEBRASKA
Omaha
Lincoln
Missouri River
Kansas City
St. Louis
KANSAS
kansas River
Wichita
MISSOURI
Tulsa
OKLAHOMA
Oklahoma City
ARKANSAS
Little Rock
Dallas-
Fort Worth
TEXAS
LOUISIANA
Austin
San Antonio
Houston
Baton Rouge
New Orleans
ATLANTIC
OCEAN
GULF OF
MEXICO

THE OXFORD HISTORY OF

T

an

MAYNARD DIXON
1942

THE OXFORD HISTORY OF

The American West

Edited by

Clyde A. Milner II
Carol A. O'Connor
Martha A. Sandweiss

NEW YORK • OXFORD
OXFORD UNIVERSITY PRESS • 1994

Frontispiece: Maynard Dixon (1875–1946). Open Range. Oil on canvas, 1942. Courtesy of the Museum of Western Art, Denver, Colorado.

Oxford University Press

Oxford · New York · Toronto · Delhi · Bombay · Calcutta · Madras · Karachi · Kuala Lumpur · Singapore · Hong Kong · Tokyo · Nairobi · Dar es Salaam · Cape Town · Melbourne · Auckland · Madrid

and associated companies in Berlin · Ibadan

Published by Oxford University Press, Inc., 200 Madison Avenue, New York, NY 10016

Maps created by the University Cartographic Laboratory, University of Wisconsin, Madison.

Research for maps on p. 330 and p. 331 performed by Kathleen Neils Conzen.

The excerpt from the Blessingway (p.41) is reprinted from Blessingway *by Leland C. Wyman with the permission of the University of Arizona Press, copyright © 1970.*

Buffalo Poem #1 *(p. 737) is reprinted with the permission of Geary Hobson.*

Library of Congress Cataloging-in-Publication Data

The Oxford history of the American West / edited by Clyde A. Milner II, Carol A. O'Connor, and Martha A. Sandweiss.

p. cm.

Includes bibliographical references and index.

ISBN 0-19-505968-9

1. West (U.S.)—History. I. Milner, Clyde A., 1948- . II. O'Connor, Carol A., 1946- . III. Sandweiss, Martha

A.F591.095 1994 *93-38829*

978—dc20 *CIP*

9 8 7 6 5 4 3 2 1

Printed in the United States of America on acid-free paper

To our children—Catherine Carol Milner,
Charles Clyde Milner, Adam Sandweiss Horowitz,
and Sarah Sandweiss Horowitz—
who have inherited our love of the West, and to
Bob Horowitz, who shared this long trail.

Contents

SECTION III: TRANSFORMATION

SECTION IV: INTERPRETATION

List of Maps, Tables, and Color Plates

MAPS AND TABLES

COLOR PLATES

Preface

This book is both a harvest and a celebration. It is a harvest of revitalized scholarly interest in the American West that began in the 1950s and continues to grow. It is also a celebration of the renewed recognition of the significance of the West in American history—a recognition that has increased through the efforts of numerous scholars. Some of those scholars have written chapters for this volume, and many others have made vital intellectual contributions that influenced what was written.

Books are often created with idealistic intentions. This volume is no exception. It is meant to serve as a reference work for readers who desire interpretation of large topics instead of exhaustive details about all topics. Important case studies and lively vignettes are incorporated into the individual chapters to illustrate significant points of interest. The editors have encouraged the creation of chapters meant to engage an attentive reader much as a masterful speaker tries to engage a listening audience. A good speaker does not exhaust his listener with particulars, and the good writing exhibited in this volume has a similar purpose. Every chapter in this volume could be a book in its own right, but our authors have generously attuned their efforts to readers who wish to be well informed, but not overwhelmed.

The illustrations for this book serve a similar idealistic purpose. They are a distinct contribution to each chapter's contents, often presenting information not incorporated into the text. The idea was to avoid constantly repeating visually what has been presented in writing.

Through the eighteen chapters of its first three sections, our volume flows forward chronologically. We begin in the era before the West's native peoples encounter Europeans and conclude with the challenges in the late twentieth century for the region's established residents and recent immigrants. The book's final section of five chapters explores the ways in which the West has been interpreted, not only in historical studies but also in literature and art, as well as in popular culture and international comparison. These different interpretations continue to influence how much of the world understands the West, at times regardless of the region's history.

◆ ◆ ◆

The formal planning for *The Oxford History of the American West* began in the spring of 1988. Clyde Milner and Carol O'Connor outlined the organization for the book and invited the twenty-five contributing authors. Marni Sandweiss planned the book's visual contents including the maps. She wrote all the captions and coordinated all illustrations with the authors as well as helped to edit several chapters.

For each of the three editors, it was a happy opportunity to work together. We received our Ph.D.'s from the same institution, Yale University, and two of us, Clyde Milner and Marni Sandweiss, had the same dissertation director, Howard R. Lamar. In addition, Carol O'Connor and Clyde Milner are husband and wife as well as professors in the same department at Utah State University. A colleague several years earlier had jokingly insisted that a married couple so fortunately situated owed the profession at least one joint book. Perhaps this volume will fulfill that informal obligation.

The editors' most generous thanks go first to each of our contributors. These authors shared our vision of the book and made the most essential intellectual contribution. It is always humbling when such talented people will go the extra mile to complete such a large project. Thanks too should go to many people whose names do not appear as authors but who helped us greatly. At an early stage of development, William H. Goetzmann and Howard R. Lamar gave very useful advice. Peter Nabokov, Patricia Nelson Limerick, and Paul Andrew Hutton contributed ideas for three chapters that kept the book moving forward.

Colleagues, students, and friends have also given us support. At Utah State University, Clyde Milner and Carol O'Connor wish to thank the staff of the *Western Historical Quarterly* and of the Department of History. Barbara L. Stewart, Carolyn Fullmer, Ona Siporin, and Jane A. Reilly aided efforts to produce readable texts, multiple documents, and timely communications. R. Edward Glatfelter, head of the history department, supported mailing costs and telephone budget and thoughtfully avoided asking when this huge project would finally end. He also endorsed the utilization of Milner's and O'Connor's year-long sabbaticals in 1990–91 to work on this book. C. Blythe Ahlstrom, the university's assistant provost, provided aid during and after this on-campus sabbatical leave.

Utah State University has its own history of nurturing the study of the American West. The creation of this book arose in this special place where a group of talented colleagues have been very supportive. David R. Lewis, Barre Toelken, Steve Siporin, F. Ross Peterson, Thomas J. Lyon, and Anne M. Butler—the latter three authors of chapters—are all making major contributions to the understanding of the West. Until he took early retirement in 1989 and moved to far southern Utah, a good friend and mentor, Charles S. Peterson, would have been part of this group. Appropriately, the three editors also wish to thank D. Teddy Diggs, who received her master's degree in history at Utah State after working two years as a graduate editorial assistant at the *Western Historical Quarterly*. Teddy Diggs is now a successful free-lance editor in Little Rock, Arkansas. She applied her keen editorial eye to the text of this volume and helped ensure the quality of its final production.

At Amherst College, whose location in the Pioneer Valley of western Massachusetts bespeaks the existence of America's old frontiers, Marni Sandweiss thanks her colleagues in the American Studies Department and the Mead Art Museum for their good fellowship and support throughout this project. Jerry Smolin provided useful assistance with picture research, and Lois Mono helped coordinate a seemingly endless string of letters, photocopies, phone calls, and faxes. Particular thanks, however, go to Elizabeth Burke, who helped track down picture sources and patiently organized and managed all of the paperwork involved in securing illustrations from more than 150 sources around the world. The many librarians, curators, and archivists who responded to these queries deserve our gratitude.

At the Cartographic Laboratory at the University of Wisconsin-Madison, our thanks go to Onno Brouwer and his staff, particularly Daniel J. Maher, who produced the maps included in this book.

Finally, the three editors wish to thank individuals at Oxford University Press in New York City. Linda Halvorson Morse extended the original invitation to Clyde Milner, in 1988, to begin the planning of this work. She has remained our editor since that time. We thank not only Linda for her support but also Liza Ewell, and most especially, John Drexel and Liz Sonneborn, the development editors at Oxford who worked closely with us to ensure the book's completion.

We dedicate *The Oxford History of the American West* to our children—Catherine Carol Milner, Charles Clyde Milner, Adam Sandweiss Horowitz, and Sarah Sandweiss Horowitz—and to Marni Sandweiss's husband, Bob Horowitz. Our own westering adventures have been immeasurably enriched by their love and companionship.

Introduction

America Only More So

Often a dream, sometimes a metaphor, the American West is a place that millions of people can visualize. Certain landscapes of mountain and desert are instantly recognizable. So are certain residents, if they ride horses and wear broad hats or feathered headbands. In these nearly universal images, the West seems grandly conceived and easily explained. It is the West that serves as popular myth and national symbol.

The American West of historical interpretation is much more complex. The people of the West are not readily stereotyped. Lakotas, Navajos, and other Native Americans retain distinct cultural identities, as do a cornucopia of immigrants whose cultural heritages may be traced to Europe, Africa, or Asia but who sometimes arrived from farther north via Canada, from farther west via the Pacific Islands, or from farther south via Mexico and Latin America.

Like its residents, the location of the American West has changed over time. One generation's West became another generation's Midwest or Upper South. Such was the case for Andrew Jackson, who served as president of the United States from 1829 to 1837. In the late 1780s, after two years of studying law in Salisbury, North Carolina, while still in his early twenties, Jackson moved first to Martinsville in North Carolina and then to Jonesboro across the Appalachian Mountains in what would become the state of Tennessee. By the fall of 1788, he had arrived in the log-cabin village of Nashville, where he made his home and established his political career. Throughout his adult life, Jackson called himself a "westerner," and his avid supporters reveled in "Old Hickory's" regional identity.

By the late decades of the twentieth century, a western president could no longer hale from Tennessee, and his western identity needed to relate to cowboys of the open range rather than pioneers of the backwoods. Ronald Reagan fit both the image and the location. He owned a ranch, rode a horse, and wore cowboy boots. Reagan grew up in Illinois, but he found personal and political success in California. In his career as an actor, he spent four years on screen in western clothes as the announcer for the television series *Death Valley Days*. He became governor of California, which made him an important political figure. California had not been part of the United States during Jackson's lifetime, but even before Reagan's presidency (1981–89), it had the largest population of all the states, not only in the West but also in the nation.

Historical study recognizes the significance of change over time. One story of change explains how the West of the nation's political map once included Tennessee but not California and then at a later time included California but not Tennessee. This story of change emphasizes the nation's expansion westward. But there is another way to tell the historical story of the West. This story forms around the idea of place—a West firmly located beyond the Mississippi River. In this trans-Mississippi West, various historical factors, including the expansion of the United States, have shaped the story.

This volume will view the American West primarily as a distinct place whose historical interpretation follows no one master narrative and no single factor of plot. Instead many narratives, themes, and ideas, like the many peoples of the region, are brought together. In other words, the large and complex story of the West stretches across a shared historical terrain. It is a terrain containing many discrete locations, separate voices, and diverse ideas. For example, the West's story has many beginnings—first in the different origin tales of the region's native peoples and then in the tales of exploration created by the new peoples from Europe. The West's geographic boundaries are also variable. The *eastern* boundary takes form beyond the Mississippi River where the aridity of the Great Plains is clearly established, such as along the ninety-eighth meridian, which roughly coincides with the isohyetal line of less than twenty inches annual rainfall. The *western* boundary extends beyond the coast of the Pacific Ocean to the Aleutian Islands of Alaska and to the chain of islands, atolls, and reefs that make up the state of Hawai`i.

Alaska and Hawai`i represent a greater West that connects to the history of the United States in both the nineteenth and twentieth centuries. Their inclusion also recognizes the grand geographic diversity of the American West. Just as the West has no fixed set of external boundaries, it has no fixed geographic or cultural unity. Crosscut with various subregions, disparate states, and distinct peoples, the West demonstrates an intentional oxymoron, what the historian Richard Etulain has called a "fragmented unity."

In this complex and puzzling place, in this "fragmented unity" that is the West, the clearest boundaries may be those delineated by the people within the region. The historian Martin Ridge has stated: "There is a location on the plains of the West where, for some undetermined reason, people think of themselves as being westerners. . . . There is a psychological fault line that separates regions. As these people see it, they are not from the South but from the Southwest; not from the Middle West but from the West." Travelers sense this same fault line. Somewhere beyond the Mississippi, the horizon is more distant, the land more open, and the sky much larger. Well before they reach the Rocky Mountains, they know they have reached the West.

Archaeologists and historians know that crossing the plains was not the only way to reach the West. The first major migrations by native peoples came from the north, whereas the first major incursion by Europeans came from the south. These Spanish expeditions arrived in the homelands of well-established settlers. Did either group know that they were *in* the West? Of course not.

The American West is an idea that became a place. This transformation did not occur quickly. The idea developed from distinctly European origins into an American nationalistic conception. The western edge of several European empires, especially the British, moved into the hinterlands of North America. The United States inherited this westward edginess and made it the main directional thrust of its own empire. Once across the Mississippi, these American lands did not fill up with a steady progression of settlers. Overlanders and gold seekers pushed ahead to Oregon and California. The mountains, plains, and deserts would be filled in later, if at all. Throughout the nineteenth century, the United States laid claim to more and more of its West, culminating in 1898 with the annexation of Hawai`i. All of this occurred because a nation established mainly by African and European peoples created a region that replaced a world—a homeland once defined exclusively by native peoples.

Throughout the twentieth century, the West became a very American place with an increasingly diverse population. To expand on a statement about California, attributed to the writer Wallace Stegner, the West is America *only more so*. In this light, the culmination of America's history occurs in the West in the last half of the twentieth century. Writing the history of the West becomes, therefore, a vital aspect of understanding the history of the United States. Such a perspective is hardly new, but it should be carefully considered. The important distinction is whether the West is created and shaped primarily by the forces of American history both from within the region and from without, or whether the history of the West is what shaped the history of the United States. The latter proposition, in somewhat different formulation, has generated an astoundingly long-lived scholarly debate.

For several intellectual generations, a dispute has raged over the concept of the "frontier" as articulated by Frederick Jackson Turner in his 1893 essay "The Significance of the Frontier in American History." Turner first presented his famous frontier thesis on an especially warm July evening in 1893, when as a young, thirty-two-year-old historian from the University of Wisconsin, he gave the final talk at the last session on the second day of the World's Congress of Historians and Historical Students organized as part of the Columbian Exposition in Chicago. His audience did not respond with any enthusiasm, but four other speakers had preceded him. So it was a long meeting after a very hot day, yet Turner's essay would not be forgotten.

Turner's published version of his talk began with a quotation from the 1890 bulletin of the superintendent of the census. "Up to and including 1880 the country had a frontier of settlement, but at present the unsettled area has been so broken into by isolated bodies of settlement that there can hardly be said to be a frontier line." The census bureau defined the frontier as the outer margin of non-Indian settlement with a density of two persons per square mile. Turner made more of the frontier than just a line of population scarcity. He told his readers: "Up to our own day American history has been in a large degree the history of the colonization of the

Great West. The existence of an area of free land, its continuous recession, and the advance of American settlement westward, explain American development." Turner went on to claim that the frontier also explained American democracy and American character. The challenges of the wild land forged the values of the American citizen. If nothing else, Turner's thesis expressed a vital moment in the conception of the nation's history.

More than a century after its presentation, Turner's frontier and its scholarly significance have undergone extensive revision—at times to praise his ideas and at other times, such as in recent years, to bury them. Even his advocates are aware that Turner's words occasionally exhibit the ethnocentric perspective of a bygone era. Early in his essay, he refers to the frontier as "the meeting point between savagery and civilization." Of course, anthropology and ethnohistory now recognize the complexity of native life that seems more "civilized" in many locations than the behavior of the invading nonnative peoples. Other critics remain frustrated by the vague use of the term *frontier* itself. In this case, Turner candidly admitted, "The term is an elastic one, and for our purpose does not need sharp definition."

Turner ended his essay on a grandly apprehensive note by asserting, "The frontier has gone, and with its going has closed the first period of American history." He did not apply his "elastic" term into the next century, although others have done so. This volume does not ignore the concepts of Frederick Jackson Turner or his intellectual legacy, but its assembled authors make their own case for the significance of the history of the American West in all centuries under examination. It is a significance not based on one thesis but built on the authors' own thoughts and the work of numerous other scholars. It also is an attempt to free the history of the American West from what could be termed "the Mount Rushmore dilemma."

The four gigantic presidential heads carved into the granite of a prominent mountain in the Black Hills of South Dakota represent American nationalism in its most artificially monumental form. George Washington's nose alone is longer than the entire head, from the chin upward, of the Great Sphinx at Giza. Three of the four historical figures seem out of context for such a distinctly western setting: Washington, Thomas Jefferson, and Abraham Lincoln never visited the area. Theodore Roosevelt spent three years, from 1884 to 1886, ranching in the Badlands of the Dakota Territory. But Roosevelt was not chosen because of his sojourn in the West. According to Lincoln Borglum, the son of Gutzon Borglum, who designed the monument, then-president Calvin Coolidge made the decisive argument. Coolidge admired Roosevelt on the questionable assessment that he was a president who had protected the rights of working men.

In 1923 and 1924, Doane Robinson, the state historian of South Dakota, conceived the idea of a monument that depicted the great heroes of western history such as Lewis and Clark, John C. Frémont, and Red Cloud. Borglum accepted the challenge to find the appropriate mountain, but as his son reported, "He had become convinced that the proposed theme—heroes of western history—was too regionally circumscribed, too insignificant nationally." In effect, Borglum had decided that the West did not have historical figures grand enough to match its own mountains. Both Robinson and Borglum thought of heroic men and considered no representation of women. The elimination of Red Cloud only increased the irony that surrounds this monument, since his people, the Oglala Lakota Sioux, consider the Black Hills to be sacred land.

The dilemma of the Mount Rushmore monument is that it presents a visual argument for the *in*significance of western history, whereas the dilemma of Frederick Jackson Turner's frontier thesis is that it presents a scholarly argument for too much significance. No one monument and no single theory can either eradicate or explain the history of the American West. Instead, during what is now at least four decades of revitalized study, a set of significant themes has emerged. Four are worthy of some consideration here; others will become evident to the readers of this volume.

The Non-Vanishing Native Americans. The persistence of native peoples today would startle visitors from the nineteenth century. Even the best "friends" of the Indians assumed that Native Americans would die out, culturally if not physically. Native population reached its nadir in the 1890s at approximately 250,000. In the same decade, the non-Indian population of the United States grew to over 75 million. These original residents had not lived exclusively in what the United States considered its West. But governmental efforts in the first half of the nineteenth century attempted to relocate all Indians to lands beyond the Mississippi. The removal policy did not eliminate eastern Indians, but it did concentrate even more native peoples in the West, especially in the Indian Territory, which eventually became the state of Oklahoma.

A hundred years after its nadir, federally recognized Native American population is approaching 2 million. These native peoples exist in greater numbers and have greater landholdings in the American West than in any other region of the nation. Native peoples, their cultures and histories, have not been eradicated. They are living proof of the centuries of human history connected to what is now the American West.

The Impact of the Federal Government. In a region supposedly characterized by personal freedom and rugged individualism, the federal government has played an astonishingly large role. In political terms, the federal government created the territories and states of the trans-Mississippi, and it has not left them alone. From the explorations of Meriwether Lewis and William Clark in 1804–6 to the atmospheric testing of nuclear bombs in southern Nevada from 1951 to 1958, projects funded by Congress and administered by federal agencies have affected life in the West. Federal land grants underwrote the establishment of four transcontinental railroads after the Civil War, and military spending transformed the West's economy during and after World War II.

The federal government remains the greatest single landholder in the West. National parks and national forests are only two examples. In 1944, the federal government controlled 99 percent of the land in Alaska and 87 percent of the land in Nevada. Currently, the average for all twelve states from the Rocky Mountains westward, excluding Hawai`i, is over 50 percent. Not surprisingly, water, forest, and land-management programs continue to shape the West through the sometimes heavy-handed implementation of national policy.

The Exploitation of a Golden Land. The West's majestically scenic landscape has inspired not only artists but also entrepreneurs. Many of the latter sought wealth, not only on the western land but also beneath it in the mining of gold and silver and then of copper, coal, and uranium. Only a few made their fortunes. Many more toiled for little gain and often greater loss. Oil, especially in Texas, Oklahoma, and California, developed a similar story of boom and bust. Even wheat farming in the Red River valley of North Dakota had a bonanza era in the late 1870s and early 1880s.

For regional, national, and global markets, the economic development of the West has exploited a cornucopia of resources from timber in the Northwest to hydroelectric power on the Colorado River. Yet, in a term coined in the 1960s by Arizona's Stewart Udall, the "myth of super-abundance" has obscured a troubled and fragile natural environment. Water is an especially problematic resource in a predominantly arid West. The large-scale environmental cost of economic development began in the nineteenth century but became a major public issue only in the final decades of the twentieth century.

The Global Population of an International Borderland. The grand American story of mobility and immigration culminates in the American West. To some excited observers, it seemed nearly the entire world rushed toward California after 1848 to pan for gold. The influx of diverse peoples has continued to the present day, especially along the Pacific coast and in Hawai`i. The American West began as an international borderland between native peoples, the Spanish, Russians, French, and British. Today the international borderlands are even more significant with connections directly to Canada in the north, Mexico to the south, and numerous Asian nations via the Pacific. Immigrant groups who first arrived on America's east coast have representative populations in the West, as do peoples from Latin America, Asia, Australia, and the Pacific Islands. Just as the West's economy since the nineteenth century has increasingly connected to world markets, its population has increased its connections to the world's peoples.

◆ ◆ ◆

Ultimately, the case for the significance of the history of the American West moves beyond the national context to the world's stage. The world knows the American West through mass media images and occasional tourist visits. It also knows the

West in the nationalistic rhetoric of journalists and politicians. The West has been oversold and oversimplified as a vast vista of mountain, plain, and desert occupied by heroic, often male, archetypes noted for their violent actions. Through these misrepresentations, the world's peoples, and even the people of the United States, have little knowledge of the history of this distinctively American region. The authors in this volume have undertaken the important task of replacing easy assumptions about the West with thoughtful analysis of the West's past. In so doing, we desire to supplant widespread distortions with informed insight. This objective is perhaps the first, best purpose for any historical writing. We leave it to our readers to judge whether we have obtained our goal.

CLYDE A. MILNER II

Logan, Utah

SECTION I

Heritage

One West of the popular imagination is a place of spacious landscapes and few people. It is beyond the fringe of settlement on the frontier of national, and even international, expansion. It is natural wilderness, defined as a place where people are not, more often than as a place where people are. This imagined West ignores the well-established presence of native peoples and the diverse groups of nonnatives who have arrived in the region. It neglects the human heritages that have shaped the West both from within and from without. New peoples from across the Atlantic contributed to the mixture of cultures on the continent. Well before these developments, native cultures met and mixed in the West.

The European and African newcomers to North America did not settle an empty land. Some travelers, traders, and explorers may have observed a great wilderness with few native inhabitants, but what they did not understand was how the arrival of new peoples had created the depopulation that made a "wilderness" possible in many places. Over a dozen infectious diseases, led by smallpox, measles, and typhus—diseases common, or endemic, to one-half of the world in Eurasia and Africa—devastated the native peoples of the other half. The death rates from the "virgin soil epidemics," which spread among peoples with no resistance to these illnesses, are appalling. More than one of these killers could strike at the same time, driving the death rate up to as high as 90 percent.

As many as fifty million may have died in the decades after first contact. Not surprisingly, the most devastated were areas of dense population such as the Aztec and Inca empires of present-day Mexico and Peru and the Mississippi Valley of what is now the United States. After initial outbreaks, diseases often spread ahead of contact with Europeans and Africans. In areas of sparse settlement, or where the inhabitants had migratory patterns for subsistence, the onslaught might be delayed. For example, along the upper Missouri River of North America's Great Plains, major epidemics of smallpox were still occurring in the late 1830s.

The impact of disease after 1500 in the Western Hemisphere produced the greatest demographic disaster in human history. It also has produced many troubling questions. The first concerns the number of native peoples who inhabited this half of the planet before diseases took their toll. Historical demographers and other interested scholars keep revising their estimates. The geographer William Denevan compiled all the scholarly recalculations produced between 1976 and 1992 for native population. He estimated there were fifty-four million people in North and South America, with slightly less than four million north of present-day Mexico. But Denevan warned that his range of error was approximately 20 percent.

A second troubling question concerns the lack of comparable Native American diseases that could kill Europeans and Africans. Native residents migrated over a land

The Mandan chief Mató-Tópe's drawing of his slaying of a Cheyenne chief suggests the complexities of cultural exchange in the West. Its subject conveys the long history of intertribal tensions on the plains, whereas the detailed and carefully modeled drawing reveals the influence of the white artists Karl Bodmer and George Catlin on traditional plains art. A friend of both painters, Mató-Tópe died a bitter man, cursing the European-American people who had introduced the smallpox that proved fatal to him and many of his people.

Mató-Tópe (ca. 1800–1837). Mató-Tópe Battling and Killing a Cheyenne Chief with a Hatchet. *Watercolor and pencil on paper, 1834. Gift of the Enron Art Foundation. Joslyn Art Museum, Omaha, Nebraska.*

bridge from Asia thousands of years ahead of the newcomers from across the Atlantic. In this earlier passage to what was then a truly New World, the diseases of Asia may not have survived, or they may not have even existed in the migrant populations. The most deadly of the diseases that arrived after 1500 were eruptive fevers that evolved from microbes that initially may have infected various species of domestic animals and pets. For example, the microbe that produces rinderpest in cattle is closely related to the measles virus. Humans and domestic cattle have lived in proximity for eight thousand years, during which time the rinderpest virus, which does not affect humans, may have evolved into the measles virus, which does not affect cattle. Native Americans did not have domestic cattle. In fact, the only domestic animal found throughout the hemisphere was the dog.

Ironically, the domestic animals of Europe, especially horses, cattle, pigs, and sheep, flourished in the Western Hemisphere and greatly affected native societies. For example, the horse after 1500 would produce a cultural revolution for native peoples, especially on the Great Plains.

The eventual domination of Europeans in their New World differs from other stories of European expansion. After all, European empires tried to control much of Africa, India, and China, but today it is not peoples of European heritage that are the dominant population in these parts of the world. If diseases had not done their deadly work, what population today would dominate in North and South America? Indeed, would these two continents even carry such names?

And what of the heritage of the American West? Could the Spanish empire have pushed so far north? Could the Russian-American Company have established itself in Alaska and along the Pacific coast? Would only a few French traders, British explorers, and Christian missionaries have found their way across the plains and even out to the Hawaiian Islands? Would there have been sufficient room for non-Native Americans east of the Mississippi River to create a United States of America and then to acquire a western region for that nation?

Such questions are merely conjecture. They do not alter certain historical realities. Native Americans are still the first "westerners"—the original inhabitants of the region. Despite the impact of diseases and the arrival of nonnatives, many native cultures retain a vital existence. Perhaps one of the best-established groups is the Pueblo peoples of the modern-day Southwest. These villagers first dealt with the Spanish in 1540. From an estimated population of sixty thousand in the mid-1550s, the Pueblos shrank in number to a little over nine thousand residents by 1790. But they did not die out, and their population eventually increased to nearly fifty-three thousand in 1990. To the present day, the Pueblo peoples retain their cultural identity, religious traditions, and historical memories, particularly of their successful revolt against the Spanish in 1680.

Native peoples made their home on the land that became the American West, and they have not left it. Europeans began arriving in these homelands nearly five centuries ago. Many more immigrants are still arriving, and not just from Europe. The West has become a region of mixed heritages that originated both from within the region and from outside. It is a place defined initially by the histories of native peoples and by the arrival of newcomers—first from the nations of Europe and then from a new nation to the east.

— Clyde A. Milner II

CHRONOLOGY

28,000 B.C.	Clear signs of human settlement in what would become North America.
A.D. 1,000	Two oldest Pueblo communities of Acoma and Hopi in existence. Polynesians well established on Hawaiian Islands.
1200–1400	Navajos arrive in the Southwest.
1492	Christopher Columbus's landfall in the Bahamas inaugurates centuries of cross-cultural exchange.
1540–42	Francisco Vásquez de Coronado leads Spanish expedition from Arizona to Kansas.
1565	The Spanish found St. Augustine, Florida.
1580	Over five hundred European vessels fish for cod off coast of Newfoundland.
1598	Juan de Oñate establishes towns for Spain in northern New Mexico.
1607	The English plant a permanent settlement in Jamestown, Virginia.
1608	The French found Quebec.
1614	The Dutch establish Fort Nassau on the Hudson River.
1616–19	"Virgin soil epidemic" decimates native peoples in coastal New England.
1620	English Pilgrims settle at Plymouth, formerly Patuxet.
1620s	Tobacco boom in Virginia; fur trade flourishes.
1630–42	"Great Migration" from England to New England.
1664	The English take over New Netherland.
1680	The Pueblo Revolt expels the Spanish from New Mexico.
1682	La Salle reaches the mouth of the Mississippi.
1692–96	Diego de Vargas reconquers the Pueblos.
1707	Lakota pictographs show these native peoples trading for horses.
1718	The French establish New Orleans.
1741	Vitus Bering and Aleksei Chirikov explore the coast of Alaska for Russia.
1763	Treaty of Paris ends French and Indian War and cedes French Canada and lands east of the Mississippi to Great Britain. Spain holds former French lands to the west.
1778	Captain James Cook visits the Hawaiian Islands.
1781	Representatives of New Spain found El Pueblo de Nuestra Señora la Reina de los Angeles in Alta California.
1783	Treaty of Paris ends the American Revolution and extends U.S. borders to the Mississippi.
1785	Ordinance passed by Congress establishes standard survey grid for the United States.
1786	The viceroy of New Spain announces a new Indian policy.
1787	Northwest Ordinance establishes procedures by which U.S. territories can become states.
1791	Defeat of Arthur St. Clair's troops by forces led by Little Turtle on Wabash River.
1799	Aleksandr Baronov establishes a fort in southeastern Alaska near present-day Sitka.
1802–3	France regains Louisiana from Spain and then sells it to the United States.
1811	Tenskwatawa's village on Tippecanoe Creek falls to Indiana militia under William Henry Harrison.
1812–14	War with Great Britain allows the United States to fight Indians opposed to Anglo-American domination of the trans-Appalachian region.
1821	Mexico secures its independence from Spain.

Chapter One

Native Peoples and Native Histories

PETER IVERSON

Perhaps thirty thousand years ago, the first settlers arrived in the land that would come to be known as North America. These pioneers did not travel by ship, nor did they claim territory for any monarch, but they did discover America. Migrating on foot across a land bridge that spanned the narrow Bering Strait during the Ice Age, they gradually made their way throughout the continent and down into South America, all the way to the tip of present-day Argentina. They sought not empires to swell national treasuries but new hunting grounds to feed growing populations. Of course, it took generations for these first explorers to become well established in specific locations. By A.D. 1000, or even before then, some groups occupied the territory they still claim today. Others continued to migrate, to create new homes and new customs for themselves. Even after the first Europeans made it to American shores, obviously uncertain of where they were or what the land might hold, many of the people they labeled "Indians" remained in motion.

Both for relative newcomers and for more established Indian settlers in any region, the process of building communities and traditions was essentially the same. Around winter fires or on the hunt, while working in the fields or participating in religious rites, the younger members of each group learned from the older members. Their elders taught them about the creatures who shared the earth and sky with them and about the origins of their people. Through this instruction, they understood from an early age that their people belonged to the land they occupied. Creation stories often contained accounts of migration, but these stories also reassured people that the place where they lived was meant to be their home.

One such legend has been told for at least ten centuries by the people of Acoma Pueblo in present-day western New Mexico. That community traced its beginnings to Sipapu, a mythical place beneath the ground. In the legend, a spirit named Tsichtinako met up with two sisters, Iatiku and Nautsiti, who lived in the underworld. The spirit taught them to speak and prepared them to leave the darkness for the strong light of the world above ground. When they were ready, Tsichtinako gave the sisters baskets filled with seeds and with images of the animals who were to live in their new home. The sisters planted four kinds of pine trees underground according to the spirit's guidance. One of the pine trees grew faster than the others and pierced a small hole in the earth above, causing the image of the Badger to come alive.

The middle figure at the top of this painting by a young Acoma artist is Iatiku, a central figure in the Acoma Pueblo creation myth. Framed by mothers of the first-born girls, she is surrounded by symbols of the natural world and fetishes specific to Acoma culture.

Wolf Robe (Kiwa, also known as William Henry, 1905–77). Fire Society Altar. *Watercolor on paper, 1928. Smithsonian Institution, National Anthropological Archives (#45014-C), Washington, D.C.*

At the request of the sisters, Badger climbed the pine tree and enlarged the hole, then returned to the underworld. The sisters rewarded him with a happy life in a place that would be neither too hot nor too cold. Next, Locust came to life and the sisters asked him to smooth the edges of the hole with plaster. Before returning from his task, however, Locust ignored the sisters' prohibition and ventured out into the light. On his return, he lied about his transgression, and the sisters punished him. They condemned him to a short life, although they allowed him to be born again each year.

Iatiku and Nautsiti came into the world above ground through the hole prepared by Badger and Locust. When they arrived, they sprinkled plant pollen and sacred cornmeal from their baskets in prayer to the Sun. Each chose a name for the clan of her descendants, Iatiku settling on Sun and Nautsiti on Corn. The sisters then set about creating the mountains, plains, mesas, and canyons that characterize the Acoma homeland. From these, they gained a sense of the four directions. Next, they grew plants from the seeds in their baskets and animals from the images, in the process learning how they belonged to the land and all that lived upon it. Evil entered the world when one of the images came to life as a Snake. Snake tempted Nautsiti to become pregnant by the rain that fell from a rainbow. From then on, the sisters became increasingly unhappy with each other; eventually they chose to go their separate ways. After the sisters split up, signs of discord multiplied among the birds and animals: Magpie, for instance, ate the intestines of a deer that had been killed by a puma; as a result, Magpie forgot how to hunt and was reduced to scavenging.

Nautsiti left the region, but Iatiku remained behind and in time bore many children. Each child's name became the name of a different Acoma clan. Iatiku also brought to life the spirits of the different seasons as well as the katcinas, the spirits who became so prominent in the beliefs and rituals of generations to come. The katcinas would help Iatiku's descendants to be brave, to learn how to lead good lives, and to grow corn. In homage, the Acomas would imitate the katcinas in dances and ceremonies. After creating the katcinas, Iatiku decided the time had come for the people to build their homes. She told them how to construct their homes and how to design the village they would inhabit. Following Iatiku's instructions, the people built a sacred room, a kiva, in their village. The kiva symbolized Sipapu, the mythical place from which Iatiku had entered the aboveground world. For worshipers inside, its ceiling represented the Milky Way and its walls symbolized the sky. The kiva reminded the people of their origins, of their links with the earth and the sky.

Thus, the legend recounts the origins of Acoma Pueblo. Visitors to the pueblo, built high atop a mesa in a magnificent valley, can still appreciate why the Acomas believe the place to be the center of the world.

The American Indian West

By 1500, the first inhabitants of the American West had created many such centers of the world. Still more would emerge in the years leading up to the early nineteenth century as different peoples continued to migrate, to settle in new homes, and to endow important places—prominent mountains or hidden lakes—with sacred significance. Although European intruders would misunderstand the Native American peoples as a single, monolithic population, many different cultures—varying in size, ambition, and economy—inhabited North America.

In the salubrious environment of the northwestern Pacific Coast, for example, a relatively dense population of related peoples thrived. Small groups occupied compact niches where the abundant land and sea supplied everything they needed. They had no need to conduct extended hunting or gathering forays, but they enjoyed a lively trade with groups from the interior. Through that trade, they acquired items useful for ceremonial purposes and in turn provided the interior peoples with abalone shells and other items unavailable in the less bountiful mountains, valleys, and deserts.

By contrast, fewer tribal groups lived where resources proved scarce and hard to find, for people in these regions had to traverse a wide terrain in order to survive. The passage of the seasons dictated annual cycles of hunting and gathering; the people came to appreciate and even respect the plants and animals with which they shared the earth. In areas burdened with uncertain weather, hunger remained a perennial possibility, but even under challenging circumstances, many communities attempted to farm. Depending on the amount of rainfall, soil conditions, and the length of the growing season, some farmers could coax significant yields from fields of maize, squash, and beans. But they knew their success depended on more than their own diligence, so they performed rituals to call forth the rain.

Although the diversity of the many native communities of the American West makes it difficult to generalize about them, they did share certain fundamental features that set them apart from Europeans. Perhaps most notable was their dedication to community. In general, each person stayed in his or her community of origin from birth to death, never leaving it for the next village, the next tribe, or the next hemisphere. Such permanent affiliation daily reminded individuals of their obligations, values, and loyalties. Individual expression, achievement, and recognition no doubt had a place within indigenous cultures, but people never forgot their role within their group. In a world without horses or rifles to assist in the hunt or win the war, success absolutely depended on cooperation.

This rejection of individualism shaped native attitudes toward the land. Individuals did not own parcels of real estate, so status and power had nothing to do with the acquisition of acreage, as it did in Europe. Nor did groups as a whole own land, although this does not indicate an absence of territoriality. Each people claimed the right to use the land in a given area; communal priorities dictated how the resources there were employed. If their population increased or if drought or a reduction of available resources forced them to move, Indians would attempt to expand their hunting and gathering domain through war with their neighbors. Skirmish or war could often be avoided, but both before and after the Europeans arrived, Indians occasionally fought over land.

Despite this tendency to territorial disputes, Indians did not, as historians too frequently have written, have "ancient" or "traditional" enemies. Each group surely had its allies and enemies, but such relationships were neither permanent nor necessarily long-lived. Alliances changed and animosities withered or flared, both before and after European contact. Probably far more frequently than they fought with each other, different peoples learned from each other. As they moved about, Indian bands transported old ways of doing things to new places and at the same time picked up improved methods of building houses, growing crops, hunting animals, or weaving rugs. Their flexibility carried over into the postcontact period, when many groups incorporated European technology into their culture. Absorbing new influences—whatever their source—did not signal the decay or diminution of any culture. On the contrary, these additions kept cultures viable and eventually seemed as much a part of their traditions as older ways did. The people created stories to explain such augmentations: we have always had horses; the holy people gave us sheep; the Black Hills have been sacred ground from time immemorial.

But even before the introduction of European technology, the American Indian West was a place of progress and innovation. The challenges faced by the people living there—the altitude, the wind, and the aridity—highlight the economic and cultural success enjoyed by many native societies. The resourcefulness of Plains buffalo hunts, the sophistication of Northwest Coast art, and the complexity of Navajo philosophy all point to the vitality of the aboriginal peoples of North America. Imagine what the Yavapais knew about plants, what the Inuit learned about whale hunting, or what the Pueblo peoples observed in the heavens. Consider the rich symbolism of tribal stories or the vibrant imagery of native songs. Far from primitive, the first Americans had a great knowledge of the world, a wisdom that benefited those Europeans open to it.

Likewise, the Europeans possessed knowledge and goods that the Indians wanted. But problems arose between these potentially equal peoples when the Europeans tried to impose undesired new ways on the Indians or tried to achieve their own goals at the expense of those who had first settled in the American West. Because this economic, religious, and social friction did not disappear with time, echoes of confusion and misunderstanding reverberate in the West even today. At the heart of the contemporary conflicts stand the centuries-old issues of land, language, law, sovereignty, and identity. Any insight into the history of the American West must therefore start with an examination of the nature of native societies in the first centuries after Columbus sailed and of the effect the European arrival had on those communities.

The Southwest

To the casual observer, Acoma and the other Pueblo communities of the American Southwest before the European incursion may appear to have been changeless worlds, where one generation after another repeated the familiar patterns of life and where innovation and imagination were devalued. Indeed, the Pueblo peoples cultivated continuity and respect for tradition, but their ancient stories establish that they were not frozen in a quaint ethnographic tableau. These cultures maintained continuity through change. By adaptation, experimentation, and trial and error, they discovered what worked and what did not.

The history of the different Pueblo communities of present-day Arizona and New Mexico centers on change. Acoma and Hopi are probably the two oldest surviving communities, both dating back at least a thousand years; some scholars contend that the Hopi area has been occupied for several thousand years. In the centuries before the Spaniards arrived, the village populations at Acoma and Hopi swelled from immigration.

The Anasazi peoples, who had occupied such fabled sites as Mesa Verde and Chaco Canyon, dispersed, perhaps because of drought, soil depletion, or disease; this brought new residents to old villages and led to the formation of new settlements. Some of the people from Mesa Verde drifted down to Acoma while others ventured to the Rio Grande valley. At the same time, people from the Marsh Pass and Kayenta areas to the north migrated to Hopi.

By the early 1500s, these Pueblo peoples had learned a great deal from numerous contacts with other Indian groups through raiding, war, immigration, and most often, trade. Items and ideas from hundreds or even thousands of miles away arrived with the transient traders and foreign immigrants, some of whom were permitted to reside in the Pueblo villages. The arrival of newcomers probably prompted a renaissance in Hopi pottery technique at the beginning of the 1300s, which transformed the distinctive black-on-white designs to black-on-orange. At Acoma, meanwhile, migrants from the Cebollita area altered not only the local pottery but also housing construction: villagers abandoned adobe in favor of horizontal, wet-laid masonry.

Even though the Pueblo peoples had previously encountered strangers and sometimes learned to accept the changes they brought, the arrival of the Spaniards inaugurated a new era. The initial stages of Spanish-Indian contact churned with misinterpretations and violence. In 1539, for instance, the Zuni people met and killed Esteban, a North African who eight years earlier had trekked from the Gulf of Mexico to present-day Mexico City with the party of Alvar Núñez Cabeza de Vaca. Esteban arrived at Zuni as an advance scout for an expedition headed by the Franciscan padre Marcos de Niza. This group had been dispatched to investigate persistent rumors of wealthy cities in the shadowy reaches of northern New Spain. Shrouded in mystery, Esteban's untimely demise is said to have been the response of his initially hospitable Zuni hosts to Esteban's inappropriate requests and violations of proper behavior.

A frightened Marcos de Niza quickly fled to the south. Uncertain why Esteban had been killed, he reported to his superiors that he did not know whether the Zunis had any wealth. His report may have been ambiguous, but it inspired the historic expedition of Francisco Vásquez de Coronado in search for El Dorado, the "city of gold." From 1540 to 1542, Coronado's large expeditionary force lumbered around the Southwest, even stumbling as far east and north as present-day Kansas. Kansas did not look like El Dorado to the weary travelers, who in 1542 retraced their steps south and west past the Zuni village of Hawikuh, which they had fought and brought under their control in July 1540.

Coronado not only explored the Southwest but also sought to conquer the people who lived there. He occasionally prevailed, yet both New Spain and the native inhabitants of the Southwest paid a terrible price for his efforts. At the end of the century, the colonizer Juan de Oñate encountered fierce resistance at Acoma. The Indians there had no doubt heard about the fate of other Pueblo Indians along the Rio Grande, whose villages had been plundered or whose occupants had been burned at the stake by Coronado's men. Told again and again during the decades before Oñate's arrival, the stories of horror were seared into the Indians' consciousness. They did everything possible to repel Oñate, who nevertheless suppressed the Indians.

Despite Oñate's cruelty, the Acomas in time accepted the missionary Juan Ramirez, who from 1629 to 1649 brought fruit trees and livestock to the mesa. He lived in a way that inspired the people; they even helped him build an imposing, thick-walled church. But his successor, Lucas Maldonado, had the misfortune of being stationed on the mesa during the great revolt of 1680. Some of the people seized him as the embodiment of Spanish colonialism and hurled him off the mesa to his death on the rocks below. He died for the sins of his Spanish ancestors. The Spaniards did not abandon their northern outpost at Acoma in the aftermath of the revolt. Instead, in 1692 they returned under the resolute Diego de Vargas. Recognizing the risks of violence, Vargas tried to rein in his men. But he was no pacifist, and after one flare-up of resistance in the summer of 1696, he shot five Acoma captives and destroyed the Acoma cornfields.

By the turn of the eighteenth century, however, an uneasy peace appeared to prevail. Indians and Spaniards seemed to recognize that both were there to stay and that neither could be totally subjugated. Although this realization did not force the Franciscans to take a vow of perpetual pluralism, they naturally hoped to avoid a recurrence of the revolt that had killed their predecessors a generation before. As a result, they were unlikely to raid the kivas in the future. Similarly, especially in the Rio Grande area, where the Spanish population was more concentrated, the Pueblo peoples often became Catholics. They practiced their new religion while continuing to participate in traditional ceremonies that fulfilled native needs. Outside the realm of religion, the Indians found some Spanish ways useful and even pleasing. They already farmed, but the Spaniards introduced new things to grow, such as peaches, apples, and wheat. The sheep, cattle, goats, and horses brought from Europe offered food, clothing, and transportation to the Pueblo peoples; in time they could not imagine that they had ever lived without these animals.

Nonetheless, resistance to Spanish influence did not disappear. In the final two decades of the seventeenth century, various fragmented Pueblo communities reestablished themselves to avoid either absorption or control by the Spaniards. One band of refugees, from the Rio Grande valley near present-day Albuquerque, traveled hundreds of miles to the Hopi mesas to seek shelter. They established the Tewa-speaking community of Hano on First Mesa and maintained a separate community in the years to come. Other Pueblo Indian communities were absorbed into the expanding clan structure of the Navajos, and yet another group initially sought shelter at Acoma. In the very last years of the 1600s, this last group founded a new pueblo of their own, the relatively recent subdivision of Laguna.

But it was the Hopis, the westernmost Pueblo people and the most removed from contact with the Spaniards, who most staunchly rejected European influence. In 1700 they destroyed Awatovi, the one Hopi village that had tried to adopt Christianity. Six years later, they attacked the nearest Pueblo village of Zuni to protest the return of Christian influence there. The Spaniards repeatedly sent military emissaries out to chastise the Hopis for their obstinacy, but the Hopis repelled the unwanted delegations in 1701, twice in 1707, and again in 1717. Likewise, Franciscan and, later, Jesuit priests came away chagrined on several different occasions. Father Silvestre Vélez de Escalante made an unsuccessful overture in 1775, Father Francisco Garces in 1776, and Governor Juan Bautista de Anza in 1780. The case of the Hopis illustrates a general truth about the nature of intercultural contact: more frequent encounters with outsiders do not necessarily erode community identity. Indeed, the Hopis gained greater community cohesion over the years; the arrival of the Spaniards heightened Hopi identity rather than diminished it.

California

The Spaniards extended their campaign of economic, political, and social colonization beyond the contemporary borders of Arizona and New Mexico. Because of the general Spanish interest in the resources of the Pacific coast, California received considerable attention. In California, where the extraordinary topographical diversity had produced a corresponding diversity in its first peoples, the Spaniards encountered a great range of smaller native communities.

Here too, as in Acoma, native children learned about the world around them through stories told by their elders. According to the Modocs, volcanic Mount Shasta was once the home of a spirit who kept a fire in his lodge for himself and his family; the smoke from the fire came out of the hole in the top. They also believed grizzly bears were their ancestors and so should not be harmed. The Maidus, meanwhile, said the Sacramento River had been created when a hole had been poked in a mountainside, allowing the waters of a great flood to surge into the valley below.

Within the different environmental zones of California, Indian communities had a range of economic and cultural activities. Especially along the coast, ample land and water resources permitted localized economies and worldviews. Young people assumed they would live their entire lives within the boundaries of an area that would seem small to postindustrial Americans but that did not feel confining to the first Californians. Indeed, Spanish accounts testify to the richness of that world. In addition, the Indians knew how to manipulate or alter the land for their own benefit. In grassland, woodland, and chaparral regions, they employed regulated, seasonal burning to control brush, stimulate the growth of crops, and improve conditions for hunting.

Human activity within the world of the first Californians was carefully regulated by ritual. Hunters, the Indians believed, did not succeed solely by virtue of their own stealth and cunning but in proportion to their respect for their prey. Just as rituals were meant to bring rain in the Southwest, in California they brought deer to the hunter. A Pomo deer hunter, for example, applied to his body the pleasing smell of angelica and peppertree leaves, thereby showing his respect for his prey and its feelings and emotions.

As their rituals show, the natives of California believed that arrogance and insensitivity toward nature led to disaster. That disaster often took the form of drought, making many areas inhospitable to humans. Some Indians, most notably the Paiutes of the Owens Valley, diverted water for agricultural purposes, but most groups had little need for this technology. Instead, they simply chose to live wherever water was more plentiful. There they could take advantage of the easy life made possibly by an adequate water supply. They did not feel compelled to make the land more fruitful, for they did not seek to dominate nature. Proper living and respectful attitudes preserved this harmonious way of life by ensuring that the rains came. If drought came instead, the people revised their rituals or moved away. Not so the Spaniards, the Mexicans, or the

Americans, who sought to conquer nature by manipulating the water supply to create artificial environments.

These concerns, however, did not vanish. Indeed, they became magnified over time. For example, in 1925, a Wintu named Kate Luchie contrasted the actions and beliefs of whites and Indians. Whites, she said, had never cared for the land or its creatures. Disregarding the pleas of the rocks and the ground, they plowed up the earth, pulled up the trees, and killed the animals. "They blast rocks and scatter them on the earth," she wrote. "The white people dig long deep tunnels. They make roads. They dig as much as they wish. They don't care how much the earth cries out. How can the spirit of the earth like the white man?" Someday, she predicted, the water would come down from the north in retribution for all the white man had done, and all humanity would drown. When all the Indians died, the world would end. Her warnings echoed the centuries-old teachings of California's Indians.

When the Spaniards appeared on the scene, they imported other stories and other dictates. The Franciscan Junípero Serra and his compatriots established mission stations along the coast and set about informing the Indians that they had not been worshiping properly, for according to the Spaniards, humans were superior to animals. White historians until recently portrayed Serra as a kindly man who gently instructed his native charges on the error of their ways. But in the late twentieth century, the prospect of sainthood for Father Serra brought a different, though equally long-held, image to the fore. Indian history called Serra little more than a slaveholder who made prisoners rather than converts of native Californians.

Whether Serra was a saint or a scorpion, there can be little doubt that the Spanish presence adversely affected many Indian communities in California. European diseases reduced the native population by perhaps half. Indians who lived within mission walls suffered severe restrictions on their freedom and, over time, lost their culture. But Spanish domination did not extend to all the peoples of California. Those, such as the Yahis, who lived farther in the interior—in the foothills of the northern Sierra Nevada—were more than a hundred miles from the nearest mission at San Francisco and escaped the worst ravages of the period.

Moreover, some Indians resisted Spanish subjugation. For example, members of the Ipai-Tipai villages attacked Mission San Diego de Alcalá on 4 November 1775, killing Padre Luís Jayme and two other Spaniards. Thousands of southern California Indians rebelled in 1824, destroying Mission Santa Ynez, taking over Mission La Purísima Concepción, and fighting white soldiers called in to protect Mission Santa Barbara. The Spaniards quelled these and other rebellions, but Indian resistance continued. Indians often tried to run away from the missions, and other forms of more passive resistance no doubt also took place.

In the far northern part of California, the Yurok village of Tsurai on Trinidad Bay affords an instructive and well-documented glimpse of native encounters with different Europeans. After initial observation of the bay by the Portuguese sailor Sebastian Rodriguez Cermeno in 1595, the Yuroks apparently had little contact with European vessels until the arrival in 1775 of the Spanish captains Don Bruno de Hezeta and Don Juan Francisco de la Bodega y Quadra. The Spanish entrance into Trinidad Bay convinced whites and Indians alike that a new day had come to northern California.

A member of a Russian exploring party that arrived in San Francisco in 1816, Ludovik Choris became one of the few Europeans to depict Indian life in Spanish California. Here, in front of Mission San Francisco de Asís, members of an unidentified tribe dance under a towering cross, suggesting the accommodation of cultures that marked Spanish-Indian relations during the early 19th century.

Ludovik Choris (1795–1828). Danse des Californiens. *Watercolor over pencil on paper, 1816. Courtesy, The Bancroft Library, University of California, Berkeley.*

Bodega y Quadra's journal noted: "The commandant took possession of those lands with all the dignity and solemnity that the accommodations of the port afforded. Mass was celebrated, a sermon was preached and many volleys of cannon and guns were fired as an act of thanks to the creator." He observed that the Indians "were terrified by these noises," believing "that those volleys could demolish the nearby mountains." The Spaniards clearly saw the area as a land of great promise for farming and ranching and noted the "well-ordered" hunting, fishing, and gathering economy of the natives. Perceived again and again as timid and docile, the local people did not appear to threaten European or American ambitions and prerogatives.

Late in the eighteenth century, the British captain George Vancouver stopped in Trinidad Bay a few months before another Spanish ship anchored there. At that point, starting in the early 1800s, sea otters in the bay began to attract various trading ships. The Yuroks grew increasingly enmeshed in the global economy and its conflicts; in an 1806 clash, whites from a ship called the *O'Cain* killed a Yurok. Owned by Americans, the *O'Cain* fulfilled a Russian contract, carried Aleut hunters, and landed in territory still claimed by Spain. The *O'Cain* incident heralded the beginning of the end of Indian autonomy in northern California.

The Pacific Northwest Coast

Stretching fifteen hundred miles from southern Oregon to the Gulf of Alaska and spanning no more than one hundred or two hundred miles east to west, the Pacific Northwest Coast is a long narrow strip that housed native peoples in a world between mountains and ocean. Water defined the region, in the tides and waves of the sea, in the flow of the rivers and creeks, and in the frequent showers of falling rain. The skies ranged from light to dark gray, but the indigenous residents probably did not complain too much. Not only was it the only world they knew, but in it they found abundant resources and prosperity. Along nearly the entire coastline, Sitka spruce and western hemlock flourished, together with yellow cedar in the north and Port Orford cedar in the south. The natives usually selected cedar for their world-renowned woodworking, but they used a variety of woods for different implements, weapons, and ceremonial objects.

Chinook, coho, sockeye, pink, and chum salmon ran in seemingly inexhaustible numbers from fresh water to the ocean and back again to fresh water. Halibut, herring, and other saltwater fish, as well as various sea mammals, supplemented the local diet. On a seasonal cycle, the people hunted bears, beavers, and other land animals. Far more frequently than Indians in many other parts of North America, they could anticipate a surplus during the brief, leaner winter months. The region's rich resources also allowed the people to spend time nurturing a highly creative artistic culture, whose accomplishments included stylized woodworking and splendid twined basketry.

The Cascade Mountains split present-day Oregon and Washington into two regions: to the east, the land is dry, and resources are less plentiful. But some indigenous peoples say this has not always been so. A long time ago, according to Quinault legend, Ocean sent his sons and daughters, Clouds and Rain, to the dry country in answer to the pleas of its residents. Fearing the return of drought, the eastern people refused to allow Clouds and Rain to leave after their visit. A distraught Ocean beseeched the spirits to punish the people. In response, the spirits built a barrier of dirt between the eastern people and the abundant west. This partition became the Cascades; the hole from which the dirt was taken filled with water to form the Puget Sound. From that time on, the eastern peoples struggled to survive in an arid land with few resources.

West of the Cascades, though, many different communities formed along the bountiful coast, including Eyaks, Tlingits, Haidas, Tsimshians, Bellacoolas, Kwakiutls, Nootkas, Makahs, Quileutes, and Tillamooks. No single culture can be said to represent all these communities, but many of the people shared certain characteristics. Like other peoples throughout the area, the Tlingits told stories about mischievous Raven, whose exploits, travels, and travails amused and instructed listeners. Tlingit myths and texts also describe many other beings important to the people, including the orca, otter, porcupine, wolverine, beaver, halibut, clam, salmon, frog, sea lion, and bear. In these tales, the lives of animals and fish intermingled with those of humans. A woman married a frog, a man entertained the bears, a salmon chief talked to a fisherman, a devilfish married a woman and impregnated her. From childhood, Tlingits learned that other beings had feelings and impulses and that all beings were capable of good and evil.

The Tlingits' seasonal pattern of hunting and fishing brought them into contact with many different creatures of the land and sea. Because of their beliefs, Tlingit hunting and fishing assumed the character of a holy quest, in which participants carefully prepared to enact their sacred duty. Boys learned early that the proper response to success on the hunt was not arrogance but gratitude. To heighten their chances of snaring their quarry, hunters wore charms made by shamans, the Tlingit religious leaders. The charms reflected not only the artistic ability of the shaman but also his or her relationship with the spirits.

Armed with religious rituals and proper behaviors, as well as the appropriate implements and a detailed knowledge of animal and fish behavior, the Tlingit people ably harvested the resources of the region. Hunting sea lions, bears, or other imposing prey required sophisticated physical and spiritual knowledge on the part of hunters. The Tlingit people also learned the effectiveness of strategy and cooperation between fellow hunters while tracking and felling large animals. Based on aboriginal spears, bows and arrows, harpoons, and traps, their hunting techniques changed little with the introduction of European technology. They might have added a metal tip to a weapon or fashioned a tool more quickly, but their basic approach and essential equipment remained largely the same.

Although the Tlingits valued self-sufficiency, they enhanced their lives through trade. Neighboring peoples, including the Eyaks, Haidas, Tsimshians, and various Athapaskan-speaking groups, traded with them for items of utility and decoration. Tlingit communities especially sought the furs that interior tribes could provide in return for coastal products such as seal oil. Heavily reliant on canoes, they knew how to make good ones, but the Haidas appeared to make still better ones. A large Haida canoe, impressive not only in its size but in its capacity and overall beauty, was a prized acquisition.

Sometimes the exchange of material items was involuntary because in the Tlingit world, relationships rested not only on goodwill but also on power. Within a community, one clan might be strong enough to dictate trading terms to other clans. It might demand a blanket of surpassing beauty or even a human being to be used as a slave. The social stratification that arose as a result of the prosperity of the Northwest Coast peoples included slavery and turned people into a highly valued commodity. Spread out along a strategic part of the coast, the Tlingits were particularly well situated to profit from the unfortunate tradition of human bondage.

When Europeans intruded into the Tlingit world of the eighteenth century, they sailed into a complex social, economic, and cultural environment. Spanish, British, French, Russian, and American exploring parties disembarked during the next two centuries, complicating an already confusing scene. The first whites to make contact with the Tlingits, the Russians, were the most firmly established European presence in the region by the late 1700s. When Tlingits recount the arrival of the first Russians, they recall how they thought perhaps the famed Raven had paid them another visit:

> At one point one morning
> a person went outside.
> Then there was a white object that could be seen
>   way out on the sea
> bouncing on the waves
> and rocked by the waves.
> At one point it was coming closer to the people.
> "What's that?"
> "What's that, what's that?"

"It's something different!"
"It's something different!"
"It's something different!"
"Is it Raven?"
"Maybe that's what it is."
"I think that's what it is—
Raven who created the world.
He said he would come back again."

The story tells how the people heard strange sounds coming from the object:

Actually it was the sailors climbing around the mast.

Finally, two brave young men paddled out in their canoe to examine the huge boat of the white men. When the crew brought them into the ship's cabin, they saw odd and confusing things:

. . . they saw—
they saw themselves.
Actually it was a huge mirror inside there,
a huge mirror . . .

And in the ship's galley:

There they were given food.
Worms were cooked for them,
worms.
They stared at it.
White sand also.

After they hesitantly tasted the rice and sugar, they were given alcohol to sample:

They began to feel very strange . . .
to feel happiness settling through their bodies.

They then returned to tell the people all the remarkable things they had experienced, and their tale has been retold among the Tlingits ever since.

The European newcomers had just as much difficulty understanding the peoples they encountered, but their early accounts leave little doubt that certain elements of Tlingit material culture impressed them. The French visitor Jean Francois La Perouse, for one, remarked on the unique skill of Tlingit basket and hat makers. Illustrations by Tomás de Suría, a gifted artist who accompanied the Spanish explorer Alejandro Malaspina, support this view. One shows an intricate ceremonial hat woven of spruce root and perhaps worn by a leader of a Tlingit village.

The Chilkat blanket represented the peak of Tlingit creativity. Emblazoned with a stylized crest figure, it featured long strands of wool dangling below the design. Crests belonged to the different clans and portrayed the animals and places significant to those clans. Thus, Chilkat blankets not only demonstrated the artistic genius of their creators but also preserved the meaning of traditional symbols and associations. A Chilkat

In November 1788, the Spanish crown authorized Alejandro Malaspina to lead an exploring expedition to the northern Pacific. When the artists assigned to the survey became ill, Tomás du Suría, an engraver at the Mexican mint, replaced them. His pictures of the residents of Port Mulgrave documented the material culture of an exotic people unlike any the Spanish had previously encountered and became part of the report that Malaspina sent to Madrid.

Tomás du Suría (1761–?). Portrait of the Head Man, Port Mulgrave. *Ink on paper, 1791. Beinecke Rare Book and Manuscript Library, Yale University, New Haven, Connecticut.*

Framed by the spectacular house screen and carved house posts of the Whale House is a broad selection of Tlingit objects including a well-worn Chilkat dance apron, spruce-root hats, and a mask. The large woven basket on the left, known as the "Mother-Basket," was a prized possession of the Gaanaxteidi clan and was used to serve food on ceremonial occasions.

Lloyd Winter (1866–1945) and Percy Pond (ca. 1872–1943). Interior of the Whale House, Klukwan. *Photograph, 1895. Winter and Pond Collection (#PCA 87-13), Alaska State Library, Juneau.*

blanket attested to the power of a clan's religious and political leaders and strengthened the bonds of kinship within the clan, linking the generations that passed the prestigious possession among themselves.

Suría's work also indicates that by the late eighteenth century, Tlingit warriors were very well equipped. Clad in a kind of armor fashioned from wood, sinew twine, and leather, they also wore helmets of wood, copper, and shell. Through trade with the English, some Tlingit soldiers acquired copper knives to attach to their lances. European-made knives and hatchets made their way into battle and the hunt, and even muskets became part of the Tlingit arsenal. So armed, Northwest Coast peoples such as the Tlingits could not be easily controlled by outside forces.

And the Tlingits needed their weapons, for tension as well as friendship characterized early relations between whites and Indians. Three years after Aleksandr Baranov founded a fort near present-day Sitka in 1799, Tlingits raided it and took it over. In 1804, the Russians evicted the Tlingits and erected another fort, New Archangel, which soon served as the hub of Russian America. But despite such clashes, the Russian arrival was not as disruptive to Tlingit culture as it might have been. Missionaries of the Russian Orthodox church criticized traditional Tlingit religious practices and worked to convert the people to a new faith, but they were less determined to transform the Tlingit way of life. Most of them recognized that the people worked hard, spoke an advanced language, and created fine things from wood and other natural elements. The Russian missionaries understood that the Tlingits had come to terms with their environment; the people could not be faulted as hunters or as fishermen. Unlike the American missionaries, the Russians did not hope to make the Tlingits into farmers. Perhaps in part because the Northwest Coast resembled their own home, the Russians could appreciate more of Tlingit culture on its own terms.

The Russians did, however, seek to tap into the existing trading patterns along the coast. Eventually, especially after the British-chartered Hudson's Bay Company arrived on the scene, trading relationships between whites and Indians may have made the Tlingits dependent on Europeans and Americans for certain goods that they could not produce. But from the end of the eighteenth century through the first half of the nineteenth, the trade network that linked the Tlingit people to the rest of the world did not erode their tribal economy or culture. Instead, it often revitalized or expanded some elements within that economy and culture.

For instance, the Tlingits had a fully developed woodworking technique before they started trading with whites. As they acquired iron and then steel tools from the foreigners, they carved and sawed with greater speed, efficiency, and perhaps, imagination. Metal tools allowed them to refine the elaborate carving they applied to various objects, including the house posts known as "totem poles" to the whites. Enhanced by the introduction of new tools, the remarkable evolution of Tlingit art and craft was revealed in such achievements as the interior of the Whale House of the Raven in Klukwan, Alaska, constructed early in the nineteenth century. Approximately fifty feet wide and slightly longer, it boasted a central rain or raven screen and wonderfully detailed house posts. The house posts reflected the aesthetic preference of the people for emphasized eyes, eyebrows, noses, and mouths. As elsewhere in Tlingit art, Whale House artisans used curved shapes defined by black paint and filled in with red. The results were bold and striking.

The Tlingits had many neighbors who also benefited from trade with the whites. To the south, for example, the Haidas prospered as shrewd traders who sometimes declined to trade with one ship in the expectation of making a better deal with another. Similarly, the Bellacoolas played the British and the Americans off each other for many years. All along the Northwest Coast, the indigenous peoples gained much from contact with whites until about the middle of the nineteenth century. American missionaries and other intruders forced some cultural changes on many peoples, but they did not manage to destroy them for a long time.

Alongside the missionaries, the destructive force of smallpox, tuberculosis, malaria, and other diseases hit various Indian groups in epidemics that ran their course in a year or two. Native populations in the region noticeably declined during the nineteenth century, in many instances to half their original size. But even those peoples hit hardest by sickness displayed a consistent resilience that prevented the immediate collapse of their cultures. In Northwest Coast tribal populations, which were large enough to withstand the sudden loss of elders who passed along traditions, disease was less devastating than in other parts of the Americas, where smaller groups could not easily spring back. Demographers debate the precise extent and nature of population decline in the Northwest, but its peoples' survival cannot negate the horrors of the epidemics or their tragic consequences. Those cultures showed a strength and cohesion that permitted them to carry on under unprecedented pressure. Disease killed many native people, but it did not always kill their cultures.

Both in spite of and because of its implications for the Indians, the incursion of whites into the Northwest was an important historical event. Considerably to the south of Tlingit territory, near the mouth of the great river that the whites would name the Columbia, the local residents encountered a famous American exploring party. Meriwether Lewis, William Clark, and their team met the Chinookan peoples of the Oregon coast early in the nineteenth century. Obviously ethnocentric, their observations of native groups reflected white attitudes of the time. Nearly two hundred years after this expedition, the much-maligned Chinooks are still coming into focus.

That the coastal peoples of Oregon were not portrayed favorably by Lewis and Clark is not very surprising. The entourage may have been overjoyed at the sight of the Pacific Ocean, but they were less than thrilled with other dimensions of their stay at Fort Clatsop during the winter of 1805–6. Depressed by the eternally damp weather, they battled exhaustion after their long trek from St. Louis and tried to survive on dwindling supplies. They treated the local people badly, complaining about Indian thievery while rationalizing their own theft of an Indian canoe.

Nonetheless, Lewis and Clark recorded many plodding but valuable ethnographic details of Chinookan life. Unimpressed by the Indians' physical appearance, they yet admired the hats that kept them dry. "They are nearly waterproof, light and I am convinced, are much more durable than either chip or straw," wrote Lewis. Demonstrating the most sincere form of appreciation for Indian craftsmanship, Lewis and Clark bought the hats. When they did, they no doubt discovered the Chinooks' exasperating insistence on getting a good deal. Eons of native trade, as well as contact with non-Indian traders, had taught the people the fine art of bargaining.

Lewis and Clark never formally recorded the frustrations of trading with the Indians of the region, but their disdain for their temporary neighbors was obvious. Like previous white visitors, they found fault with the local people for making a home in a place the explorers found thoroughly unappealing. Fort Clatsop could never be home to Lewis and Clark. At best a stopover on a long and uncertain journey, it had in their view nothing to recommend it to the permanent settler. The Indians, meanwhile, marveled that the white men could not appreciate the environment that had for generations provided them with a comfortable home.

The Plains

In the northern plains, the arrival of Lewis and Clark marked the start of a new era. Only a few months into their westward sojourn, the American explorers had a tense encounter with Black Buffalo and his band of Brulés, a branch of the Dakotas—the Lakotas, or Western Teton Sioux. As representatives of separate and sovereign nations, the two parties tried to impress each other and establish themselves as forces to be reckoned with. Although their posturing probably made everyone apprehensive, they avoided bloodshed that autumn of 1804. The incident would eventually lead to violence, however, for Lewis and Clark's expedition journals labeled the Sioux the vilest miscreants of the savage race. Unless the U.S. government acted firmly and swiftly to uproot them from

George Catlin was deeply impressed by the Mandan villages he visited on the upper Missouri in 1832. Trying to capture the "thrilling panorama" that lay before him, he included in his painting of Mandan earth lodges a drum-like shrine in the center of the open area, a medicine lodge, the distant scaffolds of a Mandan cemetery, and many small scenes of daily life. Catlin's romanticized records of tribal life acquired greater historical value after a smallpox epidemic ravaged the upper Missouri tribes in 1837.

George Catlin (1796–1872). Bird's-eye View of the Mandan Village, 1,800 Miles above St. Louis. *Oil on canvas, 1837–39. National Museum of American Art, Washington, D.C./Art Resource, New York, New York.*

the Missouri River country, the explorers asserted, the fledgling nation would never be able to wrest control of the northern plains from the Indians. Unfortunately, there are no comparable written Brulé journals to offer a record of the 1804 encounter from the natives' perspective. No streams of invective inked in defiant Lakota give a balancing account of the event. But by the early nineteenth century, there could be no doubt that there were not one but two expanding powers on the northern plains.

During the autumn and winter of 1804–5, Lewis and Clark came across the Hidatsas, Arikaras, and Mandans, whose way of life contrasted with that of the increasingly nomadic and broadly territorial Lakotas. These village peoples of the upper Missouri River country demonstrated a remarkable ability not merely to survive but to prosper in country that would ruin the less knowledgeable. For well over a thousand years, Indians had farmed in the plains region, successfully confronting the icy winters and searing summers. People identified specifically as Mandan may be traced back perhaps nine hundred years.

The Mandans and other northern plains farmers prospered because they knew the land and saw the earth itself as a living thing. Following time-honored ways, they never forgot how to plant and harvest, hunt and preserve. As a result, despite their harsh environment, they rarely knew famine. The world they lived in demanded that they make full use of whatever they found, caught, or grew. They used more than seventy wild plants for food, ritual, and healing; they mined flint beds along the rivers for arrow- and spearpoints, scrapers, and knives. In this spare and unsparing world, children grew up knowing that their mothers, fathers, aunts, uncles, grandmothers, and grandfathers knew things that they did not.

Northern plains villagers reached maturity wanting to succeed within the natural and social boundaries of the world prescribed for them. In the early twentieth century, an old Hidatsa woman described these boundaries when she remembered the earth lodge of her childhood in the 1840s. She considered it to be alive, with "the door for a mouth" and four living posts, ten to twelve feet long, to support the roof. Inside, space was methodically allocated: the fireplace, of course, marked the center, with beds, a food-storage platform, a cache pit, a place to store sacred bundles, a place for honored guests to sit, and other specific spots lining the walls. As this orderliness suggests, northern plains villagers learned to do things in certain ways. As children, they entered and moved about their lodges in a counterclockwise fashion, sat down only when told they could do so, and ate the food provided whether or not it tasted good and whether or not they were hungry.

Most often the food tasted good, for it was gathered and prepared according to long tradition. Staple crops, such as maize and squash, were cooked in a variety of ways or dried and stored from one year to the next. Bison meat could be preserved and consumed over an extended period of time. As they farmed or hunted, the people observed ceremonial dictates to enhance their chances of success. Not to do so would be to risk hunger or starvation, but few on the northern plains would dare tempt fate in this way. To people steeped in cultural tradition and social discipline, such inattention to ritual would have seemed absurd and unacceptable.

For so long threatened only by the possibility of raids by peoples less blessed and productive than they, the villagers of the northern plains faced a more insidious danger when whites arrived. Their strength, derived from a large population based in permanent settlements, contributed to their vulnerability because they could not easily avoid the new diseases brought by the whites. Nor could they quickly abandon a way of life so proven and satisfying to take advantage of the newly arrived horse, which made a valuable contribution to other communities.

The smallpox epidemic of the late 1830s, well-known to historians because it wiped out perhaps half the region's indigenous population, represented a final blow rather than an initial onslaught. In 1795, one observer recorded that the Arikaras were devastated by a smallpox outbreak that obliterated thirty of their thirty-two villages and killed thirty-five hundred of their four thousand warriors. Fourteen years earlier, the Mandans and the Hidatsas had suffered similar catastrophes. By the time Lewis and Clark arrived, the Hidatsa population stood at about twenty-seven hundred, perhaps half the total before the importation of European disease.

The horrible experience of the sedentary farmers of the northern plains made their way of life far less attractive to migrating tribes than it once had been. Near the end of the eighteenth century, the various branches of the Lakotas and the Cheyennes had paused along their westward migration from the eastern woodlands to sample life in this seemingly secure world. For a few years, they shifted their focus away from migratory hunting and toward farming and village life. But once they acquired horses and witnessed the terrifying consequences of smallpox, the Lakotas and the Cheyennes scattered with the winds.

The horse came surprisingly late to these prototypical inhabitants of the plains. Not until the last decades of the 1700s and the first years of the 1800s did they accumulate sufficient numbers of horses to begin the transformation of their cultures. But in time, the horse changed them by giving them greater mobility. It increased the territory they could try to control and expanded their capacity to hunt bison and carry out raids. Thus, the village peoples of the upper Missouri River and other more narrowly circumscribed populations, such as the Omahas, which remained settled longer and adapted to horses later, lost their dominant status on the plains.

The popular image of the Lakota and Cheyenne plains peoples, of mounted warriors in feathered headdresses, reflects only the later stages of their evolution. Not only had they lived most of their history without the horse, but they also arrived in the region fairly late. Migration, adaptation, and culture building characterized their way of life from the time they left their original home in the woodlands to the East. Each group has its own story about the reason for that long migration from the upper Midwest. The Ojibwas, for example, say they forced the Sioux, their word for "enemy," out of Minnesota, but the Lakota people do not subscribe to this story. Instead, they speak of their imagination and initiative in following the bison and tell how they sought out opportunities for trade and expansion, which could be realized only in the West.

The Cheyennes, meanwhile, traditionally divide their history into four periods. According to their legends, they originally lived in the Northeast between the Great Lakes and Hudson Bay. An epidemic forced them southward during the second era, known as the time of the dogs. Part-wolf dogs traveled with them until the third age dawned, that of the time of the buffalo. With the buffalo as the key to tribal prosperity, they enjoyed a phase of plenty. Then, when they followed the buffalo onto the plains, they entered the fourth period, the time of the horse. The number four is sacred to the people, so their history gave special significance to this time.

Although the early experience of the Cheyennes is clouded with uncertainty, recent studies have turned up evidence of a migration similar to that recounted by the elders. Originally hunters and gatherers of wild rice, the Cheyennes also farmed and probably did know wolves in the Mississippi River country. Before 1700, they moved west to southwestern Minnesota and settled along the Minnesota River. For almost a century, they lingered in the prairie at the edge of the plains. Then the westward retreat of the buffalo, combined with the growing number of horses in their herds, drew the Cheyennes to the interior plains.

As they claimed a new homeland, the Cheyennes redefined the heart of their culture. They designated a great mountain in the western Dakotas as the most sacred place in their universe and named it Noaha-vose, "The Hill Where the People Are Taught." This solitary promontory resembled a bear to the Lakotas and came to be known to Anglo Americans as Bear Butte. In the Cheyenne belief system, "Nothing lives long except the rocks," as the historian John Stands in Timber put it. Accordingly, the Cheyennes looked to enduring Noaha-vose as the source of power—the power to heal and renew them in the years to come.

They expressed their beliefs in the story of Sweet Medicine and his Woman, who came to a cave in this sacred mountain. Ma-heo-o, the Creator, met Sweet Medicine there and gave him four arrows with special powers. To learn how to care properly for them, Sweet Medicine remained at Noaha-vose for four years. While receiving instruction from Ma-heo-o, he saw Neve-stanevoo-o, the Four Sacred Persons who bless the people from their homes at the Four Directions of the universe. Other holy beings who occupied the Cheyenne universe, such as Sun, Moon, and Thunder, revealed themselves to Sweet Medicine as well, along with the Ahtonoone-etaneo-o, the Underground People. When Sweet Medicine finally left Noaha-vose, his Woman carried Maahotse, the Sacred Arrows, down to be bestowed on the Cheyennes. These four arrows symbolized Ma-heo-o's love for and commitment to the people. Two of them granted the Cheyennes power over the buffalo, and two gave them power over human enemies. Ever since Sweet Medicine's sojourn, the Cheyennes have linked their collective well-being to these forces and entities.

According to Cheyenne legend, Sweet Medicine lived among them for many years before he died. During that time he taught them how to live properly by following the principles he had learned at Noaha-vose. In the uncertain and trying world of the plains, the people were always concerned with simple survival. As they made the transition to a less agricultural way of life, they faced the persistent possibility of not merely hunger but starvation. The vast, open plains country offered little protection, not only from the wind, the snow, and the sun but also from rival groups such as the Pawnees. The arrows bestowed by Ma-heo-o and the ideas communicated through Sweet Medicine reassured the people that they would endure in the face of adversity.

The Cheyennes say that Sweet Medicine warned them to hold fast to their values, for strange people would someday come into their country. They would recognize the newcomers by their light-colored skin and short hair. Speaking a harsh and incomprehensible language, the strangers would pose an unprecedented threat to the Cheyenne way of life. After they came, the buffalo would disappear, and in its place would appear another hoofed animal with a long tail, whose meat they would learn to eat. First, though, they would learn to ride a round-hoofed animal, which would allow them to travel quickly on the hunt or from one camp to another.

As with other indigenous Americans, the arrival of white people did not spell immediate doom for the Cheyennes. They probably encountered the first light-skinned strangers in the early 1700s. Spanish traders and French trappers and traders made their way into Cheyenne territory well before Anglo Americans such as Lewis and Clark traversed the country. Accustomed to trading with other Indians, the Cheyennes began to exchange commodities with non-Indians. Over time, through trade and raid, they acquired horses, guns, and metal implements that eventually had a significant impact on their lives.

New technology changed the way Cheyennes did things more than it altered their values and priorities. They had decided to follow the buffalo herds long before they obtained horses; horses merely allowed them to hunt the buffalo more effectively. Long before they obtained guns, they had fought their enemies; guns simply made them more lethal warriors. Likewise, they had made arrow tips of stone or bone since time immemorial; the introduction of metal only enabled them to make better points. Thus, the items they gained through trade or other means promised to make life better. The Cheyennes also avoided the worst of the epidemics of the late eighteenth and early nineteenth centuries because they were so widely scattered across the land. (Cholera may have been more of a killer than smallpox during this period.) By the second decade of the 1800s, most Cheyennes looked to the future with some measure of confidence.

The Cheyennes could afford to feel confident, because their economic and social order seemed to function well in a world that included a few whites. But they did feel some concern about blending their old values with new technology. The increasing use of horses threatened to make individuals more independent, less likely to participate in traditional, cooperative social and economic endeavors. Similarly, rifles might allow people to hunt by themselves, without working with their peers as in the old days. Given the opportunity to acquire various material possessions, families could hoard rather than share their wealth. And a more nomadic way of life, some worried, was likely to compromise the physical well-being of the elders. The new age clearly posed these and other challenges to the collective well-being of the Cheyennes.

Still, they managed to sustain the behavioral ideals that had always informed the patterns of their life. In times of peace and in times of war, the urge toward individual achievement or action was tempered by the need for group cohesion. Young men wanted to be known for their daring deeds, so they tried to count coup on their enemy in the heat of battle, to stay and fight when retreat might have been tempting, to steal horses even when the odds of a successful getaway seemed slim. Those particularly skilled in wartime exploits or on the hunt had every right to feel proud. Yet individuals did not and could not live apart from the community. Warriors had to move with the group when it moved and fight alongside others if an enemy suddenly attacked the village. In countless ways, individual life was inextricably bound up with the people's common fate. As a result, men generally adhered to the social expectations placed on them from birth. They measured their worth not by what they took or hoarded from others but by what they shared. Reciprocity permeated Cheyenne life, influencing the distribution of food, clothing, housing, and horses.

Among the Cheyennes, altruism and generosity counted for more than ambition, and these admirable traits were fostered by a reward system. Individuals who aspired to lead in the Council of Forty-Four, the organization that oversaw tribal life in times of peace, had to live by certain values and behave in a certain way. Those charged with religious duties also had to live in the proper way. To earn the honor and responsibility of leading a religious ritual, for example, a Cheyenne had to be judged a good person. Otherwise, the ritual was unlikely to achieve the desired results.

For those who made mistakes or committed undesirable or outlawed acts, the Cheyennes developed an elaborate system of law and order that emphasized rehabilitation rather than extended punishment of the individual. Obviously, the Cheyennes had no jails; other than a degree of social isolation within the camp or temporary exile, an individual could not be effectively ostracized. In the rare case of a murder within the community, the perpetrator was apprehended immediately to prevent a spiral of revenge and retribution. The Cheyennes also believed that a murderer left undiscovered in the group would smell bad to the buffalo, and the buffalo would stay away from the hunters. This example graphically illustrates how the individual's fate intertwined both with that of the community and with the workings of the natural world.

During the first half of the nineteenth century, the Cheyennes preserved their traditions while absorbing white influences. At this stage, buffalo were still fairly abundant on the plains, but already the people who hunted them had to keep moving into the interior plains and venturing farther south to find their prey. Like a number of Plains Indian groups, the Cheyennes continued to migrate. A portion of the tribe moved south of the North Platte River in Wyoming and Nebraska, down into eastern Colorado and western Kansas, and toward the Arkansas River. Indian trading networks expanded in the process and began to mesh with those of the Anglo Americans. The growing white economic presence led to the construction in 1833 of Bent's Fort, a landmark on the plains in southern Colorado. Soon afterward, the stresses of migration, tribal politics, economic changes, and war led to the division of the people into the northern and southern Cheyennes.

Other groups who made major migrations included the Kiowas and the Comanches. Formerly peoples of the North, they were linked to the traditions of the Shoshonean Great Basin area by language and economics. By the late eighteenth century, though, the Kiowas and the Comanches both made their way from Wyoming down to the southern plains and established new homelands between the Arkansas and Red rivers. They also abandoned their traditional enmity with each other in favor of greater intertribal harmony. In oral accounts, both groups preserved tribal memories of their move, in stages, to the southern plains. According to Ko-sahn, a Kiowa woman:

> There are times when I think that I am the oldest woman on earth. You know, the Kiowas came into the world through a hollow log. In my mind's eye I have seen them emerge, one by one, from the mouth of the log. I have seen them so clearly, how they were dressed, how delighted they were to see the world around them, I must have been there. And I must have taken part in that old migration of the Kiowas from the Yellowstone to the Southern Plains, for I have seen antelope bounding in the tall grass near the Big Horn River and I have seen the ghost forests in the Black Hills. Once I saw the red cliffs of Palo Duro Canyon. I was with those who were camped in the Wichita Mountains when the stars fell.

The last event, the Leonid meteor shower, occurred early on the morning of 13 November 1833, by which time the Kiowas had established themselves on the southern plains.

Even as the Kiowas and the Comanches migrated out of the Great Basin region to the southern plains, other peoples remained in the Basin. This area, which encompasses present-day eastern Oregon, southern Idaho, western Wyoming, western Colorado, the California-Nevada border territory, and all of Utah and Nevada, was home to a culture dominated by branches of the Shoshone, Bannock, Ute, and Paiute nations. In part because it was among the last areas to be penetrated by Europeans, it remains the least-known portion of the American West.

By the time of the American Revolution, Spanish officials had a nebulous concept of the northern reaches of New Spain. In the same month that Thomas Jefferson signed his name to the Declaration of Independence, two Spanish missionaries set out on an expedition at once remarkable and unrecognized. They were the Mexican-born

Francisco Atanasio Dominguez and the Spanish native Silvestre Vélez de Escalante. Their small party found much to note along an elliptical route that started and ended in Santa Fe, New Mexico. Escalante's journal described peoples who forged a living from apparently meager territory that yet offered adequate resources to those who knew how to exploit the land.

Encountering peoples and places that did not always match their expectations, the bearded travelers at one point came on men with beards so long "they looked like Capuchin or Bethlemite fathers." The priests lamented the fate "of those miserable little lambs of Christ who had strayed only for lack of the Light," but they were impressed by the natives' willingness to help them search for a missing member of their group. The expedition's artist, mapmaker, and astronomer, Don Bernardo de Miera y Pacheco, provided images of the bearded Indians, clothed in nearly knee-length shirts, wearing moccasins, and clutching animal-skin quivers, a net, and a rabbit.

These Indians, the western Utes, hunted by driving rabbits into fences of soapweed, sagebrush, or hemp cords. They netted sage grouse the same way and hunted deer and elk with particular effectiveness in the drifts of winter snow. Wearing deerskins as a disguise, the hunters crept up as close as they could to surprise their prey. In the streams and lakes of their home country, the people used spears, arrows, weirs, nets, and traps to catch fish. Seeds, fruits, nuts, and berries added variety to the Ute diet, with piñons (pine nuts) being a favorite staple. When necessary, they ate rattlesnakes, lizards, cicadas, crickets, and ants. Europeans frowned on such foods, but the Utes were a practical people in an environment that demanded pragmatism. They either ate what the land provided or went hungry.

The Utes had little in common with the Acomas, the Tlingits, or the Hidatsas. Nevertheless, in the years leading up to the period of white contact, various Indian peoples all over the West sought to build cultures and homes, often in new territory. As they made their way onto new lands, they redefined their view of the world around them. They mixed old ideas and ways with new ones, hoping to forge alliances both with other peoples and with the environment itself. In time, the horse and the gun combined with migration to produce greater competition for natural resources, which led to conflict as well as coexistence among Indians. War broke out whenever groups struggled over territory and the right to establish an economic base.

When white people began to appear in growing numbers at the end of this period, Indians saw them not only as enemies or competitors but also as potential allies or pawns. Small communities threatened by more powerful tribes sometimes perceived whites as a useful means toward survival. Certainly, the first whites who struggled onto the plains or into the Great Basin gave the indigenous groups little reason to worry. But the ultimate effect of the white incursion remained to be seen.

The Navajos

A final example of migration and adaptation in the Southwest produced the largest Indian nation of the modern American period: the Diné, known as the Navajos. Their humble beginnings, centuries before, hardly foreshadowed their eventual rise. Linguistic analysis links the Navajos to the large family of Athapaskan-speaking peoples, now

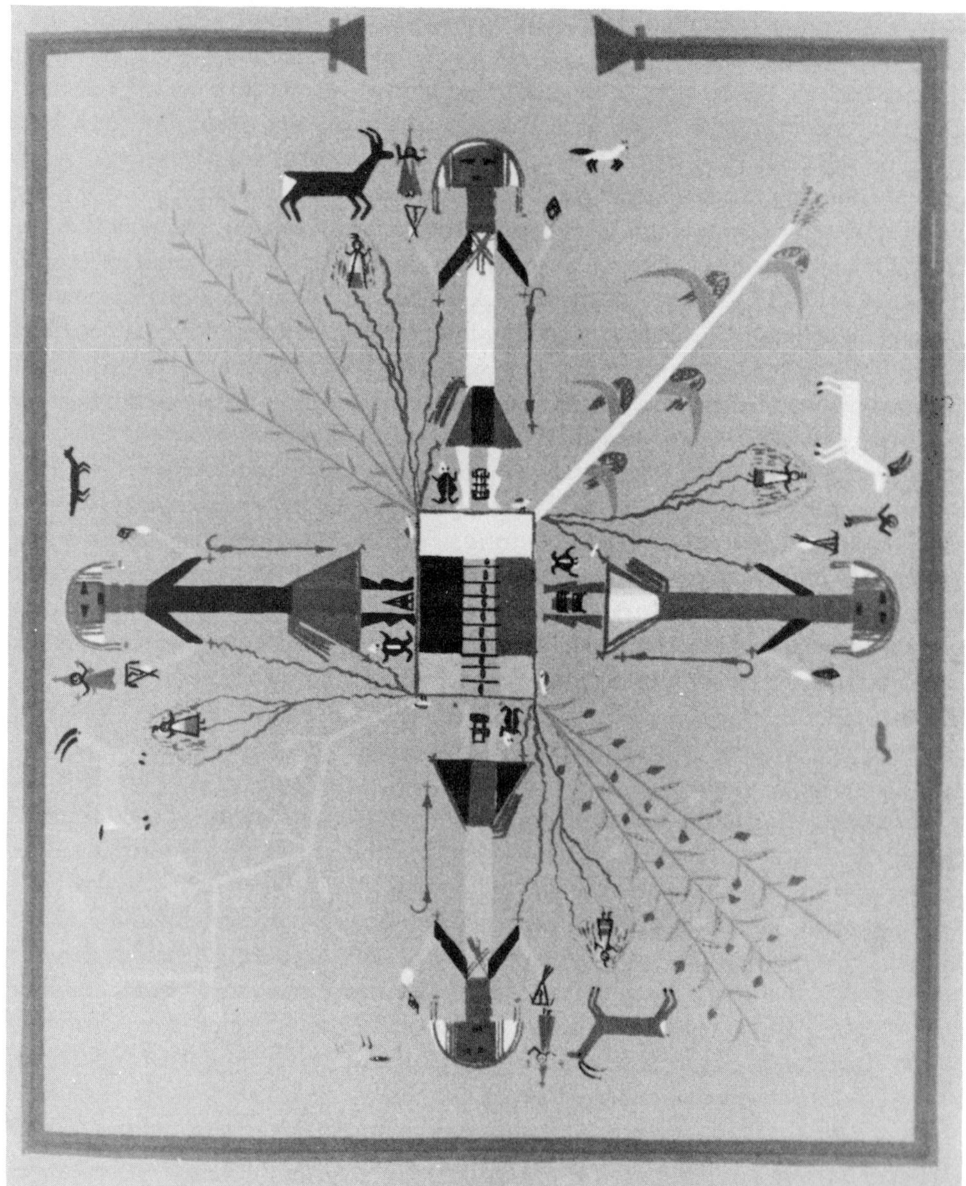

In this image representing one of the central stories of Navajo cosmology, four figures surround the place of Emergence, where, according to myth, the Navajo people migrated into this world. The figures represent the sectors of the sky and the sacred mountains that mark the cardinal points of the Navajo world. Rivers flowing from these mountains nourish the food-producing plants that grow from the Emergence ladder. Though sand paintings were widely used in Navajo ceremonials, these ephemeral images were rarely recorded until the 20th century.

Louie Ewing (1908–83). Emergence Sand Painting. *Serigraph, 1949 (after a sand painting recorded by Mrs. John Wetherell). From Mary C. Wheelwright,* Emergence Myth *(1949).*

scattered over a wide portion of northern North America. The Navajos and the Apaches are the only Indian communities of the Southwest to speak a language of this classification. But their speech clearly ties them to peoples in Alaska and western Canada, such as the Eyaks, Haidas, and Kutchins. The languages as spoken today are not identical, but even to the relatively untutored ear the cadences and accents sound quite similar.

The Navajos are also tied to the North by the stories they tell. When the anthropologist Franz Boas studied Indian societies a century ago, he discovered striking parallels between some of the tales told by Athapaskan-speaking peoples now separated from one another by thousands of miles. More than coincidence, the commonality was another important piece of evidence that the Navajos had lived in the North before making their way to the American Southwest. Why and how—and even when—the Navajos decided to make this extended trek remain uncertain. It is unlikely that they moved as one group; it is even more unlikely that they moved directly from the far North to the Southwest. What seems more probable is that theirs was a gradual, halting migration by stages, which took them through the Rocky Mountains into northern New Mexico. When they arrived, the Navajos may have contributed to the general reshuffling of southwestern populations in the thirteenth and fourteenth centuries. Internally, the Navajos underwent their own adjustment. More hunters than gatherers, and decidedly not farmers, they slowly came to terms with their new environment.

Like the Cheyennes, the Navajos in time redefined themselves. Children learned that their people had always lived in the Southwest, that they belonged in the place called Diné Bikéyah, or the Navajo country. The Indians marked the boundaries of this territory by four sacred mountains: the San Francisco Peaks (Dook'o'oosliíd) were the sacred mountains to the West, Mount Taylor (Tsoodzil) was the sacred mountain to the South, Blanca Peak (Sis Naajiní) was the sacred mountain to the North, and Hesperus Peak (Dibé Nitsaa) was the sacred mountain to the East. Adorned, respectively, in abalone, turquoise, white shell, and jet, these mountains were not the only places with sacred significance to the Navajos. In their stories, they designated as holy the part of Diné Bikéyah called Dinétah ("among the people"), where they believed they originated. Archaeological evidence corroborates Navajo legend, identifying the region as the first the people occupied when they entered the Southwest.

The Navajos' story of their beginnings is one of emergence, of long migration from one world to the present world. According to the tale, the central figures of Navajo mythology lived in Dinétah. First Man, First Woman, and the children of Changing Woman occupied Huerfano Mountain; Changing Woman resided at Gobernador Knob. The adventures of the children of Changing Woman, Monster Slayer, and Child Born for Water gave meaning to other natural features of Navajo territory. For example, the story refers to the lava flow near present-day Grants, New Mexico, as the dried blood of the creature felled by Monster Slayer. Throughout the tale, other beings, such as animals and insects, appear as an integral part of life itself.

Indeed, the similarities between the Navajo origin story and the origin stories told by the different Pueblo communities are so striking that it seems the Navajos reworked a version of another story to create their own. Such adaptation would be very much in keeping with Navajo character, for the great strength of the Navajos, the key component of their expansion and prosperity, was their ability to learn from others and to incorporate new ways of life into their culture. In time, acquisitions such as the content of stories, the crafting of silver, or the raising of sheep became part of traditional Navajo life.

Essentially a hunting people at first, they became a people who not only hunted but also farmed. In the two or three centuries before the Spaniards came, the Navajos probably learned a great deal from their more established neighbors, the Pueblo Indians, about how to exploit the southwestern environment. The Navajos, of course, do not tell the story that way. Rather, they tell of early Navajos who learned how to grow corn or how to weave from the holy people. Spider Woman, according to their legends, instructed them in the art of weaving. In some respects, it does not matter who taught the Navajos their skills. What matters is that before the Spaniards arrived in the area, the Navajos had thoroughly adapted to their new home, which lay in country remote from the main regions of Spanish incursion.

The Navajos developed an elaborate system of ceremonies designed to promote harmony within themselves and within the larger universe. Hunting remained far more important than agriculture in the different chantways, evidence that the Navajos did not discard all their old beliefs when they migrated. But at the same time, the cornfield became a primary image in the Navajo cosmos. Corn pollen and the corn plant are central to the Navajo ceremonial world.

Nearly a century and a half elapsed between the first Spanish foray into the Southwest and the revolt of 1680. During this period, the Navajos learned about the Spaniards' religion, language, and culture and gradually determined the usefulness of these things. Few Navajos saw any point in converting to Catholicism, but some Navajo names do show up in the baptismal records kept by the missions of New Mexico. Nor did Spanish replace the Navajo language, although many Indians saw the value of learning a few key words for the purpose of communication and trade. Horses, sheep, cattle, and goats, however, were another matter entirely.

Because they had adopted a more sedentary life-style and had claimed a large, relatively uncontested country, the Navajos were ideally situated to incorporate Spanish livestock into their economy. In addition to the meat that all the animals could provide, the horse also promised a new dimension in transportation, and the sheep offered wool, the ideal material for weaving. Once again, new additions to the culture became gifts from the gods, who had always possessed them and only now had seen fit to bestow them on the people.

The Navajos said that horses had originally belonged to the Sun, who kept four of the animals corralled separately to the North, South, East, and West. The coats of the horses showed different colors, one for each of the four directions. When Turquoise Boy journeyed to see the horses, he saw a white shell horse to the East, a turquoise horse to the South, a yellow abalone-shell horse to the West, and a spotted horse to the North. Different Navajo accounts give different versions of how the people acquired the horses, but they all recount the joy with which the Navajos accepted these wonderful animals. One account praises the horse of the South:

The turquoise horse prances with me.
From where we start the turquoise horse is seen.
The lightning flashes from the turquoise horse.
The turquoise horse is terrifying.
He stands on the upper circle of the rainbow.
The sunbeam is in his mouth for a bridle.
He circles around all the people of the earth
With their goods.
Today he is on my side
And I shall win with him.

As demonstrated by the successful introduction of the horse, contact between the aboriginal Navajo community and the invading Europeans did not lead instantly to decline. To be sure, some dimensions of the Spanish presence—the capture of Navajos for the regional slave trade, for example—were clearly destructive. But just as certainly, the Navajos could not have reached the cultural pinnacle they now occupy were it not for the Spanish influence. Acquisition of Spanish livestock provided them the necessary means to complete their transformation into a people not only with a rich past but also with a viable future.

Even the revolt of 1680 benefited the Navajos by increasing their numbers. In the generation following the rebellion against Spanish oppression, thousands of Pueblo peoples fled their home villages. Some of them established new communities, but others joined existing groups such as the Navajos. In turn, their children became Navajos, and the number of Navajo clans expanded. The Jemez Pueblo people, for instance, migrated to Navajo country and established the Coyote Pass Clan (Ma'ii Deeshgiizhnii), while other immigrants founded the Mescalero Apache Clan (Naashgalí Dine'é) and the Mexican Clan (Naakaii Dine'é). These new arrivals had an immediate demographic impact on the Navajos, swelling the tribal ranks in the wake of the revolt. But they also enhanced Navajo culture by contributing to the art of weaving, the skill of farming, and the technology of animal husbandry. Almost overnight, the Navajos found themselves on the brink of a century of unprecedented expansion, cultural development, and prosperity.

During the seventeenth and early eighteenth centuries, the Spaniards could not contain the growth of the Navajo people or curtail their independence. Meanwhile, through trade and war, the Navajos greatly increased their wealth. True to their time-honored traditions, the Navajos conducted warfare and raiding only after performing the appropriate ceremonies. These rituals—as well as the prospect of proving their valor and acquiring riches—inspired Navajo warriors to perform daring exploits in combat and on raids. For protection and courage, they called on the power of animals and imagined that they assumed the form of these animals. They chanted:

> I am a big black bear.
> My moccasins are black obsidian.
> My leggings are black obsidian.
> My shirt is black obsidian.
> I am girded with a gray arrowsnake.
> Black snakes project from my head.
> With zigzag lightning projecting from the ends of my
> feet I step . . .
> Black obsidian and zigzag lightning stream out from me
> in four ways.
> When they strike the earth, bad things do not like it.
> It causes the missiles to spread out.
> Long life, something frightful I am.
> Now I am.

The Navajos must sometimes have seemed frightful to other native peoples of the Southwest. Sometimes friendly and sometimes hostile to their neighbors, the Navajos absorbed other peoples, trading with them, raiding them, and learning from them. Like

In the early 20th century, many artists presented a timeless image of a romanticized Indian world that obscured the many changes that had transformed native life since European contact. Although images of Navajo sheepherders and Lakota chiefs on horseback might seem to reflect unchanging cultural patterns, they in fact underscore the theme of adaptation in Native American life by emphasizing the cultural transformations initiated by the introduction of the sheep and horse.

Laura Gilpin (1891–1979). Shepherds of the Desert. *Gelatin silver print, 1934. © 1981, Laura Gilpin Collection, Amon Carter Museum, Fort Worth, Texas.*

Edward S. Curtis (1868–1952). The Prairie Chief *(Pine Ridge Agency, South Dakota). Photogravure, 1907. From Curtis,* The North American Indian, *supplement to vol. 3. Courtesy Special Collections, Amherst College Library, Amherst, Massachusetts.*

the Lakotas of the northern plains, they were an expanding power as the American era dawned.

At the end of the second decade of the nineteenth century, the American era was about to begin in the West. In 1821, Mexico claimed its independence from Spain, only to lose a large part of its territory in the revolt in Texas and the war with the United States. In the meantime, the first wave of Anglo-American explorers, missionaries, trappers, traders, adventurers, miners, farmers, and urban entrepreneurs began to arrive in the region. No one could predict that the stream of settlers would soon turn into a flood.

Indeed, the previous experience of indigenous peoples with various Europeans in the West bore little hint of the onslaught that lay ahead. Despite the massive toll exacted by disease, Spanish imperialism, Catholic mission work, and all the other upheavals caused by contact, the many surviving Indian communities had reason to be cautiously optimistic about the future. The Spaniards, the British, the Russians, and the first Anglo Americans had many designs on them and their lands, but the native peoples had not been displaced. Their tenacity thus far suggested that they would not disappear quickly in the decades to come. Thus, although the history of the native peoples of the American West is one of migration, adaptation, and change, it is also one of settlement, persistence, and continuity. The presence of Indians in the West obviously altered the course of Anglo-American history, just as the arrival of outsiders changed the direction of Indian history.

Too often in histories of the American West and of the United States as a whole, Indians greet boats filled with whites, fight and lose to the invaders, and in the end totter on the verge of assimilation and extinction. They are always on the defensive. But from the early sixteenth century to the early nineteenth century, native peoples experienced gain as well as loss, and expansion as well as contraction. Like the history of Anglo Americans, theirs is a tale of discovery and new beginnings.

The great ceremony of the Navajos, the Blessingway, includes a song about finding yucca and corn. The words say just as much about the Indians of the West, people who belonged to the land:

When I found it I became fabrics of all kinds.
When I found it I became jewels of all kinds.
When I found it I became game of all kinds.
When I found it I became plants of all kinds.
When I found it I became corn of all kinds.
When I found it I became horses of all kinds.
Now I am long life, now happiness as now I found it.
Before me it is blessed, behind me it is blessed,
Below me it is blessed, above me it is blessed,
Around me it is blessed, my speech is blessed,
All my surroundings are blessed as I found it, I found it.

Bibliographic Note

Any attempt to synthesize the experiences of the many native peoples of the West depends heavily on the work of others. The following list of sources suggests works especially important in the field and in the writing of this essay. Readers should consult the extensive bibliographies of the Smithsonian Institution's *Handbook of North American Indians* series for a comprehensive collection of work. In addition, the D'Arcy McNickle Center for the History of the American Indian at the Newberry Library in Chicago has also sponsored a series of critical bibliographies that will prove helpful to those seeking additional sources.

The origin story of Acoma Pueblo is told in Matthew W. Stirling, *Origin Myth of Acoma and Other Records*, Bureau of American Ethnology Bulletin No. 135 (Washington, D.C., 1942). The classic overview of relations between and among the Indians, Spaniards, Mexicans, and Anglo Americans is Edward H. Spicer, *Cycles of Conquest: The Impact of Spain, Mexico, and the United States on the Indians of the Southwest, 1533–1960* (Tucson, 1962). Many authoritative articles on Indian communities and on relations between Indians and non-Indians may be found in the volumes published by the Smithsonian Institution. See in particular the *Handbook of North American Indian* series: *California* (volume 8, 1978), *Southwest* (volume 9, 1979), *Southwest* (volume 10, 1983), *Great Basin* (volume 11, 1986), and *Northwest Coast* (volume 7, 1990). Peter Nabokov and Robert Easton, *Native American Architecture* (New York, 1989), is a richly illustrated study, telling us much about buildings but also about their cultural context.

Traditional stories from California, the Pacific Northwest, and Canada may be found in Katharine Berry Judson, *Myths and Legends of California and the Old Southwest* (Chicago, 1912), Ella E. Clark, *Indian Legends of the Pacific Northwest* (Berkeley, 1963), and Ella E. Clark, *Indian Legends of Canada* (Toronto, 1960). Robert F. Heizer and John E. Milles, *The Four Ages of Tsurai: A Documentary History of the Indian Village on Trinidad Bay* (Berkeley, 1952), provides details of the different encounters between the Yuroks and the Europeans and Anglo Americans. Norris Hundley offers an authoritative examination of choices made about water in *The Great Thirst: Californians and Water, 1770s–1990s* (Berkeley, 1992).

Erna Gunther, *Indian Life on the Northwest Coast of North America, as Seen by the Early Explorers and Fur Traders during the Last Decades of the Eighteenth Century* (Chicago, 1972), is a thorough, well-illustrated overview, including a chapter on the Tlingits. The Tlingit oral historical narratives quoted in the essay are from Nora Marks Dauenhauer and Richard Dauenhauer, eds., *Haa Shuká, Our Ancestors: Tlingit Oral Narratives* (Seattle, 1987). The standard collection of Tlingit stories was recorded by John R. Swanton in *Tlingit Myths and Texts*, Bureau of American Ethnology Bulletin No. 39 (Washington, D.C., 1909). Sergei Kan provides a perceptive introduction to his translation of the Russian Orthodox priest Anatolii Kamenskii's *Tlingit Indians of Alaska* (Fairbanks, 1985).

Richard White's "The Winning of the West: The Expansion of the Western Sioux in the 18th and 19th Centuries," *Journal of American History* 65 (September 1978), is a stunning explanation of Lakota ascendancy. James P. Ronda, *Lewis and Clark among the Indians* (Lincoln, 1984), is a masterful study of the explorers' relations with both Plains and Northwest Coast Indian peoples. Carolyn Gilman and Mary Jane Schneider, eds., *The Way to Independence: Memories of a Hidatsa Indian Family, 1840–1920* (St. Paul, 1987), takes full advantage of the pioneering work of Gilbert Wilson and adds new essays about upper Missouri River native life by W. Raymond Wood, Gerard Baker, Jeffrey R. Hanson, and Alan R. Woolworth.

Peter John Powell's magnificent studies of the Northern Cheyennes are essential readings; see his *Sweet Medicine: The Continuing Role of the Sacred Arrows, the Sun Dance, and the Sacred Buffalo Hat in Northern Cheyenne History* (Norman, 1969) and *People of the Sacred Mountain: A History of the Northern Cheyenne Chiefs and Warrior Societies, 1830–1879, with an Epilogue, 1969–1974* (New York, 1981). E. Adamson Hoebel, *The Cheyennes: Indians of the Great Plains*, 2d ed. (New York, 1978), is a good introduction. George Bird Grinnell, *The Cheyenne Indians: Their History and Ways of Life* (1923; reprint, New Haven, 1973), presents a thorough examination of Cheyenne society. Margot Liberty and John Stands in Timber, *Cheyenne Memories* (New Haven, 1967), is the account by a respected tribal elder. The Kiowa writer N. Scott Momaday includes Ko-sahn in his splendid essay "The Man Made of Words," in Geary Hobson, ed., *The Remembered Earth: An Anthology of Contemporary Native American Literature* (1979; reprint, Albuquerque, 1981), and writes of the Kiowas' great migration to the southern plains in *The Way to Rainy Mountain* (Albuquerque, 1969). Herbert E. Bolton, *Pageant in the Wilderness: The Story of the Escalante Expedition to the Interior Basin, 1776* (Salt Lake City, 1950), is the standard translation and explication of the Escalante diary. Gloria Griffen Cline, *Exploring the Great Basin* (Norman, 1963), outlines the paths taken by various entrants into the region. Omer C. Stewart has written widely about the different peoples of the area; one of his introductory essays is "Ute Indians: Before and After White Contact," *Utah Historical Quarterly* 34, no. 1 (1966): 38–61. Joseph Jorgensen, *The Sun Dance Religion: Power for the Powerless* (Chicago, 1972), is an important study that brings Ute history into the modern era.

Aileen O'Bryan, *The Diné: Origin Myths of the Navaho Indians*, Bureau of American Ethnology Bulletin No. 163 (Washington, D.C., 1956), and Pliny E. Goddard, *Navajo Texts*, Anthropological Papers of the American Museum of Natural History, vol. 34 (New York, 1934), are important sources. LaVerne Harrell Clark, *They Sang for Horses: The Impact of the Horse on Navajo and Apache Folklore* (Tucson, 1966), also employs material from O'Bryan and Goddard that is quoted in this essay.

Sam Bingham and Janet Bingham, eds., *Between Sacred Mountains: Navajo Stories and Lessons from the Land* (Tucson, 1984), is a wonderful collection of material gathered by people at Rock Point School on the Navajo Nation. James F. Downs, *The Navajo* (New York, 1972), and Peter Iverson, *The Navajos* (New York, 1990), present overviews of Navajo history, society, and culture. Ruth Underhill, *The Navajos* (Norman, 1956), and Clyde Kluckhohn and Dorothea Leighton, *The Navaho* (Cambridge, Mass., 1946), remain helpful. The quotation from the Navajo Blessingway ceremonial is from the translations of Father Bernard Haile in Leland C. Wyman, ed., *Blessingway* (Tucson, 1970).

A sweeping survey of western Indians at the time of the Columbus landing may be found in Peter Iverson, "Taking Care of Earth and Sky," in Alvin M. Josephy, Jr., ed., *America in 1492: The World of the Indian Peoples before the Arrival of Columbus* (New York, 1992). Edward Spicer and his students and associates have contributed greatly to our understanding of Indian continuity and change. See, for example, a book of essays in Spicer's honor edited by George Pierre Castile and Gilbert Kushner, *Persistent Peoples: Cultural Enclaves in Perspective* (Tucson, 1981), as well as Edward Spicer and Raymond H. Thompson, eds., *Plural Society in the Southwest* (New York, 1972). Anya Peterson Royce, *Ethnic Identity: Strategies of Diversity* (Bloomington, 1982), is a persuasive study of the symbols employed by groups in their efforts to maintain flexible, working identities.

Peter Nabokov was originally assigned this chapter and was unable to complete it because of events in the Middle East that led to the war in Kuwait and Iraq. His ideas about the structure of the chapter certainly influenced my approach, and I would like to express my thanks to him.

Chapter Two

The Spanish-Mexican Rim

DAVID J. WEBER

In 1826 a Pueblo Indian appealed to New Mexico officials to stop non-Indians from acquiring land belonging to his community. As alcalde, or mayor, Rafael Aguilar claimed to represent the "principal citizens of the Pueblo of Pecos," a once-powerful town that lay astride a key pass between the Rio Grande valley and the western edge of the high plains. Writing in phonetic Spanish, Aguilar reminded authorities that Pueblo Indians enjoyed the rights of citizens, that the law guaranteed their ownership of four square leagues of land around their pueblo, and that non-Indians had no right to acquire Pueblo lands. Aguilar's petition was one of several formal complaints lodged in the 1820s by natives of Pecos to protect their farms and pastures. In legal terms, the petitions paid off. In 1829 the New Mexico legislature ordered non-Indians to vacate Pecos Pueblo lands.

Like other Pueblo Indians, the *pecoseños*, or residents of Pecos, for whom Aguilar spoke, had remained a culturally distinctive people. Nonetheless, as personified by Aguilar, over two centuries of exposure to Hispanic neighbors and missionaries had profoundly altered Pueblo culture. Like many Pueblo leaders, Alcalde Aguilar held a Hispanic office, understood how and when to appeal to Hispanic law, communicated with Hispanics in their language, and identified himself with a Spanish surname and a Christian given name. [*Please see "A Note about Language" on p. 75.*]

Moreover, Hispanic influence went beyond law, language, politics, and religion into the economic life of Pueblo communities such as Pecos. On farmlands, such as those Aguilar sought to protect, Pueblos raised crops they had not known before the Spaniards arrived in the sixteenth century. In addition to the corn, beans, squash, and cotton they had cultivated for centuries, Pueblos grew tomatoes, chiles, and new varieties of corn and squash brought by Spaniards from central Mexico, as well as exotic coriander, wine grapes, cantaloupe, watermelon, wheat, and other imports from the Old World. Pueblos also tended apricot, apple, cherry, peach, pear, and plum orchards and raised sheep, goats, cattle, horses, mules, donkeys, oxen, and flocks of chickens—all previously unknown to them.

By adopting these foreign crops and stock, Pueblos, like many other Indians, enriched their economy and diet. Changes introduced by Spaniards, however, also had deleterious effects. For over a century before Aguilar wrote his petition, the number of *pecoseños* had declined steadily. Spanish-introduced diseases had taken a toll, and so had raids by Apaches and Comanches. Those raids had apparently intensified as Old World crops and livestock made the Pueblos more productive and thus more tempting as targets for raiders well equipped to strike astride Spanish-introduced horses.

Among the earliest surviving Indian images depicting contact with Spaniards, is this pictograph of Spanish horsemen drawn high on a canyon wall in Canyon del Muerto, Arizona. The riders with long capes, broad-brimmed hats, and flintlock guns may represent the soldiers of Lt. Antonio Narbona who led a raid against the Navajos in 1805.

Attributed to Dibé Yázhí Nééz (Tall Lamb). Spanish Horsemen. *Pictograph, ca. 1805. Canyon del Muerto, Arizona. Photograph by Helga Teiwes. Arizona State Museum (#28883), University of Arizona, Tucson.*

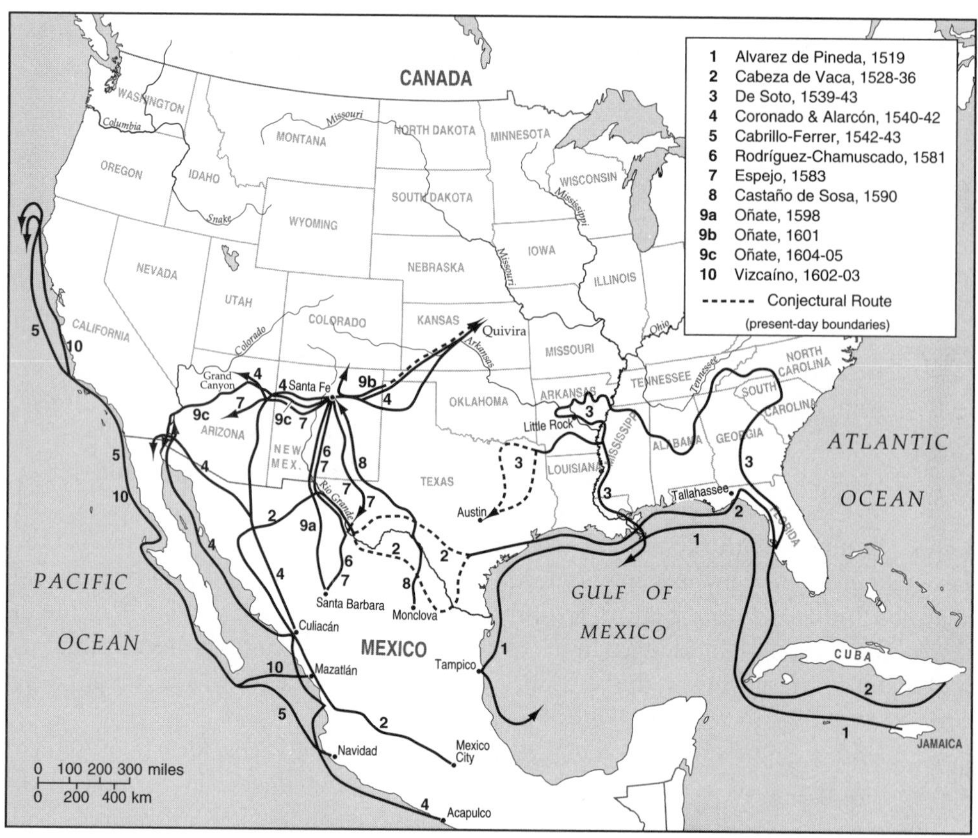

Spanish Exploration, 1519–1660

Whether from emigration or premature death, the population of Pecos fell from about 1,000 in 1700 to fewer than 150 by the end of the century. By the 1830s, it hardly mattered that Aguilar had won a legal victory against Hispanic encroachment, for there were too few *pecoseños* to prevent unscrupulous Hispanics from killing the Pueblos' stock, poisoning their water holes, and otherwise making their lives intolerable. In the late 1830s, the last residents of Pecos, numbering fewer than 20, abandoned the town and moved across the Rio Grande to Jémez Pueblo, which became their permanent home.

In general, the story of Pecos Pueblo exemplifies the Hispanic impact on the Pueblo world, but many of the details of the story are unique to Pecos. Even among Indians as seemingly similar as the Pueblos of New Mexico, Hispanic influence varied with time, place, and circumstance. Some Pueblo communities became extinct much earlier than Pecos, but others, strengthened by their adaptations to the Hispanic world, have survived to the present day. Whatever the rate or quality, the changes that Spaniards engendered in indigenous communities were remarkably pervasive. Directly or indirectly, Spanish influence extended beyond the Pueblo world across the southern rim of what is now the American West, from California to the Mississippi and northward to Oregon and Washington, the Great Basin, the Central Rockies, and the upper Missouri.

Between 1821 and 1846–47, when Americans seized the West in the war with

Mexico, much of this area had constituted the far northern frontier of independent Mexico. Before 1821, Spain had claimed the western half of the continent as the northern reaches of the viceroyalty of New Spain. Never numerous but highly influential, Spaniards had invaded and explored much of the West in the first half of the sixteenth century and had begun to settle on its southern rim before 1600. For the next two and a half centuries, Hispanics remained and expanded their presence in the region. In the process, they transformed the indigenous landscapes and peoples of what is today the American Southwest.

The story of the Hispanic transformation of southwestern North America has been poorly understood. For Hispanics themselves, the vast region lay "at the ends of the earth . . . remote beyond compare," as one conquistador wrote in 1692. In 1800, a Spanish mining engineer reported to the king that in central Mexico, "the people speak with as much ignorance about the regions immediately to the north as they might about Constantinople." Many Indians also forgot much about the changes that occurred when Hispanics invaded their lands. Over time, for example, as the historian Elizabeth John has noted, the origin of some of the new crops introduced by Spaniards to the Pueblos "faded from memory . . . [and] came to be accepted as 'Indian food,' eaten in the kivas and named in rituals and prayers." Anglo Americans, who began to push into the region toward the mid-nineteenth century, displayed little curiosity about the Hispanics who had preceded them. Blinded by anti-Spanish and anti-Mexican biases, many of the earliest Anglo Americans preferred to imagine the trans-Mississippi West as a virgin land and readily overlooked the region's long Hispanic past.

First Encounters

The transformation of the continent's Spanish-Mexican rim began in the first half of the 1500s, when Iberians introduced Native Americans to the predatory ways of Europeans. The Spaniards who invaded, explored, conquered, and settled much of the Western Hemisphere in the wake of Columbus's discovery arrived with several beliefs that suited them well for these tasks. Their society glorified and rewarded the warrior, whose skills had been hardened in the crucible of a seven-century struggle to drive Muslims out of Iberia—a task completed in 1492. This protracted war had nourished Iberians' zeal and intolerance and bolstered their self-confidence. Like other Christians, Spaniards understood that their god had given them "dominion" over all creatures on the earth, including the newly discovered infidels. Indeed, Spain's onward-moving Christian soldiers crossed the sea with the conviction that the New World belonged to them, since the Spanish pope, Alexander VI, had given it to the Crown of Castile in the famous papal donations of 1493 (the next year, Spain inadvertently granted the eastern edge of South America to Portugal, in the Treaty of Tordesillas).

Spaniards also arrived in the New World with a variety of practical advantages that enabled them to turn their fantasy of superiority into a reality. Peculiar weaponry, including steel swords and explosives, and animals strange to the New World, especially horses and greyhounds, gave Spaniards tactical and psychological advantages that helped them defeat overwhelming numbers of Indians on their native ground. Certain pathogens, foreign to the Western Hemisphere before the 1500s, traveled on Spanish ships and became the microscopic allies of the conquistadores by killing vast numbers of Native Americans, who, unlike Spaniards, lacked immunities. Then too Spaniards arrived as representatives of an emerging state society, with institutions designed to enforce social order—armies, police, and bureaucracies adept at manipulating written symbols. Still fragmented into rival kingdoms, Spain had not yet coalesced as a single nation, but in North America, Iberians found tribal societies far less unified than their own. Like other Europeans in America, Spaniards used Indian disunity to their advantage.

The Spanish reconnaissance of what would become the American West began along the Gulf of Mexico in the same year that Hernán Cortés discovered the Aztec empire. In 1519, in search of a sea lane to Asia, Alonso Alvarez de Pineda set out from Jamaica for the western coast of Florida, where Juan Ponce de León had preceded him, then continued westward to discover the northern shores of the Gulf of Mexico, including the Texas coast. Although Pineda's voyage contributed substantially to Spanish geographical knowledge, it had no direct effect on North American natives, since he did not land on North American shores.

The first Spaniards to set foot in what would become the American West arrived unintentionally, fleeing from a disastrous attempt by Pánfilo de Narváez to conquer and colonize Florida. A storm drove the refugees' makeshift craft onto the Texas coast at or near Galveston Island in the autumn of 1528 as they sailed from Florida toward sanctuary in Mexico. Four men reached Mexico to tell the tale: Alvar Núñez Cabeza de Vaca and three companions, including a black Moorish slave named Esteban. After six years of living among Indians on the Gulf coast, the four had crossed Texas and much of northern Mexico before uniting with fellow Spaniards.

Cabeza de Vaca's lost party contributed little to cartographical knowledge, but his return sparked more purposeful exploration. Despite his apparent reluctance to exaggerate what he saw, or perhaps because of that reluctance, Cabeza de Vaca's reports seemed to substantiate earlier rumors of a wealthy civilization to the north of Mexico and kindled the ambitions of powerful men. Cabeza de Vaca did nothing to dispel these rumors when he hurried to Spain to request a patent to settle Florida. He arrived too late. Hernando de Soto, fresh from enriching himself in the conquest and plunder of Peru, had received royal permission in 1537 to explore and colonize Florida, which then included all of southeastern North America. Two years later, De Soto landed in Florida with a large, costly expedition of nine vessels carrying over six hundred men and assorted livestock.

Meanwhile in Mexico, a private fleet belonging to Cortés had begun to explore Pacific waters to the northwest, and the viceroy of New Spain, Antonio de Mendoza, had laid his own plan to stave off these rivals. In the autumn of 1538, Mendoza put a peripatetic Franciscan, Marcos de Niza, in charge of a reconnaissance of the lands to the north of Mexico, reckoning that a small expedition led by a priest would draw little attention. Within a year Fray Marcos returned to Mexico City, stirring up waves of sensational rumors. He claimed to have seen, but not have entered, a city "bigger than the city of Mexico." The natives, the friar had learned, called this place Cíbola, perhaps from an Opata word meaning Zuni (Spaniards later applied the word *cíbolo* to the curious "cattle" they found on the Great Plains—the buffalo). Fray Marcos described Cíbola as just one of seven cities in a country that appeared to be "the greatest and best of the discoveries"—an extravagant recommendation from a man who knew firsthand the wealth of Mexico and Peru.

On the strength of Fray Marcos's reports, Viceroy Mendoza authorized one of the most elaborate and significant of Spain's reconnaissances of the interior of North America—one to rival De Soto's expedition to Florida. Mendoza, who entertained the idea of leaving the comforts of Mexico City to lead the expedition himself, finally entrusted it to a protégé, thirty-year-old Francisco Vázquez de Coronado. The party included three hundred Spanish adventurers (at least three of them women), six Franciscans, more than one thousand Indian "allies," and some fifteen hundred horses and pack animals. In search of precious metals and great cities, Coronado made his way into the heart of today's Southwest in 1540, establishing a base among Pueblo farmers on the Rio Grande who could supply food, clothing, and shelter. From there, Coronado's band explored parts of Arizona, New Mexico, Texas, Oklahoma, and Kansas. Meanwhile, De Soto had landed at Tampa Bay in 1539 and cut a sanguinary path across eight of the present southern U.S. states. In 1541, after De Soto had crossed the Mississippi into Arkansas and Coronado had marched into Kansas, only some three hundred miles separated the two expeditionary forces.

Neither Coronado nor De Soto succeeded in finding another Mexico or another Peru. In the spring of 1542, Coronado and his entourage retraced their route to Mexico. De Soto's party, far from the vessels that had brought them to Florida, found it more difficult to leave North America. After De Soto took sick and died in the spring of 1542, the remnants of his group tried to make their way back to Mexico by land. Somewhere in eastern Texas they grew disheartened and turned back to the Mississippi. There they built small boats and sailed for Mexico, reaching safety in the autumn of 1543.

Meanwhile, in June 1542, before the results of the Coronado or De Soto expeditions were known, another venture sponsored by Viceroy Mendoza set out for the Pacific coast of North America. Led by Juan Rodríguez Cabrillo, three small ships sailed from the Pacific port of Navidad in 1542 and beat their way up the western coast of Baja California. In 1539, Cortés had sent Francisco de Ulloa to explore that same coast, and Ulloa had made it three-quarters of the way up the peninsula. Beyond that point, Cabrillo's expedition entered waters that no European had seen before. Along the rocky shores of what is today southern California, Cabrillo staked out claims for his sovereign, Carlos I, and mapped the principal harbors—including the ports of San Diego, San Pedro, and Santa Barbara. Cabrillo died while the expedition waited out the winter on Santa Catalina Island, but the next spring his chief pilot, Bartolomé Ferrer, pushed the expedition north to the Rogue River, beyond the present California-Oregon boundary, before turning back.

By 1543, then, the reports of the De Soto, Coronado, and Cabrillo-Ferrer expeditions had opened new horizons to the north of Mexico for Europeans. As a result, the contours of the western half of the continent began to appear on European maps, and Carlos I indulged himself in the common European conceit that discovery gave a claim to lands actually held by a variety of native peoples.

In North America itself, the earliest Spanish expeditions set powerful forces into motion, altering native populations and institutions. European infectious diseases, particularly smallpox and measles, accompanied or preceded Spain's earliest explorers.

If this presidial soldier followed special regulations issued in 1772, he carried 123 pounds of accoutrements in addition to food and water. His weighty equipment included a heavy, knee-length sleeveless coat made of seven layers of buckskin. Designed to deflect Indian arrows, the coat alone weighed 18 pounds. These Spanish soldiers were ambulatory arsenals, able to defend themselves but too burdened down to effectively pursue Indian raiders.

Raymundus à Murillo. Presidial Soldier. *Watercolor on paper, 18th century. Archivo General de Indias (Uniformes 71), Seville, Spain.*

Deadly to natives, who had no previous exposure and, therefore, no immunities, these diseases took a heavy toll in certain times and places. Cabeza de Vaca reported that shortly after he and his companions arrived ill on an island off the coast of Texas, "half the natives died from a disease of the bowels and blamed us." Neither extant documents nor present archaeological evidence permits informed appraisals of the extent of decline in native populations during the early sixteenth century in what is now the West. Judging from the better-documented demographic collapses that occurred elsewhere in the hemisphere, however, it seems likely that some native communities suffered heavy losses. Wherever those occurred, societal transformations followed. With diminished populations, for example, some native groups must have found themselves suddenly weak in comparison with their neighbors and, in extreme cases, forced to abandon their communities. In the most complex societies, greatly reduced populations would not support specialized functions, undoubtedly causing some simplification.

Along with deadly microbes, the earliest Spanish expeditions also left a legacy of ill-will. Coronado's men, for example, although vastly outnumbered, had taken advantage of firearms and hard steel to seize by force what they could not gain by persuasion. Pueblo Indians who refused to cooperate learned the cost of resistance. Word of Spanish atrocities spread quickly and far. At San Diego in September 1542, Cabrillo met surprising hostility from Ipai Indians who, he learned, were alarmed because "men like us were traveling about, bearded, clothed and armed . . . killing many native Indians, and . . . for this reason they were afraid."

The violent aggression of the earliest Spanish explorers affected Spanish-Indian relations for generations thereafter. Indeed, late in the nineteenth century, Pueblos at Zuni, whose towns Coronado had seized by force, reportedly still remembered that the first Spaniards had worn "coats of iron, and warbonnets of metal, and carried for weapons short canes that spit fire and made thunder." They recalled, "These black, curl-bearded people drove our ancients about like slave creatures."

Spanish exploration and the mutual discovery that it engendered did not end, of course, in 1543. Exploration, and its intentional and unintentional consequences, continued as long as Spain and Mexico held the region. For several decades after 1543, however, Spanish exploration of western North America halted. The new vistas to the north had revealed few tangible attractions—no rich civilizations and no strait through the continent to carry Spanish mariners toward the riches of Asia. For the time being, Spain's American colonies developed around the mineral- and labor-rich realms of the vanquished Incas and the Aztecs, leaving western North America beyond the periphery of the Spanish empire.

Settlers, Friars, and Pueblos in Seventeenth-Century New Mexico

Late in the sixteenth century, as memories of previous exploration dimmed and fresh rumors of another Mexico to the north rekindled their interest, Spaniards returned to New Mexico. This time they stayed. New Mexico became the first permanent European colony in what is today the American West and the second enduring European settlement in what is now the United States (preceded only by St. Augustine, founded by Spain in Florida in 1565).

New Mexico's colonizer, Juan de Oñate, led a small contingent north from Mexico in 1598. Some of his 130 soldier-settlers traveled with their wives, children, and servants, bringing the expedition's number to over 500. All were Spanish subjects, but few were born of Spanish parents. Most of the Spaniards who settled in northern New Spain over the next two centuries were mixed-bloods (mestizos and mulattoes), and some were Mexican Indians and blacks. After pausing at what later became El Paso, where Oñate proclaimed Spanish dominion over the new land and its inhabitants, "from the leaves of the trees in the forests to the stones and sands of the river," the party moved up the Rio Grande, and across the great dry stretch that would come to be called the Jornada del Muerto, into the heartland of the Pueblo Indians. North of present-day Santa Fe, Oñate made his headquarters at the Tewa-speaking pueblo of *ohke.* He declared the pueblo a Spanish town, renamed it for San Juan, and forced the king's new vassals out of their apartments. He kept the natives close at hand, however, to provide labor, food, and clothing.

The Pueblos probably acceded to Oñate's demands to avoid the deaths and damage that Coronado, as well as more recent expeditions by Antonio de Espejo and Gaspar Castaño de Sosa, had inflicted on the Indians when they failed to offer hospitality to Spaniards. Oñate reinforced the lesson. Before his colony was a year old, Pueblos at Acoma resisted Spanish requisitions of food and clothing and killed eleven Spanish soldiers in a single surprise attack. Oñate retaliated swiftly and audaciously, sending a small force against the almost impregnable mesa-top pueblo, where Acomans enjoyed the numerical as well as logistical advantage. Oñate's men succeeded brilliantly. They destroyed Acoma, killing some five hundred men and three hundred women and children and taking perhaps eighty men and five hundred women and children captive. Tried and found guilty of murder, the adult Acoman prisoners received a punishment calculated to make them living reminders of the cost of resistance. Oñate sentenced all captives between the ages of twelve and twenty-five to twenty years of personal servitude, and he condemned males older than twenty-five to have one foot severed. The mutilations, a common punishment for miscreants in Renaissance Europe, were carried out in public.

Like Coronado, De Soto, and other leaders of legitimate Spanish expeditions, Oñate had entered North America as a private contractor, under an arrangement with the Crown that permitted him to settle New Mexico at his own expense. He planned to realize a handsome return from his investment by building a transcontinental empire that would include rich mines, abundant Indian labor, a strait through North America, and seaports on both oceans. Like many of his contemporaries, however, Oñate had underestimated the distance across the continent, and his luck was no better than Coronado's.

Exploring west to the Gulf of California and northeast to the edges of Kansas, Oñate found no wealthy Indians or mines, much less a transcontinental strait. The colonists and the king quickly grew disillusioned and prepared to abandon New Mexico, but Franciscans had baptized so many Pueblo Indians that the project received a royal reprieve. Rather than move the baptized Pueblos south, or see them revert to paganism, Felipe III permitted Franciscans to stay.

For much of the seventeenth century, New Mexico endured largely as a missionary outpost. Isolated over eight hundred miles beyond the edges of Mexico's mining frontier, it attracted few immigrants. The number of Spaniards in New Mexico in the 1600s probably never exceeded three thousand. This scanty population could support only one formal municipality, Santa Fe, established in about 1610 by Oñate's successor, Pedro de Peralta, at an altitude of seven thousand feet amidst piñon and juniper at the foot of the Sangre de Cristo Mountains. No military garrison or presidio existed in New Mexico in the 1600s, but the king expected his vassals to serve as soldiers as well as settlers—particularly those who had received a specified number of natives in trust, or in *encomienda*. Rewarded by Oñate and his successors according to their rank and services to the Crown, *encomenderos*, or holders of *encomiendas*, enjoyed the privilege of receiving tribute from one or more pueblos or fractions of pueblos.

Many of New Mexico's pioneers lived outside of the province's single urban center in clusters of fortified farmhouses or ranches too small to call towns. To be close to the province's single source of revenue, the labor of Pueblo Indians, the Hispanic population scattered up and down the Rio Grande, from Taos Pueblo to below Albuquerque (founded in 1706). Farther south along the Rio Grande, where Oñate had first entered New Mexico, a mission was built in 1659 at present-day Ciudad Juárez, across the river from today's El Paso, Texas, but no civilian community developed there until the 1680s.

In contrast to the slow growth of the civilian population, the number of missions in New Mexico expanded rapidly. By 1629, according to the enthusiastic count of Fray Alonso de Benavides, Franciscans had already overseen the construction of fifty churches and residences for priests in New Mexico. Central to Spain's enterprises in America, missions had become the dominant institution on the frontiers of the Spanish empire by the seventeenth century. Comprehensive Orders for New Discoveries, promulgated by Felipe II in 1573, had made missionaries the Crown's primary agents for exploration and pacification. "Preaching the holy gospel," Felipe II had noted, "is the principal

The work of mission building in New Mexico advanced in the 1620s, Franciscans said, due to the appearance of a Spanish mystic and nun, María de Jesús de Agreda, who journeyed miraculously to America and persuaded Indians to seek out missionaries. Although the nun later repudiated the story, it served as an inspiration to Franciscans in Texas and California in the 18th century.

Unidentified artist. María de Jesús de Agreda Preaching to Chichimecos of New Mexico. *Woodblock print, 1730. From the first printing of a letter that Fray Alonso de Benavides wrote from Madrid in 1631 to his fellow Franciscans in New Mexico,* Tanto que se sacó de una carta . . . *(Mexico, 1730).*

purpose for which we order new discoveries and settlements to be made." The regulations also prohibited the entry of unlicensed parties into new lands and prohibited the use of the word *conquest* to describe future "pacifications." Those habits proved impossible to break, for there remained an inherent tension between the Crown's imperative to save souls and its interest in exploiting new lands and peoples.

If the age of military conquest had ended, spiritual and cultural conquest had not. Through interpreters, Oñate had spelled out the new Spanish agenda when he ordered Pueblos to accept Spanish rule in 1598. Submission, Oñate said, would bring peace, justice, protection from enemies, and the benefits of new crops, livestock, and trade. Obedience to the Catholic church would bring even greater rewards: "an eternal life of great bliss" instead of "cruel and everlasting torment."

Missionaries embarked on a program to eradicate Pueblo religion and replace it with Catholicism. Using techniques they had honed over time in other areas, Franciscans, many of them skillful and zealous, persuaded Pueblos to build churches and to receive baptisms and religious instruction. In some cases, the friars probably succeeded in bringing about complete conversions; in most cases, it appears, Pueblos simply adopted those features of Catholicism that they found congenial while maintaining their traditional spiritual beliefs.

Many Spanish priests, like their English counterparts, could not imagine that a people could become Christians unless they lived like Europeans. Thus, in the ideal mission, Franciscans sought to reshape the natives' temporal lives by teaching them to dress, eat, and live like town-dwelling Spaniards. Depending on climate, soils, and local needs, Franciscans in New Mexico taught native converts the Spanish trades: to husband such European domestic animals as horses, cattle, sheep, goats, pigs, and chickens; to cultivate European crops, from watermelon to wheat; to raise European fruit trees, from peaches to plums; to use such iron tools as wheels, saws, chisels, planes, nails, and spikes; and to practice those arts and crafts that Spaniards regarded as essential for the civilization they knew. The Franciscan program for the cultural transformation of these tribal societies enjoyed the financial support of a large state apparatus, presided over by the Spanish Crown, which saw missions as a practical and pious way to advance the frontier and to increase the number of Hispanicized subjects without having to export more Spaniards.

Franciscans accomplished some of their purposes through persuasion, gifts, and the promise of spiritual rewards, but they also resorted to force and the threat of force (for Pueblos, for example, memories of Spanish atrocities at Acoma and elsewhere must have lingered). Schooled in a time and place where the good of the community prevailed over the rights of the individual, Franciscans justified the use of force as beneficial to the common good. Thus, once natives consented to receive baptism, Franciscans commonly relied on military force to prevent converts from slipping back into apostasy—a danger to the common weal as well as to the individual soul. Soldiers aided Franciscans in compelling baptized Indians to live in mission communities as Spanish law required, hunted down neophytes who fled, and administered corporal punishment to natives who continued religious practices that Spanish priests found loathsome. From the Spaniards' viewpoint (but certainly not the Indians'), whipping seemed a particularly appropriate punishment in an era when the lash was applied to Spanish miscreants, from schoolchildren to soldiers, and when Franciscans whipped themselves in atonement for their sins. As they had done since the earliest stages of the conquest of America, the Franciscans smashed, burned, or confiscated objects sacred to the natives—what one friar in New Mexico described as "idols, offerings, masks, and other things of the kind which the Indians were accustomed to use in their heathenism."

Believing they had something to gain by accepting the new religion and the material benefits that accompanied it, or too much to lose by resisting it, Pueblos initially cooperated. Like other native societies that had not been vitiated by war or disease, Pueblos adopted from the outsiders what they perceived as both useful for and compatible with their essential values and institutions. Ideally, they sought to add the new without discarding the old, or to replace elements in their culture with parallel elements from the new—as they had done long before the arrival of Europeans. In the religious sphere, for example, many Pueblos simply added Jesus, Mary, and Christian saints to their rich pantheons and welcomed the Franciscans into their communities as additional shamans. However selective neophytes might have been in adopting aspects of Christianity and Spanish culture, their decision to accept missionaries began to transform their cultures—often in ways that neither they nor the missionaries intended. By cultivating certain European crops and raising European domestic animals, for example, natives often enriched their diet, lengthened the growing season, deemphasized hunting in favor of agriculture, and made it possible for their villages to support denser populations. Their prosperity, on the other hand, also made them more attractive targets for raids by nomadic Indians and forced them to devote more resources to defense.

For the Pueblos, the bright future that Franciscans offered at the outset of the courtship lost its luster as the terms of exchange shifted. Along with gifts and access to trade goods came demands for labor and resources. For individual Indians, those demands increased as the size of their communities, their land base, and their productivity declined. Obedience to the Franciscans and their god had not stopped the spread of diseases strange to the natives. From 1598 to 1680, the Pueblo population fell by at least half, to some seventeen thousand.

Violent eruptions, which Spaniards characterized as rebellions but which Indians probably saw as armed struggles for freedom, broke out in New Mexico on a number of occasions. Spaniards suppressed them, sometimes brutally. In 1680, at a time of unusual stress precipitated by drought and starvation and aggravated by Apache raids and Spanish religious persecution, Pueblos revolted with near unanimity. In their carefully planned offensive, the Pueblos caught the Spaniards off guard. Reports of a plot had reached the Spaniards, but they could not have guessed the magnitude of this unprecedented campaign, which involved over two dozen independent towns spread out over several hundred miles and separated by at least six different languages and numerous dialects, many of them mutually unintelligible. Within two weeks, Pueblos had cleansed New Mexico of Spaniards. The natives had killed over four hundred of the province's twenty-five hundred foreigners (nearly all in the initial days of the rebellion), destroyed or sacked every Spanish building, desecrated churches and sacred objects, killed twenty-one of the province's thirty-three missionaries, and laid waste to the Spaniards' fields. The survivors and some Pueblos who remained loyal to them fled south to El Paso. There could be no mistaking the deep animosity that natives, men as well as women, held toward their former oppressors. "The heathen," wrote one Spanish officer in New Mexico, "have conceived a mortal hatred for our holy faith and enmity for the Spanish nation."

Some Pueblo leaders urged an end to all things Spanish as well as Christian and counseled against speaking Castilian or planting crops introduced by the Europeans. This nativistic resurgence succeeded only partially in reversing the cultural transformation that Spaniards had set in motion. Some reminders of Spanish rule, such as forms and motifs in pottery, seem to have disappeared, but Pueblos continued to raise Spanish-introduced livestock and to make woolen textiles. Just as they had selectively adapted certain aspects of Hispanic culture, so too did they selectively reject them.

Pueblos had carried out one of the most successful Indian rebellions against Spanish colonizers anywhere in the hemisphere, but only briefly did they enjoy freedom from

Living far from the world of portrait painters, few of the individuals who settled in New Spain left visual images of themselves for posterity. Among the exceptions was Diego de Vargas, the reconquerer of New Mexico. Before he set out for the Indies in 1672 at age 29, leaving his wife and four children behind in Madrid, the fashionable nobleman posed for this portrait, still preserved in a private chapel in Madrid.

Unidentified artist. Diego de Vargas. *Oil on canvas, ca. 1672. Capilla de San Isidro, Madrid. Photograph courtesy J. Manuel Espinosa.*

Christian strictures and obligations of labor and tribute. In 1692, Diego de Vargas began a slow and shrewd reconquest of the Pueblos—a task he completed after smashing another bloody Pueblo rebellion in 1696.

Exhausted from war, their property and population diminished, Pueblos did not launch another major offensive while under Spanish rule. To the contrary, threatened by Apaches and other common enemies, Pueblos became loosely allied with Spaniards and fought with them to defend the province. Spaniards, in turn, feared another rebellion and did not offer as much provocation. After the 1696 revolt, pragmatic

Franciscans displayed less zeal in attempting to stamp out Pueblo religious practices, and colonists and officials eased (but did not end) their demands on Pueblo laborers.

This cycle of initial acceptance, growing tension, rebellion or attempted rebellion, and mutual accommodation—the native-Spaniard relationship that manifested itself in seventeenth-century New Mexico—reoccurred in the eighteenth century as Spanish missionaries attempted conversions among other sedentary peoples in far northern New Spain, from the Caddos of East Texas to the Pimas of southern Arizona and to the coastal tribes of Alta California.

Defensive Expansion: Texas and Louisiana

Throughout most of the seventeenth century, New Mexico stood alone on the northern fringes of New Spain, the only permanent Spanish settlement in what is now the western United States. Only temporarily had Franciscans extended the province westward into what is today Arizona, to the land of the Hopis in the high, arid northeast corner of the state. But Hopis had expelled the Franciscans in the 1680 revolt and maintained their independence until the end of the Spanish era. Not until 1700 did Jesuits, building a chain of missions along the Pacific slope of New Spain, cross the future U.S.-Mexico border and establish an ongoing Hispanic presence in present-day southern Arizona. In the parched, cactus-covered desert of southern Arizona, which Spaniards called the land of the Upper Pimas, or Pimería Alta, Jesuits altered the lives of Pima farmers much as Franciscans had effected changes in the Pueblo world. After Spain expelled the Jesuits from all of its territories in 1767, Franciscans carried on their work in Pimería Alta. Meanwhile, Spanish Arizona had attracted few colonists. A modest influx of miners and ranchers began in the 1730s but ended with the bloody Pima rebellion of 1751—a widespread nativistic reaction to Spanish intrusion. As a part of Sonora rather than a separate province, Hispanic Arizona remained confined largely to the Santa Cruz Valley between the present-day border and Tucson. Its non-Indian population seldom exceeded one thousand because it had little to offer would-be colonists and, for much of the century, because it held no strategic importance to attract the attention of government officials.

By the late seventeenth century, defense rather than conversion had become Spain's foremost concern in western North America—notwithstanding the Crown's frequent assertions to the contrary. Indeed, with the exception of the Jesuit advance into Pimería Alta, geopolitics more than religious considerations lay behind all Spanish expansion into North America from the 1690s on. Even the decision of Carlos II to support Diego de Vargas's reconquest of the Pueblos was based in large part on the strategic value of the province in the face of perceived threats to northern New Spain by Frenchmen and Indians.

Beginning in the late 1600s, a growing French presence in the Gulf of Mexico and in the upper reaches of the Mississippi Valley ended Spanish hegemony on the western half of the continent and weakened Spain's ability to control Indians. French traders broke the Spanish monopoly by offering an alternative source of European trade goods and a market for furs while making no demands that Indians change their culture or religion. Moreover, French trade goods, lower priced and better made than Spanish merchandise, included arms and ammunition. In contrast, based on the belief that firearms made natives more effective adversaries, Spanish law had prohibited furnishing guns to Indians. Its enforcement would leave Spaniards at a disadvantage.

Even before French traders appeared on the eastern horizon, New Mexico was under siege by Apaches and other tribes whose cultures had been transformed by Spanish-introduced horses. (Spain had tried, but failed, to keep Indians from acquiring horses.) Mounted and highly mobile, nomadic and seminomadic tribes from the plains raided New Mexico settlements, including those of the Hispanicized Pueblos, and threatened the mining regions of northern New Spain. In the early eighteenth century, southward-moving Comanches, some armed as well as mounted, pushed Apaches farther south and west. As Comanches took possession of the high plains, they alternately raided and traded in New Mexico and Texas.

Initially, Spanish policymakers faced the growing French and Indian threats in Texas, where they relied on an antiquated solution to a new problem. Galvanized by La Salle's bold effort to plant a colony on the Texas coast in 1685, and disturbed by the penetration of the lower Mississippi Valley by Canadian-based French traders, Spain sent a small detachment of soldiers and Franciscans into Texas in 1690. Spanish officials hoped to continue the tradition of advancing and defending frontiers with peaceable and inexpensive missions. Thus, they bypassed the strategic Gulf coast, with its intractable hunting and gathering tribes, and targeted Caddo agriculturalists, the most sophisticated peoples in Texas, as having the greatest potential to become Christians and allies. In the rolling piney woodlands of the "Kingdom of Tejas" in what is now East Texas, at a site some six hundred miles from the Rio Grande over a difficult land route and inaccessible by sea, Father Damián Massanet founded two missions among the Tejas, or Hasinai Indians—the westernmost Caddo confederacy. Spain, however, failed to impose an ecclesiastical and cultural solution to a defensive and commercial problem. Decimated by European diseases, aggrieved by Christian arrogance, and uncompensated by trade goods, the Hasinai forced the defenseless and hopelessly isolated Spanish missionaries to leave three years later, in 1693.

Thereafter, as the French threat to Texas and the Gulf coast seemed to subside, so did the interest of an overextended and bankrupt Spain. In 1716, however, the specter of widening commercial influence from French Louisiana (which had grown slowly from shaky beginnings at Biloxi Bay in 1699 and Mobile Bay in 1702) led Spain again to try to settle Texas. This time the Crown succeeded. Returning to the Kingdom of Tejas, a small party of about seventy-five—soldiers, Franciscans, and colonists, including women—built four small wooden churches and a presidio in the summer of 1716. While that work proceeded, the expedition's leader, Captain Domingo Ramón, continued east, beyond the Hasinai lands, to Natchitoches, the westernmost French trading post in Louisiana. Evoking the papal donation of 1493, Spain officially denied that France had a legal right to be in Louisiana, but Captain Ramón tacitly acknowledged Natchitoches as the western limit of the French colony, and he took measures to check its growth. Just to the west of Natchitoches, he founded two more missions in nearby Caddo communities: San Miguel de los Adaes and Dolores de los Ais.

Undermanned and remote from the nearest Spanish settlements and sources of supplies, the fledgling colony represented little more than a symbol of Spain's interest in maintaining Texas. The weakness of Spain's position became painfully evident when war broke out between France and Spain in 1719. Seven Frenchmen invaded Texas from Louisiana and managed to panic the colonists, soldiers, and missionaries into abandoning the Caddo country. They fled to San Antonio, a civil, military, and ecclesiastical complex founded the year before as a way station on the trail between the Rio Grande and the East Texas missions.

The French "invasion" shocked Spanish officials into reinforcing Texas. In 1721 the wealthy marqués de San Miguel de Aguayo recaptured eastern Texas with the most imposing force Spain would send into the province—some five hundred men, four thousand horses, and assorted other livestock. Aguayo bolstered the presidios at San Antonio and East Texas and built two more. One rose at the very site of La Salle's failed enterprise on Matagorda Bay and another at Los Adaes, just twelve miles from Natchitoches. The post at Los Adaes, where Aguayo left a one-hundred-man garrison and six cannon, became the capital of Texas, a position that it held while the French flag flew over Louisiana.

If Texas succeeded in checking further French expansion, it was largely because of the corresponding weakness of Louisiana. Texas itself stagnated. Under Spain, it failed to develop significantly beyond the points reinforced by the marqués de Aguayo. The mission system, effective in promoting the expansion of earlier Hispanic frontiers, failed across much of Texas. The Hasinai, who recalled prior experience with Christians, declined to receive baptism. "They have formed the belief," one Franciscan wrote, "that the [holy] water kills them." Thanks to French traders, the Hasinai, together with mounted nomadic or seminomadic tribes such as Comanches and Apaches, had better options; they could not be induced to embrace the restricted lives of neophytes in Spanish missions. Only along the San Antonio River did missionaries have modest success at recruiting. San Antonio de Valero and the four other stone-walled missions that developed nearby between 1720 and 1731 provided refuge for small bands of linguistically diverse hunters and gatherers, who found themselves squeezed between Hispanics moving northward into Coahuila and Tamaulipas and Apaches moving southward. At their peak in mid-century, the five San Antonio missions were home to about a thousand neophytes, many of them Coahuiltecan-speaking peoples. A few other missions along the coastal plain attracted modest numbers, but none achieved the potential that the dense native population of Texas seemed to offer.

Texas also failed to attract a significant number of immigrants. In part, it lacked a large pool of tractable Indian laborers, and it suffered, as did all of Spain's frontier provinces, from remote markets and from restrictive economic policies that stifled commerce. In normal times, Spanish mercantilism limited legal trade to a few key ports in the New World that were to be supplied only by Spanish goods carried on Spanish vessels. None of the natural harbors along the Texas coast were among these ports. Fearful of encouraging smuggling, Spain kept the Texas coast closed to shipping until the end of the colonial period, despite occasional entreaties from Texas officials. These commercial restrictions jacked up transportation costs for exports and imports while stifling consumption and production across the northern frontier.

Notwithstanding the relative lack of economic opportunity, Texas attracted some immigrants because the Spanish government, eager for bodies to man the barricades against French expansion, offered incentives. Giving free passage, land, and the title of hidalgo, officials managed to sell Texas to a small number of poor farmers from Tenerife in the economically depressed and overpopulated Canary Islands. Fifteen families (fifty-five persons) arrived in San Antonio in 1731 after a year-long journey by way of Havana, Veracruz, and Saltillo. Together with the colonists who had come to Texas with earlier expeditions, the Canary Islanders brought the Hispanic population of the province to perhaps 500. By 1763, Hispanics in Texas—not counting Hispanicized mission Indians—numbered 1,850, the sparsest Hispanic population of any province in New Spain. Nonetheless, members of the small but resourceful *tejano* community managed to develop a ranching economy adapted to the harsh conditions of life on the frontier. They also found ways to circumvent trade restrictions by trading illegally with Frenchmen in Louisiana.

Spanish Texas endured primarily as a buffer colony, controlled largely by Indians and sparsely populated by Hispanics. When it no longer served imperial purposes, Spain nearly abandoned the province. After Spain acquired Louisiana from France in 1762, it withdrew its missionaries, soldiers, and settlers from East Texas by royal order and relocated them at San Antonio, which, together with nearby La Bahía, was too heavily populated to shut down. Nonetheless, *tejanos* soon drifted back to the edge of Louisiana, where they founded Nacogdoches in 1779. The new town, a center for contraband, rivaled San Antonio through the end of the colonial period.

Just as Spain expanded into Texas in response to a foreign threat, so it acquired western Louisiana for purely defensive reasons. Louisiana promised to become a financial liability for Spain, as it had been for France, but Spain needed it. An immense province, western Louisiana stretched from the mouth of the Mississippi into the Illinois country, with a vague and seemingly limitless boundary to the west. If Louisiana fell into English hands, Spain would face its most powerful imperial rival on the doorstep of New Spain. On the other hand, if Spain occupied Louisiana, it could hold England at arm's length. As one of the king's advisers argued, the Mississippi River would form a "recognizable barrier, a good distance from the population centers of New Mexico." After weighing costs and benefits, Carlos III reluctantly accepted the expensive gift of Louisiana from his French cousin Louis XV in a secret treaty signed in 1762.

After a dilatory and stormy beginning, Spain asserted its sovereignty over Louisiana's French population. The Crown built fortifications and encouraged immigration, as it had in Texas, but it did not attempt the large-scale conversion of French-influenced Indians. Compared with residents of Texas or New Mexico, Louisianians enjoyed economic prosperity under Spain, perhaps because Spanish officials adopted France's more liberal trading policies, opened Louisiana to direct trade with Spanish ports, and welcomed non-Spanish immigrants. On the other hand, even without these policy changes, a growing contraband trade with neighboring Anglo Americans had invigorated Louisiana's economy. Nonetheless, although Louisiana's colonists prospered more under Spain than they had under France, the province remained a net liability for the Spanish Crown.

Spain held title to Louisiana for nearly four decades but failed to Hispanicize it, much less profit from it. Numerically, Frenchmen simply overwhelmed Spaniards. Through incentives, Spanish officials managed to lure Canary Islanders to Louisiana, as they had to Texas. Beginning in the late 1770s, some two thousand Canary Islanders reached Louisiana, along with a trickle of immigrants from other parts of Spain. A hundred or so colonists from Málaga, for example, founded New Iberia in 1779. The number of Spanish immigrants, however, never exceeded the fifty-seven hundred Frenchmen who lived in Louisiana in 1766, when the first Spanish governor arrived to take possession of the province. Moreover, the number of Frenchmen increased dramatically under the Spanish regime as French residents of lands to the east of the Mississippi fled British rule. In addition, Acadian refugees from Nova Scotia swelled the French population of Louisiana by perhaps three thousand.

By 1800, Spaniards in Louisiana were outnumbered not only by Frenchmen but also by Anglo Americans. Eager to regain the Floridas, which it had lost to Britain in 1763, Spain had joined the Americans in 1779 in their rebellion against England and had driven British forces out of the lower Mississippi, Mobile, and Pensacola. The United States, however, with a burgeoning population and manifest designs on the trans-Mississippi West, proved more dangerous than Britain. As early as the 1780s, Spanish officials began to regard Americans as analogous to the "barbarians" who had swept into the Roman empire centuries before. To halt Anglo Americans who swarmed into Louisiana illegally, Spain permitted some to settle legitimately. The idea—that the settlers would become loyal Spanish subjects—soon seemed ill conceived. Unable to limit the number of westward-moving Americans or to ensure their loyalty, a weakened Spain yielded to pressure from Napoleon Bonaparte to return Louisiana to France. Spain still regarded Louisiana as the key to the defense of northern New Spain but would henceforth depend on France to hold the line against the Americans. Thus, the lengthy Franco-Spanish negotiations, finally completed in 1802, proceeded in secret for fear that the United States would invade Louisiana to prevent the transfer. The invasion never came. Breaking his agreement with Spain, Napoleon sold Louisiana to the United States the following year.

When Louisiana passed into the hands of the Americans in 1803, demographics and longevity had combined to ensure that French culture still predominated. Many descendants of old Spanish families stayed in American Louisiana, but in the popular imagination they became culturally indistinct from the more numerous Frenchmen, and their era was nearly forgotten. New Orleans was rebuilt in Spanish style after fires obliterated the French-built city in 1788 and 1794, but the Spanish-built heart of New Orleans remains known today as the French Quarter.

War and Peace

Even though Spain failed to stop the flow of Anglo Americans into Louisiana in the late eighteenth century, it did succeed in coming to terms with Apaches, Comanches, and other Indian nomadic tribes who had long plundered northern New Spain. With their lands threatened by the continuing expansion of the Spanish frontier and with their liberty at risk from slaving parties and missionaries, some Indian raiders had powerful motives besides loot. They fought Spaniards bitterly and successfully, initially meeting little effective resistance.

In the seventeenth century, Spain had no presidios west of the Mississippi, but the destruction of New Mexico in 1680, the Indian rebellions that rippled across northern Mexico, and the growing threat from foreigners had led Spain to increase its professional military forces on the northern fringes of its empire. Friars with crosses and small military escorts no longer filled the bill. In New Mexico, authorities constructed a presidio at El Paso del Norte in 1681, after survivors of the Pueblo Revolt had taken refuge there. As Diego de Vargas reasserted Spanish control over the Pueblo country in 1693, he began construction of a presidio in Santa Fe, authorized by the viceroy. In 1716, when Spaniards returned to Texas to counter French influence, they built presidios to protect every population center—a total of five by mid-century. In Pimería Alta, too, where threat from Apaches grew, Spain constructed a number of presidios, with Tubac (1752), Tucson (1776), and Terrenate (1776) all to the north of the present border.

As the burden of defense in these provinces shifted to presidial troops paid by the king, economic power increasingly resided with the military and its access to the government payroll. Slowly, the military supplanted the missions as the dominant institution on the Spanish rim in the eighteenth century, and soldiers and their families became the mainstay of many communities. Despite its strong presence, however, the military failed to end the chronic insecurity that characterized the lives of Hispanic frontiersmen. Underfunded and headed by an officer staff riddled with corruption, the frontier fortifications were a model of inefficiency for much of the century. Presidial soldiers—cheated of their pay by their officers, vastly outnumbered by their adversaries, and ill equipped and badly trained for either conventional or guerrilla warfare—soon became demoralized, seldom winning a victory. Nonetheless, the presidios themselves enjoyed a reputation as secure places of refuge from hostile Indians. That, however, had less to do with the effectiveness of the soldiery or the strength of the structures than Indians' reluctance to incur losses by laying siege to a fortification.

In the late eighteenth century, Spain ended this dreary chapter in its relations with hostile tribes by making two fundamental changes: it reorganized the administrative structure of the frontier provinces, and it took a fresh approach to its treatment of Indian adversaries. The administrative restructuring aimed to make the presidial soldiers more efficient, to coordinate military campaigns, and to foster immigration and economic development. These activities, the Crown recognized, required closer supervision than a distant viceroy could give. In 1776, four years after restructuring the military command, Carlos III granted semiautonomous status to the northern region of New Spain, long known as the interior provinces. This new administrative entity, the Comandancia General de las Provincias Internas, included much of the present-day northern tier of Mexican states—then known as the provinces of Baja California, Sonora, Sinaloa, Nueva Vizcaya, and Coahuila—together with today's American Southwest, from California through Texas (Louisiana fell within the general jurisdiction of the viceroy of New Spain but came under the immediate supervision of Havana and was never regarded as one of the interior provinces of New Spain). At the head of the Comandancia General was a military officer with the title of *comandante general*, or commander in chief, who had authority over governors of individual provinces and individual military posts, thus enabling Spain to construct a comprehensive military strategy and execute interprovincial campaigns.

In various permutations, the Comandancia General remained in place for the rest of the colonial era and achieved some successes. The first commander in chief, for example, Teodoro de Croix (1776–83), found imaginative ways to increase the flexibility and effectiveness of the frontier military against guerrilla forces. Croix, like most frontier commanders, hoped to win peace through force of arms, but success came only when Spain imitated its French and English competitors, put diplomacy ahead of warfare, and took Indian interests into account.

The individual most responsible for this shift, Bernardo de Gálvez, had experience with war on the Apache frontier and peace in Louisiana. In 1769, as a well-connected twenty-two-year-old officer assigned to Chihuahua, Gálvez had acquired a respectful understanding of Apaches—along with serious wounds from a lance and arrow. In 1776, as a youthful governor of Louisiana, Gálvez observed firsthand how the French and English used trade rather than war to maintain harmonious relations with natives. He noted that French and English traders, by making Indians dependent on them for "sundry conveniences," including guns and ammunition, had caused Indians to forget the use of the bow and arrow and to rely entirely on Europeans for powder to hunt and to defend themselves.

Other Spanish officers had made similar observations, but as the nephew and protégé of José de Gálvez, the powerful and dynamic minister of the Indies in charge of American policy, Bernardo de Gálvez had a unique opportunity to influence Spanish policy. In 1778 Bernardo de Gálvez urged his uncle to adopt the French-English model for the northern frontier of New Spain. Although he had no illusions that trade would rapidly alter Indian cultures, he thought the approach better than a costly, ineffective, and unwinnable war. Through trade, he argued, "the King would keep [Indians] very contented for ten years with what he now spends in one year in making war upon them."

Bernardo de Gálvez's cost-cutting recommendation met a ready reception in Madrid as Spain prepared to enter the war against England on the side of the thirteen rebellious colonies. In 1779, José de Gálvez instructed the commander in chief of the interior provinces, Teodoro de Croix, to halt plans for a large-scale offensive and to seek alliances based on trade and gifts, including firearms. Prompted by the fiscal stringencies of wartime, Gálvez's instructions of 1779 seemed temporary. Instead, they became the foundation of an enduring Indian policy when, seven years later, José de Gálvez indulged his penchant for nepotism by appointing his nephew viceroy of New Spain. As viceroy, Bernardo de Gálvez prepared a detailed exposition, his well-known *Instructions of 1786*, which reconciled conflicting practices with a three-pronged approach. First, he urged the maintenance of military pressure on Indians, to the point of exterminating Apaches if necessary. Second, he endorsed the building of alliances, such as those that Spaniards had long enjoyed with Pueblos and other tribes. "The vanquishment of the heathen," he coldly noted, "consists in obliging them to destroy one another." Third, he argued for the extensive use of trade and gifts to make Indians who sought peace dependent on Spaniards.

In his explanation of how to increase Indian dependency on Spaniards, a policy that he termed "peace by deceit," Bernardo de Gálvez surpassed his uncle in imagination and cynicism. He urged that Indians be furnished with firearms and ammunition, but he specified that guns be made of poorly tempered metal with long barrels that would make them awkward to use and easy to break. Natives, then, would depend on Spaniards for repairs or replacements. As to ammunition, Gálvez believed that Indians should be given an abundance of it. The more they used powder and shot, the less they would use arrows. Soon, they would "begin to lose their skill in handling the bow," which Gálvez correctly understood to be a more effective weapon than the firearm. In short, Gálvez planned to use tried-and-true English and French practices to destroy the basis of native culture as the first step toward turning nomadic Indians into Spaniards. By this means, the military might succeed in bringing about the cultural transformations that missionaries had failed to achieve through less violent and less cynical means.

With various modifications and embellishments, the *Instructions of 1786* governed Spanish-Indian relations on the northern frontier for the remainder of the colonial period. Spain, strapped for funds and arms for its own army, lacked sufficient resources to buy a peace entirely, and there is no evidence that Spanish agents provided significant amounts of alcohol to Indians. But Gálvez's *Instructions* did establish clear rules under which some of Spain's ablest officers could play a new game that included gifts, access to trade fairs, cooperation against mutual enemies, and more equitable and consistent treatment than Indian belligerents had received in the past. Those Indians who had resisted Spanish domination and military pressure had, in effect, forced Spanish leaders to make these concessions.

Thus, conciliation and negotiation, previously subordinate to force, became the cornerstone of a new Spanish policy in the interior provinces. Coupled with military pressure, this policy achieved remarkable results—even before Bernardo de Gálvez reformulated it in 1786. One of the most notable successes occurred in New Mexico under the leadership of Juan Bautista de Anza, a third-generation presidial officer whose father had been killed by Apaches on the Sonora frontier. As governor of New Mexico from 1778 to 1787, Anza won an enduring peace with western Comanches, who had been the scourge of the province since mid-century. Once he had secured peace with Comanches, Anza went on to lay the foundations for an alliance with Navajos, who were soon persuaded to turn on their former allies, the Gileño Apaches. In the face of such evident shifts in the balance of power, Gileños began to sue for peace and its attendant benefits. Similar successes occurred all across the northern frontier, and in many locales Apaches settled down to lives as farmers and ranchers. Near some presidios, at what Spaniards called "peace establishments," soldiers began to distribute goods and instruct Indians in the ways of Spaniards.

This turning point in Spanish-Indian relations occurred during the administration of the experienced and exceptionally able Jacobo de Ugarte, who served as commander in chief of the interior provinces from 1786 to 1790. Ugarte benefited from the groundwork that his predecessors had laid, and he continued to provide the overall coordination to prevent hostile Indians from playing off one Spanish province against the other. Peace, of course, was never absolute. Raids and occasional acts of violence continued on all sides, but minor infractions by individuals were overlooked by Spanish and Indian leaders alike. Both Spanish and Indian leaders had come to believe, as Bernardo de Gálvez hoped they would, that "a bad peace . . . would be more fruitful than the gains of a successful war."

In the interior provinces, the understandings that Spaniards had with Comanches, Apaches, Navajos, and other tribes lasted until the 1810s. Then, when rebellion in Mexico diverted resources away from the frontier, making it difficult to continue to buy friendships or offer a steady supply of trade goods, the peace establishments began to collapse and alliances weakened. On the Texas frontier in particular, the chaotic intermural quarrels between Spanish royalists and Spanish insurgents, both of whom solicited the aid of Indians, made Spaniards undependable and unpredictable allies.

To California

The reorganization of the interior provinces and the forging of a successful Indian policy during the long reign of Carlos III (1759–88) was accompanied by Spanish expansion to the Pacific coast—Spain's last defensive thrust into the western half of the continent. In 1769, the same year that Spanish troops took forcible possession of Louisiana, sea and land expeditions pushed into the northwest of New Spain to occupy the bays of San Diego and Monterey. Had they known of it, they would also have raised the cross on the magnificent Bay of San Francisco, but it had remained concealed from Spanish mariners behind the narrow, fog-shrouded Golden Gate. After a land party discovered the bay later in 1769, Spain built a presidio and mission at the tip of the San Francisco peninsula in 1776.

As in Texas and Louisiana, Spain's motives for expanding into Alta California were purely preemptive. Missionaries and mariners had long urged the Crown to occupy the Pacific coast of North America, all of which Spaniards termed "California," but not until Spain perceived the area as threatened by foreigners did it commit resources for such an enterprise.

The architect of Spanish expansion to the Pacific was José de Gálvez, who served as a nearly omnipotent royal inspector in New Spain from 1765 to 1770, before becoming minister of the Indies. Gálvez regarded both England and Russia as threats to Spanish claims. Beginning with Francis Drake in 1579, English mariners had entered California waters from the south; if England's well-publicized search for the Northwest Passage succeeded, Gálvez feared that English vessels would soon enter the Pacific from the north. At the same time, he predicted that Englishmen would continue westward from Canada and the Mississippi, finding their way to California along great rivers. "There is no doubt," Gálvez wrote in 1768, "we have the English very close to our towns of New Mexico and not very far from the west coast of this continent." The threat from Russian fur traders seemed still more immediate. In 1759 a book by a Spanish Franciscan, José Torrubia, had appeared in Italy with the alarming title *Muscovites in California.*

On his own initiative, the energetic and ambitious Gálvez began to lay the foundations for Spanish expansion to the northwest of New Spain. In 1768, when reports of large numbers of Russians settling the California coast prompted Madrid to order him to secure Monterey Bay, Gálvez was prepared to move swiftly. Strapped for resources and volunteers, he relied heavily on Franciscans, led by Fray Junípero Serra, and soldiers, commanded by Captain Gaspar de Portolá.

Missions had fallen from vogue among the Enlightened ministers of Carlos III, who had expelled the Jesuits from Spain and its colonies in 1767, and Gálvez himself believed them antiquated. Nonetheless, they flourished in Alta California even as they were being secularized in Texas, New Mexico, and Arizona—that is, the missions converted into parishes with parish-supported secular priests and their communal property divided among the remaining Indians. Using the mixture of rewards and punishments that had

A German amateur artist, Ferdinand Deppe reinforced the popular European image of California mission life as a harmonious relationship between peaceful natives and benevolent Spaniards.

Ferdinand Deppe (active in California, 1828–36). San Gabriel Mission. *Oil on canvas, 1832. Courtesy Santa Barbara Mission Archive Library, Santa Barbara, California.*

worked among the Pueblos, Pimas, and Coahuiltecans, Franciscans supervised the construction of twenty-one missions between San Diego and Sonoma, the last completed in 1823. At the height of their occupancy in the early 1820s, the California missions housed some twenty-one thousand Indians—on the average, five times as many per mission as in the San Antonio missions at their zenith. As elsewhere, Franciscans along the densely populated coast met resistance of various types, including rebellion and flight, but the linguistically diverse natives, organized into tiny units and without benefit of horses, guns, or French or British allies, proved more tractable than Apaches or Comanches. Bernardo de Gálvez had hoped in vain in 1786 that the "innocence" and "tranquility" of the California tribes might by maintained and that they might be denied "the use and handling of the horse."

Despite the populous missions and salubrious climate, Alta California attracted few colonists. The government offered material rewards to encourage immigration and sent some convicts and orphan girls as colonists, but the area remained too distant from population centers and markets to attract willing immigrants. In effect, Alta California was an island through most of the colonial period, dependent solely on occasional ships from New Spain to bring additional colonists and supplies and to provide access to markets. Baja California had provided soldiers, missionaries, and livestock for the initial settlement of Alta California in 1769, but the impoverished peninsula had spent itself. In 1774, California's isolation diminished when Juan Bautista de Anza blazed a trail from Pimería Alta to California via the critical Yuma crossing of the Colorado River. The Quechans at Yuma revolted in 1781, however, destroying two nearby missions and closing Alta California's only land connection to Mexico, which Spain never reopened. As much through natural increase as through colonization, then, the Hispanic population reached about three thousand by 1821—five hundred more persons of European descent than lived in Texas that same year and three times the number who lived in Arizona. New Mexico, the least populated of today's border states, had by far the largest population of Hispanics in 1821—some thirty thousand.

California's Hispanic population was thinly dispersed, clustered in well-watered valleys along a five-hundred-mile stretch of dry coast. The province had only three self-governing municipalities under Spain: Los Angeles (1781), San José (1777), and Branciforte (today's Santa Cruz, 1797). Civilian communities also developed near some of the missions, such as San Luis Obispo and San Juan Capistrano, and around the province's four ill-equipped presidios: San Diego, Santa Bárbara, Monterey, and San Francisco. Appropriate to the Enlightenment, Spanish scientists reconnoitered Pacific waters far into present-day Alaska, but Spain never took effective control of the coast beyond San Francisco. In 1790, Spain's short-lived attempt to occupy Nootka Sound on Vancouver Island, off the coast of present British Columbia, ruptured relations with England. A much-weakened Spain surrendered its exclusive claims to the Pacific Northwest rather than go to war with England. In 1812, while Spain was preoccupied with rebellions in its American colonies, Russian fur traders, whom Gálvez had feared, established Fort Ross on Bodega Bay, to the north of San Francisco.

Impacts and Adaptations

Small in numbers, the Hispanics who moved into California in the late eighteenth century made a powerful impact on the land and its native peoples—as they had wherever they settled along the Spanish rim. Largely because of European diseases, the native population of California fell from an estimated 300,000 in 1769 to 150,000 in 1821. Some Indian survivors fled into the interior, but those who remained along the coast found their traditional cultures eviscerated. Unlike the Pueblos, whose permanent, cohesive towns had provided a measure of refuge against forced culture change, many of California's hunting and gathering peoples had been lured into institutions whose routines, regulations, economic and social activities, and housing and sleeping arrangements differed radically from those they had known before. With the breakup of the missions in the 1820s and 1830s, most of these Indians became marginalized outsiders without a firm footing in either Hispanic or Indian culture.

As Indian populations declined in California, as elsewhere along the Spanish rim, European domestic animals proliferated. The temperate climate along the California coast and the grasslands of South Texas proved especially congenial for cattle and horses, which multiplied with few natural predators. The introduction of European quadrupeds, like the addition of any new species of flora or fauna, had a ripple effect throughout the ecosystem. Voracious, sharp-hooved grazing animals, for example, destroyed protective ground cover and compacted soils; on hillsides or gently sloping land, their well-worn trails deepened into gullies that carried rainwater off too swiftly. In the Spanish era, overgrazing and "gullying" did not turn vast areas of grassland into desert (that process did not become apparent until the late nineteenth century, hastened perhaps by a change in climate), but by the eighteenth century the effects of overgrazing began to be felt in Hispanic and Indian communities alike.

The new pastoral economy also altered the types of plant life that characterized the region because cattle, horses, and sheep (which thrived in New Mexico) exterminated native species of grasses and left a void that more aggressive European species rushed to fill. Wherever they went, Old World grazing animals effectively transported the seeds of Old World grasses—including those, such as Kentucky bluegrass, that we have come to think of as 100-percent American. For centuries, European grasses had adapted to close cropping and bare or compacted soil and had developed seeds specially equipped to travel with grazing animals. In California alone during the Spanish era, Mediterranean forage plants and weeds—bromegrasses, common foxtail, curled dock, Italian ryegrass, red-stemmed filaree, sow thistle, wild oats, and other plants—moved well beyond areas of Spanish settlement into northern California and the interior valleys of the San Joaquín and Sacramento rivers. As in much of the Southwest, the transformation of grasslands in California from native to alien species was not completed until late in the nineteenth century, but the process was well under way before the American era in California and along the Spanish rim.

Just as Spaniards made a powerful impact on the land and the natives from California to Texas, so did the land and the natives have a powerful impact on Spaniards. For example, Spaniards adapted most readily to the same temperate zones where European Spanish domestic animals flourished—along the California coast, in the high country of northern New Mexico where altitude mimics latitude, and on the well-watered, fertile, and salubrious prairies inland from the Texas coast. In contrast, much of the desert remained alien to many European species and, therefore, to Spaniards as well. Across the region, microenvironments also shaped, but did not determine, Hispanic life and institutions. In the mountainous villages of northern New Mexico, scant farmland and the need to share water put a premium on cooperation and intensified Iberian communitarian traditions, including communal ownership of land. Below Santa Fe, where the widening Rio Grande floodplain offered more farm- and rangeland, large, privately held estates became common.

Human geography also transformed the lives and institutions of Hispanics in the far North, whose societies never became mirror images of their metropolis but rather resembled other Hispanic frontiers, such as those in Chile or Argentina. For example, no place on the Spanish rim offered the large, docile, and sophisticated Indian labor force that supported the elegant institutions (universities, seminaries, libraries, and guilds) and complex hierarchical society of central New Spain. To be sure, the frontier societies had at their apex a small aristocracy whose status derived from family, racial purity (real or imagined), land, livestock, and governmental or ecclesiastical positions. At the broad base of the frontier societies one could, of course, also find exploited Indian labor, including Indian servants and slaves (who were seldom called "slaves," since enslaving Indians was prohibited). Across much of the Spanish rim, however, the demands of frontier life forced a high percentage of Hispanics, even those with a few unpaid servants, to work their own land and tend their own livestock—at times simultaneously defending themselves from hostile nomadic tribes. Such circumstances may have shaped the character of Hispanic frontier peoples, making them more

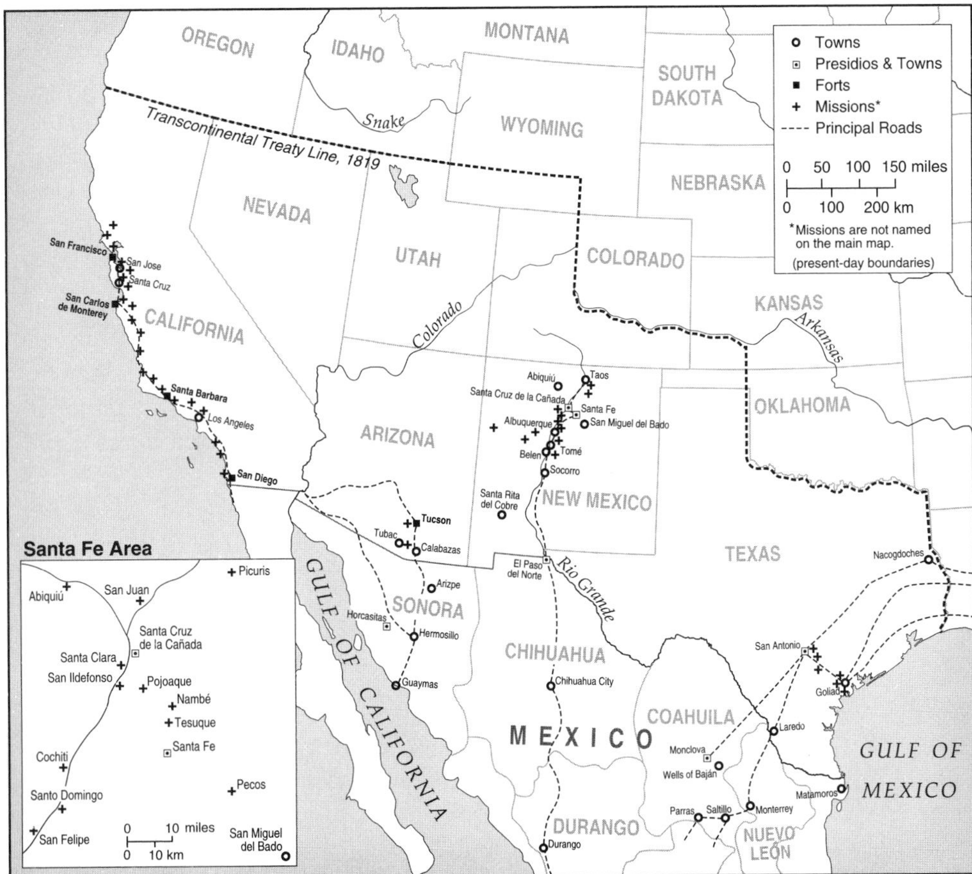

The Spanish-Mexican Rim, 1821

independent, self-reliant, and hardy, as contemporaries such as Alexander von Humboldt and Miguel Ramos Arizpe believed. More concretely, Indian raiders distorted Spain's institutions and society on the frontier, discouraging immigration, initiating the preeminence of the military in the eighteenth century, and making a disproportionate percentage of widows in places like San Antonio, where fighting was most intense.

In contrast to the English colonies, where Indian influence on the formation of Anglo-American cultures appears to have been slight and limited to initial contact, the Spanish colonies experienced ongoing and profound contact with Indians. In many parts of the frontier, the fate of Spaniards remained intertwined with that of their Indian neighbors, on whom they initially depended for subsistence and from whom they learned to adapt to the new land. From Pueblo women, for example, Spanish women learned to plaster adobe, which became women's work among Hispanics as it was among Pueblos. Some Spaniards acquired competence in Indian languages, consulted with Indian physicians, and used Indian drugs, including peyote on occasion. *Atole*, chocolate, pinole, *posole*, tortillas, and other Indian foods became integral to the diet of Spanish frontier folk.

At its deepest level, Spanish-Indian contact included the ongoing amalgamation of races as Spaniards, mestizos, blacks, and mulattoes came north from Mexico and coupled with natives along the frontier, in and out of wedlock. Racial boundaries, which blurred throughout New Spain in the waning decades of the colonial era, seemed to break down even more rapidly on the frontier, where institutions that maintained such distinctions were relatively weak.

In what California Governor Pedro Fages called a "pernicious familiarity" with Indians, Spaniards may also have adopted nonmaterial elements of Indian cultures, but such influence is difficult to assess. As early as 1631, a Franciscan in New Mexico complained to the Inquisition of the difficulty of life in New Mexico, where Spaniards were "reared from childhood subject to the customs of these Indians." Over a century and a half later, Governor Fernando de la Concha lamented that the unruly and independent New Mexicans had adopted "the liberty and slovenliness which they see . . . in their neighbors the wild Indians." Similarly, the californios earned a reputation for scattering into "remote regions without King to rule or Pope to excommunicate them," in the words of Fray José Señán.

However much Spaniards adapted to their new human and geographical environment, most sought to replicate Hispanic culture on the frontiers and to maintain allegiance to Pope and Crown. Like other Europeans, Spaniards struggled to reconstruct the family structure, social organization, communities, and modes of work, play, and worship they had known at home. The greater their means, the more easily they obtained and maintained the trappings of their civilization. The possessions, for example, of Juana Luján, a wealthy widow who died at her ranch near Santa Cruz de la Cañada in New Mexico in 1762, included, in the words of the historian Richard Ahlborn, "silver and porcelain utensils, painted chests, religious images, jewelry, and fine clothing from central Mexico, Europe, and China." When Pedro de Villasur led an expedition from Santa Fe to the Platte River in present Nebraska in search of Frenchmen in 1720, he traveled with silver dishes, cups, and spoons, a silver candlestick, an inkwell, writing paper, quills, and a saltcellar.

Spaniards, who possessed the technological, economic, and political strength of a large state society, succeeded so well in rebuilding their familiar world on the frontiers that members of tribal societies, such as Rafael Aguilar, the alcalde of Pecos Pueblo, probably borrowed more from Spaniards than Spaniards did from them. Ultimately, however, the extent of cultural borrowings depended on so many contingent circumstances, including time, place, and class, that no composite picture can be drawn for the entire Spanish rim. *Californios* differed from *arizonenses*, who, in turn, differed from *nuevo mexicanos* and they from *tejanos*. To cite graphic examples, the carved wooden saints, or *santos*, associated with late colonial New Mexico and valued today by aficionados of folk art, had no counterpart elsewhere along the Spanish-Mexican rim, and the red-tile roofs common to the missions of California toward the end of the colonial era did not exist in New Mexico.

Independent Mexico and American Manifest Destiny

In the first two decades of the nineteenth century, Spain's New World empire crumbled. In 1821, when New Spain followed other former Spanish colonies in declaring its independence, the provinces from California to Texas quietly endorsed the new order. Mexico and the United States agreed to an international boundary that put much of what

Hispanic art in New Mexico found its most creative expression in the form of naive but powerfully evocative wooden carvings of saints, or santos, *and paintings of saints on pine boards, known as* retablos. *The earliest Anglo-American visitors to New Mexico denigrated these objects as "miserable pictures of the saints," but modern-day* santeros *are now regarded as heirs to one of the few indigenous artistic traditions in the United States.*

Miguel Herrera (active 1880s). Our Lady of Sorrows. *Gesso, natural pigments on wood, ca. 1880s. Photograph by Blair Clark. Charles D. Carroll Bequest to the Museum of New Mexico, Museum of International Folk Art, Museum of New Mexico, Santa Fe.*

is now the American West, including all of Nevada, Utah, and parts of Colorado, Kansas, and Wyoming, as well as the four border states of California, Arizona, New Mexico, and Texas, under the jurisdiction of independent Mexico. But the region remained part of Mexico only briefly. Texas rebelled successfully in 1836; a decade later, in 1846, American troops invaded New Mexico, southern Arizona, and California.

In this quarter century from 1821 to 1846, Anglo Americans and other foreigners moved into the region for the first time in significant numbers as Mexico nullified Spanish restrictions against foreign residents and foreign commerce. The newcomers entered a frontier zone whose landscape and peoples often seemed to them exotic, even though it had been deeply transformed by Europeans. Moreover, foreign visitors found themselves in a region undergoing an immediate turmoil of rapid structural transformations imposed from the outside. In its first decades, independent Mexico made fundamental changes in the nation's political, economic, military, and ecclesiastical institutions. Those changes, intended to modernize the nation, came about so rapidly and imperfectly that they promoted profound discontent along the frontier.

After a brief flirtation with monarchy, Mexico attempted to convert itself into a republic, with its Constitution of 1824 modeled in part after that of the United States. But to lawmakers in Mexico City, it seemed clear that the northern frontier's sparse and relatively unschooled population could not afford to operate the institutions of state government. The compromises that ensued left the frontier populace with inadequate government and with little hope for political reform. The entire region from California to Texas lacked political autonomy at the provincial level and adequate representation or influence at the national level. The situation worsened when Mexico's political institutions collapsed in the early 1830s. The following years of chronic instability disaffected many frontiersmen and contributed to separatist movements that culminated in Texas in 1836 and that had undercurrents in California and in New Mexico from the mid-1830s through the mid-1840s. "Hopes and promises are only what [New Mexico] has received . . . from its mother country," one disenchanted New Mexican, Manuel Chávez, wrote in 1844.

At the same time that Mexico's weakening political institutions began to lose legitimacy in the eyes of some frontier oligarchs, the authority of its ecclesiastical institutions diminished. The mission system ended, and the last Franciscans departed. Like the political changes on the frontier, the demise of the missions came about in part because of efforts to modernize Mexico. The liberal view that missions represented antiquated institutions had been articulated by Spanish thinkers in the late eighteenth century and prevailed in Mexico in the 1820s and early 1830s. Among other things, liberals asserted that missions oppressed Indians by holding them forcibly and denying them the full equality accorded other Mexican citizens. However noble the philosophy, the dismantling of the missions produced a spiritual crisis on the frontier, where the secular church failed to replace the departing Franciscans, who had ministered to Indian and non-Indian alike. Many priests preferred more comfortably familiar parishes in the nation's heartland.

Meanwhile, Mexico's military position on the frontier never fully recovered from the tumultuous decade of civil war that preceded Mexico's final declaration of independence in 1821. The frontier military declined in relation to the strength of Apaches, Comanches, and other nomadic tribes partly because of the reluctance of the Mexican army's highly politicized officer corps to fight Indians in the remote North. Greater awards awaited officers whose units stood poised near Mexico City when governments tottered and fell—as they frequently did. Then too, money for gifts for Indian allies appears to have diminished in the Mexican era. As a result, the frontier presidios never reached the level of effectiveness that they enjoyed before the Mexican wars for independence, and the burden of frontier defense fell on the frontiersmen, who organized themselves into ill-equipped militia. As military effectiveness declined, the firepower of Indian adversaries increased, due largely to American gunrunners. By the

1830s, American merchants had broken into the Indian trade as far west as Arizona, and some of them traded weapons for loot that Indians stole from Mexicans.

In this and other ways, Americans who entered the borderlands became participants in the region's transformation. When independent Mexico sought to invigorate its economy by opening its borders to foreigners and foreign trade, it succeeded notably on its northern frontier. From the nearby United States in particular came manufactured goods, capital, access to markets, and colonists. The tempo of economic life increased, as Mexico hoped it would, but Hispanic frontiersmen from California to Texas grew dependent on outsiders, especially Americans, for manufactured goods, and Americans came to dominate commerce and local industry even before the United States acquired the region between 1845 and 1848.

In the first half of the nineteenth century, westering Americans represented a society that had rapidly outstripped fledgling Mexico in economic productivity and population growth. While the American economy expanded remarkably between the so-called panics of 1819 and 1837, the once robust economy of Mexico withered as a result of the dislocations that accompanied the wars of independence. After independence, Mexico's gross national product fell to less than half of its 1805 peak and would not surpass that figure again until the 1870s. A similar disparity occurred in population growth. In 1820, the population of the United States had reached 9.6 million, more than doubling since 1790; by 1840, it had nearly doubled again, to over 17 million. Meanwhile, the number of Mexicans had fallen by about 10 percent during the turbulent 1810s, to 6.1 million, and remained relatively static for several decades.

Like the encounter between the Spanish state and the North American tribal societies of the sixteenth century, then, this latest contest for control of the Spanish-Mexican rim was lopsided. The new winners enjoyed not only demographic and economic advantages but also a mercantile ethos and a certitude in what they believed to be the superiority of their race, religion, and political institutions. Those conceits lay behind the Americans' sense of their own manifest destiny and served them well as rationalizations for conquering and transforming northern Mexico, much as Spaniards' ethnocentric values had facilitated their domination of indigenous Americans several centuries earlier. But if the contest between the United States and Mexico was lopsided, the transformation that followed the American invasion of 1846 was not entirely one-sided. Just as Hispanic frontiersmen altered and enriched their culture by borrowing from vanquished Indians, so did Anglo Americans embrace aspects of the culture of the defeated Hispanics, whose influences can still be seen today in the Southwest, especially in architecture, diet, language, literature, laws, and ranching and farming techniques.

Constructing and Reconstructing the Hispanic Past

The American transformation of the Spanish-Mexican rim extended to the rewriting of its past. Initially, Anglo Americans dismissed the long Spanish-Mexican tenure in the region as a time of despotism, religious intolerance, and economic stagnation and Hispanics themselves as indolent, vicious, and superstitious (characteristics that Spaniards had often applied to Indians). Painting the Hispanic past in dark hues enabled Anglo Americans to draw a sharp contrast with the enlightened institutions that they imagined they had imposed on the region. Hispanophobic interpretations of the

Carleton E. Watkins systematically photographed all of the California missions in 1880–81, and marketed his large-scale views to buyers who valued the ruined buildings as nostalgic relics of the Spanish West. The renewed late 19th-century interest in the Spanish past was closely linked to efforts to stimulate the California tourist industry.

Carleton E. Watkins (1829–1916). Mission, San Juan Capistrano. *Albumen silver print, 1880–81. Department of Special Collections, University Research Library, University of California, Los Angeles.*

Spanish-Mexican era never died, but in the late nineteenth century a rosier view emerged. In the 1880s, as demographic changes reduced the influence of Mexicans and Mexican Americans in the region, Anglo Americans could afford to indulge in nostalgia about the Hispanic traditions they had nearly obliterated. A sanitized and quaint rendition of the Hispanic past provided a pleasant sense of place for rootless newcomers and added an aura of exoticism and romance that lured tourists and their dollars. Artifacts from the Hispanic past came to be treasured, architecture and building materials (including the once-despised adobe) emulated, and crumbling missions restored. As Charles Lummis, southern California's most energetic Hispanophile, crassly put it, "The old missions are worth more money . . . than our oil, our oranges, or even our climate."

Historians simultaneously shaped and reflected this new appreciation of America's Hispanic past—an appreciation that prevailed through much of the twentieth century and has continued to the present day in some circles. Most vigorous were those historians associated with Herbert Eugene Bolton and the so-called Borderlands school, many of whom sought to enlarge Americans' understanding of the nation's multicultural origins. To correct the Hispanophobic distortions of the past, they emphasized the heroic achievements of individual Spaniards and the positive contributions of Hispanic institutions and culture. In so doing, many lost sight of the cultural and racial blending that made the societies of northern New Spain essentially Mexican. Anglo Americans had fallen under the spell of what Carey McWilliams, in a pioneering 1949 history of Mexican Americans, called "a fantasy heritage." The fantasy, which McWilliams described as "an absurd dichotomy between things Spanish and things Mexican," enabled Anglo Americans to glorify the region's "Spanish" heritage while ignoring or discriminating against living Mexicans. So too did it allow older Mexican Americans to disassociate themselves from recent immigrants from Mexico by imagining themselves of pure Spanish ancestry. The fantasy implicitly denied Mexicans and Indians their historic roots in the region.

Although a small number of scholars decried the "fantasy heritage," their objections were largely ignored until the late 1960s. Then, the coincidental rise of the Chicano movement and a growing interest in the prosaic questions of social history gave impetus to a fuller and less poetic rendering of the past. Younger scholars, who tended to sympathize with the exploited rather than the exploiters, examined themes that resonated with the problems of contemporary Mexicans in America as well as the concerns of social historians: workers, migration, women, class, race, miscegenation, acculturation, urban life, crime, punishment, social control, family, faith, and the fortitude and adaptability of common folk who endure in times of stressful change.

Hispanophobic and hispanophilic interpretations of the past continue to have strong adherents, but the more inclusive history of today seems likely to prevail until future historians feel the need once again to reimagine the past in order to make it more useful for their generation. At present, we understand the Spanish-Mexican rim as a place where Hispanic men and women, most of them mestizos, contended for power and resources with Indians, with European rivals, and with one another. As both the exploiters and the exploited, they transformed the human and natural geography of southwestern North America and in the process were themselves transformed. Similar transformations continue in this place, which we now know as the American Southwest.

Bibliographic Note

A Note about Language. Spaniard and *Hispanic* have political overtones in certain contexts. In this essay, I usually use those words to designate peoples in the past who identified themselves as culturally more Spanish than Indian, whatever their racial background. Because ethnicity is decidedly contextual and ethnic labels are constructed in opposition to some other group, meanings change with time and place. In the sixteenth century, for example, Iberians would normally have defined themselves by region or city—that is, as Castilians, Aragonese, or *sevillanos* rather than as Spaniards or Hispanics. In the New World, however, the word *Spaniard* acquired currency and took on racial connotations, signifying a person born of Spanish parents as opposed to an Indian or a mestizo. I am not using *Spaniard* in that restricted sense.

Like *Spaniard* or *Hispanic*, the terms *Native American* and *Indian* are ideologically charged and conceal the obvious political and cultural diversity of many specific peoples. I use these words largely to distinguish Native Americans in the aggregate from non-Indians.

In this essay, *North America* refers to the continent above present-day Mexico. Although the northern rim of Spain's empire extended across the continent to Florida, I have treated only the western half of the continent.

Much of this essay draws heavily from my own work *The Spanish Frontier in North America* (New Haven, 1992), which contains citations to many of the quotes as well as guidance to more specialized literature on which I have depended.

Readers in search of a short, well-illustrated overview of the Spanish era and its aftermath will enjoy *The Spanish West* (New York, 1976), by the staff at Time-Life Books. Peter Gerhard, *The North Frontier of New Spain* (Princeton, 1982), is the best reference work. For the Mexican era from California to Texas, see David J. Weber, *The Mexican Frontier, 1821–1846: The American Southwest under Mexico* (Albuquerque, 1982). An overview that examines the Spanish impact on native peoples is the anthropologist Edward H. Spicer's *Cycles of Conquest: The Impact of Spain, Mexico, and the United States on the Indians of the Southwest, 1533–1960* (Tucson, 1962). Spicer focused on northwestern New Spain and the Pueblo world; for insight into the indigenous worlds in northeastern New Spain, from the Rio Grande to Louisiana, see the detailed narrative by the ethnohistorian Elizabeth A. H. John, *Storms Brewed in Other Men's Worlds: The Confrontation of Indians, Spanish, and French in the Southwest, 1540–1795* (College Station, Tex., 1975).

The best surveys of sixteenth-century Spanish exploration are in David B. Quinn, *North America from Earliest Discovery to First Settlements: The Norse Voyages to 1612* (New York, 1977), and Carl Ortwin Sauer, *Sixteenth Century North America: The Land and the People as Seen by the Europeans* (Berkeley, 1971). Biographies of individual explorers include Harry Kelsey, *Juan Rodríguez Cabrillo* (San Marino, Calif., 1986), and Herbert Eugene Bolton's classic *Coronado: Knight of Pueblos and Plains* (Albuquerque, 1949). See too Stewart L. Udall, *To the Inland Empire: Coronado and Our Spanish Legacy* (Garden City, N.Y., 1987), with its handsome photographs and unabashed advocacy.

If one could read only a few books on Hispanic New Mexico, several recent titles should be among them: Marc Simmons, *The Last Conquistador: Juan de Oñate and the Settling of the Far Southwest* (Norman, 1991); John L. Kessell, *Kiva, Cross, and Crown: The Pecos Indians and New Mexico, 1540–1840* (Washington, D.C., 1979); and Ramón A. Gutiérrez, *When Jesus Came, the Corn Mothers Went Away: Marriage, Sexuality, and Power in New Mexico, 1500–1846* (Stanford, 1991).

For Arizona too, the lively narratives of John L. Kessell stand out: *Mission of Sorrows: Jesuit Guevavi and the Pimas, 1691–1767* (Tucson, 1970) and *Friars, Soldiers, and Reformers: Hispanic Arizona and the Sonora Mission Frontier, 1767–1856* (Tucson, 1976). Hispanic families are the focus of the anthropologist James E. Officer's *Hispanic Arizona, 1536–1856* (Tucson, 1987).

The authoritative works of Robert S. Weddle provide fine reading on Texas: *Spanish Sea: The Gulf of Mexico in North American Discovery, 1500–1685* (College Station, Tex., 1985); *Wilderness Manhunt: The Spanish Search for La Salle* (Austin, 1973); *San Juan Bautista: Gateway to Spanish Texas* (Austin, 1968); and *The San Sabá Mission: Spanish Pivot in Texas* (Austin, 1964). Jack Jackson, *Los Mesteños: Spanish Ranching in Texas, 1721–1821* (College Station, Tex., 1986), looks at the province's most important economic activity and its participants. Donald E. Chipman, *Spanish Texas, 1519–1821* (Austin, 1992), is the best overview.

There is no recent single-volume overview of Spanish Louisiana in English, but among the books that shed light on important aspects of life in the colony are John Preston Moore, *Revolt in Louisiana: The Spanish Occupation, 1766–1770* (Baton Rouge, 1976), Jack D. L. Holmes, *Gayoso: The Life of a Spanish Governor in the Mississippi Valley, 1789–1799* (Baton Rouge, 1965), and Gilbert C. Din, *The Canary Islanders of Louisiana* (Baton Rouge, 1988).

Spanish expansion to California and the Pacific Northwest is treated in magisterial fashion in Warren L. Cook, *Flood Tide of Empire: Spain and the Pacific Northwest, 1543–1819* (New Haven, 1973). Iris H. W. Engstrand, *Spanish Scientists in the New World: The Eighteenth-Century Expeditions* (Seattle, 1981), explores one little-known aspect of that expansion. The California missions received a hardheaded look in Robert Archibald, *Economic Aspects of the California Missions* (Washington, D.C., 1978). For the secular world, C. Alan Hutchinson, *Frontier Settlement in Mexican California: The Híjar-Padrés Colony and Its Origins, 1769–1835* (New Haven, 1969), is rich in insight and detail.

The best overview of Spanish military policy and practice is Max L. Moorhead, *The Presidio: Bastion of the Spanish Borderlands* (Norman, 1975). The analogous volume on frontier society is Oakah L. Jones, Jr., *Los Paisanos: Spanish Settlers on the Northern Frontier of New Spain* (Norman, 1979). Michael C. Meyer, *Water in the Hispanic Southwest: A Social and Legal History, 1550–1850* (Tucson, 1984), explores one aspect of the reciprocal relationship between man and environment. Thomas D. Hall, *Social Change in the Southwest, 1350–1880* (Lawrence, 1989), is an unusual effort, by a sociologist, to place southwestern history in a theoretical framework.

Carey McWilliams, *North from Mexico: The Spanish-Speaking People of the United States* (Philadelphia, 1949), is a sprightly, perceptive, and influential commentary on the past by a passionate social reformer. John R. Chávez, *The Lost Land: The Chicano Image of the Southwest* (Albuquerque, 1984), reveals a new historical sensibility in the making.

The above recommendations are to secondary books. Numerous essays and articles, many of them essential to understanding the field, have appeared in scholarly journals over the years. Some are gathered together in David J. Weber, ed., *New Spain's Far Northern Frontier: Essays on Spain in the American West, 1540–1821* (Albuquerque, 1979). Still other important articles have been written specifically for a single volume, such as those that appear in David Hurst Thomas, ed., *Columbian Consequences*, 3 vols. (Washington, D.C., 1989–91).

For those inclined to the pleasure of reading original narratives and documents, an astonishing number have been published in English translation, from the classic work translated and edited by Cyclone Covey, *Cabeza de Vaca's Adventures in the Unknown Interior of America* (New York, 1961), to the recently discovered correspondence translated and edited by John L. Kessell, *Remote Beyond Compare: Letters of Don Diego de Vargas to His Family from New Spain and New Mexico, 1675–1706* (Albuquerque, 1989).

For valuable critiques of this article I am indebted to many generous colleagues: the historians Susan Deeds, Northern Arizona University, Ramón A. Gutiérrez, University of California at San Diego, Elizabeth A. H. John, of Austin, Oakah L. Jones, Jr., Purdue University, John L. Kessell, University of New Mexico, and Clyde Milner, Utah State University, as well as the anthropologists Bernard Fontana, University of Arizona, and David Hurst Thomas, Museum of Natural History in New York. Of the splendid support that I receive from my own university and the Robert and Nancy Dedman Chair in History, none is more valuable than the research assistance of the efficient and affable Jane Lenz Elder. As always, my most reliable critic, Carol Bryant Weber, Esq., went through the manuscript line by line, wringing clarity out of obscurity.

Chapter Three

Empires of Trade, Hinterlands of Settlement

JAY GITLIN

Antoine Le Page du Pratz left the French port of La Rochelle in May 1718 and arrived in America—at the entrance to Mobile Bay—on 25 August. Le Page was an educated man, a professional architect with training in mathematics and engineering. He had fought for Louis XIV in the War of the Spanish Succession. He came to Louisiana as a *concessionaire*, a man of means with a grant from John Law's Company of the West, ready to invest his energy, capital, and skills in this New World.

Several days after his arrival Le Page purchased a female Indian slave, "in order to have a person who could dress our victuals." Le Page and the young woman could not speak each other's language, but they could communicate by signs. Le Page decided to establish his settlement on Bayou St. John, half a league from New Orleans, the new capital of the colony. At the time, New Orleans was—in Le Page's words—"only marked out by a hut, covered with palmetto leaves." After choosing his land and having that choice confirmed by the resident company agent, Le Page built a hut of his own, then made a fire. Then the following incident occurred:

> It was almost night, when my slave perceived, within two yards of the fire, a young alligator, five feet long, which beheld the fire without moving. I was in the garden hard by, when she made me repeated signs to come to her, I ran with speed, and upon my arrival she shewed me the crocodile, without speaking to me; the little time that I examined it, I could see, its eyes were so fixed on the fire, that all our motions could not take them off. I ran to my cabin to look for my gun, as I am a pretty good marksman: but what was my surprize, when I came out, and saw the girl with a great stick in her hand attacking the monster! Seeing me arrive, she began to smile, and said many things, which I did not comprehend. But she made me understand, by signs, that there was no occasion for a gun to kill such a beast; for the stick she shewed me was sufficient for the purpose.

The anecdote related by Le Page raises the expectation of a dramatic confrontation between an American "monster" and a European hero, a "pretty good marksman" at that. But the drama is undercut by surprise and, finally, an ironic smile: Le Page discovers that the native girl and her stick were "sufficient for the purpose." Stephen Greenblatt, in his book on European exploration of the New World, *Marvelous Possessions*, describes

The earliest European eyewitness picture of America, this image of French explorers and Timucua Indians in Florida suggests the willful optimism of Europeans' first New World visions but masks the surprise and disappointment that often greeted settlers. The artist depicts a bountiful land of obedient and worshipful natives, but the French colony he visited was, in fact, near collapse due to the Frenchmen's inability to supply their own food or cooperate with local Indians.

Jacques Le Moyne de Morgues (1533–88). René de Laudonnière and Chief Athore. *Gouache and metallic pigments on vellum with traces of black chalk outlines; date depicted, 27 June 1564. Bequest of James Hazen Hyde, Miriam and Ira D. Wallach Division of Art, Prints and Photographs, The New York Public Library, Astor, Lenox and Tilden Foundations, New York, New York.*

the anecdote as the "principal register of the unexpected." As such, it is the perfect form for conveying the experience of discontinuity, the sense of surprise and dislocation that belongs in our histories but is too often missing. A textbook might state that John Lederer, a German physician, led three early expeditions to explore the western lands of Virginia in 1669 and 1670. Would it add that the sight of the Blue Ridge Mountains literally stopped him in his tracks, turning him back emotionally to the more familiar coastal landscape? In just such a moment, Le Page could only stop and marvel at the actions of his Indian slave. Like these early colonists and natives, we should expect the unexpected when studying these early frontiers—not because alligators were lurking behind every stump but rather because ordinary people doing ordinary tasks were caught in the act, as it were, by the strange transposition effected by the European colonization of America.

Of course, natives and colonists soon regained their composure and attempted to make sense of their experiences. Naturally, they relied on their respective cultures to inform their interpretations and guide their responses. On a personal level, the narrative of colonization, this frontier history is truly a dialogue between the familiar and the unfamiliar. The use of languages illustrates this point. Indian people learned how to speak European languages—not just Spanish, English, and French but also Dutch, Portuguese, and German—and colonists, in turn, learned a daunting array of Indian languages. In colonial Louisiana, many Europeans, Indians, and Africans spoke Mobilian, a trade language based on several Western Muskhogean languages. *Nanta shnu bana shnu chumpa*—"you want it, you buy it"—was a phrase recognized by all. Trade jargons and lingua francas sprang up all over North America, two of the more famous being the Chinook jargon in the Pacific Northwest and the Delaware jargon used in New Netherland and New Sweden. Even late in the colonial period, an Englishman in New Jersey would understand *How' s't netap* as a greeting to a friend, and a Connecticut colonist might say a word of thanks—*taubut ne*—to a Pequot neighbor.

This is also a story of—in the words of the historian Fernand Braudel—"dietary frontiers." Colonies were places where people learned to eat "other people's bread." Europeans and Africans in Louisiana learned to use bear oil for cooking and, when it hardened, for spreading over bread as a substitute for butter. A trade in food commodities developed, one that even used Indian standards of packaging and measurement in the case of "deers [deer heads] of oil" and "mococks of maple sugar." Europeans learned to eat various corn-based native dishes such as sagamité and supawn. Delaware Indians learned to raise chickens—which they named *tipas* after a Swedish word used to call domestic fowl. The Pequots came to appreciate *beesh* (peas) and *boige* (porridge). They also came to despise *beksees* (pigs), for these European animals destroyed cornfields, oyster banks, and clamming sites.

Not all such exchanges were beneficial. Europeans learned to smoke tobacco, and Indians learned to drink brandy. And exchanges were often predicated on values rooted in different worldviews, different economies, different environments. The contrast might prove amusing. Europeans obtained land—or, at least, thought they had—for mere "trinkets." The Montagnais of Canada noted, "The English have no sense; they give us twenty knives like this for one Beaver skin." Ultimately, the values or prices set

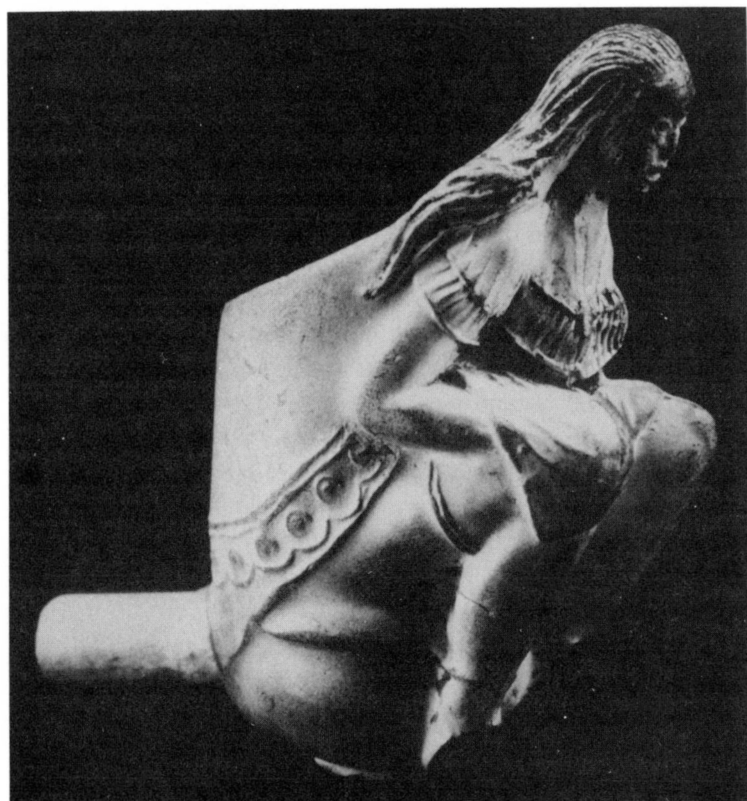

From a modern perspective, early native and European depictions of one another seem rife with ambiguity. A French pipe featuring an Indian woman sitting on a chamber pot hints at a world of complex social interactions and cultural jokes.

French (near Kaskaskia, Illinois). Clay Pipe. *Clay, 18th century. Erwin Peithman Collection, Illinois Historic Preservation Agency, Springfield.*

on North American resources or products by Europeans in the New World reflected European needs in the Old World, and the commercialization of the American landscape disrupted Indian economies. Conflicts were often accompanied by complaints of Indian duplicity and European greed.

The sad truth is that mediation and mutuality were, in the long run, overshadowed by misunderstanding and conflict. America became a contested terrain, and five hundred years after Columbus first arrived, the very history of American colonization has become the object of some struggle. From a European perspective, Columbus made a great discovery. From a Native-American perspective, he began an invasion. Historians struggling to eliminate bias from their work describe the arrival of Columbus as an encounter. Although such neutral terms have their uses, there is no disguising the fact that the arrival of Europeans and their plants, animals, and germs generated a demographic catastrophe for Native Americans. The narrative of colonization is, indeed, a story with tragic dimensions, but it is also a story of adventure and empowerment. The fact that history itself has become a contested terrain simply shows that the story set in motion five hundred years ago is an ongoing one.

Models of Colonization and Appropriation

When Columbus reached the Bahamas on 12 October 1492, a new era of surprising discoveries began; yet this landfall was hardly an accident, a mere chance encounter. The

Columbian voyages represented the summation of past European experiences as much as they generated a field of new ones. The merchants of Genoa and Venice had been expanding the commercial interests of Europe in the eastern Mediterranean for centuries. They controlled the bulk trades, importing grain from Constantinople and the cities of Egypt and Syria and exporting oil, wine, and cheese. The most lucrative ventures were the trades in luxury goods: silks from China or Cathay and Persia; linen and cotton cloths such as damasks and calicoes, whose names reveal their place of manufacture. The spice trade was the most important of all, and in the later Middle Ages, "spices" included not only condiments for food but also perfumes, dyes, and cosmetics. This trade functioned within a complex and far-flung network of merchants, cities, caravans, and shipping routes. In cities such as Alexandria and Aleppo, the Italians lived in protected merchant colonies or Latin quarters, called *fondaci*. Throughout the Levant and the Maghrib, merchants learned the lessons of cross-cultural trade. They learned how to convert currencies, how to conduct negotiations that were both diplomatic and commercial, and above all, how to extract profits while residing in a foreign land. These lessons were fundamental to the evolution of colonial culture. The fur trade in North America, especially as it was pursued by the French, resembled a net of mercantile enclaves cast over a vast, animal-rich environment. At its height, this American enterprise became—in the historian Philip Curtin's words—"a series of European trade diasporas meeting a series of native American trade diasporas at convenient sites on major waterways." The merchants at New Orleans, St. Louis, New Amsterdam, and a number of other colonial entrepôts kept one foot in their host society and one in their cultural home. It was their business to be a part of the larger world—to keep abreast of the conditions of trade and to regulate the flow of commodities they were sending and receiving. Their connections were their source of identity and prosperity.

The Mediterranean provided another model of colonial expansion, one that reached its fullest expression in lands lapped by the waters of the Atlantic. In the islands of the eastern Atlantic Ocean, colonization "avec des gros bagages"—that is, colonies based on the importation of settlers, animals, and plants—developed. The cross-fertilization of Mediterranean and Atlantic worlds had been sparked by the Arago-Catalan conquests in the Balearic Islands and the expansion of Aragon and Castile into Andalusia, Valencia, and Murcia. The island of Majorca, in particular, became a center of maritime information, ideas, and technology. This Catalan colony took the lead in what the historian Felipe Fernández-Armesto has dubbed the late medieval "space-race" into the Atlantic.

European mariners explored a new zone of navigation defined by the Atlantic wind system. The prevailing westerlies in the north, the northeasterly trade winds that produce the clockwise-moving ocean currents in the North Atlantic: here was the basis for the linking of distant coasts.

The Canary Islands off the northwestern African coast and the Madeiras and Azores to the north served as stepping-stones—way stations on the routes connecting the New World and the Old; they also became laboratories for colonial experimentation. The patterns of ecological displacement, the destruction and transfer of biota, began to take shape here. These islands also witnessed the creation of plantation societies that would eventually link Africa, Europe, and new American lands.

Sugarcane arrived in Madeira by the 1450s, after experiments with honeybees and wheat and after the destruction of the island's forests by fire, pigs, and cattle. By the end of the century Madeira had become the world's leading producer of a commodity destined to become a reliable source of wealth in the New World. In addition, an ecological epilogue to this story of Madeira would be replayed elsewhere: wood, needed as fuel for the boilers in the sugarhouse, became a scarce commodity on these once-forested islands. Regulations were issued—but too late.

The Canaries had a similar experience with sugar, but it was delayed: the Canaries had an indigenous population. The Guanches were described as being olive-colored, tall, and—when necessary—fierce. They were isolated from the mainland, lacking any nautical traditions, and were therefore subject to the ravages of virgin-soil epidemics. Although the Guanche population may have been as high as one hundred thousand, the people were divided, spoke different dialects, and were never able to unite against their invaders. The invaders, of course, exploited divisions among the natives and were aided in this task by missionaries. The Guanches barely survived the fifteenth century. As one sixteenth-century diplomat noted about European expansion, "Religion supplies the pretext . . . gold the motive."

Sugar plantations—which combined agricultural and industrial processes—required large numbers of unfree laborers. With the explosion of sugar production in the Madeiras and islands off the African coast, the traditional sources of slave labor proved insufficient. The Portuguese, searching for African gold, which they found, also found new supplies of African slaves at places much closer to the islands generating the demand. The Senegambia region soon became the focal point of African-European—and ultimately—American exchange. Senegambia not only became the center of the slave trade but also witnessed the transfer of crops: yams (possibly from the Wolof word *nyami*, "to eat"), millet, bananas, okra, and rice were carried to the Americas; peanuts, papayas, manioc, and maize came to Africa. Colonization required the uprooting of plants and people, and in the pursuit of profits, American enterprises would consume large numbers of African slaves. It was to be a tragic connection, and the linkage of color with slavery would fix European perceptions of Africans for centuries. In his second voyage to the New World, Columbus brought sugarcane from the Canary Islands to Hispaniola. African slaves soon followed.

It is worth noting that this lengthy prehistory of American colonization is not merely about precedents and patterns that shaped that colonization; it is about the consistent importance of the larger European and global context. The importance of that context goes beyond the early settlements in Virginia or Louisiana. It continues to Alaska and Hawai`i. This larger perspective sheds light on social and economic links, even on personal connections. How can we explain, for example, the presence in late-eighteenth-century Natchez on the Mississippi River of merchants such as Domenico Tevezola, a native of Genoa with a French-Creole wife and an English-speaking African slavewoman, and of Francisco Bazo, a native of Palma in Majorca? The ubiquity of such Mediterranean businessmen seems somehow less surprising when we consider the long-standing "western" interests of Majorcans and Genoese. The father of Manuel Lisa, the famous Missouri River trader and adventurer, was born in the old Catalan frontier province of Murcia. An important merchant in St. Louis, Bartolomeo Bertolla, was a partner with

Pierre Chouteau, Jr., and imported glass beads for the fur trade from his brother Alessandro in Venice. The Fort Berthold Indian Reservation in North Dakota preserves the memory of this Venetian connection. Where, indeed, does the American West begin?

In terms of exploration, efforts in North America proceeded by fits and starts throughout the sixteenth century. Members of the international community of mariners and scholars such as Giovanni Verrazzano and John Cabot Montecalunya established claims for patrons in France and England. In the process, the national and religious rivalries and conflicts of Europe were exported to the Americas. Europeans mapped the outlines of a new continent, but their searches for a Northwest Passage and for rumored places of great wealth such as Quivira, Saguenay, and Cíbola accomplished little more than the alienation of native hosts, on whom they depended for survival.

The English, who would eventually dominate North America, were just beginning to think about the process of living and thriving away from home. The idea of subduing Ireland by means of plantations—by seizing native lands, converting them into an English system of land tenure, and distributing them to English settlers—would have a formative influence on English colonies in America. English perceptions of the natives of Ireland would also have an impact on their stereotypes of Native Americans. The "Wild Irish" were seen as uncivilized; that is, they lacked proper towns, spoke a different language, were pastoral and therefore were not sedentary agriculturalists, had long hair, and wore rough clothes made of animal skins. English activities in Ireland and America were connected and roughly contemporary. The first English attempts to colonize Virginia in the 1580s were dismal failures. Promoters underestimated the effort necessary to feed the colonists. Unless the natives fed them, the Europeans starved.

It was the Dutch—together with the spectacular rise of Amsterdam after 1576—who created the preconditions for the subsequent development of North America. Improvements in shipbuilding and marine insurance contributed to Dutch supremacy as the shippers of choice. The use of printed bills of exchange and the formation of the Bank of Amsterdam in 1609 helped make Amsterdam the central money market of Europe until the nineteenth century. Although an old inclination for domination by warfare and the control of land lingered, the English looked at the Dutch, a people once on the periphery and now grown rich through trade, and began to change their tune. The English now started to look for "merchantable commodities" that they might ship, store, and trade in an emerging world-system of markets. The English had their first success at Chesapeake Bay with the cultivation of tobacco, and this success owed a direct debt to pioneer Dutch planters in Guiana. The Dutch presence in the Caribbean, due in part to their search for an alternative source of salt for the herring trade, kept the Spanish on the defensive and allowed the French and the English to gain a foothold in the Americas.

Until the advent of tobacco, only the Newfoundland codfishery made a trip to North America worthwhile. By 1580 there were over five hundred ships from La Rochelle and Bordeaux, from the ports of Normandy and Brittany, from Portugal and England, in the waters around Newfoundland. Fish was an essential part of the European diet, and the North American fishery became a major European industry, employing as many as fifteen thousand men. It fact, the fishery was the only significant European enterprise in North America during the entire sixteenth century, and it was located on the periphery of the continent. The discovery of America and the appropriation of its resources were part of the global expansion of European activities. Within that context, America—with the important exception of the silver mines of Peru and central Mexico—had proven to be a sideshow compared with the theater of the Orient. Other than one defense post in Spanish Florida, no permanent European settlement existed in North America. The French and the English had shown no ability to survive, let alone flourish, in an unfamiliar environment. Certainly the natives had little to do with this; on the whole, they had shown themselves to be gracious hosts. Although the Europeans had explored much of the Americas, in most cases they had—in Braudel's words—"left no more trace behind than a ship does in the sea." (The pathogens the Europeans brought, however, had a tremendous impact.) Only the fishermen in the North Atlantic had any continuous interaction with the North American environment, and they were a transient group living on or near their ships. On the land of North America, the Europeans were like fish out of water.

From Tsenacomoco to Virginia

In 1606 James I acquiesced to the wishes of several groups of merchant adventurers and committed the crown to the colonization of North America by creating the "King's Council of Virginia." This royal council was to oversee the efforts of the Virginia Company—in reality, two private companies centered in Plymouth and in London. The two companies divided the North American coast between them, with a curious overlap in the region between present-day New York and Washington, D.C., and each sent would-be colonists to their respective turfs in 1607. The Plymouth Company's colony along the Sagadahoc (Kennebec) River in present-day Maine lasted only one miserable and contentious winter. The London Company's colonists at Jamestown in the Chesapeake Bay region may well have wished that they too had returned home. Indeed, Sir Thomas Dale, appointed governor of the colony in 1611, reported "a generall desire in the best sort to returne for England." (In our inclination to put a positive spin on the efforts of the pioneers, we often ignore the pleas and complaints of those who regretted their actions. Two centuries after Dale's remarks, Margaret Dwight, a young Connecticut woman on her way to Ohio, wrote: "We have concluded the reason so few are willing to return from the Western country, is not that the country is so good, but because the journey is so bad.")

Ignorance of Virginia's climate prompted optimistic directives from the home company to plant oranges, lemons, sugarcane, and olive trees, and gentlemen colonists starved while frantically pursuing gold. The Europeans were still strangers in a strange land. Clearly, all European colonists came to America with preconceived notions of the situations in which they hoped to find themselves. Their projections and the realities they found were often a bad match. Throughout the history of America, the gap between speculation and occupation has produced comic and tragic episodes, from Jamestown to "swampy acres" subdivisions in twentieth-century Florida. Moreover, the search for viable situations—which usually translated into a search for marketable commodities—was

Eager to encourage further English settlement in Virginia, John White depicted the native communities as clean, prosperous towns where—the Indians' strange customs notwithstanding—fertile soil and a long growing season ensured the well-being of would-be colonists. His optimism proved unfounded. Following his first trip to Virginia in 1585, White returned in 1587 for a brief stint as governor of the Roanoke colony. When he came back to Virginia again in 1590, he found no trace of the European settlement.

Theodore De Bry after John White. The Town of Secota. *Engraving, 1590. From Thomas Hariot,* A Brief and True Report of the New Found Land of Virginia *(1590). Courtesy Virginia State Library and Archives, Richmond.*

often impeded by a predilection for conquest. Commerce and conquest would remain connected throughout the history of European expansion on this continent, but the former would increasingly shape the latter over time.

As Europeans attempted to secure—in the words of the geographer D. W. Meinig—a "point of attachment," many variables influenced their efforts. There was, of course, the land: its shape, soils, climate, plants, and animals. The site of Jamestown was chosen because of its anchorage, its view downstream, and its potential as a fortress situated on the neck of a peninsula. Unfortunately, it was a low and marshy place with an inadequate water supply. Typhoid fever, dysentery, and malaria plagued the colonists and gave the colony a reputation as a sickly place. Most natives wisely lived upstream and inland.

The historian Timothy Breen has described the European immigrants who first arrived in America as "charter groups." More so than later immigrants, the first colonists had an enormous impact on the "rules of the game": the labor systems, the patterns of behavior and transmission of European customs, the reaction to and treatment of native peoples, even the incorporation of later immigrants. Similarly, the situations that emerged in the first permanent colonies of New France, New England, Virginia, and New Netherland might be described as "charter" frontiers that had a seminal influence on later developments in America. Although every frontier situation was unique, these early colonial frontiers involved the creation of conditions that, to some degree, determined the nature of European expansion westward into the interior of North America. The first frontier settlements served as models for the creation of additional Euro-American settlements. We will focus first on two fundamentally rural frontiers, Virginia and New England.

All European colonists in America entered an occupied land. Estimates of the number of Algonquian-speaking natives living in coastal Virginia in 1607 range from fourteen thousand to twenty thousand. Both in Virginia and in New England, it took roughly forty to fifty years for the European population to approximate the indigenous population. In short, the Europeans built their new homes in the midst of Indian country. The first English adventurers in Virginia hoped that they, like the Spanish, might encounter stratified Indian societies that they could live among and exploit as an imperial elite. Failing that, they might live on plantations like their counterparts in Ireland and reap the rewards of other people's labor. But the natives of Virginia—Tsenacomoco, as they knew it—had no wish to labor for Englishmen, men they perceived to be pale and prone to disease, ugly (too hairy), and needy. (It was logical for the natives to assume that the English had come to their land to acquire maize, trees, and women.) The English, on the other hand, soon recognized that the Indian villages of Virginia lacked gold or spices, indeed lacked any commodity worth stealing. Unlike the Indian groups who lived on the upper Potomac, Susquehanna, and Delaware rivers, these natives did not even have a tradition of preserving or processing valuable beaver pelts.

The Indians of the region were primarily agriculturalists. Men cleared the land around a village by the slash-and-burn method. Women planted the crops, weeded the fields, and gathered nuts and berries. The primary crops were maize, beans, squash, and tobacco. Men went fishing. Seasonal hunting trips were often a communal affair. This economy was fairly common throughout the eastern coast. One historian has described the coastal Algonquians as "sedentary commuters" or "seasonal nomads." In 1607, many of the villages or tribes were part of a state, a chiefdom led by a paramount chief *(mamanatowick)* named Powhatan, or Wahunsonacock. Powhatan's state was surrounded by hostile Siouan and Iroqouian peoples west of the fall line. The newly arrived English represented either a potential ally or another threat. Two years of posturing and bullying on both sides decided the matter: they were a threat. More than three decades of hostilities followed, including three Anglo-Powhatan wars. In the end, Tsenacomoco was devastated. Virginia had replaced it.

A quarter century of conflict drove home many lessons. As early as 1609, the Virginia Company changed its opinion about the Powhatans and wrote, "If you make friendship with any of these [Indian] nations, as you must doe, choose to doe it with those that are farthest from you and enemies unto those amonge whom you dwell, for you shall have least occasion to have differences with them." Few English colonists were ever able to find "common ground" with their Indian neighbors. (Powhatan's daughter, Pocahontas, did marry a colonist and even visited England, where she died in 1617.) Too many factors stood in the way of peaceful coexistence: conflicting ways of managing the land and its resources; the lack of institutions for regulating cross-cultural conflict; the simple absence of anything that might promote cooperation, such as extensive intermarriage and mutually profitable trade relations. The English in Virginia had no use for the natives, and so the natives had to go. Even if the Indian communities they encountered had been able to supply valuable commodities, it is doubtful that the English would have been comfortable living as tolerated foreigners under an alien jurisdiction. Their traditions had prepared them to live as conquerors, not as intercultural brokers like the Genoese and Jewish merchants of Asia Minor. The English were only beginning their education in multiculturalism. As soon as the second Anglo-Powhatan war ended, the English constructed a six-mile palisade on the Virginia peninsula passing through Middle Plantation (Williamsburg). This tangible boundary served two purposes: it kept cattle in and Indians out. After a third war broke out in 1644, the English position became quite clear. A treaty of submission signed in 1646 forced the new Indian leader to become a vassal of the English king. The natives agreed to leave the Virginia peninsula from the fall line to the coast. Any Indian found in the area could be shot on sight, unless he was a messenger wearing a special striped coat. The defeated and marginalized "tributary" Indian communities were positioned on the edge of European settlement by colonial authorities to serve as a barrier to the aggressions of larger, independent tribes. It was now the Indians who found themselves strangers in their native land.

Emerging during these war years were several story lines that would be repeated again and again in the drama of European westward expansion. Before the second Anglo-Powhatan war of 1622, native spiritual leaders may have begun to preach purification as a prelude to resistance and rebellion. A spiritual vision—a prophecy of an Indian millennium—may have served to unite communities experiencing conflict not only from without but also from within as European goods, habits, and values upset an established social consensus. The English, for their part, began to realize the distance between the cultural hearth and the fires at home. The Virginia Company's attempts to promote interethnic harmony were not well received by colonists embroiled in total war and race hatred. It would not be the last "east-west" division over Indian policy.

Other problems separated the company back in England and its colonists in Virginia. The company, throughout its brief history, had a vision of orderly, self-sustaining Virginia settlements with company tenants producing valuable commodities on company land until their term of service expired. The reality did not even come close. The only commodity to emerge in the colony was tobacco, the one crop the company discouraged the planters from raising. The first shipment of Virginia tobacco was sent to England in 1617. The market price was an incredibly high three shillings a pound. Tobacco plants were soon being raised on every available plot of land. Far from being self-sufficient, the colonists relied on maize taken from Indian villages during hostilities or purchased from Indian allies. Tobacco exports, on the other hand, rose from sixty thousand pounds in 1619 to a half million pounds in 1628. Although prices fell drastically by 1630, production spiraled upward for the rest of the century.

The historian Edmund Morgan has described the Virginia of the 1620s as a "boom" colony. High prices for tobacco meant fast profits. Land and labor were the necessary ingredients. Although tobacco cultivation did not require an inordinate amount of either, the work was hard, and more hands meant greater profits. Our notions of the independent, hardy pioneer are certainly not sustained by the history of the tobacco colonies. Most men and women were either masters or servants. And few Englishmen came to America prepared for the backbreaking work of converting forests to fields. One observer in Bermuda noted, "[Settlers] do sigh to see how many trees they have to fell, and how their hands are blistered." Not surprisingly, Europeans were quick to seize cleared Indian lands, a fact pointed out by Indians outraged by squatters in their cornfields. Land was abundant and cheap in Virginia. Labor was harder to come by, and a labor market quickly developed—one controlled by an entrepreneurial elite. A coercive labor system was predictable, given the high demand and the long terms of service agreed to by those colonists who could not afford to pay their own way. Masters simply bought and sold men and women for the time remaining on the contracts. Some planters apparently used servants as stakes for gambling.

Life was indeed cheap in Virginia. Some seventy-two hundred settlers came to Virginia between 1607 and 1624. Around twelve hundred were still alive in the latter year. The vast majority died of diseases—usually in "seasoning" time, the dreaded Virginia summer. One colonist estimated that only one out of every six immigrants survived the year. Greed and the uncertainty of survival, combined with the absence in Virginia of English laws and courts designed to protect servants, contributed to a system many laborers found intolerable. By the time Virginia became a royal colony in 1625, it was clear that the vision of the now-defunct Virginia Company had not been realized. Like mid-nineteenth-century California during the gold rush, mid-seventeenth-century Virginia was a transient place. Planter and servant alike crowded passing ships and turned them into floating saloons. Men outnumbered women by three to one. The majority of the colonists were under thirty years old. Demographic and economic factors shaped a place that lacked a sense of permanence or order. Few substantial houses were built. Observers noted that Jamestown was a shabby little village. It was obvious that those who survived hoped to prosper and return to England.

Conditions in Virginia gradually improved. The European demand for tobacco continued to rise, and despite lower prices—tobacco was being grown in Europe in substantial quantities by the 1650s—the annual income of the colony rose from less than ten thousand pounds sterling in 1630 to over seventy thousand by 1670. Immigration increased after the 1630s. The vast majority of new arrivals were still servants, but families with resources responded to perceived opportunities in the colony. By 1660, Virginia and Maryland combined had over thirty-five thousand colonists. By the end of the century, the southern colonies—now including the Carolinas—were home to over one hundred thousand settlers.

Although Virginia was no longer a "boom" colony by the 1660s, Governor William Berkeley reported that the elite of Virginia were still "looking back on England" in hopes of retiring there on the profits made from tobacco. Virginia had become a somewhat sleepy, provincial place on the periphery of the English metropolis. Indeed, after the 1660s, many competitors of the Chesapeake region stopped producing tobacco, in part because of the superior product grown in Virginia and because soil exhaustion threatened a limited supply of land. Planters on islands such as Barbados, Jamaica, Martinique, and Guadeloupe switched from tobacco to crops such as sugar or indigo, which required greater investments in labor and equipment but were far more valuable. Tobacco planters were the poor relations among the producers of cash crops.

Virginia planters, on the other hand, began to diversify. Some planters became farmers and abandoned tobacco for wheat. Most stayed with the leaf but combined that crop with grain production, livestock, and orchards. Unlike planters in the West Indies, those in Virginia could not afford the luxury of becoming absentee proprietors. On the contrary, the ideal in Virginia became the independent gentleman living in an autonomous, self-enclosed world. When African slaves replaced white servants toward the end of the seventeenth century, the picture was complete. A successful plantation appeared to be a world easily controlled. "Economic privatism" had been from the beginning, in the words of Breen, the "colony's central value," and the landscape reflected this. "An extreme form of individualism, a value system suited to soldiers and adventurers," had been transformed into "a set of regional virtues, a love of independence, an insistence upon personal liberty, a cult of manhood, and an uncompromising loyalty to family." Indeed, in a landscape that valued distance from one's neighbors and enjoyed only sporadic contact with the wider world, who else could be trusted? Historians have also suggested that persistently high mortality rates led parents to encourage independence in their offspring. As the Virginia countryside achieved a settled, English appearance, the illusion of English country life tended to obscure the colony's ultimate dependence on global markets. Planters in the eighteenth century, dependent on Scottish factors in Virginia and commission agents back in Great Britain, forgot—to their peril—that connections based on debt and credit could not be expressed simply in personal terms. When the fluctuations of world trade rudely reminded Virginians in the 1760s that they lived on the edge of the metropolis, Thomas Jefferson captured the bitterness of the moment. Virginians, after all, were "a species of property annexed to certain mercantile houses in London."

New England, with Reservations

An eminent historian of colonial America recently wrote, "It would be difficult to imagine how any two fragments [the Chesapeake colonies and New England] from the same metropolitan culture could have been any more different." There are many reasons for this difference, including the native worlds encountered, the economic and environmental conditions, the process of migration, and the regional origins of the immigrants. The earliest English efforts in New England were of little consequence. A small group of dissenting Separatists attempted to settle the Magdalen Islands in the Gulf of St. Lawrence. In the words of the English historian K. R. Andrews, they "dissented from each other and everyone else." The project came to naught. After several expeditions and the failed colony at Sagadahoc, the *Mayflower*, carrying 101 passengers, landed at Plymouth Rock in 1620. The situation they encountered in this region was quite different from that faced by the first Virginia immigrants.

Southern New England—defined roughly as encompassing the present-day states of Connecticut, Massachusetts, and Rhode Island—had a native population before 1616 of between 70,000 and 140,000 people. The people in this region spoke a variety of dialects and languages belonging to the Eastern Algonquian group. The languages were similar, and it is said that neighbors speaking different languages were able to comprehend each other. Tribes—if we may call them that—consisted of allied villages with kinship ties and a system of episodic political, economic, and military relations. Sachems and sagamores exercised leadership based on consensus. Europeans, used to less egalitarian arrangements, tended to magnify the importance of native leaders. Three "nations" inhabited the coast from southern Maine to Cape Cod, from north to south: the Pawtuckets, the Massachusetts, and the Pokanokets or Wampanoags. The Narragansetts lived in Rhode Island, and the Pequot-Mohegan people occupied eastern Connecticut. A variety of smaller groups lived in central Connecticut and Massachusetts. (In fact, various Native-American communities are still in the region today, some of them as federally recognized tribes.)

Two important events occurred in the years immediately preceding the founding of Plymouth. First, French traders along the Massachusetts coast and Dutch traders along the Long Island Sound and up the Hudson and Connecticut rivers initiated a commerce in furs that undoubtedly resonated with existing native exchange networks. The fur trade had begun and, with it, the competition among European nations for spheres of influence in the region. Establishing jurisdiction in this area would not be a simple matter of Indian versus European. A complex series of alliances and struggles encompassing various Indian and European groups would ultimately establish the boundaries we know today.

The second event, probably related to the first, was a virgin-soil epidemic (possibly chicken pox or smallpox) that killed from 75 percent to 90 percent of the people in the coastal villages of the Pawtuckets, Massachusetts, and Wampanoags from 1616 to 1619. The tragedy that devastated the coast had many implications. Tribes that were not hit, such as the Narragansetts and the Pequots, assumed a new-found dominance in the region. Survivors from decimated villages formed new composite villages. Spiritual confidence was undermined. Traditional political and social relationships were disrupted, and even the wisdom and technical knowledge of the people—held in common and transmitted orally by elders—was, to some degree, dissipated by the enormous loss of life. Into this situation sailed the Pilgrims in November 1620, looking for the site of Patuxet (Plymouth), an Indian village they knew to be depopulated.

Although half the colonists died that first winter, the colony survived, thanks in large measure to Squanto—a native of Patuxet who had been kidnapped and brought to England during the time of the great epidemic and had returned before the *Mayflower* arrived. According to William Bradford, governor of the colony, Squanto "directed them how to set their corne, wher to take fish, and to procure other comodities, and was also their pilott to bring them to unknowne places for their profitt, and never left them till he dyed." For the Pilgrims, Squanto was a godsend. For Squanto, the Pilgrims were probably a source of power in a world turned upside down. The local Pokanokets and their sachem, Massasoit, also welcomed an alliance with the English as a bulwark against their traditional enemies, the Narragansetts.

After several seasons, the Pilgrims were producing a surplus of corn, or maize. (Some settlers had already switched to the cultivation of English grains.) The surplus was exchanged for furs from Abenaki villages farther north along the coast. This profitable relationship was overshadowed by the fur trade of the Dutch, who had discovered the value of wampum or wampumpeag—called *sewan* by the Dutch—in 1622 and had established a monopoly with those tribes who controlled its production. Fathoms of wampum were produced from the white and the purple shells of whelks and quahogs by Indians living on the shorelines of the Long Island Sound. The demand for this commodity altered relationships between tribes and facilitated the commerce in furs with inland tribes. In fact, wampum was used as currency by Europeans in New England and New York, and the number of beads in a fathom fluctuated with exchange rates in Europe. (Wampum remained legal tender in specie-starved New England until 1661 and was used as change in the more remote areas of the region into the eighteenth century.) Wampum producers, agriculturalists, and hunters were thus linked with Europe in a market economy. The Dutch and the Pilgrims worked out an arrangement that designated respective trading zones, and throughout the 1620s the colonists at Plymouth discouraged, even threatened, other Englishmen who attempted to settle around Massachusetts Bay. In so doing, they unwittingly prepared the way for a powerful new enterprise forming in England.

In 1629 the Massachusetts Bay Company, with the support of patrons such as Robert Rich (the second earl of Warwick), William Fiennes (Lord Saye and Sele), and Nathaniel Rich, received a royal charter. Although originally designed as a typical joint-stock company, the venture was quickly transformed by influential Puritans such as John Winthrop, a lawyer from Suffolk and first governor of the colony, who divided the commercial and governmental aspects of the company between trustees and colonists. A unique settlement was about to begin.

English Massachusetts was populated quickly and carefully. Over twenty thousand people arrived during the "great migration" from 1630 to 1642. In a generation, the English population doubled, so that by 1660, New England contained approximately twice as many Europeans as the Chesapeake region. And by 1660, most New Englanders were native-born; an Anglo-American colony had developed. After a rather short "starving" time, the English in Massachusetts had become self-sustaining. Birthrates were high and infant mortality was low. The first generation of colonists lived on average to the age of seventy—quite a contrast to life expectancy in Virginia. Although immigration to New England diminished significantly after 1642, in 1700 the descendants of these first New England settlers constituted 40 percent of the colonial population of North America.

Simple population figures do not tell the whole story. This Puritan migration was a family affair, a chain migration of neighbors and relatives. Over 70 percent of the first colonists of Massachusetts came as part of a family group, and the ratio of men to women was a remarkably low 1.5 to 1. The "great migration" was also a middle-class movement, the majority being farmers, artisans, and merchants with some resources at their disposal—almost 75 percent of the adults coming over paid their way. Literacy among the adult males of Massachusetts was twice as high as in old England. Harvard College was founded in 1636, and laws requiring towns to establish schools were passed in the 1640s.

What does all this mean? In part, it means that the transfer of English culture to New England was unusually successful. (This transfer, of course, was influenced by the Puritans' social and religious ideology.) Social relations were not disrupted by the move. Patterns of deference to traditional secular and clerical leaders were brought to the New World; indeed, many leaders themselves made the trip to New England. The environment initially offered no serious challenge to the transfer of English grains such as wheat, oats, rye, and barley. More important, perhaps, the environment was not hospitable to the production of "boom" crops such as sugar. English farms, not tobacco plantations, were carved out of New England soil. Land systems in individual New England townships mirrored the local customs of the English birthplaces of the colonists. Many settlers came from East Anglia (Suffolk, Norfolk, Essex, and Hertfordshire), a region of small towns, enclosed farmsteads, and an active market in land. The historian David Grayson Allen has shown that people in Massachusetts towns such as Ipswich and Watertown simply and effectively reestablished the patterns of their former regions. Origins were of great importance throughout the colonization of North America. Local allegiances and habits were more deeply felt than national ones. The merchants of Creole St. Louis, for example, came primarily from the French provinces of Aunis and Saintonge, and their patterns of family formation and distribution of wealth reflect their place of origin.

Luckily for the Puritans, the land they first settled on lacked powerful native communities. By 1633 some three thousand English settlers had gathered in communities around Massachusetts Bay. By contrast, there were only several hundred natives, and local Indian leaders almost immediately acknowledged the protection offered by the newcomers. (There were, of course, many natives living elsewhere in the region. Thirty thousand resided around Narragansett Bay.) The leaders of the colony, with a natural ethnocentric bias reinforced by a somewhat arrogant sense of mission, quickly extended their jurisdiction over the natives. The policies adopted by Massachusetts were to have a powerful impact on cross-cultural relations in North America. The leaders of the colony attempted both to assimilate and to segregate the small native communities nearby; that is, Massachusetts established reservations.

At first the magistrates were content to restrict the comings and goings of Indian neighbors. In the 1640s a committee led by the Roxbury minister John Eliot formulated a plan for "praying towns," settlements of native converts. Natick was the first, established in 1651, with thirteen others following in the next quarter century. The goal was to bring the good news of Christianity to Native Americans, and that goal was best achieved by weaning native people from their traditional habits. After all, old actions supported old ways of thinking. Besides, many Indian habits were offensive to English sensibilities. To "reduce them to civility," Eliot supervised the Natick Indians in the construction of English-style houses. Commuting to the woods could not be reconciled with the fixed habits of churchgoers. In addition, English leaders wanted to know where the Indians were at all times, both to allay fears of attack and to have it, in the words of one colonial English soldier, "more in our power to Distress them [the Indians]." (Wigwams were retained by many Indians, however, since they were warmer in the winter and cooler in the summer.) Numerous Indian offenses were listed, among them long hair, naked breasts, and gambling. The dangers of multiculturalism were suggested in a Connecticut law of 1672, which prohibited any English person from "playing with any Indian" or laying "any wager with, or for, any Indian." Clearly, Englishmen must have been doing exactly that.

Puritan leaders were quite anxious about the possibility of Englishmen "going native." Great care was taken to maintain adequate cultural distance. Early immigrants built English-style wigwams covered with Indian mats as temporary shelters—clearing fields and planting crops being the top priority. This practice was soon prohibited. Tobacco was also banished from the colony, and lobsters and clams—staples of the Indian diet—were not looked on with great favor by the English, at least not at first. The colonists retained their own ways, preferring "pease porridge"—which developed into the well-known New England baked beans—and brown bread. Indeed, the brown bread of the Puritans might serve as a symbol of cultural interaction on this frontier. Made from a mixture of wheat flour and Indian cornmeal—when a wheat rust forced a greater reliance on rye flour, the mix was known as "rye 'n' injun"—the bread was English in conception and function but borrowed from Indian culture.

In the end, the local native communities of Massachusetts were situated in small enclaves surrounded by English people, grasses, and animals. Indeed, all Native American groups in southern New England suffered the same fate after two devastating wars in the seventeenth century—what the Puritans called the Pequot War of 1637–38 and the even bloodier King Philip's War in 1675–76. Tribal communities, already weakened by death and disease, were further undermined by religious cleavages. Native leaders who served as intermediaries in land transactions and the execution of English justice lost face with their fellow villagers. By the eighteenth century, tribesmen were looking around for support in their struggles against hereditary sachems, some of whom had been corrupted by English speculators or had simply become alcoholics. One Narragansett leader, King Tom Ninigret, was educated in England and spent sixty thousand pounds on a palace designed by an English architect and built in Rhode Island. His debts mortgaged the tribal estate and provoked a declaration of independence by his tribesmen—a decade before the Narragansetts' English neighbors declared their independence from King George. The signs of cultural breakdown were all around by the eighteenth century: migration, alcoholism, pauperization. Yet, many New England natives remained, adapted, and survived. They built fences, sold wood, made baskets, worked as domestic servants, even doctored English neighbors with herbal medicines. Native and European spiritual values were combined by "new light" Indian preachers who responded to the Great Awakening of the 1740s and the new emphasis on the spoken word. On a more limited scale, they continued to use a variety of habitats to produce and gather food. But the land had changed. Hogs had ruined clam banks. Wild turkeys were disappearing. Deer hunting was first regulated in Massachusetts in 1694 (Virginia in 1699). Overhunting, competition from domestic livestock, and the

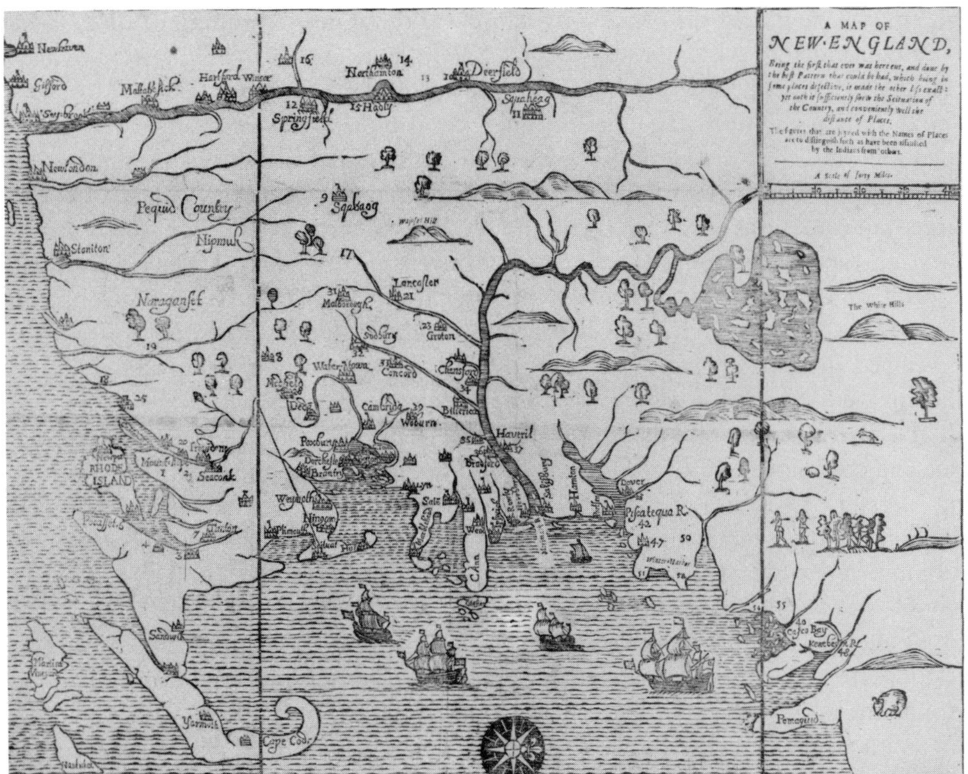

This map, widely recognized as the first printed in the present United States, appeared in a narrative of New England Indian wars written by William Hubbard and published in Boston in 1677. With the direction of west appearing at the top and Cape Cod at the lower left, the map effectively illustrates how English settlers quickly took control of strategic riverfront and harbor sites, pushing Indian communities inland.

John Foster (1648–81). A Map of New England. *Woodcut, 1677. Courtesy of the John Carter Brown Library at Brown University, Providence, Rhode Island.*

disappearance of areas of open forests with attractive grasses for deer and other herbivores—grasses created by careful Indian burning—caused the decline in animal populations.

In the first sixty years of the seventeenth century, around a quarter of a million people left Britain. Perhaps seventy-five thousand of them moved to North America; most of the rest went to Ireland and the Caribbean. Seen in this larger context, North America was hardly a strong magnet for immigrants. Nevertheless, settlements had been secured; indeed, the English in North America were no longer fish out of water. They had moved beyond their first palisades and were running all over the countryside. In many ways, by the 1670s the links with old England were becoming weaker on the farms and plantations of Virginia and Massachusetts. In New England, writers were already glorifying the achievements of the first Puritan immigrants. New England was a new homeland, one with a history. Boston merchants were developing their own carrying trade, and the colony schools were producing a homegrown elite. Observers were less certain about the achievements of Virginia. Critics, from Captain John Smith to Thomas Jefferson, were not proud of Virginia's past and preferred instead to dwell on Virginia's future. The rather thin social and political life of seventeenth-century Virginia had an improvised flavor.

The two colonies had similarities. Both had fewer institutions, fewer social classifications, and fewer rules than old England. But America had land, and land assumed a central, defining power in both regions. In New England, land was essential because it conferred the ability to define community without interference from ungodly quarters.

In Virginia, land held the promise of future wealth, of new beginnings. Although for New Englanders the vision of land retained a social frame, and for Virginians it remained essentially individualistic, for all colonists land promised mastery.

It is fair to assume that so much seemingly available land was an irresistible attraction for Europeans. But the ideal as it evolved in North America was not common to all people; the British, in particular, seemed to measure status by one's landed estate. American conditions encouraged an imported cultural inclination. The basic commodity was cheap enough. It was only a matter of converting native land to rural English land. Clearly, on these settlement frontiers the natives were mere obstacles to development, part of the "howling wilderness" that needed to be reclaimed. The brief fur trade that developed in New England was but a prelude to settlement as debts contracted by Indians were eventually—as animals disappeared—paid off in land.

Native Americans and the English developed few, if any, bonds. Intermarriage was rare and commerce limited. (This was not as true in the Carolina backcountry, where the deerskin trade assumed major importance.) Although some New Englanders developed a conqueror's interest in Indian survival, race hatred was probably the norm for most. When Eleazar Wheelock—a Yale-trained missionary and founder of an Indian school—passed his collection plate at a church in Windsor, it was returned empty "but for a bullet and flint." As James Axtell, one of the foremost scholars of European-Indian relations in the colonial period, has written, the colonists viewed the Indians as "sometime adversaries and full-time contraries."

With no links to the Indians and rather infrequent contact with Europe, most colonists settled down to a simple rural life. Seen from a European perspective, these colonies were sleepy provinces—in historians' terms, the periphery. For colonists born in America, their towns and farms were the center of the universe. Those with the money to do so tried their best to be fashionable in a European manner. Some traveled to old England and were reluctant to return to America. Increase Mather, the famous Puritan minister, was one who longed for London life. He envied his son Samuel, who had moved to the metropolis where he could "furnish himself with variety of books." For those with a cosmopolitan inclination, the problem with English America in the seventeenth century was not only its distance from Europe but also its lack of urban amenities and social distinctions. Ann Eaton, wife of the governor of New Haven, was anxious to return to England. Lady Deborah Moody, a friend of the Winthrops and a proper Puritan, left Massachusetts in part for religious reasons but also because she found the social atmosphere stifling. She chose to move to New Netherland.

Handelstijd: Dutch New Netherland

The golden age of the Dutch Republic coincided with the first wave of colonizing efforts that swept over North America in the seventeenth century. Viewing the commercial opportunities of the entire world, many Dutch merchants concluded that North America had limited possibilities; nevertheless, a few—such as Arnout Vogels and Lambert van Tweenhuysen—sent ships to the region referred to by 1614 as New Netherland. When Dutch sea captains and traders explored the Noordt Rivier (Hudson River), they looked over the country with Dutch eyes. They searched for places to trade and people to trade with, for routes into the interior and ways to connect. As the historian

Donna Merwick has reminded us, the Dutch were true townsmen who valued communication and "purposeful movement." In Holland, the countryside was dependent on the city; the city was a world in motion encompassing the activities that brought virtue and prosperity. Heroes and leaders came from a *burgerlijk* society and represented its values. They were not landed aristocrats. Jeremias van Rensselaer, sensing the need to know more about the English, who might take over New Netherland, wrote home and asked for a map of England. Dissatisfied with the first one he received, he wrote, "Send me another map . . . but the country must be set off better by cities, for . . . the other map was no good, there were no cities shown on it."

The Dutch came to North America to possess the commerce of the country. Occupying the land was, at best, a secondary consideration. The value of agricultural land had to be compared with the value of other commodities. A good yacht, a consignment of beaver pelts, or a prime location in Manhattan might be worth considerably more. New Netherland did not remain Dutch for long: only until 1664 with a brief Dutch interlude in 1673–74. Nevertheless, Dutch culture in New York had considerable staying power. The Dutch created what would become the dominant city in North America and modeled and named it after Amsterdam. Even distant Beverwijck (Albany), named for the commodity that justified its existence, had its *stadhuis* (town hall) and prominent *handelaars* (merchants). Connections mattered most to the inhabitants of a frontier trading post. Traders literally knew their world. Many Beverwijck burghers had property in New Amsterdam and business associates in old Amsterdam. Distances were deceiving: rivers were meant to be navigated, and people, their letters, and business accounts traveled back and forth. The pioneers of New Netherland, in short, were pioneer agents of trade poised to create markets in the wilderness.

This was, perhaps, a more exciting prospect than the conversion of Indian acres into European farms. How far and in which directions might a trading area extend? Which lines of trade would prove profitable? Which goods would satisfy native customers, and what might these customers produce that would sell in Amsterdam? Indeed, native North America had cultivated acres and boundaries before Europeans, but there were no stores, no merchants. Here was an opportunity: a power to shape, in a sense, virgin territory. Rural colonies, of course, had their entrepôt cities; tidewater Virginia was exceptional. Hartford, New Haven, Boston—these pioneer market towns gathered—in the words of the geographer James E. Vance, Jr.—the "periodic threads of trade" from their hinterlands into a ball of sufficient magnitude to allow for transatlantic exchange. In New Netherland and New France, where the fur trade greased the flow of transatlantic commerce, frontier entrepôts such as Detroit, Michilimackinac, St. Louis, and Beverwijck sprang up in Indian country. Moreover, these places and the larger cities they were connected to—Nouvelle Orleans, Montreal, and Nieuw Amsterdam—dominated the life of the colonies in which they were situated. Historians who have studied these places have been surprised to find merchants on the edge of the "wilderness" in regular contact with European correspondents, enjoying an unexpected array of urban amenities. If life on this frontier was fundamentally urban, even from the beginning, what difference did this make?

It made a difference in cross-cultural relations. Trading posts required profitable links with host communities. The Mahicans gave the Dutch permission to establish Fort

Nassau on the Hudson in 1614, opposite a Mahican village. There the chief trader became fluent in Mahican—an Algonquian language—and enjoyed good relations with his customers. When the Dutch abandoned Fort Nassau and established Fort Orange near present-day Albany in 1624, the Mahicans moved their village nearby on the opposite shore. Generally speaking, the Dutch got along with their Indian trading partners. (They were less diplomatic with natives who could not supply furs.) Unlike French traders, most Dutch traders did not make a regular practice of visiting, residing in, or marrying into Indian villages. They did, however, eagerly desire the natives to come to Fort Orange or Beverwijck. During the trading season—*handelstijd*—from 1 May to 1 November, hundreds of Indian people arrived. Sheds and temporary houses were erected outside the palisades of the town. Some Indian families even stayed at burghers' homes.

Trade required communication. The Unami- and Munsee-speaking Delaware people who traded with the Dutch on the South (Delaware) River and around New Amsterdam developed a trade jargon. The Munsee-speakers also borrowed Dutch words for *appel, knoop* (button), *komkommer* (cucumber), *melk, pannekoek* (pancake), *suiker* (sugar), and many other items. The Jersey Dutch, in turn, used the words *spanspak* and *tahaeim* (Munsee for cantaloupe and strawberries). These examples give us some idea of the nature of the cultural exchange occurring. (The Delawares also borrowed a Dutch slang word for penis, which suggests another kind of exchange.) Although the two peoples lived apart and maintained some distance, physical and social, between one another, good business demanded a certain degree of mutuality. The Indians themselves probably put it best in the following complaint, made several decades after the English takeover of the colony: "When the Dutch held this country long ago, we lay in their Houses; but the English have always made us lie without Doors."

It should be understood that good relations were helped immeasurably by the simple fact that trading posts and incipient cities held small numbers of Europeans and kept them in an enclosed space away from native lands. Indeed, many shareholders in the Dutch West India Company (the Westindische Compagnie, or WIC), the national joint-stock company responsible for administering the colony, wished to keep the trading-post regime as uncomplicated as possible. Unfortunately for the advocates of a pure trading-post regime, the French and the English protested against the presence of Dutch traders in the region. Therefore, in 1624, the WIC settled thirty Walloon (French-speaking) families at four locations on the Delaware and Connecticut rivers, at newly built Fort Orange and on Governor's Island in New York harbor. In short, the new colonists were sent to stake out Dutch claims in North America and thus to substantiate the limits of the proposed colony.

Settlement increased slowly at first, but friction between Indians and the Dutch grew with a series of incidents at Staten Island and Pavonia in 1640 and 1641. When Director General Willem Kieft led a massacre of over eighty Indians outside Communipauw in Pavonia in 1643, a full-scale war ensued. Kieft's policy was not supported by all the colonists: influential men such as the trader David de Vries, who negotiated a temporary peace in 1643, and Dominie Everardus Bogardus, New Amsterdam's Dutch Reformed minister, were vocal in their opposition. Officials back in the United Provinces were also displeased with Kieft and—after a combined Dutch-English force destroyed Indian villages at Hempstead, Fort Neck (Massapequa), and Pound Ridge—replaced him with Peter Stuyvesant in 1647. Stuyvesant was generally more conciliatory, but conflicts occurred again in 1655—the so-called Peach War—and in 1659 and 1663. Many of these struggles hinged on incidents typical of agricultural resource competition: the stealing of a peach or the appropriation of an Esopus Indian field. Indian villagers eventually responded by moving away and establishing refugee camps with neighbors farther inland.

The fur trade also played a role in these conflicts. Kieft had attempted to extort wampum from Indian coastal villages in 1639, but the Mahicans were more successful in extracting tribute from their Indian neighbors. The Wiechquaeskecks massacred by Kieft in 1643 were, in fact, seeking refuge from Mahican attacks. The Mahicans dominated the coastal villages and the supply of wampum and joined the attack on the Indians of western Long Island in 1655. The Mohawks, reacting to a depletion of animals in their own territory and encouraged by the Dutch, attacked northern tribes allied with the French in the 1640s and 1650s in an effort to control the flow of furs. (They also sought captives to be adopted into Iroquois villages to replace members lost through violence and disease.) From 1662 to 1675, the Mohawks, the Mahicans, and their respective Indian allies fought a series of battles for supremacy in the region. The competition between these regional rivals spread to distant areas as the search for furs and allies reached the lands of the Cherokees and the Creeks to the south and the Miamis and the Ottawas in the Great Lakes region. In short, a complex set of conditions involving trade, refugee movements, and Indian and European diplomatic and military attempts to control native decisions over a wide area had evolved by the late seventeenth century. If game depletion, alcoholism, the breakdown of traditional Indian society, and the continuing arrival of European settlers ultimately stacked the deck against many Native American communities and resulted in dependency and enclavement, it still must be said that the story was played out in large measure through Indian agency.

What can be concluded, then, about Dutch-Indian relations? Despite a great degree of cross-cultural contact and exchange, Dutch occupancy of the land resulted in the destruction or removal of Indian villages. Away from the areas of Dutch farms, Indian people retained a measure of autonomy and were necessary to Dutch interests. Many voices in the colony spoke in favor of good relations. At least after Kieft's administration, negotiations were generally favored over confrontations. Efforts to convert the natives were minimal. Measures aimed at preserving the peace were enacted. A law of 1640 ordered colonists to keep their livestock from straying into unfenced Indian fields. The best that can be said is that the Dutch had Indian allies and, therefore, endeavored to live with them.

It was clear to the Dutch by the 1640s that the trade diaspora models in Asia Minor and the Far East were not applicable to North America. A small foreign merchant community living within a host city was not possible: there were no native cities, and food surpluses were hard to extract from native villages. Therefore, the Dutch created new cities. Beverwijck came into existence in almost classic Old World fashion. The patroon of Rensselaerwyck, who owned the land around Fort Orange, had been granting trading privileges and lots to artisans who wished to live near the fort in hopes of grabbing a piece of the fur trade. Director General Stuyvesant, determined to protect the WIC's

Framed by two Indians exchanging furs, New Amsterdam appears in this Dutch print as an urban community dependent on commerce with the hinterlands, its prosperity intertwined with that of the Native Americans.

Engraved by Aldert Meijer. Nieu Amsterdam at. New York. *Colored engraving; date depicted, 1673; date issued, ca. 1700. I. N. Phelps Stokes Collection, Miriam and Ira D. Wallach Division of Art, Prints and Photographs, The New York Public Library, Astor, Lenox and Tilden Foundations, New York, New York.*

prerogatives, restated its claims to all land within a nineteen-hundred-foot radius of the fort in 1652 and declared Beverwijck to be a chartered town (*dorp*). The following year, New Amsterdam was finally granted municipal status. A city government was immediately installed, consisting of two burgomasters, a *schout* (sheriff), and five *schepens* (aldermen). The *vroedschap* (municipal council) exercised the city's privileges, which included collecting an excise tax on beer and wine and regulating bakeries, slaughterhouses, ferries, and other important urban institutions. The city also confirmed property titles within its jurisdiction. (Even Stuyvesant had to go to the city magistrates to perfect a title to land that he had granted to himself in the name of the WIC.) In all they did, the magistrates of Beverwijck and New Amsterdam followed the customs and practices of old Amsterdam back in the fatherland, or *patria*, as they called it.

The cities of New Netherland also came to resemble their counterparts in the *patria* visually and physically. Space was limited at first, and prime commercial property was at a premium. Even the initial house lots were small. In Beverwijck, they ranged from one-half to one-twentieth of an acre. Garden lots were smaller. By comparison, lots in New England market towns were rarely smaller than one acre. And in New Netherland, lots were quickly subdivided. Narrow frontage and close neighbors quickly became the norm. Merwick has calculated that house density in Beverwijck was remarkably similar to that of small towns in the Netherlands—an average of five houses to the acre—almost from the beginning, and "deeds frequently carried clauses covering damage resulting from a neighbor's downspout." Joint ownership of city property was not uncommon, and renting was a feature of New Amsterdam life early in its history. Dwellings made of wood were gradually replaced by ones made of brick and stone with stepped gables facing the street. Rather late into the colonial period, New York City still had a Dutch appearance, with tall buildings made of special red-and-yellow brick covered with red-and-black tile. One feature marked Beverwijck as different from the Dutch Old World and the English New World: an area to the west of town remained forested so that traders might "walk in the woods" with Indian customers.

Pioneer Dutch urbanites bought property outside of town; but when they did, it was often dispersed fragments of real estate—an investment, not a source of status. Status and power, on the contrary, derived from city rights—*burgerrecht.* Even in the English period, New Yorkers paid a fee for the *burgerrecht*, which brought the privilege of doing business. (In 1675, a shopkeeper paid six beavers, and an artisan paid two.) Business was the lifeblood of New Netherland's towns. Interdependency, not self-sufficiency, was the rule. Everyone specialized in at least one trade, and many inhabitants had more than one source of income. In Beverwijck, a tailor was also a half-owner of a bakery. A cordwainer also served as a notary. Capital was meant to be invested, and investments were spread around in any number of ventures. The Dutch practice of partible inheritance reflected a commitment to a dynamic economy that put resources into as many hands as possible. The hands included those of women, since Dutch sons and daughters received equal shares from the estates of their fathers and mothers. Spouses combined their property in a community of goods, which was divided in half at the death of either one. The custom of *boedelhouderschap* continued the community of goods after the death of a spouse and protected the widow's right to manage the estate. Not surprisingly, many women in New Netherland were experienced in business affairs. Many inns were kept by women, and many other women were merchants. Women also engaged in the fur trade, and one was thought to be the best interpreter in the colony.

New Netherland's commerce increased along with its population in the 1650s. With the collapse of the WIC's colony in Brazil in 1654, the passage of the first English Navigation Act in 1651—which was aimed at reducing the Dutch share of the carrying trade—and the Anglo-Dutch war of 1652–54, Dutch merchants and officials took a new and rather anxious interest in New Netherland. Propaganda and incentives produced the desired effect: a substantial migration of young families to the colony. Although the estimates are rough, the colony's population may have grown from twenty-five hundred in 1645 to around nine thousand in 1664. The population, never homogeneous, grew increasingly heterogeneous. People of Dutch ethnicity constituted less than half of the colony's population in 1664, with nearly 20 percent of the total being German, another 15 percent English, and substantial Scandinavian, French, Belgian, and African minorities. (The African minority, both free and slave, was itself ethnically diverse and came to represent a sizable segment of the population later in the colonial period.) As early as 1643, an observer in New Amsterdam noted eighteen different languages being spoken. Except for the Indians and the Africans, the colony's ethnic diversity was reminiscent of old Amsterdam. Indeed, when a group of Jewish refugees from New Holland in Brazil arrived in New Amsterdam in 1654, Stuyvesant—a confirmed anti-Semite—and the orthodox *predikanten* (ordained preachers) of the colony protested. The directors of the WIC told Stuyvesant to mind his own business and hold his tongue lest he offend the Jewish shareholders in the company. They repeated that advice some years later when Stuyvesant was involved in persecuting Quakers.

Living in or near a frontier city was different from being situated on and defined by the land. Living in New Netherland meant being connected to Europe by the Atlantic, not separated by it. Even the Dutch farmers of Brooklyn and Jersey came to the city often. On Tuesday and Saturday, the market days, the country people came to town, reenacting the customs of the Netherlands. As in the *patria*, waterways connected the countryside and the city. Traveling and communicating were crucial because the city held the keys to this culture, and in Merwick's words, the "*lantsman* who simply farmed . . . remained a person without status." But in this New World, partaking of the city meant something more, especially in New Amsterdam/New York where diversity meant that no one cultural tradition could be taken for granted. For those who lived farther away, traveling to Brooklyn or Manhattan meant renewing one's sense of identity—tasting the food of home, visiting the old church. The city was a reference point, a crucible of ethnicity. Braudel has remarked that cities served as "social amplifiers of markets." In the New World, they also amplified cultural traditions, providing the critical demographic mass necessary to reproduce social identities and form consumer communities for the products of the home country. For a colonial people who valued their sense of being connected, the city was a truly central place.

The Dutch wanted to see each other; they likened themselves to bees in a hive. English visitors always remarked on the crowded feel of places like New Amsterdam or French Detroit. The existence of "free" land clearly cannot explain everything about our American history. How the resources of the country were exploited—indeed, which resources were exploited—explains a great deal, but the cultural values Europeans cherished led them to live on the land in ways that must have seemed marvelously inappropriate to their Indian neighbors.

The irony of New Netherland seems to be that its colonists desired direct connections to both Indian country and Europe. Perhaps the irony derives from hindsight influenced by the Anglo-American perspective. In the rural settlements of the English colonies, Indians came to be regarded as alien, inhuman "others." The settlers' fear of declension, the fear that European standards of civility were disappearing among themselves, reinforced fears about the natives. In New Netherland, communication with Europe and with Indian communities assuaged anxiety over the savage within and without. In New England, captivity narratives described horrible scenes of frontier women being abducted by the "savages" and carried into the wilderness. This was the stuff of English nightmares. In New Netherland, women traders visited Indians at their villages and sold them goods for a profit.

Comptoirs: French Traders and Their Partners

The French, like the Dutch, were in North America to trade—although they did not know it at first. When Jacques Cartier crossed the Atlantic in 1534, his orders from the French crown were to search for gold and other precious metals and for a route to China (a search mockingly memorialized in Montreal, where the embarkation point above Sault-Saint-Louis is known as Lachine). Cartier discovered the Gulf of St. Lawrence, and then a group of Micmac (Mi'kmaq) Indians discovered Cartier and suggested that the

French forget China and trade for furs. The Micmacs showed great patience, waiting out several attempts by the French to frighten them off with weapons. The Micmacs returned the following day and persuaded the French to take their skins in exchange for knives, "other Iron wares, and a red hat to give unto their Captaine." The fur trade, of course, became a central, organizing factor in the life of the colony. For the remainder of the sixteenth century, seasonal cod-fishing expeditions and fur-trading fairs at Tadoussac, at the mouth of the Saguenay River, grew in importance without the benefit of permanent European establishments. Samuel de Champlain, who emerged as the first leader of New France in the first decade of the seventeenth century, finally established a habitation at Quebec in 1608. He also secured an alliance with the Hurons and through this alliance secured the Hurons' Iroquois enemies. Champlain began a policy of sending young traders to Indian villages to learn native languages and customs. The first such *hivernant*, or "winterer," was Étienne Brulé, who lived with the Hurons. According to the report of one missionary, he was familiar with too many Huron women, so his hosts killed him and ate him. W. J. Eccles—in one of the classic lines of Canadian history—observed, "He was the first Frenchman, but by no means the last, to be completely assimilated by the Indians."

Canada, like New Netherland, remained little more than a fortified warehouse, a *comptoir* in Indian country, for decades. Cardinal Richelieu formed a joint-stock company, the Compagnie des Cent-Associés, in 1627 in an attempt to build up the colony. The company got off to an inauspicious start when its ships, loaded with colonists and provisions, were captured by English privateers in 1628. The following year Champlain surrendered Quebec to an Anglo-Scottish expedition. A treaty restored French claims, and Champlain began to rebuild in 1633. The population of the colony remained low. As late as 1650, New France contained only 675 permanent settlers, one-quarter of the population of New Netherland and far less than the colonial population of New England.

Two factors complicated the task of establishing trading-post empires in North America: competition from rival European claimants and the lack of native cities and merchants. (Native Americans did, however, trade through fairly elaborate networks, traveling well-worn paths to distant communities.) In both New Netherland and New France, colonial officials recognized the need to establish and defend imperial boundaries—usually with the aid of native allies—and to encourage the growth of a European agricultural sector, though rural settlements in these colonies remained of secondary importance to the main business of trade. In the cities, *marchands* and *négociants* (the larger merchants who handled exports and imports) and the smaller *traiteurs* and *voyageurs* (the traders who handled cross-cultural exchange) made their homes. In Canada and Louisiana, as in New Netherland, cities—or villages in the process of becoming entrepôts—held the keys to development, the symbols of status, and the institutions that maintained cultural traditions. And so, the Dutch re-created Amsterdam on the Hudson, and the French left a trail of cities in North America, from Montreal to Detroit, New Orleans to St. Louis, St. Paul to Kansas City.

The fur trade provided the foundation. If the conversion of Indian land into European farms and plantations underwrote colonial development in Anglo-America, the production of furs and skins by Indian men and women and the consequent rise of native purchasing power underwrote the building of French-Canadian and Creole cities. The beaver was the main object of European affections. The soft, barbed underfur of the pelt was used in the manufacture of felt hats. *Castor gras* (greasy beaver)—a fur that had been worn for several seasons—was more valuable than *castor sec* (dry beaver) because the long guard hairs had been worn off. Markets also existed for moose hides and deerskins, worked into leather for a variety of manufactures. Finally, there were the peltries, or *pelleterie* (skins worn as furs), such as marten, raccoon, otter, and black fox.

Indian labor produced the majority of these furs. Men did the hunting and trapping; women processed the furs and hides through a variety of tasks such as scraping, stretching, rubbing, and curing. For many of the tribal peoples involved, such techniques were part of their cultural repertoire. Some groups had to learn: the Miamis, when the French first encountered them, were in the habit of roasting beavers—burning off the fur—and eating the animals. Women of the Crow tribe of Montana, on the other hand, were esteemed for their processing skills. Indian producers became Indian consumers. Axtell has described the infiltration of European goods into native societies in the seventeenth and eighteenth centuries as the "first consumer revolution." Indeed, metal tools soon became essential items. Iron axes, awls, chisels, knives, fishhooks, and kettles were more durable than their native equivalents and reduced the amount of labor required for many daily tasks. Red-and-blue woolen duffels and strouds, calico shirts, brandy and rum—these items were always in demand. Native Americans could be shrewd bargainers, playing off competing traders and demanding "good measure" and the extension of credit. They also had an eye for quality and fashion. Small paper packets of vermilion—used for body and face paint—came from distant China. Fashion-conscious Indian men counted European mirrors among their prized possessions. Glass beads desired during one trading season could easily be passé during the next.

Baron de Lahontan observed in 1690, "The trade in goods usually brings in a 700 per cent profit for the Indians get skinned" (*écorche les Sauvages*). The markup on trade goods was high, but the profits were not. The Canadian historian Louise Dechêne has calculated normal margins for traders and merchants to be around 10 to 15 percent. The trade involved many middlemen and many burdensome expenses, among them the distribution of gifts to Indian clients. The trade was not simply conducted or expressed in European terms. Indian customers usually articulated exchange in the language of reciprocity, of mutual gift-giving. For Indian villages, the arrival of European traders meant much more than commerce: it meant the establishment and maintenance of social and political relations. Of all the European groups involved in the trade, the French entered most deeply into native worlds.

At first, the French—like the Dutch—waited for native groups to come to them. Montagnais, Nipissing, and Huron canoes brought furs to Quebec in the 1630s, but Iroquois enemies to the south—eager to obtain furs to trade with the Dutch—began attacking parties of Frenchmen and their native partners. Thus began a pattern of warfare and destruction that would last until 1701 and encompass a vast area from the St. Lawrence to the Great Lakes. Iroquois actions forced French reactions. In 1634 a post was built at Trois-Rivières at the mouth of the St.-Maurice River to protect one transportation route to the north. The village quickly became an important training center and jumping-off point for the trade. The French experience began to diverge from that of the Dutch: French traders were leaving home and traveling to Indian communities. Another crucial distinction between the Dutch and the French colonies was the presence in the latter of active missionaries, eager to confront Indian societies and convert them to the true faith.

The paths to Indian country were followed not only by traders but also by clerics and soldiers. (This was true in seventeenth-century Canada and equally true in the nineteenth-century United States, when missionaries and soldiers followed fur traders from St. Louis up the Missouri River.) Recollet friars arrived first, reaching Quebec in 1615, but they had little success in Canada. Their standards of European civility and Franciscan models of reduction—based on experiences with sedentary Mexican natives—were of no use in this northern country. The Jesuits arrived in Canada in 1625 and, at first, pursued a similar plan based on the economic and social Frenchification of the natives. Failures at the reserve at Sillery, founded in 1637 several miles from Quebec, and at a *séminaire*, or boarding school, for Indian children north of the city forced the Jesuits to reconsider. By 1640, the Jesuits were ready to travel with Indian bands and take up residence in distant native villages. Depending on one's perspective, the Jesuits might be congratulated for their insights into native cultures and brave attempts to translate between worlds or condemned for their insidious efforts to undermine traditional gender roles, patterns of authority, and cultural inventories. One thing is definite: their knowledge of Indian customs and languages and their annual *Relations*, which publicized their work and the needs of the colony, were a great resource for the colonists. Nevertheless, not every Jesuit enterprise brought success. From 1647 to 1649, the Iroquois attacked the Hurons and destroyed not only their villages but also the Jesuit missions.

Ste. Marie, the largest of the missions, went up in flames in 1649. The Iroquois, searching for beaver, hunting grounds, captives to replenish their villages, and prisoners to torture, targeted the Hurons because they were the main trading partners of the French. The Iroquois then attacked several allied groups, the Eries, the Neutrals, and the Petuns, and pursued the refugees to the *pays d'en haut* (Great Lakes basin), where they attacked a number of Algonquian-speaking groups such as the Miamis, the Mascoutens, and the Ottawas. In short, for twenty years—from 1647 to 1667—the Iroquois blazed a path of destruction and created a new landscape of refugee communities in a vast area from Ontario to Illinois. The chaos disrupted the fur trade. French farms along the St. Lawrence also came under attack. An Iroquois chief claimed that the French "were not able to goe over a door to pisse."

At this critical point in the history of New France, Louis XIV and his minister, Jean-Baptiste Colbert, decided that the colonies were worth keeping and undertook an ambitious series of reforms to strengthen them. The crown's first act was to take the colony out of the hands of the Compagnie des Cent-Associés. By 1674, New France had become a royal colony. Colbert established a new form of government, designated the Coutume de Paris as the exclusive body of law in the colony, and reorganized the seigneurial system—the plan of settlement.

The seigneurial regime had the appearance of a feudal system, but in fact the seigneur's privileges were limited, and his relationship to his settlers, or *censitaires*, was contractual. In essence, the seigneur was a developer sanctioned by and responsible to the state. It was a system that guaranteed the construction of mills and provided for a minimum level of support for all newcomers. Land was granted to all in exchange for the payment of a nominal annual tax—the *cens*—and rents. A settler could sell his land, but a tax—the *lods et ventes*—had to be paid by the purchaser. The seigneur could also be replaced by the intendant if he neglected his duties. In short, it was a system that encouraged orderly development and discouraged speculation. Speculation was unlikely anyway. Land was cheap and markets were inaccessible. (Canadian ports were closed for over half the year.) Although grain surpluses were produced by Canadian farmers, West Indian markets were dominated by the English colonies. Colonial merchants had no interest in investing in land. Profits accrued from imports.

Rural Canada then was a self-sufficient, isolated world. The landscape assumed a familiar pattern: long, narrow rectangular lots fronting the river, grouped into areas, or *côtes,* that shared physiographic traits. Farmhouses situated along the river brought habitants closer together. This pattern of settlement facilitated the building and upkeep of roads and made ploughing easier—decreasing the number of times a team had to be turned around. Social life centered on the parish church. By the end of the French regime, farms stretched along the north and south banks of the St. Lawrence, giving the appearance of a continuous village. The forest remained intact behind the seigneuries.

That the *côtes* filled in at all was due in large measure to Colbert's policies. From 1665 to 1673, the crown sent approximately one thousand young women, the *filles de roi,* to Canada to help compensate for the pronounced gender imbalance—that is, to provide wives for the colonists and produce children. These young women were, for the most part, orphans and other girls without resources. The king provided dowries. In addition, grants and other incentives were provided for parents of ten children or more. The strategy seemed to work: the colony had a very high birthrate. Recent research has shown that immigration, though never comparable to the flow of people to British North America, was higher than previously thought. There were perhaps as many as sixty-seven thousand men and women emigrants to Canada during the French regime. Many, however, returned to France, giving a net migration calculated at roughly twenty thousand.

Of all those that came over, the soldiers of the Carignan-Salières regiment had the most immediate impact. This force of twelve hundred men arrived in 1665, built three forts on the Richelieu River—the route of Mohawk war parties—and took the war to Iroquois country. Although the French had limited success in battle, they burned enough cornfields to force the Iroquois—who were already fighting the Susquehannocks and the Mahicans on other fronts—to press for peace in 1667. Peace led to a new expansion of trade in the west. Although a flood of beaver depressed prices, the profits were still considerable, and engagés and unemployed soldiers headed for new fur frontiers in the *pays d'en haut.* A new governor, Louis de Buade de Frontenac, pursued an expansionist policy that ran counter to Colbert's plan to concentrate men and resources in the St. Lawrence region. Frontenac had a fort built—named after himself—on the eastern end of Lake Ontario in 1673. That same year Louis Jolliet and Father Jacques Marquette journeyed down the Mississippi. Supported by Frontenac, René-Robert Cavelier, Sieur de La Salle, explored the Illinois and Mississippi rivers and established a chain of posts. By 1680 there were reportedly eight hundred illegal traders, or *coureurs de bois*, in the western Indian lands. The trading fairs of Montreal (founded in 1642) declined and disappeared. Exchange now took place in Indian country.

The Iroquois, freed by 1677 of their other battles, responded to what they saw as an attempt to rob them of furs and hunting grounds. Other natives, however, forged tentative alliances with the French. In 1683 the French sent troops to fortify trading posts in the Great Lakes basin, thus beginning a succession of commandants who pursued private interests while—hopefully—following orders. Attacks in Illinois country in the 1680s gave way to attacks in Iroquoia in the 1690s. By 1701, Iroquoia was decimated. The fur trade at Albany had suffered during the war; convoys had continued to reach Montreal. The Grand Settlement of 1701 established a general peace. An illicit trade, known in New York as the Canada trade, developed between Albany and Montreal, with the Mohawks at Caughnawaga acting as intermediaries. (In the 1730s two Canadian sisters named Desaunier were directing smuggling activities.)

Despite a glut of furs, a severe drop in prices, and a temporary decision in 1696 to abandon the western country, the French were there to stay. The crown, faced with the threat of an English expedition from the Carolinas, ordered Pierre Lemoyne d'Iberville to establish a colony at the mouth of the Mississippi in 1698. (The ministers reasoned that Louisiana would check the expansion of the English colonies and also provide a strategic position from which to attack or defend Spanish possessions, depending on the outcome of the impending war over the Spanish succession.) Detroit, founded in 1701 by Antoine de la Mothe Cadillac, and Michilimackinac, located on the passage between Lakes Huron and Michigan, became the centers of the *pays d'en haut.* Canadian merchants opposed the establishment of Detroit because it was too close to the Iroquois and Albany and of Louisiana because a new port on the Gulf of Mexico would loosen their control over the trade and the traders who were in their debt. Imperial policymakers won the day, and their plans were carried out by a new generation of Canadian-born soldiers and entrepreneurs, many of them members of the Lemoyne clan, ready to realize profits and policy at the same time.

New France, entering its golden age in the first three decades of the eighteenth century, now had two distinct socioeconomic worlds: one was rural, self-sufficient, and rather static; the other was urban and dynamic. Canada came to be more and more a strictly European colony. In Louisiana, the Illinois country, and the Great Lakes basin, French cities and villages developed alongside Indian villages. On this frontier, described by one historian as "one cabin with two fires," Europeans and natives lived side by side. There were violent struggles between the French and certain tribes such as the Fox and the Natchez; indeed, there were problems, large and small. Nevertheless, as the historian Richard White has noted, "Their knowledge of each other's customs and their ability to live together . . . had no equivalent among the British." Here, natives and Europeans found that their different goals were complementary. The French posed no demographic threat. Although French villages in the Illinois country developed agricultural surpluses and although Louisiana, including Illinois, had a colonial—African and European—population of some ten thousand by 1750, the landscape of Indian life had not been seriously altered. The fur trade depended on the integrity of that landscape. (Of course, there came a point when game animal populations were depleted. Once that point was reached, cross-cultural relationships often deteriorated.) Exchange also depended on a variety of bridges such as language and marriage.

In the first few decades of the eighteenth century, French-Indian marriages in Illinois predominated. Intermarriage and clan adoption were ways of going beyond bridges; they formed the basis of a true middle ground. People of mixed ancestry—métis—became an important part of this shared world. Some, like Jean-Baptiste Richardville, of Indiana, assumed positions of authority within Indian society. He became the principal chief of the Miamis in 1814 and was said to be one of the richest men in Indiana at the time of his death in 1841. Others, like Antoine Leclaire, the founder of Davenport, Iowa, and the son of a Canadian trader and a Potawatomi woman, opted for status in the non-Indian world. Ultimately, when this frontier had passed into history, many métis people came to regard themselves as a distinct people. In eighteenth-century Illinois, however, intermarriage was simply one expression of peaceful coexistence.

Another intriguing expression of cultural negotiation can be found in some of the deeds from colonial Illinois. One such deed between a Frenchman and a French husband-and-wife couple advises the purchaser that the deed is valid *unless* the Indians decide to take back the land in question. Such caveats would not be found in a deed from New England, where Englishmen simply assumed that English sovereignty was final and complete. In early Illinois, the French made no such assumptions.

So Louisiana and the Illinois country—which became part of Louisiana in 1717—represented, perhaps, the logical outcome of a trading-post frontier in native North America. In a region without cities, the French established New Orleans in 1718 and St. Louis in 1764. Though they conceived the cities, French colonists relied to an extraordinary extent and for an unusually long period of time on Indian communities for food supplies. Some Indian communities relocated to the suburban districts to serve such a function. Even in the nineteenth century, Indian vendors continued to bring wild turkeys, venison, filé powder, and other products to the markets of New Orleans, Natchez, and St. Louis. African, Indian, and French cooks all contributed to the development of such regional delights as jambalaya and gumbo. Louisiana would later become famous for another unique regional contribution to our national heritage, jazz. It has been suggested that the term *jazz* derives from the French word *jaser*, "to chatter." How appropriate that this polyphonic musical form would be born on a frontier so full of cultural conversations.

Adjustment

The four different frontiers each represented an extension of European languages, laws, people, plants, and animals. Each one represented, in addition, a unique way of mediating between the Old World and the New. European permanence in Virginia and New England required the transformation of the land and the social reorganization of its management. The Dutch and the French, primarily concerned with the acquisition of commodities produced by native peoples, built urban bases. In areas shared by French and Indian peoples, a mixed culture began to develop. All colonies depended on capital

Unidentified artist. Pierre Chouteau, Jr. *(1789–1865). Oil on canvas, ca. 1820s. Art Collection (#POR-C-106A), Missouri Historical Society, St. Louis.*

Pierre Chouteau, Jr., and Jean Baptiste Richardville grew up in a world of cross-cultural exchange and conversation. Though both became wealthy businessmen, they chose different cultural paths. Chouteau thrived in the American world and passed along his social prestige to his descendents, who occupied positions of status in St. Louis. Richardville, the son of a French-Canadian man and a Miami Indian woman, abandoned white dress and stopped speaking French or English in his later years. His Miami world in Indiana was shattered by the policies and practices of the United States and its citizens.

RICHARDVILLE
The Head Chief of the Miami tribe of Indians

James Otto Lewis (1799–1858). Jean Baptiste Richardville *(ca. 1761–1841). Lithograph, ca. 1835. Marion Public Library, Indiana.*

investments—living away from home was an expensive proposition. Economic activities certainly played a major role in shaping new societies, but there were other factors. The New England landscape reflected a social and religious agenda born in England and nurtured on American soil. The crowded, narrow streets of New Amsterdam recapitulated old Amsterdam without any reference to the available space of seventeenth-century Manhattan. In Virginia, the seeds of independence were planted in American soil by a status-conscious would-be landed gentry. The Creole and Indian world of the Mississippi Valley was forged in the crucible of North American geopolitics.

Ultimately, international rivalries produced a transoceanic war—the Seven Years War—fought on two continents. In 1763, the British emerged victorious on each front, and the French nation lost its North American empire. But the French people did not leave, nor did their native allies. Frenchmen and -women continued to occupy key settlements from Mobile to Detroit, from Quebec to Montreal. Only the imperial maps had changed significantly. While British land speculators projected their visions onto

Founded as a commercial center on the edge of a frontier, St. Louis exemplified the French form of community-building in the New World. Its strategic location near the confluence of the Mississippi and Missouri rivers made it an ideal base from which merchants could operate as middlemen between the Indian communities to the west and the burgeoning American populations to the east. The trade and transportation links established during the French period later helped the city become a leading American transportation and industrial center in the late 19th century.

John Caspar Wild (ca. 1804–46). Front Street, St. Louis. *Lithograph, 1840. Missouri Historical Society (#SS 589), St. Louis.*

surveyors' grids, Creole merchants and their Indian partners continued to travel up and down rivers and continued to live in those supposedly empty spaces.

The process of adjusting frontiers, of remaking the maps, created some wonderful ironies. When George Rogers Clark recaptured Vincennes in 1779 during the American Revolution, his troops included Illinois Frenchmen. The "British" army that surrendered to the Americans was composed of French militiamen from Detroit. Some of the French held commissions from both armies. French settlers at Vincennes became quite adept at taking oaths of allegiance. This masquerade in present-day Indiana led to the extension of Virginia's jurisdiction and eventually to the inclusion of this frontier within the boundaries of the United States. Across the Mississippi River in Spanish St. Louis, French-Creole traders such as Auguste Chouteau—whose portrait suggests a self-styled western Napoleon—and their native partners continued about their business, even after the transfer of sovereignty in 1804, which necessitated a new oath of allegiance. Pierre Chouteau, Jr., Auguste's nephew, created a fur-trade empire in the trans-Mississippi West from 1813 to 1865. Chouteau's fur company realized a centuries-old French colonial dream. Company steamboats plied western waters; the Chouteau name was emblazoned on flagpoles and stamped on Indian medals. Government officials, tourists, artists—all who traveled in the vast regions united by company trading posts—relied on Chouteau's good offices and the company's hospitality. At the same time, Chouteau and his firm looked toward the settlement frontier advancing from the east. With connections to tribesmen in Indian country and to federal officials and politicians in Washington, the company—in a perfect position to assist in the process of extinguishing Indian title—pioneered the use of government funds to pay tribal debts in land-cession treaties. Lobbyists pressured Congress to pass appropriations favorable to the company. When emigrants on the Oregon Trail needed military protection, Chouteau sold Fort

Laramie to the government. Money from all these sources was reinvested in railroad stocks, steamboat lines, and iron mines. Several frontiers intersected in St. Louis. Chouteau, in many ways an old-fashioned, French-speaking *négociant* presiding over a family firm, understood this confluence and profited during this critical period of adjustment. He died a very wealthy man, and with his death—in 1865—the roar of national expansion drowned out the faint echoes of passing colonial frontiers.

Bibliographic Note

Many innovative studies of the colonial era have appeared in the last two decades. Jack P. Greene and J. R. Pole, eds., *Colonial British America: Essays in the New History of the Early Modern Era* (Baltimore, 1984), contains thematic essays by the leading historians in the field. T. H. Breen's contribution, "Creative Adaptations: Peoples and Cultures," elaborates on the notion of "charter groups" and the cultural "conversations" that generated new societies in North America. Three broad reinterpretations of colonial British America have appeared in the last decade. David Hackett Fischer, *Albion's Seed: Four British Folkways in America* (New York, 1989), and Bernard Bailyn, *The Peopling of British North America: An Introduction* (New York, 1986), both emphasize the process of transplantation, but there the resemblance ends. Jack P. Greene's *Pursuits of Happiness: The Social Development of Early Modern British Colonies and the Formation of American Culture* (Chapel Hill, 1988) offers the concept of "social development" as a key to understanding the interaction between metropolitan "inheritance" and American "experience." D. W. Meinig's *The Shaping of America: A Geographical Perspective on 500 Years of History*, vol. 1, *Atlantic America, 1492–1800* (New Haven, 1986), casts a wider spatial and temporal net in an effort to conceptualize the interactions between "three Old Worlds": Europe, America, and Africa.

Histories of Europe and the forces that led to and encompassed the colonization of America abound. Fernand Braudel's magisterial *Civilization and Capitalism, 15th-18th Century*, published in English translation in three volumes—*The Structures of Everyday Life*, *The Wheels of Commerce*, and *The Perspective of the World* (New York, 1982–84)—is comprehensive yet full of insightful analysis and rich detail. Immanuel Wallerstein, *The Modern World-System*, 3 vols. (New York, 1974–89), focuses more strictly on the history of European and global economies and the relationship between core-states and peripheries. The early patterns of European expansion are discussed in Charles Verlinden, *The Beginnings of Modern Colonization*, trans. Yvonne Freccero (Ithaca, N.Y., 1970), and Felipe Fernández-Armesto, *Before Columbus: Exploration and Colonisation from the Mediterranean to the Atlantic, 1229–1492* (Basingstoke, Hampshire, 1987). J. H. Parry, *The Age of Reconnaissance* (Cleveland, 1963), remains a lucid overview of European exploration and conquest, with a concise discussion of the means and preconditions of discovery. Carlo M. Cipolla, *Guns, Sails, and Empires: Technological Innovation and the Early Phases of European Expansion, 1400–1700* (New York, 1965), examines in greater depth the impact of developing technologies.

The Columbian quincentenary witnessed a flood of books on the Genoese mariner, the age of discovery, and the consequences of the encounters between Europeans and Native Americans. Alfred W. Crosby, Jr., *The Columbian Exchange: Biological and Cultural Consequences of 1492* (Westport, Conn., 1972), pioneered the study of "species shifting." His *Ecological Imperialism: The Biological Expansion of Europe, 900–1900* (Cambridge, Eng., 1986) contains an examination of what he terms "Neo-Europes." Crosby's work inspired a Smithsonian exhibit and companion volume to mark the Columbian anniversary: Herman J. Viola and Carolyn Margolis, eds., *Seeds of Change: A Quincentennial Commemoration* (Washington, D.C., 1991). Stephen Greenblatt's *Marvelous Possessions: The Wonder of the New World* (Chicago, 1991) is an elegant and sophisticated reading of the encounter. Those wishing to explore the sometimes depressing, often surprising, and always fascinating history of native-colonial interaction should begin with any of James Axtell's collections of essays: *The European and the Indian: Essays in the Ethnohistory of*

Colonial North America (New York, 1981); *The Invasion Within: The Contest of Cultures in Colonial North America* (New York, 1985); *After Columbus: Essays in the Ethnohistory of Colonial North America* (New York, 1988); and *Beyond 1492: Encounters in Colonial North America* (New York, 1992).

David B. Quinn, *North America from Earliest Discovery to First Settlements: The Norse Voyages to 1612* (New York, 1977), provides a comprehensive and clear overview of European exploration and settlement in this early period. Ralph Davis, *The Rise of the Atlantic Economies* (Ithaca, N.Y., 1973), and K. G. Davies, *The North Atlantic World in the Seventeenth Century* (Minneapolis, 1974), remain useful surveys of imperial systems. Two volumes by Philip D. Curtin—*Cross-Cultural Trade in World History* (Cambridge, Eng., 1984) and *The Rise and Fall of the Plantation Complex: Essays in Atlantic History* (Cambridge, Eng., 1990)—are invaluable for their global perspective and thematic focus. Kenneth R. Andrews, *Trade, Plunder, and Settlement: Maritime Enterprise and the Genesis of the British Empire, 1480–1630* (Cambridge, Eng., 1984), provides an urbane inquiry into a variety of topics pertaining to English expansion.

As for the literature on specific colonial frontiers, Edmund S. Morgan's *American Slavery American Freedom: The Ordeal of Colonial Virginia* (New York, 1975) remains the best single examination of North America's first boom-time frontier. Essays by E. Randolph Turner and Frederick J. Fausz in William W. Fitzhugh, ed., *Cultures in Contact: The Impact of European Contacts on Native American Cultural Institutions, A.D. 1000–1800* (Washington, D.C., 1985), and Helen C. Rountree, *The Powhatan Indians of Virginia: Their Traditional Culture* (Norman, 1989), round out the narrative of Indian-English relations on that frontier. T. H. Breen, *Tobacco Culture: The Mentality of the Great Tidewater Planters on the Eve of Revolution* (Princeton, 1985), is a revealing study of the colony's values and habits. Some of Breen's seminal essays on Virginia and New England are collected in *Puritans and Adventurers: Change and Persistence in Early America* (New York, 1980). The theme of continuity and the persistence of local traditions in New England is pursued in depth by David Grayson Allen, *In English Ways: The Movement of Societies and the Transferal of English Local Law and Custom to Massachusetts Bay in the Seventeenth Century* (Chapel Hill, 1981). David Cressy, *Coming Over: Migration and Communication between England and New England in the Seventeenth Century* (Cambridge, Eng., 1987), is a brilliant and lively account of the move itself and provides many insights into the self-selection process. Francis Jenning's *The Invasion of America: Indians, Colonialism, and the Cant of Conquest* (Chapel Hill, 1975) provides a thoroughly revised narrative of New England's settlement, taking into account the natives of the region and their treatment at the hands of the English. Neal Salisbury, *Manitou and Providence: Indians, Europeans, and the Making of New England, 1500–1643* (New York, 1982), provides an in-depth ethnohistorical account of the region's formative years. William Cronon's *Changes in the Land: Indians, Colonists, and the Ecology of New England* (New York, 1983) offers a brilliant and original assessment of the conflicting ways natives and colonists managed natural resources and the ecological transformation effected by the English invasion.

Donna Merwick's *Possessing Albany, 1630–1710: The Dutch and English Experiences* (Cambridge, Eng., 1990) captures the distinctiveness of the Dutch approach to the New World. It should be supplemented with Oliver A. Rink, *Holland on the Hudson: An Economic and Social History of Dutch New York* (Ithaca, N.Y., 1986). Joyce D. Goodfriend, *Before the Melting Pot: Society and Culture in Colonial New York City, 1664–1730* (Princeton, 1992), a valuable study of ethnicity and community, sheds much light on Dutch culture in colonial America. Allen W. Trelease, *Indian Affairs in Colonial New York: The Seventeenth Century* (Ithaca, N.Y., 1960), remains the standard treatment but should be used in conjunction with Daniel K. Richter, *The Ordeal of the Longhouse: The Peoples of the Iroquois League in the Era of European Colonization* (Chapel Hill, 1992). Thomas Elliot Norton, *The Fur Trade in Colonial New York, 1686–1776* (Madison, 1974), traces the activities of Dutch and British Albanians in a later period.

Historians of New France, appropriately enough, have been among the leaders in the effort to write narratives that include native peoples as significant actors. In addition to the work of

James Axtell, Cornelius J. Jaenen's *Friend and Foe: Aspects of French-Amerindian Cultural Contact in the Sixteenth and Seventeenth Centuries* (New York, 1976) explores the effects of a variety of cross-cultural relationships. Bruce G. Trigger, *Natives and Newcomers: Canada's "Heroic Age" Reconsidered* (Kingston, 1985), revises the early period of the colony's history, and Olive P. Dickason, *Canada's First Nations: A History of Founding Peoples from Earliest Times* (Norman, 1992), provides an Indian-centered narrative of Canadian history. The standard surveys are W. J. Eccles: *The Canadian Frontier, 1534–1760*, rev. ed. (Albuquerque, 1983) and *France in America* (New York, 1972). Louise Dechêne, *Habitants and Merchants in Seventeenth Century Montreal*, English ed. (Montreal, 1992), originally appeared in 1974 and has achieved the status of a classic over the years. Richard White, *The Middle Ground: Indians, Empires, and Republics in the Great Lakes Region, 1650–1815* (Cambridge, Eng., 1991), explores the French and Indian world of the *pays d'en haut* with the creativity and mastery of a novelist. Daniel H. Usner, Jr., *Indians, Settlers, and Slaves in a Frontier Exchange Economy: The Lower Mississippi Valley before 1783* (Chapel Hill, 1992), rescues Louisiana from historiographical oblivion and reveals a fascinating and colorful story. William E. Foley and C. David Rice, *The First Chouteaus: River Barons of Early St. Louis* (Urbana, 1983), examines the first generation of a remarkable Creole dynasty. Jacqueline Peterson and Jennifer S. H. Brown, eds., *The New Peoples: Being and Becoming Métis in North America* (Lincoln, 1985), provides an essential introduction to the subject. Carolyn Gilman, *Where Two Worlds Meet: The Great Lakes Fur Trade* (St. Paul, Minn., 1982), is perhaps the best short overview of the fur trade and is beautifully illustrated. Finally, R. Cole Harris, ed., *Historical Atlas of Canada*, vol. 1, *From the Beginning to 1800* (Toronto, 1987), a visual feast of charts and maps, is a textbook in and of itself.

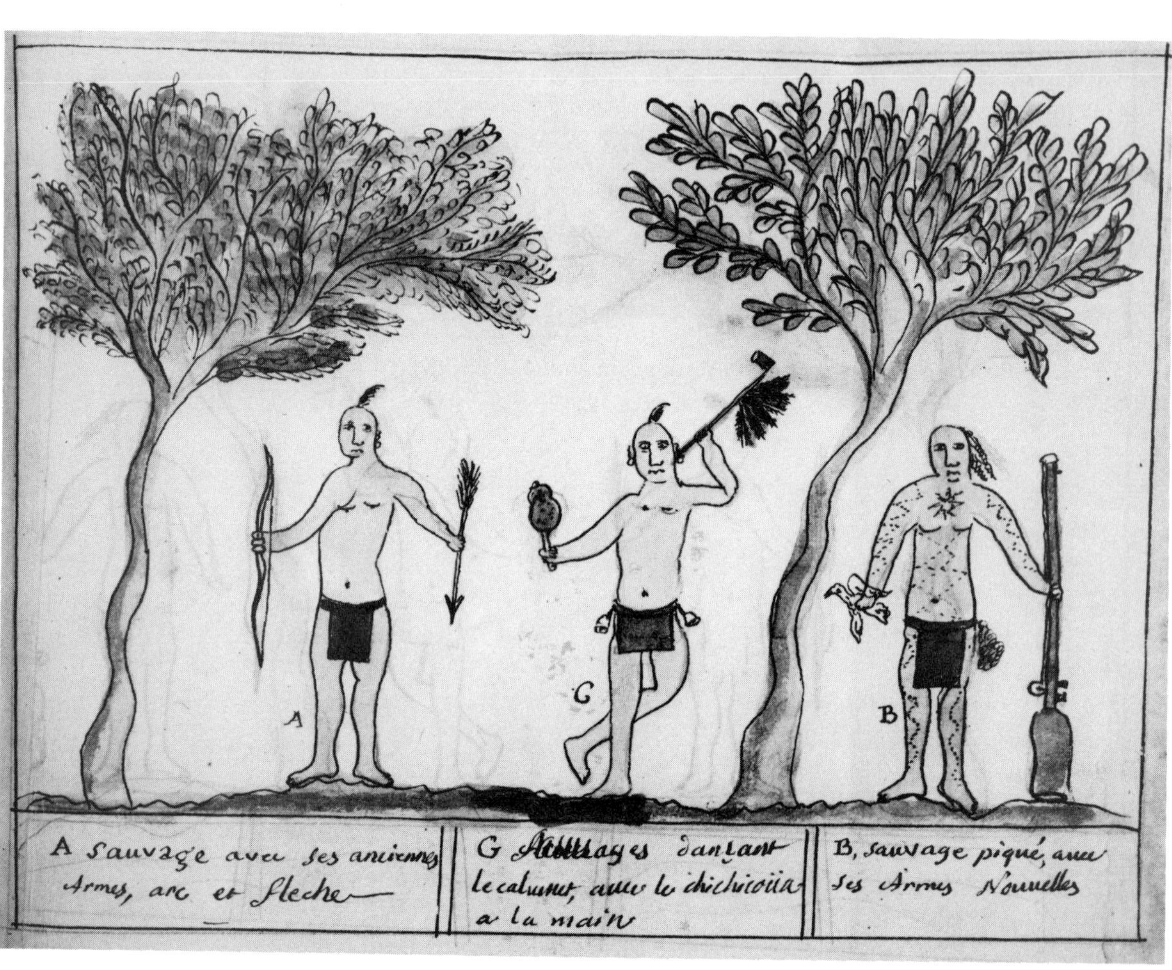
A
C
B
A sauvage avec ses anciennes Armes, arc et fleche
G Sauvages dançant le calumet, avec le chichicoüa a la main
B, sauvage piqué, avec ses Armes Nouvelles

Chapter Four

American Frontier

ELLIOTT WEST

Is any word in the American idiom more evocative and elusive than *frontier*? Few persons can agree on what the frontier was, yet few will deny that it existed and that before its passing it somehow helped shape the nation.

Trans-Appalachian America is a case in point. The frontier certainly left its stamp on this region of more than half a million square miles between the Appalachian plateaus and the Mississippi River. Through the region's standard histories walk figures—from Daniel Boone and the "men of western waters" to Tecumseh and William Henry Harrison—who became, even in their own lifetimes, icons of the "pioneer experience." That vague phrase, and particular images of this American frontier, are bound up as well with popular perceptions of the American people and their emerging character.

Just what the frontier was, however, remains something of a puzzle. It refers in some way to the emigration from the East of thousands of hunters, trappers, traders, farmers, merchants, soldiers, and other assorted floaters. This intrusion into the lands beyond the Appalachians was under way by 1763, and by 1820 its work there was largely done.

That much is clear enough. But the particulars of the frontier's meaning—its definition and its results—are slippery. It is helpful to think of the frontier as both a condition and a historical force. This frontier had certain traits that set it apart from the rest of the continent during these years. Those traits in turn shaped the land and people of trans-Appalachia, leaving marks that ever since have helped define that country as one of the nation's distinctive regions. Besides a condition and force, the frontier was also an anticipation. After 1820, as the wave of emigration rolled beyond trans-Appalachia, most of these traits would reappear in the lands between the Mississippi and the Pacific. Then the Far West, in its turn, would feel the frontier's shaping power.

Of the frontier's defining characteristics, five were paramount. First was its human diversity. Between the Appalachians and the Mississippi was a wider range of cultures than east of the mountains—more than thirty widely varied Indian groups as well as French, English, and Spanish communities, all of them joined by an influx of westering emigrants of many different cultural traditions. That, in turn, contributed to the second characteristic—an intricate set of power relationships. Native peoples exercised a high degree of autonomy. Doing so, they dealt with other centers of power, older nations in Europe and a new one on the Atlantic coast. The strength of these nations, although much greater than that of particular Indian groups, was exercised from afar. Like an expanding circle of light, that power was often felt rather dimly in trans-Appalachia, so whites and Indians were frequently on something close to an equal footing. The resulting play of authorities was arguably more complex than anywhere else in the Western world.

The trans-Appalachian frontier was a dynamic place of cultural exchange between Indians and Europeans. As this watercolor sketch by a French officer in Louisiana suggests, Native Americans quickly adapted European technology for their own purposes.

Jean Benjamin Francois Dumont dit *Montigny (1696–?).* Indians with Old and New Arms. *Watercolor on paper, finished in 1747. From Jean Benjamin Francois Dumont* dit *Montigny, "Memoire de Lxx Dxx Officier Ingenieur . . . a la Louisiane depuis 1715 jusqua'a present. . . ." Courtesy The Edward E. Ayer Collection, The Newberry Library, Chicago, Illinois.*

Behind this interplay of power was the frontier's third defining trait—a diversity of appealing resources. These were certainly not unused; Indians had flourished on the land's bounty for at least ten millennia. But newer arrivals wanted to use the land in new ways. And besides, the region's potential—its economic meaning—was changing rapidly with the stirrings of industry and the rise of a capitalist, market-oriented economic system on both sides of the Atlantic. Value, like beauty and perversion, depends on who does the looking, so the country's promise expanded with the changing economic perspectives of Indians and whites alike. This evolving economic meaning was basic to the fourth element of the frontier—its symbolic significance, most of all for the people of the youthful United States. To them, the country not only offered opportunities to individuals but also ensured a true independence and survival of the nation's distinctive virtues. The frontier became inseparable from the Republic's sense of itself.

All four elements combined to make the fifth. This American frontier was a region of extraordinarily dynamic changes, transformations that were arguably of greater depth and scope than anywhere else in the Atlantic community. The frontier was a competitive arena of a few dozen powers whose varying strengths were ebbing and surging. It was a place of rich cultural exchanges among a changing mix of different, evolving peoples. Here a new nation tested and modified its emerging institutions. The land itself, and perceptions of it, changed dramatically. So did what people wore, how they fought, and why and how they hunted deer and sought salvation.

This American frontier was only the latest of many to enter trans-Appalachia. Indian peoples had been moving into this country in successive waves for at least ten thousand years. In 1763, more than one hundred thousand Native Americans occupied the region. They included more than two dozen groups, varying in population from a few hundred to twenty thousand. In contrast to the common image of Indians rooted for eons to their home sod, most of these peoples had been in movement for generations—some migrating from the East under pressure of European expansion, some being displaced by newcomers, and others expanding their range against weaker competitors. Their cultural, religious, and artistic lives were varied, rich, and complex. Certainly this frontier was not what its most famous scholar, Frederick Jackson Turner, once called it—"the line between civilization and savagery." It was, rather, a zone of exchange and conflict among many civilizations.

Indian peoples had felt the presence of Europeans for more than two centuries before 1763. The *entrada* in the 1540s of the Spanish conquistador Hernando de Soto was the first European intrusion into the interior of the Gulf Coast. Afterward, the Spanish concentrated their attention on the Atlantic side of the Florida peninsula and only in 1698 did they establish an outpost at Pensacola to counter French influence. By then English traders among the Cherokees and Creeks were causing grave concern among both the Spanish and the French. The English also had established themselves as patrons of the powerful Iroquois of western New York, and by the 1740s their traders were among the Indians of the Ohio Valley.

The French, however, were the dominant European force in trans-Appalachia. After establishing a base at Quebec on the St. Lawrence River in 1608, the French, eager to expand their lucrative fur trade, began one of the most remarkable campaigns of exploration of the modern era. By the mid-1630s, they were west of Lake Michigan, and

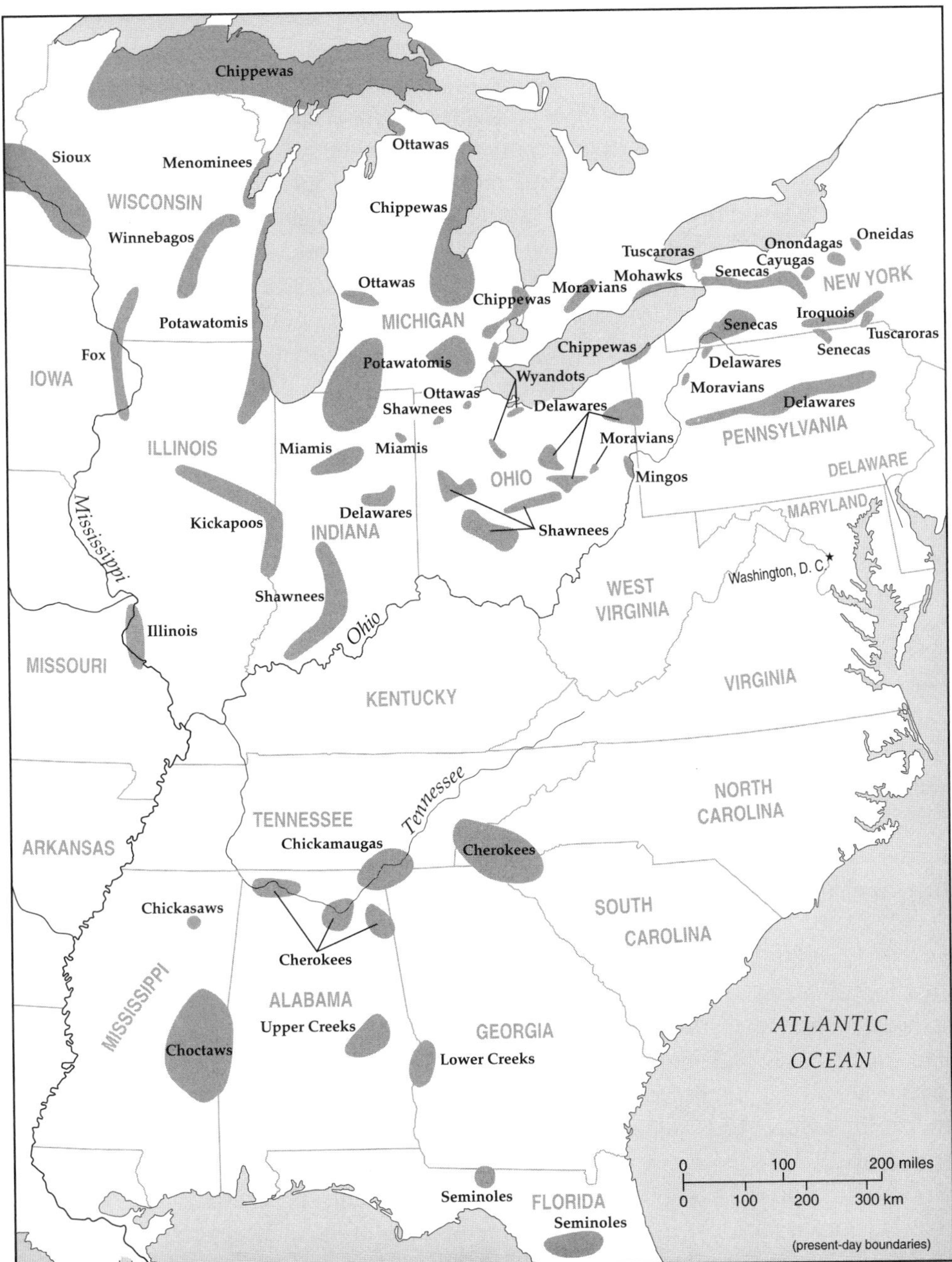

Trans-Appalachian Indian Settlement, 1760–94

by the 1670s, they were in the lower Ohio Valley. Soon they explored and founded posts down the Mississippi and along the Gulf coast. When Sieur de Bienville founded New Orleans in 1718, French traders were probing up the Missouri River and southwest toward the upper Rio Grande. Based on La Salle's voyage down the Mississippi to its mouth in 1682, the French claimed the land stretching from the Appalachian crest westward to the continental divide of the Rocky Mountains. Their presence was concentrated in a string of posts and settlements from the Great Lakes down the Mississippi Valley and eastward along the coast. But their influence radiated outward from that axis. By 1750 they had outposts in northern Alabama and the upper Ohio Valley, on the upper Missouri, and on the southern Great Plains. Their hold on this country depended on an intricate diplomacy among a score of Indian peoples. They lavished gifts and attention on natives even at the far edges of this orbit. In 1724, a delegation of Otos was escorted from their homes on the Missouri River to France, where they hunted in the Bois de Boulogne and basked in admiring gazes at the court of Louis XV. The travelers returned to report that women of the French court smelled like alligators.

Then, in 1763, France surrendered it all. In that year, the diplomatic map of North America was redrawn, at least from Europe's perspective. After nearly seventy-five years of hot and cold wars among England, France, and Spain, England emerged the clear victor. From France, Britain received eastern Canada and all French claims east of the Mississippi; Spain ceded Florida to England. With little use now for its holdings beyond the Mississippi, France transferred the enormous western half of Louisiana to its ally, Spain.

This shuffling marked the start of the frontier era considered here. The next fifty-seven years can be divided into three periods, each with its own theme. The years between 1763 and 1783 saw the first European settlers from east of the mountains and the combat and realignment of powers that came with the American Revolution. During the next three decades, as the nation born of that revolution exerted its authority over trans-Appalachia, the swelling tide of emigration brought rapid change to the land and its peoples. In the final stage, from 1812 to 1820, the United States tightened its grip on the region. But even then, as in the first two periods, the changing frontier conditions in turn changed the new arrivals, their government, and their various institutions.

A Muddling of Powers

In 1763, England suddenly found itself the master of eastern North America. Just as quickly, it began to learn how success bred difficulties. Most Indians of the Ohio Valley and Great Lakes had long been the allies of France. Understandably wary of this expansion of English power, the natives were infuriated when British commanders drastically reduced the annual gifts that were a staple of the Indians' economy. In the spring of 1763, Indians throughout the region launched ferocious assaults against the British and captured every installation except Forts Detroit, Pitt, and Niagara.

This war, traditionally attributed to the ambitions of the Ottawa leader Pontiac, was in fact a wide-ranging attempt to establish a native position of power at a time of uncertainty and change, and although the British retook their posts by 1765, the natives' point was made. In later calculations, the English would weigh Indian perceptions and demands more heavily. In the fall of 1763, as the war raged, London issued orders meant to stabilize conditions in the West to avoid giving Indians further cause for discontent. The Proclamation of 1763 recalled all settlers from west of the Appalachian crest, forbade emigration there until further notice, and authorized trade with Indians only by licensed government agents.

The success of the proclamation, or rather the lack of success, illustrated an important element in understanding the changes coming to this region. The frontier was not a place beyond the reach of authority. Rather, it was an area where many competing authorities were imperfectly exercised. Each of the three centers of settlement that appeared during the dozen years after 1763 took advantage of this situation.

The first, at the headwaters of the Ohio, already existed at the time of the proclamation. After 1758, when English and colonial forces captured France's Fort Duquesne, traders and settlers quickly moved into the area. Renamed Fort Pitt, the garrison offered protection and a market, and by 1760, an officer reported, the country was "over run by a Number of vagabonds, who under pretense of hunting, were making settlements." The second group of settlements, in eastern Tennessee, was fed by discontent in western North Carolina, where predominantly Scotch-Irish farmers complained of corruption, insensitive local officials, too many taxes, and not enough representation. After the one-battle "War of the Regulation" in 1771, many regulators crossed the mountains to join James Robertson, Arthur and John Campbell, and other early settlers, believing this country was part of Virginia. When they discovered their error, they set up a semi-independent government, the Watauga Association.

The third region, Kentucky, was doubly appealing. Besides being known as a beautiful and fertile land teeming with game, it had no permanent Indian settlements, although Shawnees, Miamis, and Cherokees ventured there often to hunt and fight. The prime movers behind Kentucky's earliest white settlement were eastern speculators, well connected and well-to-do. Since the 1740s, various groups had maneuvered to grab the region through vast but ill-defined grants and land bounties originally given to officers of the French and Indian War. By 1775, agents of several Virginia interests were surveying the area, but the most aggressive group was the Transylvania Company of Judge Richard Henderson, a North Carolinian. At Sycamore Shoals on the Tennessee River, Henderson "purchased" from Cherokee leaders about twenty million acres in central Kentucky, then he sent Daniel Boone and thirty companions to cut an emigrant road to the Kentucky River. Soon Henderson and several families were settled around the new outpost of Boonesborough.

In these first beachheads could be seen the most common patterns of settlement, both in trans-Appalachia and farther west: an uneasy partnership between settlers and an outpost of government power, a spontaneous movement of independent pioneers, and a speculative grab by wealthy interests who brought settlers with them. And in each case, the frontier's endemic confusion of interests and authorities helped open the way. Nowhere was this shown more clearly than in Kentucky. The land "Carolina Dick" Henderson hoped to colonize was claimed by Virginia and closed by the Proclamation of 1763. His right to title through Indian purchase was based on a highly dubious legal opinion, and in any case, the Cherokee signatories were giving away land that was not theirs in the first place. Henderson, the Fort Pitt "vagabonds," and the Tennessee regulators all used government claims and the military presence if their needs demanded; they all ignored prohibitions when it suited them. On this frontier, as on later ones, overlapping authorities, geographical ignorance, and ill enforcement of laws brought opportunities as well as problems.

All this was happening during the final stages of the crisis of empire. The history of the frontier and that of the American Revolution are intimately bound together, although the relationship is a bit problematical. Earlier historians, influenced by the ideas of Frederick Jackson Turner, considered the frontier the cradle of democratic ideals that ultimately fostered the discontent causing the breach between the colonies and England. This line of reasoning is far too simplistic, yet some connection between the empire's troubles and conditions on its outer rim is undeniable. For one thing, a concern with defense of the West led Parliament to pass the Stamp Act and other money-raising measures, which in turn brought the first serious colonial protests.

The frontier had a deeper, if vaguer, influence on the imperial crisis. By the middle of the century, many rural Americans had come to associate personal independence with possession of enough land to support a family—a hundred acres was generally considered a minimum. The French immigrant J. Hector St. John de Crèvecoeur wrote that as he stood on his own land, he was elated by "the bright idea of property, of exclusive right, of independence." A lack of land for family freeholds, by contrast, was linked to social corruption and eventual economic and political tyranny. By 1775, as the eastern population grew and farms were divided over and over to provide for maturing children, landholdings were shrinking alarmingly. But to the west (as long as one ignored the native inhabitants) was land to ensure a free and healthy society, especially after the victory over France in 1763. "Now behold!" wrote Nathaniel Ames, of Dedham, Massachusetts. "The farmer may have . . . land enough for himself and all his sons." To the people of his Virginia parish, the Reverend John Brown reported, the country across the mountains was like "a new Paradise." Even those who would never venture west still could see the region as the stuff of independence. It followed that any attempt to stem the emigration was a threat to a free society.

But slowing frontier expansion was just what many British officials advised. Not only would westward emigration bring war with the Indians and increase the imperial debt; the growing colonies would also draw away from England many thousands of tenants and badly needed laborers. Some opposed expansion for the very reasons that American farmers wanted it. To the earl of Hillsborough, secretary of state for the colonies, an expanding society of freeholders threatened "a just subordination to and dependence upon this kingdom." The home government consequently imposed first the Proclamation of 1763, then its modification, the line of 1768, which was to hold settlement east of the Allegheny River and in far eastern Kentucky. To dampen the speculative frenzy after 1770, London in 1773 forbade any new grants to speculators. The Quebec Act, passed the next year, closed land north of the Ohio River to all but the most limited settlement and trade.

The frontier, then, was certainly not a prime cause of the Revolution, but it was a provocation. The West was an occasion for conflict between the home government and particular economic interests, some of them quite powerful. For a far larger group of Americans, the western country was a symbol of possibilities and a hedge against baleful social tendencies. From the Republic's birth, many associated the nation's continuing independence, its unique virtues, even its very identity, with the frontier.

The phrase "War of Independence," when applied to the region beyond the

Appalachians, is both misleading and ironic. In one sense it was not an internal struggle within the British Empire but an imperial rivalry among four nations—three older ones and one just emerging—for control of the continent. In another sense the phrase is accurate, although not in the way it is usually used. American Indians were trying, once again, to salvage a measure of independence and control over their lands, and as usual that meant maneuvering among the various newcomers, lining up with the particular Euro-Americans who seemed at the moment to be the least threatening.

The fighting in the West was an extension of a conflict that had been gathering force for several years as the population around Fort Pitt grew and surveyors ventured into Kentucky. Rising tensions led to Lord Dunmore's War in 1774, after which Shawnees and Delawares allowed white settlement in Kentucky, at least for the time being. To the south, Creeks struck out in 1773 against the advancing Georgia frontier. It was hardly surprising, then, that virtually all Indian groups aligned with Great Britain against the colonists. Nonetheless, British leaders first pleaded with the natives to stay out of the fray, realizing that these people, acting out of their own motives, were thus ultimately uncontrollable. In fact, distant British and colonial commanders found their white frontier "allies" as unpredictable as the Indians, and partly for the same reasons. The changing patchwork of western Whig and Tory support depended on a complex of reasons, including speculation on which side would control land and trade after the fighting. Indians and settlers fought for their own goals, so leaders on both sides hoped to avoid relying on them.

Making matters even less predictable were the other two imperial rivals, Spain and France. Both opposed England, but neither looked favorably on colonial expansion. By 1781, the governor of Spanish Louisiana, Bernardo de Gálvez, having taken Natchez, Mobile, and Pensacola, hoped to control the coastal region as far north as the Cumberland River. French aid to the revolutionary cause was indispensable, and former French citizens sometimes gave direct aid to revolutionary forces in the West. But the French recognized no American claims west of the mountains. Few would have been surprised if the French tried to reassert themselves among their old Indian allies of the Great Lakes and Upper Mississippi.

The war in the West was fought in three theaters. Conflict began south of Kentucky when Cherokee leaders, against the pleas of the British agent John Stuart, struck white pioneer settlements along the Watauga and Nolichucky rivers in the summer of 1776. After a counterattack by white settlers destroyed Indian villages and corn supplies, Cherokees surrendered most of the Tennessee River valley. In the northern theater—upstate and far western New York and northwestern Pennsylvania—loyalists and Iroquois warriors battered colonials with hundreds of raids overseen by the British colonel John Butler and the brilliant Mohawk strategist Joseph Brant. Revolutionary Generals John Sullivan and John Clinton led a brutal retaliatory campaign into the Mohawk Valley, razing forty towns and burning 160,000 bushels of corn and other foodstuffs. By the end of the war, the region was devastated.

Between the northern and southern theaters lay the battleground of Kentucky and western Pennsylvania. Beginning in 1777, Shawnees, Delawares, and their Indian allies struck with such ferocity that "the year of three sevens" became synonymous with loss

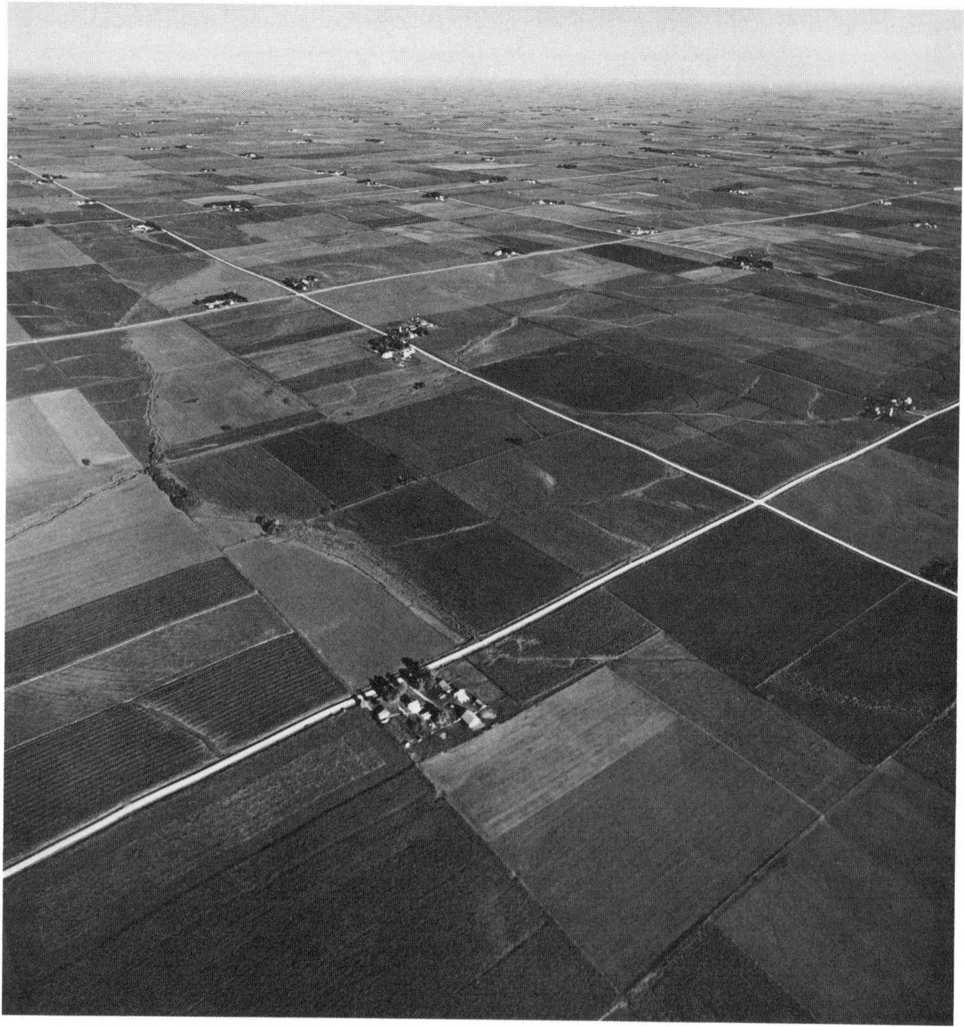

The grid system established by the Ordinance of 1785 created a rigidly precise way of marking the western lands for eventual sale and settlement. Superimposed on a wide range of topography as far west as California, the grid system continues to shape land-use patterns in vast areas of the rural West.

William Garnett (b. 1916). Farms and Section Lines in the Great Plains. *Gelatin silver print, ca. 1969. Courtesy of the photographer.*

and bloodshed in white folklore of the region. Scores, perhaps hundreds, of settlers died over the next five years. Twice Boonesborough was besieged. In 1778, George Rogers Clark with 175 men descended the Ohio and with help from French in the area took the three major British outposts in the Illinois country—Kaskaskia, Cahokia, and Vincennes. At the last of these, he captured the British western commander, Major Henry Hamilton. His audacious campaign, a tonic to colonials, eased the pressure on the upper Ohio. But within a year Clark had withdrawn, and Shawnees and Delawares lashed out with new energy. Colonials in turn swept north of the Ohio, destroying villages and food. These exchanges—bloody, destructive, but ultimately inconclusive—pushed the level of bitterness on both sides progressively higher.

The most striking result of this bloody conflict was the lack of a result. Little territory changed hands, and although white population west of the mountains had grown, Native American and British forces were, if anything, on the ascendant at the end of the war; one of their greatest victories, an ambush at Blue Licks, came in August 1782, nearly a year after the surrender of Cornwallis at Yorktown. So, on the face of it, the terms of the Treaty of Paris (1783) seem astonishing. Great Britain granted the new nation both its independence and an extraordinary western domain stretching in the north to the St. Lawrence and the Great Lakes, in the west to the Mississippi, and in the south to the thirty-first parallel. It was a great gulp of land. The addition of this territory, unwon and mostly unoccupied by colonials, was greater, as a percentage of territory actually controlled, than the expansion of the Louisiana Purchase.

British actions—military stalemate, then apparent generosity—seem a paradox but are easily resolved, for Great Britain in fact was acting out of self-interest. In the short term, London was giving up nothing. Given the new nation's pitiful military and limping economy, British soldiers, agents, and traders could stay in the region indefinitely, counseling the Indians and profiting from the fur trade. England was also guaranteeing its former colonies a host of troubles. The western country was filled with native peoples hostile to white settlement. And Spain, having reestablished itself in the lower Mississippi Valley, now claimed a band of land, roughly the southern half of modern-day Alabama and Mississippi, that the British supposedly were giving away. But what if the republican experiment survived? In the long term, an American nation with western resources and room to expand would depend less on French support and protection. By isolating the Americans from England's two prime rivals, Spain and France, Great Britain was playing a game—that of balance of power—at which it had no peers.

England's bet-hedging strategy was a fitting coda for this first era of the trans-Appalachian frontier. As it showed, this and later frontiers were unusually subject to the swings in diplomatic fortune among distant powers. Great nations were drawn to the country's resources, but their grip on the land was tenuous. In barely twenty years, three nations had passed title of the region among themselves—four if Spain's bid for the south is counted. This, with the changing allegiances among the many Indian groups and these competing nations, made for kaleidoscopic patterns of power and an extraordinary instability.

A Long Reach Westward

During the second period, from 1783 to 1812, one of these powers, the emergent United States, asserted its political control over the country beyond the mountains. Saying this, however, conceals as much as it describes. Behind this apparent grasp, which did not come unresisted, were changes of dazzling complexity. The government itself and its institutions were modified in the process of taking control. Westering emigrants adapted to demands and opportunities of the new country even as they brought their own changes. The many peoples already in trans-Appalachia influenced and were influenced by the newcomers, and all felt the effects of powerful outside economic forces.

Looking westward, the government of the nascent United States first had to unsnarl the conflicting claims of several colonies to country beyond the mountains. By 1786, Virginia, Connecticut, and Massachusetts had surrendered to the national government all claims north of the Ohio River and west of Pennsylvania. This Northwest Territory, or Old Northwest, was the first land over which the central government had exclusive jurisdiction. There, consequently, the frontier first shaped and expanded the government's power.

The national legislature now passed the two most important laws in the history of westward expansion. The Ordinance of 1785 ordered the Northwest Territory surveyed into sections (one mile square, or 640 acres), which were to be grouped into townships (six miles square, or thirty-six sections). Surveyed land was to be auctioned—part in sections, part in townships—in several eastern cities; the minimum price would be a dollar an acre. Next the Ordinance of 1787, or Northwest Ordinance, set out a plan of government for the territory that legislators assumed (incorrectly) would soon be sold and settled. The region would eventually be organized into three to five territories, which would mature into states by a metamorphosis of three stages. In the first, a new territory would be ruled by a governor, his secretary, and three judges, all appointed by the national government. Once five thousand free males had arrived, a legislature would be elected, though the governor could veto its actions. Finally, when sixty thousand settlers were counted, the territory could petition for statehood and, when admitted, enjoy the same rights and powers as its counterparts in the Republic. To establish a social and political order, the ordinance extended English common law into the Northwest Territory, guaranteed freedom of religion and other civil rights, forbade slavery, and set aside one section in each township to support public schools.

Although modified slightly over the decades, both laws were eventually extended to most of the Republic, first to the gulf coastal frontier and then to lands beyond the Mississippi. The ordinances determined the political form the growing nation would take. In principle, the first central government could have created a colonial system, with the original thirteen states as the parent country and the western settlements as perpetual children, but the Northwest Ordinance provided that the expanding nation would be one of sovereign states with equal rights. The three-stage evolution specified by the 1787 ordinance set the framework for the early political life in thirty-one of the thirty-five states west of the Appalachians.

The Ordinance of 1785 had a heritage at least as enduring as that of the Northwest Ordinance. In 1786, government surveyors laid out a baseline westward from the precise point where the Ohio River left the State of Pennsylvania. From there, they laid off the first townships and sections, then built more onto those, range after range, throughout the Old Northwest and eventually in virtually all the region southward to the Gulf of Mexico and westward to the Pacific. That first square inch of the first surveyor's stake was a kind of polestar of national development, the anchored point of reckoning for more than a billion acres. Nowhere else in the world would an area of such size be laid out in a uniform land system.

The survey's rigid predictability was meant to avoid the hopeless tangle of claims and boundaries found in the early land system of Kentucky and Tennessee. But there was far more to the ordinance than that. Adapted from the survey system used in southern New England, this vision of a gargantuan national grid also expressed that passion for reason and symmetry typical of the enlightenment mentality of the founding fathers, particularly Thomas Jefferson. And behind this dream of reason were assumptions of breathtaking audacity.

The system, first, presumed its own growth. With each range serving as the base for the next, the grid was an infinitely reproducible pattern, the perfect machine for national expansion. But the assumptions went deeper, to the very nature of the expanding society. "The art of civilization is the act of drawing lines," Oliver Wendell Holmes wrote later. He might have been referring to this ordinance, for the way the land was plotted profoundly influenced how new settlers thought of it, treated it, and lived on it. The system allowed a simple exchange of parcels of land. Squares, after all, are easily divided and combined: here was the ideal geometry for wheeling and dealing. This encouraged impulses already abroad in the new nation—a restless mobility, a search for profit through transforming a place and moving on, a tendency to see land as a commodity. The arrangement virtually dictated particular styles of farming, most obviously the devotion of one rectangle of soil to one crop, the next to another, with each plowed into straight lines among straight fences beside straight roads.

The Ordinances of 1785 and 1787 were in some ways terribly flawed. In one sense, the grid system was an illusion, one that presumed to impose a sameness on a magnificently diverse landscape. Surveyors would plot out the same legal checkerboard on mountains, canyons, prairies, pine barrens, gullies, deltas, plains, hummocks, and bogs. The rigidly linear brand of agriculture dictated by the survey was woefully unsuited to some regions, including much of the Far West. The result would contribute to later environmental calamities. The ordinances were also at odds with themselves. The provisions for representative government, education, and civil liberties spoke to the farmers of modest means who were close to Jefferson's heart. The land was surveyed not into huge tracts but into parcels that could be easily subdivided; this too seemed to presume a society of yeoman freeholders. Yet other terms clearly favored wealthy speculators. At auctions, to be held not on the frontier fringe but in eastern cities, the smallest unit offered was a section, far more than a family could use, and its minimum price of $640 was several times the annual income of a typical rural household. Besides simply trying to raise money quickly, politicians were recognizing those two groups—monied interests and common squatters and farmers—whose energies had first pushed the frontier over the Appalachians. Playing to both sets of interests, this first western policy was bound to be disharmonious.

In an even more obvious conflict, the government set down a plan of settlement for land that was already occupied. To be sure, the tens of thousands of Indians in the Northwest Territory were not ignored by the government. Its promise to the natives was unambiguous: "The utmost good faith shall always be observed towards the Indians; their lands and property shall never be taken from them without their consent; and, in their property, rights, and liberty, they shall never be invaded or disturbed, unless in just and lawful wars authorized by Congress; but laws founded in justice and humanity, shall from time to time be made for preventing wrongs being done to them, and for preserving peace and friendship with them." But what was the context of this pledge? These words composed the third article of the Northwest Ordinance. We shall protect the Indians and their ancestral lands, the government promised, and we shall lay out the means, step by step, for that land to be sliced up, sold, cleared by the ax and broken to the plow, then layered over by English common law, parliamentary republicanism, and traditions carried by white families and the common school. A policy that could make such promises, all within the same pair of documents, had moved beyond contradiction to schizophrenia.

Not surprisingly, these muddled intentions led to policies that were often at war with themselves, as seen when the new government began to dispose of its land. During the mid-1780s, speculators once more scrambled to control enormous tracts of western lands. They were eagerly received by state and national governments starved for income and groaning under huge debts. In 1787, the national government sold five million acres in southern Ohio to the Ohio Company of Associates, which in turn sold three and a half million acres to the Scioto Company, a group with several members of Congress among its principals. The terms—including discounts for swampland and payment by inflated certificates of indebtedness—came to barely nine cents an acre. Virginia interests purchased veterans' land warrants, which they used to claim great stretches of Kentucky. By 1792, New York had dispensed more than five million of its western acres at prices that averaged under twenty cents an acre. Surely the most outrageous transaction, and probably the most infamous land steal in the nation's history, came in 1795, when a thoroughly bribed Georgia legislature sold more than thirty-five million acres reaching to the Mississippi for half a million dollars, or just over a penny per acre. Georgia soon repudiated the "Yazoo fraud," which the acid-tongued Virginian John Randolph later called a "many-headed dog of Hell," but the U.S. Supreme Court eventually upheld the contract.

These land deals inevitably raised the ire of settlers, a constituency that, if not as powerful as wealthy speculators, was certainly more numerous. Some emigrants, to be sure, were settling land in hopes of selling for a profit when the country filled up; "squatters" and "speculators" sometimes differed only in degree of ambition. That said, the settlers' hostility toward speculators was both genuine and justified. The frenzied dealing naturally drove up the price of land. Many speculators, furthermore, simply held on to their grants, waiting for their value to rise, and thereby closed millions of acres to settlement. Some, like William Blount, the future territorial governor of Tennessee, engaged in outrageous chicaneries. The investors' complex dance—the trades within trades, the dubious transactions, and the outright frauds—muddled titles for generations to come.

Soon after the end of the war, Kentuckians were petitioning Congress, saying they would never be "Slaves to those Engrossers of land" who lived "at ease in the internal parts of Virginia." Some threatened to head down the Mississippi to take up land among the Spanish. Others asked to settle north of the Ohio. Those courses obviously would complicate the government's already troubled diplomacy. Some politicians nonetheless sympathized with families chafing to move into Ohio. When families went there on their own, a congressman later asked his fellows, would Congress "then raise a force to drive them off"? The answer was yes. In fact, troops were patrolling the northern bank of the Ohio as early as 1785, confiscating arms, expelling squatters, and burning their houses.

Pulled between conflicting interests, the government revised the system of land sales. A new law in 1796 opened land offices in Pittsburgh and Cincinnati, thus appealing to settlers, but then raised the minimum price at auction from one to two dollars an acre. With the Harrison Land Act of 1800, however, Congress dramatically shifted its favors. Henceforth settlers could buy land in parcels as small as 320 acres, a size more practical for the typical family. More important, land now was sold on credit; after a down payment of a fourth of the price, purchasers had four years to pay the rest.

The government's pirouette—first nodding one way, then bowing another—suggested the varied economic interests that fueled westward growth. The performance also anticipated the shifting, often contradictory policies that would become familiar as the frontier pushed farther west.

As it tried to appease its constituents, the new government also faced a collection of powerful adversaries across the Appalachians, not only a few dozen Indian groups but also two of the world's great colonial powers, England and Spain. All were determined to push back, or at least retard, the Republic's expanding frontier settlements.

Despite the deep hostility at the end of the revolutionary war, fighting among whites and Indians receded after 1783. Both sides were ready for a rest, and some Indians argued for an accommodation and a search for a middle ground. But the Treaties of Fort Stanwix (with the Iroquois in 1784), Fort McIntosh (with the Wyandots, Delawares, Ottowas, and Chippewas in 1785), and Fort Finney (with the Shawnees in 1786) proved ineffectual. The Indians received nothing for ostensibly surrendering part of western New York and western Pennsylvania and most of what is today Ohio, the fallacious reasoning being that as losers in the recent war, they were merely turning over its spoils. Besides, many Indian groups who lived or hunted in the surrendered regions were not even present at the negotiations. Meanwhile thousands of emigrants were descending the Ohio, and squatters were crossing the river to build on its northern shore. Several towns sprouted along the river in the Ohio country, most of them connected to land speculation schemes; by 1790 Marietta, the first, had been followed by Cincinnati, Steubenville, Gallipolis, Waterford, and a few others.

The uneasy peace was rapidly unraveling by 1786. English agents, still operating in posts they had promised to cede to the new nation, now provided substantial nonmilitary support to natives in the region and worked to fashion a confederation of the more than thirty Indian groups in the Ohio Valley and Great Lakes region. For several years Indians acted with a remarkable degree of unity in their dealings with the new American government, and at one point English officials suggested the formation of an Indian state in the Northwest. Buoyed by encouragement from London, native leaders were increasingly impatient with government policies. Led by the Miami chief Little Turtle, some were soon lashing out at farmsteads in northern Kentucky. "Scarcely a week has passed without some person being murdered," wrote a militia leader. These attacks were followed by retaliatory raids into the Illinois country and against the Shawnees, who so far had been mostly peaceful. By 1789, when the American government tried to pay for the surrender of the upper Ohio in the Treaty of Fort Harmar, events had raised the Indians' alarm past reconciliation. Arthur St. Clair, appointed governor of the Northwest Territory in 1788, summed up the situation with wry understatement. The natives found white intentions so clear and dreadful that there was "little probability of . . .

General Anthony Wayne led his troops in a charge against Indian opponents in the Battle of Fallen Timbers, 1794. After fighting just 40 minutes, Wayne reversed a string of recent American military defeats and gained concessions to vast areas of land north of the Ohio River, land that soon opened to white American settlement.

Frederick Kemmelmeyer (active 1788–1803). General Wayne Obtains a Complete Victory over the Miami Indians. *Oil on canvas, ca. 1794. Courtesy, Winterthur Museum, Winterthur, Delaware.*

cordiality." He added, "The idea of being ultimately obliged to abandon their country rankles in their minds."

Faced with the consequences of these ineffectual, conflicted policies, President George Washington authorized in 1790 a punitive expedition against "certain banditti of Indians from the northwest side of the Ohio." But after General Josiah Harmar's command razed a few towns and destroyed corn crops, a contingent of his troops was ambushed and badly mauled by Little Turtle's warriors. Washington then ordered Governor St. Clair himself, a revolutionary veteran who had served at Valley Forge, into Ohio. In November 1791, Little Turtle, leading a coalition of several groups, sprang a second trap, this time on St. Clair's poorly trained, ill-disciplined soldiers and their accompanying families, camp followers, and rum sellers. Caught at dawn in an open area against the Wabash River, the expedition was raked for hours by gunfire and attacked by Indians "so thick we could do nothing with them," a combatant wrote later. Of the 1,400 troopers, 630 were killed. Also lost were twelve hundred muskets and eight cannons (including two taken from Cornwallis at Yorktown). In some ways, it remains the worst defeat in the nation's military history.

After Indian leaders turned aside renewed efforts at another settlement, this one on far more favorable terms, Washington reached once again for a military solution. This time General Anthony Wayne, a seasoned commander, led a well-prepared expedition methodically northward toward Fort Miamis, recently built by the English on U.S. soil near the western end of Lake Erie. Four miles from there, on 20 August 1794, Wayne soundly defeated a force of about eight hundred Indians of the confederation on a storm-littered field called Fallen Timbers.

Paradoxically, this combination of dismal failures and a single success made a point that an initial victory would not have. The Republic's armies might often be guilty of tactical blundering and carelessness, but behind this pitiful floundering were, compared with what the Indians could muster, massive resources and, more important, the willingness to commit them in a war of attrition. Driving the point deeper was another realization by the Indians. At Fallen Timbers, the confused and panicked warriors had bolted toward Fort Miamis, calling for sanctuary as they approached its walls. But the fort's commander, unwilling to risk a crisis with the United States, had barred the gates. The Indians were devastated, and for good reason. Only a power like Britain had resources to match those of the United States. The natives, however, could never depend on such help, because Britain's stance toward the Indians was only one of dozens of factors in that nation's complex, ever-changing foreign policy. The barred gates of Fort Miamis showed the natives that their fate was ultimately subject to the veer and sway of European interests.

And those interests were changing rapidly indeed. In 1783, both England and Spain had given high priority to opposing the expansion of the frontier. The former wished to keep the profits flowing from the region's fur trade; more than half the pelts processed at the great market of Montreal came from the forests of the new nation. Spain feared American expansion would ultimately threaten Spanish Florida and Louisiana. Although neither nation gave outright military aid to its Indian allies, both provided moral support and gifts and encouraged natives to form a united front against Anglo-American expansion. In bald violation of treaty terms, English agents remained in six posts and several Indian towns around the Great Lakes and Ohio Valley. The Spanish funneled up to fifty thousand dollars in goods annually to the Creeks and in 1793 built an Indian post, aptly named Fort Confederation, on the Tombigbee River. Spain took further advantage of its control of the lower Mississippi, heavily taxing foreign commerce and periodically banning American traffic from the river that was the pioneers' only practical avenue for exporting goods. At the same time the Spanish—hoping, like the Romans, to assimilate the barbarians—offered full access to the river and generous land grants to westerners who would emigrate to Spanish territory in present-day Missouri.

For a while the English and Spanish had reasons for confidence. While the coalition of northern Indians stung the pioneer settlements and their militia forces, the southern groups held firm against further cessions. Discontented westerners seemed to flirt with the Spanish; at one point the district around Nashville was named Miro after the governor of lower Louisiana. Yet in 1795 both England and Spain conceded virtually everything the United States wanted. In Jay's Treaty, Britain agreed to remove all agents and soldiers from U.S. soil in exchange for concessions on certain maritime issues. Spain's capitulation was even more complete. The treaty of San Lorenzo (or Pinckney's Treaty) set the boundary with Florida at the thirty-first parallel, as provided in the Treaty of Paris, and guaranteed U.S. citizens the use both of the Mississippi and of New Orleans as a site to unload and reship their goods.

Why? England's and Spain's acquiescence, like the story of the Treaty of Paris, demonstrated how distant developments could have profound effects on the American frontier. Events between 1789 and 1795—the French Revolution, the onset of continental wars, and the surprising successes of the French—forced European governments to reevaluate their diplomatic postures. England, now almost wholly without allies, had to anticipate a lengthy conflict with France; Spain prudently looked toward an alliance with the French and, with that, war with England. Neither nation could afford the possibility, however remote, of serious troubles with the United States, and Spain in particular feared that England and the United States might unite to seize Florida and its Caribbean holdings. England and Spain both saw the sense of settling squabbles in a part of the world that was now on the periphery of their main concerns. Once more, an unsettled Europe had worked to the new nation's advantage.

But behind England's and Spain's abandonment of the Indians, and behind the Indians' retreat, was a more fundamental force. The population of the United States stood at more than four million, or nearly fifty times that of the Indians and Europeans to the west. Most citizens of the Republic would never see beyond the Appalachians, but as their sense of nationhood took vague shape, many of them thought of the western country as vital to the Republic's security and future possibilities. A "wall of public opinion" supported white U.S. settlement across the mountains, as a Spanish official put it. The new government, whatever its contradictions and clumsy use of power, would ultimately respond to that popular notion, as well as to the economic concerns of powerful interests.

More immediately, the number who *were* moving westward, though only a tiny portion of the larger society, far outweighed their opponents. The annual white emigration to just one part of the frontier, western New York and Pennsylvania, was at least ten times the number of warriors who fought at Fallen Timbers. During 1788, the average number of people floating on flatboats down the Ohio *each month* was greater than the entire population of the Shawnees; the total for the year nearly equaled the population of the largest southern group, the Cherokees.

Emigration to the frontier was a great weight leaning westward. Standing against that weight was a collection of interests of varying strengths and in unstable combination. They had in common a determination to hold the frontier where it was. But that coalition was vulnerable to events, whether a defeat in the field or upheavals on another continent. The westward-leaning weight, by contrast, continued to build despite military embarrassments, economic sanctions, and diplomatic opposition. Eventually the opposing coalition was bound to crack, and when it did, the force of population would fill any openings and wedge them wider.

Politicians later waxed eloquent on how a superior society, blessed by God and made strong by republican virtue, inevitably triumphed over lesser peoples in its westward march. But if this triumph was inevitable, numbers were the most obvious explanation. On this frontier, as on later ones, demography was destiny.

Land Pikes and Cotton Gins, Libraries and Taxes

The growing wave of settlers from the East, more than military engagements and diplomatic maneuvering, explains how the United States secured control of trans-Appalachia. This emigration in part told of the public's increasing infatuation with the

HERITAGE:

Encounters in the West

◆

No other aspect of the West's complex cultural heritage fascinated European and American artists during the mid-nineteenth century as much as the region's many Indian cultures. Though each artist ventured west with his own interests in mind, most depicted native peoples with a mixture of anthropological curiosity, romantic awe, and a firm conviction that Indian cultures would disappear with an increased American presence in the region.

Fearful that the western Indian cultures he observed in the 1830s would soon disappear, George Catlin placed himself squarely in the forefront of the effort to preserve a record of Indian life with his prodigious output of paintings, prints, and books documenting Native American cultures.

George Catlin (1796–1872). Catlin Painting the Portrait of Máh-to-tóh-pa—Mandan. *Oil on composition board, 1857–69. Paul Mellon Collection. National Gallery of Art, Washington, D.C.*

Karl Bodmer's detailed and lucid watercolors of Indian life on the upper Missouri reflect the widespread international interest in the West's native peoples. A Swiss-born artist who studied in Paris, Bodmer came to America in the employ of the Prussian prince Maximilian zu Wied, an ardent explorer and amateur naturalist. Many of his paintings, such as this portrait of a distinguished Minnetaree leader, were later engraved for an international audience. Bodmer combined ethnographic detail with a romantic admiration for native life.

Karl Bodmer (1809–93). Péhriska-Rúhpa, Hidatsa Man. *Watercolor on paper, 1833. Gift of the Enron Art Foundation. Joslyn Art Museum, Omaha, Nebraska.*

St. Louis–based artist Charles Deas was praised for his attention to frontier subjects and the "essentially American" themes that would attract interest from a European audience. Perhaps his most successful painting, this small narrative picture depicts a group of Sioux Indians as they pause in flight from intertribal warfare. The detailed renderings of material objects suggest the artist's general familiarity with Sioux life, though the scene is clearly one he never observed.

Charles Deas (1818–67). A Group of Sioux. *Oil on canvas, 1845. Amon Carter Museum, Fort Worth, Texas.*

The tragic but inevitable demise of Native American cultures became an increasingly popular theme in American art of the second half of the 19th century. In this allegorical painting, John Mix Stanley depicted representative figures from both eastern and western tribes at the westernmost edge of the continent as the sun—and by implication their cultures—disappeared over the horizon.

John Mix Stanley (1814–72). Last of Their Race. *Oil on canvas, 1857. Buffalo Bill Historical Center, Cody, Wyoming.*

Daniel Boone's efforts to settle Kentucky were first outlined by John Filson in 1784. The artist George Caleb Bingham appealed to popular mid-19th-century thinking about Boone by depicting the frontier hero as a latter-day Moses, leading his people through darkness into the promised land.

George Caleb Bingham (1811–79). Daniel Boone Escorting Settlers through the Cumberland Gap. *Oil on canvas, 1851–52. Gift of Nathaniel Phillips, Boston, 1890. Washington University Gallery of Art, St. Louis, Missouri.*

West. When Benjamin Franklin disembarked in France to seek aid during the American Revolution, he wore the fur cap that Europeans associated with backcountry pioneers. With characteristic insight, Franklin caught an essential truth—that on that continent and his own, the emerging nation's identity was inextricably bound to the imagined possibilities of its hinterland. Prominent frontier figures were already suffering the fate of many to come, transmogrified from flesh and bone into mythic creatures. In John Filson's *The Discovery, Settlement, and Present State of Kentucke* (1784), Daniel Boone's wilderness ordeals change him from a mere hunter into a kind of Moses who embodies a people reborn by their occupation of the new land. His passage and theirs is as much spiritual as physical. This vision, expressed in songs and fireside stories and jokes, set more and more Americans in motion. From Connecticut, an observer wrote of social gatherings where young people "marched to rude melodies which taught them to dream that toward the setting sun lay an earthly paradise with gates open to welcome them. From hill and valley the processions hurried away."

Emigrants moved across the mountains in broad, parallel streams—northeast to northwest, southeast to southwest—so that the Ohio Valley was dominated by settlers from New England, New York, and Pennsylvania, while Kentucky, Tennessee, and the Gulf frontier filled with Virginians, Carolinians, and Georgians. This settlement, however, was never an unbroken, steadily advancing line. Settlers spilled from western Pennsylvania and the Carolinas into the next closest regions—Kentucky, Tennessee, and southern Ohio. But others moved up the Tombigbee into south-central Alabama, down the Tennessee to its great bend, and into the Mississippi delta around Natchez, forming islands of Anglo-American population hundreds of miles from the closest settlement to the east.

By the turn of the century this emigrant society was supporting itself with a surprisingly diverse economic system. The majority of newcomers were farmers, mostly of the "backwoods" sort who would become, in the popular mind at least, the essential frontier common folk. Their lifeways, evolved in the woodlands east of the mountains, were well tested in this new setting.

They first faced the overwhelming fact of trees. "O the woods! the interminable woods!" wrote a visitor to Kentucky in the 1830s. Farmers opened a few acres in this thick forest by girdling or felling the largest trees. Trunks and brush then were burned to clear ground for planting and to enrich the soil with ash. Among the green stumps, families planted corn, which produced four times the food value of wheat with a tenth of the seed. Farmers often intercropped, usually with beans and squash, with the vines of the former climbing the corn stalks and the latter spreading over the soil to choke out weeds. Pioneers grated green corn to make pudding they flavored with berries and grasshoppers; they ate the culms, the soft inner lining of the stalks, like asparagus; they ground some kernels into meal for bread and unleavened johnnycake and boiled and pounded others for hominy; they washed it all down with whiskey distilled from fermented seed. In their gardens they grew potatoes, pumpkins, peas, gourds, and tobacco, and from the surrounding woods they gathered pecans, walnuts, papaws, sassafras, and medicinal herbs and roots. Backwoodsmen fished the streams and preyed on deer, bears, raccoons, squirrels, and opossums. A family kept some chickens, a dog or two, and perhaps a cow for milk, but most valuable were pigs, which roamed the forest on their own, flourishing on roots and acorns. Lean and rangy (some called them "land pikes"), a woodland porker was a walking, grunting smorgasbord. Besides hams, chops, and bacon, families ate the ribs, intestines, lungs, and brains. They made head meat into pies. These essential animals became folkloric figures; it was said they could jump through fences by turning sideways in midair and swim heroically through floods to roost safely in trees. For shelter a family usually began with a lean-to, then turned to the simple, utilitarian log house that visitors found emblematic of the backwoods frontier, then expanded that into a dogtrot (or saddlebag) house, with two rooms separated by a covered breezeway. Around the farmstead they built a sapling or split-rail fence in either the straight "post-and-rider" or, more likely, the zigzag "snake" or "worm" style.

The earliest arrivals practiced extensive and subsistence farming. They exploited the land vigorously for a maximum yield, then moved on to begin anew after a few years when that yield began to decline. And they were almost wholly self-supporting, concentrating on diverse production of things immediately usable. A surplus meant not gain but wasted effort. If the family could not eat it, wear it, drink it, or smoke it, what was the point? A second agrarian wave, overlapping the first, hoped to stay longer on the land. They too began by planting corn among the stumps of felled trees, but soon they began plowing, first around the stumps but eventually in fields cleared of the rotting remains. (Their familiar labors became political metaphors—"log rolling" for cooperation and "plowing the green stump" for avoiding a stubborn problem.) A signal of this intensive agriculture was the appearance of wheat in cleared fields whose potential was conserved by rotating crops and fertilizing with manures. These farmers looked outward, hoping to sell their products at local and distant markets. In one sense this trade was nothing new. As early as 1746, the French on the Wabash and Illinois rivers had sent flour to New Orleans, and by the late 1770s, farmers of western Pennsylvania marketed wheat as far away as the West Indies. By 1811 this trade had grown many times over. Cargoes floating past Louisville included about two hundred thousand barrels of flour, eighty thousand bushels of corn, sixteen thousand barrels of whiskey, twenty-two thousand barrels of pork, and a million pounds of bacon.

On the southern frontier, the most important commercial crop by far was cotton. The gulf coastal plain between Georgia and eastern Texas, with its long, humid summers and its bottomlands of rich, black soil, was the largest region on earth in which short-staple cotton could be grown. After the invention in 1793 of the cotton gin—with its mechanical combs easily removing seeds previously taken out laboriously by hand—cotton suddenly could be produced cheaply and sold dear. A gin was operating in Natchez only two years after Eli Whitney had built his first model. During the following twenty years, production expanded substantially around Natchez, along the lower Tombigbee River, and near the "great bend" of the Tennessee River. In one way, Natchez was at the forefront of the cotton culture; between 1805 and 1810, a local landholder experimented with a newly imported type of cotton, "Mexican upland," to develop the primary strand that would be used throughout the inland South. This cultivation brought its distinctive labor system. Slavery was well entrenched in the Mississippi delta by 1800, and by 1810, slaves were working extensive fields in the Tennessee Valley and the northern Mississippi Territory.

This frontier—and these frontiersmen—were unlike any others that had come across the Appalachians. Dedicated to commercial farming, these settlers were establishing a system of large-scale staple agriculture at the outset, with no preliminaries or transition. Many were planters of means who brought capital, equipment, and slaves. They lived with huge investments and onerous debts and succeeded or failed by shifts among complex markets and by the distant decisions of investors.

Their appearance, in turn, taught something about the dynamics of expansion. Trans-Appalachia by now was part of an expanding capitalist economy emanating from Europe and commercial centers of the Atlantic Coast. The value of this land depended on its potential, and that potential, in a world of growing markets and technological wizardry, expanded or shrank with events hundreds or thousands of miles away—in this case, when Whitney made a prototype of a cotton gin in the guest room of a Georgia plantation house. Land is, in a sense, what people see in it, and so this country, before a single tree was cut or an acre plowed, had become another place, one instantly desirable to squatters and wealthy eastern landholders alike. In the history of westward expansion, there would be no better example of how the accidents of an international economy could profoundly affect the far fringe of the frontier.

Another commercial enterprise—cattle raising—also flourished in the Old Southwest. By the early eighteenth century, Spanish colonials were raising cattle in substantial numbers on the sea islands off the Alabama coast, and after 1750, visitors reported many thousands of grazing animals on the prairies northwest of New Orleans, along the

The popular mythology of the lone frontiersman long obscured the actual importance of families in frontier settlement. After the Civil War, as an eastern interest in domestic culture spread westward, popular prints such as this one reimagined families as central to the frontier experience.

After Frances Flora Palmer (1812–76). Published by Currier & Ives. The Pioneer's Home: On the Western Frontier. *Lithograph, 1867. The Harry T. Peters Collection (#57.300.23), Museum of the City of New York.*

Mississippi in the Natchez district, and in the valleys of the Perdido, Tombigbee, and Chickasawhay rivers to the east and north of Mobile. By the War of 1812, more land was devoted to cattle raising than to any other economic activity. Cattle were driven by herders on horseback to New Orleans, Mobile, and other ports, then shipped to the West Indies and to cities along the Atlantic coast. Profits were good. Even as the cotton economy boomed, delta planters would "yearly mark thousands of calves, and send them to the prairie to feed," a visitor wrote. As the historians John Guice and Thomas Clark have suggested, many estates in the Old Southwest could have been more accurately called ranches rather than plantations.

In frontier towns, yet another economic pattern could be found. Despite the popular image of trailblazing pioneer farmers opening new regions to settlement, towns typically preceded the agricultural frontier. Their economic patterns varied according to geography and historical circumstances. Some turned to manufacturing goods that could not be profitably imported over the primitive roads across the Appalachians. For this, Pittsburgh's location was ideal. Whatever was made in the town could be floated down the country's grandest distribution system, then marketed from towns along the Ohio and Mississippi rivers. By 1815, with a flourishing iron and glass industry, Pittsburgh was the region's largest urban center. Lexington turned locally grown hemp into cloth and rope worth half a million dollars a year by 1809. Cincinnati and Louisville sold Pittsburgh-made iron goods, light manufactures from the East, and sugar, cotton, and molasses carried from the lower Mississippi Valley by keelboat to the rapidly filling Ohio and Kentucky countryside. These towns also sent downriver the harvests from the developing farms, foodstuffs produced with farm implements made in the shops of local artisans. St. Louis and Natchez did much the same, and they profited as well by exporting the frontier's furs and cotton. New Orleans, at the opposite end of the water highway from Pittsburgh, reshipped goods from upriver to the outside world, sent local cotton both upstream and off to England, and supplied the gulf coastal frontier with imported necessaries.

For all their differences, all these towns exploited the frontier's dynamic changes—its vigorous migration and its evolving rural economy. They had a vibrance that visitors found exciting, if not always appealing. Walking the streets of St. Louis, one would see trappers, merchants, and slaves, planters and their families, boatmen and Indians, loafers and thieves. By 1815, Cincinnati residents could choose among the shops of thirty merchants. Below the bluffs of Natchez was the town's shadowy brother, Natchez-under-the-Hill, where rivermen and traders mixed with murderers, prostitutes, and mountebanks in gambling houses and whiskey holes as repulsive and dangerous as in any eastern city. The respectable town above supported several physicians, artisans in two dozen trades, and perhaps the ultimate sign of a maturing economy: eight lawyers.

These economic changes were part of a wave of influences washing westward over the mountains. The new patterns in turn mingled with ones already there, making this American frontier, like those to follow, a cultural swapping-ground as well as a land of conflict and conquest.

White pioneers carried a mix of traditions. Those that served their needs survived and spread so successfully that they were often seen, incorrectly, as unique creations of the frontier. Just whose traditions were most influential, however, is a matter of dispute. Grady McWhiney and Forrest McDonald have vigorously argued that Celtic lifeways took root and flourished, particularly on the southern frontier and in the Old Northwest closest to the Ohio River. These influences, they have written, explain the most distinctive differences that emerged between the developing South and the North, which drew its traditions from England's southeastern lowlands. Others have stressed the contributions to pioneer life of the Germans who settled in western Pennsylvania. More recently, Terry Jordan and Matti Kaups have written that another group—the Finns who emigrated from the wooded eastern part of their homeland to Pennsylvania's Delaware Valley in the mid-1600s—brought methods of farming, hunting, housing, and herding that others took westward into the vast central woodlands of trans-Appalachia.

Besides these broad influences stretching back across the Atlantic, much of the dominant national culture was being transplanted. "Higher" culture was most apparent in cities, which sat along the best lines of communication with the East and which had the money and population to support cultural institutions. Subscription libraries were established early, and the private book trade could be substantial; a shipment of thirty-five hundred books was auctioned in Natchez in 1808. Even communities of modest size boasted newspapers that offered news and, as the *Missouri Gazette* (St. Louis) advertised in 1808, "Belles Lettres [and] historical and Poetical extracts." Churches appeared early and in surprising numbers. Methodists in Pittsburgh established eight congregations during the decade after 1796, and Cincinnati residents could choose among six denominations by 1815. Protestants, especially Methodists, Presbyterians, and Baptists, dominated cities of the Ohio Valley, while Catholic churches were sprinkled along the Mississippi Valley, with its French and Spanish heritage. Private schools typically were operating within a few years of the founding of a town. Nor was higher education wholly ignored. Lexington's Transylvania Seminary, founded in 1785, had evolved by 1799 into a university that soon had schools of liberal arts, law, and medicine.

Outside the towns, most settlers had only a casual acquaintance with the written word; "all were illiterate," one frontiersman recalled, "but in varyious degrees." But even this man's parents brought a small library that included the Bible, Benjamin Franklin's *Autobiography*, and John Bunyan's *Pilgrim's Progress*. Subscription, or "field," schools appeared quite early in many areas, although their terms and attendance were erratic and their quality decidedly mixed. Home schooling was probably at least as effective, especially north of the Ohio in country dominated by pioneers from New England.

In town and countryside, immigrants also brought the attitudes, values, and traditions that constituted the new nation's varied "folk culture." In this, the role of the family was paramount. A common popular impression portrays the archetypal frontiersman as a lone woodsman who savored his isolation. This image could not be more wrong. The need for labor and protection, not to mention companionship, made a family deeply desirable. Families, in fact, were more common on the frontier than in the East. Bachelors and widowers were rarely single for long. Women living on their own were almost unheard of; a recent study found only twenty among a sample of fifty thousand persons on the trans-Appalachian frontier between 1800 and 1840. In the entire region, one demographer estimates, only about two-fifths of 1 percent of the population lived alone and planned to stay that way. Like everywhere else in the nation, the frontier was dominated by nuclear families, but these often clustered near their kin, forming communities of geographically dispersed but closely related households. Families were like cells encoded with instructions for the cultural outlines of the societies into which they would grow. Through them were carried religious and ethical beliefs, standards of behavior, concepts of the proper nature of community and the respective roles of women and men, humor, songs, superstitions, games, and much more.

Local governments also nurtured transplanted ways of life. The first county courts protected and extended the new economic system, recording land titles and imposing harsh penalties for crimes against property; a Tennessee horse thief was pilloried, whipped, and branded and had both his ears cut off. Courts conscripted labor to build public roads that would encourage emigration, the growth of market economies, and cultural connections to the world beyond. An evolving tax system fed this implanting of institutions. In the earliest stage of development, taxes were levied on the raw stuff of settlement—free men, slaves, horses, and cattle. As the labor of men and beasts gave monetary value to the land, taxes came from real estate and licenses for some commerce, taverns especially. This income in turn paid for the expanding government services of a society tightening its grip on the land—judges and courts, care of orphans and paupers, bounties for wolves. Weightier civil and criminal matters were heard in courts of common pleas, quarter-session courts, and the territorial supreme courts, but most people encountered the court system, if at all, through a justice of the peace. This official heard some criminal cases but mostly domestic squalls and squabbles between neighbors. Sessions, which were usually a good show, became social as well as legal occasions. A justice's style was informal and his solutions based more on custom than anything else. Through these courts were transplanted both law and those folk traditions and social mores summed up as "common sense."

In their outward appearances, these agents of authority had the frontier's raw look. Records tell of courts interrupted by nearby tavern noise and by pigs rooting under the floor. But from the lowest level to the highest, territorial governments were both bearers of traditions and powerful reinforcers of the economic and social institutions taking root in the country.

A Swapping Ground of Cultures

Yet this intruding culture did not overwhelm and replace all existing culture—far from it. The pioneers' folk and more formal culture was itself evolving, changed through adaptation to this new setting and shaped by the extraordinary mix of other peoples in the region.

By now this was an old story. For two centuries, Indians and Euro-Americans had been changing one another. Certainly nothing remotely like a pristine native culture had existed for generations. At French, English, and Spanish military and trading outposts and in Indian villages, visitors saw an amalgamation of customs and startling juxtapositions. An Englishman wrote one evening of a gala ball at a Shawnee settlement on the Maumee River, with white and Indian dancers sporting fur caps and ostrich feathers. Soon afterward he told of a warrior proudly displaying a Kentuckian's withered heart, "like a piece of dried venison."

Plumed dancers and body parts: the combination of cross-fertilization and conflict continued after the American Revolution. Nowhere was the phenomenon more apparent than in the region's evolving language, starting with new names on the map. All but two colonies on the Atlantic coast had names drawn from the Old World. But once beyond the mountains, the white invaders, having taken the land from the Indians, then named virtually all their new states after the natives they continued to fight. Many other words, following various paths, entered the American idiom from the frontier. Some came directly from European rivals—*alligator* and *dollar* from the Spanish and *coulee* from the French. Many came from Indians: *pecan* and *caucus,* for instance. Others arrived indirectly. *Bayou* was borrowed from the French, who had taken it from the Choctaw *bayuk.* Old words took on new meanings. Because the hide of an adult male deer brought one Spanish dollar, a dollar was called a *buck. Corn,* which in England meant any grain, applied in America only to Indian maize by the revolutionary era. Then west of the mountains, where corn was the family's mainstay, the word was joined to others to form, by one count, 151 new words and phrases.

The newer arrivals absorbed many influences from the older Euro-American settlements, particularly from French communities in the Illinois country and from the Spanish farther south. Euro-Americans of all backgrounds in turn borrowed heavily from the Native American cultures. Backwoods farmers had learned about girdling and intercropping from Indians on both sides of the Appalachians. Other techniques, like burning trees and brush to increase fertility, were native practices that confirmed and reinforced similar methods brought by settlers; most foods gathered and grown, as well as the family's medicinal plants, were adopted from Indian customs. At least in how they supported themselves—hunting and gathering, cutting and burning—early farmers and Indians were more similar than not.

Hundreds of Anglo-Americans—the "white Indians"—opted to live in native societies. To replenish populations devastated by warfare and disease, Indians captured and adopted white women and children. When later given the chance to leave, many of these refused, and if children were forcibly returned, Benjamin Franklin wrote a friend, they soon fled to the woods, "from whence there is no reclaiming them." Some "white Indians" rose to leadership among their adopted cultures. A Pennsylvanian, seized during a raid at the age of four, grew up to become Old White Chief, a prominent Iroquois. Others moved back and forth between cultures. Simon Girty was captured by Delawares at fourteen, and though he returned to white society a few years later, he spent the rest of his life along the cultural borderland. His raids against Kentucky settlements made him probably the most hated figure of the revolutionary frontier era. Later, while dabbling in land speculation, he advised Wyandot, Mingo, and Delaware leaders and helped command Indian warriors in their crushing defeat of Arthur St. Clair in 1791. In 1818, besotted and crippled by rheumatism, he died in Canada, having lived a life as ambivalent and conflicted as the frontier heritage itself.

The Indians in turn found much to borrow from the white newcomers. For generations they had been exchanging animal pelts for European goods. This fur trade sent a wave of European influence over the region—then on, well beyond the Mississippi—far in advance of the farming frontier. In the 1790s, as the first white towns were appearing on the banks of the Ohio, Indians on the upper Missouri River were trading with British and French agents for a variety of goods, including corduroy trousers and peppermint. Trade goods offered obvious advantages. A metal awl or fishhook was more durable than one made of bone. Factory-made blankets were lightweight and easily maintained. For women who heated water by throwing hot rocks into baskets of tightly woven fibers, an iron pot hung above a fire was a wonder. Apart from trade, colonial governments, particularly the French, gave huge amounts of gifts to keep various groups within their alliance. The result was an often dramatic change on the surface of Indian life—in weapons, in utensils, in playthings. As whites adopted some Indian garments, natives wore linsey-woolsey shirts and painted themselves with imported vermilion. From Europeans, the Indians also acquired horses. That, in turn, brought changes in warfare, hunting, and other areas of economic life. By the early eighteenth century, southern Indians on horseback were tending cattle—another animal immigrant—and Seminoles had become acknowledged masters at breeding and raising horses.

In so much of how they passed their days, Indians and whites were moving toward one another, creating what the historian Richard White has called "the middle ground" in which the various lifeways wove together into rich, complex patterns. Did these peoples approach each other at a deeper level? There are hints. Take, for instance, bear stories. Bears—large, graceful, powerful—walked prominently among Indian legends. Typically they embodied both the threat and the bounty of the natural world. They were asked ritually to give themselves to the hunters' spears, then were thanked afterward,

Bears figured prominently as creatures of semi-divine power in the folklore of both Indian and European-American inhabitants of the trans-Appalachian and Far West. Whether in an illustration for Davy Crockett's epic tales or in an Assiniboin Indian's sketch of a European hunter, the animals appeared in 19th-century illustrations as worthy antagonists for their human pursuers.

Unidentified artist. Perilous Adventure with a Black Bear. *Wood engraving, 1838. From* Davy Crockett's Almanack, of Wild Sports in the West, Life in the Backwoods. . . . *(Vol. 1, No 4), 1838, reproduced in Franklin J. Meine, ed.,* The Crockett Almanacks: Nashville Series, *1835–38 (1955).*

Unidentified Assiniboine artist. Man Shooting a Bear *(detail). Ink on paper, 1853. Reproduced from* Bureau of American Ethnology Forty-Sixth Annual Report, *1928–29 (1930).*

with the understanding that the animals sometimes would take people in their turn. Looking on these animals as semidivine, Indians also recognized a close affinity to bears, which were, after all, the woodland creatures most closely resembling humans. Natives told of women marrying bears and of children wandering into caves and slipping easily into bear families. The line between hunter and prey, like that between man and his natural setting, was ultimately indistinguishable. Backwoods settlers too concocted tales of all-powerful bears of mysterious properties; living in the Indians' world, these people embraced stories that spoke of the same view of life. The classic distillation of these tales was Thomas Bangs Thorpe's "The Big Bear of Arkansas" (1841), in which a seasoned hunter and rip-roarer tells of stalking a giant, elusive animal, first with anger, but then with identification ("I loved him like a brother"), and finally with awe. In the end, pushing ever deeper into the forest, he confronts the bear, which looms "like a black mist." He shoots it, and it dies. But, he concludes, the death came not truly from him, for this was "a creation bear . . . an *unhuntable bear,* [which] *died when his time come.*"

So for all its bloody conflict, the frontier was also a place of accommodation and exchange, a zone of mutual influence among one of North America's richest mixes of peoples. Still, that "middle ground" was unstable and shifting. Behind those complex changes was a dynamic that worked again and again in favor of the new arrivals and against native cultures.

Initially, for instance, the Indians' interest in trade goods and gifts was conservative in intent; they wanted pots, blankets, spearpoints, and fishnets because those goods would allow them to live, more efficiently and with less effort, by a way of life that had served them well for generations. But inevitably, traditional skills fell into relative disuse, and natives came to rely on items that only Europeans could provide. Indians also felt a spiraling desire for some of the trade goods, especially alcohol. They then began hunting by a new, self-destructive strategy. To meet an escalating need for goods available only from white traders, Indians killed more and more of the very animals they would require to satisfy that need. Populations of game animals plummeted. The effect was felt first in the Ohio Valley; as early as the 1780s, an observer in Ohio estimated that a Delaware hunter typically killed between 50 and 150 deer each autumn. By 1800, areas as far west as Wisconsin and upper Michigan were severely depleted of white-tailed deer, elk, bear, and bison.

White settlers contributed even more to these ecological changes. Thousands of backwoods families stripped the trees from tens of thousands of acres. Others pulled up the stumps and pushed clearings outward into ever-larger cattle pastures and fields for cotton and commercial food crops. All remade the natural setting, disrupting game habitats and destroying the bounty of wild plants. Cotton cultivation often exhausted the fields. Lowlands became bogs that hosted thick clouds of mosquitoes; topsoil from higher ground washed into streams, killing fish and leaving the fields above scarred and brushy.

Thus Indians entered into a pernicious sequence. Trade initially made their lives easier, but time left them increasingly needful of white goods; yet season by season, Indians controlled less of the land and fewer of the animals essential to acquiring those goods and to supporting an independent existence. At the end of this progression—from advantage to reliance to dependence—the balance of power had shifted from native peoples to the whites. Native Americans saw where this process was taking them, and they responded. The result was another level of complexity in the story of cultural exchange. Whites and Indians developed some behaviors that a casual observer would find quite alike. But beneath the surface similarities, the cultural messages were dramatically different.

Nowhere was this more clearly shown than in two of the Indians' and pioneers' most notable activities—drinking and worshiping. Visitors often noted that alcoholic consumption on the frontier far outpaced that in the East. "When I was in Virginia, it was too much whiskey," a traveler noted. "In Tennessee it is too, too much whiskey!" The habit was especially pronounced among the South's Celtic frontier families, but accounts from all trans-Appalachia told of both sexes, and often children, downing draughts of whiskey and cider at all meals and social occasions. When a congregation drained two barrels of whiskey before ten o'clock one Sunday morning, a visitor wrote, "We could hear them firing, hooping and hollowing like Indians." The comparison was apt. By most accounts the natives too drank quickly, passionately, and in prodigious amounts, with spectacular and appalling results. Witnesses told of brawls and murders, of mass rapes, and of men rubbing burning logs on their heads.

Behind these common alcoholic enthusiasms, however, were different causes. White immigrants drank to relieve the tensions growing from the initial isolation and the heavy labors of pioneering. As for the Indians, there may have been some biological basis for their habits, but their drinking seems mainly a response to—and a cause of—their deteriorating position. In alcohol they found brief escape from dispossession and cultural disruption. Rampant drunkenness worsened problems of disease and sped the unraveling of social structures, which in turn gave Indians more reason to drink. It was a brutal cycle helped along by traders, who understood that the demand for alcohol, unlike that for awls or blankets, increased with the product's use. So frontier drunkenness among whites and natives told of different tensions. One habit expressed the trials of conquest, the other the despair and dependence of the conquered.

Backwoods settlers developed a distinctive style of religion as well. Sermons stressed a stark duality of good and evil and the sudden grace of salvation, when a God with strength beyond man's imagining would banish sin from a soul and, at the End Time, from a world rotten with corruption. This simple theology inspired an astounding emotionalism that found its fullest expression at camp meetings. As a sermon built, listeners would wail and weep and occasionally fall unconscious. Some got "the jerks," powerful convulsions that sent hats and bonnets flying. The evangelist Peter Cartwright claimed that five hundred persons were twitching crazily at one of his sermons. However else this religious style might be explained, it at least reflected lives of unrelenting stress and the dichotomous perspective from a cabin surrounded by a host of dangers.

During the same years new religions arose among the Indians. For all their differences, these religions shared certain traits. In fact, they seem to have been part of a far larger phenomenon, what Vittorio Lanternari called "religions of the oppressed," that appeared among native peoples in settings as varied as central Africa, Micronesia, and Latin America. These movements, typically led by a messiah figure, borrowed some Christian concepts from European conquerors; these concepts were then amended to fit natives' perspectives and needs. Besides stressing self-discipline and a rigorous moral code, cults typically demanded that members preserve precolonial skills and ways of life. Most found one Christian teaching particularly appealing: the Day of Judgment. If followers were firm in their faith and kept to the old ways, messiahs promised, a vengeful God would destroy or banish the colonial masters and restore His people to a golden age. Thus many movements became religions of liberation and cultural regeneration.

At least two such movements arose among Indians during the period studied here. The first, which helped inspire the Indian war of 1763, was led by a shadowy Delaware holy man named Neolin, who apparently demanded retention of traditional lifeways and promised divine protection in a crusade against the white invasion. The second movement, much better known, was associated with the most famous native leader of this era—Tecumseh, a Shawnee of diplomatic and oratorical brilliance who attempted to fashion a confederation of tribes north and south of the Ohio against further expansion of white settlement. But at least as important as Tecumseh was his brother. Known during his first thirty years as Lalawethika, this brother was a lazy blowhard; his name translated as "Noisemaker" or "The Rattle." One-eyed, alcoholic, and a poor hunter with no taste for fighting, he was utterly undistinguished until the spring of 1805, when he fell into a trance (or coma) and emerged from it to say that he had died and had been reborn to teach a new religion entrusted to him by the Master of Life. Its basic tenets were the unity of all Indians, preservation of traditional customs and means of living, rejection of virtually all Euro-American ways and technology, abstinence from alcohol, and monogamous marriages with only Indians. He insisted on a revival of some older rituals and the adoption of some new ones strikingly similar to the Catholic confession and rosary. If natives would follow His teachings, the Master of Life had promised, He would restore harmony to the world and "overturn the land." The prophet told his followers, "All the white people will be covered, and you alone shall inhabit the land." Now a divine messenger, Lalawethika called himself Tenskwatawa, or "The Open Door," from Christ's words, "I am the open door."

Indian and backwoods religions had much in common. All found comfort in Christian precepts and ceremonies; all embraced a chiliastic worldview in which the faithful triumphed through the intervention of a fearsome God. Behind the similarities was a fundamental difference, however. Backwoodsmen and Indians all were wrestling for control in a changing world, but whereas settlers were crying out for protection as those changes were accomplished, "The Open Door" desperately preached to reverse those same transformations. His was a theology of the last chance.

Flood Tide

During the third period of this American frontier, from 1812 to 1820, Tenskwatawa's deepest fears were confirmed. The United States solidified its military and political domination as the tide of white migration grew still greater and the market economy took tighter hold in towns, cotton country, and wheatlands. Ecological changes further eroded Indian independence.

The War of 1812 was a triumph for the expansionist forces in trans-Appalachia. This culminating conflict in fact had begun several months before the declaration of war. The movement for Indian resistance led by Tecumseh and Tenskwatawa, also called "The

Prophet," had gathered support after 1806. Incensed by the surrender of land through a series of treaties negotiated by Indiana's territorial governor, William Henry Harrison, Tecumseh devised a complex and precarious strategy. He would use the appeal of his brother's mystical, apocalyptic vision to fashion a practical policy for the here and now. The Shawnee leader knew the odds against successful military resistance. His main weapon instead would be an inclusive agreement among all Indian groups against any further surrender of land. War might be necessary, but it had to be avoided at all costs until this common front was achieved. Tecumseh traveled through the Ohio Valley, the Great Lakes country, and the Old Southwest, urging conversion to the new religion and an alliance against the expanding white frontier. Younger warriors in particular responded to the movement, and the brothers' village of Prophetstown, on Tippecanoe Creek in western Indiana, grew to a few thousand persons by 1811.

But this reach toward Indian independence failed, partly from one rash action and partly from the frontier's snarl of power relationships and the vagaries of distant diplomacy. Many Indian leaders resisted a new order that would, after all, undercut their authority. Then, while Tecumseh was traveling among the southern tribes, Harrison led a militia force to Prophetstown. When Tenskwatawa broke his brother's cardinal instruction and ordered his followers to attack, Harrison's troops drove the Indians from the village. That day—7 November 1811—was perhaps the darkest for Indian interests in the history of the region. Tenskwatawa's powers, and so his religion, were discredited, and the premature fighting crippled Tecumseh's campaign for a Native American front. With the coming of the War of 1812, Tecumseh had little choice but to link Indian interests once more to the ultimately unreliable British forces. His followers fought in at least 150 engagements, but at the end of the conflict, the United States retained control of the entire region. The Indians' greatest disaster came in October 1813, when Tecumseh, protecting British troops withdrawing into Canada, was killed at the Battle of the Thames.

The most militarily formidable group in the South, the Creeks, was divided into two increasingly hostile factions, the Upper and the Lower Creeks. The former, adamantly opposed to any accommodation with white invaders, not surprisingly were also more receptive to overtures from Tecumseh and Tenskwatawa. Many young warriors, calling themselves Red Sticks, converted to the new gospel. As tension grew during the summer of 1813, Lower Creeks and whites from the Tombigbee settlements took refuge in Fort Mims, a stockade on the Alabama River. But this proved to be not a sanctuary but a slaughter pen. On 30 August, more than seven hundred Red Sticks stormed the fort and killed more than five hundred persons. Seven months later a retaliatory force, commanded by Andrew Jackson, trapped the main body of Upper Creeks in their village of Tohopeka, wrapped on three sides by the Horseshoe Bend of the Tallapoosa River. In a withering cross fire, more than eight hundred died, their retreat cut off by Jackson's Cherokee allies.

In barely five months, Indians hoping to resist white expansion saw their shrewdest leader die and their most effective military force annihilated. In the Treaty of Fort Jackson, the Creeks (both Upper and Lower) surrendered an enormous L-shaped parcel of more than twenty-two million acres that included part of western Georgia and much of present-day Alabama. After signing treaties with the demoralized Indians of the Northwest, the United States moved to build a series of military posts among them. And in the final and most famous battle of the war, Jackson's rout of the British at New Orleans assured western farmers and merchants control of their most vital artery, the Mississippi.

So ended the last hopes of Native Americans and other nations to block the absolute control of trans-Appalachia by the United States. Small wonder, then, that these events lifted to political eminence the men most associated with those episodes. Out of the western campaigns came two presidents, Jackson and Harrison, and a vice president, Richard Johnson, of Kentucky, who after claiming to have slain the greatest Indian of the era, ran for office with the following slogan: "Rumpsey Dumpsey,/Rumpsey Dumpsey,/Colonel Johnson killed Tecumsey."

After 1815, the forces of change that had emerged during the previous thirty years gathered an extraordinary momentum. The stream of white migration became a flood. Between 1810 and 1820, Tennessee grew by more than 160,000 persons; Ohio's population more than doubled, from 230,760 to 581,434. But the most impressive growth was in the next tier of territories, those combining accessibility with the best chances of good land. Indiana's population grew by 600 percent, Alabama's by 1,420 percent. Several factors inspired this "Great Migration." The war that ended resistance from Indians and Europeans also publicized the country's promise. Soldiers once more were paid with public land. The value of that land increased as national and international markets for its products grew. With Atlantic trade lanes reopened, the price of cotton doubled within several months of the last fighting.

The territories' passage to statehood was quickly accomplished. Between 1816 and 1819, a state was admitted each year (Indiana, Mississippi, Illinois, and Alabama). Except when four Far Western states joined the union in 1889, the union of states would never again expand at such a pace. The particulars of the new constitutions suggested regional differences and the continuing pull and tug among economic and social groups. Indiana's constitution reflected the interests of middling farmers on the rise; it prohibited slavery and allowed universal manhood suffrage, popular election of governors, and annual election of the lower house. Mississippi's gave the vote to virtually all white men, but like Kentucky's, it also firmly established black slavery and leaned toward larger landowners. The economic patterns north and south of the Ohio became even more deeply entrenched: north of the river were diversified farms and small market towns; to the south was a mixture of large-scale cotton plantations in the blacklands, smaller subsistence farms in the piney uplands, and commercial cattle and pig farming in both.

Throughout trans-Appalachia, residents engaged in one economic activity—land speculation—with a common, frenzied enthusiasm. Eastern investors bought up military bounties for a pittance in hopes of future profits. An unconfirmed report told of a half section in Indiana purchased for $17.50 and sold for $5,000. As usual, squatters complained that their land was being sold from under them after they had held it against foreign and native competitors, but just as usual, settlers of modest means were playing the speculative game on a smaller scale. By the thousands they bought land on credit, pouring all resources into the down payment on the bet that rising values would allow

Set in a mountainous New Hampshire setting, Thomas Cole's painting evokes one of the central tensions of frontier settlement as it was reenacted across the trans-Appalachian West. The sunlit cabin, bountiful garden, and warm domestic scene suggest the idyllic possibilities of frontier life. Nonetheless, the raw tree stumps and fallen logs serve as the artist's warnings about the transformation of the land.

Thomas Cole (1801–48). The Hunter's Return. *Oil on canvas, 1845. Courtesy Amon Carter Museum, Fort Worth, Texas.*

them to sell off part to make installments or to dump it all for a handy profit. Sales at an Alabama land office increased from under fifteen hundred acres in 1814 to more than two hundred thousand in 1817. In just three years after 1816, indebtedness to the government for public lands increased nearly 600 percent. Rarely in the history of westward expansion would so many buy so much with so little.

Land sales were directly linked to another government encouragement—migration. As Congress admitted each state, it reserved about 3 percent of the net income from the sale of public lands to construct public roads. The first such project was authorized in 1806. The Cumberland Road, which ran between Cumberland, Maryland, and Wheeling, Virginia, was meant to ease the way across the Appalachian barrier to the upper Ohio Valley. Well before its completion, thousands of farm and freight wagons and throngs of people, cattle, dogs, and pigs filled it each year on their way west. Before the War of 1812, Congress had also paid to build the Federal Road from Milledgeville, Georgia, to St. Stephens on the Tombigbee River, and after the conflict it authorized another overland route, popularly called Jackson's Military Road, connecting New Orleans with Nashville and the upper South generally. As with the Ordinances of 1785 and 1787, the demands of the frontier allowed the national government to expand its powers and to project its shaping force into the future of the growing Republic.

These thoroughfares of migration and commerce sealed even more tightly the fate of native peoples and thus deepened the government's well-worn dilemma: how could the United States treat the Indians justly while taking their lands? Especially among the populous southern tribes, Washington now shifted its emphasis to a strategy of "education." Government-sponsored agents, most of them men of the cloth, would tutor natives, not only in Christian principles but also in the skills and virtues of agriculture—Euro-American style, of course—which would promote orderly communities and a veneration of private property. "We want to make citizens out of them," wrote Thomas L. McKenney, an architect of this policy, "and they must first be anchored to the soil." The various peoples would also be encouraged to abandon native languages for English, "the lever by which they are to elevate themselves into intellectual and moral distinction," and to adopt white concepts of family and white styles of housing, dress, and general propriety. This was thought to be a merciful policy—McKenney called it a "cup of consolation"—that would ensure natives eternal glory and a happy assimilation into the society now pressing hard upon them. Raised up in the ways of the virtuous Republic, Indians would be freed from both outside threats and their own resentments. And the government's dilemma would dissolve.

This optimistic program in one sense reflected certain Enlightenment influences embraced by its early advocate, Jefferson. But in a larger sense this policy was only a newer variation of an assimilative dream as old as the first colonies. And it was as wrongheaded as what had been tried at Jamestown. Indians *were* absorbing white culture (just as whites were absorbing Indian culture), but many of the lessons had little to do with the missionaries' goals. At work was a dynamic of infinite complexity, one that was far beyond the government's understanding and even farther beyond its control. The current policy was bound for frustration, and some already were advocating a final option. In 1817 the government set aside a portion of present-day Arkansas where Cherokees might relocate. Beyond the Mississippi, advocates of removal argued, native peoples would find sanctuary and enough time for a full and final adoption into the national family.

By 1820 the frontier had left its mark on trans-Appalachia. As a historical condition, it had moved on. Its defining traits now prevailed in the country beyond the Mississippi. There the pace of exchange among the many peoples was accelerating. Guns, hawk bells, horses, pelts, blankets, and cloth moved back and forth through a trading network stretching from southern Canada and the upper Missouri to Santa Fe. The scramble of powers had shifted westward. The Adams-Onís Treaty of 1819 rounded out the southeastern boundaries of the United States and projected the nation's imperial energies beyond the great river, where the U.S. government would contest with entrenched European rivals and another few dozen Indian groups. The arm of government was reaching farther west. As Missouri petitioned for statehood, surveyors continued laying out their checkerboard, edging now toward the eastern fringe of the Great Plains, even as Major Stephen Long was mapping part of the region farther west in the latest of several government-sponsored expeditions. The public embraced ever more fondly the idea of the opening West. In 1823 James Fenimore Cooper published *The Pioneers.* Its protagonist, Leatherstocking, would stalk through four other novels before drifting west to die on the plains. A wilderness regenerate, doomed to solitary freedom, brother to and slayer of Indians, Cooper's hero personified a people wedded to the frontier idea in all its obsessiveness, naïveté, hope, and contradictions.

Millions would move west vicariously with Leatherstocking. Of the thousands who went in person, many came from trans-Appalachia. And so the patterns and conflicts of that earlier frontier were begat again. The first serious foray of fur traders up the Missouri River was organized by William Morrison and the Canadian-born Pierre Menard, both Kaskaskia merchants, and Manuel Lisa, a Spaniard from the lower Mississippi Valley. The trapper and scout Christopher "Kit" Carson had moved as a child with his Scotch-Irish parents to Missouri from a Kentucky backwoods farm in 1812. Members of the first party of farmers to cross the plains to California were virtually all from the Ohio Valley and upper South. Guiding them part of the way was John Park, a half-Iroquois hunter.

Among those looking westward was Moses Austin, a Connecticut Yankee who had taken up lead mining in Missouri. In 1820 he won from Spanish authorities the permission to establish a colony of his countrymen in central Texas. A shrewd observer of forces around him, Austin had watched and wondered at the crowds of emigrants he had seen pushing across the Appalachians twenty-five years earlier. "Ask these Pilgrims what they expect when they git to Kentuckey the Answer is Land. have you any. No, but I expect I can git it. have you any thing to pay for land, No. did you Ever see the Country. No but Every Body says its good land . . . here is hundreds Traveling hundreds of Miles, they Know not for what Nor Whither, except its to Kentuckey."

The traits of that crowd—not so much optimism as a driven, unappeasable restlessness—had only gathered strength since then. More "Pilgrims" than ever were pressing westward, but now their promised land was not Kentucky but Texas and other places almost as improbable. As one American frontier closed, another opened.

Bibliographic Note

Anyone curious about the trans-Appalachian frontier can turn first to several excellent surveys. Besides Malcolm J. Rohrbough's *The Trans-Appalachian Frontier: People, Societies, and Institutions, 1775–1850* (New York, 1978), which remains the best and most comprehensive, the reader should consult Reginald Horsman, *The Frontier in the Formative Years, 1783–1815* (New York, 1970), Jack M. Sosin, *The Revolutionary Frontier, 1763–1783* (New York, 1967), Thomas D. Clark and John D. W. Guice, *Frontiers in Conflict: The Old Southwest, 1795–1830* (Albuquerque, 1989), and Francis S. Philbrick, *The Rise of the West, 1754–1830* (New York, 1965). Beverly W. Bond, Jr.'s *The Civilization of the Old Northwest: A Study of Political, Social, and Economic Development, 1788–1812* (New York, 1934), although thin on interpretation, remains a fine starting point for a study of the Ohio Valley frontier. To put this story in context, works on the background of imperial competition are helpful: W. J. Eccles, *The Canadian Frontier, 1534–1760* (New York, 1969); John A. Caruso, *The Mississippi Valley Frontier: The Age of French Exploration* (Indianapolis, 1966); and Jack M. Sosin, *Whitehall and the Wilderness: The Middle West in British Colonial Policy, 1760–1775* (Lincoln, 1961). For a taste of the many approaches to the study of Native Americans and their societies, see Harold E. Driver, *Indians of North America*, 2d ed. (Chicago, 1969), Charles Hudson, *The Southeastern Indians* (Knoxville, 1976), R. Douglas Hurt, *Indian Agriculture in America: Prehistory to the Present* (Lawrence, 1987), and Helen Hornbeck Tanner's superb *Atlas of Great Lakes Indian History* (Norman, 1987), useful for its text as well as its maps and illustrations. There are, as well, anthropological and historical studies of all major native groups of the region.

Solon Buck and Elizabeth H. Buck, *The Planting of Civilization in Western Pennsylvania* (Pittsburgh, 1939), still is an essential introduction to expansion into the upper Ohio; for expansion into Tennessee and Kentucky, one should begin with Thomas P. Abernethy, *From Frontier to Plantation in Tennessee* (Chapel Hill, 1932), William S. Lester, *The Transylvania Company* (Spencer, 1935), John Mack Faragher, *Daniel Boone: The Life and Legend of an American Pioneer* (New York, 1992), and Charles A. Talbert, *Benjamin Logan: Kentucky Frontiersman* (Lexington, 1962). On popular attitudes toward the frontier's possibilities and the revolutionary crisis, I found the research of Alan Taylor, soon to appear in print, especially helpful. As for the war itself, Jack Sosin's *The Revolutionary Frontier* can be supplemented by Thomas P. Abernethy's *Western Lands and the American Revolution* (New York, 1937), which stresses the role of speculation and economic interests.

Two recent studies provide fascinating, but conflicting, ethnic analyses of the roots of frontier society: Grady McWhiney, *Cracker Culture: Celtic Ways in the Old South* (Tuscaloosa, 1988), and Terry G. Jordan and Matti Kaups, *The American Backwoods Frontier: An Ethnic and Ecological Interpretation* (Baltimore, 1989). James E. Davis has given this society a close demographic look in *Frontier America, 1800–1840: A Comparative Demographic Analysis of the Settlement Process* (Glendale, Calif., 1977); to put his conclusions into a larger context, see Walter T. K. Nugent, *Structures of American Social History* (Bloomington, 1981). Richard Wade, *The Urban Frontier: The Rise of Western Cities* (Cambridge, Mass., 1959), remains the best source on the topic. Much can be learned about daily life in the countryside in Charles W. Towne and Edward N. Wentworth, *Pigs: From Cave to Cornbelt* (Norman, 1950), Jared van Wagenen, Jr., *The Golden Age of Homespun* (New York, 1953), and Nicholas P. Hardeman, *Shucks, Shocks, and Hominy Blocks: Corn as a Way of Life in Pioneer America* (Baton Rouge, 1981). Louis B. Wright's brief and highly readable *Culture on the Moving Frontier* (Bloomington, 1955) is a must as an introduction to early cultural development. A classic older study of economic development, Percy W. Bidwell and John I. Falconer's *History of Agriculture in the Northern United States, 1620–1860* (Washington, D.C., 1925), is still full of helpful information, although it should be read with more recent works, such as David C. Klingaman and Richard K. Vedder, eds., *Essays on the Economy of the Old Northwest* (Athens, 1987). On the popular image of the frontier and its complex relationship to Americans' perception of themselves, two classics—Henry Nash Smith, *Virgin Land: The American West as Symbol and Myth* (Cambridge, Mass., 1950), and Arthur K. Moore, *The Frontier Mind: A Cultural Analysis of the Kentucky Frontiersman* (Lexington, 1957)—now have a third companion, Richard Slotkin's *Regeneration through Violence: The Mythology of the American Frontier, 1600–1860* (Middletown, Conn., 1973).

John Barnhart's *Valley of Democracy: The Frontier versus the Plantation in the Ohio Valley, 1775–1818* (Bloomington, 1958) still offers much of value on politics and government, although it should be supplemented by more recent works. Among the best of these is Andrew R. L. Cayton, *The Frontier Republic: Ideology and Politics in the Ohio Country, 1780–1825* (Kent, 1986). Two of the most useful among the many works on land policy are Roy M. Robbins, *Our Landed Heritage* (Princeton, 1942), and Vernon Carstensen, ed., *The Public Lands* (Madison, 1963); Malcolm J. Rohrbough's *The Land Office Business: The Settlement and Administration of American Public Lands, 1789–1837* (New York, 1968) remains the essential book on the distribution of the public domain. Standard studies of the era's diplomacy include Arthur P. Whitaker, *The Spanish-American Frontier, 1783–1795* (Boston, 1927), Jerald A. Coombs, *The Jay Treaty: Political Battleground of the Founding Fathers* (Berkeley, 1970), E. Wilson Lyon, *Louisiana in French Diplomacy, 1759–1804* (Norman, 1934), and two works, both now revised, by Samuel F. Bemis: *Jay's Treaty: A Study in Commerce and Diplomacy*, rev. ed. (New Haven, 1962), and *Pinckney's Treaty: America's Advantage from Europe's Distress, 1783–1800*, rev. ed. (New Haven, 1960).

Indian-white relations and warfare have inspired a voluminous literature. Readers should begin with Richard White's splendid recent study, which blends environmental, cultural, and diplomatic history: *The Middle Ground: Indians, Empires, and Republics in the Great Lakes Region, 1650–1815* (Cambridge, Eng., 1991). Wiley Sword's *President Washington's Indian War: The Struggle for the Old Northwest, 1790–1795* (Norman, 1985) and Reginald Horsman's *Expansion and American Indian Policy, 1783–1812* (East Lansing, Mich., 1967) should be read with earlier standards, such as Randolph C. Downes's *Council Fires on the Upper Ohio* (Pittsburgh, 1940). On government policy and relations with Native Americans, every reader should begin with Francis Paul Prucha's magisterial *The Great Father: The United States Government and the American Indians* (Lincoln, 1984). The best introductions to Tecumseh, Tenskwatawa, and the movement they inspired are both by R. David Edmunds: *Tecumseh and the Quest for Indian Leadership* (Boston, 1984) and *The Shawnee Prophet* (Lincoln, 1983). Vittorio Lanternari's *The Religions of the Oppressed: A Study of Modern Messianic Cults*, trans. Lisa Sergio (New York, 1963), can help the reader understand Indians' new religions in a far larger context, whereas another study is full of insights into dynamics behind the natives' drinking habits: Craig MacAndrew and Robert B. Edgerton, *Drunken Comportment: A Social Explanation* (Chicago, 1969). The complex questions of white perceptions of Indians and how these images were translated into public policy are the subjects of Roy Harvey Pearce, *Savagism and Civilization: A Study of the Indian and the American Mind* (Baltimore, 1953), and Bernard W. Sheehan, *Seeds of Extinction: Jeffersonian Philanthropy and the American Indian* (New York, 1973).

Finally, readers who want to taste the flavor of the times can turn to two fine reprint series. Arno Press has republished early histories and commentaries of the region in the "Mid-American Frontier" series, and University Microfilm's "March of America Facsimile Series" includes reprints of contemporary descriptions and travel accounts.

SECTION II

Expansion

In the course of the nineteenth century, the trans-Mississippi region became the *American West.* Nationalism in all its diverse manifestations enfolded the vast lands and distinct places. The governing of the West and the peopling of the region now acquired a clearly American imprint. The growth of the West in economic and demographic terms not only demonstrated the close connection of this region to the growth of the American nation but also indicated the growing interconnection of the West and the United States to the rest of the world.

The domination of the West by the United States, its incorporation into an American empire, implies a form of political, and even military, control that obscures an important story. Total subjugation of both lands and peoples did not occur. No monolithic power took control of the West. Instead a web of interconnected relationships bound new and old residents of the West to each other, even when they saw each other as enemies. The full story is not one of great victories and clear successes. It is a story of economic boom and financial bust, of natural resources exploited for distant markets, and of agricultural expansion established with great human cost. It is a story marred by violence but shaped by important human institutions—governments appointed and elected, churches traditional and new, communities temporary and sustained, and families separated and united. The actors are not of one generation, one color, or one gender, and the story would not end with the closing of the century.

The complex interconnections of the West's many peoples and institutions can be recognized in some individual lives. Such is the case of the Native American woman whose name was most probably Sacagawea ("Bird Woman" in Hidatsa). In the fall of 1800 at age twelve or thirteen, this Lemhi Shoshoni girl had been captured by the Hidatsas; she was later sold to a French-Canadian fur trader, Toussaint Charbonneau. When the American explorers Meriwether Lewis and William Clark met Charbonneau at Fort Mandan on the Missouri River in present-day North Dakota, Sacagawea, the youngest of the three Indian women with whom he lived, was pregnant with the trader's child. Charbonneau became an interpreter for the Lewis and Clark expedition and took Sacagawea on the epic trek to the Pacific coast and back.

What role did she play in the journey? Lewis dismissed her participation. He reported, "If she has enough to eat and a few trinkets I believe she would be perfectly content anywhere." Clark felt differently. He wrote to Toussaint Charbonneau, "Your woman . . . diserved [*sic*] a greater reward for her attention and services." But what were these services? No direct accounts from Charbonneau or Sacagawea exist, but many legends have been created. The most persistent has Sacagawea guiding the expedition through dangerous lands, thus ensuring the party's survival. A careful reading of the daily accounts left by Lewis and Clark show that only twice did Sacagawea help guide the

The feelings of nationalism inspired by the American victory in the Mexican War created renewed interest in narrative paintings that supported the myth of America's "manifest destiny" as a nation stretching across the continent.

William Tylee Ranney (1813–57). Boone's First View of Kentucky. *Oil on canvas, 1849. Courtesy of The Anschutz Collection, Denver, Colorado.*

expedition. More often she aided communication with other native groups, especially with her knowledge of the Shoshoni language. Yet her most important role, as the historian James P. Ronda has explained, may well have been as a woman and a mother. Sacagawea gave birth to her son Jean Baptiste Charbonneau early in the journey on 11 February 1805. The presence of a young mother and her baby may have reassured many of the native peoples—women and men—who met the expedition. Clark, with his infamous grammar and spelling, observed: "The Wife of Shabono our interpreter We find reconsiles all the Indians, as to our friendly intentions. A woman with a party of men is a token of peace."

Other women, not as legendary, made their own journeys that interwove with stories set in the West. An African-American woman named Biddy may have started her travel west by oxcart from Georgia. She arrived in southern California in 1852 as a slave, the property of Robert Smith, a Mormon convert from Mississippi. Smith had taken Biddy and another slave woman, Hannah, to Utah in 1851 before migrating with other Mormons to San Bernardino. By 1855, Smith wanted to take Biddy, Hannah, and their twelve children and grandchildren to Texas, but Biddy and Hannah managed to take Smith to court. The judge in Los Angeles granted the mothers and the children their freedom, making special note of Robert Smith's desire to relocate four of the children, born in the free state of California, to the slave state of Texas.

Biddy took the full name of Biddy Mason and settled in Los Angeles, where she worked as a confinement nurse for $2.50 a week. She also did domestic work but soon purchased her own home and eventually bought two city lots. Thus began a successful series of investments in real estate that made Biddy Mason a wealthy woman. She used her house on South Spring Street to aid needy travelers, and she used her money to support education for African-American children. Biddy Mason, like Sacagawea, was a woman and mother who could ensure safety and a form of security for others.

Not all women's lives in the West would have such positive results. Sacagawea was able to facilitate peaceful interactions, but many other Native American women died in brutal attacks, such as the Cheyenne mothers shot down with their children by the Third Colorado Volunteers in 1864 at Sand Creek. And Biddy Mason may have found financial success in the West, but many African-American women found themselves in western prisons, well out of proportion to their numbers in the general population. As the historian Anne Butler discovered, from the 1860s to the early twentieth century, African-American women composed the majority of women in western state penitentiaries. Some were accused of extremely petty crimes, such as stealing a nightgown.

The many lives of the West's many peoples are an indication of the different ways that the history of the American West may be told. Some of these ways may appear familiar at first, examining topics such as the policies of the national government or the pitfalls of economic development. But surprises can emerge—the government was more intrusive and the economy more global than previously recognized. Other perspectives will seem fresh—the creation of animals of enterprise or the multigenerational migrations of families. All perspectives are part of a larger story that improves as much by rethinking as by retelling. It is a story that will expand to include more peoples and more ideas the more often it is told.

— Clyde A. Milner II

CHRONOLOGY

1804–6	Meriwether Lewis and William Clark explore the Louisiana Purchase.
1808	John Jacob Astor founds the American Fur Company.
1821	William Becknell opens the Santa Fe Trail.
1825	William Ashley organizes rendezvous for fur trappers at Henry's Fork on the Green River.
1829	Depletion of sandalwood in Hawaiian Islands and sea otters in southeastern Alaska.
1834	Protestants open mission in Oregon's Willamette Valley.
1836	Texas Revolution results in independence from Mexico.
1837	Epidemic diseases devastate many Plains Indian tribes.
1841	General Preemption Law acknowledges squatters' right to land.
1845	The United States annexes Texas.
1846	Treaty with Britain secures half of Oregon for the United States.
1847	Latter-day Saints (Mormons) arrive in Utah.
1848	Treaty of Guadalupe Hidalgo, ending the Mexican-American War, adds 1.2 million square miles of territory to the United States.
1848	James Marshall discovers gold at Sutter's Mill, California.
1850	Entry of California into the Union forces the most complex legislative compromise in U.S. history.
1857	One-sixth of U.S. Army dispatched to quell presumed rebellion by Latter-day Saints in Utah.
1858	John Butterfield opens overland stage route.
1859	Mining rushes occur in Nevada and Colorado.
1862	Congress passes the Homestead Act, granting 160 acres of public land to settlers after five years of residence.
1864	The Navajos are forced to take the Long Walk to Bosque Redondo. Cheyennes are massacred at Sand Creek in Colorado.
1867	35,000 Texas cattle are driven up the Chisholm trail to Abilene, Kansas. The United States purchases Alaska from Russia.
1869	Completion of the first transcontinental railroad. Wyoming Territory extends the vote to women.
1876	Battle of the Little Bighorn marks short-lived victory for Lakotas.
1877	African-American settlers, "Exodusters," establish Nicodemus in northwest Kansas.
1878	Lincoln County War in New Mexico produces legendary outlaw, Billy the Kid.
1880s	Drought and harsh winters devastate cattle herds throughout the West.
1881	Gunfight near the O.K. Corral in Tombstone, Arizona, results in three deaths.
1882	Huge deposits of copper found at Anaconda mine in Butte, Montana.
1883	North American bison hunted to near-extinction.
1885	Massacre of fifty-one Chinese by miners at Rock Springs, Wyoming.
1886	Division of Forestry formally recognized within Department of Agriculture.
1887	General Allotment Act, or Dawes Act, initiates break-up of reservation lands.
1890	U.S. Census announces that the frontier has closed. Lakota Ghost Dancers shot down at Wounded Knee, South Dakota.
1893	Great Northern Railroad reaches Seattle.
1896	William Jennings Bryan of Nebraska runs for president on both the Democratic and Populist tickets.
1898	Spain cedes Pacific possessions to the United States with treaty ending the Spanish-American War. The United States annexes Hawai`i.

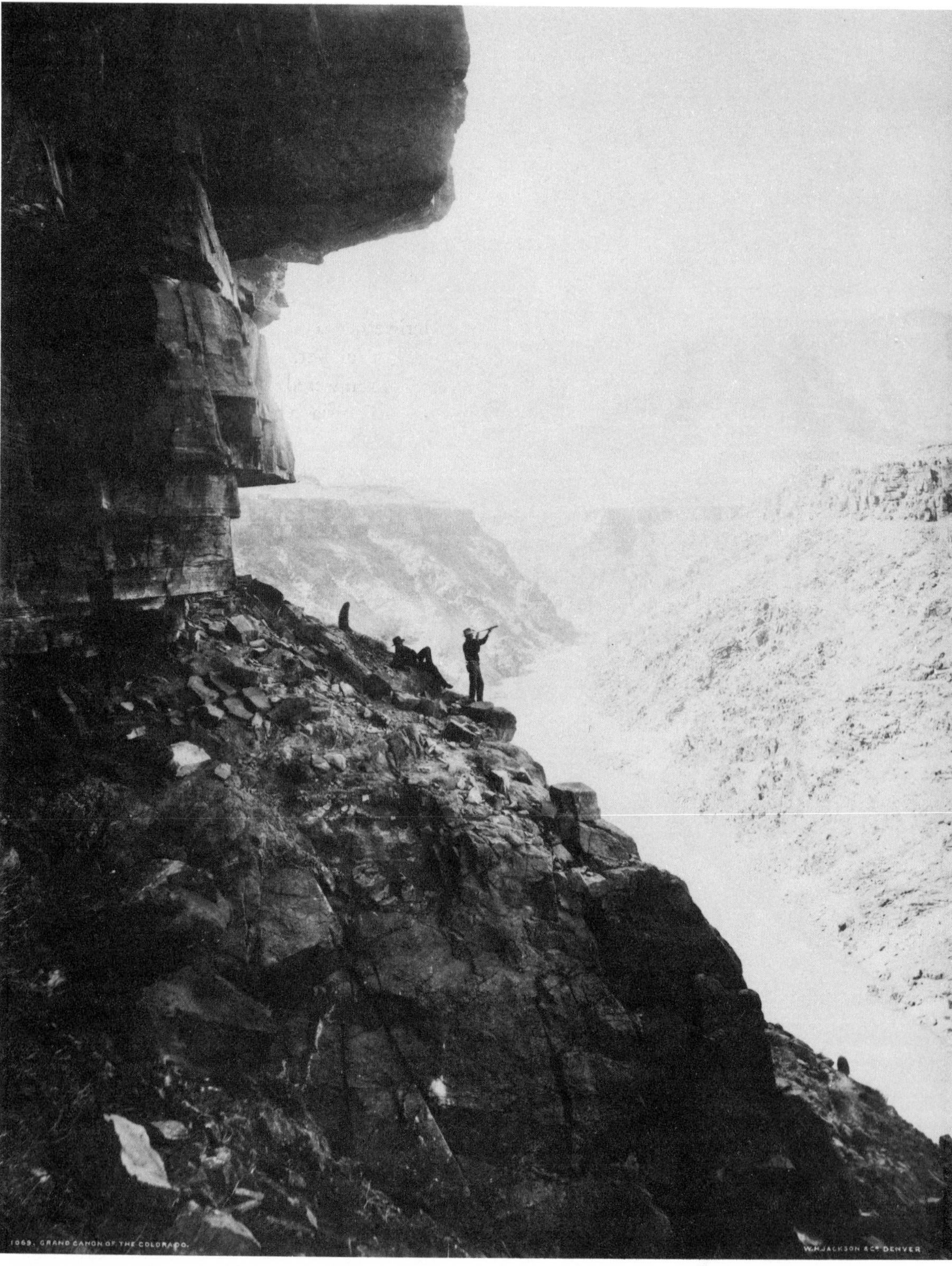
1069. GRAND CAÑON OF THE COLORADO.
W.H.JACKSON & Cº DENVER

Chapter Five

National Initiatives

CLYDE A. MILNER II

Who owned the Louisiana Territory? Meriwether Lewis and Carlos Dehault Delassus could not agree on an answer. Delassus, a Spanish official, thought that he still administered lands controlled by the French. Rumors of the sale of Louisiana had reached St. Louis in August 1803, but no formal instructions had been sent to Delassus, lieutenant governor of the territory. On 8 December 1803, Captain Lewis had crossed the Mississippi River to meet with Delassus. Lewis knew that the United States had purchased Louisiana from Napoleon's government, but he carried no official documents concerning the sale. So after presenting his credentials, Lewis informed Delassus that he planned to lead an expedition funded by the government of the United States. Perhaps the captain paraphrased the instructions that President Thomas Jefferson had sent to Lewis, his twenty-nine-year-old former secretary, in June. "The object of your mission is to explore the Missouri river, & such principal stream of it, as, by it's [*sic*] course and communication with the waters of the Pacific ocean, whether the Columbia, Oregan [*sic*], Colorado or any other river may offer the most direct & practicable water communication across this continent for the purposes of commerce."

Delassus insisted that the American party not proceed up the Missouri until he received notification of Louisiana's sale. Lewis willingly complied. He crossed back to the east side of the Mississippi and established his winter camp on easily recognized American soil. In January 1804, the residents of St. Louis officially learned that all of the Louisiana country had been purchased by the United States. In March, a formal transfer ceremony occurred in the town. In May, William Clark, who shared command of the expedition with Lewis, led the party across the Mississippi. As many as forty-six men may have begun the ascent of the Missouri. Aside from the two commanding officers, twenty-seven young, unmarried recruits served as permanent members of the "Corps of Discovery." Two nonmilitary personnel also accompanied the corps—York, an African-American slave whom Clark had inherited from his father, and George Drouillard, half French-Canadian and half Shawnee, who worked as a hunter and interpreter for twenty-five dollars a month. Over the journey of twenty-eight months and eight thousand miles, only one member of the corps lost his life. Sergeant Charles Floyd died from what may have been a ruptured appendix. York, Drouillard, Lewis, Clark, and even Lewis's black Newfoundland dog, Seaman, survived the entire transcontinental trek.

A photographer who documented the Rocky Mountain West from 1870 to 1878 as a member of Ferdinand V. Hayden's ambitious geological and geographical survey, William Henry Jackson later created images that romanticized this period as a heroic age of exploration. These two men poised confidently at the edge of a cliff evoke the sense of self-assurance that characterized 19th-century settlement of the western half of the nation.

William Henry Jackson (1843–1942). Grand Canyon of the Colorado. *Albumen silver print, ca. 1892, 21 x 17 1/16 in. Collection of the J. Paul Getty Museum, Malibu, California.*

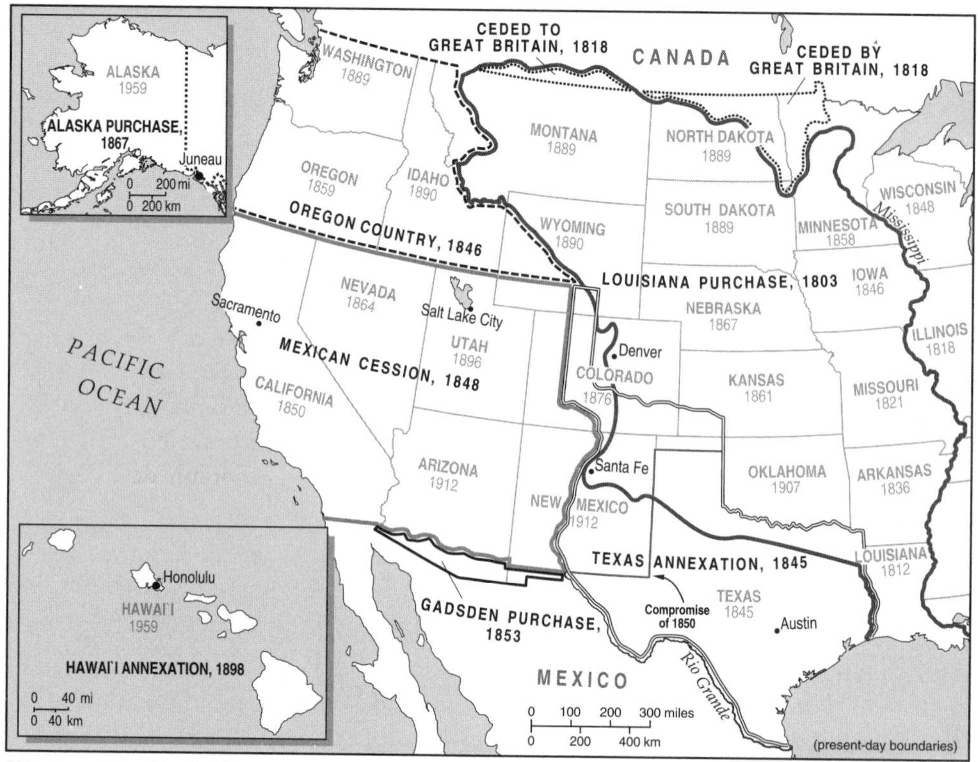

Western Territorial Expansion, 1803–1959

Three centuries earlier, an expedition of three ships under the command of Christopher Columbus had launched an era of European exploration and expansion. On 14 May 1804, an era of American exploration and expansion began when a keelboat and two pirogues carried the Corps of Discovery up the Missouri. Under instructions from the nation's president and with congressional funding, the federal employees of the Lewis and Clark Expedition initiated the incorporation of the trans-Mississippi West into the United States.

During the century that followed, the federal government continued the scientific exploration of land beyond the Mississippi. Yet, federal action did not end with these expeditions. The government not only explored these lands but also claimed new ones. Through purchase, treaty, annexation, and war, the federal government added extensive territory beyond the vague boundaries of the original Louisiana Purchase. The most prominent additions came from what had been Mexican territory with the annexation of Texas in 1845, the cession after the Mexican-American War in 1848, and the Gadsden Purchase in 1853. Britain gave up Oregon, Russia sold Alaska, and numerous Native American tribes lost lands to the growing United States.

More than a century of national expansion culminated on 14 February 1912, St. Valentine's Day, when President William Howard Taft signed the proclamation that made Arizona the forty-eighth state of the United States. Nearly five weeks earlier, his signature had admitted New Mexico to the Union. In less than eleven decades, new states had been formed out of the Louisiana Purchase and the vast lands that bordered it. The two newest additions completed the political incorporation of the contiguous trans-Mississippi West into the United States. Like the acquisition of lands, this process of political incorporation resulted from actions of the federal government and its agents.

Acquisition was simply a claim to western lands that were already occupied by an incredible variety of native peoples, a distinct population of Hispanic settlers, and a diverse representation of fur trappers. Political incorporation meant embracing people as well as lands. Often it involved helping people move to, and settle on, the land. In other words, the federal government promoted the increased occupation of the West as well as its acquisition. Occupation then produced political incorporation. The federal government aided this occupation of the West directly and indirectly. For example, it used the treaty process to reduce the lands held by American Indians and thus to increase the lands that non-Indians might settle. It published the reports of scientific exploration that indicated routes for immigration and lands for cultivation. It protected overland migration and helped supply immigrant parties. It stimulated economic development, especially by building roads, subsidizing stage and wagon lines, and constructing railroads. It also imposed the social, political, and legal models established east of the Mississippi. The U.S. Army supported most of these federal actions, whereas the U.S. Congress prolonged the political apprenticeship of much of the West before granting statehood. As expansion led to acquisition and as occupation led to incorporation, the initiatives of the federal government greatly shaped the history of the American West from 1803 to 1912—and beyond.

Search for a Usable West

The Louisiana Purchase and its aftermath form a long prelude to the more vigorous era of western expansion that began in the 1840s. By the end of that decade, the American presence in the trans-Mississippi had been transformed. Until the 1840s, federal activity in the region extended little beyond a list of scientific, military, and diplomatic initiatives that suited political and economic interests *east* of the Mississippi.

Consider Thomas Jefferson's reasons for sending the Lewis and Clark Expedition across the continent. His explanation to Congress in January 1803 had stressed support of the fur trade through improvement of relations with the Indians along the Missouri River. Jefferson believed that a small party of as few as a dozen men might "explore the whole line [of the Missouri] even to the Western ocean." He noted that improvement of commerce through conferences with the Indians fell within the constitutional powers of Congress. The national assembly accepted the president's plan and appropriated twenty-five hundred dollars for the project. By June, Meriwether Lewis had spent nearly all the initial appropriation, before receiving detailed instructions from the president and before the president had learned of the purchase of Louisiana.

The instructions, dated 20 June, came first. In his letter, Jefferson underscored the purposes that he had presented to Congress and then requested detailed scientific observations of the peoples, weather, rivers, terrain, climate, and natural resources in the lands traversed. On 4 July 1803, a day already reserved for recognition of the signing of the Declaration of Independence, the *National Intelligencer* reported news of the Louisiana Purchase. Jefferson had received notification only the day before, two months after the event. Great news indeed, but how great was the territory acquired? French officials did not know. On 1 October 1800, the French and Spanish had signed the secret Treaty of San Ildefonso, returning Louisiana to France. In the description of the ceded territory, Louisiana contained "the same extent . . . that it had when France possessed it, and such as it should be after the treaties subsequently entered into between Spain and other States." The agreement for the American purchase in 1803 contained the same ambiguous language. In fact, the French had reacquired and then sold Louisiana without establishing political control over all the territory. This explains why in St. Louis, a Spanish official, Delassus, served as lieutenant governor.

Thomas Jefferson knew that his country had acquired a vast territory with most of the new land lying beyond the Mississippi. Some critics considered it a useless wilderness, and no one knew if it contained lands suitable for the yeoman farmers that Jefferson idealized. The president could only guess at the boundaries to the north, west, and south. These would not begin to be clearly defined until 1818, when a convention with Great Britain set the northern boundary along the 49th parallel to the Rocky Mountains and established joint occupation of the Oregon country. The Transcontinental Treaty of 1819 between Spain and the United States, also known as the Adams-Onís Treaty, set the western and southwestern boundaries. Jefferson recognized that the Louisiana Purchase did not stretch to the Pacific Ocean. Yet, he understood that in 1792, an American merchant captain, Robert Gray, had first discovered the mouth of the Columbia River on Oregon's west coast. Gray gave the United States some claim to the Oregon country, so the president instructed Lewis and Clark to cross the entire continent to the Pacific. Jefferson dreamed of an easy portage from the headwaters of a far western river, such as the Columbia, back to the headwaters of the Missouri. Perhaps furs could be collected on the Pacific coast and transported back across the continent instead of across the Pacific or around South America. Of course, others had dreamed of a Northwest Passage that would not only cross the continent but also establish a base on the Pacific coast for trade with Asia, especially China. In 1803, the prospects of profits from both the fur trade and the China trade had captured the attention of commercial interests in Britain, Canada, and the United States. Two great rivals, the British-owned Hudson's Bay Company and the Canadian-controlled North West Company, already had designs on the Oregon country.

On their return in 1806, Lewis and Clark reported to the president. They had not found a Northwest Passage—an easy portage from the Columbia to the headwaters of the Missouri. The Rocky Mountains presented too formidable a barrier. Nor had they found agricultural lands within the Louisiana Territory. But they had collected an immense trove of information about the native peoples, diverse geography, and abundant resources of the trans-Mississippi West. Only a portion of their observations appeared in print. Nicholas Biddle, of Philadelphia, consulted their daily journals to write a narrative of the trip. After modest alterations by Paul Allen, it was published in 1814, along with a map produced by William Clark. Dr. Benjamin Smith Barton had agreed to produce a scientific report, but he died before fulfilling the task. Some materials in his care disappeared, including nearly all the collected plants and animals, the recorded vocabularies of Indian languages, and the astronomical observations.

Without a published scientific report, Jefferson felt that only half of the expedition's story had been told.

Although the trip did not benefit science, agriculture, or commerce in the manner that Jefferson had idealistically envisioned, some people did profit from the Lewis and Clark Expedition. For example, Manuel Lisa, one of the expedition suppliers in St. Louis, established trade with Indian tribes along the upper Missouri River and, by 1807, had built a post at the mouth of the Bighorn River in present-day Montana. In 1809, William Clark, now a brigadier general in the militia and the principal Indian agent for the Louisiana Territory, joined with Lisa, Pierre Menard, and two members of the Chouteau family to create the Missouri Fur Company, based in St. Louis. John Jacob Astor, the squat, German-born, New York merchant, had even grander commercial dreams. He envisioned a series of trading posts that would follow the route of Lewis and Clark all the way to the Pacific. He also desired some sort of relationship with the government. With friendly encouragement from Thomas Jefferson but no formal federal support, Astor forged ahead and founded the American Fur Company in 1808; with Canadian partners, he created the Pacific Fur Company in 1810. Astor launched this latter enterprise by sending both an overland party and a maritime expedition to the mouth of the Columbia where, in March 1811, the overlanders founded Astoria. After the outbreak of the War of 1812, a British warship attempted to take Astoria, but it arrived too late. The North West Company, an enterprise established in Canada under British dominion, had bought the trading post in October 1813 for the bargain price of fifty-eight thousand dollars. This sale under duress, if not quite under the gun, hastened the demise of the Pacific Fur Company, but neither event prevented Astor from becoming one of the wealthiest men in the United States. He continued to profit from the fur trade until 1834, when he wisely left the business, anticipating the decline in the international demand for beaver pelts.

Both before and after this downturn in the market, fur trappers and government explorers used each other's knowledge. For example, fur companies sent their trappers to the upper Arkansas River because of information gathered in 1806–7 by Lieutenant Zebulon Pike's expedition across the Southwest. More typically, fur trappers explored the Far West on their own. Jedediah Smith in the 1820s and Joseph R. Walker in the 1830s undertook astounding journeys. Both men crossed the arid Great Basin and forged through the Sierra Nevada. Smith demonstrated a feasible route through the Rockies at Wyoming's South Pass, which became the main passage for overland travel to Oregon. In the 1840s, Walker helped guide two of John Charles Frémont's expeditions for the U.S. Army Corps of Topographical Engineers. Kit Carson, another famous mountain man, aided three of Frémont's parties, exploring the central Rockies, the Great Basin, Oregon, and California.

Frémont's scientific explorations under military auspices followed the precedent set by the Lewis and Clark Expedition. Throughout the nineteenth century, the federal government supported such explorations with the expectation of commercial, diplomatic, or political benefits. These scientific expeditions produced vital information about the American West, but like the Lewis and Clark Expedition, they did not produce all the benefits expected by the government. Nonetheless, each expedition set out with

Between 1840 and 1863 more than 700 prints of western scenes—virtually all based on eyewitness observation—appeared in government reports documenting exploration of the West. Particularly influential were the Pacific Railroad Survey reports (1855–61), which flooded the market with more than 6.6 million copies of western views. These scenic landscapes, along with carefully drawn scientific illustrations, fueled popular interest in westward expansion.

After W.P. Blake. Mammoth Tree "Beauty of the Forest." *Lithograph, 1856. From* Reports of Explorations and Surveys to Ascertain the Most Practicable and Economical Route for a Railroad from the Mississippi River to the Pacific Ocean, Vol. 5 (Explorations in California), *1856.*

After Heinrich Balduin Mollhausen (1825–1905). Canadian River Near Camp 38 *(Oklahoma). Lithograph, 1856. From* Reports of Explorations and Surveys to Ascertain the Most Practicable and Economical Route for a Railroad from the Mississippi River to the Pacific Ocean, Vol. 3 (Thirty-fifth Parallel Survey), *1856.*

a plan—a "program"—that outlined the government's intentions about what should be found. The historian William H. Goetzmann has argued that such cultural preconditions distinguish "exploration" from "discovery." The latter may be unexpected and serendipitous, but the former demonstrates cultural values and existing knowledge.

Between 1840 and 1860, the federal government placed great value in what could be learned from scientific explorations. During these two decades, Congress supported such scientific endeavors, as a percentage of the national budget, far more extensively than at any other time in the nation's history, in some years appropriating one-third of total federal expenditures. Most of this money financed what has been called the "great reconnaissance" of the trans-Mississippi. Congress not only paid for these explorations in the West but also published sixty reports—many in multiple volumes, complete with lithographs produced by artists who accompanied the scientists. Typically the U.S. Army Corps of Topographical Engineers carried out these western expeditions. Frémont served in this elite unit, established in 1838 as a separate organization from the army's larger Corps of Engineers. The contributions of the topographical engineers reached their zenith in 1853–55 with the transcontinental railroad surveys. The reports from these surveys appeared in a magnificent set of twelve lavishly illustrated volumes. With extensive maps, detailed geology, elaborate ethnography, and exhaustive ecology, this virtual encyclopedia of the West cost the government over a million dollars to publish—twice the cost of the surveys themselves.

The railroad surveys demonstrate how rapidly the nation had changed by the 1850s. A half-century earlier, such transcontinental routes could not have been predicted. At the time of the Lewis and Clark Expedition, a major question had been left unanswered. Beyond exploration and commerce, what plans did the federal government have for its newest West? Jefferson believed that territorial expansion might not bring national expansion. Ideally, republics should remain compact and homogeneous. In August 1803, he told Senator John Breckinridge that settlement west of the Mississippi would lead to the creation of a second republic—a friendly neighbor sharing a common language and political tradition. "The future inhabitants of the Atlantic and Mississippi states will be our sons."

Later presidents did not share this vision. James Monroe wanted everything east of the Rocky Mountains to join the Union, whereas John Quincy Adams advocated a continental nation stretching from the Atlantic to the Pacific. In his later years, Jefferson too lost his fear of national bigness. In 1817, he talked of the "enlargement of territory" leading to an enduring republic, if that nation was founded "not on conquest, but [on] principles of compact & equality."

As the nation expanded westward, these Jeffersonian ideals did not prevail. Military conquests, unfair treaties, and political inequities sustained government initiatives beyond the Mississippi. For example, in the first major effort to encourage settlement, the government forced the resettlement of American Indians in the trans-Mississippi by removing them from lands east of the Mississippi. As early as 1808, some Cherokees had agreed to go west to hunt and then to settle. After the War of 1812, the Monroe administration became more committed to the idea of exchanging Indian title, and thus changing Indian residency, to lands beyond the Mississippi. Proponents of Indian removal, from Thomas Jefferson to Andrew Jackson, justified this policy as a benefit to

Native Americans. In 1829, President Jackson explained: "[This] just and humane policy . . . recommended them [the Indians] to quit their possessions on this side of the Mississippi, and go to a country to the west where there is every probability that they will always be free from the mercenary influence of White men, and undisturbed by the local authority of the states: Under such circumstances the General Government can exercise a parental control over their interests and possibly perpetuate their race."

The policy of removal clearly benefited agricultural expansion east of the Mississippi, especially after the War of 1812 with the establishment of a cotton kingdom in the Deep South. America's presidents, beginning with George Washington, had advocated European-style agriculture and conversion to Protestant Christianity as the path to social and spiritual salvation for the Indians. Removal forced Indians to emigrate to lands not coveted by American farmers, especially cotton planters. Under the protection of the federal government and under the direction of Christian missionaries, these native peoples, though not recognized as citizens of the United States, might somehow secure "civilization" through cultural assimilation. Removal treaties targeted not only the largest southern tribes—the Choctaws, Chickasaws, Creeks, Cherokees, and Seminoles—but also smaller northern tribes such as the Potawatomis, Shawnees, Ottawas, and Sauks and Foxes. In terms of the American West, the removal policy demonstrated an early variation of what the historian Alfred Runte has called the "worthless land" thesis. The same logic that relegated eastern Indians to "worthless" western lands would later be used to justify the creation of such diverse entities as national parks, nuclear bomb test sites, toxic waste dumps, and numerous Indian reservations.

West of the fertile Mississippi Valley, government explorations led by Lieutenant Zebulon Pike in 1806–7 and Major Stephen H. Long in 1820 had produced vivid descriptions of the Great Plains as a vast arid area. The map that accompanied the publication of Long's report in 1823 labeled most of present-day Kansas, Nebraska, and eastern Colorado as "The Great American Desert." Such information indicated that Indians removed beyond the Mississippi would live, at best, on the far western fringe of usable agricultural lands within the Louisiana Purchase, primarily in the eastern portions of present-day Kansas and Oklahoma.

The government continued its removal efforts into the 1850s. Not all Indians were forced across the Mississippi. After the roundup of thirteen thousand Cherokees by the military, at least two thousand died in late 1838 along the infamous "Trail of Tears" on their way to present-day eastern Oklahoma. Perhaps a thousand Cherokees avoided the military by hiding in the mountains of western North Carolina. Some Choctaws stayed behind in the state of Mississippi. The Senecas of upstate New York swapped the lands reserved for them in the West for a portion of their old homeland. In the Black Hawk War of 1832, belligerent Sauks and Foxes failed to retain their lands in Illinois and Iowa. On the other hand, some Seminoles fought a successful guerrilla war and stayed in the Everglades of Florida. Other natives managed to keep some lands in Michigan and Wisconsin. Nonetheless, most eastern Indians were removed, with the result that American Indian history after the 1840s is overwhelmingly, though not exclusively, set in the West.

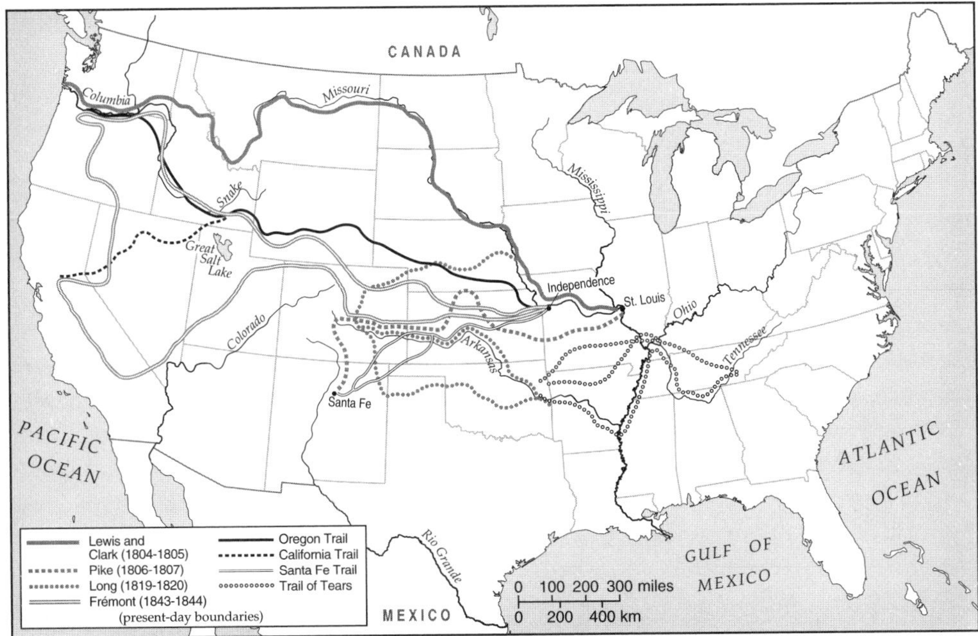

Exploration and Migration Routes, 1804–60

Although Indians in the West did not remain "free from the mercenary influence of White men" as President Jackson had promised, a flood of white settlers did not immediately follow the high tide of Indian removal in the 1830s. In fact, the bulk of the Louisiana Purchase, not just the area of present-day Oklahoma, could still be viewed as "Indian Territory" until after the Civil War. National initiatives leapfrogged beyond the borders of this first trans-Mississippi West—to Texas, the Oregon country, California, and the Great Basin. The Great Plains and the Rocky Mountains were places to pass through or go around. With no settlers in these areas, political incorporation—first with territorial governments and then with state constitutions—had to wait. American citizens were gone to Texas, bound for Oregon, or rushing to California.

Manifest Destiny

Anglo Texans made bad Mexicans. The independence of Mexico, established in 1821, did not produce a loyal citizenry in the frontier province of Texas. Anglo-American settlers had first migrated to Texas in large numbers in the 1820s under the careful direction of an *empresario* from Missouri, Stephen F. Austin. By 1836, open rebellion had created an independent Republic of Texas. Military victory did not come easily. At Goliad, not far from the Gulf of Mexico, one Texas garrison surrendered, only to see Mexican firing squads execute 342 soldiers as pirates on 27 March 1836. Earlier that month, on 6 March 1836, a superior Mexican force commanded by General Antonio López de Santa Anna had stormed the Alamo in San Antonio and killed every one of the Texas and American defenders, more than 180. The victorious Mexicans lost over 600 soldiers in the battle. Then on 21 April, Santa Anna suffered a crushing defeat—730 captured, 630 killed, and 208 wounded—after a surprise attack at siesta time by an army of 900 under the command of Sam Houston. The Battle of San Jacinto took only eighteen minutes, but it would take much longer for the now independent Texas to become part of the United States. In the new republic's first election in September 1836, Texans endorsed a constitution, elected Sam Houston president, and voted 3,277 to 91 to seek annexation to the United States.

Despite the ardor of the Texas suitor, family members of the prospective bride delayed the marriage. Many of the Anglo Texans had southern origins. They had come to Texas from such nearby states as Louisiana and Alabama and from the Arkansas Territory, and they had brought African-American slaves with them. The Mexican constitution of 1824 had abolished slavery, but the Texans maintained a charade of "contract labor" until the Republic of Texas reestablished the institution. Antislavery members of the U.S. Congress opposed the acquisition of Texas for nearly a decade. Westward settlement in the Louisiana Purchase had produced only slaveholding states: Louisiana, Missouri, and Arkansas, joining the Union in 1812, 1821, and 1836 respectively. The consideration of Missouri's statehood had precipitated a dangerous political crisis. In the resulting Missouri Compromise of 1820, Maine was admitted to the Union as a free state and Missouri as a slave state to balance the number of slave and free states. In addition, the compromise prohibited slavery in the remaining area of the Louisiana Purchase north of 36°30'—approximately the extended southern boundary of the new state of Missouri. Texas was not part of the Louisiana Purchase, and in any event, like Arkansas, it was south of the line of compromise. Nonetheless, elaborate political maneuverings, culminating with an extraordinary joint resolution by a lame-duck Congress at the behest of a lamer president, were required before Texas was annexed. Less than a year after this arranged marriage, Texas became a state, on 29 December 1845. Almost exactly one year later, on 28 December 1846, Iowa entered the Union as the first free state carved out of the Louisiana Purchase.

Like Texas, Oregon had not been part of the Louisiana Purchase. In a series of agreements beginning in 1818, the United States and Great Britain recognized their joint occupation of this far northwestern territory. During the decade that Congress by turns debated or ignored the annexation of Texas, Americans began to settle in the Oregon country, especially the fertile Willamette Valley. In 1834, Methodist missionaries from New England, led by Jason Lee, joined some of the first arrivals.

The short trip in the 1820s from the United States to the Austin colony in eastern Texas had few of the physical and psychological challenges produced by the overland trip to Oregon. The American immigrants who declared that they were "gone to Texas" did not have to go far, compared with the journey for Oregon's pioneers. At the time of independence in 1836, the population of the new Texas republic was approaching 40,000. When Texas entered the Union, this figure had soared to 142,000. By 1845, perhaps 6,000 Americans had settled in Oregon. Many were recent arrivals—900 in 1843, known as the "Great Migration." The next year 1,200 immigrants arrived, predominately from Missouri and Illinois. Traveling by covered wagon, on horseback, or, very often, by foot, these overlanders followed the Oregon Trail. Usually starting in Independence, Missouri, they crossed the plains along the North Platte River, rested at Fort Laramie, traveled through the Rockies via South Pass, stopped at Fort Bridger, and headed to Fort Hall on the Snake River. The immigrants survived more barren and mountainous terrain before reaching The Dalles for the hazardous descent of the Columbia River to the Willamette Valley. The journey stretched for two thousand miles and lasted 150 to 180 days.

The success of this overland migration supported the American claim to Oregon. Some political figures wanted all of the territory up to the northern boundary with Russian Alaska. In the 1844 presidential campaign, the Democratic candidate, James K. Polk, of Tennessee, took this position. He also advocated the annexation of Texas. His

Although the Mexican War became the first important event in American history to be documented by the new medium of photography, the American public preferred the dramatic fiction of contemporary prints that glorified the war in the name of "manifest destiny."

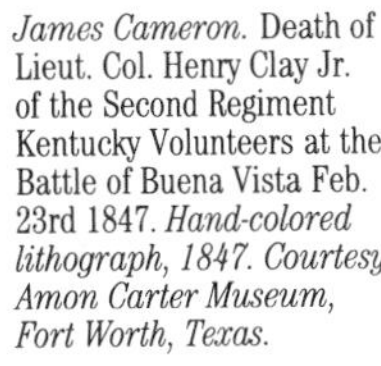

James Cameron. Death of Lieut. Col. Henry Clay Jr. of the Second Regiment Kentucky Volunteers at the Battle of Buena Vista Feb. 23rd 1847. *Hand-colored lithograph, 1847. Courtesy Amon Carter Museum, Fort Worth, Texas.*

Unidentified photographer. General Wool and Staff, Calle Real to South. *Daguerreotype, 1847. Beinecke Rare Book and Manuscript Library, Yale University, New Haven, Connecticut.*

major opponent, Henry Clay, of Kentucky, obfuscated. Polk won a narrow victory. The outgoing president, John Tyler, of Virginia, pushed through the joint resolution for annexation of Texas. After taking office, Polk addressed the "Oregon Question."

The new president articulated one of the major themes of his era—national expansion. His party, the Democrats, especially those in the Mississippi Valley, supported this ideal. In the debate over Texas annexation, an important phrase had appeared: "manifest destiny." John L. O'Sullivan, the editor of the widely circulated *Democratic Review*, placed these two words into print in July 1845. He argued that foreign interference in the acquisition of Texas was attempting to check "the fulfillment of our manifest destiny to overspread the continent alloted [*sic*] by Providence for the free development of our yearly multiplying millions."

In 1751, Benjamin Franklin had talked about a similar demographic destiny for Americans to fill up new western lands. The earlier continental consciousness of presidents like Jefferson, Monroe, and Adams also recognized a form of expansion. But what appeared in the 1840s had a more aggressive tone and nationalistic zeal. The reference to divine purpose and the claim of national superiority underlay an at times idealistic rhetoric that stressed the benefit to all humanity of America's growth. American Indians, who had been subjected to the paternalistic attitude and ethnocentric idealism of the removal policy, were already familiar with such self-serving assumptions.

The government's acquisition of Indian lands in the East closely paralleled the acquisition of western lands under manifest destiny. For its own purposes, the government used various combinations of negotiations, military actions, and treaties. In the case of the Indian removal policy, these aggressive efforts resembled a form of "domestic" foreign policy. In fact, the Supreme Court had created a special legal status for Indian tribes. Chief Justice John Marshall used the phrase "domestic dependent nations" in 1831 when he ruled that the Cherokees were not a foreign nation in relation to the federal government. Instead, the tribes came under the exclusive "protection" of the federal government, as demonstrated by numerous treaties. Marshall claimed that Indian nations occupied "a territory to which we assert a title independent of their will." In the mid-1840s, the administration of James K. Polk assumed that the British and the Mexican rights of possession to certain lands had also ceased. Manifest destiny dictated the removal of old land titles.

The dispute over British claims to the Oregon country ended without war. When James K. Polk said he wanted all of Oregon, he coveted a vast area that stretched from the northern boundary of Alta California to the southern border of Russian Alaska. This latter boundary inspired the political slogan "54°40' or Fight." But diplomatic officials already had begun to discuss the division of Oregon along the 49th parallel. Polk kept his belligerent tone in public statements that cheered expansionist Democrats. Behind the scene, compromise proceeded. Neither British nor American officials wished to push this issue to war. The British protected the fur-trade interests of the Hudson Bay Company by retaining the northern half of Oregon. The southern half had been nearly trapped out, and the company intended to move its major trading post to Vancouver Island. In Britain, the Anti-Corn Law League effectively agitated for peace as part of its campaign for free trade in food. Support for peace grew in the United States, since the government preferred to fight one war at a time and Congress had already declared war with Mexico before the proposed Oregon treaty reached Washington. Ratification quickly followed, on 18 June 1846.

Polk agreed to half of Oregon because the United States hoped to acquire all of northern Mexico. The possibility of open conflict had loomed since the formal annexation of Texas. Within a week after this joint resolution by Congress, the Mexican minister vehemently protested and left Washington. Mexico had refused to recognize the independence of Texas or the claim, made by the new republic, that its international boundary was the Rio Grande. The Nueces River had been Mexico's southern boundary for Texas. Roughly 150 miles separated these two rivers, but President Polk wanted more than this stretch of sandy soil. On his first day in office, he told one of his cabinet members that a prime objective of his administration would be the acquisition of California. He also wanted to settle the numerous claims of American citizens for financial losses in Mexico. Beginning in the 1820s, civil unrest and political turmoil had led to the loss of property by American owners. President Andrew Jackson had demanded repayment, but nothing happened until 1840, when the Mexican government accepted the ruling of a five-member international arbitration commission that awarded American claimants $2,026,000 to be paid in five annual installments. Mexico made three payments and then announced its inability to pay more.

The dispute over land and money led to war. The issue of the Texas boundary provided the flash point. In May 1845, President Polk directed military and naval forces to prepare for actions against Mexico. A few weeks later, American troops under General Zachary Taylor arrived in Corpus Christi, Texas, at the mouth of the Nueces. In November, the president sent a special envoy, John Slidell, to Mexico City. He instructed Slidell to purchase California, assume all claims against Mexico, and settle the boundary along the Rio Grande. Public hatred of the United States had grown so intense that President José de Herrera did not dare receive the U.S. envoy. Despite this anti-American gesture, Herrera's government fell in a bloodless revolution in late December. Slidell informed Polk, "Be assured that nothing is to be done with these people until they shall have been chastised."

The president received Slidell's message on 12 January 1846; the next day he ordered General Taylor to dispatch troops into the area north of the Rio Grande. This action provoked a Mexican response. On 25 April 1846, soldiers under the command of General Mariano Arista crossed the Rio Grande and attacked an American patrol, killing or wounding sixteen men. Truth became the seventeenth casualty when Polk proclaimed to Congress in his war message on 11 May, "Mexico . . . has invaded our territory and shed American blood upon the American soil."

American armies invaded Mexico across its northern frontier, south from the Rio Grande, and eventually inland from its eastern coast. This last campaign occurred in 1847, when General Winfield Scott marched his army through mountainous terrain from Veracruz to Mexico City. The fall of the capital helped Nicholas Trist, Polk's special commissioner, complete the negotiation of a final treaty. In exchange for a payment of $15,000,000, Mexico ceded the vast northern provinces of California and New Mexico to the United States and recognized the Rio Grande as the international border. The United States assumed the claims of $3,250,000 against Mexico by American citizens. Polk, the Democratic president, reluctantly supported the Treaty of

Guadalupe Hidalgo and forwarded it to the Senate. He had started to distrust Trist when the commissioner had become friendly with General Scott, a well-known Whig who, Polk reasoned, might emerge as a potent rival to the Democrats.

Some senators argued that the United States should ignore the treaty and claim all of Mexico by right of conquest. Other senators disagreed. John C. Calhoun, of South Carolina, a noted advocate of states rights and slavery, wanted only the sparsely settled northern lands. Using his best racial logic, he reasoned that Indians had not been incorporated into the Union but had "been left as independent people amongst us, or been driven into the forests." Calhoun warned: "To incorporate Mexico, would be the very first instance of . . . incorporating an Indian race; for more than half of the Mexicans are Indians, and the other is composed chiefly of mixed tribes. I protest against such a union as that! Ours . . . is the Government of a white race."

The Senate ratified the Treaty of Guadalupe Hidalgo by a vote of 38 to 14. Calhoun's rhetoric in defense of the treaty expressed the realpolitik of federal relations with American Indians and Hispanic Americans in the nineteenth century. The federal government willingly acquired new land, but it did not willingly embrace the people inhabiting that land. The remainder of Mexico had too many racially unattractive people to make the land attractive. As for the people within the Mexican cession to the United States, they were left on the margins of American society. A clear sense of racial hierarchy, based on the assumption of white cultural superiority, often led to legal, political, and social exclusion for racial minorities. Even rights guaranteed in a federal treaty could be reversed or ignored. Such was the case with the promised recognition of the title to land in the area ceded to the United States. Over time, the Californios and Hispanos of what had been northern Mexico lost most of their property. By the 1890s, often under the pretense of "clearing" the title to land that predated the Treaty of Guadalupe Hidalgo, Anglo-American lawyers and settlers controlled 80 percent of the original Spanish land grants in New Mexico. Like many American Indians, Hispanic Americans learned that the government of a white race paid little attention to the promises made to peoples of other cultures.

By force of arms, the United States acquired its second trans-Mississippi West in less than half a century. The Mexican-American War fulfilled President Polk's desires for expansion to the Pacific. Combined with the settlement of the Oregon Question in 1846, the Treaty of Guadalupe Hidalgo in 1848 provided the United States with an ample coastline in the Far West. It also increased the nation's internal empire by adding 1.2 million square miles, out of which would be carved the states of California, New Mexico, and Arizona and much of Nevada, Utah, and Colorado. With the fruits of war, the nation had grown by 66 percent.

The Mexican-American War changed the map of the United States, but it did not address the critical issues facing the nation. The apparent fulfillment of manifest destiny in continental terms amplified the need to reexamine the national agenda. Should slavery be permitted to grow westward with the nation, as it had after the Louisiana Purchase? How could a country that stretched from the Atlantic to the Pacific be kept unified? If the modest pace of migration to the West Coast had continued throughout the 1840s and 1850s, the importance of the West in national politics might also have

To St Louis. — 350 Miles
To California. 1700 Ms.
THE INDEPENDENT GOLD HUNTER ON HIS WAY TO CALIFORNIA.
I NEITHER BORROW NOR LEND

The discovery of gold in California in 1848 drew more than a quarter million immigrants to the area within five years, establishing a powerful American presence in a region only recently won from Mexico.

Unidentified artist. The Independent Gold Hunter on His Way to California. *Lithograph, ca. 1850. Courtesy Amon Carter Museum, Fort Worth, Texas.*

remained modest. But one event brought dramatic changes. News of the discovery of gold in northern California in January 1848 set off a rush of people, tying the fate of America's Far West to the fate of the nation.

The Nation's Region

More than 100,000 people streamed to California in 1849. This flood grew to over 250,000 by the end of 1852. They came from around the globe—from Peru and Chile, from Australia and China, from the British Isles and continental Europe, from the

Hawaiian Islands, Mexico, and Oregon, and from the Mississippi Valley and the eastern United States. Excluding Indians, California's population had been no more than 14,000 at the beginning of 1848. By the fall of 1849, enough people had arrived for a convention at Monterey to draw up a state constitution. This document prohibited slavery because the majority of miners did not want the unfair competition of slave labor in the diggings. The nation's Congress and new president now had to respond to the addition of a free state to the Union.

Since 1846, every northern state legislature, with the exception of Iowa, had endorsed a proviso originally authored by David Wilmot, a Democratic congressman from Pennsylvania. Wilmot advocated the prohibition of slavery in all territories acquired from Mexico. The Senate had blocked the enactment of the Wilmot Proviso into federal law, and President Zachary Taylor believed he could avoid further sectional discord if New Mexico and California immediately applied for statehood and avoided territorial status altogether.

By August 1850, Congress had fashioned an elaborate compromise that ignored New Mexico's application and balanced the interests of North and South. Most significant, a federal Fugitive Slave Law offset Southern opposition to California's admission as a free state. Yet the Compromise of 1850 merely delayed the larger crisis of civil war. During the 1850s, most slave states of the South became convinced that they could no longer stay in the Union. The South feared the subjugation of its political rights and the abolition of its peculiar institution of slavery by a Northern majority in control of the national government.

The South's anxiety expressed a widely held belief about the nature of freedom and government. Before America's Civil War, the concept of freedom often meant freedom from government. Eleven slave states took this position to its extreme and tried to free themselves from the Union as an exercise in radical states rights. The North's victory asserted the power of the federal government to ensure the nation's political unity. In addition, legislation passed during the Civil War demonstrated the government's support of the supposedly "free" enterprises of railroad development and agricultural homesteading. These economic initiatives continued after the war as federal troops and constitutional amendments attempted to politically reincorporate the former rebel states. The West, on the other hand, had come directly under federal power even before the Civil War, in ways that the South did not experience until after the war. Freedom from government did not apply so readily in the nation's newest region. In the West, the national government, aided at times by the U.S. Army, took direct action.

Immigrants appreciated the federal presence in the West. John D. Unruh, Jr., the preeminent historian of the trails to Oregon and California, observed, "Most pre-Civil War overlanders found the U.S. government, through its armed forces, military installations, Indian agents, explorers, surveyors, road builders, physicians, and mail carriers, to be an impressively potent and helpful force." For example, in 1843–44, the second of Frémont's three expeditions for the U.S. Army Corps of Topographical Engineers produced much helpful information about the route to Oregon, including a detailed map drawn by Charles Pruess. The U.S. Senate ordered the publication of ten thousand copies of Frémont's report, which contained the map. In 1846, Pruess produced a second, more elaborate map of the Oregon Trail, which became an oft-used reference.

Overlanders insisted on protection during their travels, as well as information about the route. The federal government provided military support. In fact, soldiers easily outnumbered other groups of federal employees in the West. At times during the 1850s, 90 percent of the U.S. Army was stationed at the seventy-nine posts throughout the trans-Mississippi. In 1860, on the eve of the Civil War, the total reached 7,090 enlisted men and officers. Initially, in the 1840s, the army had relied on expeditions to assert its presence in the West. Along the main immigrant route in 1845, Colonel Stephen W. Kearny took a military force of three hundred to South Pass via the fur-trading post at Fort Laramie. His dragoons wanted to impress the Plains Indians with the might of the United States and prevent any attacks against the overlanders. The next year, Congress approved legislation to establish fixed military posts along the Oregon Trail. By 1850, three had been established: Fort Kearny, on the south bank of the Platte; Fort Laramie, purchased by the government; and what became Fort Dalles, on the Columbia River. The freighting of goods, the rotation of troops, and the communication between posts improved the safety and services along the trail. Supply towns, where passing travelers could find necessities and diversions, grew up near each fort.

Government planning in the 1840s had anticipated increased migration to Oregon's Willamette Valley. By the 1850s, gold in the Sierra Nevada had made California the major destination for cross-country travel. National political leaders advocated the construction of a transcontinental railroad that could ship the riches of California to the East and deliver people and goods to the Far West. The state that contained the eastern terminus would gain a gigantic economic windfall. The railroad frenzy amplified divisions between North and South during the intense political maneuverings in Congress. Not one but four transcontinental surveys were authorized under the direction of the army, in order to determine the best route. Undertaken in 1853–54, these four expeditions produced their own gold mines of scientific information about specific cross sections of the West. Nonetheless, since a southerner—Jefferson Davis, of Mississippi—had overseen these surveys as secretary of war, few were surprised in 1855 when a preliminary report endorsed the southernmost route, along the 32nd parallel.

The North did not want the South to profit from the transcontinental railroad. One attempt in 1854 to preempt a southern route was the Kansas-Nebraska Act organizing the Kansas and Nebraska territories. With the federal government established in these areas, construction of the federally supported railroad could begin in Chicago, the largest city of the home state of the architect of the act, Senator Stephen A. Douglas. Ironically, Douglas had marshaled Southern votes for the two new territories by including in this legislation a repeal of the Missouri Compromise line. Although this meant that slavery could expand into areas that had been considered "free soil" since 1820, Douglas confidently assumed that "popular sovereignty," the political action of the citizens in the new territories, would prevent the growth of slavery. The aftermath of the Kansas-Nebraska Act is well known. Emotions ran high, especially in the North, where a new political party coalesced around the anti-Nebraskaites, who soon took another name—Republicans. The Kansas Territory, which bordered on the slave state of Missouri, plunged into violence and turmoil, with fighting between proslavery and antislavery settlers. The decision about the transcontinental railroad remained unresolved until after the outbreak of the Civil War, when all southern routes were readily eliminated.

Yet, Douglas's assumptions about the workings of popular sovereignty in the West were not impractical. In 1854, slavery was not permitted in the territory of Oregon or the state of California. New Mexico, which shared a border with the slave state of Texas, had proposed a state constitution, written in 1850 and ignored by Congress, that also denied the expansion of slavery. In fact, the Compromise of 1850, whose passage Douglas had engineered, organized the territories of New Mexico and Utah without reference to slavery. Applying the principle of popular sovereignty, Congress had empowered the territorial legislatures of New Mexico and Utah to decide on the slavery issue. Neither had established the South's peculiar institution.

Although these territories were allowed popular sovereignty on the slavery issue, the federal government retained considerable power in the West. As a result, in 1857, the Mormons in Utah Territory found themselves at war with the United States. The territory of Utah had not taken the fast road to statehood but remained under federal domination for decades. After the Mormon pioneers arrived in the Great Basin in 1847, their leader, Brigham Young, advocated the creation of the state of Deseret, which stretched to the southern California coast. Congress rejected a petition for Deseret's statehood in 1849 and instead created the territory of Utah in 1850, with greatly reduced boundaries. A paper government for the state of Deseret continued until 1872, but any possibility of political independence for the Mormons ended in 1857 when President James Buchanan dispatched one-sixth of the U.S. Army to Utah to prevent what his friends and advisers considered a potential rebellion.

What focused the fury of the nation against the Latter-day Saints was not simply Mormon ambitions to control more land. In 1852, the church's leadership had publicly announced what some people had suspected for years. As part of their religious practice, many Latter-day Saints engaged in plural marriage—that is, men had two or more wives at the same time. America now had a second peculiar institution or, as the new Republican party viewed it, a twin evil to slavery. Federal territorial appointees complained that a notorious polygamist, Brigham Young, controlled the Utah Territory through his control of the Mormon Church.

As the historian Howard R. Lamar has explained, President Buchanan's decision to send the U.S. Army into Utah contained several political messages. "The Utah crisis could divert the whole nation from its preoccupation with slavery. The sight of the United States reasserting jurisdiction over a region practically claiming independence would also give pause to Southern secessionists and delimit the more extreme demands uttered in the name of states rights." It would also reassure Northern Republicans that "to believe in popular sovereignty was not to condone polygamy."

The soldiers of the so-called Utah War saw very little fighting. In effect, the national government established military occupation in the territory. Camp Floyd, the major installation in Utah, was located forty miles south of Salt Lake City. A garrison that averaged twenty-four hundred soldiers stayed there until 1860. During 1858 and 1859,

Created by Congress in 1867, the Peace Commission sought to persuade Indians to settle on reservations in exchange for government benefits. Its name notwithstanding, the Peace Commission was backed by the federal army, which stood ready to enforce reservation boundaries if peaceable negotiations failed.

Alexander Gardner (1821–82). Peace Commission Treaty Negotiations with the Cheyenne and the Arapaho at Fort Laramie, Wyoming. *Albumen silver print, 10 May 1868. Smithsonian Institution, National Anthropological Archives (#3686), Washington, D.C.*

as a show of force, the army concentrated more troops at Camp Floyd than at any other location in the United States.

War and the White Road

During the 1850s, the federal government tried to control not only the apparently rebellious residents of Utah Territory but also the diverse native peoples of the trans-Mississippi West. The result was a series of wars that lasted into the 1890s. Sometimes warfare flared because of federal actions and sometimes because of federal inactions. During these four decades, not all Indian tribes fought American troops, but all Indians came under the federal government's policy of cultural assimilation.

After negotiated agreements, military pacification, or a combination of the two, most native peoples relocated to reservations, where federal agents attempted to lead them on the "white road." Through education and missionization, the government hoped to transform tribal peoples into independent Christian farmers. Under the federal government, both civilian and military officials implemented Indian policies. Initially, in 1824, Congress organized the Indian Bureau as part of the Department of War. In 1849, the bureau became part of the newly created Department of the Interior. For the next three decades, advocates of military control of Indian affairs argued that the bureau should be returned to the War Department, since soldiers had more experience and exerted firmer discipline in dealing with native tribes. They pointed to the corrupt financial dealings by some agents and the continued belligerence of some natives. Supporters of civilian control pointed to examples of unnecessary military violence and bloodshed. They argued that soldiers were ill prepared to deal with the Christian, humanitarian needs of the Indians on the reservations. Each side had its supporters and detractors in Congress. The Indian Bureau stayed put, but Indian policy remained a combined, if sometimes contentious, effort by both sides. For example, during the presidency of Ulysses S. Grant, Christian denominations administered specific Indian agencies. Meanwhile, the military attempted to settle all Indians on their assigned reservations, where they would be cared for by the Christian agents. General William T. Sherman described the effect on the Indians as a "double process of *peace* within their reservation and war *without.*"

The concerted commitment to a reservation policy had begun during James K. Polk's presidency. William Medill, Polk's commissioner of Indian affairs, advocated the establishment of "colonies" for the Indians native to the region beyond the Mississippi. His was not an original idea. Concentrated colonies, or reservations, had been tried in the East before the removal policy attempted to create a "permanent" Indian frontier. The reservation policy became a form of internal removal within the West. It either reduced the homeland of a native people or moved them to a new location—where they also had less land but were to be protected from evil influences and guided along the road to assimilation. Like the justification of removal, the defense of reservations fused humanitarian concern for the Indians' future welfare with a self-serving ethnocentrism. By the 1880s, in near unanimity, government officials, military officers, congressional leaders, and Christian reformers agreed that allowing tribal landholdings and promoting tribal culture should end. They believed that reservations should disappear along with Indian identity.

Reservations were now seen as obsolete in terms of assimilation. This led to the eventual breakup of many reservations after the passage, in 1887, of the General Allotment Act, often called the Dawes Act after the Massachusetts senator who sponsored the bill. The Dawes Act began a process that further reduced the lands claimed by American Indians. Initially, lands not allotted to Indians became "surplus" and could be sold to non-Indians. Eventually, after the lapse of the government's title in trust, many individual Indians sold their allotments. In 1881, the federal government recognized over 155 million acres as Indian lands. By 1900, the figure had shrunk to under 78 million acres. A few tribes, such as the Navajos and Senecas, managed to escape allotment entirely, but for those who did not, until federal law ended the process of allotment in 1934, Indians under the Dawes Act and its successors lost 60 percent of their lands, which included 66 percent of the lands allotted to individual natives.

The Dawes Act helped create more acreage for non-Indian agriculturalists, but it did not guarantee economic prosperity. In the extensively arid portions of the West, such as the Great Plains, farmers recurrently failed. In the 1870s, drought and grasshoppers doomed many efforts. In the 1880s, a major agricultural depression—tied to deflation, debt, overproduction, and world markets—ruined more farmsteads. In each decade, destitute citizens turned to state governments and federal agencies for relief. Yet, these hard times for white farmers never altered the Jeffersonian idealism that produced the Dawes Act. The government's policy continued to assume that the route to salvation for Indians followed the white road to the church, the school, and the farmstead.

Only a few decades earlier, during the 1850s, the federal government had emphasized treaty negotiations to establish reservations in places such as California and Oregon and to ensure safe passage across the Great Plains. In California, three federal commissioners hastily attempted to set aside 11,700 square miles of land in a series of reservations for 139 small tribes and bands. More than 175 other California tribes were ignored, but that did not matter to the gold-hungry Californians who protested giving to Indians any lands that might contain mineral wealth or that could be used for agriculture. The U.S. Senate refused to ratify these treaties. Many members of Congress believed that California's diverse Indians had no title to the land. By March 1852, Congress did approve California's first superintendent of Indian affairs, the well-intentioned Edward F. Beale. He believed that the old mission system could serve as a model to train the Indians in useful labor, under the guidance not of a Catholic priest but of a federal agent aided by U.S. troops.

In May 1854, Congress learned that Beale was nearly $250,000 in arrears in his official accounts. A later investigation exonerated him, but not before he had lost his appointment. The government's search for parcels of land suitable for Indian resettlement limped along, with modest results and no clear establishment of title. The reservations in California were mostly small and scattered rancherias that, individually, gave some shelter to a few Indians. Meanwhile, the impact of the gold rush had destroyed thousands of other Indians. Some California argonauts hunted Indians like wild animals, and all refused to respect Indian land use. Murder, starvation, and disease created a demographic disaster. In 1845, about 150,000 native people lived in California. By 1856, they numbered as few as 25,000.

The Indian situation in Oregon produced its own set of tragedies. President Polk appointed Joseph Lane, a hero of the Mexican-American War and a politician from Indiana, as the first governor of the Oregon Territory in 1848. As was often the case, the governor was responsible for Indian affairs as well. Lane traveled throughout Oregon to observe the Indians. He also noted the natural riches of the new territory and determined that the rush of settlers to western Oregon would doom the Indians to "poverty, want, and crime" if they were not relocated to "a district removed from the settlements." The territorial legislature followed the governor's advice and asked Congress to purchase native lands and remove the Indians. The territory's delegate to Washington, Samuel R. Thurston, proposed that the Indians be placed east of the Cascade Mountains. In 1850, three federal commissioners were appointed to carry out the extinguishment of titles and the removal of several Oregon tribes. The commissioners reported that removal might prove disastrous, since most of the Indians of the Willamette and lower Columbia valleys survived by fishing. Instead of relocation, the commissioners negotiated six treaties that reduced the Indians' lands but reserved a few sections, including some fishing grounds, for native habitation. Congress never ratified these treaties or thirteen more negotiated by the territory's superintendent of Indian affairs, Anson Dart.

Meanwhile, settlement aggressively expanded under the Oregon Donation Land Law of 1850, which ignored Indian title and granted 320 acres to any citizen or prospective citizen who cultivated the land for four years. The arrival of gold seekers in the Rogue River country, south of the Willamette Valley, created more problems. Army regulars and Oregon volunteers quelled hostilities in the summer and fall of 1853. A treaty of cession left the Rogue River Indians with a small temporary reservation. Tensions increased, and a larger war broke out in the area in 1855. Oregon volunteers and U.S. regulars this time forced the Rogue River Indians to surrender. In the summer of 1856, these Indians were removed to a reservation in the Coast Range. From 1853 to 1855, without open warfare, other Oregon tribes capitulated to federal demands and accepted treaties that left them with small reserves of land and the promise of annual government support in the form of material goods, formal education, and agricultural development. Congress ratified all these treaties by 1859.

In California and Oregon, during the 1850s, new arrivals overwhelmed—and in some places annihilated—the native peoples. On the Great Plains, the Indians' fate did not unfold so rapidly. During the first three weeks of September 1851, an estimated ten thousand American Indians gathered for a treaty council near Fort Laramie. The powerful western Dakotas, known as the Teton Sioux or Lakotas, with their Arapaho and Cheyenne allies dominated the conference. But Assiniboins, Shoshones, Arikaras, Hidatsas, Mandans, and Crows also attended. A former fur trader, David D. Mitchell, of St. Louis, led the U.S. delegation under the protection of 270 soldiers.

The Treaty of Fort Laramie did not establish reservations for its Indian signatories. This agreement simply marked boundaries for the different tribes, to promote peace on the Plains and provide safe passage for overland travelers. It also permitted the government to build military forts and construct roads on Indian lands. Beyond the immediate distribution of numerous presents, the government promised to deliver annually for fifty years useful goods, such as domestic animals and agricultural implements, with a total value of fifty thousand dollars per year. The Senate later reduced these annuities to ten years with a possible five-year extension and increased the annual value to seventy thousand dollars.

The Indians accepted the presents, but they did not necessarily recognize the boundaries imposed by the treaty. Blue Earth, of the Brulés, warned, "We claim half of all the country; but we don't care for that, for we can hunt anywhere." Another Lakota, Black Hawk, of the Oglalas, complained: "You have split the country and I don't like it. What we live upon we hunt for, and we hunt from the Platte to the Arkansas, and from here up to the Red But[t]e and the Sweet Water. . . . These lands once belonged to the Kiowas and Crows, but we whipped these nations out of them, and in this we do what the white men do when they want the lands of the Indians."

The Fort Laramie Treaty of 1851 did not end intertribal raiding or warfare. It did not prevent the eventual hostilities between the U.S. Army and the aggressively expanding Teton Sioux. And it may not even have been necessary to protect the overlanders as they crossed the Plains. The historian John Unruh's careful analysis of the California and Oregon trails from 1840 to 1860 showed that of the approximately 250,000 overlanders who took these routes, fewer than 400 were killed. Of these killings, nearly 90 percent occurred west of South Pass. Despite the great fears expressed by travelers during the journey and the fabricated accounts of Indian attacks in contemporary newspapers and later pioneer memoirs, the trip across the Plains was usually peaceful.

Conquest and Survival

Peace did not continue between the U.S. Army and the Teton Sioux. The fate of the Lakotas on the Great Plains and of another Indian society, the Navajos, in the Southwest demonstrated that military conquest could result in political subjugation but not cultural assimilation. The Lakotas and Navajos exemplify the mixed results of the government's Indian policy in the last half of the nineteenth century. These native peoples lost wars with the United States and were placed on reservations, but they retained their cultural identity and their commitment to a tribal homeland.

Warfare began, at times, from seemingly trivial causes. The first significant armed clash between Lakota warriors and U.S. regulars revealed how miscommunication could produce tragic results. During late July 1854, some four thousand Brulés, Oglalas, and their allies had gathered near Fort Laramie for the distribution of the annuities guaranteed by the treaty of 1851. By mid-August, the federal Indian agent had not arrived to supervise the distribution, and tensions had grown between the outnumbered soldiers at the fort and the Indians encamped in the valley of the North Platte. The nearly constant stream of overlanders along the valley only increased the volatility of the situation. Late on the afternoon of 17 August, a weary cow lagged behind a passing train of Mormon immigrants and wandered among some Brulé lodges, where a visiting Miniconjou Sioux killed it. The Mormon owner demanded punishment of the Miniconjou, so on the morning of 19 August, the commander at the fort sent twenty-nine of his seventy-five infantry, two cannon, and one interpreter, under the command of Brevet Second Lieutenant John L. Grattan, to make an arrest. As this force approached the Brulé camp, the interpreter, drunk with more than courage, called out that the army planned to kill all the Sioux and that he would cut out and devour their hearts. The Brulés did not attack, even when Grattan formed a battle line in the center of their camp. The band's leader, Conquering Bear, had offered many horses in payment for the dead bovine and tried to calm both Grattan and the threatened Miniconjou. The interpreter garbled Conquering Bear's words, and Grattan had his men level their guns at the superior Indian force. Whether the Miniconjou or the soldiers fired first is not clear, but after shots were fired, Conquering Bear urged his people not to shoot. A second volley killed the Brulé leader, and the Indians attacked, killing every member of Grattan's command. The Miniconjou survived. The Lakotas emptied the warehouse and distributed the annuities on their own. They did not attack the fort but split into small bands and left the area.

A year later, in August 1855, after the public outcry over the "Grattan Massacre," the army sent General William S. Harney to lead seven hundred troops out of Fort Leavenworth in Kansas. This show of force began with the annihilation of a Brulé village in western Nebraska, where more than a hundred Indian men, women, and children died. During the fall, Harney marched his men through the heart of the Lakotas' hunting grounds, from Fort Laramie northeast to Fort Pierre on the Missouri. In March 1856, a major council convened at this fort, with representatives from all bands of the Lakotas, including the aloof Hunkpapas. This gathering reaffirmed the Fort Laramie Treaty of 1851. The Lakota leaders also agreed to surrender all warriors still wanted by the army for the Grattan incident.

The federal attack on Indian self-determination during the 19th century included the forcible displacement of tribes, the creation of a reservation system, and the more subtle devaluation of Indian cultures and histories.

Removed more than 300 miles in 1864 to Fort Sumner in Bosque Redondo, New Mexico Territory, these Navajos work under the watchful eye of military guards. In 1868, the Diné were permitted to return to a reservation carved from their former homeland.

Unidentified photographer. Displaced Navajos at Fort Sumner. *Photograph, 1866. Courtesy Museum of New Mexico (#1816), Santa Fe.*

The end of the great buffalo herds aided the government's efforts to confine native peoples to reservations. By the late 19th century, many Plains Indians—like the Lakotas shown here—were increasingly dependent on food and other supplies distributed by government agents.

Unidentified photographer. Dispersement of Government Rations. *Photograph, late 19th century. Smithsonian Institution, National Anthropological Archives (#56630), Washington, D.C.*

By the close of the nineteenth century, both government schools and privately run training institutes encouraged cultural assimilation and discouraged traditional behavior. In this photograph, Native Americans share a history class with African Americans at the Hampton Institute in Virginia. Next to a bald eagle, one native poses in traditional clothing—not necessarily of his own people. The lesson appears to be that both the Indian and the bird are exotic subjects doomed to survive only in museums.

Frances Benjamin Johnston (1864–1952). Class in American History *(plate from an album of the Hampton Institute). Platinum print, 1899–1900, 7½ x 9½ in. Gift of Lincoln Kirstein, The Museum of Modern Art, New York.*

Despite Harney's apparent success, the balance of power on the plains had not clearly shifted in favor of the U.S. Army. The outbreak of the Civil War in 1861 incited military combat from the Mississippi Valley eastward and left much of the West with few regular army units. More volunteers replaced the regulars. This change in personnel did not improve Indian affairs, as demonstrated by one infamous event on Sand Creek in the Colorado Territory. At dawn on 29 November 1864, one thousand of the Third Colorado Volunteers, under the command of Colonel John M. Chivington, attacked the camp of some five hundred sleeping Cheyennes. Black Kettle and White Antelope, the Cheyenne leaders, believed that a peace treaty was in effect and had turned in their arms at Fort Lyon. The soldiers slaughtered men, women, and children indiscriminately and mutilated their bodies. At least one hundred and fifty Indians died. The volunteers returned to Denver to cheering crowds that admired the scalps and severed genitals displayed like trophies of battle. A joint congressional committee later investigated the Sand Creek Massacre and condemned Chivington, but he could not be punished, since he had left the army.

The Cheyennes retaliated by twice attacking Julesburg, Colorado, and by halting travel across the plains to Denver. The Teton Sioux, close allies of the Cheyennes, were already agitated by an uprising in 1862 of the starving Santee Sioux bands in Minnesota, who had not received their treaty annuities. Militia and volunteer troops had defeated the Santees, who were expelled from Minnesota. Many Santees fled westward into

Lakota country, but thirty-eight captives were hanged at Mankato on 26 December 1862. Events in Minnesota and Colorado enraged many Lakotas, who also condemned the opening in 1863 of a route from Julesburg, Colorado, to the new gold camps in western Montana. The Bozeman Trail cut through the heart of Lakota territory in the Powder River country, and the U.S. Army had established three forts along the trail to protect travelers and freighters. Red Cloud, of the Oglalas, led the Lakotas' resistance and trapped U.S. troops in all three forts. On 21 December 1866, outside one of these forts, eighty men under the command of Lieutenant Colonel William J. Fetterman died in an ambush skillfully planned by a force of Lakotas, Cheyennes, and Arapahos. As a result of the annihilation of Fetterman's command, along with an ineffective military expedition in 1865–66, Congress advocated negotiations to resolve the Powder River War. After elaborate efforts, the government's peace commission produced a treaty at Fort Laramie late in 1868. Red Cloud promised to keep his Oglala followers out of war, and the army agreed to abandon its three forts. Hostilities had ended, and the Lakotas held the upper hand.

By 1868, in the Southwest, the Navajos had fared much worse in their relations with the United States. War with American forces had come during the years that Americans were fighting each other. The largest Civil War battle in the Far West had been fought in New Mexico at Glorieta Pass in late March 1862. At this engagement, John M. Chivington, who would lead the slaughter at Sand Creek in 1864, helped defeat a Confederate army. The Union's victory ended any serious threat of Confederate control in the Southwest. In August 1862, Brigadier General James H. Carleton arrived in New Mexico in command of a column of California volunteers; with the Confederate force already defeated, he turned to fighting the Indians. Carleton wanted to end raids by the Mescalero Apaches and by the Navajos. He directed his old colleague, Colonel Kit Carson of the New Mexico volunteers, to invade first the lands of the Apaches and then those of the Navajos. By the end of March 1863, more than four hundred Apaches had been relocated to the new reservation at Bosque Redondo, next to the new military post of Fort Sumner. Carson next attacked the Navajos, whose population of ten thousand may have been twenty times greater than that of the Mescaleros. General Carleton had one message for the Navajos: "Go to Bosque Redondo, or we will pursue and destroy you. We will not make peace with you on any other terms."

Carson's men destroyed orchards, crops, and livestock. They marched through Canyon de Chelly, the Navajos' great citadel. To avoid starvation, six thousand Navajos surrendered by the spring of 1864. The military then organized the Navajos' Long Walk—three hundred miles southeast to Bosque Redondo. By the end of the year, eight thousand Indians had been relocated there. Those who refused to surrender hid in isolated areas of their homeland or fled west. One Navajo, Curly Tso, recounted that many of the Diné (Najavos) saw Hweeldi (Bosque Redondo) as a place "where they would be put to death eventually."

Carleton saw the new reservation as a place of cultural transformation where the Apaches and Navajos would take up farming and where their children would learn to read and write and acquire the "arts of peace" and the "truths of Christianity." The superintendent of Indian affairs for New Mexico, Michael Steck, had his doubts. He had been the agent for the Mescalero Apaches and he knew that they considered the Navajos to be "inveterate enemies." He also knew that the land at Bosque Redondo could not support such a concentration of people. Carleton's grand experiment failed, destroyed by the forces of nature as much as by the forces of culture. Drought and insects devastated the crops. The government delivered inadequate supplies. Once more the Diné faced starvation.

In 1868, the same congressionally appointed peace commission that negotiated the new treaty at Fort Laramie sent two representatives to Bosque Redondo. On 1 June the representatives, who saw the suffering of the Navajos, signed a treaty that allowed the people to return to a reservation carved out of the Indians' old homeland. The document still advocated programs such as schooling and farming for the Navajos' cultural "advancement," but it recognized the need for the Navajos to begin again on familiar ground.

What unfolded for the Diné after their return home is a remarkable story. They reestablished their pastoral life-style with herds of sheep, goats, and horses, but they did not continue to raid their neighbors. Before removal, the Navajos had been a people divided into extended families, bands, and clans. But the four bitter years at Hweeldi had increased their sense of tribal unity and expanded the Diné's familiarity with Anglo-American culture. The treaty of 1868 gave the Diné clearly defined borders for their homeland. The historian Peter Iverson has observed, "Their political boundaries had been established: the Navajo Nation had begun." It also began to grow. By 1870, the population reached fifteen thousand. By the early twentieth century, the Navajo Nation was double that figure. The reservation grew as well. From 1878 to 1886, five additions to the original 1868 boundaries quadrupled the Navajos' territory. Most significant, the Navajo reservation was never broken up into individual allotments. The Diné had escaped the deleterious results of the Dawes Act and its successors, and the Navajo population and Navajo lands continued to grow throughout the twentieth century.

The Teton Sioux followed a different road into the twentieth century. Their treaty of 1868 had created the Great Sioux Reserve, which stretched from the Missouri River to the western boundary of the Dakota Territory. In addition, the Lakotas could still hunt in the Powder River country. The agreement contained the usual provisions for promoting "advancement," such as the establishment of farms and schools. It also included an article that promised no future cessions of the reservation without approval by three-fourths of the adult male Indians. Later documents ignored this provision and set the stage, in the late twentieth century, for a legal effort to reclaim Lakota lands. This case eventually reached the U.S. Supreme Court, where in 1980 a favorable verdict for the Lakotas resulted in financial offers from the federal government but no reacquisition of territory.

Many bands among the Lakotas, such as Sitting Bull's Hunkpapas, refused to recognize the Fort Laramie Treaty of 1868. Sitting Bull taunted the apparently cooperative Lakotas who accepted annuities at the government agencies located on the Great Sioux Reserve. He said, "You are fools to make yourselves slaves to a piece of fat bacon, some hard-tack, and a little sugar and coffee." When gold was discovered in the Black Hills, miners invaded the area, destroying an already ineffective peace. Instead of removing the miners, President Ulysses S. Grant, in late 1875, ordered the Lakotas to leave their winter camps and come into the agencies. Few complied, and the Great Sioux War of 1876–77 began.

The most famous battle of this war occurred on 25 June 1876 near the valley of the Little Bighorn River. The destruction of George Armstrong Custer's command by a vastly superior force of Lakotas and Northern Cheyennes became national news on 4 July, just as the United States prepared to celebrate the centennial of its Declaration of Independence. The Indians' one dramatic victory sealed their fate for the longer war. On 7 July 1876, General Philip H. Sheridan, in command of the Division of the Missouri, assured his superior, General William T. Sherman, "I will take the campaign fully in hand, and will push it to a successful termination sending every man that can be spared."

In a massive military effort, the U.S. Army defeated the Lakotas. Even before the final victory, the federal government began to reduce the lands of the Great Sioux Reserve. The president appointed a commission that met with Indian leaders at various agencies during the fall of 1876. The commissioners insisted that the Lakotas cede the Black Hills to the United States. The Indians complied, but this acquiescence did not constitute the approval of three-fourths of the adult males as specified by the treaty of 1868. Another round of reductions and divisions began in 1889 and created six reservations: Standing Rock, Cheyenne River, Lower Brulé, Crow Creek, Pine Ridge, and Rosebud. Eventually the six reservations were to be allotted and "surplus" lands acquired by non-Indians. The Great Sioux Reserve had been shattered.

Shattered too was the Lakota way of life. The loss of the Great Sioux War had been followed by the loss of the great buffalo herds. Some Lakotas turned to cattle raising and crop farming, but most became dependent on the government for food rations. By the late 1880s, the situation was desperate. Disease decimated Lakota cattle herds. Crops failed. Measles, influenza, and whopping cough were epidemic. Crops failed again. The government reduced rations, and the Lakotas starved.

The destitute Lakotas took heart in a new religious movement. The Ghost Dance promised the return of the buffalo and the disappearance of the white people. Nervous neighbors and anxious agents feared another war and asked the army to intervene. Once more the military moved on to Lakota lands, and once more tragedy resulted. The slaughter at Wounded Knee on 29 December 1890 killed 146 Indians, including 44 women and 18 children. Much hope died with them.

Like the Navajos' Long Walk, Wounded Knee placed the Lakotas at their historical nadir. In the twentieth century, their population would grow and their cultural identity would be maintained. But with reduced lands and continued impoverishment, the Lakotas became domestic, dependent peoples—often functioning as economic wards of the federal government. The Diné made a more successful adjustment to economic life in the twentieth century, developing arts, crafts, tourism, grazing, and energy resources on their unallotted reservation. But the Diné have not escaped the droughts, diseases, displacements, and internal divisions that continue to plague the twentieth-century Lakotas. And Indian reservations, whether allotted or not, remain internal colonies controlled to a great extent by a federal bureaucracy. This political status began to improve somewhat in the mid-1930s when the government created a New Deal for America's Indians. Allotment ended. Tribal governments and tribal lands gained legal recognition. But federal supervision continued. Only in recent decades has the possibility of self-determination in economic and political terms been openly considered as national policy.

The Prolonged Territorial Era

Self-determination may not yet fully exist on Indian reservations, but it supposedly came to the federal territories of the American West when they were granted statehood. Yet, admission as a state did not wean the former territories from the agencies of the federal government. During the last half of the nineteenth century, a federal bureaucratic system became well established in the western territories through the Department of the Interior's Indian Bureau, its General Land Office, and its Geological Survey (which after the Civil War expanded on the work of the Army Corps of Topographical Engineers). These federal agencies tried to map, assess, distribute, and develop a vast public domain while limiting the Native American use of that domain. By mid-century, east of the Mississippi, the policy of Indian removal and the massive sale of public lands had resulted in a federal bureaucratic presence limited primarily to institutions like the Post Office and the Customs Service, which also existed in the West. With the exception of the Civil War and the military occupation of the South during Reconstruction, the U.S. Army throughout the nineteenth century was largely a western army, manning posts and pursuing native peoples.

Much more than in the East, the federal government directly affected the lives of the residents of the trans-Mississippi through the actions of its bureaucracies as well as its army. These federal institutions became established in the West along with, or sometimes in advance of, settlements in the region. New arrivals in the West relied on the federal government for military protection, gainful employment, title to land, and relief during disasters. The modern federal government of the twentieth century with its expanded power and bureaucratic structure began to take shape in the American West of the nineteenth century. The historian Richard White has concluded: "While the federal government shaped the West, . . . the West itself served as the kindergarten of the American state. In governing and developing the American West, the state itself grew in power and influence."

Initially, political patronage typically determined appointments within these federal agencies that controlled the West. By the 1880s, in response to the efforts of reformers such as Carl Schurz, secretary of the interior from 1877 to 1881, a professional civil service began to emerge. In that same decade, the federal government started to reserve public lands, which eventually became national forests and national parks. Unlike in the East, vast acreage in the public domain did not pass to private ownership but remained public. Thus in the West when statehood came, the federal government controlled land within the states and retained professional bureaucrats to administer these lands. But statehood itself did not come quickly. With the exceptions of Texas and California, the western territories underwent a prolonged period of apprenticeship within the federal system.

Some people profited during the decades of delay before statehood. Within each territory, self-serving elites acquired much wealth and power as part of a process that the historian Howard Lamar has labeled "the politics of development." During the same era, the impoverishment of Native Americans demonstrated what might be called "the politics of underdevelopment." The loss of reservation lands and natural resources was the result of an assimilationist policy that failed, whereas the economic growth of the West and the admission of new western states might be considered an assimilationist policy that succeeded.

Not surprisingly, the Civil War influenced the commitment to firmer federal control over the western territories. The Lincoln administration extended the territorial system over the remaining unorganized lands beyond the Mississippi. The Republican-dominated Congress also wanted to keep the West in the Union, so in 1861 it organized the Dakota, Colorado, and Nevada territories. Arizona and Idaho followed in 1863, and by 1870, Wyoming and Montana had also become federal territories. Within the contiguous trans-Mississippi, only the Indian Territory remained an exception. But its special status as a reserve for native peoples did not survive once political incorporation into the Union began with the establishment of the Oklahoma Territory in 1890.

Congress readily created territories, but it reluctantly created states. Before the logjam broke in the late 1880s, only four western states had entered the Union—two during the Civil War, Kansas in 1861 and Nevada in 1864, and two after the war, Nebraska in 1867 and Colorado in 1876. Six territories finally gained statehood in 1889 and 1890: North Dakota, South Dakota, Montana, Washington, Idaho, and Wyoming. Four more states had to await admission because of issues delaying congressional approval. In Utah, the Mormon leadership officially abandoned polygamy in 1890; statehood followed in 1896. Oklahoma was next in 1907, after fusion with the remainder of the Indian Territory. Admission came for New Mexico and Arizona in 1912, after Congress reconsidered its attempt to join these territories. In New Mexico, hostile attitudes toward the Hispanic culture and citizenry had prolonged the territorial era for sixty-two years.

The years of New Mexico's territorial status, 1850–1912, are nearly congruent with a period that one scholar has labeled "The Second United States Empire." According to Jack Eblen, the First Empire lasted from 1787 to 1848, or from the passage of the Northwest Ordinance to the admission of Wisconsin. During these sixty years, the United States incorporated all the states through the first tier west of the Mississippi, with the exception of Minnesota. The Second Empire incorporated states across the continent before giving way to a Third Empire, which more firmly established American interests in the islands of the Caribbean and Pacific as well as northward in Alaska.

In terms of territorial policy, the First Empire began with an assumption of federal control over the organization of new lands under the model established by the Northwest Ordinance. Over time, the expansion of voting rights to nearly all adult white males resulted in greater political participation for a new territory's residents, especially in the election of a legislature. By 1836, these democratic ideals reached their culmination in the Wisconsin Organic Act, which replaced the Northwest Ordinance as the model for organizing new federal territories. Of course, federalism itself came under fire during the decades before the Civil War. The democratic rights of territorial residents became part of the debate over states rights and popular sovereignty. For those promoting states rights, the states and not the federal government held title to the territories. It followed, therefore, that if an institution such as slavery was permitted in some states, it could not be excluded from any territory. For the advocates of popular sovereignty, the people of the territory, and not the states or the federal government, should decide issues such as the establishment of slavery.

At least in terms of the slavery question, this debate ended after the Union victory in the Civil War. Federal power again became the guiding force in organizing the territories of the Second Empire, and some controls over popular sovereignty appeared. For example, by 1869, Congress had limited territorial legislatures to a biennial meeting of no more than sixty days. All territorial elections had to be reported to Congress, which reserved the right to resolve disputed elections, especially if they involved the selection of a territorial delegate.

The delegate represented his territory in Washington, but federal appointees represented the national government in the territory. More accurately, these appointees often represented particular political interests and specific factions within the national political parties. Some appointees were crooks. Others were hacks. The most prominent ones were the territorial governors, secretaries, and judges. Nominated by the president, they were confirmed by the Senate. Beyond these appointments, political patronage often determined who held other positions of federal employment in the territories, from surveyor generals, land registrars, and postmasters to U.S. marshals, customs agents, and Indian agents.

Within each western territory, stories of corrupt federal officials abound. Consider the case of Victor Smith, U.S. collector of customs for Puget Sound from 1861 to 1863. An appointee of the new Lincoln administration, Smith, a former newspaper reporter from Cincinnati, was a political crony of a fellow Ohioan, Secretary of the Treasury Salmon P. Chase. Soon after arriving in Port Townsend, Washington Territory's port of entry, Smith began to conspire to move the customshouse to a new townsite, Port Angeles, some forty-five miles to the west. Smith owned twenty-five acres at the Port Angeles site. Since loss of the customshouse would destroy Port Townsend's economy, Smith lied about his intentions while bombarding the secretary of the treasury with letters detailing the attractions of Port Angeles. Smith even returned to Washington, D.C., to help Chase steer through Congress the legal transfer of the port of entry. Meanwhile, the acting customs agent, Lieutenant James H. Merryman, had already written to Chase about Smith's fraudulent use of federal funds. At least $4,354.98 was missing.

Chase did nothing, and Smith returned to Port Townsend aboard the federal revenue cutter *Shubrick*. With the transfer enacted, Smith had come for the customshouse records. The town residents protested, but Smith threatened to have the *Shubrick* shell the customshouse. The ship's commander said he would have to follow Smith's orders. An angry mob of sailors and townspeople backed down in the face of this gunboat diplomacy. The records, along with the customs safe, were delivered to the *Shubrick*. One federal appointee had made off with the town's most important economic asset.

Victor Smith may stand as an example of the corrupt official who tried to profit from the territorial system. Civil service reform only started to take hold in the 1880s and never produced a territorial service for the Second Empire comparable to the colonial service of Great Britain's overseas empire. With patronage in control, men like Victor Smith sought federal jobs in order to rake off some of the funds intended for the

After the Civil War, the federal government renewed its interest in the western territories. Four major survey teams, each accompanied by artists and photographers, scientifically charted the interior West. Led by Clarence King, Ferdinand Hayden, John Wesley Powell, and George Wheeler, these teams produced extensive reports with illustrations that underscored the dramatic features and economic possibilities of the western landscape.

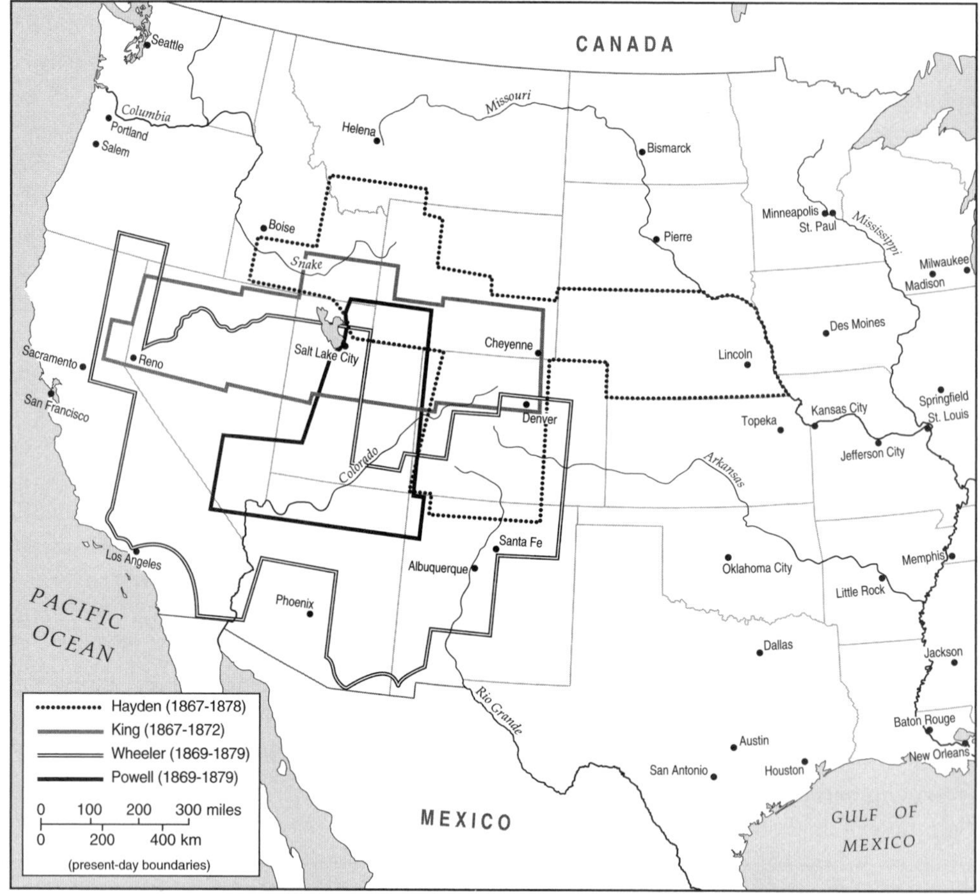

The Great Surveys, 1867–79

territories. They then participated in land investments, railroad schemes, and other business speculations—efforts not always so unwelcomed as the relocation of a customshouse.

Lamar's "politics of development" demonstrated that territorial governments functioned for the benefit of an oligarchy of federal appointees and business leaders. Simply put, the main business of territorial government was business. Mine owners, railroad developers, merchants, cattlemen, and bankers as well as lawyers, judges, and governors played an often profitable role in the politics of development. This fusion of business and government spurred economic development, as did the large federal subsidies to the territories. Along with the salaries of federal appointees and budgets to operate the territorial legislatures, federal funds maintained military posts, land offices, postal routes, and Indian agencies. Federal money also fed Indians on the reservations and built improvements within the territories, such as roads, forts, capitols, and prisons. Because of these expenditures, the federal government became the biggest business in some areas of the territorial West.

But what of the federal appointees who were themselves another form of federal investment in the West? They represented a system of patronage that did not necessarily

The wondrous images of Yellowstone created by artists with Ferdinand Hayden's survey helped persuade Congress to set the area aside as the country's first national park in 1872. For those unable to see the original photographs or watercolors, engraved reproductions in Hayden's annual reports made views of Yellowstone's marvels widely available.

Bisbing, after Thomas Moran and William Henry Jackson. Yellowstone Geysers. *Wood engraving, 1873. From F.V. Hayden,* Sixth Annual Report of the United States Geological Survey on the Territories *(1873).*

Like other survey photographers, Timothy O'Sullivan often included a human figure in his landscapes to give a sense of scale to his dramatic vistas.

Timothy O'Sullivan (1840–82). Shoshone Falls, Looking over Half of the Falls *(for the King Survey, along the Snake River in Idaho). Albumen silver print, 1868. International Museum of Photography at George Eastman House, Rochester, New York.*

produce corrupt or inept officials. Consider, for example, the collective qualifications of all territorial judges. Six hundred and eight individuals received appointments to the territorial courts between 1789 and 1959. Of this number, a majority (57.9 percent) were appointed between 1829 and 1897, from the presidential administration of Andrew Jackson through that of Grover Cleveland. These dates correspond to the years of greatest expansion for the territorial system. The judges of this era typically came from privileged, middle-class families. Well over half of them had attended or graduated from college, which placed their participation in higher education above that of members of the House of Representatives. Less than one-third of these judges had experience in any other court at any other level, a significantly lower rate of judicial service than may be found in other federal courts of the same decades. Sixty percent of the territorial judges had won election to a public office before their appointment, a level of political activism that paralleled that of Supreme Court justices and other federal judges. In other words, territorial and all other federal judges regularly established their affiliation with a political party by running for office. For some, this effort paid off with partisan appointments to the bench.

Territorial judges were well-educated but inexperienced appointees, with active political affiliations. In addition, they were outsiders, and they stayed only briefly on the bench. Between 1829 and 1896, three-quarters of the judges were not originally residents of the territories in which they served. Three-quarters also served less than four years on the territorial bench. The average time of service was 3.2 years, whereas the judges of the other lower federal courts averaged 12.4 years. Rapid turnover was endemic in the western territories. Of the 424 territorial governors, secretaries, and judges appointed between 1861 and 1890, 288 served for less than the four years of their original commission. In the case of territorial judges, most either resigned or were not reappointed. Only about 10 percent were removed from office. Party politics had created a revolving door of territorial appointments. The legal historian Kermit L. Hall notes, "Incumbents on the territorial bench after 1850 must have sensed that, if they did not resign, an incoming administration of the opposition party was likely to replace them."

The constant turnover in territorial appointees may have increased the citizenry's desire for statehood. Taking more government into their own hands could bring some stability, but first statehood needed to be established. And statehood did not come quickly. The case of Montana is typical. Political leaders in Montana, like those in other territories, saw statehood as a certification of successful growth in both economic and political terms. In the push for statehood, the politics of development had been successful enough to produce the politics of home rule. Advocates of statehood recognized that Montana would gain the right to elect its own governor, choose its own judges, and establish its own full representation to Congress. Large grants from federal lands could be used to support education, and a state government had the power to tax local corporations as well as the landholdings of railroads. Knowing the benefits of statehood, the 1883 territorial legislature called for the election of a state constitutional convention. In mid-January of 1884, the forty-one elected delegates gathered in Helena.

Like their election, the convention was generally a nonpartisan affair. Nonetheless, mining and cattle interests had effective representatives, led by the elected president of the convention, William A. Clark, of Butte, one of the wealthiest mining magnates in the United States. In twenty-eight days, the convention produced a document, which borrowed heavily from the constitutions of other states. In November, Montanans gave it their overwhelming approval by a vote of 15,506 to 4,266. They then awaited a positive response from Congress. It took four years.

The national election of 1884 produced a balance between the Democrats and the Republicans. The party of the Democratic president, Grover Cleveland, controlled the House of Representatives, but the opposition Republicans controlled the Senate. Each party blocked the admission of any state that might aid the other party. Since Montana and New Mexico were seen as potentially Democratic states, the Republican Senate stopped their admission. Likewise the Democratic House blocked the admission of Republican Washington Territory as well as a plan to divide Dakota Territory into two states and thus double the gain in Republican votes in Congress.

The election of 1888 gave Republicans control of both houses of Congress as well as the White House. Awaiting the inauguration of Benjamin Harrison, the lame-duck Democrats allowed the passage of an Omnibus Bill that enabled North and South Dakota, Washington, and Montana to proceed to statehood. Grover Cleveland signed the legislation before leaving office. New Mexico was not admitted, but Montana could now join the Union because Republican leaders had become convinced that immigration into the territory would soon produce a Republican majority in the new state. To add to the Republican forces in Congress, Idaho and Wyoming gained admission in 1890.

Fitting the American Mold

A Republican torrent had ended the logjam of admission for six western states. Such forceful partisanship also channeled the admission of the contiguous West's remaining four states. But it did not speed the process. For Utah, Oklahoma, New Mexico, and Arizona, delays continued until each territory had assured congressional leaders that it fit the American mold.

In the case of Utah, an agenda set during the era of Civil War and Reconstruction prevailed. In 1856, the newborn Republican party had denounced the "twin barbarisms" of slavery and polygamy. The Civil War dispatched one barbarism, but Republican Reconstruction ended when the South returned to domination by the Democratic party and the white race. Although Republicans lost their grip on the postwar South, they did not give up the fight against polygamy in the West. They saw Utah as a bastion for one party, the Democrats, under the domination of one church, the Latter-day Saints. Senator George Franklin Edmunds, of Vermont, took the initiative in attempting to reconstruct the Utah Territory. Edmunds had come to the Senate in 1865 as a radical Republican. He served as chair of the Senate Judiciary Committee from 1872 to 1891, with only a brief interruption from 1879 to 1881 under a Democratic majority. In the 1870s he supported legislation to dismantle the Ku Klux Klan and to guarantee civil rights in the South. These efforts had little success. In the 1880s, he pushed to end polygamy and to dismantle the Mormon church. These initiatives had more success.

The Edmunds Act of 1882 established fines and terms of imprisonment for men found guilty of formal polygamy or of informal cohabitation. Belief in plural marriage could exclude citizens from jury duty, public office, and voting. Children born after 1883 of polygamous marriages were declared illegitimate. A federal "Utah Commission" of five men registered voters and oversaw territorial elections. The work of this commission and of the federal courts had some impact, but it did end not polygamy. So by 1887, Senator Edmunds had established more radical powers for the federal government. The so-called Edmunds-Tucker Act of that year gave U.S. marshals wide latitude in arresting offenders. It empowered the Utah Commission to administer a qualifying oath for voters. It also dissolved the incorporated Church of Latter-day Saints and placed church assets in the hands of a non-Mormon receiver.

These antipolygamy laws sent hundreds of Mormon men either to jail or into hiding. The laws so disrupted the work of the Mormon church that after the death of the church president, John Taylor, in 1887, it took two years to go through the ritual of choosing the next president. After prayer and revelation, the new church leader, Wilford Woodruff, instructed his followers to abandon polygamy and to recognize the separation of church and state in the governance of Utah. Woodruff's manifesto in 1890 opened the way to statehood, which nonetheless required six more years of effort by Utahns.

Another Republican principle—homesteading—affected the Indian Territory. The Lincoln administration had produced the Homestead Act of 1862. In legislation such as the Timber Culture Act of 1873 and the Desert Land Act of 1877, Congress expanded the acreage that one settler might claim. By 1887, the homesteading ideal had been applied to Indian lands through the Dawes Act. The goal of Americanizing native peoples through small-scale farming resulted in the Americanization of "surplus" Indian lands by non-Indian farmers. In 1889, Congress authorized the opening of two million acres of "Unassigned Lands" in the middle of the Indian Territory. To claim farmsteads and townsites, fifty thousand people gathered on 22 April to participate in the first of the wild and dramatic "runs." A few months later, these lands formed the basis for the new Oklahoma Territory.

A federal commission chaired by the former governor of Michigan, David H. Jerome, aided the growth of the Oklahoma Territory. Under the principles of the Dawes Act, the Jerome Commission managed to create more surplus lands through the allotment of reservations in the western sections of the Indian Territory. Henry L. Dawes came out of retirement to head another commission that attempted to allot the remaining lands of the Indian Territory. Leaders of the Cherokees, Choctaws, Creeks, Seminoles, and Chickasaws resisted these efforts, but eventually, between 1898 and 1907, over 100,000 Indians from these five tribes were assigned lands. Surprisingly, very few acres became directly available for non-Indian settlement.

The five tribes of the Indian Territory tried to form a separate state that could not be absorbed into Oklahoma. A convention at Muskogee in the spring of 1905 produced a constitution for the proposed state of Sequoyah. Unfortunately, the chairman of the Senate Committee on Territories, Senator Albert J. Beveridge, of Indiana, had become favorably impressed with the "American" settlement of the territory of Oklahoma. The Indian Territory had a population equal to the state of Maine. It contained numerous railroads and incorporated towns. It also had more fertile land and better economic resources than the Oklahoma Territory. Beveridge, who had visited Oklahoma Territory briefly in 1902, believed that the Indian Territory should be joined to the "American" territory of Oklahoma to form one new state. His fellow Republican, President Theodore Roosevelt, agreed and recommended joint statehood in his annual message to Congress in December 1905. On 16 November 1907, the forty-sixth state officially entered the Union.

President Roosevelt and Senator Beveridge so liked "jointure" in the case of Oklahoma that they advocated a similar union for the Arizona and New Mexico territories. Both men belonged to the generation of Republicans influenced more by the Spanish-American War than by the Civil War. Another one of Beveridge's brief visits had been to one of America's newest possessions, the Philippines, which he considered a backward outpost of the former Spanish Empire. His contempt for the Spanish cultural heritage apparently extended to America's own Southwest. A quick tour of that region in 1902 convinced him that neither Arizona nor, most especially, New Mexico was advanced enough for statehood. Nonetheless, Beveridge decided that one vast state might shift the cultural balance toward the more American, but still backward, Arizona. A critic in Congress noted that Beveridge had decided "that one rotten egg is bad, but two rotten ones would make a fine omelet."

Given a chance to vote on "jointure," Arizona and New Mexico hatched different results. The referendum failed in Arizona, whereas it passed in New Mexico with the aid of political gamesmanship and much fraud. President Roosevelt eventually gave up on the proposition and announced in his last annual message to Congress in 1908 that each territory should gain statehood. Beveridge tried to persuade the new Republican president, William Howard Taft, to block admission for partisan reasons—Arizona might become Democratic, and New Mexico supported the wrong Republican faction. By 1910, the opinionated Indiana senator had reluctantly accepted the admission of both states. In 1912, Taft's signature ended this saga.

The admission of New Mexico and Arizona completed the formal political incorporation of the lands that formed the Second United States Empire. But, as some of the final disputes over statehood indicated, a far more complex process of cultural incorporation had not been completed. If anything, the admission of Utah, Oklahoma, Arizona, and New Mexico demonstrated that the American mold would have to be modified to incorporate distinctive peoples within the nation's expanded boundaries.

For decades, members of Congress had viewed the expansion of the nation as a grandly heroic adventure. Indeed, they could daily walk by a dramatic representation of national expansion—a painting that measured twenty by thirty feet on one wall of the U.S. Capitol. Congress commissioned this work in 1860 from Emanuel Leutze, the same artist who had produced the famous portrait *Washington Crossing the Delaware* (1851). In a vividly heroic tableau, Leutze filled a spectacular mountain passage with overlanders emigrating in wagons, mountain men traveling on horseback, women holding babies, men swinging axes, and even a freed slave leading a mule. Inspired by a line in a poem by the eighteenth-century Anglo-Irish philosopher George Berkeley, Leutze titled his work *Westward the Course of Empire Takes Its Way* (see color essay following p. 370). Only one figure, who waves his hat from atop a central pinnacle and resembles John Charles Frémont, hints at any role for the federal government in this great rush through the mountains. U.S. troops, Indian agents, and territorial officials are not part of the picture; neither are the native peoples and Hispanic residents of the Far West. Congressmen, in their own capitol building, could comfortably imagine that national expansion had been the manifest destiny of the American people and not the creation of the nation's government.

Although this image kept the government's cart behind the people's horse, it produced too simple of a picture. At times, the government proceeded westward more like a horseless carriage as it acquired, explored, and administered new lands. Into this vast territory, not only American emigrants but also federal agents and federal policies westward took their way. The initiatives of the nation's government sustained the country's growth, but these actions also enlarged the complexity of the nation's destiny. Thomas Jefferson's ideal of a compact and homogeneous republic could in no way stretch across the expanse of continent that, from east to west, measured itself as one nation. Nor could that nation grow westward in splendid isolation. The resources of the American West brought the country more fully into a global economy. Expansion also opened new doors for immigration from east and west, north and south. Ultimately, the incorporation of the American West multiplied the internal diversity of the United States and increased the nation's interconnections with the rest of the world. Globally the course of empire now would flow.

Bibliographic Note

For many of the topics, events, and individuals presented in this chapter, the best single-volume reference is Howard R. Lamar, ed., *The Reader's Encyclopedia of the American West* (New York, 1977). A revised second edition is in preparation for HarperCollins Publishers. Several books have surveyed the history of the American West with extensive attention to the actions of the federal government. Richard White's *"It's Your Misfortune and None of My Own": A New History of the American West* (Norman, Okla., 1991) presents a most insightful reinterpretation of how the federal government shaped the American West. Patricia Nelson Limerick's *The Legacy of Conquest: The Unbroken Past of the American West* (New York, 1987) gives witty, provocative commentary on some federal efforts. Frederick Merk's *History of the Westward Movement* (New York, 1978) is filled with useful information, whereas Ray Allen Billington and Martin Ridge's *Westward Expansion: A History of the American Frontier*, 5th ed. (New York, 1982) remains a thematically focused, grandly conceived narrative.

The best examination of Thomas Jefferson's interest in the trans-Mississippi West may be found in Donald Jackson, *Thomas Jefferson and the Stony Mountains: Exploring the West from Monticello* (Urbana, Ill., 1981). David Lavender's *The Way to the Western Sea: Lewis and Clark across the Continent* (New York, 1988) is a well-wrought narrative. Under the editorship of Gary E. Moulton, a superb multivolume edition of *The Journals of the Lewis and Clark Expedition* is being published by the University of Nebraska Press. James P. Ronda has considered the expedition in *Lewis and Clark among the Indians* (Lincoln, Neb., 1984) as well as *Astoria and Empire* (Lincoln, Neb., 1990). Three works by William H. Goetzmann have brilliantly explained the significance of American exploration: *Army Exploration in the American West, 1803–1863* (New Haven, 1959); *Exploration and Empire: The Explorer and the Scientist in the Winning of the American West* (New York, 1966), which received the Pulitzer Prize in history for 1967; and *New Lands, New Men: America and the Second Great Age of Discovery* (New York, 1986).

A reassessment of "manifest destiny" awaits a new generation of scholars. A preliminary step is Thomas R. Hietala, *Manifest Design: Anxious Aggrandizement in Late Jacksonian America*

(Ithaca, N.Y., 1985). Frederick Merk's *Manifest Destiny and Mission in American History: A Reinterpretation* (New York, 1963) is the standard work, whereas Reginald Horsman's *Race and Manifest Destiny* (Cambridge, Mass., 1981) is forcefully argued. The grand narrative of Texas is *Lone Star: A History of Texas and the Texans,* by T. R. Fehrenbach (New York, 1968). David Pletcher has considered annexation in *The Diplomacy of Annexation: Texas, Oregon, and the Mexican War* (Columbia, Missouri, 1973). The best military history of the Mexican-American War is John S. D. Eisenhower, *So Far from God: The U.S. War with Mexico, 1846–1848* (New York, 1989).

Before his untimely death in 1976, John D. Unruh, Jr., completed his monumental study of the Oregon and California trails, *The Plains Across: The Overland Emigrants and the Trans-Mississippi West, 1840–1860* (Urbana, Ill., 1979). Among its many insights, this work revised earlier scholarly assumptions about interactions between Indians and emigrants and about the role played by agents of the U.S. government. An effective overview of the political turmoil before the Civil War is provided by David Potter, *The Impending Crisis, 1848–1861* (New York, 1976). This book ignores the "Utah War" of 1857, however. A sound monograph that considers this event is Norman F. Furniss, *The Mormon Conflict, 1850–1859* (New Haven, 1960).

All the writings of Francis Paul Prucha on government Indian policy are models of masterful scholarship. Especially impressive is Prucha's *The Great Father: The United States Government and the American Indians,* 2 vols. (Lincoln, Neb., 1984). Albert Hurtado, *Indian Survival on the California Frontier* (New Haven, 1988), reveals the adaptations made by native peoples before and after the gold rush. The cultural complexities of treaty-making are considered by Raymond J. DeMallie in his essay "Touching the Pen: Plains Indian Treaty Councils in Ethnohistorical Perspective," published in Frederick C. Luebke, ed., *Ethnicity on the Great* Plains (Lincoln, Neb., 1980). Three books by Robert M. Utley are well written and informative: *Frontier Regulars: The United States Army and the Indian, 1866–1891* (New York, 1973); *The Indian Frontier of the American West, 1846–1890* (Albuquerque, N.M, 1984); and *The Last Days of the Sioux Nation* (New Haven, 1963).

Paul Andrew Hutton's *Phil Sheridan and His Army* (Lincoln, Neb., 1985) is an excellent study of the army in the West after the Civil War. The story of the Navajos' removal is told in Gerald Thompson, *The Army and the Navajo: The Bosque Redondo Reservation Experiment, 1863–1868* (Tucson, Ariz., 1976). The later history of the Navajos is presented in Peter Iverson's *The Navajo Nation* (Westport, Conn., 1981). Alfred Runte espoused his "worthless land" thesis in his book *National Parks: The American Experience* (Lincoln, Neb., 1979).

Earl S. Pomeroy, *The Territories and the United States, 1861–1890: Studies in Colonial Administration* (Philadelphia, 1947), and Jack Ericson Eblen, *The First and Second United States Empires: Governors and Territorial Government, 1784–1912* (Pittsburgh, 1968), provide solid overviews of the federal government's administration of the western territories. For the larger story of developments in the West during the territorial era, the work of Howard R. Lamar is unsurpassed. His *Dakota Territory, 1861–1889: A Study of Frontier Politics* (New Haven, 1956) and its marvelous successor *The Far Southwest, 1846–1912: A Territorial History* (New Haven, 1966) are models of definitive scholarship and graceful writing.

Before establishing his fame as a novelist, Ivan Doig wrote about Victor Smith's escapades in "Puget Sound's War within a War," *American West* 8 (May 1971): 22–27. Kermit Hall analyzed the qualifications of territorial judges in his article "Hacks and Derelicts Revisited: American Territorial Judiciary, 1789–1959," *Western Historical Quarterly* 12 (July 1981): 273–89. John Guice has carefully examined territorial courts in *The Rocky Mountain Bench: The Territorial Supreme Courts of Colorado, Montana, and Wyoming, 1861–1890* (New Haven, 1972). Full scholarly histories exist for individual states throughout the West. Two of the best are Michael P. Malone and Richard B. Roeder, *Montana: A History of Two Centuries* (Seattle, 1976), and Arrell Morgan Gibson, *Oklahoma: A History of Five Centuries* (Norman, Okla., 1965).

Chapter Six

Entering the Global Economy

KEITH L. BRYANT, JR.

James W. Marshall walked along the American River inspecting John A. Sutter's millrace on a cold January morning in 1848. As he tried to determine the water pressure needed to turn the wheel of Sutter's sawmill, Marshall's eyes detected the glint of metal in the stream, yellow metal, and he collected samples. The nuggets he found that morning produced the California gold rush. Some fifty-three years later, Captain Anthony F. Lucas pushed more drill stem into a salt dome at Spindletop, Texas, near Port Arthur. A rumbling noise, a sharp vibration, and the Lucas gusher blew in on 10 January 1901, oil shooting two hundred feet in the air for nine days. Within four years, twelve hundred nearby wells produced over thirty million barrels of petroleum. Between the discovery of gold in California and the coming of "black gold" in Texas, the economy of the trans-Mississippi West was transformed from subsistence agriculture and herding into a modernized and urbanized capitalistic economy integrated into a worldwide structure.

White settlers pressed inexorably westward from the early seventeenth century, but the conclusion of the Mexican War found only a few U.S. citizens occupying the coastal towns on the Pacific while the "Great American Desert" remained largely unoccupied. From the Mississippi River westward to the eastern reaches of the Colorado, Columbia, and American rivers, vast tracts of land beckoned a people whose pioneering spirit permeated the society. The settlement and urbanization processes moved forward vigorously, but the trans-Mississippi West would not be occupied along formal, geographical lines. The slowly moving "frontiers" of the Old Northwest and the Southeast were not replicated in the West.

Endless land, abundant natural resources, scarce labor and capital, a spirit of entrepreneurship, and social and political institutions favorable to economic growth combined to produce a western society undergoing changes that mirrored those occurring around the globe. The favorable movement of the prices of key staples and heavy infusions of capital from the East and abroad stimulated the economy, as did the growing presence of a relatively well educated and technologically sophisticated middle class. Westerners shifted easily and quickly from a rural to an urban environment within a highly mobile society. There was no refuge from change in the region, which imitated the nation and the world as it formed a capitalistic, industrial economy within half a century. Altered consumer demands in the East and in western Europe, the international mobility of both capital and labor, and the global flow of technological information combined to shape the economy of the West.

Painted to illustrate a popular travel guide, John Gast's picture presents a capsule history of western economic development as seen through the rosy guise of manifest destiny. Following a group of Indians come the symbols of American know-how: a wagon train, a horse-drawn stage, and the three lines of the transcontinental railroad. Over them floats a woman wearing the "Star of Empire" and carrying a schoolbook as a symbol of enlightenment and a telegraph line "to flash intelligence throughout the land."

John Gast (active 1870s). American Progress. *Oil on canvas, 1872. Gene Autry Western Heritage Museum, Los Angeles, California.*

Students of the nineteenth-century West have been overly concerned with a nonexistent uniqueness. Elements of the "mining frontier" in California, Colorado, and Montana could be found in South Africa and Latin America. The impact of the railroad was as significant in India as in Kansas and Texas. A legal structure based on capitalism shaped the American West as well as New Zealand and Australia. Indigenous peoples were subjugated, exploited, and exterminated on all these global frontiers. Some scholars of western American history also fail to differentiate between economic growth and economic development. Dependence on the exportation of raw materials or agricultural products may expand an economy to substantial levels, but that very success may limit diversification and the creation of an infrastructure to sustain industrial and commercial development. This has been the recurring nightmare of colonial and Third World economies. This was not the pattern in the American West before 1900, however, as external investment and internal capital formation produced a diversified regional economy.

Europeans occupied "free land" in North and South America, Asia, Africa, and the South Pacific throughout the nineteenth century. A treasure trove of minerals spurred that process. The movement of peoples into the trans-Mississippi West introduced both material progress and concepts of liberty even as it destroyed those non-European indigenous cultures that resisted. Technologically superior Europeans simply overwhelmed less-sophisticated societies. Transportation and communication accelerated this expansion as peoples of low skill levels were pushed aside. These "frontiers" quickly joined a world-capitalistic core market centered in western Europe. Maturation beyond self-sufficiency brought rising levels of participation in the world economic order to peripheral areas like the American West. As the French historian Fernand Braudel has shown, capitalism is an identifying theme for studying the modern world, for it provides structure and organization for examining relationships within a society. Capitalism emerged as the prevailing force in world history by the nineteenth century and as an all-pervasive aspect of American life, especially in the West.

The trans-Mississippi West formed a segment of several networks of productive processes and commodity chains. The various participants in these networks became interdependent as they sought to maximize capital accumulation. The creation of this world-capitalistic economy was not constant but occurred in wavelike spurts of expansion and contraction. Periods of stagnation led to class struggles, and weaker and less-efficient producers were often eliminated. Western Americans, and many of those who have written their history, saw these boom-and-bust cycles as regionally unique rather than as a reflection of a global economy that had no regard for political boundaries. World trade tied the peripheral economy of the American West to an international market system; then, as the economy of the West matured, it became a part of the core. Transportation, technology, military superiority, and medical science allowed for the inexorable expansion of capitalism, for as the economist Joseph Schumpeter has argued, "Stationary capitalism is impossible."

The economy of the trans-Mississippi West initially shared characteristics with other areas undergoing transformation in the nineteenth century. Like the occupying populations of South Africa, Australia, and portions of Latin America, the settlers of the West were young, largely male, transient, and violent. The absence of women and young children and a low birthrate reflected a restless, transitory society. Demographic characteristics soon changed, however, as the West became urbanized and the population came to reflect national norms. Industrialization produced towns and cities, in some of the most difficult conditions imaginable, in the mountains of Colorado or on the plains of eastern Montana, as westerners sought the amenities of urban life. Schools, churches, and opera houses quickly followed the railroad station, bank, and general mercantile store. If there was a unique element in the American West, it was the speed of this transformation. Mineral rushes produced long lines of wagons headed westward, followed by shining iron rails and singing telegraph wires. Freight lines and steamboats radiated from regional centers as trade routes stimulated traffic in lumber and foodstuffs. Chinese and Irish, Cornishmen and Italians, Australians and African Americans joined thousands of newcomers from the East determined to strike it rich. James Marshall's walk along the American River set in motion a process whereby throngs of settlers sought the "main chance" in the American West.

Mining, Magnates, and Miners

The mineral rushes of the nineteenth century stimulated settlement, forced the early formation of laws and government, created a demand for transportation, and lured labor and capital westward. Waves of pioneers swept into California, Nevada, Idaho, Montana, and the Dakota Territory seeking gold and silver. Ironically, many of these prospectors, miners, investors, and their auxiliaries moved not from east to west but from California eastward into the hinterland.

The forty-niners of the Golden State who flocked to Sutter's Mill found nuggets, "color," or "dust" in the rivers and streams. Placer mining—washing the dirt from the stream in a pan, leaving the heavy grains of gold in the bottom—required little labor, capital, or skill. If the pay dirt proved of sufficient quantity, a wooden box, or cradle, could be used to wash larger amounts of sand and dirt as the box was rocked to and fro. Wooden cleats in the bottom of the box held the gold as the water and earth washed away. A group of men might build a sluice, a series of long wooden boxes fitted with riffle bars across the bottom. They diverted water from the creeks through the sluice, and the flowing water carried away the dirt and sand dumped into the sluice by the miners. Nuggets and dust remained trapped in the riffle bars. Muscle and sweat produced wealth for a few and created a true cornucopia of publicity to lure thousands to the West.

The days of gold placer mining proved short-lived, however. As prospectors moved into the interior, into the Sierra Nevada and the Great Basin, gold and silver were found, but the minerals were locked in quartz lodes, or veins, buried deep in the earth. To reach this treasure, miners had to sink shafts, install timber linings, and use pumps to remove water seeping into these subterranean labyrinths. Capital for mills to crush the quartz and for vessels of mercury to dissolve the gold came not from the prospector but from investors in San Francisco, Philadelphia, and London. Mills, tunnels, machinery, and transportation brought the mining corporation to the strike, along with an army of laborers.

Sutter's Mill in 1848; Virginia City, Nevada, and Cherry Creek, Colorado, in the 1850s; Montana and Wyoming in the 1860s; and the Black Hills of South Dakota in the 1870s saw rushes of prospectors, followed by ramshackle towns filled with lawyers and gamblers, merchants and prostitutes, assayers and drifters. More lasting on the landscape were mineheads, smelters, mills, and tailings heaps, the debris produced by washing and concentrating. These rough, crude communities contributed to a national output of eighty-one million ounces of gold by 1852 and six and a half million ounces of silver by 1863. Virtually all of this treasure came from the Far West. The gold production alone transformed the role of the United States in world trade and led to rising foreign investments in the West. Even more important, the export of gold paid for the importation of steel for the railways and for machines for factories, mills, and mines. The trade surplus generated by western mining encouraged the expansion of credit and brought the United States into the mainstream of the world economy.

The mining rushes began with an initial discovery by a prospector followed by the communication of the news to miners elsewhere, who soon arrived looking for a bonanza. Within a short time, capital requirements brought in corporate investors even as the community moved from tar-paper shacks to wooden stores, stone churches, and brick opera houses. The rapid transformation of the area brought social chaos, physical change, and cultural amenities amid vigilante law, violence among the laborers, and often catastrophe in the tunnels.

All participants, whether a prospector engaging in placer mining along a creek or a mining firm having millions in capital to invest, needed the protection of the rule of law. The absence of legal precedents led westerners to borrow some of the finer points of Spanish and English mining laws as they formulated a body of statutes to protect their interests. Groups of miners elected officers at mass meetings and drafted rules based as much on common sense as common law. "Reasonable" amounts of land could be claimed if clearly marked and filed with a local recorder. The elected officers often settled disputes and allocated water rights. Workable codes, democratically formed, created an idealistic framework that originated in California and was exported to the mining camps of the interior. These spontaneous codes were subsequently incorporated into territorial, state, and federal mining laws. But the miners also conducted lynchings and vigilante actions and displayed disregard for the claims of the Native Americans and Hispanics who preceded them in the area. The mining rushes produced a demand not only for mining law but also for land offices, town marshals, and local governments. These requirements, in turn, led to premature territorial governments and eventual statehood. The mining booms initiated and hastened the arrival of government in the West.

The early Spanish settlers of the Southwest and California, unlike those who first entered Mexico and South America, did not concern themselves exclusively with the acquisition of mineral wealth, but they were not ignorant of its presence. The Catholic fathers of the California missions saw the discovery of gold not as a help in Christianizing the Indians but as a hindrance. Secular elements among the Spanish and then the Mexican settlers did not share that view and initiated modest mining operations in California and elsewhere. The Californios had productive mines long before the Americans arrived, and they taught the new settlers how to crush gold-bearing rocks and to remove the metal. Whereas some miners, known as the "Old Georgians," would come to California from previous diggings in Georgia and North Carolina and would bring their knowledge and skills with them, the Yankees also learned much from the Mexican settlers about the mining of gold.

The early days of placer mining in California were short-lived. Once the ore was mined from rivers and streams, expensive technology was required to obtain the additional gold ore embedded in quartz or buried deep in the earth. An image of easily obtained wealth drew thousands of gold seekers west, but independent miners such as these soon found the richest lodes to be controlled by heavily capitalized corporations.

Unidentified photographer. [Goldminers]. *Daguerreotype, half plate, ca. 1850. Courtesy Amon Carter Museum, Fort Worth, Texas.*

By mid-March 1848, news of Marshall's discovery at Sutter's Mill reached San Francisco, and within sixty days that small port town seemed deserted as men rushed to sites along the American River and then to other streams that flowed westward from the Sierra Nevada. President James K. Polk formally announced the discovery in his presidential address to Congress that December, and his confirmation of the rumors of vast mineral wealth initiated the rush of the forty-niners. Dozens of boats sailed from East Coast ports, headed around South America and then north through the Pacific to California. Even more intrepid were those would-be miners who took riverboats to Missouri and set out across the Great Plains for the West. In May 1849, some twelve thousand wagons crossed the Missouri River while other wagon trains departed from Fort Smith and Santa Fe. Shipwrecks in the straits of Tierra del Fuego and disease and death along the trails failed to slow the rush to El Dorado. In Dry Diggings, Hell's Delight, and Hangtown, the former shopkeepers, farmers, herders, and sailors panned, sluiced, clawed, and fought to find the illusive nuggets. They sometimes leased sites from the local Indians, but usually the prospectors simply staked claims. When Marshall found his first bits of yellow metal, California had few more than 14,000 U.S. citizens, but by 1852, there were over 250,000 people residing in the area. Immigrants from Asia, Chile, Hawaii, Great Britain, Continental Europe, and Mexico joined the Yankees in their quest.

The mining pattern that emerged in California would be replicated across the American West and in British Columbia. Along rivers and streams, the forces of erosion exposed veins of gold bearing ores and washed the nuggets or dust into the waterway. The first arrivals found easy pickings in their large metal pans and sluice boxes. A prospector might find as much as eight thousand dollars in gold in a day, but the initial discoveries played out quickly. Soon the miners braved the cold waters of upstream sites and resorted to sluices and then to mixing crushed rock with mercury to amalgamate the gold. But even as the output at the sites declined, the cost of food, equipment, and transportation rose; the merchants and freighters soon reaped much of the great bonanza.

Within three years, with dwindling yields from surface mining, corporations entered the diggings to pursue veins hidden beneath the gravel and rock of the land near the streams. Mills to crush the ore and machinery to erode vast quantities of dirt and sand displaced the individual prospectors and miners. Hydraulic mining, the use of water pressure to blast away rock and soil, devastated streambeds and created mountains of rubble. Hydraulic mining required large-scale reservoirs and elaborate flumes and ditches that delivered huge volumes of water at high pressure. Investors in such operations found laborers among the disillusioned forty-niners, who could earn more per day in the mines than with their pans.

Lonely prospectors moved along the Feather River, striking gold there, on the Stanislaus, and in the valley of the Shasta. Each new "rush" created more exaggerated publicity of finds in gulches and ravines, of instant fortunes and riches for all. The ten million dollars in gold produced in 1849 grew each year as miners spread out across the Mother Lode country, but by the end of the next decade, these laborers faced unemployment as technology and capital reduced the need for workers even as output increased. The small-scale miner disappeared, to be replaced by the daily-wage earner.

The gold rush in California created wealth for the few and labor for the many. When the output of gold reached eighty million dollars in 1852, the economy of the region had been transformed. Demands for goods and services made San Francisco a city with merchants, bankers, shipowners, freighting firms, and manufacturers competing for a share of the wealth. Clipper ships sailing round the Horn could not deliver enough clothes, shovels, nails, mercury, and other necessities. The urbanization of northern California initiated a pattern to be found across the West as mining rushes created demands for specialized products. Around the Bay, crowded docks piled high with goods alternated with small foundries and machine shops. Skilled ironworkers could earn far more there than along the streams of the Sierra Nevada or deep in the mine tunnels. The California gold rush set in motion the economic maturation of the West.

Interestingly, the geographical expansion of the mining boom came as a result of digging not by prospectors but by Mormons in the Nevada country. Probably in 1848, but definitely by the following year, members of that religious faith discovered a vein of gold-laden quartz in the Washoe Mountains, a range of hills extending east from the Sierra Nevada into the Great Basin. The Mormons soon lost faith in the area as the amount of dust they found proved negligible. Although the Mormons who made the strike lingered only briefly, miners in California soon heard the news and moved eastward, hoping to find another bonanza. Luck eluded them in "Gold Canyon" until June 1859, when Peter O'Riley and Patrick McLaughlin found the Ophir silver vein. When Henry T. P. Comstock ventured along shortly thereafter and claimed a share of their prize based on his alleged ownership of the spring they were using, the partners acquiesced. The lazy, shiftless Comstock soon had his name attached to this great bonanza, the Comstock Lode. The ore samples sent to California to be assayed contained three-fourths pure silver and one-fourth gold and were worth $3,876 per ton; the Ophir mine was the richest in history.

News of this fabulous strike spread quickly, and investors from California flocked to the Washoe country to purchase claims. Believing that the vein would play out shortly, just as others had, the original owners sold the mine. Judge James Walsh, of Nevada City, and George Hearst, of San Francisco, arrived early, purchased wisely, and prospered greatly.

In the largely treeless and waterless Washoe, the boomtown of Virginia City emerged, and thousands flocked to the forlorn site. Because the gold and silver were embedded in quartz, only crushing machinery could unlock the treasures. Prospectors owned claims, not capital, and sold or leased to others, but only a few of the three thousand sites proved profitable. The major mines—the Ophir, Central, Mexican, and the Gould and Curry—required drilling, blasting, and removal of the ore to steam-powered stamp mills. The tunnels flooded in the spring, and cave-ins marked the paths of the miners. Only heavy timbering allowed the intrepid laborers access to a vein that grew to a width of nearly two hundred feet. Cornishmen with their knowledge from the tin mines, Welshmen with their smelting experience, and Germans with their metallurgy background contributed skills and technology to this mining venture. As the population of Virginia City rose to fifteen thousand, an infrastructure developed to support the mining operations. Freighting, lumbering, and the mercantile trade became almost as profitable as owning a small claim. Yet the greatest days lay ahead, for in 1873 the Consolidated Virginia combine excavated a hole 1,167 feet into the earth and found the "Big Bonanza," a vein fifty-four feet wide filled with gold and silver. A fortune of two hundred million dollars created the "Kings of the Comstock."

Above C Street in Virginia City, the ornate homes of the merchants and bankers looked down on the gaudy, vulgar town, which had swelled to twenty thousand people by the mid-1870s. While the Irishmen, Cornishmen, Germans, Mexicans, and a polyglot of miners labored in the tunnels far below, town matrons sat on their porches eating ice cream and drinking champagne. They toasted their friends and financial allies, such as Hearst, whose capital had made their families a part of the Comstock's elite.

George Hearst had gone to California from Missouri, where he had experience with lead mines. Arriving in 1850, he tried placer and quartz mining but soon engaged in the buying and selling of claims. When news of the strike in Nevada came, he raced to the area to purchase a share of the mighty Ophir mine. Selling all his California holdings, and borrowing money, Hearst put everything into the Ophir. The "Old Californians" like Hearst knew how to register claims, form syndicates, organize stamp mills, and perhaps as important, create a legal mining district. The first thirty-eight tons of ore were packed across the mountains to San Francisco for smelting, but the high quality of the silver from that sample led to the construction of a complete mining operation in Virginia City. Flooding of the mines led to the installation of steam-powered pumps and extensive timbering; it took vast amounts of capital to produce an extraordinary profit. Hearst and his associates retreated to San Francisco, erected mansions, extended their fortunes, and prospered long after Virginia City became a virtual ghost town in the 1890s.

Almost simultaneously with the rise of Nevada as a monumental source of silver and gold, Colorado beckoned both Old Californians and Old Georgians to the foothills of the Rockies. Rumors of gold brought a handful of prospectors, led by Captain John Beck and W. Green Russell, to the headwaters of the Arkansas River in the spring of 1858. Veterans of the California goldfields, they panned along the frontal range of the Rockies. A member of the party hit pay dirt on Cherry Creek near present-day Denver. When residents of "the states" received the news, another rush of epic proportions began. Pilgrims and greenhorns struck out for Pikes Peak and Cherry Creek based on early reports that one need not travel all the way to Nevada or California to strike it rich. Nearly one hundred thousand people headed for Colorado, undaunted by six or seven hundred miles of travel across the semiarid plains of Kansas and Nebraska. Tall tales and the aftermath of the national depression of 1857 sent hundreds to Auraria and Denver City, the twin towns that sprang up along Cherry Creek. Every frontier outpost from Iowa to Arkansas dispatched wagon trains westward to Colorado. "Pikes Peak or Bust" painted on the canvas of their wagons revealed their destination and their determination.

The prospectors found little to warrant their optimism, and half were gone by the next summer. Some of those who remained were rewarded, however, when John H. Gregory hit more pay dirt on Clear Creek. Over five thousand miners swarmed into Central City, the boomtown that developed around Gregory's claim. Although the riches did not rival those of the Comstock Lode, the findings along Gregory Gulch sent hundreds of prospectors into the valleys south and west of Central City. More gold, silver, and lead were found, but Colorado yielded only twenty-five million dollars in gold in the decade after 1858. The greenhorns fled, but the veterans of Georgia, California, and Australia persevered. They had the knowledge and experience required to make the finds, whereas others had the technological skills and capital to develop the strikes.

As the true placers of Colorado played out, it became evident that only shaft mining offered hope for large-scale profits. Pyritic gold ores would not amalgamate with mercury, and thus new processes were required. Undercapitalized, the Colorado mines lacked efficient means to remove the gold, and the boom fell on hard times in the 1860s, but the introduction of smelting soon revived the mines. Smelting converted ores to a fluid state through heat and the use of chemicals, allowing the separation of the various metals. The mining promoters utilized the skills of German-trained metallurgists as well as a large amount of costly equipment.

The failure of early smelting efforts did not deter attempts to build a technologically sophisticated smelter in the mountains. An improved performance led to the removal of the smelter to Denver in 1878, where access to coal and coke made available by the newly arrived railroad allowed for both expansion and greater efficiency. Ores from throughout the Rockies soon came to several smelters in Denver, where foundries and machine shops turned out equipment and parts for the processors and for the mines. An industrial base brought prosperity to the "Mile High City."

The elite of Denver included men who owned mines or shares in mines, but the families who occupied the Cherry Creek Gothic mansions more often obtained wealth based on transportation, banking, smelting, real estate, or food processing. Merchants and bankers such as Jerome B. Chaffee and David H. Moffat used contacts in the East or Great Britain to create such enterprises. Luther Kountze, from Ohio, formed the Colorado National Bank; Amos Steck, also a Buckeye, succeeded in real estate speculation and, along with Kountze, inaugurated horse-drawn street railway service in Denver. Interior towns in the Rockies could only aspire to emulate Denver's growth and prosperity.

Some mining camps failed quickly, but others, such as Leadville and Durango, succeeded in becoming regional centers for mining and transportation. Leadville grew from a village of log huts in the late 1870s into a city of over fourteen thousand by 1880 with schools, churches, a waterworks, and several rail routes to Denver. Hard-rock miners, over one-third foreign born, brought ores containing silver, lead, copper, and zinc to the railway cars. Horace A. W. Tabor, a storekeeper, mayor, and mineowner, made enough money to divorce his wife, marry the famous "Baby Doe," and get himself elected lieutenant governor and later U.S. senator. Tabor went through his riches with dispatch, but his Leadville operations demonstrated the significance of technology and base metals to the mining economy. Gold and silver produced great wealth over short periods, but scientists and technicians showed how to make money from base metals too. Capital came to Leadville to dig new mines and erect smelters for permanent operations. The future of the western economy rested on its ability to attract investors for the long haul and to generate and reinvest its own capital.

Although only a few of the Old Californians traveled east to Colorado, thousands more headed north and west in the 1860s to British Columbia, Washington, Idaho, and Oregon in their quest for the Big Bonanza. A hearty band of prospectors pushed into British Columbia after gold had been found in 1855 near Fort Calville, an old Hudson's Bay Company outpost on the upper Columbia River. The find played out quickly, but some daring souls entered the Fraser River country farther north and discovered a new El Dorado two years later. Thousands of Californians entered British Columbia, only to be disappointed by the paucity of placer sites. While a few continued their quest northward along the Fraser, others moved south into Idaho and the Snake River valley. Elias Davidson Pierce tried his luck in British Columbia, then headed south to the Nez Percé Indian reservation in northern Idaho. He formed a trading partnership in the area, married a Nez Percé woman, and using that position, began to prospect on Indian lands. In 1860 he found gold on the Clearwater River, and despite the protests of the tribe, a rush of prospectors followed. The miners fanned out along the Clearwater and the Salmon rivers and to the south and began to stake claims in the rich placers of the Boise Basin in 1862. That summer a few of them abandoned their Idaho sites to join contemporaries in Oregon, where placers had been found on the John Day and Powder rivers. Because these placers were scattered and small in size, large-scale operations like the Comstock failed to materialize in the Inland Empire. The coming of the "Younder siders" from California did create a demand for farmers and more permanent residents, who initiated the process to organize a territorial government. By the mid-1860s, both Idaho and Montana had achieved territorial status.

As tens of thousands of miners arrived at Portland, Oregon, on their way eastward to the goldfields, that community expanded its role as a mercantile and shipping center. Merchants outfitted the prospectors, and local firms supplied horses, mules, and boats for those headed east. Within the Inland Empire, communities such as Walla Walla

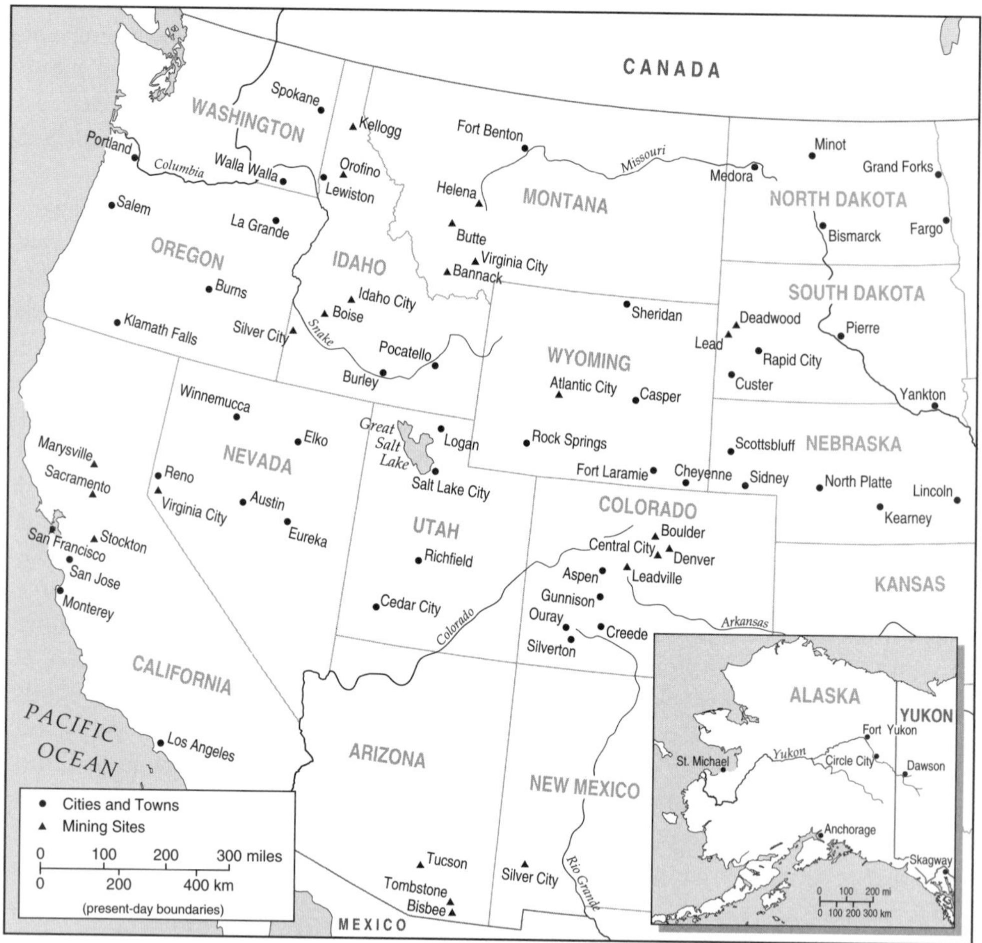

Primary Western Mining Sites, 1848–1900

flourished as more strikes were made. By 1863, Idaho City claimed six thousand residents, who flocked to its saloons, gambling parlors, and bordellos as well as to the stores, theaters, churches, and the hospital. Farmers received $5.00 for a chicken and sold butter for $1.20 a pound as Inland Empire agriculturalists tried to meet the needs of the miners. The Idaho mining communities filled with unskilled laborers, largely American born, but the merchants and skilled workers tended to be foreign-born veterans of other mining regions. The communities were ethnically diverse, and in the absence of widespread property ownership and large numbers of families, their populations often fell drastically, with some simply disappearing.

It would be hard to exaggerate the rootlessness of the miners and those who followed them. Prospectors who reached Idaho and Oregon had hardly wet their pans before news of strikes in Montana sent them headlong to the east. James and Granville Stuart found gold in the headwaters of the Missouri River on Gold Creek in 1862. Their sluice boxes produced modest dust, and the word spread even to Colorado. "Fifty-niners" from Pikes Peak roamed northward, finding a pocket of gold on a tributary of the Jefferson River.

The great Montana gold rush began the next year, with miners coming from the west along the military road from Walla Walla and others coming up the Missouri River from the east in steamboats. Gold on Alder Gulch produced Virginia City, a roaring camp of four thousand people, eight billiard halls, five gambling establishments, and many saloons and bawdy houses. The thirty million dollars in dust and nuggets found in the gulch almost doubled the riches extracted at Last Chance Gulch near Helena, but that town would outlast Virginia City. Helena lay astride the trail between Fort Benton, Bannock, and Virginia City, and the influx of merchants, bankers, and farmers gave it stability. Further, the presence of gold-bearing quartz required crushing operations leading to permanent facilities and large capital investment. Helena also benefited from strikes in Montana and Idaho in the 1870s and 1880s. As late as 1883, a rush in the Coeur d'Alene mountains of northern Idaho saw thousands of miners enter this remote region, only to depart in a few years as the gold played out. That gold rush did lead to discoveries of silver, lead, and zinc and the establishment of permanent operations for the extraction and smelting of base and precious metals in northern Idaho.

Some of the Coeur d'Alene miners, veterans of the Black Hills rush of a decade earlier, carried the scars of that bitter struggle. By the 1870s, clear evidence existed of the presence of gold in the Black Hills of South Dakota. Part of the Sioux Indian lands, the Black Hills fostered rumors and then reality as Indians came to Fort Laramie with nuggets and dust. The army sought to prevent a full-scale invasion of the Sioux reservation, but an expedition into the Black Hills, led by General George Armstrong Custer and intended to lay the rumors to rest, only confirmed what prospectors suspected: gold in extractable quantities existed throughout the area. By the spring of 1875, thousands of miners had gathered at Sioux City, Iowa, and other points around the Black Hills, and the army's thin line of troops could not deter them. Groups of miners penetrated the Black Hills, only to be removed by the soldiers. Escorted off the reservation, the prospectors reentered at other points. When the Sioux refused to relinquish their claims to the land, exasperated federal agents opened the Black Hills to the miners at their own risk. Over fifteen thousand prospectors swarmed in, and by the following April, Deadwood was home to seven thousand people. A wide-open, lawless, frontier town, Deadwood symbolized the last of the famous camps. After the placers and sluices removed the "easy" dust and nuggets, large-scale mining entered the Black Hills, and trained engineers and laborers again replaced the prospectors. The days of thousands of people racing from one place to another, having little or no hard information and enduring hunger, danger, and frequently death, had almost disappeared.

Sometimes the romantic legends of the camps reflected reality, and luck, as when the formerly penniless Irish immigrant Marcus Daly persuaded George Hearst, James Ben Ali Haggin, and Lloyd Tevis, San Francisco investors, to develop a silver mine near Butte, Montana. Armed with their mining venture capital, Daly began operations, only to discover in 1882 that the Anaconda mine had little silver but contained one of the largest copper deposits ever discovered. "The Richest Hill on Earth" would make millionaires of Daly and William A. Clark and add to the fortunes of Hearst and the others. The partners invested nearly four million dollars before producing a profit; soon nearly two thousand people occupied the company town of Anaconda. The largely immigrant laborers, guided by professional engineers and managed by Daly, extracted the copper ore, smelted it, and sent tons of refined copper east to voracious markets. Access to rail transport allowed Anaconda to develop a vast, high-grade copper deposit thousands of feet deep at the beginning of the age of electricity. In 1883, the United States became the world's leading producer of copper, with Montana being a major source of the metal. Anaconda Copper, like other mining giants in the West, depended on technology and science to increase output and reduce costs. A major factor in the rise of industrialization, copper from Montana contributed to the expansion of electrification in the United States and in western Europe.

The "Copper Kings" of Montana engaged in bitter disputes and lengthy litigation to establish control over not only copper output in Montana but also the state's economy and political structure. The copper magnates purchased their primary rival at Butte, consolidated operations, and increased profits. Output from Montana rose from 5,000 tons in 1882 to 176,000 tons in 1916. Anaconda invested huge sums in electrical generating plants, mining equipment, and smelting facilities and joined an international cartel in an effort to corner the copper market. Ultimately Anaconda came to be controlled by eastern investors linked to Standard Oil Company and the powerful National City Bank of New York City. Copper operations in Montana, and later Arizona, symbolized the shift to a highly concentrated and technologically sophisticated metals industry.

The extraction of gold and silver, as well as copper, lead, and other base metals, employed thousands of men across much of the trans-Mississippi West. For decades, mining represented the largest nonagricultural source of jobs in the region. As hard-rock mining supplanted placer operations, more miners became laborers paid by the day or week to work in gangs underground, where "human machines" with strong arms and backs harvested ore for the stamping mills. Life in the tunnels was hard, dangerous, and monotonous, characteristics that led to a highly mobile work force. Itinerants, or floating workers, became the norm as they moved from the silver mines of Nevada to the South Dakota goldfields and then to the copper ranges of Arizona. Some miners broke out of the mold and labored in the smelters or joined railroad construction crews, but many would return to the shafts and pits, only to move on again in a few months. Young, single, and mobile, living in company-owned boardinghouses or tar-paper shacks, many miners failed to settle down, form families, or develop roots. Nevertheless, the sense of adventure and the opportunity for employment attracted workers from around the globe to the mines of the West.

A key characteristic of the labor force was its ethnic and racial diversity. Mining corporations hired gangs of Chinese laborers from headmen, not to work in the shafts but to dump cars, remove surface rubble, cook food, and wash laundry. White miners refused to work with the Chinese in the tunnels, and management feared that the Chinese, unlike other groups, would quit if not paid on time. Nevertheless, the Chinese appeared in large numbers not only in California but also in Idaho, Nevada, and even Colorado. Many Chinese immigrants turned to placer mining, taking over claims abandoned by whites as unprofitable. Chinese and Hispanic miners often accepted the least-skilled and the lowest-paying jobs, the ones that white men did not want.

Working deep in the sweltering shafts of Virginia City's gold and silver mines, Timothy O'Sullivan used magnesium flares to make the first underground photographs of western mining. This image of collapsed timbers suggests the ever-present danger of the work, while the scale of the pick and boot suggest how cramped and claustrophic the shafts could be.

Timothy O'Sullivan (1840–82). Cave-in, Gould & Curry Mine, Virginia City, Nevada. *Albumen silver print, winter 1867–68. Courtesy Amon Carter Museum, Fort Worth, Texas.*

The highest levels of discrimination against the Chinese and the Hispanic workers often came from some of the immigrants who brought a sense of racial superiority and prejudice with them from Europe. Job competition simply exacerbated the racial views of those from Cornwall, Ireland, Wales, Germany, and eastern Europe. An ethnic pecking order developed in some of the mines as a result; the Cornishmen, Welshmen, Irishmen, and Americans labored in the shafts while the southern and eastern Europeans loaded and pushed the cars. Yet all who worked in the mines saw a glimpse of hell.

The tunnels reeked of blasting powder, unclean bodies, rotten timber, and human waste, since there were no toilets. The deeper the shaft, the higher was the temperature and the greater the danger of collapse or flooding. Electricity brought improved light as well as bare copper wires, huge clouds of dust from power tools, and elevators that could become death traps. Mineowners in the West, not unlike those who controlled mines in the East, the United Kingdom, and continental Europe, moved slowly, if at all, to install safety devices and open-air vents and to improve timbering. As the historian Rodman Paul has noted, the absence of concern for the welfare of workers could be found throughout capitalistic economies in the nineteenth century.

The extraordinarily threatening conditions of work within the mining industry led to the first labor action, or strike, in the Comstock Lode in 1864. The Miners' Protective Association sought higher wages, shorter hours, and better working conditions—"bread and butter unionism"—that created a sense of brotherhood among the miners. The union used monthly dues to help the injured or the families of those killed in the mines. It wanted independent hospitals, free from company control, and built union halls as alternatives to the saloons and brothels. The demands of the association, its charter and ideals, spread among the mobile work force. In 1892, miners struck at Coeur d'Alene, Idaho, demanding recognition for their union, and when state and federal troops arrived to break the strike, seven miners died for their cause. The following year the center of union activity shifted to Butte, Montana, where the Western Federation of Miners (WFM) made that city the "Gibraltar of Unionism." The WFM unified local labor organizations in an effort to stop intervention by the military in disputes. Within a decade, the WFM had over fifty thousand members across the region and would serve as a starting point for the more militant Industrial Workers of the World (IWW). The mining corporations and their managers inspired the rise of trade unionism by refusing even to discuss basic issues with their employees and by importing "scabs," calling for the national guard, and seeking injunctions to terminate strikes.

Labor-management relations varied widely from one mining center to another. Some miners joined militant anticapitalist unions, but in other camps, labor remained virtually docile. The WFM elected members to public office in the Comstock, and the miners at Silver City, Idaho, won settlements for better wages and safer working conditions under the leadership of the future militant William "Big Bill" Haywood. But in some Idaho and Colorado camps, class warfare raged. Driven to desperation by the companies, laborers retaliated with dynamite and rifles. Although few California mines unionized, the IWW would later create a virtual commune at Goldfield, Nevada. Even in Butte, the unionists were split between socialists and "bread and butter" Democrats courted by the mineowners Clark and Daly, who wanted their votes. Life for workers in the mining camps of the West differed little from that of miners in the industrial towns of Pennsylvania, the English midlands, and the Ruhr Valley. When corporate mining broke down the relationship between prospectors and townspeople, a polarized labor and management resulted. When absentee ownership collided with demands for the end of paternalism and for an improved standard of living, miners in the West unionized or responded with violence. Mining created wealth, generated significant labor problems, and laid the basis for an expanding economy, especially for a transportation network.

Freighting on the Prairies, Steamboating on Western Waters

Strikes, violence, and militant unionism could not have been further from the minds of the forty-niners headed to the California goldfield as they stood on the docks in New

York, Philadelphia, and Boston, begging ship captains to take their money. The discovery of gold created an enormous demand for transportation and placed a heavy burden on existing shipping facilities between the coasts. Before the War with Mexico, vessels from the United States plied the coast of California. The ships, largely from New England, carried coffee, sugar, metal implements, cloth, shoes, and spices around Cape Horn and into the Pacific. In California, ship captains traded for cattle hides and tallow. Californians greeted each white-sailed clipper with a fiesta, rowing small boats out into the harbor to initiate trading. Voyages of a year to eighteen months could generate profits of 300 percent, unless the straits at Cape Horn swallowed up the vessel.

Shipping companies and their captains knew the routes, and the risks, and after 1848 entered the newly found business to California with a vengeance. The U.S. government initiated mail subsidies to California in 1847, and the United States Steamship Company opened a new route to San Francisco with an overland crossing of the Isthmus of Panama. Cornelius (Commodore) Vanderbilt joined the fray with steamship lines in the Caribbean and the Pacific linked by a land route across Nicaragua. Would-be prospectors and the mail contracts produced demands for faster, and safer, transportation to California. The circuitous routes took thirty days or longer, and the ship lines charged twelve to eighty cents per ounce for letters. A railroad across Panama did open in 1855, reducing the time and dangers of the long voyages. Yet these steps still did not address the issue of cost, with passenger fares from New York to San Francisco averaging over five hundred dollars. Furtive efforts at a mail route through Mexico at the Isthmus of Tehuantepec made clear to Congress that land routes across the Great American Desert were an absolute necessity. An overland route had to be established as freight and passenger traffic to California soared.

Even as demands for an overland route grew louder, shipping firms established a triangular trade across the Pacific. After unloading passengers and cargo in California, ship captains had nothing to haul back to the East Coast. They thus loaded masts, spars, and lumber, which they carried to China and Japan. There they acquired tea, spices, rice, and Chinese goods and set sail for Cape Horn and ultimately New York, London, or Boston. As early as the 1850s, California had entered a global capitalistic economy. Trade with Asia heightened demands for an overland route to the West Coast.

Visionary publicists, political orators, and the promoters of Kansas City, Memphis, St. Joseph, Omaha, and Fort Smith lobbied for the creation of a national road to California, to begin in their town or congressional district, of course. Seventy-five thousand Californians signed a petition in 1856 demanding that Congress establish an overland mail route to the Far West. Clearly, existing services were inadequate. The Brigham Young Express and Carrying Company operated monthly from Independence, Missouri, to Salt Lake City, where a connection reached San Francisco. A military express line extended from Independence to Santa Fe, in the New Mexico Territory, and a monthly mail service linked San Antonio to San Diego. These operations were neither dependable nor inexpensive. As pressure mounted, Congress removed responsibility for road construction from the U.S. Army Topographical Engineers to civilian contractors and authorized improved wagon routes from Nebraska to California over South Pass, from El Paso to Fort Yuma, and from Albuquerque to Needles, California. The Department of the Interior assumed responsibility for the new road work, which was carried out largely by the mail contractors. Using packtrains to carry the mail or handing letters to a freight-wagon driver at St. Joseph offered no guarantee of safety or speed; service had to be improved. The federal commitment to improved transportation in the West accelerated rapidly in the 1850s.

After Congress appropriated an annual subsidy for mail service from the Mississippi River to San Francisco, the leading express firms in the East formed the Overland Mail in a united effort to meet the required delivery schedule of twenty-five days to California. In the midst of a sectional dispute over the expansion of slavery, Congress approved routes from both the North and the South, a compromise beneficial largely to westerners.

The postmaster general, a Tennessean, awarded the mail contract to John Butterfield and William G. Fargo, of New York, experienced express operators. A founder of the American Express Company, Butterfield proposed a route from a rail connection in Missouri, south and west through Fort Smith and El Paso, to Yuma and San Francisco. When howls of protests from the North greeted this decision, the postmaster general defended the southern route as "all weather." Butterfield spent a year building stations along the trail, training drivers, purchasing mules and horses and coaches, and finding agents to operate the stations. He soon had invested more than one million dollars. Service began when a heavy-wheeled stagecoach, its body swaying on leather braces, raced to the West Coast in twenty-four days, arriving on 10 October 1858. The east-bound coach to St. Louis took three fewer days. President James Buchanan telegraphed Butterfield, "It is a glorious triumph for civilization and the Union."

The Overland Mail Company linked the East to the West for three years, offering service twice each month. At the stations, passengers received meals of fried beef or pork, beans, and bread while the agent changed the four horses or six mules for the next leg of the twenty-eight-hundred-mile journey. Blizzards, hostile Indians, and the hellish ride did not deter passengers, who paid two hundred dollars for the adventure. Government subsidies, the mail contract, and passenger fares led to profits for Butterfield and his partners. The postmaster general found the service so reliable that he issued additional contracts for routes from Kansas City to Stockton, California, and from San Antonio to San Diego.

Butterfield, Fargo, and Henry Wells dominated American Express, a powerful firm in the East and Midwest, and now they had a link to California. But even before the opening of the Overland Mail, the Golden State offered prospects as well, and they had challenged the leading firm there, Adams and Company. The enterprise they formed, Wells, Fargo and Company, dramatically reduced rates, driving Adams and Company into bankruptcy in 1855. Wells, Fargo and Company dominated internal transportation in California, with lines from San Francisco to the gold camps of the Sierra Nevada and north toward Oregon. In addition, Wells, Fargo and Company entered the banking business, an increasingly profitable enterprise. The coming of the Civil War in 1861, however, shifted the mail service to the more central California Trail route.

The war led to the consolidation of western operations as the government demanded more frequent and faster service to California. The Overland Mail Company, dominated by Wells and Fargo, established a virtual monopoly in the coaching and freighting business west of the Mississippi River after the failure of two of the greatest names in overland transportation: Russell, Majors, and Waddell; and Ben Holladay.

The acquisition of the Mexican Cession in 1848 meant that the U.S. Army had to establish forts and garrisons throughout the region to protect settlers and traders. The discovery of gold filled the trails with prospectors, and the central routes soon saw caravans of canvas-topped Conestoga wagons heading to California and to the Oregon country. The needs of the military reached mammoth proportions as troop contingents increased in size to protect the trains. Flour, sugar, salt, feed grains, livestock, and military equipment had to be delivered safely and quickly. The Bureau of Indian Affairs guaranteed food, cloth, and other commodities to its wards on the reservations, and the bureau too required considerable transportation. Government alone would have generated vast freighting operations, but the discovery of gold in California, Nevada, and Colorado only exacerbated demand. The response to this need came from large freighting firms, heavily subsidized but with access to substantial capital.

Professional freighters accepted cargoes from the steamboats or railroads at Independence, St. Joseph, Kansas City, Atchison, Omaha, or Nebraska City. These border towns received goods from St. Louis or later Chicago, consigned to the Far West. An experienced trader, Alexander Majors, entered the tough, competitive freight business using oxen rather than mules. The oxen could forage on the plains, he argued, but mules could not, and six yoke of oxen could pull a wagon loaded with six thousand pounds of freight. In 1855 he formed a partnership with William H. Russell and W. B. Waddell, and within three years the firm of Russell, Majors, and Waddell operated thirty-five hundred wagons and employed four thousand men. Armed with a two-year contract to supply the military garrisons in the Southwest, the firm monopolized trade in the region. Extraordinary profits from freight tariffs and subsidies allowed the firm to expand and then to enter stagecoaching, which led to its downfall.

Russell tried to persuade his partners to operate a stage line over the central route, but they refused to join this scheme without a federal subsidy. Russell forged ahead, opening the route to Denver and then on to Salt Lake City, where his line made a connection to California. The operation failed, without a subsidy, and fearing that Russell's collapse would bring down their firm, his partners rescued him. They took over the stage line as part of Russell, Majors, and Waddell, a tragic error.

Faced with a desperate situation, Majors and Waddell followed Russell into yet another ill-conceived scheme, the Pony Express. Hoping to win a federal mail contract for their central route and to prove its superiority, the partners sought to create a means to carry the mail far more rapidly than the Butterfield route. Relays of young riders could move the mail to California in ten days, they reasoned, and they set about establishing 190 stations at ten-mile intervals from St. Joseph to San Francisco. They spread five hundred fine horses along the route, and on 3 April 1860, riders left each terminal at full gallop. Soon eighty riders were in the saddle every day, forty in each direction, tossing the mail pouch to each other at the end of a seventy-mile leg of the journey. The Pony Express was efficient, tightly organized, and very romantic, but hardly profitable, lasting only two years. The Pacific Telegraph Company and the Overland Telegraph Company set the first pole for a transcontinental line in July 1861, and on 24 October, telegraphers tapped out the initial message. Technology killed the Pony Express and helped drive Russell, Majors, and Waddell into bankruptcy.

For several years the firm had received infusions of capital from Ben Holladay, and in 1862 he foreclosed, taking over the leading freight line in the West. Although the Overland Mail Company had won the subsidy for the central mail route, Holladay, the coarse frontiersman, put together a huge freight and coach operation out of the ruins of Russell, Majors, and Waddell. Holladay possessed cunning and managerial skills and recognized the needs of the mining centers far off the main trails. He extended coach lines into Idaho, Montana, and Colorado, becoming the "Napoleon of the West." His five thousand miles of stage routes and twenty thousand wagons produced good profits as he ruled the firm in an absolute manner. The early settlers, miners, and the army depended on his system, but the rates, based on distance and weight, kept costs high and precluded substantial expansion of the regional economy. Holladay purchased surplus army wagons and animals after 1865, moved them west, and extended his freight services. Always shrewd in his business affairs, Holladay saw the effects of the expansion of the railroads and in 1866 sold out to Wells, Fargo, and Company, which had already acquired Butterfield's western interests. The coming of the railways soon relegated the coaches and the freight wagons to secondary routes, but even after 1900 some of these operations continued in remote areas in the region.

Although Wells and Fargo and other investors brought capital for freighting and coaching lines from the East, much of the investment in these transportation services initially came from the Missouri Valley. Families, longtime business associates, and independent draymen formed partnerships to enter the risky and yet often profitable activity. Independent operators, such as Madame Canutson, a female freighter in the 1880s, vied with the larger companies for mercantile traffic when they could not win subsidies or army contracts. From their terminals along the Missouri River, the freighters kept the lifelines open not only to California and Oregon but also to the merchants of Santa Fe, Tucson, and Virginia City as well as to military outposts from Canada to Mexico. This land transportation system often competed with the more efficient steamboats that penetrated western rivers and bays.

Steamboats brought goods to the freight company docks along the Missouri River, carried cargoes up the Columbia River to the Inland Empire, plied Puget Sound with timber and passengers, and transported miners across San Francisco Bay and along its tributaries. In 1859 the Chouteau family of St. Louis constructed a shallow-draft vessel, the *Chippewa*, and steamed up the Missouri almost to Fort Benton, Montana. The discovery of gold in Montana brought seventy boats in 1867, carrying freight for the miners and military goods for the army outposts. As rail lines reached the Missouri River's eastern bank, the point of departure shifted from St. Louis to Omaha and then to Sioux City. Boats on the upper Missouri successfully competed with overland freighting firms.

In California, rankled that the oceangoing ships at their docks were owned by New Yorkers, San Franciscans determined that transportation on the Bay would be locally controlled. Initially a small fleet of sloops and schooners plied the Bay and moved upriver to Sacramento, but the introduction of steamboats revolutionized shipping. The owners of the steamboats formed a monopoly in 1854, the California Steam Navigation Company, a joint-stock firm capitalized at $2,000,000. Since individual steamboats had been owned in shares by builders, merchants, captains, and investors, the new company had a wide-ranging list of stockholders. The firm owned nearly every major vessel linking San Francisco to the interior and over the next two decades earned good profits, paying more than 100-percent dividends on par value stock by 1860. Cash dividends of $225,000 in 1870 enhanced the pocketbooks of the owners and generated capital to be reinvested in mines, factories, railroads, and banks. The Bay steamboats remained a locally owned enterprise and a source of investment capital.

Similarly, the Columbia River served as a water highway into the Inland Empire. The Hudson's Bay Company had initiated river traffic with a fleet of barges transporting furs to Fort Vancouver, and steamboating came to the region in 1850 when the *Columbia* began to ply the Willamette River from Astoria to Oregon City. Transit of the Columbia would not prove so simple, given the rapids at the Cascades and at The Dalles. The solution was the construction of portage tramways around the rapids, with steamers positioned on either side. The *Jennie Clark*, a stern-wheeler, showed that a shallow-draft vessel could navigate the boiling waters of the lower Columbia. Such boats moved upriver from Portland to the Cascades, where freight would be transshipped around the rapids to an awaiting vessel, which moved on to The Dalles. Another portage, more unloading and reloading, and yet a third ship carried the cargo and passengers to a point nearly thirty miles from Walla Walla for transshipment first to wagons and later to a local railway. In 1862, John Ainsworth, Jacob Kamm, and Simeon Reed joined with R. R. Thompson to form the Oregon Steam Navigation Company. Combining their vessels, these men created a virtual monopoly on the river. They replaced the portage roads with steam railroads and added more efficient vessels. At the height of the Idaho mineral rush, the firm earned over $780,000 amid loud complaints by shippers and passengers about high rates. The alternatives, freight wagons and coaches, proved even more expensive. The answer to shipper complaints would come not through state intervention for lower rates but from the railroads, for soon iron rails extended not only up the gorge of the Columbia but also across the plains of Kansas and over the Sierra Nevada. The day of the railroader was at hand.

Railways and Railroaders

"Manifest destiny," economic ambitions, nationalism produced by the Mexican War, a romantic infatuation with "a passage to India," and the lure of the China trade captured the imaginations of many Americans in the 1840s. The New York merchant and China trader Asa Whitney proposed to Congress in 1845 that the federal government grant a strip of land sixty miles wide, from Lake Superior to the Oregon country, to a firm willing to construct a railway to the Pacific Ocean. When Congress failed to respond, Whitney organized a propaganda campaign, and for the next decade he lobbied for an "iron path" to capture the trade of the Far East and to fulfill the national destiny. Residents of California and the leaders of Chicago, St. Louis, and Duluth supported his pleas and bombarded Washington with petitions. Whitney's pleadings were echoed by ambitious politicians such as William Gilpin, who told an audience in Independence,

Documenting the construction of the Union Pacific tracks, Andrew Joseph Russell paused in Granite Canyon, Wyoming, to photograph a train loaded with covered wagons. Few images so eloquently suggest the ways in which technology, in particular the railroad, transformed life in the American West.

Andrew Joseph Russell (1830–1902). Train on Embankment, Granite Cañon. *Albumen silver print, 1868–70. Beinecke Rare Book and Manuscript Library, Yale University, New Haven, Connecticut.*

Missouri, in 1849, that the East was holding the West in bondage by its failure to support a railroad to the Pacific. Although Whitney and Gilpin might be dismissed as dreamers, Senator Thomas Hart Benton, of Missouri, could not be so categorized, and he carried the message to the corridors of power in Washington: the federal government must authorize a transcontinental railroad. The propaganda and the oratory received greater attention as California gold flowed into the banks of New York, Philadelphia, and Boston; finally Congress responded.

From the beginning of the nation, the federal, state, and local governments aided transportation developments. Funds were directed to the construction of post roads, canals, and coastal waterways. The Erie Canal and the National Road reminded the public of the role of government in creating a transportation system. Federal agencies provided surveys, mail subsidies, and road engineering while state and local governments offered subsidies and land for rights-of-way. The result was the rise of a substantial transportation system in much of the East and Midwest by the 1850s. Water transportation along the coasts carried products brought to the ports by the railroads and canal boats. In the interior, roads and canals supplemented an expanding rail network. The shift from a north-south commercial market based on the Ohio and Mississippi rivers to an east-west trade using the new trunk rail lines only heightened tensions generated by the issues of slavery and slavery expansion in the 1850s.

Members of Congress largely agreed that a transcontinental railroad had to be built, but where would the eastern terminus lie? Senator Stephen A. Douglas contended that only Chicago would do, whereas Jefferson Davis argued for Memphis or perhaps New Orleans. Acting in its time-honored fashion, Congress asked for a study of all feasible routes, thereby postponing a decision. In 1853 the army initiated the surveys leading to a thirteen-volume report two years later. Four routes were possible: Lake Superior to Portland; the Overland Trail to San Francisco through South Pass; the Red River westward to southern California; and southern Texas to San Diego via El Paso and Yuma. Thus the thorny political issue remained with alternative surveys originating in both the northern and the southern states. Although Douglas triumphed in obtaining territorial status for Kansas and Nebraska to enhance the chances of the central route, he offered a compromise: to build three transcontinentals—one from the North, one from the South, and his own favorite. The staggering sums required to finance three such schemes deterred even the most vocal proponents. It became obvious to all that this gigantic enterprise required loans, land grants, and financial guarantees of enormous magnitude.

The secession of the Confederacy in the spring of 1861, and a Republican party pledge to link the East to the West, gained passage of the Pacific Railroad Act. The catalyst for this action, Theodore D. Judah, had overwhelmed senators and representatives with charts, graphs, engineering drawings, energy, and determination. An engineer by training, Judah left San Francisco and arrived in Washington in the spring of 1861; he was an emissary of Leland Stanford, Mark Hopkins, Collis P. Huntington, and Charles Crocker, San Francisco and Sacramento merchants and bankers. They proposed to construct the Central Pacific Railroad from San Francisco across California to the Sierra Nevada, enter Nevada, and strike eastward to meet a line beginning on the Missouri River. The audacity of the promoters of the Central Pacific Railroad, the persuasiveness of Judah, and the need for transportation to the West in time of war won the day. Congress approved the proposal and incorporated the Union Pacific Railroad to build the link between the Missouri River and the Central Pacific. Lines from Sioux City, Leavenworth, Kansas City, Omaha, and St. Joseph would converge near the one-hundredth meridian before heading to the West, thus ending the furor over the location of the eastern terminus.

The act provided for seemingly substantial subsidization for the project. The railroads would receive a four-hundred-foot right-of-way, ten alternate sections of land for each mile of track, and first-mortgage loans of sixteen thousand dollars per mile in flat country, thirty-two thousand dollars in foothills, and forty-eight thousand dollars in the mountains. Yet even these inducements failed to generate private capital sufficient to initiate work on the Union Pacific. Congress responded in 1864 by doubling the land grant, reducing the government loan to second-mortgage status, and increasing the number of shares of one-hundred-dollar par value stock tenfold, to one million. Thus the scheme began with federal underpinnings in line with land grants to some canals and railroad projects in the East.

Westerners, meanwhile, enthusiastically watched the penetration of the Missouri River valley by the railroads. The Hannibal and St. Joseph reached its western terminus in 1859, and the Missouri Pacific Railroad linked St. Louis to Kansas City in 1865. In Iowa, railroad fever rose to epidemic proportions. The Chicago and North Western reached the Mississippi River in 1855 and pushed on across Iowa to Council Bluffs on the Missouri a dozen years later. The Chicago, Rock Island and Pacific threw the first span across the Mississippi and gained access to Council Bluffs in 1869. With substantial capital from New England, the Chicago, Burlington and Quincy connected its namesake cities and marched westward to Fort Kearney, Nebraska, by 1873. The race to join Chicago with the incipient Union Pacific saw the Illinois Central construct a western branch from Dubuque to Sioux City by 1870. The Union Pacific would have no shortage of connections to the East; what it initially lacked was a massive infusion of capital and material.

Engineers and surveyors plotted a line westward from Omaha across the plains of Nebraska even as company managers competed with military demands for iron rails, locomotives, rolling stock, and labor as the Civil War intensified. When the war ended, track extended only forty miles, but the termination of hostilities brought an influx of rails and equipment as well as gangs of Irish tracklayers. Materials came upriver until 1867, when the Chicago and North Western arrived in Council Bluffs across the Missouri River from Omaha. While the surveyors faced Indian raids and the challenges of the weather, loneliness, and thirst and sought the easiest grades for the roadbed and supplies of water for the locomotives, construction materials piled up in the yards at Omaha. From Boston and New York came urgent telegrams from company executives pleading for more track to be laid, for each new segment of line meant bonds sold and land acquired.

Chief Engineer Grenville Dodge depended on John Stephen (Jack) Casement and his brother Dan to get the track laid, and they, in turn, depended on Irish laborers, many of whom were veterans of the war recruited by the Casements in the East. A brigadier general when the war ended, Jack Casement knew how to mobilize troops, and he used that experience and years of work in railroad construction in New York and Ohio to organize the Union Pacific. He took the "Casement Army" into the field while brother Dan served as quartermaster general in Omaha. Crews of ex-soldiers followed the surveyors, grading the earthen roadbed in one-hundred-mile segments and building wooden bridges. Casement's tracklayers followed, pulling rails from flatcars pushed by a locomotive as crews placed ties on the roadbed. They laid four lengths of rail per minute and in the first six months pushed the end of the track 250 miles. The object was not to construct the best or most efficient route but to move quickly, at the least expense. The trackage was not ballasted with rock, the bridges were often flimsy, and the ties were generally untreated and frequently of poor quality. In 1866 and 1867 the toiling crews drove the shining iron over 500 miles, reaching Cheyenne, Wyoming, in November.

The incredible expense staggered even the most optimistic of the Union Pacific's leaders, and government loans failed to approach even minimal construction costs. Dr. Thomas C. Durant, Oliver and Oakes Ames, and their friends who controlled the line turned to a construction company to raise the necessary funds. The Credit Mobilier, secretly owned by the Union Pacific's leaders, received contracts far in excess of costs to construct the railroad. The Credit Mobilier's profits attracted investors, who ultimately earned over $20,000,000 while creating a railway whose capitalization of $110,000,000 was nearly double its value. "Watered stock," internal corruption, bribes to members of

Congress and the executive branch, and managerial clashes did not deter the Casement Army or the surveyors as crews reached Utah and looked westward. Indeed, the survey teams passed their counterparts from the Central Pacific as both firms sought to lay as much track as possible in order to obtain land and sell mortgage bonds.

The "Big Four" of the Central Pacific—Crocker, Hopkins, Huntington, and Stanford—had gotten their project off to an earlier start after securing a loan of $1,659,000 from the State of California and untruthfully convincing President Abraham Lincoln that the Sierra Nevada began almost at Sacramento, so that the federal subsidy was forty-eight thousand dollars per mile, not sixteen thousand dollars. Using their mercantile contacts in the East, they ordered equipment and rails sent around Cape Horn to San Francisco and upriver to Sacramento. Lumberjacks cut trees in the foothills for ties and trestles, and hundreds of Chinese laborers threw up embankments and graded the right-of-way. Tracklaying began in 1863, but the Central Pacific extended only 115 miles four years later as the construction gangs entered the mountains. Not even Judah anticipated the deprivations and hardships to be faced by Central Pacific employees in breaching the mountains. Seven thousand, then ten thousand, pig-tailed Chinese "coolies" struggled against the rocks, snow, and terrible winds of the Sierra Nevada. Landslides tore away work that had taken weeks or months. Blizzards buried the roadbed and many of the workers, forcing the company to build huge snowsheds to protect the line. Chinese workers swung in baskets suspended by ropes, drilling holes for dynamite in the face of sheer rock walls. They collapsed from heat and humidity deep in tunnels driven through solid granite. Hundreds died, and still they came, recruited in China, brought to California, and taken by rail to the end of the track. They persevered. In the summer of 1867 the Central Pacific crossed the crest of the Sierra Nevada, and the laborers moved swiftly downgrade toward the deserts of Nevada. As Casement's Irish gangs drove westward, the Central Pacific's Chinese raced them to a connection.

As the grading crews passed each other in northern Utah, Congress intervened and ordered the rails joined at Promontory. On 10 May 1869, workers and officials observed the placement of the last tie, laurelwood bound in silver, and the driving of the last spike, fittingly made of gold. They groaned as Stanford missed with the swing of the sledge, but the spike pierced the laurel, and the locomotives moved forward, pilots touching. A telegrapher tapped out, "It is done." The nation, so recently torn by bloody civil war, stood united, east and west.

The construction of the transcontinental railway made millionaires of a few, and some, such as Huntington, reinvested those profits in the region, further stimulating economic growth. Huntington rose from poverty in Connecticut, where he was born, leaving school at fourteen, becoming a traveling salesman, and joining several partners in a hardware store in rural New York. Gold fever took him to California in 1849. Buying and selling commodities in Panama en route gave him five thousand dollars in capital when he arrived in San Francisco. Dry goods, not mining, proved to be his game, and a partnership with Hopkins led to a highly successful hardware business. When the Big Four formed the Central Pacific in 1861, Huntington traveled to Washington to represent the firm in the halls of Congress. Huntington lobbied hard, misrepresenting

Andrew Joseph Russell's photograph of the joining of the transcontinental rail routes at Promontory, Utah, on 10 May 1869 celebrated the triumph of American labor and technology over the vast, often inhospitable, stretches of the western landscape. Excluded from the celebratory image, however, were the Chinese laborers who did the back-breaking work of building the Central Pacific's tracks through the Sierra Nevadas. Joseph Becker, a staff artist for *Leslie's Illustrated Weekly Newspaper*, traveled west aboard the first cross-Rockies Pullman train in 1869 and sketched Chinese workers during a six-week stay in California. Here, as Chinese laborers shake their fists at a train emerging from a snowshed, it is unclear whether they cheer their accomplishments or jeer at the system that subjected them to such difficult labor.

Andrew Joseph Russell (1830–1902). Dodge and Montague Shake [Golden Spike Ceremony]. *Albumen silver print, 1869. The Bettmann Archive, New York, New York.*

Joseph Hubert Becker (1841–1910). Snow Sheds on the Central Pacific Railroad in the Sierra Nevada Mountains, May 1869. *Oil on canvas, ca. 1869. From the collection of the Gilcrease Museum, Tulsa, Oklahoma.*

the length of the railroad and the terrain it occupied, generally following the "ethics" of the day. After the roads were joined at Promontory, he determined to create a rail and steamship empire in California and the West.

Huntington had already purchased independent short-line railroads in California, and he formed the Southern Pacific Railroad as his primary vehicle for expansion. He moved to make the Southern Pacific the major segment of the railroad that Congress had authorized from southern California to Texas. He pushed rails south from San Francisco to Los Angeles and then eastward into Arizona. Clashing with Thomas A. Scott, of the rival Texas and Pacific Railroad, Huntington crushed Scott and drove the Southern Pacific eastward to Yuma, Tucson, El Paso, and on to San Antonio. Even as he reached Texas, Huntington built north from San Francisco, creating a line to Portland. By 1890 his empire stretched from Oregon to Louisiana; railroads, timber, land corporations, ferries, and coastal shipping entered his grasp.

But like others in the West who accumulated vast wealth, Huntington also transferred capital to large-scale investments in the East. As early as 1869, Huntington led a syndicate that seized control of the Chesapeake and Ohio Railroad and rebuilt and expanded that line from tidewater at Newport News, Virginia, into the coalfields of West Virginia. Huntington promoted Newport News as a major port for the export of coal, and he formed a dry dock and shipyard company, making the port one of the largest on the East Coast. Millions of dollars earned in the West migrated eastward to finance railways, terminals, and shipyards. The flow of capital in nineteenth-century America was indeed two ways.

Capital from Boston financed Huntington's major rival in the Southwest, the Atchison, Topeka and Santa Fe Railroad. Founded in Kansas in 1859 by the promoter Cyrus K. Holliday, the company began as an effort to link Atchison, and then Kansas City, to Santa Fe. Financed by Boston investors, construction proceeded to Newton and then Dodge City, where trail herds from Texas created substantial traffic. The company encouraged the migration of farmers to Kansas, and after Russian immigrants introduced hard winter wheat, that commodity generated vast traffic. The Bostonians secured William Barstow Strong as the railway's president in 1877, and under his leadership the firm pushed a line north from Pueblo, Colorado, to Denver, then south toward Santa Fe. Strong and his stockholders abandoned the idea of Santa Fe as the end of the track and, together with the St. Louis and San Francisco Railroad, gained control of the Atlantic and Pacific Railroad, authorized by Congress to construct a line to California. Strong first built to Deming, New Mexico, where a connection with the Southern Pacific formed a through route to Los Angeles. Determined to control its own destiny, the Santa Fe then built a line across Arizona into southern California, reaching Los Angeles over its own tracks by 1887. The Santa Fe acquired more than thirteen million acres of land in Arizona and New Mexico through the Atlantic and Pacific's land grant, but like other carriers, it received the bulk of its capital from private investors, not land sales. Strong extended the company's lines to Chicago and built south from Kansas to connect with the Gulf, Colorado and Santa Fe Railway, giving the firm access to Fort Worth, Dallas, Houston, and Galveston. Acquisition of the Colorado Midland and the St. Louis and San Francisco, combined with headlong expansion, led to bankruptcy in 1893. Reorganization in 1895 created a new entity that proved highly profitable, becoming a major railway in the Southwest. The Bostonians lost control of the carrier to investors from New York and abroad, but both groups sought developmental rather than opportunistic results.

The early literature on capital investment in western railroads emphasized the voracious greed of "robber barons" who fleeced "widows and orphans," constructed poorly built lines across huge federal land grants financed by watered stock, and overcharged farmers and small merchants. Such a simplistic analysis ignored the facts that only a very small percentage of track was laid over land grants, that purchases of railway securities involved great risks for any investor, and that lines across barren prairies did not generate sufficient revenue for years. Arthur M. Johnson and Barry E. Supple have shown that investors had mixed motives. "Developmental" investors committed capital to the railways with the expectation of long-run economic returns as the region matured. Other investors purchased securities for the purposes of manipulation, or for "opportunistic" profits. Whereas the Credit Mobilier clearly represented the latter, the Bostonians who invested in the Santa Fe and the Chicago, Burlington and Quincy saw vast prairies filled with wheat farms; they hoped for the success of Denver's smelters, and they sent their sons and nephews to Los Angeles to purchase real estate, open banks, and construct office buildings and hotels. It has recently been argued that even Jay Gould saw his western investments as developmental in nature.

In a revisionist biography of Gould, Maury Klein contends that after Gould gained control of the Union Pacific, the Missouri, Kansas and Texas (Katy), the Wabash, and the Missouri Pacific railways, he rebuilt and revitalized the properties, sought economic expansion along their routes, and operated the carriers for long-term return rather than for the manipulation of securities. Born on a farm in Delaware County, New York, in 1836, Gould overcame a frail, sickly childhood and entered the business world in 1852. Possessing a quick mind, perseverance, an indomitable will, self-control, and a strong practical bent, he turned his talents first to tanneries and then to Wall Street. Mastering the intricacies of the financial world, he gained a seat on the board of the Erie Railroad, allying himself with Daniel Drew and James Fisk against Cornelius Vanderbilt. The famous "Erie War" won for Gould the sobriquet "The Mephistopheles of Wall Street," yet he tried to make the Erie a trunk-line carrier between New York and Chicago, Klein argues.

Similarly, when Gould gained control of the Union Pacific after the Credit Mobilier scandal and near bankruptcy in 1873, he revitalized the demoralized firm. Within two years the railroad was paying dividends as Gould revamped internal management, formulated lower rates, and labored to create new industries along its routes. Acquiring an encyclopedic knowledge of the Union Pacific, Gould petitioned the federal government to settle a substantial body of issues relating to its charter, the land grant, and the status of the second-mortgage bonds. All the while he developed the company's properties, opened coal mines, and encouraged agricultural settlements. The Union Pacific built branches into Utah and Colorado, planned lines to Idaho and Montana, and purchased the Kansas Pacific running from Denver to Kansas City. By 1878, Gould transferred his ambitions from the Union Pacific to the Wabash Railroad, the Denver and Rio Grande, and more important, the Missouri Pacific and the Missouri, Kansas and Texas. The latter two companies gave Gould a system extending from St. Louis and

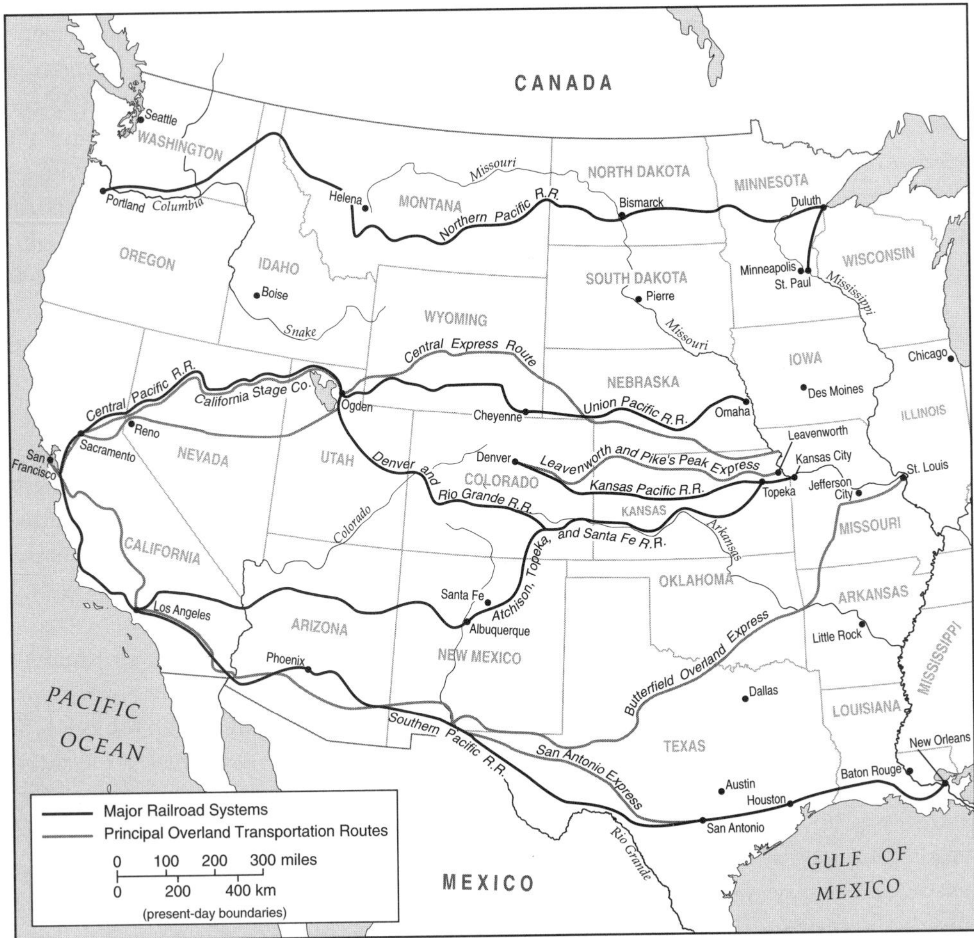

Principal Overland Transporation Routes

Kansas City south and west into Texas and Colorado. The acquisition of additional carriers took the Gould lines to Houston, San Antonio, and Galveston. Klein contends that Gould followed the same pattern with these firms: he purchased new equipment, rebuilt trackage, and modernized the carriers while encouraging economic development along their routes. Through battles with Huntington, the Santa Fe, and the federal government, Gould fought to build or acquire new routes in the West and the Southwest. Though attacked by the press as a scoundrel and seen by the public and later generations of historians as a rapacious manipulator, Gould used his railroad empire to enhance the economic development of large areas of the West.

Even as Gould created his rail system in the Southwest, a young German immigrant, Henry Villard, put together a transportation network in the Pacific Northwest. Born in Bavaria in 1835 as Ferdinand Heinrich Gustav Hilgard, Villard immigrated to the United States in 1853, changed his name, and settled in Belleville, Illinois. Learning English, Villard read law, edited a newspaper, and toured Colorado in the late 1850s and 1860s. Suffering from poor health, he returned to Germany in 1871, where he became involved in financial matters with a group of investors in the Oregon and California

Railroad (O&C). This line, formed by Ben Holladay, extended south from Portland toward California. When the O&C defaulted in 1873, Villard traveled to Oregon to represent the German bondholders. Holladay failed to reorganize the O&C successfully, and Villard persuaded him to return that property, another railroad, and the Oregon Steamship Company to the creditors. Villard assumed leadership of these firms and invested much of his own capital in them, for he saw many opportunities for economic development in the region.

Seeking to enhance Portland's position in the Pacific Northwest, Villard sought transportation links to defeat the city's major rivals, Seattle and Tacoma. The leaders of Portland feared the consequences of the Northern Pacific's goal of a transcontinental terminal at Tacoma. Villard shared that view and sought a connection between Portland and a line being built by the Union Pacific north and west from Ogden, Utah. To thwart Tacoma's ambitions, Villard bought heavily into the Northern Pacific (NP), formed a "blind pool" of eight million dollars with friends, and gained control of the incipient transcontinental. Under Villard's leadership, the NP built south along the Columbia River toward Portland before constructing a line north into Tacoma. Villard purchased additional railways and steamship lines, merged them into his Oregon Railway and Navigation Company (OR&N), and drove that line eastward to a connection with the Union Pacific. He formed a holding company, the Oregon and Transcontinental (O&T), as an umbrella for his enterprises. Alas, the empire dissolved. The NP fell into receivership because of the costs of construction and the paucity of traffic; the OR&N and the O&T lost control of the line from Portland south to California; and Villard suffered a nervous breakdown. Yet Villard had succeeded in creating a substantial rail and steamship network that contributed to the rise of three cities, the establishment of lumbering as a major industry, and the opening of vast tracts to agricultural pursuits. A person of honesty and integrity, Villard would be compared by some to the "Empire Builder," James J. Hill.

While the Northern Pacific struggled from the 1860s to the 1890s to build across its land grant from Lake Superior to Tacoma, falling into bankruptcy twice, Hill constructed the Great Northern Railway (GN) from St. Paul, Minnesota, to Seattle without benefit of a federal land grant and, in the process, created a prosperous region of farms, ranches, mines, lumber camps, and small-scale manufacturing. A Canadian born in 1838, Hill migrated to St. Paul in the 1850s. In that raw, ugly town he saw economic opportunities in many fields, especially transportation. He formed a partnership to put steamboats on the Red River to Winnipeg and organized a firm to bring anthracite coal to St. Paul. When the St. Paul and Pacific railway project to link Minneapolis with Winnipeg floundered, Hill persuaded Canadian and English investors to finance its acquisition. A Minnesota state land grant represented a significant asset for the line, which reorganized as the St. Paul, Minneapolis and Manitoba and resumed construction under Hill's direction. Supported by investors at home and abroad, Hill devoted the next decade to the expansion of the railway.

Hill proposed first to gain access to the Canadian prairie provinces with a through route to Winnipeg, but this was a temporary goal. Only a direct line to the Pacific Coast would keep his carrier independent and profitable. Further, he needed to secure service to Chicago and the Great Lakes from the Twin Cities. Hill and his allies invested in a

With the completion of the transcontinental rail lines, West Coast ports grew in importance as links in an expanded trade network between the western hinterlands and the vast markets of South America, Europe, and the Far East.

Carleton E. Watkins (1829–1916). Pacific Coast Steamship Company Building and Wharf, San Diego. *Albumen silver print, 1880. Courtesy of the California State Library, Sacramento.*

subsidiary of the Chicago, Burlington and Quincy that was building a line to St. Paul from Chicago, and they constructed trackage to Superior, Wisconsin, where a line of steamships reached eastward to Buffalo, New York. With these connections secured, Hill drove the Pacific extension of his line, called the Great Northern, across the Dakota prairies toward the Rocky Mountains. His engineer, John Stevens, found Marias Pass through the Rockies and Stevens Pass across the Cascades. As the line crept westward, Hill brought farmers to the Dakotas, urged the settlers to diversify their production, and ordered his surveyors to note the presence of resources along the route. By the time the GN reached Seattle in 1893, Hill had initiated schemes for irrigated farming and for plant breeding along the line.

Hill insisted that the GN be constructed to high standards with quality rail, easy curves, and the lowest-possible grades. Management purchased excellent equipment for the railway and operated at a high level of efficiency. Although the GN reached Seattle as the depression of 1893 began, the railway prospered and, under Hill's tutelage, continued to expand. He visualized a united railway system from the Twin Cities to the Pacific Northwest and twice tried to acquire the Northern Pacific. Failing in those efforts, he did succeed in a scheme for joint control of the Chicago, Burlington and Quincy with archrival Northern Pacific. Hill established an excellent relationship with shippers along the GN, touring the line constantly and speaking to local audiences on economic diversity and the need to develop resources. Hill won the title "Empire Builder," and yet he would often be linked to the nefarious robber barons by those in the West who felt that government should strenuously regulate, or even own, the nation's railways.

Ironically, even as the Farmers' Alliances, the People's Party, and William Jennings Bryan demanded socialization of the nation's railways, competition between the carriers drove rates lower and lower. The efforts by the railroads to form traffic associations and pools to maintain rates were rarely successful and became illegal in 1887. On the Great Northern, for example, the average rate to haul a ton of freight one mile fell from 2.88 cents in 1881 to only 0.77 cents in 1907. Long-distance rates all but collapsed for short periods, a leading cause of the bankruptcies of many lines in the 1890s. Rates charged related not to the distance freight was hauled but rather to the value of the service provided. Most of the costs of carrying the freight were fixed, regardless of how far the cargo was carried. As a result, rates were higher for manufactured products than for heavy, bulk raw materials. The idea that transportation costs should be based on what the traffic would bear did not find favor in Tacoma, Tucson, or Telluride. With Congressman John H. Reagan of Texas leading the way, the federal government initiated efforts to hold down the cost of transporting all commodities, whether the shipment was Washington timber sent to the Dakotas or California oranges dispatched to Chicago. What the critics had overlooked were the major contributions of the railroads to the economic development of the West.

By 1900, the trans-Mississippi West possessed not only an intricate rail network linking the region to all areas of the United States, Mexico, and Canada but also ocean steamship lines operating from Galveston, San Francisco, Los Angeles, Portland, Seattle, and other ports, giving the region access to the trade of Latin America, Europe, and Asia. The railways brought agricultural commodities, base metals, lumber, and manufactured goods to the harbors and returned with industrial products and foodstuffs not indigenous to the region. Along the rail system, farms sprang up like winter wheat in a Kansas spring. The carriers extended loans to farmers and provided them with seeds and free transportation in times of economic distress. Huge "bonanza" wheat farms spread across the Dakotas while irrigated vegetable farms and orchards appeared in New Mexico, Texas, and California. With the resulting agricultural bounty came flour mills, fertilizer plants, meat packers, canneries, dairies, and other food processors in the towns and cities.

Indeed, the urbanization of the West reflected the coming of the railways, with cities as diverse as Dallas, Los Angeles, Tacoma, Spokane, and Denver largely owing their existence, and their growth, to the presence of railways tapping their hinterlands. The detractors of the railroads decried their "capricious" rate making, corrupt financial activities, and political involvement, but the iron and later steel rails gave the region the transportation base necessary to form a viable segment of the world economy.

Loggers, Merchants, Bankers, and Wildcatters

The railways helped to create a vital ingredient in the West's economic growth and in its position in global trade—lumbering. In the region, only the Pacific Northwest, northern California, eastern Texas, and a few mountainous areas of the Rockies received rainfall sufficient to produce trees of a size necessary for efficient timber harvesting. From colonial times into the mid-nineteenth century, the export of logs, masts, spars,

Surrounded by the tools of their trade, these Washington lumberjacks typify the young, single men who worked as independent logging operators before the turn of the century, when large corporations began to dominate the industry.

Darius Kinsey (1869–1945). Bachelor Bolt Cutters at Home in Their Little Shake Cabin. *Modern print from original glass-plate negative, 1897. D. Kinsey Collection, #10322 (0322), Whatcom Museum of History and Art, Bellingham, Washington.*

and cut lumber helped sustain international trade. The timber industry devastated the hardwood stands of New England and moved westward into the white pine forests of Michigan and Wisconsin in the 1850s and 1860s. Removal of the trees closest to water and rail transportation led these firms to eye the timber of the Pacific Northwest. Lumbering began there with the arrival of the first Europeans; the Russians cut timber at their fur stations along the coast, and the Spanish missions in California produced some wood exports. As has been seen, Californians exported timber to Asia in the late 1840s and 1850s. Only the absence of transportation into the interior prevented large-scale lumbering, and as long as wood remained the universal American building material, a vast potential market existed.

The demand for lumber and timber generated by the mining industry and urbanization led to the opening of forestry operations in the West. New England lumbermen discovered that shipping logs from Maine around Cape Horn made no sense when huge stands of timber reached almost to the coast of the Oregon country; in the 1850s they established sawmills around Puget Sound. Loggers from New England led a lonely, isolated existence at the Sound, cutting timber for the mines of California. Operations expanded slowly into the interior as the firms used waterways to float logs to the Sound and the sawmills. An infusion of capital led to the construction of short logging railroads around Puget Sound and in Oregon's Willamette River valley. Later the invention of the steam donkey allowed the harvest of larger trees that could be pulled to a stream or to the rails. In the Oregon country, as in California, loggers simply stripped timber from the public domain and moved on to richer sites.

The forests of the Northwest held Douglas fir, red cedar, and other trees that produced bountiful yields of cut lumber that was transformed into houses, stores, fences, flumes, and even planked streets. The entrepreneurs from Maine, Michigan, and Wisconsin recognized the potential, brought in equipment and veteran employees, and initiated a giant transplanted industry.

The firm of Pope and Talbot provides a case study of the rise of lumbering in the region. In 1850, two lumbermen from East Machias, Maine, Frederick Pope and William Talbot, sent a cargo of lumber from Maine to San Francisco to test the California market. With the successful sale of that shipment, they decided to establish a sawmill at Port Gamble on Puget Sound in 1852 and began a substantial operation. As the business grew, they relocated the company headquarters to San Francisco and built docks, workshops, and living quarters for their loggers at Port Gamble. The timber cut by their employees entered the domestic California market and was exported to Hawaii, South America, and Asia. Their success attracted investors willing to risk even greater capital to build redwood empires.

By the 1880s, firms from New England, the Great Lakes, and San Francisco began to secure title to large blocks of timberland. These contiguous acreages had the capacity to produce vast quantities of lumber. The firms recruited "homeless, womanless" lumberjacks from New England, Michigan, and Wisconsin and initiated operations. The extension of the Oregon and California Railroad to Redding, California, opened the pine and redwood forests of that state to timber operators. The completion of the Northern Pacific meant that lumber from the Northwest could be shipped to the plains states to compete with the Chicago lumber market. Indeed, the harvests of timber reached such proportions that as early as the 1880s, cries for conservation could be heard. Congress gave the president the power to establish forest reserves in 1891, in response to the growing threat to western timberlands.

Conservationists indeed blanched when James J. Hill told an audience in Everett, Washington, in 1892, "Lumber, gentlemen, is your greatest resource today." Hill meant what he said and encouraged Frederick Weyerhaeuser to develop the fir and hemlock forests of the western Cascades. Weyerhaeuser followed a model established by a group of Great Lakes lumbermen who purchased eighty thousand acres of the Northern Pacific's land grant to form the St. Paul and Tacoma Lumber Company on Puget Sound. In the midst of frenzied speculation in timber holdings, Weyerhaeuser purchased nine hundred thousand acres of prime fir lands from the railroad in 1900. The days of the lumber giants began at the outset of the twentieth century.

The coming of a handful of major firms to the industry brought some order to what had been a chaotic business of many small operators, but the situation for the loggers and lumberjacks did not improve. Like those who worked in mining and railroading, the men who came to the timberlands tended to be young, single, and immigrants. There were relatively few Asians or African Americans, but large numbers of Scandinavians could be found in the lumber camps. The work proved hard and dangerous in the forests and in the mills. Companies dammed streams to form small lakes and, after filling the streambeds with logs, blew up the dams, using the cascading floodwaters to move the logs to a nearby river. Men died or lost limbs maneuvering the logs on the water. Crosscut saws ripped into fir trees and into the hands and legs of the loggers. Lumberjacks fell from trees when chains gave way, and they were crushed by trees that fell "the wrong way." Operatives in the sawmills labored over the huge, unprotected blades of circular saws and breathed air filled with sawdust six days a week, twelve to fourteen hours each day. Only a few of the men were considered skilled craftsmen—those who produced shingles, for example—and replacements were easy to find. The rate of industrial accidents in lumbering exceeded that of most other businesses in the West. Nevertheless, lumbering became a substantial contributor to the economy and international trade of the West.

The mill villages, the mining camps, the railroad towns, and even San Francisco all relied on another western business: the mercantile stores with their supplies and services. From early in the nineteenth century, the mercantile traders provided the region with tools, clothes, farm implements, sugar, coffee, and other necessities. The ubiquitous country store served the rural South and Midwest and was replicated across the West as well. Along the Mississippi and Missouri rivers, steamboat companies often established warehouses and retail stores, a pattern repeated by some of the freighting firms west of the Missouri. Local merchants formed partnerships with the boat companies and the freighters, often doubling as their agents and bankers. Western entrepreneurs moved quickly from mercantile interests to transportation, banking, and real estate, as well as town promotion, as their communities matured into towns and cities.

In many communities the merchants and bankers formed the town's elite, both financially and socially. Some were sons or nephews of eastern bankers and merchants and were sent west to open branches or to supervise local investments. An unusually large number were German immigrants, and some were German Jews. Those with European connections often had lines of credit with continental banking houses. While the merchants enjoyed access to capital and social status, the miners, farmers, ranchers, and professional people in their communities frequently resented them because of their privileges. The merchants often controlled transportation facilities, and customers blamed high prices on alleged collusion rather than on the great distances goods had to be carried to their communities. Merchants bought cheap and sold dear, but they were indispensable for economic growth.

Not all merchants were Anglos from the East Coast. In New Mexico and Arizona, Hispanics frequently dominated freighting and the mercantile trade, some families having been engaged in such activities for decades. The Hispanic merchants in New Mexico used their stores as collecting points for wool from local herds, trading that commodity for freight goods brought over the Santa Fe Trail. Some formed partnerships with Yankee traders and served as agents for firms in St. Louis or Independence. They entered the import-export trade and were bankers as well as commission agents. Felipe Chávez, for example, was born in 1835, south of Albuquerque. His grandfather served as the first governor of New Mexico under Mexican rule, and the family had excellent social and economic connections. Chávez attended St. Louis University to study liberal arts and business practices, and he returned to New Mexico to join his father's trading

With the coming of the railroad in 1868, the Mormon leader Brigham Young abandoned his emphasis on economic isolation and the local manufacture of goods. Acknowledging the inevitability of imported products, he instituted a program of church cooperatives in every settlement to keep retail businesses in Mormon hands. Built in 1878, this Lehi store operated as a typical cooperative with stockholders limited to no more than a $200 share in the business.

George Edward Anderson (1860–1928). Peoples Co-operative Mercantile Institution, Lehi, Utah. *Modern print from original glass-plate negative, ca. 1885. Harold B. Lee Library, Brigham Young University, Provo, Utah.*

company. He opened the Chávez General Merchandise Store in Belen in about 1860, and using his own wagons to transport his trade goods, Chávez flourished. He became interested in the New York Stock Exchange and purchased securities as well as New York real estate. Eventually he purchased a seat on the exchange for his son José, who moved to the East to manage the family investments. A stockholder in the Santa Fe Railway, Chávez worked diligently to have his town located on a low-gradient freight line that the carrier was building across New Mexico. His success mirrored that of Esteban Ochoa in Tucson and other Hispanic business leaders in the Southwest.

Paralleling this story were the careers of many German and Polish Jews who came to the West as merchants, traders, bankers, and incipient manufacturers. Michael Goldwater failed in several ventures in Sonora but established a successful store in Phoenix, where the Goldwater family became prominent members of the community. Aaron Meier tried his luck in the California goldfields before moving to Portland, where he created the Meier and Frank's store in 1857. From a small plant in San Francisco, Levi Strauss provided heavy cotton work clothes to the miners, and I. W. Hellman formed the Farmers and Merchants Bank in Los Angeles. Anti-Semitism could be found in the West, as elsewhere, but that did not deter Jewish families from seizing economic opportunities in the cities and towns.

Merchants have largely been ignored by scholars, as have capitalists of the second echelon whose careers were not as colorful as those of the robber barons. Critics of Montana politics and Anaconda Copper focus their ire on Daly and Clark while failing to see the contributions of men such as Simon Pepin. Neither as prominent nor as controversial as the Copper Kings, Pepin came to Montana after living in Maine and working his way westward first as a riverboat deckhand and then on a freighting outfit to the Utah Territory. A Canadian by birth, Pepin entered the freighting business in Montana by 1864, still a young man in his mid-twenties. As a wagon boss, he gained a favorable reputation and was seen as a "comer." He used his savings to buy a cattle herd, which he sold to the army and the Indian agencies, and then with several partners started a cattle ranch. When the Great Northern Railway built its track through Pepin's ranch, he established a townsite on his property, and Hill agreed to locate a division point in the community. Pepin founded Havre, built a store, hotel, and houses there, and donated land for parks and churches. A banker and developer, Pepin acquired real estate in several Montana communities before his death in 1914.

Transportation allowed towns and cities to grow, and entrepreneurs saw economic opportunities in merchandising, real estate, banking, and construction. Money could be made in transshipping goods, in accepting consignments of equipment, and in developing urban properties. The Mormons of Salt Lake City set examples as they developed freighting firms, trading posts, and stores in Utah and beyond. Brigham Young recognized the impact of the coming of the railroad and entered into construction contracts with the Union Pacific. Mormons built railways north and south out of Salt Lake City and expanded their wholesale houses and small-scale industries. Church leaders planned economic investments and stood united against "Gentile" merchants and bankers. Their concerted efforts succeeded as Salt Lake City grew from 8,236 people in 1860 to over 20,000 two decades later. Salt Lake City, like San Francisco, Portland, and Houston, grew as a result of access to capital and the creation of banking houses with connections in the East and abroad.

Miners, farmers, merchants, builders, promoters, realtors, and westerners generally wanted and needed banks and bankers. Despite the historic image of the banker and the banking system vilified by the press and the populace, in most communities in the region the banks and those who led them were held in considerable regard. Westerners desired sound banks and bankers who could win their respect. Those who formed these institutions usually did so using their own capital and that of a group of partners. Few turned to the East for capital, but they did establish close connections with eastern banks that served as clearinghouses and as sources of currency. The bankers in western communities knew their customers by name, sent their children to school with the town children, and belonged to the same civic organizations and the same churches as their customers. There was little of the venom found in Populist rhetoric directed at local bankers; rather, the ire was directed toward the "monied interests" in the East.

Banking in the West developed largely after the National Banking Act of 1863 and the Internal Revenue Act of 1864, so that western state and territorial governments never had the opportunity to charter their own note-issuing banks or to create a system of "wildcat" banks. It was no longer possible to form banking institutions solely to generate credit by issuing paper money. As a result, most banks in the West were chartered by the states to provide only basic financial services. The regional banking system expanded at an extraordinary pace. The Miners Bank opened in California in 1848, and when Seattle had only four hundred people and a steamer arrived only once a week, it had a bank. The West needed currency more than it needed credit, at least initially. No longer could the regional economy be based on bartering trade goods; an infusion of currency was necessary. This problem had existed from the days of the fur traders and had to be resolved if the economy was to grow and modernize.

In the mining camps the use of gold, gold dust, or raw silver represented a temporary form of currency, but transfer of this currency proved dangerous. As a consequence, gold brokerages, stagecoach lines, and merchants began to accept deposits of the metals and issue certificates in receipt. These certificates of deposit could be exchanged, and thus the firms or individuals issuing them became de facto bankers. One Denver firm even minted coins and put them into circulation, but the arrival of Luther Kountze from Omaha and the establishment of his bank eased the local currency problem. A similar crisis occurred in San Francisco: gold flowed in from the mines, but there was little or no currency to exchange for the dust and bullion. Merchants needed to pay cash for goods ordered in the East, and as a result, banks appeared in rapid order.

The most significant banking house in San Francisco, the Bank of California, opened in 1864 with two million dollars in capital secured by its founders, William Chapman Ralston and D. O. Mills. Born in Ohio in 1826, Ralston worked on the Ohio and Mississippi rivers before going to Panama as a merchant, banker, and shipping agent. By 1854, Ralston moved to San Francisco and joined friends from his steamboat days to form a small bank. In the absence of incorporation laws, such banks were usually partnerships, and their reputations reflected those of the partners. Ralston quickly became a civic leader and served as a director or officer of mining and transportation companies. Although he persuaded Mills to serve as titular head of the Bank of California, Ralston led the firm, and through friendships, opportunism, and sheer energy he took over and consolidated the mines of the Comstock. He built a woolen mill, a winery, and a sugar refinery and invested in a dry dock, an insurance company, and steamships. The coming of the transcontinental railway undermined some of his enterprises, and following the national depression of 1873, Ralston's investments shrank in value, as did the collateral for loans extended by his bank. His empire collapsed in 1875, and Ralston turned over all his personal assets to satisfy his debts to the bank. A lonely swim in the Bay led to an apparent suicide. The economic growth of San Francisco and much of northern California depended on the entrepreneurship of men such as Ralston.

Firms like the Bank of California established branches or corresponding banks in Virginia City and in other mining camps in the West. Whereas the Mormons of Utah issued script to meet their need for currency, bankers in the villages and towns in the region depended on the growing banks of Seattle, Denver, and San Francisco or issued their own notes. Merchants with safes often became the town bankers, and in the absence of state or territorially chartered banks, trust in the banker was the only hope for the depositor. The absence of regulation produced a free-wheeling economy, but by the 1890s, demands for banking restrictions led to new laws in the region. Failures, such as Ralston's, forced legislatures to act to provide minimum protection for banking customers. The states required that both incorporated bankers and private bankers meet certain operating standards.

The depressions of 1873 and 1893, combined with demands from the mining companies and midwestern and southern farmers for the expansion of the nation's currency supply, placed western bankers in a dilemma. While the mining firms and their workers, as well as many other customers, cried out for "free silver," the banking community nationally opposed any movement away from the gold standard. The collapse of small-town banks in farming areas led to louder cries for currency inflation and lower interest rates. From the 1870s into the twentieth century, bankers in the West became more than suppliers of currency; certainly by the late 1880s, they were the primary sources of credit as the regional economy matured. Demand for credit came from many sectors of the economy, but particularly from manufacturing.

The construction of large, elaborate bank buildings in the major cities and the larger towns indicated economic maturation and the development of a commercial and industrial base beyond the extraction of minerals and the harvesting of forests. Although only San Francisco could be described as a city in 1860, by 1900 Houston, Dallas, Denver, Seattle, Portland, Los Angeles, and other urban areas boasted large populations and complex economies. Mining created demands for a wide range of products, from chemicals to candles and even ice, and small firms emerged to supply such items. Denver and San Francisco met some of the demands of miners for consumer goods by opening breweries and small garment plants. More significant, foundries and machine shops began to manufacture equipment for the mines, and over thirteen hundred men labored in such facilities in San Francisco by 1870. Denver and San Francisco attracted the greatest concentration of manufacturing facilities, but other cities also acquired a wide range of industries. Portland, for example, milled flour, produced pig iron, and developed woolen mills. Yet the coming of the railroad hurt these high-cost, high-wage factories; goods from the East or Europe sold for much less in many instances. Some cities that had previously suffered from a lack of labor found large numbers of unemployed men and women filling the streets. Only in the 1880s would the effort to industrialize vigorously advance as wage and transportation differentials achieved some balance and as western urban markets created greater demand. The economy of the region remained urban-centered in 1900, just as it had been since the arrival of the first European settlers; the towns and cities were the vanguards of the settlement process.

At the turn of the century, some western cities stood poised to take advantage of yet another regional resource, petroleum. After the successful development of the oil fields of northwestern Pennsylvania in the 1860s, wildcatting spread to West Virginia, Ohio, and Illinois. Although John D. Rockefeller's Standard Oil Company secured over 90 percent of the crude oil production of the major fields, other firms sought oil across the nation. Commercial production began in Kansas and the Indian Territory in the 1890s, and a large field was discovered south of Dallas near Corsicana, Texas, in 1894. Geologists, speculators, and drillers brought oil booms to the West, but California appeared to be the most promising area before 1900.

As early as the 1860s, oil seepages in California suggested the possibilities of commercial development. But the crude oil of California proved to be of low gravity and to contain excessive carbon, limiting its use as the basis for kerosene, the most important petroleum derivative. It could be used as boiler fuel, however, and was adopted by some industries. In 1892, E. L. Doheny found the enormous Los Angeles field, and other discoveries were made in Coalinga and the Kern River areas. These fields produced boiler fuel for railroad locomotives, steamships, and even Arizona copper smelters. California crude oil could compete with imported coal as a source of heat, and refineries, pipelines, and storage tanks soon dotted the landscape. Capital for the creation of the oil firms came largely from local investors, many of whom, like Doheny, had backgrounds in mining. As the California production grew, Standard Oil Company entered the field by purchasing the independent Pacific Coast Oil Company, soon renamed Standard Oil of California. By 1900 California ranked fifth among the states in petroleum production, and the West contributed 9 percent of the nation's output.

The development of petroleum spread slowly across the West before the discovery at Spindletop in 1901. Fields in Kansas and the Oklahoma Territory yielded good results, as did areas of Wyoming. These widespread discoveries led to the creation of independent oil firms in the region and to a substantial decline in Standard Oil's monopolistic position in crude production after 1900. The oil companies, railroads, and university scientists and geology departments cooperated to expend output, discover new fields, and streamline the refinery processes. By the time the Supreme Court ordered Standard Oil Company of New Jersey to divest itself of major subsidiaries in 1911, the West had replaced the old fields of the Northeast and Midwest as the nation's major source of petroleum. The capital created by regional oil firms contributed to the rise of many cities and towns and was frequently reinvested in other western businesses.

The End of the Beginning

In the last two decades of the nineteenth century, the West grew in population at a rate greater than the rest of the nation even as the rate for the United States as a whole exceeded that of all other industrialized countries. After 1890 the population of the region became more urbanized, and by the turn of the century almost 40 percent of westerners lived in an urban society. The sex ratio moved closer to the national norm as urbanization accelerated. The region retained its large percentage of foreign born while the number of African Americans increased slowly. Economic opportunity attracted this burgeoning populace as the discovery of gold in Alaska in 1896 and then at Tonopah, Nevada, four years later helped to sustain the boom in mining precious metals.

Westerners tended to be boastful, yet there was more than a grain of truth in their claims of economic success. A severe price had been paid in the misuse of natural resources and in the treatment of workers, but the West had functioned within a rising capitalistic system. Westerners were also optimistic, and many believed that although there were areas of poverty and although conservation measures had to be adopted, prosperity and the regulation of business would eradicate these problems in the twentieth century. They would be disappointed.

The economy of the West in the nineteenth century did not develop along the lines of Frederick Jackson Turner's rolling frontier. Rather, it grew in spasms over a widespread terrain and often from west to east. The people of the West adopted and used the system of laws and government they brought with them, but they also borrowed from other societies and cultures. In making use of the natural resources of the region, they incorporated the ideas and experiences of peoples from several other continents. The economy they forged in the West included features common not only to the rest of the United States but also to the world capitalistic economy. Westerners could not lay claim to uniqueness in the creation of a complex regional economy, but they could persuasively argue that few other areas of the world had witnessed such remarkable growth in such a brief span of time.

Bibliographic Note

In the absence of a general survey of the economic history of the trans-Mississippi West, readers should initially turn to the relevant chapters of several histories of the region. Ray Allen Billington and Martin Ridge, *Westward Expansion: A History of the American Frontier*, 5th ed. (New York, 1982), Robert V. Hine, *The American West: An Interpretive History*, 2d ed. (Boston, 1984), and Rodman W. Paul, *The Far West and the Great Plains in Transition, 1859-1900* (New York, 1988), are helpful and suggestive. To place the economic development of the West in a global perspective, William H. McNeill's *The Great Frontier: Freedom and Hierarchy in Modern Times* (Princeton, 1983) is indispensable. Although the jargon can be almost overwhelming, Leften Stavrianos, *Global Rift: The Third World Comes of Age* (New York, 1981), Theodore H. Von Laue, *The World Revolution of Westernization: The Twentieth Century in Global Perspective* (New York, 1987), and Imanuel Wallerstein, *The Politics of the World-Economy: The States, the Movements, and the Civilizations* (Cambridge, Eng., 1984), are critical to placing the economic history of the West in a broader framework. The largely discredited "plundered province" thesis first espoused by Bernard DeVoto in "The West: A Plundered Province," *Harper's Magazine* 149 (August 1934): 355–64, is examined most recently by William G. Robbins in "The 'Plundered Province' Thesis and the Recent Historiography of the American West," *Pacific Historical Review* 55 (November 1986): 577–98. Regional studies are also helpful, such as Carlos A. Schwantes, *The Pacific Northwest: An Interpretive History* (Lincoln, 1989), and Earl Pomeroy, *The Pacific Slope* (New York, 1965).

The mining industry in the West has attracted the interest of many scholars, but the reader should start with Rodman W. Paul's *Mining Frontiers of the Far West, 1848–1880* (New York, 1963), and William Greever's *Bonanza West: The Story of the Western Mining Rushes, 1848–1900* (Norman, 1963). Also important are two works by Clark C. Spence, *British Investments and the American Mining Frontier, 1860–1901* (Ithaca, N.Y., 1958), and *Mining Engineers and the American West: The Lace-Boot Brigade, 1849–1933* (New Haven, 1970), and Richard H. Peterson's *The Bonanza Kings: The Social Origins and Business Behavior of Western Mining Entrepreneurs, 1870–1900* (Lincoln, 1977). For the question of mining law, see Gordon Morris Bakken's *The Development of Law on the Rocky Mountain Frontier: Civil Law and Society, 1850–1912* (Westport, 1983). Lewis Atherton addresses the role of the mining entrepreneur in "The Mining Promoter in the Trans-Mississippi West," *Western Historical Quarterly* 1 (January 1970): 35–50. For mining in California, the best survey is Rodman W. Paul's *California Gold: The Beginning of Mining in the Far West* (Cambridge, Mass., 1947). See also Rudolph M. Lapp, *Blacks in Gold Rush California* (New Haven, 1977). A splendid case study of mining in Nevada can be found in *Treasure Hill: A Portrait of a Mining Camp* (Tucson, 1963) by W. Turrentine Jackson. The emergence of mining in Colorado is traced in Joseph E. King, *A Mine to Make a Mine: Financing the Colorado Mining Industry, 1859–1902* (College Station, Tex., 1977), and Duane A. Smith, *Rocky Mountain Mining Camps: The Urban Frontier* (Bloomington, 1967). Aspects of mining in Washington and Idaho are addressed by John W. Fahey in *The Ballyhoo Bonanza: Charles Sweeny and the Idaho Mines* (Seattle, 1971). For Montana, see Michael P. Malone, *Battle for Butte: Mining and Politics on the Northern Frontier, 1864–1906* (Seattle, 1981). Gold mining in South Dakota is described by Watson Parker in two works: *Deadwood: The Golden Years* (Lincoln, 1981) and *Gold in the Black Hills* (Norman, 1966).

The "new" labor history has produced excellent studies of miners in the West. The lives of these laborers are portrayed by Ronald C. Brown, *Hard-Rock Miners: The Intermountain West, 1860–1920* (College Station, Tex., 1979), Richard E. Lingenfelter, *The Hardrock Miners: A History of the Mining Labor Movement in the American West, 1863–1893* (Berkeley, 1974), Mark

Wyman, *Hard Rock Epic: Western Miners and the Industrial Revolution, 1860–1910* (Berkeley, 1979), and Ping Chiu, *Chinese Labor in California, 1850–1880: An Economic Study* (Madison, 1963).

Prerailroad land transportation in the West is surveyed by Oscar O. Winther in *The Transportation Frontier: Trans-Mississippi West, 1865–1920* (New York, 1964). Studies of various aspects of freighting and stagecoach operations can be found in Raymond W. Settle and Mary L. Settle, *War Drums and Wagon Wheels: The Story of Russell, Majors and Waddell* (Lincoln, 1966), W. Turrentine Jackson, *Wagon Roads West* (New Haven, 1965), Henry Pickering Walker, *The Wagonmasters: High Plains Freighting from the Earliest Days of the Santa Fe Trail to 1880* (Norman, 1966), and Oscar O. Winther, *The Old Oregon Country* (Bloomington, 1950).

No modern survey of railroad expansion into the West has replaced Robert Reigel's *The Story of the Western Railroads* (New York, 1926). Robert W. Fogel, an economist, challenges the significance of the railroads in the modernization process in *The Union Pacific Railroad: A Case in Premature Enterprise* (Baltimore, 1960). Arthur M. Johnson and Barry E. Supple provide insight into the motives of investors in western railroads in *Boston Capitalists and Western Railroads* (Cambridge, Mass., 1967).

Corporate biographies for the period include Maury Klein, *Union Pacific*, vol. 1, *The Birth of a Railroad, 1862–1893* (Garden City, N.Y., 1987), Keith L. Bryant, Jr., *History of the Atchison, Topeka and Santa Fe Railway* (New York, 1974), Ralph W. Hidy et al., *The Great Northern Railway: A History* (Boston, 1988), Richard C. Overton, *Burlington Route: A History of the Burlington Lines* (New York, 1965), Robert G. Athearn, *Rebel of the Rockies: A History of the Denver and Rio Grande Western Railroad* (New Haven, 1962), and John Hoyt Williams, *A Great and Shining Road: The Epic Story of the Transcontinental Railroad* (New York, 1988).

Important biographies of railroad leaders include Maury Klein, *The Life and Legend of Jay Gould* (Baltimore, 1986), Albro Martin, *James J. Hill and the Opening of the Northwest* (New York, 1976), and James B. Hedges, *Henry Villard and the Railways of the Northwest* (New York, 1930). A fine case study of the railway workers is James H. Ducker's *Men of the Steel Rails: Workers on the Atchison, Topeka and Santa Fe Railroad, 1869–1900* (Lincoln, 1983).

Steamboat operations on the Missouri and Columbia rivers as well as in San Francisco Bay and Puget Sound have been well documented. Some of the best studies are Louis C. Hunter, *Steamboats on the Western Rivers: An Economic and Technological History* (Cambridge, Mass., 1949), Jerry MacMullen, *Paddle Wheel Days in California* (Palo Alto, Calif., 1944), William E. Lass, *A History of Steamboating on the Upper Missouri River* (Lincoln, 1962), and Randall V. Mills, *Stern-Wheelers up the Columbia: A Century of Steamboating in the Oregon Country* (Palo Alto, Calif., 1947). Coastal shipping has received far less scholarly interest, but *The Panama Route, 1848–1869* (Berkeley, 1943), by John H. Kemble, is helpful.

The lumber industry and its laborers have been studied by Thomas R. Cox, *Mills and Markets: A History of the Pacific Coast Lumber Industry to 1900* (Seattle, 1974), William H. Hutchinson, *California Heritage: A History of Northern California Lumbering* (Santa Cruz, 1974), William G. Robbins, *American Forestry: A History of National, State, and Private Cooperation* (Lincoln, 1985), Robert S. Maxwell and Robert D. Baker, *Sawdust Empire: The Texas Lumber Industry, 1830–1940* (College Station, Tex., 1983); Ralph Hidy, Frank E. Hill, and Allan Nevins, *Timber and Men: The Weyerhaeuser Story* (New York, 1963), Edwin Truman Coman, Jr., and Helen M. Gibbs, *Time, Tide, and Timber: A Century of Pope and Talbot* (Palo Alto, Calif., 1949), Andrew Mason Prouty, *More Deadly Than War: Pacific Coast Logging, 1827–1981* (New York, 1985), and Daniel A. Cornford, *Workers and Dissent in the Redwood Empire* (Philadelphia, 1987).

One of the important aspects of the western economy is the mercantile trade, but it is also one of the most neglected fields. Only in a few articles and monographs can the role of the merchant be delineated. See, for examples, the following: Lewis Atherton, "The Santa Fe Trader as Mercantile Capitalist," *Missouri Historical Review* 77 (October 1982): 1–12; Richard Griswold del Castillo, "Tucsonenses and Angelenos: A Socio-Economic Study of Two Mexican

American Barrios, 1860–1880," *Journal of the West* 18 (July 1979): 58–66; Peter R. Decker, *Fortunes and Failures: White-Collar Mobility in Nineteenth-Century San Francisco* (Cambridge, Mass., 1978); Floyd S. Fierman, *Guts and Ruts: The Jewish Pioneer on the Trail in the American Southwest* (New York, 1985); and William J. Parish, *The Charles Ilfeld Company: A Study of the Rise and Decline of Mercantile Capitalism in New Mexico* (Cambridge, Mass., 1961).

Bankers fueled the economic engine of the western economy, and a study by Lynne Pierson Doti and Larry Schweikart, *Banking in the American West: From the Gold Rush to Deregulation* (Norman, 1991), surveys the role of banking in the region. A history of Houston's Texas Commerce Bank, by Walter L. Buenger and Joseph A. Pratt, *But Also Good Business: Texas Commerce Banks and the Financing of Houston and Texas, 1886–1986* (College Station, Tex., 1986), is informative, as are Larry Schweikart, *A History of Banking in Arizona* (Tucson, 1982), and L. Milton Woods, *Sometimes the Books Froze: Wyoming's Economy and Its Banks* (Boulder, 1985).

Also neglected by scholars is the story of manufacturing in the West before 1900. Leonard J. Arrington's works on Utah—*Beet Sugar in the West: A History of the Utah-Idaho Sugar Company, 1891–1966* (Seattle, 1966), *David Eccles: Pioneer Western Industrialist* (Logan, 1975), and *Great Basin Kingdom: An Economic History of the Latter-day Saints, 1830–1900* (Cambridge, Mass., 1958)—and H. Lee Scamehorn's *Pioneer Steelmaker in the West: The Colorado Fuel and Iron Company, 1872–1903* (Boulder, 1976) need to be emulated.

The early days of petroleum in the West are related in the following works: Ralph Andreano, "The Structure of the California Petroleum Industry, 1895–1911," *Pacific Historical Review* 39 (May 1970): 171–92; Henrietta M. Larson and Kenneth Wiggins Porter, *History of Humble Oil and Refining Company* (New York, 1959); and Gerald T. White, *Formative Years in the Far West: A History of Standard Oil Company of California and Predecessors through 1919* (New York, 1962).

Chapter Seven

Animals and Enterprise

RICHARD WHITE

If a roof had been built over the southern plains in the early 1870s, the American zoologist William Hornaday wrote, it would have been "one vast charnel-house." During the fall of 1873 the corpses, stinking and rotting in the sun, lay in a line for forty miles along the north bank of the Arkansas River. William Blackmore, an English traveler, counted sixty-seven bodies in a space not covering four acres. The bodies were those of bison. They died in such numbers and so close together because if a hunter got properly downwind, if a bluff partially concealed him, and if his luck held, he might shoot, reload, and shoot, again and again and again, without the animals stampeding.

In 1872 George Reighard had left Dodge City for the Texas Panhandle with "a buffalo-hunting outfit." He killed, or so he remembered, an average of one hundred bison a day. He killed the bison for the "hide and the money it would bring." Asked, years later, whether he felt pity for the animals as day after day he dropped his hundred, he replied no, he did not. "It was a business with me. I had my money invested in that outfit . . . I killed all that I could."

Money and pity, these are the words that mark a great divide in the history of the American West. Reighard stood at a point where animals were only dollars on a hoof; those who later asked him about pity regarded animals as being worthy of concern within a human moral universe. But for all their differences, those who saw animals as commodities and those who saw them as objects of sentiment stood on the same side of a cultural divide. On the other side was a world in which animals were persons and pity was the sentiment that animals felt toward humans. This earlier West appears to us now at once recognizable and utterly strange. Remembering it, we may feel like Dorothy remembering Oz. Because once, when animals were persons, the West was a biological republic.

For many Native Americans, animals formed the bridge by which humans tapped supernatural powers. Working at Fort Union in 1832, Pennsylvania-born artist George Catlin painted this Blackfoot medicine man cloaked in the skin of a grizzly bear, whose strength gave power to the healing ceremonies.

George Catlin (1796–1872). Medicine Man Performing His Mysteries Over a Dying Man. *Oil on canvas, 1832. National Museum of American Art, Washington, D.C./Art Resource, New York, New York.*

It was Indian peoples who had made animals other-than-human persons with whom relationships were social and religious instead of purely instrumental. Indians, like all other humans, certainly sought to order and control the natural world, but the order they constructed was a social order, and control partially came through what amounted to religious negotiation. Indian religions made hunting holy and gave human-animal relations a depth and complexity largely lacking among Europeans. In hunting, some persons died so that others might live. Ceremonies preceded the kill. Animals consented to die; they, or more powerful beings—holy people, keepers of game, or other supernaturals—pitied the hunter and instructed him in the rules and rituals necessary to kill them. Indians killed game as much by prayer, pleading, and reverence as by the arrow or spear. They recognized the obvious wariness of game and the reluctance of animals to die, but they explained it in terms of previous ritual abuse by humans or even supernaturals. The difficulty of obtaining the consent of animals only made strict observance of hunting rituals all the more necessary.

Europeans brought to the West a far more instrumental view of animals, and they brought new animal species that had evolved as instruments of human purposes. Europeans regarded their domesticated animals as sentient tools, and on the basis of utility, they classified them as superior to the native wild fauna. They recognized that their horses, sheep, goats, cattle, burros, and smaller stock gave them an advantage over Indians, for the native North American biota (the continent's native animal and plant life) lacked suitable candidates for domestication. Western Indians had domesticated only the dog and, in the Southwest, the turkey.

Europeans and Indians both linked the success of human endeavors to the biological success of at least some animals, but it was Europeans who made the measure of success commercial and who, like Reighard, could see their own commercial success as tied to the virtual destruction of entire species. Animal persons yielded to animals of enterprise, which gleaned the energy of western ecosystems (the interacting species and the nonbiological environments of the West) to produce hides, meat, and wool that found markets all over the world. Animals ceased to be creatures of a limited set of adjoining ecosystems and became instead, as the historian Alfred Crosby has noted, movable creatures of the biosphere (the entire planetary system that sustains life). By the close of the nineteenth century, human masters took tribute from subject animals and determined their fate. Animals not subject to humans had become, in effect, enemies.

The transition did not come easily or smoothly, and the animals of enterprise themselves, in a sense, rebelled. Without domesticated animals, Europeans would have neither survived nor conquered, but their animals often proved fickle allies. Indian peoples too enlisted the aid of horses, sheep, and cattle. And some domestic stock, disregarding both Indians and Europeans, went wild, turning against their masters' purposes. All of this greatly complicated the history and the mental and physical landscape of the West.

The horse, more than any other animal, exemplified the unpredictability that domesticated livestock introduced into the West. Horses became tools of Indian resistance as well as of European conquest. Horses became persons who entered the visions and dreams of Plains Indians. Among native peoples, the horse greatly increased the efficiency of nomadic hunters and the mobility of raiders. Along with the exotic diseases brought by Europeans, horses shifted the balance of power away from settled horticulturalists—both Indian and Hispanic—and toward nomadic hunters. The horse helped create the flourishing nomadic culture of the Great Plains.

Indians spread horses rapidly and widely across North America. West of the Rockies, they transported the animal to the Snake River valley by 1700 and the Columbia Plateau by 1730. East of the Rockies, the horse reached the central Great Plains by the 1720s and western Canada by the 1730s. Its distribution among Indians of the Pacific Coast was spottier, but by the 1770s the Yokuts of the central valley in modern-day California had adopted horses, and in the Pacific Northwest, Yakimas introduced horses among villagers living in the small prairies on the western foothills of the Cascades. Indians used horses for transport, war, hunting, and more rarely, food. For most groups, a life without horses became unimaginable.

Cattle and sheep did not spread so far or so fast. The Papagos moved from hunting semiwild cattle to herding them; refugee Indians from the East brought livestock with them; and in the early nineteenth century, some groups on the Columbia Plateau became cattle raisers. Before the reservation period, however, most western Indians remained raiders, not ranchers. They plundered Spanish, Mexican, and later American herds. The Apaches, in particular, regarded Spanish and later Mexican herds as virtual supply depots to be tapped at will.

Only the Pueblos and Navajos quickly adopted sheep and cattle. The Pueblos began raising sheep, horses, and cattle in the seventeenth century. The Navajos started as did conventional Apache raiders, preying on the Spanish and the Pueblos. But they quickly became livestock raisers, looking after their sheep, in the words of one Spaniard, "with the greatest care and diligence for their increase." At the end of the eighteenth century the Spanish regarded Navajo herds as "innumerable," and the Navajos found themselves targets of Spanish and Ute raids. As sheep and goats became critical to Navajo subsistence, these animals, like horses, bridged the gap between wild animal persons and domestic animal tools. Navajos regarded sheep and goats as metaphorical mothers and fathers because they sustained life. They were thus loved and identified with the family even as they were killed and regarded as wealth.

It was through raiding, which marked the movement of animals from European to Indian control and back again, that some animals escaped human control entirely. Feral horses and cattle appeared on the margins of human settlements. Not all species of introduced livestock could seize the opportunity to go wild. Domestication had given sheep, even the *churros*—the tough sheep of the southwestern borderlands—too great an inventiveness in finding ways to die. Although hardy and able to withstand drought, sheep could not live without shepherds. By themselves, they could not find water or feed. They died from poisonous weeds, and they could not resist even the most inept predators.

Cattle and horses did go feral, but their populations remained localized, growing as it were from seed animals that Indians and white migrants planted. The mustangs of Texas and other wild horses of the West developed from strays left by raids and from animals abandoned as lame. Wild horses were most numerous in Texas, but there were also significant wild horse populations along the Snake and Columbia rivers and in the San Joaquin Valley of California. There were relatively few wild horses on the northern plains or in the Rocky Mountains before 1800. They did not appear in Nevada, their present-day stronghold, until the transcontinental migrations of the 1840s. At their nineteenth-century peak, only an estimated two million horses grazed between the Rio Grande and the Arkansas River, with perhaps another million in the rest of the West. True wild cattle, as distinct from unbranded stock, had an even more confined range and fewer numbers. Into the nineteenth century, feral Criollo cattle from mission herds remained largely limited to East Texas and the lands around the Arizona missions. After several generations in the wild, they could not be redomesticated. They could only be hunted.

The introduction of new species and their spread created sweeping and significant changes, but these changes still lay on the other side of the great nineteenth-century divide. Western animals had not yet become animals of enterprise. Animals of enterprise are commodities; their value is exchange rather than use. For both Navajos and *nuevo mexicanos*, for example, the *churro* remained primarily a subsistence animal. A full-grown *churro* dressed out to only seventeen pounds of meat, less than a modern lamb, and yielded only about a pound of wool. Such small yields were commercially unacceptable, but they were not liabilities when fresh meat had to be consumed quickly to avoid spoiling. Similarly, *churro* wool (actually closer to hair) was meager but ideally suited to hand spinning.

In a world where social relationships took precedence over economic relationships, commerce took a backseat to other forms of exchange, and such exchanges dominated the transfer of livestock. Indians stole livestock from their enemies, but they gave livestock freely as gifts to friends or relatives. Trade lay in the originally limited region between these extremes; it was a way to seek advantage among those who were not friends but also not clearly enemies. Commerce was thus a form of sublimated theft, and as such, it often shaded easily into outright theft. Plains Indians traded horses at the Mandan and other Missouri River villages, but given the opportunity, traders would quickly turn into raiders. The subtleties of an exchange where the goal was to accumulate wealth at the expense of others and to profit by others' need, but where outright theft was forbidden, were lost on many Indians. And livestock, by its very nature, invited theft. Large domesticated grazers were a plunderer's dream, for unlike all other forms of property, these would gallop off with an insistent thief. Raided by Indians, the Spanish and later the Mexicans raided in turn.

Livestock raiding and its companion, slave raiding, became deeply entrenched in the Southwest and on the Great Plains. The Spanish and later the Mexicans raided and traded for Navajo, Apache, and Pawnee slaves; the Navajos and Apaches took Spaniards and later Mexicans as slaves in turn. In Texas the herds of the missions on the Rio Grande and farther north around La Bahia and San Antonio de Béxar provided inviting targets; first the Apaches, then northern villagers ("los norteños"), and finally the Comanches preyed on these herds. Ute, Apache, Comanche, and Navajo raiders struck deep into the heart of New Mexico, repeatedly devastating sheep herds; in the 1760s the Comanches virtually stripped New Mexico of horses. By the 1840s, Yokut and Miwok horse raiders from the interior had crippled many Mexican ranchos in California, aided by the Indians living on the ranchos.

Under such conditions, cattle, horses, and sheep only slowly emerged as animals of enterprise. The first step in this direction came when they began to serve as capital among both Indians and Hispanics. In eighteenth- and nineteenth-century New Mexico, sheep raising took place on the *partido* system. Sheep essentially became capital lent at interest. The owner (banker) of the sheep turned over a certain number of ewes (capital) to the *partidario* (borrower), who agreed to make set annual payments of wool and lambs, usually 20 percent of the original head count (interest). Eventually, the contracts specified responsibility in case of losses to disease, lightning, and Indian raids. If the *partidario* could realize a return greater than the interest that he paid the owner, he

By the late 19th century, the animal skulls holy to Plains Indian tribes had become for American military troops mere monuments to slaughter.

Karl Bodmer (1809–93). Assiniboin Medicine Sign. *Watercolor on paper, 1833. Gift of the Enron Art Foundation, Joslyn Art Museum, Omaha, Nebraska.*

William Henry Jackson (1843–1942). A Collection of Buffalo, Elk, Deer, Mountain-Sheep, and Wolf Skulls and Bones Near Fort Saunders. *Photograph, 1870. United States Geological Survey (Jackson, W.H. 348), Denver, Colorado.*

profited and could establish his own flock. If he failed, he sank into debt and lost his collateral—land, if he had any, or if not, his own labor. *Partidarios* hoped the arrangement would lead to wealth and freedom, but it often led to poverty and peonage, a form of debt slavery. Sheep owners, on the other hand, succeeded in transferring part of the risks to the *partidarios*, solved their own labor problems in a cash-short economy, and freed themselves from active management of the herds.

Partidario arrangements involved little buying and selling of animals, but livestock gradually became commodities used to acquire other commodities. Comanche and Apache raiders turned easily stolen livestock into more difficult to steal European manufactures by trading what they had stolen. In the nineteenth century, the Yokuts and Miwoks traded stolen California horses to American mountain men and New Mexican adventurers, who drove the horses east to New Mexico and Missouri.

Until the late eighteenth and early nineteenth century, commerce per se remained more an incidental outcome of livestock raising in New Mexico, Texas, and California than a rationale. Indeed, in eighteenth-century New Mexico the governors, fearful of losing breeding stock, placed substantial restrictions on trade in sheep, ordering that only wethers (neutered rams) be sold in the mining towns of Nueva Vizcaya and occasionally forbidding the export of any live animals. Textiles woven from the wool of *churros* remained more important than the direct sale of sheep.

Only in the late eighteenth century did reforms in the Spanish empire and relative peace with the Indians create a commercial boom in sheep raising. By then the *partido* system had concentrated the wealth in the hands of relatively few interrelated families. Large sheep owners became *comerciantes*, rancher-merchants who developed a trade with Chihuahua, Nueva Vizcaya's leading commercial and mining center. By the 1790s, sheep drives had become an annual event as New Mexicans sent roughly twenty-five thousand head of sheep south each year in large caravans guarded by soldiers against the Apaches. Both farmers and *partidarios* were in debt to the *comerciantes*, and the *comerciantes*, in turn, were usually in debt to the merchants of Chihuahua. Only a few *comerciantes* escaped this chain of debt and became *ricos*—the richest tier of New Mexican landowners and merchants. Clemente Gutierréz, born in Aragon and married well in New Mexico, used the *partido* system and the Chihuahua trade to make himself the most powerful of the New Mexican *ricos*. Sheep delivered wealth and power into his hands; he and the other *ricos* dominated New Mexico's economy and politics.

This commerce in sheep, however, remained at the mercy of war and theft. It collapsed in the years of turmoil that led to Mexico's independence. After the revolution, exports resumed, and by 1835, New Mexico was sending eighty thousand sheep south. New Mexican sheep, marketed in Durango, found their way deep into the interior of Mexico, until the Mexican War and the American conquest of New Mexico finally aborted this trade.

The California gold rush redirected New Mexican sheep away from Old Mexico and to the new markets provided by the California mines. Even after the early boom prices in the mines declined, sheep drives continued. In 1858 one hundred thousand New Mexican sheep went west into California. Only the growth of California flocks, to nearly three million head by 1870, closed the California market.

In Texas, the early development of commercial livestock raising was far slower and less successful. Between the 1750s and 1810, ranchos developed in three distinct areas of Texas: between the Rio Grande and the Nueces River (then part of Coahuila and Nuevo Santander), on the San Antonio River, and finally, on the Louisiana border near Nacogdoches. It was on these ranches and on the missions that many of the practices of later western cattle raising evolved. The first Texas cowboys were Indians. When not restricted by raiders, these vaqueros conducted annual roundups to gather and brand free-ranging cattle of the mission herds.

But Texas cattle never quite became full-fledged animals of enterprise during the Spanish era. Indian raids often kept missionaries and ranchers from conducting roundups to brand their cattle. Ownership of these unbranded cattle remained in constant dispute, and they became an irresistible temptation to meat hunters (*carneadores*), soldiers, and, later, soap makers, all of whom slaughtered the unbranded cattle with incredible waste. What was commerce and what was theft remained a matter of considerable disagreement. To make matters worse, legal markets remained limited largely to sales to the king's presidios in Texas and to the south. Burdensome regulations made the transport of cattle to markets in Coahuila and Louisiana difficult and often illegal. This taint of smuggling, plus the endless disputes over the ownership of the unbranded animals, continued to give ranching an aura of criminality. As the slaughter of unbranded stock reduced the once abundant herds, the rancheros turned to gathering and exporting the wild mustangs that also abounded on the range. By 1796, mustangs had supplanted cattle as the province's chief export, until the bloody struggles in the province before the Mexican Revolution stymied the growth of all commerce.

In terms of cattle raising, *californios* succeeded where *tejanos* failed. California missionaries had the good fortune of having a ready market appear virtually at their doors in the form of American and British trading vessels, which arrived to collect hides and tallow—the rendered fat used for candles. Neophytes—the baptized Indians of the missions—and vaqueros working for private ranchos slaughtered the cattle, left the meat to rot, and took away the hides and tallow. After the Mexican government secularized the California missions between 1836 and 1842, much of the mission property intended for the Indian neophytes ended up in the hands of rancheros. The mission herds that had once numbered 150,000 head of horses, sheep, and cattle dwindled to 50,000. The remainder had been slaughtered and sold. The rancheros no longer had to share the trade with the missions.

Fur Trade

By the time New Mexican sheep and California cattle succeeded in becoming full-fledged animals of enterprise, a select group of wild animals had also fallen victim to commodification. To be sure, a small trade in products of the hunt had long existed out of Taos and Pecos. Some hides and buffalo robes traded there continued south, finding an ultimate market in Mexico. Much farther north, at the Mandan villages and Wichita villages, Plains Indians had also bartered deerskins, antelope hides, buffalo robes, and dried meat for corn, beans, and other produce. By and large, these remained local exchanges that disposed of small surpluses gained during normal subsistence activities.

It was not production for the market per se. The large-scale commodification of wild animals emerged only in the late eighteenth century. Even then it affected few species—buffalo, beaver, and sea otters. For these species, however, the results were so dramatic that within a relatively few years, each faced extinction.

In the commodification and killing of wild animals for their furs, two different cyclical patterns met. One was natural. The turning of the seasons made beaver fur and bison wool grow thick in the winter. As long as consumers desired beaver for felt to make hats and bison for wool to make robes, the seasons shaped the hunt. The other pattern was economic and far less regular. On the world markets of emergent capitalism, prices fluctuated, and this cycle too shaped the fate of the animals selected for enterprise. Yet even as human beings conformed to the patterns that nature and the market imposed, they also struggled against them. Producers found new technologies that allowed hatmakers to use lower grades of beaver, and tanners found uses for summer-killed buffalo. The large companies that came to dominate the fur trade unsuccessfully sought monopolies that would allow them to avoid, as much as possible, the cycles of the market.

No animal suffered as rapid a subjection to the market and as rapid a destruction as the sea otter. Sea otters congregated in rafts or schools of up to one hundred animals along the northern Pacific rim from Hokkaido to Baja California. Russian fur traders, lacking the skill to take the animals themselves, virtually enslaved Aleut and Kodiak Indians to hunt the otter during the spring and early summer. They shipped the fur to China, where mandarins prized it for hats and as a trim on their garments. The British somewhat accidentally opened their own direct trade with China in 1778 when British sailors of Captain James Cook's voyage of discovery purchased fifteen hundred otter skins from Indians at Nootka Sound on Vancouver Island. The sailors intended to use the pelts for clothing on their northern voyage, but in China they discovered that, as one sailor said, "skins which did not cost the purchaser six-pence sterling sold . . . for 100 dollars."

The promise of fortunes to be made in the sea otter trade brought English merchantmen sailing to the coast in Cook's wake. The Englishmen, the Russians pushing south from Sitka, and the even more numerous Americans who traded with the Indians for skins from the Pacific Northwest south to California brought the animal to the brink of extinction. As guns replaced bows and arrows in the hunt, otters became increasingly vulnerable. The number of sea otters traded in Canton began dropping early in the nineteenth century. The otter population was badly diminished by the 1820s; by the 1850s, the sea otters had nearly vanished.

It took time for the mainland fur trade to duplicate this pattern of overhunting and near extinction. The western fur trade in beaver and buffalo was an extension of a much older trade that had begun in the East and that had depended on Indians to procure the furs. The French pushed this trade west of the Missouri during the eighteenth century. The dependence of the traders on free Indian labor initially acted to slow the destruction of furbearers, since the Indians of the prairies and plains gave primacy to subsistence hunting over fur trapping. Indeed, many mounted Indians disdained beaver trapping and would engage only in buffalo hunting.

The reluctance of Indians to become wholehearted partners in the slaughter that the traders contemplated had as much to do with the Indians' ideas of a proper economy as

Without the labor of Blackfeet women, the buffalo hides secured by Blackfeet hunters had no value.

Walter McClintock (1870–1949). [Blackfeet Woman] Tanning a Skin—Fleshing a Hide. *Photograph, ca. 1900. Courtesy of the Southwest Museum (#N.36397 MCC.334), Los Angeles, California.*

with their ethical beliefs about animals. For Indians, the proper ritual treatment of animals remained essential; as long as hunters killed animals in an appropriate manner, ritually honored the animals, and took game only to fulfill their own legitimate needs, Indians believed animals would return. Because most Indians regarded killing for exchange as a legitimate activity, religion did not offer a significant ideological obstacle to the trade. Indians often interpreted diminishing game in terms of improper treatment rather than overhunting. Thus when the buffalo finally disappeared from the plains, the Sioux did not view their demise as permanent nor the cause as biological. The buffalo had, they believed, gone underground because whites had killed the animals with disrespect. The buffalo would return when Indians could ensure that the animals would receive proper ritual treatment.

Indians could thus, in good conscience, hunt animals even as those animals grew less and less numerous. Yet most Indians did not kill as many animals as they could have because their wants remained limited. The demand for goods and wealth is cultural, and wealth among western Indians took quite specific terms. Horses or medicine bundles were forms of wealth sought by members of many Indian groups. Both the Northwest Coast Indians and the Navajos equated wealth and status, but maintaining status involved the redistribution of accumulated wealth. Most Indians, however, sought only limited amounts of white trade goods. Having met their immediate and limited necessities, they often refused to engage in further hunting. And when they did face shortages, "begging" (asking aid of friends or kinspeople with more than enough to share) or stealing (seizing goods from enemies) seemed to them just as appropriate as engaging in further hunting.

This common Indian refusal to accept European premises of a proper economy led white traders to complain constantly of the capriciousness of Indian hunters, the "laziness" of Indians, and the difficulty in collecting debts. The huge paper profits the trade generated often dwindled beneath the heavy overhead of gifts and the losses to accident and theft. Traders compensated by resorting to alcohol to stimulate greater demand and thus more hunting. Traders themselves dealt in stolen furs, and they favored those Indians who did seek to accumulate wealth, for whatever reason.

The trade thus brought economic and social changes in Indian societies, and these changes increased the impact of Indian hunters on game populations. Among the Blackfeet, a complex brew of material, social, and ideological factors increased their involvement in the British (but not the American) fur trade. In the nineteenth century, Blackfeet chiefs, already distinguished by their wealth in horses, became notable for the number of their wives. A man with many horses could use them to acquire wives or female slaves taken in raids. Because of the gendered labor system of the plains, multiple wives were necessary to process the buffalo hides into robes. A woman could, on average, process eighteen to twenty robes in a winter, and the more wives a man possessed, the more finished robes he accumulated, since a single mounted hunter could kill more buffalo than a single woman could process. Indeed, buffalo hides were useful to a man only if he had wives to process them. Robes, in turn, became critical for chieftainship, for robes obtained the trade goods that chiefs increasingly needed for the gift giving that demonstrated their generosity and maintained their following.

It was women rather than men who made the production of buffalo robes a virtual Indian monopoly before the Civil War. White men could slaughter buffalo, but they could not process robes. By 1840, commercial production had reached about ninety thousand robes a year on the northern plains, and trade robes represented about 25 percent of the total buffalo kill of the plains. The work of Indian women also allowed Indians on the Canadian plains to control the production of pemmican—the mixture of buffalo meat, fat, and berries that formed a basic foodstuff for trappers in the beaver trade.

The production of pemmican linked the work of women to the trade in beaver pelts, but by the early nineteenth century, the work of western Indians, both men and women, was becoming less and less critical to the trade. With western Indian hunters unwilling to supply sufficient furs, white and eastern Indian hunters and trappers replaced them. Operating from fixed posts, American fur-trading companies on the Missouri River had begun to dispatch parties of white men to trap beaver. On the Pacific coast, the Hudson's Bay Company, which had incorporated the earlier North West Company, also used white and Iroquois trappers organized in fur brigades to supplement the Indian trade. A major innovation in the trapping came in the mid-1820s when William Ashley, driven off the Missouri by Arikara attacks and unable to persuade Indians in the central plains to trap on the scale he desired, persuaded his white trappers to stay permanently in the mountains. Ashley sent a supply train to annual rendezvous, where trappers and Indians exchanged furs for supplies and engaged in an extended bacchanal. Because processing a beaver pelt demanded far less skill and labor than processing a buffalo hide, even white trappers who lacked Indian wives or lovers could ready their own catch for shipment. This eliminated the production bottleneck that limited the trade in buffalo robes.

The Rocky Mountain trappers who gathered at the annual rendezvous fell into three broad categories: *engagés*, men supplied and salaried by a fur-trading company; skin trappers, the sharecroppers of the fur trade who operated on credit advanced by a company; and finally free trappers, the small entrepreneurs of the trade who sold their furs to the highest bidder. For all of them, trapping was hard and extraordinarily dangerous work. Trappers may have had the common entrepreneurial ambitions of Jacksonian America, but few other Jacksonians so regularly risked their lives as well as their capital. The lucky ones lived, but they often ended up, in the words of Nathaniel Wyeth, who organized an unsuccessful fur company, "mere slaves to catch beavers for others." The real profits went to the large companies that organized production or brought goods west. These companies would maintain the so-called Rocky Mountain Trapping System, in one form or another, until 1840.

The southern Rockies spawned a final variant on the trade. This was the domain of free trappers who operated, usually illegally, out of Santa Fe and Taos. Kit Carson and other Anglo-American, Franco-American, and Mexican trappers methodically moved through the southern Rockies, stripping the streams of beaver and shipping their furs east along the Santa Fe Trail. After depleting the southern Rockies of beaver, they found themselves unable to compete with the more efficient rendezvous system of the central Rockies.

Between 1820 and 1840, these branches of the trade battled each other, and their contest destroyed the beaver over much of the West. At the height of the beaver trade in 1832, the Missouri and Rocky Mountain systems deployed perhaps one thousand trappers, the Hudson's Bay Company provided another six hundred, and a smaller number of free trappers worked out of Taos. Two large companies came to dominate the fray: the American Fur Company and the Hudson's Bay Company. Each combated its competition as ruthlessly as it slaughtered beaver. The Hudson's Bay Company created a fur desert along the Snake River to stop the progress of American trappers into its stronghold, the Oregon country. The American Fur Company sent its parties to tail those of its competition in the northern and central Rockies. Each side rushed to trap out the streams. Only the Blackfeet country, where the Indians maintained their long-standing animosity to American trappers, retained significant beaver populations.

Sea otters and opium, in an odd way, saved the surviving beaver from slaughter. Sea otter pelts and opium allowed American and European merchants to acquire Chinese silk without expending valuable silver, and by 1833, silk hats had begun to replace beaver hats on fashionable heads in both the United States and Europe. The demand for beaver dropped, prices fell, and the beaver trade went into a precipitous decline. By 1840, it had largely ended. Although substantial, the destruction brought by the mountain men was limited in an important sense: they had destroyed only animals and not habitat. Given a respite, beaver populations staged a recovery in the 1840s and 1850s, only to decline again later in the century from part-time trapping by miners and other western workers and from habitat destruction.

The buffalo, or more properly the North American bison, was the last furbearer to suffer near extinction, and its fate is at once the most instructive and the most mysterious. It is instructive because the buffalo died as an industrial animal rather than as an animal of fashion, like the beaver or otter. The discovery that buffalo hides could be turned into a cheap leather suitable for making machine belts, together with the expansion of the railroad network across the West after the Civil War, sealed the bison's fate. No longer did hunters have to hunt bison in the late fall or winter when robes were full; no longer did Indian women have to painstakingly process hides; and no longer did traders have

Albert Bierstadt's monumental painting reflected a scientific interest in the fate of the buffalo, as well as a widespread belief that the animal—like his Indian pursuer—was a symbol of a vanishing West.

Albert Bierstadt (1830–1902). The Last of the Buffalo. *Oil on canvas, ca. 1888. Gertrude Vanderbilt Whitney Trust Fund Purchase, Buffalo Bill Historical Center, Cody, Wyoming.*

to rely on river transport or wagons to move robes to market. Both the seasonal and the Indian bottlenecks on production vanished. Now white hunters could kill buffalo at any season and transport the hides to market on the railroads.

Against Indian resistance, professional buffalo hunters moved onto the southern plains in the early 1870s. The southern hunt peaked between 1872 and 1874. In all, the hide hunters took an estimated 4,374,000 buffalo during these years. To this has to be added the Indian kill of approximately 1,215,000 on the southern plains during this same period, as well as the smaller number of bison killed by settlers and sportsmen. In the 1870s, Congress passed a bill protecting the bison, but President Ulysses Grant vetoed it.

The efficiency of the killing was coupled with a staggeringly inefficient use of the carcass. Virtually all the meat rotted. Some hunters initially did not even know how to skin the animals properly, and they thus wasted the hides. Other hunters killed more bison than they could skin. Contemporary sources estimated that at the peak of the hunt in 1872, three to five buffalo were killed for each hide that reached market. By 1875, the southern herd had largely ceased to exist.

The destruction of the smaller northern herd came later. In 1876 the Northern Pacific Railway reached Bismarck, North Dakota, and began pushing its tracks west into the buffalo country. That same year, the American army began the campaigns that broke Sioux control of the northern plains. In 1880 the assault on the northern herd began in earnest. By 1882, there were an estimated five thousand white hunters and skinners at work on the northern plains; by the end of 1883, the herd had vanished. The slaughter was so thorough and so quick that not even the hunters could believe what they had done. In the fall of 1883, many outfitted themselves as usual. But there was nothing to hunt except piles of bones bleaching in the sun and wind.

The causes and mechanisms of the final slaughter are thus obvious and instructive.

What is mysterious about the fate of the buffalo is that a precipitous drop in bison numbers appears to have occurred in the decades *before* this slaughter, and neither the number killed by Indians nor the early fur trade can account for it. Only quite recently have historians begun to unravel this mystery. Bison numbers on the plains probably peaked at about twenty-five million, well under older estimates. And it now appears that bison were in trouble by the 1840s not so much from overhunting, although this was increasingly a factor, as from a combination of drought, habitat destruction, competition from exotic species, and introduced diseases. During droughts, such as the one that struck the southern Great Plains in 1849, bison had to compete with Indian horse herds and wild mustangs for food and water in critical riverine habitats. At the same time, livestock taken by Indian raiders and cattle driven across the plains by white migrants introduced tuberculosis and brucellosis to the buffalo herds. The expansion of trails through the plains and the settlement of whites along the edges of the region drove the bison from the peripheral habitat on which they had depended as refuges from drought and hunting. The result was a buffalo population already unable to maintain its numbers when the white hunters struck.

The wholesale slaughter of the buffalo eventually abated from a lack of targets, but killing continued for a while on a retail basis with hunters, like the young Theodore Roosevelt, who rushed off to get a trophy before it was too late. Others, however, sought to save the pitiful remnants of the species. Some ranchers started private herds; William Hornaday, of the New York Zoo, organized the Bison Society; and George Bird Grinnell, the editor of *Forest and Stream*, worked to protect the small group of bison in Yellowstone from poachers. The buffalo moved from being a commodity to a symbol of the American West, gracing American coins and exhibited in Wild West shows. Such symbolic status became the last refuge from extinction open to some species whose uniqueness, size, and power allowed Americans to endow them with a special national meaning.

Animals and Energy

The slaughter of the buffalo opened the West for an expansion of the domesticated grazers that would more efficiently provide food, clothing, and energy for humans. Twentieth-century Americans readily think of animals as food, but they forget that for millennia, animals were also energy. Whale oil and tallow provided light. Animals moved people and freight. Before the coming of the railroads, animal-drawn freight wagons hauled the commerce of the prairies, plains, and mountains. Even after the building of the transcontinental railroads, stagecoaches and freight wagons served those areas away from the tracks. Until displaced by the automobile, horses and buggies provided local transportation. Animals pulled the plows that broke the prairies and much of the plains. Oxen dragged logs from the woods; the limits of animal power determined how far from navigable streams early loggers in the Pacific Northwest could go.

In providing energy, the preeminent animals in western enterprises were oxen and mules rather than horses. East of the Mississippi, where improved forage, water, and shelter were readily available, the big, grain-fed Conestoga horses provided the traction for freight wagons. Under the harsher conditions of the West, mules, an infertile cross between a donkey and a horse, stood up better to the rigors of freighting. Although, as the saying went, "without pride of ancestry or hope of progeny," they had the longest working life (eighteen years) of any draft animal, were plagued by few diseases, resisted saddle and harness sores, and could, freighters claimed, do as much work as a horse on one-third the food. There were few mules in Missouri when William Becknell loaded three farm wagons with trade goods, hitched them to horses, and opened the Santa Fe Trail in 1821. But thanks to the Santa Fe trade, mules became a major export of the Mexican borderlands to the American borderlands.

Oxen were slower than mules and more liable to disease, but they had other offsetting advantages. They were much cheaper than other draft animals, and so was their harness equipment. Indians were less likely to steal them, and unlike mules and horses, oxen could live and work on grass alone. The oxen that initially pulled western wagons before the Civil War were not the massive draft animals of the Northeast; they were Texas or Cherokee range cattle. In 1860, what appears to be a partial census counted an estimated 65,950 oxen, valued at $35 each, pulling freight across the high plains; only 7,574 mules, at $125 each, were in harness.

Oxen powered the western freighting industry, but horses and mules supplied the power for western stagecoaches. Transporting freight and people demanded tremendous numbers of animals. Ben Holladay, who in the 1860s briefly reigned as the "Stagecoach King" of the West, supposedly employed 15,000 men on his stage and freight lines and owned 20,000 wagons and 150,000 draft animals. In such operations, freight counted for more than passengers. The Stagecoach King and his competitors relied on federal mail contracts for their essential revenue and tended to regard passengers as so much paying ballast for the coaches. Except in California, passengers did not initially ride in the familiar horse-pulled Concord coaches of western movies. Between 1850 and 1864, a stagecoach trip from Missouri to Santa Fe meant a ride on mule-pulled, watertight wagons that could be dismantled and turned into scows on swollen streams. And since stage stations on this route were often few and far between, passengers and mail shared the wagons with fodder for the mule. Even with the Concord coach, a western stage ticket bought, in the words of a Denver passenger, "fifteen inches of seat, with a fat man on one side, a poor widow on the other, a baby in your lap, a bandbox over your head, and three or four more persons immediately in front, leaning against your knees." Nightfall meant sleeping on "the sand floor of a one-story sod or abobe hut, without a chance to wash, with miserable food, [and] uncongenial companionship." It was a jolting, pounding trip of crying children, swearing drivers, angry passengers, abominable whiskey, and brackish water. Dust was everywhere. Holladay was an ideal man to run such a crude and primitive enterprise. According to the railroad promoter Henry Villard, he was a "genuine specimen" of the successful pioneers of western enterprise: "illiterate, coarse, pretentious, boastful, false, and cunning."

Mules, oxen, and horses had pulled him into great wealth, but Holladay recognized that in long-distance transportation, the days of animal power were numbered. After 1866 he withdrew from freighting and stagecoaching and invested his money in coastal shipping and steamboats on the rivers of the West Coast and, finally, in railroads. Animals, as Holladay realized, were fleeing before steam, and this new source of power, divorced from living ligament and muscle, drove draft animals into more and more

The sinew and muscle of animals powered agricultural production in the West well into the 20th century.

J.E. Stimson (1870–1952). Harvesting Oats near Dallas, South Dakota. *Photograph, 1908. Wyoming State Museum, Cheyenne.*

remote sections of the West. In these corners of the West, their capacities continued to mark the limits of enterprise. As long as animal power hauled ore, for example, transportation was expensive and only high-grade ore was worth processing. Low-grade ore or less valuable minerals like copper had to await the railroads. As long as bull teams hauled logs, logging could not penetrate much farther than a mile or two from tidewater or navigable rivers and remain profitable.

Long after steam displaced animals from long-distance transportation, draft animals continued to provide power in the fields. They were in a sense the largest of barnyard stock: animals that lived in close association with humans. There was a surprisingly obvious gendering of the work involved with such domestic animals. Those that produced milk or eggs for consumption or local sale were the domain of women; those animals that produced power, meat, wool, or hides were the domain of men. This changed only when chickens or dairy cattle became fully commercialized. At that point, they too became the domain of men.

The hold of draft animals on American farms proved quite tenacious. Because of their rapid gait, horses were the traction animals of choice over most of the West, but the mule's tolerance for heat gave it the advantage in Oklahoma and Texas. Oxen were less common, although farmers used large oxteams to break raw prairie, particularly wet prairie, until the late 1860s. By the late nineteenth century, steam power had replaced animal power on a large scale only in threshing. Even after the invention of the tractor, many farmers kept their horses and mules because they were far superior for cultivating row crops. Not until 1920 did mechanical power drive most draft animals from the field. By the 1930s, the horse and mule population of the United States had dropped ten million from its peak. As horses and mules vanished, so too did the patchwork of irregularly shaped fields, with their hedgerows and windbreaks that were suited to horse-drawn plows but not to tractors. Thirty million acres of hay, barley, and oats, previously needed to feed traction animals, now became available for the production of other crops. Profits from such crops were necessary, for with the dependence on machinery, the farmers' need for capital rose.

Industrialism, with its creation of new sources of power and products, narrowed the spectrum of animals of enterprise to those that were good to eat or those with skin or wool that was good to wear. Americans were increasingly urban dwellers: 40.3 percent of the people in the Northeast and 20.8 percent in the Midwest lived in cities or towns by 1870, and this growing urban population wanted meat, hides, and hair in quantities greater than ever before. Above all, they wanted beef.

Eastern Tables and Western Beef

Why late-nineteenth-century Americans wanted such vast quantities of beef is not immediately obvious; Americans have not always chosen beef over other meats. If late-nineteenth-century consumers had wanted chicken or pork, or had decided not to eat meat at all, the history of the West would have been considerably different. Before the Civil War, urbanites ate preserved pork, but even then it appears that their desires were turning to fresh beef. Pork may have been on their tables, but beef was on their minds. It was the food that managed to denote both high status and down-to-earth Americanism. The managers of William Henry Harrison's 1840 campaign for the presidency captured the potent symbolism of beef when they coupled Harrison's Indian fighting and log cabin with his diet of "raw beef and salt" to create defining symbols of a rude but democratic American. His opponent, Martin Van Buren, with his weakness for French food and his supposed aristocratic tendencies, seemed effete and elitist by comparison.

Plentiful grass for cattle in the West and eastern consumers hungry for beef were necessary but not sufficient conditions for a prosperous western cattle industry. Range-fed cattle from the grasslands initially provided tough and stringy beef that captured only the bottom end of the growing beef trade: the cheaper cuts of meat in American cities and cured beef for the trade with England. Middle-class Americans, desiring fatter, tenderer meat, bought corn-fed midwestern beef.

Texas Longhorns provided the first inroads into eastern markets. Butchers derided the Longhorn as "eight pounds of hamburger on 800 pounds of bone and horn," but the Longhorns had assets more apparent on the western plains than on the eastern table. Texas Longhorns were a distinct breed of cattle and not merely the description of the horns of a steer. The Spanish cattle of Texas, the Criollos, had long horns, but they were not the Texas Longhorns. True Texas Longhorns did not appear until the early nineteenth century. A backcountry cattle industry had long existed in the American South, and American migrants brought some of these cattle with them into Texas. Criollos thus interbred with American cattle, some descended from a special English breed—the English Longhorns. The crossing of Criollos and Anglo-American stock, like so much of the history of western animals of enterprise, resulted from violent acts. In the wake of the Texas Revolution, raiders from both sides of the border drove off cattle and thus mixed Mexican and Anglo-American animals. Between 1836 and 1865, fertile, long-lived, pugnacious, multicolored cattle with long legs, long tails, long bodies, and long horns (a trait derived from both parent stocks) developed. These cattle were resistant to the tick-carried Texas fever, and they thrived on grass without supplemental feeding.

As southern migrants into Texas abandoned their earlier custom of penning cattle and adopted the Mexican practice of allowing cattle to roam free, the mixing of breeds (including wild Criollos) increased. When the Civil War eliminated both the markets for cattle and many of the men who raised cattle, untended Longhorns bred promiscuously with other range cattle. By the end of the Civil War, Longhorns formed the majority of the roughly five million cattle in Texas. Most of these cattle were unbranded and thus free for the taking. Unlike the feral Criollos, they were manageable, and so they would not remain mere wild cows, the prey of human hunters.

The very biology that gave Texas Longhorns their advantage, however, seemed doomed to limit their commercial prospects. In the arithmetic of enterprise, cattle were only commodities, but cattle continued to demonstrate that they were animals—complicated living things. Texas cattle, for example, carried Texas fever. Although Longhorns had an immunity to Texas fever, when driven out of Texas, the cattle carried the ticks that transmitted the disease. The ticks dropped off and found new hosts, and the new hosts died. Northern farmers did not know how the disease was communicated; they knew only that when Texas cattle came in contact with their domestic stock, their animals died. Northern cattle sometimes sickened simply after crossing a trail used by Texas cattle. The Texans, who never saw their livestock die of the disease, denied that it existed or, at least, that it came from Texas cattle, but they had to face the reality of northern attempts to ban their cattle.

Ticks, as much as markets, determined the initial course of the plains cattle industry. When Texas cattle raisers tried to drive herds north after the Civil War (as they had before the war), Missouri and Kansas farmers violently resisted their passage, shutting off direct access to northern markets. Texas cattle had to stay west of the line of agricultural settlement; thus, except for a small trade to New Orleans, the initial postwar markets for Texas cattle were the mines, military posts, and Indian reservations of the West.

Providing beef to hungry miners in the Rockies had arisen almost as a side effect of transcontinental travel. A percentage of the oxen hauling migrants to Oregon and California and pulling freight wagons within the West inevitably wore out and went lame. J. W. Iliff, who as a failed miner in the 1859 gold rush to Pikes Peak was a sort of human equivalent to these failed oxen, realized that their misfortune might cancel out his own. Iliff established a little store near what is now Cheyenne, Wyoming; in addition to cash sales, he began bartering with passing migrants for lame, footsore, and emaciated cattle and oxen.

Like other early cattlemen in Colorado, Wyoming, and Montana who began roughly similar operations, Iliff discovered that cattle could overwinter with little care on the plains. To easterners, the bunchgrasses of late summer looked scanty and sorry when compared with the big bluestem of the prairies or a bluegrass pasture of the East, but the bunchgrass plains were a storehouse of surprises. Nature annually turned bunchgrasses into hay on the stem. Unlike big bluestem or introduced grasses, bunchgrass retained much of the original plant protein in the dry leaves. And this natural hay remained accessible. Despite cold winters, the light dry snows of the region usually did not crust, thus allowing cattle to push their way down to the grass. On the south-facing slopes, cattle did not have to dig at all, since the winter sun soon melted the snow. In short, it appeared that in the West, animals did not require winter feeding. Footsore cattle, bunchgrass, and hungry miners combined to make Iliff the first of the cattle kings of Colorado and Wyoming.

Iliff's kingdom eventually expanded to thirty-five thousand head; he had cattle in Colorado ranging from Julesburg to Greeley. Whatever the regenerative properties of bunchgrass, however, it could not make oxen reproduce. Iliff built his herds from cattle driven north from Texas. In 1866 Iliff bought cows and bulls that two peripatetic Texans, Charles Goodnight and Oliver Loving, had brought into the Arkansas Valley.

Goodnight, although born in Illinois, was raised in Texas, and he was one of many restless, ambitious Texas cattlemen seeking an outlet for the Longhorns. In 1866 he trailed a cattle herd west from the Cross Timbers to Fort Sumner, New Mexico, where the military needed beef for themselves and for the Navajos and Apaches confined at Bosque Redondo. Goodnight later established a ranch, neither his first nor his last, forty miles south of the fort. By 1870, the federal distribution of cattle to reservation Indians, whose livelihood the Americans had destroyed, required fifty to sixty thousand head a year.

When Iliff began collecting footsore oxen, the mines west of the Rockies had already supported a fledgling cattle industry for a decade. The California gold rush provided the *californio* ranchers with markets virtually at their doorsteps, and cattle once valued for their tallow and hides became beef animals. The *californios* sold cattle to the very miners who had banished them from the goldfields, and they sold cattle in prodigious numbers. Thinking the boom permanent, they spent freely and mortgaged their land at usurious rates to pay the taxes on it and to hire lawyers to defend their ranchos from Anglo squatters. But by 1861, the boom was over. Then came floods, followed by the worst drought southern Californians had endured. Between 1862 and 1864, roughly three million cattle starved to death on the range. Disease administered the coup de grace to the cattle and the ranchos; by 1869, only thirteen thousand cattle remained in southern California. Sheep had replaced them. The ranchos passed into Anglo hands.

By driving down prices, cattle from the Pacific Northwest played their part in destroying the California cattle industry. Some of the Oregon steers that went south to the mines were, in a sense, returning home. They were descendants of California cattle imported by the Puget Sound Agriculture Company, a Hudson's Bay Company subsidiary that had earlier sought to produce tallow, beef, and hides to replace dwindling fur returns as a company export. But most of the cattle for the mines were domestic stock from the eastern United States, stock that in the 1840s had accompanied American migrants west along the Oregon and California trails. More docile, better milkers, and carrying more flesh than the California cattle, they became the preferred stock in Oregon and Washington.

In the 1850s and 1860s, Oregon cattle added much of the interior grass and sagebrush lands of the Great Basin to the bovine empire. They finished their conquest of the region by the 1870s. In Nevada they met herds from the central valley of California fleeing first the drought and then the conversion of interior cattle ranges into wheat fields and orchards. Only Utah resisted the open-range system of Oregon and

California. The Mormons had adapted to the arid Great Basin by creating a unique village-based stock-raising system that combined hay production on irrigated fields, the winter feeding typical of midwestern livestock systems, and summer drives to mountain pastures.

Feeding miners, soldiers, and Indians maintained a small western cattle industry, but as long as Texas fever kept Longhorns from eastern markets, a beef bonanza on the scale envisioned by promoters remained a mirage. At the end of the Civil War, cattle were a glut on the market in Texas, selling for three to four dollars a head, with a fat beeve bringing only five to six dollars. In New York, a three-year-old steer (admittedly of better quality than a Longhorn) brought eighty dollars; in Illinois, a steer brought forty dollars; and even in relatively nearby Kansas, the steer cost thirty-eight dollars.

A tick held all these profits hostage, and they might have remained hostage if the tick could have endured cold. But it could not; a hard freeze killed it. Drovers discovered that when cattle were held on a northern range until after a hard frost, the cattle could be safely shipped. And cattle that spent a year or more fattening on the central and northern plains before shipping were free of fever. Cattlemen did not as yet know how, but the isolation and hard winters of the plains were freeing the profits held hostage by Texas fever.

Joseph G. McCoy claimed credit for being the first to realize the financial possibilities that the convergence of Texas fever, Texas cattle, and the railroads presented, but McCoy was as lucky as shrewd. He was only one among several entrepreneurs who sought out likely spots along the Kansas Pacific and solicited Texas cattle. McCoy helped found Abilene, the first of the cattle towns that enterprise, railroads, and ticks created. Jesse Chisholm, a Texas cattleman of Cherokee ancestry, pushed a trail north from Texas across the Indian Territory to Abilene. During the summer of 1867, an estimated thirty-five thousand Texas cattle came up the Chisholm Trail.

These first drives were not profitable, but McCoy persisted and advertised for more cattle. Over the next twenty years, two million cattle would come up the Chisholm trail and its successors. Traveling ten to fifteen miles a day in herds numbering two to three thousand head, the cattle wore troughs as deep as shallow canals, which remained visible for years. The alliance of ticks and Texas cattle forced a steady westward march of cow towns and the trails that fed them, for once farmers surrounded a town, they banned Texas cattle, from fear of Texas fever. And so the trails led first to Abilene, then later to Ellsworth, Dodge City, and Hays.

McCoy and other cattle-town entrepreneurs solved the problem of moving Longhorns east, but they had not made the Longhorns any more palatable. Northern tastes demanded marbled beef, and Longhorns did not fatten easily when fed grains or, later, cottonseed by midwestern farmers. To get better-quality beef, cattlemen began to interbreed the Longhorns with improved stock from the East. This hybrid stock more easily put on weight on the farm feedlots of Iowa and Illinois. By the late 1870s and early 1880s, midwestern farmers were selling improved cattle in the West and rebuying the steers they produced as feeders. To service these cows, western ranchers began importing purebred Hereford and Shorthorn bulls. Longhorns gradually disappeared as feedlot buyers demanded cattle that could turn a minimum amount of corn into a maximum amount of beef in the minimum amount of time.

Abilene, Kansas, became the first of the cattle towns, where cattle driven north from Texas were loaded on railroad cars for shipment east.

After Henry Worrall (1825–1902). Abilene in 1867—Celebrating the Shipment of the First Train-load of Cattle. *Wood engraving, 1874. From Joseph G. McCoy,* Historic Sketches of the Cattle Trade of the West and Southwest *(1874). Courtesy Amon Carter Museum, Fort Worth, Texas.*

ABILENE IN 1867—CELEBRATING THE SHIPMENT OF THE FIRST TRAIN-LOAD OF CATTLE.

Quality was one obstacle the western cattle industry faced; price was another. For beef to be cheap, it needed to be mass-produced, and this could not happen until it could be preserved. Packers had been working on refrigerating beef since the 1860s, but technical problems stymied them until Gustavus Swift created a fleet of refrigerator railroad cars; by 1882 he was successfully marketing refrigerated Chicago beef in New York. Refrigeration allowed shippers to avoid paying freight on the inedible portions of cattle. It cost half as much to ship a dressed carcass as a live animal, and centralized, industrial slaughter allowed a fuller use of by-products. By the 1880s, refrigerated beef undersold fresh beef in the East. The price of prime cuts dropped 40 percent between 1883 and 1889. Because the production of refrigerated beef demanded industrial slaughterhouses, a fleet of refrigerator cars, and a coordinated marketing network, capital costs were high. A few Chicago packinghouses—Swift, Armour, Morris, and Hammond—quickly created an oligopoly that forced smaller regional packers out of business. By the early twentieth century, they, along with Cudahy and Company, produced 82 percent of the beef in the United States. Other meat-packing centers—Kansas City, Omaha, and later Fort Worth—arose as outposts in the dominion of the "Meat Trust."

Americans came to think that they were living in the "Golden Age of American Beef." With cookbooks and magazines dismissing pork as difficult to digest, unwholesome, and unhealthy, fat beef became a health food. The urban Northeast became a world of beef and potatoes; the older hog and hominy diet retreated south. For Anglo Americans, Irish Americans, and German Americans in particular, beef reigned as the symbol of the good life, and the good life became attainable at least at the dinner table. According to the German sociologist Werner Sombart, it was on the "reefs of roast beef and apple pie" that the dreams of American socialists were dashed. And the newer immigrants from eastern and southern Europe soon adopted this preference for beef.

American per capita beef consumption reached a peak in the early twentieth century, only to dip after World War I, before rising to what would be all-time highs after World War II and then falling again.

Making the West Comfortable for Cattle

In the promotion of western stock raising, ideologues were in a sense as significant as the packinghouse magnates. The ideologues were a varied lot: a general of the U.S. Army, James S. Brisbin; a German aristocrat who came as a tourist and stayed as a speculator, Walter Baron Von Richthofen; a surgeon for the Union Pacific, Dr. Hiram Latham; a publicity man for the Union Pacific Railroad, Robert Strahorn; and the entrepreneur and townsite promoter Joseph McCoy, of Illinois, who generously credited himself with creating the cattle industry. It was they who worked for and proclaimed the rise of the cattle kingdom. It was they who urged and celebrated the transformation of the plains, deserts, and mountains from a biological republic to a biological monarchy where humans reigned, where uselessness among lesser living things was a crime punishable by death, and where enterprise was the reigning virtue.

The importance of these boosters was not the example they set; rather, it was the new order they perceived. McCoy soon left the western cattle business; Latham's ranch investments failed; and Richthofen's major livestock experience seems to have been operating a dairy barn for consumptives who rested on second-story porches while sipping fresh milk brought up from purebred cows on the floor below. What the boosters, for all their individual limits, realized was that stock raising was not a romantic retreat from industrial America; stock raising was part of its foundation and its base. Hiram Latham, in particular, conceived of the enterprise in grand, nationalistic terms. He was intent on harnessing useful animals to their rightful task of helping create an industrial power. Industry, in return, provided the railroads, refrigeration, modern packing plants, and urban populations that made beef production practical.

In his 1871 pamphlet *Trans-Missouri Stock Raising*, Latham presented a series of what he regarded as largely self-evident propositions. To maintain a large population of laborers who would sell their products in the competitive markets of the world, the United States must furnish them with cheap food and clothing. The cheaper the food and clothing, the cheaper every article they would produce, and "therefore, in proportion to the abundance and cheapness of our food and clothing, will be our success as a manufacturing nation." This food must include meat, for a solely vegetable diet brought degeneration "to the condition of the Macaroni Eaters of Italy." Cheap meat and cheap wool for clothing demanded cheap lands, but since land prices in the East were rising, the future of stock production was in the West, where there lay "a billion acres . . . boundless, endless, gateless, and all of it furnishing winter grazing."

Cattle and sheep were so many machines to turn grass into meat, hide, or wool, all of which could be readily turned into dollars. Latham's ambitions for sheep and cattle knew no bounds. In a speech to the Colorado Stockgrowers Association in 1873, he said stock raising should not cease until "every acre of grass in Colorado is eaten annually." Latham and the other boomers mastered the arithmetic of optimism. There is, Richthofen assured his readers, "not the slightest uncertainty in cattle raising." In the Spanish proverb that Latham loved to quote, "Whatever the foot of the sheep touches turned to gold." Ideologues and stock raisers alike conceived of sheep and cattle as biological dollars that could mate and produce little dollars with the same regularity with which interest compounded. In their books are tables of investment and yield that have cattle endlessly procreating and cows, heifers, and steers relentlessly transforming themselves into money. At the end of six years, Richthofen estimated a profit of 156 percent on the original investment. Experienced cattlemen, he said, thought it too low.

There was no possibility too remote for this arithmetic of enterprise. For example, a limitless demand for cheese would supposedly push dairy cows into the grasslands. To provide the market, Latham happily imagined the Chinese and Japanese to be cheese eaters—allocating them fifty thousand tons annually—and spiraled the resulting figures into wonderful castles of profits. The Platte River valley alone could support twelve hundred cheese factories and five hundred thousand milk cows and would yield $33,295,000 in profit. As for the scope of such enterprises, Joseph Nimmo, in his *Report in Regard to the Range and Ranch Cattle Business*, prepared for Congress, estimated that in 1885, 44 percent of the United States, exclusive of Alaska, was devoted to grazing. All of it was in the West.

This fascination with the free, apparently limitless grass of the West relegated the old pioneering eastern animal of enterprise—the pig—to the western barnyard, for grass formed only a minor part of a hog's diet. Pigs had accompanied and sustained Anglo-American farmers in the wooded, humid lands of the East, but in the West they prospered only on the prairie margins where farmers grew corn and in rare places such as the western parts of Washington Territory or eastern Texas. Most of the West was inhospitable to the pig. Pigs do not like direct sun and heat; they need access to shade and water. They do not like extreme cold. Pigs became mere animal tourists in the West. They clustered on the edges of Solomon Butcher's famous photographs of Great Plains farms, pampered and peripheral. Unless human beings made them comfortable, pigs wished they had never come west.

The tendency of American universities to name their athletic teams after animals nicely reveals the boundaries between the kingdoms of the cow and the hog. The athletic teams of the University of Arkansas are the Razorbacks. Razorbacks are feral pigs that once abounded in the woods of Arkansas, and University of Arkansas athletes presumably share the better qualities of wild hogs. One state over, to the West, you encounter the Longhorns of the University of Texas. Although eastern Texas proved hospitable to pigs, western Texas welcomed only cattle, and Texas football players today derive inspiration from their identification with neutered cattle.

As pigs lost their pioneering role in the West, they regained it in another sense in the East, for it was on their bodies that entrepreneurs perfected the mechanisms of industrial slaughter. The industrial slaughter of pigs paved the way for the industrial slaughter of cattle. Chicago emerged from the Civil War as the major meat-packing center of the nation. The massive new Union Stock Yards, underwritten by the railroads in 1865, became the central distributing point from which butchers elsewhere bought live cattle. In the 1870s, before refrigeration, most cattle sold in Chicago were still shipped east for butchering, but Chicago was "Hog Butcher to the World." Hogs retained two major advantages over cattle. They more efficiently turned corn into meat, and hogs—whether as bacon, ham, salt pork, or lard—took to preservation better than

Systematically photographing the people of Custer County, Nebraska, for a regional history, Solomon Butcher created a visual inventory of his subjects' material possessions. Here the settlers' domestic animals and draft horses are displayed as symbols of prosperity while the pigs—animals that never adjusted well to the western climate—remain on the periphery.

Solomon Butcher (1856–1927). East Custer County, 1887. *Photograph, 1887. Solomon D. Butcher Collection, Nebraska State Historical Society, Lincoln.*

beef in the days before refrigeration. Although there was some market for dried and salted beef, most cattle were slaughtered near the point of consumption and sold quickly as fresh beef; pigs could be killed far away and held for sale.

Meat packers transformed slaughterhouses into year-round factories that turned hogs into meat and an increasingly formidable line of by-products. Chicago packinghouses learned to escape the seasonal constraints nature had previously imposed on their enterprise. Using ice, they began, on a small scale, to extend the packing season from the cold-weather months into the spring and summer. The slaughter of hogs increased fivefold between 1872 and 1877. In killing pigs, the packers created a centralized infrastructure that could be expanded to killing and marketing cattle once the packers had perfected a way to preserve the meat.

The western expansion of the railroad network, industrialized slaughter, and the development of refrigeration all allowed western cattle to penetrate more and more deeply into eastern markets. And once cattle found their markets, they became creatures of those markets. When the economy boomed, as it did in the early 1870s and again in the early 1880s, the cattle industry prospered and cattle expanded their numbers and range. When the economy slumped, as it did in the mid-1870s and late 1880s, the cattle industry contracted. The booms of the early 1870s and early 1880s differed, however. The first involved stocking the plains; the second involved stocking butcher shops once cattle became the leading animal of industrial capitalism. During both booms, and in the years that followed, Americans seemed to have no higher ambition than making the awesome western landscape comfortable for cows.

Stocking the Great Plains and the Southwest absorbed most of the cattle produced in Texas. An estimated two-thirds of the cattle driven north out of Texas in the 1870s and early 1880s were yearlings and two-year-olds. Such cattle were too young to market. They, along with a smaller number of cows and calves, were sent to stock the northern grasslands. A second, subsidiary drive, usually numbering 20–25 percent of the northern drive, took cattle into New Mexico and Arizona. The Sioux on the northern plains and the Apaches in the Southwest opposed this expansion, but financial markets proved more deadly than Lakota and Athapaskan warriors.

The optimism of Texas cattle raisers—made tangible in the 600,000 head of cattle they drove north in 1871, the largest total for any year—foundered when one-half the cattle driven north remained unsold. Placed on already overstocked ranges near the cattle towns, many of the cattle died in the harsh winter of 1871–72. The industry was barely recovering from this disaster when the Panic of 1873 cut off the drovers' access to credit. When the banks refused to extend loans, the herders had to ship their young and thin cattle to market, further depressing prices and driving many to ruin. Markets remained depressed, and by 1875 only about 150,000 cattle were going up the trails from Texas. West of the Rockies, the mining boom ended, and the cattle of the Oregon country lost three-quarters and more of their value. This combination of bad weather and bad markets signaled the end of the first boom on the open range.

Although cattle prices remained low until 1881, the industry began to recover in the late 1870s. In 1876 over 300,000 cattle came up the trail from Texas, twice the total of the previous year. For the rest of the decade, the number fluctuated between 200,000 and 250,000 head annually. In the Far West, cattle from southeastern Oregon and southwestern Nevada headed to Winnemucca, Nevada, on the Central Pacific Railroad for shipment to San Francisco. Stocking the central and northern plains quickened with cattle from Oregon joining the herds coming up from Texas.

Oregon and Texas cattle moved onto the plains by their own power; they left the plains in cattle cars provided by the expanding railroads. In 1880–81, Denver became a major sales center with the creation of the Denver stockyards. Farther north, Cheyenne, Wyoming, the headquarters of the Wyoming Stock Growers Association, became one of the capitals of the cattle kingdom. Ogallala, Nebraska, at "the end of the Texas Trail," joined the Kansas cattle towns as a shipping center, as did Miles City, Montana, which went from serving the military sent to subdue the Sioux to serving the cattle raisers who succeeded the buffalo hunters. The eastern demand for western beef grew great enough by the mid-1880s to reach across the Rockies. Ontario, Oregon, replaced Winnemucca as the shipping point in the Great Basin as Chicago superseded San Francisco as the destination of Oregon and Nevada cattle.

In this second cattle boom, large railroad corporations shipped the cattle, and the beef trust slaughtered them. Ranching corporations in the West sought to match this corporate dominance of shipping and slaughtering stride for stride. By the 1880s, western cattle had become creatures of the world capitalist market; the greed of investors became as essential to the propagation of cattle as grass on the western plains. Western banks, particularly those in Denver and Kansas City, had financed much of the early cattle industry at rates of interest ranging from 10 percent to 25 percent, but now eastern and European investors, who commanded much more abundant and cheaper capital, took over the industry.

Cattle corporations were organized in Boston and New York; a large number arose in Great Britain, particularly in Scotland. John Clay, a Scotsman who transplanted to the West, marveled at "this love of money making, enterprise you might call it," that pulled first Scottish investors and then English investors into cattle raising. American investors took money made from mining, railroads, and merchandising and invested it in cattle. In Dundee, Scotland, profits from the jute trade financed western cattle corporations. The thrifty Scot, as Clay remarked, was also the speculative Scot, and this financial schizophrenia found full play in the grasslands. Having gambled on cattle, the Scots tried to enforce precise managerial and accounting techniques in cattle companies that neither knew how many cattle they possessed nor owned most of the land their cattle grazed. The imaginary arithmetic of the cattle business reached new heights in the book count that applied presumed fertility rates to estimates of existing herds and passed off the results as actual cattle. Skeptics in the British financial press complained that purchasing shares of such operations was only playing "poker on joint stock principles." And the players in this particular game—American promoters such as "Uncle Rufus" Hatch, a New York speculator—were not above dealing off the bottom of the deck.

Between 1882 and 1886, as many capitalists as cowboys seemed to be chasing cattle in the West. Much of the total British cattle investment of forty-five million dollars entered the West during these years, and eastern investors outspent the British. Wyoming boasted nearly 100 new cattle companies, New Mexico over 100 (with, however, less funds behind them), Montana 66, and Colorado 176. Some of these corporations grew to gargantuan proportions because the larger the operation, the cheaper it was to raise cattle. Expenses did not increase as a ratio of the size of the herd. In southeastern Oregon, for example, it took three cowboys to care for a herd of one thousand head, but eight thousand head demanded only twelve cowboys.

The holdings of the new cattle companies rivaled European principalities in size. The XIT ranch in Texas arose from the offer in 1879 by a syndicate of Chicago investors to build the Texas state capitol in exchange for the three million acres the legislature intended to sell to pay for the building. The XIT (an abbreviation for "Ten in Texas," referring to ten counties) was unusual not because it was so large but because it owned its land. Other large companies did not hold title. The greatest of them, the Prairie Cattle Company, straddled New Mexico, Texas, and Colorado and contained three divisions, each named after the major river that watered it. Its Arkansas Division in Colorado contained thirty-five hundred square miles; its Cimarron Division in New Mexico had over four thousand square miles. By comparison, its Canadian Division in Texas was relatively tiny: it held only four hundred square miles.

Operations of this size, like other corporations in the West, tried to leave little to chance. They sought a rationalized industry. They wanted to take living animals—cattle—and make the lives of these cattle, from birth to slaughterhouse, as predictable as possible. The companies wanted to control everything about the cattle: their genes, what they should eat and when, how much it should cost to get them to market, and the price they should bring. The cattle corporations wanted to keep labor costs low, and they wanted a dependable, malleable work force.

Cheap labor was one of the appeals of cattle raising, but the corporations wanted to make it even cheaper and far more malleable. The companies' peak labor needs came at roundup and during cattle drives; during the rest of the year, "line riding" cowboys traveled the high ground between watersheds, drifting cattle back toward the ranges claimed by their respective owners. For their labor, these cowboys received only from twenty-five to forty dollars a month plus room and board, and the room was usually as simple as a dugout or a board shack. When cowboys struck for higher wages in the Texas Panhandle in 1883, the Panhandle Cattleman's Association ruthlessly broke their strike. Corporations did not seek romantic heroes for employees; they wanted men who worked cheaply, did as told, and did not get drunk and shoot each other or the cattle. Into the 1880s, the majority of these cowboys were white, with a considerable minority of African Americans working in Texas and Oklahoma, but gradually cattle companies in southern and western Texas began to replace Anglo cowboys with *tejanos* and Mexicans, who drew only one-half to two-thirds the pay of Anglos. Elsewhere, companies began to forbid their cowboys to gamble, drink, or carry six-shooters. R. G. Head, of the Prairie Cattle Company, issued a lengthy circular to his cowboys, reminding them of their moral responsibilities. They took his remonstrances well. Like good company men, the cowboys gave him a silver service when he lost his position as manager.

Cattle raising was becoming an increasingly specialized business by the 1880s. The arithmetic of enterprise, so deceptive in many respects, had accurately revealed that a division of effort best served the industry. A greater percentage of calves survived in Texas than in the cattle country to the north, but young steers fattened more quickly on the northern ranges. The Texas cattleman George B. Loving claimed in 1880 that a Texas steer removed to Nebraska at one or two years old would weigh 1,100 to 1,300 pounds at four years. That same steer, left in Texas, would weigh only 850 to 950 pounds at the same age. Thus, although Texas continued to produce cattle that went directly to market and northern cattle raisers continued to breed their own stock, a rough specialization between southern breeding grounds and northern fattening grounds developed. Many northern cattle raisers annually imported Texas yearlings to fatten on their grasslands.

As specialization increased, a Texas calf (or for that matter a Wyoming calf) was likely to know several owners before its head met a hammer in a Chicago slaughterhouse. Born in Texas and fattened for several years on the northern plains, it might then be shipped to the Flint Hills of eastern Kansas. There it would feed for several months on the lush bluestem grasslands before those grasses began to decline in protein in early July. Now two to three hundred pounds fatter than when it began to gorge on bluestem, the steer would either go directly to slaughter or be sold to a midwestern farmer for further fattening on corn before being killed. By 1883, the Swan Cattle Company had established its own fattening pens near Omaha, where it prepared cattle for the early spring market.

To broker the buying and selling of cattle and the procuring of the grazing land and capital that cattle required, a host of commission merchants acted as middlemen in the business. Most firms were located in Chicago. John Clay was a Wyoming cattleman, but he located his commission firm in Chicago, as did his successful competitor Joseph Rosenbaum.

Cattlemen cultivated an aura of individualism, but the big companies sought oligopoly and a safe market. They organized themselves into powerful local and state associations, which by the 1880s flourished all over the West. The associations held cooperative roundups in the spring to gather and brand newborn calves; they helped establish registry of brands; they hired stock detectives to track down cattle thieves. All attained political influence, but none so great as the Wyoming Stock Growers Association, which faced no major economic rivals within the territory. It represented a concentration of private wealth flaunted in the famous Cheyenne Club, where an eastern visitor reported watching a member simultaneously play tennis and carry on a chess game at the side of the court while periodically refreshing himself with bourbon.

The cattle raisers had trouble extending this regional influence into the national arena. A national organization of cattlemen did not appear until 1883, and it almost immediately split over measures to be taken against the cattle diseases pleuropneumonia and the still prevalent Texas fever. Pleuropneumonia had led to a British embargo of American cattle imports, and northern cattlemen backed an Animal Industry Bill for federal inspection. Southern plains cattlemen, who feared bans on the movement of southern cattle, were already at odds with northern plains cattlemen in the old dispute over Texas fever. As northern cattle raisers turned to improved eastern stock, many came to fear the continued importation of Texas cattle because of both Texas fever and competition for the diminishing grass. In 1885, cattle raisers and farmers in Kansas and Colorado succeeded in getting their legislatures to pass laws that virtually prohibited the driving of Texas cattle into those states. Texans responded with demands that Congress create a national cattle trail so that their herds could move north.

Settling these disputes provided an occasion for the expansion of federal power. Congress established a Bureau of Animal Husbandry but refused to establish a national cattle trail. The Supreme Court ruled that the laws of Kansas and Colorado represented an unconstitutional restraint of commerce by the states, and the federal Bureau of Animal Husbandry took over quarantine inspection and regulation. By the late 1880s, the days of the long drive and the open range were numbered. When Texas yearlings went north, they increasingly went by rail.

Cattle raisers also resorted to soliciting federal intervention to settle disputes with the packers and railroads. Cattle raisers resented the control they believed Chicago packinghouses exerted over prices and the rates the railroads charged. But here they faced powerful enemies, and their victories were minimal. In the 1890s, the Interstate Commerce Commission did secure somewhat lower railroad rates, but cattle raisers failed to get effective federal aid against the packers.

The public domain represented the greatest failure of the cattle corporations to turn regional economic power into national political power. The "free grass" of the public domain that had created the cattle kingdom also presented the greatest danger to the corporations. For if the grass was free to the first cattle raiser to come across it, it was also free to the second, third, and fourth. And not even the wonderful arithmetic of the cattle business could feed an indefinite number of cattle on the same blades of grass. The first or second cattle raisers in an area saw themselves as hardy pioneers and men of enterprise; they saw the third and fourth cattle raisers in an area as "range pirates" stealing "their" grass. To deny competitors access to the ranges already claimed by members, the associations denied range pirates the right to participate in the roundups and refused their herds the protection of stock detectives.

With infinite ingenuity, cattle raisers sought ways to keep grass on public lands for themselves while denying it to competitors. There were legal ways to do this in some

Like many ranch workers of the late 19th century, these men were actually corporate cowboys, employed by a far-distant company.

Unidentified photographer. Home Ranch of the Prairie Cattle Co. Ltd., Scotland *(New Mexico). Photograph, ca. 1888. The Denver Public Library, Western History Department, Denver, Colorado.*

areas, but staying within the law was usually expensive or impractical. In Texas, cattle companies could and did lease state lands. Elsewhere they could purchase railroad lands or school lands. But a company that paid for land was obviously at a disadvantage when competing with a company that got its land for free. And so cattle raisers perfected the technique of obtaining large amounts of grazing land with few purchases. They turned aridity to their advantage by securing water rights and then enforcing extralegal customary rights by which those controlling a stream possessed range rights on its watershed. Controlling the lands bordering a stream thus meant the exclusive use of a much larger area of rangeland lying around the stream. This system of range rights originated in Texas and flourished into the 1880s. In 1882, for example, the Matador cattle company of Texas claimed a range of over a million and half acres on the basis of one hundred thousand acres of waterfront held in fee simple.

When purchasing land or enforcing extralegal rights proved impossible, the cattle corporations resorted to fraud. On occasion, cattle raisers could buy out legitimate homesteaders on riparian lands, but more often they resorted to "dummying" land along the streams. They used their cowboys as "dummy" homesteaders to file fraudulent claims on the land; the cowboys then transferred the land to the cattle raisers. Similarly, cattle raisers filed false claims under the Timber and Stone Act, the Desert Land Act, and the Timber Culture Act.

But barbed wire proved to be the cattle raisers' best friend. In the early 1880s, cattle raisers eagerly used the new technology to defend "their" grass. Before the invention of barbed wire, cattle raisers had bitterly opposed fencing laws designed to confine cattle, but inexpensive barbed wire changed their minds. In the early 1880s they fenced vast acreages of the public domain, even public roads, to keep out competitors, to cut labor costs by preventing their cattle from drifting, and to ensure that their improved bulls did not waste valuable semen on neighbors' cattle. When smaller cattle raisers violently objected, the result was fence-cutting wars such as the one that erupted in Texas in the mid-1880s.

To protect their claims to the public domain, the cattle companies launched a battle for the leasing of federal lands. In doing so, they challenged deep-seated beliefs that agriculture was the proper ultimate use of all land. They also faced popular resentment, for it angered much of the public that foreign corporations could control public lands and exclude American settlers. President Grover Cleveland refused to allow leasing, and in 1885 he issued an executive order for the removal of fences on public land. He evicted herds of cattle trespassing on land in the Indian Territory, and Congress in 1887 passed laws banning foreign land purchases in the territories.

Congress and the White House were the decisive theaters of the battle over land, but a much nastier and often bloody conflict was waged in the West itself. In defense of customary rights, which were sometimes enshrined in state or territorial law, cattle raisers sought to drive competitors—whether sheep raisers, small stock owners (often ex-cowboys), or farmers—from the public domain. The associations redefined legitimate enterprise so that what had once been regarded as the natural progress of enterprise was now seen as theft.

In the early Texas cattle industry, enterprise alone could turn a cowboy into a cattleman. A cowboy who assisted at branding received a portion of the cattle in return. Cowboys too could, with little trouble, take up mavericks (motherless calves), whose ownership could not be determined, and thus become cattlemen. Indeed the gendered terms of *cowboy* and *cattleman* themselves seemed to embody a natural progression that both connected men (but not women) to the raising of large meat animals and correlated the progression from caring for animals to owning them with the growth from boyhood to manhood.

As cattle grew more valuable and range scarcer, large stock raisers sought to stop the mavericking that cowboys had come to regard as their rightful route to independence. Mavericking became an object of dispute in part because mavericks could be created as well as found, and the bigger operations sought to end a practice that at once drained their own herds and created competitors on the public domain. In Colorado, Wyoming, and elsewhere, mavericks were declared the possession of the stock associations and were annually auctioned off. Mavericking—a way to begin a career of enterprise—became rustling—a way to begin a career of crime. Small stockmen did not always agree with this redefinition of enterprise, particularly when the men who claimed higher virtue were themselves appropriating millions of acres of public land while protesting the actions of those who appropriated a few stray cattle. Such differences of opinion lay behind the Johnson County War of 1892, when members of the Wyoming Stock Growers Association hired gunmen to clear out small ranchers whom they accused of rustling.

The Johnson County War was the act of desperate men, for the days of the large cattle corporations were already numbered. They had engineered their own demise. As cattle prices rose in the early 1880s, cattle raisers had crammed animals onto the ranges. Roughly twenty million cattle grazed the American West by 1884, and although both federal studies and some stock raisers denied it, there were fears that there was no longer either enough grass to feed the cattle or enough consumers with the money to buy the beef. When the country once more became mired in depression, cattle, as commodities, went into decline. In 1885 the price for young steers fell dramatically. Too young for prime beef, and with no ranges open to them, they were a glut on the market.

Cattle could not feel their decline in exchange value, but as living things that experienced hunger, cold, and pain, they could feel the consequences of the human greed that had pushed vast numbers of them onto grasslands that could no longer feed them in climates where bad winters could bring enormous suffering. In the 1880s, drought and bad winters combined to inflict damage on western cattle, sickening those who saw it. The winter of 1880–81 brought devastating losses to the herds of the Columbia Basin, reaching 50 percent or more in some counties. Blizzards and bitter cold on the southern plains in the winter of 1884–85 forced hungry cattle to drift before the storms. But the animals' instinctual reaction came up against the new barbed-wire fences, where the cattle piled up and died. Hungry cows aborted their calves. The ones that gave birth were too weak to feed their offspring, and both cow and calf died. The calf crop plummeted. These bad winters moved like serial killers across the West. In 1886–87, a bitter winter followed a hot, dry summer on the northern plains. White arctic owls, for the first time in nearly a generation, appeared in Montana; the chinooks, the warm winds off the Pacific, never arrived, and the result was, as John Clay remembered, "simple murder." Cattle died as they had on the southern plains. Because the cattle companies operated on book counts, no one could be sure how many cattle died, since no one was quite sure how many there had been to begin with. Many companies took the opportunity to claim huge losses and thus remove the discrepancy between their book counts and their actual herds. Probably about 15 percent of the herds died, many more in the hardest-hit areas. Most of the survivors were weak, emaciated, and disfigured from the cold. To meet their loans, cattle raisers rushed many of these steers onto a declining market, pushing prices down farther. In Chicago, cattle worth $9.35 a hundred weight in 1882 brought $1 in 1887. In the Great Basin, 1889–90 was the terrible winter.

The suffering of livestock in these winters affected even hardened cattlemen; they could not, it turned out, fully commodify their animals. Confronted with carcasses of dead cattle and with living animals so weakened that they could not move from the mudholes that mired them, Granville Stuart found what had been a fascinating business suddenly distasteful. "I wanted no more of it. I never wanted to own again an animal that I could not feed and shelter." A concern for the humane treatment of animals destined for slaughter had begun to infringe on the concerns of enterprise well before Stuart's revulsion. The American Humane Society had obtained legislation governing the shipment of cattle as early as 1873, and their disputes with cattle raisers continued for the remainder of the century. Stuart's reaction only revealed the ambivalence of the cattlemen themselves.

Falling markets and the losses of these terrible winters hurt the industry in general, but some companies did better than others. Those companies that, in the derisive words of an English journal, were mere "hunters of wild cattle," although they styled themselves "graziers on a princely scale," did not survive. Better-run companies, such as the Matador of Texas, rode out the bad times, for they had managed to bureaucratize and centralize the business. They endured losses in the 1880s and in the early 1890s, but they bought land, upgraded their stock, leased and purchased northern grazing lands for fattening, and raised hay. Their cowboys became company men who were as likely to dig irrigation ditches, put up fences, or cut hay as ride herd.

But even for the better-run cattle companies, the remainder of the century was more respite than reprieve. There would be another speculative boom, and another bitter collapse after the depression of the early 1890s, but the direction of stock raising had changed. The need to winter feed improved stock changed cattle raising into ranching and reduced its scale. The future lay with small operations of two hundred head or less. These cattle were still seasonally run on the public lands, but they relied on alfalfa and sorghum for the majority of their winter feed. Ranchers also marketed animals more quickly. Because young steers put on weight more rapidly than older cattle, ranchers realized the greatest return on investment by selling steers as soon as their ability to turn feed to fat began to slacken. Eventually, with selective breeding, they could market a steer after only two years. A relatively small herd of improved cattle with a rapid turnover of animals could yield more beef and more profit than the larger herds of the old free-range systems.

As the twentieth century progressed, these reduced operations came to depend on federal permits to run cattle, first in the new national forests and then, after the Taylor Grazing Act of the 1930s, on the remainder of the public domain. Ranchers depended on water from federal projects to fill their irrigation ditches. They raised improved breeds—mainly Herefords—and sent them for fattening to farmers in the corn belt. Corporate control did not vanish; it just became less obvious to outsiders. By the late 1890s, the packinghouses had largely taken over the financing of the cattle industry. The big packinghouses loaned money to commission agents, who loaned it to farmers seeking to buy feeders from the western range. The agents held a mortgage on both the corn and the steer. Long before the steer ever reached Chicago, Kansas City, or Fort Worth, the packers controlled it.

In the late nineteenth and early twentieth centuries, sheep invaded the shrinking domain of cattle. Some of the early boosters had been undifferentiated enthusiasts of grazing animals, and in Montana and elsewhere there were men who ran both sheep and cattle and their range. But on the whole, sheep raisers and cattle raisers competed. Cattle raisers and homesteaders, who resented the presence of tramp sheepmen whose herds consumed grass they wanted for their own cattle and horses, complained that cattle hated the smell of sheep and would not prosper where sheep had grazed. Sharpening this competition was the status of the sheep as an ethnic animal. The owners of sheep were often either immigrants or nonwhites; shepherds were usually Scotsmen, Basques, or Mexican Americans. The Mormons too were often sheep owners; not only were a high proportion of Mormons immigrants, but in the nineteenth-century West, all Mormons were regarded as un-American. To many Anglo-American cattlemen and cowboys, sheep were inferior animals herded by inferior men. The constant care that sheep demanded, the hostility they often created, and their pervasive smell and sound made sheep raising a life utterly without romance.

Yet cattle raisers could not stop the rising tide of sheep. Cattlemen constantly threatened violence; they occasionally slaughtered sheep or burned the haystacks of farmers who sold hay to shepherds; they more rarely killed shepherds. But the numbers of sheep inexorably increased. In New Mexico, the old heartland of sheep, sheep raisers crossbred the *churros* with Merinos, raising the wool yield to four or five pounds, as they found a profitable market for wool in the East. The number of sheep in New Mexico rose to nearly five million in the late 1880s, and their ownership grew even more concentrated. On the upper Rio Grande, Frank Bond and Edward Sargent acquired grazing rights on so much public and private land that many formerly independent *nuevo mexicano* sheep ranchers could not find grazing land and had to sign on as *partidarios* with Bond or Sargent. In Arizona, sheep spilling over from California joined Navajo sheep and Mormon sheep from southern Utah. In northern Arizona, sheep outnumbered cattle ten to one as early as the late 1870s. On the Columbia Plateau, sheep outnumbered cattle four to one in 1890, and by 1900 the more than one million sheep outnumbered cattle eight to one. These sheep were Spanish Merinos and French Rambouillets.

Although cattle raisers did not realize it, their own cattle formed a fifth column paving the way for sheep. As cattle overstocked the ranges and overgrazed the bunchgrasses, they opened up the land to invasion by exotic forbs (that is, nongrasslike herbs). Cattle did not thrive on this weedy growth, but sheep did; in addition, sheep could crop short grasses left behind by cattle. The growth of weedy forbs peaked early, but sheep raisers compensated by adopting a pattern of transhumance, a system of grazing in which herds grazed lowlands in winter and spring when the mountains were covered with snow and then moved into the mountains in the summer as the snow retreated. In the mountains, the sheep, or hoofed locusts as John Muir called them, devastated the mountain pastures.

In the 1880s, sheep raising outside of New Mexico offered poor people the shot at enterprise that the larger cattle companies sought to deny small cattle raisers. In the Columbia Plateau, poor immigrants could work herds on shares, taking minimal pay plus a claim on part of the lamb crop. Sheep required a much smaller initial investment than cattle, and since they matured more quickly, provided quicker returns. They could survive on lands that cattle could not, and they could go much longer without water, thus opening range closed to cattle. The very docility that made them so tempting to predators meant that a single herder and a pair of dogs could control a band of fifteen hundred to three thousand sheep. Their wool was a commodity that could be stored indefinitely, was protected by tariff from foreign competition, and could be harvested annually with little harm to the animal.

But sheep also exacerbated the problems of the open range. Tramp sheep raisers competed with each other, and the only way to keep outfits off a range was, as an observer remarked, "to strip it utterly naked," a practice known as "sheeping off" the range. With the expansion of farming into both the high plains and the Columbia Plateau, and the decision of railroads to ban grazing on their land grants without leases, sheep raisers found their winter ranges constricted; with the creation of national forests, the government began to regulate their use of summer ranges as well. Like cattle raisers before them, sheep raisers found they could not control the open range. Those operations that survived had to turn to leasing lands from the railroads, obtaining grazing permits in the national forests, and purchasing other land to produce hay for winter feed. The capital demands for sheep raising necessarily increased; only those who could meet them would remain in business. And sheep too, as animals of enterprise, were creatures of the market. The Panic of 1893 and the depression that followed sent wool and mutton prices plummeting to their lowest levels since the Civil War.

When sheep battled cattle for the land, the fight was between two industrial animals—animals whose very bodies, whose genetic makeup, humans had altered through selective breeding to fit their needs and whose every part humans processed into a product. Reshaped, these animals, in turn, reshaped the land. The results of their relentless overgrazing differed from place to place, but everywhere they opened up the land to invasion by other exotic species, everywhere they changed the composition of plant communities, and virtually everywhere they brought increased erosion. The changes varied with the local environments that the animals grazed and with the use or suppression of fire. On the Great Plains, the exotic Russian thistle—the tumbling tumbleweed of the cowboy songs—became such a familiar mark of overgrazing that people came to think of it as a native. In the California mountains, light burning and overgrazing by sheep restricted forest regeneration. In the Southwest, overgrazing by cattle and the banishment of fire led to the expansion of juniper and piñon into what had been grasslands. In the Great Basin, overgrazing eliminated native bunchgrasses, stripping the landscape to stark sage-dominated communities whose missing understory provided a vacuum into which exotics like cheatgrass could expand. In the Sangre de Cristos of New Mexico, native grasses such as Thurber fescue and alpine timothy became mere remnants amid the invading bluegrass and forbs such as yarrow and fleabane.

Industrial Animals and Animals of Leisure

The introduction of large herds of cattle and sheep also threatened wild animals, which now became either competitors with or predators on the domestic livestock. Americans, it is true, regarded game animals as useful in the sense that they provided food, but only temporarily useful. These animals were a resource meant to be used up and replaced by domestic stock. Elk, deer, mountain sheep, and any animal that might make a meal fell before the rifle, the victims of a remarkable slaughter.

The slaughter extended beyond animals that pleased the human palate. The western devotion to making the land as comfortable as possible for cattle and sheep led to a relentless campaign against animals that might make a meal of domestic stock. Stockmen shot wolves on sight, but since wolves prudently learned to avoid the sight of stockmen, ranchers and cowboys resorted to putting strychnine in animal carcasses. In a campaign that continues in some parts of the West to this day, bears, mountain lions, wolves, coyotes, wildcats, and lynxes were poisoned, trapped, and shot. Eventually grizzly bears disappeared over most of the contiguous American West, mountain lions over much of it, and wolves over all of it.

Other animals died not because they ate domestic animals but because they ate grass that domestic animals might eat. In Colorado, between 1 August 1885, and January 1886, the Bartholf brothers killed 1,080 antelope that made the mistake of grazing on "their" range. Ranchers also killed elk, which they claimed made their cattle wild. This kind of slaughter continues today on the ranges of Arizona. The carnage extended even to ground squirrels, pocket gophers, jackrabbits, skunks, hawks, and, above all, prairie dogs. Employing the kind of arithmetic that ranchers, U.S. Biological Survey employees, and later agricultural scientists and Forest Service personnel found irresistible in rearranging the western landscape, Frank Benton calculated that the elimination of one huge prairie dog town in Wyoming could alone provide a home and meals for 180,000 more cattle. Wyoming was shortgrass country, but in tallgrass environments, at least, overgrazing by cattle promoted the expansion of the very prairie dogs that cattlemen tried to eradicate. When grazing was restricted, the prairie dog towns shrank. Neither enterprise nor slaughter always yielded the desired results.

Finally, animals died as an unintentional result of livestock raising. Pronghorn antelopes, for example, could easily move through conventional barbed-wire fences, but close-woven sheep fences confined them. They died cornered in blizzards or confined to forage-depleted ranges.

If the citizens of the newly industrialized Republic had retained a strictly instrumental attitude toward animals, species after species would have yielded to the animals of enterprise. But industrialism created not only an intense pride in what human ingenuity had produced but also an increasing nostalgia and fear over what disappeared with that triumph. The loss was cultural. If game animals vanished, Americans could easily replace their meat, but could they, some began to ask, replace the masculine virtue cultivated by killing game?

Sportsmen, as distinct from hunters, began to argue that a particular kind of virtue—hardiness, bravery, self-reliance—impossible to cultivate in an urban, industrial environment was the true product of the hunt. This attitude arose among an eastern elite, who themselves tended to ape the English, but it diffused downward from elite groups like the Boone and Crockett Club to other hunters. Because such killing sought virtue rather than profit, sportsmen had to engage in a "fair chase" and abide by the "sportsman's code." Sportsmen disdained market or subsistence hunters as mere pothunters, and they struggled to institute and enforce state game laws and bag limits. By the early twentieth century, they had clearly won their battle to protect game populations. By dying so that American males could maintain their virility and virtue, game animals achieved a symbolic utility and a protected status.

Relatively few animals, east or west, qualified as game, but other animals benefited from the rise of middle-class nature appreciation. Nature appreciation usually operated within the same middle- and upper-class milieu as sport hunting but on the opposite side of the gender division. Nature appreciation was thought to be appropriate for women and children, who came to treasure the common experiences of wood and field that urbanization and industrialism made rare. Coupled with nature appreciation was the rise of "animal psychology," which sentimentalized and anthropomorphized wild animals. Animals became lovable—"our friends in fur and feathers," as nature stories put it. During the early 1890s, Ernest Thompson Seton and Charles G. D. Roberts developed the animal story into a special genre for which there was a huge popular appetite. Although nature lovers and hunters arose from common roots and appealed to the same classes, there was a latent conflict between the two groups. Both wanted to save Bambi, but the sportsmen wanted to shoot Bambi when he grew up.

Although a product of industrial society, sport hunting and nature appreciation in one sense opposed that society's instrumentalist tendencies; in another sense, however, they opened up a new basis for the commodification of wild animals. Displaced by

Fellow creatures of leisure, a bear and a tourist meet in Yellowstone National Park, where elk, bear, and buffalo populations reflect deliberate wildlife management policies.

Ellen Todd. Yellowstone Bear and Tourist. *Gelatin silver print, ca. 1935. The Denver Public Library, Western History Department, Denver, Colorado.*

animals of enterprise, wild animals became animals of leisure. And it turned out that deer and other game animals could yield revenues as people consumed them while at play. There were the obvious revenues that the state obtained through license fees, but also the revenues that hunters (and sport hunting became a mass sport in the twentieth century) produced in buying supplies and equipment, in travel, and in obtaining food and lodging. Hunters were armed tourists; keeping them happy meant keeping game populations high enough that they had a likely chance of killing something. All of this gave game animals a commercial value, so that even those skeptical about the cultivation of virtue could see convincing financial reasons for maintaining their populations and introducing new and exotic things for people to shoot at. Hunters, and those who profited from hunters, became the main political and economic supporters of wildlife preservation programs, programs whose results became more apparent as the twentieth century wore on.

Some domestic animals too became animals of leisure. The dog and cat had long ago become leisured, and in the twentieth century the horse, for all practical purposes, joined them. These species were almost wholly consumers, creatures that absorbed wealth and did not produce it. Horses, because of their size and the expense of maintaining them, became a special symbol of status, and in the West, part of that status, and part of their perceived value, came from their connection to the Old West of cowboys, Indians, and overland emigrants. The horse's status as an animal of leisure was far more exalted than that of wild game, since horses were usually neither killed nor consumed.

Only predators and rodents initially resisted classification as either lovable or leisured animals. Before Mickey Mouse, rodents failed to be lovable animals, but despite mass poisoning campaigns by the U.S. Biological Survey, they survived. Large predators were not so lucky. Sport hunters disliked predators because they killed game. Ranchers disliked them because they killed domestic stock. Nature lovers disliked them because they killed cute animals. They were, in the early nature-appreciation literature, cruel and murderous. At the turn of the century, even the Audubon Society recommended the killing of hawks, owls, and foxes. In 1915 the federal government took over the war on predators. The U.S. Biological Survey eliminated the last breeding wolf packs in the Dakotas, Wyoming, Colorado, New Mexico, and Arizona by the 1920s. The ecological effect was to make humans the only predator on some large-game species.

These changes in attitude occurred nationally, but they had particular consequences in the West, which contained the largest amount of remaining animal habitat and the largest reservoirs of remaining wild species. Yellowstone National Park represented perhaps the most extreme version of these consequences. It does not appear that Yellowstone was particularly rich in wild animals before white settlement, but the combination of habitat destruction and overhunting outside the park made it an unintentional haven for game. As the twentieth century wore on, elk, bears, and buffalo began to overshadow the geysers and waterfalls as tourist attractions. Government hunters eliminated wolves within the park, and a ban on hunting eliminated human predation. Elk, in particular, increased and thrived even though the park could not provide the whole herd with a winter range. Although presented as a salvaged remnant of aboriginal America, the park by the late twentieth century came more to resemble a petting zoo with a highway running through it. In Yellowstone the commodification of wild animals was everywhere apparent, and a closer look revealed the damage that elk and bison could do by overgrazing a habitat artificially shaped by park boundaries.

Animals of enterprise and animals of leisure were the two sides of the coin of industrialization in the West of the late nineteenth and early twentieth centuries. There was no escape from a logic of commodification that eventually even commandeered the memories of an older West. Animals became symbols of the lost West, symbols of a freedom and wildness that could, advertisers promised, be acquired if one bought an automobile that was also a Mustang, or a Bronco, or an Eagle. Even manitous—the other-than-human persons, often in the form of animals, that Indians conceived of as giving them the qualities they otherwise lacked—could be commodified and mass-produced. Motorized manitous could deliver other-than-human aid without the necessity of the fasting and self-mutilation required to appeal to the older manitous. These new manitous, of course, were not persons; they were only machines. They evoked the old to expand the new. They, as much as the giant feedlots, genetically engineered cattle, or Yellowstone, represented the logical culmination of the triumph of animals of enterprise in the West.

Dances with Wolves, a movie that reached theaters as the twentieth century entered its last decade, reflected even in its title the symbolic burden that animals had come to bear when Americans thought about the West. The first environmentalist Western, the film reversed all of the verities of enterprise. In the film, white soldiers and pioneers were "bestial" in an older sense: they were filthy and greedy, befouling themselves and everything they touched. Their enterprise was mere slaughter and destruction. Animals, especially horses and of course a wolf, were noble and "human": loyal and self-sacrificing. These were animals from the pages of Ernest Thompson Seton—the great sentimentalist of nature and outdoor life. In its depiction and condemnation of the slaughter of bison and its portrayal of a human-animal relationship that went beyond utilitarianism, the film seemed to reject the commodification of animals that had so marked the history of enterprise in the American West. And yet, the film itself was a commodity; the audience paid for the sentiments, and the animals were highly trained. Regarded this way, the film was also what it condemned: yet another stage in the evolution of animals of enterprise in the West.

Bibliographic Note

An interested reader can re-create most of the sources for this article from the bibliographies of recent publications, so I will mention these recent works as well as those sources that readers might not normally consult.

For a wider perspective on these issues, see Alfred W. Crosby *Ecological Imperialism: The Biological Expansion of Europe, 900–1900* (Cambridge, Eng., 1986). For the organization of the livestock industry, see Mary Yeager, *Competition and Regulation: The Development of Oligopoly in the Meat Packing Industry* (Greenwich, Conn., 1981), J'Nell L. Pate, *Livestock Legacy: The Fort Worth Stockyards, 1887–1987* (College Station, 1988), Gene M. Gressley, *Bankers and Cattlemen* (New York, 1966), Margaret Walsh, *The Rise of the Midwestern Meat Packing Industry* (Lexington, 1982), and Sigfried Giedion, *Mechanization Takes Command: A Contribution to Anonymous History* (New York, 1948), for a start. For American dietary habits, see Harvey A. Levenstein, *Revolution at the Table: The Transformation of the American Diet* (New York, 1988).

Works on the plains cattle industry are too numerous and uneven to cite here, but along with the older, classic works, Don Worcester's *The Texas Longhorn: Relic of the Past, Asset for the Future* (College Station, 1987), is recent and worthy of note. Maurice Frink et al., *When Grass Was King: Contributions to the Western Range Cattle Industry Study* (Boulder, 1956), Charles L. Wood, *The Kansas Beef Industry* (Lawrence, 1980), J. Orin Oliphant, *On the Cattle Ranges of the Oregon Country* (Seattle, 1968), W. M. Pearce, *The Matador Land and Cattle Company* (Norman, 1964), Harmon Ross Mothershead, *The Swan Land and Cattle Company, Ltd.* (Norman, 1971), James A. Young and B. Abbott Sparks, *Cattle in the Cold Desert* (Logan, Utah, 1985), and John T. Schlebecker, *Cattle Raising on the Plains, 1900–1961* (Lincoln, 1963), provide regional or corporate studies. The chapter on meat in William Cronon, *Nature's Metropolis: Chicago and the Great West* (New York, 1991), is essential reading.

For sheep and culture in the Southwest, see William deBuys, *Enchantment and Exploitation: The Life and Hard Times of a New Mexico Mountain Range* (Albuquerque, 1985), and John O. Baxter, *Las Carneradas: Sheep Trade in New Mexico, 1700–1860* (Albuquerque, 1987). For the Northwest, see Alexander Campbell McGregor, *Counting Sheep: From Open Range to Agribusiness on the Columbia Plateau* (Seattle, 1982)

For animal power, see Henry P. Walker, *The Wagonmasters: High Plains Freighting from the Earliest Days of the Santa Fe Trail to 1880* (Norman, 1966), Morris F. Taylor, *First Mail West: Stagecoach Lines on the Santa Fe Trail* (Albuquerque, 1971), and Oscar Winther, *The Transportation Frontier: Trans-Mississippi West, 1865–1890* (New York, 1964), as well as W. Turrentine Jackson, *Wells Fargo in Colorado Territory* (Denver, 1982), and Jackson's other works on the Wells Fargo Company.

For the fur trade, see David Wishart, *The Fur Trade of the American West, 1807–1840: A Geographical Synthesis* (Lincoln, 1979). For buffalo, the classic study is still Frank G. Roe, *The North American Buffalo: A Critical Study of the Species in Its Wild State*, 2d ed. (Toronto, 1970), but Dan Flores's recent "Bison Ecology and Bison Diplomacy: The Southern Plains from 1825 to 1850," *Journal of American History* 78 (September 1991), is an extremely important new study. For changing attitudes toward animals, see Thomas R. Dunlap, *Saving America's Wildlife* (Princeton, 1988), and Roderick Nash, *The Rights of Nature: A History of Environmental Ethics* (Madison, 1989).

Chapter Eight

An Agricultural Empire

ALLAN G. BOGUE

From the early nineteenth century through the first decades of the twentieth, settlers created farms in the American trans-Mississippi West. Each life story in this process was unique, but similar elements might be found in all of them. Collectively, these biographies compose the history of American agriculture and settlement west of the Mississippi. The settlers directed torrents of grain and mountains of meat, hides, wool, and cotton to the markets of the older states and of the world and by their example drew millions of hopeful compatriots and immigrants into the American West to share in the adventure of planting European-American settlements.

Among these hopefuls was Joseph Fish's Yankee grandfather, who pushed across the Vermont boundary line, without realizing it, and settled in lower Canada. A generation later, Joseph's mother converted to the religion of the Latter-day Saints (Mormons), and the family moved with other church members to Will County, Illinois, where her husband, Horace Fish, also accepted the faith. In the fall of 1840, when young Joseph was a few months old, the Fish family moved to Nauvoo. During the persecution of the Saints and the ensuing exodus across Iowa, the family lost most of its possessions. In 1850, after three years in Council Point, the family continued to Deseret. Horace Fish prospered at Circleville on a twenty-five-acre holding, but in 1852 he moved to Parowan to join in the development of the southwestern Utah Territory. The family set out in a heavy snowstorm, but twelve-year old Joseph cared not; he was wearing his first pair of new boots.

The forted village of Parowan, where Joseph Fish lived until 1878, lay on Center Creek upstream from Little Salt Lake. Since his father was both miller and tanner, much of the agricultural labor fell to young Joseph. The Fishes had a holding in the community's common field, and there they planted potatoes and grain, with several neighbors typically exchanging work in the cultivation and seeding of their plots. To prepare the community field, the settlers pulled clumps of sagebrush with oxen and chains and used wooden plows reinforced with iron from the rims of wagon wheels. Irrigation was essential; ditching, as well as fencing, had to be done, and water needed to be diverted from the ditches to the crops. Joseph learned to cradle two acres of wheat or oats in a day during harvest. Sometimes he helped guard the community herd or helped drive livestock to new range. In winter, Fish often cut timber for the village homestead or for making pine tar.

The Parowan settlers feared hostile Indians and rustlers; the community militia drilled frequently, and males were armed. Joseph helped build roads and hunted wolves

Though engaged in a common sort of labor, every western farm family had its own story to tell. Like his neighbors, Ether Blanchard relied on family assistance at harvesttime. But Blanchard, who was also a poet, insisted on harvesting the grain on his 13-acre Utah farm with an outmoded cradle scythe.

George Edward Anderson (1860–1928). Ether Blanchard Farm, Mapleton, Utah. *Gelatin silver print, 1902. Harold B. Lee Library, Brigham Young University, Provo, Utah.*

or waterfowl. However busy they were, residents found time for the religious observances typical of Mormon settlements, for sessions of lay school, and for theatrical entertainment—in which Joseph joined—as well as for frequent games of shinny or horseshoes. This was Parowan during the 1850s and early 1860s.

Farther west, the enumerators of the federal census in California in 1850 recorded several hundred rancheros. Among them was Don Carlos Antonio Carrillo, a prominent resident of Santa Barbara and owner of the Rancho Sespe. Born in 1783, the son of a Spanish Army captain, Carrillo enlisted as a private at fourteen and left the service as a sergeant while still relatively young. Thereafter, he served in the California assembly, as well as representing his region in the National Congress of Mexico. His political fortunes climaxed with his appointment as governor of California, an honor that was contested by a rival who pressed his cause in Mexico City and in California. In 1838, California residents on both sides of the dispute engaged in a short period of comic-opera warfare. When the Mexican government recognized Carrillo's rival, Don Carlos accepted the decision philosophically and returned to the life of a "pobre ranchero," regretting only that his political career had been expensive.

In 1829, Carrillo petitioned the California authorities for six leagues (twenty-four thousand acres) of land along the Sespe, a tributary of the Santa Clara. At the time, Carrillo owned five hundred head of cattle and two hundred horses and mules. Carrillo established his ranch on the Sespe, but he did not perform the ceremonial crossing of his boundaries until 1842, when, according to the rancho's historian, he "pulled up grass, scattered handfuls of earth, broke off branches of trees, and performed other acts and demonstrations of possession." By 1845, he was running approximately three thousand cattle, droves of horses and mules, and flocks of sheep. A two-story adobe house served as headquarters for the rancho, which included two vineyards and two cultivated fields. In these days of the hide and tallow trade with the United States, Carrillo's herds provided a comfortable life marked by visiting, lavish hospitality, fiestas, and rodeos featuring brilliant horsemanship.

The historian of Rancho Sespe pictured Carrillo and his sons leading the leisurely annual expedition of family and servants to the ranch, all astride spirited Arabian horses, saddles and bridles sparkling with silver. The men were clad in "short breeches extending to the knee, ornamented with gold or silver lace at the bottom . . . soft deer skin [leggings] well tanned, richly colored and stamped with beautiful devices . . . tied at the knee with a silk cord . . . with heavy gold or silver tassels . . . long vests, with filigree buttons of gold or silver . . . [jackets] of dark blue cloth . . . long serape or poncho made in Mexico . . . [vicuña] hats imported from Mexico and Peru." Behind rode the women and children in *carretas*, great squeaking oxcarts. At the ranch was the commotion of the annual rodeo: the sounds of bawling calves and pounding hooves mixed with the shouts of vaqueros and with the thick dust that accompanied roundup. The calves were branded and perhaps a herd selected for slaughter.

Hard work was balanced by leisure; the Carrillo men hunted grizzly bears, attended bullfights, and enjoyed the fiesta. Initially, Carrillo benefited under American rule; goldfield markets raised the value of his herds to unimagined levels. In 1852, at age sixty-nine, the old soldier died. The United States Land Commission approved the Carrillos' title to Rancho Sespe in April 1853, and in September, the administrator of the estate auctioned the land. Three brothers named Moore, cattle buyers from Ohio, purchased the rancho for $18,500, about seventy cents per acre.

In Iowa, we find a different story. "Our land is beautiful, though there are few trees . . . . We have good spring water near by. Best of all, the land is good meadowland and easily plowed and cultivated." So wrote Gro Svendsen from Emmet County, northern Iowa, in November 1863. This young woman had accompanied her husband, Ole, from Norway the previous year. After time with friends and relatives in eastern Iowa, the couple established their "new home" in the prairie country drained by the upper Des Moines River. Gro died in 1878 after the birth of her tenth child. The daughter of a respected teacher in Hallingdal, Gro had recorded the details of her life in letters to her parents and siblings in the old country.

Ole Svendsen had hoped to homestead his land and erroneously believed that he must be naturalized to do so. Drafted in 1864, he served in the South with General William Sherman, returning unscathed. Home again, Ole purchased a timber lot to provide boards, rails, and firewood; he tilled more acreage, harvested good crops of wheat, and saw his livestock increase. Meanwhile, Gro made and sold butter. Aided occasionally by work and money from Ole's relatives, the Svendsens never knew abject poverty. But the 1870s were hard years—agricultural prices were low, grasshoppers damaged crops, and the resolution of real estate titles in the area left the Svendsens indebted to a railroad.

The school system was rudimentary in Emmet County; Gro taught her children in both Norwegian and English. She read assiduously, and on summer evenings she liked to sit in the yard and play the alpenhorn. She and Ole were active in the local Lutheran congregation. As the years passed, Gro corresponded less frequently, but her longing for her family remained strong. The Svendsens sent family portraits to Norway and affectionately scanned the pictures of their relatives. However, the recurrent strain of childbearing, her duty as scribe for less-educated neighbors, and constant work wore Gro down. In her thirties, Gro examined a photo of herself and believed she saw an old woman.

The experiences of the Fish, Carrillo, and Svendsen families illustrate the diversity of agricultural settlement in the West. Other examples of the varied backgrounds of western farmers abound: peoples of the southwestern pueblos, residents of the plazas of Hispanic New Mexico, Swedish settlers in Minnesota and Washington, Willamette Valley pioneers, participants in the Oklahoma land runs, and reclamation homesteaders waiting for that first rush of water in "the ditch." In each new environment, settlers were challenged to create productive farms, to modify old ways, to make do with inadequate resources, and to reconstruct the relationships and boundaries of the community.

Of course, European Americans were not the first to utilize the agricultural potential of the North American continent. Earlier migrants, the Indians, had adopted farming practices long before the era of European expansion. Natives of the northeastern woodlands were raising corn, beans, and squash when the Puritans arrived, and centuries earlier, Hohokam Indians had tended irrigation canals where Arizonans today practice a more elaborate water husbandry. Apache Indians channeled water to their crops in New Mexico during the eighteenth century. Although some Indian groups depended on hunting, the natives had identified a range of seeds, nuts, roots, tubers, bulbs, leaves, and

The transformation of open range into farmland struck many artists as a metaphor for the shape of western history. For Charlie Russell, devoted artist of the cowpuncher's West, "trails plowed under" signaled the end of the exhilarating freedom of western life and the emergence of a new, more constricted, agricultural society.

Charles M. Russell (1864–1926). Trails Plowed Under. *Pen-and-ink on paper, ca. 1926. From* Trails Plowed Under *by Charles M. Russell. Copyright 1927 by Doubleday, a division of Bantam Doubleday Dell Publishing Group, Inc. Used by permission of Doubleday, a division of Bantam Doubleday Dell Publishing Group, Inc.*

barks to use for food, fiber, and medicine. They sometimes fostered particular flora by planting seeds or otherwise manipulating the environment. As American settlements expanded, the frontiersmen planted Indian corn, potatoes, squash, and beans side by side with the wheat, oats, barley, and peas of Europe.

Among the European Americans who went West, most dreamed of fertile fields and the bounty that cultivation provided. But there is a no-man's-land between farming and pastoralism in American history. Was the young Hamlin Garland, tending herd on the Iowa prairies in the 1870s, a cowboy or a farm boy? Garland herded cattle on grasslands, but he also knew that he, his pony, and the cattle were an integral part of a nearby farm enterprise. Garland's cowboy days illustrate the tendency of frontier agriculturists to use the free range adjacent to settlements. Access to market was difficult for the pioneer farmer, and livestock provided an ideal crop; cattle, sheep, and hogs could be driven to market by their owners or sold to traveling drovers.

For a time in the plains country of the trans-Mississippi West, the cattle and sheepherders ranged far ahead of the pioneer farmers, and the scale of their operations exceeded such activity east of the Mississippi. Even here, however, the herders built on the foundations of other livestock frontiers—northern, southern, and Hispanic—and found markets for feeder animals and supplies of young range stock or breeding animals in the western farm regions. The agricultural settlers of the trans-Mississippi West failed to oust the graziers from all of their domain, but they followed them doggedly. One of

Charles M. Russell's most poignant sketches shows a cow pony sniffing curiously at the edge of plowed land, its grizzled rider dejectedly contemplating the constricted horizons of his future.

Who Were the Settlers?

For over three centuries, Americans expanded the geographic limits of farming. Many of the new farms were maintained by succeeding generations of the settlers' families, but children and grandchildren also established their own farms, and immigrants joined the native-born. There were about 450,000 European-American farmers by 1800. Fifty years later, the number was almost 1.5 million, and by 1910, it was 6.4 million. By the mid-1930s, the total of farm operators had peaked, well short of 7 million.

Of those 1.5 million American farmers in 1850, only 119,000 were located beyond the Mississippi, most of them in the states along the river from Iowa to the Gulf. Sixteen thousand of them farmed in Texas and in the New Mexico Territory. The census takers reported fewer than 1,000 farmers in Utah and California and 1,200 in Oregon. This distribution foretold the future. Farmers moved west in the hundreds of thousands after 1850, but most trans-Mississippi farmers tilled their fields in the two tiers of states adjacent to the river. Thus, in 1890, the 201,903 farms in Iowa more than doubled the total in Washington, Oregon, and California combined. Moreover, the number of farms in the older states west of the great river continued to grow. The 16,552 farms added in Iowa during the 1880s exceeded the combined total in Montana, Wyoming, and Idaho in 1890 and almost equaled the number in Colorado. Although true pioneering had vanished in the Hawkeye State, farm makers continued to develop unimproved lands, draining wetlands and subdividing older units.

The best-known historian of the West, Frederick Jackson Turner, loved maps. In those the federal census cartographers were drawing during the late nineteenth century depicting American population growth, he found the basic patterns of the westward movement across the Mississippi. From various staging points in the central Mississippi Valley, settlers fanned out to form an area of contiguous settlement that reached the western border of Missouri by 1850 and then advanced into the plains country in a great ellipse, its northern perimeter anchored in Wisconsin and Minnesota, its southern in Texas, with the Indian lands that were to be Oklahoma breaking the pattern until the end of the century. The overleaping movement of pioneer farmers, Mormons, and gold seekers to Oregon, California, and the promised land of the Latter-day Saints during the 1840s and thereafter provided the cartographers with a different configuration—one of vertical bands in the great valleys of the Pacific rim and of strips and islands in the basin, plateau, and mountain country. Here the maps show the tendency for population to move not only westward but also north or south, or even west to east, as farmers followed the miners but were more rigorously confined to river valleys and irrigable areas.

Even on the central plains, settlement did not always surge simply westward. Local settlement jutted out along transportation routes and sometimes proceeded from north to south or vice versa or filled upland gaps between previously settled valleys. The outer margins of the farming frontier reflected the pulsations of the general economy and the alternation of years of little or of generous rainfall. In good times, the agricultural

Based on data from federal census reports, these population-density maps graphically document the westward movement of the American people. In 1890, the superintendent of the census noted, "Up to and including 1880 the country had a frontier of settlement, but at present the unsettled area has been so broken into by isolated bodies of settlement that there can hardly be said to be a frontier line." This observation, vividly supported by the 1890 map, helped inspire the historian Frederick Jackson Turner's famous remarks about the disappearance of the American frontier.

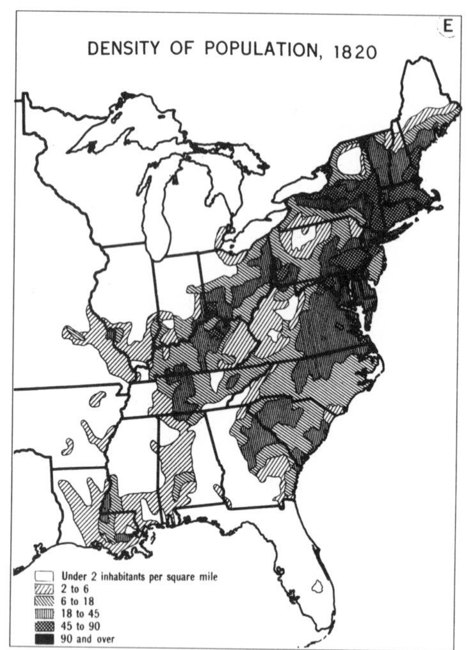

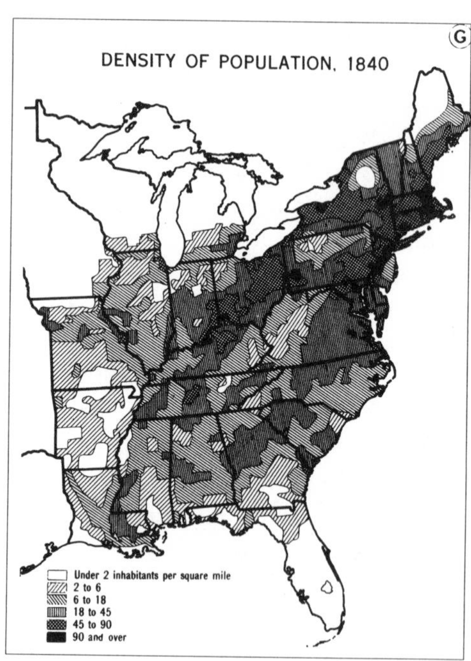

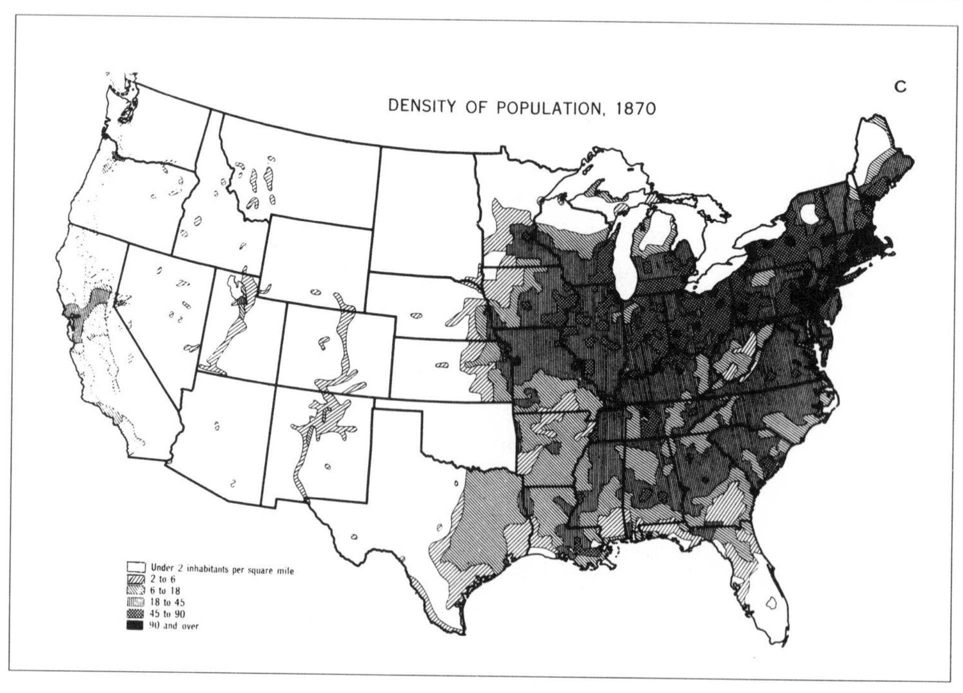

Maps reproduced from Charles O. Paullin, Atlas of the Historical Geography of the United States *(1932).*

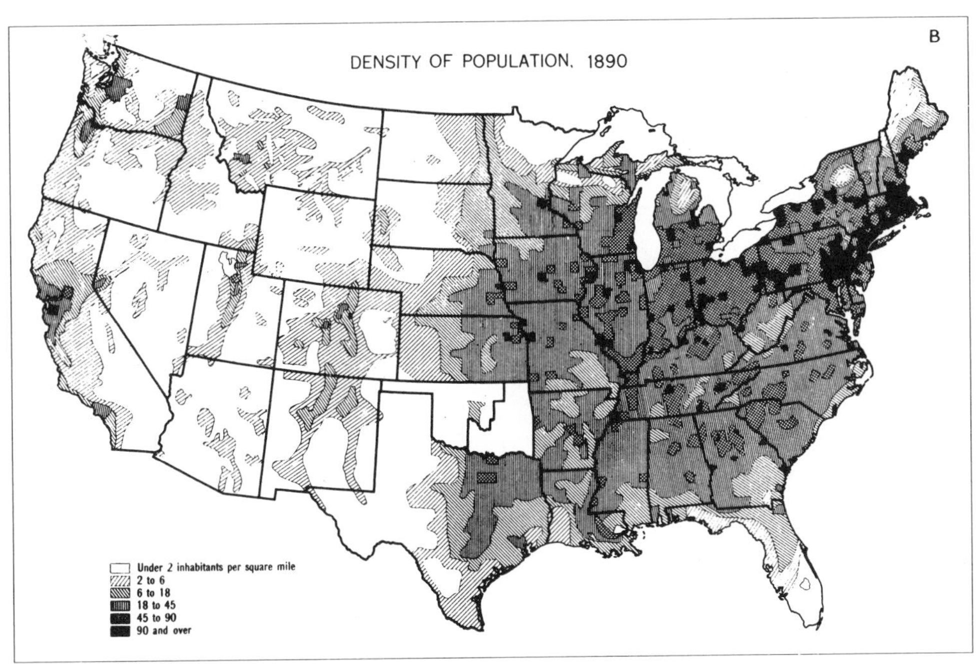

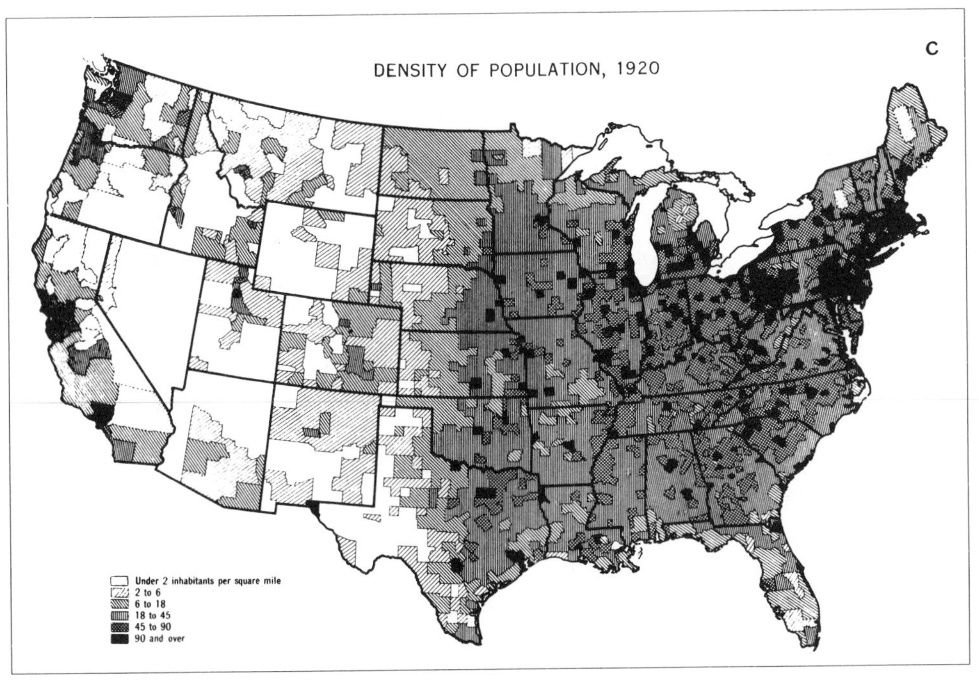

frontier expanded; it remained stationary, or retreated—as in eastern Colorado and western Kansas—during the depression and drought years of the late nineteenth century. In part, such contraction involved the departures of those who failed, those whose will, health, or resources proved inadequate—"in God we trusted, in Kansas we busted" read the wry doggerel. But even in years of prosperity, the population was mobile as new settlers came and residents left.

For the most part, the treaty makers and the soldiers swept the Native Americans from the path of the settlers; relatively few trans-Mississippi pioneers lived close to Indians for long. When the overlanders of the 1840s began to arrive, they found French-Canadian settlers, former employees of the Hudson's Bay Company, living in the lower Columbia-Willamette region. The American settlers drew on this outpost of habitant-métis culture for agricultural knowledge and supplies, but the fur trade left a minimal legacy of agricultural settlement.

Our inheritance from Spanish colonization was a different matter. Within a generation in American California, the argonauts and American businessmen and agriculturists overran a culture of rancho and petty irrigation agriculture, penetrating the world of the thirteen thousand Californios by purchase and intermarriage and transforming it by an expansion of large- and small-scale commercial agriculture on undeveloped Spanish or Mexican grants and federal and state lands. The legacy of Hispanic institutions and landholdings influenced the later development of California, but we cannot distinguish a Hispanic frontier there after 1848.

In New Mexico the situation was different. Although most of the population would have been categorized as mestizo rather than Spanish, some seventy thousand Hispanos lived in New Mexico during the 1840s, near five to ten thousand Pueblo Indians. During the early nineteenth century, the Hispano population of New Mexico clustered along the upper Rio Grande del Norte from Socorro to Taos and extended for short distances up tributary rivers. From Santa Fe, the population branched eastward into the upper valley of the Pecos. Islands of population existed on two tributaries of the Little Colorado, while a feeble settlement struggled for existence on the Santa Cruz River below Tucson. Isolated by the arid country stretching from El Paso to Socorro, pressured by the Comanche, Apache, Navajo, and Ute Indians, riven with the tension of Hispano versus Pueblo, and with a population unschooled and religiously insular, Spanish New Mexico has been termed "a forlorn gateway to imaginary cities of gold on the arid plains." One author wrote that New Mexico in about 1815, "stalemated physically, culturally, and economically by the conditions of the land and by the Indian menace, was an arrested frontier society." It was not, however, arrested in terms of population expansion after 1820.

During the years 1820–70, the Hispanos had their own agrarian frontier. Settlers filtered out from the old communities, developing holdings on streams where they could irrigate patches of grain, corn, beans, peppers, and other vegetables or in canyons that served as bases for grazing flocks. To the east of the old settlements, the New Mexicans pushed their ranchos, or plazas, "crude clusters of adobe and stone houses in a labyrinth of pole and brush corrals" (according to one scholar), north along the margins of the Sangre de Cristo Mountains and east into Canadian headwaters to Tascosa, south along the Pecos almost to Fort Sumner, and southeast to the Penasco River and neighboring streams. To the west, the Hispano pioneers penetrated the San Luis Valley, pushed along various of the Rio Grande's tributaries beyond previous boundaries of settlement, and washed into the headwaters of the San Juan and Little Colorado rivers. When American rule began, this frontier expansion was under way, reflecting population growth and the vitalization of the New Mexican economy that occurred after trade relationships developed with the lower Missouri River valley. After the Mexican War, the American army's efforts to curb the southwestern tribes encouraged the spread of Hispano settlement. By the early 1870s, Spanish-speaking settlers were confronting Texas cattlemen and Mormon farmers. As the Hispanos struggled to hold their lands against these aggressive competitors and to contend with small-scale irrigation agriculture, this frontier thrust lost both its distinctive character and its vitality.

When the American colonists began their great republican experiment, the union was one of cultures as well as one of economic and political institutions. The northeastern states formed a reservoir of Yankees, children both of the sea and the field, who were energetic and commercially oriented and who nurtured a Puritan sense of mission. The people of the middle states were of Dutch and British origins, leavened during the eighteenth century with German and Scotch-Irish settlers who moved into the interior of Pennsylvania and followed the Great Valley into the backcountry of Maryland, Virginia, and the Carolinas. A few representatives of other cultural groups were scattered within this matrix of population. Of the white population of the United States in 1790, some 60 percent were English in origin, and the Scots, Irish, and Germans by birth or descent each made up 5 to 10 percent. Those of Dutch, French, Swedish, or Spanish origins totaled 3 percent or less. From the 1630s onward, on the basis of importation and natural increase, there was a growing African-American population, held, for the most part, in slavery. That institution survived into the late colonial period in the northern colonies and residually into the early national period of some northern states, but it thrived in the southern colonies. The census takers of 1790 estimated that within a U.S. population of 3,929,000 were 697,624 slaves and 59,557 free blacks.

After 1790, a stream of migrants from the northeastern and Middle States flowed westward, occupying western New York and even parts of southern Canada; from these places they moved into the upper areas of the Old Northwest. Simultaneously, settlers from the Virginia and Carolina backcountries flooded into Kentucky and Tennessee and the southern regions of Ohio, Indiana, and Illinois. In turn, the southern border states and the Old Northwest were sources of migrants for the trans-Mississippi Middle West. While some of the population in the southern backcountry moved north and west through the Appalachian chain, another current of southerners trekked along the Gulf, creating a pool of potential settlers for Louisiana, Arkansas, and Texas.

Modest during the first thirty years of the American Republic, immigration quickened in the 1830s, when more than half a million new residents arrived. By the 1850s, the number had risen to almost three million, and five and one-quarter million entered the country during the 1880s. Initially, most of the immigrants in this period were Irish, but by the 1850s, Germans were arriving in comparable numbers. Many newcomers came from the United Kingdom, Scandinavia, and Canada, and almost every European state or province was represented. One historian of American immigration argued that the newcomers shunned the frontier—they were "fillers in." Many did establish themselves in towns or cities or in older farming communities, but immigrants also shared in the agricultural settlement of the trans-Mississippi West. Foreign-born Kansans never totaled more than 13 percent of the population in the period of most rapid settlement in the state, 1860–90, and the proportion in Texas was smaller. But Nebraskans of foreign extraction made up more than 20 percent of the state's residents between 1860 and 1880, dropping to 19 percent in 1890. At that time, 32 percent of Dakotans had been born under other flags. A magnet for gold seekers, California counted 39 percent of its residents as foreign-born in 1860, with approximately 30 percent remaining in that category thirty years later. These facts of population composition and growth are basic to understanding agricultural settlement in the trans-Mississippi West because there, the agrarian culture bearers of the older American North and South mingled with settlers of different heritage, and together they faced new challenges of adaptation.

According to one of the songs of the Mormon handcart migration, "Some may push and some may pull." So it was when individuals considered moving into or within western America. Inability to find farms close to parents or relatives, inhospitable social or religious environments, nagging loads of debt, the conviction that one's locality was ague-ridden or otherwise unhealthy, and clouded land titles—these were factors that pushed people westward. On the other hand, settlers were lured by opportunities for cheap land, tales of rising land values and fantastic crop yields, the urging of relatives in the new country, and newspaper stories or letters singing the praises of a frontier region.

Some westward migrants believed themselves to be agents of destiny. Southern frontiersmen flocking to Mexico and, later, independent Texas considered themselves apostles of republican institutions and were eager to expand their domain. Others who pioneered the overland trail to Oregon or Hispanic California fused the ideology of imperial republicanism with an ambition to prosper in a rich new country. Marching into the Great Salt Lake basin in the wake of Brigham Young and his pioneers, the Latter-day Saints sought not only a religious utopia but also economic security. John Brown and his sons waged war on the slave power in Kansas but staked out land claims there as well.

Some migrants impulsively decided to emigrate; others deliberated at length, push and pull commingled in their minds. Sometimes the embittered bachelor, the bereaved widower, or the independent young woman resolved, in solitary contemplation, to leave his or her home; but usually the decision was a product of family, community, or congregational interaction. Many wives went unwillingly, dreading the disruption of family, the dangers of the trail, and the primitive conditions of pioneering. Others welcomed the adventure and shared their spouses' hopes for success in the West. "A setting hen never gains any feathers," one wife pertly responded when importuned by family members to let her husband proceed her.

The rural householders who left midwestern communities during the nineteenth century were younger than the community mean, possessed less property than those who stayed, and had smaller families. In mid-nineteenth-century Appanoose County, Iowa, for example, the mean total wealth of persisting householders, age thirty to thirty-nine in 1860, was $2,348, whereas those of the same age group who left the county during the next decade possessed assets averaging only $1,401. Such differences were not always present and often were not great. Although family heads in every adult age group migrated, the typical nonpersisters were in their thirties and were not impoverished. Indeed, some members of the initial migration to Oregon and California possessed means beyond the average of the communities from which they came.

For most westering Americans, the decision to emigrate involved the vision of a verdant farm and a thriving family. Land was the lure. Settlers learned of the agricultural possibilities of the West in myriad ways. Land speculators advertised their western holdings with handbills, pamphlets, and newspaper stories. In the early nineteenth century, the American travel account was a popular literary form, and many publications described parts of the American frontier. Although few literary travelers penetrated the land beyond the Mississippi during the first half of the nineteenth century, more of them did so as transportation expanded into the western grasslands and mountain regions. Eastern newspapers featured letters from the West, sometimes from former residents happily reestablished on the new frontiers and on other occasions from eastern newspapermen exploring the region. Early in the nineteenth century, guidebooks became available, describing the natural resources and surface features of western regions and providing digests of the federal land laws and state laws of interest to prospective emigrants, as well as depicting appropriate routes and transportation agencies. Josiah T. Marshall's *Farmer's and Emigrant's Hand-Book: Being a Full and Complete Guide for the Farmer and the Emigrant*, of 1845, was almost five hundred pages long, with chapters on the "Naturalization and Preemption Laws" and a "Miscellany—Containing a vast variety of Recipes, Hints, Tables, Facts, etc. etc., to aid the Emigrant, whether male or female . . . in daily life."

Westerners were keenly aware of the importance of attracting settlers. If others joined them, western economies would thrive; emigrants would be not only comrades in state building but also consumers and producers. Western lawmakers, officials, and community leaders subscribed to the gospel of development. Minnesota established a State Board of Immigration in 1855, and other western states followed suit. Such agencies hired representatives who distributed pamphlets and other materials, trumpeting the abundance of cheap farmlands. By 1864, Kansas was sending emissaries abroad, and after 1872, the thick *Biennial Reports* of the Kansas Board of Agriculture informed readers about crop production and the availability of farmland. In 1887, the governor of the Wyoming Territory promised wonderful futures to "practical, every-day farmers, who will put their hands to the plow and not look back."

The federal government granted public lands to the new western states for the support of education, for the erection of public buildings, and after 1820, for the improvement of transportation. If revenue was to be derived from their sale, the lands must be advertised. By the 1840s, railroads had begun obtaining land grants from the states, and with the advent of the Illinois Central project in 1850, the great era of federal land-grant railroad advertising dawned. Companies like the Union Pacific and the Northern Pacific energetically advertised their imperial domains. Railroad land departments organized excursions for newspapermen and land lookers, sent agents abroad, and printed handbills and pamphlets describing their fertile lands and favorable terms of purchase. The Burlington and Missouri River Railroad Company advertised "millions of acres of Iowa and Nebraska Lands" for sale—on ten years time, at 6 percent interest,

Railroad companies energetically promoted settlement along their western routes, luring prospective homesteaders with pictures of bountiful fields, prosperous farmers, and easy rewards. Enthusiastic (if self-interested) boosters of western settlement into the 20th century, they simply altered their images to keep up with changing times.

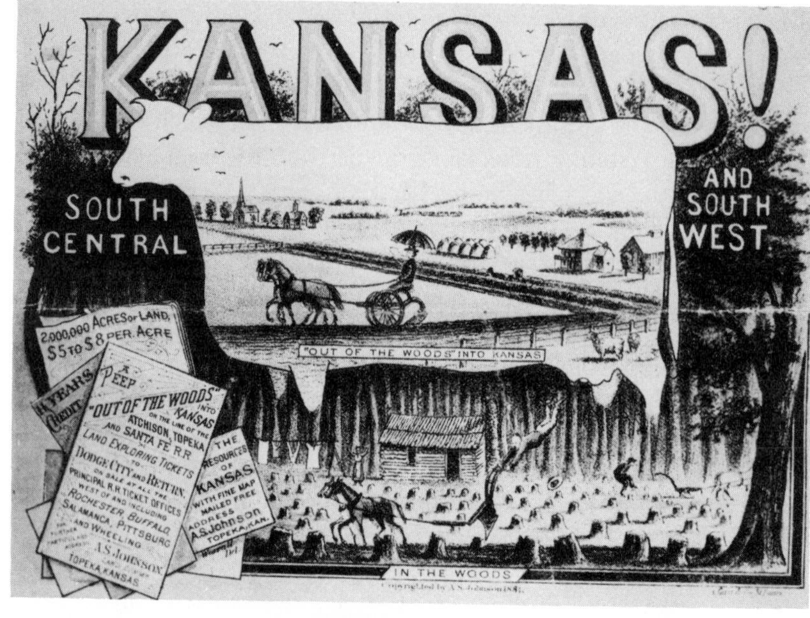

A. S. Johnson, publisher. Kansas! *Chromolithograph, 1881. Kansas State Historical Society, Topeka.*

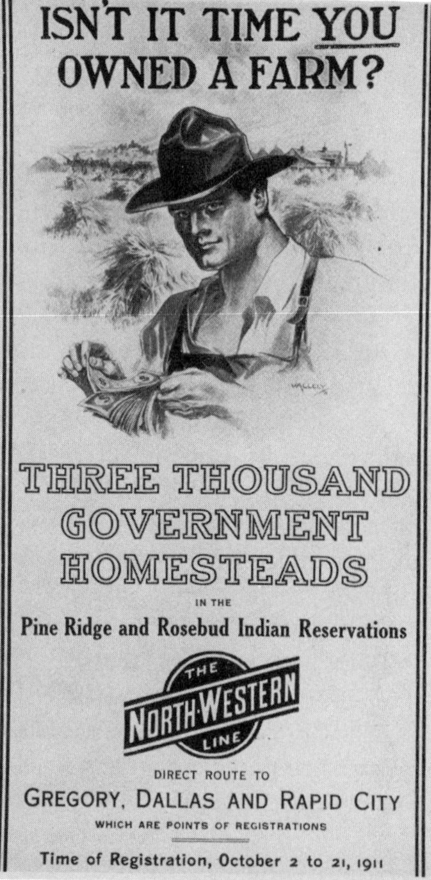

Isn't It Time You Owned a Farm? *Poster, 1911. South Dakota State Historical Society, Pierre.*

payments of principal to begin after two years, and with the cost of a land excursion ticket to be deducted from the first payment if the purchase was made within thirty days. While the settler sought his new farm, his family might wait in "free rooms" in Burlington, Iowa, or Lincoln, Nebraska.

Western community leaders were also active. Local lawyers and real estate men persuaded easterners with capital to invest in western land, mortgages, and tax certificates and to entrust the supervision of such investments to them. In the process, they sang the praises of the country. On the postbellum frontiers, realtors thrived first as claim locators, guiding land seekers to lands that were available under federal laws. Some attempted to attract settlers from the older regions.

The community of black settlers at Nicodemus, Kansas, owed its existence to such activity. "All Colored People that want to GO TO KANSAS, on September 5th, 1877, Can do so for $5.00." So read the handbills that announced the formation of the colony (membership, one dollar) in Lexington, Kentucky, during the summer of 1877. Among those who paid their dollar was the freedman Thomas Johnson. He was part of a group of three hundred who reached the townsite of Nicodemus in Graham County, Kansas, in mid-September, only to find a settlement of bankside dugouts. Arriving later, the wife of an African-American minister saw "various smokes coming out of the ground." "The scenery," she wrote, "was not at all inviting, and I began to cry." A white Kansan from Indiana, W. R. Hill, provided the initial impetus for the establishment of this community of blacks in the valley of the South Fork of the Solomon River, 240 miles west of Topeka. Hill was a townsite promoter and claim locator, and of the five dollars mentioned in the handbill, two dollars was paid to him for his services in locating the colony families on claims, two dollars covered the filing fee at the land office, and one dollar went into the colony treasury.

Western urban centers harboring pretensions (and most did) organized chambers of commerce or boards of trade whose advertising and reports were designed to attract emigrants. By the last third of the nineteenth century, community leaders viewed "booming" editors, such as Marsh Murdock, of the *Wichita Eagle*, as prizes beyond price. In the spirit of the game, the *Kirwin* (Kansas) *Chief* printed Lucy Larcom's "A Call to Kansas" in 1876: "Yeoman strong; hither throng! / Nature's honest men, / We will make the wilderness / Bud and bloom again."

"Boom" went "bust" when the feverish settlement activity of the 1880s in the plains country collapsed amid dust, low prices, and foreclosures. But as western frontier communities prospered in the early twentieth century, community leaders once more sang of opportunity, only now they were "boosters." Their intent was the same.

Although the journalistic scribbling and advertising had an impact, personal contact with western family members or neighbors who returned for visits often provided the incentive to go West. And letters from those in the West were passed from hand to hand within families. Ephraim G. Fairchild, a resident of Jones County, Iowa, in 1857 wrote to a relative, "I think that I can plough and harrow out hear without being nocked and jerked about with the stones as I allways have been in Jersey . . . if Father and Mother and the rest of the family was out here . . . they would make a living easier than they can in Jersey." So worked the process of chain migration. In Scandinavia such letters were

"America Letters" and were read by family networks. The same process occurred in church congregations, spreading migration fever.

Colonies of former neighbors developed in the West, by specific plan or by the accretion of family members, former neighbors, or acquaintances from the same eastern locality or foreign country. Of the organized colonies, none was more exotic than Bethlehem Yehuda, a corporate colony based on socialistic principles established during the early 1880s by idealistic young unmarried Ukranian Jews near Mount Vernon, South Dakota. But as the writer Irving Howe noted, "The leap from a Ukranian *shtetl* to . . . South Dakota—the cultural leap, the economic leap—was simply too great." Bethlehem Yehuda failed.

The last generation of American social historians has shown that nineteenth-century Americans were highly mobile. The federal census takers of one enumeration found less than half of the same pioneer farm households recorded by their predecessors a decade earlier. In the interim, settlers had come and gone. Given that there were fifty-five live births per one thousand of population in the United States in 1820—a figure that stood at thirty in 1910 (as compared with eighteen in 1970)—we might have expected such results. Once fully settled, the rural communities of the nineteenth century produced more children than could be accommodated without drastic subdivision of farms. Barring local urban or industrial opportunities, individuals had to leave, and some went West. That region did not draw equally from the older states and foreign countries; among American migrant residents in western regions, a predominance of settlers came from the nearer states. This held true even in the transcontinental migration to Oregon, where Missourians were most common in the early days. Settlers from more than a few hundred miles away became more common on the plains during the late nineteenth century, when the emigrant's boxcar provided an attractive alternative to the covered wagon.

Although many western farmers moved several times during the course of their lives, the early generations west of the Mississippi blended stability with movement. Most rural communities developed a core of longtime resident farmers; other settlers came and went, and some maturing children sought opportunity elsewhere. Thirty or forty years after the settlement of western rural communities, farmers bearing the surnames of original settlers could still be found, whereas their siblings were farming on new frontiers. In other cases, particularly among some foreign-born settlers, sons and sons-in-law acquired the adjacent lands of aging farmers and established family enclaves within the community.

Acquiring Land

In early June 1873, Henry Ise and his bride, Rosie, reached Henry's homestead and dirt-roofed log cabin in Osborne County, Kansas. Frank and Sam, their team of horses, had pulled the canvas-covered wagon, loaded with supplies, and the husband and wife had alternated between sitting on the wagon seat and driving their small herd of cattle. That day, the sun shone, prairie flowers bloomed in profusion, meadowlarks filled the air with birdsong, and Rosie wore calico and a pink sunbonnet. Although threatened by a flood in Dry Creek at the corner of Henry's claim, the couple arrived at the farmstead safely. The cabin measured eighteen by fifteen feet and was flanked by a straw-roofed stable, a chicken house of sod, a well with wheel and buckets, and a small fenced corral. Within the cabin were a bed, a tiny stove, boxes for chairs, and a nail-box washstand positioned on a floor of cottonwood boards. Nearby patches of sturdy corn, wheat, and oats promised good things to come. Spartan though this was, other homesteader wives found less-developed claims awaiting them.

When the pioneer farmers pushed across the Mississippi during the early nineteenth century, they understood the major features of the federal land-disposal system. Federal surveyors had preceded them, or were soon to follow, leaving a grid of baselines and prime meridians in which were set ranges of townships six miles square, subdivided by survey crews into thirty-six mile-square sections, each further divisible into smaller units. Within the framework of the federal survey, the settlers sought land to their liking. Initially, they used the land auction with its minimum price of $1.25 per acre (after 1820) and the right of purchase by private entry of offered, but unsold, land. Due in part to the political power of the pioneers, Congress liberalized the system, passing preemption, donation, and homestead laws, as well as implementing environmental adaptations like the timber culture laws and the Desert Land Act of the 1870s. But the settlers also found many fertile western lands available only by purchase from railroad companies or western territories or states, as well as from the reviled land speculators.

Congress approved thousands of laws relating to the nation's public domain, but the Homestead Act (1862) symbolized the process of western settlement. Thomas Jefferson wrote, "Those who labor in the earth are the chosen people of God, if ever he had a chosen people . . . whose breasts He has made His peculiar deposit for substantial and genuine virtue." Jefferson hoped that the lands of the United States would be used to foster the development of this class of citizens. The Homestead Act seemingly epitomized such thought. In reality, Congress established a system of land disposal that also served the interests of capitalists and developers. For this reason, as well as the feeling that the law did not meet western needs, historians have sometimes belittled the importance of the Homestead Act and the amendments that increased the size of the grant from 160 to 640 acres and shortened the period of compulsory residence. The facts of farm making between 1860 and 1920 contradict this belittlement. During those years, the number of farms in the United States increased by some 4.4 million. At the same time, 1.4 million homesteaders or their heirs received final patents, equivalent to 32 percent of the increase in farm numbers. So, the Homestead Act probably accounted for a substantial proportion of the new farms opened during this country's greatest period of agricultural expansion. Also, many pioneers filed claims under the law and relinquished them for a sum proffered by other settlers or commuted them to cash purchases. Sometimes commutation reflected poverty and distress; on other occasions the relinquisher or the commuter expected a quick profit. Still, for seventy years the Homestead Act gave legions of Americans inexpensive access to the land market and farm ownership.

Acquiring a farm was not always easy, and over time, it became more difficult. When Congress passed the Homestead Act in 1862, the farm frontier of the prairies and plains still included much of Minnesota, some of northwestern Iowa, most of Nebraska, and two-thirds of Kansas. For over a generation, 160 acres had been an ample unit through much of the unsettled portion of these states and in California, Oregon, and the Washington Territory. But settlers moving into the high plains and the dry plateau and basin country beyond the cordillera found 160 unirrigated acres inadequate to support a family.

The busiest homesteading areas of the nineteenth century were Kansas, Nebraska, and the Dakotas, where more than 430,000 settlers had filed homestead claims by the end of 1895. The most spectacular burst of settlement occurred in the "Great Dakota Land Boom" between 1881 and 1885, when 67,000 settlers took up homesteads in the territory. Between 1896 and 1920, homesteaders were most common in Montana, North Dakota, Colorado, South Dakota, Oklahoma, and New Mexico, each attracting at least 125,000 land entrants. Almost 200,000 settlers poured into James J. Hill's high plains railroad empire in Montana, the flood peaking in the years 1906–10. Of these states, only the Dakotas had seen comparable activity during the nineteenth century. Homesteading also increased after 1900 in the plateau and basin states as settlers moved into the cold desert of southern Oregon and into interior Washington, California to the east of the Sierras, and Arizona. Despite its size and the fact that homesteading began there in 1863, only 114,000 settlers filed homestead claims in California. Nowhere was Jefferson's dream so obscured as in that state.

Gro Svendsen's husband, Ole, was in error when he believed that he had to be naturalized in order to homestead land; he needed only to have declared his intention of becoming a citizen. In this respect, the Homestead Act was a generous one. Its application to the American Indians was ironic; few of them could utilize its provisions. But there were exceptions. During the summer of 1862, warfare blazed in the Minnesota-Dakota borderlands, where starving Santee Sioux rebelled against the exactions of traders and a corrupt Indian service. After a bloody uprising, thirty-eight Indians were adjudged guilty of rape or murder and were hanged. Others went to prison, and the tribe was moved into the Dakota Territory. In 1869, some twenty-five families from the tribe trailed into an unsettled area along the Big Sioux River in South Dakota, determined to live like white farmers. Others joined them; thus emerged the Flandreau Sioux colony, whose members asked to enter individual holdings under the Homestead Act. Friendly government officials and missionaries helped them do so, although the Flandreaus were required at first to surrender their claims on the tribal assets.

Steven Arrow, Big Eagle, David Faribault, their fellows, and their families slowly developed a settlement of log cabins. Initially, they hired oxen from white neighbors, tilled with spades and hoes, and supplemented their husbandry with hunting and trapping. A Sioux Falls merchant reported that they "gave more indications of civilization and industry and 'a show of living like white people than the same number of Norwegian families located a few miles below.'" Like other pioneer farmers, the Flandreau homesteaders faced killing frosts, grasshoppers, and a general shortage of livestock, equipment, and tools; in addition, the Indians confronted defections, factionalism, burdensome taxes, the threat of alcoholism, and the temptation to sell out to white men. But an Indian agent of the early 1880s wrote, " [They] pay their taxes promptly, their word can be relied upon, and they make good neighbors." Few American Indians, however, were able to use the Homestead Act, although an amendment to the statute of the early 1870s was intended to make it generally possible.

The above only outlines the ways in which westering farmers acquired virgin land.

The options of any individual might be limited, but the federal system was varied when viewed in broader perspective. Although the rectangular units within the basic survey grid did not always allow an effective or conserving use of land, settlers found the federal survey easy to understand and, therefore, less a source of boundary disagreements with neighbors than were the haphazard metes-and-bounds surveys found in some older states. Still there were problems. During the early period, when purchase at federal land auction or by private entry prevailed, squatters pushed onto the public domain before land surveyors had completed their tasks. Although such incursions were criticized in Congress and were contrary to law, squatters complained about rapacious speculators who thronged to the ensuing land sales and threatened to bid on claims that the settlers had improved. Among the frontier settlers there were also rascals ready to jump claims. To protect each other's holdings, squatters formed claim clubs or associations. They either ousted claim jumpers or forced them to purchase the claim in question. Club members attended land sales en masse, carrying sturdy walking sticks and threatening unwary capitalists. The squatters' clubs also allowed early settlers to engross land beyond their farming needs and to force latecomers to buy squatters' titles.

With the passage of the general preemption law in 1841, settlers no longer needed to fear speculators. But often thereafter, when land titles were uncertain, claim clubs appeared. In the upper valley of the Des Moines River during the 1850s and 1860s, the assignees of a river-improvement grant found their titles challenged by both a land-grant railroad and settlers hoping to acquire title through the preemption or homestead laws. A settlers' association denounced the other claimants. When the courts ruled on the title to these Des Moines River lands, the squatters learned, to their dismay, that they had to buy their lands from the river lands assignees or the railroad. During the 1850s, settlers poured into eastern Kansas before the federal land surveys, and free state and proslavery settlers used claim clubs to reinforce their positions. Settlers' associations on former Indian reserves and railroad lands in Kansas battled for preemptive rights into the 1870s, and such organizations were common in Nebraska and present in Colorado.

In California, conflict over land titles was complex and confusing. Although farmers were responsible for the first trickle of American overland immigration into California, the gold strikes brought migrants hoping for quick riches. Meanwhile, Congress grappled with the slavery issues attendant on the Mexican Cession and neglected California's land problems. The United States did not establish a California land office until 1853, and the first federal land sales did not occur until 1858. Settling the status of the lands that the Spanish or the Mexican government had granted to individuals was difficult. There were some eight hundred such grants, embracing eleven million acres, and the United States had pledged protection of the property rights of the grantees. Typically, the grants lay in attractive agricultural areas, and in many cases, their boundaries were vague.

The U.S. Congress established a claims commission in 1851 to determine the validity of the Mexican titles. The commission was generous in approving claims, even some of dubious validity; but evaluation proceeded slowly and was further complicated by federal court rulings. In the meantime, squatters settled on undeveloped parts of the grants—some in ignorance of the pending Mexican title, others either believing that the claims were fraudulent or hoping to win preemptive rights to their holdings. California officials complicated the situation by failing to identify state swamplands clearly and by claiming tracts that settlers were already trying to homestead or preempt. The locations of the state agricultural college lands further constricted the acreage available to settlers wishing to use the federal land laws. The allocation of railroad grant lands in areas already occupied by pioneers exacerbated the situation. The Californios were ill prepared to cope with the aggressive, Anglo-American businessmen, legal costs, and tax burdens associated with U.S. sovereignty. Anglos with capital acquired title to all or portions of many of the Mexican grants while others accumulated large holdings of state or railroad lands or of federal lands subject to private entry. Such individuals often refused to acknowledge that squatters had any rights, and the courts became mired in land litigation. In self-defense, the settlers formed claim associations, retained lawyers, and threatened violence against law officers seeking to evict them. Near San Jose in 1861, more than five hundred settlers confronted a posse seeking to evict squatters from the Chabolla grant. Some California farmers were still living on lands with clouded titles in the 1870s.

The New Mexican story of land disposal is even more depressing. Here were hundreds of Hispanic grants, most lying in the corridor of settlement along the Rio Grande and its tributaries but some extending into latter-day Colorado and Arizona. Under these titles a few rich families claimed ownership of large domains, and the Hispano masses held some form of title to modest acreages, with rights to share adjacent range with their neighbors. In New Mexico, grantees submitted their claims to the territorial surveyor general for approval. His rulings, however, could be challenged in the courts, and individuals with resources could beseech the U.S. Congress to confirm titles. Predatory Anglo lawyers and businessmen purchased claims, obtained shares in grants in payment for legal services, and tried to expand the boundaries of their holdings. Exemplified in Stephen B. Elkins and Thomas B. Catron of the "Santa Fe Ring," the Anglo wheeler-dealers found rich pickings, despite charges of corruption. Meanwhile, stockmen used dummy entrymen to obtain control of great stretches of New Mexico range. When Grover Cleveland's advisers considered the administration of public lands, they found widespread belief that "swindling cattle kings, surrounded by a gang of swindling herders, all of whom are in collusion with swindling surveyors, have swallowed our Western acres as a gourmand swallows oysters." New Mexico was the cesspool of the public land system.

With their range rights disregarded and the parent grants within which they lived frequently discredited, Hispano smallholders often hesitated to submit evidence of title and so lost their lands, surviving as hired laborers or renters. Anglo ranchers and settlers manipulated the federal homesteading system more successfully than did the Hispanos. When Congress finally established the Court of Private Land Claims in 1891, the Hispanic land base had already been subjected to four decades of plundering. The court confirmed title to only 2 million of the 37.5 million acres in claims presented.

Spanish Americans did not always suffer the assaults on their land base passively. After fence-building Texas cattlemen invaded the Las Vegas community grant, Hispanic residents, enshrouded in white, responded with night raids. Directed by Juan José

Following the Mexican War, Hispanic landowners in California filed illustrated plats to prove ownership of their property to the federal government.

José Rafael Gonzales. Rancho San Miguelito, California. *Watercolor and ink on paper, 1852. National Archives and Records Administration (Record Group 49, Records of the General Land Office, California Private Land Claims, Diseno #27), Washington, D.C.*

Herrera, "Las Gorras Blancas" (The White Caps) cut fences and burned Texan improvements. The movement was the seedbed of New Mexican populism, and although unsuccessful, this protest contributed to the incorporation of the Las Vegas grant and its continuing occupation by longtime residents.

Federal land disposal in Utah also had unique features. Most farmers here practiced irrigation and lived in villages from which they tended neighboring fields or livestock. In federal eyes, however, the Mormons were, initially, squatters. Not until 1869 did American officials open a land office in the Utah Territory. Meanwhile, church officers supervised settlement and land distribution. With the federal system in place, the Saints moved rapidly to ensure possession of the lands they had occupied, utilizing the Preemption Act, the Homestead Act, the Timber Culture Act, and the Desert Land Act. Many Mormon holdings were less than 160 acres in size, and although the practice was technically illegal, church officers often acquired the acreages allowable under the acts and apportioned them to the settlers occupying them. Wives in polygamous relationships sometimes entered homesteads as family heads and transferred title to their spouses, and villagers evaded the residency requirements of the Homestead Act by periodically camping on their claims. Thus did social realities overcome the technicalities of federal land statutes.

In much of the trans-Mississippi West today, roads, crops, and field boundaries delineate for airborne travelers the checkerboard squares of the federal surveys. Land records in hundreds of county courthouses still show evidence of large holdings amassed in the settlement period, as well as the tendency of ethnic groups to keep and to extend family holdings. In general, the government patent was only one transfer of title on any piece of rural property; there might be many more thereafter. Indeed, in some areas, a quarter-section homestead might be sold several times by relinquishment before the patentee occupied it. Many frontier settlers never acquired land from the government, obtaining their farms from original owners or the latter's assignees. Some of this churning represented the failure of settlers to become independent owners, but in other cases, it tells us that pioneers believed better opportunities lay elsewhere.

Some came to the frontier without means to acquire land. They found landholders willing to rent land to them, and hence, tenancy settled on the frontier along with the freehold farmer. Tenancy also developed because aging pioneers or widows used it in transferring farms to the next generation. For some young farmers, tenant status was a step in their progress toward debt-free farm ownership. But tenancy also indicated that not everyone could enjoy full fellowship in the smallholder's republic lauded by Thomas Jefferson. When the federal enumerators of 1880 counted farm tenants for the first time, they discovered that 24 percent of Iowa farmers were tenants, as were 16 percent of Kansas farmers; in California the rate was also 16 percent. Twenty years later the numbers had risen to 35, 35, and 23 percent.

Mechanization

When American farmers moved into the trans-Mississippi West in the early nineteenth century, their implements were much the same as those used by colonial farmers. Two-wheeled carts or crude wagons, mold-board plows of wood with an iron point and—perhaps—share, drags of heavy planks or tree trunks, harrows of wooden spikes set in triangular plank frames, double-shovel plows for row crops, harnesses, ox yokes, and fittings—these were the major items, along with smaller tools such as scythes, sickles, forks (often of wood), shovels, and coopered pails. The husbandman of the early nineteenth century broadcast his seed from a bag slung over his shoulders; he cut the ripened crop with a sickle or a scythe while others raked it into sheaves and bound them with knotted grain stalks. Eventually, the farmer flailed the grain, taken from barn mow or stack, and winnowed it on a canvas. The settler planted Indian corn with a hoe or, if the crop was sod corn on new breaking, sometimes used an ax. Gathering techniques varied by region; but however it was gathered, ripened corn was stored in cribs while the stalks and leaves provided fodder.

As settlement spread into the prairie states, mechanization accelerated. By about 1820, Jethro Wood had developed the prototype of the walking plow, with all of its earth-turning parts made of iron. Though Wood's plow was an improvement, the dense, matted prairie grass and forb roots defied conventional plows and, once turned, revealed soils that clogged iron-mold boards. Prairie blacksmiths developed massive beamed plows to break the sod, and during the 1830s and 1840s, John Deere and William Oliver led in manufacturing plows with mold boards of steel or polished chilled iron to which these soils did not adhere.

The agricultural inventors worked wonders in reducing work hours for the harvest of wheat and other small grains, hitherto cut with scythe or cradle and bound into sheaves by hand. Cyrus H. McCormick and other machinist-inventors mechanized and melded several processes. For the scythe's short cutting blade, they substituted a long cutting bar with serrated edges set within a framework of metal guide teeth, or fingers, attached to the front of a platform on wheels and pulled by draft animals. In forward motion, gearing motivated by a drive wheel at one end of this platform caused the cutting bar to whicker-snicker within its frame and cut the stalks of grain a few inches above the ground as an elevated reel and guide teeth steered them to the blade. A worker with a rake followed McCormick's contraption of the 1830s, periodically clearing the table of cut grain; but soon this worker was replaced by automatic devices.

The Marsh harvester of the 1870s revolutionized the sheafing and binding process by introducing moving canvases that elevated the cut grain over the drive wheel and dropped it on a shelf for sheaf binders standing on a step attached to the machine. These workers vanished with the introduction of an automatic twine knotter. The resulting grain binder was found in many harvest fields of the trans-Mississippi West by the 1890s. Such binders reduced field labor still more when manufacturers attached sheaf carriers that accumulated the bound bundles. When the driver judged that there were enough to make a stook of grain, he tripped the device.

The final process of separating the grain from the straw had long since left the flail or tramping-floor stage. By the 1840s, crude grain separators that flailed the grain and blew away the chaff and straw with internal fans were in use. Initially, horse treadmills and circular horsepowers drove such machines, but by the 1880s, Pitts or Case threshing machines were trundling down western roads behind steam engines, followed in turn by a horse-drawn water wagon to keep the engine puffing. In the Far West, imaginative rural inventors consolidated the stages of grain harvest; by the 1870s, huge, cumbersome combines drawn by twenty or thirty horses were rattling through the grainfields in the central valley of California, leaving a trail of filled grain sacks behind them. At the turn of the century, combines were conquering the hilly wheat fields of interior Washington. The binder-driver of the plains country encouraged his three or four horses with imprecations and a long whip; the driver of the combine's multiple hitch hung a pail of rocks under his elevated seat and threw them at lazy members of his huge team.

Meanwhile, mechanization proceeded in other areas of husbandry. Practical, if not always reliable, horse-drawn broadcast seeders, grain drills, mowing machines, rakes, and hay loaders had appeared by the 1850s. Plows and cultivators were enlarged and made available as riding machines, and rollers and other tilling machinery also appeared in larger sizes that were stronger and more efficient. The Indian corn crop seems, at first glance, to have benefited less from mechanization than the small grain harvest. Although mechanical planters and twin-row riding cultivators entered the cornfield in the 1860s, handpicking into a horse- or mule-drawn wagon was the general practice until the 1920s. In John Herbert Quick's *The Hawkeye* (1923), Fremont McConkey described the process during the late 1870s:

> Forget the sky, the clouds, the blood drawn by rosebrier or the sharp tips of the kernels of the "hackberry" ears; forget everything but the economy of movement, the making of every second count. Make sure that you do not fail to tear the ear from the stalk and throw it into the wagon by a single movement of the muscles; see to it that

> when the right hand returns from the throw, the body has moved forward if necessary to another proper position, and that the left hand has seized another ear and holds it ready for the husking peg; and do not fail to remember that if you husk your hundred bushels in a day, the steady "clump, clump, clump"' against the throw-board must continue hour after hour, even while the trained horses are making the turn at the end of the field.

The settlement era was past when mechanical corn pickers became common in the Middle West during the 1920s.

Mechanization was not the only factor contributing to the efficiency of agricultural production west of the Mississippi during the nineteenth century. The fertility of western soils, improvement of hand skills in tasks such as corn picking, and the tendency of farmers to specialize also enhanced production. But the decline, for example, in the number of work hours expended per hundred bushels in the production of wheat between 1840 and 1900—from 233 to 108 hours—and in the production of corn—from 276 to 135—reflected the use of more sophisticated machinery. By the latter date also, steam, the electric generator, and gasoline engines had begun supplying the energy hitherto provided by Buck and Bright, Fan and Pomp, and Sam and Jenny—the oxen, horse, and mule teams that powered animal-drawn farm machinery.

Climatic Adaptation

The pioneer farmer of the Genesee Valley chopped his farm from a mixed deciduous forest, extending the ramparts of stump and rail by a few acres each year. His son or nephew in central Illinois came to understand the fertility of the tallgrass prairies but preferred a farmstead amid the trees along streams or in the prairie groves, where he found timber for firewood, rails, and lumber. If he moved into the prairie interiors, he might buy a woodlot beside a stream. A generation or two later, settlers on the central or high plains drove their teams as far as forty miles to cut wood. But by this time the railroads were aiding farm making by transporting lumber, milled from logs originating in the pineries of the Great Lakes states—raw material for grassland houses, barns, and fences.

Other western pioneers experienced drastic revisions of their childhood experiences. In the Great Basin of the Mormons, mountainsides to the east were forested, mountain crests glistened white for much of the year, and springtime streams ran full with snowmelt. But the Saints discovered that their accustomed crops of the Mississippi Valley needed supplemental moisture or would fail in all but the most unusual years. Those pioneers who pressed into the coastal regions of upper Washington experienced annual precipitation of seventy inches or more—what the Kansas farmer might expect in several years. In the Central Valley of California, the situation seemed even stranger. One historian imagined the newcomer's first reactions: "The absence of summer rains seemed to prove that the valley was unfit for agricultural purposes. How could crops be raised where there was only a wet and a dry season? How could one farm where the grass turned green in the winter and died in the summer?"

For the farmer who left the Mississippi Valley, a transformation from forest man to grassland man was necessary. The Texan Walter P. Webb dramatized this story: "The whole technique of pioneering and the ways of living which had become habitual with the people and had proved so effective as to become standardized broke down completely when carried from the Eastern Woodland region into the Great Plains." Webb argued that only after waiting from 1825 to 1860 for industrial America to develop such inventions as the six-shooter, barbed wire, and the windmill did pioneer farmers establish themselves beyond the ninety-eighth meridian in the plains country.

No demography demonstrates Webb's claim. The westward march did not stall at the central grassland until the army was equipped with six-shooters and a new technology of settlement adopted. During the 1850s, settlement was incomplete in the states adjacent to the Mississippi. In that decade, preparations for opening the lands beyond the Missouri began—midwestern Indian tribes were moved, formal relations with the Plains Indians were initiated, and territorial governments were established west of Iowa and Missouri. Still, in part, Webb spoke the truth. The pioneer farmers adapted to subhumid environments as they moved westward. Forest man, however, first accepted the challenge of becoming Webb's "new man" on the bluestem prairies of the "prairie triangle," that lazy V of grassland, striped with wooded watercourses, that flares from its apex on the Wabash prairies through northern and central Illinois to include much of southern Wisconsin, more of Minnesota and Missouri, and all of Iowa before merging into the plains proper.

The problem of aridity was manifest in various ways, including home construction. The dominant feature of any new farmstead was the house. The first dwelling of the pioneer was typically the log cabin, which settlers in the grasslands preferred. During the first months of settlement, however, some families lived in lean-tos, or even canvas-topped carts or wagon boxes. Once families could build, cultural influences became apparent, such as the differences between the Yankee and the southern tides of settlement. The well-built, southern-style log cabin had a breezeway through the center. Henry Ise took his bride, Rosie, to a simpler structure near the ninety-ninth meridian. But the dwellings that Rosie saw in the last stages of the journey were mostly dugouts, scooped from the sides of draws or creeks, with front walls of sod, though there were some frame or stone buildings. Elsewhere on the plains, the genuine sod house was common, with walls and usually the roof constructed from rectangles of tough, matted sod.

If many settlers used soddies, few lived in them for long. As the Ises prospered and the family grew, they added a wing to their cabin and built a new house of local stone. More commonly, as railroad lines laced the postbellum West, settlers obtained scantlings, planks, and boards to build claim shanties or the second house that denoted improved circumstances. The new houses were "sawed" rather than "sod" dwellings, one historian notes. When Elinore Plaisted's mother, Em, took her children to the Dakota wheat country late in the century, her husband proudly drove them to a two-story, wooden, framed house where a few years earlier there had been only waving grass. Less-affluent folk lived in wooden shacks, sometimes with an additional coating of tar paper nailed to the board walls, slight shelter against blizzards. Within, the farm wife might stoke her stove with buffalo chips, twisted hay, or stalks of sunflowers or other woody plants if creek-bed wood or coal was unavailable.

On washdays, Em Plaisted carried water from a nearby slough to fill her tubs, but few farms could continue without a good well. Under much of the trans-Mississippi West lay aquifers, but the depth of the water table varied. Sometimes the well-digging

The idealized agrarian image of Currier & Ives' "western farmer's home" reflected eastern sensibilities, not western reality. In the trans-Mississippi West, settlers such as this young Nebraska mother or these African-American homesteaders often started out in simple temporary structures. But dugout homes and claim shanties were not easily incorporated into the popular images that romanticized life on the western farmers' frontier.

Currier & Ives, publishers. The Western Farmers Home. *Chromolithograph, 1871. The Harry T. Peters Collection (#56.300.1448), Museum of the City of New York.*

Unidentified photographer. Our Home *(near McCook, Nebraska). Photograph, 1890. Nebraska State Historical Society, Lincoln.*

A. P. Swearingen (?–1931). Homestead of a Family near Guthrie, Oklahoma Territory. *Photograph, ca. 1889. Western History Collections, University of Oklahoma Library, Norman.*

settler might reach water in less than thirty feet; other times he labored with pick and shovel to greater depths, running the risk that the windlass rope might break or that he might be overcome by well gas.

Loessial in origin and fortified with humus, grassland soils were highly productive if adequate moisture was available. But since the rainfall diminished in amount and increased in variability meridian by meridian, settlers obtained lower average yields per acre as they moved farther west and were compelled to cultivate larger units. The uncertainties of plains country weather also dictated that work be done when conditions were favorable. Larger and stronger farm machinery, illustrated by two and three bottom plows and wider harrows, provided a partial solution to this problem. The trend culminated in the big hitches in which several sturdy teams were combined to pull massive arrays of tillage machinery and in the introduction of the tractor on the high plains wheat frontier of the early twentieth century.

In their eagerness to produce, pioneer farmers managed to raise a half crop of corn on fresh breaking by hacking holes for the seed in the turned sod. During the 1880s, settlers in the Dakotas found that flax grew well as a first-year crop. Beyond the line of the Missouri, however, was a belt of territory where corn thrived during the moist years but where wheat was more dependable in dry periods. When the spellbinding Populist orator Mary Elizabeth Lease exhorted farmers of this region to "grow less corn and raise more hell," she was at least half right on the basis of agronomic principles. During the 1870s and early 1880s, farmers in "The Golden Belt" of central Kansas learned that the hard, red winter wheat that Mennonite settlers had brought from the Russian plains was their most dependable cash crop. Improved milling techniques expanded the market for this grain, which came to dominate a winter wheat region on the central plains. To the north, farmers discovered that spring wheat was more satisfactory. During the dry years of the late nineteenth century, farmers on the high plains found that in their area, sorghums provided a drought-resistant crop.

There were other illustrations of climatic adaptation in the mechanization process. To some Kansas and Nebraska farmers, the droughts of the late nineteenth century revealed the beneficial effects of using lister drills, which allowed the planting of wheat in trenches that ran at right angles to the direction of the prevailing winds. Many plains country farmers adopted the grain header, which cut the small grains just below the seed heads and elevated them into an accompanying wagon for transport to a threshing machine. Thus, they took advantage of the fact that the moisture content of the harvested grain was low and did not need to dry in the shock.

As native timber was consumed, pioneer farmers sought alternatives to rail fences. Herd laws, restricting the stock owner's right to allow his animals to roam at will, were only a partial solution. The prickly Osage orange tree (bois d'arc) provided a popular substitute for fences. If planted thickly and pruned properly, it made a formidable hedge. During the 1850s and 1860s, farmers planted miles of hedge throughout the Middle West but later adopted the more effective barbed wire that northern Illinois inventors had developed. Along with less-punitive wire-mesh fences, barbed wire solved the grassland farmer's fencing problem, but in a way that required an outlay of cash rather than labor.

Real estate agents, officers of land-grant railroads, state officials, and immigration agents assured the farmers who moved into the plains country during the 1870s and 1880s that "rain followed the plow." During some years of those decades this appeared to be true. But the late 1880s and the early 1890s revealed the cruel variability of the grassland climate. Hardy Webster Campbell's gospel of dryland farming sustained the thronging homesteaders after 1900. Publicized by the railroads during the first decade of the twentieth century, the Campbell system dictated that fields be fallowed while moisture accumulated in packed undersoil, topped with a dust mulch. Dubbed "scientific farming," at the urging of railroad executives, dry farming was not complete nonsense. Properly managed, fallowing conserved moisture, but Campbell's system could not withstand a succession of dry years. After 1900, settlers attempted to build a dry-farming empire on the northern high plains served by the Great Northern and the Northern Pacific railroads. From 1917 onward, they suffered recurrent droughts, and many failed, bitterly repeating the aphorism "Dry farming works best in wet years."

Yet, some western farmers grew wheat successfully in areas of low rainfall, unaided by either irrigation or the charlatan rainmakers who preyed on gullible husbandmen during the 1890s. During California's transition from ranching to a varied commercial agriculture, both operators with large mechanized operations and small-scale farmers grew bumper wheat crops on a dryland basis in the Central Valley. After 1900, the farmers of central Washington successfully grew wheat in a region where rainfall was less than twelve inches per year. The timing of the precipitation explains such successes. Although the mountain ranges of the West drained eastward-moving cloud masses and created tributary areas of low rainfall, they also accumulated magnificent snowpacks that created rushing rivers in the spring and early summer. Irrigation, thought visionaries, might create a western oasis. Looking across the heat-shimmering reaches of their kingdom, the Mormons had little choice; if the faithful were to be gathered and to survive, man needed to assist the heavens. In 1847, Mormon farmers in the Salt Lake Valley diverted water to their crops from City Creek. Although some initially tried to harvest crops without irrigation, the Saints built their first highline canal in 1850.

In Deseret, community structure was also church structure, simplifying the mobilization of community effort essential to building an irrigation system. Church officers supervised the construction of the stream diversions and canals and apportioned water quotas. The Saints were, however, human. One pioneer Mormon lurked in the dark near the ditch that provided water for his acres, rock in hand, poised to repulse any neighbor from down the valley who might try to drop the headgate before watering was completed. For a generation and more, Utahns advised residents in other water-scarce regions of the West. There was much to learn about laying out canal systems, constructing impoundments and gates, caring for particular crops, and distributing water evenly. But Utah's irrigation system was a relatively simple one with few elaborate reservoirs or lengthy canals.

In the pursuit of precious minerals, the miners of western regions used water in great quantities. Their activities encouraged the development of agriculture, as onetime farmers realized that eggs at fifty cents apiece and meat and flour at equivalent prices were as rewarding as flakes of gold. With the water rules of the miners in mind, western irrigationists overturned centuries of European and eastern precedent, enunciating a doctrine of prior appropriation in place of riparian rights. The first user was entitled to the amount taken prior to the arrival of others, rather than sharing a right in common with adjacent users and being obligated to return diverted water to streams. During the 1870s, the lawmakers of Colorado and other western states wrote prior appropriation into constitutional and statutory law, although elements of riparian rights lingered, particularly in California.

John Wesley Powell, the intrepid runner of the Colorado canyons and director of the U.S. Geological Survey, was the first government scientist to understand fully the climatic challenges of the West. At the close of the 1870s, he suggested that federal land-disposal laws be revamped—the irrigation farmer needed less than a quarter section to make a good living, whereas the stockman needed more. In the summer of 1889, he told the North Dakota Constitutional Convention, "One year with another, you need a little more [rainfall] than you get." After western developers and politicians drove Powell from the Geological Survey because of his conservative approach to irrigation issues, he continued to perfect his ideas for the development of the West's irrigable lands under a system of self-governing hydrographic basins. As for promoters who dreamed of irrigation as a quick fix, he wrote: "Terpsichorean', sacrificial, and fiducial agencies fail to change the desert into the garden, or transform the flood-storm into a refreshing shower. Years of drought and famine come and years of flood and famine come, and the climate is not changed with dance, libation or prayer." Still, western community builders were convinced that reclamation could help them build populous states. In response, the federal government passed the Carey Act of 1894, providing for the transfer of lands to western states for irrigation purposes. The law was ineffectual.

For the first time in 1890, the federal census takers noted irrigation farmers. They found 54,136 in sixteen western states and territories, one-quarter of them in California. Despite such figures and the success of private developments like the Greeley Colony in Colorado, private or state efforts to develop irrigation were marred by the failures of optimistic entrepreneurs and by land speculators who monopolized reservoir sites or irrigable land. Obviously, argued westerners, the federal government should play a larger role. When Congress passed the National Reclamation Act of 1902, creating the Bureau of Reclamation, the western settler had a new option—homesteading on reclamation developments, as in Arizona's Salt River Project or the Truckee-Carson district of Nevada. In reclamation districts, settlers paid water-user fees that were expected to return the project costs to a revolving reclamation fund from which other developments might be funded. Even federal homestead land was never free, given the farm-making costs involved. Land was certainly not free in the early reclamation projects. By 1923, there were 34,276 farms on twenty-eight western reclamation districts, and in only one district had the settlers met all of their payments.

Commercialization

We can look to the American past and imagine the subsistence farmer, a sturdy fellow living on an isolated farm from which he and his family satisfied all their needs. This individual never existed. But the colonial farmer resembled him more than did the agriculturist of 1850, and the colonial frontier farmers were much more subsistence-oriented than were the farmers of the older settlements adjacent to the markets of colonial ports. In contrast to the self-sufficient husbandman, we can also visualize the commercial farmer, who concentrated on producing the crops that had the greatest market value and who purchased all other necessities.

By the early nineteenth century, American farmers had moved significantly toward the commercial end of the continuum between subsistence and commercial farmers. But even where capitalism had most influenced agricultural production—on the slave plantation—field crops were grown, gardens were tended, and some clothing was made at home. Before the 1820s, in the West, homespun was still common, buckskin was not unknown, and farmers cultivated numerous crops and raised various animals and fowl. Beginning in the 1820s, the transportation system of the United States was revolutionized, expanding the markets for farm produce and making a broad range of manufactured articles generally available. Although nineteenth-century depressions left a residue of failed or financially embarrassed farmers in the less-established communities of the West, agriculturists in general seized market opportunities provided by the expanding economy. In 1863, California's most eminent authority on agriculture reported, "The farmers generally are anxious to make as much money as possible, and as soon as possible." When the railroad vitalized the economy of the hitherto isolated Cache Valley community in Utah, a settler reported, "The jingle of coins was stimulating." Through the last forty years of the nineteenth century, farmers in many western townships or counties supported the purchase of railroad bonds by their local governments in order to amplify that stimulating jingle.

Some historians believe that the early nineteenth century was an era of small-scale farms tilled by hardy republicans who lived a life of community marked by trading of labor and by barter until a subversive capitalism dragged them unwillingly into the marketplace. Trading butter and eggs and other farm products for store goods and bartering labor, goods, and services within the agricultural community *was* common during the early nineteenth century and continued into the twentieth century; but accounting was usually done in monetary units. Whether it was wheat, as on many frontiers, or oats in the Walla Walla Valley, or flax in the Dakotas, the pioneers sought a cash crop that produced a maximum return; they were perennially short of capital and forced to invest much labor in farm improvements from which benefits were derived only in the long run. The United States was born capitalist, and the attendant values did not change, though transformations in transportation, industry, and marketing substantially altered the pioneering process during the nineteenth century. And as farmers learned the peculiarities of their areas and the most remunerative combinations of crops, livestock, capital, and labor—sometimes involving practices that observers called wasteful or exploitive—agricultural regions and subregions emerged.

American institutions encouraged the frontier farmer to think in terms of dollars and cents. The Founding Fathers had established a federal land system that assumed the existence of a market economy. When squatters poured into the Black Hawk Purchase of eastern Iowa during the 1830s, they established claim clubs and purchased and sold claims. They understood also that they must pay the federal government for their claims in money. At the federal land auctions in Iowa, eastern capitalists or their agents made loans to them at 50 percent or more interest. Thus, commerce in land, down payments, and interest charges made even squatters on the Iowa prairie think in terms of markets and cash crops. Once settlers had title to their lands, there were taxes to pay, and western counties did not take payment in kind, although road levies were typically worked off. Before the railroads, some firstcomers were far in advance of organized markets. A few served military posts or Indian agencies. Others satisfied their needs for cash, for iron products, for salt, and for "luxuries" like coffee or cane sugar by selling surplus products to new settlers, by selling livestock to traveling drovers, or by transporting wagonloads of wheat or oats to distant market points.

What did it cost to establish a farm during the nineteenth century? The economist Clarence Danhof estimated the costs of developing a 40-acre farm on the midwestern frontier during the 1850s as "\$50-\$400 for land, \$60–800 for breaking the sod or clearing woodland, \$112–320 or more for fencing, \$100 for implements, \$150–200 for livestock, \$40-\$80 for the first year's seed, and \$25-\$450 for housing." On the average, Danhof believed, the midwestern farmer of this era needed about \$1,000. Labor shortages during the Civil War stimulated the mechanization process. In 1862, a farm editor published a list of equipment appropriate to a 150-acre holding; the value of the items amounted to \$968.

Farmers seldom outfitted themselves completely at one time; some brought secondhand machinery, and others shared the cost of reapers or other more expensive machines with relatives or neighbors. The owners of custom breaking plows did a thriving business, and commercial threshers were a feature of rural life until the 1940s. But the average value of machinery on farms in Johnson County, Iowa, was four times greater in 1890 in constant dollars than it had been in 1860, and in the Ise family's Osborne County, Kansas, the average investment in machinery per farm increased two and one-half times between 1870 and 1890. Some settlers, however, did not initially own draft animals and hoped to work off the claim and thus accumulate farming equipment and stock. Others deferred investment in land by becoming tenants. As mechanization proceeded, the agents of implement manufacturers sold machinery on short-term credit, taking the farmer's note with a chattel mortgage as security.

Throughout the period of our study, the family farmer typified the rural community of the West. This does not mean that all western farmers had equal resources or status. Older farmers had had more time to accumulate property than had younger farmers. Luck, inheritance, business skills, education, good or ill health, and the number of family members all might be reflected in an individual's property holdings, as were the land-disposal systems, land-use patterns, and regional economic development in general. The most common size of farm in Iowa in 1860 was fifty to one hundred acres; in Kansas at that census date the typical holding was twenty to fifty acres; and in California the most common farm was one hundred to five hundred acres. In that year, the census reported seventy-six Iowa farmers who had holdings of five hundred or more acres, with eight such farmers listed in Kansas and eight hundred in California.

Californians developed an agriculture characterized by both small irrigated holdings under intensive cultivation and huge ranches and grain farms, but they were not alone in developing spectacular farm operations. During the 1870s, investors in the Northern Pacific Railroad Company exchanged company bonds on favorable terms for acreage in

In the Guthrie, Oklahoma, land rush of 1889, as in most other federal land sales, those who profited most were often not the actual settlers but the entrepreneurs—like these lawyers—who serviced the settlers' needs.

C. P. Rich. Guthrie, Oklahoma Territory, April 22, 1889, Open Air Law Offices. *Photograph, 1889. Western History Collections, University of Oklahoma Library, Norman.*

the railroad's land grant in the Red River Valley of the north, and they developed bonanza grain farms on these holdings. Oliver Dalyrymple managed a seven-thousand-acre wheat operation there in 1877. He used "eighty horses, twenty-six breaking plows, forty cross plows, twenty-one seeders, sixty harrows," and harvesting equipment in proportion. But over time the owners of the bonanza farms found it worthwhile to sell out to small farmers.

When settlers first crossed the Mississippi, men of capital had long been investing in raw western lands, expecting to earn large returns as the country developed. By the mid-nineteenth century, well-informed capitalists knew that investments in western mortgages and tax titles were also remunerative. Western lawyers, bankers, and real estate men eagerly served as western agents of such investors. As the tide of settlement flowed into the West after the Civil War, a less visible stream of capital accompanied it. Increasingly in the western county-seat towns and supply centers, loan agents, sometimes also lawyers or real estate men, offered funds to settlers who wished to preempt land, take advantage of the commutation clause of the Homestead Act, or buy supplies, equipment, or livestock. Mortgage agents recruited savings from eastern banks and insurance companies as well as from professionals and small businessmen who were eager to double or triple the 5 or 6 percent paid by eastern savings institutions. Successful western loan men incorporated their businesses, set up branch offices in New York or abroad, offered "guaranteed" mortgages for sale, and began to sell debentures against mortgage paper. Such companies helped finance the "Great Dakota Land Boom" of the 1880s and the flow of farmers into other western states and territories during the same period.

Entrapped in the collapse of the boom, many western settlers failed to meet their mortgage interest or capital payments and abandoned their land. Caught between penniless borrowers and importunate lenders, the mortgage companies foreclosed. But the corporations could not resell the land, and most of them failed also. There had been earlier "wring-outs" in new settlement areas, but none so devastating as that of the late nineteenth century. Populist orators castigated mortgage men, along with railroad companies and land monopolists, for being responsible for the misery of the western farmer during the early 1890s. When Congress passed the Federal Farm Loan Act of 1916, memories of those years of discontent inspired advocates.

On every western agricultural frontier, the commercial production of farmers stimulated the development of an urban superstructure. The villages, towns, and cities of the West provided skills, services, capital, machinery, markets—or gateways to markets—and a good deal of political leadership for rural America. In return, the food, feed, and fiber surpluses of the farmers and their needs for manufactured goods, equipment, and supplies invigorated western trade centers. Above the local level, an urban hierarchy of subregional, regional, and national entrepôts and manufacturing centers developed, sustained by the movement and processing of agricultural products and the needs of hinterlands for manufactured goods, supplies, and services.

Glance at a listing of American millionaires compiled in 1892. In Moline, Illinois, Charles H. Deere controlled the fortune amassed by his father in the plow business. Nearby in Quincy, Edward Wells prospered by packing midwestern hogs; in Chicago, three members of the Armour family became rich in the same business. Thomas Lynch was a millionaire distiller, and Cyrus H. McCormick made millions from the sale of patented mowers and reapers. In Minneapolis-St. Paul, one finds the Pillsburys, enriched by western wheat. Frederick Weyerhaeuser provided thousands of western farmers with milled lumber, and James J. Hill's railroads served the farmers of the northern plains and Pacific Northwest. In St. Louis, D. R. Frances attributed his wealth to dealing in grain options, and Adolphus Busch used western grains to brew "fine beer." In San Francisco, L. L. Baker prospered in the agricultural machinery business, and the millionaire Henry Pierce was a grain dealer. These men exemplify urban individuals or firms that accumulated vast assets during the agricultural settlement of the trans-Mississippi West. Other businessmen derived part of their substantial wealth from rural land speculation. William Jennings Bryan echoed the belief of millions of rural westerners when, at the convention of the Democratic party in 1896, he thundered, "The great cities rest upon our broad and fertile prairies. Burn down your cities and leave our farms, and your cities will spring up again as if by magic; but destroy our farms and the grass will grow in the streets of every city in the country."

Cultural Adaptation

Did cultural groups adapt similarly to the economic challenges of agriculture while they created a vast social mosaic in the West? Did the eagerness of some of these groups to develop commercial agriculture assist their adjustments to western conditions, or did cultural predispositions sometimes obscure economic opportunities? We have seen that two powerful American subcultures were involved—northeastern Yankee and upland southern. Studying early-nineteenth-century regional cultures in Indiana in areas that would soon be sending populations westward, Richard L. Power contrasted hay-gathering, orchard-planting, dairy-oriented, energetic Yankee farmers with corn-growing and hog-raising southern upland or Hoosier agriculturists, who lacked the sense that leisure was sinful. The one ate white bread and chicken "fixens"; the other subsisted on corn bread and "common doins." One used Virginia rail fences; the other preferred posts and straight boards. Yankee and Hoosier farmers even laid out their farmsteads differently. To the cultural options that settlers from such backgrounds brought across the Mississippi, the foreign born added the preferences and practices that they had known. In facing problems, people first consider accustomed solutions—so it was in the trans-Mississippi West.

Cultural proclivity, however, was modified by western realities. Swedish settlers from Upper Dalarna might prefer to make barley their major small-grain crop in Isanti County, Minnesota, as at home in Sweden. But when the local market and natural environment suggested that wheat and oats were more profitable, the immigrants accepted the cue. Had they followed their cultural inclinations, Russian Germans in eastern Nebraska might have concentrated on the small grains of the Russian plains, but a major part of their acreage was soon in Indian corn. The agricultural patterns of ethnic groups did sometimes appear to differ from those of nearby American-born pioneers. These differences tended to disappear in the first generation of settlement. Still, some preferences in farm management continued into the second generation. And sometimes, practices that had been dysfunctional during settlement reappeared. German-Russian settlers in eastern Nebraska were once more growing their homeland crops, flax and rye, during the early 1890s, illustrating cultural rebound. Another adaptation pattern involved the transfer of crops or farming practices from the homeland without any perceptible break in continuity, such as when Mennonites continued to emphasize the wheat crop in the central grasslands.

The manuscript censuses of the late nineteenth century frequently show that ethnic groups differed from colonial-stock Americans in their choice of combinations of crops and livestock. But differences in age of farm operators, in the degree of farm development, in access to capital, in the soils that the groups farmed, in the availability of labor, in the distances to markets, and perhaps in other factors must be considered in such comparisons, and historians have not always done so. Nor did members of particular national ethnic groups invariably behave in the same way. That cultural background might serve as either cue or brake in the immigrant's adaptation to American practice, we cannot doubt. We are tempted to conclude that ethnic differences were less noticeable in making a living than were contrasts in household economy, food preference, work routine, language, religion, neighboring, and family values. Some members of ethnic groups were more strongly committed to passing their farms to the next generation and establishing family members in proximity than were those of Yankee or southern upland descent. Historians have particularly related such behavior to the German settler, but none of the numerous Ise children stayed on the lands painfully accumulated by Henry and Rosie, both impeccably German in origin.

Sometimes the immigrant farmers brought much more to their new country than labor and the will to succeed. The Mennonites' contribution of hard winter wheat is the best known of such legacies. The alfalfa seed that Wendelin Grimm brought from Bavaria was crucial to the later development of American agriculture. Agoston Haraszthy's importations of European vine cuttings made Sonoma County, California, the center of the West Coast wine industry. Thus, immigrant settlers broadened the options open to American farmers, although sometimes their contributions were controversial. When hundreds of industrious Chinese truck gardeners purchased or leased small holdings in California during the 1850s, supplying countrymen, miners, and urban dwellers with vegetables, neighbors criticized their use of night soil in fertilization and urine in pest control as dangerous to the public health and offensive to the nostrils.

Family Ventures

Critics maintain that historians of agricultural settlement have accentuated production to the neglect of cultural values and have emphasized the role of the male head of the farm household while ignoring the farm wife. Such criticism is justified, although the earlier emphases are understandable—if settlers failed to make a living, all else failed as well. But the farm frontier was a family frontier, even though sex ratios were sometimes skewed by the presence of large numbers of bachelors and men improving claims before they brought out families. On the other hand, the censuses of the nineteenth century show that farm households headed by women were present in some numbers, and it is estimated that women represented 10 to 15 percent of the homesteaders in some states during the early twentieth century.

After losing her husband in a railroad accident, Elinore Pruitt moved with her small daughter to Denver, where Elinore worked as a domestic. Hoping to become a homesteader, she advertised her availability as a housekeeper in a region where government land was available. Thus, in 1909, she moved to the ranch of Clyde Stewart near Burnt River in southwestern Wyoming. Soon Stewart proposed marriage, and Elinore accepted.

Elinore Pruitt Stewart filed her homestead claim on land adjacent to her husband's holding, allowing them to build her log cabin as an extension of the Stewart home. Meanwhile, she milked cows, tended poultry and pigs, gardened, cooked, put up preserves, and made clothes for herself and her daughter. She had other children, and she wrote, in 1912, that she wanted "a great many things" that she lacked, but she was neither discontented nor forgetful of her many blessings:

> I have my home among the blue mountains, my healthy, well-formed children, my clean, honest husband, my kind gentle milk cows, my garden which I make myself. I have loads and loads of flowers which I tend. There are lots of chickens, turkeys, and pigs which are my own special care. I have some slow old gentle horses and an old wagon. I can load up the kiddies and go where I please any time. I have the best, kindest neighbors.

With real pleasure, Elinore Stewart proved that she knew how to operate the hay mower.

Frontier wives and mothers did what most wives and mothers were doing elsewhere in America; they tried to provide "proper" homes and give their children a good upbringing. When Anne E. Bingham, a seminary-trained former schoolteacher, journeyed west with her husband in 1869 to a Kansas farm, a woman's suffrage worker in Leavenworth asked for her views on women and the vote. She responded that "voting had better be left to the men, for a woman's place was in the home." Many farm wives of the time would probably have agreed. Still, sixteen of the seventeen states in which women already had the vote before passage of the Nineteenth Amendment were western states. If most of the leaders in winning woman's suffrage were urbanites, rural women had swelled their battalions.

Farms on the frontier were family enterprises. The wife cleaned, cooked, prepared meals, washed dishes and clothes, and mended. At threshing or hog-slaughtering time she fed the extra men. She preserved food—from wild plums to headcheese. During the early nineteenth century, small flocks of sheep were usual, and carding, spinning, and the knitting of socks, mittens, and other apparel were common female tasks. Some women owned looms. Home manufactures were waning by the 1850s, but at this time the treadle sewing machine became available, and many farm women found it indispensable. Although mechanization rapidly changed the character of external farm work after 1850, America's industrialization did less for the farm wife than for her husband. True, the cookstove replaced the fireplace, and the sewing machine was a boon, but the washing machine of the late nineteenth century was a crude device, and few frontier women had running water or an indoor toilet.

The farm woman's work did not stop when she crossed the threshold of her house. Almost certainly, she worked in the garden, she sometimes cared for the barnyard fowl, and she often milked and churned the cream into butter. The dribble of money or store credit from sales of eggs and butter sometimes made the difference between success and failure in the new settlements. Women helped with the binding and shocking of grain in prebinder days and often drove the team or "built the load" at haying time or grain harvest.

In matters of domestic economy there were probably differences depending not only on the material and cultural circumstances but also on the personal inclinations of the women. Some pioneering women found the landscape hostile and the outdoor tasks wearing; others enjoyed the companionship involved in such labor and thrilled to the sights, sounds, and smells of the new land. On large, prosperous farms, the farm wife might be supplied with hired help and never feel the need to venture beyond the household perimeters in her work. Within some cultural groups, outdoor work was an accustomed part of the woman's heritage. On the other hand, Anne Bingham wrote: "My work . . . would have been much harder if I had not had the very best kind of a good husband. Before he went to work morning and noon he saw to it that there was wood and water in the house. I never did any milking and never took care of the poultry, and seldom did the churning. Many women I knew did all these things besides their own housework."

To merely contrast the roles of husband and wife is to oversimplify the social contours of farm life. A demographic analysis of the frontier areas during the years 1800–1840 showed that children constituted at least half of the household in 72 percent of northern frontier households and 83 percent on the southern frontiers. Having "suffered reverses," the Indiana carpenter William Dorsey established a preemption

Unidentified photographer. Farm Women at Work. *Photograph, date unknown. The Denver Public Library, Western History Department, Denver, Colorado.*

The farm frontier was a family frontier, where women and children performed essential labor. In addition to producing most of the foods and household goods consumed within the home, women might assist with the care of animals and the planting and harvesting of crops while children performed domestic chores. Their labor would help generate much-needed cash, since the added income brought in from the sale of eggs and dairy products could be critical to a small farm's success.

Solomon Butcher (1856–1927). Nebraska Lassie *(Custer County, Nebraska). Photograph, date unknown. Solomon D. Butcher Collection, Nebraska State Historical Society, Lincoln.*

claim on the Little Nemaha River in Nebraska in 1857. So lively was the cabin, wrote his diarist daughter Mollie, and so recurrent the demands for hospitality from visitors and the older daughters' beaus that Mother Dorsey often "took the babies and fled to her retreat in the woods . . . to gain her equilibrium." The frontier farm family is best understood when viewed in the context of family interaction and changing family goals as well as from the standpoint of gender polarities.

The full history of American farm labor is unwritten. We know that there were kin and neighborhood work rings, that farmers sometimes hired help at harvest and threshing time, and that there were full-time laborers on large holdings. We also believe that the farmer and his family provided most of the labor on frontier farms. Western children joined the work force at an early age. Seven- or eight-year-old boys could herd livestock, drop corn into waiting hills, and pick potato bugs from infested vines. Increasingly thereafter they shared in farm tasks in the field and at the farmstead. In larger families, boys might hire out to neighbors, with the wages sometimes destined for family coffers. Within the house, the farm wife might move from the position of domestic worker to manager, over daughters who minded younger siblings, learned to cook and sew at an early age, and worked outdoors when needed. On the Little Nemaha it was Mollie, or her sixteen-year-old sister, who milked the cow, rather than their mother, and Mollie also considered herself the chief cook.

One of Rosie's sons wrote a sensitive account, *Sod and Stubble,* of the Ise family homestead and family life on the plains frontier. He depicts the dangers faced by the pioneers: family privations, tragedies, comedy, and victory; the first-born who died and the crippled child among the healthy ones. He writes of the church picnics, the celebration when the mortgage was paid off, the relationships of neighboring and community, the road routing that sparked local conflict, the cultural heterogeneity (leavened with one Democrat), the disdain for "little Dutchies," and the movers and the stayers. In time, the railroad came, and rural-urban differences sharpened. Meanwhile, the Ises developed their farm through times of good returns and periods of drought, grasshoppers, or low prices, slowly accumulating equipment and additional land. Rosie was the dominant force in the Ise family, but the history of the Ise farm is a family story.

Pioneers in the trans-Mississippi West found the first years to be the most difficult. There were exceptions; some came to the new land with ample resources, strapping sons, hired help, strong teams, implements and machines, and the funds with which to buy sufficient land and supplies to bridge the gap before tilled fields began to produce. But for most, the scarcities, hardships, and dangers of pioneering weighed heavily. Drought, grasshoppers, prairie fires, hailstorms, the wheat-attacking Hessian fly, and even jackrabbits on the high-desert homesteads of the twentieth century could turn a year of hope into one of disillusion, or famine. And there were other dangers: poisonous snakes, rabid animals, wandering Longhorns that could tumble through a dugout's sod roof, runaway teams, well-digging accidents, and the errant ax or pitchfork. Medical practice was rudimentary. "Out here," wrote Mrs. Stewart, "we have to dope ourselves." Malaria was common on the nineteenth-century frontier and tetanus always a threat. Other diseases, later rendered minor by vaccine or antibiotic, shattered frontier farm families.

We can make too much of the isolation faced by members of the farm family. The native born were children of an open-country, rural culture and were unused to living in each other's pockets. Remembering a more densely settled countryside, the foreign born sometimes found the absence of close neighbors or the wind of the high plains distressing, even psychologically unsettling. But most pioneer families soon had neighbors, sometimes more than they would have preferred, and among these neighbors almost invariably were relatives, former neighbors, or members of the same religious persuasion. Farm wives visited, quilted, and were active in church activities. Still, Anne Bingham wrote in retrospect of their farm near Junction City, Kansas: "In all the years spent there we never could see a neighbor's light in the evening. I did wish so much we could, to relieve the aloneness."

In the last stage of their epic trek to Oregon in 1843, the Applegate families lost young Warren and Edward in the turbulent waters of the Columbia. In contrast to such tragedy, Herbert Quick's memorable character "Cow" Vandemark wrote: "The prairies took me, an ignorant, orphaned canal hand, and made me something much better.... The best prayer I can utter now is that it may do as well with my children and grandchildren." On the other hand, the year 1888 was one of drought in Kansas, and in the fall, the wagons of the "defeated legion" began to roll eastward past the Ise homestead, "grizzled, dejected, and surly men; sick, tired and hopeless women," along with children often unaware of the family tragedy of which they were a part. Given shelter overnight, some left lice, bedbugs, and pools of tobacco juice behind; others tried to steal chickens or pigs. The Ises, therefore, reluctantly agreed when the Hutson family asked for shelter for the sick woman traveling with them. The Hutsons proved, however, to be cultivated people. The sick traveler had lost her husband in an Indian raid in Decatur County, where Hutson had, in two years, lost all of his money. The family was returning to Iowa to begin again. Farmers who enjoyed success comparable to that of "Cow" Vandemark are well represented in the county histories and old settlers' accounts, but such publications seldom tell the stories of victims like Warren Applegate and the Hutsons. Was the settler's story a tale of ruthless and uncaring conquest over man and nature? Sometimes it was that, but more frequently it was a chronicle of hope, new homes, adaptation, family success, and sometimes, tragedy or failure.

Bibliographic Note

Those wishing to carry their reading or research beyond suggestions here should consult the articles, reviews, and indexes in *Agricultural History*, edited at the Agricultural History Center, University of California/Davis, and published by the University of California Press, Berkeley. During the 1960s and 1970s, the center also published a series of helpful regional bibliographies. Critiques of the recent literature dealing with frontier agriculture are provided by Gilbert C. Fite, "The American West of Farmers and Stockmen," in Michael P. Malone, ed., *Historians and the American West* (Lincoln, 1983), 209–33, and James W. Whitaker, "Agriculture and Livestock Production," in Roger L. Nichols, ed., *American Frontier and Western Issues: A Historiographical Review* (New York, 1986), 51–67. Unfortunately, space limitations require that this note focus on the monographic literature without specific mention of relevant articles in the periodical literature.

During the last generation, a number of scholars have written histories of American agriculture that treat the frontier experience from a variety of viewpoints: John T. Schlebecker, *Whereby We Thrive: A History of American Farming, 1607–1972* (Ames, 1975); John L. Shover, *First Majority, Last Minority: The Transforming of Rural Life in America* (DeKalb, Ill., 1976); Willard W. Cochrane (an agricultural economist), *The Development of American Agriculture: A Historical Analysis* (Minneapolis, 1979); and Walter Ebeling (an entomologist), *The Fruited Plain: The Story of American Agriculture* (Berkeley, 1979). Nor should one ignore William Parker's brilliant essay "Agriculture" in Lance E. Davis et al., *American Economic Growth: An Economist's History of the United States* (New York, 1972), 369–417. Several more narrowly focused surveys of agricultural development or settlement are useful: Fred A. Shannon, *The Farmer's Last Frontier: Agriculture, 1860–1897* (New York, 1945); Paul W. Gates, *The Farmer's Age: Agriculture, 1815–1860* (New York, 1960); and Gilbert C. Fite, *The Farmers' Frontier, 1865–1900* (New York, 1966). More broadly conceived and recently published is Rodman W.

Paul, *The Far West and the Great Plains in Transition, 1859–1900* (New York, 1988), which contains relevant chapters.

A number of regional or subregional studies deal with agricultural settlement in the trans-Mississippi West or special aspects of the subject, although the coverage as a whole is incomplete. Despite its age, conceptual weaknesses, and substantive thinness, Walter P. Webb's *The Great Plains* (Boston, 1931) cannot be ignored by anyone who wishes to comprehend the history of the plains country. Basic to understanding the development of agriculture on the prairies, though by no means comprehensive, is Allan G. Bogue, *From Prairie to Corn Belt: Farming on the Illinois and Iowa Prairies in the Nineteenth Century* (Chicago, 1963). The unique developments in the Red River Valley of the North are covered in Stanley N. Murray, *The Valley Comes of Age: A History of Agriculture in the Valley of the Red River of the North, 1812–1920* (Fargo, 1967), and Herman M. Drache, *The Day of the Bonanza: A History of Bonanza Farming in the Red River Valley of the North* (Fargo, 1964). Robert Ostergren, *A Community Transplanted: The Trans-Atlantic Experience of a Swedish Immigrant Settlement in the Upper Middle West, 1835–1915* (Madison, 1988), is an excellent illustration of current approaches to the settlement experience of immigrant groups. Richard L. Power, *Planting Corn Belt Culture: The Impress of the Upland Southerner and Yankee in the Old Northwest* (Indianapolis, 1953), however, is still a classic introduction to American agricultural subcultures.

One of the most talented of western historians, James C. Malin, made the central grasslands his special interest; of his many contributions, *Winter Wheat in the Golden Belt of Kansas: A Study in Adaption to Subhumid Geographical Environment* (Lawrence, 1944) and *The Grassland of North America: Prolegomena to Its History, with Addenda and Postscript* (1947; reprint, Gloucester, Mass., 1967) are central to the concerns of this chapter. Richard G. Bremer develops some of Malin's methods still further in *Agricultural Change in an Urban Age: The Loup Country of Nebraska, 1910–1970* (Lincoln, 1976). Two fine collections from the Center for Great Plains Studies include various relevant papers: Brian W. Blouet and Frederick C. Luebke, eds., *The Great Plains: Environment and Culture* (Lincoln, 1979), and Frederick C. Luebke, ed., *Ethnicity on the Great Plains* (Lincoln, 1980). Mary W. M. Hargreaves, *Dry Farming in the Northern Great Plains, 1900–1925* (Cambridge, Mass., 1957), ably describes the development of that system of agriculture. Terry G. Jordan, *German Seed in Texan Soil: Immigrant Farmers in Nineteenth-Century Texas* (Austin, 1966), and Richard G. Lowe and Randolph B. Campbell, *Planters and Plain Folk: Agriculture in Antebellum Texas* (Dallas, 1987), provide a good introduction to the Texas end of the plains country spectrum.

Several monographs have been basic to our understanding of Mormon settlement: Leonard Arrington, *Great Basin Kingdom: An Economic History of the Latter-Day Saints, 1830–1900* (Cambridge, Mass., 1958); Lowry Nelson, *The Mormon Village: A Pattern and Technique of Land Settlement* (Salt Lake City, 1952); and Nels Anderson, *Desert Saints: The Mormon Frontier in Utah*, 2d ed. (Chicago, 1966).

Turning to the Pacific Northwest, Donald W. Meinig, *The Columbia Plain: A Historical Geography, 1805–1910* (Seattle, 1968), is extremely useful, as is William A. Bowen. *The Willamette Valley: Migration and Settlement on the Oregon Frontier* (Seattle, 1978). Several recent monographs skillfully develop important themes: James R. Gibson, *Farming the Frontier: The Agricultural Opening of the Oregon Country, 1786–1846* (Seattle, 1985); John Fahey, *The Inland Empire: Unfolding Years, 1879–1929* (Seattle, 1986); and Barbara Allen, *Homesteading the High Desert* (Salt Lake City, 1987). Alexander C. McGregor, *Counting Sheep: From Open Range to Agribusiness on the Columbia Plateau* (Seattle, 1982), describes the transition from pastoralism to wheat raising in interior Washington.

Despite California's later position as one of the nation's major agricultural producers, the agricultural history of the state still has many gaps. Various articles in James H. Shideler, ed., *Agriculture in the Development of the Far West: A Symposium* (Berkeley, 1975), are very helpful. See also Rodman W. Paul, "The Beginnings of Agriculture in California: Innovation *vs.* Continuity," *California Historical Quarterly* 52 (Spring 1973): 16–27. Paul W. Gates, ed.,

California Ranchos and Farms, 1846–1862, Including the Letters of John Quincy Adams Warren . . . (Madison, 1967), provides the contemporary observations of an agricultural authority. Since irrigated agriculture is such a large part of the California story, Donald J. Pisani, *From the Family Farm to Agribusiness: The Irrigation Crusade in California and the West, 1850–1931* (Berkeley, 1984), is invaluable.

Donald W. Meinig, *Southwest: Three Peoples in Geographical Change, 1600–1970* (New York, 1971), is a fine, if brief, introduction to the settlement history of the Southwest. Henry C. Dethloff and Irvin M. May, Jr., eds., *Southwestern Agriculture: Pre-Columbian to Modern* (College Station, Tex., 1982), contains some useful articles. Water has been the key to the location of cropland agriculture in the Southwest, and Michael C. Meyer, *Water in the Hispanic Southwest: A Social and Legal History, 1550–1850* (Tucson, 1984), contains much of interest.

Much of the monographic literature cited thus far touches on several of the major themes of this chapter. Something should be said of more specialized treatments. Oscar Handlin et al., *Harvard Guide to American History* (Cambridge, Mass., 1954), 152–61, lists many of the best books of travel and description. Two surveys of regional booming activity are David M. Emmons, *Garden in the Grasslands: Boomer Literature of the Central Great Plains* (Lincoln, 1971), and Jan Blodgett, *Land of Bright Promise: Advertising the Texas Panhandle and South Plains, 1870–1917* (Austin, 1988). Paul W. Gates has dominated scholarly study of the disposal of federal lands. He summarized much of his work in *History of Public Land Law Development* (Washington, D.C., 1968). Two of his many articles are particularly relevant: "The Homestead Act: Free Land Policy in Operation, 1862–1935," in Howard W. Ottoson, ed., *Land Use Policy and Problems in the United States* (Lincoln, 1963), 28–46, and "California's Embattled Settlers," *California Historical Society Quarterly* 41 (June 1962): 99–130. John Opie, *The Law of the Land: Two Hundred Years of American Farmland Policy* (Lincoln, 1987), sets the settlement period into chronological perspective.

Almost ignored until 1960, the role of women in western settlement has attracted many scholars during the last generation. Any sampling should include Sandra L. Myres, *Westering Women and the Frontier Experience, 1800–1915* (Albuquerque, 1982), Glenda Riley, *The Female Frontier: A Comparative View of Women on the Prairie and the Plains* (Lawrence, 1988), and Sarah Deutsch, *No Separate Refuge: Culture, Class, and Gender on an Anglo-Hispanic Frontier in the American Southwest, 1880–1940* (New York, 1987). So far, the family in general has been of less interest to scholars than women, but note Elliott West, *Growing Up with the Country: Childhood on the Far-Western Frontier* (Albuquerque, 1989). James E. Davis analyzes the demographic structure of the early-nineteenth-century frontier population in *Frontier America, 1800–1840: A Comparative Demographic Analysis of the Frontier Process* (Glendale, Calif., 1977), but later frontiers still need attention. Also still seeking additional historians are the subjects of frontier agricultural credit and finance and frontier labor, although Allan G. Bogue, *Money at Interest: The Farm Mortgage on the Middle Border* (Ithaca, 1955), David Schob, *Hired Hands and Plowboys: Farm Labor in the Midwest, 1815–1860* (Urbana, 1975), and Thomas D. Isern, *Bull Threshers and Bindlestiffs: Harvesting and Threshing on the North American Plains* (Lawrence, 1990), have demonstrated the promise in these fields.

BOARDING HOUSE

Chapter Nine

A Saga of Families

KATHLEEN NEILS CONZEN

A family story lies at the heart of American western history. Through the oft-told tales of western conquest and resistance, settlement and development, wind subtle, insistent themes of family, kinship, and community. Consider, for example, one extraordinarily evocative scene from *My Darling Clementine*, John Ford's classic 1946 western film starring Henry Fonda as Wyatt Earp. Silhouetted against the bright southwestern sky rises the skeletal frame of a church tower, a cross at its peak, a joyous bell pealing from its topmost rafters. A lowering mesa broods in the background, nature dwarfing the raw town at its base. In the foreground, two American flags whip bravely in the brisk wind. Men, women, and children, all dressed in their Sunday best, crowd onto the plank floor of the unfinished meetinghouse. And in their midst, knees awkwardly lifted in cautious celebration of the new church, dances the town's tough, gunfighting, poker-playing marshal, tamed by the genteel Bostonian whom he holds gingerly in his arms.

Ford's haunting imagery encapsulates one of the primal themes in the history and mythology of the nineteenth-century American West. Time and again, in memoirs and novels, folk songs and films, political speeches and academic histories, Americans have insisted that the story of western settlement is a story of the conquest of nature and the taming of human nature in the name of the family and of the community that families together form. In this familiar saga, the trajectory from savagery to civilization both defines and legitimates the westward expansion of the American people, and the essence of American civilization lies in its institutions of family and community life. The western drama may seem to be a violent, masculine one, its main protagonists almost exclusively male. Mountain men and miners, cowboys and speculators, native warriors and the U.S. Cavalry, all crowd onto center stage. But sooner or later even the most relentlessly masculine conquest narrative yields to the developmental logic of domestication, and the spotlight shifts to those other archetypal figures who have been waiting in the wings—families trekking westward in covered wagons, wives working alongside husbands to erect a log cabin, warriors setting their hands to the plow, children trudging to a one-room schoolhouse, a community building a church. In popular perception and scholarly interpretation alike, the final integration of the western saga into the nation's ongoing history always seems to turn on those pregnant moments when family and community finally take root, just as Ford's turbulent film pivots on the gentle scene at the church dance.

The fictionalized Tombstone marshal Wyatt Earp danced across an unfinished meetinghouse floor in My Darling Clementine *(1946) with a schoolteacher from Boston, signaling the real and metaphoric centrality of women and the family in the building of western communities. When Earp left town at the conclusion of John Ford's classic film, Clementine remained to preside over the community's transformed domestic order.*

Film still from My Darling Clementine *(1946), directed by John Ford. Courtesy of the Academy of Motion Picture Arts and Sciences, Beverly Hills, California.*

Thus when Ford's boomtown marshal, the fictionalized Wyatt Earp, symbolically embraces the communal and familial values represented by the newly formed Tombstone congregation and by Clementine, the lady from Boston, his feud with the vicious Clantons becomes a crusade for civilization and his success at the O.K. Corral a bittersweet victory that brings to an end the only way of life in which he, with his shooter's skills, could flourish. Ford establishes the essential savagery of both nature and man at the opening of the film with stark camera shots of a pair of menacing gunmen who observe cattle and cowboys in a parched desert setting. But at the end of the film it is a woman, Clementine, the town's new schoolteacher, standing by a tamed, fenced-in landscape, who takes the observer's role as the shooters depart.

The three Earps, seeking to avenge their murdered youngest brother, initially encounter a darkly lit Tombstone of saloons, dance halls, and poker games, a world where Shakespearean tragedians can only degenerate into farce. "Wide-awake, wide-open town, Tombstone! You can get anything you want there." Its denizens are rowdy miners and cowboys, gamblers and whores, its boss a consumptive, alcoholic physician who has turned his back on the civilization of the East. Clementine, his former fiancée, proves unable to reclaim Doc Holliday for Boston and all it stands for. But she begins the taming of Wyatt Earp, and as he primps for her in the local barbershop, Ford lets us glimpse for the first time a different, brighter Tombstone. Beyond the dark of the hotel porch the street floods with sunshine. Wagons and buggies purposefully stream past, families greet one another on the sidewalk, and serving girls in their best hats bustle out of the hotel. "If I wasn't in the territory," Virgil Earp observes, "I'd swear we were back home on a Sunday morning." But the Sabbath has indeed come to the territory. "You know," says Morgan Earp, "there's probably a lot of nice people around here. We just ain't met 'em." Then Wyatt accepts Clementine's ladylike challenge to escort her to church; he slowly leaves the shadows to step with her out into the Sabbath sun. Thus when the Earps subsequently destroy the Clantons at the O.K. Corral, they not only have avenged their brother; they have also, as Wyatt promised, left behind a country where young kids like him "will be able to grow up and live safe," where law and culture are free from ridicule, where the schoolteacher literally replaces the gunslinger. Female sexuality yields to feminine sensibility, and the madam shows herself to be a sensitive nurse. Even Doc Holliday, although too corrupted to be allowed to survive into the new era, is redeemed in death by his belated decision to join Wyatt's crusade.

At the heart of the western saga, Ford would appear to be telling us, lies the civilizing process. The Wild West is an individualistic, solipsistic male world untrammeled by law, morality, higher culture, or feminine domesticity. Its women are deceiving whores and half-breeds, fully deserving the symbolic cleansing in the horse trough that Wyatt administers to Chihuahua, Doc's paramour. Domesticity is as doomed as young James Earp and his plans to marry his sweetheart. "Mac, you ever been in love?" Wyatt asks one of Doc's employees. "No," he replies, "I been a bartender all me life." Women, the family, and the bonds of community can thrive, it seems, only when the West has been tamed. They are the motive that invests the history of western violence with virtue, just as their ultimate triumph signals the end of the uniquely western experience and the final integration of the frontier into the ongoing history of the settled, civilized nation. And in the process they are themselves changed, strengthened, made more robust. A formerly dependent woman like Clementine can now stand alone and shape a role for herself in constructing the new community; the community, like the deacon's womenfolk, can reject the artificial conventions of the East for the wholesome naturalness of western life. Family and community, fresh and reinvigorated, are the rewards that lie at the end of the trail, the ends but not the means of western conquest.

Or are they? Ford's story on one level is indeed the familiar parable of family, community, and the civilizing process that has provided so much of western legend and history with its standard plot. But on another level it charts what can be understood only as a far more complex transformation from one regime of family and community to another. The Clantons, after all, are in Ford's telling also a family of a kind, a dynasty of four brothers held together by their father's will and whip, and the Earps are yet another set of apparently motherless sons, bound in duty and affection to the family economy headed by their Pa back in California. Powerful motives of family honor and vengeance fuel the actions of both clans. Not even Doc Holliday has fully shaken off the claims of family and community. Rather, he is a renegade who prolongs his exile to avoid bringing shame to those who claim him as their own, and in his resistance to the new regime that Clementine represents he even contemplates making his liaison with Chihuahua a permanent one.

Nor is Ford's raucous Tombstone without its own communal structures. It has a mayor and a marshal and the forms of law; when Wyatt Earp marches to the O.K. Corral, he does so with an arrest warrant in his pocket and with the blessings of both the mayor and the deacon of the new church. Untamed Tombstone can constitute itself a moral community in quest of common goals, whether simple entertainment or support for a wounded singer. And it is linked to the wider national community by everything from accounts in the Lordsburg bank to newfangled barber chairs imported from Kansas City. It may not share the moral code of the domesticated, cultured East, but it has its own communal morality nonetheless, a masculine morality that proscribes equally stealing cattle, playing eight-handed poker games, cheating on one's lover, and drawing on a man who is not carrying a gun. Ford's ambivalence about the passing of this order may not be as palpable here as in a later film like *The Man Who Shot Liberty Valance* (1962), with its overt acknowledgment of the falsity of the myths upon which the civilizing process rested. But when Wyatt Earp rides away from Tombstone and its new teacher, it is not just Clementine who is "lost and gone forever" to him. In remaining true to the dictates of one kind of family and community, he has midwifed another, different familial, communal order in which a man like himself can find no place. It is not so much that family and community triumph as that one kind of family order replaces another.

Ford's story, of course, bears only a tenuous relationship to historical reality. It is rather difficult to turn the actual saga of the Earps, the Clantons, Doc Holliday, and the O.K. Corral into a comparable drama of domestication, no matter how ambiguous. The historical Earps were hardly the virtuous retired lawmen and cowboys of Ford's fable. The best recent accounts make it clear that the Earps, though probably not the stage robbers and horse thieves of the revisionist counterlegend, were definitely gamblers, bartenders, brothel keepers, and small-time speculators, as well as sometime lawmen and farmers. The famous shoot-out—not at, but near, the O.K. Corral—had less to do with family vengeance than with tensions arising from economic and political rivalry. And rather than nobly sacrificing his life, Doc Holliday—a Georgian, not a Bostonian—probably initiated the slaughter by drawing first, and lived to trade on his notoriety. Tombstone indeed had a young, single schoolteacher, twenty-four-year-old Lucy McFarland who lived with her sister and her lawyer brother-in-law, but she was from West Virginia, not Boston, and it was not she but the undoubtedly less virtuous Josephine Sarah Marcus, an "artiste" raised in a prosperous German Jewish mercantile family of San Francisco, who attracted Wyatt Earp's wandering eye.

Nevertheless, like their filmic counterfeits, the real Earps inhabited a West defined by distinctive bonds of family and community. They were the offspring of an agrarian Kentucky clan who migrated westward as a family in chainwise fashion, first to Illinois and Iowa, then to Missouri, Kansas, and California, as the various brothers peeled off on a series of continually intersecting trajectories that carried them singly or together through mining camps, railheads, and cow towns from Montana to Kansas and Texas before reuniting them all in 1880 in the new Arizona boomtown of Tombstone. They were, in fact, all married men. Each arrived in Tombstone with his wife, common-law or otherwise, firmly in tow, though Wyatt would replace his partner with Josie during his Tombstone stay. Their women labored for them and moved with them. The Earp brothers were on occasion founding members of church congregations and candidates for public office. They supported and leaned on one another in their efforts for economic advancement, and outside the O.K. Corral they joined to affirm their reputations against not one but two pairs of brothers—McLaurys as well as Clantons—who like them were striving to make it family-fashion in the West. The Clantons, like the Earps, were members of a southern clan seeking to reproduce a familiar pastoral lifestyle in the new territory. The McLaurys, by contrast, were young New York–born, Iowa-raised entrepreneurs who cut loose from their parents to find the fortune that would enable them to found new families of their own. Doc Holliday's initial move west can be seen as a similar attempt by the professional son of well-to-do urban, middle-class parents to establish a home for himself in a healthier climate. The real Tombstone, like its movie version, was overwhelmingly a community of unattached miners leavened by a sprinkling of families like these, to be found mainly within the town's small entrepreneurial class. But the surrounding countryside was punctuated with smaller, family-centered communities—one of them Mormon, several Mexican—that testified to the area's older, deeper domestic roots. And although the only Native American in Ford's film was the drunkard whose literal removal signaled the beginning of Wyatt Earp's crusade to civilize Tombstone, the constant threat of Apache attack was a disturbing reminder to the historical Tombstone of yet another familial tradition in the region.

Thinking about the Family in the American West

In fact as in film, it would seem, the American West was not a domestic tabula rasa. As Ford intuitively sensed, there is indeed a family story at the core of America's western history, but it is something other and more complicated than our familiar fable of domestication and civilization. Many different kinds of family arrangements were to be found in nineteenth-century America's successive Wests, before, during, and after their incorporation into the American nation. Kinship ties could extend across great distances, and family logic could influence even seemingly unattached men and women. Families performed a range of different functions, for their members and for the broader society, that varied with time and place, culture and class. The family in any society is necessarily a cultural and legal construct that invests the basic biological relationship of parents and children with social meaning. Societies determine for themselves what defines a family, who constitutes its members and what their responsibilities are to one another, how far the family identity extends through marriage and consanguinity and across generations, on what basis a family is to be formed, and what societal functions it is meant to fulfill. Tombstone in 1880, no less than Texas in 1820 or Iowa in 1850, was a family frontier in the most literal sense: an arena in which culturally variant constructs of the family, carried by individuals pursuing varying family strategies, met, intermingled, and clashed. These frontier family dynamics played a crucial role in shaping the new societies that emerged, and in turn the struggle to define these new societies helped define the range of regional family models that would survive, the directions in which they would evolve, and their broader influence on American family construction.

Though scholars long neglected the contours and implications of this family frontier, to nineteenth-century Americans the link between western lands and the family was a self-evident though complicated one. "It is in the very philosophy of things, in a country like ours, whose free institutions awaken and bear up the spirit of aspiration from a humble hut, as well as the lofty palace, that the poor man, surrounded by his wife and his children, and animated by a holy love of those endeared objects, with a pure conscience and a resolved purpose, relying upon his own unassisted arm, should go forth to the wilds of the far West to improve his fortunes, and confirm his personal independence," enthused one senator in 1841. To provide for their children, families moved west. In moving west, they escaped the corrupting influences that nineteenth-century American thought attributed to urban life, thereby preserving for the nation its body of independent, freedom-loving householders and, not coincidentally, preserving the West for the nation. As an 1846 observer noted, "All we had to do was to let our women and children go [to the Oregon region] and, without assistance from any one, they would take possession of the country."

Thus easy access by settlers to public land was defended in terms of both its benefits to families and their benefits to the nation. "I know the character of the pioneer," insisted a territorial delegate to Congress in 1852, "and of the men who even now are on their way to the West, and I speak understandingly when I say that it is in such homes as this bill, if adopted, will create, which will forever remain the nurseries of that love of freedom by which alone our present happy form of government can be perpetuated." But by the same token the West could endanger the American family, should it lure away too many of its children or should the family in the West become too isolated from uplifting moral influences, too exposed to the risks of excessive speculative gain. The nation needed strong families; strong families required the independence that western lands could give them; family settlement was the best and cheapest insurance that the West would be tied firmly to the national culture; but familiar models of the family might also prove vulnerable in the West.

Nowhere was the complex circular interplay of family and western opportunity in nineteenth-century American thought better explained than by that acute analyst of Jacksonian America, Alexis de Tocqueville. Without a law of primogeniture, Tocqueville insisted, American families had developed few strong attachments to place or patrimonial land. The resources of the West preserved the promise of economic progress and equality for each new generation on which the belief in American democracy rested, while encouraging the family to reshape itself in democracy's mold. Patriarchy waned, relations of dominance between father and son, and brother and brother, yielded to companionable cooperation, and women, in Tocqueville's reading, achieved and accepted an equal though separate status. Such democratic families raised children appropriately prepared to leave home to make their fortunes; at the same time, these families became a central reason for the stability of American democracy. Order and community in a society lacking strong central authority depended not only on the intersecting self-interests of its self-governing citizens but also on shared mores inculcated by religion and cultivated within the family. But this meant that if the West was the source of the American family's strength, it could also become the nation's weakness, should the bonds of family become too distended in the course of settlement. High levels of speculative profit in the West, Tocqueville feared, too often attracted westwardly mobile men unhampered by kith and kin and thus undisciplined by morality or public opinion. Western opportunity shaped the domesticity on which American democracy depended, but opportunity also placed the family in peril.

American artists in the middle decades of the nineteenth century similarly constructed for their viewers a domestic West, a West peopled by families, a nurturing West for the homes that would cradle the nation's future. The archetypal images were the Madonna-like women whom a stalwart Daniel Boone leads through the Cumberland Gap in George Caleb Bingham's 1851 painting, the three-generational family alongside its covered wagon that listens to a rugged trapper in William Ranney's 1853 *Advice on the Prairie*, and most especially the log cabin domesticity depicted in Thomas Cole's 1845 *The Hunter's Return* (see p. 145). Cole's cozy vine-trellised cabin nestles beneath sublime mountain scenery. A broom and a washtub flank its door, laundry dries beside neat rows of cabbages in the garden, one dog sniffs at the dinner meat airing on a bench and another playfully nuzzles a small child, while smoke drifts invitingly from the cabin's stone chimney and two women eagerly hail their menfolk returning laden from the hunt. No matter that these were in good part images explicitly crafted to celebrate and encourage American expansionism. They worked because they spoke to mid-century Americans' perception of the West as a family resource, the same perception that buttressed antebellum demands for a liberalized homestead law and fueled the attack on slavery in the territories as a threat to family settlement. When the West was portrayed as violent and dangerous, as it often was in paintings of Indian attack, abduction, and rape, it was the threat to the domesticity of the settlers' West that was particularly emphasized, the threat to the purity of its women and the sanctity of its homes. But even the ostensibly untrammeled West of the Native Americans and the fur trappers who

In William Ranney's painting, the men of the family listen intently to an old trapper's tales and absorb the historical lore of the plains, while the Madonna-like mother and child engage the viewer more directly and affirm the family strategies that motivated western settlement.

William Tylee Ranney (1813–57). Advice on the Prairie. *Oil on canvas, 1853. Courtesy Buffalo Bill Historical Center, Cody, Wyoming.*

lived among them was often domesticated by the artists of the 1820s, 1830s, and 1840s, who produced a compelling iconography of Indian daily life and trapper-Indian family unions.

But by the eve of the Civil War, a more somber reassessment was beginning to inform the log cabins that Jasper Cropsey and Sanford Robinson Gifford depicted as inappropriate and unlovely intrusions in the natural landscape, and soon a different kind of artist's West began to take center stage. Families still moved west, their journey now hastened by the railroad, in the popular lithographs of Currier and Ives, but increasingly in western art the domestic middle landscape faded before, on the one hand, a masculine West of soldiers, Indian warriors, and cowboys—Frederic Remington, Charles Schreyvogel, and Charles Russell country—and, on the other, the tourist's delight of unpeopled vistas and valleys, as cultivated by Albert Bierstadt and Thomas Moran. The semiarid lands and boom-and-bust mining camps and cow towns of the post–Civil War West were chancy environments in which to seek family security. The new suburbs of the burgeoning cities back East seemed to offer a safer domestic haven, and in art as in literature a wilder West now shaped the public perception.

It is no accident that it was in the context of this dominant new masculinized and nature-bound western vision that Frederick Jackson Turner in 1893 fashioned his lyric argument for the significance of the West in the nation's development. For the earlier generation of Tocqueville, mores—culture—had taken precedence over environment, and the centrality of the western family seemed self-evident. But by Turner's time, many American reformers had come to doubt the ability of the family to withstand the pressures of the new urban environment. What role, then, could such a feeble institution hope to play in the face of the even greater savagery of the wild? It is little wonder that

"What we want in California," this popular print of mid-century implied, was the civilizing influence of eastern families whose domestic culture would replace the less desirable influence of the single male miners or native California Indians they would inevitably displace.

Britton & Rey, lithographers (active 1852–58). What We Want in California. *Lithograph, ca. 1855. Courtesy, The Bancroft Library, University of California, Berkeley.*

consideration of the family was overwhelmed in Turner's analysis by his emphasis on the competitive individualism of the frontier and that his stage theory of frontier settlement could readily be taken to imply that families culminated rather than coordinated the process. It was left to others, most notably Arthur W. Calhoun in his pioneering 1917 history of the American family, to work out the more specific implications of Turnerian thinking for the relationship between the family and the West. Calhoun noted that the pioneer's necessary focus on "home building and home protection" meant that "the psychology of domesticity was supreme," that a strong "clan-spirit" often developed, and that "the family was the one substantial social institution" on the frontier. But he did not explore the structuring role played by the family in the West, a role that these insights implied. Instead, he focused on the effects of the West on the family, and here his Turnerianism emerged most clearly. The most important formative influence on the American family in the decades after the Revolution, he insisted, was "pioneering and the frontier." Frontier dispersion, hardship, and democracy combined to foster early marriage, high fecundity, easy divorce, the emancipation of children, improved status for women, and insensitivity to the claims of lineage. The frontier, in short, encouraged a fundamental transformation from a regime of "familism" to the "parentalism" that he saw dominating his own era.

By the time that historians began to develop a systematic interest in family history in the 1960s, however, Turnerian approaches were so thoroughly discredited that the question of a specific western or frontier influence on the American family was barely raised. Family historians uncovered a rich and finely textured account of the fundamental transformation of American family patterns and ideology in the course of industrialization, urbanization, national centralization, political change, and ethnic and racial pluralism but made little effort to incorporate geographical expansion and western regional development into their central interpretation. Only as western historians belatedly began to explore the role of women—and children—in the West did family-related issues again receive much scholarly attention. We can now draw on a diverse and growing body of research to sketch, in broad outlines, answers to the four basic questions that structure this essay. How did family auspices influence the processes of migration, settlement, resistance, and integration that peopled the West? What were the resulting variations in western family patterns? Can we identify a distinctive western influence on the development of the American family? And how did the particular values of nineteenth-century American family life shape western regional development?

Family and community were not a simple culmination of western conquest. Nor did new kinds of conditions create a new breed of family or community in the American West. As European and African Americans in the course of the nineteenth century pushed westward over the Appalachians, through the great valley of the Mississippi, across the plains and the Rockies and the deserts to the Pacific and back into the intermountain plateaus and parks, and as Asians moved eastward and Mexicans north, they all carried with them familiar and diverse assumptions about domestic life and firmly established patterns of domestic structures, functions, and relationships. They encountered similarly deep-seated domestic patterns among the Native Americans and French, Spanish, and mixed-descent colonists already inhabiting the land. Bonds of family and community fundamentally shaped the processes of both conquest and resistance, and those processes in turn molded the domestic landscapes that westerners would inhabit by the end of the century. There would be not one form of the western family, one kind of western community, but many. Changing American family and community norms, selective migration, local resource variation, differential incorporation of indigenous peoples, and locally varying degrees of economic integration with the national order all combined to construct a complex and continually evolving domestic landscape in the American West that affected every aspect of western life and influenced evolving conceptions of the American family. Family threads were woven deeply into the fabric of western development from the outset, and western history cannot fully be understood without tracing the patterns they formed. The O.K. Corral showdown was "strictly a family affair," insisted John Ford's Wyatt Earp. The same, we shall see, might almost be said of the West itself.

The Family Logics of Western Settlement

When young Millard Fillmore moved west to Buffalo in 1822, he was both following and breaking a family tradition: following, because he was repeating a pattern set by his father and his grandfather before him; breaking, because whereas they had sought new land on the frontier, it was the new opportunity of a western city that drew the young lawyer who later would become the nation's thirteenth president. Westward migration was certainly an established family habit. In 1765 Millard's grandfather, after seeing his two older brothers move to new settlement areas in Nova Scotia and northwestern Connecticut respectively, set out for frontier Vermont, leaving his youngest brother behind to farm the family land in Norwich, Connecticut. As his sons came of age, they repeated their father's pattern. The two youngest remained behind to take over the Vermont homestead while the oldest moved with his young family to Oneida County, New York, and then on to the Buffalo area as his children approached adulthood.

Then the second and third sons—the second was Millard's father—in their turn joined forces to purchase wild land in Cayuga County, New York. But luck turned against them. Their title proved defective, and they were forced to lease poor land elsewhere in the county. Disgusted with farming under such circumstances, Millard's father, as Millard later recalled, became "anxious that his sons should follow some other occupation" and schemed to apprentice them to such standard rural trades as carding and cloth finishing, carpentry, and masonry. But he soon realized that his talented oldest son could set his sights still higher. Born with the century, in a primitive log cabin, Millard came of age as the pace of local life was quickening in upstate New York. Lending libraries and academies were bringing the culture of a wider world into the backwoods, and Millard arduously scraped together a sort of education for himself in the time he could steal from farmwork, apprenticeship, and winter schoolteaching. So Fillmore senior finally approached his landlord, a wealthy judge, to beg—successfully—that he give young Millard a chance to read law. When Millard was ready to move on a couple years later, it was Buffalo, portal to the vast western wilderness about to be opened by the Erie Canal, that inevitably attracted—and amply rewarded—his ambitions. Two of his younger brothers would attempt to follow the same path (both died young), and two of his Buffalo-area cousins became local ministers. But the rest of this generation of Fillmores continued to trod the accustomed family paths. One of the Buffalo cousins stayed behind on his father's land while the remaining cousin, and Millard's three other brothers and their families, followed the Great Lakes westward to the new frontier farmlands of northern Indiana, Michigan, and later Minnesota.

Few families produce a future president, distinguished or otherwise. But in other respects the Fillmores typified a fundamental family process that peopled much of the American West. Westward migration for them, as for so many rural American families, was a generationally recurring event, as expected and regular a stage in the normal life course as marriage or retirement. It formed a central element in the family's strategy for providing livelihoods for the coming generation; family considerations governed the timing and composition of migration, and extended family groups sought common destinations. Individualistic loners were probably more common in western legend than in life. No matter how tenuous family ties might seem when a young scapegrace like Millard's cousin Henry Glezen Fillmore drifted into the free and easy life of the Indian trader on Minnesota's northern frontier in the late 1840s, it was no accident that brothers and cousins soon followed in his wake. And when he subsequently married, it

As the three generations depicted in this 1854 field sketch of western immigrants suggest, extended families often moved together, preserving a labor force that might be essential in developing a new home.

W. C. Weisel (active 1854). Camp of the Covered Wagon Train. *Watercolor on paper, 28 December 1854. M. and M. Karolik Collection. Courtesy, Museum of Fine Arts, Boston, Massachusetts.*

was to a young woman with a similarly extensive local family connection of her own. The story of the family in the West necessarily begins with its varied influences on the peopling of the region itself.

It often seemed to nineteenth-century Americans that there was something inexplicable, even irrational, about the pace and intensity of westward migration. They referred to it in terms suggestive of irresistible natural phenomena. They spoke of "Oregon fever" breaking out, of migrants "swarming" like a hive of bees, of stampedes and epidemics, waves and floods. For the *Cleveland Herald* in 1839, migration was a "tide" that in "the past season [had] been setting toward the west stronger than ever." Americans seemed a breed of restless wanderers, ever seeking elbow room or speculative profits over the next horizon. What but irrational, individualistic ambition or the equivalent of a force of nature could pull so many away from well-tended domestic hearths and the ample opportunities to be found in the more developed sectors of the nation's economy?

Scholars have tended to seek more rational explanations in "push factors" peculiar to a particular time and place—the worn-out fields of New England that by the 1830s could no longer compete with the productivity of new western lands, for example, or the fevers, floods, and low crop prices with which Missourians had to contend in the 1840s—or in uprooting personal crises like bankruptcy or family tragedy. Or they have stressed the "pull" of western opportunity for a striving, entrepreneurial nation—the special "one time only" profits that could accrue to those who reaped the first crops from virgin soil, to those who pastured the first cattle on prairie grasses, trapped the first beaver, mined the first ores. But there were always crises somewhere, and migration was not the only possible response; western opportunity always beckoned, but only some Americans responded. Underlying the complex interactions of specific push and pull factors that help account for the peopling of any given frontier was a family regime that made westward migration for countless American families like the Fillmores—and the Earps—a viable, sometimes even an inevitable, option.

The family, Brazil's noted anthropologist Gilberto Freyre asserted several decades ago, was, for his country, "the great colonizing factor." The same can be said for the United States. The successive Wests that were integrated into the American nation in the course of the nineteenth century were shaped from the outset by the needs and strategies of American families. A number of different and changing family cultures coexisted in the older settled areas of the United States—and in Europe, China, and Mexico—during the nineteenth century. Their varying assumptions about family function, size, and relationships fundamentally influenced who chose to migrate to new frontier areas, when, and with what goals. The varying family patterns of the native peoples already resident in these areas were similarly influential in shaping their ability to resist, adapt, or succumb to the newcomers. Equally significant was the ideological colonization of the West by a powerful new familial ideal that, once embedded in institutions and law, proved able to play an independent role in the shaping of western life.

It was the family cultures of the westward-migrating European-American populations that would be the dominant vectors of change in the new region. Historians have identified the nineteenth century as a time of fundamental transformation in American family life, when a family regime suitable for life in a traditional agrarian society yielded to a new kind of family better adapted to the new industrial order. Various labels have been proposed to capture this transformation—from the "traditional" to the "modern," from the "patriarchal" to the "companionate," from the "instrumental" to the "affective," from the "household" to the "domestic" family—with many recognizing in the nineteenth-century "Victorian" family a style distinctive both from what came before and from the regime that would become dominant in the twentieth century. Whatever the terminology, a new set of assumptions about family life came to be widely shared among nineteenth-century Americans: that the ideal family rested on an affectionate union of a man and a woman who provided one another with love, companionship, and comfort and who carefully and lovingly nurtured their children to be well-developed individuals and productive members of society. Each parent had a separate, distinctive sphere within this family ideal: the man was to be its head, its public representative, its wage earner, and its ultimate source of authority; the woman, its heart and its conscience, kept the home, trained the children, and deferred to the husband while providing gentle moral counsel. No longer a basic unit of production, this new family ideal offered its members a private, sentimental retreat from the rigors of the new, vigorously competitive, capitalist economy. This new style began appearing in prescriptive popular literature discussing family life by the 1810s, and within a generation it overwhelmed the previous emphasis on large, patriarchally governed and productive families with well-defined public roles.

But too much emphasis on the ideal family model defined by the ideology of the times ignores the actual variety of family styles within the nineteenth-century American family regime, and a periodization based on the replacement of one family ideal by another obscures the extent to which older family styles continued to coexist with newer.

If we abandon familiar efforts to classify family styles on the basis of the emotive character of familial relationships or the source of power within the family and turn instead to the economic basis of family organization, we can identify at a very general level four distinctive, coexisting nineteenth-century American family styles that shaped the westward migration process. We can term these the patrimonial, the proletarian, the entrepreneurial, and the slave family styles. Each style had its own distinctive logic of migration.

To understand those logics, we have to recall in very general terms how these styles evolved. American families of all stripes were heirs to what has been termed the western European family system. Any traditional people, dependent on relatively unproductive agriculture, limited craft production, and inefficient trade, necessarily faces the classic Malthusian dilemma of balancing resources and population. With population multiplying over successive generations, land and hence food will run short, resulting, if nature is left to run its course, in famine, disease, and war until the population is again reduced to a level that can be supported by available resources. Of course, nature does not have to be left to run its course, and in traditional western Europe, the most important regulator proved to be the cultural assumption that procreation should occur only within marriage and that marriage should occur only when the new couple were capable of supporting their own household. This meant that the proportions of the population who married, the age at marriage, and consequently the number of children produced by the marriage even in the absense of effective contraception could vary greatly with time, place, and class, depending on the productivity of the local economy and a particular family's place within it. When times were hard, marriages were postponed, birthrates declined, and family size grew smaller. High death rates, particularly among children, also kept family size in bounds. The average western European age at marriage, consequently, tended to be relatively high. The average family was never large, nor were extended multigenerational families residing in the same household very common.

Central to the survival of such a traditional family was the patrimony on which it depended—the land or craft from which it earned its living and which it transmitted from generation to generation. The family passed through a recurrent cycle, from when a young couple married and took over its patrimony, to the maximum earning years when children worked alongside their parents to amass the resources needed for their start in life, to the point when the parents retired to make way for the next generation. In some European regions, the patrimony was divided among all children; in others it remained with one heir while the rest had to find other niches in society. But whatever the specific strategy, the basic principle remained: it was through the family that the individual gained access to a livelihood, and families thus depended on maintaining the patrimony from generation to generation to ensure the livings of their members. Because families were the fundamental units of local society, performing most of its basic functions of reproduction, production, education, welfare, and even governance, they had a fundamentally public character, and local society—formally through the agencies of government and religion, informally through gossip and social sanctions on behavior—paid close attention to the regulation of life within individual families. Wider kinship networks were also important assets. In an age when few institutions existed to guarantee the behavior of strangers, shared blood was one of the best ways of securing trust, and kinship ties could be invoked to ensure everything from emergency assistance to business partnerships, making marriage as much an alliance of two kin networks and two patrimonies as of two people.

Family limitation through delayed marriage was not, of course, the only way that resources and population could be kept in balance. Over the long term, improved productivity released gradually increasing proportions of the western European population from agriculture, paving the way for the quickening of manufacture and trade that would usher in the industrial revolution of the nineteenth century. In the process, two other family styles took shape. One was the proletarian family. Wage earners, urban or rural, lacking either rights in land or a trade that could be passed down to their children, had little reason to think of their families in cross-generational, patrimonial terms. For them, the family economy could be only a pooling of wages for mutual survival. There was little reason to postpone marriage, husband and wife shared the necessity of labor, and additional children were more often regarded in terms of the wages that they would earn rather than the drain on scarce resources that they might represent. Near the other end of the economic spectrum emerged what might best be termed the entrepreneurial family, a family style focused less on the preservation and transmission of the patrimony than on enterprising investment in the constantly changing opportunity offered by the emerging capitalist order. For those with the means to sustain it, the family ceased being the main unit of production. The husband followed his work out of the home into the male world of business, leaving his wife behind to cultivate domesticity, rear their children in the entrepreneurial virtues, and through her tasteful consumption, affirm the gentility of the striving family—in sum, the family ideal that would be propagated as the desired national norm in nineteenth-century America. These families remained enmeshed in the web of kinship and continued to enshrine the ideal of patriarchal authority, but the productive unit of husband, wife, and children—the core of the patrimonial family—was gone, and the number of children declined. Success in the new economy favored those endowed with familial gifts of education and capital but unhampered by constraining obligations to people or place; for such men on the make, the new domestic ideology served as a substitute tether to society's need for the support of women and children.

Accompanying these changes was yet another resolution to the Malthusian dilemma: migration. Even in traditional Europe, those who could not find a niche in rural society often sought opportunity in a nearby town or moved to new lands on the margins of European cultivation. From the early seventeenth century onward, they could also move to America. Their emigration was necessarily influenced by the family economies of which they were a part and which shaped their hopes for life in America. Thus the historian Bernard Bailyn has documented the presence of two very distinct migration streams in the heavy emigration from Britain to America in the mid-eighteenth century. One stream he terms "provincial," made up of mature families with growing children who moved from the agricultural areas of northern England and Scotland with the hope of reestablishing what we have called a patrimonial agrarian family economy on American land. The other "metropolitan" stream consisted mainly of young men, and some young women, who moved as individuals from England's largest cities to seek laboring or entrepreneurial employment in the more urbanized areas of America in particular.

By the time of national independence, some version of the patrimonial family ordered life for the vast majority of Americans who lived in the new nation's rural areas—some 95 percent, according to the 1790 census—and for many in the few but growing cities. The rigorous logic of the proletarian family was, however, also spreading in the cities, and the entrepreneurial family style was beginning to flex its metaphorical wings. Equally in evidence was a fourth generalized family style, formed under conditions of slavery and hence dominant among the roughly 20 percent of the 1790 population who were of African descent. The slave family, as historians have described it in its mature form in the decades before the Civil War, generally centered on a stable two-parent household wherever possible. But parents had little control over either their own work or the fates of their children, and families were constantly at risk of disruption through sale and forced migration. Their situation encouraged general status equality between the sexes and embedded each nuclear family deeply in a broad local net of kin and community that could be called on for the support that the nuclear family was never able to ensure.

The vast expanses of land lying to the west, by European standards lightly used and often lightly defended, almost inevitably played a particularly prominent role in the American family calculus of survival and intergenerational succession. Why delay marriage, why limit children, why divide a patrimony, why cling to the old family hearth, when the limitless resources of the West seemed there for the taking? "To better their condition in an unknown land our forefathers left all that was dear in earthly objects," President Andrew Jackson declared in 1830. "Our children by thousands yearly leave the land of their birth to seek new homes in distant regions. Does Humanity weep at these painful separations from everything, animate and inanimate, with which the young heart has become entwined? Far from it. It is rather a source of joy that our country affords scope where our young population may range unconstrained in body or in mind, developing the power and faculties of man in their highest perfection." Future historians would note that the expenses of establishing a western home could be substantial, precluding this option for many, but European-American families would prove surprisingly capable of marshaling the resources needed for migration when it fit within the logic of their long-term strategies. And because that logic varied with family style, so too did the influence of family on the timing, composition, and goals of migration and hence also on the character of the new communities created in the West.

Two Families Move West

There may be no better way to illustrate this point than to take a closer look at two representative American families as they moved across the continent from 1700 to 1900. The Fillmores can serve as our New England example, whereas the Maxeys, who arrived in Virginia at about the same time as the first Fillmore set foot in New England, provide a southern counterpart, necessary because both slavery and kinship systems created distinctive regional differences. These two families were chosen almost at random from

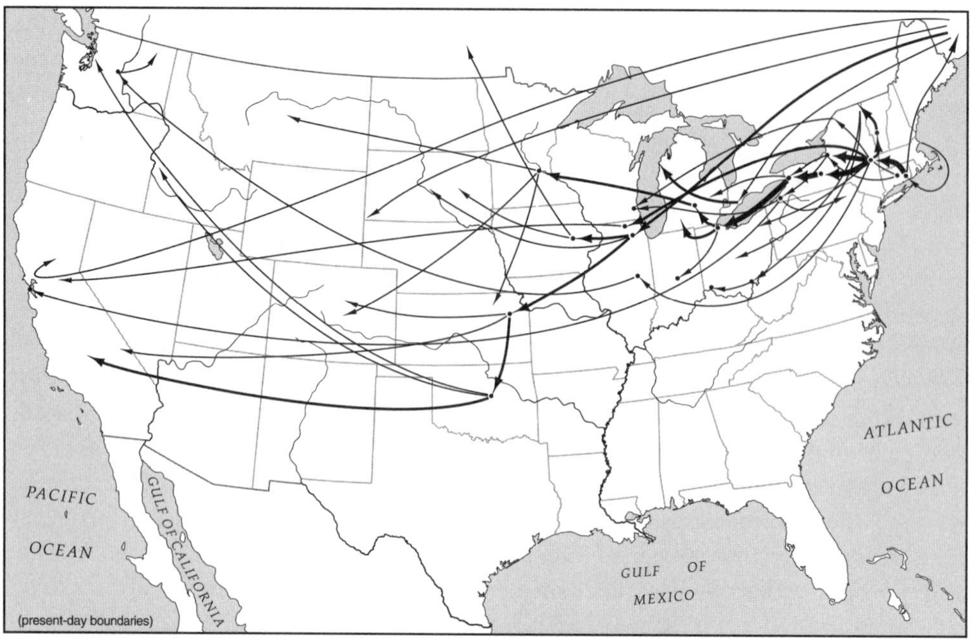

Generalized Fillmore Family Western Migration Paths, 1700–1900

among the plethora of families for whom well-researched genealogies have been published in recent decades. Despite gaps inevitable in even the best-researched record, the basic information about places and dates of birth, marriage, death, landownership, occupation, and the like compiled in Charles L. Fillmore's and Edythe Maxey Clark's authoritative family genealogies yields a surprisingly rich and consistent account of family migration strategies when set within the broader context provided by family and settlement history. Though the Maxey family, unlike the Fillmores, never produced a president, it did number among its members both the wife of Abraham Lincoln's law partner and a Confederate general who later became a Texas senator. More important for present purposes, however, is the very anonymity of most of the more than 1,400 Maxeys and 650 Fillmores whose eight generations of westward migration illuminate the strategies that they shared with countless other European-American families of the time. Like most genealogies, these accounts permit us to follow only the male lines through extended generational sequences, but the daughters of each successive generation are included. It is also important to note that our generalized analysis necessarily skips over many intermediate family moves, side paths, and backtrackings in the interests of interpretive economy.

Either Fillmores or Maxeys participated in virtually every stage of the European-American penetration of the successive Wests of the nineteenth century. Most of the Maxeys, like the Earps and the Clantons in Tombstone, formed part of that vast procession of upland southerners who trailed Daniel Boone across the upper South and lower Midwest to Missouri and points west, while other lines of Maxeys moved, like Doc Holliday's family, through the plantation lands of the lower South and Texas. By

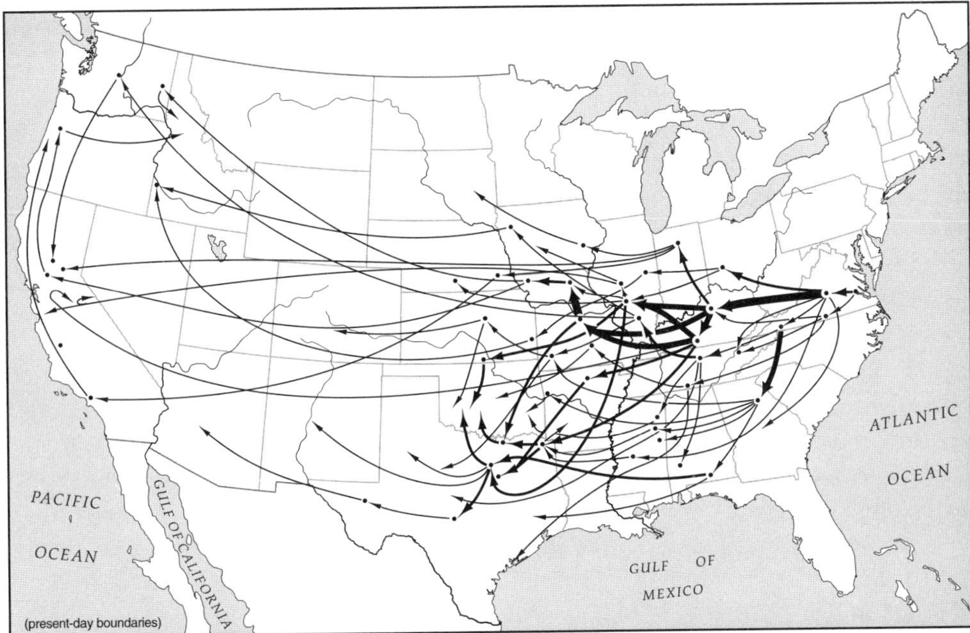

Generalized Maxey Family Western Migration Paths, 1700–1900

contrast, the Fillmores were, like the McLaurys, part of the Yankee diaspora that fanned out to northern New England and upstate New York before and after the Revolution and then pushed westward to the Great Lakes and—barring the occasional southern diversion—California and the Pacific Northwest. Many members of each family would stop and take root at each stage along the westering march, and at each stage urban opportunity soon began to exert a counterpull, but as long as new frontier resources beckoned, there was a Fillmore or a Maxey to answer the call.

The first American Fillmore, John, was a mariner who married in New Hampshire in 1701 and established his family in Beverly, Massachusetts. After his early death at sea, his wife remarried and moved with the children to Norwich, Connecticut. John's younger son remained on the family's Norwich land, but the older son, John junior, began his career as a seaman like his father. John junior's children seemingly shared his broader horizons, and began the family's march westward when service in the 1755-63 war with France exposed them to the opportunity that lay beyond New England's northern and western borders. This time too, it was one of the youngest of the five brothers who remained behind to cultivate his father's Norwich farm, though some of his offspring would subsequently join their uncles and cousins in Vermont or New York and move on westward with them. The oldest of the five brothers settled in New Brunswick, where his progeny participated generation by generation in the slow opening of Canada's Maritime frontier. One line of these Canadian Fillmores, however, followed the logging frontier westward into Ontario and Michigan while another moved to the farmlands of northern Illinois and then on to Iowa, South Dakota, and Manitoba, on the one hand, and to Kansas, Oklahoma, Colorado, and the Pacific Northwest, on the other—a family trek that culminated only when some of the Oklahoma Fillmores joined that last great westward pilgrimage of the Okies to California in the 1930s. Two of the other prerevolutionary Fillmore brothers, along with one of their sisters and her family, settled in the Bennington, Vermont, area. Millard Fillmore stemmed from this migrant branch, and his relatives pushed the family westward from New York through the Great Lakes states to Montana, Colorado, and Kansas. The remaining brother's line effectively halted its frontier advance in western New York State.

The Maxeys' parallel procession westward began early in the eighteenth century when Edward and Susannah Maxey arrived from England to take up plantation land in Tidewater Virginia. Here the oldest of their six sons remained while the rest moved west to plantations in Virginia's Piedmont. Once the frontier beyond Virginia's boundaries opened in the turbulent years during and after the American Revolution, the next generation of Maxeys began to cross the mountains, encouraged in several instances by land grants received for revolutionary service. The first to leave Virginia was Jesse, one of the original Edward's youngest grandsons. His father had settled twenty years earlier in southwestern Virginia under the shadow of the Blue Ridge, and now, in the early 1770s, Jesse with his young wife and infant son trekked down the Great Valley and across Virginia's southwestern border to help create the first permanent colonial settlement in Tennessee. When, during the Revolution, settlers pushed farther west to the Cumberland Valley, forming a local government for themselves at the site of Nashville, Jesse was again among their number. Surviving a severe wound at the hands of the region's native defenders, the "old Indian fighter" and revolutionary veteran lived to see his family well established as slaveholding planters on the rich tobacco and cotton lands of the Cumberland that he had fought to conquer.

It was the Georgia wilderness farther to the south that attracted Jesse's younger brother Walter soon after the Revolution. Choosing to move within his wife's kin group rather than follow his brother, the revolutionary veteran was able to exploit Georgia's generous land lotteries to embed his growing family firmly among the plantation aristocracy of that new cotton frontier. About the same time, other Maxey cousins established a third family foothold across the mountains when they joined the great postrevolutionary Virginia migration that flowed into the Kentucky Bluegrass and then spilled over into the Barrens of southern Kentucky. And in 1804, the son of yet another of Jesse's cousins chose yet another frontier, this one lying to the north, in the Virginia Military District of Ohio. It was probably no accident that Horatio Maxey moved his large family into free territory. Horatio's father, converting to Methodism in the religious revival of the revolutionary years, had emancipated his slaves in 1788, and although three of his sons remained slaveholders, Horatio and his younger brother Bennett, along with a nephew and a niece, sought nonslaveholding lives for their families on the Ohio frontier.

No sooner did one Maxey clan secure a foothold somewhere in the West than they could expect to be joined by siblings, nephews, nieces, cousins, and second cousins. Since most Maxey lines retained some representatives in Virginia, news of settlement opportunities passed quickly along the family grapevine. Kentucky proved particularly enticing. By about 1800, children or grandchildren of all of Edward and Susannah's sons had gravitated to one or the other of the Maxey settlement areas there. For many,

"ALLITO FARM" PROPERTY OF U.S. NYE.
16 MILES NORTHWEST OF WILLOWS COLUSA CO. CAL.

In the many county atlases and histories published from the 1860s to the 1880s, Americans could find idealized images celebrating the virtues of domestic life in the rural West. The owner of this California farm conveyed his own happy prosperity with images of his prize animals, contented children, and farm and domestic laborers and signaled his social connections by including an image of visitors stopping to call at his neat frame house.

Unidentified artist. "Allito Farm": Property of U.S. Nye. *Lithograph, ca. 1880. From Will S. Green,* Colusa County, California *(1880; reprinted 1950, Sacramento Lithograph Company).*

however, Kentucky proved more a staging area than a stopping place, as did Tennessee. By 1815, the first of the Kentucky Maxeys were following their second cousins north to free soil—in this case, Indiana. A couple years later, Jesse's oldest son, like his uncle a Methodist convert, emancipated his slaves and moved his large family from Tennessee to southern Illinois. Within a year or two, one of the Ohio Maxeys was also exploring Illinois opportunity, and soon numerous other Kentucky Maxeys followed, part of the great surge of migration from the upper South into free territory in the decades after 1820. Their children and grandchildren pressed forward with the free frontier to Iowa, Kansas, Nebraska, and the Pacific Northwest or rejoined Maxey cousins farther south. Other Kentucky Maxeys, retaining their commitment to the "peculiar institution," trekked to Missouri as early as 1819; a few of their offspring took the Overland Trail to California and Oregon while others flowed south to Arkansas, Texas, and Oklahoma. The speculative plantation frontier of the antebellum period lured Georgia and Tennessee Maxeys westward to Alabama, Mississippi, Arkansas, and Texas while a later, wilder Texas offered numerous Maxeys a refuge from the war-torn South. And finally, like the Fillmores, Maxeys of all stripes in the 1890s dreamt the frontier dream one last time in the Indian lands of Oklahoma.

A closer look at these migration histories suggests, first of all, the dominant role of patrimonial logic in so many of the Maxey and Fillmore moves. They migrated as family groups, and they migrated for land. The farm that Millard Fillmore's grandfather pioneered in Vermont, for example, could provide a living for only the two youngest of his five sons, one of whom never married; the other three could hope to replicate their father's success only by moving to a new frontier. By moving after they had married and when their children were still young, they could expand cultivation as their family's needs and labor resources increased and with luck—the luck that eluded Millard's father and uncle—jointly earn farms for both generations. Moving together, or in chainwise fashion, permitted siblings to draw on one another for aid. Fillmore daughters and their husbands and children were an integral part of these sibling chains; a move west for Fillmore women seldom meant leaving all members of their parental family behind.

Indeed, the older generation itself often traveled west with their children and grandchildren, so that their wisdom and labor could aid the family in establishing its new farm and so that their children could assist them in their old age. Thus one of Millard's uncles first moved his young family from Vermont to Oneida County, New York, in 1790 and then, twenty-one years later when his oldest children were approaching marriageable age, transplanted the whole family again to new land in the western part of the state. This land proved able to support one son and his family; two other sons became ministers, and the fourth moved on to Michigan, where he was soon joined by the families of the three sons of his farmer brother, assorted other in-laws, and at least five cousins, siblings of the future president, some of whom then extended the chain to Minnesota.

It was family auspices like these that established the diverging lines of agrarian Fillmores on one frontier after another in the century and a half after 1750. The move from upstate New York to southern Ohio around 1818, the two moves from Vermont to different parts of Illinois in the 1830s, the extended series of frontier moves within the Maritimes, and most particularly the lengthy migration chain that led from New Brunswick in the 1840s all the way to California during the Great Depression—all exhibited the same patrimonial logic as the migrations within Millard's line. The Fillmores who followed the logging frontier from New Brunswick to Ontario to western Michigan in the 1840s and 1850s were engaged in a similarly structured cross-generational family project. Virtually all of these migrants grew up in large families of seven, eleven, or even fourteen children, and virtually all would raise similarly large families on the next frontier.

The Maxeys replicated these basic patrimonial patterns, though not without some distinctive southern accents. For one thing, the generous size of southern land grants gave many early Maxeys, both in Virginia and across the mountains, access to a level of wealth unmatched by most Fillmores, affording parents and some of their offspring the luxury of remaining behind in settled territory while other siblings, able to substitute slave or hired labor for the family labor they lacked, migrated on their own. Thus, for example, many of the initial Maxey moves from Virginia were made by young married couples rather than by older established families, as was common among the Fillmores.

Such moves were facilitated by a second distinctive southern feature of Maxey family strategy: the comparative complexity of the kinship system on which they could draw. Pairs of siblings married other pairs of siblings, and cousins often married cousins; kin relationships constructed in one generation were reaffirmed by new marriages in the next. Migrating Maxeys did not have to rely only on their immediate family for support; they could move within their far-flung networks of kin. Thus, although Edward Maxey remained behind in Virginia when most of his siblings and numerous other kin headed west in the decades after the Revolution, the options for his maturing children in the 1830s and early 1840s were not confined to the weary lands of the Piedmont. Two of his sons joined first cousins in Mississippi, one sought out second cousins in Illinois, and three linked up with third cousins in Missouri; indeed, Patrick, the youngest of the brothers, married one of those Missouri Maxeys and moved on with his in-laws to new land on Missouri's western border.

"I think any of you would be much better satisfied in this country," wrote Walter Maxey from Alabama to his "Dear Brothers and Sisters" in Illinois in 1820. "I think that I can make five dollars here easier than I or you can make one in your country. Trade is so much better. . . . There is a good deal of land in our country to enter and it would not be a hard matter to move from your country here. . . . I would assist any of the connection if they see proper to move to this country. . . . [Brother John] yet lives [in a nearby Alabama county]. He is building a mill but very probable the next time you hear from him he will be living here and they all appear willing." Although Walter's hopes were doomed to double disappointment—not only did the Illinois kin remain in Illinois but John and his family soon left to join them—both his sentiments and his practical offer of assistance demonstrate the kinds of family solder that fused the Maxey migration chains. Thus the same interrelated communities of Maxeys, Bondurants, Fords, and others who moved together from Virginia to Kentucky in the 1780s were to be found in southern Illinois or in Missouri a generation later.

Slavery gave antebellum Maxey migrations a third distinctive twist. There was one kind of Maxey move that almost invariably involved the more mature, often three-generation patrimonial family, even when other kin were already waiting at the destination: the move from slave to free territory. Perhaps so fundamental a transformation in the character of the family itself demanded the emotional support of its most intimate unit. More prosaically, slaves formed part of the patrimony, and any decision to strip the family of that patrimony depended on familial consensus. And once bereft of slave labor, migrating Maxeys were forced to rely on themselves to carve out their new homes on the new frontier and undoubtedly, like the Fillmores, timed their migrations to take maximum advantage of the labor of a maturing family. That certainly became the pattern for subsequent westward migrations among the northern Maxeys, in contrast to many among their southern kin. For example, William Maxey, the son of Jesse, the Tennessee pioneer, after converting to Methodism in his late forties, freed his slaves, sold his cotton mill, and moved his maturing family to Illinois in 1818. Here his eleven children prospered, but many among his numerous grandchildren repeated the family pattern after the Civil War, moving with growing or grown children and grandchildren to new farmlands in western Missouri, Iowa, Arkansas, Kansas, Texas, Oklahoma, Colorado, Idaho, or Washington. Simeon's experience is perhaps representative of this generation. Using borrowed money to purchase forty acres of government land in Illinois in 1853, he set out an orchard and began taking prizes for his fruit at agricultural fairs, his success interrupted only by Civil War service in his brother's company. But with five teenage children to establish by the late 1870s, the arguments for re-investing family resources on new land in the West must have seemed compelling, and he set out on an exploring trip west. He chose the flat Kittitas Valley in the center of the Washington Territory, where the approaching railroad and a series of bad winters that had devastated the cattle herds were forecasting the end of the open range. Here his family began homesteading in 1882, here they established the first of Washington's famed commercial apple orchards, and here they were soon joined by other relatives, friends, and former wartime comrades.

Finally, the incredibly speculative character of the southern market in western lands in the decades before the Civil War encouraged among some Maxey lines one other distinctively southern pattern: frequent joint mobility shared by parents and their married children as the clan moved restlessly from one frontier to the next, using their joint labor less, perhaps, to construct a stable home than to marshal the capital to grubstake a speculative bonanza on the next frontier. Among the less successful players in this game was Simeon's great-uncle Walter, the Alabama correspondent noted earlier. Among the most successful was Walter's cousin William, born in Virginia in 1783, who moved with his parents, siblings, and mother's kin in a complex, intertwining dance that took family members through the land lotteries of Georgia and the federal and Indian land booms of Alabama and Mississippi, prospering as planters and slaveholders seemingly innoculated against the Methodist conscience that plagued others of their kin. The dance culminated for William in 1840, when he sold his large eastern Mississippi plantation, cotton gin, and thirty slaves to his brother in a complicated transaction for twenty-six thousand dollars. Transferring his family and his business to Texas, he carved out, with his children, rich cotton plantations from the tangled lands of the Big Thicket and inspired other relatives to join them on the Texas trail.

But northern or southern, all of these Maxeys exemplified, just as did the Fillmores, the predominant pattern of migration that peopled so much of the American West: the patrimonial migration of kin-linked groups of large rural families, farmfolk or artisans, seeking to preserve a customary family economy by endowing the next generation with the bounty of newly opened public lands. They could, of course, have made other choices, as did Fillmores and Maxeys who remained behind when their kinfolk moved west. If a pioneering venture succeeded, a migrant's children could reasonably expect to live out their lives in the new settlement, but some, at least, of the grandchildren of all but the most prosperous settlers would clearly face the renewed necessity of western migration if they were to maintain both living standards and the customary patrimonial strategy. But pioneering, it seems, was a habit or skill that had to be passed on from parents to children, an option that seldom seemed feasible or desirable unless there was immediate contact with previous migrant generations or relatives already in the West. And as former frontier areas were drawn into the modernizing national economy, more attractive resolutions of the old rural dilemma of balancing population and resources was resolved with attractive options closer to hand. Nearby cities offered market incentives to improve the productivity of local farming. They diffused new ideas about family life and gender roles into their developing hinterlands, encouraging smaller families that placed fewer pressures on available land, and urban opportunity itself proved a potent lure to farmers' children.

Thus the younger of the first John Fillmore's two sons and his male descendants would remain in the Norwich, Connecticut, area as farmers and skilled workmen for five generations. Family sizes soon dropped to fewer than four children, and out-migration, when it finally began, was to the growing cities of the region rather than to the West. The same generational pattern of fertility reduction and subsequent urban migration often in family groups appeared among the other branch of Connecticut Fillmores and among the Fillmores who were subsequently to remain behind in Vermont, New York, Michigan, Indiana, Illinois, and Iowa. Some Fillmores, to be sure, found their way from Vermont and upstate New York to San Francisco and California in the boom years of the 1880s, several generations after their westering kin had departed, but this was new-style city-to-city relocation rather than a late revival of long-forgotten habits of pioneering. The federal patronage that relocated one Nantucket-born Fillmore to a position in the San Francisco Mint in 1863, and one of the Cincinnati branch to an Indian agent's appointment in Oklahoma a generation later, created variants of the same pattern.

Indeed, by the middle of the nineteenth century, when the maturing of any new frontier was becoming a matter of years rather than decades, the logic of patrimonial family migration seemed to be fading even for the most migratory of the Fillmore lines. The march westward halted; the urban drift began. Where westering continued, it now generally took a different pattern. Young people exposed to urban influences might still choose to move west, and families—brothers and sisters, in particular—might still move together or in chains, but they were more apt to move before marriage, and it was urban rather than agrarian opportunity that they seemed to seek on the frontier. Millard provides one early Fillmore example of this alternate strategy. Then there was his cousin's son who grew up on a farm near Buffalo, trained as a physician, and located in a Minnesota frontier town in the 1850s, and whose son, in turn, established his own practice in Kansas in the early 1880s; or the two sons of the Lake Champlain resort hotel owner, one of whom moved to Buffalo and then to newly bustling Zanesville, Ohio, in the 1830s and the other to Milwaukee a decade later; or even the farmer's son from Fillmore Corners near Syracuse who found his urban frontier in the late 1850s in the new coal and iron boomtown of Scranton in the Pennsylvania mountains. Their aspirations were entrepreneurial rather than patrimonial, and the small families that they would establish in the West were testimony to the decidedness of their break with the past.

Northern Maxeys exhibited similar changes in strategy. Various stay-at-home Maxey lines in Missouri, Ohio, Indiana, and central Illinois all responded to new local opportunity with both reduced family size and urban migration. The Illinois land that Joel Maxey chose when he left Kentucky in 1827, for example, happened to be located near the future state capital of Springfield. The agrarian frontier, to be sure, lured one son, Nelson, west in 1847 to join a Kentucky cousin who had settled along the Arkansas River fourteen years earlier. Soon marrying one of the cousin's daughters, he managed to resist the California gold that lured so many along the trail west from Fort Smith and instead turned south to Texas, to the booming cotton lands of the Brazos, where as a deputy sheriff he met a gunfighter's death attempting to make an arrest in the tempestuous years after the Civil War. But remaining at home in Illinois brought both prosperity and a brush with history to his less adventurous brothers, one as an entrepreneurial farmer, the other as a Springfield politician and father-in-law of

Abraham Lincoln's law partner and biographer. One grandson would briefly try his hand at Nebraska farming, and another would die in California in 1866, but most remained in the new middle-class mold that the family had fashioned for itself as Illinois emerged from the frontier.

Southern Maxeys, by contrast, tended to persist much longer in the patrimonial mode. Fewer Maxey lines halted their westward march quite so decisively as did many of the Fillmores. Both the slower southern pace of antebellum modernization and the disruptions and destruction of the Civil War and its aftermath held out fewer incentives to southern Maxeys to abandon familiar family strategies. Families remained large, and well-cultivated kinship connections ensured that westering habits seldom died out completely in any Maxey line. Where they did, the predictable consequence of thinning patrimonial resources was a species of rural proletarianization that underlined just what was at stake in continued family access to ever newer Wests. For southwestern Virginia Maxeys reduced to coal-mine or stone-quarry labor by the late nineteenth century, for example, the West might still offer hope, but for protelarianized families like these it was now the workman's hope of jobs at quarries in Nebraska or at new coal mines in Arkansas or northern New Mexico, where one Maxey died in a mine accident in 1913.

Well-to-do planting families could prove receptive to new kinds of domestic and educational aspirations, and the occasional Maxey planter's son was able to realize an entrepreneurial family style in a nearby town one, for example, became a leading Nashville manufacturer of tin and sheet iron and the city's mayor in the early 1840s, whereas another, after trying an Illinois law practice and the California goldfields, settled down to storekeeping, law, and politics back home in Kentucky. But more often, the limits of the local economy ensured that entrepreneurially inclined southern Maxeys, like their more patrimonial kin, chose to relocate to the frontier as they came of age, drawing on the family network to assist them in their search for urban opportunity in the West. Perhaps the clearest examples occurred in the family of William Maxey, one of those revolutionary veterans who settled in Kentucky in the 1780s. One of his older sons early moved his young family to the Boons Lick area of Missouri and then late in life removed with them farther west in the classic patrimonial pattern; several of William's other children settled into prosperous local farming back in Kentucky. But two of his sons took up the law, one went into local politics, and another became a cabinetmaker, and the next generation continued the family march into the professions and other urban occupations. Not even advantageous marriages into other planter families could secure this family's economic aspirations in Kentucky's Pennyrile, however, and by the late 1840s one young lawyer was on his way to relatives in the Arkansas Valley, and his brother, a former teacher, was in the California goldfields. But it would be Texas where the family's entrepreneurial energies would particularly focus. Samuel Bell Maxey knew the area from service in the Mexican War and from glowing reports by earlier migrating in-laws. So in the summer of 1857, he and his father—William's youngest son—sold their joint Kentucky law practice and, with their households, set out by wagon through Arkansas and Oklahoma to the new boomtown of Paris in the Blacklands of northern Texas. Here Samuel Bell prospered in a legal and political career that brought him to a Confederate generalship and the U.S. Senate. An uncle soon arrived, then an aunt's family left their Tennessee plantation to buy two thousand acres nearby; they were quickly followed by other young cousins—a lawyer, a doctor, a merchant—in an entrepreneurial migration that continued to dispatch young Maxeys westward along the kinship chain to Texas and Oklahoma for more than three decades after the Civil War. When James, a Civil War veteran three years out of law school, left Tennessee for Texas, a law partnership with his father's second cousin, General Maxey, was waiting; his distant Tennessee cousin Napoleon drew on kin in Illinois, where he read law, attended the old University of Chicago, and after a decade of Illinois practice moved with his family to Muskogee to become one of the first two lawyers admitted to the new Oklahoma bar.

Among the Fillmores it was the New Brunswickers, confined like many of the southern Maxeys to a slowly developing periphery of the new economic order, who most resembled them in their resistance to newer family strategies. The two families' migration streams would finally meet as the Canadian line projected the Fillmore patrimonial tradition of large farm families and westering habits through America's final farm frontiers—Kansas, Oklahoma, Idaho—to its sad end in the Dust Bowl trek to California. After three generations of internal colonization in the Maritimes, restlessness—probably induced by family involvement in the logging boom—seemed to strike several of the area's large Fillmore clans after the late 1830s, first sending one group of families to Ontario and later Michigan, then another to northern Illinois and, after the Civil War, points west. Perhaps their northern isolation had left them less prepared to take advantage of newer kinds of opportunity than were their distant cousins reared within the increasingly entrepreneurial communities of the Yankee diaspora. Indeed, at the very time that the first Fillmore moved west from the Maritimes, one of his brothers blazed the trail south to the milltowns of Massachusetts, a trail that many among the increasingly proletarianized subsequent generations would be forced to follow.

Finally, one other pattern of westward migration also becomes more visible in the historical record of both families by the middle of the nineteenth century: the feckless drift or adventuresome quest of the young man who seemingly cuts loose from the family and lights out for the territory, on his own or with others of his kind. The frontier had always tempted the young in this fashion, though such rebels and free spirits are the most difficult for the genealogist to track. Jesse Maxey must have been such a one in his prerevolutionary youth, as surely was Elisha Maxey who, orphaned early, found his way in his early twenties from Tennessee to Stephen B. Austin's young Texas colony on the Brazos. But now there were added inducements. For one thing, the constant series of mineral discoveries that began in 1849 proved a powerful solvent of family bonds, extracting young men from their families' patrimonial strategies and seemingly rewarding most those least hampered by the inflexibility of immediate family responsibility. Gold could be integrated as a resource into the family economy, to be sure, whether by using a grubstake from the goldfields to re-endow the patrimony back East or by establishing a new family line in the West. Two young Maritimers from separate Fillmore branches returned home from California prosperous enough to establish substantial patrimonial families of their own; entrepreneurial Maxeys from Illinois (two), Kentucky, and Tennessee established urban families on similar gold rush foundations.

Another drifted back from California to family farming in Missouri, and yet another made the transition from miner to prosperous commercial farmer and family man in California's Santa Clara County. But others, like the California-bound Maxey, recently widowed, who left his young children with various relatives back in Indiana and was never heard from again, or like the Fillmore father and son of New Brunswick stock who disappeared into the goldfields of the Black Hills in the 1870s, document the disruptive potential of the lure of western wealth.

Other factors also conspired to weaken or supplement family direction of westward migration in the latter decades of the century. Civil War service diverted northerners as well as southerners from planned life paths, and postwar offers of soldiers' homesteads beckoned veterans west to the chancy fortunes of the semiarid plains. The ever more rapid pace of settlement and development compromised the ability of many families to adapt old patrimonial strategies to new circumstances at the same time that the expansion of railroads and natural resource extraction created western wage-earning opportunities largely independent of the agricultural sector to which family pioneering had proven so well-adapted. Even the farming frontier found it easier to dispense with a family work force in this new capitalist era of machinery, hired labor, and speculative homesteading for future sale, as the phenomenon of the single female homesteader attests.

Thus when John Jefferson Maxey left Missouri for the West Coast in 1862 after the death of his father, family planning undoubtedly guided his initial steps. He probably traveled with his sister's family and in Oregon found a home with relatives who had emigrated a couple years earlier. But a political quarrel with his uncle forced him out to earn a living on his own as a farm laborer and miner, drifting down to California and marrying along the way. Quarrels with their stepfather probably launched two of his second cousins, ages twenty-one and twelve, on a similar odyssey from Missouri to the wage earners' frontier in the Pacific Northwest a couple decades later. Yet family still had a way of reasserting its claims. Recall Millard Fillmore's black-sheep cousin, Henry Glezen, who lived a scandalous life in the Minnesota Indian trade but nevertheless drew siblings and cousins to settle nearby once the area opened for white settlement. Or consider Henry's sons, reared by their divorced mother in near poverty on the Minnesota frontier. One ran away early, later surfacing in Montana, but the other, after drifting across the plains from Minnesota to Texas to Colorado and working here as a railcar checker, there as a mule driver or miner, finally married, entered the real estate business, sent for his mother, and put down roots in Kansas City to found the still thriving Unity School of Christianity and to pioneer in radio evangelism. Even the young Illinois Fillmore who rode with the U.S. Cavalry in the Indian wars of the 1870s later settled down to raise a family in the Washington Territory.

For eight generations the frontier was an integral part of Maxey and Fillmore family strategy. Maxeys and Fillmores were loggers in Michigan and fur traders in Minnesota, ranchers in Texas, New Mexico, and Colorado, miners in Arkansas and California, fruit growers in Washington, planters throughout the South, doctors and lawyers in boomtowns across half the nation, and farmers everywhere. Their histories, and those of countless other westering old-stock American families, illustrate the complex ways in which family logic influenced who moved west, with whom, at which stages in their life cycles, where they chose to settle, and what kinds of communities they formed, as well as how the opportunity they encountered in the West shaped subsequent family choices.

Patrimonial strategies undoubtedly peopled most agrarian frontiers, demanding successive generational migrations of young, growing families moving within chains of neighbors and kin in the hope of maintaining the linkage between land that could support the family and the large family that alone could provide labor to work the land. Their dominant form of movement was what might best be termed *colonization*, the uprooting and transplantation of significant segments of a family line, often in conjunction with a larger group of relatives and kin, from an older agrarian region to a newly opened one—the functional equivalent of Bailyn's "provincial" trans-Atlantic migration stream. *Colony*, it might be noted, was a term employed by nineteenth-century frontier folk to describe both highly organized settlement ventures and clustered family settlements. Formal colonization schemes were a feature of virtually every newly developing western region throughout the century, whether organized by land or transportation companies, by religious groups, or by voluntary organizations. But even they relied on colonizing families for their main recruits.

Consider the patrimonial logic of colonization. A farm family with growing children could sell its farm, developed with their common labor, perhaps to an older child who would remain behind, perhaps to a stranger. The family could then invest the profits and its own joint labor—its greatest resource—in cheaper western land, providing farms for the remaining children as they came of age. The migration might occur all at once, or different family members might migrate in stages, some staying behind to earn money and others going ahead to prepare the way. It often made sense for the older as well as the younger generation to head west. Parental wisdom and labor were a critical part of the human capital that the family economy could muster; the parental presence helped keep the children within the cooperative circle of the family, and the children embodied the parents' best hope for a secure retirement. Friends and neighbors often migrated together, turning the wider community into a portable resource that could be carried along rather than left behind and creating chains of linked communities stretching across the nation.

In theory such colonizing families and friends could continue together westward in stages generation after generation. An infant might make her first move with her parents and grandparents when she was still in her mother's arms; as a young bride she might move again with her husband, child, parents, and siblings and move one final time when herself a grandparent. "I had been reared to a belief and faith in the pleasure of a frequent change of country," one southern autobiographer recalled. A move could fail, of course, and all hope of perpetuating the patrimony vanish. Individual children might drift away on their own. The family might be fortunate enough to acquire land enough for more than one generation, permitting members to break the cycle of colonization. Or—and this occurred increasingly over the course of the nineteenth century, as we have seen—the character of local opportunity could change, tempting families to abandon the logic of patrimonial conservation and venture into the world of enterprising individualism. Parents with sufficient resources might decide to remain in place to enjoy the fruits of a developed society; children might take their inheritance in the form of an education preparing them for entrepreneurial or professional life or in the form of capital or credit that would enable them to relocate—on the frontier as readily as in the more developed parts of the country—without the necessity of relying on the pooled labor of the family or the support of a colony. And the West itself changed. By the 1830s, the quickened pace of western development was multiplying urban opportunities for those of an entrepreneurial bent. In the South, frontiersmen with capital could substitute slave for family labor, and everywhere it became increasingly easy for an individual with funds but no family to hire the labor and buy the machinery necessary to establish a farm or a business outside the context of formal family colonization.

The smaller families that broke free of patrimonial logic to take entrepreneurial advantage of the new modernizing economy thus had less dependence on the frontier, but its burgeoning cities and resource speculation could still beckon both young folk leaving home to establish new families of their own and more mature families relocating for a better life. And for those trapped in the hard logic of the proletarian family, the West increasingly in the course of the nineteenth century meant jobs, particularly transient jobs for single migrants whose remitted wages could support family members left behind or free young people from the burden of family demands, even if at the price of long years isolated from the world of the family itself.

The Diversity of Domestic Landscapes in the West

For all their regional and temporal variation, these family strategies for western settlement still rested upon a basic domestic consensus derived from the western European tradition. But as young Volley Maxey, for one, tragically discovered on a summer day in 1872, there were other familial traditions, other family logics, that also played major roles in shaping the nineteenth-century West. Born and raised in a small kin-linked settlement on the northwestern Texas frontier, the six-year-old was playing with his toddler sister and a couple of other children around the woodpile where his widowed Indiana-born grandfather was chopping wood when suddenly his playmates' mother urgently called them back to the cabin. She had seen a raiding party of Comanches, who for generations had supported the family bands in which they lived through raiding and trade in horses, cattle, and captives. Her warning came too late. Volley's grandfather and his two playmates were killed, his baby brother died in his mother's arms, and he and his sister were abducted by the raiding party. When his sister tired and began to cry, she was soon killed. Volley, known to his captors as "Topish," lived three years with them, until the final stages of Comanche resistance to reservation confinement. Horseback, the chief of one of the Quahada bands, traded him and several other young captives to the Quaker agent at the Fort Sill reservation in return for captured Comanche women whom the army was holding. Volley's reintegration into white society was not easy: he had forgotten all his English; he ran from the soldiers when he first saw them; he was rude and, by his family's lights, uncivilized. As an adult he drifted beyond the family's ken.

Maxeys and Fillmores were seldom forced to experience quite so immediately the clash between their family strategies and other family traditions they encountered in the West. Theirs was, after all, the dominant pattern in the region's evolving domestic landscape. Their family system, as Congress had early recognized, ensured the motive and the demographic means to maintain a continual frontier flow, and along with their herds and their household effects, they carried westward their domestic ideology and the power of the American state, which they used to shape the new institutions and law that their family regime required. Family practices different from their own were often central to the critiques they leveled against other groups: to their perceptions of Indians as savage, Chinese as depraved, Mexicans as loose and lazy. Thus family life was irrevocably transformed for the Native Americans who encountered Maxey or Fillmore Indian fighters in Tennessee or Texas or Wyoming and for the African Americans whom Maxeys took west as slaves or sold before their own departure. But it was similarly if more subtly influenced also by the Fillmore Indian agent who sought to impose federally dictated family policy in Oklahoma and by the Maxey judge who coped with the Mormon family system in Utah. In Minnesota Henry G. Fillmore lived and worked within an Ojibwa-French mixed-blood community that soon acquired a German cast; his son made his first money in a Colorado community similarly shaped both by mixed-bloods and by expansionist Hispanic settlers from farther south. Maxey and Fillmore forty-niners undoubtedly confronted Chinese in the California mines. A Fillmore in Cincinnati married a German pioneer's daughter; a San Francisco kinsman took an Irish wife.

Such alternative family traditions were inevitably forced to adapt as they confronted the economic changes and institutional pressures set in motion by the main currents of American westward penetration. But they also proved surprisingly successful in using family as a defensive instrument to ward off some of the consequences of unwanted change and to preserve enclaved cultures, particularly in cases where their family strategies rested on an agrarian patrimonialism strongly defended by religion or geographical isolation.

Thus the rural Hispanic families of northern New Mexico after the American takeover in 1846 drew strength from the communal economies of the extended family plazas that constituted the basic units in their settlement system. They could evade many of the consequences of American governmental and ecclesiastical pressures by retreating into the networks of kin that defined their daily world, effectively converting local justice of the peace courts, communal land rights and community water-control associations, and religious sodalities into alternative forms of local ordering, and hiving off new colonies northward into Colorado as population grew or land was lost to Yankee claimants. For old elites throughout the former Mexican territories, intermarriage with incoming Yankees proved an important tactic in maintaining status and at least partially assimilating the newcomers to accustomed ways. Over the longer term there were real limits to the autonomy of the family. Anglo inheritance laws, Anglo control of land claim courts, and Anglo access to the expanding sectors of the national economy, for example, steadily subverted Hispanic landownership throughout the Southwest, dissolving the vital link between the family and its patrimony. The proletarian niche that increasing numbers of native Tejanos and immigrant Mexicans occupied in later-nineteenth-century Texas involved such family adjustments as more households headed by single persons, more children in the work force, and smaller household size. Nevertheless, even when, at the end of the century, the land of the New Mexico and

Inspired by an encounter with a wagon train of German immigrants moving west through Nebraska, Albert Bierstadt's painting becomes an allegory for the future of the West. As the pioneers move onward, the Indians fade into the sunset.

Albert Bierstadt (1830–1902). Emigrants Crossing the Plains. *Oil on canvas, 1867. National Cowboy Hall of Fame and Western Heritage Center, Oklahoma City, Oklahoma.*

Colorado plazas was proving inadequate for many a family's support, creative family strategies of seasonal male labor migration channeled the adaptation and helped preserve the distinctive core of community culture.

Mormons migrating from the East similarly constructed a distinctive and enduring society for themselves in frontier Utah resting on the family practices that lay at the core of their religious belief. The Mormon family system was in many ways a logical extension of the patrimonial American family in which it was rooted, and if patrimonial logic yielded to religious fervor in fueling the initial exodus to Utah, it probably played a coequal role in supporting subsequent Mormon colonization throughout the intermountain West. It was doctrine and not economic logic, however, that saw in marriage and childbirth the divine plan for hastening the millennium and in polygamy a path to heavenly exaltation. Life in the Mormon West, accordingly, was fundamentally shaped by early and almost universal marriage, often undertaken without a secure economic base, by very high fertility, and by the practice of plural marriage, which by one recent estimate framed about a third of the men's lives, two-thirds of the women's, and half of the children's. By 1890 this self-confident society could no longer resist federal pressure against polygamy, but its religiously sanctioned patriarchal familism, marshaling family and community resources for both agriculture and trade, remained at the core of its expanding settlement region.

The commitment of various European immigrant groups to their own versions of patrimonial logic planted yet other sets of enduring family cultures in the West. It was long a cliché of western history that immigrants, in contrast to native-born Americans, lacked the skills and resources needed for frontier settlement and that if they entered farming, they did so as "fillers-in" on settled land vacated by the westward-moving native-born. Apparently, as the case of the Irish canal-diggers seemed to suggest, only as proletarian migrants in work gangs of single men could immigrants readily make their way to the West. Certainly the mining towns, lumber camps, and cities of the West always drew more than their share of foreign-born families as well as single workers. But most nonsouthern agrarian frontiers settled after the onset of mass immigration in the 1830s likewise attracted disproportionate numbers of immigrants, either directly from Europe or from older-settled areas to the east. In 1850, for example, when 9.7 percent of the American population was foreign-born, newly settled Wisconsin with 36 percent of its population born abroad was the most heavily immigrant state in the nation; forty years later, almost 43 percent of recently settled North Dakota's population was foreign-born.

The immigrant pioneers came mainly from Germany, Scandinavia, the Netherlands, Britain, the old German colonies in Russia, and even parts of Ireland and Slavic central Europe, products of a predominantly peasant emigration rooted in the old dilemma of growing families and limited land. They were confronted with the same local options as their American counterparts—entrepreneurial transformation for some, proletarianization for most—and when improved trans-Atlantic communication and transportation brought the new lands of the American West within reach, many made the same conservative choice: to move in order to preserve the family system, rather than to change the family in order to stay. The artist Albert Bierstadt encountered a "very picturesque" train of such immigrant families lumbering westward across Nebraska in 1863. "They had," his companion noted, "a large herd of cattle, and fifty wagons, mostly drawn by oxen. . . . The people themselves represented the better class of Prussian or North German peasantry. A number of strapping teamsters, in gay costumes, appeared like Westphalians. . . . All the women and children had some positive color about them, if it only amounted to a knot of ribbons, or the glimpse of a petticoat. . . . Several old women, of less than the usual anile hideousness of the German Bauerinn, were trudging along the road with the teamsters, in short blue petticoats and everlasting shoes. . . . In the wagons all manner of domestic bliss was going on . . . . Many mothers were on front seats, nursing their babies in the innocent unconsciousness of Eve . . . . Every wagon was a gem of an interior such as no Fleming ever put on canvas, and every group a *genre* piece." There is no little irony in the recognition that Bierstadt's luminous archetypal painting of American wagons moving west probably owes much to this immigrant encounter.

The same familial tradition that motivated their emigration lent such immigrant pioneers their greatest strength. With more restricted access to information than the American-born, they tended to migrate within even tighter chains of neighbors and kin, forming clustered communities whose self-segregation within the social walls of linguistic, religious, and cultural difference, buttressed by their control of local schools and government, encouraged the perpetuation of patrimonial values long after they had waned within other western family traditions. As Yankee families like the Fillmores brought their westward march to a halt, immigrants took up the slack, and by 1880, when the foreign-born comprised just over 14 percent of all farmers nationwide, their proportions among the farmers in the predominantly northern-settled states of the post-1840 frontier ranged from 65.5 percent in Minnesota, 60.5 percent in Wisconsin, and 58.4 percent in Dakota to 26 percent in Washington. Even in southern-settled and relatively immigrant-poor states like Kansas, Colorado, and Oregon, they significantly exceeded the national average. Immigrants substituted family labor for the capital they lacked, they transplanted familiar peasant inheritance customs to help bind the next generation to the farm, they used the power of religion to ward off the blandishments of outside opportunity, and when land grew scarce, they exported daughter colonies to the next frontier.

Among those who may have been on the Oregon wagon train that Bierstadt encountered, for example, or one like it, was Franz Nibler, one of Henry G. Fillmore's Minnesota neighbors, a "good solid Backwoods man" of Bavarian birth recently discharged for illness from the Union army. Four years later he returned to lead seventeen families of his German Catholic neighbors and kin in a twenty-six-wagon train from Minnesota across the northern trail to Oregon. His companions were, by and large, married couples in their thirties and forties with six to eight growing children each—Bavarians, Eifelers, Alsatians, and Westphalians who had migrated a decade earlier to the new lands of Minnesota from older German frontiers in Ohio, Michigan, Missouri, and Wisconsin. The extended Schultheis family, for one, had emigrated from Bavaria to Missouri as early as 1836, from whence they trekked in a large family caravan of ox-drawn wagons to Minnesota in 1854. Once in Oregon's Willamette Valley, they became fillers-in for a time, establishing on land abandoned by early American settlers the nucleus of an enduring German Catholic settlement sheltered, like its Minnesota progenitor, beneath the towers of its Benedictine abbey. But when the undulating hills of eastern Washington's Palouse opened to settlement in the mid-1870s, younger members of the Schultheis family headed east to found yet another expansive German Catholic community on yet another frontier.

Family processes like these stretched archipelagos of ethnically defined European settlement islands across successive nineteenth-century Wests. Their family cultures varied from one another in detail and inevitably changed with time. Thus Norwegian families learned that they had to rely more on the labor of the nuclear family than they ever had in Europe; more of their children married, at younger ages, and fertility increased. Women's domestic duties thereby increased as well, while new kinds of farming altered the gendered division of labor in field and barn, and gradually, with time and increasing prosperity, old habits of prenuptial conception disappeared, and bourgeois fertility reduction began. Sharing a familiar patrimonial familism, posing no fundamental challenge to American domestic ideologies, European immigrants like these were able to carve out distinctive and enduring western domestic landscapes.

The pressures of prejudice helped deny other western family cultures the luxury of such relatively autonomous evolution, however, and weakened their ability to directly mediate the settlement process. For the Chinese, for example, it was the absence of family that defined their situation. They left their families behind when they responded to the lure of California gold, seeking as entrepreneurs or laborers the stake that would enable them to maintain or advance family interests back home. In this they differed little from their Yankee counterparts in the mining West or from later labor migrants like the Italians, who would show even less tendency to remain in America than the Chinese. The small numbers of Chinese women who arrived on America's western shore tended to come as prostitutes, perhaps after being sold to relieve their family of the burden of a female mouth or to help finance the advance of more favored members. No more than 5 percent of the region's Chinese population before 1880 was female. Wives remained behind to serve the interests of the family at home, and single men who wished to marry usually had to recross the ocean to do so. The price of these long-distance family strategies was an American life lived in single-sex households and community institutions geared to the needs of single men. Kinship ties, however, forged familiar chains of migration and settlement, and the gradual establishment of local families—by 1880 a third or more of the women in most communities were now wives, though often considerably younger than their husbands, reflecting their relative scarcity—suggested that some of the Chinese were finding motive and means to remain. But from the beginning, they encountered virulent racism that culminated in 1882 in a series of federal exclusion acts effectively curtailing their ability not only to form families in America but even to return from visits to families back in China. It would be well into the twentieth century before a real second generation and a supportive family culture could emerge.

There was even less room for family autonomy among African Americans brought west as slaves before the Civil War. Recent studies suggest that migrant slaves were more youthful than the slave population as a whole, with as many as half still in their teens, but perhaps more evenly divided between males and females than westward-moving whites. "I do not wish families and only desire to purchase those who are YOUNG and likely," advertised a Missouri dealer in 1856. Regardless of whether slaves were sent west by traders or migrated with their owners, such westering almost by definition meant family disruption. Although some masters may have sought to keep slave families together, most apparently did not or could not. Both debt and death could force slave sales; when a master's estate was divided among the heirs, so too were the slaves, and when some family members moved west, so too might their share of the family slaves. Many slaves were hired out on the frontier, as elsewhere, and many frontiersmen owned or rented only a slave or two. Most slaves thus did their pioneering with few of the even minimal supports of family that they may have enjoyed farther east; westering required of them the arduous construction of new domestic ties.

Free blacks, however, shared many of the patrimonial values of white society and were able to participate to a limited extent in the agrarian settlement process. Thus in 1855, two decades after North Carolina–born Walden Stewart and his family settled in Illinois, he pulled up stakes once again at the age of sixty and moved with them north to the Wisconsin frontier, undoubtedly seeking not only more land for his growing family but also freedom from the increasing restrictions that Illinois was placing on people of color. Here they were soon joined by at least ten other black families, a number of them also from North Carolina, and here they prospered and remained, sharing school, church, and ultimately intermarriage with their immigrant neighbors. Similar patrimonial aspirations among relatively self-sufficient African-American farm families from the border South drew increasing numbers to colonies on the Kansas frontier after the Civil War, culminating in the 1879 "Kansas Fever" that for a brief period also spread to the more desperately poor of the lower South. The numbers of westering black families were always small, however, in comparison with those who remained within the South or later moved north. These families had fewer resources of their own than their

Seth Eastman's paintings of Indian life focus particularly on domestic genre scenes. Based on observations made during his tour of duty as an officer at Fort Snelling (Minneapolis) in the mid-1840s, this watercolor sketch of a Winnebago sugar camp suggests the important role played by women in domestic labor and production.

Seth Eastman (1808–75). Indian Sugar Camp. *Watercolor on paper, ca. 1850. James J. Hill Reference Library, St. Paul, Minnesota.*

white counterparts and drew little institutional support from local society in the West; black homesteaders in the Nicodemus, Kansas, area, for example, were unable to find a local surveyor willing to survey their claims. Their communities, where they survived, could seldom prosper. The substantial African-American presence in the later-nineteenth-century West undoubtedly owed far more to labor-seeking migrants drifting from or with other members of their essentially proletarian families and living in households like the one in Tombstone in 1880 shared by a thirty-year-old South Carolina laborer, his young California-born wife, and eight other African-American men in their thirties and forties—blacksmiths, waiters, porters, laborers, cooks—all, like so many others in Tombstone at the time, without families of their own. The family trajectories of pioneers like these remain largely unstudied.

It was undoubtedly Native Americans, however, who felt the most concerted pressure on their family systems. It is difficult to generalize about the variety of family systems that Native Americans evolved before white penetration into the West. Several points are clear, however. First, few bore much relationship to the patrimonial logic rooted in family rights to privately owned land from which the European-American family system derived. Second, as Volley Maxey learned only too well, systems of kinship, marriage, and adoption, often more expansive, complex, and flexible than those of most European Americans, functioned as effective demographic and economic adaptive mechanisms to the uncertainties of low-technology lives. Third, common practices like polyandry, serial marriage, and women's physical labor, whatever their logic within Indian family systems, struck nineteenth-century European Americans not only as alien but immoral, and inevitably provoked repression. And fourth, like their European-derived counterparts, these family systems were in a constant process of change and adaptation.

Although research has addressed the specific historical contours of such familial transformation, it has tended to focus much more on how various Indian family systems changed or resisted change than on how the family variably influenced adaptation. Often the first consequences of white contact for many Indian peoples, experienced at long distance, were new techniques of hunting and transportation made possible by the horse, or higher standards of living through trade, and family systems were frequently reshaped to take advantage of them. The woodland Winnebagos of eighteenth-century Wisconsin, for example, broke their large agrarian community into small and mobile family bands, the better to hunt for the furs desired by their white trading partners. Polygynous familes organized around groups of co-wives emerged among the early-nineteenth-century Cheyennes as they abandoned agriculture and moved out onto the plains to hunt and trade, an adaptation both to the surplus of females that warfare created and to the need for more women to process the increased products of the hunt. The status of women changed with their exclusion from the hunt, and extrafamilial male sodalities took on a greater societal role. As Cherokees on the trans-Appalachian frontier adapted to closer contact with white settlers and markets in the period before federal removal policies, many exchanged their communalism, clans, and extended family life for agriculture and nuclear families while others, reflecting the same sort of logic that propelled white families to invade their lands, migrated westward across the Mississippi in an effort to preserve the old familial order. Indians in California who survived the initial American onslaught, with its particularly severe attacks on women through rape, prostitution, and disease, frequently had to take what refuge they could in proletarian strategies that left them working for Americans in nonfamily settings with little chance for family life or biological and cultural reproduction.

Native American family systems faced intensified pressures with coerced concentration on Indian reservations in the latter half of the century and again with the subsequent breakup of many reservations into individual land allotments toward the end of the century. For one thing, both federal policy and missionary persuasion aimed at forcing Native Americans into the nineteenth-century American mold of male-supported nuclear agrarian families and female domesticity, using both formal means like schools, supervision by matrons and agents, and Courts of Indian Offenses and even more effective tactics like excluding women from public discussion, channeling annuity payments only through heads of household, designing housing for nuclear families, providing men with wage labor on the agency, and implementing family-based allotment itself. Under such demands, for example, the extended, kin-linked Comanche bands broke up into more numerous, smaller residential clusters often composed of lineally extended family households, and the salience of family and kin increased over that of larger tribal groupings; the gender balance of power within the families and kin groups of the Teton Sioux shifted further in the male direction. The very ability of the family to reproduce itself was endangered when nutrition was compromised, as it so often was, through federal agents' efforts to force behavioral change by withholding

William Fuller, a carpenter and laborer at Crow Creek Agency, depicted local Indian life in the 1880s as heading inevitably toward a European-American model of domesticity. Tidy, fenced frame houses stand as a goal toward which tipi-dwellers might strive; Native Americans dressed in suits become the ideal for traditionally clothed men; and neat agricultural fields suggest a new future for a people whose livelihood had depended on hunting.

William Fuller (active 1860s–80s, South Dakota). Crow Creek Agency, Dakota Territory. *Oil on canvas, 1884. Acquisition in memory of René d'Harnoncourt, Trustee, Amon Carter Museum, 1961–68. Amon Carter Museum, Fort Worth, Texas.*

rations or banning traditional medicine or through the sheer incompetence and corruption that plagued the reservation system. But federal policy also worked against intended familial consequences. Thus individual land allotment, in theory designed to anchor Indians irrevocably in the logic of the nuclear family, more often in practice proved a means of stripping their land from them altogether, lending new rationality to still vital traditions of extended family communal living and kin support as a strategy for shared survival. And increasing reliance on cash income from land leases or welfare payments that, thanks to American inheritance laws and domestic ideology, passed equally—or in the case of welfare, primarily to women—could mean that family balance shifted away from the male head.

Indeed, despite severe constraints, Native American family systems, like other western family traditions, were able to influence as well as be influenced by the currents of change, though we know much less than we should about the implications of differing family cultures for Indian negotiations of life in the modernizing West. Thus customs of fictive kinship, adoption, and marriage alliance proved potent means of integrating early traders into Indian society on essentially Indian terms. Through Indian wives, European Americans acquired both access to Indian trade and processors of the pelts and skins for which they traded. Their mixed-race families mediated relations between their two peoples in many areas of the West for generations, with family strategies that sometimes promoted the native relatives' gradual adaptation to newer ways but equally often depended on their ruthless exploitation; in contrast to the case in Canada, however, American circumstances and racial mores ultimately prevented them from consolidating an enduring peoplehood of their own.

But interracial alliances were not the only way that Native Americans could use the family to structure adaptation. Many Winnebagos, for example, converted their small family bands into successful instruments for resistance to removal, repeatedly slipping away from successive reservations in Iowa, Minnesota, and Nebraska to old camps in the forests of Wisconsin, until the government in the 1870s was finally forced to acknowledge their right to claim Wisconsin homesteads of their own. Hopis conserved traditional, sedentary family forms in their densely settled high pueblos at the price of disease, high mortality, and ultimately labor migration and fertility reduction; their more flexible Navajo neighbors adopted pastoralism, then home manufacture and wage labor, maintaining high fertility and high population growth and, like their Hispanic and Mormon neighbors, supporting territorial expansion in the process. Band allegiances often structured settlement location and later allotment choices on Plains Indian reservations, and among the Crows, for example, the clan system provided valuable continuity in marriage regulation despite the unfamiliar reservation setting. Many Indians clearly found their flexible systems of marriage commitment a more rational response to the uncertain vagaries of modernizing life than the lifetime commitment that white society tried to impose and was indeed already in the process of abandoning for itself. Reservation life in effect lent Native American family cultures some of the protective isolation that other enduring western family traditions had also managed to find, but such protection was purchased at a daunting price.

Family Life in the West

There was, then, no single style of family life in the West. America's successive nineteenth-century Wests were shared by a variety of indigenous and colonizing family cultures. Nor could the meanings, functions, and roles that these cultures assigned to the family and to individuals by virtue of their family membership ever be assumed. In the West, as elsewhere, the contours of family culture were constantly evolving through the negotiation, contestation, and unconscious accommodation of family members, in their work sharing, at social gatherings, in bedrooms, even—as in the case of the Fillmores of St. Cloud, Minnesota—on the kitchen floor and in the courtroom.

Mary and Henry Fillmore must have entered their marriage in the early winter of 1853 with conflicting and confused sets of assumptions and aspirations. Mary Georgiana Stone was just sixteen, newly arrived on Minnesota's logging frontier with her family and other relatives from Wisconsin and before that Maine and the Maritimes. Perhaps part of the attraction of Henry Glezen Fillmore, the twenty-five-year-old trader whom she married at a little millsite on the Mississippi, was the éclat of his kinship with the nation's recently retired president. She clearly aspired to much of the life-style promised by the new domestic ideology of the middle class. She had taken dancing lessons in Wisconsin. She enjoyed fine clothes, furniture, and entertaining and was resentful when her husband's business reverses forced her to do heavy housework. Producing a son almost exactly nine months after her marriage, she soon induced her husband to conduct their sexual intercourse so as "not to beget her with child," though "through a mistake on her part of the effect of free and full sexual cohabitation immediately after her menstrual course" she bore a second son three years later. Not only did she seek to restrict her fertility but, Henry would later charge, she gossiped about her methods with her female friends. She was also, it seemed, careless enough in her general demeanor to lend credence to charges of adultery with the housepainter who lived two doors away.

If Mary Georgiana was not fully adept in the practice of the new domesticity, Henry was even less so. His family expectations seemed more old-fashioned; he wanted purity and probity of his wife, of course, but he also wanted more children and more work and expressed his objections to her life-style in a decidedly ungenteel fashion derived more, perhaps, from the rough world of the Indian trade than his patrimonial parents' home. Not only did he beat and curse her often, but once (he admitted) or twice (she charged) he even attempted to rape her on the kitchen floor in front of her children and his young female cousin, declaring, she said, "that he . . . had no other desire to gratify by such intercourse than the desire to abuse . . . and disgrace [her] in the eyes of his . . . family, and bring her . . . to scorn, and loathing." She was, he charged, "a damned whore" who had committed adultery, admitted to being raped by that dancing master in Wisconsin, and bragged that another man had fathered her second child. She resisted him, she said, because he had "a loathsome venereal disease," and once she found out (from her female support network, perhaps?) how such a disease was contracted, she left him in the autumn of 1860 and sought out the judge. After her divorce, Mary succeeded in raising at least one son in the entrepreneurial values to which she aspired and through him achieved a secure old age in the middle-class world of Kansas City. Henry became a horse doctor and lived out his life bereft of family in the male world of St. Cloud boarding houses.

The confusion of family models and practice within the Fillmore household mirrored the mosaic within their broader community. One young neighbor reveled in her companionate marriage, doting over her children, contributing sentimental prose and verse extolling frontier domesticity to a Philadelphia ladies' magazine, and cherishing the sensuously loving letters she received from her attorney husband when he was away. Another was a tough divorced feminist who edited an abolitionist newspaper. Still another was a former schoolteacher who worked with her pioneer husband to carve out a frontier farm for the family and a gracious, busy domestic sphere for herself and who reared their three sons to professional and business occupations. Yet just a few miles down the road were wooden-shoed German immigrant women in thatched-roof log cabins, carefully bargaining with time-honored peasant logic about dowries and family bonds of maintenance for aged parents, and mixed-blood women negotiating the exchange of one partner for another.

The varieties of family cultures to be found in the nineteenth-century West were all subject to the currents of economic and ideological change that were transforming the nation itself. They were influenced also in regionally distinctive fashion by the opportunities and constraints of the western environment, by the self-selection inherent in their migration or local persistence, and by the mutual need to coexist. The result was both regional variation and temporal change. The older agrarian Wests of the first half of the century were the locus classicus of the patrimonial family, with relatively little extractive industry to attract a permanent nonfamily work force, fairly even sex ratios from an early date, and few remaining indigenous families. In the South, slavery created the main variant family style; in the North, prosperity encouraged rapid acceptance of new entrepreneurial family styles for those who chose to remain. Frontiers incorporated after mid-century offered more varied opportunities to attract more varying kinds of families. There were greater numbers of entrepreneurial and proletarian familes to be attracted; new family ideals had filtered even into farming families, and new machinery and labor availability permitted farming without large families; more immigrants were arriving; and indigenous peoples had nowhere left to go. The result was a more variegated domestic landscape in the later Wests, more enclaves of variant family types, and more rapid shifts when, as we saw in Tombstone, one kind of economy replaced another and brought with it a new mix of family styles. Differing family economies were as often symbiotic as competing, as in the case of the farm families who clustered around the edges of mining, ranching, and logging regions to provide them with food and seasonal labor or the entrepreneurial and proletarian families of the towns that serviced the countryside. Yet their cultural incompatibilities were often real and have become the stuff of western legend.

Nineteenth-century observers were indeed more astute than early-twentieth-century commentators: families easternized the region more than they were westernized by it. Men, women, and children brought westward with them habits and aspirations nurtured in the East; like the women who struggled to preserve accustomed domestic ideals in their wagons on the Oregon trail, the settlers used these aspirations to filter and focus their adjustment to the west. Family life changed as people moved west, but it probably would have changed for them had they stayed home. The western women who complained most about loneliness, about being torn from relatives and kin, one suspects, were women who were migrating precisely because their husbands had already broken free of patrimonial ties. For so many others, migration in extended processions of kith and kin was a way of avoiding precisely what, like one Iowa pioneer, so many feared from the changes of the modernizing world: "a drifted family . . . a broken, distorted chain."

Perhaps what was most distinctive about family life in the West, then, was the new lease on life that the region offered a patrimonial logic that was rapidly becoming obsolescent in more settled areas. Fertility ratios were consistently higher on new frontiers than elsewhere, less perhaps because opportunity there encouraged families to have more children, as scholars have often argued, than because families enmeshed in strategies that rested on large numbers of children knew they had to seek the frontier. The West in this sense proved not so much the source of all that was new in American family life as a haven for much that was old, and it was the demographic inexorability of those old forms as much as the speculative entrepreneurship of the new that drove Americans westward. But the region also shared the nation's public aspirations for the new ideals of the middle-class companionate family, however much difficulty westerners like the Fillmores might have had in putting them into practice, and by the end of the century the accelerating pace of economic and cultural transformation was rapidly marginalizing other family styles.

Thus there is a final irony to be wrung from the John Ford saga of western domestication with which this chapter began. In the film, as in the real world of the late-nineteenth-century West, what was really conquered and banished was not so much savagery and anomic individualism as it was the domestic tradition of the extended, patrimonial families whose energies had engulfed and reshaped frontier after frontier in the American mold. It was, after all, Wyatt Earp who drove off into the sunset to rejoin his patrimonial kin group, while entrepreneurial Clementine—single, cultured, now career-oriented, and far from her family of origin—remained to inherit the new community of strangers that foreshadowed the social organization of the twentieth-century West. It was "My Darling Clementine" who claimed both the movie's ending and its title.

Bibliographic Note

At the core of this essay lies an analysis of the thousands of pieces of family information contained in two fine published genealogies: Charles L. Fillmore, *So Soon Forgotten: Three Thousand Fillmores* (Halifax, Nova Scotia, 1984), and Edythe Maxey Clark, *The Maxeys of Virginia: A Genealogical History of the Descendants of Edward and Susannah Maxey* (Baltimore, 1980). The story was supplemented through further research in manuscript census schedules, local histories, and for the Fillmores, "Mary Georgiana Fillmore vs. Henry G. Fillmore," St. Cloud District Court, Civil Case File No. 44, Minnesota Historical Society, and Henry G. Fillmore's obituary, *St. Cloud Journal-Press*, November 15, 1899.

Family history in general is a relatively new field of academic study. Charles Tilly, "Family History, Social History, and Social Change," *Journal of Family History* 12 (1987): 319–30, and Tamara K. Hareven, "The History of the Family and the Complexity of Social Change," *American Historical Review* 96 (1991): 95–124, provide orientation. In "Family Strategy: A Dialogue," *Historical Methods* 20 (1987): 113–25, Leslie Page Moch et al. debate a central concept for this essay. Accessible introductions to the history of western European family systems are Martine Segalen, *Historical Anthropology of the Family*, trans. J. C. Whitehouse and Sarah Matthews (Cambridge, Eng., 1986), David Levine, *Reproducing Families: The Political Economy of English Population History* (Cambridge, Eng., 1987), and James Casey, *The History of the Family* (Oxford, Eng., 1989). Important for conceptualizing nineteenth-century change is Leonore Davidoff and Catherine Hall, *Family Fortunes: Men and Women of the English Middle Class, 1780–1850* (Chicago, 1987).

American family historiography begins with Arthur W. Calhoun, *A Social History of the American Family from Colonial Times to the Present*, 2 vols. (Cleveland, 1917). For its evolution, consult Susan M. Juster and Maris A. Vinovskis, "Changing Perspectives on the American Family in the Past," *Annual Reviews in Sociology* 13 (1987): 193–216. Modern surveys include Carl Degler, *At Odds: Women and the Family in America from the Revolution to the Present* (New York, 1980), Stephanie Coontz, *The Social Origins of Private Life: A History of American Families, 1600–1900* (New York, 1988), and Steven Mintz and Susan Kellogg, *Domestic Revolutions: A Social History of American Family Life* (New York, 1988). John Demos, "Images of the Family, Then and Now," *Past, Present, and Personal: The Family and the Life Course in American History* (New York, 1986), offers an important periodization.

A pathbreaking approach to the family in the American West can be found in Elliott West, *Growing Up with the Country: Childhood on the Far-Western Frontier* (Albuquerque, 1989). James E. Davis, *Frontier America, 1800–1840: A Comparative Demographic Analysis of the Settlement Process* (Glendale, Calif., 1977), explores frontier demography. For the fertility issue, see Richard A. Easterlin, "Population Change and Farm Settlement in the Northern United States," *Journal of Economic History* 36 (1976): 45–75, and Daniel Scott Smith, "'Early' Fertility Decline in America: A Problem in Family History," *Journal of Family History* 12 (1987): 73–84. The history of western women is much better developed than western family history per se; see Susan Armitage, "Women and Men in Western History: A Stereoptical Vision," *Western Historical Quarterly* 16 (1985): 381–95. Overviews include the following: Julie Roy Jeffrey, *Frontier Women: The Trans-Mississippi West, 1840–1880* (New York, 1979); Sandra L. Myres, *Westering Women and the Frontier Experience, 1800–1915* (Albuquerque, 1982); Glenda Riley, *The Female Frontier: A Comparative View of Women on the Prairie and the Plains* (Lawrence, 1988); and Lillian Schlissel, Vicki L. Ruiz, and Janice Monk, eds., *Western Women: Their Land, Their Lives* (Albuquerque, 1988).

Classic narratives of westward migration include Lois Kimball Mathews, *The Expansion of New England: The Spread of New England Settlement and Institutions to the Mississippi River,*

1620–1865 (Boston, 1909), and Stewart Holbrook, *Yankee Exodus* (New York, 1950). For modern approaches, see John D. Unruh, Jr., *The Plains Across: The Overland Emigrants and the Trans-Mississippi West, 1840–60* (Urbana, Ill., 1979), and John C. Hudson, "North American Origins of Middlewestern Frontier Populations," *Annals of the Association of American Geographers* 78 (1988): 395–413. William A. Bowen, *The Willamette Valley: Migration and Settlement on the Oregon Frontier* (Seattle, 1978), analyzes family selectivity and settlement, whereas John M. Faragher, *Women and Men on the Overland Trail* (New Haven, 1979) is a pioneering interpretation of gendered responses to migration; see also Lillian Schlissel, *Women's Diaries of the Westward Journey* (New York, 1982). James N. Gregory, *American Exodus: The Dust Bowl Migration and Okie Culture in California* (New York, 1989), charts the culmination of the family's westward march.

But social historians have tended to pay more attention than western historians to family migration strategies: Gordon Darroch, "Migrants in the Nineteenth Century: Fugitives or Families in Motion?" *Journal of Family History* 6 (1981): 257–77; Allan Kulikoff, *The Agrarian Origins of American Capitalism* (Charlottesville, 1992); Richard L. Bushman, "Family Security in the Transition from Farm to City, 1750–1850," *Journal of Family History* 6 (1981): 238–56; and John Solomon Otto, "The Migration of the Southern Plain Folk: An Interdisciplinary Synthesis," *Journal of Southern History* 51 (1985): 183–200. Studies using genealogies include the following: John W. Adams and Alice Bee Kasakoff, "Migration and the Family in Colonial New England: The View from Genealogies," *Journal of Family History* 9 (1984): 24–42; Russell M. Reid, "Church Membership, Consanguineous Marriage, and Migration in a Scotch-Irish Frontier Population," *Journal of Family History* 13 (1988): 397–414; and Randy W. Widdis, "Generations, Mobility and Persistence: A View From Genealogies," *Histoire Sociale/Social History* 25 (1992): 125–50. Important for methodological and interpretive comparison are Bruce S. Elliott, *Irish Migrants in the Canadas: A New Approach* (Kingston, Ontario, 1988), and Alida C. Metcalf, *Family and Frontier in Colonial Brazil: Santana de Parnaiba, 1580–1822* (Berkeley, 1992).

Gerald McFarland, *A Scattered People: An American Family Moves West* (New York, 1985), offers an evocative case study of generational migration; John Mack Faragher, *Sugar Creek: Life on the Illinois Prairie* (New Haven, 1986), remains the most satisfying explication of family life in a frontier community. Christopher Clark, *The Roots of Rural Capitalism: Western Massachusetts, 1780–1860* (Ithaca, N.Y., 1990), and Hal S. Barron, *Those Who Stayed Behind: Rural Society in Nineteenth-Century New England* (New York, 1984), examine family change within older rural regions. Even when not explicitly family-focused, western social histories suggest how varying family styles interacted in different local contexts: Peter K. Simpson, *The Community of Cattlemen: A Social History of the Cattle Industry in Southeastern Oregon, 1869–1912* (Moscow, Id., 1987); Kathleen Underwood, *Town Building on the Colorado Frontier* (Albuquerque, 1987); Ralph Mann, *After the Gold Rush: Society in Grass Valley and Nevada City, California, 1849–1870* (Stanford, Calif., 1982); and Paula M. Nelson, *After the West Was Won: Homesteaders and Town-Builders in Western South Dakota, 1900–1917* (Iowa City, 1986).

Allan Kulikoff, cited above, provides a valuable introduction to slave migration. Data have been drawn from Peter D. McClelland and Richard J. Zeckhauser, *Demographic Dimensions of the New Republic: American Interregional Migration, Vital Statistics, and Manumissions, 1800–1860* (Cambridge, Eng., 1982), and James William McGettigan, Jr., "Boone County Slaves: Sales, Estate Divisions and Families, 1820–1865," *Missouri Historical Review* 70 (1978): 271–95. Zachary Cooper tells the story of *Black Settlers in Rural Wisconsin* (Madison, 1977), and William Cohen, *At Freedom's Edge: Black Mobility and the Southern White Quest for Racial Control, 1861–1915* (Baton Rouge, 1991), furnishes a starting place for understanding postbellum black family migration.

Larry M. Logue, *A Sermon in the Desert: Belief and Behavior in Early St. George, Utah* (Urbana, 1988), probes Mormon family patterns in a community context, as do essays in Jessie L. Embry and Howard A. Christy, eds., *Community Development in the American West: Past and Present Nineteenth and Twentieth Century Frontiers* (Provo, Utah, 1985). See also Lee L. Bean, Geraldine

P. Mineau, and Douglas L. Anderton, *Fertility Change on the American Frontier: Adaptation and Innovation* (Berkeley, 1990).

An interpretive review of research on immigrant family patterns is sketched in Kathleen Neils Conzen, "Immigrants in Nineteenth-Century Agricultural History," in Lou Ferleger, ed., *Agriculture and National Development: Views on the Nineteenth Century* (Ames, Iowa, 1990). The Nibler and Schultheis story rests on records consulted for a larger study introduced in Kathleen Neils Conzen, "Peasant Pioneers: Generational Succession among German Farmers in Frontier Minnesota," in Steven Hahn and Jonathan Prude, eds., *The Countryside in the Age of Capitalist Transformation* (Chapel Hill, 1985), 259–92; their journey west can be followed in C. S. Kingston, "The Northern Overland Route in 1867: Journal of Henry Lueg," *Pacific Northwest Quarterly* 41 (1950): 234–53. The German wagon train that artist Albert Bierstadt saw is described in Fitz Hugh Ludlow, *The Heart of the Continent* (New York, 1870). For the Norwegian case, see Jon Gjerde, *From Peasants to Farmers: The Migration from Balestrand, Norway, to the Upper Middle West* (Cambridge, Eng., 1985); other examples include Robert C. Ostergren, *A Community Transplanted: The Trans-Atlantic Experience of a Swedish Immigrant Settlement in the Upper Middle West, 1835–1915* (Madison, 1988), and Rob Kroes, *The Persistence of Ethnicity: Dutch Calvinist Pioneers in Amsterdam, Montana* (Urbana, 1992).

For the Chinese family experience, consult Sucheng Chan's *This Bitter-sweet Soil: The Chinese in California Agriculture, 1860–1910* (Berkeley, 1986) and her "European and Asian Immigration into the United States in Comparative Perspective, 1820s to 1920s," in Virginia Yans-McLaughlin, ed., *Immigration Reconsidered: History, Sociology, and Politics* (New York, 1990), 37–75; see also Lucie Cheng Hirata, "Chinese Immigrant Women in Nineteenth-Century California," in Carol Ruth Berkin and Mary Beth Norton, eds., *Women of America: A History* (Boston, 1979), 223–44. Frances Leon Swadesh, *Los Primeros Pobladores: Hispanic Americans of the Ute Frontier* (Notre Dame, 1974), David Montejano, *Anglos and Mexicans in the Making of Texas, 1836–1986* (Austin, 1987), Arnoldo De Leon and Kenneth L. Stewart, *Tejanos and the Numbers Game: A Socio-Historical Interpretation from the Federal Censuses, 1850–1900* (Albuquerque, 1989), and Sarah Deutsch, *No Separate Refuge: Culture, Class, and Gender on an Anglo-Hispanic Frontier in the American Southwest, 1880–1940* (New York, 1987), provide similar access to Hispanic family processes in the West.

The complexities of Native American family history can be approached through S. Ryan Johansson, "The Demographic History of the Native Peoples of North America: A Selective Bibliography," *Yearbook of Physical Anthropology* 25 (1982): 133–52, and Nancy Shoemaker, "Native American Families," in Joseph M. Hawes and Elizabeth I. Nybakken, eds., *American Families: A Research Guide and Historical Handbook* (New York, 1991). Significant articles are collected in Patricia Albers and Beatrice Medicine, *The Hidden Half: Studies of Plains Indian Women* (Lanham, Md., 1983), and volume 15 of *American Indian Quarterly* (1991). Useful case studies include the following: John H. Moore, *The Cheyenne Nation: A Social and Demographic History* (Lincoln, 1987); Nancy Oestrich Lurie, "The Winnebago Indians: A Study in Cultural Change" (Ph.D. diss., Northwestern University, 1952); William G. McLaughlin and Walter H. Conser, Jr., "The Cherokee Transition: A Statistical Analysis of the Federal Cherokee Census of 1835," *Journal of American History* 64 (1977): 678–703; Albert L. Hurtado, *Indian Survival on the California Frontier* (New Haven, 1988); Morris W. Foster, *Being Comanche: A Social History of an American Indian Community* (Tucson, 1991); and S. Ryan Johansson and S. H. Preston, "Tribal Demography: The Hopi and Navaho Populations as Seen through Manuscripts from the 1900 U.S. Census," *Social Science History* 3 (1978): 1–33. For the development of mixed-race families, see Jacqueline Peterson, "Many Roads to Red River: Métis Genesis in the Great Lakes Region, 1680–1815," in Jacqueline Peterson and Jennifer S. H. Brown, eds., *The New Peoples: Being and Becoming Métis in North America* (Lincoln, 1985), 37–71, and Gary Clayton Anderson, *Kinsmen of Another Kind: Dakota-White Relations in the Upper Mississippi Valley, 1650–1862* (Lincoln, 1984).

Finally, those interested in Wyatt Earp can find the film's script in Robert Lyons, ed., *My Darling Clementine: John Ford, Director* (New Brunswick, N.J., 1984). The best recent interpretation is Paula Mitchell Marks, *And Die in the West: The Story of the O.K. Corral Gunfight* (New York, 1989), supplemented for present purposes with older accounts as well as manuscript censuses, county histories, and other local sources; see also Patricia Jahns, *The Frontier World of Doc Holliday: Faro Dealer from Dallas to Deadwood* (1957; reprint, Lincoln, 1979). For the family in western art, consult Dawn Glanz, *How the West Was Drawn: American Art and the Settling of the Frontier* (Ann Arbor, 1982) and Ron Tyler et al., *American Frontier Life: Early Western Painting and Prints* (Fort Worth, 1987). Worry about a "drifted family" can be found in Alfred B. McCown, *Down on the Ridge: Reminiscences of the Old Days in Coalport and Down on the Ridge; Marion County, Iowa* (Des Moines, 1909), 124.

Chapter Ten

Religion and Spirituality

FERENC M. SZASZ AND
MARGARET CONNELL SZASZ

In the spring and summer of 1788, a number of eastern cities staged celebrations in honor of the new Constitution of the United States. The most impressive of these "federal processions" occurred in Philadelphia, where, on 4 July 1788, a crowd of about seventeen thousand watched five hundred people file past in a mammoth parade. According to the eyewitness Francis Hopkinson, the marchers grouped themselves by guild or profession, and eighty-fifth in line (after the lawyers but before the doctors) strolled "the clergy of the different Christian denominations, with the rabbi of the Jews, walking arm in arm." This public display of "charity and brotherly love" by Philadelphia's clergy proved a first, not only for America but probably for the entire world. It pointed to the fact that religion in the new federal Republic would play a vastly different role from anything that had gone before.

The clerics' optimism drew heavily from the political theory of James Madison, the American Enlightenment figure who thought most deeply about church-state relations. Acknowledging that a person's faith could never be determined by reason alone, Madison placed religious belief as the foremost of all natural rights. Since the state existed to protect these rights, it should never unnecessarily interfere with the realm of faith. The Philadelphia Convention of 1787 incorporated Madison's ideas into the Constitution; in 1791, these ideas formed the heart of the First Amendment. Unlike those nations with established churches, which included most of Europe, the United States would never develop any official church. Except for nineteenth-century denominational schools and missions among American Indians, no American church could rely on state support. Rather, each denomination *voluntarily* had to convince others that its position was the correct one. Almost every religious group accepted these boundaries. Each faith would set forth its position as best it could; "the people" would then choose their own religion.

The eminent twentieth-century theologian Paul Tillich once observed, "Religion is the substance of culture and culture the form of religion." Certainly this proved true for the trans-Mississippi West. The religious history of the West is all-embracing. It cannot be limited simply to kivas or churches, ceremonies or sermons, medicine men or clerics. Rather, western religion permeated the realms of politics, culture, and society. Perhaps the key to understanding religion in the West was the land. The vastness of this immense territory, with its many ecological subregions, provided a multitude of homes for native belief systems, as well as for the diverse faiths brought by European, African, and Asian immigrants. In the Great Plains, Rockies, Southwest, Plateau, Great Basin, and Pacific

A region marked by both religious pluralism and tolerance, the West is home to a wide variety of faiths, including the folk Catholicism that flourishes in the Hispanic communities of northern New Mexico.

Eliot Porter (1901–90). Church. Placita de Taos, New Mexico. *Gelatin silver print, 1940. Copyright Amon Carter Museum. Eliot Porter Collection, Amon Carter Museum, Fort Worth, Texas.*

Coast regions, a variety of religious subcultures flourished. With a few notable exceptions, tolerance and openness characterized the world of western faiths. In the generations encompassed by our story, the West initiated a pattern of religious pluralism in American society—often without a culture-shaping mainstream—that anticipated many developments of the late twentieth century.

We begin with the 1840s, a pivotal decade for both the religious and the political fortunes of the nation. By this time, the main outlines of American religious history had been generally sketched out. The Roman Catholic church had become the nation's largest single denomination, a position it would sustain to the present day. With growth fueled largely by immigration, the church wrestled with multiethnic congregations and a "foreign" image for over a century. The same stream of immigration brought over 250,000 German Jews, who soon scattered across the land. These Jews played vital entrepreneurial roles in the West, and some, such as the clothier Levi Strauss, rapidly rose to the realm of legend. In 1844, when the Latter-day Saints prophet Joseph Smith, Jr., died at the hands of an Illinois mob, the Saints numbered only about 14,000. The pundits of the day predicted their imminent collapse, but their subsequent move to the Great Basin region of Utah and Idaho gave the church new life. The mainline Protestant churches (Methodists, Baptists, Congregationalists, Presbyterians, Episcopalians, Lutherans) congratulated themselves that they had saved the trans-Appalachian region from "barbarism" through their "benevolent empire" of Bible, tract, Sunday school, and education societies. All were looking for new fields to conquer.

Simultaneously, in an era dominated by ideas of "manifest destiny," many Americans pushed across the Mississippi to claim Indian lands in Oregon country or Mexican California. Integral to this mass emigration, the Christian clergy joined the exodus in a race both to convert the Indians and retain the emigrating church members. During the antebellum era, the mainline Protestant denominations wielded the most influence in national affairs. Together, these groups composed what has been termed a "voluntary" religious establishment. While they disputed among themselves over theology and church polity, they agreed on essentials: Christianity had broken into "denominations," each of which had a distinct mission; Protestantism and democratic republicanism were forever intertwined; America had become God's "New Israel"; and the churches felt compelled to carry their mission to both whites and Indians *west* of the Mississippi.

The religious diversity that the European Americans brought west met an equal diversity among the indigenous faiths of the Native Americans. When the historian Robert F. Berkhofer, Jr., spoke of the "multiplicity of [the Indians'] specific histories," he referred primarily to their means of warfare, hunting, fishing, and social organization. But the Native Americans' varied ceremonial life and relationship to the supernatural shared a similar "multiplicity." Thus, in the nineteenth-century West, heterogeneous European-American religions interacted with equally heterogeneous native religions. The resulting blends, as seen in the Pueblo-Roman Catholic, Sioux-Episcopal, and Pima-Presbyterian amalgamations, proved unique in the history of American faith.

Long before the voyages of Columbus, American Indians had engaged in "religious borrowing and synthesis." Thus, when they began to graft European Christianity onto their own faiths, this was, as the anthropologist Robert Brightman has noted, "simply one more instance of a traditional receptivity to religious innovation." A major part of the history of native religion in the West is the story of its interaction with this imported Christianity.

From the 1760s, native groups of southern California had encountered the highly motivated Franciscans, who forced them into mission enclaves stretching from San Diego to San Francisco. The Franciscans retained their hold over thousands of native Californians until the Mexican government secularized the missions in the 1830s. In other regions of the Southwest, including present-day southern Arizona and parts of Texas, natives had also been influenced by Catholicism through missions founded in the late seventeenth and early eighteenth centuries. In the late 1500s, along the Rio Grande valley in what is now New Mexico, Tanoan and Keresan speakers, as well as the Zuni, had come under the control of these Hispanic Catholics, who occupied the region for eight decades—an era dominated by bitter church-state rivalry—before the natives drove them out in 1680. Don Diego de Vargas's *reconquista* of 1692 acknowledged native rights and marked the beginning of a rich blending of native ceremonies and worldview with those of Hispanic Catholicism, a blending that continues into the present. East of the Llano Estacado, crossed by Coronado in the 1540s, former Southeast Woodland tribes—Cherokees, Creeks, Choctaws, Chickasaws, and Seminoles—were settling in. Even before the era of removal forced their emigration, most of these groups had met Protestant missionaries. In general, the Christian messages were well received, especially by the Cherokees, whose leadership, epitomized by the mixed-blood John Ross, welcomed change and the incorporation of European ways. Christianity and traditional values blended among these Indians, historically known as the "Five Civilized Tribes," during their early decades in the Indian Territory.

Elsewhere in the West, however, native religions had remained beyond the thrust of Christian missionaries. In the Northwest Coast and Columbia River Plateau regions, Salishan, Sahaptian, Chinookian, and other linguistic groups had begun extensive cultural borrowing with the opening of the sea otter trade in the late eighteenth century, the startling visit of Lewis and Clark, and the intense international rivalry for beaver. Bargaining for iron pots, metal fishhooks, weapons, or the much desired blue beads had changed their cultures. They had incorporated the epithets of the Boston men into the Chinook trade jargon, and they had sharpened their shrewd trading skills in the vast exchange network that stretched east via the Nez Percés. Moreover, they had been weakened by European disease. But with the exception of a band of Catholic Iroquois, who settled among the Salishan-speaking Flatheads around 1820, and the quasi-religious influence of the Hudson's Bay Company, this cultural borrowing had generally excluded Christianity. Not until the 1830s and 1840s, with the arrival of Oblate and Jesuit priests, plus missionaries from various Protestant denominations, did the Northwest Coast and Plateau people begin to address the many messages of Christianity. In the central Rockies, much of the Great Plains, and the western Great Basin, these missionaries arrived even later.

The Intertwining of Politics and Religion

In the mid-1840s, the Utes, Paiutes, and other natives living in the eastern Great Basin met one of the most unusual religious groups in nineteenth-century America. In no other area of the West were politics and religion more closely intertwined, for this region is forever linked with the saga of the Church of Jesus Christ of Latter-day Saints (the Mormons). Western Protestant-Catholic and Christian-Jewish tensions generally remained confined to harsh words and editorials. Only Mormon-Gentile (i.e., non-Mormon) relations crossed the line into mob violence. For many mid-nineteenth-century contemporaries, the Latter-day Saints pushed beyond the limits of America's famed religious toleration.

The story of the angel who led an upstate New York farm boy, Joseph Smith, Jr., to the buried golden plates on Hill Cumorah is well-known. Seated behind a curtain, Smith translated these plates to form *The Book of Mormon*, first printed in 1830. Read literally, *The Book of Mormon* tells the story of ancient Near Eastern peoples who migrated to the Americas: the Jaredites, the Nephites, and the Lamanites (the latter designated as ancestors of the American Indians). The account culminates with the visit of Jesus Christ, shortly after His resurrection, to the Nephites. Read metaphorically, the book depicts the success of those civilizations that follow the Commandments of the Lord and the collapse of those that become filled with pride and arrogance. In either case, *The Book of Mormon* was America's first indigenous holy scripture.

The Mormons invoked controversy wherever they settled. Their new scripture, Smith's 130 special revelations from the Lord—especially those concerning polygamy (an open secret, fueled by rumor, from the late 1830s until officially proclaimed in 1852), Mormon "bloc voting," and their alleged violation of the church-state separation—all played on Gentile fears. The culmination came on 29 January 1844, when Joseph Smith, Jr., announced that he was a candidate for the presidency of the United States.

Consequently, what the novelist William Dean Howells once termed "the foolish mob which helps to establish each new religion" proved a major factor in early Mormon history. Many church leaders, including Smith, were either tarred and feathered or thrown in jail on trumped-up charges. Their northern origins made them especially suspect in slaveholding Missouri, where proslavery settlers and politicians persecuted them mercilessly. As a Mormon hymn writer put it: "Missouri/Like a whirlwind in its fury,/And without a judge or jury,/Drove the Saints and spilled their blood."

When the Saints established the Mississippi town of Nauvoo, Illinois—a well-run prototype for the later Mormon communities in the Great Basin—local outrage could no longer be contained. On 27 June 1844, an angry mob stormed the jail at Carthage, Illinois, to martyr both Joseph Smith, Jr., and his brother Hyrum.

Virtually all observers expected the Saints to collapse with the death of the prophet. Indeed, several schisms weakened them considerably. Sidney Rigdon led a fragment to Pittsburgh, Pennsylvania; James J. Strang headed a larger remnant that thrived in a communal setting on Beaver Island in Lake Michigan, until his assassination; and Joseph Smith III, the prophet's son by his first wife, Emma Hale Smith, rejected polygamy to lead a group that became the Reorganized Church of Jesus Christ of Latter Day Saints, with headquarters in Independence, Missouri. That the entire body of Saints did not similarly fracture may be credited to the skills of the newly appointed prophet, Brigham Young, and his decision to move to the West.

Painted by a Mormon artist, this idealized portrait of the prophet Brigham Young places Mormon family values squarely within the realm of Victorian virtues. Surrounded by traditional symbols that belie the actual western environment, the Youngs appear as prosperous members of Utah society.

William Warner Major (1804–54). Brigham Young's Family. *Oil on canvas, ca. 1847–53. The Church of Jesus Christ of Latter-day Saints, Salt Lake City, Utah.*

The historian Jan Shipps has argued that the great trek from Missouri and Illinois to Utah formed the central event in Mormon history. The journey to the Great Basin carried the Saints not simply to the promised land of Deseret but also "backward" into a primordial sacred time. From this journey, Shipps has suggested, the Mormons emerged as a distinctly new religious faith, as different from Christianity as Christianity was from Judaism.

Both Mormon social practices and theology proved unique. The Saints rejected the Christian trinity and downplayed the concept of original sin. Their communalism, polygamy, and authoritarian church polity formed a sharp contrast to the romantic individualism that dominated contemporary American Protestantism. Believing that God "was once as we are now," the Mormons taught that most devout male Saints would eventually hold similar dominion over future worlds of their own. Their maxim phrased it thus: "As God is at present Man may become." Essentially universalists, the Saints maintained that all of humanity would achieve salvation but that Mormon believers would reach a higher degree of glory. The King James translation of Scripture, *The Book of Mormon* (written in the King James idiom), and Smith's subsequent revelations were accorded equal divine status. The head of the church was assigned the mantle of contemporary prophet.

The evolving Mormon folk religion transcended even the official pronouncements from church leaders. The Saints celebrated special holidays: Joseph Smith's birthday,

Brigham Young's birthday, the birthday of the church; the day of arrival in the Salt Lake Valley (still observed in Utah on 24 July as Pioneer Day). They wove heroic legends of the "Great Trek" west and the suffering of the later emigrants, some of whom pushed handcarts over twelve hundred miles to their new home. They commemorated the sego lily, whose roots the early pioneers ate to avoid starvation, and seagulls, which arrived to devour a plague of crickets that threatened to consume the Saints' first wheat crop. They danced and sang with vigor. When their hymns spoke of "Israel" or the "Camp of Israel," they claimed these concepts for themselves, and thus the term *Gentiles* took on new meaning in the Mountain West. Like the ancient Hebrews, the Saints forged a separate concept of "peoplehood" that persists up to the present day.

The federal government, however, viewed the rise of a semi-independent kingdom in the Great Basin with considerable suspicion. In the mid-1850s, Congress accused Brigham Young of complicity in the harassment of Utah's federal officials. Spurred on by exaggerated coverage by the eastern press, President James Buchanan ordered federal troops to Utah in 1857 to bring the Saints into line.

The Saints viewed the arrival of the federal army as reminiscent of their persecution in Missouri and Illinois. The Mormon leaders seriously considered relocating to Central America or elsewhere. Eventually cooler heads prevailed, and the "Mormon War" ended without direct confrontation. But the tension caused by the war did lead to bloodshed. In August 1857, a wagon train of Missouri and Arkansas settlers crossed southern Utah, where they were attacked by a band of Mormons and their Indian allies. This raid, in which 130 people died, ranks as one of the worst examples of religious violence in American history. The Mountain Meadows Massacre, as it is known, assumed a symbolic role in defining Mormon-Gentile relations.

Politics and religion were equally intertwined in the story of religious expansion into the Pacific Northwest. In 1833, four Flathead and Nez Percé Indians journeyed to St. Louis to inquire about Christian missionaries. This seemingly inconsequential request would help to determine the course of the history of the Northwest. It opened the door for missionaries and migrants and thus became the basis for America's claim to the Oregon Country.

The native appeal for "white religion" probably implied a desire for increased knowledge of a general, all-defusing cultural power. In 1833 and 1837, other groups of Salishan and Sahaptian natives traveled the same path to St. Louis. The retelling of the story created one of the most famous legends of nineteenth-century western religious history. Catholic journals broadcast the Indian journey as a call for "Black Robes" who said "Great Prayers" (the Mass). Protestants declared that the Indians had requested the "white man's book of heaven." Within a few years, both Catholic and Protestant missionaries had begun the arduous trek to the Columbia River Plateau and the Northwest Coast.

In June 1840, the Jesuit Pierre Jean De Smet made the journey from St. Louis to the Flatheads and Pend d'Oreilles. The next year he returned with two more Jesuits, Nicholas Point and Gregory Mengarini, thus inaugurating what a later Jesuit termed "the grandest missionary work of the nineteenth century in its religious, social, economical and political aspect."

De Smet and his fellow Jesuits hoped to encourage the Indians to abandon their nomadic life and adopt a settled agrarian existence. In September 1841, De Smet began St. Mary's Mission in the Bitter Root Valley of Montana. The next year he helped create the Coeur d'Alene Mission of the Sacred Heart on the St. Joe River. The St. Ignatius mission to the Flatheads, St. Paul's to the San Poils, and St. Michael's to the Spokans soon followed.

Generally speaking, the Jesuits looked to their own history, especially their "holy experiment" in the Central Highlands of South America, as a model for this endeavor. During the seventeenth and eighteenth centuries, the Jesuits had established a string of over thirty settlements (called *reducciones*, from the Spanish *reducir*, "to bring together") in the region that is now largely Paraguay. Centered around a market square and a plaza, these communities consisted of several thousand Indians managed by only a handful of clerics. The Jesuits taught the Natives European forms of agriculture, music, architecture, and religion during an experiment that lasted over a century.

Although De Smet's dream of establishing "a new Paraguay," never occurred, these Northwest missions did serve many functions similar to those of their earlier counterparts. St. Ignatius provided a hospital, sawmill, flour mill, and printing press. All missions boasted schools that taught theology, English, and other skills. Rumor had it that every Jesuit mission contained at least one resident genius. Father Anthony Ravalli certainly qualified. During his career at St. Mary's he served as doctor, architect, sculptor, linguist, and expert manager. De Smet himself also proved a skilled negotiator. His peacekeeping efforts on the northern plains saved hundreds of lives, and many regional native leaders held him in esteem.

De Smet also drew on the romantic appeal of the American West to encourage numerous European novices and priests to follow his footsteps. Over the course of the century, perhaps two hundred Jesuits crossed the ocean to serve missions in the northern Rockies and Plateau regions. In spite of this effort, however, the string of Jesuit missions never fulfilled their founders' hopes. The harsh climate of the region proved unsuitable for extensive agriculture, and the Indians preferred their traditional hunting, fishing, and gathering cycle to a settled mission life. (To follow the tribe, for example, Sacred Heart Mission moved three times in thirty-six years.)

Some of these Jesuit missions remain modest tourist attractions today, such as St. Ignatius in Montana or the Cataldo Mission (Sacred Heart) in Idaho. As an entity, however, these missions are not well-known outside the region, and they pale when compared with their internationally known California counterparts. The life of De Smet is respected, but it has never engendered the romance that surrounds California's mission founder, the Franciscan Junípero Serra.

The Methodists were the first Protestant denomination to respond to the Indian journey to St. Louis. In 1834, Rev. Jason Lee, his nephew Rev. Daniel Lee, and three lay associates traveled to the Northwest Coast, settling in the Willamette Valley. Within a few years the Presbyterians sent out Revs. Elkanah Walker and Cushing Eells and their wives, Dr. Marcus and Narcissa Whitman, and Rev. Henry and Eliza Spalding. Narcissa and Eliza were the first European-American women to cross the Rockies into the Columbia River Plateau. Unlike Jason Lee, these missionaries were drawn to the Plateau tribes: Walker and Eells to the Spokans at Tshimakain; the Spaldings to the Nez Percés at Lapwai; and the Whitmans to the Walla Wallas and Cayuses at Waiilatpu. Like the

Jesuits, the Whitmans built a gristmill, sawmill, blacksmith shop, and school; their mission also served as an "emigrant house" for Oregon Trail travelers.

In 1842, when the American Board of Commissioners for Foreign Missions determined to close these missions to the Plateau tribes, an equally determined Whitman traveled east in a dangerous mid-winter trek to argue their case. Like the Nez Percé-Flathead trip to St. Louis, Whitman's dramatic journey to the East has also ballooned into legend. Those who argue that Whitman "saved Oregon" through his travels neglect the fact that by the 1840s, midwesterners with "Oregon fever" were already beginning the migration that led to the resolution of the Oregon boundary issue. The Whitmans' contribution to the American cause may have come later. When Congress learned of the November 1847 native uprising against the Waiilatpu Mission and of the deaths of Marcus, Narcissa, and others, it responded by creating a government for the Oregon Territory, the first official American government established west of the Rockies.

As Protestant and Catholic missionaries competed among the tribes living in the Northwest Coast, Plateau, and northern Rockies, they carried out in microcosm the most persistent American religious theme of the century: Protestant-Catholic hostility. This theme echoed and reechoed throughout the West, where it affected both native and immigrant. The Protestant and Catholic "ladders" developed in the Northwest Coast and the Plateau reflected this antagonism. Borrowing from the Salishan concept of a *sahale* stick ("wood from above"), the French-Canadian father François Norbert Blanchet created a large (six-feet-by-two-feet) paper chart with a time line portraying the life of Christ and basic Christian principles. One version of the "Catholic ladder" depicted Martin Luther as branching off on a road that led to hell. By contrast, Spalding's "Protestant ladder" for the Nez Percés peopled the road to hell with worldly popes and immoral priests.

It is far easier to count missions, sawmills, gristmills, printing presses, and even Christian "ladders" than it is to evaluate the results of the missions among the Indians of the Northwest Coast, Plateau, and northern Rockies. A number of Nez Percés became Presbyterians, and in 1890 a writer pointed out that approximately seven thousand out of ten thousand Montana Indians were Catholics. But counting converts, a popular pastime among nineteenth-century missionary groups, proved as inaccurate as the conclusion, made by some late-twentieth-century historians, that the missions were a general failure. From church buildings to native clergy, the results of both Catholic and Protestant missionary endeavors of the nineteenth century are much in evidence today. Their legacies remain as diverse as the cultures of the native peoples themselves.

Many European immigrants to the Pacific Northwest, like the natives, responded to the missionaries with indifference. By the late twentieth century, this area was widely acknowledged as "the least churched region" of the nation. The nineteenth-century boasts "the Sabbath shall never cross the Missouri" and "no Sunday west of St. Louis" proved prescient. They pointed to the fact that the eastern religious institutions would have difficulty establishing themselves in the wide-open society of the trans-Mississippi West.

Nowhere was the secular image of the new West more pronounced than in California. In 1849, the cry of "Gold, Gold, from the American River" drew thousands

Missionary work in the rural West presented particular challenges to those serving widely scattered populations.

Unidentified photographer. John Jasper Methvin, a Methodist Missionary among Kiowa and Comanche Girls. *Photograph, ca. 1894. Western History Collections, University of Oklahoma Library, Norman.*

Unidentified photographer. Fording River on a Congregational Missionary Tour *(Wyoming). Photograph, date unknown. American Heritage Center, University of Wyoming, Laramie.*

around the Horn, across Panama, or over the trail to San Francisco. The chief goal of forty-niners was seldom that of the spirit. "The Americans," complained a visiting Catholic priest, "think only of dollars, talk only of dollars, seek nothing but dollars."

Nevertheless, a group of clerical forty-niners did their best to stem the tide. By one estimate, four denominations had established about fifty small churches throughout the early "Mother Lode" country. A Unitarian pulpit orator, Thomas Starr King, tried to replicate Boston's values in San Francisco during the 1850s and early 1860s while the Congregationalist Timothy Dwight Hunt attempted to "make California the Massachusetts of the Pacific."

Such was not to be. The historian Kevin Starr has noted that the tumultuous nature of California life could never be confined within traditional religious norms, be they New England parish, Virginia plantation, or Mexican village. California manifested a religious "openness" from its earliest days.

California life also muted all the traditional religious antagonisms. The fact that the territory's first American governor, Peter H. Burnett, was a Catholic convert played absolutely no role in his political career. As a Catholic archbishop noted in 1864, his church "did not face the prejudice which is encountered elsewhere." A generation later, California's small Seventh-Day Adventist community led a successful fight to repeal the state's Sunday regulations. In the cities, the African Methodist Episcopal and African Methodist Episcopal Zion churches provided strong voices for racial equality. John Muir's "religion of nature," a transcendental appreciation for the magnificence of Creation (with little or no role for a redeemer), also drew a number of followers. Worship services by Asian faiths generally went unmolested. In religion, as in so many other areas, California became "the great exception."

Politics and religion were equally intertwined in the American Southwest. In Texas, the nineteenth century was a postmission era. The Franciscan missions, especially those among the Caddo, established in the early 1700s in part to counteract French movement in the lower Mississippi Valley, were defunct, and in the 1840s only a handful of priests still served the Texas Catholic community. After the independence movement established freedom of religion, Jean Marie Odin, the first bishop of Galveston, oversaw the rejuvenation of Texas Catholicism. In addition to the Mexicans, his diocese consisted largely of European immigrants. For example, a band of German Catholics settled the hill country during the mid-1840s, and the Polish Franciscan Leopold Moczygemba led a group of Silesian Poles to Panna Maria in 1856. By the 1850s, however, American immigration had thrust the Baptists, Methodists, and Disciples of Christ into dominance. These evangelical groups have played a major role in Texas religious history to the present day.

The political-religious connection was even more sensitive in the lands taken from Mexico in 1848. All of the Hispanos of the American Southwest were titular Catholics, but everywhere the faithful had long suffered from want of clerical attention. In southern Texas, Arizona, California, and especially New Mexico, the Hispanic settlers had responded to the dearth of priests by creating their own version of folk Catholicism.

This included an intense respect for local patron saints, many of whom were credited with frequent miracles, and a strong Mariolatry, represented by devotion to the Virgin of Guadalupe. The Hispanic communities of the borderlands celebrated a steady round of religious holidays: 17 January, the feast of San Antonio, a day for the blessing of the animals; 24 June, San Juan's Day, which became associated with the first fruits and vegetables of the season; the feast of Corpus Christi, celebrated in the seventh week after Easter; the solemn 1 November, All Saints' Day, and 2 November, All Souls' Day. December was the climax month of celebration, with *Los Pastores*, a Spanish medieval miracle play, plus a reenactment of the nine days that Mary and Joseph wandered in search of shelter in Bethlehem before the birth of Jesus. The historian Arnoldo De Leon has argued that the faith of the Rio Grande borderlands expressed "an attitude consonant more with life experience than theology."

Folk Catholicism permeated the territory of New Mexico. The healing skills of *curanderas*, the lay brotherhood of Penitentes, and the folk carvings of *Santos*, *bultos*, and *retablos* reflected a deeply held cultural faith. From the early nineteenth century forward, the little chapel at Chimayo, New Mexico, known as "The Lourdes of the Southwest," has drawn those seeking healing. This pervasive New Mexico folk Catholicism proved remarkably tolerant of the influx of Anglo Protestants.

The same basic toleration may be seen in the story of western Judaism. From the 1850s, Jews composed perhaps 10 percent of San Francisco's merchant community. Relying on a credit network that included family members and coreligionists, Jewish families provided vital economic services, both in rural areas, such as New Mexico, and urban centers, such as San Francisco, Portland, Los Angeles, Denver, and Seattle.

Contemporary visitors marveled at how well the western Jews had succeeded. In the Los Angeles 1876 centennial celebration, a young Jewish woman portrayed the "Spirit of Liberty" while a rabbi helped preside over the festivities. In San Francisco's first *Elite Directory* (1870–79), Jews composed over one-fifth of the city's "elite." The historians Harriet Rochlin and Fred Rochlin have counted over thirty nineteenth-century western Jewish mayors, plus countless sheriffs, police chiefs, and other elected officials. Although one can find traces of anti-Semitism, it played a much smaller role in western life than in the contemporary South or Northeast. The historian Eldon Ernst has concluded that California's failure to produce a "religious mainstream" allowed all faiths to flourish on a roughly equal basis. The same could be said for many other subregions in the trans-Mississippi West.

Eastern Churches Move West

In the decades after the Civil War, the American churches moved into the Far West with increasing enthusiasm. Clerics from all denominations followed the recently completed railroad lines in hopes of securing a "first strike" among the mobile frontier settlers. The railroad corporations and various booster organizations aided the church-building frenzy by gladly giving away free lots to virtually every denomination that asked. In their eyes, the presence of church buildings connoted "stability."

Although stability may have been the ultimate goal, both clerics and parishioners remained highly mobile for over half a century. The fluid environment of the early western towns created a unique situation: a brief period of genuine interdenominational cooperation. Initially, frontier clerics sought out not only members of their own denomination but all "interested parties." Thus, the first churches frequently contained people from several different denominations. In these early years, virtually all the Sunday schools on the Great Plains were "union" (multidenominational). Music proved especially ecumenical. The Episcopal choir at Bland, New Mexico, for example, consisted of two Episcopalians, one Catholic, one Mormon, one Presbyterian, and one Congregationalist. A Jew joined the first Episcopal choir in Helena, Montana, and Gentiles sang in the Los Angeles Congregation B'nai B'rith for several years. Westward-migrating Jews occasionally attended Unitarian services, and several Jewish merchants donated to local Catholic and Protestant churches and schools. This "ecumenicism of necessity" diminished when each group gathered enough members to form a church or synagogue of its own. Such cooperation, however short-lived, was seldom duplicated in the annals of American religious life.

The frontier clergy have rarely been given credit for their accomplishments. In addition to building churches and supplying the ordinances of their denominations, they provided the basic institutional infrastructure for the western states and territories. The local and state governments lacked funds, and the federal government was primarily interested in railroad and military affairs. Consequently, the clergy and churches took the lead in providing hospitals, orphanages, old-age homes, and schools.

Every denomination viewed health care as part of its mission. These efforts ranged from the modest "Industrial School and Hygienic Home for Friendless Persons," founded by the Kansas Mennonites in 1890, to the substantial network of Catholic and Episcopal hospitals that graced most major western cities. Catholic bishops spent a great deal of time coaxing women's religious orders west to staff these enterprises. In the last quarter of the nineteenth century, as many as fifteen orders of nuns were active on the northern plains. In addition to founding over twenty schools, the Presentation Sisters established three hospitals in South Dakota, plus one in Montana. The Grey Nuns of Montreal played a similar role in the Pacific Northwest, as did the Sisters of St. Joseph of Carondelet and the Sisters of Loretto in the Southwest. When the Sisters of the Holy Cross walked down the streets of Salt Lake City, curious crowds of Mormon children followed their every move.

The churches and clergy also played a prominent part in establishing and/or teaching in the public school systems. Records from the constitutional conventions of the various western states show that the fiercest arguments frequently revolved around the issue of state funding for parochial schools. Some areas, such as San Francisco, initially divided public funds between private and state-run schools, but this arrangement seldom endured. Protestant clergy frequently taught in the early school systems. The Freethinkers of Cottage Grove, Oregon, even asked the Cumberland Presbyterian minister Will Magee to establish a school. A religious school, these atheists reluctantly admitted, was better than none at all.

Since the fledgling public schools could not accommodate all of the children, many western denominations established parallel educational systems on the primary, secondary, and university levels. San Francisco's first Catholic church served for both worship and education. Many Protestant ministers, or their wives, also taught. The Presbyterians and Congregationalists established elaborate parochial school systems in both Utah and New Mexico. Everywhere, "church" and "school" overlapped.

EXPANSION:
Building an American Nation

◆

In the mid-nineteenth century, as the West acquired a decidedly American imprint, the region became the focus of a particularly nationalistic school of painting that celebrated the region and its recent American settlers as an expression of the country's democratic values and boundless possibilities. Not surprisingly, such art was generally created by those for whom western history seemed a triumph of American culture. Although art by Native Americans conveyed a different story, the incorporation of European-American motifs into traditional art forms also signaled the joined future of the West's different peoples.

Widely praised as a western artist whose paintings captured a distinctively western type of character, the Missouri artist George Caleb Bingham helped bring regional subjects into the mainstream of American art.

George Caleb Bingham (1811–79). The Jolly Flatboatmen in Port. *Oil on canvas, 1857. Museum Purchase, the Saint Louis Art Museum, Missouri.*

This study for a 20-by-30-foot mural in the U.S. Capitol presented familiar icons of westward migration in a triumphant narrative of American expansionism. Painted at the time of the Civil War, the picture reaffirmed the importance of the West as a place of optimism, opportunity, and national unity.

Emanuel Leutze (1816–68). Westward the Course of Empire Takes Its Way *(mural study). Oil on canvas, 1861. National Museum of American Art, Washington, D.C./Art Resource, New York.*

Purchased by Congress in 1872 and put on public view in the Capitol, Thomas Moran's massive painting of Yellowstone (7 x 12 feet) became the first American landscape by an American artist to be purchased by the federal government. Acquired just three months after Yellowstone was designated a national park, the painting epitomized popular interest in the park as a romantic wonderland.

Thomas Moran (1837–1926). The Grand Cañon of the Yellowstone. *Oil on canvas, 1872. Lent by U.S. Department of the Interior, Office of the Secretary, National Museum of American Art, Washington, D.C./Art Resource, New York.*

After the 1870s, when many Plains peoples were confined to reservations, new motifs in tribal art emerged to reflect the changed relationship between Native Americans and the federal government. In this Lakota shirt a beaded hand, the traditional symbol of a warrior, is combined with the symbol of another sort of power, the flag of the United States.

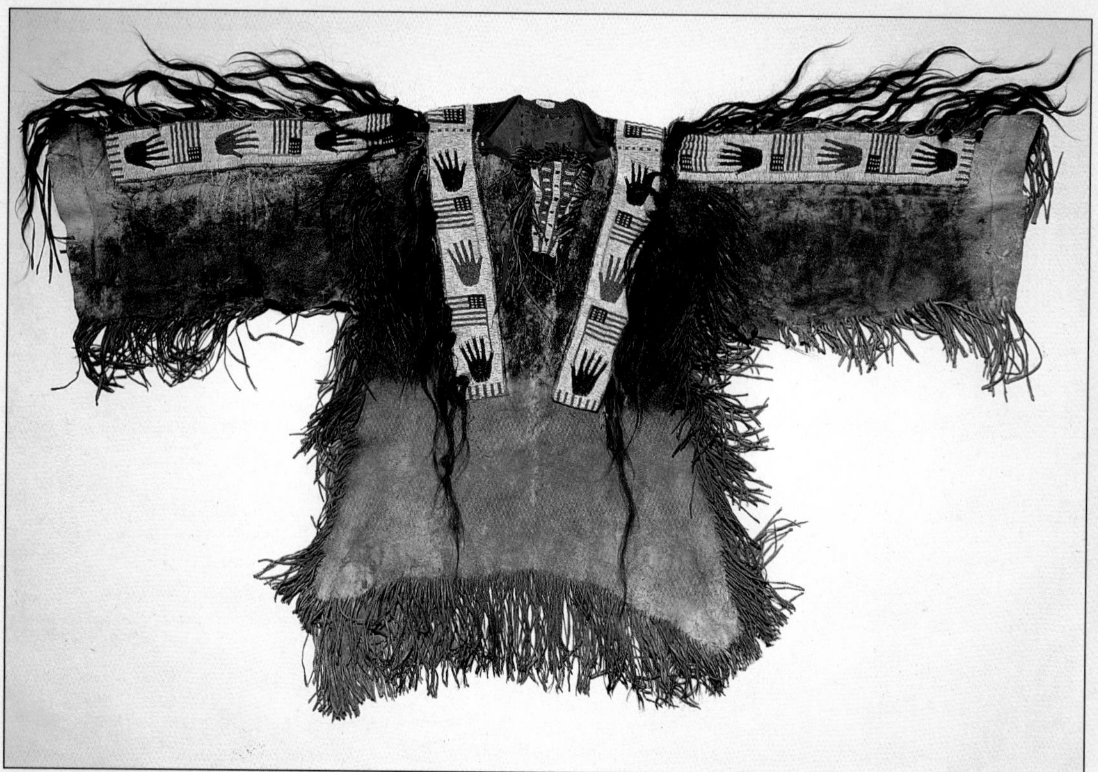

Unidentified artist (Oglala Lakota). Warrior's Shirt. *Native-tanned hide, hair, glass and metal beads, sinew, and blue-green and yellow pigment, ca. 1890. Bern Historical Museum, Ethnography Department, Bern, Switzerland.*

The various denominations devoted much attention to founding colleges. Well into the twentieth century, most western higher education maintained strong denominational links. The oldest institution of higher learning in California is Santa Clara College (now University). St. Ignatius, which later became the University of San Francisco, and St. Mary's, founded by the Christian Brothers, soon followed. Northern Baptists established McMinnville College (Oregon), the Lutherans organized Pacific Lutheran University (Washington), and the Presbyterians founded Occidental College (California), just to name a few. The Methodists established Willamette University (Oregon) and the University of Southern California, whose athletic teams were proudly known as the "Methodists" until 1912, when they became the "Trojans."

The race to found church-related colleges can largely be traced to interdenominational rivalry, but many churches overextended themselves in the effort. Lack of money, interest, and denominational commitment ensured that many of the schools would not survive. Virtually all western states are dotted with "ghost colleges." In California, about fifty church colleges collapsed, and Washington and Oregon lost about fifteen each. The Great Plains are littered with similar ghosts, bearing long-forgotten names such as Mallalico College (Nebraska) or Redfield College (South Dakota). Too many western towns wanted a college, complained a writer in 1891, but it was "immaterial with them whether the institution be a college proper or a normal school or a business college or a school for the feeble minded." By the 1890s, the college-founding boom had largely collapsed.

The clergy from the various denominations also played central roles in the establishment of most western state universities. New England Congregationalists were instrumental in founding the University of California, the University of Colorado, and the entire educational system of South Dakota. Presbyterians helped establish the University of Kansas and the University of Tulsa. The Episcopal St. Margaret's Girls School eventually grew into Boise State University. It was a rare university that did not boast a cleric on the board of trustees. Initially, the University of California had several. The churches saw their educational mission as an integral part of the settlement of the West.

Like the rest of the nation, the West also exhibited its share of free thought: Bohemian and German rationalists in Nebraska and Texas; the prolific, acid pen of William Cowper Brann's *Iconoclast* from Waco, Texas; the Liberal League in Kansas; and the Oregon State Secular Union in the Pacific Northwest. Great Plains freethinkers actually formed a town (Liberal, Kansas), and their Pacific Coast counterparts established a short-lived "Liberal University" in Silverton, Oregon. Despite its often vigorous rhetoric, western anticlericalism had little lasting significance.

The western churches frequently provided the focal point of a community's social life. Revivals and camp meetings, especially on the southern plains, offered an opportunity for isolated ranchers to gather for song and fellowship. Whereas the organizers of these gatherings usually measured "success" by the number of converts, the people were more pragmatic. "However great may have been the need for salvation," one plains woman recalled, "the need for recreation was given preference." In addition, these gatherings helped "democratize" the faith of western Protestantism. Drawing on the persistent tradition of the Great Awakenings, the western evangelists proclaimed a

The oldest cathedral west of the Mississippi, this building (constructed 1831–34) served as an urban monument to the importance of Catholicism in St. Louis, where believers of French and, later, Irish, German, and Italian descent played a major role in local affairs. The completion of an impressive new cathedral in 1914 reconfirmed the central role that the Roman Catholic church played in the religious life of the American West.

After George Morton and Joseph Laveille, lithograph by Leon Pomarede. Front View of the Cathedral of St. Louis. *Lithograph, 1835. Missouri Historical Society Photograph and Print Collection (#PB 984), St. Louis.*

simple message: mankind was a sinner, but God had redeemed the race through the gift of his only son, Jesus Christ. Humanity was saved through grace, not works, and people had only to open their hearts to the Savior. The truth of this story lay in the Scriptures, easily understood by all. Thus, a simple, democratic, Arminian biblicism formed the heart of western Protestantism. Theology was left to the theologians. The evangelical folk culture of the southern plains still bears the mark of this tradition.

All through the West, churchwomen utilized their denominations for a variety of activities. Excluded from the franchise in all states except Wyoming and Utah, many nineteenth-century women contributed to the social order through church work rather than through politics. Since the local churches were constantly short of money, churchwomen spent considerable energy in fund-raising. Naturally, these gatherings focused on the entire community. An 1882 Catholic church fair in Cheyenne lasted six days and featured dancing. Episcopal churches in Laramie, Cheyenne, and El Paso all sponsored formal dances, as did the Mormon church, but most mainline Protestants considered dancing beyond the pale. Instead, they staged raffles, "beauty contests" (twenty-five cents a vote), box lunches, and bake sales. African-American churches usually sponsored an Emancipation Day supper, with such luxuries as oysters, pound cake, and sweet potato pie. For almost forty years, the Episcopal Charity Ball opened Denver's fall social season, with funds going to St. Luke's Hospital.

Finally, the churches served yet another role on the Gilded Age western frontier: they functioned as training schools for political democracy. The numerous church gatherings introduced citizens to basic democratic principles: the conduct of public meetings via accepted rules of order, the need to speak persuasively to the issue at hand, and (usually) the realization that the majority rules. The discussions also reinforced the virtue of listening, freedom of expression, respect for others' views, and the necessity for compromise. Thus the countless church and political meetings of the era overlapped and reinforced each other.

The clerical accomplishments, however, have never penetrated the popular myth of the American West. Whereas virtually everyone recognizes the names of Billy the Kid, Sitting Bull, Annie Oakley, and Wyatt Earp, only specialists recall Rev. Sheldon Jackson, Father Ravalli, or Rabbi William Friedman. Unlike figures who have been mythologized, the western clergy suffer from the restrictions of denominationalism. Because they can be identified only by denomination, they cannot be "universalized." Thus, their particularity has excluded them from the mythological West and diminished their stature in the public mind.

The pioneer phase of western European-American religious history ended with the first years of the twentieth century. In general, it had been a "brick and mortar" era for most denominations. Religious leaders built schools, hospitals, orphanages, and local churches as fast as their finances would allow. They left an impressive architectural legacy. Denver's Trinity Methodist Church still anchors the city's downtown; when Portland's Beth Israel was built, it reflected the eclectic architecture characteristic of most Reform temples of the era. Virtually every visitor to Salt Lake City acknowledged the Mormon Temple as the most impressive building in the intermountain West. Perhaps the crowning glory came with the new Catholic Cathedral in St. Louis, "the gateway to the West." Completed in 1914, it was an architectural statement of the role that Catholicism had played in shaping the religious life of the American West.

The vastness of the American West also provided space for numerous experiments in communal living. After the Civil War, thousands of immigrants sought alternative living arrangements. The majority of these drew from a shared religious (and often ethnic) framework, and the West provided the setting. The Aurora Colony near Portland (1855–83) reflected the German pietistic roots of its mother organization, Bethel, in Missouri. During the economic crisis of the late nineteenth century, both the Salvation Army and Jewish benevolent societies experimented with religious agriculture communities in the Dakotas and Colorado. In 1867, William Davies, an ex-Mormon, achieved local notoriety when he formed the Kingdom of Heaven colony in Walla Walla and announced that his son was Christ reincarnated (the "Walla Walla Jesus"). The Land of Shalam, in southern New Mexico, was based on a spirit-delivered, second American scripture, the *Oahspe Bible*. Less well-known than *The Book of Mormon*, the *Oahspe Bible* combined convoluted verbiage with practical advice; the community lasted from 1884 to 1901.

California and the Puget Sound area provided homes for more successful utopian colonies. Alturia, Llano del Rio, and Icaria Sporanza were founded in California. Perhaps the most colorful was the Point Loma Theosophical Colony, presided over by Katherine Tingley ("The Purple Mother") with magisterial splendor. The Puget Sound region had fewer strictly "religious" communities, but it was the location of at least five

"Brotherhood of Man" socialist utopias, including Equality, Burley, and Home. Kevin Starr has suggested that for many easterners, the state of California itself seemed to be "utopia" writ small.

The numerous German and Russian-German communal settlements on the Great Plains proved the most successful of all. Descendants of the Reformation Anabaptists, these German peasants moved into Russia in the late eighteenth century. A century later, when the czar introduced a policy of forced Russification, he ordered the Hutterites and Mennonites to move or join the Russian mainstream. All of the Hutterites and over a third of the Mennonites chose to migrate, and a significant number found their way to the American Great Plains.

Both groups initially settled in communal arrangements. In 1874, a group of Mennonite Brethren organized Gnadenau (Meadow of Grace) in Marion, Kansas, a classic model of a communal village. But within three years, the leaders discovered that banding together was less necessary in America than in Russia. Although some branches of Mennonites retained the communal emphasis, the majority became individual farmers.

The Hutterites, however, did not meld into the American mainstream. They have retained their communal style of life up to the present day. Arriving from the Ukraine in the mid-1870s, they settled in the northern plains. Hutterite theology, especially the concept of *Gelassenheit*, demanded that the individual self be subordinated to the will of God. The goal of Hutterite education was to replace the individual will with a group will, bending the emotions (rather than the intellect) through indoctrination. Pacifist, communal, and deeply suspicious of the prevailing *weltgeist*, the Hutterites were also progressive farmers. Unlike their Old Order Amish "relations" (who usually failed when they ventured from the East onto the Great Plains), the Hutterites bought the latest farm machinery and welcomed technical improvements. Their centralization and capital-heavy organization, however, soon brought them into conflict with their South Dakota immigrant neighbors, who termed them "a nepotic corporation." Eventually, South Dakota passed laws restricting Hutterite land purchases, and the anti-Communist agitation of the 1950s drove several colonies into Canada.

In addition to the Russian-Germans, the Great Plains provided a home for numerous other religious-ethnic groups. German Lutherans, Danish Lutherans, Volga German Catholics, and the Bohemians who figure so prominently in Willa Cather's novels all settled on the northern plains. The villages founded by these immigrants often had an obvious ethnoreligious composition. As late as 1926, for example, the only non-Bohemian in Prague, Nebraska, was the depot agent.

Duplicating their ancestral homes in Europe, these immigrants reestablished the church at the heart of their communities. Immigrants who landed in the nation's larger cities could create a variety of institutions for mutual support: foreign-language presses, restaurants, mutual benefit societies, and bakeries. On the plains, however, the ethnic church, the easiest institution to create, had to assume all of these roles. Even those who had been indifferent churchgoers in Europe (such as most males) often became active after they crossed the Atlantic. The plains priest or pastor was always a community leader. The ethnic church played a crucial role in life on the Great Plains: it held the community together.

Native Revitalization Movements and the Challenge of Ethnic Diversity

In the 1870s, a group of Paiutes at Walker River, Nevada, who were becoming anxious about whites moving onto their lands began to focus on a ceremonial dance under the leadership of a prophet known as Wodziwob. The prophet told his people that if they danced, they could communicate with the spirits of their ancestors; he forecast that the world would soon end, the whites would be swept aside, and the Paiutes' ancestors would return to join them on a renewed land. As word of Wodziwob's message spread west to the Monos and Yokuts of northern California and east to the Shoshones, Bannocks, and Utes, word also spread of another prophet, whose dance was performed by the native people of the Columbia River Plateau. This news fueled Wodziwob's movement: the 1870 Ghost Dance. The network that had linked so many native groups in pre-Columbian America continued to hold into the late nineteenth century, binding Indians of the Columbia River Plateau to their counterparts in the Great Basin. A decade later, the tie would reach across the Rockies to encompass the Plains tribes.

The ceremonial cycle of the Sahaptian and Salishan people of the Plateau had long included a Prophet Dance. When native dancers (*Washani*) performed this ceremony, they were acting on their belief in world destruction and renewal, a concept deeply rooted in both Plateau and Northwest Coast cultures. However, after mid-century, this ceremony had incorporated Christian influences, such as observation of a Sabbath. From this time forward, Plateau native groups experienced intensive white pressures, all of which led to anxieties similar to those of the Walker River Paiutes.

In the nineteenth century, a number of Plateau prophets rose to meet these crises. The most important of these leaders was a Wanapum (*river people*) named Smohalla, whose band lived along the Columbia River between Priest Rapids and the mouth of the Snake River. A traditionalist, Smohalla appealed to the conservative groups among the Plateau tribes. During the 1870s, at the height of his influence, he may have had as many as two thousand followers, including Palouses, Nez Percés, Bannocks, and Northern Paiutes. Like his predecessors, Smohalla merged some Christian elements, such as the use of bells, into the ceremonies, but he discouraged his young men from farming, claiming that those who farmed did not have time to dream. He cautioned his followers "to adhere strictly to native dress and custom," and he refused to allow the Wanapum to move from their lands to the Yakima Indian Nation reservation. A dreamer himself, Smohalla's religion became known as the Washani faith or the Pom Pom (Dreamer) religion.

West of the Cascades another native revitalization movement began to grow in the early 1880s. Founded by John Slocum, a Squaxin of lower Puget Sound, the Indian Shakers, so-called because of the nervous twitching that accompanied their prayers and songs, combined shamanistic performances with Catholic ritual and Presbyterian doctrine. Possibly motivated by Smohalla, Slocum's group did not retain the antiwhite sentiment of the Dreamers, but it spread throughout native groups in Washington, Oregon, British Columbia, and northern California.

The widespread religious borrowing of Coast, Plateau, and Great Basin that characterized these native revitalization movements found its way through the Rockies and onto the plains in the late 1880s when Wovoka, the son of an apostle of Wodziwob, revived the 1870 Ghost Dance religion. Because of the catastrophic impact of Wovoka's teaching, his reputation has endured. His message, however, was part of a continuum. Like his predecessors, Wovoka preached that a cataclysmic event would banish the whites and return the ancestors of native people, an event that would be hastened by Indians dancing the Ghost Dance, treating each other as brothers and sisters, and following the old ways. The difference between Wovoka and his predecessors lay in the timing and the circumstances. Smohalla's Washani faith spread in an era of intense anxiety for perhaps two thousand Plateau followers, and it influenced Joseph's band of Nez Percés as well as Northern Paiute Dreamers, who joined the Bannocks in the Bannock-Paiute War of 1878. By contrast, Wovoka's Ghost Dance faith spoke to perhaps twenty thousand Plains Indian followers, whose world was shattering all about them. In the two decades between the Battle of the Little Bighorn and the rise of the Ghost Dance, the Lakotas saw their lives reduced from proud Plains warriors to starving, disease-stricken reservation dwellers. At the same time, their relation to the supernatural had come under attack when the Great Father in Washington had issued new rules prohibiting the old ceremonial dances, including the Sun Dance, and forbidding any interference of the medicine men with schooling or Christianizing. With the news of Wovoka's promised millennium, forecast for the spring of 1891, as well as both the Lakota despair and the Great Father's incredible bungling, conditions were in place for the Seventh Cavalry's massacre of a large number of Lakotas on Wounded Knee Creek in December 1890.

Although this event destroyed the momentum of the Ghost Dance, it also led indirectly to the rise of another pan-Indian revitalization movement that had already captured converts across the southern plains. Peyote, a small spineless cactus with hallucinogenic qualities that was used as a sacred medicine in ancient Mexico, was introduced in the late 1870s and 1880s by border groups, especially the Lipan Apaches, to southern Plains tribes confined to the Indian territory. Like the Ghost Dance, the use of peyote spread rapidly through, in part, European-American devices: the federal boarding school, which inadvertently spurred the growth of pan-Indianism and provided a lingua franca for students in its enforced English-language rule; and the railroads, which eased travel for Indians and aided in transporting hundreds of peyote buttons from the lower Rio Grande to the Indian Territory and the northern plains.

Peyote offered a very different solution from that of the prophet movements. Whereas the Ghost Dance had anticipated a millennium and a renewal, the peyote religion taught the individual Indian how to deal with problems of life here and now. Consequently, it had a wide appeal that ranged from returned boarding school students to Indians with no formal schooling. In the Indian Territory, young, schooled Indians, who found in it a version of Indian Christianity, often became its leaders. As it spread north, some peyotists adopted Big Moon or Cross Fire ceremonies, which were introduced in the Caddo-Oto rituals and included strong Christian elements. Others established the Half-Moon ceremony, adapted from the Kiowas and Comanches and known at the time as the "Quanah Parker Way." This ceremony emphasized the Great Spirit and Mother Earth and included the use of tobacco. As the anthropologist Omer C. Stewart has pointed out, despite these differences, both forms opposed the use of alcohol, and both retained "the ancient persistent belief in the supernatural power of the peyote plant."

The Ghost Dance revival of 1890 swept across the plains, finding wide support among tribes struggling to cope with the declining numbers of buffalo, the encroaching white settlement, and the increasing restrictiveness of federal Indian policy. This drawing by an Indian observer depicts Cheyenne and Arapaho participants in a Ghost Dance of the northern plains.

Yellow Nose (Ute captive living with the Cheyennes). Ghost Dance Painting. *Colored inks on buckskin, 1891. Reproduced from* Fourteenth Annual Report of the Bureau of Ethnology, 1892–1893 *(1896).*

By the 1920s, the movement had swept into the Great Lakes and Midwest, as well as Canada. Some tribes accepted it quickly; some rejected it completely; and some were bitterly divided over it. At the same time, peyotists encountered vehement hostility and legal action by many non-Indians, including European-American churches, the Bureau of Indian Affairs (BIA), and the U.S. Congress. The feud over peyote in the early twentieth century is a story in itself. In 1918, after a narrow victory over a congressional bill designed to make peyote illegal, a group of Oklahoma peyotists incorporated the Native American Church to bring their belief system under the First Amendment to the Constitution. However, their action did not resolve all legal problems for the peyote religion.

As Native Americans struggled with the numerous challenges to their faiths, settlers from Europe and elsewhere were meeting other religious difficulties. Within the Judeo-Christian groups, the Catholic hierarchy faced a dilemma unique among the churches: a multiethnic membership. Although the Mass was conducted in Latin, the European immigrants naturally sought a priest who spoke their language. Every bishop wrestled with this question. In Texas, Bishop Odin labored for over two decades to provide ethnic priests for his Belgian, Irish, German, Polish, Swiss, Czech, Alsatian, and Mexican-American enclaves. When the bishop of St. Paul finally sent a Bohemian priest to Tabor, Nebraska, in 1877, the Bohemian congregation openly rejoiced. In urban areas such as Denver, bishops confronted potential "secession" as French congregations refused to be "ruled" by Irish priests, or vice versa.

From the bishops' point of view, the Hispanic Catholics formed perhaps the most perplexing of all the ethnic groups. Recent European immigrants generally shared the same assumptions about the faith. But generations of isolation and poverty had forced

Hispanic Catholics to evolve their own religious folk culture. The first bishops of California, Joseph Alemany and Thaddeus Amat, confronted this dilemma. Amat attacked the Mexican practice of selling burial shrouds and actually suspended some Mexican Franciscans for what he termed "fomenting superstition."

This clash between European-American Catholicism and Hispanic folk Catholicism was highlighted in the territory of New Mexico, where the key player in the drama was Archbishop Jean-Baptiste Lamy. The novelist Willa Cather has immortalized Lamy's efforts to bring Hispanic Catholicism into the Catholic mainstream. But her romanticized portrait of "Bishop Latour" in *Death Comes for the Archbishop* (1927) glosses over Lamy's scorn for Hispanic folk art (which he considered primitive), as well as his attempt to replace it with French imagery. Lamy also had negligible appreciation for the strength of the Penitente lay order (which he tried to suppress) or for the healing shrine at Chimayo. Overall, it is a marvel that the Catholic church survived without ethnic schism.

From their stronghold in the Great Basin, the Mormons confronted a different set of challenges, most related to continuing Gentile hostility. A generation after the Republican Platform of 1856 referred to the "twin relics of barbarism" (slavery and polygamy), Congress finally decided to move against the Saints' "peculiar institution." In 1882, Congress passed the Edmunds Act and five years later the Edmunds-Tucker Act, which forced many Mormon polygamous leaders into jail or exile. In 1890, the Saints agreed to give up the practice of polygamy, a decision that led directly to Utah statehood in 1896. But the "Mormon Question," as it was called, remained a steady theme in federal, denominational, and neighboring state politics until World War II.

Hoping to wean Mormon youth from the faith, several denominations, especially the Presbyterians and Congregationalists, opened a vast network of parochial schools in Utah during the latter decades of the nineteenth century. "To educate the children and youth is to emancipate them," wrote a Congregational woman missionary in 1876. "They can be drawn into the school when the church would fail to reach them." These hopes were never realized. The Protestant parochial schools, located in the heart of Mormon country, provided needed social services, but they garnered relatively few converts. From 1870 to 1930, the Gentile population of Utah grew only in the same proportion as Gentile immigration. Nonetheless, these parochial schools did introduce national holidays and, consequently, helped "Americanize" the Saints' young people. In these decades, the Mormons were a people in transition. The maturing of state politics, the end of polygamy, and the influence of Protestant schooling all helped the Saints move closer to the American mainstream.

The Ferment at the Turn of the Century

In 1871, the Philadelphia Presbyterian Herrick Johnson preached a sermon, "The American City: What Shall We Do with It?" Johnson's plea was the opening shot of a national "education and evangelism" campaign that grew rapidly over the next half century. During these years most of the mainline American churches focused on the problems of urban life: immigration, poverty, urbanization, social services, and political corruption. Their response became known as the "social gospel."

Every western region witnessed a social gospel program. Ministers erected "institutional churches" that provided numerous social services and remained open seven days a week. Episcopal priests led city mission work in Omaha, a meat-packing city whose 1919 population was over half immigrant. Turn-of-the-century clergy voiced their opinions on a wide variety of social issues. In the Southwest, a number of churches teamed with physicians to establish hospitals or sanitoria for people with tuberculosis. These institutional combinations of medicine and faith were widely praised at the time. Such actions for the public welfare formed the "left wing" of the social gospel movement.

But even conservative clergymen, the "right wing," devoted time to reform programs during the fin de siècle years. A Seattle Presbyterian evangelist, Mark A. Matthews, spent over a decade denouncing corrupt mayors, police chiefs, and bootleggers. In nearby Tacoma, Mrs. Birgitte Funnemark and her daughter Christine established a nondenominational Evangelical Seamen's Rest as an alternative to the fleshpots of the region. From 1897 to 1903, they simultaneously ministered to sailors and tried to clean up the worst aspects—including shanghaiing—of Tacoma life. In rural Oklahoma, the Anti-Saloon League billed itself as "the church in action." In all western cities, the Salvation Army and the Volunteers of America reached out to an impoverished segment of society that eluded most other religious groups.

During this period, many western clerics revealed a social concern that extended far beyond their own denominations. They campaigned for clean government and civic responsibility. In Denver, for example, the Catholic priest William O'Ryan led the movement to coordinate the city charities (both Catholic and non-Catholic) into an umbrella Charity Organization, which eventually grew into the Community Chest. The Portland First Congregational Church relief fund eventually became the City Relief Fund. The entire city had become the "parish" for these social gospel clerics.

A second distinct regional manifestation of social gospel concerns was Chinese mission work. The nation's clergy had expressed interest in the West Coast Chinese beginning in the 1850s, when there were perhaps twenty-five thousand Chinese living in California. The clergy argued that the presence of so many Chinese in California was obviously part of God's grand design for Asia. They hoped to convert the Chinese miners, who would return and spread the Christian message throughout China. From the clerics' point of view, the conversion of the Far East lay in the offing. These events never materialized, but their absence did not diminish church support for a wide range of Chinese missions.

Another dimension of the social gospel was the work of eastern Christian reformers who led the fight for the assimilation of western Native Americans. Throughout the late nineteenth century these reformers, who included many ministers and lay church members, had the ear of the Great Father. Their efforts only increased after the demise of President Ulysses Grant's Peace Policy, which had sought to establish church influence over federal Indian programs by assigning specific denominations to each Indian reservation. Christian reformers soon tasted victory with the passage of the disastrous Dawes Act, or General Allotment Act, of 1887, which began to break up the reservations and attempted to end tribal ownership of lands. Two years later the Christian reformers gained a sympathetic commissioner of Indian affairs in Thomas

Jefferson Morgan, a former Baptist minister and educator, who strengthened the 1883 regulations enforcing Indian assimilation. Working directly with the BIA and Congress, the Christian reformers exerted control on Indian religions throughout this era. However, Native Americans demonstrated their resilience. They practiced their ceremonies in secret, and they embraced revitalization movements that brought hope to the despairing.

During this same era, the Southwest began to rediscover its Roman Catholic mission heritage, largely through the efforts of a former Massachusetts Yankee, Charles Fletcher Lummis. The Protestant Lummis spent a lifetime trying to convince California's recently arrived midwestern Protestants that Spanish missions and iconography should become central to their new self-image. He likened the Franciscan missions of California to the Puritan churches of Massachusetts, arguing that one need not be Roman Catholic to appreciate their symbolic power. "The old missions," he wrote in 1918, "are worth more money, are a greater asset to Southern California than our oil, our oranges, or even our climate." Simultaneously, the Catholic bishop of Tucson began to restore nearby Mission San Xavier del Bac, an effort that prevented San Xavier from joining the nearby presidio at Tubac and mission at Tumacácori as crumbling adobe ruins.

Thus, the Southwest began to forge a romantic, mildly nondenominational saga of Catholic missions and missionaries. The fledgling tourist industry soon discovered that "mission tours" were highly profitable. At the center of this revival stood Father Junipero Serra, an eighteenth-century figure whose reputation continued to grow: the state of California eventually placed his statue in the rotunda of the Capitol in Washington, D.C., and the Catholic church inaugurated his canonization process. During the late 1980s, however, several California Native American groups vigorously opposed Serra's proposed canonization on the grounds that the Franciscan missions had enslaved their ancestors. In reply, Catholic defenders argued that Serra should be judged as a man of his times. This controversy, still under way, shows how difficult it is for any religious figure to achieve transdenominational acclaim.

The fin de siècle years also witnessed a bewildering variety of new religious currents. The highly publicized World's Parliament of Religions of 1893 in Chicago focused the nation's attention on the wide range of new faiths, such as the Baha'i movement, Theosophy, the Vendanta Society, and Christian Science. Two of these movements forged deep roots in the American West: New Thought and Pentecostalism.

New Thought, a complex system of metaphysics, maintains that one may control both the physical and the mental circumstances of life by consciously cultivating a "positive" attitude toward one's surroundings. An offshoot of Mary Baker Eddy's Christian Science, New Thought teachers stressed that one should listen to the "voice of the indwelling Presence which is our source of Inspiration, Power, Health and Prosperity." New Thought advocates also insisted that practical results flowed from holding these views: "health" and "success." Both proved capable of endless variations.

The three foremost New Thought organizations—Divine Science, Unity, and Religious Science—were all based in the West. In the late 1880s, the three Brooks sisters of Pueblo, Colorado, began Divine Science in Denver. Nona L. Brooks served as pastor of the church for thirty-one years and shepherded it into the forefront of Denver's faiths.

The Unity movement of Kansas City proved even more successful. Founded by Myrtle and Charles Fillmore, the Unity School of Practical Christianity is now recognized as the most successful of the New Thought groups. California proved especially congenial to metaphysical thinking, and by 1907 boasted more New Thought centers than any other state. The leading California group was Ernest Holmes's United Church of Religious Science, headquartered in Los Angeles. Holmes's *Science of Mind* textbook (1926) has become a minor classic in the field, and his faith gained fame as "the religion of the Hollywood stars." Holmes maintained that the realization that individuals could consciously direct and control "the law of creative force" for their own purposes was "the greatest discovery of all time."

The flexibility of western faiths and the flexibility of New Thought proved a good match. The movement's significance lay less with its actual membership—never large by any count—than with its endless stream of "positive thinking" publications and its eventual incorporation into mainline Protestantism, Catholicism, and Judaism. By the 1990s perhaps the foremost western proponent was Rev. Robert Schuller. With his striking Crystal Cathedral located in Garden Grove, California, his popular television series, and a myriad of best-selling books, Schuller shepherded New Thought from its roots in the West to a prominent role in contemporary American culture.

Few people connect New Thought with the Pentecostal-Holiness movement, but the two have a good deal in common. Emerging simultaneously, they offered spiritual healing to many Americans who, for various reasons, were affected by the professionalization of medicine. Although both these movements fragmented into numerous small groups, their impact reached well beyond their numbers.

The turn-of-the-century Pentecostal movement had tangled roots in both the West and the South and in two interrelated beliefs. One belief emphasized "Christian perfection" or "entire sanctification," a second grace that cleansed the believer from the tendency to sin. The second belief derived from events in the book of Acts and emphasized "speaking in tongues," or glossolalia.

Although historians of Pentecostalism have uncovered scattered references to glossolalia during the nineteenth century, they agree that the 1906 Azusa Street (Los Angeles) revival began modern Pentecostalism. Led in part by William J. Seymour, a one-eyed black minister from the South, the rise of Pentecostalism is a little-known black contribution to white religious life. In the early days, most Pentecostal churches were integrated, but by the 1920s they had separated into primarily white or black congregations.

From Azusa Street, the Pentecostal "full Gospel" revival spread up and down the West Coast and into the rural areas of Oklahoma, Texas, and Missouri. There the Pentecostals also created yet another religiosocial subculture. Their world emphasized spiritual healing, religious ecstasy, glossolalia, and general renewal. The faith demanded a strict personal morality (no cards, jewelry, cosmetics, or bodily ornamentation and minimal amusements). Their musical imagery, which would later influence early rock and roll (for example, the song "Great Balls of Fire"), called for a high degree of participation and emotional release. For many, the profession of the ministry proved a popular road to success. The democracy of the message was obvious. As one minister

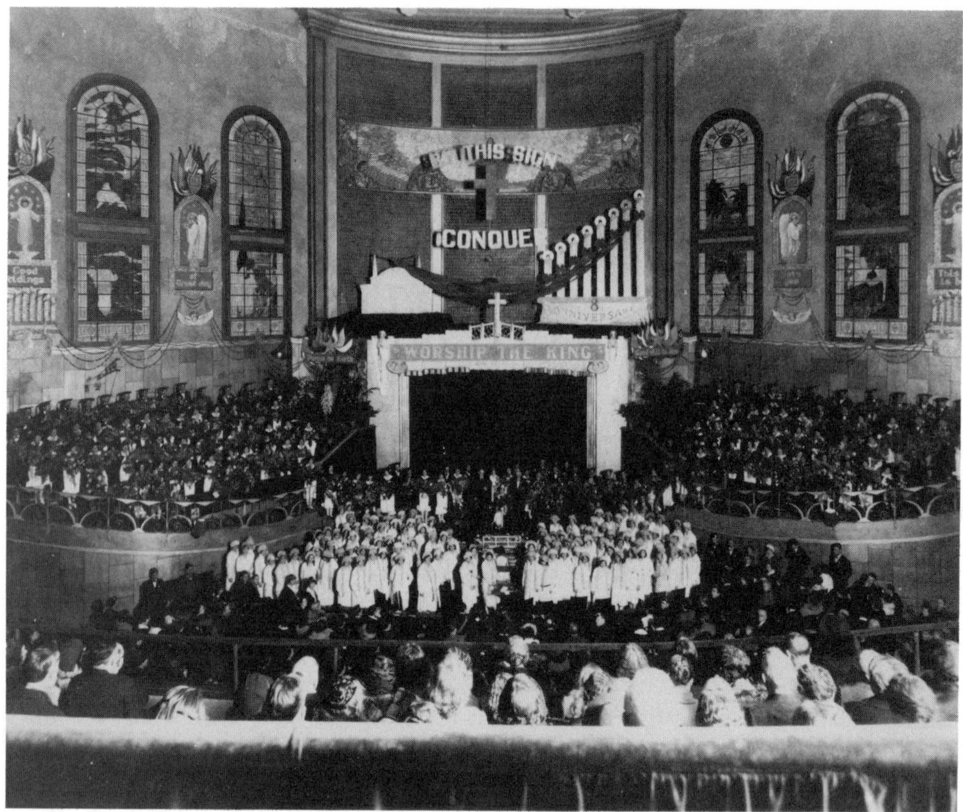

With personal dynamism and extravagant staging borrowed from Hollywood, the evangelist Aimee Semple McPherson used her Four Square Gospel Church as a pulpit from which to build a national following for her Pentecostal faith.

H. Harold Fisher. Temple Angelus, Graduation Exercises. *Gelatin silver print, 1930. Department of Special Collections, University Research Library, University of California, Los Angeles.*

stated: "We did not honor men for their advantage, in means or education, but rather for their God-given gifts." In this sense, the Pentecostals reached out to the religious needs of the common people. Not surprisingly, they manifested great strength in old Populist or socialist areas of the West.

The most effective publicist of the Pentecostal movement was Aimee Semple McPherson. Reared in a Salvation Army family, Aimee arrived in Los Angeles in 1918 to establish the Four Square Gospel Church, revealed to her in a vision. A strikingly beautiful woman, she utilized the Hollywood atmosphere to turn her worship services into media productions. "Sister Aimee" also established a religious radio station (KFSG, Kall Four Square Gospel) to spread her message. From the mid-1920s to the mid-1930s, she appeared on the front page of the *Los Angeles Times* approximately three times a week.

McPherson's most notorious escapade occurred in 1926 when she disappeared, probably for a brief romantic tryst with her radio station operator. On her return, however, she claimed to have been kidnapped by two outlaws, "Jake" and "Mexicali Rose." The endless publicity from this event dramatically increased the size of her Sunday audiences and gave her a nationwide reputation.

Beneath the hype and extravagance, McPherson emerged as America's first "superstar" media evangelist. She provided a national platform for the Pentecostal message, one that it would not regain until the 1970s and 1980s. By that time, the charismatic dimension of Pentecostalism had spread into Roman Catholicism, the Episcopal church, and a number of Native American communities and among many televangelists. By the 1990s, Pentecostalism was growing most rapidly in Latin America and Africa.

The Twentieth Century: Pluralism Expands

A revival of religious conservatism occurred in the decades between the two World Wars. The rapid growth of the Ku Klux Klan, which claimed a tenuous link with right-wing Protestantism, provided the most extreme example of this religiosocial backlash. Another form of repression came in the attacks on Native American faiths. During the 1920s, the fundamentalist-modernist controversy split the mainline Protestant churches into two warring camps. All of these national movements affected the trans-Mississippi West.

The Klan proved exceptionally strong in several western states, especially Colorado, Texas, and Oregon. Their anti-immigrant and anti-Catholic message drove some Catholics out of Oregon and Colorado and soured Protestant-Catholic relations in El Paso for a decade. In Oregon, militant nativists introduced legislation in 1922 that would have required all children to attend public school, ostensibly for reasons of "Americanism." The real goal was to destroy the Catholic parochial school system. Catholic resistance found ready allies from the Lutherans, Seventh-Day Adventists, and the American Jewish community, as well as several liberal Episcopal and Presbyterian clergymen. In 1925, the law was overturned by the Supreme Court.

Religious suppression also found its way into the long-festering question of the First Amendment and Indian religious liberties. Through the 1920s and 1930s, Indian religious freedom was a major issue for Native Americans. In the 1920s, the debate pitted Commissioner of Indian Affairs Charles H. Burke against the reformer John Collier. In 1921, Burke issued a circular that reinforced the directive of 1883, prohibiting ceremonial dances and "celebrations" that included actions deemed improper and even harmful. As the historian Francis Paul Prucha has pointed out, this attack infuriated Collier and inspired his "campaign in support of religious liberty for Indians." The ensuing national debate climaxed in 1926 when Collier and his followers defeated a congressional bill that would have legalized Burke's position.

But the fight was not over. In the 1930s, when President Franklin Delano Roosevelt appointed Collier as commissioner of Indian affairs, Collier viewed Indian religious freedom as a cornerstone of his blueprint on Indian policy. Like Burke, he issued a directive. The 1934 circular, entitled "Indian Religious Freedom and Indian Culture," declared that Indians be granted "the fullest constitutional liberty, in all matters affecting religion, conscience, and culture," and that "no interference with Indian religious life or ceremonial expression will hereafter be tolerated." Reversing the centuries-old approach, Collier declared, "The cultural liberty of Indians is in all respects to be considered equal to that of any non-Indian group." At BIA schools, Collier prohibited compulsory attendance at religious services and permitted students to return home for ceremonies. Christian reformers and Christianized Indians saw Collier's circular as a step backward. Some tribes opposed the concept of religious freedom as a matter of principle: it would violate tribal sovereignty by interfering in internal tribal affairs. If the majority of a tribe, such as the Lakotas, opposed the Native American Church, for example, the tribe did not want Washington ordering it to legalize peyote. The issue of religious freedom for Indians was not resolved in the 1930s, but Collier's stand did begin to bring Indian religions under the constitutional guarantees granted to other citizens.

The fundamentalist-modernist controversy, which so disrupted the nation's Protestant churches, also had a strong western component. Two transplanted Pennsylvanians, Lyman and Milton Stewart, used the profits from their Los Angeles–based Union Oil Company to support a series of conservative evangelical causes from the 1890s forward. During the Progressive Era, Lyman Stewart began to attack theological liberals, especially Presbyterian Thomas F. Day, who was eventually dismissed from the San Francisco Theological Seminary in 1912 for teaching Higher Criticism. In 1907, Stewart helped found what would become the Bible Institute of Los Angeles. He also financed the publication of William E. Blackstone's millennial tract *Jesus Is Coming*, which became the most widespread premillennial piece of literature in the world. Finally, Stewart funded the publication of *The Fundamentals* (1912–16), a series of conservative pamphlets that are usually acknowledged as the opening shots of the fundamentalist-modernist controversy. The Stewart brothers helped inaugurate what became the most disruptive twentieth-century controversy among American Protestants. Over the years, the nation's Protestant churches began to divide along theological lines (liberal-conservative) rather than denominational ones. The sociologists Robert Wuthnow and James Davison Hunter have argued that this ever-widening liberal-conservative split lies at the heart of the post–World War II "restructuring of American religion."

After World War II, with the flood of people moving to the Coast and Sunbelt and the meteoric rise of electronic media, western religious culture moved into the mainstream. The evangelist Billy Graham personified this movement. A Wheaton College (Illinois) graduate, William Franklin Graham toured the revival circuits from 1943 to 1948 as merely one of many conservative evangelicals. In late 1949, when the evangelist was in Los Angeles, the publisher William Randolph Hearst allegedly told his editors to "puff Graham." The unexpected publicity brought the young revivalist and his message to the attention of the nation, propelling him into a position of prominence. Graham eventually became the foremost American cleric of the twentieth century. Although Graham could never be called a "western" figure, his national prominence began among the rootless citizens of Los Angeles. Graham's sudden rise personified the religious revival of the 1950s, which sent Americans back to their churches and synagogues in record numbers.

The religiosity of the Eisenhower years furthered the "mainstreaming" of the country's Catholics and Jews. In 1955, the sociologist Will Herberg published *Protestant-Catholic-Jew*, in which he argued that the great historic faiths formed three equally valid routes to becoming "American." Religious life in the West, however, anticipated Herberg's conclusions. By the 1950s, the western Catholic universities of Santa Clara, St. Mary's, St. Martin's, Seattle University, Gonzaga, and the University of San Francisco had long-established strong regional reputations, in both academics and athletics. The Catholic parochial school systems of San Francisco and Denver educated

Important contributors to the economic and political growth of West Coast cities, Jews in San Francisco, Los Angeles, and Portland composed solidly middle-class communities by the late 19th century.

Unidentified photographer. Children's Party at Temple Beth Israel. *Photograph, 1898. Oregon Historical Society (#OrHi 25946), Portland.*

perhaps one-third of the cities' children. Since the early 1900s, Santa Fe had relied on the romance of Hispanic folk Catholicism (*luminarias, faralitos*, the burning of Zozobra) to entice eastern tourists to its hotels and shops. The California "mission tours" struck the same chord. The "ghetto mentality" that had forced eastern Catholicism to remain on the cultural defensive for over a century was never re-created in the trans-Mississippi West.

The same proved true for western Judaism. The saga of the Jewish forty-niners and the famed success of the great Gilded Age merchandisers, such as I. and J. Magnin, Meyer and Frank, and the Goldwaters, had long been an integral part of the legend of the West. With the larger-than-life stories of twentieth-century film magnates Adolph Zukor, Jesse Lasky, Samuel Goldwyn, Carl Laemmle, Louis B. Mayer, and the Warner Brothers—"the ethnic Horatio Algers who built Hollywood," as one historian phrased it—the legend evolved into myth itself.

The Jewish communities on the urban West Coast had long been intimately involved in the growth of their cities. In 1880, Los Angeles Jews dominated the mercantile world in such fields as dry goods, clothing, and book selling. By any classification, they were middle class. For years, the San Francisco Jews worked closely with the dominant Republican party, contributing politicians who ranged from the infamous city boss Abe Ruef to long-term U.S. Congressman Julius Kahn. From 1930

Reflecting changing urban demographics and the continued vitality of western religious organizations, this former synagogue in East Los Angeles has been adapted for use as an evangelical Christian church.

Tom Vinetz (b. 1945). Church in East Los Angeles. *Gelatin silver print, 1980. Bruce Henstell Collection.*

to 1960, San Francisco's major boards and commissions were about 30 percent Jewish, with certain positions traditionally reserved as "Jewish seats." The same tale, with variations, applied to Portland, Seattle, and Denver. Idaho, Utah, New Mexico, and Oregon voters all elected Jewish governors, long before New York or Illinois, with their much larger Jewish populations. In 1913, when San Francisco's mayor, Jim Rolph, extended to Rosh Hashanah and Yom Kippur the same public recognition that attended Good Friday, he was the first big-city mayor to do so. By World War II, impressive temples dotted Los Angeles, Portland, San Francisco, Seattle, Boise, El Paso, Albuquerque, Denver, Phoenix, and numerous other western cities. Thus, the saga of western religious life anticipated Herberg's ideas by at least two generations.

Although it took yet another generation, the "Americanization" process eventually expanded to include the Latter-day Saints. After the Vietnam War and the disruptions of the 1960s and early 1970s that challenged traditional values, the Saints' insistence on conventional morality and family virtues appeared more and more "mainstream." The spread of Mormonism beyond Utah, the opening of the priesthood to African Americans, plus the Saints' success in the business world confirmed this impression. In the early 1980s, the Mormon hierarchy added the phrase "another testimony to the Gospel of Jesus Christ" to the front of *The Book of Mormon*. A decade later, the church quietly dropped the denunciation of non-Mormon clergy from secret temple rituals. Both were indications of the mainstreaming under way. Mormon missionaries continue to stress that their church is the only way to salvation, but if one judges Mormons by lifestyle and values, rather than theology, the modern Latter-day Saints might well be on their way to becoming simply another American "denomination."

The historian Sydney E. Ahlstrom has argued that the 1960s formed the most tumultuous single decade in American religious history. The Vatican II reforms liberated American Catholicism on many levels but also undermined the church's historic unity of purpose. From the mid-1960s forward, prominent lay and clerical figures publicly disagreed with the official church position on a variety of issues, including abortion, birth control, clerical celibacy, and nuclear disarmament. As the historian Garry Wills has noted, just as Catholicism was accepted in American life, it proceeded to commit hara-kiri.

The mainline Protestants fared no better. From 1960 to 1990, they lost an alarming number of members. In the tumult of the late twentieth century, these aging, liberal denominations often found it difficult to establish a meaningful theology or to retain the loyalty of their youth.

Although they found their influence waning, the mainline churches continued to provide leadership in many western communities. In the urban West, the Catholic parochial schools have proven surprisingly successful in educating children from underprivileged backgrounds. Other mainline churches have inaugurated a variety of modern social gospel programs that are reminiscent of the late nineteenth century: day-care centers, meals for the homeless, hospices, and homes for senior citizens. In polyglot Los Angeles, a downtown Methodist church advertises services in Korean, Vietnamese, Cambodian, and English. These creative adaptations illustrate the continued vitality of the mainstream groups.

From the 1960s, it again became fashionable to search for spiritual values. The quest reached its apogee on the West Coast, especially in California. For a century, California's tolerant, open atmosphere had encouraged a smorgasbord of religious organizations, but the cultural ferment dramatically broadened the offerings. Mainline Protestantism, Catholicism, and Judaism; conservative evangelicalism; a variety of "Jesus People"; saffron-robed Hare Krishnas seeking "Krishna consciousness"; L. Ron Hubbard's Scientology, with its promise to adherents that they could be psychologically "clear"; Rev. Sun Myung Moon's Unification Church, with its emphasis on a "new family"; Black Muslims, with their militant, separatist racial call; the storefront churches, with a bewildering panorama of theological and social messages—all were available in an afternoon's walk in San Francisco, San Diego, or Los Angeles.

In the 1980s, the "unchurched" Pacific Northwest became home for a variety of extreme movements: sexual enlightenment on the Big Muddy Ranch in eastern Oregon, led by the Rajneesh (until he fled to India after pleading guilty to two federal felonies); the Church Universal and Triumphant near Livingston, Montana, whose armed followers went underground to avert the predicted end of the world; and a militant, racist group of Aryan Christians based at Hayden Lake, Idaho. Similar extreme millennial views were expressed by the Branch Davidians of Waco, Texas, whose violent encounter with federal agents ended in the deaths of most members of the group in April 1993.

Amid the bewildering cacophony of voices during the late twentieth century, three larger, nationwide themes emerged: the resurgence of conservative evangelical Christianity, bolstered by its use of television; the popularization of New Age ideas; and the increased visibility of Native American religious concerns. All had strong western components.

Before the 1960s, conservative evangelicals had been somewhat peripheral to the religious mainstream. But the collapse of traditional values in the 1960s called them forth in considerable numbers, bearing their old message of strict biblicism and a call for individual repentance and conversion. Demos Shakarian, a successful California dairy farmer, had created the Full Gospel Businessmen's Fellowship in 1951; its Pentecostal message soon found considerable support within the middle-class business community. The popular Fuller Theological Seminary in Pasadena, which carefully hewed to a moderate biblicism, graduated hundreds of conservative pastors and church workers. Many of these groups began to flex their political muscle.

The conservatives' aggressive use of electronic media, especially television, fueled their growth. Pioneers in the use of radio ministry, in the 1950s and 1960s fundamentalists moved to television, bringing a combined message of folksy, old-fashioned preaching, elaborate staging, and effective fund-raising. Although television audiences are national ones, the westerners Oral Roberts (Tulsa) and Robert Schuller (Garden Grove, California) commanded some of the largest followings.

The New Age movement also attracted growing attention during the 1970s and 1980s. A combination of Eastern mysticism and eclectic folklore, New Age religion celebrates the private individual; it offers the opportunity to "create your own reality." A significant number of New Agers were drawn to Arizona, California, and Washington. The popular West Coast writer J. Z. Knight claimed that she had received messages from a mystic being, "Ramtha," in a fashion similar to the nineteenth-century Theosophist Helena Blavatsky's *Isis Unveiled.* The actress Shirley MacLaine's books, plus various celebrations of "harmonic convergences" at the Black Hills of South Dakota, the San Juan Islands of Washington State, and elsewhere, have given the movement wide publicity. Crystal shops and elixirs, with claims of energy and healing powers, have brought these ideas home to millions. Most modern bookstores have a "New Age" (formerly "Occult") section, and as usual, southern California serves as a welcome home for these questers. By 1990, however, Native American groups began to protest that the New Age movement had stolen traditional native teachings and were using them for financial gain.

Native American belief systems have remained resilient in recent decades, but theirs remains an untold story. Both revitalization movements and traditional ceremonies have retained followers. In the Coast and Plateau regions, Shaker churches and Washat Longhouses are a strong presence in native communities. Throughout Indian country, the Native American Church is ubiquitous. In the trans-Mississippi West, there may be as many as two hundred thousand peyotists. Omer C. Stewart has argued, "Except for the Indian powwow, [peyote] is the most pan-Indian institution in America." In addition, traditional systems of belief persist and continue to provide a focus of ethnic identity. Among the Navajos, many faiths compete. But in times of crisis, Navajos who are Mormons or members of the new Protestant groups may seek help from the traditional ways, incorporating healing ceremonials, such as the Blessing Way. Faced with an increasingly complex world, Navajos view their ancient Diné religion as one symbol of Navajo nationalism.

Above all, however, native belief systems continue to be affected by Christianity. The result is a syncretic blend of faiths: traditional, revitalization, and Christian. When a friend of the anthropologist John C. Ewers paid his last respects to an elderly Blackfeet, the friend noted, "There were lighted candles at his head, a Methodist Bible in his hands, and his weather-worn old medicine bundle at his feet." As the anthropologists Ray J. DeMallie and Douglas R. Parks have written of the Lakotas: "Many of the leaders of traditional ceremonies belong to the Roman Catholic or Episcopal churches. They see no conflict between traditional beliefs and ceremonies and those of Christianity."

Before Vatican II and the upheavals of the 1960s and 1970s, mainstream Christianity had been uncomfortable about accepting the validity of native faiths. Growing religious liberalism, however, combined with increased concern over the deteriorating environment, has brought renewed interest in Native American beliefs. As the Lakota author Vine Deloria, Jr., has concluded, "Many people are seeking answers in American Indian religions, which must involve some form of reconciliation with the American Indian and his lands." Since most of the federal reservations and the largest Indian populations are in the West, this has special meaning for the region.

On Thanksgiving 1987, a group of prominent church leaders in Seattle (Catholic, Lutheran, American Baptist, Disciples, Presbyterian, Methodist, and United Church of Christ) issued an astounding declaration: "a formal apology on behalf of our churches for their longstanding participation in the destruction of traditional Native American spiritual practices." In addition, these liberal churches (no conservative evangelical group participated) pledged to support the 1978 American Indian Religious Freedom Act by helping the native peoples protect their sacred sites. This federal law carried Collier's directive of the 1930s one step further by granting to the American Indians, Eskimos, Aleuts, and Native Hawaiians the right "to believe, express, and exercise traditional religions . . . including access to sites, use and possession of sacred objects and freedom to worship through ceremonials and traditional rites." Other actions included the 1973 return of Blue Lake to Taos Pueblo, the 1975 return of Mount Adams to the Yakimas, and the 1988 halting of the proposed development of Madrona Point on Washington's Orcas Island (sacred to the Lummis). All these events acknowledged the validity of native faiths. As "The Bishops' Apology" concluded, "May God of Abraham and Sarah, and the Spirit who lives in both the Cedar and Salmon People, be honored and celebrated."

What makes the story of western religions distinctive? This is not easy to answer. Clearly, the West never duplicated the theological or literary thrust of New England Puritanism; nor did it produce the umbrella of evangelism that often united the diverse peoples of the American South. Instead, western religion was molded by both the historical moment of settlement and the vast spaces of the western landscape.

The West manifested a bewildering number of faiths: the Native American belief systems; the Hispanic Catholics of the Borderlands; the Latter-day Saints of the Great Basin; German and East European Jews; the faiths of Asia; the Franciscans of California; the Jesuits of the Rocky Mountains; the Russian-German Hutterites and Mennonites, the Greek Orthodox, and other ethnic churches; the Pentecostals of Oklahoma and environs; the Missouri Synod Lutherans of the Middle Border; the evangelical subculture of Texas and the southern Great Plains; and the ubiquitous Congregationalists, Episcopalians, Presbyterians, Disciples of Christ, Methodists, Lutherans, and Baptists. In many areas, these faiths were set apart by both the configurations of the land and the timing of settlement, leading to the creation of religious folk cultures that retain their vitality today.

Thus, the religious history of the American West has flowed in a myriad of parallel currents. Although specific churches may have shaped subregional cultures, no denomination has ever achieved hegemony over the entire West. The West has not produced a "western religious establishment," voluntary or otherwise. In the future, this pluralism, accompanied by the absence of any "culture-shaping" denominational mainstream, will likely characterize religious life across the nation. The pattern of faith established in the American West, therefore, has clearly pointed the way.

Bibliographic Note

The most extensive overview of American religious history is Sydney E. Ahlstrom, *A Religious History of the American People* (New Haven, 1972), although he does not focus specifically on the West. Several recent studies have tried to fill this gap: Ferenc M. Szasz, ed., *Religion in the West* (Manhattan, Kans., 1984); Carl Guarneri and David Alvarez, eds., *Religion and Society in the American West: Historical Essays* (Lanham, Md., 1987); Jay P. Dolan, ed., *The American Catholic Parish: A History from 1850 to the Present*, vol. 2 (New York, 1987); and Ferenc M. Szasz, *The Protestant Clergy in the Great Plains and Mountain West, 1865–1915* (Albuquerque, 1988).

For the Native Americans, see Henry Warner Bowden, *American Indians and Christian Missions: Studies in Cultural Conflict* (Chicago, 1981), and Robert F. Berkhofer, Jr., *Salvation and the Savage: An Analysis of Protestant Missions and American Indian Response, 1787–1862* (1965; reprint, New York, 1972), as introductory overviews. A solid case study is Clyde A. Milner II, *With Good Intentions: Quaker Work among the Pawnees, Otos, and Omahas in the 1870s* (Lincoln, 1982). The late-nineteenth-century reform movement's impact on religion is treated in Francis Paul Prucha, *American Indian Policy in Crisis: Christian Reformers and the Indians, 1865–1900* (Norman, 1976). The revitalization movements are dealt with in a number of works. Recent studies include Robert H. Ruby and John A. Brown, *Dreamer-Prophets of the Columbia River Plateau: Smohalla and Skolaskin* (Norman, 1989). Also consult Click Relander, *Drummers and Dreamers* (Seattle, 1986). On the Ghost Dance, see the classic account by James Mooney, *The Ghost-Dance Religion and the Sioux Outbreak of 1890* (1892–93; reprint, Chicago, 1965). On peyote, consult Omer C. Stewart's works, especially *Peyote Religion: A History* (Norman, 1987). Studies that develop the concept of syncretism include Raymond J. DeMallie and Douglas R. Parks, eds., *Sioux Indian Religion: Tradition and Innovation* (Norman, 1987), Alfonso Ortiz, *The Tewa World: Space, Time, Being, and Becoming in a Pueblo Society* (Chicago, 1969), and David Aberle, "The Future of Navajo Religion," in David M. Brugge and Charlotte J. Frisbie, eds., *Navajo Religion and Culture: Selected Views* (Santa Fe, 1982), 219-31. An American Indian point of view is found in Vine Deloria, Jr., *God Is Red* (New York, 1973). For recent historiography, see Robert Brightman, "Toward a History of Indian Religion: Religious Changes in Native Societies," in Colin G. Calloway, ed., *New Directions in American Indian History* (Norman, 1988), 223–49.

The most comprehensive studies of the missionary thrust into the Pacific Northwest remain Clifford M. Drury, *Marcus and Narcissa Whitman and the Opening of Old Oregon* (Glendale, Calif., 1973), Wilfred P. Schoenberg, *A History of the Catholic Church in the Pacific Northwest, 1743–1983* (Washington, D.C., 1987), and Robert Ignatius Burns, *The Jesuits and the Indian*

Wars of the Northwest (New Haven, 1966). See also Francis Paul Prucha's provocative article "Two Roads to Conversion," *Pacific Northwest Quarterly* 79 (October 1988): 30–37.

Kevin Starr, *Americans and the California Dream, 1850–1915* (New York, 1973), has a good deal on "the great exception," as do Eldon Ernst, "Religion from a Pacific Coast Perspective," in Guarneri and Alvarez, eds., *Religion and Society in the American West*, Robert V. Hine, *California's Utopian Colonies* (1953; reprint, New York, 1966), and Sandra Sizer Frankiel, *California's Spiritual Frontiers: Religious Alternatives to Anglo-Protestantism, 1850–1910* (Berkeley, 1988). Arnoldo De Leon, *The Tejano Community, 1836–1900* (Albuquerque, 1982), Paul Horgan, *Lamy of Santa Fe: His Life and Times* (New York, 1975), and Mark T. Banker, *Presbyterian Missions and Cultural Interaction in the Far Southwest, 1850–1950* (Urbana, 1993), explore religion in the Borderlands region.

The *Western States Jewish Historical Quarterly* provides a mine of information, but one should also consult Max Vorspan and Lloyd P. Gartner, *History of the Jews of Los Angeles* (San Marino, Calif., 1970), Moses Rischin, ed., *The Jews of the American West: The Metropolitan Years* (Berkeley, 1979), and Harriet Rochlin and Fred Rochlin, *Pioneer Jews: A New Life in the Far West* (Boston, 1984).

The literature on the Latter-day Saints is enormous. One should probably start with the overview by Leonard J. Arrington and Davis Bitton, *The Mormon Experience: A History of the Latter-Day Saints* (New York, 1979). The most provocative interpretation of the Saints remains Jan Shipps, *Mormonism: The Story of a New Religious Tradition* (Urbana, 1985). See also Leonard J. Arrington, *Brigham Young: American Moses* (New York, 1985), and Thomas G. Alexander, *Mormonism in Transition: A History of the Latter-Day Saints, 1890–1930* (Urbana, 1986).

Martin E. Marty has begun a four-volume study of religion in the United States under the title *Modern American Religion*. The sociologists Robert Bellah (et al.), *Habits of the Heart: Individualism and Commitment in American Life* (New York, 1985), Robert Wuthnow, *The Restructuring of American Religion: Society and Faith since World War II* (Princeton, 1988), and James Davison Hunter, *Culture Wars: The Struggle to Define America* (New York, 1991), provide the best overviews of post–World War II religious life.

Frederic Remington

Chapter Eleven

Violence

RICHARD MAXWELL BROWN

The focus of this essay is the West from the middle of the nineteenth century to 1920—a period in which the violence of the region was not only heavy but destined to become an enduring aspect of the national mythology. First is a discussion of the values that impelled westerners of the time to be violent. Next is an extended treatment of what I have termed the Western Civil War of Incorporation, the key to so much violence from 1850 to 1920. The essay concludes with brief comments on western violence in recent decades, treats the images of western violence so deeply graven in the national consciousness, and closes by addressing two vital questions: Just how violent was the West? Is the West mainly responsible for the American heritage of pervasive violence?

Values

A cluster of beliefs mentally programmed westerners to commit violence: the doctrine of no duty to retreat; the imperative of personal self-redress; the homestead ethic; the ethic of individual enterprise; the Code of the West; and the ideology of vigilantism.

The *doctrine of no duty to retreat* emerged when the West, along with the rest of America, made a transition from the English common law of homicide and self-defense, in which flight or retreat was legally required in combat situations, to the frontier-western-American concept of no duty to retreat. Crucial to the English common law of homicide was the notion of escape: in a personal dispute that threatened to become violent, one must flee from the scene. Should it be impossible to get away, however, the common law required that one retreat as far as possible—"to the wall" at one's back—before violently resisting an antagonist in an act of lawful self-defense.

Following the westward movement of white American settlers beyond the Appalachians, the highest court in state after state canceled the English duty to retreat in favor of the American right to stand one's ground. In 1876 the top Ohio court held that a "true man" was "not obligated to fly" from an assailant. The following year the Indiana Supreme Court got to the heart of the matter: "The tendency of the American mind seems to be very strongly against the enforcement of any rule which requires a person to flee when assailed." An old folk song expressed the popular attitude:

Wake up, wake up darlin' Corrie
And go and get my gun
I ain't no hand for trouble
But I'll die before I'll run

No western gun battle over unbranded cattle or range rights claimed as many lives as Frederic Remington's painting suggests. But the picture, based on a story by Owen Wister, captures the spirit of the no-duty-to-retreat gunplay that characterized violence in the 19th-century West.

Frederic Remington (1861–1909). What an Unbranded Cow Has Cost. *Oil on canvas, 1895. Gift of Thomas M. Evans, B.A. 1931, Yale University Art Gallery, New Haven, Connecticut.*

The climax of the American renunciation of the duty to retreat came with the U.S. Supreme Court's 1921 decision in the case of *Brown v. United States*. The 7–2 majority opinion endorsing no duty to retreat was written by the noted civil libertarian Oliver Wendell Holmes, whose brisk language was a withering dismissal of the duty to retreat. The Supreme Court's decision reversed a federal-court murder conviction of a self-defending Texan who stood his ground and shot to death a knife-wielding assailant. In private correspondence about the case, Holmes noted that in its common and statute law, Texas was the strongest of all states in favor of the doctrine of no duty to retreat. In Texas, Holmes wrote approvingly, "a man is not born to run away."

Throughout the West, the *imperative of personal self-redress* of grievances was strong. In American frontier history Andrew Jackson, who was reared on the South Carolina frontier and established himself in frontier Tennessee, recounted how his mother's 1781 deathbed admonition to him as a youth of fourteen had been never "to tell a lie, nor take what is not yours, nor sue . . . for slander" but to "settle them cases for yourself"—advice by which the future president, who had killed an opponent in a duel, lived. In the West itself the gunfighting Texas-born New Mexico rancher Oliver M. Lee invoked the ethic of personal self-redress to justify the killings in his embattled career. "I never in my life willingly hurt man, woman, or child—unless they hurt me first. Then I made them pay."

Another powerful inspiration for violent behavior by westerners was, time and again, the *homestead ethic*, whose morality went back to the colonial Anglo-American frontier. This grass-roots doctrine had three key beliefs: the right to have and to hold a family-size farm, the homestead; the right to enjoy a homestead unencumbered by a ruinous economic burden such as an onerous mortgage or oppressive taxes; and the right peacefully to occupy the homestead without fear of violence (such as that by Indians or outlaws) to person or property.

Stretching to the highest realm of the American and western economy was a contrasting value: the large-property owner's *ethic of individual enterprise* in a market economy. The individual-enterprise ethic was strongly supported by the greatest capitalists of the West, including such legendary self-made men as the "Big Four" entrepreneurs who built the railroad empire of the Central and Southern Pacific lines and the "cattle kings" such as Captain Richard King of Texas and William C. Irvine of Wyoming. It was not just the big-name industrialists and agrarian magnates who subscribed so ardently to the entrepreneurial ethic but also countless others in small businesses and the professions. Throughout the West, these aggressive men-on-the-make were ever ready to use violence in allegiance to the individual-enterprise ethic and in defense of their landed and industrial property.

As the nineteenth century wore on, the civilians of the West, brandishing revolvers and rifles in the ordinary course of daily affairs, became one of the most heavily armed populations in the world. The uniquely armed and conflicted society of the West—a "legacy of conquest" in the historian Patricia Nelson Limerick's apt phrase—produced notions of western honor culminating in the *Code of the West*. Central to the Code of the West were the doctrine of no duty to retreat, the imperative of personal self-redress, and an ultrahigh value on courage, which often became, in the phrase of one historian, "reckless bravado"—a bravado that, however, was praised for its courage and not derided for its recklessness.

An Englishman who traveled across the West from California to Texas in the 1870s–1880s observed firsthand the Code of the West among his quick-to-shoot cowboy mates on a long 1880s Texas cattle drive. For readers of the British *Cornhill Magazine*, this anonymous Englishman enumerated the elements of what he termed the "somewhat primitive code of honour" of the cowboys: honesty, courage, sensitive pride, stoic indifference to pain, and, above all, a violent vengefulness against insult. With the cowboy, it was "frequently not a word and a blow but a word and a bullet," for the Code of the West was upheld by ready resort to the six-gun. Allegiance to the Code of the West produced a gunfight that claimed a life on this cattle drive. The urban West shared in the code, as the writer Rudyard Kipling found when, at about the same time, he visited the "civilized city" of Portland, Oregon. To his deep distaste Kipling observed that the jury and Portlanders at large viewed a murder case from the perspective of the western code, emphasizing the proper conditions under which a gunfight might legitimately occur and the fairness of such combat. That such prescriptions were often violated was testimony to the view that they were needed.

Nineteenth-century America was obsessed by masculine honor—North, South, East, and West. The Code of the West was a variant of the national emphasis on honor, a variant that was responsive to the particular conditions of western society in which the actuality or threat of gunplay was pervasive. Basic to the Code of the West was what President Dwight D. Eisenhower, in a nationally televised address of 1953, stressed as the essence of that code and as the code of Abilene, Kansas (Eisenhower's hometown), and its frontier marshal James Butler ("Wild Bill") Hickok: "Meet anyone face to face with whom you disagree" and "if you met him face to face and took the same risk as he did, you could get away with almost anything [killing included], as long as the bullet was in front."

Whereas western gunfighting brought the Code of the West into focus, one of the most common institutions of western violence—vigilantism—had its own set of beliefs. The *ideology of vigilantism* was regularized in the vigilante bylaws, constitutions, and oaths to which westerners frequently subscribed. Motivated by the objective of supporting the values of life and property under conditions of frontier and western disorder, vigilante bands took the law into their own hands for the paradoxical purpose of law enforcement—law as they saw it, in its substantive form of justice rather than its procedurally legal sense. Since vigilantes were almost invariably led by the elite, well-to-do members of early western communities, the ideology of vigilantism reflected the need to justify taking the law into one's own hands (in effect, committing a revolution against the State) on the part of those who were ordinarily the most zealous upholders of the legal system of law and order.

At the core of the ideology of vigilantism were three elements: self-preservation, the right of revolution, and popular sovereignty. To vigilantes, self-preservation was "the first law of nature," and thus vigilantism was necessary to preserve the community against outlaw activity. By the same token, although vigilante action was a blow against legal authority, it was justified by the right of revolution, which, in analogy to the intolerable conditions that inspired revolution against the British in 1776, justified vigilante bands, which, likewise, were seen as being like "revolutionary tribunals." By the related doctrine of popular sovereignty, vigilantes as well as Americans at large saw the people as being above the law—a law viewed as ineffective against frontier crime. To its adherents, vigilantism was but a case of the people exercising their sovereign power, in the interest of self-preservation, against the disorderly. Crucial, also, to the ideology of vigilantism was its economic rationale: vigilantism was not only often far more certain and fair than the regular system of law and order but also much cheaper. A Denver newspaper reported the popular view that an 1879 vigilante hanging in nearby Golden was not only "well merited but a positive gain to the county, saving it at least five or six thousand dollars."

Behavior

With well over two hundred vigilante movements west of the Mississippi, few states escaped the severe affliction of vigilantism. From the earliest days of the Anglo settlers, Texas was the most active vigilante state. California, with the giant San Francisco vigilante movement of 1856 (whose six to eight thousand members made it the largest in American history) and with many other movements in the gold rush era, was a prototypical state for western vigilantism. In no state, however, was the ethos of vigilantism more deeply embedded than in Montana, where the state capitol memorializes frontier vigilantes.

Prominent western senators (Leland Stanford, California; Wilbur Fisk Sanders, Montana; William J. McConnell, Idaho) and governors (Stanford, California; John E. Osborne and Fennimore Chatterton, Wyoming; Miguel A. Otero and George Curry, New Mexico) had been vigilantes, as had such members of the economic aristocracy as the capitalists Stanford and William Tell Coleman of California and the cattle king Granville Stuart of Montana. Especially in Texas and occasionally elsewhere, vigilantes terrorized entire communities and, once in a while, as Walter Van Tilburg Clark suggested in his classic antivigilante novel *The Ox-Bow Incident* (1940), punished the innocent. Yet, the offense of vigilantes was far less in violating the spirit of the law than its letter. Violations of the letter of the law, although serious, were widely acclaimed by the people and even by notables of the bench and bar.

The local campaigns of vigilantes were often aspects of a crucial pattern of violence pervading the West from the 1850s to 1920. At its core was the conservative, consolidating authority of capital—the force that was, in the scholar Alan Trachtenberg's conception, "incorporating" America during the late nineteenth century. In the West this process of incorporation was well under way by 1870 and lasted to 1920. Yet, opposing factions and individuals fought the incorporating trend politically and, often, violently.

The polarizing antagonism resulting from the trend of incorporation produced a civil war in the West—one fought in many places and on many fronts in almost all of the western territories and states from the 1850s into the 1910s. In its broadest terms, the "Western Civil War of Incorporation" pitted insurgent or resistant Indians against the political pressure and military force that concentrated them in reservations throughout the West. The Western Civil War of Incorporation also impinged economically and culturally on the traditional lifeways and livelihoods of the Hispanos of the Southwest, who fought back, for example, in northern New Mexico with the Gorras Blancas ("White Caps") and in southern Texas with the *bandidos*. The expansive western farm

Though gunfighters dominate popular imagery of western violence, the perpetrators of violence came from many walks of life. Numerous politicians and business leaders—including the California senator, governor, and railroad magnate Leland Stanford and the Montana cattleman Granville Stuart—participated in the vigilante violence that enforced the interests of conservative businesses in the late-19th-century West.

Walery studio. Leland Stanford, Wife, and Son. *Albumen silver print, ca. 1881. Stanford University Archives, Stanford, California.*

E. H. Train (1831–99). Granville Stuart. *Photograph and pen and ink on paper, 1877. Montana Historical Society, Helena.*

Mugging for a family photograph, Jesse James playfully posed on horseback with his gun drawn, ironically acknowledging his own status as a popular bandit hero.

Unidentified photographer. Jesse James on Horseback. *Tintype, ca. 1870. Everhard Collection, Amon Carter Museum, Fort Worth, Texas.*

and range country was incessantly rocked by land wars and brigandage while the propertied class curbed the disorder of chaotic boomtowns. In the mines, mills, and logging camps on the wageworkers' frontier of the West, employees resisted corporate industrialists with strikes that frequently ended in violence. An alliance of capital and government fought back with paramilitary efforts to control the far-flung workplaces of the West.

In the forefront of the Western Civil War of Incorporation were the gunfighters of the region. The best known were the two or three hundred glorified gunfighters whose fame and exploits became a part of the legend of the West—gunslingers such as Wild Bill Hickok, Jesse James, John Wesley Hardin, Billy the Kid, and Wyatt Earp. Much more obscure were the thousands of grass-roots gunfighters whose exploits became little or not at all known beyond their own localities. Although generally not as effective as the glorified gunfighters, the grass-roots gunfighters could be deadly. One of them—Walter J. Crow of California—individually exceeded the single-gunfight killings of Hickok, James, Hardin, Billy the Kid, Earp, or any of the other glorified gunfighters. In the range country and boomtowns of the pastoral and mining West, gunmen were the shock troops in the Western Civil War of Incorporation. On one side of this intraregional war were the conservative incorporation gunfighters, whose ranks included glorified gunfighters like Hickok of Kansas, Earp of Kansas and Arizona, and Frank Canton of Wyoming and Oklahoma and grass-roots gunfighters like Crow of California. The incorporation gunfighters were often northern in background and members of the Republican party. Frequently southern or Texan in their roots and Democratic in politics were the dissident resister gunfighters, some of whom, like Jesse James and Billy the Kid, were mythologized as popular heroes—as "social bandits."

Conceptualized by the British historian E. J. Hobsbawm, a social bandit is, in American terms, a notable lawbreaker widely supported, paradoxically, by the law-abiding members of society. In the West the crimes of social bandits were often approved because they expressed the discontents and grievances of those who would never dare commit such crimes on their own. The historian Richard White has traced the grass-roots admiration for social bandits in the tradition of Jesse James, whose bravery and daring was applauded as being that of "strong men who could protect and revenge themselves." Skilled gunhandlers, these social bandits often robbed banks and railroads whose steep charges were deeply resented by peaceable western farmers, ranchers, and townspeople in the post-1865 period when economic conditions caused severe hardship for those of small means. These western social bandits not only were outlaws but also were resister gunfighters in the Western Civil War of Incorporation.

African-American gunhandlers, who fought effectively on both sides of the Western Civil War of Incorporation, were fairly numerous. Among the black resisters was Isom Dart (an alias of Ned Huddleston) of the Brown's Park outlaw faction of Colorado and Wyoming. On the other side of the regional civil war was tall, tough Jim Kelly, a star gunslinger for the magnate I. P. (Print) Olive, whose embattled "gun outfit" of cowboys stormed across ranges in both Texas and Nebraska.

The Western Civil War of Incorporation coincided with a trend from 1865 to 1900 in which wealthy and powerful individuals, companies, and corporations sought either to force settlers off the land or to overcharge them for their occupancy. In effect, this was a land-enclosing movement, which in the West engendered instability and discontent comparable to that caused by the land-enclosure movements in England from the Middle Ages to the eighteenth century. Especially aggressive in the West were the big ranchers, whose gunfighting cowboys tried to exclude small ranchers and homesteading farmers from the ranges. Crucial, also, to the land-enclosing trend were some top railroads of the West, which, through congressional land grants, tied up huge acreages and set the price of land sales to settlers.

It was just such a land grant, to the Southern Pacific Railroad, that bred the Mussel Slough conflict in California. In the agriculturally rich Mussel Slough country thirty miles south of Fresno in California's Central Valley, the homestead ethic of the settlers clashed with the capitalistic entrepreneurial ethic of the "Big Four" owners of the Southern Pacific—Collis P. Huntington, Leland Stanford, Charles Crocker, and Mark Hopkins. In dispute between the settlers and the railroad were thousands of acres for which the pioneers and the railroad had conflicting land claims. The legal dispute over the land's ownership entered the federal circuit court, where in 1879 Judge Lorenzo Sawyer, a friend of Stanford and Crocker, decided in the railroad's favor. The settlers responded with night-riding vigilantism to intimidate local supporters of the railroad and, in a no-duty-to-retreat mood, prepared to defend their richly productive small farms with firearms.

The crisis exploded into the deadliest civilian gunfight in far western history on 11 May 1880, when settlers resisted eviction from their homes. With a final toll of seven deaths, the Mussel Slough shootout far exceeded the three dead of the legendary Earp battle near the O.K. Corral in Tombstone, Arizona, the following year but was entirely the work of grass-roots gunfighters—five pioneers versus two railroad supporters. The five settlers (resister gunfighters) were all killed by the two incorporation gunfighters on the Southern Pacific side, both of whom also died. In killing the five settlers, however, one of the incorporation gunslingers, Walter J. Crow, took more lives than were ever claimed on a single occasion by any of the glorified gunfighters such as Earp, Billy the Kid, or Hardin.

Public opinion in the nation and in California was strongly on the side of the settlers. The conclusion was drawn that a huge American and western corporation, the Southern Pacific, headed by a few millionaires, would not content itself with depriving industrious farmers and family men of their homes but would have them shot down in cold blood. In London, Karl Marx followed the California conflict; after the five farmers died, he wrote to an American correspondent that nowhere else in the world was class conflict—"the upheaval most shamelessly caused" by capitalist oppression—taking place "with such speed" as in California. The Mussel Slough affair and its mordant gunfight burned into the consciousness of late-nineteenth-century Americans. One of the five novels based on the Mussel Slough was Frank Norris's powerful American classic *The Octopus* (1901); its title—long applied to the Southern Pacific in California—was, in effect, a hostile metaphor for the incorporating forces of the American West.

Defeated in both their courtroom and their gunfighting battles with the Southern Pacific, the Mussel Slough dissident farmers, losers in this phase of the Western Civil War of Incorporation, had no choice but to leave their farms or pay the railroad. Most left. The resulting resentment affected an entire generation in California's Central Valley, far more than the hundreds of farmers who had been in direct conflict with the railroad. An outcome of this feeling was the popular admiration for a famous team of robbers, the social bandits Chris Evans and John Sontag, who repeatedly struck Southern Pacific trains from 1889 to 1892. The antirailroad lawbreaking of Evans and Sontag, both glorified and resister gunfighters, was a surrogate for the seething resentment toward the Southern Pacific by peaceful, law-abiding Californians. Evans and Sontag fought in two spectacular shootouts with law officers and railroad detectives, the last of which in 1893 killed Sontag and ended their criminal careers.

Indirectly related to the Mussel Slough conflict was the sensational 1889 killing of David S. Terry of California—an event in which the gunfighter tradition of the West dramatically merged with the Western Civil War of Incorporation. A potent force in the anticorporation wing of the state's Democratic party, Terry had been a strong supporter of the Mussel Slough settlers against the Southern Pacific. Meanwhile, a personal and legal dispute festered between Terry and Justice Stephen J. Field of the U.S. Supreme Court. As the leading member of the Supreme Court in the late nineteenth century and in his concurrent role as a federal circuit judge on the Pacific Coast, Field, a Californian, had spearheaded court decisions favoring the Southern Pacific and other corporations. His judicial associate and protégé, the federal circuit judge Lorenzo Sawyer, had dispossessed the Mussel Slough settlers. As the head of a powerful clique of economically conservative West Coast federal judges that included Sawyer, Field was a pillar of the establishment cause in the Western Civil War of Incorporation and, as such, a political and ideological as well as personal and legal opponent of Terry.

To protect Field from the threats of the violence-prone Terry, who had killed one man in a duel, David Neagle—a tough gunhandling lawman from Tombstone in the era of Wyatt Earp—was hired to serve as Field's bodyguard in the Golden State. When Terry slapped Field in a California railroad depot on 14 August 1889, Neagle immediately shot Terry dead in what quickly became a western and national cause célèbre. The outcome of a legal process reaching the Supreme Court (Field abstaining) found Neagle to be without fault in the killing. Unconvinced were anti-Field partisans, who saw the killing as premeditated murder in the interest of an economic and judicial order that favored incorporating millionaire industrialists.

The range-cattle industry was a major theater of war in the Western Civil War of Incorporation, and it had both urban and rural battlegrounds. In urban terms the conflict was fought in the raw towns of the Great Plains that sprang up where cattle trails met the railroad shipping points to the midwestern packinghouses. In famed boomtowns like Abilene and Dodge City, the incorporating faction of urban merchants wanted to curb the disorder and violence of the Texas cowboys who whooped into town wild for pleasure after months out on the townless trails north of Texas. To intimidate and, if need be, to arrest or even kill cowboys, the mercantile clique used its dominance of boomtown governments to employ skilled gunfighters like Wild Bill Hickok and Wyatt Earp to keep the Texans in line.

The boomtown phase of the Western Civil War of Incorporation had strong political and cultural overtones. Thus, the typical cowboy who roared into the likes of Abilene and Dodge City was a Texan, a southerner in outlook, an ex- or pro-Confederate, and a Democrat. On the other side were the merchants or entrepreneurs like Joseph G. McCoy of Abilene, a northerner who arranged for Wild Bill Hickok to keep order in Abilene. Hickok had established his gunfighting credentials as early as 1861 and became nationally known for his 1865 slaying of Dave Tutt in Springfield, Missouri, in the prototypical western showdown. Hickok, a northerner who fought for the Union in the Civil War and was reared in an Illinois abolitionist family, was a strong Republican in politics. In Abilene in 1871, Wild Bill intimidated violence-prone Texas cowboys and climaxed the season with a face-to-face killing of Phil Coe, a skilled Texas gunfighter and gambler. Incorporation gunfighters and lawmen like Hickok and the Earp brothers were in the van of the movement that safely incorporated Abilene, Ellsworth, Hays, Newton, Wichita, Dodge City, and other boomtowns into a social and economic system dominated by enterprising capital.

In the immense rural range country, the pattern in the Western Civil War of Incorporation pitted the cattle kings against small ranchers, cowboys, farmers, and rustling horse and cattle thieves who resisted the land-monopolizing thrust of the big cattlemen. In Montana, the reign of Granville Stuart and other cattle grandees (including a young Theodore Roosevelt, whose home ranch was across the territorial line in present North Dakota) was challenged by horse thieves in alliance with a motley faction of wolf hunters and ruffians whose outlaw haunts were in the wild Missouri Breaks river country of Montana. The horse-theft operations stretched from the Montana-Canada borderland down into Wyoming. Fed up with these outlaw inroads, "Stuart's Stranglers," as the vigilantes were called, embarked on a devastating campaign that burned the bandit cabins along the wooded shores of the Missouri and killed the inhabitants. Stuart, later to be idolized as "Mr. Montana" (the state's most revered pioneer), deputed a strong force of cowboys, who swept through eastern Montana and on into North Dakota, where those marked for death on a hit list, provided by Stuart, were killed. Theodore Roosevelt knew well the cattle-king leaders of Stuart's vigilante campaign, strongly approved of it until his dying day, and always regretted that Stuart and the others, fearing that the loquacious Roosevelt would talk too much, had kept him out of the triumphant campaign that, with over a hundred fatalities to its credit, was the deadliest of all western and American vigilante movements.

By the time Stuart's Stranglers disbanded in 1884, the Montana-Dakota range country was conquered territory in the Western Civil War of Incorporation. This was far from true across Montana's southern border, where in the late 1880s and early 1890s a Wyoming coalition of small ranchers, homesteading farmers, and cowboy outlaws resisted the growing aggressiveness of a powerful faction of big cattle ranchers. At the core of this faction, eastern and British aristocrats presided over their investments in the cattle country, lording it over the cowboys who toiled for them. Many of the latter struck back at the arrogant employers by rustling from them on the sly in order to break free and establish competing small spreads. By 1892, as the grandees of the Wyoming Stock Growers' Association saw it, wide areas of central and northern Wyoming were held by those who harassed and stole from them. With convictions of accused cattle thieves hard to come by from juries of local folk who were hostile to the cattle kings, the latter perfected their vigilante plans. Defiant Johnson County was marked for the strongest dose of lynch-law medicine.

Political divisions in Wyoming reflected the rising range conflict. The cattlemen tended to be Republican and, indeed, had strong support in 1892 from Wyoming's Republican governor, from its Republican party state chairman (Willis Van Devanter, who as a conservative U.S. Supreme Court member in the 1930s was a staunch opponent of Franklin D. Roosevelt's New Deal), from its two Republican U.S. senators, and as it turned out, from the Republican occupant of the White House, Benjamin Harrison. Tilting against the cattle kings in Wyoming were the insurgent Democrats and Populists. Undoubtedly inspired by the success of Granville Stuart's flawless vigilante campaign only eight years before, the Wyoming big-cattlemen vigilantes (who called themselves "Regulators" in the tradition of the first American vigilante movement, the frontier South Carolina Regulators of 1767–69) replicated Stuart's operation. Like Stuart in Montana, they compiled a victim list (seventy in Wyoming) and prepared a lightning thrust by rail and horse into the enemy country.

The cattle magnate Frank Wolcott headed the Regulators. He enlisted a mercenary band of Texas gunfighters under Frank Canton, a gunfighting ex-sheriff of Johnson County. First by special train and then by horse, this paramilitary force headed for the rustlers' domain. Along the way, however, the overconfident Regulators came to grief. After a notable first success in besieging and killing two resister gunfighters, the rustlers Nate Champion and Nick Ray, Wolcott and company rode on north to Johnson County. South of the county seat of Buffalo, the Regulators were intercepted and pinned down by a giant posse of citizens alerted to the invasion.

F. Wolcott
W.C. Irvine
Johnson County Cattle Raiders - Prisoners at Ft. D.A. Russell - 1892
A.B.Clark, E.W. Whitcomb, A.D. Adamson, C.S. Ford, W.H. Tabor, G.R. Tucker, A.R.
J.E. Booke, B.M. Morrison, W.A. Wilson, M.A. McNelly, Bob Barlin, W.S. Davis, S.Sutherland,
Alex Lowther, W.J. Clarke, J.A. Garrett, Wm. Armstrong, Buck Garrett, F.H. Laberteaux, J.
Johnson, Alex Hamilton, F.M. Canton, W.C. Irvine, J.N. Tisdale, W.B. Wallace, F. DeBilleir, H.Teck
maker, W.E. Guthrie, F.G.S. Hesse, Phil DuFran, Wm Little, D.R. Tisdale, J.D. Mynett, M. Shonsey,
Joe Elliott, C.A. Campbell, J. Barlings, L.H. Parker, S.S. Tucker, B. Wiley, J.M. Beuford, K. Rickard,
Frank Walcott, B. Schultz, - Names not in order. Copied from "Longest Rope" by Baber.

Photographed a few weeks after their failed invasion of Johnson County, Wyoming, in April 1892, the self-proclaimed "Regulators," a group of big cattlemen and hired guns, were in temporary defeat. Later set free without a trial, they eventually won their battle with smaller homesteaders and ranchers for domination of the Wyoming range cattle industry.

Charles D. Kirkland (1857–1926). "The Invaders." Johnson County Cattle War *(taken at Fort D. A. Russell). Photograph, 1892. American Heritage Center, University of Wyoming, Laramie.*

On the brink of annihilation by the Johnson Countians, the Regulators were saved only by the intervention of U.S. cavalry (called out by the Republican chain of influence, which ran from the Wyoming Stock Growers' Association to President Harrison in Washington, D.C.). The cavalry imposed a truce, no bloodshed occurred, Johnson County authorities ran out of money in their legal prosecution of the invaders, and Wolcott, Canton, and all the rest went free. Meanwhile, outraged Wyoming voters avenged the blatant invasion of Johnson County by repudiating the pro-vigilante Republican party in the fall 1892 election. Yet the Johnson County War was only a temporary setback for the big cattlemen of Wyoming in the Western Civil War of Incorporation. From the open violence of a vigilante campaign, the determined cattle barons shifted to the stealth of murderous ambushes by the bounty hunter Tom Horn, who picked off victims until his homicidal career was ended by a legal execution in 1903. The result was a triumph for the big cattlemen. By 1910, possibly even sooner, the range country of Wyoming was a part of the fully incorporated West.

In the Southwest, fence cutting was the major tactic used against the incorporating efforts of the cattle kings. Resistance surged in violence-torn central Texas during the 1880s and 1890s. In county after county, farmers and small ranchers cut the fences of the land-enclosing big cattlemen who were gradually forcing so many of the small operators off the land or on to reduced holdings. The fence-cutting property destruction peaked in 1883–84 and 1897–98 but lost the battle against the broader trend.

In its institutionalization of political violence and assassination from the late 1860s to shortly after 1900, the New Mexico Territory was unequaled in the West. The government in New Mexico lacked the credibility, power, and will to curb the violence of the territory's intricately arrayed, deeply divided elements. In conflict after conflict, the incorporating forces battled against those, like the Gorras Blancas ("White Caps"), who resisted them.

The White Caps were poor Hispanic villagers who struck back—by burning barns, cutting fences, and occasionally using sniper fire—at the aggressive Anglos and *ricos* (rich Hispanos) who used their knowledge of the law and the ways of modern urban society to seize portions of the age-old communal land of the sheepherding villagers. By day these *pobres* ("poor ones") voted Populist and streamed into Knights of Labor lodges, but at night, as White Caps, they destroyed the fences and outbuildings of the *ricos*. White Cap violence was a guerrilla struggle that for a time halted the incorporating trend and through court victories preserved the communal grazing lands. A 1960s throwback to the White Caps of the 1890s was the Alianza movement, formed by Reies Tijerina to reclaim land lost by rural Hispanos to Anglo chicanery, according to Tijerina. Ultimately failing in its objective, Tijerina's crusade came to a climax in 1967 with the violent seizure of the county courthouse in Tierra Amarilla, New Mexico, during which one person died.

In the 1870s, in the northern New Mexico county of Colfax, violence erupted against the incorporating trend. The issue was the Maxwell Land Grant Company, a giant combine of absentee lawyers and capitalists who planned to convert the Maxwell grant in Colfax County into an enormous economic empire. Here the resisters were not Hispanos but Anglo small ranchers, cowboys, and townsmen who used the law as well as the violence of vigilantism and gunfighting to defend their small land claims against the Maxwell magnates. In the Colfax County War, the dissidents rallied around one of the West's most fearsome resister gunfighters, Clay Allison. Anchored by the powerful political support of Republican nabobs in Santa Fe and Washington, D.C., the Maxwell Land Grant Company outlasted the violent resistance of Allison and others and, erecting a land, cattle, and mining empire of nearly two million acres, dominated the county until the 1960s. In central Arizona a lethal vigilante movement on behalf of incorporating big ranching and commercial interests ended the chaotic, bloody Tonto Basin War of the 1880s, which took twenty to thirty lives.

The most enduring range-country episode in the Western Civil War of Incorporation occurred in southern New Mexico and Arizona from the 1880s to 1910. The opposing alignments were similar to those elsewhere, from Texas to Montana and from the Missouri to the Pacific. On the incorporating side was a faction of big cattlemen and Republican capitalists and politicos whose citadels of power were in the growing urban centers of the region. Resisting the incorporators was a typical coalition of small ranchers and cowboy outlaws, whose dissidence was spearheaded by some notable resister gunfighters opposed, in turn, by potent incorporation gunfighters on the other side. Rustling cattle from the large herds of their opponents was a constant tactic of the anti-incorporators, who, in general, tended to be southern or Texan in origin, Democratic in politics, and premodern in their values—emphasizing family and individual loyalty, the no-duty-to-retreat syndrome of personal self-redress, and manly courage. In conflict were not just contrary claims of land and property but two opposed worldviews: one stressing modern, urban, capitalistic values and the settlement of disputes through the legal system (a ground of combat favoring the know-how and sophistication of the incorporators) and the other stressing rural values, kin and friendship loyalties, and the violent settlement of disputes face to face instead of in the courtroom.

Crucial to so much of the trouble that turned southern New Mexico and Arizona into a dark and bloody ground from 1880 to 1910 was the famous Lincoln County War in New Mexico during 1878. The Lincoln County War was a veritable university for gunfighters, with no less than nineteen of them (including Billy the Kid) honing their gunshooting skills in the 1878 conflict. Trouble came from the partnership of Lawrence G. Murphy, James J. Dolan, and John H. Riley, who in the 1870s had, in effect, incorporated Lincoln County into their own economic domain based on the store they operated in the county seat, also named Lincoln. By 1876, when the ambitious young Englishman John Henry Tunstall came into Lincoln County, the small ranchers, farmers, and cowboys were restive under the oppressive domination of town and county by "the House" (the phrase was a reference to the imposing two-story store on Lincoln's single, rambling street—that is, the mercantile house of Murphy, Dolan, and Riley). More than just a store, the House was a corrupt political and economic faction that had much in common with the Tweed Ring of New York City and many other such rings in Gilded Age America and the West. The House in Lincoln County thrived on a complex system of ill-gotten gains. Murphy-Dolan-Riley outlaw hirelings stole cattle from the ranch king, John Chisum, for beef that was sold at inflated prices to the U.S. government's Mescalero Apache Indian Reservation and to Fort Stanton, both located in Lincoln County. In all of this, the House was bolstered by its allies in politics, in the judiciary, and in law enforcement.

Grass-roots discontent with the greed of the House found no practical outlet until the appearance of Tunstall, who forged an alliance with a dissident local lawyer, Alexander McSween, and with the cattle king Chisum, whose cattle losses were illicitly enriching the House. Amply backed by capital from his father's profitable London business, Tunstall (with McSween) soon opened a store in competition with the House. Customers flocked to the new Tunstall-McSween store. No less cynical and selfish than the House, Tunstall hoped to create his own ring and monopolize the mercantile possibilities of the county. Although he hired a band of tough cowboys (including Billy the Kid) to handle his burgeoning ranch (also made possible, like the new store, by munificent loans from his father), Tunstall's English culture led him to accept the ethic of legality and to refrain from violence. The House was not so forbearing. The brutal murder—or assassination—of Tunstall by House hirelings on 18 February 1878 triggered the Lincoln County War. Tunstall's gun-wielding cowboys remained loyal to his memory and to his surviving partner McSween (and his spirited wife, Susan McSween). An all-out range war ebbed and flowed across the county and climaxed in the five-day battle fought along Lincoln's lone street on 15–19 July 1878, resulting in a bitter defeat for the McSween side. After the Tunstall store was set on fire in the battle's conclusion, McSween was shot to death when he fled from the flames while Billy the Kid made one of the most famous of his many escapes.

The Lincoln County War led to the mighty myth of Billy the Kid as a social bandit. Born Henry McCarty in New York City, the lad moved with his widowed mother and his brother through Indiana and Kansas to New Mexico. With his mother remarried to

Sheriff Pat Garrett promoted the mythic stature of his nemesis, Billy the Kid, in 1882 by publishing a popular and widely inaccurate account of the gunfighter's life. Western violence, popular culture, and commerce remain closely linked, with violence still an important theme of much western literature, art, and film.

Pat Garrett (1850–1908). An Authentic Life of Billy the Kid *(Santa Fe, 1882). Western Americana Collection, Beinecke Rare Book and Manuscript Library, Yale University, New Haven, Connecticut.*

a miner, William Antrim, the future Billy the Kid led a normal schoolboy life in Silver City, New Mexico, until 1874, when his mother died and the Antrim family fell apart. The teen-age Billy committed a petty theft, became a fugitive, killed a bully in Arizona, and, back in New Mexico under the alias of William Bonney, became a cowboy working on Tunstall's Lincoln County ranch. The Kid's strong loyalty to Tunstall and the

McSweens drew him into the Lincoln County War; by the end of the battle, he was the top gun on the anti-House side. Not until 1989—more than a century after the Kid's death—did a full, realistic biography of him, by Robert M. Utley, appear. The Kid was neither the hero of myth nor the psychopath of the antimyth but a youth quite typical of the time and place; during his "short and violent life" (Utley's phrase) of twenty-one years, he took not twenty-one lives (one victim for each one of his years, according to the legend) but no more than a far-from-negligible ten. The Kid's career in his last two years collided with the forces of incorporation and became a brief but significant episode in the Western Civil War of Incorporation.

Thrown on his own at the end of the Lincoln County War, Billy the Kid—literate and ambitious—tried and failed to find a niche in law-abiding society. (His nickname was created by newspapers and dime novels in the last year of his life; his friends and enemies in New Mexico spoke of him as "the Kid" but not "Billy the Kid.") John Chisum, the wealthy Tunstall-McSween ally, denied the Kid the combat pay Billy claimed for service in the Lincoln County War—a denial that may have been influenced by Chisum's disapproval of the budding romance between Billy and his niece, Sallie, as well as by the cattle king's famous parsimony. Earlier, the Republican territorial governor of New Mexico, Lew Wallace (a Civil War general and the future author of the best-selling novel *Ben-Hur*), reneged on a deal with the Kid. The governor had promised the Kid a pardon in return for his crucial testimony against two brutal killers. The Kid kept his part of the bargain, but Wallace faithlessly denied him the pardon. With all avenues to a peaceful civilian life closed to him, the Kid became a full-time cowboy outlaw.

Heading a gang of gunfighting veterans of the Lincoln County War, Billy rustled cattle in Lincoln County and the Texas panhandle. This brought down on him the incipient forces of incorporation in Lincoln County, now centered in the mining boomtown of White Oaks and the cattle town of Roswell. A coalition of town businessmen, professional men, and aggressive big cattlemen formed to silence the Kid's deadly guns and end his cattle thefts. But the Kid had resources in this conflict. His sunny nature had earned him a wide circle of Anglo and Hispanic friends throughout southern New Mexico. The rising Roswell entrepreneur Joseph C. Lea and his neighbor, John Chisum, headed the effort to suppress Billy and his gang. The Kid's erstwhile friend Pat Garrett, a tall Texan and former buffalo hunter, was put up for sheriff and elected despite the opposition of the popular Billy, who supported Garrett's rival. Garrett broke up the gang, cornered and arrested the Kid, and saw him tried and sentenced to death for a homicide in the Lincoln County War—the only killer in the war to be tried and convicted. With Wallace's promised pardon definitely withheld, the desperate Kid shot to death two guards and escaped from the Lincoln jail. By now, "Billy the Kid" was a famed figure whose violent career was flaunted in the nation's newspapers and ten-cent paperbacks. Garrett, assisted by the spying of the cattle-range detective John W. Poe, tracked the Kid to one of his favorite haunts: the compound of Pete Maxwell at old Fort Sumner, New Mexico, on the Pecos River. Here, in midnight darkness, Garrett found the Kid and killed him with one shot.

The triumph of Lea, Chisum, Garrett, and Poe over the hapless Billy the Kid ushered in a new incorporated era of dominant town-and-country wealth and large-landed cattle property in southern New Mexico. The new order flourished under the leadership of its rising Republican political boss, attorney, and militia colonel Albert J. Fountain, who enjoyed the crucial support of the large ranchers of the country. Fountain and the big cattlemen had close ties to the lawyer Thomas B. Catron of Santa Fe, the avaricious Republican political boss of New Mexico who came to own or directly control more land than any other American in history. Meanwhile, more and more small ranchers and cowboys from central Texas filtered into the Tularosa basin, which stretched west from the mountain heights of Lincoln County. Ambitious, blessed with incomparable cowboy skills, and proudly bearing the no-duty-to-retreat proclivity to violence of their central Texas backgrounds, these aggressive newcomers saw only one way to survive against the land-enclosing tactics of the established big cattlemen: to steal from the herds and protect themselves with six-guns and rifles. The model and leader of these anti-incorporation Texans was Oliver M. Lee, a natural-born cowman, peerless horseman, and matchless gunfighter. The big cattlemen formed a stockmen's association to fight off the interlopers and employed a most-willing Albert J. Fountain to mount an antirustling militia campaign and legal effort to end the threat of Lee and the Texans to the incorporated state of affairs. Lee formed his own alliance with an ambitious Democratic politico and gunfighting lawyer, the ex-Kentuckian Albert Bacon Fall (who, decades later, with coat turned to the Republican party, became Warren G. Harding's ill-fated secretary of the interior).

Pressed by Fountain's indictment of Lee for cattle theft, Lee, Fall, and the rustling small ranchers seemed to be on the run. That soon changed in 1896 with the disappearance of Fountain and his young son, Henry, as they traveled by buggy from the courtroom in Lincoln back to their home in Mesilla on the Rio Grande. Fountain and his son were never found. It became an open secret to many in the region, and is confirmed by historians, that Lee and two of his cowboys carried out a plot hatched by or, at least, joined in by Fall to waylay Fountain (Fall's personal rival for the political domination of southern New Mexico) and murder him and his son. The bodies were buried in the mountains away from the crime and were never found. Once more, the incorporating faction of cattle kings and powerful Republican politicos led by Thomas B. Catron turned to Pat Garrett, who, again made county sheriff, was given the mission of bringing Lee and his henchmen to justice for killing the Fountains. Garrett eventually arrested Lee, who was brought to trial in 1899. Catron came south from Santa Fe to head the trial team against Lee but turned out to be no match for the histrionics and legal skill of Lee's defense attorney, Albert Bacon Fall. It took only eight minutes for a strongly anti-incorporation, anti-big cattleman, pro-cowboy jury to find Lee not guilty.

The fiasco of Lee's trial was a dramatic but only temporary check of the incorporating trend. The ironic outcome of the Tularosa war was that Fall and Lee soon became incorporators themselves. Deeply conservative in his social and economic views, Fall increasingly felt out of place in a Democratic party dominated by the quasi-populism of William Jennings Bryan. Fall switched to the Republican party and, with his rival Fountain out of the way, succeeded to the leadership of the party in southern New Mexico. In 1912, Fall realized a longtime ambition by being elected to the U.S. Senate from the newly admitted state of New Mexico. Lee's subsequent career was similar to that of his lifelong friend Fall. Lee became one of the largest ranchers in southern New

Mexico, eventually heading the million-acre Circle Cross Ranch (the area's largest) and serving two terms in the New Mexico legislature. In effect, Fall and Lee had used the violence of gunfighting and murder to move from the losing to the victorious side of the Western Civil War of Incorporation. There is no better example of a powerful western and national economic and political career built on violence than that of Albert Bacon Fall—Republican U.S. senator, cabinet member, and power in the high councils of his party as well as heavy speculator in Mexican mining property and baronial New Mexican cattle grandee. Many westerners—and none better than Fall—exemplify the historical sociologist Charles Tilly's maxim that the history of violence is nothing less than the history and organization of power.

By the early 1880s, vast Cochise County in the extreme southeastern region of Arizona was another battleground in the Western Civil War of Incorporation, with conflicts in both the rural range country and the urban streets of the county seat, Tombstone. The opposing forces in Cochise County represented the pervasive pattern of incorporating versus anti-incorporating factions. Headed by the mine owners and managers of booming Tombstone, the incorporating element was mainly Republican in politics, northern in background, urban in culture, and modern in outlook. The opposing faction included urban Democrats of Tombstone but centered on an alliance of small ranchers (many of whom rustled cattle from large ranchers) and cowboy outlaws (including "Curly Bill" Brocius and John Ringo), who were also Democrats as well as mainly Texan or southern in their backgrounds. The cowboy outlaws dominated the backcountry village of Galeyville and periodically rode into Tombstone for boisterous good times that unnerved the Republican elite of Tombstone, which was headed by the mine magnate E. B. Gage and the editor and mayor John P. Clum. Supporting this establishment was the youthful Episcopalian minister Endicott Peabody (later to be the revered schoolmaster and White House chaplain of Franklin D. Roosevelt), who, soon after his missionary period in Tombstone, founded and for decades headed America's most exclusive private school for boys, Groton, in Massachusetts. An exponent of muscular Christianity, Peabody knew and liked Wyatt Earp, for whom he had a lifelong admiration.

Wyatt Earp and his brother Virgil (as well as their younger brothers, Morgan and Warren) were a crucial bloc in the Cochise County conflict. The modernizing Tombstone elite turned to the gunhandling talents of Wyatt and Virgil (and their brothers) in an attempt to end the killings in Tombstone and play down the city's anarchic "man for breakfast" image. In turning to the Earps (and their gunfighting colleague Doc Holliday), the Republican elite hoped to stabilize life in turbulent Tombstone and convince California and eastern investors that the boomtown was a safe field for profitable investment. In contrast to the Earps, the small ranching and rustling families of the Clantons and the McLaurys (along with their cowboy-outlaw allies) were violent protagonists for the unincorporated, premodern, traditional values of the rural cowboy coalition of Cochise County. Strong Republicans of an Illinois-Iowa family of Civil War–era unionists, the Earp brothers were right at home on the side of Tombstone's urban elite, for Wyatt, Virgil, and their brothers were enthusiastic and profitable investors and speculators in Tombstone-area mine and real estate property. Personal clashes with the Clantons and McLaurys brought the Earps (and Holliday) to a violent confrontation with them near Tombstone's O.K. Corral on 26 October 1881. When the Earps fired away at the Clantons and McLaurys in their famous gunfight of that day, they were fighting for their entrepreneurial, Republican, incorporating values as well as their lives. The triumph of the Earps and Holliday (the two McLaury brothers and the one Clanton who faced them were all mortally wounded) was followed by a series of shootings in early 1882. Credited with at least two or three more killings, Wyatt left Cochise County, as did his brothers. They were consoled by their profits in booming Tombstone but saddened by the death of Morgan Earp in a pool-hall ambush. The gunpower of the Earps won a notable victory in the Western Civil War of Incorporation, for their success was to defeat and break up the cowboy-rustler-outlaw faction headed by the Clantons, the McLaurys, Brocius, and Ringo.

Resembling the cowboy-outlaw episode of Arizona was a long-range crime wave in the four-state enclave of Missouri, Kansas, Oklahoma, and Arkansas. An outlaw dynasty flourished in this region from the 1860s to the 1930s—from the time of Jesse James to that of Pretty Boy Floyd. As the historian Paul I. Wellman has shown, Charles Arthur ("Pretty Boy") Floyd (killed in 1934) was "the lineal successor" of William Clarke Quantrill, the Confederate guerrilla leader in Missouri. The fearsome Quantrill was the Civil War mentor of the youthful guerrillas Frank and Jesse James and Cole Younger in the violent arts of riding, raiding, and shooting. Thus began a middle-border outlaw dynasty perpetuated, said Wellman, "by a long and crooked train of unbroken personal connections, and a continuing criminal heritage and tradition handed down from generation to generation." The James-Younger gang (1866–82) began the American outlaw tradition of armed bank robbery at Liberty, Missouri, on 13 February 1866. Although the first train robbery had been by the Reno brothers in Indiana in 1866, it was the James-Younger gang that, again, made this innovative act of American banditry a national tradition. Carrying on this new pattern of gunfight-punctuated bank and train robberies were the 1880s gang of Belle Starr and the 1890s gangs of the Dalton brothers, Bill Doolin, Al Jennings, and Bill Cook. A vital link in the outlaw dynasty was Belle Starr's nephew, Henry Starr, who personally bridged the gap between the nineteenth-century brigands and the likes of the 1920s–1930s gangsters Al Spencer, Frank Nash, and Pretty Boy Floyd. From the James and Younger brothers down to the Dalton and Doolin era, these daring desperadoes fit the pattern of social banditry: men whose audacious exploits won the admiration of rural people wilting under the economic and cultural pressure of a modern, industrializing, corporation-dominated society.

Even as the middle-border outlaw tradition flourished, the Western Civil War of Incorporation was by 1900 making the transition from its main nineteenth-century battle sites in the boomtowns and range country to the early-twentieth-century mining camps, mill towns, metropolises, and commodity-crop fields. The overall issue was the taming of dissident, often radical, labor unionists for toil in a West marked for domination by profit-conscious private investors. The conflict between labor and capital in the Western Civil War of Incorporation predated 1900 but became critical in the new century. The first sustained violence in this industrial phase of the Western Civil War of Incorporation was what the historian George S. McGovern has termed Colorado's "Thirty Years War"—an 1884–1914 conflict amid the state's hard-rock mines and soft-coal fields.

Gunfighting continued, but the variety of violence in the Western Civil War of Incorporation now included the riot and the use of dynamite in connection with strikes and lockouts. High points in the Colorado turbulence occurred in the Rocky Mountain mining camps, where the radical new anti-incorporation Western Federation of Miners (WFM) fought back against repressive mine owners in Leadville (1894), Telluride (1901), and Cripple Creek (1903–4). Bloodiest of all was Cripple Creek, where a typical alignment was the state militia against the mine-and-mill unionists. The WFM was a losing cause in Cripple Creek, even though the professional terrorist Harry Orchard,on behalf of the union, killed thirteen strikebreakers while dynamiting the town's railroad station.

The alliance of industrial corporation and state militia figured in one of the most violent episodes in western history, the Ludlow Massacre of 20 April 1914, which tragically concluded Colorado's "Thirty Years War." This was the climactic event in a long, bitter strike of the United Mine Workers against the Rockefeller-controlled Colorado Fuel & Iron and independent companies in the southern Colorado coalfield stretching northward from Trinidad. Evicted from their company-owned houses, the miners at Ludlow and other coal camps settled into their own tent cities and stayed on strike. Many of the union men were Greek or Hispanic and were subjected to the highly prejudiced harassment of the predominantly Anglo militia, adding an ugly ethnic dimension to the conflict.

At Ludlow on April twentieth occurred the events that shocked America: an all-day gunfight between strikers and militia, the burning of the tent city, and the death by suffocation of thirteen women and children in the "Black Hole of Ludlow"—a declivity beneath a burned-over tent in which the women and children had taken refuge. Enraged by the tragedy, hundreds of miners and their sympathizers roared across the coal-mining counties in a spasm of property destruction that ended only when federal troops were sent in by President Woodrow Wilson to restore order. The U.S. soldiers, unlike the state militia, were impartial in their preservation of peace. The result was a defeat for the union and a costly victory for the Rockefeller family. Young John D. Rockefeller, Jr., never entirely overcame the onus of his disastrous intractability against the union nor, in the eyes of many early-twentieth-century Americans, did the enormous philanthropies of the Rockefeller family fully compensate for the tragedy at Ludlow.

Enmeshed in the Western Civil War of Incorporation was what the labor historian Carlos A. Schwantes terms the "wageworkers' frontier"—the social and industrial context of Colorado's Thirty Years War and other such conflicts. The wageworkers' frontier of the West embodied an explosive combination of the deep tensions of industrialization with the combative frontier psychology of the West. The Pacific Northwest wageworkers' frontier stretched from the mining camps of Idaho and Montana to the coastal logging stands and mill towns of Washington and Oregon. Keynoting the violence was the industrial warfare in the Coeur d'Alene region of the northern Idaho panhandle in the 1890s. The trouble began in 1892 with mineowners and labor unionists trading casualties and temporary victories; soon, to protect mine property from the dynamiting of union forces, state and federal troops intervened and incarcerated hundreds of miners in the infamous "bull pens" of the towns of Wallace and Wardner. Embittered by their defeat in the 1892 struggle, alienated strikers founded the

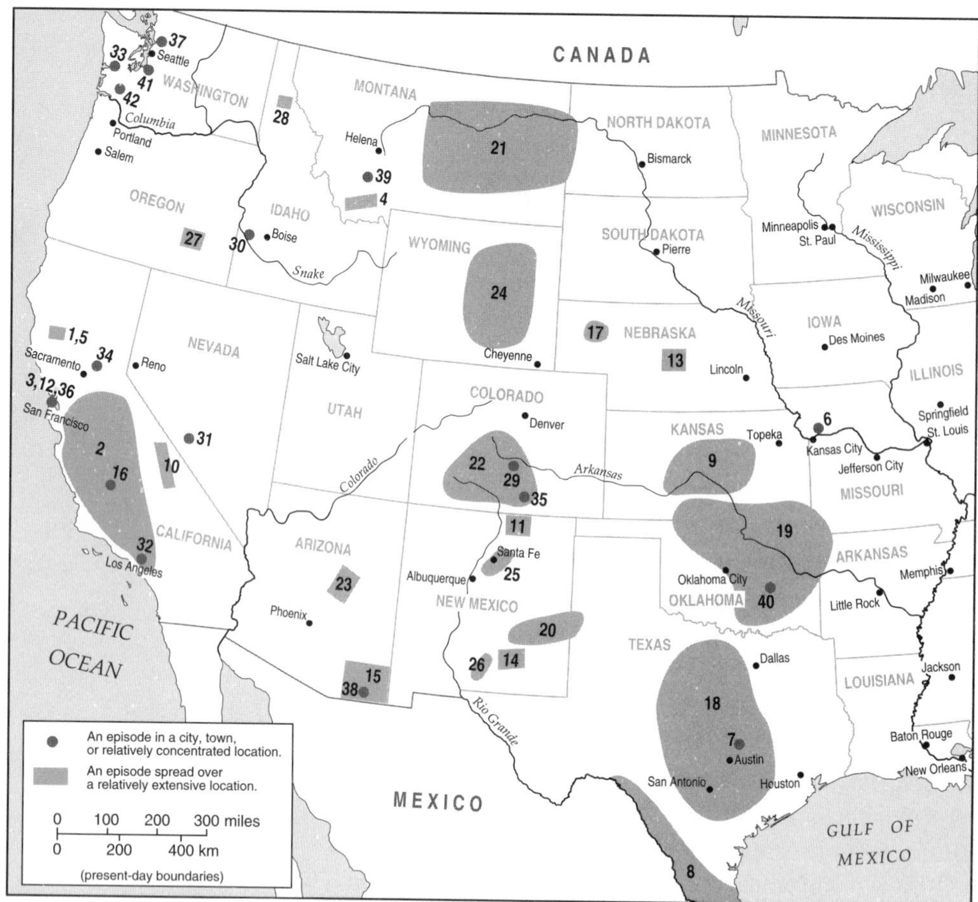

The Western Civil War of Incorporation, 1850s–1919

Below are brief descriptions of the 42 episodes in the Western Civil War of Incorporation, the numbers of which correspond to the numbers on the above map. The following abbreviations are used to designate the outcome of the episodes: V = victory for incorporating faction, D = defeat for incorporating faction, A = ambiguous outcome. (Other abbreviations used: IWW = Industrial Workers of the World, WFM = Western Federation of Miners, VIG = vigilantism was used by incorporating faction.)

1. First Round Valley War, northwest California, 1850s–1865. Incorporating white settlers carry on genocidal campaign of dispossession against local Indians. **V**

2. Mexican outlaws' activity, California, 1850s–60s. Incorporating California rangers suppress guerrilla-like insurgency of native Mexican outlaws: Joaquin Murrieta, Tiburcio Vasquez, and others. **V**

3. San Francisco vigilantes, 1856. Incorporating mercantile elite led by William T. Coleman uses vigilantism against Irish-Catholic working-class element. **VIG V**

4. Montana vigilantes, 1863–65. Incorporating faction of vigilantes of Virginia City and Bannack (headed by Wilbur Fisk Sanders, et al.) vs. an outlaw gang led by Henry Plummer. **VIG V**

5. Second Round Valley War, northwest California, 1865–1905. Land-enclosing big ranchers vs. small landholders. **V or A**

6. James-Younger outlaw gang, Missouri, 1866–82. Bank- and train-robbing outlaws led by social-bandit Jesse James and headquartered in Clay County vs. incorporating industrial, financial, commercial, and state-government forces. **V**

7. Williamson County War, Texas, 1869–76. Yegua Notch Cutters outlaw gang vs. Olive family of incorporating big ranchers. (The Olives defeat the outlaws, but their heavy losses force them to move to the more open range country of Nebraska; see #13, below.) **A**

8. *Bandido* insurgency, south Texas, 1870s–1910s. Gregorio Cortez, Juan Cortina, and other Hispanic outlaws vs. incorporating force of Texas Rangers. **V**

9. Kansas cattle towns. 1870s–80s. Incorporating merchants represented by Wild Bill Hickok and other incorporation-gunfighter law enforcers vs. Texas cowboys in Abilene, Hays, Wichita, Dodge City, and other Kansas cattle towns. **V**

10. Owens Valley, California, 1870s–80s. White settlers incorporate local Paiute Indians into labor force. **V**

11. Colfax County War, New Mexico, 1857–76, 1882, 1888. Incorporating large land company (represented by Thomas B. Catron) vs. local settlers (including resister gunfighter Clay Allison) upon whose homes the company impinged. **V**

12. San Francisco, 1877. Incorporating "Pick Handle Brigade" of establishment vigilantes (led again by William T. Coleman) vs. working-class rioters in sympathy with nationwide rail workers strike. **VIG V**

13. Custer County, Nebraska, 1877–79. Incorporating big ranchers led by Olive family from Texas (see #7, above) vs. homesteading small farmers. **D or A**

14. Lincoln County War, southern New Mexico, 1878. Unique conflict between two incorporating factions that nullified each other; the ultimate victor was New Mexico's top incorporator, Thomas B. Catron. **V**

15. Cochise County War, southeast Arizona, 1878–81. Urban, industrial elite of Tombstone (violently spearheaded by the Earp brothers) vs. a coalition of out-county small ranchers and cowboy outlaws. **V**

16. Mussel Slough conflict, California, 1878–82. Incorporating Southern Pacific Railroad headed by Collis P. Huntington, Leland Stanford, and Charles Crocker vs. small-farming settlers in dispute over land. **V**

17. Sand Hills War, northwest Nebraska, 1880s–90s. Incorporating big cattle ranchers vs. homesteaders. **A**

18. Fence-cutting conflict, central Texas, 1880–1900 (including a peak event, the Fence Cutters War, 1883–84, afflicting at least 12 counties). Land-enclosing big ranchers vs. small ranchers and farmers. **V**

19. Outlaws vs. law enforcers; Missouri, Kansas, Oklahoma, and Arkansas; 1880s–1910s. Bill Tilghman, Chris Madsen, Frank Canton, and other law enforcers represent incorporating forces vs. the Belle Star, Dalton brothers, Bill Doolin, and other outlaw gangs that often had popular support as social bandits. **V**

20. Billy the Kid outlaw activity, southeast New Mexico and west Texas, 1880–81. Incorporating big ranchers and business and professional men vs. Billy the Kid's rustling gang. **V**

21. Granville Stuart's Montana vigilante movement of 1884. Incorporating big cattle-rancher vigilantes led by Granville Stuart vs. horse-stealing outlaws. **VIG V**

22. Colorado's Thirty Years' War, 1884–1914. Incorporating mine owners and managers with state-government allies vs. organized labor. (See #29 and #35, below, for two of the major events in the Thirty Years' War.) **V**

23. Tonto Basin War, Arizona, 1886–88. Incoming incorporating commercially-minded large-landholding and ranching elite vs. traditionally-minded early settlers in a war touched off by the Pleasant Valley feud of the Grahams vs. the Tewksburys. **VIG V**

24. Wyoming range conflict, late 1880s–1901. Incorporating elite cattle ranchers and establishment allies vs. small ranchers, homesteaders, and cowboy allies. The latter won a temporary victory in the Johnson County War (1892), but the bounty-hunting kills of incorporation gunfighter Tom Horn (1894–1901) sealed the cattle kings' victory. **VIG V**

25. *Gorras Blancas* (White Caps) conflict, northern New Mexico, 1890s. *Gorras Blancas* spearheading traditional Hispanic pastoral villagers vs. incorporating land-enclosing Anglo and Hispanic elite ranchers and lawyers. **D**

26. Tularosa war, southern New Mexico, 1890s. Incorporating big cattle ranchers and business and professional allies vs. traditionalistic small ranchers and cowboys. **A**

27. Harney County conflict, Oregon, 1890s. Big cattle ranchers vs. homesteaders. **A**

28. Coeur d'Alenes War, northern Idaho, 1890s. Incorporating mine owners supported by state and federal governments and military forces vs. organized labor (including WFM) in the Coeur d'Alenes mining country. **V**

29. Cripple Creek conflict, Colorado, 1894–1904. Incorporating mine owners and managers vs. organized labor, including WFM (part of Colorado's Thirty Years' War; see #22, above). **V**

30. Caldwell, Idaho, assassination of ex-Gov. Frank Steunenberg, 1905, and its aftermath, 1906–07. Anti-incorporating miners' union (WFM) vs. incorporating forces represented by Gov. Steunenberg, who supported the incorporators in the Coeur d'Alenes war (#28, above). An important result is the unsuccessful trial of WFM leaders for the assassination of Steunenberg. **A**

31. Goldfield, Nevada, conflict, 1907. Incorporating mine owners vs. IWW. **V**

32. Los Angeles, 1910. Dynamiting of *Los Angeles Times* building by labor-union conspirators results in significant loss of life, but anti-union backlash results in victory for *Times* publisher Harrison Gray Otis as spearhead of incorporating forces in southern California vs. labor-union movement. **V**

33. Aberdeen, Washington, 1911–17. Incorporating lumber-mill magnates and town allies vs. IWW. **VIG V**

34. Wheatland, California, riot of hop pickers, 1913. Anti-incorporating IWWs and migrant workers vs. owners of Durst hop ranch and law-enforcement allies. **V**

35. Southern Colorado coal-mining conflict, 1913–14. John D. Rockefeller and other incorporating mine-owning forces vs. organized labor. Culminates in miner-families' loss of life in "Ludlow Massacre," 1914 (part of Colorado's Thirty Years' War; see #22, above). **V**

36. San Francisco, bombing of World War I Preparedness Day parade, 1916. Incorporating industrial and business forces vs. organized labor (including its radical fringe). **V**

37. Everett Massacre, Washington, 1916. Conflict between incorporating lumber-mill magnates and allies of vs. organized labor. IWW intervention results in the massacre with labor element's casualties being heaviest. **VIG V**

38. Bisbee, Arizona, conflict, 1917. Incorporating mine interests and town and law-enforcement allies vs. striking miners (including IWW). Vigilantes deport 1,186 strikers and allies. **VIG V**

39. Butte, Montana, lynching of Frank Little, 1917. Little, an anti-World War I activist and IWW organizer, fell afoul of local vigilantes. **VIG V**

40. Green Corn Rebellion, eastern Oklahoma, 1917. Uprising of anti-World War I poor farmers and tenants against incorporating landlords and townspeople.

41. Seattle General Strike, 1919. Incorporating Seattle forces (led by the mayor) defeat the general strike, an event accompanied by turbulence but not violence. **V**

42. Centralia, Washington, massacre and reprisal, 1919. Incorporating town element vs. IWW. **VIG V**

Summary and Analysis of the 42 Episodes: Thirty-four (or about 4 out of 5) of the 42 episodes resulted in clear-cut victory for the incorporators. Yet, the remaining 8 episodes (including 5 with ambiguous outcomes) underscore that, although the overall result of the Western Civil War of Incorporation was victory for the incorporating conservative forces, the resisting elements were strong. In fact, in at least 23 of the 34 episodes, the opposition or threat to the incorporators was significant. (The numbers of these are 2-4, 6-9, 11, 12, 15, 16, 18-20, 22-24, 28, 29, 32, 35, 37, 41.) The era of World War I (in this case, 1916–19) coincided with a final powerful surge of the incorporating trend, resulting in seven episodes (#36-#42). In all but two of these episodes (#37 and #41), the war was a direct factor (#36, #38, #39, #40) or an indirect factor (#42). All seven resulted in clear-cut incorporating victories. Note: The incorporation of Indian tribes through their concentration on reservations and incorporation into local labor forces was widespread in the West. Episodes #1 and #10 are examples of this type of incorporation.

radical, violence-prone Western Federation of Miners (WFM). Striking back in 1899, the WFM attacked the Bunker Hill and Sullivan mine, the leading enterprise of the Coeur d'Alenes, and destroyed its huge, costly concentrator with dynamite. This was too much for Idaho's hitherto prolabor governor Frank Steunenberg, who induced the federal government to send in troops. The latter put the area under martial law and broke the WFM strike by herding six to seven hundred miners into the hated bull pen.

Governor Steunenberg's action and the resulting repression of the WFM in the Coeur d'Alenes led to two dramatic events in the Western Civil War of Incorporation. The first was the organization of the revolutionary, anti-capitalist Industrial Workers of the World (IWW) in Chicago in 1905. The second was the 1905 assassination of Steunenberg by the WFM terrorist Harry Orchard, whose bomb killed Steunenberg (no longer governor) outside his Caldwell, Idaho, home. The aftermath of the crime was not restricted to the life sentence received by Orchard. Charges of conspiracy in the murder of Steunenberg were filed against the charismatic WFM secretary-treasurer, William D. ("Big Bill") Haywood, and two of his associates. The result was one of the greatest show trials in western history, in which the Chicago radical and criminal lawyer Clarence Darrow successfully defended the accused. The acquittal of Haywood and his colleagues was widely viewed by organized labor and its sympathizers as a vindication of the anti-incorporating militant labor movement of the West.

On behalf of the WFM, the charismatic Big Bill Haywood was one of the key founders and leaders of the IWW, whose members were frequently referred to as "Wobblies." Using class struggle as its theme, the IWW spread throughout the West. Some of its strongest support came from the loggers and sawmill workers of the Pacific Northwest. In contrast to its revolutionary rhetoric, the IWW was more often the victim than the initiator of violence, but in the spirit of no duty to retreat, it unhesitatingly fought back in a series of Pacific Northwest confrontations with capital and its supporters. These face-offs were especially acute in western Washington, where the 1916 Everett Massacre killed twelve, mostly Wobblies, and the 1919 Centralia Massacre left five dead, including one Wobbly. In Butte, Montana, in 1917, Frank Little, the IWW organizer and radical opponent of U.S. participation in World War I, died after being swung from a trestle at the end of a vigilante rope. Among many other IWW conflicts in the Western Civil War of Incorporation, Wobbly organizing among hop-field agricultural workers in Wheatland, California, in 1913 led to a strike. The ensuing riot and gunfight between sheriff's deputies and the IWW group produced five deaths, three of them on the anti-IWW side. The Wheatland episode was used by John Steinbeck as the model for the climactic act of violence in his reform novel *The Grapes of Wrath* (1939).

Early in the twentieth century, Los Angeles, in the process of supplanting San Francisco as the metropolis of the Far West, was a bastion of antiunion sentiment. On 1 October 1910, two American Federation of Labor militants, the brothers John J. and James B. McNamara, blew up the *Los Angeles Times* building, killing twenty people. The brothers were angered by the fierce open-shop policy of the *Times*' powerful publisher, Harrison Gray Otis, a prime incorporator of Los Angeles and its sun-drenched environs. Although the loss of life in the destruction of the *Times* building was a traumatic setback for Otis, the backlash of public opinion against labor and its supporters in Los Angeles was, in the long run, a triumph for the *Times* publisher and for the cause of incorporation. Otis was one of the members of the Los Angeles elite whose land speculation in the city's San Fernando Valley area greatly benefited from the early-twentieth-century construction of the Los Angeles Aqueduct, which, by bringing water hundreds of miles from the Owens Valley, gave the city the water supply needed for its booming economic development and vastly increased population. But what helped Los Angeles hurt the Owens Valley, where citizens banded together in 1924 frequently to dynamite the aqueduct in a delayed but fruitless rebellion against incorporation within the imperial outreach of Los Angeles and against the likes of Otis and his successor as *Times* publisher, Harry Chandler.

The conservative forces in the Western Civil War of Incorporation often employed the Pinkerton Detective Agency. The Pinkertons firebombed the family home of Jesse and Frank James and assisted in breaking up the bank- and train-robbing gang. Under the leadership of James McParlan (who earlier had played the key role in the Pinkertons' shattering of the Molly Maguire labor terrorists in the anthracite fields of Pennsylvania), the Denver office of the Pinkerton agency waged an effective but bitterly contested war against the nascent labor unions of the West. Wells Fargo and the Southern Pacific were among the powerful private concerns with their own detective forces.

The public enforcers of the law—local marshals and police, county sheriffs, state agencies such as the Texas Rangers and the Arizona Rangers, U.S. marshals and their deputies—were ambivalent. In numerous cases, these functionaries conducted tough but honorable—even heroic—operations to enforce the law in what was certainly an unruly region. On many other occasions, these law enforcers were willingly co-opted by the conservative side in the Western Civil War of Incorporation—for example, the gunfighting lawmen Wild Bill Hickok and Wyatt Earp and the widespread attack on the IWW. Aside from the involvement of law enforcement in the Western Civil War of Incorporation, local officers had to cope with the rampant disorder of the gunfight-prone West in the late nineteenth century. The bloodiest such confrontation occurred in 1872 in the Indian Territory (present-day eastern Oklahoma). At the community of Going Snake, an internal Cherokee feud and a jurisdictional conflict between a Cherokee court and a U.S. commissioner resulted in a gun battle that killed eleven people, eight of whom were members of a posse led by U.S. deputy marshals. This was the largest massacre in the two-century history of the federal marshal system.

The turbulent 1870s–1880s mining camp of Bodie, California, was typical of the violent pastoral and mining boomtowns where highly homicidal gunfighters were seldom condemned by law or public opinion as long as they observed the Code of the West. Away from these mining camps and prairie towns, however, the rapidly urbanizing West of the late nineteenth century resembled the rest of the United States. According to a study of Alameda County, on the east side of San Francisco Bay, these areas were becoming less violent as the criminal-justice system responded to the mounting public demand for a peaceful civic culture.

Sometimes a part of the Western Civil War of Incorporation and sometimes not were the ethnic, racial, and religious conflicts that all too often yielded massacres and murderous riots. Ethnic animosities were at times related to industrial violence in the West, whereas religious identity was frequently linked to ethnic status. In religion, one key conflict was between Mormons and the Gentiles (non-Mormons) who opposed them. After a violent expulsion from northwestern Missouri, where eighteen died in the Haun's Mill Massacre of 1838, the Mormons created the new metropolis of Nauvoo, Illinois, near which Mormonism's founder, Joseph Smith, was murdered in 1844. Moving again, the Mormons found a new refuge in the West. Established in 1847 under Brigham Young, the Mormon colony in Utah thrived, but trouble rose anew when its practice of polygamy was openly announced. Ensuing tension between the Mormons and the federal government produced a U.S. Army expedition into the Mormon country. Near-hysterical feelings of self-defense swept the Mormons of Utah in 1857 as they prepared to fight for their lives. War was averted, but in late summer a heinous act of violence—the Mountain Meadows Massacre—occurred in southwestern Utah, where, among the frontier Mormon villagers, religious frenzy in the face of the federal threat was highest. The tragic outcome was the slaughter of about one hundred men, women, and older children of a California-bound wagon train of Arkansans by a Mormon-led force of Paiute Indians and local Mormons. Only eighteen of the younger children were spared in what was the largest massacre of white civilians in western history. An order from Brigham Young came too late to save the victims.

The long-term warfare between whites and Indians sometimes passed the line from quasi-military fighting to the massacre of civilians on both the white and the Indian sides. Massacres of Indians by Indians were not unknown, with the last large episode occurring near Camp Grant, Arizona, in 1871 when some one hundred Apaches were killed by a band spearheaded by ninety-four of their traditional Papago enemies who had been incited by Apache-hating Anglos and Hispanos of Tucson. Of massacres of whites by Indians, Marcus and Narcissa Whitman and twelve others were killed by Cayuse at the Whitmans' mission near present Walla Walla, Washington, in 1847. On the Oregon Trail in Idaho were the Ward (1854) and Otter (1860) wagon train massacres, with eighteen and thirty-two lives lost, respectively, to Shoshones. These massacres were exceptions to the rule that Indians far more often aided the overland pioneers than attacked them. One hostile Indian campaign by the Apache band of Josanie (Ulzana) in New Mexico and Arizona killed forty-five civilians in 1885. Much earlier, rebel Indians killed many civilians in the Pueblo revolt of 1680 in New Mexico and in five uprisings by California Indians against Hispanic mission communities in 1775–1824.

Exacting much heavier casualties than Indian massacres of whites were white massacres of Indians: the Bear River Massacre in southeast Idaho in 1863 (90 women and children killed); the Sand Creek Massacre in eastern Colorado in 1864 (about 200 Cheyenne men, women, and children slain); and the Marias River Massacre in northern Montana in 1870 (173 Blackfeet deaths, mostly women and children). Women and children were also killed in the 1868 massacre of 103 Cheyennes on the Washita River in western Oklahoma and the 1890 slaughter of 150 or more Sioux at Wounded Knee, South Dakota. Federal soldiers conducted all of these massacres except for that by the Colorado militia at Sand Creek. Merciless were the genocidal tactics of land-grabbing white men in the fecund Round Valley region of California's northwestern mountains

The photographer George Trager recorded the mass burial of 146 Sioux gunned down by the Seventh Cavalry on 29 December 1890 at Wounded Knee, South Dakota. Though the massacre seemed to mark the end of Indian-white warfare, it was part of a continuing pattern of western violence in which federal troops were called on to police internal dissidents.

George Trager (1861–after 1892). Burial of the Dead at the Battle of Wounded Knee, S.D. *Albumen silver print, 1 January 1891. Nebraska State Historical Society, Lincoln.*

in the 1850s and 1860s. Here the population of the Yukis and other Indians fell from over 11,000 to under 1,000.

White fears of Chinese job competition inspired the West's virulent anti-Chinese movement, which was often spearheaded by radical labor reformers. The result was a long chronicle of violence and intimidation: anti-Chinese riots in Los Angeles (eighteen to nineteen Chinese dead in 1871), Seattle, and Tacoma; an 1887 slaughter of ten Chinese miners at Log Cabin Bar, Oregon, on the Snake River; and, two years earlier, a massacre of fifty-one Chinese (with the expulsion of four to five hundred others) at the coal-mining center of Rock Springs, Wyoming.

Violence between Hispanos and Indians early in the years of Spanish colonization in the Southwest was followed by Hispano conflict with the later-arriving Anglos. Violence by Hispanic outlaws and gunfighters against Anglos represented a species of resistance in the Western Civil War of Incorporation but also included the independent factor of ethnicity. This was true of the vigilante lynching of a Mexican woman, Josefa (later often called Juanita), in Downieville, California, in 1851, who had killed a drunken miner who had molested her.

Women, much less involved than men in local murders and assaults, were seldom legally executed, and Josefa was one of only several women lynched in the West. One study shows that black women inmates in the Kansas state prison were disproportionately represented in the female prison population because of racial discrimination. Recent research on physically mistreated wives in the mixed urban and rural society of

Lane County, Oregon, of the 1890s suggests that these women were unusually assertive and that their abusive husbands were economically unsuccessful and psychologically insecure.

From mid-century on, the Hispanic *bandidos* of California and Texas waged an anti-Anglo vendetta. In the Golden State, gunfighting Hispanic outlaws operated in the 1850s–1870s against the Americans who streamed in after the U.S. acquisition of California. Much of this was sheer criminal activity, but the raids and killings by the social bandit Joaquin Murrieta and others also had strong rebellious overtones. Similar animosities operated in a vast zone in southern Texas. Such notable Mexicans as Juan Cortina had the dual identity of border brigand and patriotic Hispanic nationalist. Resembling their outlaw counterparts in California were Gregorio Cortez and other resister gunfighters celebrated by Texas Hispanos in folklore and song as social bandits. Raids back and forth across the Texas-Mexico line found the *bandidos* at war with civilian law officers, Texas Rangers, and, occasionally, U.S. troops. The last such raid was a bold 1915 attack on the Norias unit of the King Ranch north of Brownsville.

Conclusion

By 1920 the Western Civil War of Incorporation was over, with the conservative side emerging strongly victorious. A final surge of seven episodes of the regional civil war had occurred in the 1916–19 era of World War I as the forces of resistance made their last stand. In the overall war, one of the episodes (the White Cap conflict in New Mexico) was a defeat for the incorporators; seven others were ambiguous or unclear in their outcomes. These eight episodes in which the anti-incorporating faction was not clearly vanquished—along with the heavy violence in most of the other episodes—show that resistance to the incorporating trend was dauntless.

Brutality and oppression were plentiful in the Western Civil War of Incorporation but should be viewed in proportion. Mitigating the harsh reality of and coexisting with much of the Western Civil War of Incorporation was a remarkably open, mobile, and expanding society in the West from the 1880s. This enabled a great many of the lower class and middle class not only to avoid the tragic battlegrounds in the regional civil war but to prosper and thrive. Nor should the popularity of the incorporating victory be overlooked and underestimated, for there was a widespread desire—by no means restricted to the elite and the affluent—for the more orderly, structured society that was one result of the Western Civil War of Incorporation. After 1890 a series of social, economic, and political reform movements and advances in popular education for the upwardly mobile softened the impact of the Western Civil War of Incorporation without diminishing the order and stability that was, in part, its legacy.

In the aftermath of the conservative triumph in the Western Civil War of Incorporation, the region, from 1920 to 1960, experienced its least violent times. But the relative calm of that era dissolved into turbulent decades of protest, riot, crime, and assassination. The anti-Hispanic Zoot Suit Riot in Los Angeles in 1943 was a portent of the West's post-1960 period of violence. In its own distinctive way, western violence of the 1960s and after mirrored the postindustrial surge of crime and disorder that afflicted all the technologically advanced democracies of the world, except for Japan. No country was more affected than the United States. Typifying the new turbulence was the unprecedented phenomenon of numerous serial murderers whose relentless violence was exemplified by Ted Bundy's killing of at least nineteen women in Washington, Utah, and Colorado in 1974–75.

Led by the massive Watts riot in Los Angeles, group violence was at its greatest in the black-ghetto uprisings of the 1960s. Watts in 1965 had much earlier precedents in the Texas riots of African-American soldiers in Brownsville in 1906 and Houston in 1917. By the 1980s, ultra-violent drug-dealing gangs of young male African Americans and Hispanos spread through the big cities of the West from their citadels in the ghettos and barrios of Los Angeles. Black gang members spearheaded the Los Angeles riot of 1992, but Hispanos, Asians, and whites were also among the rioters. The riot was a combination of protest, crime, and nihilism, a reflection of late-twentieth-century western racial diversity, and an indication of the growing anomie of western cities.

A throwback to the white-Indian conflict of a century earlier was Wounded Knee II: a fatal 1973 confrontation between Indian militants (two killed) and federal marshals at the site of the 1890 massacre in South Dakota. Most shocking of all was the assassination of President John F. Kennedy in Dallas on 22 November 1963, followed five years later by the scarcely less-unsettling assassination in Los Angeles of Senator Robert F. Kennedy, a younger brother of the president. The western setting of the Kennedy assassinations was an eerie reminder to the historically knowledgeable of a long tradition of assassinations in territorial New Mexico, a tradition that was grounded in the latter's unstable and deeply conflicted society. Aside from the western locations, however, there was nothing uniquely western about the assassinations of the Kennedy brothers.

Many of our images of recent violence in the West come not from the fiction of print or film but from the news seen by millions on television. The presidential motorcade suddenly disrupted by gunfire blasting into the body of John F. Kennedy or the flames searing the horizon of riot-torn Los Angeles are images deeply etched in the national consciousness.

Yet, the most enduring image of western violence is the "walkdown": two holstered westerners, armed with six-guns, pace toward each other down the bleak street of a frontier town, ready to draw and shoot. Climaxed by the inevitable burst of gunfire, the image focuses on one of the men, dying in the dust—dropped by the bullet of the survivor, standing tall in triumph. In this image, the victorious gunfighter is the hero and his fallen foe the villain. In the popular parlance, the hero wears a white hat—the villain a black one.

This image of the western walkdown was fixed in the American mind by its portrayal in the climactic episode of the most influential western novel—*The Virginian* (1902), by Owen Wister. In Wister's book, the heroic Wyoming cowboy, "the Virginian," slays the evil Trampas. So popular was Wister's novel (avidly read by President Theodore Roosevelt, the author's good friend, to whom the book was dedicated) that the walkdown became the central formulaic event in the western fiction of print and film. Yet, as the historian Kent Ladd Steckmesser has suggested, the fictional walkdown in *The Virginian* may have been based on a real walkdown—one at which onlookers by the town square of Springfield, Missouri, saw Wild Bill Hickok gun down his enemy, Dave

The "walkdown," fixed in the popular imagination by the climactic scene of Owen Wister's *The Virginian* (1902), remains a central visual metaphor for western culture. The enduring popularity of the image reflects a public taste for clear-cut moral distinctions between good and evil and an enduring fascination with violence as a means to enforce the rule of law.

Charles M. Russell (1864–1926). The Virginian Looks Down at Trampas. *Drawing reproduced as book illustration in Owen Wister,* The Virginian *(New York: Macmillan Publishing Co., Inc, 18th printing, 1979).*

Film still from The Good, the Bad, and the Ugly *(1968), directed by Sergio Leone. Courtesy of the Academy of Motion Picture Arts and Sciences, Beverly Hills, California.*

Roy Lichtenstein (b. 1923). Fastest Gun. *Magna on canvas, 1963. © Roy Lichtenstein. Courtesy of the artist.*

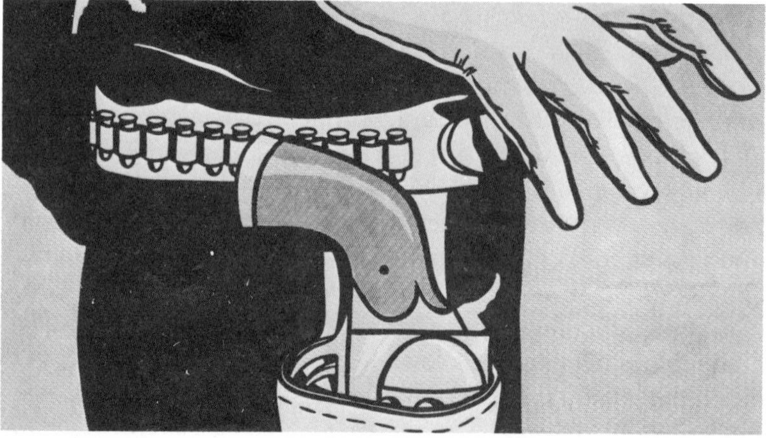

Tutt, in 1865. True to the formula, Hickok was presented as a hero to those who read the account of this prototypical no-duty-to-retreat western gunfight in *Harper's*, the nation's favorite magazine.

Scholars have devoted much attention to popular western fiction, of which *The Virginian* remains the most significant example. Henry Nash Smith traced the origins of the genre to the Leatherstocking novels of James Fenimore Cooper. Smith also emphasized the decline of the popular western into the literarily debased form of the dime novel with its mass readership in the later nineteenth century. Aside from the literary quality of popular western fiction, specialists have focused on the values upheld by it. The scholarly consensus is that popular western fiction, whether in print or film, embodies a deep formula in which the hero, according to John G. Cawelti's study *The Six-Gun Mystique* (1984), mediates between civilization and savagery (or, in the comparable terms of other scholars, between culture and nature, order and chaos). The gunfighting skill of the hero represents the savagery of violence, but his objective of besting evil is in the interest of civilization. In this deep formula the hero is also a transitional figure: one who employs the violence of the frontier West to establish the peaceable society of civilized values that should succeed it. Thus, the hero reflects deeply conservative social values aligned against the threat of anarchy.

Neither Smith nor Cawelti were impressed with the literary quality of formula western fiction, but among recent scholars who take these writings more seriously as literature are Christine Bold, Michael Denning, and Cynthia S. Hamilton. Denning holds that dime-novel authors used the western setting "not only for escapist adventure but to state social conflicts through figures of bank and train robbers," aggrieved cowboys, and range wars. Wister based *The Virginian* on the conflict between the big cattlemen of Wyoming and the rustler element, the conflict that peaked in the Johnson County War. Wister was a firm friend of some of the cattle barons of Wyoming, and *The Virginian* expresses, in literature, their conservative version of the Western Civil War of Incorporation. Wister's villain, Trampas, was modeled on a true bad man—one of the Wyoming rustlers whom Wister himself had met. The heroic Virginian rides with a vigilante band of the kind that took to the field in the Johnson County War.

Since Wister, popular western fiction in both print and film has often reflected the Western Civil War of Incorporation and the emergence of a cognitive split in the mythology of the western hero. The conservative winning side in the Western Civil War of Incorporation bred a socially conservative myth of the hero—for example, the fictional Virginian and the mythic versions of the real-life Wild Bill Hickok and Wyatt Earp. The anti-incorporating side in the regional civil war generated a dissident social-bandit myth in which the heroes were real-life outlaws like Jesse James, Billy the Kid, Joaquin Murrieta, and Gregorio Cortez. Both the conservative mythic hero and the insurgent social-bandit hero have had wide appeal because Americans are deeply ambivalent about established power and dissident protest.

Aside from mass-market formula fiction, many authentic novels of high quality have been based on episodes of western violence: *The Ox-Bow Incident*, by Walter Van Tilburg Clark, the previously mentioned antivigilante novel; *The Lady* (1957), by Conrad Richter, inspired by the tragic deaths of Judge Albert J. Fountain and his son; *A Very Small Remnant* (1963), by Michael Straight, based on the Sand Creek Massacre;

and, with its climax in the slaughter of the Blackfeet on the Marias River, the remarkable *Fools Crow* (1986), by James Welch.

There are two key questions about western violence. First, just how violent was the West? Due to the values of its people, the Western Civil War of Incorporation, and the ubiquity of ethnic, racial, and religious conflict, the West was a turbulent region. This was the result, however, of the particular historical experience of the West. Westerners were not innately more violent than people elsewhere. Leading social, economic, and political blocs freely resorted to violence to advance or defend their interests. Closely connected to key episodes of this western violence were such leading figures and men of power as Leland Stanford, Stephen J. Field, Thomas B. Catron, Albert Bacon Fall, and Granville Stuart. Yet, some qualification is in order.

Many communities and areas of the West were notably violent, but others were not. No region of the West was more violent than central Texas from 1860 to the 1890s, a huge area bounded roughly by Fort Worth, Dallas, Houston, San Antonio, and San Angelo. In this locale of multicultural convergence, Hispanos, German immigrants, and slaveholding southern whites invaded the realm of the native Indians. Violence abounded as a result of the Civil War and its aftermath and also as a result of white-Indian warfare, vigilantism, cattle-range conflict, outlaw activity, community feuds, ethnic and racial tension, agrarian discontent, and political tumult. Gunplay was common, as exemplified by John Wesley Hardin, the West's deadliest gunfighter with over twenty killings arising from his participation in post–Civil War white-black racial strife and in political conflicts and community feuds.

A central Texas culture of violence based on the spirit of no duty to retreat skewed the behavior of the people. As a contemporary wrote, self-defense was "the usual plea of the man-slayer" with "wide latitude" given to its definition. "A look may, if it have . . . sufficient of malice in it, justify resort to the pistol pocket. A touch [to the pocket] frequently justifies instant shooting." Nor was the violence self-contained in the Lone Star State. Central Texas cowboys, cattle kings, and outlaws riding the trails north, northwest, and west took their bent to violence with them, as seen in such gunfighters as Frank Canton in Wyoming and Oliver M. Lee in New Mexico as well as the outlaws John Ringo and Curly Bill Brocius in Arizona. Central Texas expatriates significantly tinctured wide western expanses with the virus of violence.

President Lyndon B. Johnson was born and reared in the heart of the violent region of central Texas. Johnson biographers and Johnson himself have averred the formative influence of his central Texas homeland in shaping his presidential attitudes and values. His relentless determination to defend militarily what he saw as the American national interest in South Vietnam was typical of his central Texas heritage of no-duty-to-retreat violence. As Johnson made his 1965 decision to commit large-scale land forces to the defense of South Vietnam, he invoked the spirit of one of his heroes, a gunfighting Texas Ranger, Captain L. H. McNelly, to admonish the American people that "courage is a man who keeps coming on."

Away from central Texas, disputes over property rights and human rights were endemic in industries like mining, timber, and cattle. Often related to the Western Civil War of Incorporation, such conflicts generated an enormous amount of violence.

Mining, mill, and cattle towns were frequently violent places, but there were also many communities in the West where violence was rare. In 1960, the homicide rate of the West was second to that of the South among the nation's regions. Contrary to this overall sectional pattern, however, two sets of western states had homicide rates in 1960 that were among America's lowest: a Northwest group of Washington, Oregon, Idaho, and Utah; and a Great Plains wedge of Kansas, Nebraska, and the Dakotas. Some basic cultural predispositions were behind the low proclivity to homicide in these states. Both groups of states had strong contingents of core settlers from the Northeast whose regional culture, as the social historian David Hackett Fischer has noted, made them averse to violence.

Governmental structure was a key factor in western violence. Comparative studies of the Canadian and the American West show that miners prone to violence and vigilantism under the loose, permissive rule of the American federal system became peaceable and law-abiding when they migrated to Canada, where the more centralized, stricter government was staunchly intolerant of violence. And yet, as violent as it was, the American West never produced anything like the hundreds of thousands of civilian casualties resulting from the anarchic political violence in the South American nation of Colombia from the 1940s to the 1960s.

Although on a far smaller scale than in Colombia, violence was a principal factor in western U.S. history. This leads to the second question about western violence: has it been mainly responsible for America's unenviable distinction as the most violent nation among its peer group of the technologically advanced democracies of the globe? The answer is no. The turbulent history and values of the West have been a major contributor to our nation's violent heritage but no more so than ethnic, racial, religious, industrial, agrarian, and political conflict or than the crime, lynch-law, and violent examples and legacies of the American Revolution and the Civil War. The West is but one example of the pluralism of American history and society that has yielded both the bane of violence and the blessings of freedom and opportunity. In spite of the incorporating trend of 1850–1920 and the excess of violence, millions of immigrants worldwide have been attracted to the open, democratic society of both America and its western region.

Bibliographic Note

A bibliographical essay is Richard Maxwell Brown, "Historiography of Violence in the American West," in Michael P. Malone, ed., *Historians and the American West* (Lincoln, 1983), 234–69. A comprehensive work on western history that perceptively treats violence is Richard White, *"It's Your Misfortune and None of My Own": A New History of the American West* (Norman, 1991). For my concepts of the Western Civil War of Incorporation and gunfighters, as well as the doctrines of no duty to retreat and the homestead ethic, see Richard Maxwell Brown, *No Duty to Retreat: Violence and Values in American History and Society* (New York, 1991) and "Western Violence: Structure, Values, Myth," *Western Historical Quarterly* 24 (February 1993): 5–20. The general concept of incorporation first appeared in Alan Trachtenberg, *The Incorporation of America: Culture and Society in the Gilded Age* (New York, 1982). The historical context of ethnic and racial violence in the West is compellingly treated in Patricia Nelson Limerick, *The Legacy of Conquest: The Unbroken Past of the American West* (New York, 1987), part 2.

An influential treatment of the frontier myth in relation to American culture and violence is the trilogy by Richard Slotkin, *Regeneration through Violence: The Mythology of the American*

Frontier, 1600–1860 (Middletown, Conn., 1973), *The Fatal Environment: The Myth of the Frontier in the Age of Industrialization, 1800–1890* (New York, 1985), and *Gunfighter Nation: The Myth of the Frontier in Twentieth-Century America* (New York, 1992). The concept of the social bandit in E. J. Hobsbawm, *Social Bandits and Primitive Rebels* (Glencoe, Ill., 1959), is applied by Richard White to "Outlaw Gangs of the Middle Border: American Social Bandits," *Western Historical Quarterly* 12 (October 1981): 387–408. On vigilantism, see Richard Maxwell Brown, *Strain of Violence: Historical Studies of American Violence and Vigilantism* (New York, 1975), Robert M. Senkewicz, *Vigilantes in Gold Rush San Francisco* (Stanford, 1985), and Richard Hogan, *Class and Community in Frontier Colorado* (Lawrence, 1990). Emphasizing the West is W. Eugene Hollon, *Frontier Violence: Another Look* (New York, 1974).

For the Mussel Slough conflict, see Brown, *No Duty to Retreat*, chap. 3. On the Johnson County War, see Helena Huntington Smith, *The War on Powder River* (New York, 1966). The following works address the Lincoln County War and Billy the Kid: Robert M. Utley, *High Noon in Lincoln: Violence on the Western Frontier* (Albuquerque, 1987) and *Billy the Kid: A Short and Violent Life* (Lincoln, 1989); Frederick W. Nolan, *The Lincoln County War: A Documentary History* (Norman, 1992); and Stephen Tatum, *Inventing Billy the Kid: Visions of the Outlaw in America, 1881–1981* (Albuquerque, 1982). C. L. Sonnichsen, *Tularosa: Last of the Frontier West* (New York, 1960), treats the Tularosa war, whereas the Cochise County War and the Earps are dealt with in Brown, *No Duty to Retreat*, chap. 2, and in Paula Mitchell Marks, *And Die in the West: The Story of the O.K. Corral Gunfight* (New York, 1989). For the James-Younger gang and the outlaw tradition initiated by them, see William A. Settle, Jr., *Jesse James Was His Name; or, Fact and Fiction Concerning the Careers of the Notorious James Brothers of Missouri* (Columbia, 1966), Paul I. Wellman, *A Dynasty of Western Outlaws* (1961; reprint, Lincoln, 1986), Glenn Shirley, *West of Hell's Fringe: Crime, Criminals, and the Federal Peace Officer in Oklahoma Territory, 1889–1907* (Norman, 1978), and White, "Outlaw Gangs."

A key article is Carlos A. Schwantes, "The Concept of the Wageworkers' Frontier: A Framework for Future Research," *Western Historical Quarterly* 18 (January 1987): 39–55. The industrial phase of the Western Civil War of Incorporation is reflected in many books dealing with violent episodes from the 1890s to the 1910s. Exemplifying this scholarship are George S. McGovern and Leonard F. Guttridge, *The Great Coalfield War* (Boston, 1972), and Zeese Papanikolas, *Buried Unsung: Louis Tikas and the Ludlow Massacre* (Salt Lake City, 1982), on the Ludlow violence. Norman H. Clark, *Mill Town: A Social History of Everett, Washington, from Its Earliest Beginnings on the Shore of Puget Sound to the Tragic and Infamous Event Known as the Everett Massacre* (Seattle, 1970), strikingly treats the Everett Massacre. John McClelland, Jr., *Wobbly War: The Centralia Story* (Tacoma, Wash., 1987), is the salient work on the Centralia Massacre. Discussing incorporation in California's Owens Valley and its violent legacy is John Walton, *Western Times and Water Wars: State, Culture, and Rebellion in California* (Berkeley, 1992).

Notable works on violent western law enforcement are Frank R. Prassel, *The Western Peace Officer: A Legacy of Law and Order* (Norman, 1972), Larry D. Ball, *The United States Marshals of New Mexico and Arizona Territories, 1846–1912* (Albuquerque, 1978) and *Desert Lawmen: The High Sheriffs of New Mexico and Arizona, 1846–1912* (Albuquerque, 1992), Leon C. Metz, *Pat Garrett: The Story of a Western Lawman* (Norman, 1974), and Robert K. DeArment, *George Scarborough: The Life and Death of a Lawman on the Closing Frontier* (Norman, 1992).

The historiography of western massacres is large. Salient studies include Juanita Brooks, *The Mountain Meadows Massacre* (Norman, 1991), Gary L. Roberts, *Sand Creek: Tragedy and Symbol* (Ann Arbor, 1984), Lynwood Carranco and Estle Beard, *Genocide and Vendetta: The Round Valley Wars of Northern California* (Norman, 1981), and Craig Storti, *Incident at Bitter Creek: The Story of the Rock Springs Chinese Massacre* (Ames, Iowa, 1991). Massacres involving Indians are cited in Robert M. Utley, *The Indian Frontier of the American West, 1846–1890* (Albuquerque, 1984). Hispanic White Caps and *bandidos* are found in Robert J. Rosenbaum, *Mexicano Resistance in the Southwest: "The Sacred Right of Self-Preservation"* (Austin, 1981). David A. Johnson is writing a book on the lynching of Josefa.

Outstanding works are Kent Ladd Steckmesser, *The Western Hero in History and Legend* (Norman, 1965), with chapters on Wild Bill Hickok and Billy the Kid; Joseph G. Rosa, *They Called Him Wild Bill: The Life and Adventures of James Butler Hickok*, 2d ed. rev. (Norman, 1974); Roger D. McGrath, *Gunfighters, Highwaymen, and Vigilantes: Violence on the Frontier* (Berkeley, 1984), on Bodie, California, and Aurora, Nevada; Don Dedera, *A Little War of Our Own: The Pleasant Valley Feud Revisited* (Flagstaff, Ariz., 1988), on the mordant 1880s Tonto Basin War in Arizona; and Gary L. Roberts, *Death Comes for the Chief Justice: The Slough-Rynerson Quarrel and Political Violence in New Mexico* (Niwot, 1990).

Henry Nash Smith, *Virgin Land: The American West as Symbol and Myth* (Cambridge, Mass., 1950), is a classic study. In addition to John G. Cawelti's seminal work *The Six-Gun Mystique*, 2d ed. rev. (Bowling Green, Ohio, 1984), formulaic western fiction is treated in Christine Bold, *Selling the Wild West: Popular Western Fiction, 1860 to 1960* (Bloomington, 1987), Michael Denning, *Mechanic Accents: Dime Novels and Working-Class Culture in America* (London, 1987), Cynthia S. Hamilton, *Western and Hard-Boiled Detective Fiction in America: From High Noon to Midnight* (Iowa City, 1987), and Jane Tompkins, *West of Everything: The Inner Life of Westerns* (New York, 1992).

For the context of late-twentieth-century Los Angeles violence, see Mike Davis, *City of Quartz: Excavating the Future in Los Angeles* (New York, 1992). On central Texas violence, see C. L. Sonnichsen, *I'll Die before I'll Run: The Story of the Great Feuds of Texas* (New York, 1962), and Brown, *Strain of Violence*, chap. 8. For homicide in the West and the peaceful impact of settlers from the Northeast, see Raymond D. Gastil, *Cultural Regions of the United States* (Seattle, 1975), chap. 3, and David Hackett Fischer, *Albion's Seed: Four British Folkways in America* (New York, 1989), 889–93.

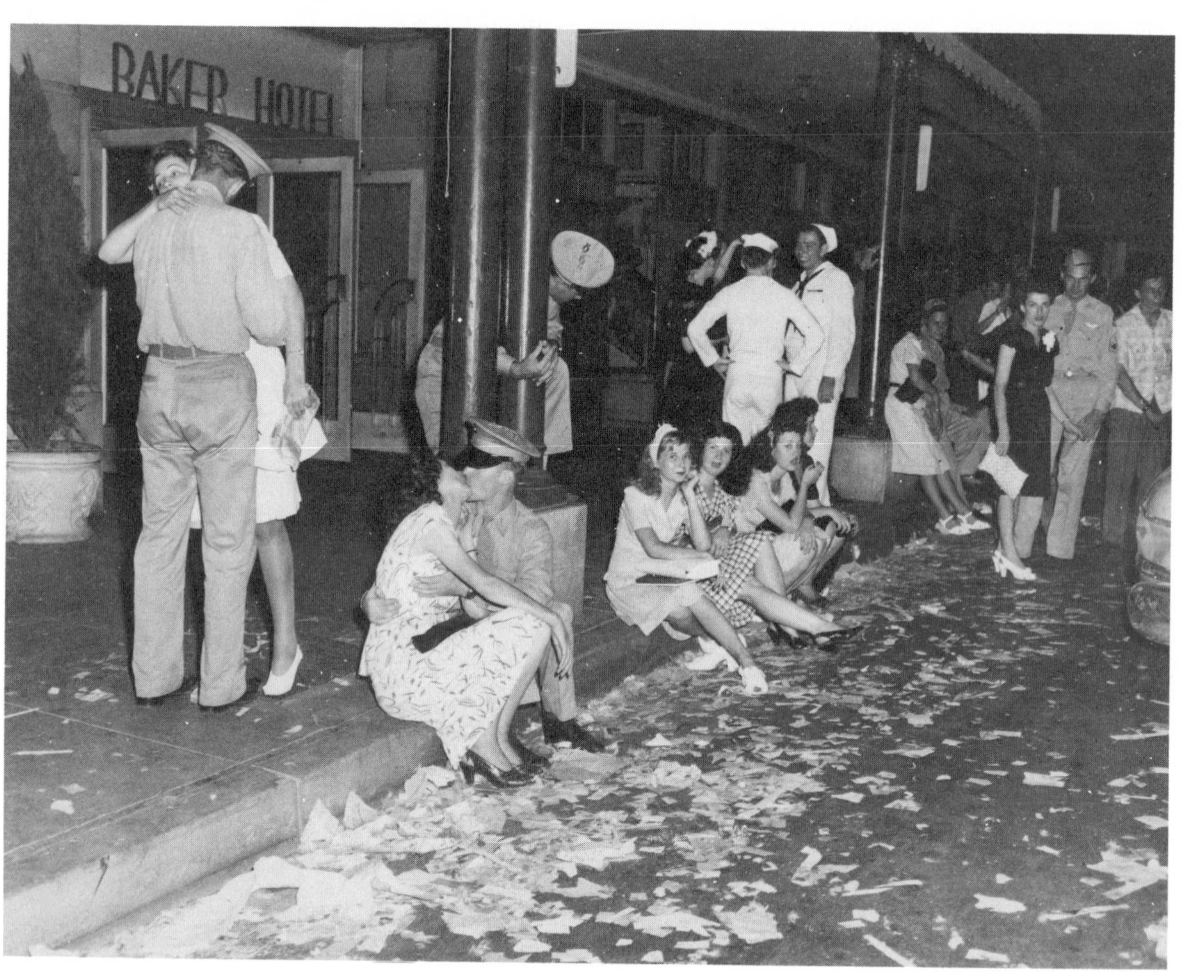
BAKER HOTEL

SECTION III
Transformation

Just as scientists map the slant of the West's lands and the flow of its streams, so historians plot the course of events that have occurred in the region. Both groups of scholars are looking for watersheds that indicate a significant shift in direction. For a time, the scholarly world assumed that the year 1890 marked the "continental divide" of western history. The closing of the frontier, as announced by the director of the U.S. Census Bureau and analyzed by the historian Frederick Jackson Turner, did not, however, have an immediate impact on settlement patterns or economic growth. Record numbers of homestead filings lay in the future, as did booms in copper, lumber, and oil.

Yet the 1890 announcement did represent a psychological turning point, for it led Americans to question how long the western resources once deemed limitless would endure. For leaders like Theodore Roosevelt, this awareness had a twofold consequence, intensifying not only his belief in conservation at home but also his commitment to imperialism overseas. Not by accident did the United States enter the twentieth century with a number of new Pacific "possessions": the Hawaiian Islands, American Samoa, Johnston Island, Wake Island, Guam, and the Philippines.

Some Americans linked the acquisition of these islands to the nation's earlier expansion. "The flag has never paused in its onward march," noted Albert J. Beveridge, the senator from Indiana. Supporting the McKinley administration's plan to withhold autonomy from the people of the Philippines, he asked, "If you deny [self-rule] to the Indian at home, how dare you grant it to the Malay abroad?" Yet even Beveridge recognized that this analogy could be distorting. The nation's latest acquisitions were not deemed destination points for Anglo-American settlement, as Indian lands had been, nor were they (including at first Hawai`i) considered candidates for statehood. Providing strategic bases as well as raw materials, they would link American manufacturers to the markets of East Asia.

The nation's new Pacific tilt excited western investors. In 1912 the streetcar magnate Henry Huntington predicted that the Los Angeles he had helped to develop would become "the most important city in this country, if not the world." Dismissing the Atlantic as "the ocean of the past," he glorified the Pacific as "the ocean of the future." "Europe can supply her own wants," he said. "We shall supply the wants of Asia."

The nation's experiment in colonialism abroad helped parts of the West to outgrow a quasi-colonial economy at home. Nevertheless, a war in the Pacific transformed the region even more dramatically than had trade across the Pacific. The two developments were not unrelated. American economic interests in East Asia and the western Pacific ran afoul of Japanese ambitions in the area. When Japanese incursions into China and Southeast Asia led to American economic reprisals, the Japanese attacked Pearl Harbor.

World War II marked a decisive moment in the history of the West. The money, jobs, and people that flowed west with the growth of defense industries helped initiate the region's emergence as a political and economic force in postwar America.

The Hayes (firm active 1930s–1960s) for the Dallas Times Herald. V-J Day, Dallas. *Gelatin silver print, 1945. From the collections of the Texas/Dallas History and Archives Division, Dallas Public Library.*

World War II involved combat on many fronts, but none loomed larger for Americans than the Pacific theater. It was here that the enemy struck without notice, all but crippling the American Pacific fleet and killing thousands of military personnel. Here too, American soldiers retreated from the Philippines. The United States suffered its most stinging reversals of World War II in the Pacific, but the nation responded with decisive action. Able to take the lead in this theater of operations, the United States pushed the Japanese out of the western Pacific and incinerated two home-island cities with atomic bombs. For both the Americans and the Japanese, this was, in the words of the historian John W. Dower, a "war without mercy."

Because the West lay closer than any other part of the nation to this theater of operations, because the region contained vast undeveloped areas ideal for training soldiers and testing bombs, and because westerners lobbied intensively to attract the business of war, the West received a disproportionate amount of U.S. military spending. Moreover, the end of World War II did not mean the end of the spending. Further international crises, including two more wars across the Pacific, helped to complete the West's economic, demographic, and political transformation.

Thus, the year 1941 is a watershed in western history, not only more profound than the artificially derived 1890 closing of the frontier but also more easily demonstrated. The post-1941 new West was a garrison state whose prosperity depended on Americans' willingness to serve as a *posse comitatus* overseas in a not-always cold war. So, in 1965, at a crucial juncture in the U.S. involvement in Vietnam, President Lyndon Johnson explained his response to a Vietcong mortar attack on U.S. Army advisers in Pleiku. "We have kept our gun over the mantel and our shells in the cupboard for a long time now. And what was the result? They are killing our men while they sleep in the night. I can't ask American soldiers out there to continue to fight with one hand tied behind their backs." A popular image of Old West readiness to rebuff Indian raids had become the rationale for initiating a long-term series of air strikes against North Vietnam.

The brutality and ineffectiveness of the Vietnam conflict shook the confidence of many Americans in the legitimacy of their cause. Two decades later, the dissolution of the Soviet Union further confused international issues. The consensus that had created the garrison state eroded, but the West had been forever changed by the decades-long mobilization.

Once a backwater of the nation, the West had grown in wealth and population to become an economic and political powerhouse. Still, its inhabitants were as diverse as ever, its internal discord just as intense. The residents of the region clashed over the development of natural resources, the extension of metropolitan areas, and the retention of the national military presence. They struggled over who belonged in the region, who had the right to name it, and who was allowed to call it home.

The West's history in the twentieth century, no less than in the nineteenth, combines elements sordid and sublime, tragic and triumphant. It is a story both more and less familiar because it has been experienced more than it has been studied. Yet so many dramatic changes have occurred in the twentieth century that this period is an essential part of the region's epic. The fullest understanding of the American West requires a historical appreciation of this watershed era.

— Carol A. O'Connor

CHRONOLOGY

1900	Frederick Weyerhaeuser opens "Sawdust Empire" in the Pacific Northwest.
1901	Major oil find at Spindletop, Texas.
1902	Newlands Reclamation Act establishes federal role in constructing dams and irrigation systems.
1906	Earthquake destroys downtown San Francisco.
1912	New Mexico and Arizona enter the Union. Center of motion picture production shifts from New York to Los Angeles.
1913–14	Strike by the United Mine Workers in Colorado culminates in the Ludlow Massacre.
1916	Congress creates the National Park Service. William Boeing begins building airplanes in Seattle.
1917–18	U.S. entry into World War I stimulates the West's economy but radical labor is suppressed.
1918	The Native American Church is incorporated in Oklahoma.
1922	Colorado River Commission Compact, involving seven states, sets a model for resource planning.
1924	Immigration Restriction Act bars Asians from migrating to the United States and establishes quotas based on "national origins" for groups outside the Western Hemisphere.
1931	Nevada legalizes gambling.
1931–40	Dust Bowl conditions hit the Great Plains.
1934	Great maritime strike ties up ports from San Diego to Seattle. Taylor Grazing Act regularizes stock growers' use of federally owned rangelands.
1939	The Hewlett-Packard electronics firm is founded in a garage in Palo Alto, California.
1941	Japanese attack Pearl Harbor; West becomes staging ground for U.S. operations in the Pacific.
1942	110,000 Japanese Americans are interned in concentration camps.
1943	Servicemen attack Mexican Americans and African Americans in "zoot suit" riots in Los Angeles.
1945	First atomic bomb is tested near Alamogordo, New Mexico. The San Francisco-based Bank of America becomes the world's largest bank.
1947	Congressional hearings on alleged Communist infiltration of the Hollywood film industry.
1955	Disneyland opens in Anaheim, California.
1959	Alaska and Hawai`i become the forty-ninth and fiftieth states.
1960	Del Webb starts selling homes in Sun City, Arizona, an entire community for retirees.
1961–62	The National Aeronautics and Space Administration's Manned Spacecraft Center opens near Houston, Texas.
1964	California passes New York to become the nation's most populous state.
1965	Riots engulf Watts district of Los Angeles. Immigration Act ends the discriminatory aspects of the 1924 law.
1970	Filipino and Hispanic workers win a five-year strike against grape growers.
1973	Members of the American Indian Movement and their Lakota supporters occupy Wounded Knee, South Dakota, for seventy-one days.
1975	Phelps Dodge halts copper production in Bisbee, Arizona. Coal towns like Gillette, Wyoming, profit from the energy boom.
1978	Dan White assassinates San Francisco Mayor George Moscone and Supervisor Harvey Milk.
1986	Immigration Reform and Control Act approved by Congress.
1989	*Exxon Valdez* runs aground in Alaska's Prince William Sound.
1992	Riots convulse metropolitan Los Angeles where members of racial "minorities" make up a majority of the population.

Chapter Twelve

Wage Earners and Wealth Makers

CARLOS A. SCHWANTES

Thursday morning, 12 July 1917, should have been like any other work-day in Bisbee, Arizona. As dawn backlit the crest of the Mule Mountains surrounding the state's premier mining city, the serpentine streets below should have echoed with the boots of several thousand men coming and going from the mines. The merchants of Brewery Gulch and other downtown streets should have been preparing for another day of brisk sales. It was wartime, and the city worked unceasingly to produce the copper so vital to communications equipment and munitions. Every rifle cartridge contained about half an ounce of the pure red metal. Copper would win the war.

Within the past decade Bisbee had briefly emerged as Arizona's largest city, and even in July 1917 it still ranked with Phoenix and Tucson. Bisbee's nearly twenty thousand residents took pride in the fact that theirs was no insubstantial mining camp. Brick buildings and paved streets, running water and enclosed sewers, and imposing public facilities like the Copper Queen Hotel clearly set Bisbee apart from nearby Tombstone, a once booming but now decaying village of wooden sidewalks and false-front stores typical of the nineteenth-century mining West.

From the mines of Bisbee came fortunes in lead, zinc, manganese, gold, and silver—more gold and silver than from any other place in Arizona—but copper above all else determined the fate of Bisbee. "That depends on the price of copper" was a refrain often heard on the town streets. In Butte, Montana, and Bingham, Utah, and dozens of Bisbee's sister settlements, the red metal equaled prosperity or adversity: a rise in price in Paris or the opening of a new mine in Chile would affect life in remote corners of the American West, and the lingo of mining was the common language of all residents of the copper kingdom regardless of economic or ethnic differences. A slump in the price of copper might close dozens of mines across the West and send hundreds, even thousands, of miners off in search of work in other camps and even to jobs in other major industries, such as logging and agriculture.

Wartime pressures drove the price of copper up dramatically in 1916 and 1917. Miners produced at a hectic pace, working around the clock in successive shifts, soon becoming frustrated by the long hours, together with wages that lagged behind inflation. In Bisbee, approximately three thousand tired miners finally walked off the job on 27 June. The strike remained peaceable, although with each passing day public attention

Staging their own portrait, these Bisbee miners pose with the lunch buckets and candles that helped alleviate the tedium of their work. The light areas on their pants, melted wax from the candles they stuck in the walls to illuminate the mine tunnels, suggest that they have just come off a shift of underground labor.

D. A. Markey. Miners, Bisbee, Arizona. *Photograph, ca. 1895. Bisbee Mining and Historical Museum (#080.73.1), Bisbee, Arizona.*

shifted from the legitimate demands of the mine workers to the radical heresies of the Industrial Workers of the World, widely regarded as the dangerous tool of enemy Germany and the real force behind the unpatriotic work stoppage.

Tension increased until on the morning of 12 July 1917, the streets of Bisbee filled with the sounds of a two-thousand-man *posse comitatus*. Their leader was Sheriff Harry Wheeler, and backing him were the bosses of local copper companies incensed by workers who dared to halt production during a national crisis. Men identified by white armbands raided the homes of known strikers and their sympathizers. Two lines of armed citizens herded twelve hundred "undesirables" aboard the boxcars and cattle cars of a special freight train that sped them east across the deserts of Arizona and New Mexico. Abruptly abandoned along a lonely stretch of track, the prisoners would have died in the July heat had they not received emergency food and water from soldiers stationed nearby to guard the isolated border area from Mexican revolutionaries. In Bisbee, strikebreakers quickly filled the mines, and the community again shouldered its patriotic duty of producing copper to make the world safe for democracy. Next to the roundup of Japanese Americans in World War II, this Bisbee deportation is the greatest mass violation of civil liberties in the twentieth-century West.

Today, most citizens of Bisbee would prefer to forget the infamous episode. Whereas nearby Tombstone attracts busloads of camera-toting tourists by glorifying its violent past, and even occasionally restaging the shootout at the O.K. Corral, Bisbee disdains the shoot-'em-up approach to western history. It retains instead the sober appearance of an industrial center not unlike a steel-mill or cotton-mill town of the East or South. Travelers do come to gaze at a mining landscape dominated by the Lavender Pit, but most Bisbee residents forswear anything that might conjure up images of a raw and immature past.

Still, it is hard to predict what Bisbee's residents will do for a living. Like many of the West's natural-resource-based communities, this one has experienced massive economic dislocation. In 1975, for the first time in nearly a century, the Phelps Dodge Corporation halted its mining operations in Bisbee. In the slump that followed, homes could be purchased for as little as one thousand dollars. By the 1980s, retirees constituted more than 40 percent of the town's eight thousand residents, and a small but visible counterculture accounted for part of the rest. At least residents can be grateful for one thing: Bisbee's mining landscape is not radioactive, as it is in some of the West's uranium mining and processing towns.

The history of business and labor in the West since 1900 is far more than a record of the troubled evolution of natural-resource communities into tourist or retirement meccas (or into ghost towns). Yet at least until World War II, economic life in cities as large as Seattle, Portland, and Denver remained closely linked to the natural-resource hinterlands beyond their doorsteps. This relationship defined the economy of the West in the early twentieth century and continues to do so in many parts of the region today.

The World of Business in the Early Twentieth Century

When the new century dawned, metropolitan centers as large as Seattle, Denver, and El Paso formed urban outposts juxtaposed with vast and lightly populated hinterlands. Many dozens of smaller population islands like Bisbee coalesced around the region's

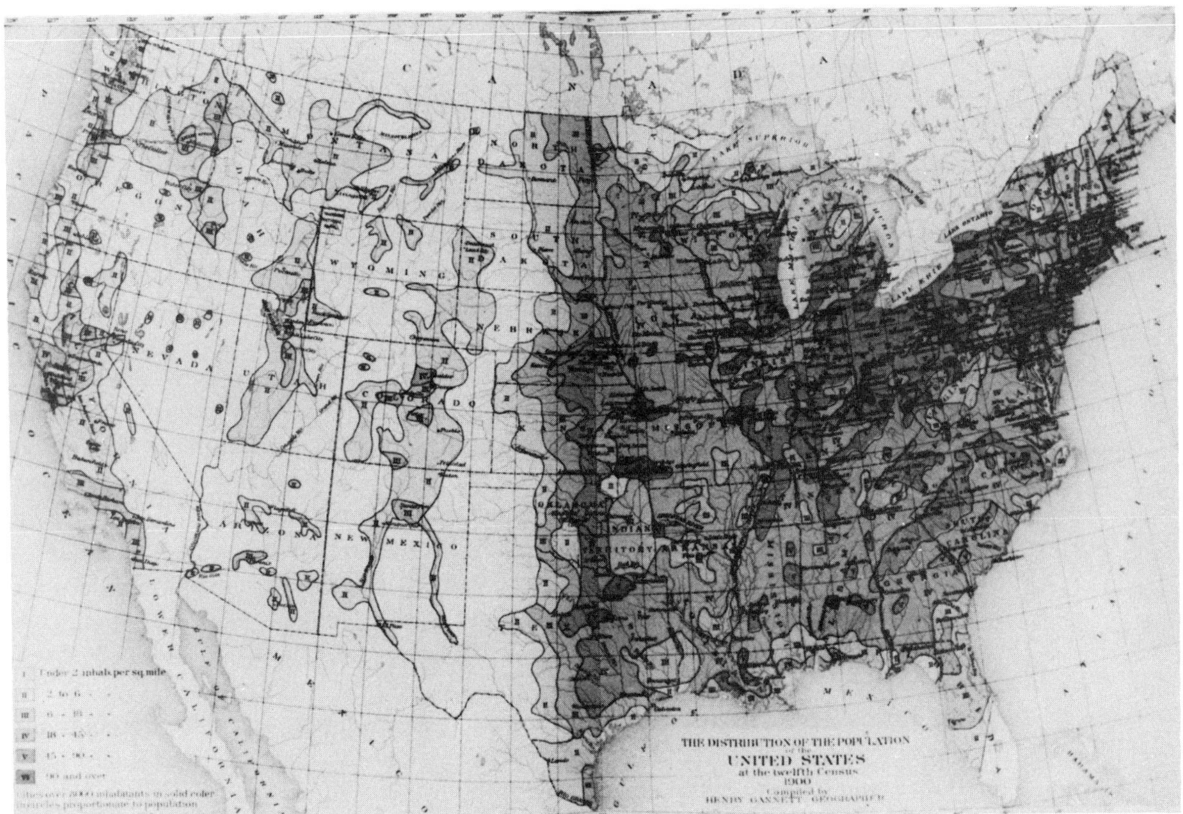

The maps drawn from the census of 1900 graphically illustrate the island pattern of settlement that characterized life in the Far West. Population density in the intermountain and Rocky Mountain regions remained tied to the availability of extractable mineral resources or to ample supplies of irrigation water.

Julius Bien & Co., lithographers. Distribution of the Population: 1900. *From the* Twelfth Census of the United States: Statistical Atlas *(1903).*

ubiquitous mineral deposits, some of them isolated at elevations of ten thousand feet or more, where a paralyzing blanket of snow often lasted from October through May, cutting them off from the rest of the world. Separating San Francisco from Salt Lake City, and Los Angeles from Denver, were hundreds of miles of nearly empty deserts, forests, and mountains, and the same imposing physical barriers that isolated so much of the West from itself also separated sizable portions of it from the rest of the United States. Separation took other forms too, notably in California where the coastal area north of San Francisco, with 5 percent of the state's population, received about 40 percent of all surface water, whereas the Los Angeles basin, with 40 percent of the state's population, received about 2 percent of the surface water. Because of the peculiar configuration of resources and settlements, particularly in the arid lands beyond the one hundredth meridian, the work of building a suitable infrastructure of canals and dams, railway lines, highways, and bridges defined the world of business and labor in the early-twentieth-century West. The construction industry necessarily loomed large in the western economy, employing thousands of workers and forming the basis for some of the region's most impressive personal fortunes.

The West's island pattern of settlement was a visual reminder that the regional economy in 1900 was still closely tied to natural resources often found in remote, nearly inaccessible places. This was especially true of the mineral industry. Some 90 percent of the nation's metal reserves were located in the West, and mining in 1900 accounted for 11 percent of the gainful employment in Colorado, 12 percent in Idaho, 13 percent in

Montana, and 15 percent in Arizona, as compared with less than 1 percent for the nation as a whole.

By the eve of World War I, the region produced the bulk of the nation's copper, gold, and silver. Gold and silver were glamour metals, but copper, so vital to the expanding electrical and telephone industries after the 1880s, created far more wealth. The West's copper kingdom centered first in Montana, which by the turn of the century had become the nation's leading copper producer, a title it had wrested from Michigan in 1892 and yielded to Arizona in 1910, which has led the country in production since that time. Utah, Nevada, and New Mexico also emerged as significant producers of the red metal.

Copper required enormous sums of money to mine, smelt, refine, and market. Such unprecedented amounts of capital were difficult, if not impossible, to raise in economically undeveloped and lightly settled parts of the United States. For that reason, Bisbee as early as the 1880s became synonymous with Phelps, Dodge and Company, a long-established New York and New England mercantile firm. Rockefeller and Guggenheim money fueled the industry in other parts of the West. In time, three giant corporations—Anaconda in Montana (backed for several years by Rockefeller money), Kennecott in Utah and Alaska (Guggenheim money), and Phelps Dodge in Arizona—dominated production. By 1940 these three eastern-controlled companies accounted for 80 percent of the West's copper output.

It would be a mistake to regard the mining giants simply as soulless corporations. For almost four decades, James Douglas personified the Phelps Dodge enterprise in Bisbee and remote parts of Arizona and northern Mexico. A broadly educated "renaissance man" from Canada, Douglas and various family members developed mine properties on both sides of the international border, arranged for building complex and expensive smelters, and knitted together a railroad system that stretched nearly eight hundred miles across the Southwest. Because of the influence of Douglas, disputes between rival mining companies in Bisbee were settled peaceably, without resort to courts, and the Arizona community was thus spared the "war of the copper kings" that enriched lawyers but wracked Butte's mineral industry. The example of Dr. Douglas's peaceable resolution of disputes was apparently lost on his son, Walter, who played a key role in the Bisbee deportation.

Although copper was synonymous with mining in many parts of the West, the production of precious metals, for which the West had been famous since the California gold rush of 1848, did not disappear. Whereas some production continued as a by-product of copper mining, the Black Hills of South Dakota emerged as a leading gold producer, and Idaho's Coeur d'Alene district did the same in silver production. Far less glamorous than silver or gold or even copper was coal, which could be found in commercial quantities in almost every western state. Coal mining dated back at least to the 1850s and 1860s in Washington. By the 1890s, Colorado alone extracted three million tons of black diamonds a year, ranking it as one of the region's chief producers of coal. Production figures climbed significantly through the first two decades of the twentieth century, but the industry experienced a long-term decline after World War I as the result of an oil glut. Coal rebounded only in the 1970s, when by the end of the decade the West accounted for approximately one-quarter of the nation's output. Much of this was produced by giant machines that stripped seams of coal in the nonunion mines in Wyoming and Montana, unlike in earlier years, when the coal had come from underground, labor-intensive, and often unionized operations.

The first intimation of future trouble for coal can be traced back to January 1901, when a wildcat driller named Captain Anthony Lucas punched through the top of a salt dome near Beaumont, Texas, to tap an enormous underground pool of oil. The black gold gushed out under such pressure that it blew away the drilling apparatus and spewed more than half a million barrels of oil high into the air before it could be controlled six days later. Hundreds of people gathered at Spindletop Field to view the black geyser, the eighth wonder of the world according to its boosters. As thousands more people rushed to form companies and drill wells, the population of Beaumont swelled from nine thousand to fifty thousand in three months. Speculative mania seized all parts of the state. Petroleum was eventually found under more than half the counties of Texas, and it replaced cotton as the primary contributor to the state's economy. Petrochemical production became the state's largest manufacturing industry. From hundreds of small companies, three eventually emerged to dominate the Texas oil industry: Gulf, Texaco, and Humble (now part of Exxon, the modern name for Standard Oil Company of New Jersey). Spindletop and its plume of black gold came to symbolize an entire industry, and oil wealth became synonymous with Texas; but during the first three decades of the twentieth century, California often produced more barrels of oil per year than did Texas.

California oil dated back to pioneer days, but it became a booming industry in the 1890s. New discoveries were made almost every year through the 1920s. A speculative frenzy gripped southern California as production in the state soared from five million barrels in 1900 to fifteen million barrels in 1914, and that was only the beginning. From the San Joaquin Valley to Santa Barbara and Los Angeles, new finds gave rise to a forest of derricks and to the beginnings of industrial giants like Union Oil of California and Standard Oil of California—now Chevron—which was originally only one more part of John D. Rockefeller's giant Standard Oil monopoly. Between 1901 and 1940, California ranked first among American oil-producing states for fourteen years, second for twenty-one years, and third for three years. Oil refining became the state's chief manufacturing industry, and the man-made harbor at San Pedro became one of the nation's chief oil-shipping ports. During the first third of the twentieth century, Oklahoma, Kansas, Colorado, New Mexico, and Montana joined the ranks of major oil producers, but nothing equaled the East Texas boom of 1931, which enabled Texas permanently to dethrone California as the nation's chief oil producer.

If Spindletop in 1901 symbolized a new era for the oil industry, something less dramatic but of similar importance occurred a year earlier to highlight profound changes in the timber industry. As the timber supply of the Great Lakes region dwindled, the extensive forests of the Pacific Northwest emerged from obscurity. Early in 1900, the Minnesota timber baron Frederick Weyerhaeuser and several associates purchased nine hundred thousand acres of the Northern Pacific Railroad land grant—mostly Douglas fir forest in Washington—for six dollars an acre. It was one of the largest private land transfers in American history. Less than fifty years earlier, Weyerhaeuser had been a

Attempts to develop oil production in California began in the 1860s. Drawing on skills and capital from their eastern oil fields, large corporations such as Standard Oil and Union Oil later helped develop California as the nation's leading oil-producing state in the first third of the 20th century.

Unidentified photographer. Union Oil Gusher, Long Beach, California. *Gelatin silver print, ca. 1925. The Huntington Library, San Marino, California.*

young immigrant from Germany beginning his career as a night fireman in a sawmill in Rock Island, Illinois. Eventually his empire totaled more than ninety affiliated companies and dominated timber production in Oregon, Washington, and Idaho.

Weyerhaeuser's show of confidence in the future of timber in the Pacific Northwest precipitated a stampede to buy timberland. Sawmills went up by the hundreds, and Idaho's scenic Lake Coeur d'Alene, for example, resembled an enormous millpond as log booms lined its shores. By 1905 Washington had grabbed the mythical title of the nation's leading timber producer, a position it would maintain for all but one year until Oregon slipped ahead in 1938. The Pacific Northwest became so dependent on the timber industry for its economic well-being that for a time it was aptly labeled the Sawdust Empire. California, however, was never far behind and usually ranked second or third. By 1940, Washington, Oregon, and California together accounted for 40 percent of the nation's lumber production.

Although during the first three decades of the twentieth century the West remained primarily a supplier of raw materials, some processing did take place within the region, usually in or near urban centers. There were smelters, sawmills, food canneries, oil refineries, foundries, and a mixed assortment of manufacturing establishments. Much of the region's manufacturing was devoted either to first-stage processing of raw materials, such as sawing logs into building materials, or to small-scale diversified manufacturing for local markets. The West's only significant steel-manufacturing facility was the Colorado Fuel and Iron Company, which produced enough steel to make Pueblo the Pittsburgh of the West. Value added through manufacturing remained very small in most of the region before World War I.

Unlike manufacturing, service industries loomed larger in the western economy than elsewhere in the U.S. economy in the early twentieth century. Banking, real estate, sales, medicine, and entertainment were all significant. Yet despite the rise of Hollywood after 1915 and even of aircraft manufacturing, which first achieved prominence during World War I, westerners continued for many more years to depend on natural-resource industries and to ride an economic roller-coaster that seemed to deny them any real control over their destiny. Given that the entire West contained a mere 15 percent of the nation's population in 1900, the region could scarcely have avoided becoming enmeshed in what some westerners complained was a colonial relationship, with the richer and much longer established East supposedly plundering the natural resources of the West.

Westerners who subscribed to the "plundered province" thesis most often cited as evidence the domination of eastern capital in Montana, where Anaconda, in addition to its huge copper mines, owned 1.7 million of acres of timberland, municipal waterworks, stores, hotels, and all but one of the seven major daily newspapers of the state. By 1900, Anaconda employed some three-fourths of the wage earners in Montana. A year earlier, Anaconda had come under the control of Rockefeller's Standard Oil and remained so until the copper trust dissolved in 1915. Rockefeller money also dominated the Colorado Fuel and Iron Company in the early 1900s, the Centennial State's largest industrial enterprise. At the opening of the twentieth century, Colorado Fuel and Iron had thirty-eight camps, rolling mills, and steelworks in Colorado, Wyoming, and New Mexico. Its thirteen thousand employees dug coal and iron ore and made coke and steel.

Many factors besides a small population and absentee ownership handicapped the West in any race to compete with manufacturers in the East and elsewhere. These included railroad-rate inequities, usurious interest charges, and tariff protection that favored eastern manufacturers over western consumers. What this "plundered province" complaint often overlooked, however, was that western business interests frequently benefited from an infusion of eastern capital and that some investment money came from the more developed parts of the West itself. For many westerners, nonetheless, the East would remain a convenient scapegoat on which to blame their recurrent economic troubles.

The World of Labor in the Early Twentieth Century

Work life in the early-twentieth-century West was not merely a pale and insignificant reflection of that in the industrial centers of the East and Midwest; instead, several elements gave it a distinctly regional character. As was true for western business generally in the years before World War I, two key concepts explain the most about work life in the West: the island pattern of settlement; and the dependence on natural-resource-based industries infamous for recurrent bouts of cyclical and seasonal unemployment. From the grainfields of the high plains to the oil rigs of Texas and southern California, an army of itinerant laborers supplied the muscle needed to harvest the products of the West's forests, fields, and coastal waters, mine its precious and base metals, drill for its oil, and load its ships. They were indispensable in constructing the West's railroads, dams, irrigation canals, and many other building projects. Before the First World War, the annual wheat harvest of the Pacific Northwest required legions of laborers, and the same was true of California's "factories in the field." In 1910 every hundred miles of railway line required an average of 156 workers for maintenance.

The physical dimensions of the itinerant laborers' domain of toil fluctuated over time. The limits expanded as new agricultural lands and timber and mining camps opened, and the boundaries contracted as natural resources were depleted or, more infrequently, as one of the raw, socially unstable communities survived and matured, as was the case of Bisbee. The domain of the itinerant laborers probably reached its greatest extent in the opening decade of the twentieth century and then diminished after World War I in the face of increased mechanization. Some form of itinerant labor, however, remained a feature of agricultural life in the West's orchards and vegetable, sugar beet, and cotton fields. As one might expect, the still developing West in the early years of the twentieth century had a much higher ratio of laborers to craftsmen and machine-tending operatives than was typical of older, more mature economic regions.

Besides the numerous laborers, the wage-earning work force of the West included printers, plumbers, painters, masons, carpenters, and other skilled workers, who could be found in every major urban center of the West. Forming yet a third group were the operatives, the men and women who ran the machines in sawmills, canneries, and other manufacturing establishments. Although they developed considerable manual dexterity in performing repetitive tasks, most operatives were not highly skilled in the same way that craftsmen were. In some cases the work had been done by craftsmen until machine production opened the way for the less skilled but highly proficient operatives. The craftsmen and operatives of El Paso, Salt Lake City, and any other western community were in many ways not much different from their counterparts in the East and Midwest, although their work lives were shaped to some degree by laborers, who seemed forever on the move throughout the West. Too many itinerant laborers in one locality might, for example, depress wage rates for operatives or tempt employers to hire the outsiders as strikebreakers.

Many operatives were young people, and tending machines was but a step on the road to some other status. Unlike craftsmen or laborers, many operatives were women, who did all types of machine work, from running sewing machines in garment factories to stuffing sausages on production lines in meat-packing plants. In 1939 approximately seventy-five thousand women, mostly Mexican and Mexican-American, worked in California canneries and packinghouses. By the end of World War II, California cannery operatives, 75 percent of them women, constituted one-quarter of the nation's food-processing work force.

Women, however, did not typically hold jobs as laborers. To be sure, women could be found in any mining camp, though not at first in large numbers. A few arrived as wives, others to manage rooming houses and eating places. An unknown number of women worked as prostitutes, especially in the predominantly male world of the laborers. If a mining camp survived to become a town or village, the ratio of males to females tended to normalize as miners married and raised families. In the pre–World War I logging camps, however, which were essentially makeshift work sites in the woods, there were seldom more than a few females. Some loggers, in fact, regarded a woman in camp as bad luck.

Laborers in the pre–World War I era usually worked in all-male gangs and supplied strong arms and backs. They functioned as "human machines" who exchanged physical labor for a daily wage. The work they did was invariably heavy and dirty and often dangerous. Their working hours seemed interminably long, often lasting from five in the morning until eight or later each evening. Some of them remained in a single industry and gained considerable proficiency as miners and loggers, even buying a house and getting married, but in most instances the work was not steady. The average job duration among laborers of the West Coast in 1914 was fifteen to thirty days in lumber camps, sixty days in mining, ten days in construction work, and seven days in harvesting. In extreme cases, an itinerant laborer might remain on the job for as few as three hours before walking off. A common saying in both the timber and the railroad-construction camps was that three crews were connected with any job: "one coming, one going, one on the job."

One scholar who studied their lives labeled them casual workers. Some were indeed casual about work, and some were hoboes and tramps. But many of those found in the ranks of itinerant laborers were at best unwilling conscripts who were ready and able to take steady jobs, if only such work could be found. These itinerant job seekers were in some ways no different from the unemployed of Chicago, New York, or Boston who, on losing one job, might trudge down the street in search of another. Their counterparts in the West, however, were more visible because the loss of a job in one of the region's natural-resource-based island communities often meant having to travel to another island, usually some distance away, to find work, often riding atop or underneath railway freight cars.

Perhaps the single most important reason for widespread itinerancy among western laborers was the cyclical nature of the region's colonial economy. Boom-and-bust cycles heightened workers' sense of dependency and encouraged their mobility. Going hand in hand with cyclical unemployment was the shorter cycle of seasonal unemployment. It became necessary to move around frequently to find work. Depending on the season, a dock worker from Seattle might be found in eastern Washington supplying the muscle needed to sack the annual harvest of wheat. During the course of the year, a metal miner might move from Bisbee to Butte or to the Coeur d'Alene district in Idaho or even to the mines of Alaska before returning to Arizona.

Some of the itinerants were, quite frankly, social misfits, and some were chronic alcoholics. Some were restless by nature and were kept on the move by a frequent surplus of labor in one locality or by reports of high wages being paid somewhere else. The very act of migrating west—which meant at least the temporary loss of the close and stabilizing ties of family, neighborhood, and church—probably contributed to a feeling of rootlessness. And because of the uncertain nature of their employment, some workers found it difficult to forge enduring social relations in new settings.

Because of the seasonal nature of their work, laborers usually spent at least part of the year in one of the West's urban enclaves of employment agencies, cheap hotels and lodging houses, soup kitchens, saloons, and brothels. Every large city had one such district, where laborers congregated and survived between jobs. These urban enclaves, like the logging and construction camps, were predominantly male societies. Seattle, as a result of its close connection with the wageworkers' frontier, in 1900 recorded the highest percentage of male population among American cities of twenty-five thousand or more people (64 percent); Butte, Montana, ranked third (60 percent). It is no accident that both cities featured large and notorious red-light districts.

The laborers of the West came from a variety of racial and national backgrounds. It was common for immigrants of one nationality to gravitate to certain types of work: Scandinavians to logging, for example, and Greeks and Italians to work on railroad construction and maintenance. In California in the early decades of the twentieth century, Japanese could often be found working in fields of sugar beets and berries, Mexicans in the citrus groves and the newly opened cotton fields, and European immigrants in fields of grapes, peas, and artichokes. American whites (sometimes called "tramps") were commonly found harvesting fruit. Few of these were farmhands like those in the East, who enjoyed year-round work and generally boarded with a single family.

There was nothing static about patterns of race and ethnicity in western industry. In 1888 an estimated four-fifths of all Colorado coal miners were English-speaking. That changed after a period of labor strife in 1903–4, when many disillusioned miners left Colorado, and Italians, Austrians, Slavs, Serbs, Poles, Montenegrins, and Greeks filled their places. Many of the first cotton pickers who went west to California and Arizona were whites and blacks from Texas and the South. The composition of cotton-field labor in California changed several times, as Mexican and Filipino laborers came to predominate in the 1920s, followed by families of displaced and desperate whites from Oklahoma, Arkansas, and Texas in the late 1930s. When the defense boom of the early 1940s attracted "Okies" and "Arkies" to urban areas, where they took jobs in shipyards and aircraft factories, Hispanics returned to the fields once more.

Mexicans and Mexican Americans constituted the largest racial or ethnic minority among the workers of the West, although at various times Asians—Chinese, Japanese, and Filipinos—were numerous in certain industries. Compared with the South, the West outside Texas and Oklahoma had relatively few African Americans among its laborers. The proximity of the Mexican border to California's factories in the field, Arizona's copper camps, and Texas's orchards was of crucial importance. The revolution that kept Mexico in turmoil from 1910 to 1920, together with the completion of railway lines north to the border, encouraged thousands of Mexicans to flee to the United States to escape violence and economic uncertainty. Moreover, by taking jobs north of the border, they might triple their wages. In time, the area of Hispanic concentration formed a rough triangle, with its base extending along the Mexican border from San Diego to San Antonio and with its apex located at Denver. Outside California, Texas, and the

Both custom and law enforced the ethnic segregation of western laborers. Italian miners probably gathered in this Bingham Canyon, Utah, boardinghouse by choice, seeking the companionship of their countrymen. But laws could also dictate housing patterns. When this photograph of an all-white Texas housing project was taken in 1942, many Texas communities had three distinct, federally funded housing projects—for whites, blacks, and Mexicans.

Unidentified photographer. Italian Boarding House, Bingham Canyon, Utah. *Photograph, date unknown. Utah State Historical Society, Salt Lake City.*

Unidentified photographer. "No Mexican Children Allowed." *Gelatin silver print, 1942. Library of Congress (#LC-USZ62-89233), Washington, D.C.*

Southwest, Mexican workers remained relatively few until the labor shortages created by World War II gave rise to the Emergency Farm Labor Program (Bracero Program), which dispatched Mexican laborers to work as far north as the fields of Idaho, Oregon, and Washington.

Racial and ethnic prejudice was pervasive. Although some of the harvest workers were fifth-generation Californians or Texans, the Mexicans and Mexican Americans were almost always perceived as cheap and temporary labor. Between 1931 and 1934, in the depths of the Great Depression, an estimated one-third of the Mexican people of the United States were either deported or "repatriated" to Mexico, even though many of them had been born in the United States. In many southwestern mining communities, strict segregation existed between Hispanic and non-Hispanic miners. Other immigrants also were often segregated into distinct ethnic neighborhoods. The copper center of Ely, Nevada, for example, had separate enclaves commonly labeled "Greek Town" and "Jap Town."

In part because of the region's ethnic diversity, labor unions found that organizing in the West was seldom easy, and they never attracted a majority of the region's wageworkers, even in Washington and Colorado, which at various times were among the most highly unionized states in the country. Neither was organized labor's influence uniform throughout the West. Its chief centers of strength were the mining towns of the interior and the Pacific seaports of San Francisco, Tacoma, and Seattle. Labor's power was especially low in Idaho, which had few urban areas and little manufacturing, and in Utah, where the Mormon church vigorously promoted the freedom to work without having to join a labor union, and in New Mexico, where the economy was essentially pastoral. Yet even in Utah, organized labor made some surprising gains, as in 1896 when the state mandated an eight-hour day for workers in mines and smelters—a law the U.S. Supreme Court upheld in the case of *Holden v. Hardy* (1898). Arizona too was friendly to organized workers until the tensions of World War I caused an abrupt about-face.

San Francisco, by far the West's largest city, emerged as perhaps the most highly organized community in the nation in the early twentieth century. Workers in many vital industries were required to join unions. But even in San Francisco, union power fluctuated between periods of strength and weakness. From the depression of the 1890s until the aftermath of World War I, the twenty-five thousand members of the building trades were a dominant force in the economic, political, and social life of San Francisco. Through the Building Trades Council, they helped govern the city, in the process gaining a reputation for corruption. A full-scale effort by local businessmen and the national turn to the right in the 1920s undercut the power of the building trades. The spectacle of San Francisco's "labor barony" frightened businessmen in Los Angeles. As a result, organized labor there confronted some of the country's most outspoken proponents of the open shop, and union weakness was the rule. Leading the battle against organized labor were the powerful Merchants and Manufacturers' Association and Harrison Gray Otis, publisher of the *Los Angeles Times* and a one-time trade unionist turned bitter foe of organized labor.

A distinguishing trait of western organized labor was the popularity of industrial or all-inclusive unions, a natural result of the inordinately large number of laborers employed in the West's natural-resource-based industries. Between 1890 and the First World War, spokesmen for the West's laborers repeatedly asserted that they had little in common with the "labor aristocrats" who belonged to the cautious, craft-conscious American Federation of Labor (AFL) unions and railway brotherhoods. The AFL, in its pursuit of bread-and-butter gains like higher wages and shorter hours, wasted little time trying to organize unskilled laborers and operatives. In the West the AFL's exclusionary philosophy in effect abandoned thousands of workers in natural-resource-based industries to any alternative form of organization that might come along. As a consequence, many were drawn to industrial, inclusive, and even militant forms of labor organization, as well as to populist, anarchist, and socialist associations and political programs.

More than any other labor organization, it was the Western Federation of Miners (WFM) that served as a rallying point for western-oriented industrial unionists in the early twentieth century. Formed in 1893 in the aftermath of a bitter dispute in Idaho's Coeur d'Alene mining district, the WFM functioned as a bridge between the Knights of Labor, an idealistic and inclusive organization popular in parts of the West in the 1880s, and the Industrial Workers of the World (IWW), probably the most famous of all labor organizations in the region. Militant metal miners played a key role in organizing the IWW; events in Colorado supplied the connection between the two unions.

Several notorious examples of industrial violence punctuate the history of work life in the American West, but no state exceeded Colorado in the overall variety, duration, and severity of conflict between labor and management during the thirty years from 1890 to 1920. Nearly one hundred people died during the great coalfield war of 1913–14 with its notorious "massacre" at Ludlow, where in a single day, ten men and a child died in a fierce gun battle between the Colorado militia and striking miners, and two women and eleven children died of suffocation when victorious militiamen finally overran the strikers' tent colony, doused it with kerosene, and set it ablaze. Yet Colorado had been bloodier earlier. Nowhere else in the United States did the level of violence remain so high for so long as in the Colorado labor war of 1903–4, a time of dynamitings, deportations, and shootings. One violent explosion, set by a union sympathizer, ripped through a railroad station near Cripple Creek on the morning of 6 June 1904, killing thirteen nonunion miners.

The WFM, which a year earlier had called 3,500 metal miners of the Cripple Creek area out on a sympathy strike in an effort to support the wage demands of smelter workers, denied responsibility for the 6 June blast. But during the uproar of antiunion hostility that followed, Governor James Peabody imposed military rule on the district. Militia officers questioned 1,569 people and ultimately banished 238 of them. They were marched to Kansas and New Mexico and abandoned, much as would happen later in Bisbee. The WFM did not officially call off the Cripple Creek strike until 1907, although the strike had run its course by the end of 1904. After Cripple Creek, the WFM was for several years but a shadow of its former self. In the camps of Colorado's mineral belt, where in 1902 the WFM could claim a third of its 165 locals and a third of its 27,154 members, the union had virtually disappeared. To get a job, workers were required to hold cards issued by the Mine Owners' Association, and no known union men could get a card. The WFM sullenly licked its wounds, and in Chicago in 1905 it spearheaded the organization of an even more militant and more broadly based union called the Industrial Workers of the World.

Although Wobblies, as members of the IWW became popularly known, did not limit their activities to the West, their union found a more congenial home among the laborers of the West than perhaps anywhere else in the United States. The IWW program seemed especially tailored to meet the needs of those most alienated from the mainstream of society by the brutal working and living conditions frequently encountered in the West's camps and mills. At a time when nationally oriented trade unions concentrated their efforts on skilled white labor, Wobblies emphasized the solidarity of all workers—men and women, whites and blacks (even Asians, who were usually shunned because of their willingness to accept low wages). The IWW kept dues low and ignored political action, which in any case made little sense to workers who seldom remained in one place long enough to qualify to vote. Wobblies refused to sign contracts with employers, whom they regarded as mortal enemies. Confronting some of the roughest working conditions in the United States, the IWW exhibited a strain of militancy that was seemingly a natural by-product of the laborers' struggle for existence.

Wobblies burst into newspaper headlines in 1909 when they launched a spectacular and unorthodox protest called a "free-speech fight," a demonstration designed to call attention to the exploitation of harvest labor in the inland Pacific Northwest. For defying Spokane's ban on street-corner speaking, hundreds of Wobblies were arrested and hauled to jail. Nearly every inbound train brought more Wobblies to Spokane until the city, faced with a lack of jail space, mounting expenses, and bad publicity, declined to pursue the struggle. One of those imprisoned was Elizabeth Gurley Flynn, the "Rebel Girl," who particularly embarrassed city authorities by publishing a description of her sexual harassment in the city jail. The resulting outcry of moral indignation caused the city to provide better treatment for women prisoners. In the end, Spokane released the jailed Wobblies.

Wobblies naturally considered the free-speech fight a success and used the technique wherever they had a grievance to dramatize. But free-speech fights also gave Wobblies a defiant reputation that aroused deep anxieties among westerners and made the union vulnerable to vigilantism and other forms of repressive action. In California in 1910, the IWW had only a thousand members in eleven locals, but the fears they aroused were sufficient to prompt Fresno, San Diego, and several other cities to ban free-speech fights from their streets. This in turn led to confrontation and violence. Free-speech fights eventually spread to Vancouver, British Columbia, and about twenty other western communities before culminating in a bloody "massacre" in the sawmill town of Everett, Washington, in November 1916. In this clash an estimated twelve people, most of them Wobblies, died.

Violence, more than any other characteristic, was attributed to Wobblies, although it must be noted that in the West, violent episodes involving labor antedated Wobblies by more than three decades. The region's sometimes turbulent labor relations owed far more to the special circumstances of its work life than to any single organization or philosophy. Rather than initiate a new and violent era of labor-management relations, the IWW merely elaborated on a strain of militancy, even radicalism, that already existed among western workers. In this way, the Wobblies earned a prominent place in modern western folklore.

Several of the most celebrated episodes of violence in the early-twentieth-century West had nothing to do with Wobblies. These included Colorado's Ludlow troubles as well as the tremendous explosion that leveled the headquarters of the militantly antiunion *Los Angeles Times* in 1910, killing twenty employees. When two union activists, John J. and James B. McNamara, admitted their guilt after first pleading innocent, the whole western labor movement suffered. Widespread public support for organized labor evaporated, and antilabor ordinances passed in several communities. Union labor was stigmatized as a bloody organization.

Another explosion that appeared to have links to organized labor occurred on San Francisco's Market Street on 22 July 1916 during a Preparedness Day parade. The blast killed ten people and wounded forty others. Police arrested Tom Mooney and Warren K. Billings, union activists and labor radicals, but unlike the McNamara brothers, the accused San Francisco murderers maintained their innocence. Labor cried "frame-up." After a long and emotional trial based on the flimsiest of evidence, Mooney was sentenced to death and Billings to life in prison in 1917. The case remained a cause célèbre until Governor Culbert Olson finally resolved it by pardoning Mooney in 1939. The violent episodes of 1916 in San Francisco and Everett foreshadowed the rising tide of antilabor activity in the West during World War I. In time, however, the industrial violence diminished, notably so by the late 1930s, when workers, managers, and government established procedures to settle disputes peaceably. But one thing that did not change in the West were the boom-and-bust cycles that made life unpredictable both for business and for labor.

Living with a Boom-and-Bust Economy

During the years from 1917 to 1941, the West experienced two major cycles of boom and bust, together with the usual vexing seasonal alternations that had long typified the regional economy. The First World War brought unprecedented demands for the resources of the West. Oil producers dramatically stepped up output between 1917 and 1919. A rapidly expanding market for lumber from California, Texas, and the Pacific Northwest brought boom times to the woods. Slogans like "Wheat Will Win the War!" and government officials' lectures to farmers that production of food was their patriotic duty helped push the output of western agriculture to record highs. Farmers planted more acres, often on borrowed money; they expanded recklessly, enjoyed the boom times, and reaped a harvest of troubles when the fighting ceased. But until the bubble burst in 1920, the rural West had never before experienced such prosperity. Manufacturers prospered too during World War I. In the Pacific Northwest, shipbuilding expanded so rapidly that seemingly overnight it ranked second in size to the huge lumber industry. The war stimulated the new aircraft industry centered in Los Angeles, San Diego, and Seattle, where in 1916 a young lumberman named William E. Boeing founded a small company that eventually became a major manufacturer of aircraft.

Jobs were plentiful, both for men and for women, but for labor organizations like the controversial Industrial Workers of the World, the Great War was a disaster. After the United States entered the conflict in April 1917, hostility toward Wobblies

Motion pictures provided both popular western images and a major western industry. Coming of age in the 1920s, the film industry generally prospered during the depression as other businesses faltered and Americans looked to movies as an escape from their economic problems.

Dick Whittington Studio. RKO Pictures. *Gelatin silver print, 1929. The Huntington Library, San Marino, California.*

increased, since people believed them to be allies of enemy Germany. It was in the context of wartime zealotry that the Bisbee deportation occurred; and the Wobblies likewise suffered harassment in the woods of the Pacific Northwest where federal officials determined to take whatever steps were necessary to harvest spruce, a light, strong wood needed for aircraft construction. The IWW suffered an even more serious blow on 5 September 1917, when agents of the Justice Department, together with state and local officials, systematically raided IWW offices across the country and arrested 160 prominent Wobblies. Wartime prosecution nearly finished the IWW as an effective labor organization.

The war ended on 11 November 1918. Allied victory over Germany and the Central Powers brought Americans joy but not peace of mind. Replacing fear of German imperialism was fear of communism, which some Americans believed had spread from the newly formed Soviet Union to the United States, perhaps even to Seattle, where in February 1919 a general strike by sixty thousand organized workers shut down the city. Although Seattle's four-day "revolution" ended without bloodshed, and radicals did not control the strike (as opponents charged), it cost organized labor popular support and contributed to mounting national hysteria, known as the Red Scare. The culminating act of violence in the immediate postwar years occurred in Centralia, Washington, where on 11 November 1919 at least five lives were lost in a bloody clash between Wobblies and American Legionnaires, the latter group celebrating the first anniversary of the end of the Great War.

The end of war also required some extremely painful economic adjustments. The federal government's precipitous cancellation of war orders threatened the financial well-being of scores of manufacturers and other businesses and cost thousands of jobs. Plummeting commodity prices devastated farmers: from 1919 through 1921, farm prices in the United States fell by 40 percent, and the wartime boom turned into a deep agricultural depression, especially for the farmers and cattlemen of the interior West. One obvious casualty of the 1921 agricultural depression was optimism about the future. Many farmers now faced a period of hardship that lasted until World War II. For most western agrarians, the 1920s were anything but the "prosperity decade" experienced in other parts of the United States.

Whether westerners would remember the 1920s as a time of adversity or prosperity depended to a large extent on where they lived and what they did for a living. Californians would be far more likely than residents of other western states to recall the era positively. Those areas most dependent on timber, mining, and agriculture continued to be buffeted by dizzying cycles of boom and bust, although for a while the region's new commercial enterprises, notably Hollywood and tourism, seemed to offer prosperous alternatives. Tourism was certainly not new to the West in 1920; railroads had lured wealthy tourists west to Colorado, California, and the national parks for decades, but the industry found new prosperity after the automobile made inexpensive family travel possible and heralded the era of the "tin can" tourist.

The rise of Hollywood produced more than an upswing in tourism. After the film industry's beginnings in the Northeast, several moviemakers headed West, lured by a climate suitable for year-round outdoor production on a varied terrain just a short distance from the indoor studios. Some independent filmmakers relocated to the far Southwest and close to the Mexican border as a way to evade Thomas Edison and the "movie trust" with their pesky lawsuits charging nonmembers with infringing on their patents. By 1912 the Los Angeles area had the nation's largest concentration of motion picture companies, with many major studios clustered in the one-time temperance village of Hollywood. Moviemaking came of age during the mid-1920s, when it ranked first among California's thirty-five top industries and grossed far more than the state's fabled gold miners ever did in their heyday. By 1929 the movie industry employed one hundred thousand people; small outfits had grown or merged into industry giants like Metro-Goldwyn-Mayer, created in 1924.

Buoyed by its solid manufacturing base and the rise of new commercial enterprises, California prospered more than any other western state; except for the depression years of the 1930s, it seemed to enjoy a perpetual boom (at least until the 1990s). Even the agricultural depression of the 1920s scarcely fazed the Golden State. Increased irrigation enabled the Central Valley to become the nation's premier fruit and vegetable producer. The decade was also California's foremost era of oil production. The refining of petroleum ranked among the state's largest businesses and for several years after 1925 replaced even food processing as California's number-one manufacturing industry.

Thanks in part to the growing importance of food processing and oil refining, California rose steadily from the nation's twelfth-ranked manufacturing state in 1900 to eighth in 1940.

The population of the Golden State nearly doubled during the decade, reaching 5.7 million people in 1930. Approximately two-thirds of the newcomers settled in the southern part of the state, which experienced another of the spectacular real estate booms that had periodically erupted there since the 1880s. Naturally the state's construction industry thrived during the 1920s, as did seemingly any business related to automobiles. The mild sunny climate that permitted year-round outdoor filming also encouraged year-round driving. By the end of the twenties, California was home to 10 percent of the nation's twenty-three million automobiles and trucks. Los Angeles had the highest automobile registration per capita in the United States. The car culture spawned some of the nation's first shopping centers, supermarkets, and gasoline service stations, in addition to a host of other roadside businesses that catered to motorists.

Some of California's new arrivals during the boom of the 1920s came from Idaho and Montana, which because of continuing economic adversity showed the largest outmigrations of any western states between 1918 and 1933. Neither the stock market crash of 1929 nor the onset of hard times the following year represented an abrupt change from the economic malaise that had plagued the northern Rocky Mountain country since shortly after World War I. When the noted journalist Mark Sullivan visited Idaho in late 1932, he observed that the state was "literally the last community in the United States to feel the depression." He did not mean that as a compliment.

The Great Depression of the 1930s exacted an especially heavy toll in a region still heavily dependent on the production of lumber, minerals, and food for distant markets. The slowdown of manufacturing in the East devastated the natural-resource-based economy of the West. Not surprisingly, during the depression years of the 1930s the West experienced the slowest rate of growth for the entire twentieth century, and some states actually lost population as residents left in search of work elsewhere.

Despite California's enduring image as a land of opportunity, it too suffered along with the rest of the nation during the depression years, although Hollywood seemed to offer a conspicuous exception to the decline. Theater attendance around the nation remained strong, providing the dollars that kept cameras rolling in southern California. Yet even this glamour industry lost $83 million in 1932 and another $40 million in 1933. A worsening oil glut caused by major new discoveries in Texas cost Californians thousands of petrochemical jobs; still more were lost when the state's agricultural revenues plummeted from $750 million a year in 1929 to less than half that amount in 1932. In San Francisco, the unemployment rate shot up to almost 25 percent in 1932 and remained high for several more years. By 1934 there were 1.25 million people on relief in California, almost 20 percent of the state's population. Indigent people concentrated in the area of warm climate in southern California, where they attempted to survive on relief payments of $16.20 a month.

Throughout the West the unemployment crisis elicited a variety of responses. In heavily timbered portions of Idaho, some of the jobless ignited forest fires and then sought work putting them out. Washington apple growers created an enduring symbol

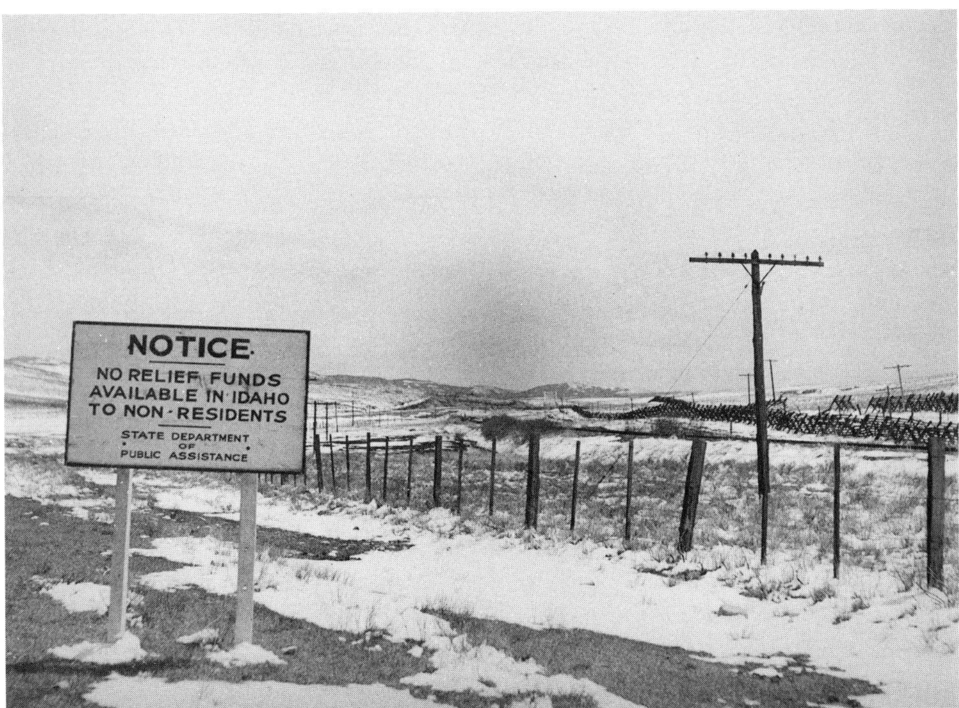

During the depression, overburdened state and local agencies posted warning signs to keep away displaced laborers seeking financial relief.

Unidentified photographer. "No Relief Funds" *(eastern Idaho). Gelatin silver print, ca. 1935. Library of Congress (#LC-USF34-65540-D), Washington, D.C.*

of the Great Depression when they sold fruit at $1.75 a crate to the unemployed, who then resold the apples, for a nickel apiece, on metropolitan street corners. Hoards of people, reminiscent of the armies of itinerant laborers common in the early part of the century, took to the roads and rails in search of jobs. Their numbers became such a burden in Wyoming that some towns required itinerants to work for room and board in special camps set up for them. The governor of Colorado declared martial law along the southern border and dispatched the National Guard to bar entry to indigent Mexicans, who he feared might take jobs from American citizens. The farce lasted only a few days. California authorities acted similarly after their state was inundated by whole families moving west in a desperate search for work: for several weeks in 1936, Los Angeles officials attempted to erect "bum barricades" on highways leading into the state from the east to turn back the unemployed. Courts ruled the procedure unconstitutional.

A primary response of the federal government to the Great Depression was the New Deal inaugurated in 1933 during the first months of Franklin D. Roosevelt's presidency. Yet even earlier, federal funds had been used to launch the construction of Boulder (now Hoover) Dam and provide loans to start work on the Oakland–San Francisco Bay Bridge. The New Deal dramatically increased the westward flow of federal dollars. Its programs ranged from building roads, dams, sidewalks, courthouses, and other public structures to cutting trails and seeking to eradicate blister rust fungus in Idaho's forests, all activities that created new job opportunities during the dark days of the 1930s. New Deal ameliorative programs pumped more dollars into Rocky

Mountain states per capita than into any others, yet that part of the West still had the highest percentage of unemployed workers in the country in 1940. At that time, 24 percent of New Mexico's work force was unemployed, the highest of any state.

The hard times of the 1930s initially meant continued trouble for organized labor in the West. The unions had endured a series of lean years during the 1920s, when business was clearly in the driver's seat and widespread and continued hostility toward organized labor had forced union members to give up many gains of previous years. The venerable Western Federation of Miners had become the International Union of Mine, Mill, and Smelterworkers in 1916, but like many of its counterparts during the 1920s, it lost important closed-shop agreements. Even in San Francisco, long the bastion of organized labor in the West, the once powerful Building Trades Council gave up ground to the businessmen who brought the open shop to the city. Throughout the West, organized labor in the 1920s showed less strength than at any other time since the 1870s and 1880s.

The 1930s depression brought massive unemployment, but the economic cataclysm also renewed the interest of workers in unions and other forms of collective action. Moribund organizations sprang back to life and made significant gains reminiscent of the opening years of the century. In the Rocky Mountain country, the Mine, Mill Union regained the closed shop in several generally nonviolent strikes. Even Wobblies, thought by many observers to be dead, briefly returned to the woods of northern Idaho and the fields and orchards of Washington's Yakima Valley. Actors like Groucho Marx and Ronald Reagan established a closed shop in Hollywood under the Screen Actors' Guild. Union membership in California tripled between 1933 and 1938, although open-shop sentiment remained strong in the southern part of the state until the rapid industrialization of World War II.

The ranks of organized labor ballooned as a result of federal legislation that encouraged collective bargaining. But the nearly uncontrolled growth of membership worsened friction between advocates of craft unionism and those of industrial unionism, the latter always numerous in the West, even in some AFL unions. When the American Federation of Labor expelled industrial unionists in 1937, dissidents formed the Congress of Industrial Organizations (CIO) the following year. Labor rivalry grew especially bitter on the Pacific Coast, where the aggressive CIO in many ways embodied the fiery militancy of the Industrial Workers of the World. The strength of the CIO lay in the timber and mining industries, in fish canneries, and on the waterfront.

The single most important labor disturbance on the West Coast during the 1930s was the great maritime strike of 1934. It occurred when the International Longshoremen's Association challenged the hated "shape-up" whereby longshoremen arrived on the docks early in the morning and waited for company foremen to select a fortunate few for the day's work, often receiving a kickback from the men chosen. When some 3,500 longshoremen quit work, they were joined by members of various marine unions. Together, they tied up ports from Seattle to San Diego for almost four months. Frightened residents came to believe that the strike was a Communist plot and that the ranks of subversives included the strike leader, Harry Bridges. In San Francisco in July 1934, the dispute erupted into a pitched battle between strikers and police. This in turn led to a three-day general strike that idled 127,000 workers. Shortly after this show of force, the walkout ended; through arbitration, the longshoremen won most of their demands, including an end to the hated shape-up.

Not only strikes but also schism marked this time in labor history. The craft-oriented AFL vigorously warred against the upstart CIO and labor radicalism. Spearheading the AFL opposition on the West Coast was Dave Beck, a fast-rising star in the Teamsters' Union. Capitalizing on the revolution in motor transport, Beck rapidly moved through Teamster ranks—from organizing laundry drivers in Seattle to being elected national head of the Teamsters' Union in 1952–62. In the mid-1930s Beck was easily the Teamsters' most powerful regional leader and the dominant personality in Northwest labor. For example, in the late 1930s when the forces of Harry Bridges, a labor radical from Australia who headed the Longshoremen's Union of the CIO, attempted to "march inland" from the docks to protect their flanks by organizing warehousemen, Beck countered with club-swinging squads of hired thugs. In time the economy improved, and both the strikes and labor's internecine warfare diminished. Beck even came to be regarded as a respectable citizen of Washington, where he served on numerous state boards and committees (although in the 1960s he served thirty months in prison for income-tax evasion). Bridges remained under a cloud of suspicion despite his repeated denials of being a Communist.

During the same years that Bridges and Beck gained national prominence as labor leaders, two western entrepreneurs gained similar recognition in the fields of banking and construction. Amadeo Peter Giannini and Henry J. Kaiser emerged in the popular mind as visionary leaders of an informal crusade to liberate the region from overdependence on eastern capital and business leadership. Their actions and public pronouncements illustrate how the western complaint of eastern economic colonialism, whether valid or not, might be used to advantage by astute westerners.

Starting in 1904 at the age of thirty-four, with no formal education in banking, Giannini launched his Bank of Italy with modest capital and an equally modest ambition of meeting the financial needs of the common folk of San Francisco. The city's money establishment initially dismissed the upstart as nothing more than a "dago bank" serving the "foreign colony" of the North Beach area. But after the 1906 earthquake, Giannini rapidly expanded and soon became a pioneer in the field of branch banking. In this way he enlarged the Bank of Italy into a giant that by 1930 controlled 40 percent of California's banking capital. Renamed that year as the Bank of America, it grew until in October 1945 its seventy-five-year-old founder proudly announced that the San Francisco–based firm had surpassed New York's Chase National Bank to become the world's largest bank. Throughout his life, Giannini chafed at what he perceived to be the West's subservient status and took inordinate pride whenever western businessmen surpassed their eastern counterparts. During World War II the Bank of America was a leader in financing the West's rapidly expanding industrial base, particularly the California aircraft industry, long dominated by eastern banks.

Another celebrated attempt to declare economic independence from the East was the formation of the Six Companies that built the sixty-story-high Boulder Dam across the Colorado River. Henry J. Kaiser, a highway contractor who would one day control an industrial empire from his Oakland, California, headquarters, regarded the chance to build the massive dam as a key to large-scale industrial development of the West by westerners. He believed that if western contractors combined their talents to build the dam—the world's largest construction venture to that time—they would pick up the lucrative contracts sure to follow. This would end long-standing eastern domination of the construction industry and help shift economic power from the East to the West.

The Six Companies, a consortium mainly of little-known western construction companies, was incorporated in 1931 and won the largest single contract let by the United States to that time (fifty million dollars). The Six Companies successfully completed Boulder Dam four years later, and true to Kaiser's prophecy, individual contractors went on to become multinational giants in their own right. Kaiser himself had a hand in constructing both Bonneville and Grand Coulee dams on the Columbia River and in time became a major player in cement, magnesium, steel, aluminum, and housing. Along the way, an enormous line of credit (forty-five million dollars) from the Bank of America plus money from Uncle Sam helped Kaiser to launch a career in shipbuilding during World War II.

According to Carey McWilliams in *California: The Great Exception*, who sounded a cautionary note, the antiestablishment rhetoric of men like Giannini and Kaiser could be used as a "smokescreen" to conceal certain aspects of California's growing dominance over other western states. California, no less than Wall Street, could become a colonizer of less populous parts of the region. Well before World War II, California had emerged as the colossus of the Far West.

From World War to Cold War

No event in the twentieth century, not even the Great Depression or the New Deal, brought more dramatic economic changes to the West than World War II. In much of the region it touched off an economic boom and significantly altered the traditional reliance on a natural-resource-based economy. On the eve of war, the West was still mainly a producer of raw materials shipped east to be made into finished products. That quickly changed as billions of federal dollars flooded the West to fund construction of a host of new military training facilities and war industries. California alone received 10 percent of all federal spending during the war. Some of those dollars took the form of huge aircraft factories that arose from the pea fields and orange groves of southern California. Here was the West's greatest concentration of aircraft factories, and airplane manufacturing became one of California's chief industries. Lockheed, for example, which built only thirty-seven planes in 1937, produced eighteen thousand between 1941 and 1945. When Uncle Sam poured more than five billion dollars into California for ships, employment in the shipyards of San Francisco Bay expanded from 4,000 to 260,000 jobs. Kaiser put together a shipbuilding empire that stretched from San Francisco Bay to the Columbia River. At the peak of production, his yards launched a new ship every ten hours and earned Kaiser the sobriquet "Sir Launchalot."

Kaiser, who regarded the war as one more chance to promote what he called "higher industrialization," became a major player in the West's new aluminum industry, which began in Vancouver, Washington, in 1940 with the opening of an Alcoa plant. Kaiser eventually built and operated five aluminum plants in the Pacific Northwest. By 1956 his Kaiser Aluminum Company was responsible for 25 percent of the nation's output. Kaiser also wanted to produce steel, but the federal government financed the construction of a big steel plant near Provo, Utah, and leased it instead to a subsidiary of U.S. Steel. Backed by Giannini and his banking muscle, Kaiser persisted and eventually won federal funding for a steel mill at Fontana, California, a site fifty miles east of Los Angeles. When his Fontana plant came on-line, Kaiser boosted the state's output of steel by 70 percent.

Oil production once again boomed in California, Texas, and other parts of the West during World War II. These were also the first prosperous years in two decades for lead and silver in Idaho and for zinc and molybdenum in Colorado. Copper boomed too, bringing newfound prosperity to Arizona, Nevada, Montana, and Utah. The huge open-pit mine in Bingham Canyon alone employed seven thousand of Utah's one hundred thousand workers.

As part of a deliberate effort to decentralize the nation's production of vital war materials, Tucson, Phoenix, Albuquerque, and Denver became sites of new defense-related industries. At the Denver Arms plant, which the federal government built in 1941, Remington Rand and later Henry Kaiser employed as many as twenty thousand workers, about 40 percent of the city's factory personnel, to make armaments such as cartridges, shells, and fuses. Even ships were built in land-locked Denver, such as the USS *Mountain Maid* completed in the spring of 1942. Near Denver, the Rocky Mountain Arsenal and its work force of fifteen thousand made chemical weapons. After the war, Denver's war industries spawned a host of defense-related businesses that included the manufacture of ballistic weapons and space hardware.

During the early years of the war, new production techniques transformed both aircraft and shipbuilding from craft to mass-production industries. This decreased the need for skilled labor and opened the doors to a variety of people who had never before been inside a shipyard or an aircraft factory. Filling thousands of jobs at Lockheed, Douglas, and a number of other aircraft companies were many Okies who had first found employment in California agriculture.

The number of lines in the "help wanted" section of the *Seattle Times* jumped from 28,631 during the first nine months of 1940 to 225,515 during the same period in 1943. Employers paid good wages and competed with one another for workers. Said one Boeing official, "We hired anybody who had a warm body and could walk through the gate." At the peak of wartime aircraft production, Boeing hired fifty-five thousand workers, 46 percent of whom were women. Nearby at the sprawling Puget Sound Naval Shipyard, women formed 21 percent of the thirty thousand workers in 1943. As recently as 1941, the facility had not employed any women outside the offices. During the war, nearly a quarter million blacks, many from rural backwaters of the South, moved to California, where most took jobs in the shipbuilding industries of Los Angeles, Oakland, San Francisco, and Richmond. Thousands more came after the war when California's economy continued to boom.

To deal with the shortage of housing for workers at his Portland and Vancouver shipyards, Kaiser built an entirely new town on the low-lying banks of the Columbia

During World War II, women moved into the work force in unprecedented numbers. Like countless others, these "timberettes" unloading lumber at a sawmill east of Seattle found employment in jobs previously held by men.

Unidentified photographer. Timberettes, Snoqualmie Falls, Lumber Co. *(Washington). Gelatin silver print, ca. 1943–44. A Weyerhaeuser Company photograph. Courtesy of the Forest History Society, Durham, North Carolina.*

River just north of Portland. Called Kaiserville and then Vanport, it became one of the largest housing projects in the world. By the end of the war, the instant community of 35,000 people had become the second-largest city in Oregon. The population of Richmond, California, the location of another major Kaiser shipyard, grew from 23,000 to 115,000. San Diego's population doubled. Los Angeles ended the war as America's third-largest city; it also climbed from the seventh-largest manufacturing center before the war to the nation's second largest after Detroit.

Much of the mountain West served as a labor reservoir for booming war industries on the coast. Idaho, except for Camp Farragut and a large naval ordnance plant in Pocatello, contributed mainly its manpower and the traditional products of its mines, forests, and fields to the war effort. The state lost 15,000 residents between 1940 and 1945. Most of them apparently moved to Oregon and Washington, which gained 194,000 and 533,000 new residents respectively. Montanans also headed west to Seattle in record numbers; by late 1943 that state had lost 88,000 people while Seattle gained almost the same number. Pacific Coast shipyards and aircraft factories drew Anglo workers from both Arizona and New Mexico while Hispanics and Indians remained behind to take jobs on railroads, farms, and construction projects.

Organized labor experienced a great influx of new members during the war, but life in the burgeoning union ranks was not harmonious. Industrial workers who were recruited from rural regions and small towns typically regarded unions with apathy or even antagonism and often resented paying any dues. The addition of females and blacks to the industrial work force further disrupted several unions. When blacks in the aircraft and shipbuilding industries applied for union membership, racial antipathy became evident. At Boeing, the union refused to admit blacks and in effect contributed to shaping company policy: Boeing hired only a small number of blacks during the war. Kaiser opposed racial discrimination, but nearly all his employees were members of metal trades unions that had no desire to admit blacks to membership. The Boilermakers' Union, for example, segregated black shipyard workers into auxiliary unions and blocked them from holding skilled jobs. Blacks paid dues but had no voice in union matters, and when the war ended, they found that their classification as temporary members gained them no seniority in the scramble for jobs.

Many westerners feared a return to hard times when World War II ended. In reality, however, the passage from war to peace was surprisingly smooth, especially when compared with the troubled aftermath of World War I. Wartime savings, terminal-leave pay, and unemployment compensation discouraged the kind of desperate search for jobs that had followed the 1918 armistice. Some of the West's war plants closed briefly only to reopen as other industries. Soldiers who had first glimpsed the West in training camps often returned after the war and added to a population boom. For a time, bulldozers ripped out three thousand acres of orange groves a day in Los Angeles County to make way for hundreds of new housing developments. In Arizona, where industrial output fell from eighty-five million dollars in 1945 to fifty-three million dollars a year later, a booming construction industry created many new jobs. An inadequate number of houses built during the Great Depression and war years, coupled with the nation's move to suburbia and an increased rate of family formation, kept the West's forest products industry producing at record levels for several years after the war, although at the cost of further depleting the region's timber supply.

In a real sense, World War II never ended. The infusion of federal dollars into the western economy, which had become so noticeable during the New Deal and World War II years, continued with the cold war. From the 1940s through the 1980s, federal investments in science, defense, reclamation, and highways all combined to further industrialization and urbanization of the West. The rising tide of federal investment in sophisticated military and aerospace technology was especially pronounced in southern California, in the urban Southwest and Texas, and around Puget Sound. "Air power is peace power," declared William Allen, president of Seattle's Boeing Aircraft, which prospered once again from defense orders after a brief postwar slump. Cold war fears kept many of the West's defense plants busy, notably those of the aerospace giants Boeing, Lockheed, and Douglas, all of which manufactured jet aircraft, intercontinental ballistic missiles, and space equipment. In the 1960s, shortly after the National Aeronautics and Space Administration was organized, the new federal agency spent 50 percent of its funds in California.

By 1958, Boeing alone employed seventy-three thousand people and added thousands more to assemble its model 707, the world's first successful jet passenger airplane, as well as to build military equipment. In the process, the state of Washington traded dependence on the forest products industry for dependence on the manufacture of transportation equipment. Defense-based industries also became a major employer in postwar Utah, where as recently as 1941 the economy had been almost entirely dependent on agriculture and mining. The aerospace industry gave Utah's economy a major boost, with the biggest facility being the Thiokol Chemical Company west of Brigham City, which produced the first solid fuel for rockets (and later, as Morton-Thiokol, built booster rockets for the ill-fated space shuttle *Challenger*).

Construction and operation of vast nuclear weapons facilities spurred economic growth in several parts of the West. These facilities ranged from the Hanford Engineer Works in central Washington, where plutonium for atomic bombs was produced, to the Nevada Test Site, where federal authorities conducted 105 atmospheric tests of atomic weapons between 1951 and 1962. Albuquerque boomed as a center of atomic age research. Seventy-five miles north of the city by air was the atomic physics laboratory at Los Alamos, and closer at hand was the Sandia complex that designed bomb cases and components. The number of employees at the Sandia Corporation increased from thirty-eight hundred in 1951 to seventy-eight hundred ten years later, and its payroll became the largest in New Mexico. For a time, Albuquerque proudly billed itself as the nation's "Atomic City." The atomic West also extended into a remote part of Idaho, where federal scientists developed the first atomic reactor to power submarines and trained crewmen for the nation's nuclear navy. They even made plans to use the atom to power airplanes, though the idea was never implemented. By 1990 the Idaho National Engineering Laboratory provided about 5 percent of the state's jobs and employed military and civilian scientists in a variety of research projects involving atomic energy, lasers, and biotechnology. For a time, it seemed as if the cold war would buoy the western economy forever.

The New Economic Equation

Military hardware, aerospace, and electronics, together with tourism, became key elements in the West's new economic equation after World War II. One unlikely pioneer of the high-tech West was Howard Hughes, the playboy son of a Texas oil-bit manufacturer who got his start in the Spindletop boom of 1901. Young Hughes came to Los Angeles in the 1920s and initially used his father's money to distinguish himself as a filmmaker and aviation enthusiast. By the mid-1930s he had established Hughes Aircraft in Burbank, which eventually specialized in electronics. After World War II, Hughes brought thousands of scientists, technicians, and other skilled workers to his southern California facility and cashed in on the first peacetime defense boom in the nation's history. With his holdings in Hughes Aircraft of California and Hughes Tool of Texas, he was probably the wealthiest man in America, although as Hughes grew older and richer he became more eccentric.

William Hewlett and David Packard launched a modest electronics business in a small garage in Palo Alto, California, in 1939. Their first breakthrough product was an audio oscillator. From this seed grew an industry that made both men very rich and transformed the agricultural Santa Clara Valley into the center of high technology now known as Silicon Valley. The name became a byword for California's computer industry, although several similar concentrations of technology emerged in the 1970s and 1980s in Orange County, on Puget Sound, and in cities as diverse as Boise and Phoenix. In Bellevue and Redmond, Washington, the rapid growth of the software giant Microsoft made its chairman, William Henry Gates III, the world's youngest billionaire.

Entrepreneurial westerners also made fortunes from less glamorous endeavors, as John Richard Simplot did from pigs and a small plot of Idaho potatoes. His big boost came when World War II created an enormous demand for dehydrated potatoes, onions, and other food products needed to feed the troops. During the war, Simplot became a supplier of dehydrated food and the world's "biggest potato farmer." The modern J. R. Simplot Company of Boise, a constellation of enterprises that range from food processing and chemical fertilizers to mining and ranching, supplies the McDonald's restaurant business with over half its french fries. In the October 1990 issue, *Forbes* magazine listed Simplot and another of Idaho's homegrown entrepreneurs, the supermarket king Joseph A. Albertson, among the four hundred richest people in America.

During the decades after 1945, several cities of the interior West, including Denver and Phoenix, reached sufficient size to put together much of the financing needed to fuel the postwar boom. The explosive growth of Arizona was financed in part by Walter Bimson and his Valley National Bank, which in their aggressiveness resembled Giannini and the Bank of America. As a result of the economic impact of World War II, Giannini himself bragged, "[The] West has all the money to finance whatever it wants to; we no longer have to go to New York for financing, and we are not at its mercy." Los Angeles and San Francisco banks continued to grow in power with the creation of new giants like Los Angeles–based First Interstate and the expansion of Bank of America through the purchase of Seattle's Seafirst bank. In banking, as in numerous other areas, Los Angeles came to rival San Francisco as a center of economic power on the Pacific Rim, home to a multitude of multinational corporations, banks, and insurance companies. California ranked second only to New York as America's banking and financial power. Long the nation's leading agricultural state, California grew in importance as a manufacturing center, with employment in that sector increasing between 1970 and 1980 at a rate more than five times that of the rest of the country. California, and to a lesser extent other western states, also benefited from the booming trade with Pacific Rim nations, especially Japan. Not without reason would Governor Jerry Brown in the late 1970s refer to his state as "the Pacific Republic of California."

In California, as in several other western states, tourism boomed after the war. When gasoline rationing ended, tourists withdrew money from their bulging savings accounts and swarmed across the "Golden West" in record numbers. The number of visitors to Washington's Mount Rainier in 1946 broke all previous records. National parks had long been recognized as tourist draws, especially so as the region's network of all-weather roads expanded and more Americans acquired the automobiles, money, and leisure time that allowed them to explore all corners of the West. One of the best-known natural attractions was Yellowstone, the nation's oldest national park, which during summer seasons in the mid-1950s lured four times as many people through its gates as lived in the entire state of Wyoming. Glacier National Park attracted more visitors each season than lived in Montana. Interior Secretary Harold Ickes correctly prophesied of Olympic

National Park, which was established in 1938, "In the long run it will mean more for the State of Washington to have a real national park on the Olympic Peninsula than it will be to log this area, either selectively or otherwise."

Another of the West's increasingly popular tourist draws was snow skiing. This industry essentially dates from the mid-1930s when the Union Pacific Railroad opened its posh Sun Valley, Idaho, resort complex, the West's first major ski facility. The demands of World War II temporarily blocked expansion of this promising new form of tourism, and Sun Valley itself was converted into a navy convalescent center. When the conflict ended, just two ski areas in Colorado were operating, with their business confined mainly to weekends. But what Sun Valley began continued after World War II with the development of new winter vacation facilities throughout the West. In Colorado, the pacesetter for winter sports activities, some of the most popular facilities were located in refurbished mining towns like Aspen and Telluride. Still others were in Vail, a new, planned community.

By 1980 tourism had become the largest industry in California. Tourism was actually an old business in the Golden State, but it evolved in a new direction after 1955 when Disneyland opened amid the orange groves of Anaheim. Before that, amusement parks catered mainly to local populations, but Disneyland was a national attraction. It drew four million visitors the first year and in time became one of America's most popular tourist stops. Disneyland also helped to transform the Anaheim–Santa Ana area into a regional rival of downtown Los Angeles. One of Disneyland's popular sections was a fantasy frontier complete with a Mark Twain–era steamboat and a runaway mine train. Outside the park, vestiges of the West's real frontier like Virginia City, both in Montana and in Nevada, discovered how to lure tourists by refurbishing and emphasizing the look of mining boom days, unconsciously, perhaps, borrowing from Disney. In dozens of such communities scattered throughout the West, from Tombstone, Arizona, to Jacksonville, Oregon, look-alike shops sold tourists much the same assortment of trinkets, T-shirts, and taffy.

An unusual form of tourism developed in Nevada, where gambling became a major contributor to the state's economy after World War II. The state had legalized gambling as early as 1931, but old ways of doing business persisted for several more years. Gambling took place in dark, secluded halls where dealers worked in shirtsleeves and wore green eyeshades. Raymond Smith took a different approach at his Harold's Club, which was located on Reno's main street, its glass front revealing brightly lit bars and gaming tables. By the end of the 1930s, Smith's advertisements reading "Harolds Club or Bust" were on billboards scattered across the country and helped launch gambling as a tourist draw.

Until World War II, gambling centered in Reno. Its chief rival, Las Vegas, remained little more than the dusty, railroad division town it had been for the previous forty years. The war, however, changed the character of the town. By 1946 Las Vegas had a population of approximately twenty thousand and was already on its way to becoming the neon-and-chrome oasis it is today. By the 1950 census, it had sixty-five thousand residents, and that was only the beginning of a time of spectacular growth.

The city's first modern casino was the Gold Nugget, which opened in 1946. It was located downtown, but the center of action quickly moved south of Las Vegas to "the

Strip," where a string of new casinos competed with one another in lavishness of decor and Hollywood entertainment. The colorful crime figure Benjamin "Bugsy" Siegel built the Flamingo Hotel, among the first and most lavish casinos on the Strip, with one million dollars of his own money and another six million of borrowed money. When the venture did poorly, Siegel became a victim of a gangland slaying at his home in Los Angeles and drew national attention to the problem of gangster infiltration of Nevada's lucrative new industry. It took a series of unsavory revelations about organized crime before Nevada attempted to clean house. Not until 1955 was any serious effort made to bring the state's gambling industry under government regulation, and nothing substantial happened until the late 1960s. Some problems with organized crime persisted through the 1980s. In any case, gambling accounted for 32 percent of all Nevada jobs in the 1980s, with another 25 percent in related employment.

Overall, the West's new service and high-tech industries enjoyed reasonably good health during the postwar decades, but even they were not immune to the region's old nemesis of boom-and-bust cycles, as the Boeing company demonstrated time and again in western Washington. By the mid-1950s, aircraft manufacturing employed more people in the Evergreen State than logging and lumbering, as compared with less than one-seventh as many before the war. Boeing was by far Washington's largest private employer. Boeing's economic well-being was tied to two markets—civilian aircraft and military weaponry. In 1970, when the company was hard hit by the failure of Congress to fund a supersonic transport plane, employment dropped from a little over one hundred thousand to thirty-eight thousand, and about fifty-five thousand people left the state in search of work. In the Seattle and Everett metropolitan area, unemployment reached 13 percent in 1971, the highest in the United States. The real estate market virtually collapsed, a fact hard to recall twenty years later when the Puget Sound area experienced soaring home prices and a superheated economy. California, like Washington, suffered greatly when recession hit the aerospace industry in the 1970s. Between 1967 and 1972, the number of aerospace jobs in California declined from 616,000 to 450,000. The state's unemployment rate climbed to more than 7 percent, compared with 6 percent for the United States as a whole.

The West's high-tech industries also suffered competitive pressures from Japan, Korea, and Singapore and thus provided no real antidote to boom and bust. For all their glamour, many high-tech industries proved little better than highly sophisticated sweatshops, employing large work forces of unskilled, low-income, nonunion labor. That was only one of the trends that did not bode well for organized workers in the postwar decades. For example, right-to-work laws weakened the power of labor unions in several western states. Arizona passed the first of these in the late 1940s by popular referendum; Nevada, Utah, and Idaho established similar laws during the following decades.

Nationwide, the number of manufacturing workers who belonged to unions fell noticeably in the latter half of the 1980s. The result in Arizona, for instance, was that less than 4 percent of the state's manufacturing workers were unionized. For unskilled workers, the future looked especially grim in the desert Southwest. There the trend toward jobs that required more education and higher levels of skill brought social dislocation to the poorest, least-prepared members of the work force. Still another disconcerting development that began in the 1960s was the rise of *maquiladora* ("twin plant") industrial parks along the Mexican border. Expensive U.S. components were imported duty free into Mexico, where they were assembled in these factories with low-cost Mexican labor and shipped back to the United States, with manufacturers paying duty only for the cost of assembly.

The *maquiladora* system rapidly expanded after devaluation of the Mexican peso; from 620 plants in 1980, the number increased to more than 1,400 by 1988, and the number of jobs the factories provided in Mexico went from 100,000 to 400,000. In Nogales, Arizona, and Mexican Sonora, some fifty companies, including some of the largest in the United States, established twin plants. In 1988 they employed 22,000 people in Mexico and another 1,250 on the Arizona side. Every day one hundred truckloads of *maquila* goods were shipped across the border at Nogales into the United States. This use of low-cost Mexican labor was one way that American automotive and electronics companies sought to meet Japanese competition.

An additional omen of a grim future for western labor was the increasing number of jobs in the tourist industry, which often paid near minimum wages, balanced against a decreasing number of jobs in industries like logging and mining, where the pay was much higher. In an affluent tourist community like Sun Valley, Idaho, which evolved in recent years from a winter sports mecca to a year-round resort, many service workers could not afford the skyrocketing price of local real estate and had to drive to work from homes located as far away as Twin Falls, seventy miles distant. The story was much the same in Jackson Hole, Wyoming, and numerous other upscale resort communities in the West.

Affluence and Anxiety

During the years from 1945 to 1990, it became increasingly clear that there were really two Wests, economically speaking. One was oriented toward sophisticated technology and service industries like tourism, whereas the other was still dependent on the production of natural resources. Timber and mining boomed after World War II and into the 1970s and then, along with agriculture and mining, suffered a collapse that lasted well into the 1980s. The globe seemed awash in natural resources, and this glut hit the West hard. The region suffered its worst economic crisis since the 1930s.

Arizona accounted for 60 percent of the copper produced in the United States until foreign competition flooded the country with copper in the 1970s and caused mines all over the West to close. In Montana, where even the capitol dome is made of copper, the red metal seemed to suffer an especially slow and agonizing death, at least when mirrored in the declining fortunes of the Anaconda Company. The copper giant relinquished control of its several newspapers in 1959; during the next decade Anaconda and its ally Montana Power relaxed their hold on the state legislature. In 1975 Anaconda became a subsidiary of the Atlantic-Richfield Company, but this did not arrest the decline. Five years later, and little more than a century after commercial quantities of copper were first discovered in Montana, the company's closure of its giant smelter at Anaconda cost the jobs of two thousand Montanans. In 1983 the company closed its remaining copper pit in Butte: the great Berkeley pit that Anaconda had opened in 1955 slowly filled with tainted water. Metals mines reopened in Montana in the late 1980s, but when

The western uranium industry, once seen as a boon to Navajo miners and others in the Southwest, is now viewed as an environmental and health disaster responsible for water and soil contamination and widespread illness and death among the miners.

Laura Gilpin (1891–1979). Navajo Uranium Miner *(Lukachukai Mountains, Arizona). Gelatin silver print, 1953. Copyright Amon Carter Museum. Laura Gilpin Collection, Amon Carter Museum, Fort Worth, Texas.*

production resumed in natural-resource-based industries like mining and timber, facilities were invariably more highly automated, less labor-intensive, and often nonunion.

In the early 1980s Nevada was another of the states beset by the worst slump in mining in forty years. Ironically, however, even as copper became an insignificant factor in the state's economy, Nevada experienced a new gold rush, centered in the Battle

Mountain area. Like the low-grade copper deposits of an earlier era, all the new gold mines were open-pit. Microscopic particles of gold could be recovered even from the tailings of earlier mine operations.

Not just metals but also the energy resources of coal, oil, and uranium experienced boom-and-bust cycles in the decades after World War II. For a time in the early 1950s, action centered in uranium deposits on the Colorado Plateau in the Four Corners area. Resembling a new Klondike rush, the search for uranium embodied the American dream of sudden wealth. Any person could stake a uranium claim on public land for $1.00 and, armed with a Geiger counter, might strike it rich. One of the lucky ones was Charles Steen, who made a $150-million uranium find near Moab, Utah, and seemingly overnight transformed the sleepy cattle and ranch center. Moab eagerly embraced the title of "The Uranium Capital of the World." The frenzied hunt was yet another facet of the cold war in the West and the belief that uranium was needed to maintain nuclear weapons superiority over the Soviet Union. Uranium finders thus had a ready customer in the federal government, which was desperate for a domestic source of the yellow metal.

During the uranium boom of the 1950s, Grants, New Mexico, grew from a ranch supply town of five hundred people to become the nation's largest uranium-milling center, with a population of ten thousand. In 1958 the six uranium mills in Grants produced ore worth $115 million. The pace slackened in the early 1950s when the Atomic Energy Commission decreased its uranium purchases, but the yellow metal boomed again in the 1960s and 1970s when the nuclear power industry expanded. The cycle again reversed itself in the early 1980s when the price of uranium plummeted from $43 to $17 a pound amid growing public skepticism about the safety of nuclear power. After the collapse, one thousand men lost their jobs in Wyoming, where Jeffrey City became a uranium ghost town. Of some two thousand uranium workers in Utah in 1960, fewer than four hundred remained employed after the bust. The story was much the same in New Mexico.

Oil, coal, and gas fared better. In the mid-1950s the discovery of oil and gas in the San Juan Basin around Farmington, New Mexico, led to the construction of numerous pipelines, one of which extended sixteen hundred miles to carry natural gas to the Pacific Northwest. In 1949, Montana's oil output for the first time exceeded the value of its copper production, and the city of Billings rode the crest of several successive oil booms. Moving to the forefront of energy development in the late 1970s was the "overthrust belt" in western Colorado and Wyoming. Here California oil giants like Getty, Arco, Champlin, Chevron, and Occidental became new colonizers of the western hinterlands, along with eastern corporations like Exxon. The latter company issued rosy predictions that by the year 2010, production from oil shale in the geologically distinctive overthrust belt would employ 480,000 people in mining, 390,000 in processing plants, and 250,000 in construction during peak years.

Encouraged by rising oil prices and predicted global shortages, Congress in 1980 approved an ambitious goal of producing two million barrels a day from coal and oil shale (synfuel) by 1992. That would mean the construction of thirty to forty mammoth coal and oil shale plants, each costing from one to six million dollars. Residents of Rifle,

Meeker, and other small towns in northwestern Colorado braced for a boom, all the while hoping to avoid becoming another Rock Springs, a Wyoming community that became a byword for social dislocation caused by too rapid development. In the end, the promise of oil from shale proved to be a mirage. By the winter of 1981–82, Occidental Petroleum had closed its operations in the face of economic change, notably the slumping price of conventional oil. Exxon followed by shutting down its Colony Project in May 1982. Once again, the tired refrain of boom and bust led to jobs, good wages, and plenty of overtime followed by massive unemployment and social problems.

In the late 1970s, while energy jobs were plentiful and money abounded, there was an outburst of anger from people who feared another "rape of the West" by "outside" capital, some of which would actually come from places inside the region like Denver and Los Angeles. But by the mid-1980s, oil optimism had run its course, grass grew in the streets of some of the energy boomtowns, and the "sagebrush rebellion" had quieted. *Newsweek* magazine in 1989 evoked the ire of many westerners when it described the northern tier states from Washington to the Dakotas as "America's Outback."

Of all the energy minerals, coal staged the most dramatic and longest-lasting comeback. As late as the 1950s, the western coal industry remained in a declining or, at best, stagnant condition. In 1956 the Northern Pacific Railway liquidated its subsidiary, the Northwestern Improvement Company, which had provided mining jobs at Roslyn, Washington, and Colstrip, Montana. Coal was simply not competitive when the price of oil was three dollars a barrel. But then came a series of sharp price increases for oil, and Colstrip made a remarkable comeback. When the first dedicated coal train left Colstrip for an electric-generating plant in Minnesota in 1969, it marked the dawn of a new era. As in open-pit metal mining, however, the reborn coal industry employed remarkably far fewer people than it had earlier. In Colstrip in 1972, strip-mining methods used only 150 men to produce 5.5 million tons of coal.

In October 1973, the Arab-Israeli War sent the price of oil up to thirteen dollars a barrel and further spurred coal production in Wyoming and Montana. The energy boom of the 1970s and 1980s probably affected no western state more than Wyoming. For a time coal production increased 20 percent annually. The state's Powder River basin, where forty billion tons of subbituminous coal lay buried beneath the soil of Campbell County, became the nation's new energy frontier. During the energy boom, the town of Gillette grew from four thousand people in 1969 to fourteen thousand in 1980. Just to haul the black diamonds to market, railroad construction crews laid 226 miles of new track in the 1970s. A decade later, coal from the Powder River country accounted for 35 percent of the receipts of the Burlington Northern, then the longest railroad in the United States.

Even as the West's coal industry enjoyed its newfound prosperity, the timber industry slumped once again. A variety of economic problems closed sawmills across the Pacific Northwest in the early 1980s. Chief causes of the trouble in timber country were the revival of southern pine forests (which Pacific northwesterners had once cavalierly dismissed as having the quality of weeds), competition from Canadian imports, a housing slump caused by high interest rates, and a shifting market. When the timber industry rebounded, as it did in the late 1980s, innovative technology that included computers and lasers enabled modern sawmills to employ far fewer hands than they had only a decade earlier. These were permanent changes, as they were in the mineral industries, and they clearly illustrated that riding the cycle of boom and bust did not mean that with the return of prosperity would come the return of all the former jobs. In sawmill towns, ranging from Garibaldi on the coast of Oregon to Potlatch in the pine forests of northern Idaho, an era had ended forever. A grassy field replaced the sprawling mill complex in Potlatch, which survived as a bedroom community; Garibaldi's mill became a storage facility for yachts and other pleasure boats. Never again would it be possible to describe the Pacific Northwest as a Sawdust Empire.

A similarly painful and perhaps irony-filled future awaited communities that had once prospered from the cold war. The American economy did not demilitarize after the Allied victory in World War II, so for forty-five years, defense-related spending buoyed the economies of nearly every western state. The apparent end of the cold war in the early 1990s thus forced the West to make some painful adjustments. There was much talk about a "peace dividend" that was supposed to translate not just into the shutdown of redundant military bases but also into massive cutbacks for defense contractors. In mid-1990, for example, the McDonnell Douglas Company trimmed expenses by laying off eight hundred employees at its Mesa, Arizona, and Culver City, California, plants. For the nation's number-one defense contractor, it was the third major set of layoffs since 1989. One study estimated that the closing of the Davis-Monthan air force base in Tucson would cost an estimated ten thousand jobs, produce an exodus of some fifteen thousand people, and cut local spending by about $250 million a year. "Welcome to peace," commented Congressman Jim Kolbe, who represented the Tucson area. "This is going to be a tough road for everybody." Ironically for the Arizona metropolis, when Hughes Aircraft Company moved production of Tomahawk missiles from San Diego to its plant in Tucson in late 1992, it was California that lost twelve hundred defense jobs. For California, which as recently as 1989 had received almost one-fifth of all federal defense contracts, the road to peace would obviously require some very painful adjustments.

Defense contractors talked of using their expertise in military technology for other purposes. Boeing believed that its skill in making satellite-blasting lasers would enable it to design medical equipment for vaporizing cancer cells. Hughes Aircraft, now a subsidiary of General Motors, hoped to transfer its expertise in flight gadgetry to automobiles. In some remote locations, abandoned defense plants and military bases might well have joined ghost towns in symbolizing the West's changing economic fortunes.

In the late twentieth century, the region's natural-resource-based economy left a distinctive signature on the land in the form of abandoned mines, sawmills, and once bustling towns that had withered or disappeared. Many such communities pinned their future hopes on tourism. Some, like Cokedale, Colorado, and Jerome, Arizona, for all practical purposes became living museums of the resource frontier. Cokedale, founded in 1906 by the American Smelting and Refining Company, was placed on the National Historic Register in 1984; but apart from industrial archaeology specialists and Colorado history buffs, not many tourists came to see its huge slag heap or its abandoned coke ovens.

Dependence on natural resources took a toll on the workers of the West, especially in timber and mining, which were two of the most dangerous occupations in the United States. In recent times, legions of agricultural workers have been exposed to a variety of chemicals, which individually or in combination are likely to have a deleterious effect on human health. Probably no group paid a higher price than uranium workers: one especially sinister aspect of the uranium bust of the 1980s was that it not only cost jobs but also left behind radioactive tailings and prospects of a rising toll of lung cancer among workers in the mines and processing plants. Some of the fifteen thousand people once employed as miners believe that uranium companies and the former Atomic Energy Commission conspired to hide the danger of radiation from workers. According to figures supplied by the Centers for Disease Control, 350 of 4,146 uranium miners studied since the early 1950s have died of lung cancer—five times the expected rate. Among the victims were hundreds of Navajos who were sent into the mines without any protective clothing, face masks, or respirators.

Despite the glamour of Nevada gambling and Hollywood films, the computer wizardry of Silicon Valley, and the burgeoning trade of the Pacific Rim, the tragedy of the uranium miners vividly illustrates how the history of business and labor in the twentieth-century West has so often been a tale of relatively short-term prosperity and technological triumphs accompanied by dire long-term consequences. One of the West's growth industries in the years to come will certainly be that of cleaning up the nuclear and chemical messes created during years of cold war prosperity. The West's freewheeling "no-deposit-no-return" post–World War II economy may have left numerous hidden costs.

Bibliographic Note

Individual state histories and historical quarterlies are a rich source of information about business and labor in the twentieth-century West. The *Utah Historical Quarterly*, for example, devoted an entire issue to the state's mining industry (Summer 1963), whereas *Idaho Yesterdays* provided a bibliography of articles on Pacific Northwest labor history (Winter 1985). I make no attempt to provide an exhaustive list of books about even a single topic. The books on western railroads, even for the twentieth century alone, would fill several shelves, and those on the movie industry would occupy many additional shelves. I note only that one of the best summaries of the vast literature on Hollywood is in Kevin Starr's *Inventing the Dream: California through the Progressive Era* (New York, 1985).

What follows is a brief description of books that address aspects of the larger subject. There is, unfortunately, no single book that synthesizes the business and labor history of the twentieth-century West, although two volumes do help to fill the gap: Michael P. Malone and Richard W. Etulain, *The American West: A Twentieth-Century History* (Lincoln, 1989); and Gerald D. Nash, *The American West in the Twentieth Century: A Short History of an Urban Oasis* (1973; reprint, Albuquerque, 1977).

The Bisbee story that was used to introduce this chapter is from Carlos A. Schwantes, ed., *Bisbee: Urban Outpost on the Frontier* (Tucson, 1992). A detailed account of the Bisbee deportation is in James W. Byrkit, *Forging the Copper Collar: Arizona's Labor-Management War of 1901–1921* (Tucson, 1982). Byrkit summarizes his research in James C. Foster, ed., *American Labor in the Southwest: The First One Hundred Years* (Tucson, 1982), a volume that provides an excellent overview of labor in that subregion. In a similar category is Hugh T. Lovin, ed., *Labor in the West* (Manhattan, Kans., 1986). Labor in the Pacific Northwest is the subject of Carlos A. Schwantes, *Radical Heritage: Labor, Socialism, and Reform in Washington and British Columbia, 1885–1917* (Seattle, 1979). David F. Selvin, *A Place in the Sun: A History of California Labor*

(San Francisco, 1981), offers a brief overview of labor developments in a single state. Also useful on California labor and business is Carey McWilliams, *California: The Great Exception* (New York, 1949).

Far more common than state or regional studies of western labor are books on individual strikes, episodes of violence, union organizations, and labor in specific industries. Among the best of these are Robert L. Friedheim, *The Seattle General Strike* (Seattle, 1964), Michael Kazin, *Barons of Labor: The San Francisco Building Trades and Union Power in the Progressive Era* (Urbana, 1987), and Bruce Nelson, *Workers on the Waterfront: Seamen, Longshoremen, and Unionism in the 1930s* (Urbana, 1988), which focuses on Pacific Coast ports.

Labor in the West's mining industry has been an especially popular topic. Among the best studies are the following: Ronald C. Brown, *Hard-Rock Miners: The Intermountain West, 1860–1920* (College Station, Tex., 1979); Howard M. Gitelman, *Legacy of the Ludlow Massacre: A Chapter in American Industrial Relations* (Philadelphia, 1988); George S. McGovern and Leonard F. Guttridge, *The Great Coalfield War* (Boston, 1972), on the troubles at Ludlow, Colorado; George G. Suggs, Jr., *Colorado's War on Militant Unionism: James H. Peabody and the Western Federation of Miners* (Detroit, 1972); James Whiteside, *Regulating Danger: The Struggle for Mine Safety in the Rocky Mountain Coal Industry* (Lincoln, 1990); and Mark Wyman, *Hard-Rock Epic: Western Miners and the Industrial Revolution, 1860–1910* (Berkeley, 1979).

The Industrial Workers of the World has been the subject of numerous studies including Melvyn Dubofsky, *We Shall Be All: A History of the Industrial Workers of the World* (Chicago, 1969), and Robert L. Tyler, *Rebels of the Woods: The I.W.W. in the Pacific Northwest* (Eugene, 1967). A basic research tool is the massive compilation by Dione Miles, *Something in Common: An IWW Bibliography* (Detroit, 1986).

Studies of agricultural and itinerant laborers include Cletus E. Daniel, *Bitter Harvest: A History of California Farmworkers, 1870–1941* (Ithaca, N.Y., 1981), Lawrence J. Jelinek, *Harvest Empire: A History of California Agriculture* (San Francisco, 1979), James N. Gregory, *American Exodus: The Dust Bowl Migration and Okie Culture in California* (New York, 1989), Thomas D. Isern, *Bull Threshers and Bindlestiffs: Harvesting and Threshing on the North American Plains* (Lawrence, 1990), Carey McWilliams, *Factories in the Field: The Story of Migratory Farm Labor in California* (Boston, 1939), Vicki L. Ruiz, *Cannery Women, Cannery Lives: Mexican Women, Unionization, and the California Food Processing Industry, 1930–1950* (Albuquerque, 1987), and Carleton H. Parker, *The Casual Laborer and Other Essays* (1920; reprint, Seattle, 1972), an old but still valuable classic study. On the world of workers, see James B. Allen, *The Company Town in the American West* (Norman, 1966), Norman H. Clark, *Mill Town: A Social History of Everett, Washington, from Its Earliest Beginnings on the Shores of Puget Sound to the Tragic and Infamous Event Known as the Everett Massacre* (Seattle, 1970), and William G. Robbins, *Hard Times in Paradise: Coos Bay, Oregon, 1850–1986* (Seattle, 1988).

Among the general business histories of the West, notably California, are Joel Kotkin and Paul Grabowicz, *California, Inc.* (New York, 1982), and Peter Wiley and Robert Gottlieb, *Empires in the Sun: The Rise of the New American West* (1982; reprint, Tucson, 1985). On transportation, travel, and tourism in the twentieth-century West, see Warren James Belasco, *Americans on the Road: From Autocamp to Motel, 1910–1945* (Cambridge, Mass., 1979), Lawrence R. Borne, *Dude Ranching: A Complete History* (Albuquerque, 1983), Don L. Hofsommer, *The Southern Pacific, 1901–1985* (College Station, Tex., 1986), Ralph W. Hidy, Muriel E. Hidy, and Roy V. Scott, with Don L. Hofsommer, *The Great Northern Railway: A History* (Boston, 1988), Maury Klein, *Union Pacific*, vol. 2, *The Rebirth, 1894–1969* (New York, 1989), Carlos A. Schwantes, *Railroad Signatures across the Pacific Northwest* (Seattle, 1993), and Earl Pomeroy, *In Search of the Golden West: The Tourist in Western America* (New York, 1957).

Studies of important entrepreneurs of the West include Donald L. Barlett and James B. Steele, *Empire: The Life, Legend, and Madness of Howard Hughes* (New York, 1979), Mark S. Foster, *Henry J. Kaiser: Builder in the Modern American West* (Austin, 1989), Marquis James and Bessie R. James, *Biography of a Bank: The Story of Bank of America* (New York, 1954), and Gerald D. Nash, *A. P. Giannini and the Bank of America* (Norman, 1992).

Many of the West's basic business and industrial enterprises have been the subjects of books: Leonard J. Arrington, *Beet Sugar in the West: A History of the Utah-Idaho Sugar Company, 1891–1966* (Seattle, 1966); Ralph W. Hidy, Frank Ernest Hill, and Allan Nevins, *Timber and Men: The Weyerhaeuser Story* (New York, 1963); Harold Mansfield, *Vision: A Saga of the Sky* (New York, 1956), the story of Boeing; Alexander Campbell McGregor, *Counting Sheep: From Open Range to Agribusiness on the Columbia Plateau* (Seattle, 1982); Gerald T. White, *Formative Years in the Far West: A History of Standard Oil Company of California and Its Predecessors through 1919* (New York, 1962); and Carl Coke Rister, *Oil! Titan of the Southwest* (Norman, 1949).

Among the more general industrial studies are Russell R. Elliott, *Nevada's Twentieth-Century Mining Boom: Tonopah, Goldfield, Ely* (Reno, 1966), Robert E. Ficken, *The Forested Land: A History of Lumbering in Western Washington* (Seattle, 1987), A. Dudley Gardner and Verla R. Flores, *Forgotten Frontier: A History of Wyoming Coal Mining* (Boulder, 1989), Andrew Gulliford, *Boomtown Blues: Colorado Oil Shale, 1885–1985* (Niwot, Colo., 1989), and Raye C. Ringholz, *Uranium Frenzy: Boom and Bust on the Colorado Plateau* (New York, 1989).

World War II, which played such a key role in the western economy, is the subject of two books by Gerald D. Nash: *The American West Transformed: The Impact of the Second World War* (Bloomington, 1985) and *World War II and the West: Reshaping the Economy* (Lincoln, 1990). Curiously, no one has yet authored a comparable volume for World War I.

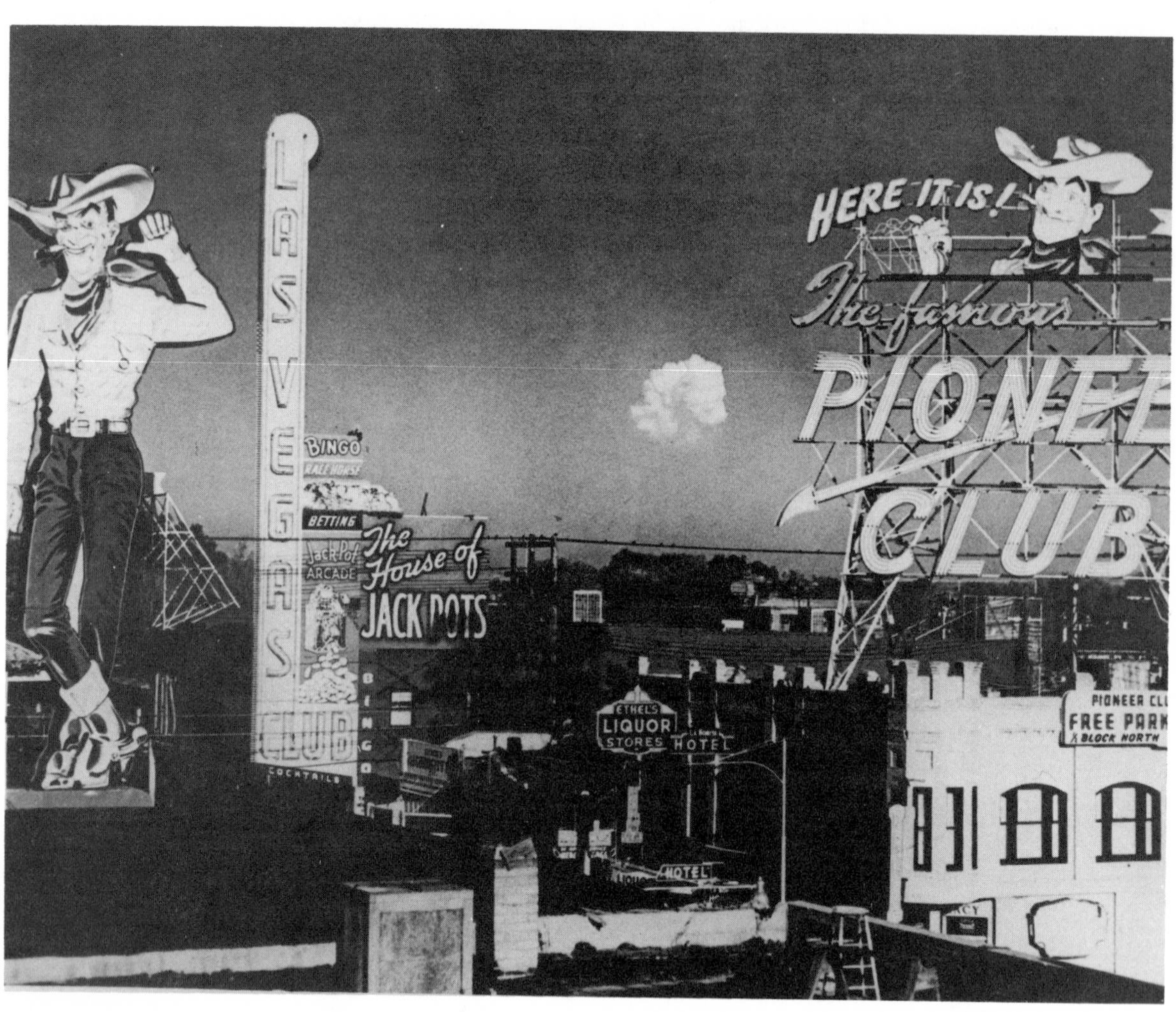

LAS VEGAS
CLUB
COCKTAILS
BINGO
BETTING
Jack Pot
ARCADE
The House of
JACK POTS
BINGO
HERE IT IS!
The famous
CLUB
ETHEL'S
LIQUOR
STORES
HOTEL

Chapter Thirteen

The Federal Presence

CARL ABBOTT

The federal government pervades the contemporary American West. Federal lands stretch from the fires of Kilauea to the ice cliffs of Glacier Bay, from the nation's highest peak in Alaska to its lowest point in Death Valley. Federal properties range in size and complexity from roadside picnic tables to Hoover Dam. Federal funds protect the ancient homes of the Anasazi on the Colorado Plateau, pay for the scientific research at the Los Alamos National Laboratory, and support the search for nuclear-waste depositories in the sparsely settled hinterlands of Nevada and New Mexico. Westerners are still known to proclaim their independence of the governmental octopus, but they now live in a region whose every corner is linked into networks of federal programs, regulations, spending, and employment.

At the opening of the twentieth century, it *was* possible to think of the federal government as something outside the West. *Honey in the Horn*, the Pulitzer prize-winning novel by H. L. Davis, depicted frontier Oregon in the first years of the twentieth century. Davis's characters move from the foothills of the Cascade Mountains to the Willamette Valley to the Pacific coast and back to the edge of the high desert. They try their hand at sheepherding, hop picking, lumbering, horse trading, and wheat farming. The federal government enters the story as a passive purveyor of cheap homesteads. It figures less in the everyday world of westerners than it does in their fading memories of the Civil War.

When Davis published this work in 1935, a generation after the setting for his story, the federal government had begun to move to the foreground of western life. In the novel's home ground of the Northwest, contractors for the Reclamation Service were at work on the first of the great Columbia River dams. The Agricultural Adjustment Administration was propping up the wheat ranchers of the Palouse and John Day river basins. Federally aided highways had replaced private toll roads. Youthful workers for the Civilian Conservation Corps (CCC) were grading hiking trails through the forests of the Cascades and the Olympics. Regional artists and writers like Davis were beginning to find paying work decorating public buildings and compiling local histories for the Works Progress Administration (WPA).

Forty years later, Edward Abbey published a different sort of best-selling "western." *The Monkey-Wrench Gang* follows an increasingly violent quartet of environmental activists who set out in the 1970s to block the further transformation of the canyonlands of northern Arizona and southern Utah. In Abbey's radical nostalgia, economic development is itself the enemy. The Forest Service, Reclamation Service, Park Service, and Bureau of Land Management, as fronts for the forces of growth, are all part of the problem.

Federal funding helped Las Vegas grow from a town of 5,000 in 1930 to a metropolis of 750,000 in the 1990s. Federal dollars built the dam that generated power for the city's neon lights and constructed the highways that brought in tourists. More federal money poured into the community via the Nevada Test Site and other defense installations. Like many areas of the West, a region of self-proclaimed independence, Las Vegas depends on this outside support.

Unidentified photographer. Las Vegas. *Gelatin silver print, 1951. Las Vegas News Bureau Archives, Las Vegas, Nevada.*

Even as they fight to reverse the effects of western resource development, however, George Hayduke, Doc Savas, and the other ecoraiders cannot avoid utilizing the same federal investments and institutions that they hate. They drive on U.S. Highways 66 and 89, 160 and 163. They move back and forth from Natural Bridges National Monument to Grand Canyon National Park, from Kaibab National Forest to Glen Canyon National Recreation Area, from the Navajo Indian Reservation to Canyonlands National Park. They sleep in federal campgrounds and cruise on federal reservoirs. They shop in the "neat green government town of Page" near the Glen Canyon Dam. In a story about the vast spaces of the Four Corners in the last quarter of the twentieth century, half the pages would be blank if the federal government were left out.

The difference between the fictionalized Oregon of 1908 and the fictionalized Utah of 1975 captures something of the growing prominence of the national government in the daily life of the West. During the nineteenth century, the federal government was essentially a facilitator of western growth. It explored the region and helped map the terrain. It sold or gave land to individuals or corporations. It confirmed the farm and mining claims of pioneers. It maintained the basis of public order through Indian wars and territorial government until Anglo-American settlers were judged ready to take local affairs into their own hands. In short, it made resources available for private appropriation, investment, and development.

Over the course of the twentieth century, the balance has changed. Uncle Sam is now more active and omnipresent than nineteenth-century pioneers could have imagined. The federal government directly manages the regional resources of grass, timber, oil, and recreational space. It is a primary customer for many of the region's leading businesses and the chief engine behind the area's massive urbanization. Since the Great Depression, it has become an active partner in previously local decisions.

This new federal presence has accumulated in generational layers like the strata of silt laid down along the banks of western rivers such as the Yellowstone or the Platte. The equivalent to periodic floods has been a succession of economic eras of boom and bust and new boom. Each period has involved a new set of federal activities and imperatives. We can follow the expanding federal presence from the era of the federally assisted frontier (1900–1918) through the long regional depression (1919–40), the Pacific war and the rise of the military economy (1940–60), and finally to the era of the West's emergence (1960–90), during which the region increasingly met the world with introductions by Uncle Sam.

A largely inadvertent effect of federal actions has been the transformation of the West from a backwater of the national and world economies of 1900 to a central player at the end of the century, bringing an economic power that westerners only dreamed of in the first generations of American settlement. Policies on resource development, defense, and international relations have moved the southwestern and Pacific states into a prominent position within the contemporary world system. With a series of strong boosts from Uncle Sam, cities such as Seattle, San Francisco, Houston, Dallas, and especially Los Angeles have become contact points and focal points for the world economy in much the same way that Boston, Chicago, and New York grew in the nineteenth century.

Federal actions have also nationalized the West by overriding or eroding regional differences. Almost by definition, federal programs place individual choices, local interests, and regional change within a framework of national requirements and expectations. Federal funds have helped to pave the highways, string the power lines, and build the airports that tie the West to the rest of the nation. At the same time, the organizational state of the twentieth century has had some of its most powerful effects on the "independent" and "individualistic" West that finds itself dependent on farm supports, subsidized water, defense contracts, and mass-transit grants. Although many of the region's public voices hate to admit the obvious, today's West would be far poorer and far more sparsely developed without a century of federal initiatives.

The Federally Sustained Frontier, 1900–18

The first two decades of the twentieth century were the era of the federally sustained frontier. In the nineteenth century, the federal government had created settlement and development opportunities that benefited individual settlers and corporations. After the drought and depression of the 1890s, federal agencies increasingly initiated development on their own. Indeed, visitors to the federal pavilions at the world's fairs that ushered in the new century at Portland in 1905, Seattle in 1909, and San Francisco in 1915 came away with a single impression. What they found on display was a federal government that was taking the lead in the development of western resources through water projects, national parks, and a canal through the Isthmus of Panama. Such activities sustained the development of the American frontier for a full generation after Frederick Jackson Turner had proclaimed its disappearance. At least until 1918, farmers, ranchers, miners, and lumbermen continued to draw on federal help to push into areas previously unsettled or unused by Anglo Americans.

What was perhaps the key legislation in this expansion was authored by Representative Francis Newlands of Nevada. The Newlands Reclamation Act of 1902 established the Reclamation Service as a new agency within the Department of Interior. Led by Frederick Newell, the new bureau took the lead in the construction of irrigation projects throughout the West. In the 1870s and 1880s, many irrigation facilities had been community projects, like the canals that watered the Mormon settlements in Utah or the agricultural colony at Greeley, Colorado. Other projects had been profit-seeking business ventures, like the Highline Canal winding through the suburbs of Denver or the unsuccessful Boise River Project, the setting for the latter chapters of *Angle of Repose* (1971), Wallace Stegner's novelized version of the lives of the writer Mary Hallock Foote and the engineer Arthur Foote.

Under the Newlands Act, profits from public land sales in the sixteen states west of the one hundredth meridian were to be deposited in an "arid land reclamation fund" to pay for major irrigation projects. Users of the water were to repay the U.S. Treasury over ten years, generating a revolving fund for new systems. The Reclamation Service quickly undertook projects beyond the capabilities of profit-oriented corporations. The Roosevelt Dam on the Salt River of Arizona allowed the development of Phoenix. Residents of Reno placed high hopes on tapping the waters of the Truckee River. Orchards blossomed around Delta and Grand Junction in Colorado and Yakima in Washington. Arrowrock Dam on the Boise River, the world's tallest when completed in 1915, finally realized the dream on which Arthur Foote had spent seven years of unrelenting work.

The dam and related projects transformed southern Idaho from desert to farmland as older cities like Boise competed with new towns like Twin Falls for hundreds of thousands of new Idahoans. As Carlos Schwantes has commented, "a whole new Idaho grew out of the sagebrush plains" of the upper Snake River drainage with the irrigation boom of 1900–1917.

The same era of activist government brought the final giveaways of federal lands. Increasingly the federal government acted like a storekeeper trying to move backed-up inventory with price cuts and premiums. The Enlarged Homestead Act of 1909 raised the homestead grant from 160 to 320 acres and reduced the time period before transfer of title from five to three years. The Stock Raising Homestead Act of 1916 raised the grant of rangeland to 640 acres. The General Land Office took on the tone of a railroad immigration department, publishing circulars that listed available homestead acreage in each western county and explained the simple procedures for homestead entry. The result was the extension of settlement into marginal lands. Encouraged by wet years in the 1910s and high food prices caused by war in Europe, farmers pushed into plains lands to plant wheat and sugar beets on acreage better suited to grazing than cultivation. In the first two decades of the twentieth century, eager agriculturalists took up more homestead acreage than in the entire nineteenth century. The "second boom" on the northern plains doubled the mileage of railroads to serve new towns and farmers. North Dakota in 1915, for example, had more towns, railroad mileage, churches, schools, and elected officials per capita than any other state.

The federal government also took the lead in promoting the Anglo-American settlement of Alaska. Interior Secretaries Franklin K. Lane and John Barton Payne published glowing articles on the attractions of Alaska. The failure of private rail companies to penetrate the interior of the territory led Congress to authorize a federal railroad from the Gulf of Alaska to the Yukon River. Woodrow Wilson chose the route from Seward to Fairbanks in 1915; Warren Harding drove the symbolic golden spike in 1923. The Alaska Engineering Commission (AEC) supervised forty-five hundred workers on the federal payroll at the height of construction.

The AEC also found itself in the town-building business, for railroad building required a construction and supply base. In the nineteenth century, towns like Cheyenne on the Union Pacific, Reno on the Central Pacific, and Billings on the Northern Pacific were private promotions. Profit-motivated rail companies or their subsidiaries surveyed the site, sold the lots, and invested in community facilities to encourage settlement. In 1915, in contrast, the AEC planned, built, and managed the new city of Anchorage from 1915 to 1920, anticipating a series of federal construction and science cities in the 1930s and 1940s. Residents resembled those of the earlier generation of private railroad towns, but until the city incorporated in 1920 it was the federally funded AEC that supplied the town manager, graded the streets, laid the water and sewer pipes, constructed the school and the hospital, bought the fire-fighting equipment, and even organized the YMCA to improve the moral tone of four thousand typical frontier townspeople.

The same development impulse also laid the foundations of professional land management by the federal government. The Progressive Era conservation movement, as the historian Samuel Hays has shown, was "an effort on the part of leaders in science,

Unlike the railroad towns of the 19th century, Anchorage was planned and developed by the federal government, which needed a town to supply workers and materials for the construction of a railway through the Alaskan interior. In July 1915, the government auctioned off 655 lots, with strict rules forbidding their development for "immoral" purposes. Just a few months later, Anchorage already resembled a small town, with retail stores, restaurants, and lawyers.

Sydney M. Laurence (1865–1940). Anchorage, 4th Avenue Looking West between G & H Streets, October 8, 1915. *Gelatin silver print, 1915. Anchorage Museum of History and Art (#B63.16.6), Anchorage, Alaska.*

technology, and government to bring about more efficient development of physical resources." As an attempt to sustain and extend the resource development of initial settlement, it was "an aspect of the history of production" whose gospel of efficiency was spread from the top down by dedicated professionals within the Agriculture and Interior departments. The model was the U.S. Forest Service, established under the direction of Gifford Pinchot in 1905 to manage a growing inventory of federal forest reservations that dated to 1891. Theodore Roosevelt entered office in 1901 with 41 million acres in reserves and left in 1909 with 151 million in the rechristened national forests. Pinchot's goal was scientific management to ensure a sustained yield of timber as a lasting contributor to national growth and the stability of local economies. In his view, national forests could protect water supplies for irrigation and western cities, provide cheap grazing for stock raisers, and repay the U.S. Treasury with timber sales.

The National Park Service as both a peer and a rival of the Forest Service was established in 1916 to provide consistent management for an assortment of federally protected parks and monuments. Director Stephen Mather and his chief lieutenant, Horace Albright, were dedicated to preserving the parks for utilization by the public. They constructed tourist facilities, lobbied for paved roads into the parks, granted concessions to private resort operators, and defined the role of the park naturalist as educative entertainment. They hired Robert S. Yard to publicize the parks as tourist destinations with tasteful books on their scenery and attractions. Their work helped to achieve a 900-percent increase in automobile visits from 1919 to 1931. Parks such as Rocky Mountain National Park (established 1915) were originally promoted in terms of local economic development. The rising tide of visitors, however, gave the Park

Service the justification to override local interests in favor of uniform management directives originating in Washington, D.C.

The other federal construction enterprise that spanned the administrations of Roosevelt, William Howard Taft, and Wilson was the Panama Canal, begun in 1905 and opened in 1914 at a cost of $365 million. As with the Alaska Railroad or the Reclamation Service dams, the federal government did what private capital could not. The canal confirmed a Pacific orientation to America's new international power. It improved access to Hawai`i, Guam, American Samoa, and the Philippines, all acquired between 1898 and 1900. It also raised expectations of world trade along the Pacific Coast. Seattle, Portland, and San Francisco all looked forward to what the journalist Wolf von Schierabend called "The Coming Supremacy of the Pacific." The Los Angeles Chamber of Commerce sponsored the book *Los Angeles: A Maritime City* in 1912. Although most Americans thought of Los Angeles as citrus groves and retirees, the author John McGroarty pointed out that the great circle route from Panama to the Far East ran seventy miles from the harbor at San Pedro. West Coast expositions like San Francisco's Panama-Pacific International Exposition and Seattle's Alaska-Yukon-Pacific Exposition were part of the same desire to capitalize on new opportunities served up by the federal treasury by publicizing the attractions of West Coast ports.

The climax of the federally sustained frontier came with World War I. Even more than the new canal, the war linked the West to a warring world with its insatiable appetite for food, lumber, and minerals. As Michael Malone and Richard Etulain have written, "The great wartime boom, with its huge export markets and artificially high prices, climaxed the three-century frontier expansion of American agriculture." The U.S. Food Administration defined markets and guaranteed base prices. Western miners found new profits in industrial metals like zinc, molybdenum, tungsten, and vanadium to make up for depressed silver prices and played-out goldfields. The Emergency Fleet Corporation made Puget Sound and San Francisco Bay into major shipbuilding centers, with thirty thousand workers learning to make steel ships in Seattle's largest yard. By 1919, the federal government had helped to advance American enterprise further than ever before—more deeply into mountain valleys and plateaus, further onto the arid plains, and back from the coastal fringe into the heart of Alaska. It had staked a lasting claim to the Pacific as a field of American enterprise and influence. In the process, western Americans had improved eighty million additional acres of farmland, more than doubling the total in nine states. The western share of national petroleum production had jumped from 29 percent to 68 percent and the share of timber production from 10 percent to 35 percent. Cities like Seattle, Portland, Los Angeles, Denver, Dallas, and San Antonio had doubled and tripled in population. It was the flood tide of the federally assisted frontier.

Regional Planning and the Long Depression, 1919–40

One of the public heroes of America's first crusade in Europe was the westerner Herbert Hoover. Born in Iowa, raised in Oregon, and educated in California, Hoover managed wartime and postwar relief efforts in Europe with American practicality. He was also known in the American West as the director of the wartime Food Administration, which had helped to urge farmers to higher and higher levels of production. As secretary of commerce from 1921 to 1928, he embodied the impulse of progressive efficiency. Given the reputation that he later earned as an advocate of federal inaction during the onset of the Great Depression, it is ironic that, especially before his presidency, Hoover helped to change the federal role in the West from promoter to planner.

The Colorado River Commission Compact of 1922 emerged with Hoover as facilitator; in his role as the congressionally authorized federal representative, he met with delegates of the seven western states along the Colorado River. His goal was to work out compromises in uses and claims on the limited flow of the Colorado River and to provide federal sanction for multistate agreements. Hoover and the delegates met half a dozen times in Washington, in Phoenix, in Denver, and finally for more than two weeks in Santa Fe before hammering out an agreement to divide the river equally between the upper basin states (Wyoming, Colorado, Utah, New Mexico) and the lower basin states (Nevada, Arizona, California). The decisions of 1922 were flawed in detail because of inexact knowledge of the river, but the federally mandated framework for planning the allocation of natural resources stood for the rest of the century.

The Colorado River Compact epitomized the progressive impulse toward "scientific" and efficient development of western resources for the greatest economic return to the nation over time. Hoover filled his very comfortable role as progressive engineer. At the same time, the meetings in Santa Fe anticipated the era of Franklin Roosevelt and Harry Truman, for the compact was a regional planning response to a slowly growing sense that the development of the West could rub up against limits. Over the ensuing quarter century, that growing recognition put a new twist on the idea of efficient use by extending the progressive idea of planning for specific resource types into a broader concern for the comprehensive planning of development by regions.

The federal government in the early 1920s also took part in another effort at regional planning through the creation of a national highway system. Responses to the automobile in the 1910s and early 1920s had followed the western tradition of local community boosterism and private entrepreneurship. Venturesome motorists traveled state to state on a series of privately named highways, bouncing back and forth from paved segments to gravel to graded dirt and even worse. Local groups and governments maintained individual road segments, which national associations grouped into as many as 250 continuous and sometimes logical "routes." Signs were usually a set of color-coded stripes on telegraph poles and fence posts. The best known was the Lincoln Highway, whose red-white-and-blue colors led from Philadelphia to Omaha, Cheyenne, Salt Lake City, Reno, and San Francisco—roughly the route of the first transcontinental railroad west of the Missouri River. The Yellowstone Trail led from Chicago to Seattle. The Old Spanish Trail (red, white, and yellow) started in Jacksonville, Florida, and ended in San Diego. The federal government began to subsidize state highway building with the Federal Aid Highway Act of 1916, but most states allocated their aid funds by logrolling and chance. As one observer noted in the 1910s, "The highways of America are built chiefly of politics, whereas the proper material is crushed rock, or concrete." New legislation in 1921 required states to designate a maximum of 7 percent of their rural roads as primary routes qualifying for federal aid, but planning remained a captive of

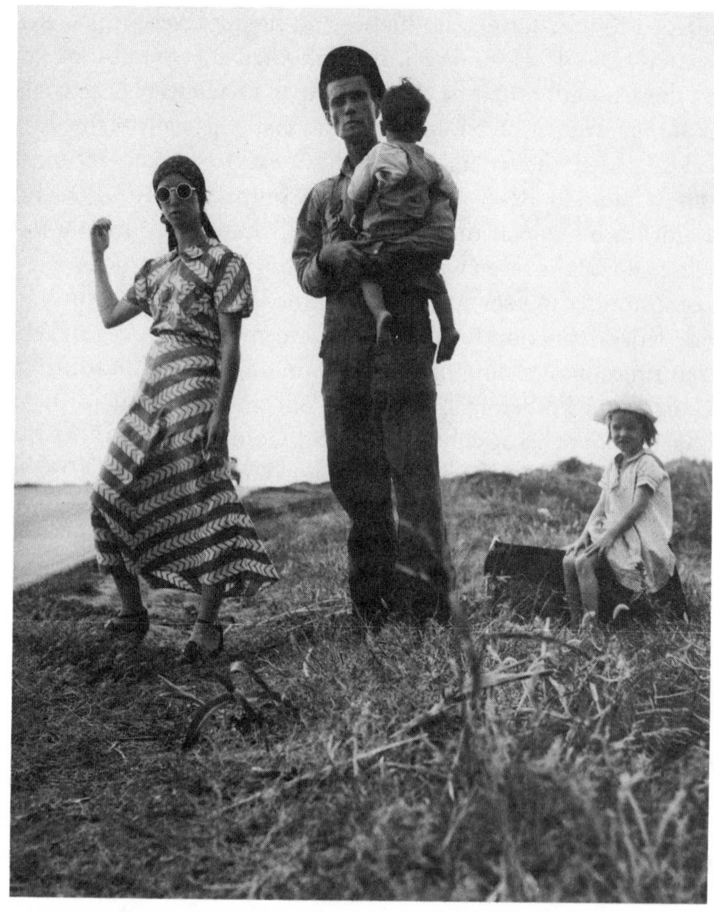

Under the aegis of the government's Farm Security Administration, leading photographers traveled the country to document the plight of depression-era America. Pausing in western Oklahoma along Route 66, a popular migration route for displaced farm workers, Dorothea Lange photographed this young Missouri family, hitchhiking their way west in search of work and a better life.

Dorothea Lange (1895–1965). Family on the Road, Oklahoma. *Gelatin silver print, 1938. Gift of Paul S. Taylor, The Dorothea Lange Collection, The Oakland Museum, Oakland, California.*

competitive politics. *Sunset* magazine, for example, reported to its travel-oriented readers that Wyoming had shouldered past Montana as the gateway to Yellowstone Park by using federal dollars to grade and gravel the entrance roads from Casper and Gillette.

The third step toward a planned highway system came when the Bureau of Public Roads within the U.S. Department of Agriculture convened a Joint Board on Interstate Highways. Under the auspices of the Joint Board and its secretary, E. W. James, state and federal highway officials selected and designated "a comprehensive system of through interstate routes" with systematic identification. The outcome amounted to a vast regional planning program. The Joint Board pared a possible 200,000 miles to 75,000 miles of U.S. highways identified by number rather than name. Even-numbered routes supplemented the great trunk railroads—U.S. 10, 30, 40, 90, and the others. Odd-numbered highways paralleled the West's valleys and ranges from north to south—U.S. 99 for the old Pacific Highway, U.S. 89 and 93 through the intermountain West, and U.S. 85 traversing the high plains from Williston, North Dakota, to El Paso, Texas.

The new highway that embedded itself most deeply in American memory through song and television illustrates the shift from local initiative to federally coordinated planning. U.S. 66 originated in the efforts of the Tulsa booster Cyrus Avery. The Joint

Board incorporated pieces of the Ozark Trail into a road that cut diagonally across the national grain, leading from Chicago and St. Louis through the Oklahoma and Texas "dust bowl" region to southern California. Like the other federal highways, it helped to tie the West into a single economic unit. During the 1930s, the fictional Joad family traveled westward on U.S. 66 in *The Grapes of Wrath*. So too did the very real Haggard family follow friends and relatives to California in a 1926 Chevrolet, two years before the birth of their son Merle. A few years later, an even larger migration traveled the same route—war workers, soldiers on assignment, and family members hoping for a few days with departing servicemen in San Diego or Los Angeles.

Whether they left home like the Joads or tried to cling to rural communities, millions of westerners found themselves dependent on the federal government for simple survival during the depression of the 1930s. Through much of the mining and farming country of the Rockies and Great Plains, falling resource prices in the 1920s had already brought individual bankruptcies and bank failures (five hundred in North Dakota alone during the twenties). In Colorado, for example, war-inflated grain and livestock prices fell 60 percent from 1919 to 1921. Unable to expand production to generate more income, small farmers were squeezed between falling receipts and fixed mortgages. The number of tenant farmers who rented rather than owned their land in the state's dry farming areas rose from 23 to 35 percent. Whereas the climatic disasters and market collapse of the 1930s simply deepened existing problems of wheat growers and ranchers, they crippled West Slope orchardists who had protected themselves in the 1920s by cooperative marketing associations. Agricultural statistics for the 1930s told a grim story: population down in all but one plains county; one million acres taken out of cultivation; thirty thousand agricultural jobs lost for Colorado as a whole.

Despite the anti-New Deal posturing of many western politicians, necessity opened the door for a grudging and often limited acceptance of federal aid. Few states had leaders who were prepared to deal with the crisis of the early 1930s. Strict economy in government sat higher on many state agendas than did effective relief. Governors such as Arthur Seligman of New Mexico or Julius Meier and Charles Martin of Oregon believed in the gospel of balanced budgets. By repeatedly striking down new taxes that would have funded state matching money for the Federal Emergency Relief Administration, the Colorado courts forced the termination of needed assistance at the end of 1933. As the historian Robert Athearn has commented: "The West was inclined to bite the hand that fed it. . . . what it amounted to was the unwillingness or inability of the states to reconcile what they conceived to be their rugged individualism, born of the frontier, with the planned society that was implied by the New Deal."

Nevertheless, westerners took what the federal government had to offer, especially after the inauguration of President Franklin Roosevelt and the onset of the New Deal in 1933. New Deal grant and loan programs had a greater relative impact in the West than in any other part of the United States. Figures compiled by the historian Leonard Arrington show that the greater the collapse of a state's level of personal income from 1929 to 1933, the more it received in New Deal assistance from 1933 to 1939. Nevada, with a small population and huge federal construction projects, led the nation with $1,499 received per capita. Next in line were Montana, Wyoming, and Arizona, followed by ten other western states before Minnesota finally broke into the list.

In the rural communities and the city neighborhoods, the New Deal farm support and work relief programs of 1933–35 held crumbling communities together. The longer-lasting federal programs of the later 1930s helped to put the same communities back on their feet. By one estimate, two of every three North Dakotans received federal assistance by 1936 through farm-support programs, business loans, emergency relief, and public works jobs. Just across the state line in Otter Tail County, Minnesota, the people of Fergus Falls, Pelican Rapids, and other small towns found that the new federal agencies kept alive the dream of a normal life, helping one family to hold on to its farm, another to think about college for the children. The CCC enrolled seventy-five hundred workers on Arizona Indian reservations and another five thousand in twenty-nine Indian Conservation Camps in Idaho. The Rural Electrification program, which began in 1935 and continued after the 1940s, raised the number of Idaho farms with power from 30 percent to 54 percent. Beyond direct relief, the federal government helped to reshape the West on paper and on the ground. The New Deal built on the regional initiatives of the 1920s to introduce the idea of systematic planning for regional recovery and future growth. As the historian Richard Lowitt has argued: "Depression, drought, and dust undermined dependence on the marketplace as an arbiter of economic activities. . . . Animating the New Deal in the West was concern for rational planning of resource use. Reports of the National Resources Planning Board, several presidential committees, various regional, state and local boards, all provided guidelines designed to encourage more meaningful regional economic development."

The Taylor Grazing Act of 1934 marked the effective end of the era in which the goal of the federal government was to transfer public lands to private ownership and development. Within two years, the Taylor Act removed 142 million acres of western lands from potential sale and reserved them for grazing under federal control. President Roosevelt reinforced the Taylor Act by officially removing the majority of the public domain from private land claims at the end of 1934. The new Grazing Service of the Department of Interior was to lease grazing rights at reasonable rates while controlling excessive use of the fragile rangelands. The Grazing Service (merged into a new Bureau of Land Management in 1946) has been severely criticized as a captive of local advisory boards that give preferred access and cheap grass to big stock growers at the expense of resource conservation. Nevertheless, the Taylor Act established the principle that virtually the entire remaining public domain should be set aside for federal management rather than sale.

The Soil Conservation Service (expanded from the Soil Erosion Service in 1935) was also part of the new layer of federally mandated regional planning. In the 1910s, the federal government had run a sort of land office fire sale. Twenty years later, H. H. Bennett proposed a regional planning solution to the agricultural crisis of the Great Plains. The Soil Conservation Service tried to work cooperatively with local and state officials, but it set its district boundaries at watersheds rather than county lines. Conservation districts were an experiment in federalism—local government units, established under state laws, implementing a federally defined agenda with the help of federal expertise. The first thirty-seven districts covered nineteen million acres where improved cropping practices, soil retention, and soil restoration became a new credo to replace the soil mining of the frontier generations.

More explicit planning came under the umbrella of an agency known variously as the National Planning Board (1934–35), National Resources Committee (1935–39), and National Resources Planning Board (1939–43). Established in 1934 to advise the Public Works Administration (PWA), the NRPB took on a life of its own. It promoted state planning agencies and established regional planning commissions for the Pacific Northwest, Alaska, Southwest, Intermountain-Great Plains, and Missouri Valley. Planners in Washington, D.C., provided money, manpower, information, and advice for the regional and state agencies. In turn, the regional and state agencies performed first-rate resource inventories and prepared dozens of reports with a focus on natural resources and public works. Reports dealt with individual towns like Sitka, Alaska, with small watersheds like the Upper Gila Basin in Arizona or the Willamette Valley in Oregon, with single states, and with entire multistate regions. The NRPB was perhaps at its strongest in what Lowitt has called the "planned promised land" of the Northwest. Throughout the West, however, it introduced the idea of comprehensive planning for resource conservation and development in areas facing the first stages of a massive economic transition that has now lasted for half a century.

More obvious to most Americans was federal involvement in reshaping the western landscape. The PWA and the WPA built 246 public buildings in the state of Washington, 227 in Montana, and thousands more throughout the West. The PWA and WPA facilities in every state ranged from courthouses to culverts, water-treatment plants to swimming pools, bridges to golf courses. Much of the basis for the postwar boom in tourism was laid in the 1930s as WPA workers built Mount McKinley Lodge in Alaska and Timberline Lodge on Mount Hood and as CCC workers opened trails, leveled campgrounds, and erected rest areas.

The 1930s also brought an engineering marvel for each of the great rivers of the West. Congress had authorized a dam at Black Canyon on the Colorado River in 1928, but the timing of construction made Boulder (now Hoover) Dam a depression-era public works project. The erection of the huge wedge of concrete in the blazing heat that vibrated between the canyon walls captured the national imagination. The current name of the dam is appropriate, for the project represented Hoover's approach to regional growth through carefully planned engineering projects. It gave the Bureau of Reclamation its first large multipurpose project designed for hydroelectric power as well as irrigation. Indirectly, the need to support the dam's thousands of workers and the growing streams of tourists gave birth to the modern city of Las Vegas in place of the small, struggling mining and railroad town of the 1910s and 1920s.

With a span of 3.7 miles, Fort Peck Dam on the Missouri River was the largest earth-fill dam on the continent. The PWA supplied the funds and the Army Corps of Engineers did the building. Drawing on normal construction experience, the Corps built the planned town of Fort Peck City with 280 units for families and 3,200 barracks spaces for single men. In accord with Montana relief regulations, however, three-quarters of the dam builders were married men with dependents, who had to crowd into improvised shelters of barn siding, cardboard, and tenting in instant towns like Wheeler, New Deal, and Delano Heights. The photographer Margaret Bourke-White caught the tone of the towns in a photo essay on "Mr. Roosevelt's New Wild West" for the first issue of *Life* magazine in November 1936. The thousands of "veterans, parched farmers, and plain unemployed parents" who labored on the dam provided "extracurricular work for a shack-town population of barkeepers, quack doctors, hash dispensers, radio mechanics, filling station operators, and light-roving ladies."

Grand Coulee Dam on the Columbia River was the best publicized, from initial PWA funding in 1933 to the generation of the first electricity in 1941. The journalist Richard Neuberger, later to serve in the U.S. Senate from Oregon, proclaimed it "Man's Greatest Structure" and the "Biggest Thing on Earth." New Deal officials were glad to use the awestruck publicity to justify an entire generation of federal activity in the West. In the words of the singer Woody Guthrie, hired to immortalize all the Columbia River dams in 1941, "From the rising of the river to the setting of the sun / the Coulee is the biggest thing that man has ever done." Although its size was what caught the public's attention—550 feet from bedrock to guardrail, with a 15-acre spillway—Grand Coulee would soon prove more important for the electricity it furnished during World War II for Oregon and Washington aluminum plants and for the production of plutonium at Hanford, Washington.

The three pioneer projects introduced a generation of massive western water projects funded with federal taxes. Bonneville, McNary, Chief Joseph, Hungry Horse, and Libby are a few of the dams that tapped the Columbia and its tributaries for irrigation water and electricity. Oahe and Garrison dams on the Missouri, Shasta Dam on the Sacramento River, and Glen Canyon Dam on the Colorado were other key projects. The Colorado-Big Thompson and Fryingpan-Arkansas water diversions took water from the western slope of the Colorado Rockies for the farms and cities of the eastern slope, the former via huge tunnels that pass directly beneath Rocky Mountain National Park.

Big projects required intricate coordination. Congress created the Bonneville Power Administration (BPA) in 1939 to market power from the entire set of Columbia River dams in accord with a rational plan. The BPA's choice was a triangular grid of power lines that marched up and down over the Cascade range to connect Seattle, Spokane, and Portland with essentially equal-price power. Its marketing decision was also a land-use decision, for it favored established centers of economic activity at the expense of new industrial development near the dams. The utilization of Hoover Dam to help implement the Colorado River Compact required coordination of water sales to Imperial Valley and Coachella Valley agribusiness, power sales to private utilities, and both water and power sales to the Los Angeles Water District (the power was used to pump the water west across the desert).

The biggest dam builders offered competing visions and plans for federal development of the West. The Corps of Engineers argued for a water program oriented to flood control and navigation, whereas the Bureau of Reclamation stressed electric power and irrigation. The Soil Conservation Service worked on the sidelines to build small dams for flood prevention and erosion control. Hells Canyon on the Snake River was left undammed in part because a standoff between the Corps of Engineers and the Bureau of Reclamation delayed plans until the 1960s, when a preservationist ethic would begin to alter national policy. On the Missouri River, Colonel Lewis A. Pick of the Corps proposed fifteen hundred miles of levees and six main-stem dams for navigation and flood control, whereas William G. Sloan from the Bureau proposed damming tributaries for hydropower and irrigation. The two men met in Billings in 1944 and hammered out the Pick-Sloan agreement to parcel out development of the Missouri much like European statesmen carving up the map of Africa.

The next logical step of regional river basin development agencies, following the model of the Tennessee Valley Authority, was never taken. Colorado politicians and business leaders shot down the idea of an Arkansas Valley Authority in 1941 as another unwanted intrusion of federal authority that might reduce local control of water resources. A Columbia Valley Authority fared little better. A Missouri Valley Authority, backed by Franklin Roosevelt in 1944, at least carried the New Deal impulse into the 1950s. Advocates like Senator James Murray of Montana hoped to follow wartime prosperity with a coordinated economic transition to prevent postwar depression. Successful opposition came from the power companies, which feared cheap federal electricity, from the railroads, which resented the competition of barge transportation, and from both the Corps and the Bureau of Reclamation, which had already divided up the dam-making job with the Pick-Sloan memorandum. The regional planning idea as a federal initiative reappeared in the 1960s and 1970s with regional commissions to coordinate economic development for the Four Corners (Arizona, Colorado, New Mexico, Utah), the Old West (Montana, Nebraska, Wyoming, the Dakotas), and the Pacific Northwest (Idaho, Oregon, Washington) and with federally mandated River Basin Commissions for the Columbia and Missouri rivers.

The first impression to take away from the 1920s and 1930s is the contrasting styles of project planning and regional planning. The nation's memory of the long depression in the West is one of streams of concrete pouring into wooden forms—the core of a dam, a stretch of federal-aid highway, steps to a new courthouse built by the WPA. Just as important, however, are the hundreds of offices and conference rooms in which federal officials sat down, sometimes with state and local representatives, to plan out a framework for future growth. On the dry and desolated plains, the chief actors often reported to U.S. Secretary of Agriculture Henry Wallace, who oversaw such programs as the Agricultural Adjustment Administration, Soil Conservation Service, Farm Security Administration, and Resettlement Administration, which were designed to move the region beyond wasteful practices of agriculture. In the mountain and Pacific states, the final arbiter was most often Interior Secretary Harold Ickes, an old-line believer in the progressive gospel of efficiency whose PWA and other construction programs were designed to redevelop regional resources.

The decision to make the federal government an active planner of western development also left it a permanent partner with western residents. The land retained in permanent federal ownership as national forests, parks and monuments, grazing lands, Indian reservations, and military installations totaled 99 percent of Alaska in 1944 and 87 percent of Nevada. Federal ownership exceeded 50 percent in five other adjacent states in the heartland of western mountains and plateaus—Arizona, Utah, Wyoming, Idaho, and Oregon. Surrounding this federal core were five additional states where federal ownership ranged between 35 and 46 percent. The only significant reductions of federal lands have come in Alaska, where the Native Claims Settlement Act and the National Interest Lands Conservation Act defined the process for transferring 44 million acres to native corporations and 105 million acres to the state by the mid-1990s (leaving 215 million acres in federal ownership). Only in the plains states and Hawai`i was the federal government a relatively minor player, with ownership levels between 1 and 18 percent.

The federal agenda of the 1920s and 1930s hastened the nationalization of American life. Tangible and intangible federal networks—highways, power lines, relief payments, social insurance—helped to incorporate isolated communities into the economic and social mainstream. Ethnic islands that had retained their distinctive cultures through the long decades of Anglo-American conquest found themselves reshaped by contradictory national expectations. The New Deal offered Native Americans and Hispanos a tension between a romantic valuation of preindustrial cultures and a desire to provide upgraded and modernized services. In northern New Mexico, for example, federal officials built roads and improved health care but also tried to "freeze" the styles of folk art and organize production for the tourist trade. The Soil Conservation Service and the Bureau of Indian Affairs forced the Navajos to reduce their herds of sheep and horses in the interest of scientific conservation of the reservation grasslands. At the same time, however, vital infusions of cash and services provided the foundations on which the Navajo people would achieve significant self-determination after 1945.

For westerners who already shared the national values, the regional planning of the 1920s and 1930s was an unambiguous success. Whether residents liked to admit it or not, the federal initiatives saved the West from collapse. There were fewer farmers and miners in 1940 than in 1920, and there were more abandoned towns on the high plains and plateaus. By and large, however, the West in 1940 had new resources in place for an economic takeoff—new electric power, new expertise in large-scale construction, workers with new skills for an industrial economy, and a renewed commitment to the progressive agenda of efficient resource development. "The program of the New Deal," wrote Richard Neuberger in *Our Promised Land* in 1938, "represents the first conscious attempt of government to utilize for all the people the vast, untapped resources of the frontier. Whatever else Mr. Roosevelt may have done to or for our country, that much he has accomplished in the Columbia River basin." Although Neuberger wrote about the Pacific Northwest, his enthusiasm was relevant to a far wider region. "Never again can the natural riches of the hinterlands be left undeveloped as they were in the years before the New Deal."

The West at War, 1940–60

The federal government in 1939 was the leading landlord and largest general contractor in the West. By 1943 and 1944, it was also the dominant employer. In an era when one thousand dollars could buy a very good car, Houston, Fort Worth, Wichita, Seattle, Portland, San Francisco, Los Angeles, and San Diego all received more than one billion dollars in war-supply contracts from 1940 to 1945. The relative impact can be seen in estimates of the proportion of California's personal income derived from the federal government—5 percent in 1930, 10 percent in 1940 after a decade of relief and public works programs, and 45 percent in 1945 after half a decade of a war economy.

Between 1941 and 1964, the United States not only fought its seven-thousand-mile war against Japan but also entered wars against North Korea and China and against North Vietnam. It came to the brink of war at least twice more in the mid-1950s in Indochina and the Formosa Straits. World War II set in motion changes whose impacts

Through traveling exhibitions and direct grants to artists, the Federal Arts Project initiated a revival of traditional Hispanic crafts in depression-era New Mexico. Even as other government agencies sought to deliver modern services to remote rural areas, the FAP sought to preserve the distinctive arts of these isolated communities.

T. Harmon Parkhurst (1884–1952). George López, Córdova, New Mexico. *Gelatin silver print, ca. 1940. Courtesy Museum of New Mexico (#9973), Santa Fe.*

are still being felt half a century later. Taken together, the Pacific war and its Asian follow-ups marked the final transition from the old to the new West and made the federal budget the essential drive wheel of western growth.

The armed forces were no strangers in the West in the nineteenth century, but the region's military establishment had faded in the new century. Indeed, boosters in cities throughout the West labored long and hard during the 1920s and 1930s to secure new military bases to prop up local economies. Cities in the San Francisco Bay area vigorously argued among themselves about the best sites for navy bases but presented a united front in Washington. The San Antonio Chamber of Commerce embraced military aviation as the winning card in that city's rivalry with Dallas and Houston. Kelly Field and Brooks Air Base date to 1916–18 and Randolph Air Base to the 1920s, all to be vigorously defended by San Antonio's congressmen and businessmen. Prewar mobilization in 1940 brought a new wave of booster opportunities. The Dallas Chamber of Commerce and the Citizens Council (representing the city's preeminent movers and shakers) campaigned in Washington to secure a naval reserve aviation base and a North American Aviation Company plant whose payroll would total forty-three thousand by the midpoint of the war. Nearby Fort Worth secured Tarrant Field and a Consolidated Vultee Aircraft (Convair) plant, the latter through the newspaper editor Amon Carter's direct lobbying with Franklin Roosevelt. Similar combinations of local promotion, political influence, and bureaucratic criteria brought Tinker Air Base to Oklahoma City and Hill Air Base and Clearfield Naval Depot to the Salt Lake City–Ogden area.

Anticipating shortages of nonferrous metals, the federal government built a huge plant near Las Vegas to produce magnesium with the help of Hoover Dam electricity and a series of aluminum plants at Spokane, Vancouver, Longview, and other sites in Washington and Oregon served by hydropower from the Bonneville Power Administration.

The Japanese attack on Pearl Harbor and the sustained fighting of 1942 brought the reality of war to American territory. In addition to the attack on Hawai`i and the famous naval battle off Midway Island, Japanese submarines briefly shelled the Oregon and California coasts in 1942, and a large Japanese force invaded the Aleutian Islands in June 1942. The Japanese bombed Dutch Harbor and occupied Attu and Kiska islands until driven off in August 1943. The invasion was preemptive and defensive, for Attu and Kiska were more than one thousand miles closer to the home islands of Japan than to the lower forty-eight states. Nevertheless, the attack helped to trigger a military buildup in Alaska and the assignment of seven regiments from the Corps of Engineers to build the Alaska Highway from Dawson Creek, British Columbia, to Fairbanks. The road was neither necessary nor very useful for the military, but it symbolized the war as a shaper of the western environment.

On 19 February 1942, President Roosevelt issued Executive Order 9066, which authorized the secretary of war to define restricted areas and remove civilian residents. In Alaska, the Japanese attacks and landings caused the military to evacuate native Aleuts from the outlying islands of Alaska. They spent much of the war in substandard conditions in southeastern Alaska and returned to gutted environments. In many ways, their experience mirrored that of more than one hundred thousand Japanese Americans who were removed from California and large parts of Washington, Oregon, and Arizona in the spring of 1942. Those who had not voluntarily moved from the coastal states found themselves in any of ten relocation camps scattered from the mountain West as far east as Arkansas. The political expediency of the removal satisfied a generation of anti-Japanese sentiment kindled into hatred by the war. Most evacuees lost businesses and property. The United States officially recognized its liability with the Japanese Claims Act of 1948 and acknowledged its broader moral responsibility in 1988 when Congress approved redress payments of twenty thousand dollars to the sixty thousand surviving evacuees.

In numbers, of course, voluntary migrants who flocked to war production centers far outnumbered the deportees. The expansion of defense production in 1940–41 and the organization of a huge military enterprise in 1942–43 shifted the national economic balance toward the South Atlantic, Gulf, and Pacific coasts. Small cities like Phoenix and Albuquerque became important urban centers within half a decade. Wartime booms accelerated the long-term development of larger cities like Denver and Salt Lake City. From 1940 to 1943, the states with the highest rates of population growth were California, Oregon, Washington, Nevada, Utah, and Arizona (along with Florida, Virginia, and Maryland). The boom touched every major city in Texas, Oklahoma, the Pacific states, and the Southwest as well as selected production centers in the interior West.

The boom cities of the West suffered a common cycle of problems in which overpriced and insufficient housing forced new workers to locate haphazardly through the metropolitan area and further overburden already overcrowded transit systems. Workers in Seattle's shipyards and Boeing plant scrounged for living space in offices,

Pausing to enjoy their last meal eaten in freedom before their forced evacuation to a relocation camp in the spring of 1942, members of a Japanese-American citizens' league protest their fate with a meal of hot dogs and apple pie, deliberately chosen to reassert their status as American citizens.

Russell Lee (1903–1986). Picnic Lunch for Members of the Japanese-American Citizens' League Just Before Their Evacuation *(San Benito County, California). Gelatin silver print, 1942. Farm Security Administration Collection, Library of Congress (#LC-USF34-T01-72516D), Washington, D.C.*

tents, and chicken coops. A well-publicized visit by the U.S. House Committee on National Defense Migration in the summer of 1941 found San Diego swamped by ninety thousand defense industry workers and thirty-five thousand military personnel. Operating on their own hurried schedules, federal agencies exacerbated the city's problems by locating defense plants and emergency housing on isolated sites, thus forcing extra service costs and creating painful traffic problems. *Life, Business Week, Fortune*, and the *Saturday Evening Post* all described the crowded schools, overpacked hotels, makeshift trailer parks, and raucous nightlife in the "rip-roaringest coast boom town." San Diego's inadequate and poorly located housing, said one expert, was the "core of every problem and controversy."

After diligently working to secure naval facilities, the cities of the San Francisco Bay area found themselves with more than they had bargained for in 1940 and 1941. The federal government expanded the Mare Island and Hunters Point shipyards, Moffett Field, the Naval Operating Base and new Naval Air Station at Alameda, naval supply depots, and new facilities on Treasure Island. Federal contracts also funded a half-dozen huge private shipyards—General Engineering and Drydock in San Francisco, Western Pipe and Steel in South San Francisco, Bethlehem Shipbuilding in Alameda, Moore Drydock in Oakland, Todd-Kaiser in Richmond, and Marinship in Sausalito. San Francisco itself became a huge dormitory housing war workers, servicemen between assignments, and their dependents. The population of Vallejo and adjoining areas tripled, and workers at Mare Island commuted as far as fifty miles on rationed gasoline. Richmond's job total increased from 15,000 to 130,000. The city imported old elevated railroad cars from New York for a new rail line to Oakland as a way of coping with its transportation problem.

The war also brought entirely new towns. Under the Lanham Act and related programs, the federal government financed one million temporary housing units for wartime use nationwide. Many of the apartments were in the Bay Area at Richmond and the model community of Marin City. Many more—enough for forty thousand residents—were built at the instant "city" of Vanport, Oregon. Rising on the floodplain of the Columbia River, halfway between the Kaiser Corporation shipyards in Portland, Oregon, and Vancouver, Washington, the first of nine thousand apartments in six hundred wooden buildings were ready in December 1942. Painted dull gray to blend into the cold winter rains of the Northwest, the buildings of Vanport included schools, community centers, a day-care program, a post office, cafeterias, a fire district, playgrounds, shops, a 150-bed hospital, and a movie theater that ran three double-bills per week. New "science cities" in the western interior also contributed to the American war effort. Richland, Washington, on a dry benchland along the Columbia River, burgeoned from approximately three hundred people growing peaches and asparagus in 1940 to a community of fifteen thousand technicians supporting the manufacture of plutonium at the Hanford Engineering Works. Los Alamos, New Mexico, on a mesa that faced the sunrise over the valley of the Rio Grande, was built to an equally careful plan. The nuclear physicists and engineers who designed atomic bombs enjoyed gracefully curved streets and public facilities laid out according to the best modern standards of urban design.

The war meant new people as well as planned towns for the Far West. African Americans from Texas, Oklahoma, and the lower Mississippi Valley followed news of labor shortages and high wages to defense production centers. Seattle's black population jumped from four thousand to forty thousand, Portland's from two thousand to fifteen thousand. Within the cities, the newcomers crowded into a few neighborhoods and housing projects carefully set aside in the style of established eastern ghettos. Many found the only affordable housing to be in neighborhoods forcibly vacated by Japanese residents, such as the Western Addition in San Francisco and Little Tokyo east of downtown Los Angeles. Whether in a federal community like Vanport, a housing project like Hunter's Point in San Francisco, or a federally backed subdivision in the East Bay, new housing preserved the distinction between white and black neighborhoods and set a pattern of suburban segregation for the postwar generation.

Tensions also rose between Mexican Americans and Anglo Americans in Los Angeles. During the 1930s, tens of thousands of Mexican Americans had permanently settled in western cities such as Denver when their rural jobs as railroad hands and farm laborers disappeared. New migration in the early 1940s swelled the Mexican community in Los Angeles to an estimated four hundred thousand, prompting discrimination and a steady stream of anti-Mexican articles in the major newspapers. On 3 June 1943, off-duty sailors and soldiers led several thousand Anglos to attack Hispanics on downtown streets and invade Mexican-American neighborhoods. Blacks and Filipinos were incidental targets. The "zoot suit" riots, named for the flamboyant clothes of some young Chicanos, dragged on for a week.

The war was a powerful force for the assimilation of Native Americans. Military service and war production jobs greatly accelerated the process of modernization started by the New Deal. Twenty-five thousand Native Americans served in the armed forces. Another forty thousand moved for off-reservation jobs, utilizing industrial and mechanical skills learned in the Indian CCC and in programs of the National Defense Vocational Training Act. The average cash income of Indian households tripled during the war, and many Native Americans stayed in Los Angeles, Seattle, Phoenix, Denver, and other western cities at the end of the war.

War also opened new opportunities for women, who made up roughly one-quarter of West Coast shipyard workers at the peak of employment, two-fifths of Los Angeles aircraft workers, and nearly one-half of Dallas aircraft workers. The shipyards turned first to women who were already in the labor force in jobs ranging from shop clerk to farm worker. Many of the "housewives" who responded to recruitment campaigns in 1943 and 1944 were women who needed jobs to support their families. As one of the workers recalled of herself and a friend, "We both had to work, we both had children, so we became welders, and if I might say so, damn good ones." The most common shipyard jobs were clerks and general helpers, but the acute shortage of welders opened more than five thousand journeyman positions in the Portland operations. A few women even found positions as electricians and crane operators—far more interesting work than waitressing or sewing in a clothing factory. Aircraft companies, whose labor shortages were compounded by stubborn "whites only" hiring, developed new power tools and production techniques to accommodate the smaller average size of women workers, increasing efficiency for everyone along the production line.

Victory in 1945 forced women out of the factories, but it left the American military extended worldwide. As the "American Century" quickly faded into the realities of the cold war, the United States prepared for the indefinite projection of power in the Pacific and East Asia and for continental defense through nuclear deterrence. The North Korean invasion of South Korea in June 1950 and the Chinese entry into the Korean conflict in December confirmed the American commitment to the dual strategy of advanced and strategic defense. In response, American military planning during the early 1950s created four strategic "layers" that overlapped the western states and territories.

The nation's forward presence in the Far East rested on bases in the Philippines, Central Pacific, Okinawa, Korea, and Japan. The need to anchor this first line of defense in Alaska led to a new surge of federal investment greater than that in either the 1910s or the 1940s. Federal agencies and contractors built the Distant Early Warning Line of radar stations across northern Alaska. They expanded military bases. They upgraded ports, highways, railroads, and airports in the territory's Anchorage-Valdez-Fairbanks core. Alaska's 1950 census counted 1 uniformed member of the armed services for every 4.3 civilians.

Supply depots, training bases, shipyards, and aircraft maintenance bases constituted the second layer in the western military system. Admirals presided over navy commands headquartered at Seattle, San Francisco, San Diego, and Honolulu. Each city lay in a cluster of operating bases, shipyards, and air stations. Army and air force training facilities were especially prominent in Washington, Colorado, Texas, and Oklahoma. The siting of the new Air Force Academy in Colorado Springs to balance the historic

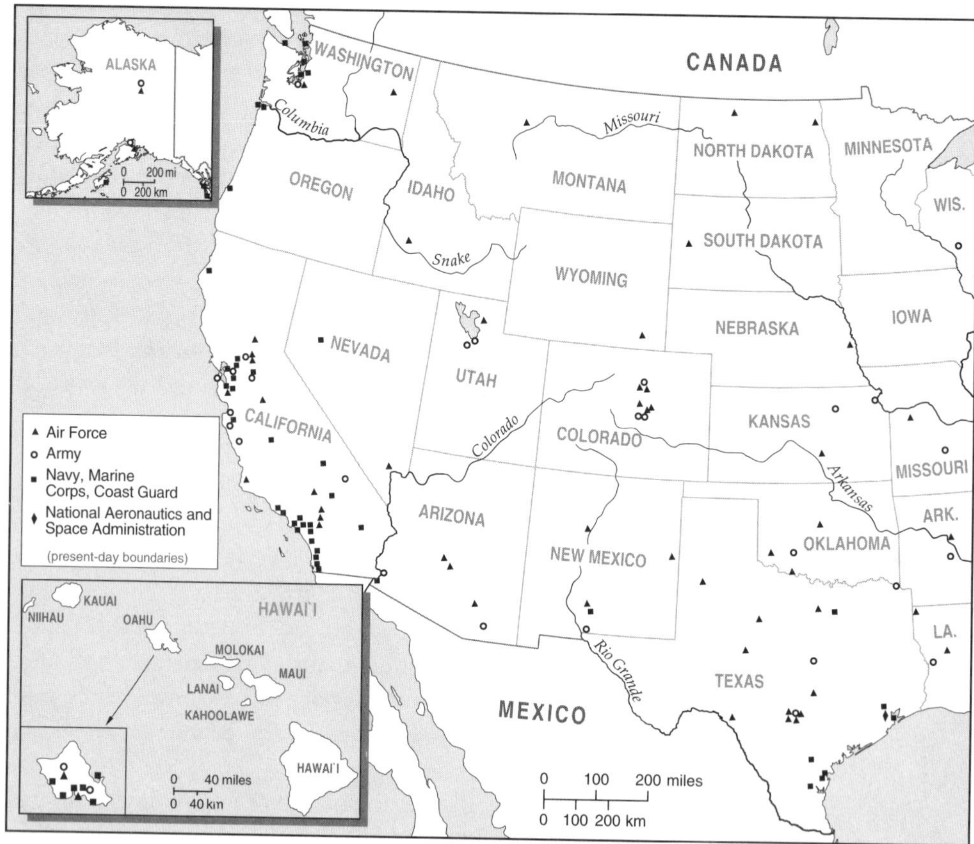

Major Military Installations and Related Institutions, 1992

service academies in Maryland and New York recognized the prominence of the West as a home base.

The strategic strike and defense forces that formed the heart of America's cold war strategy were commanded from the national heartland of the Great Plains. North American air defense was coordinated from a command post sunk deep beneath Cheyenne Mountain at the foot of the Rockies behind Colorado Springs. General Curtis LeMay directed the B-47s and B-52s of the Strategic Air Command from Omaha. By the 1970s, solid-fueled minuteman missiles were targeted for Moscow and Beijing from unobtrusive silos dotting the rolling landscapes of Montana and the Dakotas.

The fourth "layer" of the defense system was strategic weapons production. The Manhattan Project had tested the first atomic bomb at the Trinity site near Alamogordo, New Mexico. After the war, the armed forces experimented with missiles, atomic warheads, and lethal chemicals in the empty reaches of New Mexico, Nevada, and Utah. The Hanford reservation continued to produce plutonium for nuclear weapons, and the Rocky Flats facility north of Denver turned plutonium into triggers for thermonuclear bombs.

Military buildup during the Korean War brought a federally supported uranium boom to the Four Corners region of the Colorado Plateau. In the words of the *National Geographic*, it was "the land of the weekend prospector." Eager amateurs sent fifty-five cents to the Government Printing Office for a how-to pamphlet on uranium mining, stocked up on Geiger counters and USGS maps in Grand Junction, Colorado, and Moab, Utah, and joined Atomic Energy Commission and Geological Survey scientists in the search for bright-yellow carnotite ore. The federal government bought any commercial-grade ore and offered a special finding bonus of thirty-five thousand dollars for each new discovery. Although the prospecting mania died down by mid-decade, radioactive tailings from processing plants in towns like Uravan and Grand Junction, Colorado, and Monticello, Utah, remained serious health hazards for decades.

A city like San Diego showed the results of the peacetime mobilization. By one estimate, the U.S. Navy alone was responsible for adding 215,000 people to the San Diego area in 1957. Weapons makers added tens of thousands more. Most residents seemed to believe that what was good for Convair was good for the country. Cruising the harbor, as the historian Roger Lotchin has pointed out, one would pass a submarine base, a carrier anchorage, dry docks, shipbuilding firms, ammunition bunkers, and barracks ships. "In the bay, ships ply the waters and retired admirals cruise their sailing vessels; overhead, naval helicopters buzz and jets scream; and downtown, sailors throng the streets." The military was a constant presence in social life and civic activities. Scores of community organizations depended on military surplus goods to support their programs. Initiated in 1953, Air Power Day became a local equivalent of the Pasadena rose festival.

The West ended the 1950s deeply dependent on the military budget. Examining defense spending by states for 1952–62, Roger Bolton found a heavy dependence on defense for 25 percent or more of income from out-of-state sources in Washington, California, Hawai`i, Alaska, Utah, and Colorado. He found substantial dependence for 15–24 percent of outside income in Nevada, Arizona, New Mexico, Kansas, Oklahoma, and Texas. A slightly different measurement of defense spending as a stimulus to growth added Montana, South Dakota, and Wyoming to the list of defense-dependent states. Twenty years later there was little change. Six of the ten cities with the largest civilian Department of Defense employment in the 1980s were western. Of the contracts for President Ronald Reagan's Strategic Defense Initiative for 1983–86, 70 percent went to five western states.

The era of the garrison state left cultural as well as economic imprints. Military retirees make up at least 1 percent of the population in Washington, Colorado, Nevada, Arizona, New Mexico, and Hawai`i. They are an inescapable presence in San Diego, San Antonio, and other cities of California and Texas. A retired astronaut would make no sense as a central character in a story set in Buffalo or Detroit but seems perfectly at home in the Houston of Larry McMurtry's *Terms of Endearment* (1975). The West was close to the front lines of our three Pacific wars, serving as a place of embarkation, rest, and reentry. By the 1970s, it was also the new home for three hundred thousand refugees from Indochina. Whether they liked it or not, westerners found that proximity brought them face to face with growing moral dilemmas of American military action. In Herman Wouk's *The Caine Mutiny* (1951), Yosemite National Park is a healing counterpoint to the naval war against Japan. Twenty-three years later, Robert Stone's *Dog Soldiers* offered a compelling analysis of the effects of the Vietnam War on the American character, with the trans-Pacific traffic in drugs as the metaphor for the loss of national innocence. With

Vietnam only a few hours away, the most ordinary streets and ordinary residents of Oakland are pulled into direct confrontation with the moral ambiguities of Saigon. Neither a backwater nor a refuge, the contemporary West in the middle decades of the twentieth century became a central participant in the drama of war and peace.

The West in the World, 1960–90

In 1983, *Time* magazine declared that Los Angeles had become the "new Ellis Island," the destination for immigrants to the United States. A few years later, the Port of Los Angeles renamed itself "Worldport L.A." Security Pacific Bank in 1988 proclaimed that the Los Angeles regional economy was larger than that of Brazil, Australia, or India. As a "focal point of international trade and finance," greater Los Angeles ranked as one of a handful of control points in the world economy. This transformation of a western city into what geographers call a "world city" symbolizes the newest dimension of the federal impact on the West.

In the 1940s and 1950s, emergency mobilization and the entrenchment of the "garrison state" made the American West a staging ground for the projection of national power into world affairs. In the 1960s, a series of federal initiatives reopened American society and economy to the countercurrents of international trade and migration. By the 1970s and 1980s, the West had assumed a new importance in the middle of growing global networks of people, products, and ideas. In substantial part, this regional change was an inadvertent consequence of federal policies devised to handle national needs within world economic and political systems.

The reopening of American borders with the Immigration Reform Act of 1965, for example, responded directly to national economic and humanitarian needs and to the symbolic role of the United States in the world. The removal of Europocentric quotas for immigration, however, opened the door for new migrants from Latin America and Asia, who made the southwestern and Pacific states their primary rather than secondary destinations. Total legal migration to the United States grew from 3.3 million in 1961–70 to 4.5 million for 1971–80 and 6 million for 1981–90. Nearly 5 million of the total for the 1980s were Asians or Latin Americans. Indeed, the migration of the 1980s was large enough to trigger the Immigration Reform and Control Act of 1986, which offered retroactive legalization to many undocumented immigrants while trying to raise new barriers against further illegal immigration.

The liberalization of foreign trade dated to the General Agreement on Tariffs and Trade of 1947, followed by the Trade Expansion Act of 1962. Both measures looked first to the Atlantic economy, but their liberalized trade regulations provided the basis for a shift of foreign commerce to Asian and Pacific partners and ports. Imports and exports totaled 6.8 percent of the gross national product in 1960 and 14.6 percent by 1987, the heaviest orientation of the American economy to overseas trade since 1918. The accompaniment was an unplanned shift of trade to Texas and Pacific ports. By the mid-1980s, western boosters could cite figures that showed U.S. trade across the Pacific to be more valuable than the trans-Atlantic trade.

The West also gained a central role in national policy when the energy industry took center stage in world politics in the 1970s. The crisis of the OPEC embargo and world price increases brought efforts at a national energy independence that depended heavily on the West. Oil leases on federal lands on the North Slope of Alaska triggered a new Alaskan boom. Federal tax credits and subsidies encouraged attempts to extract oil from western Colorado shale deposits. The search for alternatives to fossil fuels looked to western wind and sunlight. Federal funds created the Solar Energy Research Institute in Golden, Colorado, while solar collector fields and windmills blossomed on western bluffs and hillsides. The oil glut of the 1980s resulted in a declining sense of urgency, however, allowing budget-trimming bureaucrats in the Reagan administration to cut off support for most of the experiments. National policy encouraged agricultural exports as a counterbalance to energy imports. Until halted by President Jimmy Carter in reaction to the Russian invasion of Afghanistan, grain sales to the Soviet Union were the most visible sign of a new farm boom, with some painful similarities to the 1910s. Wheat exports increased 2.4 times between 1970 and their peak in 1981. Land prices in the Missouri Valley spiraled upward in the 1970s, only to drop by 40 or 50 percent in the 1980s, with rural bankruptcies and business failures.

If the federal government has remained an ambiguous friend of resource production industries, it supplied the key market for the "fourth wave" industries that led the nation's economic expansion after World War II. Steam engines and the mass production of textiles had led the first wave of worldwide industrial growth from 1790 to 1845. Steel production and railroads had led the second wave from 1845 to 1890. Chemicals, electrical equipment, and automobiles had led a third wave from 1890 until the Great Depression. After the war, the new leaders were the electronics, communication, and aerospace industries. Whereas the first three waves had worked to concentrate industrial power in the manufacturing belt running from Boston to St. Louis, the postwar boom had its most profound impacts on the American West.

It was chiefly in the West, for example, that the aircraft industry of World War I and World War II evolved into the far more extensive aerospace business. By the 1950s, the six cities of Seattle, Los Angeles, San Diego, Dallas, Fort Worth, and Wichita accounted for nearly all of the country's airframe production and assembly. Military contracts sustained Lockheed, General Dynamics, McDonnell-Douglas, Northrop, and scores of related firms. The location of the Manned Spacecraft Center of the National Aeronautics and Space Administration (NASA) on ten thousand acres of donated land brought scores of spinoff businesses to Houston. The national space program also drew on the Jet Propulsion Laboratory in Pasadena, as well as Edwards and Vandenberg air force bases in California.

Federal contracts have been the basic support for the development and utilization of new electronics and information technologies in the newly high-tech cities of the West. Stanford Industrial Park in 1951 was the first planned effort to link the science and engineering faculties of major universities to the design and production of new products. It was the first step in the evolution of Silicon Valley, between San Francisco and San Jose, as a center of the electronics industry. Federal contracts, especially from the Department of Defense, have been a mainstay of Silicon Valley. A broader definition of high-tech, based on a high ratio of research and development to net sales, includes such industries as aircraft, guided missiles and space vehicles, computing machines, communication equipment, electronic components, and drugs. The federal government has been a primary customer for all but the last category. During the 1970s,

Arizona, California, Washington, Kansas, Utah, and Colorado ranked among the top ten states for high-tech jobs as a proportion of total employment.

The promotion of scientific research and applications became another federal industry, especially in the wake of the successful Russian launch of the first *Sputnik* in 1957. The federal subsidy of physical science utilized a combination of federal laboratories, private research-and-development contractors, and universities. One example is Albuquerque, which ranked ninth among all metropolitan areas in its federal research-and-development contracts in 1977. Scientists on the federal payroll have continued to undertake basic research at the Los Alamos National Laboratory. Western Electric operated Sandia National Laboratories under federal contract after 1949. Successful grant applications by faculty at the University of New Mexico made the state a leader in federal research and development in proportion to population. Colorado's pattern has been similar. Martin Marietta Corporation decided in 1956 to build a plant for titan missiles in the Denver suburb of Littleton. Hewlett-Packard, Honeywell, Sundstrand, and Ball Brothers Research were a few of the other high-technology firms attracted by life near the mountains. It was a short step from defense industries and science-oriented corporations to the research division of the University of Colorado and the federal research agencies located in Boulder—the National Bureau of Standards, the National Center for Atmospheric Research, and the National Oceanic and Atmospheric Administration.

The newest era of "big science" continues the western tilt. NASA's controversial manned space station is coordinated from the renamed Johnson Space Center in Houston. Astronomers use the country's biggest telescopes in California and Hawai`i and anticipate new facilities in the mountains above Tucson, Arizona. The end of the cold war has meant an end to plans for an eight-billion-dollar Superconducting Super Collider, slated until 1993 for the Texas prairies south of Dallas. Still, the nuclear weapons laboratories at Los Alamos, New Mexico, and Livermore, California, continue to function, though rapidly diversifying the range of their research.

Big science, it turns out, means big cleanups. The Department of Energy has made the western states the location of choice for disposal of nuclear wastes from military production and private reactors. New Mexico's salt deposits are scheduled to receive contaminated materials from Rocky Flats and other sites. In a multistate contest from which the Energy Department had eliminated all eastern states, Texas and Washington managed to fight off a nuclear waste repository now destined for Yucca Mountain, Nevada. At least nine billion dollars will be required to entomb high-level radioactive materials under the dry basin lands one hundred miles northwest of Las Vegas. At Hanford, Washington, tens of billions of tax dollars will be required for the safe disposal of radioactive liquids that have been sitting for decades in rusting tanks, ponds, and trenches within a few miles of the Columbia River.

In the process of building American scientific capacity, federal initiatives helped to transform western university systems. Research grants and graduate student aid, especially after the passage of the National Defense Education Act (1958), provided vital funding at a time when regional campuses were struggling to cope with the first arrivals from the postwar baby boom. The University of Washington grew from thirteen thousand to thirty thousand students by 1966 with vigorous pursuit of federal grants that equaled direct support from the state general fund. Federal funds had similar effects on university systems in Texas, California, and other western states. The University of California–San Diego was explicitly created to support high-technology research with a hefty gift and strong lobbying from the giant defense contractor General Dynamics. When federal research-and-development funding for universities is compared on a per capita basis, Alaska, Utah, Hawai`i, New Mexico, Washington, California, and Colorado all exceed the national average.

Direct federal funding and federal markets for science-intensive production have helped to make western metro areas some of the best-educated in the country. Sixteen of the metro areas that had five hundred thousand or more residents in 1980 reported that 20 percent or more of their adult population (over twenty-five years old) had completed at least four years of college. Ten of these cities were western: Austin at 28 percent; Denver at 26 percent; San Francisco–Oakland–San Jose at 25 percent; and Seattle–Tacoma, Honolulu, Houston, San Diego, Tucson, Salt Lake City, and Dallas–Fort Worth all at between 20 and 24 percent. The West claimed twenty-eight of the fifty-four metro areas of all sizes with the same high education level—including not only such likely candidates as Colorado Springs and Santa Barbara but also less obvious cities such as Grand Forks, North Dakota; Boise, Idaho; and Midland, Texas.

To gauge the global orientation of the new western metropolis, we can develop a comparative index utilizing data on foreign-born population, foreign trade, foreign banks, foreign investment, importance of foreign markets, and role as international information center. San Francisco–Oakland, the clear leader as an international city in 1960, had yielded first place to Los Angeles by the 1980s. Indeed, Los Angeles is now clearly a "world city" in both its boardrooms and its streets, tied into international flows of people, goods, and data. At one level, it is the New York City of the Pacific Rim. The volume of imports and exports through Los Angeles and Long Beach tripled from 1970 to 1990. Observers such as the *Los Angeles Times* noticed an upturn in foreign investment after 1975. Canadian, Japanese, and other foreign investors have become major speculators in downtown real estate. Several major banks passed into the hands of Japanese and British firms during the 1980s. In turn, the foreign presence has attracted U.S. banks and corporate headquarters.

Los Angeles is very much an immigrant city. Nearly 20 percent of the people in the metropolitan area were foreign-born in 1980, 27 percent in the city of Los Angeles. These new ethnic residents fill the full range of economic roles—low-skill service workers, low-wage garment workers, skilled electronics assemblers, small entrepreneurs in retailing and manufacturing, scientists, and professionals. Specific Asian and Latino enclaves dot Los Angeles and Orange counties. Downtown Los Angeles divides between its English-speaking corporate towers on the west side and its Spanish-speaking theaters and stores east of Broadway. Behind Los Angeles and the Bay Area, San Diego, Honolulu, Seattle, Houston, and Dallas have been pushing forward as second-level centers for international contacts. The landlocked Dallas–Fort Worth "metroplex," for example, has adopted an aggressive global development strategy. The Dallas partnership at the start of the 1990s talked about a future as "a preeminent center of world commerce in the twenty-first century." The adamantly American city of Dallas now cooperates with Fort Worth ("where the West begins") to publish maps showing the metroplex as the navel of the world. In fact, Dallas–Fort Worth has become a second-level center for international banking. Foreign corporations employ thirty-five thousand workers. The city is trying to acquire the critical mass of foreign firms, trade offices, and business agents necessary to rival Mexico City, Chicago, and Los Angeles rather than New Orleans and Houston.

The successful cities of the "New West" have learned how to use federal funds to shape a cityscape appropriate for the information age. Westerners have received far more than their "fair" share of federal dollars for transportation facilities to overcome the added costs and delays of great distances. With a quarter of the national population in the 1970s, the West had received 35 percent of airport construction aid (1946–72) and 42 percent of interstate highway mileage. Only three thousand westerners have to share each mile of interstate within the region, compared with six thousand easterners for each mile in their half of the country.

In contrast to the image of western cities as freeway capitals, many of them showed an initial reluctance to use federal funds that became available for urban revitalization in the later 1940s and the 1950s. Los Angeles and Portland grudgingly accepted public housing as a wartime necessity but backed away from new construction in the 1950s. Tucson rejected federal urban-renewal funds in a noisy political controversy. Politicians and newspaper editors fended off the federal octopus in Phoenix. Fort Worth, Dallas, and Houston declined to hold the local elections required by Texas law. The state of Washington failed even to adopt enabling legislation until 1957, held back, according to Seattle Mayor Gordon Clinton, by the taint of socialism.

Beginning in 1959 in several West Coast cities and in the early 1960s in many of the Rocky Mountain and southern plains cities, new urban-renewal agencies did start up the "federal bulldozer" to clear deteriorated fringes of central business districts. Civic centers, office buildings, and university campuses were the first steps in adapting cities for a global information economy. High-rise cores in Los Angeles, San Francisco, Dallas, Houston, Denver, and Seattle provide the "natural" setting for the information processing and command transactions that dominate the modern world economy. Cities resuscitated failing private bus systems by purchase and expansion with Urban Mass Transit Administration dollars. Downtown streets dedicated to bus service and new rail lines now smooth the commute of the workers who serve the metropolitan decision makers.

Even cities like Phoenix and Houston, whose community ideologies adamantly proclaim the virtues of untrammeled free enterprise, owe much of their growth to federal favor. Federal dollars helped to dredge the Houston Ship Channel that made the city a port and petrochemical center, just as they built Roosevelt Dam outside Phoenix to make the Salt River Valley a major farming center and enable it to support intensive settlement. During World War II, Phoenix acquired three large air bases. Houston ranked sixth in the nation in federal investment in factory facilities, and the federally financed Big Inch and Little Inch pipelines gave Houston a national market for its natural gas. Oil import quotas, depletion allowances, foreign tax credits, and other federal tax subsidies supported the domestic petroleum industry. The expertise of Houston corporations in petroleum research, exploration, production, and refining helped to define and secure American interests in overseas oil fields. Even the unzoned subdivisions of Houston and the sprawling suburbs of Phoenix were built with federally insured or guaranteed mortgages. Like earlier westerners, Houstonians and Phoenicians may have fulminated against the influence of federal bureaucrats, but they were happy to bank the benefits of federal assistance.

The Networked West

The western states and territories entered the twentieth century as a set of isolated resource regions organized around provincial trading centers. Copper miners in southern Arizona, stock raisers in central Wyoming, cotton farmers in Oklahoma, and wheat ranchers in eastern Washington all depended directly on decisions made in New York and Chicago and transmitted through small cities like Tucson, Casper, Tulsa, and Spokane. In a pattern reminiscent of British North America before 1775, most subregions had stronger ties to the East than to each other. In the phrase of the historian Robert Wiebe, the West was an archipelago of "island communities."

Three generations later, the national government's immigration policies, trade policies, resource development policies, infrastructure investments, and defense spending had transformed much of the West. The northern plains and northern Rockies communities found themselves repeating the events of the 1920s, losing old primary industries and scrambling to replace them with the modern resource industry of tourism. The overwhelming majority of westerners, in contrast, lived in the cities of the southwestern and Pacific states—economically diversified communities with the capacity to generate much of their own growth through innovation and import substitution. With Los Angeles–San Diego and San Francisco–Oakland–San Jose as its twin capitals, greater California embraced Arizona, Nevada, Hawai`i, and the northern coastal states within its sphere of trade, finance, and communication. Even in the midst of an energy industry depression, the metropolitan axis of Houston-Dallas-Denver organized a half dozen other states into another economic unit driven by resource production, foreign business connections, high-tech industry, and federal spending.

The political influence of the West rose with its population, making it more and more an initiator of federal policies rather than a recipient. The West added ten U.S. senators after 1900 with the admission of Oklahoma, New Mexico, Arizona, Hawai`i, and Alaska as states. The number of westerners in the House of Representatives rose from 60 in 1900 to 127 in 1980. Theodore Roosevelt in 1901 and Herbert Hoover in 1929 were the first presidents with real western connections. Between 1952 and 1992, genuine or honorary westerners—Eisenhower, Johnson, Nixon, Reagan, and Bush—accounted for thirty-one of forty possible years in the White House.

By the 1980s, western issues had become national issues. Western relations with Japan, OPEC nations, and the newly industrialized countries of East Asia defined much of national economic policy. In the 1990s, conflicting versions of a "sagebrush rebellion" are likely to define long-lasting national choices about the future of the American landscape. The free enterprise rebellion fears that the federal government is preserving and conserving natural resources at the expense of local investment and employment opportunities. The environmental rebels, especially strong in the Pacific states and northern Rockies, fear that the federal government has failed to prevent environmental degradation of lands under its control. Their agenda includes opposition

With almost one-half of the total land area of the eleven westernmost states in the lower 48—including 86 percent of Nevada—still owned or administered by various federal agencies, the region remains inextricably tied to the federal government.

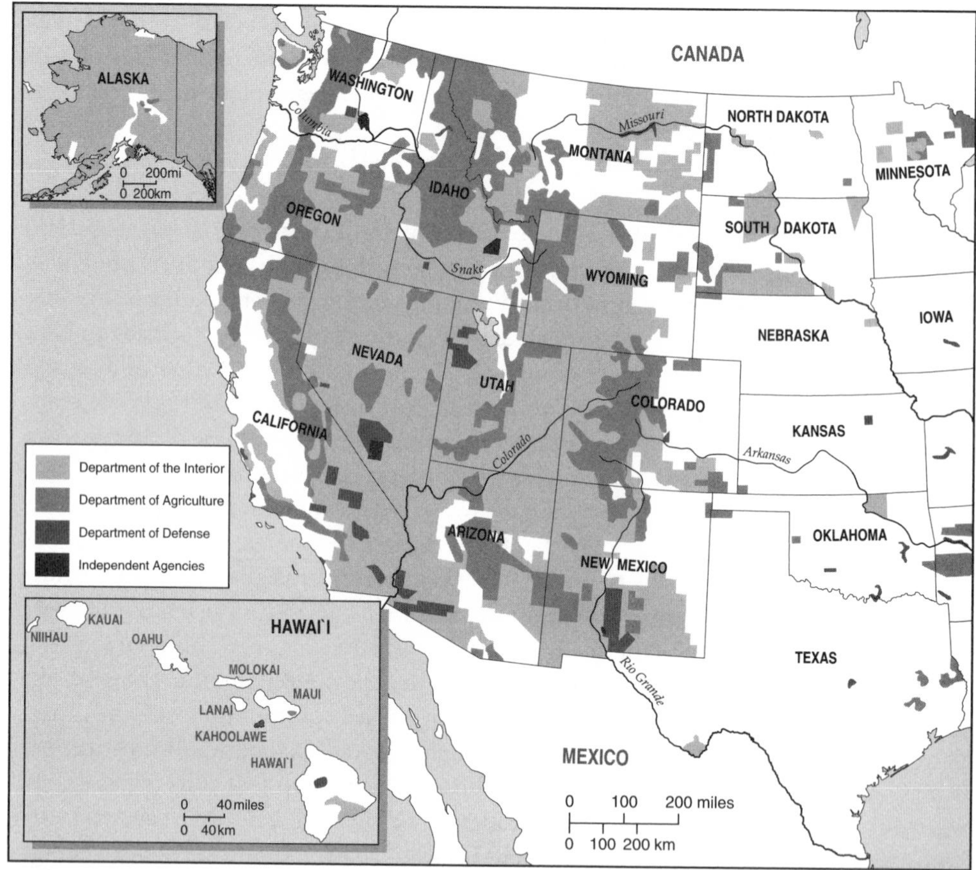

Federal Presence in the Contemporary West

to offshore oil leases and drilling without state regulation and to expansion of military training and testing ranges, concern about the western role as a toxic waste dump, and commitment to the enforcement of endangered species legislation.

At the same time, the West remains inextricably bound to the federal government. From Houston to Honolulu, the region functions within vast networks owned, operated, or defined by the national government. A number of historians and social theorists have argued that a fundamental trend of the last century has been an "organizational revolution" in American life. Individual Americans and their communities have been increasingly caught up in large and extensive organizations—large corporations, big labor, mass communications, and, of course, big government. The effect has been the nationalization of western America, the reduction of differences as subregions and cultures have been incorporated within national systems and found themselves participants in national programs.

In 1893, Frederick Jackson Turner invited his readers to imagine themselves standing at Cumberland Gap to watch the procession of trappers, traders, and pioneer settlers moving inexorably westward. In 1990, we can place ourselves instead to overlook the great gorge of the Columbia River where it cuts to the sea between Mount Adams and Mount Hood. Where Meriwether Lewis and William Clark once forged Turner's path of exploration, Japanese pickups and Korean sedans now weave among the Peterbilts and Winnebagos on Interstate 84. High-voltage lines from the dams at Bonneville and The Dalles cut straight swaths across the timbered ridges to deliver power to Portland and the California intertie. Barges loaded in Idaho pass through federally operated locks in the federal dams, on their way to rendezvous with oceangoing ships that have followed a federally dredged channel upriver from Astoria to Portland. Hikers enjoy trails and campgrounds first cleared by CCC workers and maintained by college students on the temporary federal payroll.

Less visible is the transmission of information and management decisions through the institutional networks of federal land and resource agencies. Fish and Wildlife Service workers watch over research stations and hatcheries on streams that tumble down the steep sides of the gorge. The Corps of Engineers angers local sport fishermen by identifying fishing sites for Native Americans to replace ancestral fishing grounds flooded by the Columbia River dams. Bureau of Land Management and Forest Service managers worry about balancing the demands of stock raisers, timber workers, and tourists. Other Forest Service workers try to administer the Columbia River Gorge National Scenic Area, a 1986 experiment in preserving scenic resources in the midst of continued farming, grazing, and logging.

Apple and pear growers of the Hood River valley work their own land but operate within another set of federal institutions. They draw power through the Hood River Electric Cooperative, which dates from the Rural Electrification Act of the New Deal. They work with representatives of the Soil Conservation Service in the local conservation district. Some of the orchardists are Japanese Americans whose parents were forcibly removed from the county by federal actions in the wartime panic of 1942. All of them worry about the impact of the Immigration Reform and Control Act on the valley's Spanish-heritage labor force.

Indeed, much of the economic life of the Columbia Gorge operates within the international trade rules set by the federal government. Regional lumber mills are squeezed between Japanese demand for raw logs and increased competition from Canada following the free trade agreement of 1989. The trains and barges that pass down the gorge are filled with potatoes for McDonald's outlets in Tokyo, grain for Chinese noodle factories, and bentonite for the oil wells of Indonesia. The fate of returning salmon runs is mired in multinational negotiations over the regulation of Asian fishing fleets.

As much as any private corporation or capitalist, it has been federal programs and federal dollars that have responded to the needs and opportunities of the American West. From the start of the twentieth century, the West has been the vast proving ground for an activist federal government. Thanks to those federal efforts, the West is no longer a region apart. The sparsely settled resource frontiers of 1900 have given way to a new West that lies at the center of national growth and global relations.

Bibliographic Note

The ambiguity in western attitudes toward the federal government is explored in Robert Athearn, *The Mythic West in Twentieth Century America* (Lawrence, 1986), Patricia Nelson

Limerick, *The Legacy of Conquest: The Unbroken Past of the American West* (New York, 1987), and Gene M. Gressley, "Regionalism and the Twentieth Century West," in Jerome O. Steffen, ed., *The American West: New Perspectives, New Dimensions* (Norman, 1979).

Western world's fairs are discussed in Robert Rydell, *All the World's a Fair: Visions of Empire at American International Expositions, 1876–1916* (Chicago, 1984), Burton Benedict, *The Anthropology of World's Fairs: San Francisco's Panama Pacific International Exposition of 1915* (Berkeley, 1983), and Carl Abbott, *The Great Extravaganza: Portland and the Lewis and Clark Exposition* (Portland, 1981).

Federal land policy is summarized in Marion Clawson, *The Federal Lands Revisited* (Baltimore, 1983). The evolution of irrigation policy and the Reclamation Service is treated in Donald Worster, *Rivers of Empire: Water, Aridity, and the Growth of the American West* (New York, 1985). The boom of the early 1900s is most easily approached through state histories and regional studies such as Donald Meinig, *The Great Columbia Plain: A Historical Geography, 1805–1910* (Seattle, 1968). Alaska as a federal frontier is the topic of William H. Wilson, *Railroad in the Clouds: The Alaska Railroad in the Age of Steam, 1914–1945* (Boulder, 1977).

The progressive conservation agenda is delineated in Samuel Hays, *Conservation and the Gospel of Efficiency: The Progressive Conservation Movement, 1890–1920* (Cambridge, Mass., 1959). Also see G. Michael McCarthy, *Hour of Trial: The Conservation Conflict in Colorado and the West, 1891–1907* (Norman, 1977), Elmo Richardson, *The Politics of Conservation: Crusades and Controversies, 1897–1913* (Berkeley, 1962), Harold Steen, *The U.S. Forest Service: A History* (Seattle, 1976), John Ise, *Our National Park Policy: A Critical History* (Baltimore, 1961), and Donald Swain, *Wilderness Defender: Horace M. Albright and Conservation* (Chicago, 1970). The Taylor Act and changing agricultural land policies are treated in John Opie, *The Law of the Land: Two Hundred Years of American Farmland Policy* (Lincoln, 1987).

The starting points for an analysis of the rise of federal planning are Norris Hundley, *Water and the West: The Colorado River Compact and the Politics of Water in the American West* (Berkeley, 1975), Richard Lowitt, *The New Deal and the West* (Bloomington, 1984), and Charles McKinley, *Uncle Sam in the Pacific Northwest: Federal Management of Natural Resources in the Columbia River Valley* (Berkeley, 1952).

A series of pictorial histories traces the evolution of the first federal-aid highway system: George Stewart's classic *U.S. 40* (Boston, 1953); Drake Hokanson, *The Lincoln Highway: Main Street across America* (Iowa City, 1988); Quinta Scott and Susan Croce Kelly, *Route 66: The Highway and Its People* (Norman, 1988). The standard discussion of the adoption of the second federal-aid system in 1956 is Mark Rose, *Interstate: Express Highway Politics, 1941–1956* (Lawrence, 1979).

Primary sources for analyzing the effects of economic depression and the New Deal are the reports of the National Resources Planning Board and its regional and state affiliates, discussed and inventoried in Marion Clawson, *New Deal Planning: The National Resources Planning Board* (Baltimore, 1981). The New Deal at the regional and state levels can be followed in sets of state-level studies published in the *Pacific Historical Review* 38 (August 1969) and in John Braeman, Robert Bremner, and David Brody, eds., *The New Deal: The State and Local Levels* (Columbus, 1975). The comprehensive effects of federal funds are detailed in two articles by Leonard Arrington: "The New Deal in the West: A Preliminary Statistical Inquiry," *Pacific Historical Review* 38 (August 1969): 331–36, and "The Sagebrush Resurrection: New Deal Expenditures in the Western States, 1933–39," *Pacific Historical Review* 52 (February 1983): 1–16.

The great dams are treated in Joseph Stevens, *Hoover Dam: An American Adventure* (Norman, 1988), Murray Morgan, *The Dam* (New York, 1954), Richard Neuberger, *Our Promised Land* (New York, 1938), and Mark Foster, *Henry J. Kaiser: Builder in the Modern American West* (Austin, 1989).

The relations between the New Deal and regional subcultures are discussed in Donald Parman, *The Navajos and the New Deal* (New Haven, 1976), Sarah Deutsch, *No Separate Refuge: Culture, Class, and Gender on an Anglo-Hispanic Frontier in the American Southwest, 1880–1940* (New York, 1987), and Suzanne Forrest, *The Preservation of the Village: New Mexico's Hispanics and the New Deal* (Albuquerque, 1989).

The baseline studies of World War II in the West are by Gerald Nash: *The American West Transformed: The Impact of the Second World War* (Bloomington, 1985) and *World War II and the West: Reshaping the Economy* (Lincoln, 1990).

The military strategy in western city building is discussed in Roger Lotchin, *Fortress California, 1910–1961: From Warfare to Welfare* (New York, 1992), 996–1020. Wartime boom cities are discussed in Carl Abbott, *The New Urban America: Growth and Politics in Sunbelt Cities* (Chapel Hill, 1981), Martin Schiesl, "City Planning and the Federal Government in World War II: The Los Angeles Experience," *California History* 58 (April 1979): 127–43, and Philip Funigiello, *The Challenge to Urban Liberalism: Federal-City Relations during World War II* (Knoxville, 1978). The atomic cities of Richland and Los Alamos can be studied through magazine reports and the postwar hearings of the Joint Congressional Committee on Atomic Energy, as well as in Hal Rothman, *On Rims and Ridges: The Los Alamos Area since 1880* (Lincoln, 1992). Western women as war workers are the topic of Amy Kesselman, *Fleeting Opportunities: Women Shipyard Workers in Portland and Vancouver during World War II and Reconversion* (Albany, 1990).

Aggregate data on the military role in the postwar West are presented in the following: James Clayton, "The Impact of the Cold War on the Economies of California and Utah, 1946–65," *Pacific Historical Review* 36 (November 1967): 449–53; Roger Bolton, *Defense Purchases and Regional Growth* (Washington, D.C., 1966); Rosy Nimroody, *Star Wars: The Economic Fallout* (Cambridge, Mass., 1988); and Ann Markusen and Robin Bloch, "Defensive Cities: Military Spending, High Technology, and Human Settlements," in Manuel Castells, ed., *High Technology, Space, and Society* (Beverly Hills, Calif., 1985). Urban growth in the context of the world energy business is the topic of Joe Feagin, *Free Enterprise City: Houston in Political-Economic Perspective* (New Brunswick, N.J., 1988), and Andrew Gulliford, *Boomtown Blues: Colorado Oil Shale, 1885–1985* (Niwot, Colo., 1989).

The concept of the fourth wave of industrial growth is discussed in Peter Hall and Paschal Preston, *The Carrier Wave: New Information Technology and the Geography of Innovation, 1846–2003* (London, 1988). The aircraft industry is profiled in William G. Cunningham, *The Aircraft Industry: A Study in Industrial Location* (Los Angeles, 1951). Defense spending in southern California is described in Martin Schiesl, "Airplanes to Aerospace: Defense Spending and Economic Growth in the Los Angeles Region," in Roger Lotchin, *The Martial Metropolis: U.S. Cities in War and Peace* (New York, 1984), and David Clark, "Improbable Los Angeles," in Richard Bernard and Bradley Rice, eds., *Sunbelt Cities: Politics and Growth since World War II* (Austin, 1983). The western concentrations of high-tech industries are detailed in Ann Markusen, Peter Hall, and Amy Glasmeier, *High Tech America: The What, How, Where, and Why of the Sunrise Industries* (Boston, 1986).

Data on federal science spending have been summarized by Edward J. Malecki, "High Technology and Local Economic Development," *Journal of the American Planning Association* 50 (Summer 1984): 262–69. The growth of a western science establishment is discussed in Nuel Pharr Davis, *Lawrence and Oppenheimer* (New York, 1968), and Clayton Koppes, *JPL and the American Space Program: A History of the Jet Propulsion Laboratory* (New Haven, 1982).

The emergence of a metropolitan society is discussed in case studies in Bernard and Rice, *Sunbelt Cities*, and in Abbott, *The New Urban America*. An interpretive essay that emphasizes the federally assisted sources of growth is Carl Abbott, "The Metropolitan Region," in Gerald Nash and Richard Etulain, eds., *The Twentieth Century West: Historical Interpretations* (Albuquerque, 1989). The rise of the information-processing city is the focus of Rob Kling, Spencer Olin, and Mark Poster, eds., *Postsuburban California: The Transformation of Orange County since World War II* (Berkeley, 1991), and Allen J. Scott, *Metropolis: From the Division of Labor to Urban Form* (Berkeley, 1988).

Conflicting western expectations about federal resource policy in the past twenty years are described in Thomas J. Gallagher and Anthony F. Gasbarro, "The Battle for Alaska: Planning in America's Last Wilderness," *Journal of the American Planning Association* 55 (Autumn 1989): 433–44, and Samuel Hays, *Beauty, Health, and Permanence: Environmental Politics in the United States, 1955–1985* (New York, 1987).

Chapter Fourteen

Politics and Protests

MICHAEL P. MALONE AND
F. ROSS PETERSON

On an autumn evening along the Pedernales River in the hill country of central Texas, Lyndon Baines Johnson, president of the United States and Texas native, mounted his horse and began riding through his herd of cattle. Dressed in western attire, with ever-present Stetson, the tall Texan rode among the cattle with reckless abandon as Secret Service men in business suits jumped in jeeps and cars in order to stay close to Johnson as he exhibited his riding skills.

Twenty years later, on his ranch near Santa Barbara, California, Ronald Reagan, the transplanted midwesterner, swung into the saddle of his horse and slowly led a small entourage along the hilly trails overlooking one of America's most affluent communities. The planned activity allowed the Secret Service agents to don jeans and boots. The actor-turned-politician became a westerner because of vocation, and his acquired horsemanship came from movie roles, not boyhood necessity.

Different as they were, Lyndon Johnson and Ronald Reagan shared a connection to the American West and celebrated that connection in the course of their presidencies. Moreover, their administrations underscored the western element in American politics in ways that were more than symbolic. The enlargement of federal programs during Johnson's term reflected the West's long experience with the positive impact of government spending, whereas the deregulation and cutbacks initiated by Reagan (with no curtailment of the defense expenditures on which the region was so dependent) grew out of westerners' mounting impatience with federal interference. Though Johnson was the consummate Washington insider, he possessed an individualistic style that marked him as a maverick and strengthened his ties to the people of his home region. Reagan's demeanor was more soothing, but by presenting himself as a political outsider, he too could claim maverick status. That only twelve years separated the end of Johnson's liberal presidential tenure from the beginning of Reagan's conservative one showed how swiftly the West could shift its political direction.

From the 1890s to the 1990s, the West experienced a number of swings from one end of the political spectrum to the other. Yet some concerns remained constant. Primary among these were issues related to the development of natural resources, to the dispersal of federal largesse, and to the expansion of democratic processes. As the West itself changed from a hinterland of the United States to a major population center, the issues that concerned its residents drew increasing national attention.

President Lyndon Johnson used his Texas hill-country ranch to project the image of a quintessentially American leader rooted in the rugged independence associated with the nation's western past.

Bill Hudson. President Lyndon B. Johnson Rounding up a Hereford Yearling on the LBJ Ranch. *Gelatin silver print, 4 November 1964. AP/Wide World Photos, New York.*

Politics in the Era of Closing Frontiers, 1890–1920

During the thirty years from 1890 until the close of World War I, the American West witnessed the last significant land-takings, marking the final closing of the frontier epoch. These were years of economic boom and bust and of consequent social and political turmoil, turmoil that involved the most powerful surges of leftist protest the region has ever seen but that ended with a triumphant rightist reaction. On the Right and especially on the Left, the ideological residues of these formative years lingered long into the future.

In politics, as in other aspects of national life, the tumultuous 1890s formed a watershed decade. Six new states, all of them western, entered the Union in 1889–90: Washington, Montana, North and South Dakota, Idaho, and Wyoming. These entries were heralded at the time as signaling both the governmental maturation of the region and the close of the frontier. In reality, they represented only the breaking of the congressional logjam that had held back western statehood since the entry of Colorado in 1876.

By this time, most of the West had passed beyond the raw, formative stages of territorial politics, beyond the struggles of local economic interests to find their places in the sun and the accompanying clashes between elected legislatures and appointed federal officials. By 1891, only four territories remained in the contiguous United States: Utah, which would gain statehood in 1896, Oklahoma (1907), and Arizona and New Mexico (1912).

As was true throughout the country, politics in the West took shape through the shifting interactions of inherited partisan loyalties and the region's uniqueness. The shadows of post–Civil War politics loomed over the Southwest well into the twentieth century. Thus the Democratic party of the Old South commanded the "borderland" areas—the State of Texas and the Territory of Oklahoma—and even dominated the desert territories of New Mexico and, to a degree, Arizona. Democrats could also contest or even control other states such as Montana, Nevada, or Utah, where Mormondom still shunned the "Grand Old" Republican party, which had deprecated the Mormon religion. Throughout most of the West, though, the Republican party—the party of the North, the Union, and expansion—usually prevailed.

During their territorial years, the western commonwealths had frequently displayed the trait that the historian Kenneth Owens has aptly described as "chaotic factionalism." Inherited party loyalties became confused and convoluted in the new western settings, and rampant factionalism and highly personalized politics were the rule rather than the exception. These characteristics have proven to be abiding. To this day, as the political historian Paul Kleppner has noted, party affiliations have been less stable and the power of personalities and factions stronger in the West than in the older regions to the east. This has made, over the years, for frustrated party organizers and a frequently colorful and hectic political climate.

Each of the western states evolved its own political culture, a product of its peculiar history; and in all of them, the community's distinctive economic and social makeup counted far more than party loyalties in defining behavior. Often, in these years of unblushing governmental manipulation, economic interests unabashedly manhandled the political process. For instance, in the mining states of California, Nevada, Utah,

Colorado, and Montana, millionaires such as William Andrews Clark in Montana or Thomas Kearns in Utah openly "influenced" legislatures in order to be elected to the U.S. Senate. In both Montana and Arizona, years of such abuses finally resulted in thoroughgoing corporate domination. This led some eastern critics to dismiss the more lightly populated western states as "rotten boroughs" and even to question the wisdom of having admitted them to the Union. Other extractive corporations, such as lumber companies, soon exercised similar power in Washington and Oregon.

By far the greatest corporate power lay in the hands of railroads, as Oregon Senator John Mitchell revealed in his famous quote about the owner of the Oregon Central Railroad: "Ben Holladay's politics are my politics and what Ben Holladay wants I want." For years, Nebraska's two major railroads, the Union Pacific and the Burlington, reportedly divided the state's two senate seats between them. In North Dakota, the notorious "machine" headed by "Big Alex" McKenzie acted as a political broker for railroads and other corporations based in Minneapolis and St. Paul. Under the political tutelage of William F. Herrin, the Southern Pacific Railroad exercised such a dominant influence in California that the novelist Frank Norris likened it to an octopus.

It was easy one hundred years ago to assume that the western political order was simply a creature of special interests and a colony of eastern capitalism. The truth was far more complex. Western conservatism, for instance, drew heavily not only on major vested outside interests but also on a wide range of locally based businesses and on the instinctive standpattism of wealthy farmers and ranchers. Similarly, western liberals and radicals also had their natural constituencies. Sizable elements of the region's small middle class and business community chafed at the political ineptitude of the time and at perceived mistreatment by eastern-based railroads and other corporations. Strong regional unions, including four well-organized railroad brotherhoods, were highly active.

However, it was agriculture that set the cadence of western political life in the 1890s, constituting by far the largest social and economic group in most states and territories of the region. Like other westerners, farmers and ranchers varied considerably in their political persuasions. But many of them came from liberal and socialistic backgrounds in Scandinavia and Germany; and many more resented their vulnerability to overcharging by monopolistic railroads, grain storage facilities, stockyards, and other corporations. As events of the 1890s demonstrated, westerners were ready for revolt. The Farmers' Alliance, which claimed four million members nationwide, evolved into the Populist or People's party, the most powerful third-party movement in modern American history. This revolt out of the South and West against eastern domination shook the nation.

Populist parties emerged in Kansas and then in other plains states in the early 1890s. By mid-1892, the Populist movement had grown into a national political party, which held its nominating convention in Omaha. The famous Omaha Platform passed by this gathering signaled the genuine radicalism of the movement. It called for an inflated currency, a subtreasury system of warehouses whereby farmers could secure better credit using stored crops as collateral, the nationalization of railroads and telephone and telegraph lines, a graduated income tax, and the forced forfeiture of "excess" lands granted to railroads. The Populists' political demands for direct election of U.S. senators and for democratic reforms such as the initiative, referendum, and recall substantiated the western theme of bringing power to the people.

In the presidential election of 1892, the Populist candidate James B. Weaver of Iowa did well in the West, carrying Kansas, Colorado, Idaho, and Nevada. Five of Kansas's eight congressional seats went to Populists, who joined the feisty William Peffer, already in the Senate. Populists also gained control of the Kansas State Senate. In Colorado, Populists elected Davis Waite governor and sent a sizable minority, thirty-nine members, to the legislature. These early Populist state governments, like later ones such as that of Governor John Rogers in Washington, faced severe contention because of their unstable alliances with Democrats and the enmity of Republicans. Conversely, in the Farmers' Alliance heartland of Texas, truly dedicated reformers chafed at Populist-style Democrats like Governor James Hogg, politicians who usually preached more reform than they actually achieved.

Populists and their allies in the two established parties gained strength as the hardships of the Panic of 1893 breathed fire into their radical demands. When the Cleveland administration persuaded Congress to repeal the Sherman Silver Purchase Act in November, the resulting closure of mines and smelters wrought havoc throughout the mountain West. As a result, the demand for "free silver," or the resumption of a bimetallic currency, now became the dominant theme of the reformers. It promised both rebirth to the mining industry and a beneficial inflation for debtors. Thus, while the Populists remained agrarian in the plains states and other farming regions, mining and labor interests supported the party in the mountain and Great Basin states.

As the Populist movement gained momentum in the mid-1890s, so too did the arguments of those who reasoned that it could succeed nationally only through "fusion" with an established party. The main force of political marriage was the advocacy of free silver, which appealed to indebted workers as well as farmers. This meant, practically speaking, fusion with the Democrats, as was already common not only in the South but in Nebraska as well. Although there were numerous "Silver Republicans," such as prominent Senators Henry Teller of Colorado and Fred T. Dubois of Idaho, the national Republicans remained overwhelmingly devoted to the gold standard.

During the presidential campaign of 1896, the agrarian crusade crested and then ebbed. The Democratic party, badly divided along regional lines under Grover Cleveland's stumbling direction, nominated the Nebraska congressman and orator William Jennings Bryan after he had carefully campaigned for the post and then captivated the convention with his famous "Cross of Gold" speech decrying conservative, eastern exploitation of the poor, the West, and the South. His nomination rested overwhelmingly on the silver issue; and, to the dismay of many of the more committed reformers, the Populists proceeded to nominate him also, even though this meant that they sacrificed their more substantial reform proposals to silver and its broader appeal. But to have chosen another candidate would have divided the reform-protest forces and ensured the victory of the solid-gold Republican candidate, William McKinley. Nonetheless, McKinley's massive financial support, the comparative narrowness of Bryan's appeal, and the resurgence of prosperity all combined for a GOP victory. McKinley won by an electoral vote margin of 271 to 176 and by a popular vote spread

During his second try for the presidency in 1900, Democrat William Jennings Bryan again crusaded against the eastern exploitation of western workers, linking this opposition to domestic oppression with opposition to President William McKinley's imperialist policy toward the Philippines. For many western farmers, the foreign policy issues remained too abstract, and their votes went to the conservative Republican party.

Neville Williams, publisher. William Jennings Bryan Campaign Poster. *Chromolithograph, 1900. Library of Congress (#LC-US262-2144), Washington, D.C.*

of almost 600,000, the largest in a generation. In the West and South, however, Bryan and silver nearly swept the field. Only the western states of California, Oregon, and North Dakota went Republican.

Historians have hotly disagreed about the true nature of the Populist party. Writing from a Turnerian vantage, John D. Hicks long ago portrayed the farmer-radicals as genuine frontier democrats who well understood a system that oppressed them and who valiantly tried to change it. The renowned historian Richard Hofstadter, projecting an urban-eastern bias, countered that they were poorly informed cranks and xenophobes who simply lashed out at a system they only vaguely comprehended. Contemporary historians, such as Walter Nugent and O. Gene Clanton, conclude that the Populists were in fact cogent and determined regionalists and reformers.

Clearly, the tumultuous 1890s were a seedtime for the political culture of the modern West. On the one hand, many regional conservatives often demonstrated an exaggerated individualism denoting their frontier origins and also a pronounced belief in unrestrained property rights. They saw little conflict between their dependence on federal largesse on the one hand and their condemnation of federal regulation and ownership on the other. The reformers too were highly individualistic and quite capable of savoring their federal cake even while deploring it. At the heart of their ideology lay the essence of the Populist persuasion: an enmity toward extractive corporations and a government often in league with them, a demand for progressive taxation and effective government regulation, and an encouragement of political participation that included women well before the rest of the country followed suit. For all their rancor, the Populists were true democrats and western regionalists.

Ironically, it was left to another generation of westerners to achieve the political reforms demanded by the Populists. The Progressive movement captured many Populist ideas, revised them, and gave them respectability before a national audience.

In contrast to the 1890s, the early years of the twentieth century were a time of prosperity, calm, and an optimistic belief in the inevitability of progress. Unlike the more narrowly focused and radical phenomenon of the Populist movement, the primary reform thrust of these years took a broader and more orderly approach. Known as "progressivism," an impossibly broad and imprecise term, this new movement was characterized by the urge to curb corporate and political manipulation of the public order by means of expanded democracy, moral regeneration, and increased government regulation of the economy and of society. Progressives drew support from all sectors of society, particularly the rising middle classes; they found effective presidential leadership in Republican Theodore Roosevelt (1901–9) and Democrat Woodrow Wilson (1913–21).

The West responded enthusiastically to progressivism, particularly to the call for tough regulation of railroads and other corporations and to increased participatory democracy. Direct democracy took various forms: the initiative and referendum, which allowed the public to enact and nullify laws when legislatures proved unresponsive; electoral primaries that allowed the voters to nominate party candidates; the direct election of senators by the people; and the establishment of suffrage for women. The progressives' belief that direct democracy would result in moral reform appealed strongly to frontier individualism and to the still smoldering legacy of the Populist party. As the historians Arthur Link and Richard McCormick note, "The most distinctive aspect of western progressivism was its passion for the more democratic, anti-institutional political reforms . . . they were more common there than anywhere else in the nation."

Throughout the country, progressivism energized all levels of government—local, state, and federal—during the years after 1900. In the West, the first wave of reform often came at the municipal level and usually involved the eradication of a corrupt "machine" government, in league with utilities and other businesses contracting with the city, and its replacement with a reform mayor. This is what occurred in Seattle under Reginald Thompson and in Denver, where the reform coalition included the wealthy activist Josephine Roche and two men who went on to national prominence, Judge Ben Lindsey and George Creel. A similar alliance in San Francisco, including the journalist Fremont Older and the millionaire Rudolph Spreckels, scored the region's most spectacular victory by deposing corrupt Mayor Eugene Schmitz and sending the machine boss Abraham Ruef to prison. Of all western cities, however, it was Galveston, Texas, that made the greatest contribution to urban progressivism. After a terrible hurricane and tidal wave ravaged the city in 1900, killing at least six thousand people,

Galveston instituted the businesslike "city commission" form of government, which soon became popular throughout the land.

However, progressives made their greatest mark at the state level of government. Historians have argued at length over whether progressivism derived from the earlier Populist movement or was a new phenomenon that arose primarily in middle-class society. A look at the West offers a commonsense answer. In those states where populism had flourished, progressivism bore a strong agrarian-radical flavor; in those where it had not, the new reformers came more often from urban and middle-class origins.

Two states where populism had a lingering impact were Texas and Kansas. These states passed the first antitrust laws in the nation, and both created strong railroad regulatory commissions. Kansas, under the capable Republican administrations of the reform governors Edward Hoch and Walter Stubbs, and Texas, under the Democrat Thomas Campbell, simply carried forward the efforts begun under Populists in the preceding decade.

The two leading western progressive states were Oregon and California, neither of which had figured largely in the Populist revolt. Both states patterned reforms after the model in Wisconsin. In Oregon, a comparatively middle-class state, a group of clever reformers led by the remarkable William U'Ren and the cagey politico Jonathan Bourne gained sway over the legislature through the Non-Partisan Direct Legislation League. After securing the nation's first general initiative and referendum laws in 1902, the reformers used these weapons to enact the far-reaching "Oregon System," which included the direct primary, the "recall" of unsatisfactory public officials, a corrupt practices act, and a system whereby voters could indicate to legislators their choice for U.S. senator. This constituted a major step toward the constitutional amendment of 1913 allowing popular direct election of senators. A number of basic social reforms followed, such as the defining of maximum workweeks and minimum wages for women and children. By 1912, mass democracy in Oregon had progressed to the point that puzzled voters faced a "bedquilt ballot" nearly a yard long, with 136 candidates and 37 issues to be considered.

In neighboring California, progressivism took a similar middle-class urban profile. Utilizing the new direct-primary law in 1910, the caustic and brilliant Republican Hiram Johnson captured the governorship on a campaign promise "to kick the Southern Pacific Railroad out of politics." Beginning in the 1911 legislature, the Johnson administration enacted a full panoply of progressive laws, including toughened regulation of railroads and utilities and creation of worker's compensation. Both Theodore Roosevelt and Herbert Croly, the main political theorist of progressivism, judged California to be the leading progressive state in the nation.

In national politics, the West stood in the vanguard of progressivism. Theodore Roosevelt carried every western state except Texas in his presidential bid of 1904. In the three-way presidential election of 1912, the reformist Democrat Woodrow Wilson defeated Theodore Roosevelt, who headed an independent Progressive party ticket, and the decidedly unflamboyant Republican William Howard Taft. Taft managed to carry only Utah in the West. In 1916 Wilson, the self-proclaimed reform and peace candidate, carried all but two western states, Oregon and South Dakota. The region produced a number of progressive leaders on the national stage, most of them famed for their

The West led the nation in granting voting rights to women, but the path to full suffrage was not smooth. After winning suffrage in the Washington Territory in 1883, women there were denied the vote in 1887. Not until 1910 did a well-organized campaign result in the restoration of suffrage rights.

A. Curtis. Women in Washington Campaign for Suffrage. *Photograph, 1910. Special Collections Division, University of Washington Libraries (#19943), Seattle.*

individualism and their suspicion of big business. Among them were California's Johnson, the Kansas journalist William Allen White, and Senators George Norris of Nebraska, William Borah of Idaho, and Francis Newlands of Nevada.

The West figured largely in three major facets of national progressivism: woman suffrage, Prohibition, and conservation. Beyond dispute, the region led the nation in the democratic crusade to enfranchise females. Before 1917, the only states that granted women the vote lay in the West. In fact, by the following year—two years before the Nineteenth Amendment made woman suffrage the law of the land—the only western states that had not already done so were New Mexico, Texas, Nebraska, and North Dakota. Naturally, the region also pioneered in females holding political office. Mary Howard of Kanab, Utah, became the first female mayor in the country, Bertha Landes of Seattle the first mayor of a large city, and Jeannette Rankin of Montana the first congresswoman. Miriam "Ma" Ferguson of Texas and Nellie Tayloe Ross of Wyoming became the first governors in 1925. Unlike Ma Ferguson, who fronted for her husband, Jim, Ross really did govern, and govern well.

The sociologist Edward A. Ross voiced the conventional wisdom when he explained, "In the inter-mountain states, where there are at least two suitors for every woman, the sex becomes an upper caste to which nothing will be denied from street-car seats to ballots and public offices." Scarcity was, indeed, a factor, but not the only one. Frontier egalitarianism and individualism affected women as well as men. Promoters often featured the vote to lure women and families westward, and established groups like the Mormons saw the female vote as a counterweight to untrusted immigrant ethnic groups, which tended to be heavily male. Thus, for a variety of reasons—some commendable and others less so—the West truly pioneered a basic political right for America's women.

Again, the conventional wisdom was partly right in linking woman suffrage to the companion progressive effort for the "prohibition" of alcoholic beverages. As the historian Norman Clark has written, "In those states where women could vote on such issues before 1919 (Wyoming, Colorado, Utah, Idaho, Washington, California, Kansas, Oregon, Arizona, Montana, Nevada, New York), all but two—California and New York—adopted by popular vote a state law prohibiting the saloon." In fact, women's organizations did attack the saloon, and various brewers' associations naturally opposed woman suffrage. But it is also true that some suffragists, such as Oregon's realistic Abigail Scott Duniway, viewed Prohibition as a quixotic diversion and a danger to the suffrage campaign.

Nevertheless, the West and South forced national Prohibition on the "wet" cities of the Northeast and upper Midwest. First by "local" or county option and then by statewide vote, one state after another went "dry," often with considerable turmoil, as in closely divided Texas. Among the earliest was Kansas, where Carrie Nation first captured national headlines in 1900 by wading into a Wichita saloon wielding an iron rod and throwing rocks at a painting of "Cleopatra at the Bath." The West soon provided an excellent case study of how an unpopular law cannot be enforced; those westerners who wanted alcohol seldom had trouble getting it.

Of all Progressive Era issues, it was "conservation"—federal protection, regulation, and preservation of natural resources—that most affected the West. On this issue, the region appeared somewhat self-serving and provincial, particularly in its sharp criticism of forest and other land set-asides that threatened to impede development. The Seattle Chamber of Commerce viewed the sequestrations as a "galling insult" to the region. To Colorado Senator Henry Teller, they seemed to create "a system that threatened to reduce western people to a class of servile peons"; to Utah Senator Joseph Rawlins, they were "as gross an outrage as was committed by William the Conqueror." The West strongly supported Interior Secretary Richard Ballinger in 1909–10 when he removed certain lands from the reserves and then came under sharp attack from Gifford Pinchot, head of the Forest Service, for doing so. National progressive sentiment forced Ballinger out of office. Hyperbole aside, such regional reactions were natural enough: westerners were losing their time-honored right of free access to western resources. When eastern conservationists like Pinchot—confident that they were eternally correct—dismissed such western critics as "locals" and "sagebrushers," western anger was understandable, if not endearing. The conflict arising between western property owners and developers on the one hand and conservationists and environmentalists on the other continued as a major political issue throughout the twentieth century.

However, in the area of federally funded reclamation projects, the arid West adamantly supported Roosevelt and his Newlands Act, which created the Bureau of Reclamation. Although numerous water-diversion projects existed throughout the West before massive federal involvement, the progressive engineers were the first to promise water as a consistent resource. From the construction of the Roosevelt Dam in Arizona (1911) to the building of the Jordanelle Dam in Utah (1992), the bureau has played a major part in western politics. Water as a scarce western natural resource became and remains a complicated political issue. Federal and state governments compete with agricultural irrigation districts, municipalities, and recreation interests for a limited resource. From the beginning, the progressive concept of using federal dollars to guarantee water for agrarian users has been viewed as a blessing. Once again the westerner found an internal conflict over progressive programs. On the one hand, most western property owners opposed regulation of the timber and mining industries; yet the same westerners welcomed government programs that created new sources of water.

The progressives left behind a positive, enduring heritage of direct democracy, hope, and reform. However, they also had their limitations, many of which belied their middle- and upper-class origins. For instance, California progressives pushed through a 1913 law prohibiting Japanese from owning land. Western progressives could also have done more to help the region's many workers. Their moderation in this regard left an open field for left-wing radicals during the years before World War I.

Some of the western radicals, like John Reed of Oregon and Bill Haywood of Utah, became Marxists (both are buried in the Kremlin), but most were authentic American leftists who eschewed the Marxian labor theory of value. These individuals came from the ranks of immigrant laborers and small-scale farmers and spoke the language of neopopulism. The most significant group was the Western Federation of Miners (WFM), founded in 1893. This miners' union engaged in a series of violent collisions with exploitive corporations, particularly in Idaho and Colorado. In 1905, members of the WFM led the creation of the Industrial Workers of the World (IWW), America's most famous radical union, which was always strongest in the West.

Overwhelmingly western, the IWW "Wobblies" aimed to meld all workers, regardless of skill, into "one big union." Their class-conscious appeals to exploited farm, forest, and mine workers and their "free speech" campaigns galvanized a harsh conservative reaction. So did their outspokenly anticapitalist leaders, such as Elizabeth Gurley Flynn and Haywood, who spent his final years in the Soviet Union. Joe Hill, a Swedish immigrant, was the poet laureate for the Wobblies, and his songs were sung throughout the West. Hill became a martyr when convicted of a murder-robbery in Utah, where he was executed in 1915.

As in the 1890s, radical support continued to flow heavily from the millions of family farmers who believed "the system" oppressed them. A prime instance was the Nonpartisan League, which exploded into power in North Dakota in 1915 and by the following year counted thirty thousand members there alone. Led by a shrewd ex-farmer named Arthur Townley, the league railed at Twin Cities railroad exploitation of farmers and called for a program of tax reform, state-subsidized insurance, and state-owned banks and grain elevators. As Townley noted, "If you put a lawyer, a banker, and an industrialist in a barrel and roll it down hill, there'll always be a son-of-a-bitch on top." Using nonpartisan tactics, the league gained control of the state GOP and swept the 1918 elections in North Dakota, enacting much of its program in the 1919 legislature. Meanwhile, to the dismay of conservatives and business interests, it swept into other grain states of the Northwest and into Canada.

Drawing sustenance from rural radicals like these, as well as from urban labor, western socialists gained power during the years before World War I. Exemplifying the surge of farm radicalism on the southern plains, Oklahoma was arguably the strongest socialist state in the nation. Although it had only 2 percent of the American population in 1910, it claimed a full 10 percent of national Socialist party membership. And in the elections of 1908 and 1916, it cast the highest vote percentages of all states for the

Socialist party presidential candidate. Socialists won a number of legislative and mayoralty races and in 1911 nearly captured the mayoralty of Los Angeles. A Nevada Socialist party candidate for the U.S. Senate gained over 25 percent of the vote in 1914.

The wide range of political attitudes espoused by westerners, along with their varying ethnic backgrounds, added to the region's confusion over whether to enter World War I. Generally speaking, support for intervention on the side of the Anglo-French "Allies" against the Germans was strongest in the eastern United States, more moderate in the Midwest, and weakest in the West. Seeing the war as an argument among European imperialists, many of the West's labor and agrarian radicals opposed intervention. Obviously, the West's contingents of Germans and Irish resented an English alliance; most Scandinavians, like their homelands, favored neutrality. In 1915 Nebraska's William Jennings Bryan, a friend of the German element at home, resigned as secretary of state rather than support President Wilson's truculent policy toward Germany. Bryan subsequently became a leader of the peace movement. As an example of the power of the hyphenated Americans, the entire three-member North Dakota delegation voted against the war when Congress made its formal declaration in the spring of 1917. On the other hand, representatives from the Southwest, concerned about German meddling in revolutionary Mexico, supported intervention.

Once the nation was at war, an intense atmosphere of patriotic fervor set in. In the West, much of the worst of the repression was aimed at leftist critics. Federally sponsored state "councils of defense" joined state governments and local vigilantes in denouncing anything remotely pro-German or less than "100 percent American." The Colorado progressive George Creel chaired the Committee on Public Information, which produced anti-German material. Governor Will Hobby vetoed the legislative appropriation for the German language department at the University of Texas. An editor in a town nearby opined that, although flogging might be all right for some who declined to volunteer for Red Cross work, others might have to be shot. In Nebraska, regents forced the resignations of two faculty members who were deemed too pro-German. A number of states, including Texas, passed "criminal syndicalism" and "sedition" laws that aimed at stifling dissent. Montana's Sedition Act of 1918 became the model for the federal law passed the same year, which is often considered the most sweeping abrogation of civil liberties in modern American history. Ironically, these attitudes contrasted with the fact that Utah's Simon Bamberger and Idaho's Moses Alexander, both German-born Jews, were elected governors of their states only a few months before U.S. intervention.

Leaders of the Industrial Workers of the World spoke out hotly against the war, but by the end of the war, their organization was a shambles. The public denounced them as "Imperial Wilhelm's Warriors"; and one of their most outspoken leaders, Frank Little, was murdered at Butte, Montana. At Bisbee, Arizona, an army of vigilantes seized hundreds of IWWs and other workers, illegally hauled them out of the city, and stranded them in the New Mexico desert. When the Wobblies called a mass lumber strike in the summer of 1917, the army went into the lumber business; the U.S. Justice Department joined private employers in forming a government union, the Loyal Legion of Loggers and Lumbermen. Federal agents rounded up IWW leaders with mass arrests that autumn, and many received long jail sentences.

Radical farmers faced a similar fate. When motley gatherings of sharecroppers in Oklahoma began protesting the draft by destroying property, in what became known as the "Green Corn Rebellion," they were severely suppressed. At a convention in Walla Walla, Washington, vigilantes ran five hundred members of the Grange movement, a farm-cooperative organization, out of town when they would not disclaim the Nonpartisan League. The league itself came under the severe indictment of superpatriots like ex-president Theodore Roosevelt. Eventually, in 1921, North Dakota Governor Lynn Frazier and Attorney General William Lemke became the first victims of a recall election of state officials in U.S. history as the league, like the IWW, reeled under attacks from the Right.

The national mood of reaction, nastiness, and intolerance, now intensified by fear of the Communist revolution in Russia, carried with full force into the immediate postwar years of 1919–22. In February 1919, a general strike shut down the city of Seattle for four days and deeply alienated public opinion. Miles Poindexter, Washington's formerly progressive senator, turned sharply to the Right and denounced the strikers as "reds." Kansas typified national resentment of militant labor in 1920 when its legislature declared strikes illegal. Severe race riots hit Longview, Texas, in 1919 and Tulsa in 1921; and a pitched battle between veterans and IWWs at Centralia, Washington, in 1919 left at least five dead.

More wary than the nation at large of joining the Wilsonian crusade for war, the West now seemed especially eager to isolate itself from modernity. Idaho actually repealed its 1909 direct primary law in 1919; and its prominent senator William Borah, like Hiram Johnson and other regional progressives, led the effort to defeat America's participation in Wilson's idealistic League of Nations. The U.S. involvement in the Wilsonian crusades of entry into the war, the League of Nations debate, and the oppression of dissenters damaged progressive goals and aspiration. In the West, politicians became more conservative and more concerned with domestic tranquility and foreign isolation. The region's leadership in reform measures was lost to an attitude of internal reclusiveness.

The Era of Transition, 1920–45

Just as the eras of populism and progressivism had given the West a liberal profile during the years before the war, so did the frustrations and xenophobia of postwar reaction turn the region sharply rightward during the early 1920s. However, in a broader perspective, the quarter century following 1920 witnessed a far more significant political transition, to a new and more stable order in which rising federal assertiveness became the major shaping factor. Obviously, the Great Depression and World War II altered the western political framework.

Much of the political negativism of the 1920s derived simply from reaction to postwar frenzy. It also stemmed from the wartime migration into the West of conservative southerners and midwesterners. For example, in the early twenties, Lincoln Steffens concluded that California "today is not a western, it is a middle western state." These newcomers, so different from their predecessors, made southern California dry on Prohibition and generally conservative, in contrast to wet and liberal San Francisco. Southern Californians flocked to radio evangelists like the Methodist minister "Fighting

Bob" Shuler and his sometimes scandalous female counterpart Aimee Semple McPherson. Nevertheless, of all the manifestations of postwar nativism, the most frightening was the reborn Ku Klux Klan (KKK). In contrast to its post–Civil War predecessor, the Klan of the twenties reached far beyond the Old South and not only lashed out at blacks but also violently attacked Catholics, Jews, and other "un-Americans." In one sense, it provided an extension of the immediate postwar superpatriotism.

The Klan's western manifestations mirrored national trends, and the politics of paranoia and hatred lasted until mid-decade. A few specific examples illustrate the Klan's impact on the West. In Texas, Oklahoma, and Kansas, the Klan rocked the established order. William Allen White, the grand old Kansas progressive, took on the KKK in a 1924 gubernatorial race—and lost. The hooded order mushroomed in Oklahoma, leading to the election in 1922 of anti-Klan Governor John Walton. Walton's anti-Klan zeal led to some incredible excesses, culminating in his putting the whole state under martial law. Consequently, the Klan-influenced legislature impeached Walton and removed him from office. Oklahoma's Klan days continued. In 1926, pro-KKK Henry Johnston was elected governor, only to be impeached three years later. Texas too was a cockpit of Klan activity and the homeland of its "Imperial Wizard," Hiram Evans of Dallas. By the early 1920s, the Texas Klan boasted eighty thousand members and a U.S. senator, Earle Mayfield. On the other hand, Miriam Amanda Ferguson's opposition to the KKK contributed to her election as governor in 1924.

The Klan reached into every corner of the Far West. In 1924 it attempted, unsuccessfully, to unseat Montana Senator Thomas Walsh, a devout Catholic. In Colorado, it claimed the election of a senator and a governor, as well as the mayor and chief of police in Denver, before it collapsed in an internal financial scandal in 1925. In the former progressive bastion of Oregon, an overwhelmingly Protestant state, Klansmen helped to elect Harvard-educated Kaspar K. Kubli as Speaker of the House. The state's governor, Walter Pierce, had Klan support, as did the law, enacted by initiative in 1922, that forced all children to attend public schools. The latter measure, intended to destroy Catholic schools, fell before federal district and U.S. Supreme Court rulings in 1924–25. Such were the heights and depths of Klan power before it declined in the mid-1920s, amid the shame of the national leader David Stephenson's disgrace (he was sentenced to life imprisonment for rape and second-degree murder) and the soothing balm of economic prosperity.

The West also mirrored another reality of 1920s politics—the collapse of the Democratic party under Woodrow Wilson's frustrated leadership and the dominance of conservative Republicanism. Every western state except Texas cast its electoral votes for Republican Warren Harding in 1920, all but Texas and Oklahoma voted for Calvin Coolidge in 1924, and then the entire region supported Herbert Hoover in 1928. In addition, numerous western conservative Republicans headed off to Washington, D.C., to participate in the probusiness GOP rule of the era. Utah, for instance, boasted not only of arch-conservative Senator Reed Smoot but also of equally conservative Supreme Court Justice George Sutherland and J. Reuben Clark, the influential undersecretary of state.

Beneath this surface of conservative calm, the old forces of progressivism continued to smolder. Reform governors like Pat Neff in Texas, William Sweet in Colorado, and

Joseph Dixon in Montana got elected in the 1920s, although conservative legislatures usually obstructed their programs. The West led in instituting state pension plans, beginning with Montana and Nevada in 1923, and in the unsuccessful attempt to ban child labor by constitutional amendment. Northwest progressives joined Nebraska's George Norris and pushed for public power. The region suffered from an agriculture depression throughout the decade. In Washington, D.C., angry representatives of the depressed interior West, sometimes dubbed the "Sons of the Wild Jackass" by the eastern press, joined in the "Farm Bloc" to advance the interests of agriculture in the face of solidly probusiness administrations. The McNary-Haugen Bill, which authorized the federal government to purchase and export surpluses, was vetoed twice by Calvin Coolidge.

The problem of the western progressives was that they could neither master a Republican party controlled by conservatives nor stomach a moribund Democratic party controlled by the wet machines of the East. Party irregularity again flared, as in New Mexico, where the maverick reformer Bronson Cutting temporarily switched from the GOP to the Democrats, or in Idaho, which for a time housed the nation's only active Progressive party. The height of irregularity came in the 1924 presidential campaign, which pitted conservative Republican incumbent Calvin Coolidge against John W. Davis—the nominee of a Democratic convention that had deadlocked between the forces of the East and those of the West and South over Prohibition and other issues. When an independent Progressive party ticket of Robert LaFollette of Wisconsin for president and Senator Burton K. Wheeler of Montana for vice president entered the fray, both the weakness of the Democrats and the disorganization of the Progressives became clear. In most states of the West, the Progressives actually outpolled the Democrats. In California, Davis took only 8 percent of the vote. All of this meant that, even in a region sharing only marginally in the vaunted GOP prosperity of the 1920s, the Republicans easily held sway into the 1930s.

By the later 1920s, political turmoil began to abate, and a classic mood of conservatism set in. Although some of the conservative governors of the time were rancorous, like Roland Hartley, the labor-baiting GOP lumberman from Washington, most simply advocated an efficient, low-taxation, and minimal-service government. The political cultures of the western states during the 1920s seemed to be evolving toward acceptance of the new economic stability. Unfortunately, these economic foundations were about to topple, taking with them as they fell the fleeting political calm.

The stock market crash of October 1929 had little initial effect on the West. Agriculture had been depressed throughout the 1920s, and since eastern factories tended to feel the downturn earlier than did the farther-removed sources of raw materials, the West felt the impact somewhat belatedly. However, when the full force of the Great Depression hit in 1930–32, it devastated the western economy. State and local governments, heavily reliant on property taxes, saw their revenue sources dry up with property devaluation. Most westerners demurred at taking on new obligations to help the needy and simply cut back what few services they offered. In effect, they shunted such responsibilities off to an equally hesitant federal government.

Some states did try new approaches to deal with the hard times. In Nebraska, advocates of a unicameral (one-house) legislature had been busy since 1915. Now, with

TRANSFORMATION:
The Emergence of the Modern West

◆

Contemporary artists depicting the West have developed their own sets of symbols and images to comment on the changed circumstances of modern life. While some have made ironic use of the iconic images of nineteenth-century painting, others have turned for inspiration to distinctly contemporary themes, such as the urban West, or have drawn ideas from the cultural traditions of their ancestors.

The English-born artist David Hockney created this armchair traveler's view of the West during a residency at the University of Colorado at Boulder. Combining generic mountains and cacti with a Northwest Coast Indian carving and highly stylized Indian figures, he plays on the viewer's general familiarity with time-worn western symbols to comment on the continuing influence of popular myth on visitors' perceptions of the West.

David Hockney (b. 1937). Rocky Mountains and Tired Indians. *Acrylic on canvas, 67 x 99½", 1965. © David Hockney.*

Arthur Amiotte draws from his Lakota heritage to construct critical commentaries on the white appropriation of Indian imagery. At the lower left is the artist's great-grandfather, Standing Bear, who toured with Buffalo Bill's Wild West show and performed before Queen Victoria. The queen's consort, Prince Albert, provides a link to the turn-of-the-century tobacco advertisement that serves as the centerpiece of this ironic statement on the commercialization of Native American symbols.

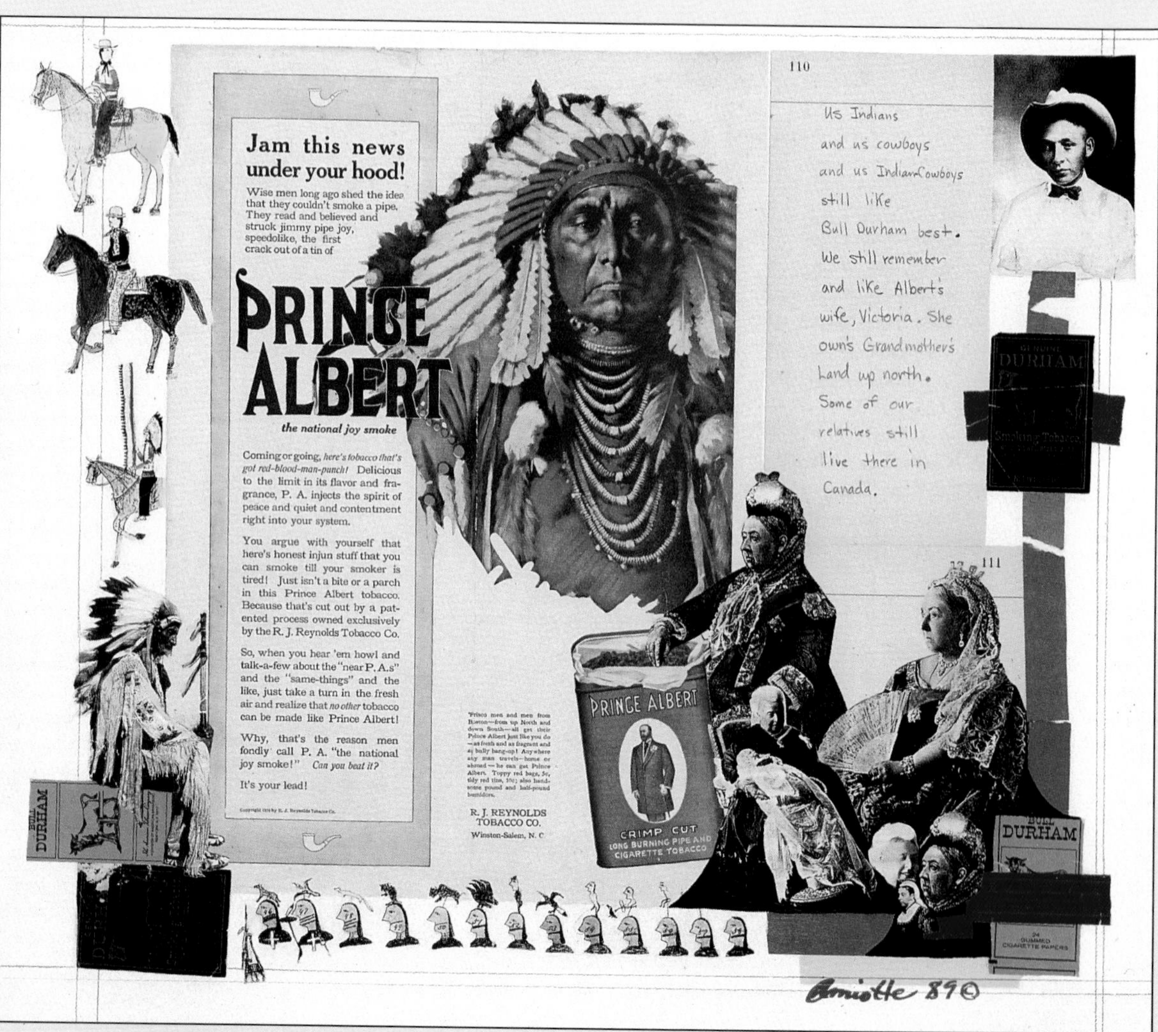

Arthur Amiotte (b. 1942). Prince Albert, 1989. *Collage and acrylic on canvas, 1989. The Ethel Morrison Van Derlip Fund, the Minneapolis Institute of Arts, Minnesota.*

Drawing on memories of her childhood in a Kingsville, Texas, barrio and the colorful distinctive style of Mexican folk painting, Carmen Lomas Garza creates narrative images that celebrate everyday life in the Chicano communities of southern Texas. Her pictures help create a visual history for a people whose presence was often ignored by earlier artists in the region.

Carmen Lomas Garza (b. 1948). Cakewalk. *Acrylic on canvas, 1987. © 1987 Carmen Lomas Garza. Photograph by M. Lee Fatheree. Collection of Paula Maciel-Benecke and Norbert Benecke.*

Born and raised in the West, Wayne Thiebaud first turned his attention to urban themes when he moved to San Francisco in the early 1970s. One of the few artists of either the 19th or 20th century to find western cities a compelling subject for his work, Thiebaud uses the unique topography of San Francisco as a vehicle for exploring ways to convey space and perspective in his painting. The results are images of a city that appears at once inviting and precarious as it perches at the edge of precipitous hills.

Wayne Thiebaud (b. 1920). Corner Apartments *(study). Oil on masonite, 1979. Private Collection.*

shrinking budgets, their arguments won favor, and the state adopted this still unique form of government in 1934. The Nevada legislature of 1931 took two controversial steps, both of which turned up new revenue sources while also raising the eyebrows of moralists. The state garnered a larger share of the growing divorce business by lowering the required legal residency period from three months to six weeks; and, in a provocative decision that paid off later, it once again legalized gambling after a twenty-year ban. Warned that legalized gambling might attract organized crime, the sheriff of Reno replied, "Al Capone is welcome in Reno as long as he behaves himself." In reality, most state governments drifted and awaited action from the federal level. The elections of 1930 sent many Democrats into local, state, and congressional office and set the stage for Franklin Roosevelt's 1932 national mandate.

The Roosevelt New Deal program of 1933–39 revolutionized the federal role in the West through a vast array of spending and regulatory innovations. Some of these programs, such as dam construction, the Civilian Conservation Corps, the Taylor Grazing Act, and the Agricultural Adjustment Administration, brought Uncle Sam directly into resource management. Other legislation, such as the omnibus Social Security program and the Federal Emergency Relief Administration, required dollar-matching and other efforts from the states. These latter programs required the states to find new revenue sources from income and sales taxes. In many different ways, therefore, the 1930s became a time of political dynamism at all levels of western government. The voters responded by electing a variety of mavericks as well as effective politicians.

The political upheaval and the rising public demand for drastic action brought a number of eccentrics and publicity seekers into candidacy for office. Kansas's John "Goat Glands" Brinkley, well known for his peculiar cures for infertility, twice ran for governor; and Washington's Vic Meyers, a former orchestra leader who once ran for the mayoralty of Seattle promising stewardesses on streetcars, became lieutenant governor. Oregon's reactionary governor Charles Martin genially advised police to "beat hell out of" labor leaders and called for the chloroforming of demented and elderly folks. Oklahoma's picturesque governor "Alfalfa Bill" Murray gained fame for wearing filthy socks and no shoes, for offering guests tea strained through his handkerchief, and for castigating the rich. The wealthy were also targets of W. Lee "Pappy" O'Daniel, a radio personality who captured the Texas governorship. O'Daniel fronted a band that included the founder of "country swing," Bob Wills. Glen Taylor, a former vaudeville actor and itinerant musician, won one of Idaho's senate seats. For all their haranguing rhetoric, though, politicians like Murray, O'Daniel, and Taylor proved loyal in practice to established economic interests.

Other depression-era radicals were more serious. Governor William Langer of North Dakota issued a general moratorium on hated farm foreclosures, then used the National Guard to halt sheriff's sales on foreclosed land. Langer also issued patently illegal embargoes on wheat and beef, in an attempt to raise prices. After the state supreme court removed him from office following conviction for forcing payments from government employees, he came back to win reelection to the governorship and later to a senate seat. The avowedly socialistic Washington Commonwealth Federation took over the Democratic party of that state for a time in 1936. And in California, the utopian novelist Upton Sinclair captured the Democratic nomination for governor in 1934 on a socialistic EPIC (End Poverty in California) program but lost the election when establishment Democrats opposed him. California, fast becoming a mecca for retirees, also produced the Townsend movement, an impractical scheme featuring monthly pensions for the elderly. Dr. Francis Townsend, a retired doctor, lost his life's savings when his bank failed, and his organization pressured Congress toward Social Security. Townsendism swept the country, but most of its twelve hundred locals were in the West. Sixteen of California's twenty congressmen supported the unsuccessful attempt to write Townsendism into federal law.

Behind all the sound and fury, much of it superficial, worked a profound revolution in federal-state relations throughout the 1930s, a revolution that dramatically affected the West. As James T. Patterson, a student of the New Deal, has commented, the New Deal was primarily a southern-western movement: its heaviest support came from these regions of low per capita income and sparse industry. Roosevelt always carried the Far West, and a significant reason for this stark shift in allegiance from Republicans to Democrats lay in the simple fact that the West benefited inordinately from the new federal dynamism and spending. Near-vacant Nevada ranked first among states in total per capita New Deal spending, at $1,130; Montana ranked second and Wyoming third. The presence of so much federal forest, park, mineral, and public land dictated that numerous reclamation, conservation, and construction projects flourished in the West.

Influential westerners in Washington, D.C., played major roles in forging this new federalism that, through investment and regulation, began the end of the old economic colonialism. Powerful lawmakers like Senators Joseph O'Mahoney of Wyoming, Burton Wheeler of Montana, and Key Pittman of Nevada proved far more able than had their predecessors in channeling new monies to home-state constituents, as they did with the Silver Purchase Act of 1934 and the Agricultural Adjustment Act of 1938. Westerners played key roles among Roosevelt's appointees. Jesse "Jesus" Jones of Houston funneled big investments westward as head of the Reconstruction Finance Corporation; and as governor of the Federal Reserve Board, the Utah banker Marriner Eccles, along with Jones, supported western allies like the California banking titan A. P. Giannini. William O. Douglas, a Washington environmentalist, became an appointee to the Supreme Court.

No western state exercised more federal power, nor better exemplified its results, than did Texas. The Lone Star State had not only Jones at the helm but also Vice President John Nance Garner; Senator Tom Connally, chairman of the Foreign Relations Committee; and a bloc-voting cadre of solidly Democratic congressmen that included no fewer than five house committee chairmen. The master legislator Sam Rayburn rose to become House majority leader in 1937 and then in 1940 became Speaker of the House, a post he held for the next two decades. In 1937 an astute young Texan named Lyndon Johnson won a special House election as a defender of the president and soon became known as FDR's "pet congressman." Allied with the soon-to-be-mammoth construction firm of Brown and Root, he carefully secured funding to complete that company's contract to build the Marshall Ford (later Mansfield) Dam on the Colorado River in Texas. In so doing, he launched the career that would one day make him the region's, and the nation's, most powerful politician.

The inflow of federal dollars and imposition of federal regulation also brought considerable upheaval to the western states. Though welcoming the money, some western elected officials bridled at federal intervention, lusted after and wallowed in the patronage of federal jobs, and railed against federal programs that stipulated the states "match" federal dollars. In Colorado, where a group of Communist-led unemployed actually stormed and occupied the capitol in 1935, anti–New Deal Democratic Governor "Big Ed" Johnson refused to raise matching dollars and openly fought with FDR's relief czar, Harry Hopkins. Other western governors, like Tom Berry in South Dakota and "Cowboy Ben" Ross in Idaho, behaved similarly—taking the money and manipulating the patronage even while yelping about it. When Ross reluctantly forced through a new sales tax to raise the matching funds, the voters turned against him. In the worst cases, as in Langer's North Dakota and Murray's Oklahoma, Hopkins finally took over state relief organizations because of irresponsible governors.

Throughout the United States, the reforming impulse and political popularity of the New Deal Democrats declined sharply in the late 1930s as a gradual economic recovery diminished the sense of urgency. This trend was especially marked in the individualistic West, where traditions of welfare liberalism had been lacking in the first place and where bitter factional struggles broke up Democratic unity in a number of states. Prominent Democratic senators like O'Mahoney of Wyoming and Wheeler of Montana led the effort to defeat the president's ill-fated attempt to "pack" the Supreme Court in 1937. And, typically, FDR failed in his attempt to intervene in the 1938 western primaries to "save" liberals like Congressman Maury Maverick in Texas or Senator James Pope in Idaho.

The rising chorus of protest against New Deal federal activism seemed reminiscent of territorial days. In truth, it smacked of considerable hypocrisy, since these federal investments—particularly those in dams and water systems, in agricultural subsidies, and in Social Security and other federal projects—proved to be enduring and underwrote a new prosperity and economic stability for the entire region. Even larger federal investments soon followed during World War II; but these investments, primarily in military bases and in huge aviation and shipbuilding industries, would be less controversial.

The West was richly rewarded both because of its loyalty to FDR and because of its geography. The arid, underdeveloped West proved perfect for numerous New Deal reclamation, conservation, and agricultural projects. Regardless of whether one's vantage was 1940 or 1945, the political West looked markedly different from that of 1930. Then, Washington had seemed remote, and what little heed one paid to government was usually to sleepy state capitals like Phoenix and Pierre. Now both the state capital and especially the national capital seemed near and vitally important. A new federalism arrived, a new world in which "government" would matter more than the baronial, eastern-based corporations of yesterday.

The bombs that fell on Pearl Harbor in December 1941 brought an abrupt silencing of the political acrimony of the New Deal years. Patriotism and its natural accompaniment, conservatism, now replaced the fervent reformism of the prior decade. Although many western politicians questioned Roosevelt programs like Lend-Lease and "Cash and Carry," every congressional member except Jeannette Rankin of Montana voted to support the war.

The conclusion of World War II left westerners anxious about continued federal support for the many defense-related projects that had fueled the regional economy. But the cold war ensured an ongoing flow of government money that made the success of regional politicians dependent on their ability to deliver lucrative contracts to their home districts.

Esther Bubley (b. 1921). V-E Day. Mayor Cecil Faris Proclaims V-E Day to Be An Official Holiday in Tomball, Texas. *Gelatin silver print, 1945. Standard Oil (New Jersey) Collection, University of Louisville Archives, Louisville, Kentucky.*

More to the point, though, the patriotic mood of wartime discouraged political debate. Even such a mass violation of civil rights as the incarceration of more than one hundred thousand Japanese Americans in "relocation centers" throughout the West provoked little protest. A surface harmony prevailed. In California, moderate Republican Earl Warren easily defeated the embattled liberal Democrat Culbert Olson in 1942 for governor. Taking advantage of wartime prosperity and California's unique system of "nonpartisan" cross-filing in both parties, Warren so soothed the electorate that in 1946 he won the nomination of both parties and coasted to a landslide reelection victory. Washington Governor Arthur Langlie successfully employed a similar bipartisan approach. Other state chief executives, like Coke Stephenson in Texas and John Moses in North Dakota, also proved able to cash in on war-born prosperity to stabilize taxes, accrue surpluses, and simultaneously enjoy political peace—the perfect world for most aspiring politicians.

Beneath the calm, however, powerful forces of change were working, driven by massive economic shifts. In Nebraska, venerable Senator George Norris fell to defeat in 1942 at the hands of reactionary Republican Kenneth Wherry, a death knell for the older progressivism. In Oklahoma, the oil magnate Robert Kerr took the governor's chair in 1943, symbolizing the new rich who were grasping political power. As he told the voters, "I'm just like you, only I struck oil." In Nevada, conservative Democratic Senator Pat McCarran routed the brother of his deceased foe, Key Pittman, in his 1944 Senate bid, an intraparty fracas that paved the way for postwar Republican victories. And in the 1944 decision of *Smith v. Allwright*, the U.S. Supreme Court ruled that blacks could no longer be excluded from Texas primary elections, a clear omen that the days of political racial discrimination were numbered.

Amid the rejoicing at the end of war in mid-1945, as westerners nervously looked ahead at what the future might hold, their main fear was that the end of massive wartime military spending might trigger a return to 1930s-style depression. They need not have worried, for quite the opposite happened. The emerging "cold war" soon fueled new rounds of defense spending, much of which flowed westward, particularly to California. Few western military bases were closed, and politicians fought to keep defense dollars moving west. These federal dollars, and many others flowing to such expenditures as reclamation, agriculture subsidies, maintenance of the public lands, and highways, harbors, and airports, would not only underwrite a new regional prosperity but also liberate the West from a dying colonialism. These dollars would, in fact, produce a new regional politics.

Politics of Pork and Protest, 1945–68

The system of federal spending that continued after the war created a new political reality. Because a politician's success depended largely on his ability to deliver federal projects into his home state (few women held national posts in the quarter century after World War II), members of both parties had to work together. But although there were incentives for bipartisan cooperation, there were also reasons for sharp attacks. First from the Right, then from the Left, the politics of the West were rocked by division.

It is easy to see why western politicians rushed to tap the federal coffers. The economic hardships of the past were still sharply remembered; the risks of overdevelopment seemed too remote to be considered. Politicians from both parties joined in and worked together.

This was especially true in matters related to defense. Among the powerful state senatorial teams who combined forces to acquire defense contracts for state-based companies and to expand military facilities within their state were the Democrats Henry Jackson and Warren Magnuson. Known as "the senators from Boeing," they represented Washington for nearly three decades. Representing Texas for a time, in contrast, were such polar opposites as Ralph Yarborough, a Democrat, and John Tower, a Republican. They battled for General Dynamics and for the state's many air bases. Meanwhile, all the legislators from California fought to obtain contracts for Lockheed and McDonnell-Douglas and to secure defense installations, especially along the Pacific coast. Even arch-conservatives like Barry Goldwater of Arizona, contemptuous of federal programs in many other areas, had no qualms pushing for federal dollars for military bases and defense.

The first major questioning of this arrangement occurred in 1961. It came from the out-going president and Kansas-born general who had led the Allied armies to victory in Europe during World War II. In his farewell address to the American people, Dwight D. Eisenhower spoke of "the total influence . . . felt in every city" of "an immense military establishment and a large arms industry." In time, politicians with closer ties to western constituencies would come to question aspects of what "the military-industrial complex"—to use Ike's phrase—was recommending. But in the early 1960s few heeded the former president's warning. Federal expenditures for defense meant unparalleled prosperity for the region.

The defense industry was not the only example of western pork-barrel politics. Federal research contracts for nuclear energy came to Washington, Idaho, and New Mexico. Western geography played a major role in this phase of federal contracts; yet politicians charted the fate of nuclear research based on economic prosperity for their respective states. Simultaneously, internal defense needs were utilized as a reason to create the gigantic interstate highway system. This huge undertaking was especially vital to the West, with its enormous acreage, for it provided up to 95 percent of federal funding for highway construction in the most rural states.

Another area of considerable pork-barrel activity focused on western reclamation projects. Ever since the creation of the Bureau of Reclamation, federally funded multipurpose dams were an accepted way of life in the West. Long before environmentalist organizations fought western projects, almost all politicians sought federal dollars for expansive water projects. The alliance among bureaucrats, politicians, and contractors such as Idaho's Morrison-Knudsen, Nebraska's Peter Kiewit, and the Utah Construction Company produced numerous dams.

Harry Truman originally sought Tennessee Valley Authority–type systems on the Missouri and Columbia rivers. Congress defeated those proposals as it came to rebel against federal government domination and regulation. Instead, numerous dams were constructed on the Missouri, Columbia, Snake, and Colorado river systems. The private power companies, municipalities, and irrigation districts combined with recreation and tourism interests to profit from federally funded projects. With such diverse backing through the 1960s, it is no wonder that western politicians continued to deliver the projects to their states.

While politicians worked together to funnel federal dollars westward, there were times when the political discourse among them attained a high level of vitriol. Initially the sharpest attacks came from the Right. Well before the name of Wisconsin Senator Joseph McCarthy became equated with the phenomenon, western politicians were ferreting out reds and smearing their opponents for being "soft on communism." Under the leadership of Martin Dies of Texas, the House Un-American Activities Committee (HUAC), since its inception in 1938, had been accusing the federal government of harboring radicals. By 1947 HUAC had gained a new member, a young veteran from California who had won his congressional seat through red-baiting tactics and who soon rose to prominence for his role in the Alger Hiss–Whittaker Chambers hearings. Thereafter, Richard Nixon ascended to the Senate in 1950 in part by branding his opponent, Helen Gahagan Douglas, as the "Pink Lady." To placate the McCarthy wing of the party in 1952, Republican kingmakers persuaded Dwight Eisenhower to include Nixon on the national ticket.

Other prominent westerners involved in the virulent anticommunism of the era included Senators Herman Welker ("Little Joe from Idaho"), Kenneth Wherry of Nebraska, and Karl Mundt of South Dakota. Although Republican politicians were most adamant in their charges, anticommunism was bipartisan. Democratic Senator Pat McCarran of Nevada, author of the Internal Security Act, and former Senator D. Worth Clark of Idaho also used McCarthyite techniques.

The search for subversives in the West amounted to the same blend of farce and tragedy as it did throughout the United States. A hunt for Communists among sixty thousand Texas schoolteachers turned up exactly one. In 1948, an investigation by the Washington State Legislature's Canwell Committee forced the dismissal of three University of Washington professors; and at the University of California's Berkeley campus, the Board of Regents fired thirty-two faculty members who refused to take loyalty oaths. After enormous controversy, the state supreme court mandated their reinstatement in 1952.

Perhaps the most celebrated aspect of western McCarthyism occurred when the House Un-American Activities Committee chose to investigate Hollywood's writers, directors, and actors. The California attorney general prepared a list of suspected Communist sympathizers, and the HUAC investigators focused on the "Hollywood Ten." Most of the ten who pleaded the fifth amendment were convicted of contempt of Congress, and all were "blacklisted" by the movie industry. The "fifth amendment" Communists became constant targets of McCarthy and his followers.

Some of the worst excesses of political McCarthyism occurred in the western states during the 1950s. In Idaho, Montana, and Utah, incumbents, usually Democrats, were attacked for being soft on communism. Glen Taylor in Idaho, who ran for vice president as a Progressive in 1948, was victimized by an attack called "The Red Record of Glen Taylor." Elbert Thomas of Utah suffered a media blitz of charges and vicious accusations of Communist sympathy. Some insiders speculated that the suicide of Senator Lester Hunt of Wyoming was precipitated by proposed personal smear tactics. Hunt's death in his Senate office sobered many western politicians. After the famous Army–McCarthy Hearings, the Senate leadership chose Utah's senior Republican senator, Arthur V. Watkins, to chair the special committee that recommended the censure of Senator McCarthy in 1954. Although McCarthy himself faded quickly from public view, the political impact remained for many years.

Indeed, a decade later, the Republicans chose as their presidential candidate a westerner whose intense anticommunism provoked parallels to McCarthy. Barry Goldwater, a department store magnate and senator from the State of Arizona, was hostile to Social Security and the federal role in civil rights. He wanted to abolish the graduated income tax, to sell the Tennessee Valley Authority, and to reduce federal functions at the rate of 10 percent a year. Moreover, he wanted an out-and-out victory over communism and said he was willing to risk war to achieve it.

With Goldwater trumpeting that "extremism in the defense of liberty is no vice," it was easy for the incumbent to present himself as a moderate. After twenty-three years in Congress—first in the House and then in the Senate, including five years as Senate majority leader—Lyndon Baines Johnson had joined the 1960 Democratic ticket as John F. Kennedy's running mate. With their campaign's success, Johnson had discovered that Kennedy and his advisers, eastern-establishment types, looked down on the lowly graduate of Southwest Texas State Teachers College. When Kennedy died on 22 November 1963, in Dallas, the once-disdained vice president became head of state.

Blending the nation's sense of mourning together with his considerable legislative talents, Johnson succeeded in gaining congressional approval for a large number of important measures. By the time he squared off against Goldwater in the 1964 race, the president had to his credit a civil rights bill, a tax cut, and an antipoverty program. Easily defeating Goldwater, he carried every western state except Arizona. Moreover, his party had won large majorities in the House and Senate.

Now president in his own right, Johnson continued to gain enactments of major legislation. In 1965 Congress approved federal aid to education, health care for the elderly, and voting rights for racial minorities. Still, if Johnson had looked to his home region in 1964 and 1965, he might have realized how fragile was the coalition that had elected him.

Two western senators, Wayne Morse of Oregon and Ernest Gruening of Alaska, alone among the members of Congress had voted against the Gulf of Tonkin Resolution in the summer of 1964. This resolution, passed under pretenses the Johnson administration had not bothered to verify, laid the groundwork both for the massive bombing of North Vietnam and for the buildup of U.S. troops in South Vietnam. The nation's actions overseas, as well as the grievances of African Americans at home, were among the issues student protesters wanted to highlight when the Free Speech movement erupted at the University of California at Berkeley in the fall of 1964. From that point through the early 1970s, the West would stand at the forefront of New Left activism.

The early civil rights movement was southern based, but the urban ghetto uprisings of 1964–65, notably the Los Angeles Watts riot of 1965, focused attention on the difficulties of urban discrimination. Thirty-four people died in that prolonged incident, and property worth millions of dollars went up in flames. The militant Black Panthers, created by Huey Newton and Bobby Seale, grew up in the East Bay area during the late 1960s. Claiming to be the godchildren of Malcolm X, the Panthers' violent rhetoric frightened mainstream America. The 1968 Olympic boycott, organized by the San Jose State sociologist Harry Edwards, had western origins. National attention also focused on Angela Davis, then a UCLA philosophy instructor and an announced Communist, when her pro–civil rights and antiwar sentiments, together with her sympathy for black prison inmates, led California Governor Ronald Reagan to attempt to dismiss her.

The more mainstream civil rights political activism centered on Cesar Chavez's National Farm Workers Association. Chavez brought national attention to the plight of Hispanic migrant workers with a series of strikes accompanied by national grape and lettuce boycotts. Chavez's movement spread throughout the Southwest, and although victories were short-lived and opposition was fierce, the national AFL-CIO gave him support.

Native Americans had their own reasons for activism. The 1950s had seen a federal effort to relocate Indians from reservations to cities to improve job opportunities. The national government also attempted to terminate its support of some Native American tribes. This policy of termination created an economic disaster for the targeted tribes such as the Klamaths in Oregon and the Menominees in Wisconsin. Before long termination itself was terminated, but during the unrest of the late sixties and early seventies, both urban and reservation Indians engaged in protests. Russell Means, Dennis Banks, and the American Indian Movement sought "Red Power" or Indian self-determination and economic self-reliance. Native activists seized headlines when they occupied, in protest, Alcatraz Island in 1969 and Wounded Knee in South Dakota in 1973.

The western civil rights movement reached a climax during the primary presidential campaign of Robert Kennedy in 1968. Kennedy campaigned among Chavez's farm workers in Oakland, among African Americans living in Los Angeles, and among the Navajos of the Southwest. His strategy of involvement, concern, care, and support paid

The American Indian Movement attracted national attention to contemporary federal Indian policy with its 71-day occupation of Wounded Knee, South Dakota, in 1973. A negotiated settlement ended the standoff with federal officials, but many of the underlying problems that precipitated the siege on the Pine Ridge Indian Reservation, including poverty and unemployment, remained unresolved.

Unidentified photographer. Armed Escort for Federal Official at Wounded Knee. *Gelatin silver print, 1973. UPI/ Bettmann, New York, New York.*

off with a major victory in California's June primary. However, his assassination that night led to despair and frustration among his followers.

In the end, the victor in the presidential election of 1968 proved to be the Californian who two decades earlier had smeared his opponents as "pinkos." Politically, the West had come full circle. Still, the region that helped elect Richard Nixon to the presidency differed from the place it had been before. No longer poor or sparsely settled, the West had entered the national mainstream. Increasingly, it would command respect in major party councils.

State Patterns of Partisan Politics

During the quarter century after 1945, the political culture of each state had been evolving in patterns at once similar yet different, both from each other and from the inherited characteristics of the postfrontier era. Four states of the interior West, North Dakota and Montana in the north and New Mexico and Arizona in the south, provide striking examples of political mainstreaming, of shedding peculiar, postfrontier cultures. The two former states represented the slow-growth political economy of the northern plains, the two latter the fast-changing Southwest.

The unique feature of North Dakota politics after the war was the old Nonpartisan League, still a major force even though postwar voters no longer favored its blend of isolationism and radicalism. In 1956, the minority Democratic party arranged a merger with the league, a move that ended its independence but also made for a genuine two-party rivalry in the state. In neighboring Montana, the corresponding anachronism was the arcane prevalence of the Anaconda Company, which finally sold its heavy-handed chain of newspapers in 1959 but did not collapse until 1983. The state had meanwhile evolved a classic, closely competitive political profile. Liberal forces, drawing on the support of both organized labor and the well-established neo-Populist Farmers' Union, managed to elect progressive congressmen, such as Senator Mike Mansfield, who were lauded for their ability to "deliver" federal pork. Conservative groups, more broadly based and relying on ranchers, large dryland farmers, and extractive corporations, usually dominated state government. Local folks called it "political schizophrenia," but this was simply a dramatic rendering of a trend typical of the interior West.

New Mexico's political culture was especially distinctive, since the Hispanic population not only dominated much of the state but also was assertive. Headed by Democratic Senator Dennis Chavez from the mid-1930s until his death in 1963, the "patron" system of rural Hispanic machines led to many abuses. Nevertheless, the rapid inflow of federal employees and defense workers during and after the war began eroding the power of this fiefdom structure. Although Democratic Senator Clinton Anderson served as the typically western deliverer of largesse, Republicans steadily rose toward parity in state government, exemplified by Edwin Mechem, who served as governor for much of the 1950s and early 1960s.

Of all the western states, Arizona witnessed the most abrupt, revolutionary political upheaval. Traditionally, Arizona embodied the classic southwestern politics: southern-style Democratic dominance, in league with the "3 C's" of copper, cotton, and cattle. Senator Carl Hayden, in Congress from 1912 until 1968, personified this tradition. But the arrival of new corporate giants like Motorola, with hordes of well-paid employees, changed everything after the war. In 1946 a business-agricultural alliance pushed through an antiunion right-to-work law; and with the triumph in 1952 of the conservative Barry Goldwater in a key Senate contest came a clear indication that Republicans were laying claim to the future.

Oklahoma and Texas, the two southern border states of the West, moved from one-party Democratic cultures more slowly. In each state, the ruling Democrats had a lesser liberal wing and a more dominant, oil-allied conservative majority. In Oklahoma the most prestigious of the liberals was the dignified and thoughtful Congressman Carl Albert, who eventually became House Speaker. The dominant conservative, serving first as governor and then as senator, was Robert Kerr, who rose craftily during the 1950s to become one of the strongmen of the Senate. Kerr used his political leverage to advance his Kerr-McGee oil concern and to channel a massive $1.2 billion into making the Arkansas River navigable all the way to Tulsa, one of the greatest pork-barrel forays in American history. By the 1960s, though, business-style Republicans were also winning statewide elections, as did Governors Henry Bellmon and Dewey Bartlett. In other ways, this southern-borderlands state changed only slowly. A large Baptist vote kept the fifty-two-year-old constitutional Prohibition law in effect until 1959.

Texas moved similarly. A vibrant, Populist strain of liberalism lived on here, best articulated by men like Senator Ralph Yarborough and Congressman Maury Maverick. But the Texas "establishment" rested firmly on the foundations of big oil, booming finance and manufacturing, and one-party southern Democracy. Conservative Governor Allan Shivers, whose "Shivercrats" preferred Eisenhower to liberal Democratic leaders like Truman and Adlai Stevenson, represented this class. So too did, in disguised form, Senator Lyndon Johnson, the new Texas strongman. Before his elevation to the Democratic national ticket in 1960, Johnson, along with his venerable mentor House Speaker Sam Rayburn, steered the national Democratic party to smooth cooperation with the Eisenhower administration, meanwhile gathering numerous benefits for Texas and Texas oil.

The silver-haired John Connally, a close associate of Johnson's, governed the state from 1963 to 1969. Connally epitomized the marriage between Texas wealth and the Democratic party, and during his years, new aerospace and other investments transformed the state. But here too, the old order was rapidly changing. By 1958 over one-third of eligible African Americans were registered to vote, and in 1966, federal courts downed the arcane Texas poll-tax law. The genuine power of San Antonio Congressman Henry B. Gonzales symbolized the awakening of Hispanic strength. And GOP Senator John Tower, who garnered the votes of many liberals angry at the Johnson-Connally "machine," signaled both the gradual rise of Republican strength and the aging of one-party Democracy.

Across the dry interior West, a fairly clear pattern emerged during these years of more or less close party competition, with an increasing slant toward conservatism and Republicanism as suburbia grew and the 1970s approached. For instance, in booming Nevada, where the gaming industry increasingly set the pace, the older Democratic organizations led by conservative Senators Key Pittman and Pat McCarran gave way in the 1960s to business Republicans led by Governor Paul Laxalt, even as reports of mafia and teamster influence over state politics raised increasing alarms.

In centrally located Colorado, with its representative mix of urban and rural constituencies, Democrats fretted at the erosion of their blue-collar and Hispanic base in Pueblo and the southeast, and Republicans fattened on the sprawling urban corridor running north and south from Denver. Typically conservative GOP senators like Gordon Allott and Peter Dominick railed at big-spending federal programs, even as Wayne Aspinall, the big-spending Democratic congressman and chairman of the House Interior Committee, worked with them to pour federal investments into Colorado and especially into Denver, America's second-largest center of federal employment.

To the north, thinly populated Wyoming had always been dominated by Republicans and stock growers. With only about twenty-six hundred members, the Wyoming Stock Growers' Association could sometimes lay claim to over half the members of the state senate. As in Colorado, the Democrats saw that their southern power base—along the Union Pacific rail line—was slowly shrinking. Similarly, in neighboring Idaho, Democrats watched the decline of their traditional power bases in the northern Panhandle mining and lumber regions and in the railroad towns, making it increasingly difficult to elect liberals like prominent Senator Frank Church. The booming cities of southern Idaho tended to vote for conservatives like Republican Governor Len Jordan and the exceptionally able Governor Robert Smylie.

In Utah, the political power of the Mormon church was truly awesome: over 90 percent of state officeholders were Latter-day Saints, whereas church members made up only 70 percent of the population; but the power was not as absolute or monolithic as some believed. As in other conservative western states, factional battles between regular Republicans, such as Senator Arthur Watkins, and tempestuous mavericks, such as Governor J. Bracken Lee, who had wooed many conservative Democrats, opened the way for a thoughtful Democrat like Frank Moss to capture a U.S. Senate seat. Farmer-labor Democrats could still win in liberal years or when Republicans became divided.

The strongest Republican heartland in all the West lay in the central Great Plains, where the party had long ago been implanted by homesteading settlers. In South Dakota, Republicans in the early seventies claimed nearly 90 percent of all political contests since statehood. Here, as in Utah, however, a liberal Democrat like the future presidential candidate George McGovern could win by campaigning against the Eisenhower farm policies and later move from the House to the Senate.

Nebraska had a strong Democratic base in Omaha and along the towns of the Union Pacific and Burlington railroads; but as in South Dakota, agriculture set the pace, and set it firmly toward Republicanism and conservatism. For years, the Senate team of Carl Curtis and Roman Hruska ruled as one of the most rightist in America. Nebraska indicated its essential conservatism by maintaining its state government until the mid-1960s with neither a sales nor an income tax. Kansas remained even more of a Republican bastion. Fires of populism and progressivism seem to have banked long ago with the departure of Senator Arthur Capper after the war, and Kansas did not get around to disposing of its antiquated temperance law until as recently as 1986. The only time since the war that the state has voted Democratic for president was in the Johnson landslide of 1964.

Whereas these eastern rim states of the region reflected the Republican conservatism of the adjoining Midwest, the Pacific Coast states were, as earlier, generally more progressive. They seemed more akin to the urban Northeast than to the aridland states that adjoined them. Historically these states, particularly northern California and western Washington, have been oriented toward maritime trade and organized labor, which tilted them leftward. Now, in all three states, sprawling suburbs have recast the political order to the cutting edge of American middle-class rule.

Washington offers a classic example of a state divided politically by geography. The populous Puget Sound area, characterized by burgeoning Seattle, aerospace and maritime industry, and union Democrats, usually outvoted the central and eastern reaches of the state, where Spokane anchored a more conservative political culture. Oregon resembled Washington in its west-east geographic split; but its sociopolitical profile was less industrial-labor and more middle class in nature. Its New England–style politics were typically so Republican yet moderate-progressive that it was sometimes called the "Vermont of the West."

As California evolved into "a nation within a nation," the surface calm of consensual, progressive-Republican governance that Governor Earl Warren had introduced in wartime continued long into the postwar era—through Warren's own nearly three terms, until he was appointed chief justice of the U.S. Supreme Court by Eisenhower, and then under his successor, Goodwyn Knight. By the 1950s, however, the middle road grew bumpy as both liberal Democrats and conservative Republicans moved away from it; and the end of the peculiar cross-filing system in 1959 further undermined the middle. The administration of Edmund "Pat" Brown (1959–67) marked a high-water point for liberal Democracy in California as the legislature enacted most of his reform proposals and a major new water-development plan.

But during the later 1960s, this imperial state spearheaded a major regional and national trend to the Right, with the triumph of arch-conservative actor George Murphy for the Senate in 1964, followed by that of another rightist actor, Ronald Reagan, over Brown for the governorship in 1966. As governor, Reagan won popular support by hacking at budgets, especially for education, and by defying radical students; but in 1967 he was forced to sign into law a record five-billion-dollar budget, with 20 percent of that being tax increases. Still, his show-business style and rightist rhetoric were fast turning him into a national political force.

The West since 1969: A Republican Stronghold?

Ronald Reagan would secure the presidency in 1980, but in 1968 another California Republican, Richard Nixon, squeaked into the White House. With only a fraction of a percentage point separating Nixon's share of the popular vote from that of the Democratic candidate Hubert Humphrey, Republican leaders wondered how they could secure a stronger popular mandate in the future. A book that appeared in 1969, *The Emerging Republican Majority*, provided the answer. The president should ignore the Northeast and Midwest and concentrate instead on winning the South and West. "From space-center Florida across the booming Texas plains to the Los Angeles-San Diego suburban corridor, the nation's fastest-growing areas" were, in the words of the book's author, Kevin P. Phillips, "strongly conservative and Republican."

Phillips's analysis was based on changing demographic patterns. In the West, rural areas were atrophying while urban complexes were quickly rising, particularly in the arc from Houston and Dallas, across the Southwest, to the boom cities of California. The new westerners were neither farmers nor blue-collar workers: they were well-paid middle-class employees, especially of new service, high-technology, and government organizations. They lived in sprawling, sunny suburbs; and if not ideologically rightist, they were inherently middle-of-the-road to conservative, disliking radicals and unions and challenging both with antilabor right-to-work laws. If the two parties had competed relatively evenly in the past, the Republicans were certain to gain the upper hand in the future—especially if the party consciously appealed to this, its natural constituency.

A look at general election trends of the 1970s and 1980s seems to suggest that Phillips was correct: the West tilted to the Right and toward the Republican party. In the Nixon landslide of 1972, the entire West—and nearly the entire country—voted Republican. More tellingly, in the closely fought campaign of 1976, which went to Democrat Jimmy Carter, the only western state to back the winner was Texas. In the two GOP landslides posted by Ronald Reagan in 1980 and 1984, the majority in every western state voted for the Republican presidential ticket. And finally, in the 1988 victory of Texas Republican George Bush, all of the western states except Washington and Oregon voted for the winner.

On closer analysis, however, a more complex reality emerges. The West had, in fact, voted Republican in *every* presidential race between 1952 and 1988 with the exception of the Johnson-Goldwater election. During that thirty-six-year span, the West had voted for the loser only twice, Nixon in 1960 and Gerald Ford in 1976. In other words, the regional voting pattern typically reflected the trend in the nation. Moreover, the Republicans had courted the region with a series of candidates who called the West home (Eisenhower in 1952 and 1956, Goldwater in 1964, Nixon in 1960, 1968, and 1972,

Reagan in 1980 and 1984, and Bush in 1988). Although a number of western Democrats had sought their party's nomination, such as Washington's Henry Jackson in 1972, Idaho's Frank Church in 1976, Arizona's Bruce Babbitt in 1988, and Colorado's Gary Hart in 1984 and 1988, only two western Democrats had succeeded in winning it—Lyndon Johnson in 1964 and George McGovern in 1972. Indeed, even though Bill Clinton was not a westerner, his campaign in 1992 showed that a Democrat who paid attention to the West could crack the Republican stronghold on the region. Clinton won 96 electoral votes from eight western states while Bush garnered 81 electoral votes from eleven states in the region.

Nevertheless, the best evidence that the Republican party's hold on the region was not inexorable came from the discrepancy between the Democrats' poor showing in presidential contests and their success in congressional, state, and local races. The truth was that, in most of the West during the second half of the twentieth century, the two-party system was thriving. Even as presidential candidates like Walter Mondale and Michael Dukakis failed miserably in the region, western Democrats as a whole did well. These Democrats succeeded because they knew how to compete for votes on their own turf. Only in a few locales like western Washington and the San Francisco Bay area could westerners win as unabashed liberals, espousing concern for the poor and needy, equality for minorities, women, and gays, protection of abortion rights, conservation of the environment, and opposition to incursions overseas. Nevertheless, Democrats could get elected in even the most conservative states if they learned to tack their sails, leaning to the Right on some issues like abortion and gays while bearing to the Left on others like the environment.

By making pragmatic choices, two highly skilled Democrats, Calvin Rampton and Scott Matheson, held down the governorship of Utah—arguably the nation's most conservative state—from the mid-1960s until the early 1980s. In Republican Arizona, moderate Democrats like Senator Dennis DeConcini, Governor Bruce Babbitt, and Congressman Morris Udall all demonstrated their popularity. Liberal Senator Alan Cranston, without benefit of any particular charisma, managed to buck the conservative trend in California. And buffeted by farm depression, conservative Nebraskans elected the liberal Democratic Vietnam Medal of Honor winner Robert Kerrey as governor, then senator, and in 1982 enacted the toughest curb on corporate farming in the country.

Yet western politics in the closing decades of the twentieth century were more than a catalogue of winners and losers. The region had to contend with a series of complex issues. On none were the lines more tightly drawn than between those who wanted to protect the West as wilderness and those who wanted to develop it for private business interests.

The modern environmentalist movement, rooted in the conservationism of the Progressive Era, rose to prominence due both to a concern for a deteriorating environment and to the assumption that American affluence could and should afford care for the land and its future. It centered on the West, and on Alaska, the location of most of the public lands.

In the eleven states of the Far West, federal ownership ranged from 29.6 percent in Montana to over 85 percent in Nevada. The managerial agencies that administered huge swaths of federally or publicly owned western land served as custodians of the people's inheritance. The Forest Service, Park Service, and Bureau of Land Management touched every aspect of western life. The governmental policies of fee collection, timber allotment, and development became more and more controversial. The West became the battleground for numerous discussions, debates, forums, and demonstrations. Westerners traditionally sought exploitation and development, but by the 1960s, militant and dedicated environmentalists demanded conservation, preservation, and limited use. Western developers and promoters felt that the federal bureaucrats and environmental allies had thwarted natural economic growth.

A clear beginning occurred in the mid-1950s, with the victory of the Sierra Club and its allies in prohibiting construction of Echo Park Dam in Utah's spectacular canyonlands. More victories followed in the 1960s, such as the creation of Utah's Canyonlands and California's Redwood national parks, even though Governor Ronald Reagan won notoriety for his comment that, having seen one redwood, he had seen them all. A series of major federal environmental enactments called both for effectively using federal lands and for setting aside pristine wilderness areas: the Wilderness Act, Wild and Scenic Rivers Act, National Recreation Area Act, the Federal Lands Policy and Management Act, and the Roadless Area Review and Evaluation Act.

Mainly in the Pacific Coast and Rocky Mountain states, westerners began in the late 1960s to manifest a new ecological concern. Oregon, with its 1971 "bottle bill" and many other measures, became a national leader. Colorado, an especially intense battleground between developers and environmentalists, caused a sensation by voting down the option to host the 1976 Olympic games. Other leaders were California, where a nasty oil spill polluted the Santa Barbara Channel in 1969, and Washington, where Governor John Spellman outraged promoters of the Northern Tier oil pipeline by refusing to allow the siting of its terminal at Port Angeles. In North Dakota, Montana, and Wyoming, ecological "resource councils" formed unlikely alliances in opposition to massive coal strip mining. Throughout the 1980s and into the 1990s, western environmentalists and wilderness advocates seemed steadily to expand their political strength and often to put developers on the defensive. The era of blatant logrolling for economic benefit had ended.

But the developers and users of federal property also mounted their own crusades, epitomized by the "Sagebrush Rebellion." Their main issues were increased regulation and grazing fees on the public land. Beginning in 1979 in Nevada, a state nearly 86 percent federally owned, the sagebrush rebels demanded the cession of federal lands to the states, which would either sell them or otherwise make them more easily accessible to stockmen and other users. Wyoming, which had already defied the federal bureaucrats by refusing to mandate a fifty-five-mile-per-hour speed limit, followed Nevada's lead, as did local politicians in New Mexico, Arizona, and Utah. Senator Orrin Hatch of Utah introduced a bill to transfer 544 million federally owned acres to the states. But other states, like Colorado and Montana, failed to support Hatch's extreme request, and critics like Governors Bruce Babbitt of Arizona and Cecil Andrus of Idaho pointed out the dubious legality, ethics, and benefits of the idea. The fact that most of the land was marginal, could not be sold, and would be expensive to administer forced the states to reconsider. In the end, the rebellion fizzled, even as the Republican presidential

With a straight-talking, straight-shooting reputation won from his career as an actor, Ronald Reagan parlayed his western persona into political capital to become governor of California in 1966. As president from 1981 to 1989, he continued to draw on his western image as a source of popular appeal.

Film still from Tennessee's Partner *(1955), directed by Allan Dwan. Archive Photos, New York, New York.*

candidate Reagan applauded it. Many of its advocates joined only to voice their frustrations, and in that they succeeded while inadvertently stirring up criticism of agencies like the Forest Service and the Bureau of Land Management for their cozy relationships with land users. Controversial environmental issues continue to pervade western politics. Conflict over the endangered spotted owl in Pacific Coast forests became an issue during the 1992 campaign; and coal-fired power plants as well as nuclear waste disposal sites became the subjects of lengthy legal, legislative, and judicial hassles among developers, federal administrators, and conservationists.

The growing complexity, even unpredictability, of western politics was likewise demonstrated in evolving relationships with Washington, D.C. On the one hand, westerners wanted to control the land and their destiny, but on the other, they still wanted federal dollars for special projects. The administration of President Jimmy Carter (1977–81) brought considerable contention between the executive branch and the West. Early in 1977, Carter demonstrated his lack of western political knowledge when he announced his intention to cut funding of eighteen large western water projects with a total price tag of just over five billion dollars. The outrage of western officialdom, from the Right to the Left, demonstrated the extent to which these giant water-reclamation projects had become the lodestone of regional politics, regardless of ideology or partisan ties. Carter beat a retreat, and in the end, all but four of the targeted projects were funded.

Carter once again ran into western frustration when, in 1979–80, he proposed to base the enormous MX (Missile Experimental) program in the deserts of Nevada and Utah. If built, the MX project would have been the greatest public program in history—a system of missile launchers on underground rail carriers covering up to forty-six thousand square miles and costing up to one hundred billion dollars. The outcry from the affected states, and especially from Utah Governor Scott Matheson, showed that even in these prodevelopment areas, which had long accepted nuclear testing, there were new limits to how much "development" was acceptable. In the autumn of 1981, the Reagan administration announced the scrapping of the MX plan and the decision instead to place the missiles in existing silos.

While westerners were beginning to question long-held assumptions, a profound leavening of western politics was occurring during the closing decades of the twentieth century. Women and minorities, hitherto largely excluded from the heart of the political process, now moved to center stage. Sandra Day O'Connor from Arizona became the first female Supreme Court justice when appointed by Ronald Reagan. The only female U.S. senator for years was Republican Nancy Landon Kassebaum of Kansas. The 1992 Senate election dramatically changed that figure as California Democrats Dianne Feinstein and Barbara Boxer, along with Patty Murray of Washington, won seats. Dixie Lee Ray, Kay Orr, Ann Richards, and Barbara Roberts had meanwhile claimed the governorships, respectively, of Washington, Nebraska, Texas, and Oregon. Numerous women succeeded in congressional, mayoralty, and state legislative elections. This trend continues as a proud reminder that western states offered the franchise to women before their eastern counterparts. True to its progressive past, the West bettered the national average in voting for the ill-fated Equal Rights Amendment as only four of its states—Oklahoma, Utah, Arizona, and Nevada—failed to vote for this assurance of political equality for women.

Native and Hispanic Americans were also making their presence known in western politics. Ben Nighthorse Campbell won a Colorado U.S. Senate seat in 1992, and Larry Echohawk became Idaho's attorney general in 1990. Manuel Luhan, a Republican from New Mexico, became the state's governor and then went to Washington as the secretary of interior. Similarly, Democrats Henry Cisneros and Federico Peña served as the mayors of San Antonio and Denver, becoming President Clinton's secretary of housing and urban development and secretary of transportation.

By the 1990s western politics were no longer the exclusive domain of a particular race, class, or gender. Nor could the Republican party claim to hold hegemony. Rocked by a long agricultural depression and by new problems in the once-booming urban states, the loyalties of western voters were clearly up for grabs. Indeed, as major party politicians seemed unable to break the deadlock between business and environmental interests or to wean the West from expensive federal subsidies, the region seemed poised to respond to new voices.

Such an appeal came in 1992 when a Texas billionaire, Ross Perot, mounted a third-party candidacy. As westerners had been doing since territorial days, they again demonstrated their weak party ties and their taste for mavericks. States such as Utah, Idaho, and Kansas gave Perot 29 percent, 28 percent, and 27 percent of their totals respectively. Even in Oregon, Washington, and California, he secured 25 percent, 24 percent, and 21 percent of the vote.

In the aftermath of the 1992 election, the Perot candidacy seemed to contain long-term implications. Whether he himself would again challenge the two-party system or whether another independent candidate would emerge, the West seemed certain to figure in the outcome. With the states from the plains westward accounting for one-third of the Electoral College vote, the West deserved the attention it was getting. Although no candidate could expect to sweep so diverse a region, neither could anyone afford to ignore it.

Bibliographic Note

Although there is no comprehensive, book-length study of politics and government in the modern West, two general histories give considerable attention to the subject: Gerald D. Nash, *The American West in the Twentieth Century: A Short History of an Urban Oasis* (Albuquerque, 1973); and Michael P. Malone and Richard W. Etulain, *The American West: A Twentieth-Century History* (Lincoln, 1989). An important general assessment is Paul Kleppner, "Politics without Parties: The Western States, 1900–1984," in G. D. Nash and R. W. Etulain, eds., *The Twentieth-Century West: Historical Interpretations* (Albuquerque, 1989), 295–338. A bibliographic overview is provided by F. Alan Coombs, "Twentieth-Century Politics," in M. P. Malone, ed., *Historians and the American West* (Lincoln, 1983), 300–322.

State and regional histories are especially helpful. Among those that pay heed to politics and government are Carl Abbott, Stephen J. Leonard, and David McComb, *Colorado: A History of the Centennial State*, rev. ed. (Boulder, 1982), Russell R. Elliott and William D. Rowley, *History of Nevada*, 2d ed. rev. (Lincoln, 1987), T. A. Larson, *History of Wyoming*, 2d ed. rev. (Lincoln, 1978), Michael P. Malone, Richard B. Roeder, and William L. Lang, *Montana: A History of Two Centuries*, rev. ed. (Seattle, 1991), Elwyn B. Robinson, *History of North Dakota* (Lincoln, 1966), Earl Pomeroy, *The Pacific Slope: A History of California, Oregon, Washington, Idaho, Utah, and Nevada* (New York, 1965), and Carlos A. Schwantes, *The Pacific Northwest: An Interpretive History* (Lincoln, 1989).

Also valuable are political studies of states and subregions, such as Robert E. Burton, *Democrats of Oregon: The Pattern of Minority Politics, 1900–1956* (Eugene, 1970), George N. Green, *The Establishment in Texas Politics: The Primitive Years, 1938–1957* (Westport, Conn., 1979), James R. Green, *Grass-Roots Socialism: Radical Movements in the Southwest, 1895–1943* (Baton Rouge, 1978), Jack E. Holmes, *Politics in New Mexico* (Albuquerque, 1967), Roger M. Olien, *From Token to Triumph: The Texas Republicans since 1920* (Dallas, 1982), Jackson K. Putnam, *Modern California Politics, 1917–1980* (San Francisco, 1980), Michael P. Rogin and John L. Shover, *Political Change in California: Critical Elections and Social Movements, 1890–1966* (Westport, Conn., 1970), James R. Scales and Danney Goble, *Oklahoma Politics: A History* (Norman, 1982), and Randy Stapilus, *Paradox Politics: People and Power in Idaho* (Boise, 1988).

Populism is one of the most debated subjects in American history. John D. Hicks's *The Populist Revolt: A History of the Farmers' Alliance and the People's Party* (Minneapolis, 1931) is the classic, appreciative study. The most influential revisionist work is Lawrence Goodwyn's *Democratic Promise: The Populist Moment in America* (New York, 1976). But perhaps the best perspective is offered in Walter T. K. Nugent, *The Tolerant Populists: Kansas Populism and Nativism* (Chicago, 1963), and in O. Gene Clanton, *Populism: The Humane Preference in America, 1890–1900* (Boston, 1991).

There is no general study of western progressivism, but see these state studies: George E. Mowry's classic *The California Progressives* (Berkeley, 1951); Robert W. Cherny, *Populism, Progressivism, and the Transformation of Nebraska Politics, 1885–1915* (Lincoln, 1981); Danney Goble, *Progressive Oklahoma: The Making of a New Kind of State* (Norman, 1980); and Lewis L. Gould, *Progressives and Prohibitionists: Texas Democrats in the Wilson Era* (Austin, 1973).

Prewar radicalism is discussed in Green, *Grass-roots Socialism*, Garin Burbank, *When Farmers Voted Red: The Gospel of Socialism in the Oklahoma Countryside, 1910–1924* (Westport, Conn., 1976), Robert L. Morlan, *Political Prairie Fire: The Nonpartisan League, 1915–1922* (Minneapolis, 1955), and Carlos A. Schwantes, *Radical Heritage: Labor, Socialism, and Reform in Washington and British Columbia, 1885–1917* (Seattle, 1979).

Historians have barely begun to study the West in the 1920s, but see Norman D. Brown, *Hood, Bonnet, and Little Brown Jug: Texas Politics, 1921–1928* (College Station, Tex., 1984), Charles C. Alexander, *The Ku Klux Klan in the Southwest* (Lexington, 1965), and Robert A. Goldberg, *Hooded Empire: The Ku Klux Klan in Colorado* (Urbana, 1981).

Richard Lowitt's pathbreaking *The New Deal and the West* (Bloomington, 1984) does not directly focus on politics, but consult the essays on western states in John Braeman, Robert H. Bremner, and David Brody, eds., *The New Deal*, vol. 2, *The State and Local Levels* (Columbus, 1975), as well as the following state studies: Robert E. Burke, *Olson's New Deal for California* (Berkeley, 1953); Michael P. Malone, *C. Ben Ross and the New Deal in Idaho* (Seattle, 1970); Francis W. Schruben, *Kansas in Turmoil, 1930–1936* (Columbia, 1969); and James F. Wickens, *Colorado in the Great Depression* (New York, 1979).

The political history of the West during World War II remains to be written, but Gerald D. Nash makes a start in *The American West Transformed: The Impact of the Second World War* (Bloomington, 1985). Also valuable is T. A. Larson, *Wyoming's War Years, 1941–1945* (Laramie, 1954).

Historians are only beginning to study the modern, post–World War II West, as Mary Ellen Glass has in *Nevada's Turbulent '50s: Decade of Political and Economic Change* (Reno, 1981). A number of contemporary regional assessments, however, are helpful: Thomas C. Donnelly, ed., *Rocky Mountain Politics* (Albuquerque, 1940); the western chapters in John Gunther, *Inside U.S.A.* (New York, 1947); Neil Morgan, *Westward Tilt: The American West Today* (New York, 1961); Frank H. Jonas, ed., *Western Politics* (Salt Lake City, 1961); and Frank H. Jonas, ed., *Politics in the American West* (Salt Lake City, 1969).

Neal R. Peirce is America's foremost political regionalist. See his *The Pacific States of America: People, Politics, and Power in the Five Pacific Basin States* (New York, 1972), *The Mountain States of America: People, Politics, and Power in the Eight Rocky Mountain States* (New York, 1972), and *The Great Plains States of America: People, Politics, and Power in the Nine Great Plains States* (New York, 1973). For contemporary assessments, consult Neal R. Peirce and Jerry Hagstrom's *The Book of America: Inside 50 States Today* (New York, 1983). Also valuable for contemporary insights are Peter Wiley and Robert Gottlieb, *Empires in the Sun: The Rise of the New American West* (New York, 1982), Richard D. Lamm and Michael McCarthy, *The Angry West: A Vulnerable Land and Its Future* (Boston, 1982), and the regular publications of the *Western Political Quarterly* and the *Almanac of American Politics*.

Finally, although space constraints allow only a brief sampling here, political biographies are also illuminating, for instance: Paolo E. Coletta, *William Jennings Bryan*, 3 vols. (Lincoln, 1964–69); Richard Lowitt's three volumes, *George W. Norris: The Making of a Progressive, 1861–1912* (Syracuse, 1963), *George W. Norris: The Persistence of a Progressive, 1913–1933* (Urbana, 1971), and *George W. Norris: The Triumph of a Progressive, 1933–1944* (Urbana, 1978); Robert A. Caro, *The Years of Lyndon Johnson*, 2 vols. (New York, 1981, 1990); Anne Hodges Morgan, *Robert S. Kerr: The Senate Years* (Norman, 1977); F. Ross Peterson, *Prophet without Honor: Glen H. Taylor and the Fight for American Liberalism* (Lexington, 1974); G. Edward White, *Earl Warren, A Public Life* (New York, 1982); Garry Wills, *Reagan's America: Innocents at Home* (Garden City, N.Y., 1987); and D. B. Hardeman and Donald C. Bacon, *Rayburn: A Biography* (Austin, 1987).

Chapter Fifteen

A Region of Cities

CAROL A. O'CONNOR

On 8 October 1957 the newspapers confirmed what the public suspected. The National League baseball team the Brooklyn Dodgers was moving to Los Angeles. Falling gate receipts at Ebbets Field, which suburban fans found inaccessible, together with the failure of squabbling borough chiefs to agree on a new site for the team, prompted the Dodgers to look elsewhere. Los Angeles, already home to the football Rams, beckoned with an audience of proven sports fans eager for a crack at the national pastime. The advent of nationwide jet passenger service enhanced the city's chances, and bold action on the part of city leaders clinched the deal. Pointing to land formerly slated for "communistic" public housing, they forced out impoverished squatters and offered the Dodgers a site for their stadium at the heart of the metropolitan network of freeways.

The Dodgers's decision, together with the news that the New York Giants were also moving west, endowed Los Angeles and San Francisco with the mantle of major-league status. In the opinion of many Americans, the cities of the West had finally arrived. To some extent such a perception was accurate. With the passing of each year after 1940, the impact of technology and the flow of government dollars furthered the integration of western cities into the national economy and culture. But to the extent that such a perception implied that cities were new to the West, it was wrong. Western history had had a significant urban dimension from the beginnings of nonnative settlement. What was more, in the nineteenth century and especially in the twentieth, western cities contributed to the shape of the American metropolis, helping to determine what modern-day living is all about.

Enmeshed in the network of freeways that helped define the Los Angeles landscape, Dodger Stadium presented a distinctive alternative to the older ball parks embedded in the densely settled neighborhoods of eastern cities. Its opening in 1962, along with that of San Francisco's Candlestick Park in 1960, signaled a new prominence for western cities in national life.

Unidentified photographer. Dodger Stadium. *Gelatin silver print, 1962. Bruce Henstell Collection.*

Eastern Models, Western Variations

"Show me a rich country and I'll soon give you a large town." By 1836 this phrase was being cited as a "pioneer proverb," and no wonder. The generation that had witnessed the rise of Cincinnati, Louisville, and Natchez—not to mention the boom then in progress in Chicago—considered the link between frontiers and cities self-evident. With every newly opened region in need of a center for trade, government, and culture, the wise investor could make a fortune. The trick was selecting a viable townsite and then working frantically to promote it.

The trans-Appalachian experience prepared the public for the competitive nature of frontier town building. In other ways, however, the rules of the game changed out West. Who would have anticipated, for example, that so many westerners would inhabit large cities? By 1890, the proportion of residents of the western states who lived in

From the start, an impulse toward town building marked American settlement of the West, where nearly every community had urban ambitions. A significant proportion of the region's population has lived in cities since the mid-19th century.

Unidentified photographer. Hogeland, Montana. *Photograph, ca. 1930. From the American Geographical Society Collection, University of Wisconsin-Milwaukee Library.*

towns of ten thousand or more was larger than that of any other U.S. region except the Northeast.

At first this information startled the statistician Adna Ferrin Weber. But by the time he published *The Growth of Cities in the Nineteenth Century* (1899), Weber had come up with a plausible explanation. In his view, the West owed its urban demographics partly to geography. The forbidding terrain and harsh climate that prevailed in vast sections of the region meant that very few people inhabited much of its acreage, whereas large numbers lived close together in its cities.

The timing of the settlement of the West also contributed to its skewered statistics. As Weber explained, a technological revolution of unparalleled magnitude was occurring—freeing the mass of human beings from the need to produce their own food but not from the need to work in order to purchase the means of subsistence. Most gainful employment now involved the production of goods and services for the purpose of exchange—activities that demanded a concentration of population. Thus, the existing stage of the world economy dictated a West that was relatively urban.

The significance of this shift was not lost on others. In *The Winning of the West* (1896), Theodore Roosevelt contrasted the Kentucky of the 1790s to the Colorado of the 1870s. When Kentucky became a state, not more than 1 percent of its population lived in its largest town and capital city, Lexington; when Colorado joined the Union, fully one-third of its people resided in Denver. "A hundred years ago there was practically no urban population at all in a new country." The future president said, with obvious pride, "Nowadays when new States are formed the urban population tends to grow in them as rapidly as in the old."

But if advances in technology helped to make the West urban, they also made possible a new kind of city. That potential would reach its fullest realization in the Los Angeles of the early-to-mid-twentieth century. Before that time, those involved in shaping western cities seemed to follow an ambivalent course. On the one hand, they wanted their cities to equal those of the East. On the other hand, they wanted their cities to be different and better.

Eastern models had a particularly strong influence on the development of Portland, Oregon. Named in 1844 for the leading city in Maine, it was located on the Willamette River, twenty miles south of the juncture with the Columbia. According to its founders, Portland stood at the head of navigation, as far inland as an oceangoing vessel could travel. The leaders of other would-be harbors disputed that assertion, but by the early 1850s the issue was moot. With a plank road across the western hills to the rich Tualatin Valley and with locally owned steamships running direct to San Francisco, Portland subdued its immediate rivals. Later it succeeded in attracting two transcontinental railroads and fended off, at least for a time, threats from Tacoma and Seattle to its regional supremacy.

Nevertheless, the leading town in the Northwest looked as if it belonged in the Northeast. With its downtown streets sixty feet wide and its blocks two hundred feet on a side, Portland was comparatively compact. Moreover, its architecture resembled that of eastern cities. Indeed, the eastern firm of McKim, Mead and White designed the Portland Public Library as a smaller-scale version of the library it had built for Boston.

Portland's similarity to eastern towns was discernible early. In the 1860s, as the city's population approached seven thousand, a visitor from Massachusetts wrote, "Portland has the air and fact of a prosperous, energetic town with a good deal of eastern leadership and tone to business and society and morals." Forty years later, another easterner affirmed that Portland was living up to its promise. Rather than "a new, crude Western town," Portland, whose population exceeded ninety thousand, was "a fine old city" with the proper "signs of conservatism and solid respectability." Even Portlanders compared their city to those in the East. The editor of an illustrated magazine, the *West Shore*, boasted, "Invoke the spirit of the lamp and transport a resident of some Eastern city and put him down in the streets of Portland, and he would observe little difference between his new surroundings and those he beheld but a moment before in his native city."

According to this type of thinking, what mattered in the Portland of the late nineteenth century were, as indicated by the *West Shore*, "the well-paved and graded streets, the lines of street railway, the mass of telegraph and telephone wires, the numerous electric lights and street lamps, the fire-plugs and water hydrants, the beautiful private residences surrounded by lawns and shade trees suggesting years of careful culture, the long lines of wharves and warehouses on the river front, and the innumerable other features common to every prosperous Eastern city and commercial port." It did not seem to matter that Portland lay within view of three spectacular snow-covered mountains or that an early town father had reserved as green space a long strip of blocks near the heart of the city.

Why was Portland judged so exclusively by eastern standards? As the historian David Hamer has pointed out, the future of urban places in the West depended in large measure on the good opinion of the East and of Europe. After all, thousands of new towns—not just in the United States—were competing for capital and settlers. Most easterners and Europeans judged the West by eastern standards; it was natural that westerners would do so as well.

Yet, even as they adhered to eastern patterns in some ways, the founders of a number of western cities deviated in other, equally telling ways. For example, nineteenth-century Americans must have assumed that God himself, instead of the government land office, prescribed straight streets to cross at right angles, for new townsites followed the gridiron with monotonous devotion. Still, the scale of the grid in the West usually differed from that in the East. Except for towns set in the sides of mountains and narrow gulches (like Virginia City, Nevada, and Deadwood, South Dakota) and in places laid out by the Spanish a century or two before (like San Antonio, Texas, and Santa Fe, New Mexico), western cities tended to have broad streets and large blocks and to spread out in all directions.

Consider the question of street widths. Early in the nineteenth century, the commissioners of the nation's most populous city had established generous standards to accommodate New York's rapid growth. North of 23rd Street, that city was to have 60-foot-wide crosstown streets and 100-foot-wide north-south avenues. Other eastern cities were laid out on smaller dimensions. In Charleston, South Carolina, the broadest streets (those bordering the town square) measured 60 feet across, whereas the standard street width in Cambridge, Massachusetts, was a mere 30 feet.

Out West, however, widths of 80 feet and more were considered the norm. Sacramento, California, platted in 1848, had streets that broad, as did Cheyenne, Wyoming, laid out twenty years later. The standard street in Omaha measured 100 feet across, whereas the city's two major avenues were 20 feet wider. The plan for Topeka, Kansas, called for more variation. Most of the east-west streets were 80 feet wide; most of the north-south streets were 100 feet wide; and the eight major avenues measured an impressive 130 feet from side to side.

In one respect, such broad streets did not make sense. Too expensive for a young municipality to pave, they became, in the words of a European visitor, George Augustus Sala, "a dusty desert" in the summer and "a Slough of Despond" in rain. Yet broad streets tended to typify big plans. The same individual who deplored "the monstrous breadth of the streets" approved of "the resolve characteristic of Americans to make their towns, even in the inception thereof, big things." Coining the phrase "Cities of the Future" to refer to the infant cities of the region, Sala observed, "Every town in the West is laid out on a plan as vast as though it were destined, at no distant date, to contain a million of inhabitants."

From the platting of Houston, Texas, in 1836 to that of Great Falls, Montana, a half century later, western town builders revealed their ambitions time and again. Yet if the nineteenth century had awarded prizes in urban design, one would surely have gone to the Latter-day Saints for Salt Lake City. In founding their Great Basin refuge in the summer of 1847, the Mormons looked to the ideas of their martyred prophet Joseph Smith. His plan for "a City of Zion" had familiar elements. In setting aside land at the center for church structures and land on the outskirts for pastures and farms, it incorporated aspects of the old New England town. Yet Smith's plan also involved some innovations—very broad streets, huge blocks, a required minimum setback for residential structures, and an unusual pattern of changing the direction of the lots on consecutive blocks.

As carried out under the leadership of Smith's successor, Brigham Young, the design

Barely twenty years old when this photograph was made for a local business directory, Topeka presents a prosperous front with its characteristically wide western streets and dense concentration of commercial establishments.

Leonard and Martin. Kansas Avenue, Topeka, Kansas. *Albumen silver print, 1876. Kansas State Historical Society, Topeka.*

of Salt Lake City followed the prophet's ideas in spirit, though not necessarily in specifics. The streets—132 feet wide with 20-foot sidewalks on each side—were said to be broad enough to allow wagons drawn by teams of oxen to turn around. A quarter of a century later, these same streets easily accommodated mass transit. Nevertheless, their breadth was probably based less on practicality than on a sense of the monumental. This was to be the temple city, the religious capital of the restored true faith. Set beneath majestic mountains on a fertile and well-watered plain, the site suited the city's purpose. Together they called for something special.

The dimensions of Salt Lake City's blocks complemented the breadth of its streets. Measuring 660 feet on a side, each block was divided into eight lots of one and one-quarter acres apiece. Strikingly, this religious group that so emphasized community designed its city in a way that maximized privacy. With one dwelling to a lot, a minimum setback of 20 feet, and, according to a Mormon pioneer, "no houses fronting each other on the opposite side of the street," a family could feel detached from its neighbors. More to the point, in a town that began in the middle of nowhere, an acre and a quarter enabled each family to put in a garden and orchard.

In subsequent years church fathers laid out additional tracts in accordance with the original plan. Modifications, nevertheless, set in. The pattern of alternating frontages, which proved impractical in commercial districts, was gradually jettisoned throughout the city; building lots were subdivided; and new neighborhoods, beginning as early as "The Avenues" section in 1855, were platted on a smaller scale.

Still, Salt Lake City retained many of its original characteristics. One writer, in 1855, dismissed it by saying, "Salt Lake City does not fulfill our ideas of a city, but is rather a gigantic village, or a collection of suburbs." Yet another, in 1886, described it as "one of the most beautiful of all modernly built cities." In the latter's view, what

As urban communities in the late 19th century attempted to incorporate mass-transit systems into existing city grids, western cities, with their wide streets, were well equipped to adapt. The streets of Salt Lake City, once valued because they were wide enough for a team of oxen to turn a wagon around, later provided commodious routes for the city's streetcar system.

Unidentified photographer. Near Main and Second, Salt Lake City. *Photograph, date unknown. Utah State Historical Society, Salt Lake City.*

Mormons, Gentiles, and visitors alike admired about Salt Lake City had less to do with its impressive buildings and more to do with its site and plan. "There is not yet a single narrow street, not a house wherein the sunlight cannot find its way, nor one whereby a stream does not run, and from which perpetual snow is not visible." From early in its history, Salt Lake's cityscape anticipated the low-density metropolis of the twentieth century. Salt Lake was truly a city—the religious, economic, and political center of a vast territory; but it resembled a suburb—its residents had land around their houses and space in which to move about.

The oldest Anglo-American city in the Rocky Mountain region and visually the most striking, Salt Lake was for a time also the most populous. By 1850 it claimed over 6,000 residents and by 1870 more than 12,000. Thereafter the city acquired railroad connections and a flourishing mining industry, but it did not grow as rapidly as it might have. The continued interference of federal officials in religious matters and of church leaders in economic and political matters contributed to social tensions until the 1890s. By that time Salt Lake City had 45,000 residents; but Denver, Salt Lake's rival to the east, had a population of 105,000, justifying its sobriquet as "Queen City of the Plains and the Rockies."

Founded eleven years after Salt Lake City, the Colorado town could not have been more different. Whereas unified action and utopian vision created the former, the latter grew almost entirely out of the forces of competitive capitalism. From the beginning, its history was chaotic. Within months of the discovery of traces of gold at the mouth of Cherry Creek, three companies had staked out townsites—St. Charles, Auraria, and Highland. Two of the three would not survive. Off scouting for buyers, the founders of St. Charles left too few guards to protect their claim. Soon the members of another company seized the land, replatted it, and named it after the Kansas territorial governor, James W. Denver, who presumably held jurisdiction. Seventeen months later, Denver City absorbed Auraria on the other side of Cherry Creek. Only Highland, which lay across the south fork of the Platte River, maintained its independence.

Despite the aggressiveness of its leaders and a mining bonanza in its hills, Denver's first decade proved shaky. Fire devastated the community in 1863; a disastrous flood followed. The Indian-white tensions that culminated in the massacre of Cheyennes and Arapahos at Sand Creek added to the Denverites' sense of instability. But the town's most fundamental problem resulted from the nature of the clientele it served.

Until 1870 Denver existed to outfit miners headed for the goldfields and to fulfill their needs and desires during slack times and weeks off. During much of the preceding decade, an estimated 100,000 to 150,000 people passed through the town each year. Unattached men for the most part, they brought to the community both money and mayhem. The saloons, gaming parlors, and bordellos they frequented created livelihoods for some Denver residents, but these institutions and the streets outside them were often the scenes of violent crimes.

Some individuals questioned whether such a town was fit to live in. When Lavina Porter arrived in 1860 with her husband and little boy, she found herself "utterly disgusted with Denver." In her view, its society consisted of "the roughest class of all states and nations." She noted, "Drunkenness and rioting . . . existed everywhere." The Porters pressed on to California. Meanwhile, William N. Byers elected to stay. A family man, the founder of the *Rocky Mountain News*, and himself the victim of a kidnapping and attempted murder, Byers agitated for the hiring of a police force and rented to the city, at a discount, a building to use as a jail. Those legal means of crime control were in place by 1862. Before then, local vigilance committees executed a swift if sporadic brand of justice. Wrote Byers in 1867, "That capital punishment had been the salvation of our western communities in their earlier periods, the history of the times proves beyond controversy."

Be that as it may, both the crimes and the retribution exacted for them earned Denver a reputation for violence. One widely read account, William Hepworth Dixon's *New America* (1866), described the town in apocalyptic terms: "As you wander about these hot and dirty streets, you seem to be walking in a city of demons. Every fifth house appears to be a bar, a whisky-shop, a lager-beer saloon; every tenth house appears to be either a brothel or a gaming house; very often both in one. In these horrible dens a man's life is of no more worth than a dog's." Dixon added: "The Vigilance Committee, a secret irresponsible board, . . . has to keep things going by means of the revolver and the rope. . . . Sometimes, when the store-keepers open their doors in Main Street, they find a corpse dangling on a branch."

Denverites responded indignantly to Dixon's portrayal. "You must remember," said one, "Denver only dates from '59; and all beginnings are a little rough." Residents feared that the book had "scared men from coming here." They failed to realize that those who tempted fate would be pleasantly surprised by what they encountered.

For in 1870, Denver entered a new phase of development. The completion of the Denver Pacific and Kansas Pacific railroads secured its position as the leading city on the eastern slope and enabled it to diversify its economy. After a decade in which vast throngs of people had passed through the town with no intention of staying, Denver was about to experience a boom. Its 1870 population of 4,759 people (only 10 more than in 1860) would grow more than seven times over by 1880 and triple again by 1890. At that time, it trailed only San Francisco and Omaha to rank as the West's third-largest city.

No wonder, then, that visitors to Denver were beginning to compare it to older, more established places. An Ohio editor in the late 1870s found "a general absence of frontier ruggedness" and "more of the repose of a settled city than . . . anticipated." A woman from England described it as "a busy place, the *entrepôt* and distributing-point for an immense district, with good shops, some factories, fair hotels, and the usual deformities and refinements of civilization." One easterner, writing in 1881, was especially flattering: "Except that the town is . . . not close knit, it might belong in Ohio, or even in New England. There are shops that would do credit to Broadway, and houses that would fit in our oldest towns."

Yet even while likening Denver to eastern towns, observers recognized a crucial difference. Denver was "not close knit" but was spread out and decentralized. With blocks that varied in size and with sections laid out on different axes, Denver lacked the symmetry of Salt Lake City. Nevertheless, when compared with urban places in the East, Denver, Salt Lake, and many other western cities seemed (like all of Montana) high, wide, and handsome.

Just as the region's geography helped to explain why so many westerners lived in cities, it also played a part in determining why those cities took the form they did. On the plains and between the mountains, the land was flat and, except along the Northwest coast, virtually treeless. Laying out streets with widths of eighty feet and more would have been an arduous task in Portland, Maine, or Portland, Oregon. In much of the West, however, the obstacles to such a scale of development were few.

Technology also affected the form that western cities took. Less than a decade after Denver's first tracts were platted, businessmen laid the groundwork for the first of several horse-drawn railways. By the late 1880s, cable cars were running on thirty-eight miles of city streets, and in 1891 the electric trolley debuted locally. Within two decades, the Denver Traction Company was averaging 240,000 riders a day on the two hundred miles of electrified track that crisscrossed the city.

Some western cities acquired a system of mass transportation earlier in their histories than others. Billings, Montana, for example, was barely one year old when its horsecars started running in 1883. Los Angeles, by way of contrast, had been around for ninety-four years before a street railway opened in 1875. Nevertheless, what mattered was not how old a city was when it acquired public transit but rather how dominant was its core and how much crowding had occurred there. In almost every case, western cities availed themselves of the technology of mass transportation before they entered the most important phases of their growth. As a result, distinct business, governmental, and industrial areas, as well as residential areas distinguished by class, emerged at dispersed locations. A low-density form of development became the norm; the suburbanization of the city had begun.

Thus, the cities of the late-nineteenth-century West proved to be "Cities of the Future" in ways that contemporaries might not have recognized. "It is curious to note," said a British visitor of Denver, "how its future vast proportions seem to exist already in the minds of its projectors. Instead of its new streets and buildings being huddled together as with us in our urban beginnings, they are placed here and there at suitable points, with a confidence that the connecting links will soon be established." What was true of Denver in 1885 was true of cities throughout the western United States. Moreover, the spread-out quality that this observer found so "curious" would increasingly affect cities in other regions and other nations.

The Metropolis on the Bay

San Francisco, the preeminent city of the nineteenth-century West, provided an exception to the regional pattern of low-density urban development. Because it grew early and almost instantaneously, because its raison d'être depended on its docks, and because it stood on a peninsula dotted with steep hills, San Francisco crowded a large population into limited confines. But although its physical growth took a form different from that of other western cities, economically San Francisco proved enormously influential. Writing in 1891, James Lord Bryce compared it to "a New York that has got no Boston on one side of it and no shrewd and orderly rural population on the other to keep it in order." San Francisco, in short, dwarfed the rest of the region.

How did the city come to assert such dominance? Location alone could not explain it, for San Francisco's site was a mixed blessing. On the one hand, the city stood at the entrance to a huge bay offering the best deep-water harbor on the west coast as well as access to the interior by way of the Sacramento and San Joaquin rivers. On the other hand, the location lacked timber, fresh water, and level ground and was wracked by unpredictable winds and fog. Surely some other spot along the bay was better suited to large-scale development.

Still, at the historical moment that mattered, San Francisco, with 800 residents, was the biggest settlement in the area. Moreover, it had recently been rechristened. As part of a ploy by its leaders to monopolize trade on the bay, the former Yerba Buena now bore the name of that famous body of water. Thus, in 1848, when the discovery at Sutter's Mill changed the course of western history, the town of San Francisco, some ninety miles distant, stood poised to profit. It was here that oceangoing clipper ships weighed anchor and unloaded goods and passengers bound for the goldfields. This break in the journey, necessitated by the technology of the day, made San Francisco an instant city. In July 1849 it had 5,000 residents, in 1856 approximately 50,000, in 1870 nearly 150,000, and in 1890 close to 300,000. By this time it ranked eighth in population among the cities in the nation.

As San Francisco grew, its economy broadened; but at first, much like any frontier outpost, it resembled a giant fulcrum. From its hinterland in northern California and Nevada, a chain of merchants collected raw materials—in this case the gold and silver gleaned from the region's streams and mountains—and exchanged them for goods unavailable locally. For a time, these goods included just about everything—from food,

A sprawling, decentralized city that grew rapidly after 1870, when newly completed rail links assured its commercial prosperity, Denver epitomized the low-density, spread-out form of most 19th-century western cities. San Francisco, in contrast, presented a more compact model of urban development that reflected the city's hilly, water-bound topography and its early commercial orientation toward a port.

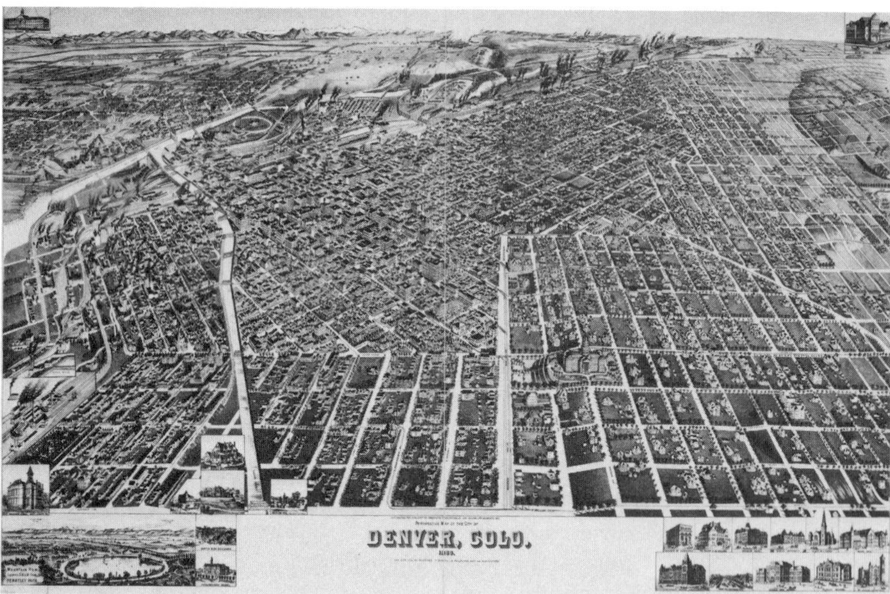

H. Wellge. Perspective Map of the City of Denver, Colorado. *Lithograph published by the American Publishing Co., 1889. Geography and Map Division, Library of Congress, Washington, D.C.*

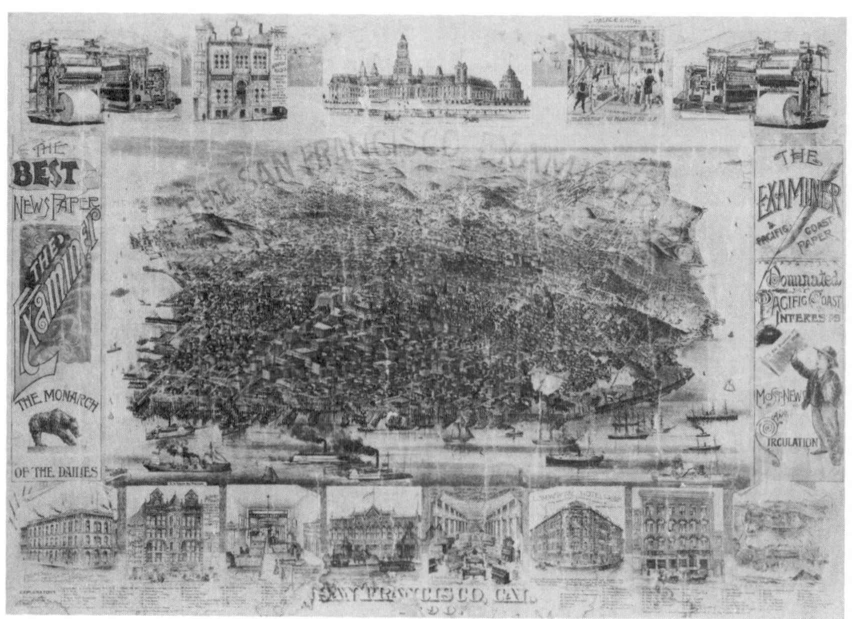

Unidentified artist. San Francisco, Cal. *Lithograph published by the* San Francisco Examiner, *1890. Arizona Historical Society, Tucson.*

clothing, and building materials to the equipment needed for mining. Soon, however, some erstwhile miners turned to farming as a surer way to make a living, and by the late 1860s, agricultural products had joined the list of the city's exports.

While agriculture grew in importance in the areas surrounding the city, manufacturing developed within it. By disrupting the flow of finished products from the eastern United States, the Civil War stimulated the demand for local production. By 1870 the city had almost six times more factories and workshops than it had had in 1860 and seven times more people employed in manufacturing. Although it dominated the Far West in both rail and water transport, San Francisco was no longer simply a hinge of commerce. Rather, its workers processed many of the goods—such as foodstuffs, shoes, and clothing—that its merchants distributed. Less than a quarter of a century had passed since the start of the gold rush, yet the number of residents employed in manufacturing exceeded the figure employed in trade and transportation. The economy was on its way to achieving breadth and balance.

San Francisco's rapid rise generated the expectation, especially strong in the 1850s, that those who moved there would prosper. Needless to say, some did. Among those who arrived in the city at mid-century were Mifflin W. Gibbs, an African American from Philadelphia, and August Helbing, a German-Jewish immigrant. While still in their twenties, both established successful business partnerships, Gibbs as a retailer of fine boots and shoes and Helbing as a wholesaler specializing in crockery. But far more common in the ranks of the city's upper and middle classes were native-born whites like Albert Dibblee. The son of a New York merchant and alderman, Dibblee used his commercial and credit connections in the East to become one of the wealthiest commission merchants in the West.

Yet for every male migrant who fulfilled his dreams in San Francisco during the 1850s, many did not. Two year-and-a-half-long booms, two far-longer busts, eight major fires, and a chronic shortage of capital meant that no one could be certain of anything. When S. B. Throckmorton's thriving business failed, he wrote a friend, "I thought . . . that I had at least brought myself to a condition that insured comfort and competency to my family; but I see now I was mistaken." He believed he had made no errors of judgment. "I simply overstaid my tide, and the waters fell and left me aground." In such unstable circumstances, it was not surprising that three-fourths of the men who held jobs in San Francisco in 1852 had left by 1860. That number included the shoe merchant Gibbs, who, denied the right to vote and testify in court, abandoned the city for British Columbia.

In subsequent decades the turnover rate of San Francisco's population began to decline. By 1880 about half the people who had held jobs there ten years earlier remained. With a population more stable than Omaha's but less stable than Boston's, the San Francisco of 1880 suggested a plausible hypothesis. Nineteenth-century Americans moved frequently, but the population of a new town was even more fluid than that of an established center.

While movement in and out of San Francisco in the late 1800s was less intense than at mid-century, movement up and down the social ladder was also less chaotic. As a result of the economic revolution that was occurring internationally, the proportion of higher-status jobs was expanding. In San Francisco specifically, the white-collar share of the work force grew almost 36 percent between 1870 and 1900, whereas at the other end of the scale, the percentage of jobs for unskilled laborers declined 50 percent. This change in the occupational structure, which was as true of Boston as it was of San Francisco, meant that many individuals moved up a notch on the social ladder and that their children rose even higher.

Still, native-born whites had greater opportunities for advancement. The city's many immigrants too often worked at semiskilled or unskilled jobs. This was surprising because San Francisco figured among the nation's most ethnic cities. Between 1850 and 1870, fully half of the population came from abroad. By 1900 that figure had fallen to 36 percent, but another 50 percent consisted of people with one or two immigrant parents.

Within the ranks of the foreign-born, some groups moved ahead more readily than others. The English and the Germans, especially German Jews, rose faster economically than the city's largest immigrant group, the Irish. Held back by a lack of skills and capital, the latter were disproportionately represented on the lower rungs of the social ladder. Nevertheless, Irish immigrants headed the Union Iron Works and the San Francisco Gas Company; others made fortunes in real estate and banking.

Many Irish sought opportunities in politics rather than commerce. In this area they rose early. Led by a second-generation Irish American who learned his politics in New York City, the Irish dominated San Francisco's government in the early 1850s. Inevitably the backlash set in. By mid-decade, David C. Broderick's machine stood accused of "stuffing the ballot-box" and "plunder[ing] the Treasury." Worse, the Irish, as a group, were characterized as "ignorant," "debased," and "dangerous." When a beneficiary of the machine killed an editor who opposed it, the city's elite made new use of a familiar form of frontier justice.

Five years earlier, San Francisco's first Vigilance Committee had hanged four men, whipped another, and banished fourteen. Like similar organizations in Denver, Colorado, and Virginia City, Montana, the committee of 1851 aimed to check a wave of violent crime. The Vigilance Committee of 1856, however, was largely political in nature. Its eight thousand members, drawn almost entirely from the upper and middle classes, focused on corruption more than violence. After executing four men, deporting twenty-five, and causing many others to flee for their lives, the ad hoc committee established an ongoing body that assured elite rule in San Francisco for a decade.

Although the Vigilance Committee targeted the Irish out of proportion to their numbers in the population, they did not disappear from local politics. San Francisco had an Irish-born mayor from 1867 to 1869, a second-generation Irish-American "boss" from 1882 to 1892, and a reforming mayor of wealthy Irish stock from 1897 to 1901. These men—Frank McCoppin, Christopher Buckley, and James D. Phelan—represented a small part of the range of Irish political activity in the city. Additionally the Irish helped to lead San Francisco's labor movement.

The members of another large immigrant group, the Chinese, exercised leadership only in the confines of their district. Inside Chinatown the wealthy merchants of the Chinese Consolidated Benevolent Association, also known as the Six Companies, acted as a quasi-government. Outside this most heavily populated section of the city, the Chinese wandered at their peril.

Chinese immigrants met with hostility in San Francisco almost as soon as they began to arrive in large numbers during the gold rush. The artist of this 1853 book illustration portrayed Chinese cultural traditions and the poor conditions of San Francisco's streets as both deserving of ridicule and derision.

Alonzo Delano (1806–74). San Francisco Scene. *Print, 1853. From Delano,* Pen Knife Sketches; or, Chips of the Old Block *(1853). Rare Book and Manuscript Division, The New York Public Library, Astor, Lenox and Tilden Foundations, New York, New York.*

The Chinese were segregated not only residentially but also occupationally. Elsewhere in the nation, women dominated many of the jobs held by this 90-percent-male group. In 1900 the Chinese (then only 4 percent of the city's population) composed 68 percent of its cigar makers, 52 percent of its laundry workers, 31 percent of its garment and textile workers, and 21 percent of its domestic servants and janitorial workers. Earlier many Chinese had fished, but Italians increasingly dominated this occupation. The Chinese had also peddled foods—usually from baskets carried on poles across their shoulders. Yet this method of hawking products became a misdemeanor in the early 1870s, and before long the city government had found other ways to harass the Chinese population.

The government passed anti-Chinese measures because, unlike European immigrants, Asians were not eligible for naturalization and because the nationality groups that could vote considered the Chinese the source of their troubles. Working-class whites accused them of driving down wages and taking away jobs from other workers. In 1877, a year of high unemployment and numerous strikes, the Irish-born head of the Workingmen's Party, Denis Kearney, tied hatred of the Chinese to hatred of the elite that employed them. Ending his speeches—"The Chinese must go!"—Kearney provoked outbursts of violence by blue-collar whites and more vigilantism from the propertied classes. Although many of its protests were forcibly quelled, the Workingmen's Party enjoyed some short-term successes. As for the Chinese, they began a long-term migration back across the Pacific.

During such times of crisis, the city's compact physical layout may have intensified its social divisions. The protesters of 1877 converged on the sandlots near city hall from their boardinghouses south of Market Street and their rowhouses in the Mission District. After hearing the likes of Kearney speak, they could wander up Nob Hill to agitate outside the mansions of San Francisco's leading industrialists or head over to Chinatown to threaten its residents with arson and violence. Though San Francisco was no longer "the walking city" it had been at mid-century, with rich and poor occupying the same district, it was still a place where a range of class-segregated districts lay within short distances of one another.

Why did San Francisco develop on such a compact pattern? The city did not lack the land: in 1856 it laid claim to the forty-seven square miles that still define its boundaries. And it did not turn its back on mass transit; on the contrary, few other cities in the American West made use of such a wide range of technologies. Horsecars, cable cars, trolleys, commuter trains, and ferries—all operated within or to and from late-nineteenth-century San Francisco. Nevertheless, legal complications resulting from the takeover of Mexican territory delayed the development of parts of the city for a couple of decades. And when that hurdle was cleared, the unusual costs of building mass transit in a hilly terrain—for example, the cost of tunnel construction—further postponed the opening of new districts. Only in 1912, when the city agreed to underwrite the expenses, did streetcar lines start to extend into the southern and westernmost parts of the city.

While the population spread slowly into the outlying districts of San Francisco proper, intensive development occurred along the bay. Those hills that could be leveled were. The resulting fill was then dumped in the water to create more of the flat land that merchants and industrialists desperately needed. As for the hills that could not be leveled, they underwent a transformation in 1873. With the invention of the cable car, the formerly spurned land atop Nob, Russian, and Telegraph hills became highly desirable for residential purposes. The result of this development of inlying locations was surprising: on average, San Franciscans lived closer to their workplaces in the 1870s than they had in the 1850s.

As San Francisco's population continued to grow, the crowding and congestion of its core worsened. By the turn of the century, many of its leading citizens believed that its streets and neighborhoods needed changes. They commissioned Daniel H. Burnham, the best-known American city planner of the day, to draw up a comprehensive urban design. Burnham reported in September 1905. Seven months later a major earthquake and three days of fire destroyed the densely populated core of the city. The opportunity existed to build San Francisco anew.

Yet instead of implementing Burnham's plan, the people of San Francisco reconstructed the city much as it had stood before. Citing the need to act as quickly as possible and at minimal cost, they refused to follow through with the planner's ideas for broader streets and more numerous parks. Yet more than practicality guided their efforts, which were based also on their own and others' affection for the city as it had been. A week after its destruction, Henry Adams, the writer, described San Francisco as "the most interesting city west of the Mississippi" and "more styl[ish] than any town in the east." He added, "I was fond of it and my generation made it."

In rebuilding the San Francisco they remembered, residents honored a city in which heterogeneous groups lived close together. Increasingly, the pattern of urban living that San Francisco represented would become anomalous in the West and in the nation.

The Rise of Los Angeles

During the second decade of the twentieth century, Los Angeles shot ahead of San Francisco to become the West's leading city. Its growth took many observers by surprise.

In their view, San Francisco looked the part of a major city; Los Angeles did not. Indeed, the small downtown, combined with the retail, industrial, and residential sprawl of Los Angeles, drew barbs. "Forty suburbs in search of a city" was one line used to describe it. "There is no there there" (initially coined by the poet Gertrude Stein to describe northern California's Oakland) was another.

Even those whose inquiries were more serious found Los Angeles a puzzle. In 1932 the writer Morris Markey concluded that the types of enterprise touted by the chamber of commerce—such as the branch factories of major industries and the high levels of fruit production—were not "the cause of a city." Rather, they were "the effect rising from an inexplicable accumulation of people." He found it "odd" that "here, alone of all the cities in America, there was no plausible answer to the question, 'Why did a town spring up here and why has it grown so big?'"

Los Angeles baffled men like Markey because it was the prototype of a new kind of city. Recent advances in technology had altered the human relationship to nature. No longer was it necessary for a city to have a natural harbor or its own supply of water. Now a settlement built on arid land in a less than ideal commercial location could become a verdant paradise and the hub of the Pacific. The sunshine, the mountains, and the ocean were what mattered; the rest could be engineered.

Also essential to the rise of Los Angeles was the existence of a class of people more interested in improving their quality of life than in maximizing their financial resources. In this regard the migration to Los Angeles differed from the various rushes for gold. Rather than consisting primarily of single young men eager to make a fortune, the area's residents included retirees and established families who already possessed a comfortable income. Coming largely from the Midwest, these people wished to duplicate, in a more beautiful place and a more benign climate, the small-town atmosphere that they knew.

None of this could have been foreseen in the first sixty-six years of the city's existence. From the time of its founding by the Spanish imperial government in 1781 until the surrender of its Mexican leaders to American forces in 1847, El Pueblo de Nuestra Señora la Reina de los Angeles existed to raise crops and livestock for the forts and missions of the region. Despite the breezes that blew in from the ocean some twenty miles distant, the land was essentially a desert. Only by tapping the waters of the river that flowed from the foothills onto a large plain could the settlers grow something more promising than sagebrush. Still, the pueblo became the center for a thriving cattle trade. On the brink of the American takeover, the population of Los Angeles numbered close to 1,500.

The arrival of the Anglo Americans changed life for the Californios. In 1850, Hispanics composed more than 75 percent of the city's population. Their numbers included rancheros, professionals, merchants, artisans, and skilled and unskilled laborers. By 1880, Hispanics made up less than 20 percent of the city's 11,200 residents. With the rancheros ruined in the 1860s by high property taxes, low cattle prices, and years of drought, Hispanics fell out of the ranks of the major landowners. Still, they maintained a toehold on the professional classes and increased their numbers in skilled laboring positions. Although four-fifths of those employed in 1880 worked as laborers, the same had been true in 1840. What was discouraging was that Mexican Americans were not advancing as quickly as changes in the region's economy might have allowed.

More distressing were the group's encounters with American justice. The state legislature not only restricted such traditional Hispanic entertainments as cockfights, bullfights, and horse races but also subjected unemployed Mexican Americans to fines and imprisonment as vagrants. Even when the laws were not discriminatory, their prosecution could be. Thus, an Anglo found guilty in the 1870s of murdering a Hispanic served only seventy days of a one-year sentence, whereas a Mexican American convicted of disorderly conduct received a ninety-day term. Hispanics (or for that matter Indians) accused of murdering Anglos could count on little mercy from the court. Most of the forty executions held between 1854 and 1870 involved such circumstances, as did the thirty-seven lynchings.

Finding themselves the victims of pervasive prejudice, many Hispanics decided to leave Los Angeles. Some went to Mexico; others stayed in the American Southwest. Meanwhile, new migrants from Mexico and from its former territory kept the Spanish-speaking population of Los Angeles at a few thousand in the late nineteenth century. Though small in numbers and fluid in composition, this segment of the city's population increasingly developed a sense of shared identity. Abetted by several Spanish-language newspapers (none of which had existed before the Anglo-American takeover) as well as by clubs and mutual-aid societies, the Hispanics celebrated their achievements and criticized their persecutors. Distinguishing themselves from the "Anglo-Sajones" or "norte-americanos," they took pride in their membership in "La Raza." As the editor of *El Democrata* noted, theirs was "a glorious race," "embodying the hopes of a race" and "possess[ing] all the secrets of all the mysteries."

As the city's population grew from 102,000 in 1900 to 1,238,000 in 1930, large numbers of blacks and Asians joined the whites and increasing numbers of Hispanics to complicate the racial makeup of the city. In 1930, when racial minorities composed less than 8 percent of the population in Chicago and less than 5 percent in New York, they accounted for more than 14 percent of the population in Los Angeles. At the same time, however, the number of foreign-born whites in the city was comparatively small. Thus, as the historian Robert Fogelson has argued, Los Angeles differed demographically from most eastern and midwestern cities. Whereas native-born Americans and European immigrants divided the latter, in Los Angeles a heavy majority of native-born whites faced a sizable minority of people of color.

The native-born population in the city was remarkable as well. Relatively few Angelenos had begun their lives in California. More came from the Midwest than the Far West; about as many migrated from the South as from the Northeast. These people had made a deliberate decision to move to Los Angeles. Having done so, they considered the city "the choicest part of the earth," and they were resolved, in the words of an early observer, that no one would "have it in his power to point out wherein" their city was wanting. From such basic emotions stemmed the legendary boosterism of the people of Los Angeles.

The city's first generation of promoters made transportation their chief concern. By 1869 the city had a rail link to the coast at Wilmington–San Pedro. In 1876 it acquired ties to northern California and the nation through the Southern Pacific Railroad, and in 1886 the completion of the Santa Fe gave Los Angeles a second cross-country

The image of Los Angeles as a tropical paradise, an image ardently prized and promoted by early civic boosters, continues to influence downtown urban development in the nation's second-largest city.

Douglas Muir (b. 1940). Landscaping at the Sheraton Grande Hotel, Los Angeles, California. *Dye coupler print, 1983. Courtesy of the photographer.*

connection. Only a short time earlier, most forecasters had predicted that San Diego's deep-water harbor would make that city the natural hub of the region. They had not counted on the determination of the citizens of Los Angeles, who taxed themselves to subsidize the railroads, or on the greed of rail officials, who deemed San Diego the greater threat to the value of their San Francisco holdings.

Ironically, then, the inadequacies of the harbor at San Pedro helped Los Angeles secure the Southern Pacific. Yet its citizens did not want their city to remain a tributary of San Francisco. Agreed that the smooth contours of its coastline should not determine the city's destiny, they decided to build an artificial harbor. The project almost collapsed when powerful interests favored Santa Monica as the site, yet the scheme's promoters persevered. In 1896 Congress set aside the first of several multimillion-dollar appropriations for the construction of a major harbor at Wilmington–San Pedro. By 1912 a rock-filled breakwater jutted two miles seaward. By 1932 it extended more than four miles into the ocean and by 1980 more than eight. Meanwhile, the city and private interests devoted comparable amounts of money to dockside improvements. By the mid-1920s, the harbor at Los Angeles was handling more tons of freight than the San Francisco harbor. Despite unpromising beginnings, Los Angeles now boasted the leading port on the west coast of the Americas.

Los Angeles acquired not only an artificial harbor but also an artificial landscape. Neither the royal palm trees that became a symbol of the city nor the citrus groves that bespoke its abundance were native to the region. These, along with the flowers whose January blossoms inspired Pasadena's rose parade, required something the area was short of—water. At first, local residents relied on the Los Angeles River as well as underground aquifers. Then they turned to distant sources. In 1907 Angelenos voted to bring in water from the Owens Valley, 233 miles away. In 1930 they looked to the waters of Mono Lake, another 80 miles to the northeast. By the time World War II broke out, they were tapping the Colorado River, about 250 miles to the east. In the 1970s they began acquiring water from the Feather River near Sacramento, and by the 1990s they had turned their sights to Utah. The sustenance of the city's artificial landscape required the ruin of farming districts elsewhere.

As destructive as these aggressive water policies were to some distant communities, they proved so attractive to neighbors of Los Angeles that many voted to join the city. Between 1915 and 1930, Los Angeles (already swollen by its annexation of Wilmington, San Pedro, and the narrow Shoestring District that connected the harbor to downtown) expanded more than four times over, from 108 to 440 square miles. In terms of area, it was the largest city in the United States in 1930; in terms of population, it was the fifth largest, behind New York, Chicago, Philadelphia, and Detroit.

Yet big as it was, Los Angeles still had a small-town atmosphere. It had this ambiance because that was what the people who were moving there wanted and because the technology existed to make their dream come true. After all, the first great migration to Los Angeles occurred in the late 1880s just as a new form of mass transit was becoming available. In 1887, the very year Julian Sprague built an electric streetcar system for Richmond, Virginia, one was running in Los Angeles. Within three decades, the southern California metropolis had the most extensive system in the world—with nearly four hundred miles of light track in operation in the city and more than eleven hundred miles of track in the area surrounding it.

Here, as in cities across the United States, businessmen financed the building of streetcar lines with the profits they made subdividing adjacent land. But the developers of greater Los Angeles broke from eastern models in one important aspect. Instead of laying out building lots all along the route so that only those at the end of the line enjoyed the beauties of nature, Henry Huntington and others pushed their tracks into the empty countryside—skirting hills, crossing fields, and placing many of their subdivisions in attractive natural settings.

The goal of developers and residents alike was to build a city of single-family, detached houses. They wanted to create what the historian Robert Fishman has called "a suburban metropolis." The squat, spread-out character of Los Angeles was therefore no accident but was the product of a conscious preference on the part of its inhabitants. In 1907 one of them boasted, "Here the tendency is to open and not crowded quarters. . . . Here even the pauper lives in surroundings fit for a king."

Not long after the electric streetcar made possible the spatial expansion of the city, yet another innovation in transportation intensified the process. Angelenos may have enjoyed one of the best light-rail systems in the world, yet they preferred the freedom of movement provided by the automobile. More than the residents of other U.S. cities, they could afford to indulge their preference. There were eight residents per auto in Los Angeles in 1915, four per auto in 1920, two in 1925, and one and one-half in 1930.

If the Los Angeles of the streetcar era seemed decentralized and spread out, it was a model of controlled growth compared with what followed. The automobile opened up for development the land that lay between the streetcars' radial routes. Fields that had formerly produced fruits, grains, or vegetables turned into factories, stores, or block after

By 1888, more than 44 miles of cable and electric railways in Los Angeles facilitated the city's sprawl outward from a small downtown. Despite the concurrent growth of the city and a rapidly expanding mass-transit system, Los Angeles would soon become a metropolis devoted to the automobile.

Unidentified photographer. Second Street Cable Railway, Looking West. *Photograph, 1888. Department of Special Collections, University Research Library, University of California, Los Angeles.*

block of low-density housing. Moreover, as long as there were streetcars, the city had a hub, a center, where the radial lines converged. Los Angeles, without streetcars, was more free-form and sprawling.

The new shape of the metropolis affected the everyday lives of its inhabitants. As an article in *Westways* magazine in 1937 made clear, Angelenos did not use their automobiles for commuting in the traditional sense of trips to and from the downtown business district. Instead, individual family members set out in a variety of directions, throughout greater Los Angeles, to do a variety of things—to work, learn, shop, and play. By the end of the day, they could easily have covered hundreds of miles of roadway as they crossed each other's paths and those of "countless of thousands of [other people] bent on similar missions." Under such circumstances, the true center of the city was "not in some downtown business district but," to quote Fishman, "in each residential unit." "Each family was its own 'core' in a decentralized city."

Before World War II and the cold war turned Los Angeles into a center for aerospace production and research, at a time when the city's dream makers created the weekend's movies but not the weeknight's television shows, back when the oil rigs of greater Los Angeles produced so much crude it was sold overseas, the city provided a preview of urban living in the future. In the Los Angeles of the 1930s and soon in cities across the United States, the automobile was the vehicle to the kind of life-style people wanted. By allowing individuals to travel on their own schedules and by their own routes, either alone or in company of their own choosing, the automobile provided maximum access to an entire metropolis's opportunities for employment and leisure and minimum exposure to its diverse population groups.

Experimental Cityscapes

Before 1940, western cities were important to the people living in the West. After 1940 they achieved national stature. That stature could be measured by statistics of population

Although the Census Bureau's definition of a metropolitan area remained essentially the same (a core city of at least 50,000 inhabitants plus its contiguous suburbs), the number of metropolitan areas in the West nearly tripled between 1940 and 1990.

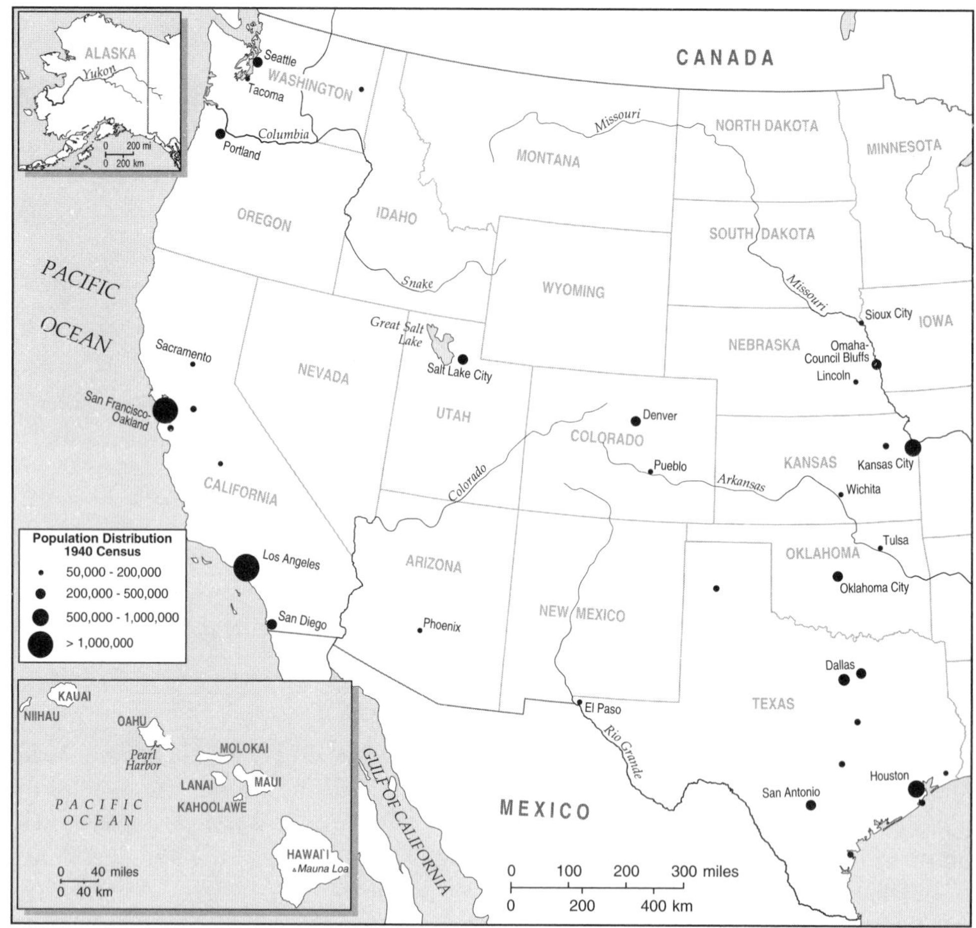

Metropolitan Areas in the West, 1940

growth and economic expansion. But it was also indicated by the frequency with which the rest of the nation took up western innovations in urban living and design. Thus the rise of the urban West did not spring entirely from decisions made in congressional cloakrooms or New York City boardrooms. Rather, it resulted from westerners' own success in creating built environments that appealed to their fellow Americans.

In 1900 only San Francisco, of all the cities in the West, ranked among the nation's top ten; in 1940 only Los Angeles could claim that distinction. By 1960 two of the nation's most populous cities stood west of the states that adjoined the Mississippi. By 1990 Los Angeles, Houston, San Diego, Dallas, Phoenix, and San Antonio all ranked among the nation's ten largest cities; of the nation's fifty largest cities, the West accounted for twenty-three.

Western cities grew in population partly because they grew in size. Between 1950 and 1990, San Antonio added 264 square miles to its municipal boundaries, Houston 380, Phoenix 402, and Oklahoma City 557. These cities annexed peripheral areas to prevent the unwieldy crosshatch of governments that typified so many older metropolises. Because they were expanding into newly developed areas whose residents had much

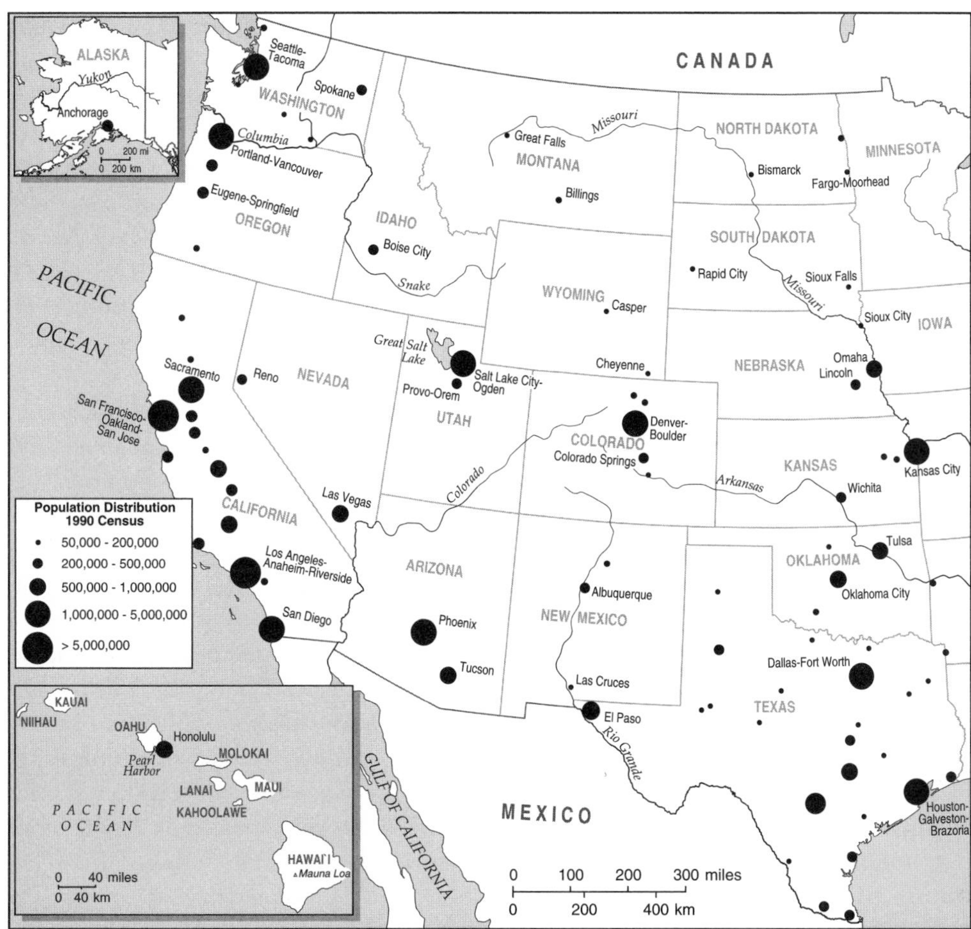

Metropolitan Areas in the West, 1990

in common with those of the city proper, annexation worked in most western cities. The notable exception was San Francisco, which had not added to its territory since the 1850s.

But if San Francisco resembled Boston, Pittsburgh, and St. Louis in the size of its municipal boundaries and in the fact that, for much of the post–World War II era, it was losing population from within those boundaries, the metropolitan area that San Francisco belonged to boomed along with much of the West. Indeed, more significant than the presence of western cities among the nation's top ten was the ranking of the West's metropolitan areas. In 1990 greater Los Angeles, San Francisco–Oakland–San Jose, Dallas–Fort Worth, and greater Houston ranked second, fourth, ninth, and tenth in the nation in population. Economically, as measured by the number of corporate headquarters and the assets of their commercial banks, these four metropolitan areas figured in the nation's top seven.

For each of these metropolitan areas, as for other urban areas in the West, World War II marked a decisive turning point. Federal dollars streamed into the West, changing economies formerly dependent on the light processing of raw materials to greater reliance on the more profitable activity of heavy manufacturing. Long associated with the lumber business, Seattle became a center for aircraft manufacturing and shipbuilding. Denver retained its stockyards but added a steel plant and factories that produced ammunition and firearms. The cotton fields that bloomed in the irrigated environs of Phoenix now competed with airfields, army bases, and defense plants for land and workers. The flow of federal dollars to the West certainly did not begin in 1940 (Phoenix owed its agricultural base to the first dam built under the Newlands Act of 1902), but government expenditures greatly accelerated at that time. Nor did the war's conclusion bring an end to the federal handouts. Cold war tensions and conflicts in Korea and Vietnam meant continued defense contracts and military bases for the cities of the American West.

Locating a defense plant or a military base in the West made good logistical sense. Facilities there were remote enough from the enemy to decrease the risk of attack but close enough to the fighting front to speed the engagement of men and supplies. The owners of non-war-related businesses had completely different matters to consider when they contemplated a move westward, but by the 1960s the situation was becoming clear to them as well. The benefits of staying in the northeastern and midwestern centers of population had declined; the benefits of moving west had risen.

Earlier, corporate managers had placed their headquarters in cities like New York to maximize their access to markets, money, and information. In the days before air travel became safe and reliable, when long-distance phone calls were costly and full of interference, the tens of thousands of experts directly available to businesses operating in Manhattan gave that city a decided advantage. This situation changed after World War II. The postwar decades brought not only improvements in air service and telephone reception but a whole new computer technology that ushered in "The Information Age." Now an executive living in Seattle or Dallas could have access to stock market prices and commodities quotations as quickly as anyone anywhere. Indeed, the non-northeasterner was likely to encounter fewer frustrations like jammed circuits, power outages, and transportation delays.

The question of relocation became pressing to different companies at different times. In the case of Shell Oil, the issue came to the fore in the latter half of the 1960s, when the company's mid-Manhattan offices could no longer accommodate its growing staff. Instead of building anew near Rockefeller Center, where construction costs were astronomical, or in suburban Connecticut, which lacked a large and diversified work force, Shell's board of directors moved its operating headquarters over sixteen hundred miles to Houston, Texas. The move, which involved twenty-two hundred jobs, cost the company twenty-five million dollars. Now ensconced in the state's tallest building, company executives considered the price tag a bargain. In Houston their paychecks went further, since housing and tax costs were lower; meanwhile, their staffs worked longer—a full forty-hour week instead of thirty-five. Additionally, all levels of employees found the journey to work much less of a hassle. Certainly the dollars and cents of the matter were all to Houston's advantage.

Nevertheless, such decisions were not made solely on the basis of economics. Other factors—amenities, they were called—helped to lure Shell Oil to Houston: a fine symphony and opera, respected universities, world-renowned medical facilities, and fashionable shops and stores. Among the latter was the Galleria, on the southwest side of Houston. It included a multistoried, glass-roofed shopping arcade, anchored by a Neiman-Marcus department store and boasting—of all things for the steamy flatlands of Texas—an ice-skating rink in the middle.

The rink represented an exaggerated version of the climate control that was ubiquitous in Houston. Formerly, the city's hot and humid weather had led the British Foreign Service to dub it a hardship post. By 1965, however, Houstonians enjoyed air conditioning not only in their houses, cars, plants, offices, theaters, and shopping malls but also at their sports stadium. "Big-as-all-indoors," the Harris County Sports Dome enabled fans to watch major-league baseball and football in air-cooled comfort. The Astrodome may have helped local residents survive the summer and early fall, but during the other months of the year, Houston, Dallas, Phoenix, and a host of other southwestern cities offered a climate preferable to that of the Northeast or Midwest. Such a consideration took on added value in the latter half of the twentieth century as people lived longer lives and worked a shorter proportion of them.

It was just such an interest in the quality of life that had sparked the growth of Los Angeles in the late 1800s. For decades, citizens of Los Angeles had held on to their vision of creating a new kind of metropolis. "It must not be a second congested London or New York," said John Anson Ford, a Los Angeles County supervisor in 1961, ". . . but a population center with many new characteristics adjusted to the outdoor life of the region, and to the era of greater leisure, greater mobility, and a wider distribution of the skills and culture of modern society." This ideal was seconded by civic leaders elsewhere in the West who accepted the automobile as the primary mode of urban transport and who wanted their cities to "spread widely rather than reach high."

In the very year that Supervisor Ford presented his formula for the future, an associate editor of *Architectural Forum* offered an opposing view. In *The Death and Life of Great American Cities*, Jane Jacobs argued that crowding was part of the essence of cities. The presence of large numbers of people at all times of the day and night made cities economically and intellectually vibrant. It also helped to make them safe. Assaults, rapes, and other major crimes occurred at higher rates, she said, in decentralized cities like Los Angeles than in congested ones like Chicago and New York. Anticipating the looting, arson, and sniper fire that would convulse the Watts district of Los Angeles in 1965, and much of the city in 1992, she called attention to the extent of its racial and economic segregation. In her view the suburbanization of the metropolis meant the addition of yet another barrier to understanding among diverse social groups—namely, physical distance.

In arguing that low-density districts were inherently dangerous, Jacobs relied on limited evidence. True, Los Angeles, along with Dallas and Houston, regularly posted a murder rate that exceeded the big-city average. But a different pattern prevailed in Phoenix, San Antonio, and San Diego. Here the rate of violent offenses (the number of murders, rapes, assaults, and robberies known to the police per one hundred thousand residents) was low; the stunning incidence of crimes against property (such as burglary, larceny, and motor-vehicle theft) drove up the overall crime rate. Moreover, the West boasted at least one large city with an enviable record in both categories. In 1990, of the seventy-four most populous cities in the United States, eleventh-ranked San Jose had the lowest total crime index. Indeed, this sprawling, suburbanized city had less than one-half the rate of property crime and one-third the incidence of violence of its more compact neighbor, San Francisco.

Although Jacobs's book lacked statistical grounding, it certainly had an impact, especially on the field of urban planning. Calling on planners to get people out of their cars and onto the sidewalks, she deserved part of the credit for the revitalization of a number of downtown areas from the 1960s on. Such projects as Pioneer Square in Seattle and Larimer Square in Denver incorporated Jacobs's notion that mixed-use districts were beneficial. Attracting different people for different reasons, such districts allowed cities to be cities. Here, in theory at least, the interactions among a multiplicity of individuals generated an atmosphere that was at once exciting and safe.

Yet influential as Jacobs's ideas proved, Supervisor Ford more nearly captured the trend of the times in his prescription that Los Angeles "must not be a second congested London or New York." By the mid-twentieth century, most Americans equated the good life with owning a single-family home and an automobile. So deeply was this ideal ingrained in the American psyche that the federal government subsidized the suburbanization of the nation. By allowing tax deductions for mortgage interest but not for rent, by providing long-term, low-interest house loans, and by financing the construction of an elaborate system of high-speed, limited-access roadways, the federal government encouraged city dwellers to move miles away from the urban core. When factories, stores, and offices followed, it was clear that other metropolitan areas were coming to resemble greater Los Angeles.

These sprawling metropolises existed not only in the West but also in the East. Analyzing data from more than fifty metropolitan areas in the East and Southwest, Carl Abbott found, in 1990, that the two groups of cities were "not markedly different." The latter *looked* distinctive because their streets were longer and broader and their architecture, with some exceptions, clung to the ground; but in terms of where their residents worked, shopped, and lived, the two had much in common. Both were experiencing the economic and demographic decline of their central business districts (some successful renewal projects notwithstanding) and the rise of peripheral areas in so-called edge cities.

But if the spread-out, multicentered metropolis had become a national phenomenon, it was nonetheless true that the West had helped to pioneer some of the land uses characteristic of the metropolis. During the immediate postwar decades, as the historian John Findlay has shown, the West developed a new type of workplace, living space, and playground. Built at the outer limits of existing metropolitan areas, these new cultural landscapes stimulated growth in their environs. They also set precedents that would be built upon across the nation and around the world.

Featured in an exhibit at the Brussels World's Fair in 1958 and visited by French President Charles De Gaulle in 1960, the Stanford Industrial Park represented the work environment of the future. Stanford University could have used the several-hundred-acre tract for housing for San Francisco–bound commuters. Instead, it laid out a vast, landscaped expanse, making parcels available to businesses in such fields as computers, aerospace, microwave technology, and pharmaceuticals. With low-slung buildings, a youthful work force, and an informal atmosphere, the Stanford Industrial Park had the look and feel of a college campus. In reality, it was the business center for Santa Clara County, whose agricultural economy was undergoing a rapid transformation. By 1980 the county, better known as Silicon Valley, ranked ninth in the nation for the value of its industrial products. By that time it was hard to remember that the term *industrial park* would once have been thought an oxymoron. By turning a contradiction into a reality, Stanford University created one of the characteristic landscapes of the new American metropolis.

Like *industrial park*, the phrase *active retirement* paired words once viewed as opposites. Adopted by the Del E. Webb Corporation as a slogan for its latest development project, the idea appealed to the emerging market of older Americans who enjoyed good health, a steady income, and no job. In Sun City, Arizona, such people would find what its advertisers termed "a complete community." Indeed, Sun City did have shopping malls, golf courses, recreation centers, and a five-story hospital, but it had no tax-consuming schools (since there were no children) and hardly any minorities. In view of what this walled community had as well as what it lacked, its proximity to Phoenix was important. With the downtown district a mere twelve miles away (and looming ever closer), citizens of Sun City enjoyed access to the amenities of a large city—stores, restaurants, theaters, and museums—without having to confront its realities. In Sun City, retirees discovered a community complete enough that forays to Phoenix could be a matter of choice rather than necessity.

Eventually Sun City drew forty-five thousand residents to its nine thousand acres, and the Stanford Industrial Park employed twenty-six thousand workers on seven hundred acres. By way of contrast, another of the West's new cultural landscapes consisted of a mere seventy acres exclusive of parking, yet it attracted as many as ten million people a year. Without a doubt, Disneyland was the West's single most important contribution to the built environment in the postwar era.

Walt Disney designed Disneyland with certain regional assumptions in mind. Whereas he wanted the amusement park to "be a place for California to be at home, to bring its guests, [and] to demonstrate its faith in the future," he rejected the model of existing parks and especially of Brooklyn's Coney Island. "Dirty grounds," "tawdry rides," "rude patrons," and "hostile employees" were not what he wanted for his brainchild.

In addition, Disney grasped an essential truth. If handled correctly, this venture in outdoor recreation would help to market his films and television shows and vice versa. For this reason Disney hired Hollywood scriptwriters to plan his park; subordinate to them were the engineers, architects, and city planners. Thus emerged in 1955 the world's first theme park. None of the attractions stood on their own. They were all part of an integrated layout that played on the guests' familiarity with Disney characters and stories. Billed as family entertainment, it appealed as much to adults as to children.

Indeed, attendance exceeded all expectations, especially in the park's first fifteen years. To entertain such swarms and keep them coming, Walt Disney Corporation

With Disneyland and the Las Vegas Strip, developers created distinctive new landscapes for the postwar West that capitalized on Americans' increased mobility and expanded leisure time. These new urban sites would become as important to western tourism as the national parks.

Unidentified photographer. Walt Disney Introducing Plans for Disneyland. *Gelatin silver print, 1954. © The Walt Disney Company. The Disney Publishing Group, Burbank, California.*

Unidentified photographer. Floating Craps Game, Sands Hotel. *Gelatin silver print, ca. 1950s. Las Vegas News Bureau Archives, Las Vegas, Nevada.*

doubled the number of attractions, building both above- and underground. The company enlarged the size of the work force five times over and coached employees in how they should look, talk, and act. It also varied the cost structure of attendance. Instead of charging for every ride in addition to general admission, it gradually moved to a single-fee passport. Such permutations helped give the park a positive billing from reviewers. The science fiction writer Ray Bradbury was especially enthusiastic. In his view, Disneyland proved "that the first function of architecture is to make men over, make them wish to go on living, feed them fresh oxygen, grow them tall, delight their eyes, make them kind. . . . Disneyland liberates men to their better selves." A less prominent observer put the matter more succinctly, "[Disneyland]'s as different from the Coney Island type of park as Walt could make it."

All three of these cultural landscapes—the Stanford Industrial Park, Sun City, and Disneyland—showed, according to Findlay, how a carefully planned fragment of the metropolis could lend a sense of order to the postwar sprawl. Yet few development projects were as well thought out as these. A more common feature of the emerging cityscape was the hodgepodge of signs, styles, and services typified by the commercial strip. Here too, the West offered a conspicuous model—the four-mile stretch of Las Vegas Boulevard known simply as the Strip.

The Strip represented the most spectacular part of a remarkable city. Fueled by federal funding, Las Vegas grew from a town of 5,165 in 1930 to a metropolis of 741,000 in 1990. The nation's taxpayers may have subsidized its water, power, highways, airport, and defense jobs, but private investors—at times with ties to the underworld—developed the resorts along the Strip. Beginning in the 1940s with El Rancho, the Last Frontier, the Flamingo, and the Thunderbird, the Strip turned Las Vegas into a major tourist attraction. By the time the Mirage and the Excalibur opened in 1989 and 1990, Las Vegas was welcoming 1.5 million convention-goers a year. In this regard it trailed only New York City, Chicago, Dallas, and Atlanta.

A success with the general public, the Strip was initially panned by critics who presumed to know what a city ought to look like. Nevertheless, in 1972, a trio of professors from the Yale School of Art and Architecture wrote a book extolling it. In *Learning from Las Vegas*, Robert Venturi, Denise Scott Brown, and Steven Izenour contrasted "the deadness . . . of present-day modern architecture" to "the vitality [of the Strip]." In their opinion the former showed the result of "too great a preoccupation with tastefulness and total design." The latter, boasting such resorts as Caesar's Palace and the Sahara, proved, on the other hand, that "people . . . have fun with architecture that reminds them of something else." Mixing allusions to the past and the present, the sacred and the profane, the Strip combined aspects of the appeal of Disneyland, the thrill of gambling, and the glitz of Coney Island. With the neon of its signs set high above the roadway (the sign for the Dunes measured twenty-two stories), the Strip represented "a new landscape of big spaces [and] high speeds." Meant to be traveled by automobile, it offered an architecture for the deconcentrated metropolis.

As much as the West pioneered forms of the late-twentieth-century cultural landscape, it did not monopolize them. Industrial parks, retirement communities, theme parks, and commercial strips soon emerged elsewhere. At times the later examples (such as Florida's Walt Disney World) clearly eclipsed the earlier models. Thus, to some extent, the metropolises of the American East could come to resemble those of the West and even to improve on them.

But in another sense the appeal of western cities did remain distinctive. For, although they could export aspects of their built environment, they could not export their natural environment. Mountains, beaches, oceans, and deserts played a part in the appeal of their settings. After all, where was the eastern metropolis that could rival a Denver, Seattle, or San Diego in offering opportunities for outdoor recreation?

Nevertheless, the lure of western cities rested on an increasingly fragile foundation. As houses grew too expensive to buy and freeways too congested to drive, residents faced a predicament that stemmed from their region's knack for attracting newcomers. Moreover, another characteristic of the region—the westerners' attachment to and dependence on the automobile—was itself fouling the healthfulness and beauty of the places they loved.

By the closing decade of the twentieth century, the day seemed close at hand when western urbanites would live under a near-constant cloud of pollution. On occasion the wind would blow, the air would clear, the landscape would come into view, and people would remember why they lived there. The suburban metropolises of the American West had succeeded too well. The "Cities of the Future" were mired in the present.

Bibliographic Note

From Richard C. Wade's classic study *The Urban Frontier: The Rise of Western Cities, 1790–1830* (Cambridge, Mass., 1959) to William Cronon's prize-winning book *Nature's Metropolis: Chicago and the Great West* (New York, 1991), scholars have stressed the role of cities in transforming rural hinterlands. They have also focused on the impact of new technologies, especially transportation technologies, in reordering the hierarchy among cities and reshaping the space within them.

A comprehensive illustrated account of the settlement of western cities appears in John W. Reps's massive volume *Cities of the American West: A History of Frontier Urban Planning* (Princeton, 1979). David Hamer's *New Towns in the New World: Images and Perceptions of the Nineteenth-Century Urban Frontier* (New York, 1990) adds valuable interpretative insights. His discussion considers locales in Canada, Australia, and New Zealand as well as the United States. Lionel Frost also provides a comparative perspective in *The New Urban Frontier: Urbanisation and City-Building in Australasia and the American West* (Kensington, 1991). Lawrence H. Larsen's *The Urban West at the End of the Frontier* (Lawrence, 1978) examines the twenty-four largest cities in the western United States circa 1880.

For perceptive sketches of San Francisco, Seattle, Portland, Salt Lake City, and Los Angeles, see "The Power of the Metropolis" in Earl Pomeroy, *The Pacific Slope: A History of California, Oregon, Washington, Idaho, Utah, and Nevada* (New York, 1965), 120–64. In *Instant Cities: Urbanization and the Rise of San Francisco and Denver* (New York, 1975), Gunther Barth treats not only the title cities but also the early history of the "City of the Saints." Mary Lou Locke looks at Portland, San Francisco, and Los Angeles to explore a neglected topic in "Out of the Shadows and into the Western Sun: Working Women of the Late Nineteenth-Century Urban Far West," *Journal of Urban History* 16 (February 1990): 175–204. Thomas G. Alexander and James B. Allen, *Mormons and Gentiles: A History of Salt Lake City* (Boulder, 1984), Lyle W. Dorsett, *The Queen City: A History of Denver* (Boulder, 1977), and Eugene P. Moehring, *Resort City in the Sunbelt: Las Vegas, 1930–1970* (Reno, 1989), provide useful information. Thomas J. Noel explores a specific feature of urban life in his highly readable monograph *The City and the Saloon: Denver, 1858–1916* (Lincoln, 1982).

No western city has been more thoroughly studied than San Francisco. Roger W. Lotchin considers the city's beginnings in *San Francisco, 1846–1856: From Hamlet to City* (1974; reprint, Lincoln, 1979). William Issel and Robert W. Cherny pick up the story in *San Francisco, 1865–1932: Politics, Power, and Urban Development* (Berkeley, 1986). Peter R. Decker, *Fortunes and Failures: White-Collar Mobility in Nineteenth-Century San Francisco* (Cambridge, Mass., 1978), Douglas Henry Daniels, *Pioneer Urbanites: A Social and Cultural History of Black San Francisco* (1980; reprint, Berkeley, 1990), and Robert A. Burchell, *The San Francisco Irish, 1848–1880* (Berkeley, 1980), discuss the experiences of specific population groups. William A. Bullough's *The Blind Boss and His City: Christopher Augustine Buckley and Nineteenth-Century San Francisco* (Berkeley, 1979) takes a biographical look at the city's government, whereas Terrence J. McDonald's *The Parameters of Urban Fiscal Policy: Socioeconomic Change and Political Culture in San Francisco, 1860–1906* (Berkeley, 1986) offers a social-scientific analysis. Judd Kahn discusses Burnham's plan for San Francisco in *Imperial San Francisco: Politics and Planning in an American City, 1897–1906* (Lincoln, 1979).

The number of works that focus on Los Angeles is beginning to look impressive as well. At the head of the list is Robert M. Fogelson's pathbreaking study, *The Fragmented Metropolis: Los Angeles, 1850–1930* (1967; reprint, Berkeley, 1993). Two provocative journalistic accounts are Carey McWilliams, *Southern California: An Island on the Land* (1946; reprint, Salt Lake City, 1983), and Mike Davis, *City of Quartz: Excavating the Future in Los Angeles* (London, 1990). The city's early history is examined in Leonard Pitt, *The Decline of the Californios: A Social History of the Spanish-Speaking Californians, 1846–1890* (Berkeley, 1966), and Richard Griswold del Castillo, *The Los Angeles Barrio, 1850–1890: A Social History* (Berkeley, 1979). Los Angeles is one of the cities Mark S. Foster discusses in "The Western Response to Urban Transportation: A Tale of Three Cities, 1900–1945," in Gerald D. Nash, ed., *The Urban West* (Manhattan, Kans., 1979), 31–39. Scott L. Bottles provides an in-depth analysis in *Los Angeles and the Automobile: The Making of the Modern City* (Berkeley, 1987). Robert Fishman's *Bourgeois Utopias: The Rise and Fall of Suburbia* (New York, 1987) contains a perceptive chapter on Los Angeles. Reyner Banham, an architectural historian, and David Brodsly, a city planner, both offer provocative insights in their respective works: *Los Angeles: The Architecture of Four Ecologies* (Middlesex, Eng., 1971) and *L. A. Freeway: An Appreciative Essay* (Berkeley, 1981). Sports fans will not want to miss Neil J. Sullivan's *The Dodgers Move West* (New York, 1987).

A brief introduction to western cities since 1940 can be found in Raymond A. Mohl, "The Transformation of Urban America since the Second World War," in Robert B. Fairbanks and Kathleen Underwood, eds., *Essays on Sunbelt Cities and Recent Urban America* (College Station, Tex., 1990), 8–32. The most comprehensive work on the subject is Carl Abbott's *The Metropolitan Frontier: Cities in the Modern American West* (Tucson, 1993). Abbott's *The New Urban America: Growth and Politics in Sunbelt Cities*, rev. ed. (Chapel Hill, 1987) also remains useful, as does his essay "Southwestern Cityscapes: Approaches to an American Urban Environment," in Fairbanks and Underwood, *Essays on Sunbelt Cities*, 59–86. For more on the Stanford Industrial Park, Sun City, and Disneyland, see John M. Findlay, *Magic Lands: Western Cityscapes and American Culture after 1940* (Berkeley, 1992), a book that is transforming how scholars think about the West and the cities it contains. Findlay analyzes the Las Vegas Strip in *People of Chance: Gambling in American Society from Jamestown to Las Vegas* (New York, 1986). The national context for many of the developments emphasized in this chapter is presented in Kenneth T. Jackson, *Crabgrass Frontier: The Suburbanization of the United States* (New York, 1985), and Joel Garreau, *Edge City: Life on the New Frontier* (New York, 1991).

Readers seeking further sources may consult the following bibliographic essays: Carl Abbott, "The Metropolitan Region: Western Cities in the New Urban Era," in Gerald D. Nash and Richard W. Etulain, eds., *The Twentieth-Century West: Historical Interpretations* (Albuquerque, 1989), 71–98; Lawrence H. Larsen, "Frontier Urbanization," in Roger L. Nichols, ed., *American Frontier and Western Issues: A Historiographical Review* (Westport, Conn., 1986), 69–88; and Bradford Luckingham, "The Urban Dimension of Western History," in Michael P. Malone, ed., *Historians and the American West* (Lincoln, 1983), 323–43.

Holland America Line
A TRADITION OF EXCELLENCE
Spring and Fall 1994
HAWAII CRUISES
ss Rotterdam

WORLD EXPLORER CRUISES
ALASKA
A 14-DAY ADVENTURE FOR THE MIND, HEART AND SOUL.
1994 PREVIEW

Chapter Sixteen

Alaska and Hawai`i

VICTORIA WYATT

At first blush Hawai`i and Alaska may seem to bear little relation to each other—or to the American West. Americans commonly think of Hawai`i less as part of the American West than as part of the American vacation: an island separated from the stresses of real life by an ocean. Yale University's alumni magazine recently advertised a tour of Alaska under the heading of travel opportunities "abroad"; and according to a common joke, Americans think Alaska is a Pacific island, floating somewhere to the southwest of California, where it basks in warm climes on maps of the nation. Alaska often *is* perceived as a metaphorical island—distanced from the "mainland" by Canada rather than by an ocean, but distanced all the same. To outsiders, Alaska symbolizes a sanitized last frontier, where travelers go to experience spiritual awakening akin to religious conversion. Hawai`i represents a tropical paradise where workaholics can relax and recharge before returning to the stresses of modern American life.

These popular perceptions mask the essential experiences of Alaskans and Hawaiians and their long-standing ties with the contiguous United States. Since Alaska and Hawai`i did not become states until 1959, they seem like new cast members in American history. In fact, many themes central to the development of the American West appeared in Alaska and Hawai`i as early as—and often earlier than—in the contiguous western frontier. These themes include the convergence of indigenous peoples, European newcomers, and non-European immigrants; dispossession of native peoples; economic enterprises based on eastern U.S. or foreign capital; dependence on natural resources for both industry and tourism; and tensions generated by a substantial federal presence in regions far from the center of federal government.

Despite these common themes, Alaska and Hawai`i have tended to remain very much on the periphery in popular concepts of America. World War II was an exception. When both regions were attacked, they suddenly seemed very much a part of America—and certain residents of both felt the impact of the federal government in similar fashion. Thus, examining two World War II experiences seems a fitting way to begin a discussion of Alaska, Hawai`i, and the American West.

Promoted as exotic sites with dramatic scenery and distinctive indigenous cultures, Alaska and Hawai`i are often viewed as peripheral to the central stories of the western American past. Nonetheless, their histories echo many of the central themes and tensions relating to natural resources, native peoples, and economic development in the contiguous western states.

Tour brochures. *1994. Courtesy Holland America Line and World Explorer Cruises.*

Hawai`i and Alaska at War: Two Vignettes

On the morning of Sunday, 7 December 1941, in Honolulu, Hawai`i, Usaburo Katamoto went to the Kokusai Theater to graduate from a first-aid class. A boat builder by trade, he and some of his Japanese friends had been studying first aid to help prepare Hawai`i for war. As they waited for commencement exercises to begin, they heard firing, and a plane came into view. Straining to see it, they finally discerned a Japanese flag painted on its body. Their graduation exercises never took place.

The people who gathered at the Kokusai Theater represented the largest single ethnic group in the islands, for Japanese immigrants and their descendants—including American citizens and aliens—constituted a third of the population of Hawai`i. On that December morning, each Hawaiian resident of Japanese descent instantly became suspect. U.S. Army General Walter Short, supported by President Franklin Delano Roosevelt, called for martial law to help prevent sabotage by Japanese Hawaiians. In an action unprecedented since the U.S. Civil War—when martial law was applied to Southern states that were overtly in rebellion—the Hawaiian people came under U.S. military rule. Although no instances of sabotage were ever found, martial law remained in effect for almost four years.

Fed by the media, rumors spread fast and far. By Monday morning, newspapers and radio stations in Hawai`i and Los Angeles reported that Hawaiian Japanese had blocked access roads to Pearl Harbor during the attack, had cut gigantic arrow shapes in sugarcane fields to point enemy planes toward Pearl Harbor, had poisoned the water supply, and had even shot at American troops. By 8 December, over one hundred local Japanese had been impounded under suspicion of disloyalty. Some were fishermen with radios. Others were suspected solely for their cultural ties and their religious or political leadership positions in local Japanese communities. Twelve Japanese newspapers and three magazines were shut down. Loyalty boards were quickly established on all islands to observe and question thousands of Japanese residents. The expression of Japanese language or culture and the observance of religious holidays were labeled implicitly—and sometimes explicitly—as unpatriotic and suspect.

Almost daily, Katamoto saw friends called in to be interrogated by the FBI. He himself was detailed from his employer to help the U.S. Army Corps of Engineers repair naval ships at Pearl Harbor. His loyalty was not at issue when his boat-building skills were urgently needed. After the urgency slackened and Katamoto returned to his regular job, the FBI abruptly arrested him and dispatched him to an internment camp in Santa Fe, New Mexico. He was suspect, they said, because before the war he sometimes entertained his brother, who was serving in the Japanese navy. Katamoto was held in the camp for almost four years.

Over the course of the war, almost fifteen hundred Japanese Hawaiians were interned. About 37 percent were already U.S. citizens. Some one thousand women and children also left Hawai`i to join the men in the internment camps. Meanwhile, on the West Coast of the mainland, the entire local Japanese population was targeted for internment and relocation. There were some calls for similar actions in Hawai`i, but the Japanese there constituted one-third of the population and a large sector of the skilled labor force. Mass internment was considered essential to national security in West Coast mainland regions, where Japanese represented only 2 percent of the population; ironically, in Hawai`i, where they were the largest demographic ethnic group, selective removal was deemed sufficient.

By June 1942, Usaburo Katamoto from Hawai`i shared the same fate as the 110,000 Japanese residents from the West Coast of the mainland who were being sent to internment camps. Oddly, the Aleut residents of Alaska's Pribilof Islands—two isolated and little-known islands in the Bering Sea—were soon to have much in common with these so-called "enemy aliens." These Aleuts provided the labor for a seal-harvesting

Aleut people from the Aleutian Islands and the Pribilof Islands were forcibly evacuated from their homes during the Second World War and relocated some fifteen hundred miles to camps in southeastern Alaska, where they lived under strict government controls. Only half would return home. Like the Japanese-American internees from mainland western states, they faced a long and difficult process of rebuilding their lives after the war.

Unidentified photographer. Evacuation of Aleut People. *Gelatin silver print, 1942. National Archives, Washington, D.C.*

operation managed by the U.S. Fish and Wildlife Service. In mid-June 1942, a U.S. Navy vessel arrived with an order to evacuate immediately. The unexpected order surprised many. Although the Japanese had attacked the Aleutian Islands the previous week, the tiny Pribilofs lay two hundred miles to the north, had no U.S. military base, and seemed an extremely unlikely target. In fact, military correspondence suggests that the navy wanted to occupy houses on the Pribilofs.

The 290 residents were shipped fifteen hundred miles east to Funter Bay, a small cove sixty miles from Juneau where a few buildings remained from an abandoned cannery and a mine site. The buildings had poor sanitation, no heat, and insufficient floor space for everyone to sleep. The U.S. Fish and Wildlife Service agents in charge protested, and in October 1942 the Pribilof women petitioned the federal government, stating, "This place is no place for a living creature." Their appeals went unheeded, and they spent the winter in these conditions.

The Aleuts at Funter Bay were not under arrest, and their loyalty to the United States was never at question. Yet when Aleut men tried to leave the camp to work in Juneau, the camp superintendent refused to release them. Such control over their movements was not new to Pribilof Islanders. They had first harvested furs for the Russians, who treated them like serfs and demanded work as taxation. When Alaska was transferred to the United States in 1867, the federal government agents stationed at the Pribilofs quickly claimed dictatorial authority. Government agents could deny Aleuts permission to travel off the islands. They selected chiefs, forced labor, controlled wages, threatened banishment to get submission, and—in efforts to increase the population—even engineered marriages.

Thus, when the superintendent at Funter Bay refused to allow men to leave the camp, he was simply assuming authority the government had exercised in the Pribilofs for decades. His motivation for the policy was also the same: to keep the government's Pribilof work force intact. Restricting the movements of free men proved more difficult

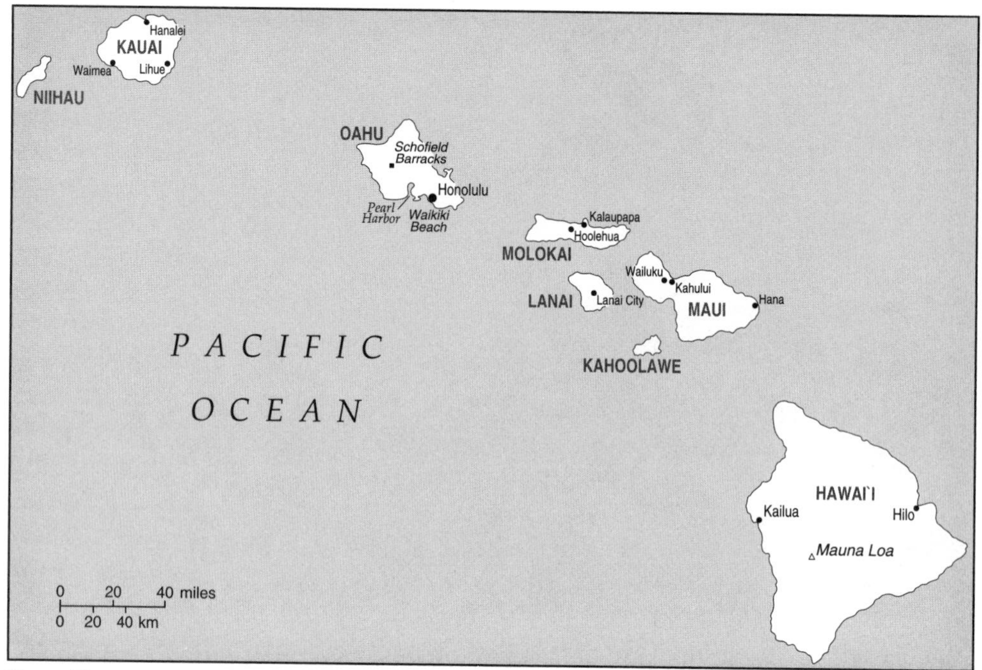

Hawai`i

away from the isolation of the Pribilofs. Other government offices intervened, and some men found work outside the camps.

In May 1944, with the Japanese gone from the Aleutians, the government sent the Pribilof Islanders home. They arrived to find that their houses and other buildings had been vandalized and looted. Personal possessions of value had been stolen, storerooms had been stripped bare, and tools had disappeared. Like the Japanese along the West Coast, the Pribilof Islanders faced a long process of rebuilding and recovery.

Ironically, the evacuation ultimately weakened the federal government's totalitarian control over the Pribilof Aleuts. After the islanders had seen free communities and met native political leaders in Juneau, they resisted the restrictions they had suffered before the war. The unthinkable conditions in Funter Bay had also brought the islanders to the public eye. The government could not reinstate the forced isolation that had made dictatorial policies possible.

Alaska's history, like that of Hawai`i, spotlights paradoxes in the development of the American West. That the government enforced a totalitarian regime in the Pribilofs for three-quarters of a century—that some federal agents sought to maintain this system throughout the disruptions of a world war fought against totalitarianism—is one of those paradoxes. In the Pribilofs, isolation from the rest of the country had made such oppression possible. In Hawai`i, ironically, isolation may actually have prevented the mass deportations suffered by Japanese residents on the mainland's West Coast. Separated from the rest of the United States geographically and no less by popular perception, Alaska and Hawai`i share an island status that fosters historical anomalies. Yet at the same time, as the lives of the Pribilof Aleuts and Katamoto attest, these states are inextricably woven to the rest of the nation. Their histories, like those of colonized islands everywhere, have been informed by continuous dynamics between isolation from the mainland and compelling, unshakable ties to it.

Indigenous Populations and First Contacts with Europeans

Of course, to define Hawai`i and Alaska as "isolated" is to define them in relation to another point of reference, the mainland United States. In fact both Hawai`i and Alaska have long histories of human settlement before U.S. involvement. To the indigenous peoples of both states, being part of a larger government outside the immediate region is a relatively new and culturally artificial phenomenon.

The Hawaiian Islands were formed over hundreds of thousands of years by volcanic eruptions on the ocean floor. Molten rock pushed upward until the tops of some volcanoes rose above sea level. The islands today are composed of volcanoes and their lava flows. The islands stretch over fifteen hundred miles. Their closest neighbors to the north are the Aleutian Islands of Alaska; the North American mainland lies about twenty-one hundred miles away. To the south, their nearest large neighbor is the chain of the Marquesas Islands, two thousand miles distant.

Hawai`i was first settled by Polynesians, who had already developed distinctive cultures on other South Pacific islands. The exact date of their arrival in Hawai`i is not clear. Native Hawaiian legends record that ancestors traveled in double-hulled canoes back and forth between the Society Islands and Hawai`i. These voyages apparently ceased about one thousand years ago, and Hawaiians had no more contact with people from other regions until the arrival of Europeans in 1778. The culture that Hawaiians developed reflects influences of their Polynesian origin, but it also has distinctive qualities born of centuries of isolation.

The first peoples who arrived in Alaska probably crossed from Siberia by land. At various times during the ice ages, the sea levels dropped sufficiently to expose a land mass—known now as Beringia—which lies beneath the Bering Sea today. Some archaeologists believe human beings were living on Beringia forty thousand years ago or even earlier. According to more conservative estimates, the first migration began about fifteen thousand years ago. Whenever they arrived, the first peoples in Alaska were also the first in North America and probably populated both North and South America. Later movements across Beringia followed, separated by thousands of years.

Unlike Hawai`i, Alaska is home to many different indigenous ethnic groups. The Tlingit and Haida peoples of southeastern Alaska have much in common culturally with native peoples of the adjacent Canadian coast. The Déné, or Athapaskan, people of the interior of Alaska share cultural traits with peoples in the Yukon. The Aleut and Eskimo peoples of south-central, southwestern, and Arctic Alaska originate from an entirely different ethnic background from the Tlingit, Haida, and Déné peoples and probably came to North America in a later migration.

The state is composed of 586,000 square miles, stretching 2,400 miles from east to west and 1,420 miles from north to south. It is surrounded on three sides by water and has a coastline of 33,000 miles; thus, many of the Alaskan native cultures adapted to the sea. Long before the arrival of Europeans, native Alaskans developed extensive trading networks along the coastline and rivers.

Because of Alaska's size and the diverse ethnic groups there, native peoples organized

Alaska

themselves locally, usually with different political leaders for each community or village. In contrast, Hawai`i's indigenous social and political system was quite uniform throughout the islands. Social rank was strictly delineated, with many hereditary high chiefs and a complex system of laws and taboos, known as *kapus*. Commoners owed a feudal relationship to chiefs.

The differences between political organization in Alaska and Hawai`i would have an immense impact on native peoples' interactions with Europeans. When British Captain James Cook arrived in Hawai`i in 1778, there were four centers of political power in the islands—four high-ranking chiefs competing for territory and influence. Their battles may have escalated in the decades after Cook's visit as the European traders introduced firearms and explosives. By the turn of the century, one leader, Kamehameha I, controlled all of the islands except Kauai, and Kauai joined his kingdom through diplomatic agreement in 1810. From that time onward, native Hawaiians dealt with foreign nations as one unified monarchy. Europeans and Americans were well accustomed to diplomatic relations with monarchies and could fit Hawai`i easily into their existing political worldviews.

In Alaska, on the other hand, there were many nations of native peoples, each divided into very local political bands. Europeans and Americans had more difficulty recognizing native bands and tribal units as nations. There was no counterpart to Kamehameha—no one monarch whom they could approach and who ruled over all native Alaskans. In the absence of such a monarch, Europeans and Americans often conveniently chose to approach no one, usurping lands and resources without even the pretense of diplomacy between nations.

Eventually, a long history of such injustice led native Alaskans to organize across cultural and geographic boundaries—a monumental task considering the distances and differences involved. In 1966 native leaders throughout the state founded the Alaska Federation of Natives. It became a driving force behind the Alaska Native Claims Settlement Act of 1971, meant to compensate native Alaskans for lands taken without treaty. For decades, however, the colonizers—first Russia and then the United States—simply operated as if there were no native title to land.

Native Hawaiians first saw Europeans in January 1778, when Captain James Cook's expedition sailed past the island of Oahu into Waimea Bay on Kauai. More than two centuries had passed since Hopi and Zuni peoples in the American Southwest had met Francisco Vásquez de Coronado, but native peoples in other parts of the American West would not encounter the expedition of Meriwether Lewis and William Clark for a quarter century more.

In 1778 Cook was on his third major voyage of exploration. He was traveling northward to explore Alaskan waters, seeking a northwest passage. He had learned some of the Tahitian language from his travels in the South Pacific, and he discovered that he could understand portions of the Hawaiian language. The Hawaiians treated him as a high chief; since such chiefs were associated with divinity, Cook may have been considered to have divine origins.

After staying in Hawai`i two weeks, Cook continued his trip. His two ships, the *Discovery* and the *Resolution*, explored parts of what is now the coast of British Columbia, southeastern Alaska, and Prince William Sound and continued on through the Bering Sea to the Arctic Ocean. Cook was certainly not the first European to arrive in Alaska. In 1741 Vitus Bering and Aleksei Chirikov, sailing from Okhotsk, crossed the Bering Sea to explore for Russia. Bering's ship, the *St. Peter*, reached as far east as Kayak Island in the Gulf of Alaska. Sailing back westward, they met and exchanged gifts with some native men in the Shumagin Islands off the Alaskan Peninsula. The parties tried to talk with each other, but without interpreters the obstacles were insurmountable.

Bad weather had diverted Chirikov's ship, the *St. Paul*, early in the voyage. Sailing independently, Chirikov reached as far east as present-day Sitka. He sent men ashore for provisions—none returned. It has often been assumed that they were killed on shore by Tlingit peoples. However, according to Tlingit oral tradition, Chirikov's men voluntarily deserted, and Tlingit peoples welcomed them into their village. There are several documented instances in later years of European seamen who deserted and were sheltered by Northwest Coast native peoples, and the Tlingit account deserves serious consideration.

In any case, after a month of fruitless waiting, Chirikov turned homeward. He traded with Aleut peoples in the Aleutian Chain and reached Kamchatka in October. Bering never returned; he died while his crew wintered on the Commander Islands. Survivors from the *St. Peter* eventually made it back to Kamchatka; like their compatriots on the *St. Paul*, they described the richness of fur-bearing sea mammals they had seen in Alaska.

The Russians already had an active fur trade with China and were eager to exploit the resources in Alaska. They wasted no time, sending a hunting expedition to Bering Island by 1743. In the next two decades Russian hunting expeditions became very active in the Aleutians. The sea otter was the prized fur because, unlike other sea mammals, the sea otter lacks a thick layer of blubber under its skin and must grow a denser fur for warmth. Russian traders wintered in the Aleutian Islands for two or three years, obtaining sea otter pelts from Aleut hunters. They sometimes took Aleuts as hostages to ensure the safety of Russians.

Aleuts and Pacific Coast Eskimos made several unsuccessful attempts to resist the Russian advance. In 1784, the Russian trader Gregorii Shelikhov established a permanent trading settlement on Kodiak Island. By 1799, Russian traders had expanded eastward through south-central Alaska into southeastern Alaska, bringing Aleut hunters with them. Among their establishments in southeastern Alaska was Fort Archangel Saint Michael, built in 1799 near the site of present-day Sitka. In 1802 Tlingit peoples destroyed the fort, but the Russian trader Aleksandr Baranov returned in 1804 with a large military force and reestablished the Russian settlement. This fortified town, called New Archangel, became the Russian capital of Alaska four years later.

Meanwhile, Spain had become concerned about Russia's advancement eastward and sent expeditions up the northwest coastline in 1774 and 1775. Juan Pérez led the first expedition with only one ship; the second was led by Bruno de Hezeta and Juan Francisco de la Bodega y Quadra. The Spaniards traded with coastal native peoples. When James Cook arrived in 1778, he did not know that Spaniards had preceded him to the Northwest Coast, but this soon became evident from the trade goods native peoples showed him.

To prepare for cold weather farther north, Cook and his men obtained sea otter furs from native traders. Later, Cook's crew discovered what Russians had known for three decades: these pelts sold in China for astounding profits. Each pelt was worth a seaman's annual pay. This news spread when Cook's journals were published in Europe in 1784, and very quickly British and American merchants sent trading expeditions to the Northwest Coast. Many reached southeastern Alaska and competed with Russian traders there. Tlingit hunters skillfully used this competition to their best advantage.

Captain Cook did not live to see this sea otter trade. He returned to Hawai`i in November 1778, having failed to find a northwest passage. One night in February 1779, islanders quietly took a British cutter. Cook armed himself the next day and went ashore to invite King Kalaniopuu, the high chief of the island, to his ship. He planned to hold the chief on board until the cutter was returned. However, Cook had not counted on the brashness of his own crew. While he was on shore talking to Kalaniopuu, sailors shot at a canoe and killed a high-ranking chief. News of his death traveled quickly to shore, where fighting broke out. Cook was killed as he retreated into the water. Over the next few days his crew took revenge, burning villages and killing indiscriminately. On 22 February 1779, the *Discovery* and the *Resolution* departed northward toward Alaskan waters, where they again sought a northwest passage.

For the next four decades, Tlingit peoples in southeastern Alaska carried on an active trade in sea otter pelts with Russian, British, and American traders. The Russians conducted their trading in southeastern Alaska from their permanent settlement of New Archangel. They brought Aleut workers from southwestern Alaska to hunt there, ignoring the fact that hunting and fishing areas were traditionally owned by Tlingit families. The British and American traders had no permanent presence in southeastern Alaska and imported no workers; they relied entirely on the Tlingit peoples, and on native peoples in present-day British Columbia, to supply the furs. En route to China, many of the ships stopped in Hawai`i to trade with Hawaiians for provisions. The traders eventually began wintering in Hawai`i, returning to the Northwest Coast for a second summer of trading before heading to China.

The firearms these traders brought to Hawai`i may have helped Kamehameha consolidate the islands under one monarchy. The trading ships also carried European diseases: smallpox, measles, influenza, venereal diseases, and more. Previously unknown in southeastern Alaska or in Hawai`i, these diseases caused devastating epidemics in both places. Social and political stability suffered. High-ranking leaders died, children lost their families, villages were decimated. Survivors faced the rest of their lives with the pain of their tremendous loss and with the ever-present danger of new epidemics.

Hawaiians frequently joined the crews of trading ships and visited other lands, including the Northwest Coast. There are also reports of Northwest Coast native peoples visiting Hawai`i. Artwork and material culture traveled back and forth on the trading ships between the South Pacific and southeastern Alaska. A Tlingit clan in Angoon, Alaska, incorporated a feather collar from the Society Islands into its clan ceremonial regalia. Thus the fur trade led to some exchanges of ideas between indigenous peoples of the two regions. Hawai`i and southeastern Alaska were both far from European and American centers, but they were no longer isolated from each other.

Many British and American seamen left their ships to live in Hawai`i. By 1810, near Honolulu alone, there were several dozen foreigners in residence; by 1819 some two hundred Americans and Europeans lived in the islands. After 1819, Kamehameha and other chiefs allowed foreigners who had skilled trades to own land. These foreigners married Hawaiian women and accumulated estates through gifts from chiefs, for whom they often served as advisers. Hawaiians learned foreign craft skills from them and no longer depended on trading ships for blacksmithing and other craft items. Initially, Europeans and Americans who chose to live in Hawai`i prospered as long as they supported native leaders and joined and contributed to indigenous communities.

The foreigners living in Alaska at this time—the Russians—had a very different relationship with the native peoples. They appropriated land for forts and hunting resources without permission from native leaders. In southwestern Alaska, they forced Aleuts to become serfs who owed labor and tariffs to the Russian tsar. In southeastern Alaska, they established New Archangel through military force and lived behind a barricade. Some Russians married or cohabited with Aleut and Tlingit women, and the progeny of these unions—Creoles—became part of the work force. Otherwise, Russians segregated themselves from native communities. Their presence in Alaska was founded on subjugation and military threat.

Far from behaving as visitors in other peoples' lands, Russians claimed Alaska for their own country. They had no monopoly on ignoring native title, however: Spain and

England claimed the Northwest Coast in similar fashion and almost came to war in the 1790s before Spain backed down. Throughout North America, colonizing nations have tended to claim land first and deal with native title later—if at all. Alaska's history provides no exception.

Nineteenth-Century Commerce

American businessmen and investors became involved in Hawai`i long before they began to develop most parts of the American West. When the first American trading vessel stopped in Hawai`i in 1790, the United States was a new nation, and its "West" was very east. The Northwest Ordinance, providing for governance of lands northwest of the Ohio River, was only three years old. Yet from 1790 onward, Americans were actively involved in commerce in Hawai`i. They quickly became the major actors in the maritime fur trade that linked southeastern Alaska, Hawai`i, and China. About twenty ships from New England stopped in Hawai`i each year to replenish supplies en route to Alaska.

The commercial success of American merchants in Hawai`i was not lost on Russian traders. In 1816 a Russian agent asked a Hawaiian king, Kaumalii, for permission to start Russian trading posts on Kauai and Oahu. Russians felt that here—unlike Alaska—the consent of the local leaders was necessary. Kaumalii granted permission, but Kamehameha I quickly overruled him and forced Russians to leave the following year. Russians did not threaten the dominance of American traders in Hawai`i again.

By 1810 the sea otter population was decreasing on the Northwest Coast, and merchants sought to supplement sea otter pelts with other trade items. American merchants discovered that sandalwood sold well in China, where it was used for wood carving, medicinal and cosmetic oils, and incense. Hawaiian chiefs and New England merchants were both eager to capitalize on the trade. Chiefs sent thousands of commoners deeper and deeper into the mountains to cut sandalwood. Furniture, clothes, carriages, and a myriad of other items flooded from New England to Hawai`i in trade. As the trade escalated, merchants paid chiefs in advance, and chiefs accumulated debts that further increased the pressure to harvest wood. Kamehameha had had the foresight to institute conservation measures early in the trade, but after his death in 1819 these measures broke down. By 1829 the supply of marketable sandalwood was virtually exhausted—leaving many chiefs with large debts from advances on orders they could not fill.

The fate of sandalwood in Hawai`i parallels that of sea otters in southeastern Alaska and the Northwest Coast. The sea otter slaughter was similarly unencumbered by conservation measures, and few sea otter remained in southeastern Alaska by 1829. Realizing the opportunity for immediate profits from sandalwood and sea otter, harvesters and merchants destroyed the very resource on which their commerce depended. Their choice to place short-term gain over long-term sustainable yield is no historical anomaly. The same short-term focus informs the rapid destruction of old-growth forests in Alaska and the Northwest today.

The commercial whaling industry faces similar issues today. Contemporary conservation measures are complicated by the need for international cooperation; and in northern Alaska, subsistence whaling has important cultural implications that must be acknowledged. In 1819, when whaling grounds were discovered off the coast of Japan, the world dealt with no such complexities. Hawai`i quickly became the major supply center for whaling vessels in the Pacific, a great number of which were American ships from New England. They stopped in Hawai`i twice a year to gather provisions and to take oil back to New England. When the sea otter and sandalwood commerce dwindled, whaling was already filling the commercial gap for American maritime businessmen. By 1829 over one hundred whaling ships called in to Hawai`i, and this number grew to four or five hundred a year in the next two decades.

As whaling grounds farther south became depleted, American whaling ships began testing Alaskan waters. In 1835 American ships explored the Gulf of Alaska; by 1852 over two hundred whaling ships were active in the Bering Sea. These whalers traded with coastal Eskimos, hired Eskimo men to work on ship, and commissioned clothing from Eskimo women.

Between 1830 and 1850, the whaling industry generated some six million to ten million dollars' worth of products each year—huge sums in those days. In Hawai`i, mercantile businesses and commercial agriculture grew quickly to meet the needs of the whaling vessels. Americans dominated this commerce, establishing and investing in the enterprises and selling to American fishing fleets. During the sandalwood trade, Hawaiian chiefs had controlled the access to the resource in demand and so had been actively involved in the commerce. They had no analogous role in the commerce that grew from whaling. Americans gained more and more power over the economy in Hawai`i, weakening the power of chiefs.

In Alaska, fur trading remained the focus of commercial activities while affording far more modest profits than whaling. As sea otter populations declined, the Russian-American Company, which enjoyed a monopoly through a Russian government charter, shifted its focus to other sea and land mammals. It also tried to diversify its economic activities in Alaska, without great success. Ice—a plentiful resource in Alaska—could be exported if there was a demand for it in warmer climates. The California gold rush of 1849 created the demand. In southwestern Alaska, the Russian-American Company had their Aleut and Creole workers cut ice; in New Archangel, they paid Tlingit peoples to do so. Up to ten thousand tons of ice were exported annually. Ultimately, though, this industry depended on California demand and declined as the gold rush there subsided. In the 1850s, the company experimented with coal mining, also to serve a California market. However, furs remained the only viable economic product of the Russian-American Company, and fur trading was not a growth industry.

In the first half of the nineteenth century, the economy in Hawai`i flourished and diversified. Hawai`i's isolation actually increased its attractiveness as a wintering and reprovisioning center for ships active in the Pacific, since Hawai`i had no nearby competitors for this commerce. As Americans dominated economic activities there, ties with New England strengthened. By 1845, four hundred Americans were living in Hawai`i—representing two-thirds of the foreign population there.

Politically these Americans were foreigners living in the kingdom of another people. However, unlike the first Europeans and Americans in Hawai`i, their prosperity did not depend on supporting the existing Hawaiian rulers and systems. They did not aspire to assimilate into the local society. Rather, they sought to transplant their own society to Hawai`i. As their economic dominance grew, spurred by investments from New England, they began to resemble colonizers rather than visitors.

In Alaska, the Russian foreigners had been colonizers from the start. From their earliest arrival, they had imposed their rulers and their institutions on local peoples, taxing them in furs for living in their own land. In 1766 Tsarina Catherine the Great declared the native peoples of the Aleutian Islands and the Alaska Peninsula to be Russian subjects. She decreed that they should be treated well, but this was not well enforced. In contrast, tax collecting did become much more systematic after this declaration as the Russian government sent tax collectors along on fur-hunting expeditions. These expeditions competed with one another, often forcing native hunters to work for them.

In 1799 Tsar Paul I granted sole right to trade in Alaska to the Russian-American Company. High-ranking Russian officials were shareholders. Chief Manager Baranov became the director of all Russians in Alaska, a position he held until 1818. Although he married a native woman from Kenai, this did not deter him from abusing Alaskan native peoples. Many of the worst Russian offenses occurred under Baranov's direction. By order of the Russian government, Alaskan natives would not pay taxes, but each year half of the Aleut and Koniag men from age eighteen to age fifty could be forced to hunt furs for the company. These native men could not leave their villages without permission from the company. Baranov also emptied villages of able-bodied men, forcing them to hunt sea otter for the Russians in other regions. These men were unable to hunt for their own families, and women and children starved.

Ironically, Baranov developed a long-distance friendship with Kamehameha, the native ruler of Hawai`i, at the same time that he abused Alaskan native peoples. Baranov and Kamehameha sent gifts and messages to each other through the American fur-trading vessels that visited southeastern Alaska and Hawai`i. Baranov even considered retiring to Hawai`i. He cultivated no such friendship with native leaders in Alaska, viewing them as entirely different from the head of a kingdom.

As colonizers, Russians brought many elements of their own society to Alaska. In Kodiak, the first Russian capital, and later in New Archangel, Russians established a museum, a library, and natural history and scientific collections. The Russian population was never large—in 1839, at its largest, there were only 832 Russians in Alaska, not counting the children of unions between Russian men and native women. However, New Archangel, the Russian-American capital, was a bustling place, with a grand residence for the manager of the Russian-American Company, two churches, a school, barracks, a foundry, a flour mill, a sawmill, a hospital, and a shipyard. Russian officials stationed there enjoyed an active—even gala—social life. In Europe, New Archangel gained a reputation as an oasis of European culture and society.

Missionary Activities

In European and American history, economic and political colonization has been accompanied by social and spiritual colonization. The experiences of indigenous

Alaskans and Hawaiians are no exception. Russians rapidly extended their society's institutions to native populations in Alaska. As early as 1784, Shelikhov and his wife, Natal'ia, founded a school in Kodiak, teaching Russian language and religion to Kodiak children they held as hostages. This school, and others, were soon operated by Russian Orthodox missionaries. Informal missionary activity had started with the earliest Russian fur traders, who began to baptize Aleut hunters. The first recorded baptism took place in 1759, just eighteen years after the Bering-Chirikov voyage. In 1794 ten Russian Orthodox monks went to southwestern Alaska. A Russian Orthodox church was built in Kodiak two years later. Missionaries taught religious, academic, and vocational subjects. When missionaries began complaining to the Russian government about the fur traders' abuse of Alaskan natives, relations between missionaries and fur traders became strained.

After Chief Manager Baranov retired in 1818, the Russian-American Company gave more support to missionaries. Its second charter, issued by the Russian government in 1821, directed the company to maintain "sufficient numbers" of missionaries in Alaska. This charter reflected an upsurge in missionary activity in the Russian Orthodox church, and many Russian missionaries came to Alaska in the 1820s. They began to train native people for positions in the priesthood and especially for service as lay workers. They wanted native lay workers to practice in remote villages where ordained priests were not stationed. They also ran schools for native children, teaching them the Russian language, European academic subjects, and vocational skills.

One of the most influential Russian Orthodox missionaries in Alaska was Ioann Veniaminov, who came to the town of Unalaska in 1824. Missionary work was his first focus, but he also wrote about ethnography and scientific subjects and sent collections back to Russia. He and an Aleut leader, Ivan Pankov, designed an alphabet for Fox Island Aleuts so that they could have a written language—to be used, among other things, for reading religious texts. After ten years in the Aleutians, Veniaminov moved to Sitka, where he worked for another eighteen years. In 1840 he was made a bishop, one of four Russian Orthodox priests in Alaska overseeing churches at Sitka in southeastern Alaska and at Atka, Kodiak, and Unalaska in southwestern Alaska. In addition there were chapels on eight other Aleutian islands.

Russian Orthodox activities continued to expand among native peoples in Alaska. By 1867, when the Russian government withdrew from Alaska, the religion was represented by over thirty Russian Orthodox clergy, with nine churches and thirty-five chapels. Many of the clergy remained active after the Americans occupied Alaska, and they competed with American missionaries for native congregations. In southwestern and southeastern Alaska, Russian Orthodox churches are still vital today.

Religious beliefs are very personal, and few native voices exist from Russian-American times to discuss reasons for native conversions. Some conversions may have been coerced. Certain Russian military personnel were skeptical of native conversions, suspecting that many native converts simply wanted the material benefits that the missionaries bestowed. Nevertheless, a great many native people embraced Christianity through religious conviction. This act did not necessarily imply a rejection of their own spiritual traditions and values. They were accustomed to a range of experiences in

Religious organizations and secular philanthropic groups played a significant role in introducing the native peoples of both Alaska and Hawai`i to the political, cultural, and material values of mainland American society.

Vincent Soboleff (ca. 1880–1950). Schoolchildren at Killisnoo [Angoon, Alaska]. *Photograph, ca. 1900. Alaska State Library (#PCA 1-178), Juneau.*

L. E. Edgeworth. Free Kindergarten, Children's Aid Association. *Photograph, 1913. Bishop Museum, Honolulu, Hawai`i.*

spiritual communication and did not perceive spiritual systems as mutually exclusive constructs. In Alaska, as elsewhere, many native peoples sought syntheses and relationships between their own beliefs and the colonizers' teachings.

As Hawai`i became increasingly like a colonized country, native peoples there faced similar religious issues. Before the first American missionaries arrived in Hawai`i, native Hawaiians had overtly challenged their own system of *kapus*, the strict laws that governed dietary restrictions, gender relationships, proper behavior toward chiefs, rituals at births, marriages, and deaths, and a range of other aspects of daily and ceremonial life. Violating a *kapu* was sacrilege and warranted severe punishment from human or divine sources.

However, the presence of foreigners—who operated outside *kapus* and broke them with impunity—severely weakened the *kapu* system. The maritime fur traders disrupted existing social balances by making material goods more widely available, by successfully encouraging Hawaiians to break *kapus*, and by introducing diseases. Kamehameha's political centralization of Hawai`i may also have removed some focus from local chiefs and the deities with whom they were associated. For a complex of reasons, some Hawaiians began to question the contemporary relevance of the *kapu* system and the religious system from which it sprang.

In May 1819, Kamehameha died, and rule passed to his son Liholiho (Kamehameha II) and to one of Kamehameha I's wives, Kaahumanu. Six months later—strongly influenced by Kaahumanu—Liholiho broke a *kapu* in public. Through this action he declared the end of the *kapu* system. To enforce that decision, he ordered the destruction of religious images. Many leaders opposed his action, and one of his cousins, Kukailimoku, challenged him in battle over the issue. Kukailimoku and his wife were killed; the remaining opposition went underground. Traditional religion survived in private, but people no longer practiced it openly.

Meanwhile, Hawaiians had started working on American sailing vessels and at least as early as 1790 were visiting New England. Hawaiian visitors began receiving Christian educations there. One Hawaiian, Obookiah, went to Connecticut in 1809 and trained at a school for missionary teachers run by the American Board of Commissioners for Foreign Missions. The school's leaders sought to extend its influence beyond American shores, and Obookiah's teachers hoped he would return to Hawai`i and preach to his people. In 1818 he died in Connecticut of typhus, but his life and death inspired the American Board to send a Protestant missionary expedition to Hawai`i. The first small group arrived in 1820, followed by reinforcements eight years later. By 1850, 153 missionaries—Presbyterians and Congregationalists—had worked in Hawai`i for the American Board of Foreign Missions.

For a variety of reasons, the American Board felt strongly that missionaries should go as couples. This presented no small hurdle, for between 1819 and 1850, only two of the men accepted by the board for missionary work in Hawai`i were already married. Often a man had to find a wife suitable to the American Board very quickly, since his sailing date had already been scheduled. Women who wanted to be missionaries faced an even tougher predicament: without formal channels to train for missionary work, they often discreetly notified ministers or relatives in theological schools that they would marry aspiring missionaries.

Lucy Goodale's experience was fairly typical. In autumn 1819 she was a twenty-three-year-old schoolteacher in rural Massachusetts when her cousin suggested she marry Asa Thurston—a man she had never met—who was sailing in six weeks to found a mission in Hawai`i. She had dinner with Thurston once and agreed to marry him. They held the ceremony three weeks later and sailed for Hawai`i eleven days after that. For many couples like the Thurstons, moving to Hawai`i meant a total transformation of their personal lives—leaving family and friends, perhaps forever, and taking a stranger as their lifelong spouse.

When missionaries left their country and their families, they did not leave their allegiance to a New England way of life. They came not to assimilate with the Hawaiian people and adopt native customs and values but to transplant American customs and values to Hawai`i. Missionaries' activities went far beyond religious services. They ran schools, administered medicine, and taught Western crafts, farming techniques, and domestic skills. The women, already overburdened with establishing households in settings they considered spartan, sewed European-style dresses for high-ranking Hawaiian women. The missionaries made one concession to their new location—they studied Hawaiian and preached in that language to reach a broad population. But their goal was to make Hawaiians resemble New Englanders, not vice versa.

Missionaries in Hawai`i did face a complication unknown to their counterparts in Alaska: they relied on the permission of the king and the local chiefs to work in the islands. For Protestants—more often than for Catholics—permission was generally granted, provided that the missionaries established missions where the chiefs wanted them.

The missionaries gained congregations quickly, for local populations were eager to explore what the newcomers offered. When Abner and Lucy Wilcox arrived in 1837, just seventeen years after the first mission was established, they noted that the Hawaiians "appeared to be rejoiced to see us." The Wilcoxes reported that 278 Hawaiians had already been accepted as church members and that this was not done lightly—they were made to "stand trial" a long time to ensure that they were truly converted. Chiefs and even commoners gave money and time to build new churches for the missionaries. The Wilcoxes estimated that along a one-hundred-mile stretch of seacoast in the Hilo and Puna districts of the island of Hawai`i, there were already 122 schools with about 4,537 students. Abner Wilcox was impressed with the students' caliber. He stated that their "capacities" equaled those of children in America. He declared, "Give them the advantages of children in our own land and when grown they, I think, will not be a whit inferior." In other words, he felt sure they could learn to live like Americans.

Like missionaries in Alaska, those in Hawai`i faced the challenge of reaching people in remote villages where missionaries were not stationed. They arrived at the same solution: training native teachers to work among their own people. In 1837, Abner Wilcox was delighted to find that native teachers already assisted at the mission in Honolulu and that about one hundred more were training in an adult teachers' school. Missionaries even hoped Hawaiians would provide manpower for missions to native peoples in the Rockies, where the Presbyterian Board of Home Missions was active. The missionaries' enthusiasm was not always matched by solid information about Native Americans: when Wilcox learned that one Hawaiian volunteer had spent three years on the Northwest Coast and could "speak a little of the Indian language," he assumed this meant the man would understand native peoples in the Rocky Mountains.

Missionaries played a strong role in developing American political power in Hawai`i. They quickly gained influential positions within the government and used those positions to lobby for American interests. Two missionaries, Reverend William Richards and Dr. Gerrit P. Judd, became advisers to the king. Missionaries influenced Kamehameha III to replace his feudal government in 1840 with a constitutional monarchy modeled on the monarchies in Europe. This government included an elected House of Commons, a hereditary House of Nobles, and a prime minister. Dr. Judd served as prime minister from 1842 to 1854, during which time he promoted many major changes that shifted power from Hawaiians to whites—who had come to be known as *haoles*.

Despite their eagerness to entrust Hawaiians with church duties, American missionaries in Hawai`i maintained a marked distance from Hawaiian peoples. Their goal was to teach native Hawaiians the American religion and customs, but they did not seek or expect to develop integrated settlements of equals. Although dispersed in missions throughout the islands, the missionaries relied on each other for friendships. They kept their own children away from Hawaiians to protect them from "heathen" influences. They developed their own political and economic base, influencing the king to adopt reforms that benefited *haoles*—and particularly Americans. They often sent their sons to New England to be educated. These sons returned to Hawai`i with their ties to America made all the stronger by firsthand experience. Many of them turned very successfully to politics and commerce. Although born in the islands, they were—like their parents—very much American colonizers.

The Formalization of American Dominance

America's religious colonization of Hawai`i went hand in hand with the growing political and economic dominance of Americans there. This American colonization started considerably earlier in Hawai`i than in some other places in the American West; by the 1840s, when wagon trains started crossing the plains on their way to Oregon, American merchants had been influential settlers in Hawai`i for some fifty years. In Alaska also, Americans—although not permanent settlers—had been active maritime traders for half a century. By the 1840s, inspired by the philosophy of "manifest destiny," American politicians began to consider colonization of Alaska.

From a military viewpoint, American control of North America depended on naval outposts in the North Pacific. Acquiring Alaska from Russia would at once create these strategic bases and eliminate a potential enemy from North America. It would also facilitate U.S. designs on British Columbia. From an economic viewpoint, Alaskan furs were an attractive resource to American companies. Russians already had harvesting systems firmly in place—based on exploitation of native peoples in southwestern Alaska—and American companies could take over the operation without great changes. U.S. exploration expeditions in the 1850s and 1860s revealed vast fur resources that were still untapped. Alaskan settlements would also help develop American markets in Asia.

The American Civil War put a temporary stop to discussions about acquiring Alaska, but these talks quickly resumed after the war. They coincided with a Russian crisis of confidence. The tsar had become concerned that Russia could not defend Alaska if the United States or Britain started a war. In an interesting parallel to events one century later, Russia's foreign affairs closer to home already strained its military budget. The Russian-American Company profits were not great and did not seem to justify the potential costs of overextending the Russian Empire. In spring of 1867, Russia and the United States agreed on a treaty to transfer Alaska to American control. The purchase price was $7.2 million. Those who believed Alaska was one gigantic icebox referred to the purchase as "Seward's folly"—after William Seward, President Abraham Lincoln's secretary of state, who actively promoted the treaty. In fact, the treaty was very popular in the Senate; only two senators voted against it. On 18 October 1867, Alaska officially became part of the United States. By the terms of the treaty, the Russian Orthodox church members owned all Russian churches in Alaska and remained an active presence there. Russians had the option of returning to Russia or becoming U.S. citizens.

Although the transfer of Alaska from Russia to the United States is often called the "Alaska Purchase," Russia did not own Alaska, for the empire had never established agreements with native Alaskans. The United States also moved in without negotiations with native peoples, and the treaty specified that "uncivilized tribes" would be subject to whatever laws the United States adopted toward them. The status of native Alaskans and native rights remained vague for over a century. In some places, such as the Pribilofs, the United States continued Russia's oppressive policies toward native Alaskans.

In 1867 these concerns were far from the minds of Americans, many of whom were eager to explore what Alaska had to offer. Most Americans who came to Alaska immediately after purchase settled in the old Russian capital of New Archangel, renamed Sitka. Some sought adventure and new horizons; others were military and government officials. The army built six posts but quickly eliminated all except the one in Sitka, which also was abandoned in 1877. Civil government was established in 1884, making Alaska officially a U.S. district with a governor appointed by the president.

Many Americans who came to Alaska did not stay long. The economy in Sitka, which boomed right after transfer, soon slowed. Americans there and in the few other settlements dotting the Alaska coast generally relied on fur trading, an industry with declining revenues. More than the Russians, however, Americans explored other industries and rapidly developed fish processing. In 1868 the first American-run fish saltery in Alaska opened in Klawock. Ten years later 2 fish canneries opened in southeastern Alaska; by 1900, 42 salmon canneries in Alaska packed 1.5 million cases annually. The numbers continued to grow, and by 1917 the 118 salmon canneries in Alaska produced half the canned salmon in the world.

Early fishing and canning operations in Alaska were connected to the mainland despite their geographic isolation. Most were operated by absentee companies based in San Francisco, Seattle, and Portland. These companies ensured that the Alaskan fishing operations had an international focus—not only in their markets but also in their labor forces. Then, as now, white Americans tended to be on the upper end of the pay scales, so the companies recruited workers elsewhere. Many native Alaskans moved in the summers to temporary company housing and worked as laborers, increasingly depending on wages as their traditional fishing grounds were appropriated by American industries. Canneries also imported foreign laborers, including workers from China, the Philippines, Japan, and Mexico, who quickly outnumbered native workers. Americans in Alaska did not embrace this ethnic mix, and incidents of violence and racism occurred.

Meanwhile, in Hawai`i, international economic development brought similar demographic shifts. As whale populations decreased and whaling seasons became less consistent, American businessmen in Hawai`i sought a more stable economic product: sugar. American-owned sugar plantations began to gain attention in the 1840s. Initially Hawaiians were hired as laborers, but with the California gold rush in 1848, demands for sugar skyrocketed. Plantation owners formed the Royal Hawaiian Agricultural Society in 1850, to seek cheap labor elsewhere. They brought some 180 Chinese workers over on contract in 1852. Before the end of the century, more than 46,000 Chinese

Brought to Hawai`i in the late 19th century to work in the burgeoning sugar plantations, foreign workers—such as these Japanese men—often remained on the islands when their contracts expired, becoming part of the racially and culturally mixed population that continues to distinguish Hawaiian society.

Unidentified photographer. Japanese Sugar Workers. *Photograph, 1906. Bishop Museum, Honolulu, Hawai`i.*

workers came to sugar plantations in Hawai`i—despite the Chinese Exclusion Act passed by the kingdom in 1886. Also in 1886, Japan agreed to allow Japanese workers to go to Hawai`i, and many came immediately as contract laborers. Between 1852 and 1900, some 15,000 Japanese, 13,000 Portuguese, 1,400 Germans, 600 Scandinavians, and 400 Galicians came to Hawai`i to work in agriculture.

These laborers faced strict contract terms and hard working conditions on the plantations. Most immigrant workers planned to return home to their families after their contracts expired. Some eventually sent for their families and settled in Hawai`i. Contracts in Hawai`i were generally for much longer terms than those in Alaska, where immigrant workers were shipped out at the end of each summer. Thus, contract laborers were more likely to settle permanently in Hawai`i than in Alaska.

Businessmen who focused on agriculture confronted a problem that had not affected whaling: to secure investments, they needed clear title to land, but in Hawai`i, land was traditionally controlled by the king and chiefs. These leaders often allowed foreigners to use land in return for services, but they did not relinquish title. In the 1840s, *haoles* continually pressed for changes in land tenure. With Dr. Judd as prime minister, they had a strong ally in the government, and by 1850 the legislature had abolished the feudal land system. Lands previously owned by the king were divided among the king, the government, and commoners; and aliens were permitted to purchase land. The system was tailor-made for foreign entrepreneurs with capital to invest in land. Influenced by *haoles*, the government was more than willing to sell its land to alien speculators; by 1886, more than two-thirds of the government allocation was owned by *haoles*. Chiefs and commoners were not accustomed to real estate transactions, and many sold their lands to *haoles* at low prices.

A small group of American investors and managers controlled the sugar plantations and dominated Hawai`i's economic development. Some of the investors in sugar plantations were American merchants; many others were American missionaries or their sons. Abner and Lucy Wilcox's son George studied engineering at Yale to prepare for an agricultural career. He devoted his life to the Grove Farm Plantation near Lihue, Kauai, using his engineering skills to introduce innovations in irrigation and agriculture. He also found places for his siblings at the plantation. Like many other missionary families, the Wilcoxes felt that economic power carried obligations; and they made generous donations to public and private charities and established foundations. Their philanthropic contributions were extremely important and testify to their commitment to their adopted land. Thus they returned much of their profit to Hawai`i but on American terms and through American institutions.

With *haoles*—primarily Americans—gaining economic power in Hawai`i, pressures for political reform continued. In 1886, a group of *haoles*, joined by a few part-Hawaiians, formed the Hawaiian League, an alliance designed to remove power from the monarchy. Some members also wanted American annexation of Hawai`i. The Hawaiian League had the support of the Honolulu Rifles, a volunteer militia with more loyalty to the league than to the monarchy. In June 1887, Lorrin A. Thurston, the grandson of the American missionary Asa Thurston, presented a set of league resolutions calling for a new constitution and less involvement by the monarchy in government.

Most Hawaiians supported King Kalakaua and did not want *haoles* to gain more power over the government. Nevertheless, Kalakaua sought to avoid armed confrontation at all costs, and he acceded to the demands. A new cabinet was established, including Lorrin Thurston, who became the primary author of a new constitution. This constitution turned the monarchy into a largely ceremonial institution subordinate to the legislature. Although the king could veto legislation, a two-thirds vote of the legislature would override the veto. Further, the constitution systematically ensured that *haoles* could control the House of Nobles. American and European male aliens could now vote in elections if they swore to support the constitution. Property requirements greatly restricted the number of Hawaiians voting in House of Nobles elections, and Asians were not permitted to vote.

The Constitution of 1887 was an explicit maneuver to make *haoles* the major power in government. However, the *haoles* had more trouble than they anticipated in controlling the legislature, and many wanted further reforms to secure their influence. When King Kalakaua died in 1891, his sister, Queen Liliuokalani, succeeded him. As she made it clear that she would take an active role in government, reformers began to talk increasingly of American annexation.

In January 1893, Liliuokalani announced her intention to revise the constitution and to return powers to the monarchy. Thurston and other reformers decided the time was right for revolution. Gathering arms, the reformers marched on the government building and took control of it, firing only a single shot and injuring one person. Neither policemen nor government loyalists resisted. This may have been because U.S. troops were stationed in Hawai`i, and it was rumored that they would intervene if loyalists resisted. The United States recognized the new provisional government even before the queen capitulated. Perhaps she hoped that the United States would reject the actions of its countrymen. Great Britain had reinstated the monarchy in 1843 when British citizens, acting without authorization, had claimed to cede Hawai`i for Great Britain.

A ceremonial disbanding of Queen Liliuokalani's palace guards marked the usurpation of power by a provisional haole *government in Hawai`i in January 1893. The guards' American-style military dress suggests the extent to which even native Hawaiian institutions had adopted the conventions of mainland culture.*

James J. Williams (1853–1926). Disbanding Royal Household Guards. *Photograph, 18 January 1893. Smithsonian Institution, National Anthropological Archives (#75-11016), Washington, D.C.*

In fact, President Grover Cleveland did feel it was ethically wrong to take the islands, and far from supporting annexation, he asked the Provisional Government to reinstate the monarchy. The Provisional Government refused, Cleveland did not back his request with troops, and the debate over annexation continued for some years. Sugar refiners in the United States opposed annexation, fearing that they would lose their monopoly. In 1896, William McKinley was elected president. Much more than Cleveland, he supported American expansion. When the Spanish-American War broke out, the United States needed a Pacific base. Despite significant opposition, an annexation resolution passed both houses of Congress by 6 July 1898, and the Republic of Hawai`i became a U.S. territory on 12 August 1898.

Hawai`i's annexation was the culmination of a long process in which American residents gathered economic power and political influence. With the help of European aliens, they gradually eroded the monarchy's power. France and England also had interests in Hawai`i, but the early dominance of Americans there left little serious opportunity for them to step in. Although Hawai`i was not technically a U.S. colony, it was certainly colonized by Americans long before it was finally annexed; when it became part of the United States, a substantial and influential group of Americans already controlled the local government.

The opposite was true when Alaska joined the United States. There was no significant American population or American government there in 1867. The United States needed to import a population of colonizers and gradually establish American political institutions. Alaska was organized into a U.S. district, with an Organic Act and an appointed governor, in 1884—a full seventeen years after the United States claimed possession. For the first quarter century after transfer, American populations grew very slowly in Alaska.

American missionaries, however, quickly became active in Alaska. Like their counterparts in Hawai`i, they did much to spread American institutions, and some also gained political power. In 1877 the Reverend Sheldon Jackson, Presbyterian superintendent of the Home Missions of the Territories, brought a lay worker, Amanda McFarland, to establish a mission for native people in Wrangell. The following year, the Presbyterian Board of Home Missions sent the Reverend John Green Brady as a missionary to native people in Sitka, where there was already a Russian Orthodox church. Over several decades, other missionaries from various denominations established missions to work with native Alaskans throughout the district. They also established churches for American immigrants.

Though he never lived for long periods in Alaska, Jackson gained much political influence in shaping the district's religious and political development. Since Alaska government officials were appointed in Washington, D.C., eastern ties were important for political fortunes; Jackson, like some of his counterparts in Hawai`i, had strong eastern ties. Raised in New York State, he was educated at Union College and Princeton Theological Seminary. In 1884, he was appointed general agent for education in Alaska, a position he held until 1908.

As a missionary leader, Jackson took a strong interest in native peoples and focused his attention on native education, health, and welfare. He encouraged various church denominations to open missions in Alaska, partially supporting them with federal funds. He also spent quantities of time in Washington, D.C., lobbying for funds for Alaskan programs. His critics—of which he had quite a few—felt that administering education through missions violated the separation of church and state. They also charged that he put too much effort into native programs and neglected schools for nonnative students. By 1892, the federal government partially supported twenty schools in Alaska, including three for nonnative students in southeastern Alaska and seventeen for native students in church missions scattered throughout the district.

Jackson's friend Brady also gained political influence in Alaska. Although raised in Indiana, Brady had strong eastern ties, for he attended Yale University and Union Theological Seminary. He quickly resigned from his missionary post in Alaska after a policy disagreement with the Presbyterian Board of Home Missions but stayed in Sitka as a merchant. He was appointed a U.S. commissioner in Sitka after the passage of the Organic Act. His proven record of government service—and likely his friendship with the influential Jackson—led to his appointment as governor of Alaska in 1897. He held that office for nine years, coinciding with a period of extremely rapid population growth. As governor, he often served as an advocate for native rights, helping native people lobby for such reforms as native citizenship. Critics felt he was too closely associated with Jackson and Presbyterians in Washington, D.C., whom they felt had excessive influence over Alaskan affairs.

Without doubt, both Jackson and Brady did mix politics and missionary goals. The tremendous influence they both exercised in Alaska's development can be traced in part to their skill in balancing and synthesizing the two. Missionaries in the American West often did not restrict their activities to religion, and neither Alaska nor Hawai`i is an anomaly in this regard. In both places, missionaries strengthened ties between a seemingly "remote" region and seats of power in the eastern United States.

Image and Economics

Since the nineteenth century, many regions of the American West have been strongly affected by popular perceptions. The image of Oregon's Willamette Valley as a promised land urged wagon trains across the Great Plains. The California gold rush attracted new populations spurred by dreams of wealth thought to be attainable—with a little luck—by anyone. In these cases and others, popular image, economic growth, and demographic trends were all integrally connected. Alaska and Hawai`i were no exception. In both places, the American public's perceptions—often greatly exaggerated—have long influenced economic and political developments.

While the year 1898 brought annexation to Hawai`i, changes in Alaska were more sudden and equally far-reaching. This year marked the height of the Klondike gold rush. Thousands of Americans, suffering from economic depression, rushed to Alaska on steamships to start the overland journey to the Klondike. Although their final destination was in Canada, they supported the development of supply towns such as Skagway in southeastern Alaska. Gold was discovered on the Seward Peninsula in 1897, and frustrated Klondike prospectors turned to Alaska. By the summer of 1900, twenty thousand people were camped in the new city of Nome.

In the decade from 1897 to 1907, over fifty new gold-mining camps operated in various parts of Alaska. Most became ghost towns, but some, such as Nome and Fairbanks, survived. Although only a small percentage of prospectors actually became rich, a significant number of American prospectors and suppliers made Alaska their permanent home. Mining corporations quickly replaced independent prospectors, bringing new technologies and company towns. In 1897 Congress authorized railroad building, and many railroad lines sprang up in the opening years of the twentieth century to serve Alaskan mines and ports. Between 1915 and 1923 the Alaska Railroad was built, and in the process Anchorage was founded—now by far the largest city in Alaska.

Alaska's population surge followed similar trends in other parts of the Northwest, which the Klondike gold rush bolstered. Between 1880 and 1900, Seattle exploded from a town of 3,533 to a city of 80,671 residents. Many of the people who came to Alaska had stayed for a time in Seattle, experiencing Washington's rapid growth and seeing the territory gain statehood in 1889. Few American immigrants to Alaska were satisfied with its political status as a district. Seeking more self-government and more representation in the federal government, Alaskans faced the issue of whether to become a territory—the usual next step—or leap to immediate statehood. Proponents of territorial status felt they needed to expand their transportation and communication infrastructure before becoming a state; they feared statehood would eliminate some federal funding. Advocates of immediate statehood, such as Governor Brady, thought that greater home rule and equality with other states in Congress would offset those problems. Pointing to the experiences of Arizona and New Mexico, still territories after some fifty years, Brady feared that if Alaska became a territory, statehood would be a long way off.

To promote immediate statehood, Alaska needed a larger nonnative population. To recruit newcomers, the state had to shake the image of a frozen northern island. That image appealed to the growing tourist traffic but did little to support a true appreciation of Alaska's needs or potential. Many Alaskans felt that federal policymakers knew little about Alaskan conditions and made decisions based on image instead of facts. Successive

As on earlier American mining frontiers, the discovery of gold transformed the Alaskan landscape. A rough tent community in 1900, Nome was by 1908 an established city serving the outlying mines.

B. B. Dobbs (1868–?). Nome, Alaska. *Photograph, July 1900. Alaska State Library (#PCA 12-157), Juneau.*

B. B. Dobbs (1868–?). Front Street [Nome] from Grandstand, Starting of the Dog Teams, All Alaska Sweepstake Race. *Photograph, 1 April 1908. Alaska State Library (#PCA 12-103), Juneau.*

governors traveled to Washington, D.C., correcting congressional misconceptions and striving to provide more accurate information about Alaska's resources. Sometimes this backfired, and they were accused of wintering in Washington to escape Alaska's frigid climate. One frustrated governor published Sitka's average seasonal temperatures in his annual report to prove that he would have been much more comfortable wintering at home than shivering in Washington, D.C.

Overcoming this "icebox" image was a challenge, especially without modern television, film, and transportation. However, Alaska would attract neither population growth nor outside investment until the world knew more about it. Governor Brady took advantage of the St. Louis Exposition of 1904, staging a lavish Alaska Pavilion with attractions such as the huge cabbages that grew in the Matanuska Valley, where twenty-four-hour sunlight offset the relatively short growing season. Photography also helped spread information about Alaskan resources. Some satirists could not resist targeting Alaskan promoters, and one cartoon shows a map of Alaska with a single (Hawaiian?) palm tree rising from the vicinity of Fairbanks. Predictably, neither population nor investments in infrastructure grew fast enough to support a strong statehood movement in the opening years of the century, and in 1912 Alaska became a territory. It would remain a territory for the next forty-seven years.

Today, however, the results of population growth in other parts of the American West have given reason for pause. It is not uncommon in Alaska to hear the comment that if Alaska's environments and quality of living were properly understood, some parts of the state would have the density and pollution of Los Angeles. This may be an exaggeration, but it points to an important irony: many of the people who currently migrate to Alaska go to escape living conditions farther south, and the more people who make that choice, the less of an escape it will be. Other western states face similar dilemmas, as Seattle's skyrocketing population and urban gridlock testify.

If Alaska's exotic image hindered most forms of growth, it spurred the economy in one way. An active tourist industry developed along Alaska's Inside Passage in the late nineteenth and early twentieth centuries. Enjoying deluxe accommodations, tourists dined and socialized on board as their ship glided through narrow channels past rain forests and glaciers. The steamships called in to port cities such as Wrangell, Sitka, and Juneau, where native women lined the docks to offer baskets and carvings for sale. For many travelers, the apogee of their trip was a walk on a glacier, an adventure that was more difficult then than now—especially for women—considering the dress of the day. The protected nature of steamship travel made it acceptable for women to travel on their own, and several women wrote travel journals publicizing the pleasures of an Alaskan expedition. These journals generally emphasized the comforts of traveling on steamships to Alaska, and some explicitly sought to counter misconceptions about the region.

A nascent tourist industry also developed in Hawai`i, although before air travel, the trip was a long one. The first tourist guidebook to Hawai`i appeared in 1875, and by the time of annexation, tourism was bringing in some five hundred thousand dollars annually. The first hotel on Waikiki beach was completed in 1901. However, services for tourists would grow only in response to demand, and demand was hampered partly by the lack of luxury services. This was less of a dilemma in Alaska, where tourists spent most of their time on steamships. In Hawai`i, it meant that the tourist industry grew relatively slowly, a surprising fact considering Hawai`i's beauty and the economic significance that tourism would later assume.

However, Hawaiians were developing other industries that helped diversify their sugarcane economy. In 1903 the Dole plantation experimented with growing pineapples. Under the leadership of James D. Dole, the Hawaiian Pineapple Company purchased the entire island of Lanai in 1922. By the 1930s, pineapples would be Hawai`i's second most important crop. As their opportunities expanded, plantation owners also faced problems. Increasingly, legislation and public pressure regulated treatment of workers on plantations. Workers were also beginning to express resentment of pay scales that were tied directly to racial background. In 1909, some five thousand

By 1920, when this casual snapshot was taken, hula dancing was being promoted to a growing tourist trade as a stereotypical Hawaiian activity.

Ernest Moses. Hawaiian Children in Backyard. *Photograph, ca. 1920. Bishop Museum, Honolulu, Hawai`i.*

Japanese plantation workers went on strike, protesting that they were paid four or five dollars less per month than Portuguese and Puerto Ricans for the same jobs. Plantation owners hired other workers and broke the strike, but they would face more challenges after World War I as workers unionized.

Meanwhile, industrialists in Alaska also faced labor issues. As the number of mines grew there, the Western Federation of Miners became active in Alaska. In 1905 it organized workers at the Treadwell Mines near Juneau, drawing on racial tensions as justification. The miners held a strike in 1905, and federal troops were called in to ensure the mine was not blown up. The Treadwell company met some, but not all, of the workers' demands, and the miners held another strike the next season. This time, though, the company was more prepared and called in workers to break the strike.

The labor tensions in Alaska and Hawai`i were similar to tensions faced elsewhere. Alaska's image as the "last frontier" and Hawai`i's image as a tropical "paradise" made both seem exotic to outsiders. In fact, the people actually living and working in Alaska and Hawai`i struggled with the same economic issues challenging workers in other parts of the American West, the nation as a whole—and the world. As international politics erupted into World War I, Alaska and Hawai`i were by no means isolated.

War and Statehood

The American West has been affected by international ties throughout its history. From the earliest exploration, it was shaped by national rivalries, and it developed under the strong influence of multinational investments and labor forces. In the twentieth century, increasing globalization and advances in technology kept national defense a priority. This emphasis, and the two world wars, changed all parts of the American West, but none more than Alaska and Hawai`i.

One of the arguments for annexing Hawai`i had been to develop a U.S. naval presence in the Pacific. Work began on Pearl Harbor in 1908, and it officially opened in 1911. The navy also built a wireless station on Oahu that enabled communication with ships in the Pacific. Army bases in existence since the Spanish-American War were bolstered, and a large base, Schofield Barracks, opened in 1908 to help protect Pearl Harbor.

World War I affected Hawai`i immediately and intimately, for Hawaiians feared actual attack. When the war broke out, ships in the Pacific, including several German vessels, rushed to Hawai`i to dock in safety. As tensions grew between Germany and the United States, Hawaiians feared that Germany would explode the vessels to destroy Hawaiian harbors. The United States seized the German ships on 5 April 1917, the day Congress passed a war resolution.

Otherwise, Hawaiians responded to the war much as U.S. residents elsewhere. Some islanders went overseas in the military and in support services; those at home bought liberty bonds and worked in relief agencies. Hawai`i experienced the anti-German sentiments prevalent on the mainland. A vigilance corps was formed, teachers had to take loyalty oaths, German language classes were stopped, and some people lost jobs when they were accused of pro-German feelings. The largest firm in Hawai`i, H. Hackfeld & Company, was controlled by German shareholders at the outset of the war and had been the agent for the German ships in Hawai`i. Rumors linked firm officials with pro-German activities. Under the Trading with the Enemy Act, the U.S. government seized most of the shares and sold them to American businessmen in Hawai`i. H. Hackfeld became American Factors, Limited, and remained in American hands after the war, reinforcing American economic dominance in Hawai`i.

Alaska's economy was suddenly and significantly affected by World War I. The tourist industry sagged, since many Americans believed that extended pleasure travel was frivolous. Federal appropriations for transportation and for communication networks slumped during the war, and the development of the infrastructure slowed. In contrast, the major Alaskan industries—fishing, mining, and timber—were tied to international markets for exported resources, and these markets boomed. New salmon canneries and cod-processing plants opened, commercial clam fishing expanded, and crab fishing was undertaken. The war also elevated prices of minerals, and copper mining prospered.

Neither Alaska nor Hawai`i was attacked in World War I. However, at the time—with men and women going overseas, with the economic implications, and with the fear of sabotage in the harbors of Hawai`i—the impact seemed great. Despite their geographic locations, both regions had long been part of global political and economic dynamics.

Like the rest of the country, Alaska and Hawai`i faced significant adjustments in the postwar period. The sudden wartime increases in world demand for Alaskan products were temporary. Some industries, such as fishing, remained active and viable. Others, such as copper mining, suffered more, and many large mines closed in the 1930s. The timber industry had benefited least from World War I, although World War II would bring a much larger boom.

In Hawai`i, owners of large plantations had long recognized the need to cooperate with each other in order to compete with sugar interests on the mainland. By 1933, five firms—known as the "Big Five"—controlled 96 percent of the sugar crop and were also gaining control of the Hawaiian Pineapple Company's operations. The Big Five controlled large shares in transportation, banking, and merchandising.

Many positions of power in the Big Five and its subsidiaries were held by descendants of nineteenth-century American missionary families. These businessmen controlled the Republican party in Hawai`i, and Republicans dominated the territorial legislature. The Big Five also exerted influence over appointments to political and judicial offices. Initially, their Hawaiian Sugar Planters' Association opposed statehood, which would reduce their influence by allowing voters to elect territorial officials. However, in the 1930s, Congress reduced the amount of sugar that Hawai`i could market and increased the quota marketed by states. The association protested, but federal courts ruled that Congress had the right to discriminate against territories when setting such quotas. Thus, when Hawai`i's nonvoting delegate to Congress, Samuel Wilder King, introduced the first Hawaiian statehood bill in 1935, he had the full support of the Hawaiian Sugar Planters' Association.

Racism complicated the statehood issue. Chinese and Japanese immigrants were leaving the plantations and establishing their own businesses in towns. By 1932, about one hundred thousand Japanese lived off the plantations, and their birthrate was high. Their Hawaiian-born children, known as nisei, were U.S. citizens eligible to vote. Some whites feared that if Hawai`i became a state, a Japanese governor would be elected—a prospect that seemed dangerous because Japan was becoming a Pacific power. These concerns about the loyalty of the Japanese in Hawai`i foreshadowed Katamoto's experience in World War II.

Unlike Hawai`i, Alaska was not yet in a position to lobby seriously for statehood. In 1937, only seven cities had over one thousand residents. With just twenty-five hundred miles of public highway built, no road connected Alaska with the contiguous states. In 1940, the territory's population was about seventy-four thousand, a drop in the bucket considering its vast size.

Japan's growing power also caused concern in Alaska. In 1933, Anthony J. Dimond, Alaska's delegate to Congress, warned that Japan was a threat. He requested that Congress bolster military establishments in Alaska and construct a highway between the territory and the rest of the United States to facilitate defense. Aware of Hawai`i's strategic importance in the Pacific, he argued that Alaska was similarly important and should receive similar fortifications. In 1937 he claimed that Japanese fishermen off the coast of Alaska were actually Japanese military spies. In 1940, Congress authorized $29,108,285 for military bases throughout the territory, but construction progressed slowly.

Despite their geographic isolation, it was clear before World War II that both Alaska and Hawai`i were integrally tied to the rest of the United States. Caucasian Americans—as distinct from indigenous peoples or immigrants—dominated both territories politically. Leaders in both places were ambivalent toward the federal government; welcoming federal spending, they also felt threatened by some federal policies and wanted more representation in Congress.

The American public, however, did not yet view the two territories as essential to the United States. In 1938, retired Major General Smedley D. Butler had recommended that in the event of an attack on Alaska, the United States should give up the territory without trying to defend it. In January 1940, a poll of Americans in the contiguous states revealed that only 55 percent of respondents felt Hawai`i should be defended if attacked. Ironically, a full 75 percent felt that the United States should help defend Canada in an attack. Clearly, at the brink of World War II, there were major differences of opinion about the importance of Alaska and Hawai`i to the United States.

The American military response to the Japanese invasion of the Aleutian Islands during the Second World War fostered national interest in the region and enhanced Americans' interest in Alaska as a part of their own country.

Unidentified photographer. Japanese Officers, Aleutian Campaign. *Gelatin silver print, 1942–43. National Archives, Washington, D.C.*

Japanese planes bombed Pearl Harbor on 7 December 1941. Hawai`i remained under martial law for almost four years, and the territory was considered a combat zone until April 1944. Perhaps because formal U.S. involvement in World War II started with the attack on Pearl Harbor, awareness of these events remains high today. However, Alaska was actually invaded during World War II—the first time an enemy military had held American soil since the War of 1812.

Japanese aircraft and ships attacked the Aleutian Islands in June 1942. U.S. aircraft fought them off from Dutch Harbor, but the Japanese occupied the islands of Attu and Kiska. From there, they posed a potential threat to military establishments in the Seattle area. The United States and Canada sent bombing raids to Kiska, gradually moving closer and closer as the United States completed air bases on the Aleutians. Some bombers engaged in dogfights with Japanese aircraft. Between combat and treacherous weather conditions, both sides suffered relatively high losses. On 11 May 1943, U.S. forces—infantry, naval ships, and one aircraft carrier—attacked Attu, finally succeeding in retaking it on 29 May. The Japanese evacuated their troops from Kiska before American and Canadian forces came ashore in August.

The war transformed perceptions of Alaska and Hawai`i as the United States rallied to defend them. Residents of both territories fought overseas in the war, contributed to relief work locally, learned first aid, and coped with shortages and rationing. Americans began to think of Alaska and Hawai`i as part of their country—still "exotic islands," to be sure, but islands their neighbor's son had helped protect and islands they might even visit some day.

This shift encouraged a post–World War II boom in tourism. Technological advances in air travel brought both territories closer to the mainland, and the spectacular Alcan Highway made Alaska accessible by car. By the mid-1960s, the number of tourists visiting Hawai`i each year outnumbered the permanent residents. The same would soon be true of Alaska, where in 1975 tourism was already a fifteen-billion-dollar industry.

World War II also brought major demographic changes to both Alaska and Hawai`i, further decreasing isolation. Civilians followed military personnel from the contiguous states, providing support services during the war while seizing the chance to combine patriotism with adventure and economic opportunities. Alaska's civilian population skyrocketed between 1940 and 1950, growing from 74,000 to 112,000 and thus advancing arguments for statehood.

Statehood movements in both Hawai`i and Alaska gained force after the war. Alaskans and Hawaiians still debated the pros and cons of statehood, but by and large residents wanted equal status with states, and new residents from the contiguous states wanted the same representation they had previously enjoyed. The statehood issues became a part of national politics, with both Democratic and Republican parties adopting promises of statehood into their party platforms. However, Alaska and Hawai`i—like many territories before them—found that politics was not a simple proposition. In the early 1950s, the Eisenhower administration supported statehood for Hawai`i—which tended to vote Republican—but was less enthusiastic about Democratic-leaning Alaska. Conservative congressmen from the southern states raised concerns about both territories.

Opponents to statehood worried that "noncontiguity" would destroy the foundation of America's union. With the cold war highlighting national defense, many Americans feared overextension—and saw Alaska and Hawai`i as precedents that would later lead to statehood for places such as the Philippines, Guam, and Okinawa. In American politics, the names of the Founding Fathers are often invoked, and this issue was no exception. Speaking about Hawai`i, Representative Kenneth M. Regan of Texas proclaimed: "I fear for the future of the country if we start taking in areas far from our own shores that we will have to protect with our money, our guns, and our men, instead of staying here and looking after the heritage we were left by George Washington, who told us to beware of any foreign entanglements. I think he had this outpost in the Pacific Islands in mind at that time."

In the conservative climate of the cold war, racism again played a role. According to the 1950 census, Hawai`i's population of about 500,000 consisted of 183,000 Japanese, 114,000 whites, 87,000 native Hawaiians, 33,000 Chinese, and 88,000 people of other ethnic backgrounds including Filipino, Korean, and Portuguese. Alaska's population of 128,643 included some 33,000 native Alaskans. Some conservative Congressmen were loath to admit states that had such ethnic diversity. Reportedly the question cropping up in discussions of Hawaiian statehood was, "How would you like to have a United States senator called Moto?"

For these and other reasons, Alaska and Hawai`i faced a prolonged fight for statehood; leaders from both territories lobbied vociferously throughout the 1950s. Finally, in 1959, Congress voted to admit Alaska as the forty-ninth state and Hawai`i as the fiftieth state.

Hawai`i's economy boomed after statehood, continuing the postwar trend. As in Alaska, the military establishment was a major source of revenue. The cold war ensured that military installations would remain active. The influx of population after the war, and the increase in tourism, supported a major building and industrial surge. Many residents today—especially native Hawaiians and long-term residents who remember the "old" Hawai`i—watch sadly as more and more of the coastline is paved with tourist parking lots and resort hotels. Many parts of the islands are now off-limits to residents who lack the wealth and time to stay at these resorts. Like many other parts of the American West today, Hawai`i is paying the price of popularity—facing the recognition that continued growth may obscure what attracted people in the first place.

In Alaska as well, the debate between development and wilderness preservation continues, pitting economic developers against conservationists. In December 1980, Congress passed the Alaska National Interest Lands Conservation Act, preserving over one hundred million acres of Alaskan lands. Most were closed to development and resource extraction, but this protection can be altered as the economy and political climate dictate. Further concessions were made to developers before the passage of the act, granting subsidies and benefits to help develop other parts of the state. As a result, regions such as Tongass National Forest in southeastern Alaska have been heavily exploited.

With the drop in oil prices in the 1980s, Alaska faced a sudden recession. Its economy had become very dependent on oil revenues; since then, Alaskans have attempted to diversify the economy. The reliance on oil and resource extraction has recently highlighted environmental concerns as disasters such as oil spills clash with pressures to open up the Arctic National Wildlife Refuge for resource exploration. Tourism has been heralded in Alaska, as in the Pacific Northwest, as a way to gain revenue from wilderness lands without destroying them. As Hawai`i's experience demonstrates, however, tourism can bring problems as well as rewards.

Alaska and Hawai`i may still be widely regarded as places to escape to, but they have long been essential participants in the history of the United States and the American West. In the 1990s, Americans increasingly focused on worldwide relationships as well as on national boundaries. No less than the rest of the country, Alaska and Hawai`i are influenced by global events, markets, and ecosystems. The breakup of the Soviet Union revealed tremendous environmental devastation in Siberia, where nuclear residue had been dumped; Alaskans and other Arctic neighbors fear contamination. Problems such as these require broad international cooperation. Alaska's and Hawai`i's isolation has for some time existed only in the popular mind.

Ethnic Diversity and Civil Rights

Throughout the history of the American West, different ethnic groups have mingled in the western landscape. Long before the arrival of nonnative peoples, various Native American groups came together in commerce and in warfare. The American settlement of the West dispossessed Native Americans, and today many of their grievances remain unresolved by the federal government and the courts. Other ethnic groups also faced discrimination. In the twentieth century, issues of civil rights and economic opportunity came to the forefront throughout the American West.

The years following World War I brought changes in native rights in Alaska. Like Native Americans elsewhere, native Alaskans had gained U.S. citizenship under the Citizenship Act of 1924, and this included the right to vote. Some native people in southeastern Alaska had successfully fought for the right to vote a couple of years earlier, but the act of 1924 made their right explicit. That year, William Paul, a Tlingit attorney, was elected to the territorial House of Representatives. One vote in a forty-member House was not substantial, but Paul's election as early as 1924 made a major statement about the potential power of native voters.

The Alaska Native Brotherhood's newspaper, the *Alaska Fisherman*, which circulated in southeastern Alaska, encouraged native people to vote according to their economic interests. Since many were fishermen struggling to remain independent, voting for their economic interests often meant voting against candidates supported by large fishing corporations and other big businesses. Some native peoples in southeastern Alaska were asked to pass tests at the polls—such as reciting the Preamble to the Constitution—before they were allowed to vote. Like African Americans in the American South, they struggled with discrimination in various forms and fought many battles in the succeeding decades.

In Hawai`i, Japanese-born immigrants who had served in the U.S. military returned to discover that they were still ineligible for American citizenship. However, their Hawaiian-born children were U.S. citizens, as were sons and daughters of immigrants of other nationalities. World War I accentuated the strong "melting-pot" sentiments in the United States, and immigrants were encouraged to forget their old country and concentrate on becoming "American." This message was particularly strong in Hawai`i, with its large immigrant population. Here again, Hawai`i's isolation from the mainland United States ended with geography. Immigrants in Hawai`i faced a pressure to assimilate just as did immigrants in the mainland United States.

In the grand tradition of political slogans then and now, the definition of "Americanism" was left somewhat vague. Promoters of "Americanism" insinuated that the country faced a dire moral threat that could be thwarted only by limiting individual liberties. Such sloganism has since proven to be a fundamental part of American politics. During the cold war, similar tactics were used to promote McCarthyism, and in the 1990s the vague but wholesome "family values" slogan spearheaded an assault on liberties deemed dangerous to America's moral fabric. Likewise, in Hawai`i immediately after World War I, "Americanism" actually meant assimilation—the eradication of cultural ties that differentiated immigrants from white Anglo-Saxon Americans.

Champions of "Americanism" targeted foreign-language schools. They viewed these institutions as obstacles to the assimilation of immigrants—and "unassimilated" immigrants, they argued, posed a dire threat to patriotism and national security. The Aloha Chapter of the Daughters of the American Revolution expressed this perception when it declared that its members were "unequivocally opposed to all practices within the borders of the United States of America subversive to the peace and order of our Nation and the undivided allegiance of our people, and unalterably opposed to all foreign-language schools of whatever nationality." Less subtly still, the chapter emphasized that it took a "firm stand for Americanism in its truest and loftiest form, and for one language—that of our heroic Revolutionary ancestors who gave their fortunes and their lives that the United States might live and prosper, and one flag—'Old Glory'!"

With such statements in the air, people who supported foreign-language schools for immigrant children could easily be labeled as subversives. In 1920, the legislature of Hawai`i passed a law entitled "An Act Relating to Foreign Language Schools and Teachers Thereof." As amended in 1923 and 1925, the act imposed special fees on these schools, controlled their curricula, and regulated selection of teachers and textbooks.

The primary targets were 147 Japanese-run schools, which successfully challenged the law in federal courts. The territory then appealed to the U.S. Supreme Court, which upheld the decision in favor of the Japanese schools on the grounds that the law abridged fundamental individual rights guaranteed by the U.S. Constitution. This particular effort to force assimilation was thwarted, but similar language issues are very much evident in the American West today—notably in California, where the use of the Spanish language in schools has become a volatile issue.

The 1930s saw changes in federal policy toward native peoples throughout the United States, including Alaska. In 1934 Congress passed the Indian Reorganization Act (Wheeler-Howard Act) and in 1936 extended the provisions of the act to Alaska, authorizing the U.S. Department of the Interior to establish reservations for native Alaskans. The reservation lands would not belong to native peoples but would be held "in trust" by the U.S. government.

The act raised fears and controversy among both native and nonnative Alaskans. Nonnative developers feared land would be tied up in reservations. Native peoples were well aware that reservations often meant dependency, with too little land and resources to support the communities. At the same time, they had no other way to protect their lands against the encroachment of the growing white population. By 1946, a total of seven reservations had been established—one of which was later ruled illegal—and three villages had voted down reservations. By 1950, some eighty other villages had requested reservations, but by that time the federal government was again encouraging assimilation of native peoples, and the requests were not granted.

The 1930s also saw another movement to protect native Alaskan lands, this one initiated in Alaska by native peoples. In 1929 at the Alaska Native Brotherhood annual convention in Haines, Tlingit and Haida leaders decided to press the federal government for compensation for lands taken when Alaska was transferred from Russia. Before they could do so, it was necessary for Congress to pass a law to authorize them to sue the U.S. government. This law was passed in 1935, but it was 1959 before a decision was handed down, the courts eventually ruling in favor of the Tlingit and Haida plaintiffs. This lawsuit helped lay the foundation for a statewide struggle for native land settlements, a struggle that culminated in the Alaska Native Claims Settlement Act of 1971.

Meanwhile, World War II had an immense impact on intangible attitudes of and toward ethnic groups who had long suffered discrimination. People from Alaska and Hawai`i who had served in the military returned to civilian life with knowledge of how people lived in other parts of the world. They also had demonstrated that they were willing to give their lives for their country. Those who experienced prejudice while in the service, or who faced it when they returned home, were ready for a change. For some time during the war, servicemen in Alaska were prohibited from speaking with native women—and this prohibition extended to native servicemen, who were not permitted to talk to their own sisters. Many Hawaiians who were targets of oppression also served in the military—Japanese, native Hawaiians, Chinese, Portuguese, Filipinos, and Koreans—and returned to Hawai`i knowing that they still faced a domestic struggle. As one Japanese serviceman, Katsumi Komentani, put it: "We have helped win the war on the battlefront but we have not yet won the war on the homefront. We shall have won only when we attain those things for which our country is dedicated, namely, equality of opportunity and the dignity of man." Just as the Pribilof Islanders had returned home from their internment camp with new political ties and a determination to change their status, so too military personnel from Alaska and Hawai`i returned home with much the same resolve. All over North America, the fight for civil rights and equal opportunity escalated after World War II.

In Alaska and Hawai`i, this struggle continued to involve access to land and natural resources. Alaskan statehood made it necessary to distribute land between the state and the federal government, and this process quickly became even more complicated than anticipated. The statehood act entitled the state of Alaska to control 104 million acres of land. Native Alaskans protested the state government's land selections, and in 1966 U.S. Secretary of the Interior Stewart Udall froze selections in an effort to encourage the fair settlement of native claims. The issue—always urgent in the eyes of native peoples—became urgent in the eyes of developers when oil was discovered on the Alaskan Arctic coast in 1969. To build a pipeline, developers needed access to the land Udall had frozen, and they joined those pressing for a solution. Unlike the earlier Tlingit-Haida settlement of 1959, the issue was decided through passage of congressional legislation rather than through the courts. The Alaska Native Claims Settlement Act of 1971 (ANCSA) established twelve regional corporations in Alaska—with land and funds—and a thirteenth corporation, without land, for native Alaskans living outside Alaska. Village corporations were also established. Alaskans born before December 1971 with at least one-quarter native background became shareholders of these corporations. The settlement has proven extremely complex and problematic, and some provisions are under revision today. ANCSA's corporate alternative to reservations has not become a model for native land settlements in the United States or Canada. Nevertheless, native groups and governments in both countries continue to pursue resolution of native land issues through negotiation and legislation rather than through litigation alone.

In Hawai`i, the focus on land distribution intensified after statehood. Due to the centralized power of the Big Five, control of Hawaiian lands was less than democratic. Thirty percent of the land in Hawai`i was owned by twelve families, and 27 percent belonged to much smaller property owners. Most people of native Hawaiian and Asian descent could not afford to purchase land. Many homeowners, then and now, own structures on land that is only leased.

Issues of land distribution, resource control, environment, and development remain central in both Alaska and Hawai`i and promise to spark controversy for some time. Both places are still mingling important grounds for a number of ethnic groups, and both must continue to address questions of civil rights and discrimination. Both retain sometimes tense relationships with the federal government as they balance local autonomy with federal involvement. Tourism continues to grow, with visitors flocking from North America, Europe, and Asia.

These developments echo similar phenomena in other parts of the American West. Alaska and Hawai`i—and the rest of the American West—have long been influenced by international economies and politics. As the twenty-first century opens, there is a growing recognition that political boundaries are purely artificial as far as issues of human welfare, economic prosperity, and biological and cultural diversity are concerned—that the welfare of all countries is vitally connected and must be played out on a global stage. With this shift in perceptions, the apparent isolation of regions such as Alaska and Hawai`i will be deemphasized in view of their integral ties to the United States and the world. Like other areas of the American West, Alaska and Hawai`i have had much experience confronting local concerns in the context of government from afar. Their history will prove valuable as they approach the global challenges of the twenty-first century.

Bibliographic Note

Several general histories of Hawai`i have contributed to this chapter and are useful sources of additional information: Gavan Daws, *Shoal of Time: A History of the Hawaiian Islands* (New York, 1968); Lawrence H. Fuchs, *Hawaii Pono: A Social History* (New York, 1961); Edward Joesting, *Hawaii: An Uncommon History* (New York, 1972); Noel J. Kent, *Hawaii: Islands under the Influence* (New York, 1983); and a three-volume work by Ralph Kuykendall, *The Hawaiian Kingdom* (Honolulu, 1938–67). John Whitehead's article "Hawai`i: The First and Last Far West?" *Western Historical Quarterly* 23, no. 2 (May 1992): 153–77, persuasively points out that many developments in the history of the American West occurred first in Hawai`i.

Eighteenth-century journals of European visitors to Hawai`i reflect early contact between native Hawaiians and Europeans, albeit through European eyes. The journal of Captain James Cook is perhaps the best known. Sketches that explorers made of native Hawaiians, and records of artworks they collected, can lend some insight into early interactions. Sketches are reproduced in Bernard Smith, *European Vision and the South Pacific*, 2d ed. (New Haven, 1985); Adrienne L. Kaeppler's *Artificial Curiosities: Being an Exposition of Native Manufactures Collected on the Three Pacific Voyages of Captain James Cook, R.N. . . .* (Honolulu, 1978), discusses both explorers' sketches and native artwork collected in the South Pacific during Cook's voyages.

Nineteenth-century writings by missionaries and colonists provide information about colonization and also reflect attitudes of the foreign newcomers. One useful account is that of the American missionary Hiram Bingham, who resided in Hawai`i from 1820 to 1840. He recorded both his own observations and his account of Hawaiian history in *A Residence of Twenty-One Years in the Sandwich Islands* (Rutland, Vt., 1981). Alfons L. Korn, ed., *The Victorian Visitors* (Honolulu, 1958), focuses on the period from 1861 to 1866 and quotes from letters of Sophia Cracroft and Lady Franklin and from the diaries and letters of Queen Emma of Hawai`i. Portions of the journal of Mary Chipman, who accompanied her husband on a whaling voyage to Hawai`i in 1856, are reproduced in Stanton Garner, ed., *The Captain's Best Mate: The Journal of Mary Chipman Lawrence on the Whaler* Addison, *1856–1860* (Hanover, N.H., 1966). Readers who prefer biography will be interested in Marjorie Sinclair's *Nahi'ena'ena: Sacred Daughter of Hawaii* (Honolulu, 1976), an account of the life of Kamehameha I's daughter, who was influenced by missionaries and the changes brought by colonization but also remained committed to her heritage.

Quite a few other published works and archival sources also reflect activities of American missionaries in Hawai`i. Patricia Grimshaw's *Paths of Duty: American Missionary Wives in Nineteenth-Century Hawaii* (Honolulu, 1989), is particularly useful; it reflects women's perspectives and offers an excellent and detailed impression of missionaries' daily lives and concerns. The information about Lucy Goodale is drawn from this insightful book, which also has an extensive bibliography of published primary and secondary sources. Barnes Riznik, *Waioli Mission House, Hanalei, Kauai* (Kauai, Hawai`i, 1987), provides a very useful case study of an American Protestant mission founded in 1834 and now open to the public as a historical museum. Bob Krauss, with W. P. Alexander, *Grove Farm Plantation: The Biography of a Hawaiian Sugar Plantation,* 2nd ed. (Palo Alto, Calif., 1965), is a biography of George Wilcox, son of the American Protestant missionaries Abner and Lucy Wilcox. It contains much information about their mission work and also documents how the children of missionaries often became active in business, politics, and philanthropy. The Grove Farm Plantation, still a working plantation, is open to the public; its archives include some journals by Abner and Lucy Wilcox as well as plantation records (including pay scales) and photographs. This chapter drew on research conducted in those archives, and I am grateful to the superintendent, Barnes Riznik, and his staff for their welcome and assistance.

Numerous publications explore the mixing of peoples in Hawai`i and/or focus on one ethnic group. A native Hawaiian author, Samuel Kamakau, recorded information about Hawaiian culture and mythology. D. Barrere has edited two volumes of Kamakau's writings: *Ka Po'e Kahiko: The People of Old* (Honolulu, 1964) and *The Works of the People of Old* (Honolulu, 1976). Linda S. Parker, *Native American Estate: The Struggle over Indian and Hawaiian Lands* (Honolulu, 1989), includes information on American appropriation of native Hawaiian lands. Michi Kodama-Nishimoto, Warren S. Nishimoto, and Cynthia A. Oshiro, eds., *Hanahana: An Oral History Anthology of Hawaii's Working People* (Honolulu, 1984), gives accounts by speakers from several different ethnic groups. The experiences of Usaburo Katamoto are presented in this work. Dorothy Ochiai Hazama and Jane Okamoto Komeiji, *Okage Sama De: The Japanese in Hawai`i, 1885–1985* (Honolulu, 1986), Dennis M. Ogawa, *Kodomo No Tame Ni = For the Sake of the Children: The Japanese American Experience in Hawaii* (Honolulu, 1978), and Alan Takeo Moriyama, *Imingaisha: Japanese Emigration Companies and Hawaii, 1894–1908* (Honolulu, 1985), discuss people of Japanese background in Hawai`i. Gary Y. Okihiro, *Cane Fires* (Philadelphia, 1991), considers the anti-Japanese movement in Hawai`i from 1865 to 1945. Works on Chinese immigrants include Tin-Yuke Char, ed., *The Sandalwood Mountains: Readings and Stories of the Early Chinese in Hawaii* (Honolulu, 1975), and Clarence E. Glick, *Sojourners and Settlers: Chinese Migrants in Hawaii* (Honolulu, 1980). Edward D. Beechert, *Working in Hawaii: A Labor History* (Honolulu, 1985), discusses the contract labor system, the development of a labor movement in Hawai`i, and the conditions under which many immigrants work. Another useful book on labor conditions affecting immigrants is Ronald Takaki's *Pau Hana: Plantation Life and Labor in Hawaii, 1835–1920* (Honolulu, 1983). John F. McDermott, Jr., Wen-Shing Tseng, and Thomas W. Maretzki, eds., *People and Cultures of Hawaii: A Psychocultural Profile* (Honolulu, 1980), discusses contemporary circumstances of various populations in Hawai`i, as does Andrew W. Lind, *Hawaii's People*, 4th ed. (Honolulu, 1980). Works of fiction also illuminate ethnic experiences. I gained insights from Milton Murayama's *All I asking for is my body*, new ed. (Honolulu, 1988), about a young Japanese American growing up on a plantation, and from various stories in Frank Stewart, ed., *Passages to the Dream Shore: Short Stories of Contemporary Hawaii* (Honolulu, 1987).

A particularly helpful general introduction to Alaskan history is Joan M. Antonson and William S. Hanable, *Alaska's Heritage* (Anchorage, 1985). Designed as a textbook, this highly readable work is also an excellent source for adult audiences, and it made many valuable contributions to this chapter. Another engaging survey, with a long bibliographical essay, is Claus-M. Naske and Herman E. Slotnick, *Alaska: A History of the 49th State*, 2d ed. (Norman, 1987).

There are many published primary sources from the periods of maritime exploration, the maritime fur trade, and the Russian occupation of Alaska. Journals of maritime explorers such as Captain James Cook and Captain George Vancouver are available in published form. Excerpts from some of these journals are reproduced in Robert N. De Armond, ed., *Early Visitors to Southeastern Alaska: Nine Accounts* (Anchorage, 1978). Sketches made by early explorers to Alaska and the Northwest Coast are reproduced and insightfully discussed by John Frazier Henry in *Early Maritime Artists of the Pacific Northwest Coast, 1741–1841* (Seattle, 1984).

Many important translations of Russian-American writings have been edited by such scholars as Richard Pierce, Lydia Black and Basil Dmytryshyn, and E.A.P. Crownhart-Vaughan. Too numerous to list comprehensively here, they are excellent sources for readers who want more detailed information about the Russian-American period. Petr Aleksandrovich Tikhmenev's *A History of the Russian-American Company*, originally published in 1861–63, has been translated and edited by Richard A. Pierce and Alton S. Donnelly (Seattle, 1978) and is a vital source of information about Russian-American activities. Excerpts from writings of Russian Orthodox missionaries in Alaska can be found in Michael Oleksa, ed., *Alaskan Missionary Spirituality* (New York, 1987). Secondary sources also illuminate the Russian colonization of Alaska and interactions between Russians and native Alaskans. These include, but are not limited to, articles in a very useful anthology edited by S. Frederick Starr, *Russia's American Colony* (Durham, N.C., 1987).

Many secondary sources focus on Alaska after transfer from Russia to the United States. Ted Hinckley's *The Americanization of Alaska, 1867–1897* (Palo Alto, Calif., 1972) is very helpful, as is his biography *Alaskan John G. Brady: Missionary, Businessman, Judge, and Governor, 1878–1918* (Columbus, Ohio, 1982). Missionary writings and travel accounts from the late nineteenth and early twentieth centuries help reveal native-white relations, albeit from nonnative perspectives. Writings by missionaries include the following: Sheldon Jackson's *Alaska and Missions on the North Pacific Coast* (New York, 1880); Eva McClintock, ed., *Life in Alaska: Letters of Mrs. Eugene S. Willard* (Philadelphia, Pa., 1884); *Hall Young of Alaska, "The Mushing Parson": The Autobiography of S. Hall Young* (New York, 1927); and a work by a Quaker missionary, Charles Replogle, *Among the Indians of Alaska* (London, 1904).

Alfred P. Swineford, the second district governor of Alaska, wrote *Alaska: Its History, Climate, and Natural Resources* (Chicago, 1898). Engaging travel accounts of visitors to Alaska, some of whom traveled as tourists on steamships, include Oskar Teichmann, ed., *A Journey to Alaska in the Year 1868, Being a Diary of the Late Emil Teichmann* (New York, 1963), E. Ruhamah Scidmore, *Journeys in Alaska* (Boston, 1885), Septima M. Collis, *A Woman's Trip to Alaska* (New York, 1890), and Ella Rhoads Higginson, *Alaska: The Great Country* (New York, 1912).

The newspaper *Tundra Times* is a useful source of native voices. Some recent publications reflect native perspectives on contemporary developments such as the Alaska Native Claims Settlement Act. These include Robert D. Arnold et al., *Alaska Native Land Claims* (Anchorage, 1976), Frederick Seagayuk Bigjim and James Ito-Adler, *Letters to Howard: An Interpretation of the Alaska Native Land Claims* (Anchorage, 1974), Lael Morgan, *And the Land Provides: Alaskan Natives in a Year of Transition* (Garden City, N.Y., 1974), and Thomas R. Berger, *Village Journey: The Report of the Alaska Native Review Commission* (New York, 1985). Dorothy Knee Jones's excellent and concise work *A Century of Servitude: Pribilof Aleuts under U.S. Rule* (Washington, D.C., 1980) contributed the information about the Pribilof Aleuts found in this chapter.

Several works focus on political history and Alaska's attempt to define a mutually satisfactory relationship with the federal government. Ernest Gruening's *The State of Alaska* (New York, 1968) discusses these themes as seen through the eyes of a prominent Alaskan politician. Ernest Gruening also wrote a short but detailed work on the Alaskan statehood movement, *The Battle for Alaska Statehood* (College, Alaska, 1967). Another concise and helpful book on the statehood movement is Claus-M. Naske, *An Interpretative History of Alaskan Statehood* (Anchorage, 1973).

The Alaska Geographic Society publishes a well-illustrated, highly readable quarterly, *Alaska Geographic*. Some titles in this series focus on regions of Alaska; others address various cultures; still others discuss historical themes or industries. Readers will find more details about material covered in this chapter in *Alaska's Native People* (Anchorage, 1979) and *The Aleutians* (Anchorage, 1980), both edited by Lael Morgan. Likewise, the *Alaska Journal* and *Alaska History* are excellent sources for additional historical information on Alaska.

Many people assisted me in my research for this chapter, and I wish space permitted me to thank them all individually here. I especially want to express my gratitude to the staff of the Grove Farm Plantation and the Bishop Museum for their generous attention during my research visits; to John Whitehead for helpful bibliographic recommendations; to Donald Worster for advice about research in Hawai`i and for recommending research at the Grove Farm Plantation; and to the University of Victoria for a Faculty Research and Travel Grant that made the research trip to Hawai`i possible. I have greatly appreciated the goodwill and good humor of Clyde Milner, Carol O'Connor, and Marni Sandweiss, with whom it has been a pleasure to work. Finally, I thank Geoff Wyatt and Muriel Wyatt for their valuable editorial assistance.

REBECCA

Chapter Seventeen

Landscapes of Abundance and Scarcity

WILLIAM CRONON

Just after midnight on 24 March 1989, the *Exxon Valdez* ran aground on Bligh Reef in Alaska's Prince William Sound. Its captain, his judgment clouded by alcohol, had maneuvered the supertanker out of ordinary shipping lanes in an effort to avoid icebergs floating south from the great Columbia Glacier. Accelerating into a mile-wide gap between the ice and the reef, and apparently forgetting that the ship (itself nearly one-fifth of a mile long) required over half a mile to turn, he had already left the bridge when the tanker shuddered to a halt, its hull ripped open as if by an enormous can-opener. So began the worst oil spill in American history.

Over the next two weeks, eleven million barrels of oil flooded into the sound while crews desperately worked to pump the tanker dry. If they had failed in this task, or if the ship had sunk before they completed their work, an additional forty-two million barrels might have spilled into the sea. As it was, the oil formed an enormous slick on the surface of the frigid waters and began to come ashore on islands and beaches in many parts of Prince William Sound. Ultimately, it would drift hundreds of miles to the southwest, traveling the same distance that separates Cape Cod and Cape Hatteras on the Atlantic seaboard. It would coat over twelve hundred miles of shoreline with a black, evil-smelling slime and kill hundreds of thousands of marine birds and mammals, in addition to countless fish and shellfish.

The wreck of the Exxon Valdez *on 24 March 1989, which sent 11 million barrels of oil into Prince William Sound, dramatically illustrated the environmental costs of the West's extractive industries and underscored the fragility of the great wilderness landscapes of Alaska.*

Nick Didlick. Exxon Clean-up of Valdez Oil Spill. *Color photograph, 1989. Reuters/Bettmann Newsphotos, New York, New York.*

Responses to the catastrophe often seemed frustratingly incompetent. The contingency plans that oil corporations and government agencies had claimed would handle such an event proved grossly inadequate. What was worse, precious time was wasted in the hours immediately after the ship ran aground—the weather stayed calm and the oil dispersed little for almost three days—as officials worried about the legal and political liabilities they might incur by moving too aggressively and thereby perhaps acknowledging their responsibility for the spill. Those who acted first were the owners of small fishing vessels; they instantly realized that the oil threatened their very livelihoods, but they had none of the special equipment or training needed to handle the challenge. Only after the dimensions of the disaster had become clear and public outcry was beginning to swell nationwide were full-scale relief efforts finally mounted. In the end, over ten thousand workers labored for months to clean beaches, save animals, and wash rocky coastlines with heated seawater. Ironically, some of their efforts probably did more harm than good by disrupting fragile ecosystems even further. The final cost of the cleanup would be measured in the hundreds of millions of dollars, and Exxon's legal liability would be more than one billion dollars. Many experts believed that coastal and marine ecosystems would need decades to recover.

The *Exxon Valdez* story is dramatic enough in its own right, but it can also stand as a symbol for much broader processes that have characterized the environmental history of the West as a whole. The oil spill in Prince William Sound came just three decades after Alaska entered the union as America's forty-ninth state, and the calamity marked a turning point in Alaskan history. Before statehood, much of Alaska had been terra incognita, a vast expanse of land known mainly to its native inhabitants and only lightly touched by development. Like earlier Wests, its very obscurity tempted those who hoped to make a quick fortune from the untapped wealth of nature. Alaska held out the promise of great natural abundance, whether in the goldfields that brought tens of thousands to the Klondike in the 1890s, in the fisheries that annually produced millions of cases of canned salmon in the early decades of the new century, or in the oil fields that were first discovered near Cook Inlet just as Alaska became a state. During World War II, Anchorage emerged as a classic western boomtown, exploding in population as the federal government poured immense sums of money into its military bases and as would-be entrepreneurs began to speculate about the economic potential of its vast hinterland.

Then came the discovery of petroleum at Prudhoe Bay in 1968, and the entire state boomed in an orgy of real and anticipated oil revenues. Along the way, Alaskan natives negotiated the largest (albeit still problematic) land deal in American history, environmentalists mounted a losing but innovative legal battle against the pipeline that would carry North Slope oil to market, and Americans suddenly awoke to the fact that their northernmost state was rich not just in oil but also in wilderness. And so the development of the young state went hand in hand with its *un*development: in 1979, President Jimmy Carter would set aside well over one hundred million acres of national parks, forests, and wildlife refuges in Alaska to make sure that its natural legacy would not be lost as oil lands were tapped. Only in this context can one see that the *Exxon Valdez* represented more than just an economic or ecological calamity; it also threatened the much less tangible spiritual values that America's "last great wilderness" had come to represent.

In the minds of many Americans, Alaska in the twentieth century moved from being a frontier of nearly unlimited natural abundance and exuberant economic promise to being a region that was both fragile ecologically and vulnerable economically. The same might be said of the West as a whole. America's many "Wests" have all begun as frontiers of real or perceived abundance whose regional identities have eventually been shaped by the experience of emerging scarcity. Alaska's identity flowed as much from the failure of the Klondike as from the initial golden dream, as much from the collapse of the canneries as from the early flood tide of salmon, as much from the wreck of Exxon's supertanker as from the extraordinary boom years that followed the discoveries at Prudhoe Bay. In much the same way, most western communities were born in the promise of plenty but did not come into their own until westerners had tested the limits of that promise to forge a new way of life on the land.

In tracing the environmental history of the western landscape, one must carefully distinguish a number of competing narrative trajectories. The story of frontier migration leads eventually to the story of emerging western regions, each with its particular cultural adaptations to the local environment. It is a long tale of people moving to frontier areas, seizing abundance, encountering scarcity, and remaking the land and themselves in the process. The result is the West as we know it today: not one single region but many smaller regions with distinctive environments and cultures. The Great Plains, Texas, the Rocky Mountains, the Colorado Plateau, the Basin and Range, the Desert Southwest, California, the Pacific Northwest, Alaska, and the Hawaiian Islands: each is a region in its own right, with its own smaller subregions like California's Sierra Nevada, Central Valley, Coast Ranges, and the sprawling urban worlds of the Los Angeles Basin and San Francisco Bay. A full environmental history of the West would have to explore the special cultural and ecological landscapes of all these places.

Set against this narrative are stories about institutional forces that have undermined regional diversity and autonomy over the course of western history. Cities and hinterlands have become linked by common markets and have grown to be more like one another. Energy resources like oil and electricity have enabled westerners to ignore the scarcities of their local environments in order to build communities of apparent abundance even in the midst of former deserts. The managerial hierarchies of the modern corporation have brought their own brands of homogeneity to the West, as have the bureaucracies of federal, state, and local governments. These narratives are all about regional (and national) homogeneity, the ability of people to use their technologies and institutions to remake the lands around them so that deserts and forests, mountains and valleys, eventually come to share common cultural forms.

These different stories are all entangled, of course, as the *Exxon Valdez* itself demonstrates. The setting for its special drama was one of the most beautiful and challenging environments of the modern West, a landscape that requires all who live there to change their habits and assumptions to meet the expectations of the land. The ship cargo was profoundly tied to the resource economy that sustains not just Alaska but also the places for which that oil was originally destined. The owners and managers of the ship, even its troubled captain, perfectly represented the corporate institutions that called the ship into being. The legal context of the oil spill, and the political responses it evoked, embody a long tradition of government involvement with interstate commerce, environmental regulation, and western lands. And its effects on the Alaskan "wilderness" lie at the very heart of recent controversies about the future of the western environment. In the long dialectic between scarcity and abundance that has shaped the landscapes of the American West, the ship on Bligh Reef embodies most major themes of the region's environmental history.

Fearing the End of Abundance

An environmental history of the modern West can begin, predictably enough, with the 1890 census pronouncement that Frederick Jackson Turner made famous in his 1893 essay "The Significance of the Frontier in American History." By declaring that "free land" could no longer be the wellspring from which the nation drew its democratic promise, Turner articulated what many Americans were already beginning to fear. America's frontier era was drawing to a close. Ideologues like Frederic Remington and Theodore Roosevelt joined Turner in worrying that the loss of the frontier would sap the nation's virility, exacerbate its class tensions, and undermine the dominance of its white races. Democracy itself might be threatened as a result.

Such fears eventually proved to be mistaken or groundless, but they rested on an even deeper anxiety that has had more lasting consequences. As the historian David Potter has noted, Turner and his compatriots, in declaring that the frontier was coming to an end, were expressing a more general concern that American abundance was giving way to scarcity. Their fears that good farmland would no longer be so easily available for would-be homesteaders suggested that other resources might also disappear from the American landscape. The forests that put roofs over American heads might vanish. The rivers that brought water to American cities might run dry. The coal mines that fueled American factories and heated American homes might give out. If these things happened, the nation's prosperity would surely erode and, with it, the political and personal freedoms that depended on prosperity for their survival. To escape such a fate, Americans must take serious steps to preserve the natural abundance that from the start had been the foundation of their nation's greatness.

Fears about resource exhaustion became increasingly common during the second half of the nineteenth century. In 1864, George Perkins Marsh published *Man and Nature*, a sprawling survey of the role that forests and other resources had played in the rise and fall of civilization. In it, he argued that Greece, Rome, and other Mediterranean civilizations had grown to greatness on the products of the forest and had collapsed when deforestation led to fuel scarcity, soil erosion, and desertification. In particular, he believed that forests at the heads of large watersheds were critical to maintaining the flow of water in navigable rivers: without their regulating effect on runoff, floods and droughts would become more common. Casting his eye on the heavily lumbered forests of his beloved northern New England, he warned that the United States was following the same path as Rome and would reach the same tragic destination if its citizens did not curb their reckless destruction of the woodlands.

In the decades that followed, Marsh's plea would be echoed by growing numbers of scientists, politicians, and corporate leaders. In 1876, an obscure rider to a congressional appropriations bill called on the Department of Agriculture to survey the nation's timber resources to determine "the probable supply for future wants" and "the means best adapted for their preservation and renewal." The result was Franklin B. Hough's *Report upon Forestry*, published in 1878, which looked to Marsh's book for its inspiration. Following Hough's lead, the 1880 census included a massive volume, authored by Charles Sprague Sargent of Harvard, surveying the forestlands of the United States. Although its statistics were roundly attacked by the lumber press, they confirmed Marsh's warnings that deforestation was occurring in many parts of the United States. Partly in response, Congress in 1886 formally recognized the Division of Forestry in the Department of Agriculture, where the German-born forester Bernhard Eduard Fernow began to conduct systematic investigations of the nation's timber resources.

Concerns about resource exhaustion and environmental degradation were not limited to woodlands. In 1878, John Wesley Powell offered to Congress his prescient *Report on the Lands of the Arid Region of the United States*. In it, he argued that the original land survey and homestead laws were radically inappropriate west of the Mississippi

River and would lead to environmental degradation unless significantly modified. Arid land settlement, Powell said, would require either small irrigated farms or large ranches, neither of which could be successfully conducted in the 160-acre units that the Homestead Act mandated. Taking a lesson from Mormon settlements in Utah, Powell urged Congress to revise existing land laws to make western settlement a more collective and regulated process. His views were supported by a report of the Public Land Commission that same year, but Congress failed to act on either set of recommendations. Powell's report joined Marsh's book as a classic of nineteenth-century conservation thought, but half a century would pass before its viewpoint would be fully embodied in government land policy in the West.

Ironically, one of the most important early responses to fears about deforestation came not in the West but in upstate New York—which would in fact become a model for subsequent conservation efforts in the West. In 1883, the New York legislature forbade any new sales on three-quarters of a million acres of public woodland in the Adirondack Mountains. Two years later, all state lands in the Adirondacks were declared to be a "forest preserve," and in 1894 their protected status was written into the state constitution. Henceforth, New York declared, the Adirondacks should remain "forever wild." In making this decision, the legislature was responding to the appeals of wealthy hunters and tourists who had flocked to the mountains in the years following the Civil War, but it was also responding to Marsh's prophecies. Among the most effective arguments on behalf of the Adirondack forest preserve were Marshian claims that deforestation in the mountains would endanger the water supplies of New York City and the Erie Canal by promoting irregular seasonal runoff and altering regional climate patterns. New York might then go the way of Rome: floods and droughts, it was said, would ravage the state's economy if its citizens failed to look after their timbered watersheds.

Despite New York's pioneering role, concerns about deforestation—about the transformation of wooded abundance into treeless scarcity—would have their greatest effect in the American West. Responding to the same public pressures that had created the Adirondack forest preserve, the U.S. Congress in 1891 passed a new statute revising many of the nation's existing land laws. Section 24 of that act, added almost as an afterthought, authorized the president to withdraw from settlement any tract of public land "wholly or in part covered with timber or undergrowth, whether of commercial value or not, as public reservations." As in the case of the Adirondacks, the apparent intent was to protect the heads of navigable rivers, particularly those that flowed past major urban centers, from the disruptions that Marsh had predicted might follow deforestation. Over the next two years, President Benjamin Harrison responded by setting aside fifteen separate reserves totaling over thirteen million acres, all in the trans-Mississippi West. Several contained precious little woodland but were requested by towns and cities whose residents feared their water supplies were threatened by overgrazing.

The 1891 Forest Reserves Act marked a turning point in federal involvement with the American West. Henceforth, there would be a steady reduction in public lands that were still available for private sale and settlement and a corresponding increase in public lands that were permanently reserved for government use. Moreover, the act laid the foundation for an entirely new federal relationship with the American environment. No one at the time had any way of knowing that Section 24 would have these consequences, and in fact it passed with little debate or public comment. But as the number of forest reserves grew, they posed a new problem for the government. Formerly, the chief task of government bureaucrats relative to the public domain had been to sell off land to yield large cash flows for the U.S. Treasury (which still had no income tax as a source of revenue) and for the rapid development of frontier areas in the American West. Now they faced an altogether different task: to oversee the proper use of land that would never pass into private hands and to make sure that government-owned resources were properly managed for the public good. This transition from public land disposal to public land management marked the real start of federal conservation efforts in the West and would become a dominant theme of the region's environmental history for at least the next half century.

Over the next fifteen years, these implications of the 1891 act became abundantly clear to all Americans. The revolution in federal land policy was spearheaded by an elite cadre of professional bureaucrats that Theodore Roosevelt appointed to key posts in his administration. Among these, the most prominent and influential was Gifford Pinchot. Born into a wealthy New York family, Pinchot had been educated to a life of privilege at elite eastern schools, preparing himself for the unlikely career his father had helped him choose: forestry. In 1896, he served on the National Forest Commission of the National Academy of Sciences, which issued recommendations about how best to manage the new western forest reserves. The result was the Forest Management Act of 1897, which would serve for the next sixty years as the fundamental law governing forest policy in the United States. Partly as a result of his experience with the commission, the next year Pinchot succeeded Fernow as chief of the Department of Agriculture's Division of Forestry, thus gaining the platform from which he would launch one of the most ambitious and successful careers in American conservation history.

The Division of Forestry initially seemed a rather unlikely place from which to direct a revolution. It had only eleven employees when Pinchot took office in 1898 and a budget of just $28,520. Worse, since the new reserves were all located in the Department of Interior, the division was without forests to manage. For these reasons, Pinchot's immediate goal was to expand the size of his operation and ultimately wrest control of the reserves from the Interior Department. Within three years, his budget had increased fivefold and his staff more than fifteenfold. By 1902, 179 people were working for Pinchot, many of them young men serving apprenticeships as part of their forestry education.

This remarkable expansion was the indirect product of a tragedy that changed the course of history: on 6 September 1901, an assassin's bullet killed William McKinley just half a year into his second term of office. His death catapulted Theodore Roosevelt into the presidency. Roosevelt had already revealed his fascination for America's frontier past and feared that the "closing frontier" might endanger the nation's democratic heritage. Now he was in a position to protect what he saw as the frontier legacy by conserving America's natural resources and by preserving the remnants of a vanishing

Theodore Roosevelt and John Muir, both passionate outdoorsmen, represented different approaches to land management. Roosevelt believed in the efficient use of forest lands—for both commerce and recreation—through careful government management. Muir, who at first supported such federal control, later repudiated the government's utilitarian aims and turned his attention to promoting the spiritual value of the wilderness.

Unidentified photographer. John Muir and Theodore Roosevelt, Glacier Point, Yosemite. *Photograph, 1903. Courtesy Research Library, Yosemite National Park, California.*

wilderness landscape where one could recover the "vigorous manhood" that the "rough riders" of an earlier day had enjoyed as their birthright. Pinchot immediately gravitated toward Roosevelt, and the two men found much in common with each other. Pinchot soon emerged as the master conservation strategist of the Roosevelt White House, wielding great influence in Washington despite his unexalted bureaucratic position.

The most striking proof of Pinchot's influence came in 1905, when Roosevelt transferred sixty-three million acres of forest reserves—the great majority of them in the West—from the Interior Department to the Agriculture Department and placed them under Pinchot's control as the head of the renamed U.S. Forest Service. But the more important innovation had less to do with who controlled those acres than with how they were managed. More than any other agency, the Forest Service epitomized Progressive Era conservation. Pinchot and his followers committed themselves to promoting professional management, believing that only those with scientific expertise should decide how best to use forest resources. A hunger for quick profits might tempt corporations and private landowners to cut the forest more rapidly than it could replenish itself. Politicians might be wooed too easily by local constituencies eager for rapid development no matter what its cost. Only a scientific forester—so the argument ran—could know enough and be disinterested enough to look after the long-term interests of people and forests alike.

To produce this new style of government manager, Pinchot relied on the new schools of forestry—all more or less inspired by German traditions—that were appearing at Cornell, Ann Arbor, Biltmore, and Yale (the latter financed by a gift from Pinchot's father). Young men—and they were all young men in the early years—who hoped to become foresters got their training from these schools and then made their way into the Forest Service to be inculcated with the values it represented. Energized by an elite esprit de corps and a vision of disinterested public service, the young foresters fanned out across the western landscape with a goal of managing public forests so that frontier abundance could be saved from scarcity and could last forever.

Like other progressives, Pinchot, Roosevelt, and their followers strongly believed in what the historian Samuel P. Hays has called "the gospel of efficiency." For progressives, the greatest villain was the waste of resources, so that "the people" could not enjoy their fullest use. Pinchot liked to borrow and extend Jeremy Bentham's famous utilitarian principle as the central goal of conservation: "the greatest good for the greatest number for the longest time." To waste resources, to use them inefficiently, was to steal from future generations. The correctness of this principle seemed so self-evident that it was hard for conservationists to see their opponents as anything other than venal and corrupt. Short-sighted landowners, dishonest bureaucrats, domineering monopolies, and craven officeholders all had bad motives for putting their own interests above the public good.

There was, inevitably, a darker side to this vision of scientific management. Despite their democratic rhetoric—their apparent defense of "the people" and "democracy" against "monopoly" and "corruption"—the progressive conservationists were suspicious of many democratic institutions. They tended to look more toward executive authority than toward the legislature to enact their reforms, and they saw the good of "the whole" (by which they often meant well-to-do middle-class easterners like themselves) as being more important than the special concerns of individual constituencies. Perhaps because of this, they were attracted to strong leaders who projected a slightly messianic air—Roosevelt and Pinchot being good examples of the type. Progressives preferred expert knowledge to the messier judgments of public debate. They generally preferred centralized authority and decision-making to local control. Pinchot's Forest Service was notable for the decentralized organization of its district system but ultimately derived its authority from Washington rather than from local communities. In their pursuit of what they saw as democratic ends, the progressives sometimes thought it necessary to circumvent democratic means.

And so it was perhaps inevitable that Roosevelt and Pinchot should come into conflict with people who did not share their vision. Among those who opposed the expanding system of national forests were senators and representatives from the western states, who saw more and more of their local landscape being removed from development and placed under Forest Service control. The conflict came to a head in a famous confrontation in 1907. Congress sought to limit Roosevelt's ability to withdraw western land from settlement by passing an appropriations bill that required the president to have congressional permission before creating any new national forests in Colorado, Idaho, Montana, Oregon, Washington, and Wyoming. The list included the most heavily timbered states in the nation and the ones most hostile to Washington's control; all were in the West. Roosevelt had no choice but to sign the bill, but the night before doing so, he ordered the creation of new national forests on sixteen million acres of western lands. These "midnight forests" enraged western congressmen and perfectly express the mingled idealism and arrogance that typified conservation during the Roosevelt years.

Federal conservation efforts at the turn of the century were by no means confined to forests. As politicians identified new environmental problems, they created new laws and bureaucratic institutions to address those problems. For instance, the first two decades of the twentieth century saw significant efforts to diminish threats to key endangered animal species by expanding government regulation of hunting. In 1900, the Lacey Act banned the interstate transport of mammals and birds that had been killed in violation of state law, thereby lending federal support to state efforts at wildlife protection. Recognizing that the reproductive cycles of many species did not respect state or national boundaries, the federal government declared, in the Migratory Bird Act of 1913, that its own jurisdiction took precedence over state laws relating to migratory game and insectivorous birds. This in turn paved the way for the landmark Migratory Bird Treaty with Canada in 1916, establishing for the first time a durable framework for regulating the critical flyways of North America. Free access to an unrestricted hunt, which had been among the defining experiences of American frontier settlement since at least the days of Daniel Boone, would all but vanish in the years ahead, giving way to far more restricted hunting conducted under the watchful eyes of scientists and managers working in the service of a regulatory state. Conflicts over who should have access to traditional hunting grounds often erupted between local communities and newly professionalized game wardens, with representatives of state and national governments working to redefine and expand the concept of poaching. In the process, Indians and other minority ethnic groups were often forced to retreat from hunting and fishing grounds on which they had long depended, sometimes for generations.

Similar efforts at fish and game regulation occurred elsewhere in the government as well. In 1905, the new Bureau of Biological Survey consolidated earlier federal programs designed to research and regulate the effects of animals on agricultural crops. Starting in 1915, it would lay the foundation for a systematic campaign to destroy predators and other "vermin" species—coyotes, wolves, grizzly bears, mountain lions—to protect livestock and to promote the increase of game species such as deer, elk, and moose, which were highly prized by hunters. The Forest Service would also join these efforts. To oversee animal species in less terrestrial environments, the Office of the U.S. Commissioner of Fish and Fisheries moved in 1903 to the Department of Commerce and Labor and became the Bureau of Fisheries, bringing increasingly coordinated oversight to the nation's fish populations, many of the most important of which spawned in western rivers and grew to adulthood off the west coast. Separate agencies overseeing Alaskan fish and fur seal resources would be transferred to the new bureau over the next half decade.

At the same time, the United States joined Canada, Russia, and Japan in seeking to avert the extinction of Alaskan and Siberian seals, eventually signing the Fur Seal Treaty of 1911, which banned pelagic (open-sea) sealing and also tried to regulate hunting at key seal rookeries like Alaska's Pribilof Islands.

Although efforts such as these did not often succeed in stabilizing animal populations at former levels, they did establish the principle that government had an essential role to play in trying to protect—or, in the case of predators, exterminate—wild animal species. More important from the point of view of western environmental history, these early efforts at federal regulation also furnish a classic example of the early abundance of a natural resource giving way to real scarcity, indeed, nearly to extinction. Attempts to respond to such changes created new regulatory institutions that helped alter not just western regional attitudes toward the environment but national ones as well. Increasingly, natural areas and animal populations would be seen not as landscapes of freedom, not as frontiers of endless natural abundance, but as endangered landscapes of scarcity, so fragile in the face of human destructiveness that only careful management could ensure their survival.

This sense of fragility lay behind the Antiquities Act of 1906, which enabled the president to establish national monuments to protect areas of special archaeological, historical, or scientific importance. Despite its apparent emphasis on antiquities—by which its authors generally meant endangered Indian ruins—the act soon became a tool for Roosevelt and his successors to set aside any area of natural or historical value. Thus, within a few years Roosevelt was able to use the act not only to protect places like Montezuma Castle in Arizona and El Morro in New Mexico—both legitimate "antiquities"—but also Arizona's Grand Canyon and a large section of Washington's Olympic peninsula, whose claims to protection clearly rested on their unusual scenic beauty. Roosevelt had already in 1904 stretched the meaning of the 1891 Forest Reserves Act to set aside the first federal wildlife reservation at Florida's Pelican Island—a wetland nesting area whose chief value had far more to do with birds than with trees. It was soon followed by more than fifty other such bird reservations—most of them not especially forested—from Florida to Alaska. When it came to scenic wonders and nesting grounds, at least, the days of limitless abundance, easy exploitation, and unregulated hunting were apparently at an end.

It is important not to misunderstand Roosevelt's interest in protecting game species and areas of extraordinary natural beauty. The goal of setting aside tens of millions of western acres as national forests, of withdrawing them from sale under the public land laws, was not to protect them as permanent natural areas; rather, it was to prevent their destruction so that they could be managed and harvested in perpetuity as a resource for future generations of Americans. The same was true of game refuges and protected nesting grounds: even though hunting was illegal within their boundaries, their purpose was not to abolish hunting but to ensure its perpetuation by protecting the reproductive cycles of key game species. What Roosevelt saw as the uncontrolled freedom of earlier frontier landscapes might have to be abandoned, but the consequence would be to protect the way of life and cultural values that the frontier had supposedly nurtured in America. The progressive conservationists saw no conflict between intelligent exploitation of natural resources and the long-term survival of those resources. Despite their sometimes apocalyptic rhetoric about what might happen if Americans failed to conserve natural resources, conservationists like Roosevelt and Pinchot were fundamentally optimistic about the ability of their own reform agenda to set the country on a course that would ensure permanent national prosperity.

The Reclamation Dream

Nowhere was this optimism more obvious—or more important to the West—than in water policy. By the time Roosevelt became president, westerners had for more than two decades been working on their own and seeking federal support to promote the construction of dams and irrigation systems on rivers and streams throughout the region. The classic dilemma of the western environment was that much of the terrain received far too little rain or snow to permit successful farming. Annual precipitation of much less than twenty inches was the norm almost everywhere except in the mountains and in the Pacific Northwest; in many areas—including the San Joaquin and the Imperial valleys of California—annual precipitation fell below ten inches. No ordinary crop could survive such conditions, though westerners did show considerable ingenuity in pushing the limits of certain crops, notably wheat, in what came to be called "dryland farming." Worse, when rain did fall or the winter snows melted, the resulting floodwaters too often raced down canyons and arroyos in muddy torrents that left little but destruction in their wake. If only these floods could be stayed in their journey and delivered to an otherwise parched earth, then natural scarcity could give way to artificial abundance.

It would be hard to exaggerate the compelling power of this bountiful vision for most Americans during the late nineteenth and early twentieth centuries. The ability of water to transform the arid West seemed wondrous, an unambiguous blessing that tempted irrigationists into flights of impassioned rhetoric that resorted sooner or later to biblical metaphors. "Irrigation," wrote the indefatigable booster William Ellsworth Smythe, "is a miracle." Just as the Nile had once watered ancient Egypt, just as "a river went out of Eden to water the Garden," so might the dry soils of the West burst into flower if the rivers could be made to share their liquid bounty with the land. Anyone who doubted this fact had only to look at what the Mormons had accomplished in Utah since the 1850s. With only the simplest of tools, their own skill, and a powerful religious hierarchy to aid them, they had captured the waters of the Wasatch Range and had turned the valley of the Great Salt Lake into astonishingly fertile farmland. Powell had appealed to the Mormon example in his *Report on the Lands of the Arid Region*, and other irrigationists did likewise. The Rio Grande Western Railroad even went so far as to produce a bird's-eye map of Mormon Utah, which drew direct parallels between its geography and that of the Holy Land. In this striking if rather distorted piece of cartography, Utah's Jordan River flowed toward the Great Salt Lake to water the Mormon Zion in much the same way that Palestine's Jordan River flowed toward the Dead Sea to water the land of Canaan. The title of the map suggests how powerfully its prophetic vision resonated with the American imagination: water would make of the West a "promised land."

Early federal efforts to help fulfill this vision tended, like most nineteenth-century resource policies, to rely on the public land laws. In 1877, Congress passed the Desert Land Act, which was modeled on an experimental effort two years earlier to promote

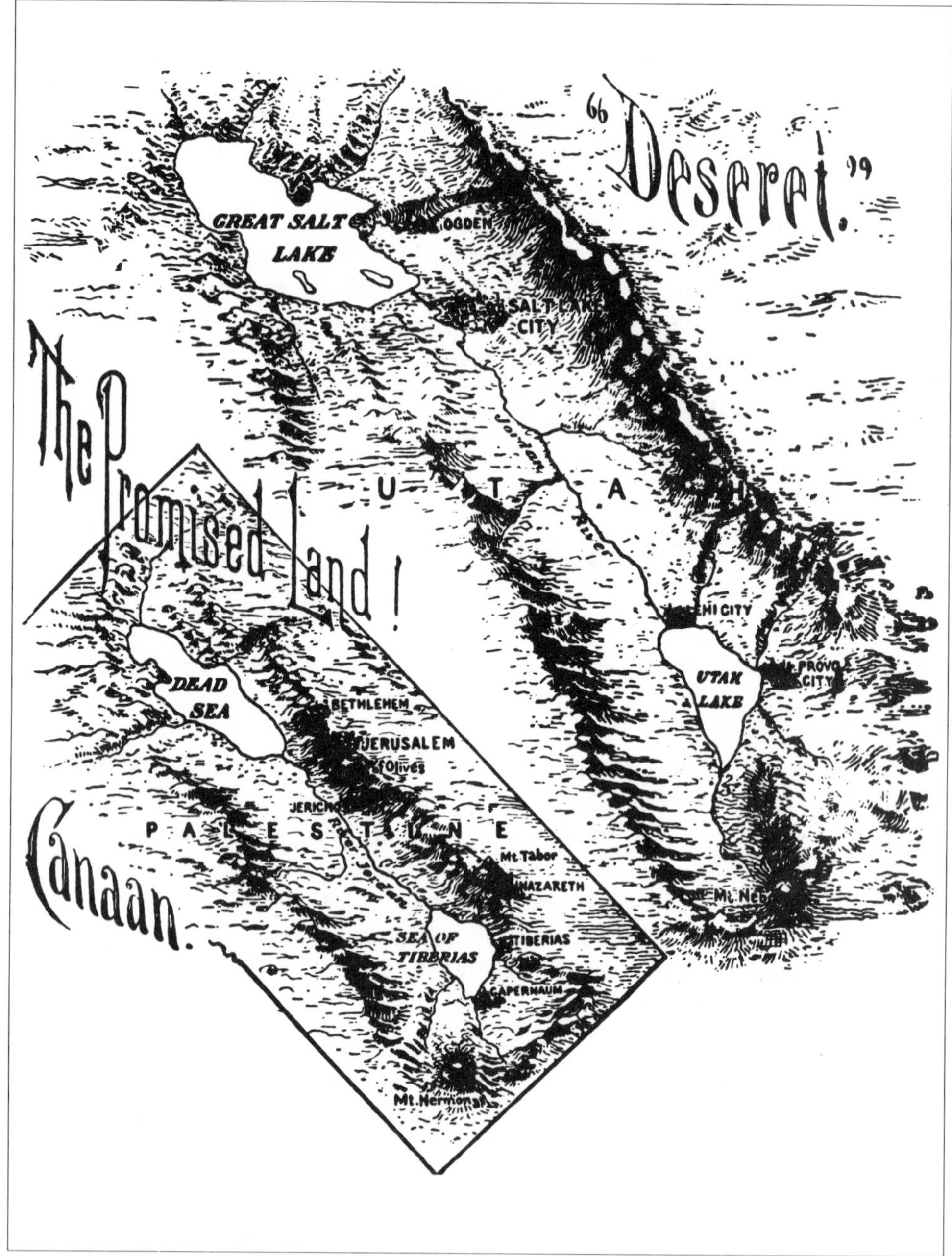

With high hopes, but little regard for geography or climate, the Rio Grande Western Railroad promoted Mormon Utah as a fertile Garden of Eden in the Promised Land.

Rio Grande Western Railroad, Publisher. Map Showing the Striking Similarity between Palestine and Salt Lake Valley, Utah. *Print, ca. 1899. From William Ellsworth Smythe,* The Conquest of Arid America *(1899).*

irrigated farming in Lassen County, California. Under the act, which applied to all western states except Colorado, individuals could purchase up to 640 acres of land (one square mile, four times the amount available under the Homestead Act) at a price of $1.25 per acre. They were required to pay only 25¢ per acre at the time they filed and then had three years to "make satisfactory proof of reclamation" before paying the balance of their debt to obtain clear title. Land was to go only to those individuals who genuinely intended to settle and improve their land; it was not intended for speculators whose only goal was to resell it after its price had risen. The act's basic intent was to promote irrigation not by direct federal investment but by making land available at relatively low prices to individuals who promised to make private investments in water technology.

Like most such land laws, the Desert Land Act was loosely drafted and poorly enforced, so that abuses of its original intentions were soon common. Would-be owners poured a few buckets of water onto their land and swore they had irrigated it. Dummy entrymen were used by large investors to amass much more than the intended square mile of land. To control water and land for their cattle, ranchers claimed narrow snake-like strips along the banks of streams or dug irrigation ditches that were promptly abandoned as soon as clear title had been obtained. The Desert Land Act undoubtedly encouraged a modest increase in irrigated acreage in parts of the West, but not on anything like the scale that its promoters had hoped.

More important, the Desert Land Act helped identify two questions that would bedevil western water policy until at least the 1920s. The first had to do with who should benefit from federal efforts to promote irrigation. Should irrigation laws, like the Homestead Act before them, encourage the development of a Jeffersonian landscape dominated by small farmers? Or should they permit farms of whatever size, no matter how large, as long as the owners could successfully irrigate the land? Even in this 1877 act, the twentieth-century tension between family farm and agribusiness was implicit. The second question was a corollary of the first: who should provide the capital that would make irrigation possible? Immediately after the Desert Land Act was passed, early critics (including Powell in his *Report on the Lands of the Arid Region*) were arguing that most irrigation districts could not be successfully organized without capital investments of one million dollars or more. From this perspective, it seemed naive to expect individuals who owned only a single square mile to be able to afford the investments in dams, canals, and other water-handling technologies without which farming would be impossible. Wealthy individuals and large corporations would have to supply the needed capital—in which case the limit of 640 acres would almost surely have to be abandoned—or else the government itself would have to make the basic investment. How to resolve these two questions in actual law and practice would be the major challenge of water policy for the next half century.

Not until Roosevelt became president would the federal government abandon its general policy of avoiding direct investment in western irrigation. In the meantime, water policy was left largely to the states and to private individuals. During the second half of the nineteenth century, western water law at the state level acquired the foundations it retains to this day. Under English common law, the use of water in a river or stream was attached as a right of ownership to the land along its banks. One could do whatever one wanted with that water as long as one did not diminish its flow, alter its course, or degrade its purity. Yet this common-law practice had already begun to change in eastern parts of the United States as the owners of canals and water-powered factories began to build dams that flooded riparian lands and fundamentally altered the flow of rivers. So long as such changes constituted "productive use," the courts were willing to tolerate them even though they violated traditional riparian rights. A new body of "appropriative rights" therefore began to emerge in American law.

The arid West posed an even greater challenge to common-law traditions. In desert areas, it was simply not possible to withdraw water from a river, use it for irrigation or other purposes, and then return it without diminishing its quantity or quality. And so westerners, starting with miners who needed water for sluicing gravel in the California goldfields, began to embrace the doctrine that whoever first put water to productive use acquired a permanent right to it. That right was not absolute, since one would lose it if one did not continue "beneficial use," but the divergence from common-law riparian rights was still very great indeed. During the 1880s, Colorado gave its name to this new water doctrine by laying claim to all surface waters within its boundaries, nullifying riparian rights to their use, and taking upon itself the task of enforcing water rights acquired by prior appropriation. If early settlers used up all the water in a stream that flowed through their lands, then later settlers had no claim whatsoever to water for their own use. And so the principle of prior appropriation came to be known as the Colorado Doctrine, which was quickly adopted by most states in the Rocky Mountain region.

Elsewhere, the triumph of appropriative rights was less complete. In California, the famous *Lux v. Haggin* court decision of 1886 held that riparian rights to water did indeed accompany the government sale of lands along a river unless that river's water was already being used by some other landowner at the time the original sale occurred. Under this "California Doctrine," the balance between appropriative and riparian rights depended on which came first, water use or land sale. Riparian rights tended to be favored by large cattle ranchers, whose use of water was relatively passive, whereas appropriative rights were favored by miners, irrigators, and factory owners, whose water use was more active and interventionist. *Lux v. Haggin*, although not a popular decision with the many Californians who believed that riparian rights favored large landowners, did establish the not unreasonable principle that different users of water had different needs, which might in turn entail different ways of conceiving legal property rights to the resource. This principle was in some ways carried to its logical extreme in yet a third body of water law, the Wyoming Doctrine, in which the Wyoming constitution claimed title to all water within the state's boundaries and retained the power to alter any existing private rights or appropriations that did not serve the public interest. This more collectivist definition of water use became the basis for water law in most of the Great Plains states.

In a region where water was at once extremely scarce and absolutely essential for most development, it is hardly surprising that none of these doctrines prevented persistent conflict—and endless litigation—over who should have the right to use this most precious of resources. And yet however problematic these doctrines might be, they did begin to establish the legal framework that made it possible for westerners to exploit the water on which regional economic growth depended. Scarce water was the key that could unlock the hidden abundance of an arid land. Throughout the West, water law rested on the utilitarian premise that both the unused water and the land through which it flowed would be "wasted" unless people intervened to ensure their "reclamation." This usage of the word *reclamation* was fairly new at the time, having appeared in the English language only in the middle decades of the nineteenth century. In England, John Stuart Mill spoke of "the reclamation of waste lands"; in the United States, lands worthy of reclamation were labeled "waste," or "alkali," or "arid," or "new." Whatever the adjective applied to them, they all existed in a state of nature that prevented their exploitation until human ingenuity could "reclaim" their potential—a potential that people tended to see as "natural" even though it served human desires and cultural values far more than the needs of existing ecosystems. So powerful was this sense of a "wasted" "natural" potential waiting to be "reclaimed"—as if a prior *un*wasted landscape had once been lost, Eden-like, and now needed to be claimed again—that by the 1890s virtually all discussion of irrigation and watershed manipulation occurred under the heading of "reclamation." Congress had a Committee on Irrigation and Reclamation of Arid Lands, and the federal agency that would be given responsibility for promoting irrigation would eventually be known simply as the Bureau of Reclamation.

The creation of that agency could not occur until the federal government was willing to embrace the mission of reclamation in a more direct way. That did not happen until it seemed clear that state and private irrigation efforts had reached an impasse. In 1887, California had implemented the Wright Act, creating new irrigation districts that could take collective control of their water rights wherever two-thirds of the electorate so voted; bond issues could then be underwritten by the tax revenues of the district and used to support investments in irrigation. Although the act helped double irrigated acreage in California by the end of the century, and although such districts would play an increasingly important role in promoting the growth of agribusiness in the early twentieth century (after large landowners gained greater control over the bonding process), their initial successes were relatively modest. The same was true in other parts of the West as well. By the 1890s, roughly seven million acres had been "reclaimed" with irrigation, an amount that seemed low relative to the "wasted" potential. Moreover, a large number of the private irrigation enterprises already under way were hardly prospering; by the start of the new century, the majority would be in or near bankruptcy.

And so westerners began to lobby for greater federal intervention. Starting in 1891, a series of "Irrigation Congresses" organized by William E. Smythe brought together engineers, lawyers, journalists, corporate leaders, government representatives, and others interested in promoting irrigation. Their initial goal was to persuade the federal government to extract from the remaining public domain all irrigable land so that it could be turned over to the states for improvement—a policy known as "cession." Washington was not ready for so dramatic a gesture, but in 1894 the Carey Act made available to each western state a grant of one million acres that could be developed by irrigation companies and then sold to farmers. Although this new act encouraged irrigation in some states—most notably Wyoming and Idaho—it shared with the 1877 act a failure to identify adequate sources of capital to "reclaim" the millions of acres it contemplated developing. Neither private corporations nor the states were willing to risk large sums on such risky investments, and so by the end of the century less than twelve thousand acres had been patented under the Carey Act. The facts that it was passed in the midst of a major economic depression and that states differed widely in their ability and willingness to take advantage of it also help explain its apparent failure.

By the time Roosevelt became president in 1901, the modest progress achieved by state, local, and private efforts to develop western irrigation seemed incommensurate with the grand prophetic vision that the word *reclamation* conjured in the minds of westerners and most other Americans. In his first State of the Union message in 1901, Roosevelt made clear his belief that only the federal government was up to the task at hand. "Great storage works," he said, "are necessary to equalize the flow of streams and to save the floodwaters. Their construction has been conclusively shown to be an undertaking too vast for private effort. Nor can it be best accomplished by the individual States acting alone. . . . It is properly a national function, at least in some of its features." Despite the apparent logic of Roosevelt's argument, one might reasonably speculate that the underlying problem with irrigation projects had more to do with the rate at which investment was occurring than with the intrinsic inability of private capital or state governments to provide funding. As with the transcontinental railroads, federal law may have been trying to accelerate developments whose time had not yet come; as with the railroads, an unsurprising consequence was that such ventures initially proved only marginally viable, heading into bankruptcy at an alarming rate. And so one should be careful not to take at face value Roosevelt's conclusion about the necessity or the proper form of federal intervention. The fact that his prophecy came true may say more about the seductive power of the reclamation dream than about the inevitability of the government's role in realizing it.

Roosevelt's prophecy did indeed prove self-fulfilling. When Representative Francis G. Newlands, of Nevada, proposed legislation dramatically expanding the federal role in western reclamation, he had little trouble moving it through Congress, supported as it was by the president and by the vigorous lobbying efforts of the California lawyer George Maxwell's National Irrigation Association. The resulting Reclamation Act of 1902—more familiarly known as the Newlands Act—clearly ranks with the 1785 Ordinance and the Homestead Act as one of the most important American land laws ever passed. The act created a revolving Reclamation Fund, which would receive revenues from the sales of all public lands in sixteen western states. Money from the fund was to be used for constructing dams and irrigation systems in those states. Individuals who purchased the resulting water were to pay high enough fees to replenish the fund within a specified number of years, thereby making it a perennial source of investment capital for reclamation projects throughout the West. (One should note that even if this fund had been fully repaid on schedule—which it was not—it would still have represented a significant federal subsidy, since water users were not expected to pay any interest on the money they had been loaned. As repayment periods stretched from ten to twenty to forty years, the invisible subsidy implied by these interest-free loans ballooned accordingly.) To make sure that the benefits of government-funded irrigation went to small farmers and not to large speculators or corporate farm operations, the act specified that no single landowner could purchase water to irrigate more than 160 acres of land. The act thus sought to continue a long tradition in American land law, dating back to the Homestead Act and before, of trying to promote agrarian communities of single-family farms as the proper foundation for American democracy.

The responsibility for managing the revolving fund was assigned to the new Reclamation Service (renamed the Bureau of Reclamation in 1923). Appealing to scientific expertise and centralized executive authority to legitimate its work on behalf of "the people," the office paralleled Pinchot's Forest Service as an expression of Progressive Era conservation values. From 1902 to 1914, the new agency was headed by Frederick H. Newell, a leading civil engineer who had worked closely with Newlands in framing the Reclamation Act. Work first began on the Salt River Project in Arizona and the Truckee River Project in Newlands' Nevada district, but by 1906 almost all of the western states had projects under way. Newell assembled a superb team of engineers who designed and built not just dams and reservoirs but also the infrastructure—roads, rail lines, cement factories, construction camps—necessary to erect them. Numerous technical problems arose almost from the outset—destructive floods, leaky geological formations, defective designs, faulty construction—but in general the engineering achievements of the act's first two decades were its most impressive legacy.

Measured in economic terms, though, the government's accomplishments under the Newlands Act were more ambiguous. In part because the Reclamation Service had no formal affiliation with the Department of Agriculture, its engineers were less attentive than they might have been to the soil and drainage conditions that farmers would face in adopting irrigation. Speculators tended to buy up land in the vicinity of new dam projects, selling it to farmers at high enough prices that little was left over to pay back the revolving fund. Settlers often moved into an area long before water was available, but they had to stay on their as yet unproductive land for up to five years if they hoped to acquire title to it. Many were undercapitalized—a classic problem of small farmers under the Homestead Act as well—and thus lacked the tools and equipment to make proper use of their government-supplied water once it arrived.

By 1910, it was already clear that farmers were having difficulty meeting their obligations to the Reclamation Fund, so Congress stepped in with the first of what would eventually be a long series of extensions delaying deadlines for making payments. As time went on, the Bureau of Reclamation more or less ignored the rule against sales of water to farm units larger than 160 acres, and large agribusiness operations came to control an increasing share of irrigated acreage in the West. By the early 1920s, federal reclamation projects had produced not much more than an additional one million acres of irrigated land, though the infrastructure was in place for significant expansion in the future. At the same time, local and private efforts based on the model of the California irrigation district had increased their acreage significantly, so that federal projects accounted for only about one-tenth of the regional whole. Throughout the West, whether constructed with private or federal capital, the irrigated rural landscape was dominated by federal bureaucracies and large landowners, with the government showing a clear tendency to favor large long-term water users over small ones. Contrary to the original promise, the society that emerged in "reclaimed" desert areas was a far cry from the Jeffersonian vision of agrarian democracy and single-family farms.

The Power of a Dam

Ironically, it was the Great Depression of the 1930s and the rising urban-industrial demand for electricity that would help secure federal reclamation efforts in the West. In 1928, Congress authorized the construction of a 726-foot-high dam in Black Canyon on the Colorado River. Once complete, it would be the tallest dam in the world. Unlike earlier projects, in which the generation of hydropower had been an incidental by-product of water impoundment for irrigation and flood control, Boulder Dam—so called because the initially proposed construction site was the one the public remembered—was designed to earn back a significant share of its cost from the sale of electricity. Its generators would be owned and operated by the Southern California

Edison Company, the Los Angeles Metropolitan Water District, and the City of Los Angeles—all of which suggested how much the reclamation agenda was shifting from its original rural focus.

Elwood Mead, then serving as head of the Bureau of Reclamation, promoted the Boulder Canyon Project as an example of multipurpose river-basin development. If the bureau conceived of its mandate in an integrated fashion, he argued, it could facilitate regional economic growth by providing construction jobs for unemployed workers, irrigation for farms, flood control for residents of river valleys, water for urban drinking supplies, and electricity for farms, factories, and cities alike. Perhaps most important, sales of electricity to urban and industrial consumers were a far more promising way to pay off deficits in the Reclamation Fund than trying to extract payments from small farmers who were perennially in arrears. In short, Boulder (later Hoover) Dam was the answer to a reclamationist's prayers. Its success would foster a series of integrated projects—Grand Coulee and associated dams on the Columbia, the Central Valley Project in California, the Colorado-Big Thompson Project in Colorado, and other equally grandiose initiatives—each of which would contribute to the Bureau of Reclamation's growing influence throughout the West. More important still, Boulder Dam and the projects that followed it also signified the increasing role that metropolitan institutions and urban political-economic power would henceforth play in reshaping the western environment.

For the Bureau of Reclamation was by no means the only entity promoting major water projects in the West. By 1900, the growing cities of San Francisco and Los Angeles were seeking guaranteed access to water to ensure their growth well into the new century. Civil engineers and politicians in both cities cast their eyes east toward the Sierra Nevada, where cold mountain peaks captured much of the state's precipitation in their winter snowpack. Los Angeles found its "river out of Eden" in the Owens Valley at the southern end of the mountains; San Francisco located its in the Hetch Hetchy Valley just north of Yosemite. Although the struggle to seize these areas for urban water supplies met with severe political resistance—from farmers and boosters who had hoped to develop the Owens Valley for agriculture and from preservationists and nature lovers who sought to protect Yosemite National Park from invasion—in the end the metropolitan demand for water proved irresistible. The Los Angeles Aqueduct was completed in 1913, carrying water 233 miles from the mountains, and the Hetch Hetchy Aqueduct—155 miles long—was finished two decades later. By then, both systems were producing electricity as well as water in such abundance that neither city could use anything close to the full supply. As a result, both cities were in the curious position of being able to sell water to farms and suburban districts in the desert environs around them, yielding important sources of revenue on which both soon became dependent. No matter whether the original source was local, state, or federal, the capital that had been invested to flood Hetch Hetchy, the Owens Valley, and Boulder Canyon made it possible for Californians to enjoy astonishingly cheap water even in the midst of desert landscapes that had never before known such abundance.

One can see in these massive projects and in the other bureaucratic legacies of Progressive Era conservation the compromises that westerners and other Americans

To supply the water and electric power that would ensure the continued growth of Los Angeles, federal and local authorities initiated massive engineering projects that fundamentally altered the landscape of more distant parts of the West. The first Los Angeles aqueduct, completed in 1913, brought water from the Owens Valley 233 miles away and destroyed the economic future of that rural valley on the eastern side of the Sierra Nevada. The construction of Boulder Dam, completed in 1935, led to the creation of Lake Mead in the middle of the Nevada desert.

James Bledsoe. The Alabama Gates of the First Los Angeles Aqueduct. *Gelatin silver print, ca. 1913. Photograph courtesy of the Los Angeles Department of Water and Power.*

Unidentified photographer. Hoover Dam *(later* Boulder Dam*). Gelatin silver print, ca. 1960. Las Vegas News Bureau Archives Photograph. Courtesy of the Special Collections, University of Nevada-Reno Library.*

were willing to make in the service of the reclamation dream. Turner had argued that the frontier had been a landscape of freedom, in which Americans had discovered the liberty and independence that characterized them as a nation. As embodied in the Homestead Act and in the land-limitation clause of the Newlands Act, frontier settlement was supposed to be for brave individuals, for yeoman farmers in the Jeffersonian mold, for the "little guy" hoping to seize opportunities unavailable in more settled and constricted lands. So said the myth, and for all its distortions it had exercised great influence over Americans' notions of themselves. But the deeper environmental reality was that the frontier had served first and foremost as a landscape of abundance—as mythical as it was material—and this abundance had made possible the way of life Americans considered essential to their national identity. If some measure of personal freedom had to be sacrificed to protect this more fundamental good, then the price might be worth paying. When Pinchot defended "the greatest good for the greatest number for the longest time," he was arguing for constraints on individual freedom—the freedom of people to enter the public domain to cut timber, graze cattle, plant crops, use water in whatever ways they liked—in order to defend a longer lasting, better managed, more collective abundance. Whether confronting the natural scarcities of the western desert or the artificial scarcities of the cutover forest, the conservationist agenda was to manage these scarcities so as to "reclaim" their usurped abundance.

In one way or another, this imperative to manage scarcity in the service of a reclaimed abundance underpinned most western environmental politics in the first half of the twentieth century. Certainly the water reclamation projects derived their emotional and political force from it. So did the famous Colorado River Compact of 1922, in which the states of the upper and lower Colorado River Basin, feuding over the anticipated consequences of the Boulder Canyon Project and finally turning to the federal government to arbitrate between them, agreed to divide the waters of the river into two halves as a not entirely successful way of preventing California from receiving the lion's share of the division. (The compact relied for its statistics on exaggerated estimates of the river's flow, thereby allowing California—the earliest effective user as a result of Boulder Dam—to appropriate more than its proper share of the actual total.) But the imperative to manage scarcity to protect abundance reached well beyond water. Its fullest embodiment was almost surely in the national forests, with their elite cadre of disinterested, scientifically trained foresters managing a natural resource for the national good. Beyond the forests, it was still possible until 1934 for individual homesteaders to stake their claims as farmers or ranchers on the public domain, but in that year the Taylor Grazing Act finally withdrew this right for most of the remaining public lands. Homesteading continued but fell from thousands of new claims per year to a few hundred. Symbolically, 1934 thus marked the culmination of the long process whereby management and regulation replaced open access (or theoretical open access) to the public domain. For all practical purposes, the era of free land was finally at an end.

Arguably, the greatest environmental disaster of that era was the exception to prove the rule. Starting in 1931, the rains failed on the southern plains, so that farms that normally received eighteen inches per year—the minimum for many crops—received as little as eleven or twelve. Crops died as the parched soil cracked in the sun and temperatures soared above one hundred degrees for weeks at a time. The drought would last for a decade, but what really made this climatic event a disaster was the fact that so many plains farmers had expanded their acreage in response to the boom years of World War I, investing in tractors and other new equipment to work farms that were significantly larger than before. The heavy debts they took on to finance this expansion bore heavily on them during the depressed agricultural conditions of the 1920s, so that by the time the rain failed in the early thirties, vast stretches of the southern plains lay open to the blistering heat and windstorms. With no ground cover to hold down the desiccated soil, dozens of extraordinarily severe dust storms occurred every year throughout the 1930s, giving the region a new name: the Dust Bowl. A single famous storm in May 1934 blew three hundred million tons of dust into the air, some of it traveling all the way to New York and Washington and finally landing on ships far out in the Atlantic Ocean. Thousands of families abandoned their homes, took to the road, and headed to California as "Okies," the refugees from Oklahoma and elsewhere whose ordeals were so movingly described in John Steinbeck's *The Grapes of Wrath* (1939).

In the face of such stark proof that the frontier dream of abundance had given way to scarcity and despair, government officials stepped in to offer expert technical assistance with their new techniques of integrated regional planning. The Forest Service organized the planting of over two hundred million trees to form "shelter belts" that would supposedly discourage soil from blowing. The Soil Conservation Service (a New Deal creation of the Franklin Roosevelt administration) promoted new techniques of contour plowing and dry tillage to discourage erosion and retain soil moisture more effectively. The Civilian Conservation Corps provided labor for these and other initiatives. Most striking, the Resettlement Administration declared that in some of the worst-hit areas, farming was simply not viable: the original frontier settlements had been a mistake. In such areas, abandoned farms were bought by the government to be converted to rangeland and pasture. And yet even this symbol of apparent failure and defeat had an optimistic underpinning. In the eyes of the planners, scarcity had been caused by the ignorant and ill-conceived use of natural resources. If settlement patterns and land-use practices could be rationalized, abundance and prosperity could be restored even to so troubled a landscape as the Dust Bowl. In the end, the return of the rains in the early 1940s—and the increasing use of fossil groundwater from the immense Ogallala Aquifer for irrigation—made the dust storms seem like a passing nightmare that technology and better management could prevent in the future. Few bothered to speculate what might happen when the aquifer itself eventually began to give out.

The multiagency assault on the problems of the southern plains was characteristic of the 1930s and carried the progressive conservation agenda to a new level of complexity. For many Americans, it seemed that economic and ecological problems were reinforcing each other in ways that required the integrated perspectives embodied in the Boulder Canyon Project. The result was a new commitment to regional planning and an effort to link conservation initiatives with government programs promoting unemployment relief, investment in economic infrastructure, and social reform. In the East, the most famous example of this new integrated planning approach was the Tennessee Valley Authority (TVA), whose dams, electrical generators, highways, and agricultural reforms transformed the economic life of an entire region. For a time, the TVA seemed to prove that regional planning could solve virtually all the nation's ills.

The nearest parallel in the West was the Columbia River. There, efforts to develop the watershed in an integrated way had to contend with fierce competition between the Bureau of Reclamation and the Army Corps of Engineers over who should develop prime dam sites. Until the appearance of the bureau, the corps had been the government's chief dam builder, and its members were not at all happy about the bureau's growing prominence in the West. In 1937, the corps succeeded in completing the celebrated Bonneville Dam, but in 1941, the bureau finished the even more celebrated Grand Coulee Dam, which at that time was the largest concrete structure in the world, backing up a lake 150 miles long. In the end, the tensions between the two agencies resulted in the creation of a third entity, the Bonneville Power Administration, charged with selling electricity generated from the new dams.

Despite this interagency rivalry over dam construction, the reclamation projects on the Columbia River brought to the Pacific Northwest environmental and economic changes similar to those in California. The actual construction of the dams provided an infusion of relatively stable, high-paying jobs in an otherwise depressed economy. The new water supplies for irrigation had the usual effect of encouraging larger farms and more intensive forms of agriculture. But the most dramatic effects by now clearly concerned electricity. The Bonneville Power Administration was committed to distributing power to municipalities and public utilities so as to foster the widest possible use at the lowest possible rates. By the 1940s, the region had an immense surplus of electricity selling more cheaply than anywhere else in the nation. As a result, it attracted new industries that had especially heavy demands for power. In particular, aluminum production, one of the most voracious industrial users of electricity, concentrated in the region. With the coming of World War II, the availability of this aluminum enabled Seattle's Boeing Airplane Company to enjoy an unprecedented boom, becoming one of the region's most important employers in the postwar era.

It was cheap electricity that linked the Pacific Northwest to one of the most spectacular examples of government-sponsored interregional integration in the 1940s: the effort to develop the atomic bomb. Cheap electricity from the TVA permitted Oak Ridge, Tennessee, to produce uranium-235 via the extremely power-intensive gaseous diffusion process; cheap electricity from the Columbia River helped Hanford, Washington, become the site for a series of nuclear reactors and machine-shop facilities to manufacture plutonium. Both were tied to a network of other sites, the activities of which—collectively known as the Manhattan Project—had to be intricately coordinated in order to make the bomb a reality. Uranium was initially acquired from sources in the Belgian Congo and from the Eldorado mine on the Great Bear Lake in Canada; uncertainty about these foreign sources would soon fuel a uranium mining boom in the West's Four Corners region in the years immediately following the war. The University of California at Berkeley, already a center for nuclear research under the leadership of the physicist Ernest Lawrence, made major intellectual contributions to the bomb-building effort, while the California Institute of Technology contributed its growing expertise in explosives and jet propulsion. Much of the technical work on the actual bomb was conducted at the government's secret laboratory at Los Alamos in New Mexico. And the first nuclear test would occur in New Mexico at a site called Trinity on the desert sands near Alamogordo (postwar nuclear tests would be conducted at an even more isolated location north of Las Vegas, the Nevada Test Site). The extraordinary effort to produce the bomb centered in the West because of its new abundance of power, its large blocks of government-owned land, its isolation, and—paradoxically—its increasing integration with the rest of the nation. The military projects that came to fruition during the war would remain critical to the western economy at least until the end of the cold war era, establishing a partnership among the federal government, the corporate sector, and many western communities—a partnership that would profoundly shape regional development far into the future.

From one point of view, the dramatic changes that transformed the western landscape in the first half of the twentieth century represented unmitigated progress: new communities had sprung into being, parts of the region seemed prosperous as never before, and the desert had indeed bloomed. And yet all had unexpected social and environmental consequences. Residents of the Owens Valley more or less had to abandon their agrarian dreams once their river was tapped to serve Los Angeles; later, the city's growing demand would decrease water levels upstream at Mono Lake, endangering brine shrimp populations and the waterfowl that depended on them. Irrigated farms eventually began to suffer from the salts that accumulated in their soils, necessitating their abandonment or expensive purification techniques. The immense agricultural operations of the Far West required the use of migrant labor for planting and harvesting, creating an underclass of workers who were exposed to any number of toxic substances as growers increasingly came to depend on pesticides to protect their crops. Mine tailings from the uranium boom of the 1950s became a serious health hazard for residents of the Navajo reservation and other communities in the vicinity. Nuclear wastes dumped at Hanford in the years during and after the war created a long-term radiation hazard at that site. Fallout from the above-ground nuclear explosions at the Nevada Test Site killed livestock and elevated human cancer rates in the region lying downwind. The list could go on and on.

But perhaps the most suggestive example of deleterious environmental change occurred on the Columbia River, where the new dams prevented salmon and other anadromous fish from making their annual spawning runs. Salmon had once traveled hundreds of miles upstream to lay their eggs in freshwater locations, where the young fry, once hatched, could grow to maturity without being threatened by the predators they would face in the open sea. Since the spawning runs sustained a large fishing fleet in the waters of Puget Sound and on the open ocean, various efforts were made to protect them. Engineers added fish ladders to the relatively low Bonneville Dam so that adult salmon could still make their way upstream. Unfortunately, young salmon heading back downstream could not find these ladders and had to make the often lethal journey through the dam's pipes and turbines; moreover, the altered temperatures and still water of the new reservoirs made the fishes' journey in both directions more hazardous. On the upper river, high dams like Grand Coulee could not possibly be traversed even by fish ladders, and so heroic airlifts were attempted to capture fish at the base of the dam when its gates were first closed. Above Grand Coulee, salmon would henceforth disappear altogether, permanently ending the spawning runs for hundreds of miles upriver.

The present-day site of Bonneville Dam, this stretch of the Columbia River once supported a thriving salmon population that played an important part in regional Indian cultures. The construction of dams along the river disrupted fish spawning cycles and necessitated an equally elaborate effort to protect and maintain fish populations.

Russell Lee (1903–86). Native Americans Fishing at Celilo Falls, Columbia River, Oregon. *Gelatin silver print, 1941. Library of Congress (#LC-USF34-70154), Washington, D.C.*

To replace those runs, both the fish and the fishing fleet would henceforth have to rely more and more heavily on the artificial output of the fish hatcheries that Washington, Oregon, and the U.S. Fish and Wildlife Service would operate along the Pacific coast. The bitter irony was that efforts to manage the river toward human ends had necessitated an equally elaborate effort—reorchestrating salmon reproductive cycles over thousands of square miles—to manage fish populations that not long before had sustained themselves without any human intervention at all. The long-term success of this massive ecological experiment remains as yet unclear; recent evidence suggests that the genetic uniformity of hatchery fish populations may eventually threaten their viability.

The Urban Wilderness

The newly integrated western landscape that had come into being by mid-century had been mandated by the state, managed by corporations and federal bureaucracies, consolidated by war, and—perhaps most important—concentrated by cities. A handful of metropolitan centers had become foci for the regional economy and homes for most of the West's inhabitants. Los Angeles had emerged as the largest city in the West and before long would surpass Chicago as the nation's second-largest metropolitan area. Joining it were Houston, Dallas, Kansas City, Denver, Salt Lake City, Phoenix, San Francisco, Seattle, and Anchorage, each pulling in the resources of a broad hinterland region—especially water and energy—to sustain itself. Contrary to the frontier myth, the West had for a long time almost led the nation in the percentage of its population living in urban areas; in 1870, only the Northeast had been more urban than the Far West. By the mid-twentieth century, the rest of the nation had caught up, so that the one-half to two-thirds of western citizens who lived in urban places roughly matched the national average of 59 percent. Moreover, western cities had become tied to markets, transportation networks, and administrative systems that were truly national in scope. Like the rest of the United States and much of the world as well, the West had become a metropolitan region by mid-century, with urban and suburban residents consuming and reworking the resources of immense rural districts as part of a fully integrated economy.

As such, they faced many of the same environmental problems and challenges as urban dwellers elsewhere. By 1950, Los Angeles was widely regarded as the extreme example of a new urban landscape peculiar to western cities that had grown to maturity in the twentieth century. Dependent on the automobile for transport, it had sprawled in all directions as its residents had sought to fulfill the suburban dream of isolating themselves from the ills of an urban downtown. In the process, they created a multicentered city with no true focus, in which residents were forced to make long car trips—for work, for shopping, for school, for recreation, for everything—on freeways that became the arteries of the city. But this in fact was the spatial landscape that was developing on the margins of even much older American cities, a fact that became more apparent as the downtown centers of those cities began having more and more trouble competing with their own suburbs. Remove Manhattan, peel away the northeastern forest, and the basic spatial arrangement of Connecticut and New Jersey is not so different from that of Los Angeles. The California metropolis had pointed a way toward the future, and eastern cities soon followed its lead. The long journey between residence and workplace, the endemic traffic jams, the smog, the highway strips, the fast-food joints, and the shopping centers, in addition to the almost invisible systems that deliver water, gas, and electricity while removing sewage and solid waste—these are found to varying degrees in all American cities, not just in Los Angeles or the West. At the same time, the spread of highway systems, rural electrification, household appliances, telephone networks, radio and television broadcasting, and national distribution networks for goods have all enabled Americans from similar socioeconomic classes to enjoy similar life-styles whether they reside in cities, suburbs, or rural districts. One of the most important trends of the twentieth-century West, then, has been its steady convergence with the rest of the nation toward a common material life. In a sense, it is precisely this shared material life that has been the real fulfillment of the reclamation dream.

And yet there remains something distinctive about this newly integrated American West, for in fact the western landscape is *not* like the rest of the country. Its aridity makes the competition for water a far more compelling struggle than in the East and much more of a zero-sum game between different places and users. The heat of its desert summers has led it to join hands with the South in adopting air conditioning as a prerequisite for continuing growth. The scale of things often does seem larger in the West than elsewhere. The long distances—between far-flung metropolises, between sprawling suburbs, between city and country, between one state and another—impose greater travel times and energy costs on even the simplest journeys, making places seem isolated from each other even when in most other ways they are not. One gets used to big things in the West, whether they be mountains or mines or dams or farms. And the enclaved nature of western settlement makes it easy for places of great poverty—Indian reservations, barrios, depressed farming districts, dying towns—to share the regional landscape with the far more affluent inhabitants of well-to-do suburbs and cities. All these qualities mark the West as a special place.

But perhaps the most distinctive feature of the modern western environment has as much to do with the way people *think* about the regional landscape as the way they *use* it. The West has become the nation's greatest repository of "wilderness," a sacred space that for many Americans embodies both the simple virtues of the frontier past and the sublime wonders of pristine nature. Wilderness marks the paradoxical fulfillment of western urbanization. Whereas earlier generations looked to the West and dreamed of a working rural landscape—a place of farms and mines and lumber camps and small towns—urban Americans over the course of the twentieth century have increasingly preferred to think of it in *non*working terms, as a recreational place for escape or play. A similar shift occurred at the same time in national environmental politics: the earlier conservation movement had concerned itself most of all with questions of production and the efficient use of natural resources, whereas the postwar environmental movement—more urban in its basic outlook—became much more interested in problems of consumption, pollution, and the protection of natural systems. As the nonrural population of both the region and the country has grown, the cultural meaning of the western landscape has shifted to reflect predominantly nonrural values. Although wilderness might seem on the surface to represent the least urban of places, the way of viewing the natural environment that it reflects—as a land with no human inhabitants—has in fact come much more easily to urban people than rural ones. Traditional rural inhabitants of the West have been accustomed to earning their living in one way or another from the land and its creatures, activities that by definition involve mingling the natural and the human in ways that one would not ordinarily call "wild." Many modern westerners, on the other hand, accustomed to earning their living from urban markets, do not obtain food or shelter or basic income by working the land. And so, unsurprisingly, when fleeing the city they have often been drawn to *un*worked land: wilderness.

The West has been a destination for leisure-class travelers since the Civil War and before. At the same time that wealthy New Yorkers were first discovering the pleasures of hunting camps in the Adirondacks, Congress was setting aside Yosemite and Yellowstone as the first nonurban American parks. The initial impulse to create national parks came from several sources: a feeling of inferiority relative to the classical monuments of Europe; a desire to preserve the most "scenic" and "picturesque" elements of the American landscape; and a sense that the nation's "natural wonders" were among its most distinguishing features. In addition, powerful lobbies saw possible benefits for themselves in establishing parks. In the earliest years, the major western railroads sought to promote parks along their routes as a way of encouraging transcontinental passenger traffic. Railroad tourism characteristically involved delivering hundreds of well-to-do travelers to a single passenger depot, where they then traveled by

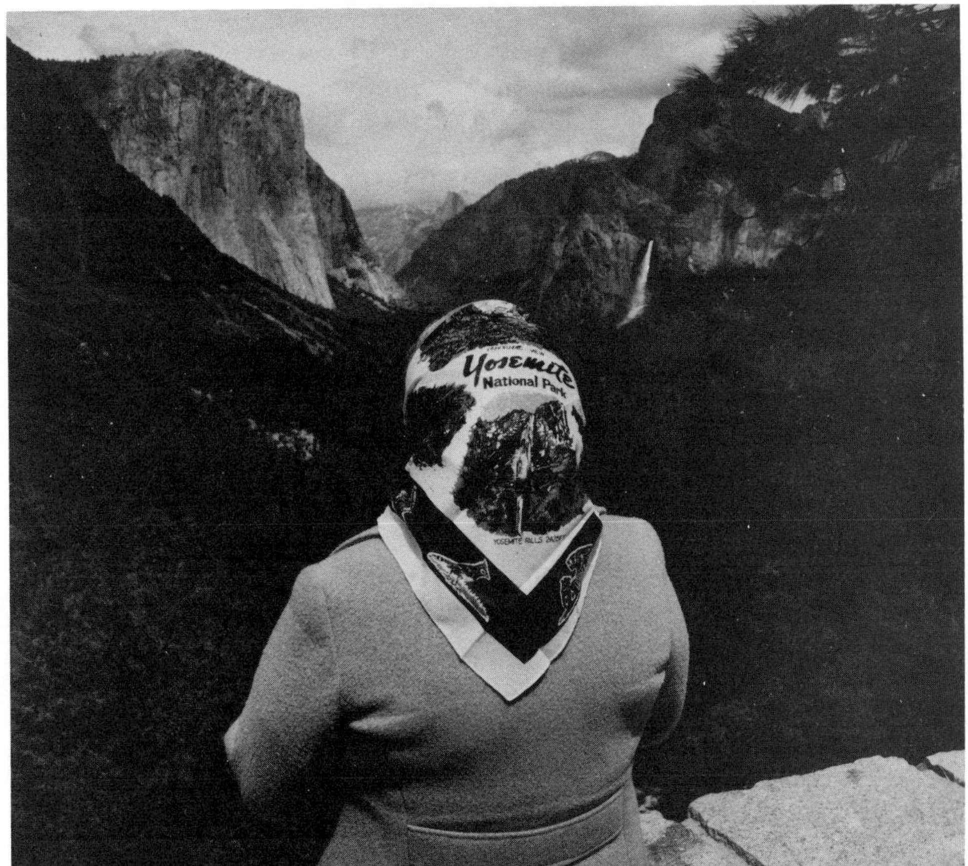

The roads that traverse the national parks not only bring tourists to these sites but also shape the ways in which visitors experience the landscape. For many, the parks are appreciated only from the well-established scenic viewing spots adjacent to the roads.

Roger Minick (b. 1944). Yosemite National Park, California, *from Sightseer Series. Ektacolor print, 1980. © Roger Minick 1980. Courtesy of the photographer.*

horse and stage to a large hotel capable of catering to their every need. Park hotels had urban amenities and were at best rustic—not "wild"—in their appurtenances. Given the need for destinations that could attract the well-heeled crowds that such hotels required, it made sense for the railroad companies to spend large sums on advertising and other promotional packages to educate Americans about the wonders of the western landscape. Such familiar places as Yellowstone, Yosemite, the Grand Canyon, and most other western parks were introduced to the public in just this way.

By the early decades of the twentieth century, as automobiles became more common and travelers began to consider them as alternatives to the railroads, new forms of western tourism started to emerge. Westerners, and Californians in particular, were prominent among supporters of the "good roads" movement, encouraging state and federal governments to invest significant sums in rural highway improvement. With the passage of the 1916 Federal Aid Road Act, states were given federal money to build and improve a system of "U.S. highways," many of which were soon being designated as preferred tourist routes. As Americans took to the roads in greater numbers, the national parks were among their favorite destinations. In 1916, Congress created the National Park Service to oversee the growing number of parks and monuments. Under the leadership of its first two superintendents, Stephen T. Mather and Horace M. Albright, the Park

Service constructed roads, built tourist facilities, designed brochures, and did everything possible to promote the success of the parks.

By the 1920s, the Forest Service, fearing that the national parks might overtake the national forests in popular affection, was also beginning to promote recreational opportunities on its lands, designating a growing number of sites as "Primitive Areas," with wilder amenities than the parks typically sought to offer. Both bureaucracies increasingly catered to the auto-based tourist, so that small motor campgrounds started to appear as alternatives to the immense railroad hotels. On the outskirts of the parks and elsewhere, "motels"—motor hotels—and other tourist facilities became more common. By the 1930s, the earliest western ski resorts were appearing as winter destinations for western travelers. Averell Harriman persuaded the Union Pacific Railroad to construct a large ski slope at Sun Valley, Idaho, in 1936 as a way of increasing passenger traffic at a time of year that had not previously attracted much of a tourist market. Aspen, Colorado, acquired its first ski lodge the same year. Throughout the West, tourism helped reorient a growing number of local economies, providing important employment alternatives for locals and outsiders alike. One striking example of a traditional western institution shifting from a rural to urban orientation was the dude ranch; throughout the West, marginal grazing operations increasingly turned to wealthy tourists as their chief source of income. Although such ranches billed themselves as places where urban visitors could get back to the basics and experience rustic living firsthand, what they marketed was less a genuine rural way of life than a carefully crafted fantasy for leisure-class consumers. Whatever value such experiences may have had for those who purchased them—and many testified to the considerable benefits for young and old alike—dude ranching was hardly "traditional" life on the land.

With the postwar era came an explosion in tourist travel throughout the West, fueled in part by baby-boom parents seeking to take their offspring to national parks, ski resorts, and even newer destinations like Disneyland, which opened in Anaheim, California, in 1955. In 1956, the popular demand for recreational travel helped secure passage of the Interstate and Defense Highway Act, which provided 90 percent of the funds for an extraordinary new network of divided highways, the largest public works project in American history. The new interstate highways, largely completed during the 1960s and 1970s, reinforced the existing urban system by linking major metropolitan centers and encouraging further growth at these key transportation nodes. At the same time, they made it easier for urban residents to escape the city for the country. The predictable result can be seen in National Park usage. In 1920, total visits to all national parks had only just reached 1 million. By 1950, this number had risen to 33 million; by 1970, it stood at 172 million. Other tourist destinations experienced comparable increases during the same period. The growth in recreational travel far exceeded national population increase.

Not coincidentally, the growing American love affair with western travel fostered new political pressure for the protection of wilderness lands. The United States had been setting aside parks in the West and elsewhere since Yosemite in 1864 and Yellowstone in 1872 but not specifically as "wilderness." Indeed, the act establishing Yellowstone described it as a "public park or pleasuring ground," suggesting the extent to which recreation was intended as its chief use. Only gradually did Americans begin to speak of the "wildness" of parks as one of their special values. Among those who led the preservationist struggle to protect wildlands, John Muir undoubtedly deserves special mention, since his writings about Yosemite, the Sierra Nevada, and the national parks in general were among the principal nineteenth-century texts convincing Americans—many of them elite inhabitants of eastern cities—of the need to set aside wild areas in the West.

In the first decade of the twentieth century, Muir and the San Francisco–based Sierra Club, which he helped found, conducted a nationwide publicity campaign to prevent the construction of the Hetch Hetchy reservoir in Yosemite National Park. Although Muir and his allies were ultimately defeated and the dam was built, Hetch Hetchy would become a battle cry in all subsequent struggles to protect western wilderness areas. More interesting, it also symbolized the growing contradictions between the material and the moral foundations of western American life. On the one hand, the people who sought to defend Hetch Hetchy were principally well-to-do inhabitants of cities—San Francisco chief among them—who viewed Yosemite National Park as a sacred icon of the western landscape as it had appeared before the coming of urban civilization. On the other hand, the people who sought to dam Hetch Hetchy were trying to defend San Francisco from a repetition of the terrible fires that followed the 1906 earthquake, while simultaneously guaranteeing that residents would continue to have water as their city grew. In the decades to come, many of those who would be most committed to saving western wilderness areas would also be drinkers of Hetch Hetchy water.

By mid-century, public enthusiasm for wildland recreation had grown to the point that the Sierra Club's next major battle against a dam would have a very different outcome. When the Bureau of Reclamation proposed building a dam in Echo Park, located within the borders of Dinosaur National Monument, preservationists mounted a major national campaign to lobby against it. This time, they won. Although part of the appropriations bill that protected Dinosaur National Monument also included funds to build a high dam in Glen Canyon—later lamented by preservationists as "the place no one knew"—it was a serious defeat for the bureau, putting it on notice that a significant portion of the American public no longer viewed dams as an unmitigated good. When plans were unveiled in the 1960s for a major reclamation project within the boundaries of Grand Canyon National Park, public outcry swelled to unprecedented levels and once again prevented the dams from being built. Never before had earlier conservation goals and emerging environmentalist values been more starkly contrasted.

The preservationist movement to protect wild areas had its greatest triumph in 1964, with the passage of the Wilderness Act. Under its terms, large tracts of land (most of them initially located in the national forests) were designated as roadless areas in a national system of wilderness preserves. Motorized traffic was not permitted, and—in the absence of new mineral discoveries—any development that might undermine the wild status was forbidden. The vast majority of these new wilderness areas were west of the Mississippi River, though a new law in 1975 mandated the reestablishment of wilderness—restored "virgin land," as it were—on eastern sites that had once been lumbered or pastured. By 1979, roughly twenty million acres had received wilderness protection, and that same year the size of the system quadrupled with the addition of new wilderness preserves in Alaska. Although the administration of President Ronald Reagan sought to cut back on the wilderness system, the resulting public outcry suggested how clearly the wilderness areas embodied environmental values that many Americans held dear.

And yet wilderness also represents one of the deepest paradoxes of the western environment. Standing in a kind of love-hate counterpoint to the urban West, it is quintessentially an urban cultural space if measured by the majority of people who visit it, defend it, and hold it dear. If one plots periods of peak visitation for western parks and wilderness areas, they coincide perfectly with weekends and major holidays as defined by urban workplaces. By the 1960s, Yosemite had acquired smog (and crime) problems that on a smaller, more symbolic scale paralleled those of Los Angeles. Crowding in most parks and backcountry areas had become so severe by the 1980s that permits and reservations—often made months in advance—were necessary if one wanted to experience the "freedom" of the hills. When a major coal-fired power plant was constructed in the Four Corners region to provide electricity to Los Angeles without compounding its pollution problems, the resulting haze would become a perennial problem at the Grand Canyon. Increasingly, those who sought escape from the city by fleeing to the wilderness found the city harder to leave than they thought, for they carried its baggage with them.

Among the most striking examples of the contradictions of wilderness occurred in the country's oldest national park, Yellowstone. Following a policy laid down in the early 1960s, rangers tried to manage the park as if it were a completely natural system, with as little human intervention as possible. As the critic Alston Chase has argued, not all of the results turned out as intended. Bear populations, which may have been kept artificially high by garbage dumps and intentional human feeding dating from at least the 1930s, plummeted disastrously when "natural"-style management was instituted. Even more striking, a decision to reverse the decades-old practice of preventing forest fires and to allow "natural" ones to burn led in 1988 to the largest and most devastating fires in the park's history. In such cases, it was not at all clear that one could just "let wilderness be" so that nature could "take care of itself." The modification of park ecosystems by intense visitor pressure, by fragmentation of habitats, by loss of native species, and by global environmental change suggested that active intervention would be necessary if wilderness was to survive.

In 1977, the Forest Service published a textbook whose title said it all: *Wilderness Management.* A quarter century before, when lobbying for a national wilderness act was still in its early stages, the notion that one could "manage" wilderness would probably have seemed antithetical to the whole concept. Americans in general, and westerners in particular, had sought to preserve wilderness because it stood for natural beauty and frontier freedom, both of which seemed seriously at risk in the modern world. Men and women who no longer earned their livings on the land, whose homes and workplaces were located in immense metropolitan districts, saw in the western wilderness a much-loved alternative to the complicated lives of quiet desperation that they both cherished and maligned. The irony was that those complicated lives—supported by dams and highways and energy resources that had made the desert bloom and had conjured cities in an arid waste to fulfill the reclamation dream—those same lives were themselves the

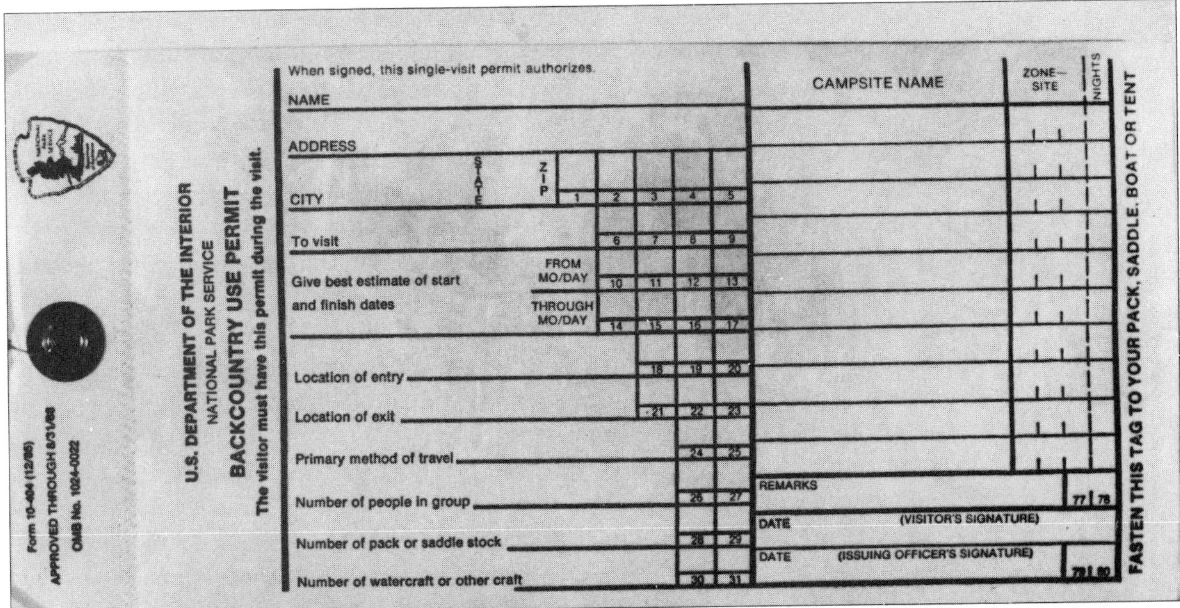
Form 10-404 (12/85)
APPROVED THROUGH 8/31/88
OMB No. 1024-0022

U.S. DEPARTMENT OF THE INTERIOR
NATIONAL PARK SERVICE
BACKCOUNTRY USE PERMIT
The visitor must have this permit during the visit.

When signed, this single-visit permit authorizes.
NAME
ADDRESS
CITY STATE ZIP
To visit
Give best estimate of start and finish dates FROM MO/DAY THROUGH MO/DAY
Location of entry
Location of exit
Primary method of travel
Number of people in group
Number of pack or saddle stock
Number of watercraft or other craft

CAMPSITE NAME ZONE—SITE NIGHTS
REMARKS
DATE (VISITOR'S SIGNATURE)
DATE (ISSUING OFFICER'S SIGNATURE)

FASTEN THIS TAG TO YOUR PACK, SADDLE, BOAT OR TENT

The "wilderness management" practiced by the U.S. Forest Service and the National Park Service suggests the complexity of contemporary American attitudes toward the land. Once valued for the limitless wildness that seemed to set them beyond the reach of civilization, wilderness lands are now viewed as dwindling resources to be shaped and controlled by human agency.

Backcountry Use Permit, *1993. National Park Service.*

principal challenge to wilderness in the modern world, calling its future into question. Furthermore, the alienated way of thinking about nature embodied in the very concept of wilderness—as a special place where nature could be experienced "pure," isolated from the "artificial" human world that surrounded it—was itself an artifact of American cultural attitudes toward the western frontier. In a very real sense, the wilderness was as much an urban cultural invention as were the dry hillsides of the Owens Valley, the flooded canyons of the Colorado River, or the suburban tracts of Los Angeles.

Furthermore, wilderness was also the ultimate symbol of abundance giving way to scarcity in the modern West. In the nineteenth century, the frontier had stood in the minds of many Americans for the unworked abundance of a savage or prehuman landscape awaiting the touch of human hands to become the site of prosperous farms, mines, factories, and cities. (Indians, of course, were shamelessly ignored or patronized by such thinking.) The progress of the nation was measured by its success in transforming wilderness into fertile countryside. The reclamation dream had extended this vision of frontier plenty even into the drylands of the arid West, making it possible to discover abundance even in the face of seemingly irrefutable scarcity. The growth of modern Los Angeles and other western cities suggests just how triumphant these nineteenth-century dreams eventually proved to be. But in the course of their fulfillment, they also undermined their original promise. The material resources of the West were *not* endless—far from it. The growth of one city could all too easily mean the stagnation of another. The damming of each new canyon left one less source of water and power to be tapped in the future. Perceived abundance gradually gave way to perceived scarcity, and nothing better represented this transition than the well-bounded, carefully managed, fragile western landscapes designated as "wilderness." By the late twentieth century, the wilderness, which had once stood as America's most potent icon of limitless abundance, seemed, in the eyes of many, to be its scarcest resource. Rarely has the reversal of a cultural symbol been more complete.

And that was why the oil that poured from the ruptured hull of the *Exxon Valdez*, coating the Alaskan coastline with black tar and devastating its marine life, was such a compelling source of outrage even for Americans who had never seen Prince William Sound or visited the far north. By the 1970s, Alaska had come to seem a near-perfect embodiment of America's frontier traditions. Its immense reserves of petroleum had fostered a boom to rival anything one could have found in the nineteenth-century West, creating a smaller, wetter, more northern version of Los Angeles in the city of Anchorage, whose supply lines reached around the world to feed a large urban population that could never have sustained itself with local agricultural resources alone. At the same time, Americans who now cherished "wilderness" as a cultural icon viewed Alaska as the last opportunity to preserve natural areas in their pristine state. Despite its immensity and harshness, the northern environment could thus seem all too small and fragile a remnant of the wonder that had once been North America. In the post-1968 struggles over whether and how to construct the Alaskan pipeline, over whether and how much to protect the Alaskan wilderness, Americans had tried once again to resolve the long-standing tension between abundance and scarcity, between the seductive dream of progress and the sublime icon of wilderness, between American national myth and material life.

Remote as it may have seemed from the day-to-day lives of most Americans, the wrecked Exxon supertanker could not have been more intimately entangled with these central questions of western environmental history. Before running aground, its ultimate destination had been Long Beach, California, where its cargo would have been refined into gasoline to help keep the wheels of Los Angeles turning. The great ship had been sent on its mission by the demands of this and other urban markets and would not have wreaked its havoc without their impetus. The connections were not always easy to trace, but the oily pollution on the shores of Prince William Sound had more than a little in common with the smoggy haze that obscured the waterfalls of Yosemite, the mesas of the Grand Canyon, and the basin of Los Angeles itself. Public outrage may have been directed against the corporations and governmental agencies that failed to prevent the spill, but in a deeper sense its ultimate causes were far too close to home for comfort. In much the same way, public anger in the wake of the spill expressed not just a legitimate outrage about serious ecological damage but an older American resentment about the fading of a pristine dream. Not even Alaska was safe, not even Alaska was far enough away to remain an unsullied landscape of frontier freedom and wilderness escape. Even there, abundance could give way to scarcity, forcing those who had counted on the promise of plenty to confront the consequences of its loss. It was not the first such failed promise in western history, and would surely not be the last.

Bibliographic Note

On the history of the *Exxon Valdez* oil spill, a competent journalistic survey can be found in Art Davidson, *In the Wake of the Exxon Valdez: The Devastating Impact of the Alaska Oil Spill* (San Francisco, 1990); for good photographic coverage, see the special report "Wreck of the Exxon Valdez" in *Audubon* (September 1989).

There is no general environmental history of the West, but Richard White's *"It's Your Misfortune and None of My Own": A New History of the American West* (Norman, 1991) does a better job of surveying environmental issues than any other available western history textbook.

For historiographical essays reviewing relevant literature, see William L. Lang, "Using and Abusing Abundance: The Western Resource Economy and the Environment," in Michael P. Malone, ed., *Historians and the American West* (Lincoln, 1983), 270–99, and John Opie, "The Environment and the Frontier," in Roger L. Nichols, ed., *American Frontier and Western Issues: A Historiographical Review* (Westport, Conn., 1986), 7–25, as well as various essays in Gerald D. Nash and Richard W. Etulain, eds., *The Twentieth-Century West: Historical Interpretations* (Albuquerque, 1989). Although tightly monographic in focus, Richard White's *Land Use, Environment, and Social Change: The Shaping of Island County, Washington* (Seattle, 1980) exemplifies the major themes of western environmental history as well as any available text; useful in much the same way are William deBuys, *Enchantment and Exploitation: The Life and Hard Times of a New Mexico Mountain Range* (Albuquerque, 1985), and Donald Worster, *Dust Bowl: The Southern Plains in the 1930s* (New York, 1979). Also suggestive are the essays in Donald Worster, *Under Western Skies: Nature and History in the American West* (New York, 1992), and my own "Kennecott Journey: The Paths out of Town" in William Cronon, George Miles, and Jay Gitlin, eds., *Under an Open Sky: Rethinking America's Western Past* (New York, 1992).

On the history of American public lands and the laws dealing with them, the standard work remains Paul Wallace Gates, *History of Public Land Law Development* (Washington, D.C., 1968). A more succinct and up-to-date survey can be found in Samuel Trask Dana and Sally K. Fairfax, *Forest and Range Policy: Its Development in the United States*, 2d ed. (New York, 1980). Quite dated but still useful is the anthology edited by Vernon Carstensen, *The Public Lands: Studies in the History of the Public Domain* (Madison, 1962).

Much of the literature on progressive conservation mingles with that on the history of forestry and lumbering in the United States. Overviews of American forest history can be found in Thomas R. Cox, Robert S. Maxwell, Phillip Drennon Thomas, and Joseph J. Malone, *This Well-Wooded Land: Americans and Their Forests from Colonial Times to the Present* (Lincoln, 1985), and Michael Williams's compendious *Americans and Their Forests: A Historical Geography* (Cambridge, Eng., 1989). On the Forest Service and its policies, see Harold K. Steen, *The U.S. Forest Service: A History* (Seattle, 1976), David A. Clary, *Timber and the Forest Service* (Lawrence, 1986), and William G. Robbins, *American Forestry: A History of National, State, and Private Cooperation* (Lincoln, 1985). On the early history of lumbering in the West proper, see Thomas R. Cox, *Mills and Markets: A History of the Pacific Coast Lumber Industry to 1900* (Seattle, 1974).

The classic history of the progressive conservation movement remains Samuel P. Hays, *Conservation and the Gospel of Efficiency: The Progressive Conservation Movement, 1890–1920* (Cambridge, Mass., 1959); for its successor on more recent environmental politics, see Samuel P. Hays, *Beauty, Health, and Permanence: Environmental Politics in the United States, 1955–1985* (Cambridge, Eng., 1987). For an invaluable survey of the predecessors of progressive conservation, see Donald J. Pisani, "Forests and Conservation, 1865–1890," *Journal of American History* 72 (September 1985): 340–59. The best edition of George Perkins Marsh's classic 1864 work is *Man and Nature; or, Physical Geography as Modified by Human Action*, ed. David Lowenthal (Cambridge, Mass., 1965). John Wesley Powell's *Report on the Lands of the Arid Region of the United States* (1878) is available in a reprint edition edited by Wallace Stegner (Cambridge, Mass., 1962).

On wildlife, the best legal history is Michael J. Bean, *The Evolution of National Wildlife Law*, rev. ed. (New York, 1983). General surveys can be found in Thomas R. Dunlap, *Saving America's Wildlife* (Princeton, 1988), and James B. Trefethen, *An American Crusade for Wildlife* (New York, 1975). More monographic works that are quite valuable include John F. Reiger, *American Sportsmen and the Origins of Conservation* (New York, 1975), James A. Tober, *Who Owns the Wildlife? The Political Economy of Conservation in Nineteenth-Century America* (Westport, Conn., 1981), and Theodore Whaley Cart, "The Struggle for Wildlife Protection in the United States, 1870–1900: Attitudes and Events Leading to the Lacey Act" (Ph.D. diss., University of North Carolina at Chapel Hill, 1971). Robin W. Doughty, *Wildlife and Man in Texas: Environmental Change and Conservation* (College Station, Tex., 1983), gives an interesting perspective on wildlife in one important western state. The forthcoming Yale University doctoral dissertations by Jennifer Price and Louis Warren will make important contributions to the history of wildlife in American popular culture and to the history of cross-class and multiethnic local conflicts over new wildlife-management regimes. As for fisheries, the story of western salmon is well traced in Anthony Netboy, *The Columbia River Salmon and Steelhead Trout: Their Fight for Survival* (Seattle, 1980). Finally, on marine resources, Arthur F. McEvoy, *The Fisherman's Problem: Ecology and Law in the California Fisheries, 1850–1980* (Cambridge, Eng., 1986), is in a class by itself as one of the finest environmental histories we have on any western subject.

Western water historiography is immense, and no brief discussion can do it justice. The most important one-volume synthesis of California water history, and the best starting point for anyone seeking a general overview of water in the West, is Norris Hundley, Jr., *The Great Thirst: Californians and Water, 1770s-1990s* (Berkeley, 1992). Broader in coverage and more provocative in argument is Donald Worster, *Rivers of Empire: Water, Aridity, and the Growth of the American West* (New York, 1985); more popularly written and polemical is Marc Reisner, *Cadillac Desert: The American West and Its Disappearing Water* (New York, 1986). For reclamation efforts in the pre–Newlands Act era, the standard work is Donald J. Pisani, *To Reclaim a Divided West: Water, Law, and Public Policy, 1848–1902* (Albuquerque, 1992). On California irrigation, see Donald J. Pisani, *From the Family Farm to Agribusiness: The Irrigation Crusade in California and the West, 1850–1931* (Berkeley, 1984), and Robert Kelley, *Battling the Inland Sea: American Political Culture, Public Policy, and the Sacramento Valley, 1850–1986* (Berkeley, 1989). On the history of individual states and reclamation projects, exemplary monographs include William L. Kahrl, *Water and Power: The Conflict over Los Angeles' Water Supply in the Owens Valley* (Berkeley, 1982), Joseph E. Stevens, *Hoover Dam: An American Adventure* (Norman, 1988), Ira G. Clark, *Water in New Mexico: A History of Its Management and Use* (Albuquerque, 1987), James Sherow, *Watering the Valley: Development along the High Plains Arkansas River, 1870–1950* (Lawrence, 1990), and Daniel Tyler, *The Last Water Hole in the West: The Colorado-Big Thompson Project and the Northern Colorado Water Conservancy District* (Niwot, Colo., 1992). A uniquely valuable reference work is William L. Kahrl, ed., *The California Water Atlas* (Sacramento, Calif., 1979). A superb bibliography and overview of the water literature published before 1980 is Lawrence B. Lee, *Reclaiming the American West: An Historiography and Guide* (Santa Barbara, 1980). Anyone wishing to understand what in this essay I call "the reclamation dream" would do well to read William E. Smythe, *The Conquest of Arid America* (1899), ed. Lawrence B. Lee (Seattle, 1969).

On the role of wilderness in the West and in American culture generally, the classic survey remains Roderick Nash, *Wilderness and the American Mind*, 3d ed. (New Haven, 1982), though one should read its arguments critically; also useful is Craig W. Allin, *The Politics of Wilderness Preservation* (Westport, Conn., 1982). Connections of wilderness to frontier ideology can be explored by reading Henry Nash Smith, *Virgin Land: The American West as Symbol and Myth* (Cambridge, Mass., 1950), G. Edward White, *The Eastern Establishment and the Western Experience: The West of Frederic Remington, Theodore Roosevelt, and Owen Wister* (New Haven, 1968), Richard Slotkin, *The Fatal Environment: The Myth of the Frontier in the Age of Industrialization, 1800–1890* (New York, 1985), and Robert G. Athearn, *The Mythic West in Twentieth-Century America* (Lawrence, 1986). On national park history, the standard history is Alfred Runte, *National Parks: The American Experience*, 2d ed. (Lincoln, 1987). A valuable journalistic survey is Dyan Zaslowsky, *These American Lands: Parks, Wilderness, and the Public Lands* (New York, 1986). The Forest Service textbook mentioned in the text is John C. Hendee, George H. Stankey, and Robert C. Lucas, *Wilderness Management*, Forest Service, U.S.D.A., Miscellaneous Publication No. 1365, (Washington, D.C., 1977). Earl Pomeroy, *In Search of the Golden West: The Tourist in Western America* (New York, 1957), remains one of the best sources on western tourism in the preautomobile era; on auto tourism, see Warren James Belasco, *Americans on the Road: From Autocamp to Motel, 1910–1945* (Cambridge, Mass., 1979), and James J. Flink, *The Automobile Age* (Cambridge, Mass., 1988). Excellent studies of the two best-known western parks are Richard A. Bartlett, *Yellowstone: A Wilderness Besieged* (Tucson, 1985), and Alfred Runte, *Yosemite: The Embattled Wilderness* (Lincoln, 1990). Alston Chase's acute, but not always generous, criticisms of Park Service policies in Yellowstone can be found in his *Playing God in Yellowstone: The Destruction of America's First National Park* (Boston, 1986). On environmentalist battles on behalf of wilderness preservation, see Stephen Fox, *John Muir and His Legacy: The American Conservation Movement* (Boston, 1981), Susan R. Schrepfer, *The Fight to Save the Redwoods: A History of Environmental Reform, 1917–1978* (Madison, 1983), and Michael P. Cohen, *The History of the Sierra Club, 1892–1970* (San Francisco, 1988).

Finally, on the related themes of abundance, scarcity, and a "metropolitan" interpretation of western environmental history, see the following: David M. Potter, *People of Plenty: Economic Abundance and the American Character* (Chicago, 1954); Earl Pomeroy, *The Pacific Slope: A History of California, Oregon, Washington, Idaho, Utah, and Nevada* (New York, 1965); Gerald D. Nash, *The American West in the Twentieth Century: A Short History of an Urban Oasis* (Englewood Cliffs, N.J., 1973); Peter Wiley and Robert Gottlieb, *Empires in the Sun: The Rise of the New American West* (New York, 1982); Carl Abbott, *The New Urban America: Growth and Politics in Sunbelt Cities* (Chapel Hill, 1981); and William Cronon, *Nature's Metropolis: Chicago and the Great West* (New York, 1991).

Chapter Eighteen

Contemporary Peoples/ Contested Places

SARAH DEUTSCH, GEORGE J. SÁNCHEZ, AND GARY Y. OKIHIRO

Boyle Heights

Toward the end of the summer of 1924, the *Los Angeles Times Illustrated Magazine* ran a one-page article extolling the virtues of southern California. In "Where Folks Are Folks," the author characterized Los Angeles as "the most American of American cities" because in Los Angeles, in contrast to the East, "the native American" (i.e., the nonimmigrant) was still "in the great majority." Like other important metropolises, Los Angeles had high buildings and high culture. But whereas New York was "overrun with Europeans," its ill-natured crowds "a nightmare to the traveler," Los Angeles was still the archetypical "middle-class American city, a metropolis with small-town ways—American ways." In the mind of this booster, Los Angeles was nothing less than "America regenerated, more eastern than the East," which had "become continental European."

The many promoters of the urban American West in the early twentieth century who extolled the absence of teeming immigrant masses ignored that the industrial and agricultural expansion that generated urban growth had, in fact, attracted large numbers of immigrants. The downtown region of Los Angeles and the areas just south and east of it held heavy concentrations of foreign-born and nonwhite newcomers. Ignored by promoters and largely excluded from the city's ruling elite, the Southside and the Eastside made up the "other" Los Angeles. Boyle Heights, an Eastside community, became the city's most heterogeneous region from the 1920s through the 1950s.

Among the first to settle in Boyle Heights were Mexican workers displaced from the central Plaza area of Los Angeles by rising housing prices and the building of offices during World War I. They shared "the Flats"—a low-lying region next to the river, close to employment by the Southern Pacific Railroad and garment and food-packing factories—with a sizable population of European immigrants, most notably Russian Molokans or "White Russians," a Christian sect that had fled Russia to escape harassment and impressment during the Russo-Japanese War. On the other side of Boyle Heights, a second expanding Mexican community joined a tiny African-American population to form the easternmost neighborhood in the city. Close to the city cemetery, this area—described as "a cheap land area" surrounded "by brick yards, railroad yards,

As this 1939 photograph of a Boyle Heights, Los Angeles elementary school suggests, the West has long been an ethnically diverse region, reflecting a legacy of Native American and Mexican residence and continued migration from Latin America, Asia, Europe, and the eastern United States.

Unidentified photographer. Breed Street School. *Gelatin silver print, 1939. Jewish Historical Society of Southern California, Los Angeles.*

and manufacturing plants"—was among the least desirable in Boyle Heights. Only housing in the Flats was in poorer condition.

Jewish laborers from the East had also begun to migrate to Boyle Heights during World War I, attracted by the congenial climate and the area's growing needle trades and other industries. They brought a tradition of radical politics and enthusiastic trade unionism. Their militancy made 1920s Boyle Heights home to local chapters of the Workmen's Circle and the hatters, carpenters, and garment workers unions. They also created in Boyle Heights and adjacent City Terrace a bustling, if paler, version of the Yiddish cultural life so typical of Jewish immigrants in New York. According to two historians of Jews in Los Angeles, on Brooklyn Avenue, "Jews bought and sold, Yiddish was freely used, and Saturdays and Jewish holidays were marked by festive appearances and many closed businesses."

Finally, although an enterprising Japanese minister had built a Buddhist temple in Boyle Heights as early as 1904, it was in the 1920s that a large number of Japanese settled there, dotting the central region of Boyle Heights, less concentrated than the Jewish or Mexican communities. Residences, community institutions, and a few businesses lined both sides of First Street. They became an extension of Little Tokyo, the largest center of Japanese life on the West Coast, located directly to the west. Initially the Japanese community consisted primarily of businessmen who preferred "suburban" living in Boyle Heights to the urban congestion of Little Tokyo. But as the 1920s progressed, more working-class Japanese families also ventured across the river, renting apartments and small houses on the Eastside.

The Great Depression pulled these disparate communities closer together. Repatriation campaigns aimed at encouraging Mexican and Filipino immigrants to return to their native countries affected the fringes of Boyle Heights by persuading unemployed Mexicans in the Flats to leave Los Angeles. The Mexicans who remained, largely skilled workers and their families, became steadfast in their commitment to the neighborhood. Some areas of Boyle Heights even grew during the 1930s. Indeed, to depression refugees from other parts of the country, Boyle Heights appeared stable and secure. The Zimmerman family of Cleveland, for example, headed west after losing the family produce business. In Boyle Heights, they could live near other Jewish families and seek employment for adults and older children in local stores and shops. Japanese Americans similarly solidified the ethnic economy they dominated in fruit, produce, and gardening. Dorothy Tomer remembered of her adolescence in Boyle Heights during the 1930s: "Everyone was poor. We pulled together. . . . If you didn't have education and money to back you, you just did anything to make a living."

This situation profoundly affected young people's intergroup perception. Enrolled in the same schools, they valued education as a luxury. Since everyone had to work, it came as no surprise when children dropped out to help the family economy. Jewish and Mexican women, many of them young adults, banded together to form the backbone of the garment and food-packing unions, which grew stronger in the decade.

Local government officials and real estate companies showed some hostility toward this struggling diverse community. For example, in 1939, the Federal Housing Authority gave its lowest possible rating to Boyle Heights precisely because, the agency explained, it was "hopelessly heterogeneous" and its "diverse and subversive racial elements" supposedly made it a bad risk for housing assistance. By 1940 the Jewish population of Boyle Heights totaled about thirty-five thousand, the Mexican population fifteen thousand, and the Japanese population five thousand.

If the depression had pulled the community together, World War II made it clear to most residents that Boyle Heights was not immune to forces out of its control. Roosevelt High School, the only secondary school in Boyle Heights, lost one-third of its population in a matter of months when Japanese-American students were interned. One English teacher encouraged her remaining students to write to interned classmates through round-robin letters kept up until graduation. Internment seemed unjustified to most in Boyle Heights, especially to the youth who had grown accustomed to this mixed environment. Indeed, the experience of living through the forced removal of Japanese-American residents during World War II seemed to leave a distinct impression on non-Japanese in the community, an impression that later translated into efforts on behalf of civil rights for all. The successful race of Edward Roybal, a Roosevelt High School graduate, for the Los Angeles City Council in 1947 was organized by an interracial group out of Boyle Heights.

The end of the war failed to bring normalcy to Boyle Heights; in many ways, peace was as disruptive as war. The severe housing shortage throughout Los Angeles contributed to an instability that left Boyle Heights vulnerable to exploitation under the guise of "development." By 1947, three public housing projects had been built within Boyle Heights by the Los Angeles Housing Authority, an "experiment" that led to the displacement of hundreds of families from the area. Despite the ten thousand residents, money for maintenance and improvement of these facilities evaporated. And between 1943 and 1960, five freeways were built through Boyle Heights, providing pathways for suburban commuters while destroying the cohesiveness of the community and displacing another ten thousand people.

These postwar developments rapidly altered the community's demographics as Boyle Heights became a less desirable place to live. By 1952, the Mexican-American population of Boyle Heights had grown to over forty thousand, forming close to half the population, while the Jewish population had shrunk to fourteen thousand. The Japanese population rested at about sixty-five hundred. For the remaining families, life in Boyle Heights had been transformed. One Jewish family, for example, found it necessary to pay a tough Mexican boy to protect their son through junior high school in the 1950s. Boyle Heights was well on its way to becoming a classic barrio with a population of poorer Chicano residents physically separated by freeways and a river and subject to racial housing restrictions and economic discrimination.

Pearl Harbor

Boyle Heights provided one model of an interethnic community in a contested landscape. World War II wrought others. In 1991, on a clear, bright Hawaiian morning, under the warming sun and with a gentle tropical breeze descending from the Koolau mountains, President George Bush faced the *U.S.S. Arizona* wreckage and memorial at Pearl Harbor. He told the assembled survivors of Japan's attack and the World War II veterans and their families: "No, I have no rancor in my heart. I can still see the faces of my fallen comrades, and I'll bet you can see the faces of your fallen comrades, too. But don't you think they're saying: 'Fifty years have passed. Our country is the undisputed leader of the free world. We are at peace.' Don't you think each one is saying, 'I did not die in vain?'" Although the president posed the question to elicit an affirmative reply, many listeners believed that although America might have won the war, it was being defeated by Japan in the global contest for economic dominance. Like the droplets of oil that still rose to the surface every twenty seconds from the *Arizona*'s tanks, the battle begun on 7 December 1941 continued to be waged on other fronts.

Months before the fiftieth anniversary services that brought President Bush to Pearl Harbor with his message of reconciliation, the Pearl Harbor Survivors Association (PHSA) had lobbied to exclude Japanese representatives from participating in the commemoration. "Would you expect the Jews to invite the Nazis to an event where they were talking about the Holocaust?" asked the national president of the fourteen-thousand-member PHSA. James G. Driscoll, in the *Fort Lauderdale Sun-Sentinel*, editorialized on the government's eventual decision to bar all foreign dignitaries from the fiftieth anniversary proceedings. "The Japanese are not invited to the ceremonies, nor should they be. Their presence would be an affront to those Americans who died there, to those who survived and who will attend and to the families of all."

Thurston Clarke, the author of one of the fifteen books on Pearl Harbor published in America in 1991, explained the legacy of hatred spawned by the Japanese attack. "Since Pearl Harbor was uniquely shocking and humiliating," reasoned Clarke, "it is only logical that American attitudes toward Japan should be uniquely sensitive, and that today's economic disputes should be haunted by Pearl Harbor ghosts. And so Japan's trade policies are likened to a form of 'treachery,' [and] a U.S. senator describes the export of Japanese cars to America as 'an economic Pearl Harbor.'" Japanese purchases of symbolic American properties such as Rockefeller Center, Pebble Beach, and Columbia Pictures, noted Clarke, stimulated among Americans an increased dislike and distrust of the Japanese. Hawai`i, where Japanese yen bought up choice businesses and homes as "minor spoils" of a trade war victory, had become "an economic colony of Japan," charged Clarke, representing "a Japanese victory more enduring than Pearl Harbor." Even the hotel in which the Pearl Harbor and World War II American veterans stayed during the fiftieth anniversary services, observed a *New York Times* reporter, was owned by Japanese. Describing C. A. Murray, a Marine corporal in 1941, who recalled that fateful Sunday, the reporter wrote: "Today he stood on the second floor of the Sheraton-Waikiki Hotel, owned by a Japanese company, and watched Japanese tourists walk by. 'I'm sorry I didn't take Japanese in school,' he said sarcastically, referring to all the signs in Japanese around him. 'I don't have any animosities, but it rubs me the wrong way.'"

Little noticed in the hoopla surrounding Pearl Harbor's fiftieth anniversary was the fact that although Japanese bombs had sparked the conflagration in Asia and the Pacific, America's presence in Hawai`i owed itself to an equally imperialist, expansive thrust during the late nineteenth century. In 1898, the United States had plucked Hawai`i like a ripe pear, the imagery employed by the American minister in Honolulu at the time.

Expansion once again stretched our boundaries and diversified our population. Forty-three years later, the invader became the invaded.

President Bush insisted that Japan had mistaken "our diversity, our nation's diversity, for weakness." He added: "But Pearl Harbor became a rallying cry for men and women from all walks of life, all colors and creeds. And in the end, this unity of purpose made us invincible in war and now makes us secure in peace." Yet Pearl Harbor instead had offered "a golden opportunity," in the words of a leading official of California's Grower-Shipper Vegetable Association in 1942, "to get rid of all Japs, sending them back to Japan either before or after the war is won." Indeed, while smoke still rose from the wreckage that was America's Pacific fleet, Hawai`i's governor proclaimed martial law in the territory, and in Hawai`i and on the mainland, army military police and Federal Bureau of Investigation agents began arresting Japanese aliens and citizens alike. By 19 February 1942, the day President Franklin D. Roosevelt signed Executive Order 9066 authorizing the mass removal and detention of all Japanese Americans along the West Coast, the army and the Justice Department had rounded up several thousand enemy aliens—including Japanese, Germans, and Italians—considered "dangerous" to the nation's security. Before the end of the year, over 110,000 Japanese Americans had been confined to concentration camps in Hawai`i and the American West.

An army leave policy that permitted Japanese-American soldiers to return to the West Coast prompted an "orgy of Jap baiting" in the press during April and May 1943, according to the writer Carey McWilliams. Despite all evidence that Japanese Americans were not the enemy, an editorial appearing in the *Los Angeles Times* on 22 April 1943 argued against the leave policy. "As a race, the Japanese have made for themselves a record for conscienceless treachery unsurpassed in history." Commenting on the leave policy, Lt. Gen. John L. De Witt, head of the Western Defense Command, told the House Naval Affairs subcommittee in San Francisco: "A Jap's a Jap. They are a dangerous element, whether loyal or not. . . . It makes no difference whether he is an American; theoretically he is still a Japanese. . . . You can't change him by giving him a piece of paper." Two California congressmen added, "If you send any Japs back here we're going to bury them." That atmosphere of race hatred and mob violence, wrote McWilliams, "contributed to kindling the fires of racial antagonism in the community."

One of those fires was the "zoot suit" race riot in Los Angeles during the anti-Japanese drive and a police and newspaper campaign against "Mexican crime" in which Mexican and African-American youth were subjected to intimidation and summary arrest. Lt. Edward Duran Ayres of the Los Angeles sheriff's department had testified before a 1942 grand jury that Mexican-American youth were genetically predisposed to criminal behavior. Beginning on 3 June 1943, and lasting for about a week, mobs of white soldiers, sailors, and civilians attacked, beat, kicked, and tore the clothing of Mexican and African Americans in the streets, bars, and streetcars of Los Angeles with apparent impunity. Similar mob violence occurred in Pasadena, Long Beach, and San Diego. During the same summer, race riots also flamed in Philadelphia, Chicago, Evansville, Detroit, Harlem, and Beaumont, Texas.

The war had not produced racial unity. It did, however, provide new employment opportunities for women, Chicanos, African Americans, and Asian Americans. And some of its legacies, notably the GI Bill, which made college education possible for many minorities, boosted the economic and political prospects for these western residents. The war also helped darken the urban West's complexion as California's booming defense industries drew African Americans at the rate of ten thousand a month during the peak wartime migration and as Native Americans too flocked to the urban worksites of the West. At the same time, while U.S.-born Mexicans moved to more lucrative jobs in the cities, the region's factories in the field beckoned braceros from Mexico to plant and harvest the crops. Having repatriated thousands of Mexican nationals back across the border to alleviate relief rolls in the 1930s, including one-third of the Mexican population in Los Angeles County or nearly thirty-five thousand people, by World War II western employers were clamoring for renewed legal immigration from Mexico to provide low-wage labor, particularly in agriculture. The Bracero Program, initiated as a wartime emergency measure, continued until 1964, supplying agricultural employers with a steady labor force whose strict regulation left few options for protest and whose availability could suppress the demand for higher agricultural wages. Even when the Immigration Service launched "Operation Wetback" in the 1950s to return illegal aliens across the border, it often ended up deporting those it had first brought to the United States as braceros. Indeed, potential deportees could be immediately "legalized" if they agreed to work under the contracts provided by the Bracero Program.

Disneyland

Braceros were not the only dream-seeking migrants in the 1950s West. On 17 July 1955, an enterprising Anglo-American transplant to the West welcomed the first visitors to what would become the most recognized site in the region for the rest of the century. As Walt Disney unveiled his uniquely American creation, Disneyland, he also exposed the unbridled optimism in technological progress, industrial expansion, and romantic saga that so marked Anglo-American thinking in the twentieth-century American West. The plaque that was laid that day in Disneyland's Town Square came to symbolize the melding of the western image with that of the nation's future. "To all who come to this happy place: Welcome. Disneyland is your land. Here age relives fond memories of the past . . . and here youth savor the challenge and promise of the future. Disneyland is dedicated to the ideals, the dreams, and the hard facts that have created America . . . with the hope that it will be a source of joy and inspiration to all the world."

Disney's fantasies, as exemplified in Disneyland, were characteristic of an entire migrant generation who made their way to California and other parts of the West in the first half of the twentieth century. Disney had grown up in Marceline, Missouri, a small town northeast of Kansas City, before breaking away from his family and moving west in 1923. Life had not been as idyllic for him as he later depicted on the mythical Main Street of his dream park. Disney's father failed in several business ventures, which perhaps caused the harsh beatings he meted out to his children. The family relied extensively on the income of young Walt and his brother, Roy. For Disney, as for an entire generation of Anglo-American migrants, the move west allowed him to wipe out unpleasant memories and reconstruct his history. That Disney's newfound vision of his past played itself out in a theme park only made his transformation more public than those of others.

Disneyland also represented a frustration with the growing sprawl and diverse population of the urban West and the associated problems of traffic congestion, pollution, overcrowding, and alienation after World War II. Disney saw his park as a leisurely retreat from that civic confusion as well as a platform to show what life in a metropolis could be like if planned correctly. His obsession with eradicating disorder translated into politics as well. A staunch Republican, he was a consistent advocate of combating "internal subversives." As a founder of the Motion Picture Alliance for the Preservation of American Ideals in 1944, Disney argued, "The American motion picture industry is, and will continue to be, held by Americans for the American people, in the interests of America and dedicated to the preservation and continuance of the American scene and the American way of life." This group invited the House Un-American Activities Committee (HUAC) to investigate Hollywood and ferret out supposed Communists who were corrupting the American mind. It was Disney and other conservative southern California businessmen who encouraged Ronald Reagan to abandon acting and turn his attention to politics, particularly after Reagan's role as a star witness identifying Communists to HUAC. Concern with "internal subversives" led Disney to hire only a handful of Jews, Asians, Mexicans, or African Americans at his new park.

Indeed, the vast Disney empire, including movies, television, marketing, and theme parks, began as an attempt by Disney to wrest American entertainment away from the largely Jewish immigrant community that had come to dominate Hollywood by the 1920s. He often equated being Jewish and a movie "mogul" with being a Communist sympathizer, and his vision was to return Hollywood entertainment to "wholesome American ideals." Yet, Adolph Zukor, Louis B. Mayer, and the Warner brothers had all distanced themselves from their immigrant pasts and any affiliation with the Left. Most, like Disney, were staunch Republican party members in the age of Roosevelt. These studio heads willingly fired suspected Communists from their studios to please HUAC in the 1940s and 1950s, even when those fired were talented Jewish writers and directors that they had recruited to Hollywood just a few years before.

A move west had allowed these entrepreneurs, like Disney, to remake themselves into a new elite to take advantage of the economic opportunities in the West, even though they needed to rely on ethnic affiliations and cultural projections brought from the East. While Walt Disney provided Americans with a sanitized version of a simpler midwestern past, the Jewish studio executives were busy appealing to immigrant notions of survival and opportunity. Both Disney and the Jewish movie magnates built institutions that relied on particularistic visions and appealed to a larger, more amorphous public. Ironically, it was their worlds of fantasy and leisure, rather than their own lives, that came to dominate representations of America. Each vision held out the possibility of remaking oneself in the migration to the American West, often by distorting one's own history.

Borders, 1965

Even as Hollywood veered toward film and fantasy, the nation took a different turn. In 1965, Congress enacted sweeping changes in immigration law, altering the racial and ethnic composition of the West and the nation. The new rules emerged in a context of profound national upheaval. Through the 1965 Voting Rights Act, four hundred thousand African Americans registered to vote in the Deep South within three years. The civil rights movement entered a period of transition marked by the full emergence of the Black Power movement, the assassination of Malcolm X, and a profound refocusing on poverty in the wake of the Watts riots, which rocked Los Angeles. Lyndon Johnson launched the Great Society in earnest, introducing the Medicare/Medicaid programs for the elderly and poor as well as creating the Department of Housing and Urban Development. The Free Speech movement had emerged in the fall of 1964 at Berkeley and would eventually spread to other college campuses. At the University of Michigan, the first antiwar teach-in was held to protest the escalation of the Vietnam War and the bombing of North Vietnam.

Amid these transformations, the nation's immigration laws appeared antiquated and racist. The myriad laws that defined who would be welcomed on U.S. shores were products of late-nineteenth- and early-twentieth-century perspectives on race and nationality. The 1924 National Origins Act practically barred legal immigration from Asia while establishing the Border Patrol to manage movement from Mexico and the rest of Latin America. In addition, the act aimed to place severe limitations on immigration from southern and eastern Europe in favor of other parts of that continent, and Europe continued to supply the vast majority of immigrants to the United States well into the 1960s. With victories by the civil rights movement, Third World liberation movements, and a cold war, the U.S. quota system seemed anachronistic at best. As Attorney General Robert Kennedy told Congress in 1964, "Everywhere else in our national life, we have eliminated discrimination based on national origins, yet this system is still the foundation of our immigration law." When the same southern conservatives who fought civil rights legislation fought the new immigration bill in Congress, it was clear that the act would signify a turn away from racial discrimination in immigration.

Recognizing white fears, the bill's proponents argued that it would *not* bring about sweeping changes in the racial makeup of the nation's immigrant patterns. They pointed to the fact that the proposed law did not do away with quotas but rather replaced quotas based on the 1890 national origins of the American population with hemispheric targets and equal allocations among countries. Within the quotas, preferences would be given to professionals, skilled workers, and refugees. Moreover, a special category outside the quota would be set up for immediate family and relatives of U.S. citizens and permanent residents. The *Wall Street Journal* predicted that the family-preference system "insured that the new immigration pattern would not stray radically from the old one." What was impossible to predict, of course, was the massive movement of refugees from southeast Asia and Central America, the deterioration of the Mexican economy, and the rise of illegal immigration.

These developments led to a profound transformation in the racial makeup of the immigrant population of the United States and created new dynamics of race throughout the American West. The national Asian-American population increased fivefold, to five million, in the twenty years following the 1965 Immigration Act while the Latino population increased to fifteen million. Moreover, a new diversity within these groups emerged. Whereas the 1965 Asian-American population had a majority Japanese-

American contingent, two decades later the Chinese and Filipino populations would vie for numerical plurality. Vietnamese, Korean, Asian Indian, Laotian, and Cambodian communities would emerge in areas previously devoid of immigrants from these parts of the world. Although the Latino population remained over half Mexican-origin, new concentrations of Cuban, Salvadoran, Guatemalan, Nicaraguan, Dominican, and Colombian migrants would take their place alongside existing Mexican and Puerto Rican communities. These changes had profound implications for a nation accustomed to thinking of race in black and white terms. The 1990 census, for the first time, recorded that African Americans constituted less than 50 percent of those traditionally considered "minority." With three major umbrella populations (Asian American, African American, and Latino) emerging in the late twentieth century to form almost one-fourth of the U.S. population, in many western localities minorities became the majority.

Nowhere was this transformation more evident than in California. After 1965, one out of every four legal immigrants settled in California, and Los Angeles International Airport replaced Ellis Island as the major port of entry into the nation. In the city of Los Angeles, "minorities" of color by 1990 outnumbered the Anglo "majority," and both the Latino and Asian communities, profoundly augmented by recent immigrants, were larger than the local African-American population. In fact, by 1990, Los Angeles had a larger proportion of foreign-born than New York City, and this proportion was nearly as large as that of New York at the height of European immigration in the early twentieth century.

Probably no other topic so clearly focused the fears of "new" immigration as did language and the "threat" of foreign tongues. California remained in the forefront of the ensuing conflict. In 1971, the U.S. Supreme Court's *Lau vs. Nichols* decision initiated instruction in the native tongue of Chinese-speaking students in San Francisco, mandating bilingual education. And as elsewhere in the United States, new laws resulting from the Supreme Court decision required that bilingual ballots allow non-English-speaking citizens to participate in the electoral process.

Such services produced a backlash from those who considered English to be one facet of the definition of an "American." In 1986, California voters passed Proposition 63, an "English-only Amendment" to the state constitution, designating English as the official language of the state. Though largely symbolic, it spawned a flurry of activity that led to the passage of similar legislation in over half the states in the country. Despite all the calls for increased foreign-language acquisition by the American population in order to compete in the international economy of the Pacific Rim, non-English-speaking immigrants from the Pacific Rim were seen as liabilities. According to the bill's proponents, "Our American heritage is now threatened by language conflicts and ethnic separatism."

Ironically, almost all studies showed that foreign-born adults and children desperately wanted to learn English but that English acquisition was often difficult for older immigrants not in school. State legislators and voters seemed unwilling to provide the services that would allow for successful English language acquisition by the foreign-born. In Los Angeles County in 1986, four hundred thousand adults were on waiting lists to enter the meager number of English-language classes offered by the state's educational system. When State Senator Richard Alatorre introduced a bill to fund more classes the day after Proposition 63 passed, it was roundly defeated by those same legislators most vocal in support of English-only.

The contest over language signified a fundamental ambivalence toward immigration, particularly focused on the West. President Reagan, who emerged from California's multiethnic society, could invoke one perspective when he spoke at the one-hundredth anniversary of the Statue of Liberty. "Call it mysticism if you will, I have always believed there was some divine providence that placed this great land here between the two great oceans, to be found by a special kind of people from every corner of the world, who had a special love for freedom and a special courage that enabled them to leave their own land, leave their friends and their countrymen, and come to this new and strange land to build a new world of peace and freedom and hope." In the same year, former Governor Richard Lamm of Colorado would provoke a different vision of newcomers in his bestseller *The Immigration Time Bomb*. "Today, we Americans must lose the dream of unrestricted immigration and must face the reality behind the dream. We dreamed of America as a country of immigration, and we identified open immigration with freedom. We believed that someone poor, someone downtrodden, someone persecuted—from any other country in the world—could pull up stakes, come to America, and have another chance. . . . It is not a dream we can keep today."

Delano

In 1965, the very year that the immigration law so dramatically changed the racial and ethnic composition of the nation's people, cross-racial alliances in stubborn pursuit of that dream were already emerging in the fields near Delano, California. California's fields were cultivated by a succession of workers, including American Indians, Chinese, Japanese, Koreans, South Asians, Mexicans, Filipinos, and Europeans, who were drawn to the state by opportunities but also by labor recruiters and shippers who plied the Pacific trade. At the locus of production, in the field, these laborers acted as individuals, as women and men, as racial and ethnic groups, and as a class of workers, depending on their individual and collective sensibilities and initiatives and on the conditions under which they labored. At times, racial and ethnic affiliation enabled them to unite for higher wages and better working conditions, but at other times, racial and ethnic affiliation divided them—a condition encouraged by some planters, who employed a mixed labor force, paid differential wages for the same task, and rotated workers in a system of migratory labor.

Racial politics mirrored the racial division of labor. In the late nineteenth century, for example, California's politicians rallied white workers under the banner, "The Chinese Must Go!" And in 1903, when the Japanese-Mexican Labor Association (JMLA) of Oxnard, California, petitioned for a charter from the American Federation of Labor, AFL president Samuel Gompers replied that the agricultural union would be admitted only if it refused membership to Chinese and Japanese. J. M. Lizarras, the Mexican secretary of the JMLA, wrote to Gompers, "We would be false [to the Japanese] and to ourselves and to the cause of Unionism if we . . . accepted privileges for ourselves which are not accorded to them." Workers must unite, he urged, "without regard to their

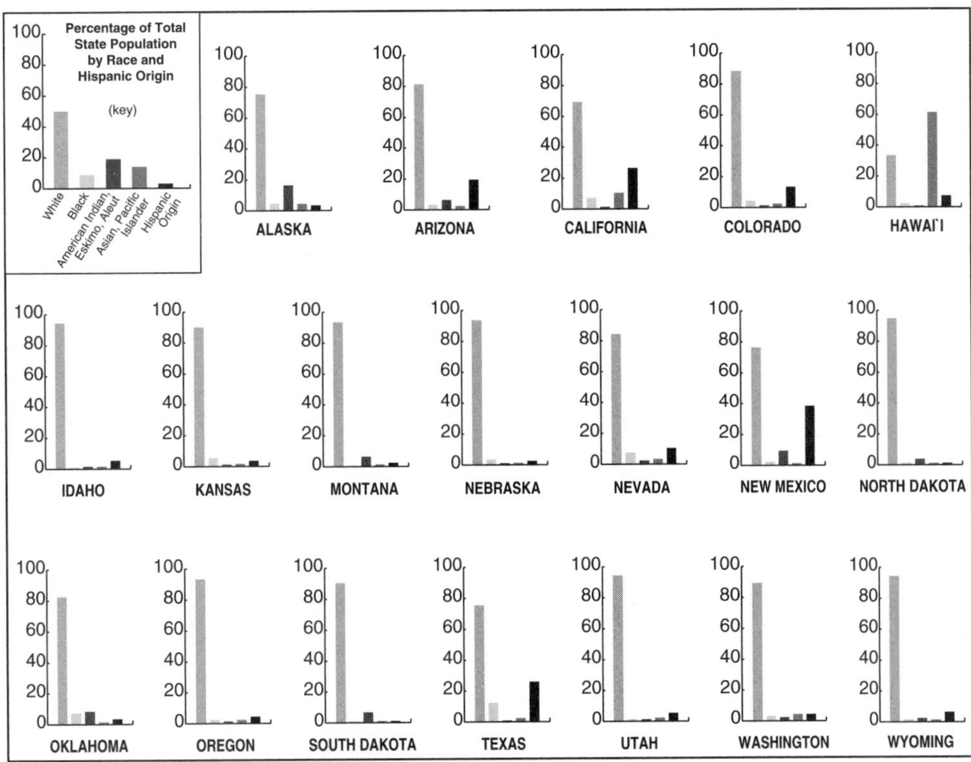

Resident Population by Race and Hispanic Origin, 1990

color or race." The Industrial Workers of the World, in contrast to the AFL, organized farm laborers "without regard to their color or race," including Europeans, Japanese, South Asians, and Puerto Ricans between 1905 and 1917. Asian workers, like whites, acted against their class interest by scabbing on and undercutting fellow Asian laborers. Class and even racial solidarity was clearly imperfect.

But the lessons learned in the fields helped to mold a new generation of leaders, who secured greater equality in the workplace and paved the way for the post–World War II drive for civil rights. When, on the evening of 7 September 1965, Filipino workers sat down in the fields near Delano, California, against the advice of their leaders in the Agricultural Workers Organizing Committee (AWOC), they followed in the footsteps of their early-twentieth-century forebears—Asian, Mexican, and European migratory farm workers who had shown the possibility of organizing and who were, in the words of a Chicago journalist, "the lowliest of workers . . . paid the least and work[ing] the hardest."

The next day, AWOC, with its Filipino organizers Larry Itliong, Ben Gines, and Pete Manuel and its Mexican organizer Dolores Huerta, struck against thirty-three grape growers. Cesar Chavez's National Farm Workers Association (NFWA) joined AWOC eight days later. In walking off his job, a Filipino farm worker in his sixties noted: "For more than thirty years I have been in strikes in the fields. I think we are going to

Emma Tenayuca helped lead a group of Mexican-American pecan shellers concerned about the advent of mechanization in a strike in San Antonio, Texas, in 1938. The workers sought to preserve jobs that paid three dollars for a 54-hour week. The widespread militancy among urban and rural workers in the 1930s laid the foundation for the civil rights and farmworkers' movements in the 1960s.

Unidentified photographer. Emma Tenayuca Leading Workers during the Pecan Shellers' Strike. *Photograph, 1938.* The San Antonio Light *Collection, U. T. The Institute of Texan Cultures, San Antonio, Texas.*

win this one, but whether or not we win, the growers will know they have been in one hell of a fight."

The five-year Delano strike was notable for the gains achieved by farm workers, for Cesar Chavez's rise to national prominence, and for the convergence of issues of labor, civil rights, religion, and education. Philip Vera Cruz, an AWOC member at the time, observed that NFWA was "a mixture of many things," including church and student groups such as the California Migrant Ministry and the Student Non-violent Coordinating Committee, African-American groups such as the Congress of Racial Equality, civil rights workers, and youth volunteers. NFWA, added Vera Cruz, was more than a union bent on raising wages; it was a social movement, "fighting . . . for the rights of people and the dignity of human beings."

Six months into the strike, AWOC and NFWA discussed the need for improved communication and a coordinated effort between the two groups. Both the Teamsters and the AFL-CIO sought to consolidate the strikers under their respective unions. The Teamsters, reported Vera Cruz, told AWOC's Filipinos that NFWA's Mexicans, being more numerous, would soon take their jobs, and both AWOC's and NFWA's members feared that unity would reduce their ethnic autonomy. Yet the groups merged in August 1966 as the United Farm Workers Organizing Committee, despite the misgivings.

The strike became a cause—for a people, for a class, for a nation. Cesar Chavez mused, "We have to find some cross between being a movement and being a union."

As a member of the Senate Farm Labor Subcommittee, Robert F. Kennedy supported the strikers in well-publicized hearings held in Sacramento, Visalia, and Delano in March 1966, joined the picket lines, and with his wife, Ethel, helped to raise funds for the farm workers. When Cesar Chavez ended a twenty-five-day fast on 10 March 1968, Martin Luther King, Jr., telegraphed him: "My colleagues and I commend you for your bravery, salute you for your indefatigable work against poverty and injustice, and pray for your health and continuing service as one of the outstanding men of America. The plight of your people and ours is so grave that we all desperately need the inspiring example and effective leadership you have given."

The strikers' persistence and the grape boycott that affected over one hundred U.S. and Canadian cities led to settlements with the growers in the summer of 1970. The great strike was over, but the farm workers' modern-day battle had just begun. "This is a war," declared the woman warrior Dolores Huerta in 1973, "this is a real war—all of the growers and right-wing elements . . . are trying to crush the farm workers . . . we have to act like it's a real war."

San Francisco City Hall

In the face of new waves of immigration and shifting power relations in the hinterlands, the meaning of the West's vaunted opportunities seemed up for grabs. Other groups who migrated to the region confronted a white population that was feeling increasingly bemused and beleaguered. On 27 November 1978, thirty-two-year old Dan White, a San Francisco native, the son of a San Francisco fireman, a Vietnam veteran, a former member of the city's Board of Supervisors, and a man with an image so clean-cut that an acquaintance told *Time Magazine* that if White " had been a breakfast cereal . . . he would have had to be Wheaties," snuck into San Francisco City Hall and shot Mayor George Moscone and Supervisor Harvey Milk to death.

The *Time Magazine* headline proclaimed it "Another Day of Death." *Newsweek* called it "one more eruption of the San Francisco syndrome, a mindless streak of politically tinged violence that has afflicted the Bay Area with nearly 100 bomb blasts and more than a dozen murder victims in the past ten years." Yet this "day of death" resulted in a gathering of thirty thousand for a torchlight ceremony at city hall, an Oscar-winning documentary on Harvey Milk, a television docudrama, and a stage play. By these vehicles, the public rejected the notion that the shootings had been "mindless" and instead struggled to make sense of the murders. Part of the sense they made was regional. Despite any number of equally lawless incidents across the country, many analysts saw in the incident something peculiarly western.

Moscone's death faded from view as attention centered on the openly gay Milk, the growth of San Francisco's gay population (approximately one-eighth of the city and one-fifth of the city's adults at the time), and the creation of a "gay city" in the Castro, the district that had elected Milk. To these observers, the Castro and the conservative vigilante White echoed images of a West both open and violent.

The story is more complicated, however, than westering gay utopians and lone gun-toters defending "law and order." It also involved San Francisco's shift from a manufacturing to a service city with corporate headquarters and deracinated yuppies. It involved the remnants of blue-collar Irish neighborhoods, an Irish police force, and Irish and Italian political machines. It involved, in short, the postwar, postindustrial transformation of the United States. In this brave new world, "homesteading" was urban, not rural. The gay men flocking to San Francisco in the early 1970s homesteaded in the mixed ethnic blue-collar and increasingly unemployed Castro. Urban or rural, homesteading still recreated the landscape.

Although stories of gay and lesbian life in San Francisco go back as far as the nineteenth century, the Castro as a gay district had two sets of more recent roots, one in World War II and the other in the 1960s. During World War II, the enormous need for men and women in the military had led to a retreat from the military's antigay policies. According to the historian Lillian Faderman, when General Dwight Eisenhower gave Sgt. Johnnie Phelps an order to ferret out the lesbians in her battalion, she replied: "Yessir. If the General pleases I will be happy to do this investigation. . . . But, sir, it would be unfair of me not to tell you, my name is going to head the list . . . . You should also be aware that you're going to have to replace all the file clerks, the section heads, most of the commanders, and the motor pool." Eisenhower responded, "Forget the order." For many gays and lesbians, particularly those from small towns, the war was their first experience meeting sizable numbers of homosexuals. Few of these people wanted to return to the isolation of their former life. They tended to stay in the cities where they had landed on their return to the United States. Moreover, after the war, when the military again adopted repressive measures toward homosexuals, it sent thousands of "undesirable" discharges to the nearest U.S. port. All of these factors helped to produce homosexual enclaves in New York, Boston, Los Angeles, San Francisco, and other cities. Businesses arose to serve a specifically homosexual clientele and, for the first time, could survive.

This was an evolution more urban than western. Indeed, many of the migrants came from a distinctly less open West, from rural Nebraska or Iowa. And yet it was in the West and not in the East that the first gay and lesbian civil rights associations began in the 1950s: the male Mattachine Society, which began in Los Angeles and moved to San Francisco, and the Daughters of Bilitis, which began in San Francisco. Unlike the counterculture simultaneously launched among the beat poets, including the gay Allen Ginsberg, these two societies were relatively conservative. According to the writer Frances Fitzgerald, they aimed primarily to prove the respectability of homosexuals. By 1961, other homosexuals gathered in the city's thirty homosexual bars, and Jose Sarrio, a drag entertainer protesting police harassment, won six thousand votes in his bid for city supervisor.

But this fluorescence was brief. In that same year, 1961, in response to election campaign accusations that he had made San Francisco the national headquarters for sex deviates, the mayor, George Christopher, allegedly launched a police crackdown. Hundreds of men were arrested simply for dancing together or holding hands. Police confiscated copies of Ginsberg's *Howl.* Bars were closed. After 1965, although other gay organizations remained in San Francisco, Mattachine's headquarters moved to New York.

In the late 1960s and early 1970s, Mayor Joseph Alioto ruled San Francisco as an old-time, big-city boss. He relied heavily on ward bosses, who kept his machine running smoothly in the city's Irish and Italian Catholic neighborhoods. But at the end of the decade, the manufacturers and shippers who employed his constituency began to falter. To bolster the city's weakening economy, Alioto lured real estate developers and corporate headquarters to San Francisco. Yet these sectors provided few jobs for blue-collar workers, and Alioto's supporters were leaving town. By the 1970s, only one of San Francisco's eleven voting districts had a blue-collar majority.

While the city's overall population declined in these years, the proportion of people age twenty-five to thirty-four rose by over 25 percent. These young, largely white professionals were lured by San Francisco's new economy. They may have come from blue-collar families, but they were college-educated and did not identify primarily by ethnic heritage. Among them were the gay men who began to gentrify the decaying blue-collar Irish streets of the Castro.

In the 1950s and 1960s gay bars had largely been in the Tenderloin and the decorators' district and, slightly later, in the warehouse district. What made the Castro different was that it was a residential neighborhood. It had shops and homes. It could cultivate a whole gay way of living, twenty-four hours a day. It arose as a result of the changing economy and demographics of San Francisco in the late 1960s and early 1970s.

Highly mobile socially, economically, and geographically, gay or straight, the new city residents were poor fodder for a city machine. The politics of the new West, at least in San Francisco, seemed to call for more accommodation. Richard Hongisto, a civil rights, antiwar activist elected sheriff of San Francisco County in 1971, tried to improve relations between gays and the police. George Moscone, elected mayor in 1977, appointed minorities, women, and gays to city offices. He exemplified the blending of the old and the new. The son of an alcoholic milk-wagon driver (another account has his father as a guard at San Quentin), Moscone had put himself through college and Hastings Law School and then rose through San Francisco's Italian-American political establishment to serve in the state senate. As the majority leader in the state senate, Moscone, along with Assemblyman Willie Brown, who would later serve as speaker of the California state assembly, had helped secure the repeal of state statutes proscribing forms of consensual sex in 1975.

Also elected in 1977 were Harvey Milk and Dan White. Milk was the city's first openly gay man to serve in high public office. Milk and thousands like him had flocked to the cities during the most assertive gay liberation movement to date, which had begun when police raided a New York gay bar, the Stonewall Inn, in 1969. San Francisco had the special attraction of a relatively large and open homosexual community already in place.

But it was not simply that Milk was gay. He was part of San Francisco's yuppie revolution. Born into a middle-class Jewish family on Long Island, Milk had attended teacher's college, served in the navy, taught high school history and math, and worked as a financial analyst for insurance and Wall Street investment companies in New York before coming to San Francisco in 1969 to run a camera store. When Milk first entered politics, in 1973, the Castro was still largely Irish and Catholic. Milk learned how to build coalitions. When, in 1977, a change in election laws meant that city supervisor elections were no longer citywide, Milk was ready for his first success. Because of Milk, gay bars boycotted Coors beer for antilabor practices. With a new haircut and three-piece suits, Milk spoke on issues of education, street cleaning and lighting, and libraries. Labor unions, housewives, teachers, drag queens, gays, and lesbians all supported Milk. Milk had proven to be not only a gay activist but also a good ward boss.

The same year, Dan White was elected from the only district in the city that voted for the Briggs Initiative to drive openly gay teachers out of the classroom. Unlike Milk, White was a native of San Francisco. In high school, he captained the baseball and football teams and was a Golden Gloves boxer. After serving in Vietnam, he worked three and a half years as a policeman before joining the fire department in 1973. While running for office, he was cited for a heroic rescue of a mother and a child from the seventeenth floor of a burning building. White ran as the law-and-order candidate, the hope of a conservative policemen's association. He was the youngest supervisor ever elected in San Francisco.

In June 1978, between 240,000 and 375,000 people turned out for the annual Gay Freedom Day Parade in San Francisco, a city of 700,000 people. Marchers included Gay American Indians, Disabled Gay People and Friends, gay political leaders, the Gay Latino Alliance, a gay Jewish synagogue, a Marilyn Monroe look-alike, Dykes on Bikes, the local Lesbian Association Kazoo Marching Band, and many others. It was, after all, a city with approximately 90 gay bars, 9 gay newspapers, 150 gay organizations, and a gay yellow pages for the clothing stores, stockbrokerages, realtors, churches, and other institutions of the Castro. Estimates of the city's gay population ranged from 75,000 to 150,000.

Were the Milk and Moscone assassinations in late November of the same year the result of a clash over defining the city's landscape? Perhaps, but White was a problematic hero for the old guard. Despite his squeaky-clean image, he had a history of gambling and expensive trips to Reno. On his policeman's salary he had bought first a Jaguar and then a Porsche before taking a leave of absence to hitchhike throughout the United States. After serving for ten months of his term as city supervisor, he had resigned for financial reasons. The potato stand at Fisherman's Wharf he had hoped would support his new wife and infant son and pay for his new house was floundering. White soon changed his mind and asked for his Board of Supervisors job back. In the meantime, it became clear that White's behavior on the board had alienated his fellow supervisors. On a closely balanced board, where narrow victories were common, White was prone to hold grudges and to be a difficult colleague. Moscone decided instead to appoint a symbol of the new accommodating West, Don Horanzy, a real estate loan officer of the federal H.U.D. Horanzy was not an office-seeker but had founded a voluntary "All People's Coalition" in White's lower-middle-class mixed-race and ethnic district. The coalition fought crime and spruced up the neighborhood.

From a radio journalist, White learned he would not be reappointed, and the next day he murdered Moscone and Milk. The trial centered on what became known as the Twinkie Defense—the impact of junk food on depressed people—rather than on city politics, race, and homophobia. When the jury, largely white, working-class Catholics, many of whom lived in White's district, returned a verdict of voluntary manslaughter rather than premeditated murder, five thousand marchers faced down the police at city hall. The police chief, Charles Gain, a Moscone appointee seen as an outsider by the policemen, particularly when he removed the American flag from his office and replaced it with plants, prevented the police from storming the crowd until the crowd began to disperse after three hours. Frustrated police supporters of White tore off their shields to avoid identification and headed for the Castro, where they invaded bars and savagely beat patrons. Sixty-one police and one hundred gay men were hospitalized.

After that violent catharsis, coalition politics reigned in San Francisco. Dianne Feinstein became mayor with 80 percent of the gay vote and appointed a lesbian as police commissioner. But she also removed Police Chief Gain and replaced him with the Irish-Catholic Cornelius Murphy, who had come up through the ranks of the department.

Florence and Normandie

In the mythic West of dime novels and movies, the law enforcers—the marshals and the sheriffs—most often symbolized the heroic, self-denying male who could single-handedly impose order on a chaotic landscape in the best interests of all. The ability of a police force to stand above the fray rather than in it, however, had been laid open to question by the violence in San Francisco, as it had in the race riots of the 1940s. The idea forcefully exploded again in Los Angeles in 1992.

Probably no other event better captured the complexity of the western experience of race than the three days of riots from 29 April to 1 May 1992. On the surface, the Los Angeles unrest was a direct result of the "not guilty" verdicts in the trial of four white police officers for the beating of a black motorist, Rodney King. Yet the reality of racial conflict can never be so easily characterized in the American West, where "race" has always had a multiplicity of meanings. Here, the historical legacy of Native American, Mexican, and Latino populations had confounded the images of race brought by scores of American migrants to the region—white and black. Moreover, Asian immigration, be it Korean, Japanese, Chinese, or Filipino, had consistently been a crucial factor in representations of race in the West, from as far back as the mid-nineteenth century. The meanings of race in the West have barely begun to be explored by eastern commentators—including those who have brought the East with them as migrants. Most Americans still believe that the diversity of the American West is solely of recent origin and can easily be reframed as simply another version of a bipolar racial model: white and "other."

For most who experienced the 1992 disturbances through the media, the most searing image of the Los Angeles riots was taken by a local television helicopter above the corner of Florence and Normandie avenues in South Central Los Angeles on April 30. The image of Reginald Denny, a white truck driver, being pulled from his cab, beaten, and spat on by a group of young African-American males quickly became a counterimage to the inhumane police beating of the black motorist King a year earlier. These two events, both captured on videotape, dominated representations of the rage in Los Angeles, a city haunted by poverty, racism, and police brutality. So powerful, they pushed out other, equally vivid and telling representations, allowing almost all commentators to explain the riot by invoking long-standing notions of racial conflict that speak exclusively of white-black tensions.

A closer look at the victims of violence at the corner of Florence and Normandie reveals not only the complexity of the Los Angeles riots but also the way many academics and social commentators avoided the implications of these events and surrendered to the simplistic notions of race fed to the public by the mass media. Most people outside of Los Angeles were surprised to hear that Denny was not the only person injured on that corner. Mesmerized by video images of a single beating of one white man, people found it difficult to imagine that thirty other individuals were injured at that corner, most pulled from their cars and some requiring extensive hospitalization. Tellingly, almost all the other victims were people of color, including a Mexican couple and their one-year-old child hit with rocks and bottles; a Japanese-American man, stripped, beaten, and kicked after being mistaken for Korean; a Vietnamese manicurist left stunned and bloodied after being robbed; and a Latino family with two five-year-old twin girls, who each suffered shattered glass wounds in the face and upper body. All of these acts of violence occurred before Denny entered the intersection.

Even more complex was the case of Byron Bowers, the adopted son of two African Americans (one of whom was an interracial communications professor), who had only recently learned that he was a product of an interracial union. A mob yelled, "He's white!" Bowers yelled back, "I'm half-black!" His jaw was broken and his car pummeled with crowbars. A twenty-four-year-old Belizian immigrant woman, already suffering a black eye for her efforts, came to the rescue.

The decisions made by young, angry African Americans at that corner as they chose whom to hurt speak volumes on the complexities of racial and ethnic identity in late-twentieth-century America. For some, the decision was not about who was white but about who was not black. For others, it centered on how Latinos and Asians had "invaded the territory" they claimed as their own turf. South Central Los Angeles, in fact, had been demographically transformed since the 1965 immigration law, so that 51 percent of the population was Latino by 1992. According to videotapes, some rioters shouted, "Let the Mexicans go!" But they added, "Show the Koreans who rules!" Whereas the famous 1965 Watts uprising could be read as an African-American community's response to poverty and police brutality, even defining "community" by 1992 became difficult.

Although the violence began as a response to a verdict passed by an almost all-white jury against an almost all-white set of police officers, other people of color were quickly engaged in the deadly discourse. The meaning of racial identities was consistently contested, as at the corner of Florence and Normandie. Over the next three days, the dynamics of racial and class tensions, rage against the police, and antiforeign sentiment came together in violent, unpredictable fashion. The mayhem spread to engulf the city, creating the worst modern race riot in American history. More than fifty people died, residences and businesses suffered about one billion dollars worth of damage, and police made over fourteen thousand arrests. In the first three days of rioting, over four thousand fires were set, and eighteen hundred people were treated for gunshot wounds. The destruction occurred throughout the Los Angeles Basin, and the participants and victims were indeed multiethnic.

Underlying much of the frustration of the riot participants was the collapse of the inner-city economy, the negative flip side of the new "Pacific Rim." Like the transformation of San Francisco's economy, which had set the stage for tensions in the 1970s, Los Angeles had lost 150,000 manufacturing jobs in the previous three years, and each of these jobs was estimated to take another three associated jobs with it. The new jobs

The violence directed at Korean-American merchants by Latinos and African Americans in Los Angeles during the riots of 1992 suggested the complexity of late 20th-century race relations and the demographic patterns of the West's largest urban center.

Jean-Marc Giboux. A Korean-American Stands Guard during the Los Angeles Riots. *Photograph, 1992. © Jean-Marc Giboux. Gamma Liaison, New York, New York.*

created were disproportionately low-wage and dead-end forms of employment; 40 percent of all jobs created in Los Angeles from 1979 to 1989 paid less than $15,000 a year. Most of these jobs were taken by recent immigrants, leaving African Americans few viable options for secure employment. The average earnings of employed black men fell 24 percent from 1973 to 1989, and unemployment swelled to record levels in the inner city. Middle-income Los Angeles was rapidly disappearing, leaving little opportunity for anyone to move up the economic ladder. This inequality was highly racialized; the median household net worth for Anglos in the city in 1991 was $31,904 but was only $1,353 for non-Anglos.

Clearly, one obvious target of the frustrated residents in the inner city were the Korean merchants in South Central. In 1990, 145,000 Koreans lived in Los Angeles County, a 142 percent increase from ten years earlier and a phenomenal growth from only 9,000 in 1970. Unable to transfer their education and skills to the U.S. labor market, many Korean immigrants had pooled their funds to start small businesses in ethnic communities throughout the city. They had replaced Jews, who had departed after the 1965 Watts riots. They now saw their businesses burn to the ground and suffer widespread looting. These small merchants had filled a vacuum created by the abandonment of the inner city by large retail businesses and by discrimination against African-American entrepreneurs. As Korean immigrants became targets for racial attack, other Asian Americans, particularly American-born Chinese and Japanese, worried that they might be mistaken for Korean and thus distanced themselves from recent immigrants.

Yet much of the damage to Korean businesses occurred in Koreatown itself, where one-third of that community's businesses were located. The residential population of Koreatown was overwhelmingly Latino, and it was this ethnic group that primarily engaged in looting these stores. In fact, 43 percent of those arrested during the riots were Latino; only 34 percent were African American, contradicting the notion that this was simply a black-Korean conflict. At the same time, the Immigration and Naturalization Service took advantage of arrests for curfew violations and deported over two thousand Latino noncitizens.

The situation remained charged in the weeks and months following the riots. Overt citywide battles diminished, but one week in October 1992 witnessed the eruption of racial hostility between black and Latino students in three high schools and one junior high, leading to injuries and arrests. Korean merchants who sought to rebuild their businesses were stymied by community efforts to keep liquor stores and other unwanted enterprises out of inner-city neighborhoods, causing a second financial disaster on top of their burned or looted merchandise. A summer campaign to ensure jobs for African Americans in rebuilding Los Angeles led to confrontations with Latino workers on construction sites. An effort to replace the departing Latino school superintendent with another Latino set Mexican-American organizations squarely against the hiring of a qualified African American. In the wake of the disturbances Rodney King asked, "Can we all get along?" The question seemed directed not toward the city's white population but toward its communities of color.

Zuni Salt Lake

To some of its residents, the contested landscape of the West is hardly new, and the stakes have always been high. The region's various indigenous peoples struggled over the land before European Americans, African Americans, and Asian Americans arrived. Although those disputes continued, in a sense like the battles at the corner of Florence and Normandie avenues, the enormity of the European-American challenge to native rights eventually dwarfed them.

Zuni Salt Lake covers more than eleven hundred acres in western New Mexico. Surrounded by piñon, saltbush, greasewood, and gama grass, the lake lies in a high valley, where only nine inches of rain falls annually. Fresh and salt waters stream and seep into the lake. Long before the advent of Europeans, the Zuni, Hopi, Navajo, and Acoma Pueblo peoples gathered salt there in sacred rituals. In 1540, Coronado called it "the best and whitest" he had ever seen.

In the 1930s, the indigenous peoples still gathered salt at the lake, maintaining the ancient rituals, but they no longer walked their hot, long, and dusty trails and prayed at the shrines along the way. Instead, in pickup trucks, they gratefully headed for the lake along modern roads. Yet, they had not exactly abandoned the old trails. The Hopis still referred to the trails in rituals and prayers. They had once been sacred; they remained sacred.

In 1990, when the Salt River Project, an Arizona Utility Company, developed an 18,612-acre coal mine near Zuni Salt Lake with a forty-mile transportation corridor to a power plant in Arizona, little in this practice had changed. The Hopi's sacred landscape was the geologist's map of resources. One hundred years earlier, the outcome of that conjunction would have been clear. In 1874, after all, the discovery of gold in the Black Hills had ultimately led the U.S. Supreme Court tacitly to condone the violation of a Lakota Sioux–U.S. treaty that had guaranteed that the Lakota alone would possess the,

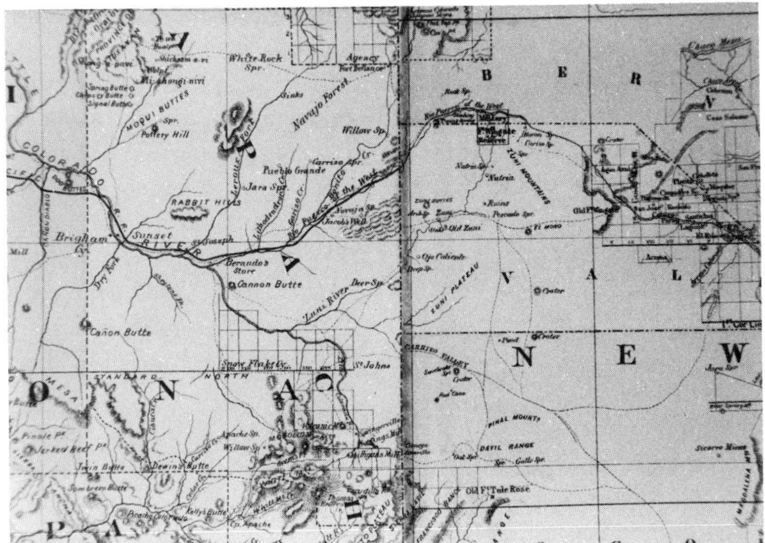

The diverse ways of depicting the physical terrain around Zuni Salt Lake represented in a 19th-century commercial map and a 20th-century Hopi artist's mural, suggest the vast cultural differences to be accommodated in the joint-use venture between European-American and Hopi residents of the Southwest.

S. Augustus Mitchell, publisher. County and Township Map of Arizona and New Mexico *(detail). Lithograph, 1881. Courtesy E. Richard Hart.*

Fred Kabotie (b. 1900). Hopi Salt Pilgrimage. *Mural (located in the Painted Desert Inn, Petrified Forest National Park, Arizona), 1948. Photograph courtesy of the Western Archaeological Center, National Park Service, Tucson, Arizona.*

to them, sacred territory. The settlement offered them money instead. The Lakota rejected the settlement, to no avail. In 1990, Indians owned none of the land involved in the Salt River Project's development. But other factors had changed. As recently as 1965, fewer than a dozen Indians held law degrees; in 1990, between five hundred and six hundred did. This small army of lawyers not only had experienced more success in direct engagements with adversaries but also had become more adept at using laws for their own ends. For example, they had used endangered species legislation to protect sacred lands and traditional uses. In the case of the Salt River Project, they could, if need be, have used historic preservation legislation for the same purposes.

When the research required for the environmental impact statement revealed the concerns of the various tribal groups, however, the Salt River Project pursued the matter cooperatively. Despite the fact that no one alive could remember exactly where the trails were, the Salt River Project asked the tribes to recommend ethnohistorians with whom they felt comfortable working, hired those ethnohistorians, asked each tribe how it would like to proceed, and worked with the subsequent tribal-created research teams to identify the pilgrimage trails and other sites of concern. The composition of the research teams varied with the tribe; some included members of the tribal council, and all included elders.

Such a collaboration was not easy for either side. Each of the thirty-three Hopi clans and each of the ten villages had a private history of its origins, carefully guarded not only from non-Hopis but also from other Hopi clans and villagers. European-American scholars accustomed to academic freedom had to accommodate tribal and individual decisions about confidentiality.

Even beyond notions of history and historical research, this cross-cultural collaboration required extra effort. The degree to which the landscapes of the indigenous peoples and the Salt River Project diverged can be seen in the dramatically different ways each group represented time and space, that is to say, the differences in the maps each produced. The Painted Desert Inn at Petrified Forest National Park displays a mural, *Hopi Salt Pilgrimage*, painted by Fred Kabotie. To European-American eyes, it looks like a nice, geometric picture, with a squiggly line running around the edge of the frame, with some people walking, cooking, bathing, sleeping, and hunting, and with some animals, one rectangle of water, and one plaza. To Hopis, the squiggly line represents the pilgrimage trail, with appropriate symbols as to distance, landmarks such as springs, and behavior appropriate at each site. As early as 1881, a European-American company run by Augustus Mitchell also made a map that included the trail. To European Americans, his map represents "reality." Euro-Americans can "read" this map. Yet it too is full of squiggly lines, dots, grids, and symbols (including writing) that no one sees when walking through the countryside. Each group's map represents a way of thinking about the land, of knowing the land, of possessing the land. The maps are of such different systems that simply superimposing one on the other makes the landscape harder, not easier, to read. In the past, the solution was to erase one map. For Zuni Salt Lake, despite all the difficulties, the two maps would coexist.

The tribal peoples who inhabit this region of the Southwest have had the strength to create an environment in which companies such as the Salt River Project value their contributions in constructing the landscape. But these tribal peoples' power is distinctly limited. They could save their pilgrimage trails. They could not, though they wanted to, prevent the opening of the strip mine.

Since the energy crunch of the early 1970s, increased pressure has been brought to bear on resources within the nation's boundaries. Though the Salt River Project's mine lay outside reservation lands, many of the nation's energy resources do not. With the management of both the nation's resources and the Indian reservations in the same department (Interior), it is not surprising that for years the federal government let leases to develop reservation resources at below market price to non-Indians. In 1976, twenty-five western tribes joined in self-protection to create the Council of Energy Resource Tribes (CERT). These tribes hold approximately half of the U.S. private uranium reserves, 15 percent of all coal in the United States, and 30 percent of all low-sulfur strippable coal in the U.S. West, as well as smaller percentages of oil and natural gas.

Although CERT was able to garner a larger share of the profits from resource extraction for its members, it did not make Indians wealthy. Oil and gas production on CERT reservations earned $169 million in 1980, before energy prices fell, but that still amounted to only $422.50 for each of the four hundred thousand inhabitants. Indeed, the seemingly intractable poverty of reservations despite numerous development and welfare programs mystifies some and exasperates others. At one end of the wide range of explanations lies world systems theory, which argues that expansive, relatively wealthy economies like those of European Americans, in order to sustain their own growth, systematically drain peripheral areas, such as Indian reservations, of resources, thereby not simply neglecting to develop them but rendering them unsustainable. At the other end, theorists blame the lack of entrepreneurial spirit among the communally oriented indigenous peoples. Small business funds, pencil factories, industrial parks, and resorts have all been tried and have failed on these usually remote and often arid lands with little infrastructure. The most successful enterprises seem to involve gambling. For example, bingo cut unemployment among California's Morongo Indians from 64 percent to virtually zero by 1987.

Whatever the cause, on the nation's roughly 270 Indian reservations in 1980, where just over one-half of the 1.4 million Indians lived, unemployment stood at twice the national average and on some reservations hovered near 80 percent. Roughly one-third of rural Native Americans in that year lived below the poverty line, and with this bleak picture, over 40 percent of Indian students entering high school dropped out before finishing. Indians ranked first of all groups in deaths caused by suicide and alcohol consumption. In the cities, Indians fared only slightly better. Just under one-quarter there lived below the poverty line. Most still had close ties to the reservation and planned to retire there. They visited often and left children to grow up in a countryside they hoped was more wholesome than the parts of the cities in which they could afford housing.

In 1984, Indians received a total of $2.6 billion from a wide range of federal agencies; if evenly distributed, the sum amounted to $1,900 per Indian. The Reagan administration and Congress in the 1980s systematically reduced expenditures on Indians, by 18 percent between 1982 and 1984 alone, hurting not only relief recipients but also the almost one-third of employed Native Americans who held government jobs. (Since the 1960s, the percentage of Bureau of Indian Affairs employees who are Native Americans rose steadily, reaching over three-quarters in 1982.) Dwindling resources heightened tensions, even within tribes, about who qualified for these funds, what constituted a Native American, and who should decide.

Despite increasing delegation of responsibilities to tribes, including contracting health and welfare services previously provided directly by the government, the federal government still held the trump cards in both legal and financial resources. It is no accident that one of the most militant Indian rights groups to emerge from the 1960s, the American Indian Movement, continually attacked federally held entities, from Alcatraz in San Francisco Bay to the Bureau of Indian Affairs headquarters in Washington, D.C. In turn, the American Indian Movement became the target of constant federal harassment and infiltration.

As national resources grew scarcer, including water and fish as well as coal, oil, and lumber, conflict sharpened between seemingly remote populations (Hopis and Los Angeles developers, for example) and between intimately contiguous populations (such as Hawaiians and European Americans) who had different notions about the best use of land and water and who still used different maps. Conditions varied tremendously from reservation to reservation, and for most Indians, tribal identities were still far stronger than pantribal ones; yet like many western issues, relations between Native Americans and European Americans became international. The International Indian Treaty Council lobbied the United Nations, and the U.N. Working Group on Indigenous Populations in 1990 rejected any planned triumphal celebration of the quincentenary of Columbus's voyage. As the group stated, "It negates our existence, our systems of government, our cultures, and our pre-columbian and pre-colonial history."

Borders, 1993

Borders, whether Indian reservations or international, whether racial or physical, do not exist in nature. They must be constructed. In the West, those constructions and their human costs have been particularly evident and particularly, perhaps predictably, imperfect.

In April 1993, police in San Diego, California, responded quickly to a report of a kidnapping of a blond preschool boy by a dark-haired Latino at a cafe in a posh neighborhood. Two women dining in the cafe had seen a "suspicious-looking" couple bribing the youngster in Spanish with promises of toys before the man left with the boy in a yellow cab. Police tracked down the taxi and, with guns drawn, entered the house where the two had been dropped off. They feared the kidnapper would quickly take his victim south and disappear into Mexico. Finding the house empty, they began a citywide search before they found the two walking back from a neighborhood park. After several tense moments of interrogation, it became clear that the police had made a terrible mistake. The supposed kidnapper was Guillermo Gómez-Peña, a recipient of one of the coveted MacArthur Fellowships awarded to a few extraordinary individuals each year. He was a world-renowned performance artist whose work highlights life in the hybrid world of the U.S.-Mexico border region. The boy he had been suspected of kidnapping was his own son. At the restaurant, his Anglo ex-wife had transferred the boy to him for an extended weekend visit.

This story contains many themes ever present along the border with Mexico in the late twentieth century. In these borderlands, the terms *alien* and *native* have come to capture the poles of historical and contemporary discussion surrounding social legitimacy in the American West. *Alien* embodies the notion of an outsider to society, an interloper without a claim to an area's resources or history. *Native* paints a picture of one who belongs—an insider, a genuine participant in the society with full legal, historical, almost "natural" rights—one to be counted as part of a particular community. The process by which those once considered native to a region came to be seen as alien and by which those once clearly alien came to be seen as native has been intricately tied to issues of race, colonization, immigration, and law enforcement throughout the century.

A result of the 1846–48 Mexican-American War, the southern border of the United

As much a gate as a barrier, the long border between the United States and Mexico has become a focal point of contemporary debate about the future of U.S. immigration and trade policy.

Douglas Kent Hall (b. 1938). The Border Fence: Tijuana/San Ysidro. *Gelatin silver print. © Douglas Kent Hall. Courtesy of the photographer.*

States remains one of the longest continuous borders shared by two nations, spanning a distance of over two thousand miles. In addition, in no other place does such a rich nation share such a long boundary with such a relatively poor country. Though Mexican culture continued to dominate both sides of the border until after the turn of the century, Anglo Americans quickly assumed positions of political and economic power along the border after 1848, and cultural definitions of insider and outsider remained drawn in largely racial terms. In the early twentieth century, the increased migration of Mexicans to work on the railroads, in mines, and in agricultural fields in the Southwest began to concern American public officials. As one worried Labor Department official put it in 1922: "The psychology of the average Mexican alien unskilled worker from Mexico is that when he enters in any manner into the United States that he is only upon a visit to an unknown portion of his own country. . . . To him there is no real or imaginary line." The presence of a strong border culture in which passage had been largely unregulated mitigated against stringent enforcement of immigration regulations already in place by the turn of the century.

In 1917, the U.S. Congress passed an immigration act that placed new restrictions on immigration from Europe, South Asia, and Mexico. Southwestern border officials realized that these new restrictions, which included a literacy test, a medical examination, and a head tax, would lead to increased violations of the law. They were right. For the first time, significant numbers of aliens illegally crossed into the United States to avoid the head tax. The Immigration Service also realized that labor recruiters and southwestern employers cared little how their prospective employees had come to the United States. These conditions led the chief inspector in El Paso to report that supervision of the border was so lax "that practically any alien desirous of entering the

United States and possessed of ordinary intelligence and persistence could readily find the means of so doing without fear of detection." Border officials realized that they were almost completely unable to stem the rising tide of undocumented immigration from Mexico.

Yet, unquestionably, the Border Patrol, established in 1924, was crucial in defining the Mexican as "the other," "the alien," in the region. J. C. Machuca, who worked for the El Paso Department of Immigration in the late 1920s, recalled that some of the early immigration inspectors were members of the Ku Klux Klan, a leading organization in the El Paso region at the time. Those Mexicans who had been long-term border residents could continue to cross in a casual fashion if, and only if, they were granted this special privilege by some Anglo benefactor. Officials would consistently denigrate others who crossed at the bridge, even if their papers were perfectly legal. Eventually, crossing the border became a painful and abrupt event permeated by an atmosphere of racism and control—an event that clearly demarcated one society from another.

Most early immigration officials were, in fact, newcomers to the region. Clifford Alan Perkins, who arrived in El Paso, Texas, in 1908 to begin a fifty-year career as an immigration inspector, remembered his initial impressions. "To a young man from a small Wisconsin farming community with a suspected case of tuberculosis, disappointed in his hopes for a college education and a career in professional baseball, it was a strange and wonderful place. I knew nothing about the people of Mexico, whose history, language and customs were to be so deeply interwoven into the fabric of my life and I was totally unprepared for some of the experiences that were ahead of me, but I was fascinated by what I saw." Perkins must be viewed as a cultural immigrant to the West and, as such, reflects the majority experience of European Americans in the region, much like the European Americans who decided to settle in Hawai`i in the twentieth century. Indeed, the cultural transformation of internal migrants was just as powerful an experience as that of a transnational immigrant, since Perkins arrived in El Paso with little knowledge of the area except a general optimistic portrait of the West.

Still, as a U.S.-born Anglo American, Perkins would become an immigration official within six months of his arrival. He was entrusted with the power, born of the U.S. conquest of the region, to administer the regulations of passage into American society. Because of the state's growing role in the twentieth century, the job of immigration officer came to mark a new cultural dynamic in the region. The administration of national policies made two thousand miles away in Washington, D.C., was handled by a newcomer whose authority was wielded over a population more experienced and at home than he. Though local concerns often encouraged the bending of the immigration laws, the die was cast toward greater restriction of movement. El Paso was defined as a "port-of-entry"—a term developed for seaports where movement implied travel over long distances and few interlocking points of contact between nations. Increasingly, areas hundreds of miles from the nearest ocean—regions defined by common land boundaries—would be asked to serve as politicized points of passage and cultural lines of separation.

By the late twentieth century, the border began to imply a sharp demarcation not only between the United States and Mexico but also between Mexican-origin people

The ongoing tensions over illegal immigration across the Mexico-U.S. border pose a particular problem for U.S. federal agents of Mexican descent, who can find their own identities and loyalties challenged by co-workers as well as immigrants.

Jay Dusard (b. 1937). Agents Larry Dalton, Elvin Harmon, Rogelio Martínez, Domingo Sánchez, and Fabián Casas, U. S. Border Patrol, Rio Grande City, Texas. *Gelatin silver print, 1985. © Jay Dusard 1985. Courtesy of the photographer.*

based on their nativity. In 1993, over 40 percent of all Border Patrol officers were Latino, primarily because of the need for Spanish-speaking personnel. As with the Bureau of Indian Affairs, the position led to tensions about identity. One Mexican-American agent in Texas described those he arrested: "They're not my people. I'm an American. They're here illegally." Having overcome nearly a half-century of questioning whether they were competent to police the border, Latinos found participation in the Border Patrol to be a lucrative, if controversial, avenue for upward mobility in the impoverished border region. Many Latino officials believed that racial bias continued to dominate the Border Patrol. One seventeen-year veteran who filed suit, alleging that he was passed over for promotion because of his Mexican ancestry, also claimed that Latino agents were consistently put under added pressure. He noted: "For Hispanics, there's no in-between. You're either a hard-ass or a bleeding heart."

The new immigration statutes and their administration on the border heightened the significance of the boundary line between Mexico and the United States. Indeed, the modern version of the border implied a rigid line of separation even while the intricate economic relationship between Mexican labor and American capital was perpetuated through labor recruitment agents, government contracts, and specialized exemptions. Here immigration officials, by inspecting new arrivals and border residents and enforcing laws barring illegal entry, made it clear who was alien and who native in this region once part of Mexico. The central role of the immigration inspector was duly noted by an El Paso attorney as early as 1912 when he wrote to Washington, D.C., "His business has brought him in contact with the poor, the ignorant, the friendless and the foreigner, over whom he has practically almost limitless power." This power over the dreams of the individual immigrant became increasingly evident at the border crossing.

To an ever greater extent, that "almost limitless power" has been translated into relationships of violence and intimidation. In December 1992, a U.S. judge in El Paso ruled that the Border Patrol had committed "wholesale violations" of the rights of citizens and noncitizens alike, including unjustified shootings, sexual misconduct, beatings, stealing money from prisoners, drug trafficking, embezzlement, perjury, and indecent exposure. One of the most tragic cases involved the death of twenty-six-year-old Dario Miranda Valenzuela, shot in the back by a U.S. Border Patrol agent, Michael Elmer, near Nogales, Arizona, on 12 June 1992. Rather than calling an ambulance after the shooting, Elmer dragged Miranda's body 175 feet in an attempt to hide the corpse, then threatened his partner if the other agent would not help him cover up the crime. In a landmark trial, Elmer and his colleagues admitted to practices contrary to Border Patrol regulations and to the cover-up, calling the border a "war zone" and the immigrants crossing it "the enemy." The Arizona jury, imbued with a similar notion of the border, refused to convict Elmer even on reduced charges, even for those crimes to which he had readily admitted. Unlike the reaction to the Rodney King trial in Simi Valley, which set off riots in Los Angeles, no one took notice of a verdict that seemed to fulfill the expectations of the power of the native over the alien in the American West.

◆ ◆ ◆

The United States, since its inception, has always looked west and south with the same intensity as, but with more covetousness than, it has looked east. In recent years, relations with Asia and with Mexico and other southern neighbors have increasingly defined the nation's economy, demography, and even culture. These relations are not new, but they are newly powerful, and they are within as well as outside national borders. They have spurred backlashes: Buddhist temple burnings and English as a First Language movements. They have raised the stakes on how the United States, as a nation, defines its western territory and whom it allows to participate in that defining. And these relations have made it crucial that we understand the foundations on which this contested landscape, this multifaceted frontier, of the late twentieth century stands. Most crucial to understand is that the foundation is not a single story but many overlapping stories of invaders who become invaded, of dreams, of histories revised, of identities invented, reinvented, and contested. From that fragmented complexity of individual stories and group experiences emerges a multifaceted history organized around the extremely unequal relations of power that have always marked the region. The West is built as well as riven by that multiplicity.

Bibliographic Note

As more historians turn their attention to the modern West (defined here as post-1930) and its diverse peoples, the field is changing rapidly. Journalistic accounts, such as Carey McWilliams, *Factories in the Field* (Santa Barbara, Calif., 1939) and *North from Mexico* (Philadelphia, 1949),

Peter Wiley and Robert Gottlieb, *Empires in the Sun: The Rise of the New American West* (New York, 1982), and Frances FitzGerald, *Cities on a Hill: A Journey Through Contemporary American Cultures* (New York, 1986), remain crucial but have been supplemented by surveys attempting to incorporate the diversity of the twentieth-century West as a central theme, such as Richard White, *"It's Your Misfortune and None of My Own": A New History of the American West* (Norman, 1991), Patricia Nelson Limerick, *The Legacy of Conquest: The Unbroken Past of the American West* (New York, 1987), and Ronald Takaki, *A Different Mirror: A History of Multicultural America* (Boston, 1993). In addition, surveys of particular ethnic and racial groups in the twentieth-century West include Rodolfo Acuña, *Occupied America: A History of Chicanos*, 3d ed. (New York, 1988), Sucheng Chan, *Asian Americans: An Interpretive History* (Boston, 1991), Ronald Takaki, *Strangers from a Different Shore: A History of Asian Americans* (Boston, 1989), and James S. Olson and Raymond Wilson, *Native Americans in the Twentieth Century* (Provo, 1984). For a brief survey that links gender and racial issues, see Sarah Deutsch, "Landscape of Enclaves: Race Relations in the West, 1865–1990," in William Cronon, George Miles, and Jay Gitlin, eds., *Under an Open Sky: Rethinking America's Western Past* (New York, 1992), 110–31. Peggy Pascoe's pathbreaking work on intermarriage in the West, "Race, Gender, and Intercultural Relations: The Case of Interracial Marriage," *Frontiers* 12 (1991): 5–18, leads a special issue focusing on writing twentieth-century multicultural women's history. This issue also includes Valerie Matsumoto, "Desperately Seeking 'Dierdre': Gender Roles, Multicultural Relations, and Nisei Women Writers of the 1930s," pp. 19–32.

The monographic literature includes regional works on Mexican Americans: David Montejano's award-winning *Anglos and Mexicans in the Making of Texas, 1836–1986* (Austin, 1987), which traces the transformation of race relations and definitions as the result of economic changes; Julia Kirk Blackwelder, *Women of the Depression: Caste and Culture in San Antonio, 1929–1939* (College Station, Tex., 1984), which also includes African-American and Anglo women, and Richard A. García's *Rise of the Mexican American Middle Class: San Antonio, 1929–1941* (College Station, Tex., 1991). On Arizona, see Thomas E. Sheridan, *Los Tucsonenses: The Mexican Community in Tucson, 1854–1941* (Tucson, 1986). On Colorado and New Mexico, see Suzanne Forrest, *The Preservation of the Village: New Mexico's Hispanics and the New Deal* (Albuquerque, 1989), and Sarah Deutsch, *No Separate Refuge: Culture, Class, and Gender on an Anglo-Hispanic Frontier in the American Southwest, 1880–1940* (New York, 1987). On California, see Vicki Ruiz, *Cannery Women, Cannery Lives: Mexican Women, Unionization, and the California Food Processing Industry, 1930–1950* (Albuquerque, 1987), George J. Sánchez, *Becoming Mexican American: Ethnicity, Culture and Identity in Chicano Los Angeles, 1900–1945* (New York, 1993), and Camille Guerin-Gonzales, *Mexican Workers and American Dreams: Immigration, Repatriation, and California Farm Labor, 1900–1939* (New Brunswick, N.J., 1994), which sets Mexican immigrants in the context of the agrarian ideals of American farming and the industrial reality of California agriculture. On the rise of the United Farm Workers in California, see Sam Kushner, *Long Road to Delano: A Century of Farmworkers' Struggle* (New York, 1975), and Margaret Rose, "From the Fields to the Picket Line: Huelga Women and the Boycott, 1965–1975," *Labor History* 31 (Summer 1990): 271–93, and "Traditional and Nontraditional Patterns of Female Activism in the United Farm Workers of America, 1962–1980," *Frontiers* 11 (March 1990): 26–32.

Crossing regional boundaries are Adela de la Torre and Beatríz M. Pesquera, *Building with Our Hands: New Directions in Chicana Studies* (Berkeley, 1993), and Vicki L. Ruiz and Susan Tiano, eds., *Women on the U.S.-Mexico Border: Responses to Change* (Boston, 1987), on women's cross-border experience in *maquilas* and as domestic workers. Border studies are beginning to describe the significant twentieth-century shift toward restriction and control. The best works include Kitty Calavita, *Inside the State: The Bracero Program, Immigration, and the I.N.S.* (New York, 1992), and Raul A. Fernandez, *The United States–Mexico Border: A Politico-Economic Profile* (Notre Dame, Ind., 1977). Leo R. Chavez has written a scholarly analysis of the fate of "illegal aliens" in *Shadowed Lives: Undocumented Immigrants in American Society* (Fort Worth,

1992). First-person accounts include Luis Alberto Urrea, *Across the Wire: Life and Hard Times on the Mexican Border* (New York, 1993), and Marilyn P. Davis, *Mexican Voices/American Dreams: An Oral History of Mexican Immigration to the United States* (New York, 1990). On the dynamics of modern cross-border cultural borrowings, see Rubén Martínez, *The Other Side: Notes from the New L.A., Mexico City, and Beyond* (New York, 1993).

On Asian migrant labor, see Gary Y. Okihiro, *Cane Fires: The Anti-Japanese Movement in Hawaii, 1865–1945*, (Philadelphia, 1991); and Lucie Cheng and Edna Bonacich, eds., *Labor Immigration under Capitalism: Asian Workers in the United States before World War II* (Berkeley, 1984). There are several books that deal with Japanese internment, including Roger Daniels, *Prisoners without Trial: Japanese Americans in World War II* (New York, 1993), and Sandra C. Taylor, *Jewel of the Desert: Japanese American Internment at Topaz* (Berkeley, 1993). On generational change and ethnic identity, see Evelyn Nakano Glenn, *Issei, Nisei, War Bride: Three Generations of Japanese American Women in Domestic Service* (Philadelphia, 1986), Karen Isaksen Leonard, *Making Ethnic Choices: California's Punjabi Mexican Americans* (Philadelphia, 1992), Victor G. Nee and Brett de Bary Nee, *Longtime Californ': A Documentary Study of an American Chinatown* (New York, 1973), Judy Yung, *Chinese Women of America: A Pictorial History* (Seattle, 1986), and Richard Chalfen, *Turning Leaves: The Photograph Collections of Two Japanese American Families* (Albuquerque, 1991), on self-definitions of ethnicity. For the more recent period, see Asian Women United of California, ed., *Making Waves: An Anthology of Writings by and about Asian American Women* (Boston, 1989), Mary Paik Lee, *Quiet Odyssey: A Pioneer Korean Woman in America* (Seattle, 1990), and Craig Scharlin and Lilia V. Villanueva, *Philip Ver Cruz: A Personal History of Filipino Immigrants and the Farmworkers Movement* (Los Angeles, 1992). On Hawai`i see Elizabeth Buck, *Paradise Remade: The Politics of Culture and History in Hawai`i* (Philadelphia, 1993) for a stimulating account of power and culture, and Haunani-Kay Trask, *From a Native Daughter: Colonialism and Sovereignty in Hawai`i* (Monroe, Maine, 1993), for a trenchant case for Hawaiian sovereignty.

For Native Americans in the twentieth century, Kenneth R. Philp, *John Collier's Crusade for Indian Reform, 1920–1954* (Tucson, 1977), presents a policy-centered version of the impact of the New Deal. Very few authors center their work on Native American women's history in this period; Patricia Albers and Beatrice Medicine, eds., *The Hidden Half: Studies of Plains Indian Women* (Washington, D.C., 1983), is a fascinating and useful account. There is, on the other hand, a small industry in books on land claims. A useful introduction is Imre Sutton, et al., eds., *Irredeemable America: The Indians' Estate and Land Claims* (Albuquerque, 1985). For studies more centered on tribal governance and economic development, see Peter Iverson, *The Navajo Nation* (Westport, Conn., 1981), Kenneth R. Philp, ed., *Indian Self-Rule: First-Hand Accounts of Indian-White Relations from Roosevelt to Reagan* (Salt Lake City, 1986), Stephen Cornell, *The Return of the Native: American Indian Political Resurgence* (New York, 1988), and Larry Burt, "Western Tribes and Balance Sheets: Business Development Programs in the 1960s and 1970s," *Western Historical Quarterly* 23 (November 1992): 475–95. Also useful are Alvin M. Josephy, Jr., *Now That the Buffalo's Gone: A Study of Today's American Indians* (New York, 1982), and Peter Iverson, ed., *The Plains Indians of the Twentieth Century* (Norman, 1985). On the Columbian quincentenary, see Margaret Connell Szasz, "American Indians and Outsiders: A Crucial Dialogue of the Columbian Quincentenary," *Montana, The Magazine of Western History* 42 (Autumn 1992): 53–62. On the Zuni Salt Lake, T. J. Ferguson and Richard Hart, both of the Institute of the NorthAmerican West, and Judy Bruncon, of the Salt River Project, all presented research papers that dealt partly with this topic at the 1992 Western History Association Conference in New Haven.

Material is particularly thin for the nonurban twentieth-century African-American experience: see Era Bell Thompson, *American Daughter* (1946; reprint, St. Paul, 1986), on homesteading. Material on urban blacks includes Blackwelder, *Women of the Depression*, and Shirley Ann Moore, "Getting There, Being There: African-American Migration to Richmond, California, 1910–1945," in Joe William Trotter, Jr., ed., *The Great Migration in Historical Perspective: New*

Dimensions of Race, Class, and Gender (Bloomington, 1991), and the last chapter of Douglas Henry Daniels, *Pioneer Urbanites: A Social and Cultural History of Black San Francisco* (Berkeley, 1980).

On Rodney King and the 1992 riots, see Mike Davis, *City of Quartz: Excavating the Future in Los Angeles* (London, 1990), for a Marxist foreshadowing, and David Rieff, *Los Angeles: Capital of the Third World* (New York, 1991), for the recent upsurge in ethnic diversity. Raphael J. Sonenshein, *Politics in Black and White: Race and Power in Los Angeles* (Princeton, 1993), traces the politics of race in the city, and the *Los Angeles Times* staff, *Understanding the Riots: Los Angeles before and after the Rodney King Case* (Los Angeles, 1992), provides a centrist coverage. A more leftist approach is taken by a variety of journalists in *Inside the L.A. Riots: What Really Happened—and Why It Will Happen Again* (Los Angeles, 1992). For more scholarly analyses, see Haki R. Madhubuti, ed., *Why L.A. Happened: Implications of the '92 Los Angeles Rebellion* (Chicago, 1993), and Robert Gooding-Williams, ed., *Reading Rodney King, Reading Urban Uprising* (New York, 1993).

Sarah Deutsch would like to acknowledge the delightfully good nature of her coauthors and the life-saving energy and creativity of her research assistant, Melissa Walker. George J. Sánchez thanks his two coauthors for their congeniality and commitment to collaboration and especially Sarah Deutsch for her persistent yet gentle prodding to get the job done well.

SECTION IV

Interpretation

Just west of downtown Cody, Wyoming, on the well-traveled tourist route to Yellowstone National Park, lies Trail Town, a collection of old wooden buildings moved from remote sites in the Rocky Mountain West and reconstructed as the imagined Main Street of a frontier town. Strung along raised sidewalks in a kind of convivial proximity to one another are homes and schools, shops and workplaces, most transported from considerably more spacious plots. At the end of the single broad dirt street is a small graveyard, surrounded by a Victorian wrought-iron fence, which contains the remains of John "Liver-eating" Johnson (1824–1900), the mountain man and trapper renamed and made famous by Robert Redford in the film *Jeremiah Johnson* (1973). Like most trappers, Johnson was a solitary fellow, but in his lifetime he earned wide renown for his reputed vendetta against the Crow Indians who, in May 1847, allegedly killed and scalped his pregnant Flathead wife. As an early biographer wrote: "For many years thereafter he killed and scalped Crow Indians. Then he ate their livers, raw. He ate them not for hunger's sake but upon principle."

The rugged country stretching from Cody to Yellowstone would have been familiar to Johnson, but he did not die there. After spending his final days in the Veterans Hospital in Los Angeles, he was buried in a military cemetery not too far from present-day Beverly Hills. In 1974, a group of Lancaster, California, schoolchildren, as part of a history project, petitioned the Veterans Administration to have Johnson's body removed and reburied in the mountain West. The schoolchildren raised some money, an airline donated its services, and Trail Town, a private commercial enterprise, paid to reinter Johnson's body and erect the new marker that stands over his grave. More than two thousand people attended the reburial services.

The story of Johnson's ultimate fate presents many ironies, not the least of which is that this self-reliant, violent, and not altogether admirable man should end up being rescued by schoolchildren and corporate largess and transformed into a tourist attraction in a town named for Buffalo Bill Cody, the great western showman Johnson reportedly despised. But this story also serves as an instructive metaphor for the way the West is popularly imagined in American culture: as a place with a particular story line and fixed set of heroes that stubbornly persist despite considerable evidence to the contrary.

Western history has long been a kind of participation sport—as seen in the current popularity of mountain man rendezvous, Louis L'Amour novels, pay-as-you-go cattle drives, Ralph Lauren's ranch clothes, or the Buffalo Bill Historical Center down the road from Trail Town, which draws up to three thousand visitors on a busy summer day. The schoolchildren who initiated the relocation of Johnson's grave were behaving like millions of other Americans of diverse backgrounds who feel empowered to claim western history as a story with particular relevance to their own lives. This is the great strength of a popular western history that celebrates the exploits of mountain men like

The West of myth and memory has long proved an alluring theme for artists, writers, and other interpreters of regional culture who have found their subject matter in an imagined past.

Unidentified photographer. Charles Schreyvogel Painting on the Roof of His Apartment Building in Hoboken, New Jersey. *Photograph, 1903. National Cowboy Hall of Fame and Western Heritage Center, Oklahoma City, Oklahoma.*

Liver-eating Johnson or extravagant self-proclaimed heroes like Buffalo Bill and constructs a vast region of imagined places like Trail Town, locked forever in a particular stage of community development. It serves as a widely appealing national myth. But the very popularity and accessibility of this myth suggest the problem with this sort of history. As the writer Wallace Stegner succinctly put it, "The western culture and western character with which it is easiest to identify exist largely in the West of make-believe, where they can be kept simple."

The very word *western*, an adjective that has come to serve as a noun, carries popular connotations that suggest the simplicity and limitations of the popular western history celebrated in films and novels, paintings and prints, theme parks and local summer pageants. The word suggests a particular time, place, and cast of characters—a sparsely populated part of the West during the mid-to-late nineteenth century, inhabited by cowboys and Indians, mountain men and cavalry troops, most of whom adhere to a particular moral code of honor. It excludes more people and places in the West than it includes.

Life often imitates art. One of the reasons that the old western myths remain so compelling is that they continue to shape contemporary behavior, as they have ever since Buffalo Bill began dressing up, old cowhands started reading western novels, and tobacco companies began using pictures of ranchers to sell cigarettes. "This is the West, sir," said a small-town newspaper editor in John Ford's classic film *The Man Who Shot Liberty Valance* (1962). "When the legend becomes fact, print the legend." Thus Ford, who understood the popular appeal of the West as well as anyone, ironically acknowledged the power of myth to acquire the trappings of truth and reshape not only the past but the present as well. The popular legend of the West should properly be discounted as good history, but it should not be disregarded as a good story that continues to have a direct impact on American behavior and national beliefs.

Frederick Jackson Turner delivered his seminal 1893 address "The Significance of the Frontier in American History" to an academic gathering in Chicago just a few blocks from a production of Cody's Wild West show. As the historian Ann Fabian argues, "The scholarly pursuit of the western past was joined immediately by an impish popular double." One might make a comparable argument about serious painting in the West, or serious writing. A good history of western creative endeavors must be broad enough to embrace the serious while acknowledging the popular; it must be loose enough to include Georgia O'Keeffe and Robinson Jeffers as well as the modern-day Charlie Russells and cowboy poets.

In the little modern cemetery in Trail Town and through the persona of Robert Redford, Liver-eating Johnson has engaged the imagination of an international audience of tourists and filmgoers who would otherwise remain ignorant of his reputed deeds in the Rockies. Johnson's story thus suggests one final lesson about the history of the American West. Its vitality lies in its continual discovery and reinvention by the artists and writers, historians and schoolchildren, tourists and citizens who come to it with fresh questions. When the West is reimagined from an international perspective, from a contemporary viewpoint, from the perspective of the vast variety of peoples who call it home, it becomes a place even grander than the place of myth—bigger in space and time and infinitely more interesting.

— Martha A. Sandweiss

CHRONOLOGY

1810	Zebulon Pike's record of his expedition to the Southwest sparks public interest.
1814	Extensively edited, the journals of Lewis and Clark appear in published form.
1823	Novelist James Fenimore Cooper introduces the frontier hero Natty Bumppo.
1830s	Painters George Catlin, Karl Bodmer, and Alfred Jacob Miller journey to the interior of the American West.
1849	Francis Parkman taps the mood of "manifest destiny" in his work *The Oregon Trail.*
1859	Painter Albert Bierstadt travels through the Rockies.
1869	Edward Judson, also known as Ned Buntline, discovers William F. Cody in Nebraska.
1871	Photographer William Henry Jackson and painter Thomas Moran accompany the government survey of the Yellowstone country.
1872	Mark Twain recalls his western years in *Roughing It.*
1874–90	Thirty-nine volumes of historical works appear under the byline of Hubert Howe Bancroft.
1887	Reuben Gold Thwaites begins his career at the Wisconsin Historical Society.
1889–1909	Frederic Remington creates more than twenty-seven hundred paintings and drawings and twenty-four editions of bronze sculptures.
1893	Frederick Jackson Turner delivers his address "The Significance of the Frontier in American History."
	Former cowboy Charles M. Russell takes up art full-time.
1894	Naturalist John Muir writes *The Mountains of California.*
1902	Long a subject of dime novels, the cowboy receives artistic treatment in Owen Wister's *The Virginian.*
1903	Arguably the first Western, *The Great Train Robbery* is filmed in New Jersey.
1914	Twenty-five thousand attend the dedication of the new five-story Kansas State Historical Society.
1917	Herbert Eugene Bolton calls attention to Hispanic contributions in "The Mission as a Frontier Institution in the Spanish-American Colonies."
1925	*The Professor's House,* by novelist Willa Cather, addresses postfrontier themes.
1929	Painter Georgia O'Keeffe begins painting in New Mexico.
1931	Walter Prescott Webb publishes *The Great Plains.*
1936	Dorothea Lange's stop at a pea-pickers' camp results in the searing depression photograph "Migrant Mother."
1939	John Ford and John Wayne team up to make *Stagecoach.*
1940	"Gene Autry's Melody Ranch" premieres on radio.
1950	*Westward the Tide,* the first of Louis L'Amour's ninety novels, is published.
1959	Seven of the top-ten television shows in the United States are Westerns.
1962	John Steinbeck receives the Nobel Prize for Literature.
1965	*The Pacific Slope* by Earl Pomeroy presents the twentieth century as integral to western history.
1968	The publication of N. Scott Momaday's *House Made of Dawn* signals a renaissance in Native American writing.
1969	Merle Haggard records "Okie from Muskogee."
1971	Inspired by the life of Mary Hallock Foote, Wallace Stegner writes *Angle of Repose.*
1987	Patricia Nelson Limerick interprets the western past in *The Legacy of Conquest.*
1989	Larry McMurtry's novel *Lonesome Dove* (1985) is re-created as a television mini-series.
1990s	The films *Dances with Wolves* (1990) and *Unforgiven* (1992) gain acclaim, reinterpreting themes of earlier Westerns.
1991	*The West as America* exhibit at the National Museum of American Art stirs controversy.

Chapter Nineteen

The Visual West

BRIAN W. DIPPIE

On 25 June 1876, Lieutenant Colonel, Brevet Major General, George Armstrong Custer and 209 men with the Seventh U.S. Cavalry died in a battle against Lakota and Cheyenne Indians on the Little Bighorn River in south-central Montana. There were no white survivors of what has gone down in American history as "Custer's Last Stand." The battle did in fact happen, but Custer's Last Stand is a cultural myth created mostly by painters, poets, dramatists, and novelists who found in the news of military disaster out West a heroic lesson. Defeat was victory. It affirmed the pioneering American spirit, the willingness to pay any price for progress, including the sacrifice of self. Today, we reject the presumption of racial and cultural superiority on which the heroic myth of Custer's Last Stand rested. We wonder, with Henry Wadsworth Longfellow, "Whose was the right and the wrong?" We doubt the imperatives of "manifest destiny" and sympathize with the dispossessed, but we do not doubt Custer's Last Stand. Instead, we read its lesson differently. It now seems a cautionary tale about the price of personal ambition and national greed. But we can still visualize Custer's Last Stand, and that is a tribute to the enduring influence of artists who, through the power of imagination, fashioned a reality all their own. Their achievement is what western American art is all about.

Western art commonly refers to representational paintings, drawings, and sculptures showing men (and it is essentially an art by and about men) and animals in unspoiled natural settings. The men are mostly Indians and whites—trappers, miners, plainsmen, soldiers, cowboys—and the viewpoint is exclusively white. The content evokes a historical period, the nineteenth century, but it is historical art of a special kind. Certain storied events like Custer's Last Stand have become standard subjects, but the history in western art is usually generalized. Mountain men frolic at rendezvous, settlers guide their covered wagons westward, Indians and cavalry clash in battle, cowboys tend cattle on the unfenced plains and shoot it out in clapboard towns that squat in the open land suspended between wilderness and civilization. Such images are timeless rather than historical, except in the sense that they constitute a kind of tribal history for white Americans. They seem typical of a time and place because artists, by repeating them, have implanted them in the public's mind. Successive images affirm one another, verifying the truth of the scenes portrayed. Collectively, the artists have created a version of the West, a romantic West that vanished long ago, never to be forgotten. Western art sees to that.

The embattled white heroes of Frederic Remington's painting reiterate one of the central themes and favorite myths of western art—the conquest of the land and its native peoples by Anglo-American frontiersmen.

Frederic Remington (1861–1909). The Last Stand. *Oil on canvas, 1890. Woolaroc Museum, Bartlesville, Oklahoma.*

Invention, repetition, and refinement define the western art tradition, as Custer's Last Stand can demonstrate. On 19 July 1876, the *New York Daily Graphic* featured a dramatic, full-page illustration by William de la Montagne Cary (1840–1922), *The Battle on the Little Big Horn River—The Death Struggle of General Custer.* It showed Custer and his troopers bravely resisting an overwhelming, savage foe. Though outnumbered and doomed, each soldier still struggles to get off a parting shot as the Indians, armed with rifles and bows and arrows and war clubs, and some wielding scalping knives, close in for the kill. The situation is hopeless, but there, in the midst of chaos, stands Custer, tall and calm, a peculiar light falling on him through the heavy clouds and dust, illuminating his moment of imperishable glory. Saber drawn back, blazing away with a pistol, he scrambles over a fallen horse to get at the enemy and defy death itself. Civilization, never more glorious than in defeat, was never more certain of ultimate victory. That was the message Cary's illustration conveyed. Drawing on his personal experiences among the Missouri River tribes and on his acquaintance with Custer—the credentials that established his authority—Cary had distilled the news, gossip, and rumors then flooding the nation's newspapers into a single compelling image. William Cary had just invented Custer's Last Stand.

Could the battle have been depicted in any other way? In fact, it was. The *Illustrated Police News* a week earlier devoted its cover to a drawing of two Indians rushing at Custer as he reels from a mortal wound. It showed his death, but not why his death mattered. The same was true of the cover of the *New York Illustrated Weekly* that August, which featured a buckskin-clad Custer leading his men in a charge over the corpse of the only Indian in view. What kind of "last stand" was this?

Cary's seminal version of Custer's end might have been lost in a welter of competing images in 1876, denying the battle its distinctive visual form and gutting it of mythic appeal. But that autumn, only months after Custer's death, a full-length biography was rushed into print, Frederick Whittaker's *Complete Life of Gen. George A. Custer*, and among its illustrations was *Custer's Last Fight.* The work of another experienced military artist with western credentials, Alfred R. Waud's (1828–91), painting pared the battle down to essentials and triumphantly reaffirmed Cary's heroic conception. A tight pyramid of soldiers, with Custer at the exact center, parts a surging sea of Indians. It speaks of sacrifice and courage, grace under pressure, good form in dying, and the moral superiority of civilization. Thereafter, artists who essayed the Custer theme—and hundreds have—did their homework in Whittaker and studied in the school of Waud.

That Custer's Last Stand was a cultural creation becomes evident when we consider that a battle renowned for having no survivors actually had hundreds of survivors—the Indian victors. And they viewed things differently. Instead of defiant last stands and military stoicism in the face of death, they remembered panicked soldiers fleeing in disarray, voices crying out and dissolving into sobs, weapons dropped in wild-eyed terror by men desperately seeking a way out. They talked about whiskey or madness to account for the sudden collapse of resistance. And they praised a few who fought hard that day. Custer's Last Stand did not exist for the Indians until time and the persistent demand of white questioners for stories about the heroic death struggle of the chief with yellow hair taught them how they were supposed to respond.

The early depictions of Custer's death, quickly rushed into print after the fateful battle at the Little Bighorn on 25 June 1876, show the rapid evolution of the myth of Custer's Last Stand. Alfred Waud's image, published in the fall of 1876, established Custer as a quintessentially American hero fiercely fighting to defend his nation from an alien threat. The triangular composition of the image, highlighting Custer's own actions, would often be repeated by artists seeking to find, in this moment of white defeat, a redemptive moral for the American people.

THE ILLUSTRATED POLICE NEWS
LAW-COURTS AND WEEKLY RECORD

Unidentified artist. The Indian War—Death of General George A. Custer, at the Battle of the Little Big Horn River, Montana Territory, June 25. *Engraving, 1876. From the* Illustrated Police News, *13 July 1876.*

William de la Montagne Cary (1840–1922). The Battle on the Little Big Horn River—The Death Struggle of General Custer. *Engraving, 1876. From the* New York Daily Graphic, *19 July 1876. Library of Congress (#LC-US262-60706), Washington, D.C.*

After Alfred R. Waud (1828–91). Custer's Last Fight. *Engraving, 1876. From Frederick Whittaker,* A Complete Life of General George A. Custer *(1876). Brian W. Dippie Collection.*

Commissioned by the artist Frederic Remington to create an account of the Custer battle, the Lakota artist Kicking Bear produced an image that gave a central role to the battle's Indian protagonists. But his work was influenced by earlier white renditions as well as memory. Lying just outside the main group of Indians, Custer is recognizable by the buckskins and long hair made familiar through other artists' pictures of the scene.

Kicking Bear (1848–?). Battle of Little Big Horn. *Muslin painting, 1890–1900. Courtesy of the Southwest Museum (#N 37788), Los Angeles, California.*

An unforgettable event in the tribal history of white Americans barely registered in the tribal histories of the Sioux victors. It did, however, take its place among the exploits of individual Indians who recorded their deeds that day on hides and muslin and sheets of paper. Few of these pictographic accounts, a traditional form of Plains Indian art devoted to battles and hunts, give much comfort to heroic myth. Some, showing overviews of the battle, define defensive positions and support the idea of a final stand. Most portray uniformed soldiers, anonymous in their alikeness, being chased and killed by distinctively costumed Indians. They show an indistinguishable jumble of white corpses and honor fallen warriors identified individually. Custer was conspicuous by his absence in early pictographic accounts, but native artists, under white influence, began including a Custer figure, firing his pistols or lying among the dead. And so a white cultural myth became an Indian myth as well, a process worth tracing.

Frederic S. Remington (1861–1909), the most influential western illustrator in the late nineteenth century, was obsessed with military subjects from the time of his boyhood in upstate New York. His father, a cavalry officer in the Civil War, was his hero, and he filled school notebooks and readers with sketches of soldiers from armies everywhere. He doted on fancy-dress uniforms, the stiff martial bearing, and scenes of combat and carnage. Naturally the sensation created by Custer's defeat in 1876 was irresistible for Remington, an impressionable fourteen-year-old about to enter the Highland Military Academy in Worcester, Massachusetts. That fall, presumably, relying directly on Waud, Cadet Remington drew and colored his own version of Custer's Last Fight. Waud's heroic triangle of soldiers, with Custer at its center, planted itself in the mind of the artist most responsible, in turn, for planting a heroic image of the Indian-fighting army in the public's mind. In 1890 Remington painted a full-scale oil, *The Last Stand*. By then, besides enhanced skill, he brought to the subject a personal acquaintance with the West. After visiting Montana in 1881, he had owned a sheep ranch in Kansas and had patrolled with the U.S. cavalry in Apache country. During the winter of 1890–91 he was on assignment in South Dakota as artist-correspondent covering the Ghost Dance War, though he managed to be absent on 29 December when the Seventh Cavalry took belated revenge on the Lakotas at Wounded Knee. The battle was a last stand for the Indians—more than 150 died—but Remington's painting, reproduced in *Harper's Weekly* twelve days later, remained faithful to myth. Here was Waud's pyramid of troopers, white civilization at bay, still bravely facing the savage foe. The accompanying text wondered, "How many scenes of which this is typical have been enacted on this continent, who can say?" Through repetition, the atypical had become the typical. Custer's Last Stand, a startling exception to the rule of white victory and Indian defeat, by 1890 represented the routine self-sacrifice of countless white Americans in the winning of the West.

Just days after *Harper's Weekly* published Remington's painting, Kicking Bear (1848–19?), a leading Lakota Ghost Dance apostle, formally surrendered to U.S. troops, ending the last major Indian war in American history. Facing exile, Kicking Bear and other Lakota leaders instead chose the option of accompanying Buffalo Bill Cody's Wild West show on a European tour. Their punishment, presumably, was being routed by Cody at each performance. Show business did not reconcile Kicking Bear to the white man's ways—as late as 1902 he was still preaching Ghost Dance gospel on the Fort Peck Reservation in Montana. But he had partly assimilated the white man's view of Custer's Last Stand. In about 1897 Remington commissioned him to paint a pictographic account of the Custer battle, and over the winter he produced a work that rivaled some of the grander nineteenth-century conceptions. Its center was reserved for Indian leaders, including Kicking Bear himself, framed by scenes of battle and dead soldiers. Most prominent among the soldiers, just outside the main grouping of Indians, was the long-haired Custer, sprawled in his buckskins. In Kicking Bear's painting, Custer was not the center of attention, but he was instantly recognizable, looking pretty much as Waud and then Remington had shown him. Artistic contrivance was triumphant—Custer's Last Stand had become a myth for all Americans.

Once established, Last Stand imagery has proven immensely adaptable. Western movies repeatedly borrowed Waud's heroic triangle of soldiers: *The Scarlet West* (1925), *The Plainsman* (1937), and *They Died With Their Boots On* (1941) to honor and praise; *Sitting Bull* (1954) and *Little Big Man* (1970) to vilify and mock. Imagery capable of any refinement, any interpretation, naturally invites parody, and Custer's Last Stand has attracted its share.

Edgar S. Paxson's solemn and gargantuan battle piece *Custer's Last Stand* (1899) derived its impact from sheer size (six feet by nine feet) and its astonishing detail. Paxson (1852–1919) arrived in Montana in 1877 and based his subsequent career on the fact that he was Montana's pioneer painter. His long residency was thought to lend authenticity to his work, in keeping with a cherished tenet of western art: personal experience is everything, validating even the entirely imaginary. In approaching the Custer theme, Paxson augmented what he "knew" with research, since another tenet of western art holds that truth depends on getting the details right, however preposterous the basic conception. Paxson corresponded with officers present at the Little Bighorn, interviewed Indian participants, visited the battlefield, and then, over a five-year period, painted a picture that is almost zanily implausible, so crammed with men, horses, and paraphernalia that, spacious as the canvas is, it feels cramped.

Parodists have paid Paxson tribute. Fritz Scholder (1937–), a Minnesotan of part-native descent and an inspiration for a whole generation of Indian artists, made a reputation in the 1960s and 1970s by pointedly reworking white images of the Indian, exposing their inherent values and reinvigorating them in the process. Playing off stereotypes from photographs and paintings—vanishing Indians, tourist Indians, drunken Indians—he uses brilliant color, simplification, and distortion to jolt the eye into reexamining the familiar and unquestioned. His painting *The Last Stand*, only slightly smaller in size than Paxson's oil, focuses attention on the figure of Custer by reducing the busy details of the Paxson original and unifying the composition through a simplified color scheme. Custer, rendered in buckskin yellow with hair to match, is a golden cloud floating over a turbulent purple sea—and is a ham actor hogging the spotlight. Peter Saul's *Custer's Last Stand* (1972–73), in contrast, is a twisted jumble of figures in cartoon colors, but it too found its inspiration in Paxson's painting. Saul, born fifty years after the Custer battle, was intrigued by Paxson's clutter. Even if his picture parodies Paxson's theme, it pays tribute to a hallmark of western art: surface detail supporting the appearance of truth. The Custer's Last Stand parodies make an additional point crucial to understanding western art. Over time, it has become an art about itself, repeating certain themes and images until they have become enshrined in the collective memory. Western art constitutes its own reality.

First Impressions

Western art as we know it began soon after the United States acquired the Louisiana Territory from France. No artists accompanied Meriwether Lewis and William Clark on their epic exploring trip (1804–6). But while they were still in the field, eastern artists were already portraying Indian visitors from west of the Mississippi River. Later, between 1822 and 1842, through the Office of Indian Affairs, the federal government commissioned nearly 150 portraits, most the work of Charles King Bird (1785–1862). Reproduced to illustrate *A History of the Indian Tribes of North America*, issued in parts between 1837 and 1844, the portraits were among the most familiar likenesses of Native Americans available before the Civil War. But they were studio productions, painted in the comforts of the nation's capital.

The first artists to actually visit the West and make field studies of Indian life accompanied the third of the government-sponsored explorations in the Louisiana Territory. In 1819, Samuel Seymour (active ca. 1797–1822) was appointed painter to Major Stephen H. Long's expedition up the Missouri to the Yellowstone River, and Titian Ramsay Peale (1799–1885), from a prominent Philadelphia family of artists, was hired as assistant naturalist with responsibility for natural history illustration. Aborted by financial constraints, Long's expedition instead traced the Platte River west to within sight of the Rocky Mountains before turning south. Seymour accompanied one party down the Arkansas River, Peale the other down the Canadian (the group was supposed to be on the Red). Disappointed in its original objectives, the Long expedition repaid its frustrations by declaring the region traversed "almost wholly unfit for cultivation, and

of course uninhabitable by a people depending on agriculture for their subsistence"—the Great American Desert. The two artists produced nearly three hundred sketches. Though only a small selection illustrated the official reports, Seymour and Peale had pioneered several themes that would dominate western art. They showed the Rocky Mountains, Plains Indians, and buffalo hunts and set a precedent for artists' participation in the government exploring expeditions that fanned out across the West in the middle of the nineteenth century. John Mix Stanley (1814–72), James W. Abert (1820–71), Richard H. Kern (1821–53), Edward M. Kern (1823–63), Gustavus Sohon (1825–1903), H. Balduin Möllhausen (1825–1905), and Arthur Schott (1813–75) would contribute to the fund of information about the flora, fauna, topography, and native peoples of America's expanding hinterland, firmly establishing the documentary tradition that nourishes western art to this day.

Titian Peale and his companion, the painter Samuel Seymour, were expected to be accurate reporters whose work would match in importance the scientific findings of the other members of Major Stephen H. Long's exploring expedition (1819–20). Though only a few of their small sketches of western Indians, wildlife, and scenery were reproduced in the official expedition report, their work initiated a long tradition of visual documentation of western exploration.

Titian Ramsay Peale (1799–1885). Sunset on the Missouri. *Watercolor on paper, 1819. American Philosophical Society, Philadelphia, Pennsylvania.*

Naturally, the documentarians had their own preconceptions. The historian William H. Goetzmann has argued that explorers are always culturally programmed; the same applies to the artists who accompanied them. The plates by Stanley and Schott illustrating southwestern subjects in Major William H. Emory's official reports of 1848 and 1857 resemble those illustrating the eighteenth-century voyages of Captains James Cook and George Vancouver. All were intended to record faithfully what was seen in order to satisfy the demand for factual information about faraway places and the peoples who inhabited them. The work of the expeditionary artists was a kind of visual naming. It brought to public consciousness what was there but previously unknown, transforming it in the process of showing it. Such is the nature of two-dimensional representation: art creates a separate reality reflecting the artist's cultural values. American exploration artists, Goetzmann notes, shifted from the exotic and pastoral to the sublime and picturesque, keeping apace with the spirit of different ages of discovery. The pastoral mode was appropriate to the earliest phase of exploration, when the West seemed a remote wonderland full of possibilities; the sublime reinvigorated interest in exploration, an activity that had become predictable by the middle of the nineteenth century.

In 1832, when the Missouri River was still redolent of mystery for most Americans, George Catlin (1796–1872) staked his claim as one of the century's great visionaries. The West was a phantom, he said, "travelling on its tireless wing": "it flies before us as we travel, and our way is continually gilded, before us, as we approach the setting sun." Catlin was after an idea, not a place, and in 1832 he headed upriver on the first steamboat to ascend the Missouri to Fort Union near the mouth of the Yellowstone, bound to see unspoiled Indians and paint them "in all their grace and beauty." The eighteen-hundred-mile journey introduced him to the Blackfoot, Crow, and Lakota Indians and all the river tribes, notably the ill-fated Mandans. Catlin subsequently toured the southern plains, the upper Mississippi, and the Great Lakes region. In 1836 he visited the red pipestone quarry on the Coteau du Prairie, concluding his travels in Indian country. His ambition was to win fame and fortune through his enterprise by preserving a pictorial record, "Catlin's Indian Gallery," of all the uncivilized tribes within the borders of the United States. It was an ambition in keeping with an expansionist age.

Catlin's supreme achievement was the gallery itself. Consisting of particulars, all factual he claimed, it was in conception and realization a great act of the imagination, an artistic equivalent of exploration itself. The individual portraits and camp scenes in his gallery—some five hundred paintings—were unified by the sense of purpose that first sent him West to paint Indians. Something vast was happening in America. The new, as ever, was displacing the old; wilderness was yielding to an expanding white population, savagery to civilization. The process was already nearly two centuries old when Catlin was born, but in his lifetime the pace of change had dramatically accelerated. The Old Northwest had appeared on the map of American ambition in the early 1800s and then was engulfed in a wave of speculation, settlement, and statehood. Where Indians had hunted not twenty years before, cities thrived. Only by venturing west could one see "real" Indians—Indians still relatively independent of the white man, still practicing their own rites and garbed in their own costumes, unspoiled by civilization's vices. Catlin was not the most technically accomplished artist to paint Indian portraits, nor was he the first. He was capable of pandering to popular prejudices, and self-interest was never far beneath the surface of his humanitarian professions. He was more often a showman trying to pry coins from a crowd than the serious student of Indian life he had aspired to be. But George Catlin was a magnificent dreamer. Where others saw miniatures, he saw the larger picture, and he created his gallery to mirror it. His generation, he was convinced, was witness to the last act in a great human drama. A race, America's native race, was being extinguished by another, yet his countrymen were mostly oblivious to the fact. His mission was to preserve a visual record of the western tribes before civilization came calling and ruined them forever. In a few years what he saw and painted would be seen no more. Indeed, Catlin counted on this. An artist-entrepreneur, he proposed to invest time and money in the creation of his Indian Gallery and then to retire on the proceeds when his countrymen awakened to the fact that the great drama was already over and that no other record of it survived. It was the power of Catlin's imagination, his ability to harness raw data to a commanding

On a trip up the upper Missouri River in 1832, George Catlin worked at a feverish pace, on some days painting half a dozen works for his contemplated Indian Gallery. In this portrait of the Pawnee warrior Buffalo Bull, he carefully painted the buffalo head sign that was the subject's special "medicine" but used broad strokes to sketch the lower torso and the background, which could be filled in later.

George Catlin (1796–1872). Buffalo Bull, a Grand Pawnee Warrior. *Oil on canvas, 1832. National Museum of American Art, Washington, D.C./Art Resource, New York, New York.*

vision of their meaning, that makes his work collectively the original masterpiece of western art.

Others might disparage Catlin's paintings, pointing to flaws in observation and execution, but his example influenced even his detractors. Two artists followed on his heels in the 1830s, Karl Bodmer (1809–93) and Alfred Jacob Miller (1810–74). Unlike Catlin, they accompanied private patrons, Bodmer traveling with a German prince who toured up the Missouri to the Yellowstone in 1833–34 and Miller with a Scottish nobleman bound for the site of the 1837 fur trade rendezvous on the Green River. With the luxury of time for close observation, Bodmer, a polished draftsman, produced a series of meticulously detailed images of western Indian life, though his portraits, clinical and exact, lacked the human qualities Catlin was able to impart. Miller, who painted some wonderfully evocative watercolors—and several comparatively stilted oils—of mountain men and Indians, thought Catlin a humbug, though literalism was not his forte either. Rather, he created a spacious natural paradise in which time was suspended while his children of nature, white and Indian alike, frolicked to their hearts' content, oblivious to the progress that would outmode them both.

Bodmer and Miller paralleled Catlin on the extremes of scientific observation and romantic indulgence. Artists who painted Indians thereafter usually joined Catlin

Based on his own eyewitness observations, Alfred Jacob Miller's large (69 x 99 inches) painting of the grand entry of the Snake Indians at the annual fur trade rendezvous of 1837 conveyed the exuberant romanticism of the artist's vision. Sir William Drummond Stewart, Miller's patron, acquired the picture for his Scottish castle as a nostalgic memento of his western adventures.

Alfred Jacob Miller (1810–74). Cavalcade. *Oil on canvas, 1839. State Museum of History, Oklahoma Historical Society, Oklahoma City.*

somewhere in between. Paul Kane (1810–71), Charles Deas (1818–67), Seth Eastman (1808–75), and Stanley each aspired to form an Indian gallery to rival Catlin's and receive some of the recognition and more of the compensation he had been seeking from governments and learned bodies at home and abroad since 1838. Kane, a Canadian, succeeded beyond any of the others. The patronage of the Hudson's Bay Company permitted him wide latitude to roam in the company's vast western empire in the 1840s, and after his return from a tour that had taken him to Forts Victoria and Vancouver on the West Coast, he received a substantial commission for one hundred oil paintings of native life, which constituted an Indian gallery of sorts. Deas was a St. Louis–based painter known in the 1850s for his scenes of frontier trappers. But his plan never got off the ground, since his mental health failed him shortly after he conceived the scheme of painting a gallery of his own. Eastman, an army officer whose postings in the West from the 1820s through the 1850s introduced him to a variety of tribes, came no nearer to realizing his ambition of forming an Indian gallery than serving as the illustrator of Henry R. Schoolcraft's multivolume government-sponsored work *Historical and Statistical Information Respecting the History, Condition, and Prospects of the Indian Tribes of the United States* (1851–57) and as the painter of nine Indian scenes in oil commissioned for the House Indian Affairs Committee room in Washington (1867–70). Stanley was another matter. Western tours, including two as draftsman on official explorations in 1846 and 1853, exposed him to tribes in the Indian Territory, the Southwest, and the

Oregon Territory and along the upper Missouri River, a range that exceeded even Catlin's. Stanley, unlike Deas and Eastman, actually completed and in 1850 began exhibiting a collection of 154 oil paintings of Indians. They were hung at the Smithsonian Institution in 1865, awaiting congressional action on purchase, when fire destroyed the collection, along with Charles Bird King's Indian portraits. Only five of Stanley's paintings were saved, leaving Catlin's gallery unrivaled as a pictorial record of the North American Indian.

Romantic Realism

Because the subject matter of western art is distinctive—Indians, mountain men, cowboys, and the like—there is a tendency to consider the art sui generis. But the obvious links between the work of the American expeditionary artists and their European counterparts alert us to other artistic parallels. Catlin, for example, originally aspired to be a history painter, and the poses in his Indian portraits reflect a conscious classicism. "The native grace—simplicity, and dignity of these natural people," he wrote of the Indians, "so much resemble the ancient marbles, that one is irresistibly led to believe that the Grecian Sculptors had similar models to study from and their costumes and weapons . . . the Toga—the Tunique & Manteau—the Bow—the shield & the Lance, so precisely similar to those of ancient times, convince us that a second (and *last*) strictly classic era is passing from the world."

Catlin's fusion of classicism and romanticism points to a European art movement in the nineteenth century that shared striking parallels with western American art. Orientalism offered colorful Bedouins on horses and camels in lieu of Indians, "these Arabs of the Prairie" as one army officer called them, and harem girls instead of languorous Indian maidens lolling in buckskin dresses, swinging half nude from branches, or performing their toilets by the river. European and American "odalisques" satisfied a taste for the exotic and forbidden. The dry, clear atmosphere, piercing light, and immense expanses of plains and desert linked outdoor scenes, but most of all, western and Orientalist art were linked by the painters' common treatment of their subjects, a supercharged realism steeped in romanticism. Orientalism went out of vogue after the 1860s, though examples lingered into the next century. By then, an emphasis on American exceptionalism (the core of Frederick Jackson Turner's frontier thesis) stressed the Americanness of western art, obscuring the Orientalist connection. But before the Civil War, critics often recognized European parallels. Viewing Miller's painting *Cavalcade*, showing a party of Snake Indians arriving at the rendezvous, a journalist in 1839 wrote: "This is a work which would not discredit Horace Vernet himself, if he had painted it; and by the novelty of the subject and the skilful execution; its evident truth to nature, and the mastery of its difficulties; we fully agree with the uniform opinion expressed by the artists and amateurs who daily congregate around it, that it will attract great attention abroad, and make a favorable impression there of the progress of American art."

Vernet (1789–1863) was one of the most admired French academicians of his day. When Miller visited him at his studio in Rome in 1833 Vernet was working on a battle piece and had yet to discover Arabia. Subsequently, he made Orientalist subjects part of his repertoire, directly influencing some of the American artists who trained abroad. Carl Wimar (1828–62), a St. Louis painter who studied at Düsseldorf, Germany, in the 1850s and later twice journeyed up the Missouri to see for himself the Indians and buffalo and strange rock formations that had so impressed Catlin and Bodmer, translated Vernet's *Arabs Travelling in the Desert* (1843) into an Indian subject, *The Lost Trail* (ca. 1856), while he was still a student in Germany. Years later, Charles M. Russell (1864–1926), born in St. Louis two years after Wimar died, painted a homage also titled *The Lost Trail* (ca. 1915). The trail from Arabia to the American West was rarely so direct. Had Wimar lived longer or been better known in his time, he might today be recognized as a major transitional figure in western art. He went on to paint buffalo herds on the move, buffalo hunts, and panoramic vistas set along the Missouri in the twilight of the Indians' independence; he also painted bloody skirmishes between Indians and whites—a running battle with the cavalry, a war party fleeing with captured horses, an attack on a wagon train, the stout defense of the besieged emigrants—anticipating themes that would preoccupy the next generation of western artists.

Whereas Wimar bridged generations *and* art traditions, the parallels between Orientalist and western art are usually more suggestive than exact. For example, Vernet's first desert scene, *The Lion Hunt* (1836), belonged to the venerable tradition of hunting pictures. It depicted a frenzied melee of horses, lions, men, and a camel and presaged a group of related works on that quintessentially American theme, the buffalo hunt. Eastman in 1848, Miller perhaps two years later, and Stanley in 1853 painted buffalo hunts showing the herd racing away in the left background while the central action in the foreground to the right of center is dominated by an Indian preparing to spear a buffalo that, in the Eastman and Stanley versions, is goring a fallen horse and rider. These buffalo hunts echo Vernet's picture, which showed a spearman, a downed horse and rider, and in lieu of the buffalo, a pouncing lion. In the 1890s Remington and Russell both painted buffalo hunts in which a wounded buffalo vents its fury on a fallen rider while his companions struggle to bring it down. So a lion hunt translated into a buffalo hunt and one buffalo hunt into another, creating through conscious repetition enduring images of the American West long before Custer died on the Little Bighorn and artists invented his glorious Last Stand.

The subjective realism of the painter was challenged in the 1840s by a new technology with apparently unlimited potential for visual documentation. In the very decade that America's "manifest destiny" realized its continental ambitions, photography appeared to offer an objective means of illustrating the Far West. There were drawbacks. The technology was in its infancy, photographic apparatuses were cumbersome, materials were fragile, and the process was time-consuming. Nevertheless, beginning in 1842, artists like Stanley, Wimar, and Albert Bierstadt (1830–1902) made photographs, presumably as aides-mémoire for paintings, since no examples survive. The formal use of photography as part of the record-keeping on surveys, an idea floated during John C. Frémont's Rocky Mountain, Great Basin, and California explorations in the early 1840s, was accomplished on an expedition commanded by Lieutenant Lorenzo Sitgreaves in the Indian Territory in 1850. But three years later Stanley, as expeditionary draftsman with the Pacific Railroad Survey, Northern Route, was still

Though western art, almost by definition, deals with distinctively western themes, its treatment of these subjects is often informed by other artistic traditions. The composition of the French painter E. J.H. Vernet's *Lion Hunt*, for example, presaged a group of paintings on the buffalo hunt, a favorite subject for western artists at midcentury.

E.J.H. Vernet (1789–1863). Chasse au Lion *(The Lion Hunt). Oil on canvas, 1836. Reproduced by permission of the Trustees of the Wallace Collection, London, England.*

John Mix Stanley (1814–72). The Buffalo Hunt. *Oil on canvas, 1853. Photograph courtesy of The Gerald Peters Gallery, Santa Fe, New Mexico.*

using photography to service his art, not to augment the permanent visual record. "I have just returned from a trip of twelve days among the Black Foot Indians," he wrote from Fort Benton on the Missouri. "I have made many interesting sketches of their customs—and daguerretypes [*sic*] of their chiefs—Thus far my trip has been very successful in subjects." But none of his photographs exists today. We do have a small but significant batch of daguerreotypes made during the Mexican War, however, anticipating the role that field photography would play in the Civil War and its role in documenting western expansion after the war.

Photography was first used in a systematic way on government-sponsored surveys of the West in the period 1866–79. Its advantage over art was still thought to be its unblinking factuality, its absolute fidelity to things as they are. It promised to eliminate even the hint of contrivance. Ferdinand V. Hayden, geologist-in-charge of the U.S. Geological and Geographical Survey of the Territories, 1869–78, praised the camera's "uncompromising lens"—an opinion that held sway even after the photographers themselves had asserted artistic aspirations. Preeminent among the western photographers were Carleton E. Watkins (1829–1916), with his almost three-dimensional views of Yosemite; Timothy H. O'Sullivan (1840–82), with his often haunting images, notable for their geological precision and raking light, of the barren lands of the Southwest; and William Henry Jackson (1843–1942), with his compelling pictures of the wonders of what became the nation's first national park, Yellowstone. Each accompanied official expeditions, enjoying freedom from the constraints that commercial considerations imposed on photographers who made their living taking portraits, group pictures, and town views and recording the local attractions, including natives, that would appeal to travelers passing through. Jackson, for example, located in Omaha in 1867 and with a brother operated a successful studio. He did the outdoor work, photographing nearby Indians and scenes along the newly completed Union Pacific Railroad, before he was taken on by Hayden in 1870, beginning an eight-year association that took him all over the West, from the badlands of Wyoming to the cliff dwellings of Arizona and New Mexico. Among expeditionary photographers, Jackson was on the prosaic end of the scale when it came to the aesthetic possibilities of the medium. In his vast body of work, he concentrated on the subject rather than its presentation or its philosophical implications. But he had a natural flair for composition, a talent that was enhanced through contact in 1871 with a gifted landscape painter praised by Hayden as an exquisite colorist, Thomas Moran.

Moran (1837–1926), English-born but American-raised, was highly trained in other aspects of his art as well. Even though he espoused close outdoor study and natural representation, he devoutly admired the English painter J.M.W. Turner. "Literally speaking, his landscapes are false," Moran wrote, "but they contain his impressions of Nature." This was a fundamental distinction, raising again the issue of the artist's subjective input into his art, his *impression* of reality. The Moran-Jackson relationship began on Hayden's 1871 survey of the Yellowstone, and the two proved kindred spirits in their commitment to western landscape as a distinctively American theme. Indeed, Jackson's photograph of Moran at Mammoth Hot Springs calls to mind Asher B. Durand's celebrated 1849 painting *Kindred Spirits*, showing the poet William Cullen Bryant and the painter Thomas Cole viewing the Catskill scenery. Here, photographer and painter contemplate western scenery, one directly, one through the camera's viewfinder. Even as Jackson became more self-conscious about the art in his photography, about selecting vantage points and framing his subjects, he also increasingly manipulated his images through retouching and extensive overpainting. Photography, it turned out, was not nature replicated. Like painting, it was one person's impression of reality.

Moran was not the only western landscape artist to discover his calling on an official

Accompanying Ferdinand V. Hayden's government-sponsored expedition to Yellowstone during the summer of 1871, the photographer William Henry Jackson and the painter Thomas Moran produced compelling views of the region's dramatic scenery that proved useful in enlisting eastern support for the movement to make Yellowstone the first national park.

William Henry Jackson (1843–1942). Thomas Moran at Mammoth Hot Springs, Yellowstone. *Albumen silver print, 1871. National Park Service, Yellowstone National Park, California.*

expedition. Bierstadt, born in Germany two years before Catlin ascended the Missouri, grew up in America but returned to Germany in the 1850s to polish his craft at Düsseldorf and to hone an ambition, equal to Catlin's, to achieve fame and fortune from his art. His often huge, theatrical paintings of western scenery were long out of favor by the end of his life, victims of changing tastes in art. But in the 1860s Bierstadt was at the pinnacle of popular renown. With a letter of introduction from the secretary of war, he had accompanied an 1859 expedition that was plotting a wagon road through the Rockies, and thus he found his life's work. Moran was always something of a topographical draftsman, despite his enthusiasm for Turner, and his Yellowstone watercolors, brilliant as jewels in their colors and craftsmanship, provided a visual catalog of the major geological wonders. But for Bierstadt, nature was the starting point, not an end in itself, and rearrangement was the artist's role. He gave Americans a western Eden just a little grander than reality. Miniature bear, deer, and Indians cavorted in the foreground while behind them, glimmer-glass lakes, shafts of light, and silver threads of plunging water carried the eye up peaks that soared to extravagant, cloud-piercing heights. He made unspoiled western nature his equivalent of Catlin's unspoiled western natives, staking out a domain (*his* Rockies, *his* Sierra Nevada) separate from that of Frederic Edwin Church, a much-admired contemporary who claimed the rest of the hemisphere.

The West belonged to Bierstadt. A friendly reviewer in 1868 joked that he had "copyrighted nearly all the principal mountains." The next year that noted art critic

George Armstrong Custer praised nature's wonderful ability to imitate the painters. "We are now in the Wichita Mountains . . . a high level plateau, with streams of clear water, and surrounded by a distant belt of forest trees," Custer wrote. "Tom [his brother] and I sat on our horses as the view spread before us, worthy the brush of a Church, a Bierstadt, the structure of the mountains reminding one of paintings of the Yosemite Valley, in the blending of colors—sombre purple, deep blue, to rich crimson tinged with gold." Shortly before departing on his final campaign, Custer lunched in Bierstadt's New York studio; subsequently, Bierstadt contributed fifty dollars to a monument fund for the fallen hero, resisting the temptation to add his version of Custer's Last Stand to the body of American historical art.

Bierstadt did, however, paint the alpha and the omega of the Indian's decline: the landing of Columbus and the last of the buffalo. His most effective statement on progress and change was an allegorical painting whose didacticism was muted by his bravura in handling landscape and light, *Emigrants Crossing the Plains* (1867; see p. 344). A passing caravan of pioneers dissolves into the setting sun; almost invisible behind the wagon train is a tipi village, bearing silent witness to civilization's inexorable advance. Back in 1819, Peale had painted a sunset on the Missouri; the West had then seemed all possibility. But by 1867, history had overtaken the future, and the setting sun, a beacon for the white pioneers, had become instead a symbol of the end of the native way of life. The years after the Civil War marked the final phase in the conquest of Indian America as the government's "Peace Policy," with its humanitarian rationale resting on military force, proved lethally effective. The Piegan, Modoc, Comanche, Kiowa, Lakota, Cheyenne, Nez Percé, Ute, and Apache Indians all felt the brunt of "conquest by kindness" after 1869. Catlin, living abroad in self-imposed exile, fretted over the "spectral picture . . . of well fed soldiers entering the wigwams of these starving and unsuspecting people, with sabres in hand splitting down the heads and mangling the bodies of women and children, crying and imploring for mercy." Now all could see what he had long ago predicted, "the reality of *Extermination* . . . the going down of the sun (and its last glimmering rays) of the North American Indians."

Remington, Russell, and Their Legacy

Born in the decade in which Bierstadt reached the heights of popular acclaim, both Frederic Remington and Charles M. Russell grew up entranced by stories and pictures of Indian-white conflict. Remington took the military side; Russell, while thrilling to the daring antics of dime-novel frontiersmen and real-life counterparts such as Buffalo Bill Cody, shied away from uniforms and favored the Indians. Both early showed their inclinations, which they faithfully followed through careers that spanned the closing of the frontier and the nostalgia of the twentieth century. Remington championed the winning of the West, whereas Russell lamented all that was lost. Each fashioned a distinctive body of work around his commanding theme. They were opposites in every way but one: both were certain they had witnessed the end of something profound. "I knew the wild riders and the vacant land were about to vanish forever, and the more I considered the subject the bigger the Forever loomed," Remington wrote in 1905. Russell said simply, "The West is dead! You may lose a sweetheart, but you won't forget her." It was this conscious awareness of the end of things that impressed their work so indelibly on future generations. Today, Remington and Russell still define western art.

There was never much doubt about Remington—the twig was bent early, and he was still painting and modeling military subjects the last year of his life. Indeed, his natural inclination was as a military, not a western, artist. He became the painter of the Indian-fighting army because Indian wars were all there was. When a foreign war did come along, he champed at the bit to take it in and eagerly awaited a "big murdering" in Cuba in 1898 so that he could be there to see it all. He was there, but it was too late by then. Real war proved messy and horrible, not splendid and glorious, and Remington was too old to be playing soldier. Anyway, time and circumstances had made him primarily a western artist, and the public held him to his task. As a prolific illustrator in the 1880s and 1890s he had relished "men with the bark on"—soldiers, certainly, and mountain men, cowboys, and their ilk. The Indian was for him the necessary foe—brutal, unfathomable, dangerous. Remington had neither empathy nor insight to offer: his was the Indian seen down a rifle barrel.

The Wyoming journalist Bill Nye set the terms for this brand of "western realism." "A dead Indian is a pleasing picture," he observed in 1881. "The picture of a wild free Indian chasing the buffalo may suit some, but I like still life in art. I like the picture of a broad-shouldered, well-formed brave as he lies with his nerveless hand across a large hole in the pit of his stomach." Artists had often depicted Indian atrocities. Paintings of the capture and murder of white women, a venerable tradition in American art, served as visual arguments for retaliation. Massacre scenes—not battles but furious slaughters by monsters in paint and feathers—were a nineteenth-century staple. Illustrations of both subjects filled popular histories of the Indian wars and even school readers. Portrayals of Custer's Last Stand were in the second tradition, and so was Remington. What distinguished his work was its concentration not on Indian savagery but on white heroism in opposing it. Other illustrators portrayed the Indian-fighting army after the Civil War, including Theodore R. Davis (1840–94) and Rufus F. Zogbaum (1849–1925). But Remington saw an epic playing itself out, the winning of the West, and his theme transcended the particulars. He prided himself on a knowledge derived from personal experience. But his imagination was his real strength. He conjured up convincing scenes of cavalry charges and close-quarter combat because he believed in them, and he made others believers too. In his time he was praised not for *creating* much of the imagery still fundamental to western art but for reporting, accurately and unimaginatively, what was. Remington knew better. He did not go to West Point himself, he once explained, because it "might have spoiled it all—it might have paralyzed the sentiment of the thing." And it was the sentiment that counted; facts were simply stepping-stones to larger truths.

Shortly after his return from Cuba, in the fall of 1900, Remington made a dispiriting trip to Colorado and New Mexico. "Shall never come west again," he wrote his wife. "It is all brick buildings—derby hats and blue overhauls—it spoils my early illusions—and they are my capital." Indeed, Remington's priorities changed in the twentieth century. He wanted to be a great artist, not merely an acclaimed illustrator. He wanted to make paintings freed from the demands of the literal. The Old West was mostly an idea, after all, and in increasingly impressionistic oils he gave free rein to his imagination. Color and light, not subject matter, were now his obsessions. Moonlight was a particular challenge, and Remington turned to nocturnes in which dark shadows cloaked a West become mundane, restoring its mystery and allure. Action was often minimized; a hush fell over figures poised expectantly, immobile, as though after three raucous centuries of westering, "of smoke and dust and sweat," only stillness remained. Remington even discovered a late-life regard for the old-time Indian, conquered more by hunger than fair combat and, in retrospect, possessed of a certain "nobility of purpose." But for all his artistic experimentation, Remington never deserted his theme. To the end he painted thundering cavalry charges and believed absolutely in the Anglo-Saxon mastery they represented. And he still defended his original turf against trespassers—artists like the Hoboken, New Jersey, painter Charles Schreyvogel (1861–1912), who in 1900 won a coveted prize from the National Academy of Design for a cavalry picture, *Wounded Bunkie*, and thereafter competed against Remington for popular favor as the painter-historian of the Indian-fighting army.

In truth, Schreyvogel was better equipped to carry on the old, heroic tradition. His West was derived from Buffalo Bill's Wild West show. It was all whooping Indians, dashing cavalry, and plunging horses, pure melodrama untroubled by subtlety or fact. Nevertheless, when Remington decided to expose the pretender, he chose factual accuracy as his grounds. The opportunity came in 1903 when Schreyvogel unveiled *Custer's Demand*, a quiet set piece for once showing a specific historical event—Custer again, though this time parleying with Kiowas on the southern plains seven years before his Last Stand. Remington could not find enough bad to say about it, but he tried, sparking a lively newspaper controversy full of irony as an artist who had disavowed documentary concerns grappled with another who had never embraced them over the issue of "historical correctness." It was a revealing note on which to introduce western art to the twentieth century.

Charles Russell represented an alternative tradition. His work was consistently commemorative, his heart in the past, not the present. He never changed. He fell in love with the wide-open spaces and rough-hewn ways of Montana as a sixteen-year-old in 1880, and he was still in love when he died in Great Falls forty-six years later, though it was the old Montana that commanded his loyalty. Russell came by his title as "the cowboy artist" honestly. He wrangled horses on the roundup most springs and falls from 1882 to 1892, sketching, modeling, and painting on the side. A close observer of what passed before him, he stored up the impressions of roping, riding, and Indian life that became his artistic stock-in-trade. In 1887 his small watercolor *Waiting for a Chinook* caught the public's fancy. It was a plain-spoken document showing a starving cow surrounded by wolves during the terrible winter of 1886–87, yet it had symbolic resonance. That winter marked the beginning of the end for the open-range cattle industry and the cowboy life Russell cherished. By 1890 the outline of a buffalo skull was part of his signature, expressing what he could not yet verbalize: the passing of a distinctive way of life. Soon roving Indians and cowboys, stagecoach drivers and bullwhackers, professional hunters and the rest would be fading memories. Documentation *was* commemoration, so rapidly was change transforming the land. Remington's work speaks to the heroism called forth by the winning of the West, whereas Russell's speaks only to the certainty, and pain, of loss.

Entranced by the romance of the Old West, Charles M. Russell painted a world that existed mainly in memory and imagination. His heroic men of the open range were figures of the past, relics of the days before barbed wire and changing business practices altered the cattle industry that Russell cherished.

Charles M. Russell (1864–1926). Men of the Open Range. *Oil on canvas, 1923. Mackay Collection, Montana Historical Society, Helena.*

Russell took up his art full time in 1893 and settled down for good a few years later when he married Nancy Cooper. She proved a shrewd, tough-minded businesswoman, adept at managing her husband and his career. Success was her goal, and together they achieved it, spectacularly. Russell advanced rapidly in technical sophistication in the twentieth century, but he never advanced an inch in his views. "I often think of you and the good old times we had," he wrote a cowboy friend soon after his marriage, and the years only intensified his nostalgia. "My me[m]ory often takes me back to the range, and camps we knew so well," he wrote another cowboy friend in 1917. "Theres not maney of the old bunch left . . . thirty seven years Iv lived in Mantana, but Im among strangers now." And just months before he died, calling himself a has-been horse wrangler, he mused that he would trade all the canned music in the world "to here the bells of a saddel band like most men of my age my harte lives back on trails that have been plowed under." Art for Russell was the shortest path back to yesterday.

Beginning with a good eye for detail and a wonderful grasp of human and animal anatomy, Russell had a rare knack for portraying action but little knowledge of how to compose a painting. His work through the 1890s was uneven as he learned his craft. At his peak, between 1905 and 1920, everything came together in glowing canvases that integrated figures and landscape to tell stories in paint that still speak of a magic time and place. Personal experience was important to his art. But some of Russell's favorite themes predated his arrival in Montana—Lewis and Clark, intertribal battles, mountain men, even buffalo hunts. His West was a compound of memory, imagination, and research.

He made it more real than reality. And he never forgot the glue that held it all together, a romanticism that spurned the mundane for the picturesque. "Sinch your saddle on romance," he advised a western writer. "Hes a high headed hoss with plenty of blemishes but keep him moovin and theres fiew that can call the leg he limps on and most folks like prancers." In the last years of his life, nostalgia, not the West, was Russell's real subject. His power—and his enduring legacy to western art—was the ability to make his yearning universal.

Remington and Russell are commonly lumped together as the progenitors of a "Wild West school" of art. But others anticipated them in some respects, and both drew on the older generation of western artists for information and inspiration—notably Catlin, Bodmer, and in Russell's case, Wimar. Still, the pair's influence does constitute a school of sorts. Remington offered advice to artists as diverse as Maynard Dixon (1875–1946), best known for his desert landscapes, and Carl Rungius (1869–1959), a gifted wildlife painter, and his influence was pervasive. It would be safe to say that no western artist after 1890 was entirely unaffected by Remington's example. Certainly Edwin W. Deming (1860–1942), W.H.D. Koerner (1878–1938), Philip R. Goodwin (1882–1935), Frank Tenney Johnson (1874–1939), Schreyvogel, and Russell himself owed Remington a debt. Russell actually had a protégé, the painter Joe De Yong (1894–1975), who from 1914 on sometimes worked beside him in his log cabin studio in Great Falls. And he encouraged a range of cowboy artists who found in him their model, including Will James (1892–1942), E. W. Gollings (1878–1932), and the sculptor Charles A. Beil (1894–1976). Edward Borein (1872–1945), Russell's California counterpart as a cowboy artist, was a close friend, and Russell's influence touched several of the outdoor illustrators of his day, including John N. Marchand (1875–1921), W. Herbert Dunton (1878–1936), and Goodwin (whose technical skills and palette, in turn, influenced Russell's own). The Great Falls painter Olaf C. Seltzer (1877–1957) was a direct imitator, indeed almost a Russell clone. Through calendar reproductions and cheap color prints, Russell's example was broadcast everywhere. Even William R. Leigh (1866–1955), a European-trained and highly polished craftsman given to a Schreyvogel kind of Wild West, named Russell as an inspiration, and members of the Cowboy Artists of America to this day are still proudly in his tradition.

Ageless Cowboys and Sunset Indians

The twin legacies of Remington and Russell have long since passed into the popular culture. John Ford's cavalry western movies relied on Remington for everything from camera angles and color to costume and action bits, and the director Howard Hawks recalled trying to capture Remington's light in a night scene set outside a saloon. The phenomenally successful Marlboro Man advertising campaign, begun in 1954 and still going strong, has played effectively on the notion of the rugged, self-reliant cowboy hero riding free under the "Big Sky" and not about to take advice from Washington bureaucrats like the surgeon general. Imagery and attitude alike are straight out of Remington and Russell. They set the terms for the Old West in the twentieth century. But although the two can be linked as progenitors of a Wild West school, it is equally important to understand their differences. If subject matter connects them, it also distinguishes them from one another. A Montana newspaper in 1902 drew the conventional distinction: "Russell is the acknowledged superior in handling subjects on the ranges and in the cattle country, while Remington's cavalrymen and scouts are the best types of his western work."

The cowboy became a major presence in the American imagination only in the last few decades of the nineteenth century. Artists like James Walker (1819–89) had painted the dashing Mexican vaqueros of California and Texas, but the earliest representations of Anglo cowboys appeared in the popular press shortly after the Civil War, the work of such illustrators as Waud, of later Custer fame. As the mountain man departed the scene, a new individualistic western hero was in order. Plainsmen-scouts in buckskin costumes kept the trappings of the mountain man before the public, but the cowboy was more current. His appeal, like that of the mountain man, stemmed from his free-roaming, transient occupation. One day the beaver would be trapped out, the range would be fenced in, and both mountain men and cowboys, their purpose gone, would be rendered obsolete, mourned as part of the nation's vanished youth. Indeed, their actual youthfulness, their profligate ways—the big blowout at rendezvous or end-of-the-trail cowtowns—and their devil-may-care disregard for convention gave them poignant stature. They would not be here tomorrow. As alternatives to sober-sided earnestness, acquisitiveness, and workaday reality in an ever more urbanized America, cowboys passed through life scornful of money, danger, and the serious business of growing up.

The cowboy story has been characterized as a male escapist fantasy, and the western myth that exalts the cowboy remains overwhelmingly a male myth. Women are plugged into set roles (sunbonneted pioneer mothers, golden-hearted prostitutes, the occasional female outlaw), but their presence is minimal and correspondingly minimal in western art. Again, this reaches to the crux of the matter: western art is about cultural assumptions, not historical realities. It evokes cleaner air and purer water—the greener grass on the other side of the fence or over the next hill. The combination of fiction, showmanship, and art that planted the cowboy's image in the public's mind came at a time when nostalgia was already cloaking the West in the romance of faded youth and better yesterdays. The cowboy promised men they need never grow old; they could be boys forever, leaving behind the cares and duties of ordinary folk and riding off into a "land without end, a space across which Noah and Adam might have come straight from Genesis," as Owen Wister put it in *The Virginian* (1902). In western fiction, the cowboy's work rarely intrudes on the action; indeed, as the Texas writer J. Frank Dobie observed, cowboy stories are notable for the absence of cows. Artists were more faithful to the cowboy's working life. Russell, for one, drew on his rangeland experience to record the technical aspects of cowboying, though his themes were mostly action-oriented: bucking broncos and perilous moments when "ropes go wrong," horses fall, and steers turn hostile and dangerous. The message was mythic.

Besides the cowboy, a more venerable figure took center stage at the turn of the century. The cowboy conjured up a past that need never die; the Indian, decked out in feathers and fairly reeking of pathos, stood for all that was already lost. Before landing in New Mexico in 1918, the influential modernist painter Marsden Hartley (1877–1943)

Echoing contemporary politicians and policymakers, early 20th-century Anglo artists often depicted Indians as doomed members of a once noble race.

James Earle Fraser (1876–1953). The End of the Trail. *Bronze, 1918. Clara Peck Purchase Fund, Buffalo Bill Historical Center, Cody, Wyoming.*

confessed that he "wanted to be an Indian, . . . go to the west, and face the sun forever." Neoprimitivism played its part in the reemergence of the "Vanishing American," but nostalgia was the star. "If the cogs of time would slip back seventy winters," Russell wrote in 1920, "thaird be another white Injun among the Black feet Hunting hump backed cows." The cowboy artist loved dressing up in wig and blanket and playing Indian, and he filled his paintings with proud warriors riding across the open plains. It was not that he took no interest in the Indian in the present, but as he remarked, he had "always studied the wild man from his picture side," and his picture side was yesterday. The sculptor James Earle Fraser (1876–1953) achieved lasting renown for his allegorical model *The End of the Trail*, which he worked up on epic scale in 1915. It showed a bowed warrior on a drooping pony, and Fraser dreamed of having it cast in bronze to occupy a promontory above San Francisco Bay. There, horse and rider "would stand forever looking out on the waste of waters—with nought save the precipice and the ocean before them . . . , in very truth, 'The End of the Trail.'" But Fraser's most widely circulated work, the buffalo–Indian head nickel first minted in 1913, told of other trails that had ended as well. The Old West was gone, and the Indian, as its most compelling symbol, now also stood for a vanished America.

Determined to be the George Catlin for a new century, the Seattle-based photographer Edward S. Curtis (1868–1952) in 1900 began creating a comprehensive pictorial history of the western tribes. His enterprise, like Catlin's, would be a calculated fusion of science and romanticism dedicated to a simple proposition: the Indians and their cultures were vanishing. The Indian had long put the putative objectivity of photography to the test. After the Civil War, field photographers had recorded Indian camp life, while studio photographers following in the footsteps of King had made individual and group portraits of Indian delegates visiting in Washington, D.C. Jackson combined both approaches in photographing the Pawnees and Omahas before joining the Hayden survey; subsequently, his ethnological bent found expression in views of the Shoshone, Bannock, Ute, and Pueblo Indians. Other expeditionary photographers were active too, and in 1877 Jackson compiled a descriptive catalog of the government's collection of Indian photographs, over a thousand in number. Hayden, impressed by its size and comprehensiveness (twenty-five tribes were represented), echoed the claims Catlin had made for his Indian Gallery forty years before and declared the government collection unique, irreplaceable, and of inestimable scientific value. The "Vanishing American" still validated the urgent need to preserve a visual record of the native tribes; inevitably, the assumptions it embodied also shaped the record that was being preserved.

Curtis, for example, inaugurated the first volume (published in 1907) in his twenty-volume series *The North American Indian* with a symbolic view of a party of Navajos riding off into the desert haze, toward dark shadows where the light glimmering along the top of a distant bluff provided the only ray of hope for their future. Curtis titled the photograph *The Vanishing Race.* "The thought which this picture is meant to convey is that the Indians as a race, already shorn of their tribal strength and stripped of their primitive dress, are passing into the darkness of an unknown future," he explained. "Feeling that the picture expresses so much of the thought that has inspired the entire work, the author has chosen it as the first of the series." Curtis's pictorial record of the tribes was faithful to this premise. He was not simply out to show Indians; like Russell, whom he later met, he was out to show the romantic and picturesque side of Indian life. Many western towns had a resident photographer in the 1870s, and amateur camera bugs proliferated by the end of the century. They were responsible for some of the frankest, least artfully contrived pictures of Indian reservations, where life, changed as it was, went on. But photographs documenting acculturation never had the emotional appeal of those that suspended time or even rolled it back. To this day, the public prefers pictures of Indians not as they are but as Curtis showed them—colorful, noble, and doomed.

No wonder modern native artists have focused so often on what Scholder has called "Indian kitsch." By defining Indian existence, white stereotypes have defined Indians out of existence. Thus the influence of Scholder's ironic mode: it makes the white images do the work of cultural criticism. Fraser's *End of the Trail* can be turned against itself, for example, exposing the whole self-serving tradition of the "Vanishing American." And Custer's Last Stand is a natural to get at the mythology of the winning of the West.

If the disappearance of native cultures was a subject for European-American artists of an earlier date, the continued vitality of native cultures is an important theme for many contemporary Native American artists. Ironically addressing the cultural stereotypes established by mainstream western art and popular history, the Lakota artist Randy Lee White evokes the imagery of Custer's Last Stand to comment on a fight over gasoline supplies.

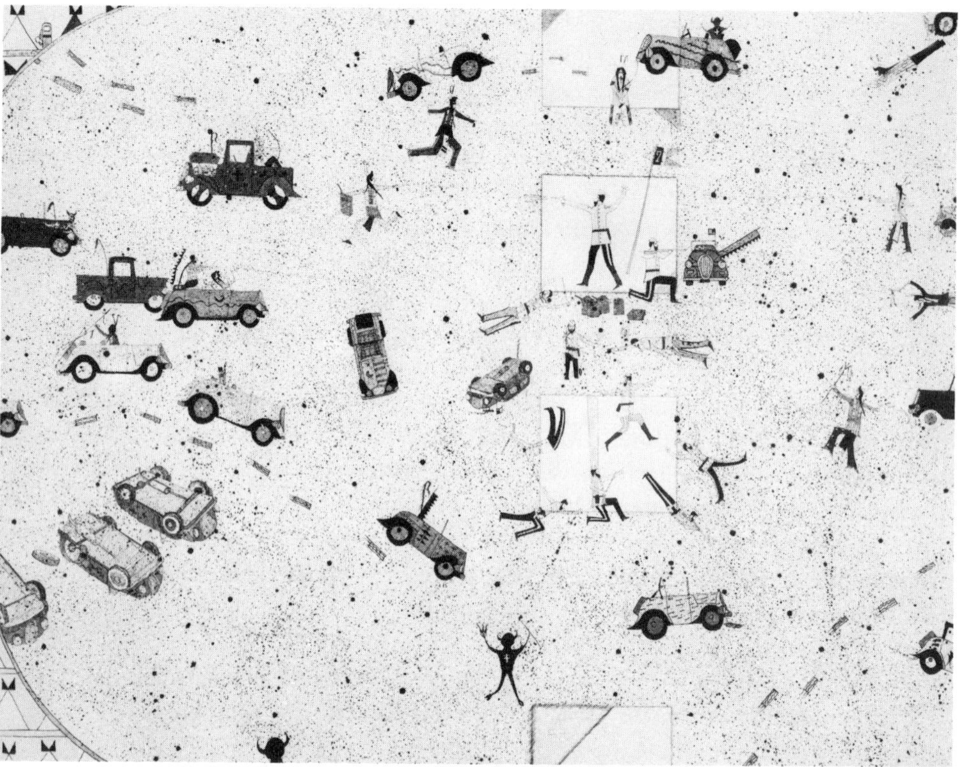

Randy Lee White (b. 1951). Custer Revised. *Mixed media, 1980. Gift of Manfred Baumgartner, 1984. Collection of the Museum of Fine Arts, Museum of New Mexico, Santa Fe.*

"Custer Wore an Arrow Shirt," a popular Red Power slogan goes; another more bitingly asserts, "Custer Died for Your Sins." These suggest the range of possibilities Custer opens for the native artist. Scholder offered a sly treatment in his amusingly titled painting *Indian, Dog, and Friend* (1973), based on a photograph of Custer at ease with an Indian scout and a hunting dog. His most celebrated student, T. C. Cannon (1946–78), mocked the heroic Custer image in portraits whose titles were also humorous but angrier: *Zero Hero, Ugh,* and *Custer, "Go Gettum."* Randy Lee White (b. 1951), a Lakota, wittily combined Custer's Last Stand and traditional pictographic form to comment on postreservation developments in *Custer Revised* (1980). It shows old-time warriors battling white salesmen, in soldiers' uniforms, who are defending their supply of gasoline, needed to fuel the automobiles that the white men have duped the Lakotas into buying and that, now useless, litter the battlefield. One warrior makes off with his trophy—not a scalp or a rifle but a gasoline can—while a "soldier," still desperately defending his culture's values (here, a fuel hoard), fires a parting shot. Native artists range freely in their choice of subjects. But their work is most clearly part of western art when it overtly comments on white preoccupations and stereotypes, taking visual revenge on the often oppressive mythology that the art sustains.

Outsiders

Cowboys and Indians dominate western art because they express so well its commanding theme of evanescence. This implies, accurately, that the artistic record of the American West is highly selective. Western art omits much in order to emphasize a little, establishing that little, through repetition, as the whole. Fundamentally, it has been a frontier art, and like frontiering, it is about a process, not a result. The West, as the novelist A. B. Guthrie, Jr., noted, "is an adventure of the spirit. . . . more than journey's end, it is the journey itself that enchants us." Repudiating the frontier as a useful historical construct today will change neither its importance for an earlier generation nor the fact that it is now entrenched in western art. As a new generation of historians challenges the inherited story of westering, reevaluation of western art is thus unavoidable, and desirable. An essentially white male escapist fantasy cannot survive unexamined (and perhaps undiscredited) into the twenty-first century. Critics who never did like western art have now found just cause for dismissing it: it is both sexist and racist. Still, the art may prove refractory material for revisionism. Its landscapes remain appealing and its action pieces exciting, and it is encrusted with layers of cultural meaning. Because it is so ideologically charged, it is an exclusionist art. It eliminates alternate visions and even much of the subject matter that a western art should logically embrace.

Urbanization is part of the story of civilization's advance into the wilderness. But a major painter like George Caleb Bingham (1811–79) of St. Louis seems most western not when he is showing political life in Missouri—politicians are still with us—but when he is showing rivermen and trappers, who are not. Western communities were well represented among the popular nineteenth-century bird's-eye views of American towns and cities. The urban intrusion on western nature is evident in prints showing tree stumps pockmarking the hills outside the town. Often, tiny foreground figures serve as aids to perspective and as commentaries on white progress: strolling ladies with parasols and gentlemen in top hats, farmers in their fields, cowboys with their herds, prospectors with their picks, hunters with their faithful dogs, fishermen in their boats, even an artist at his easel. Once in a while an Indian creeps onto the fringe of things. A print made in about 1850 showed an Indian man gazing at Sutter's Mill and Coloma, mute testimony to the devastating impact the gold rush had on California's native population. In an 1849 lithograph of Oregon City, four Indians stand with spears in hand looking across the river at a settlement from which they are both literally and figuratively barred. A view of Omaha, Nebraska, published in 1868 made a similar point. In the foreground, a small forest shelters a lake and a miniature canoe and Indians continuing their traditional pursuits, oblivious to the burgeoning, well-ordered city just beyond the treeline. Sometimes tipis would be sprinkled over the prairies outside the town limits to contrast old and new, past and present.

Despite the occasional concession to such allegorical motifs, western town views were not really concerned with the frontier past. Instead, neat rows of buildings and grids of streets welcomed the future. By the 1980s nearly 83 percent of those living in the West resided in metropolitan areas—the highest percentage in the nation. But this urban reality remains outside the western art tradition. In the nineteenth century, illustrators working for the eastern periodicals might show gambling halls and hurdy-gurdy houses, ignoring the substantial private residences and public buildings that were prominently featured whenever local pride set the artistic agenda. But in western art, towns exist primarily as settings for gunfights and cowboy high jinks. They might be saloon-centered urban oases, but they were never home. *That* was on the range.

Similarly, the farmer is outside western art. Harvey Dunn (1884–1952) gave his

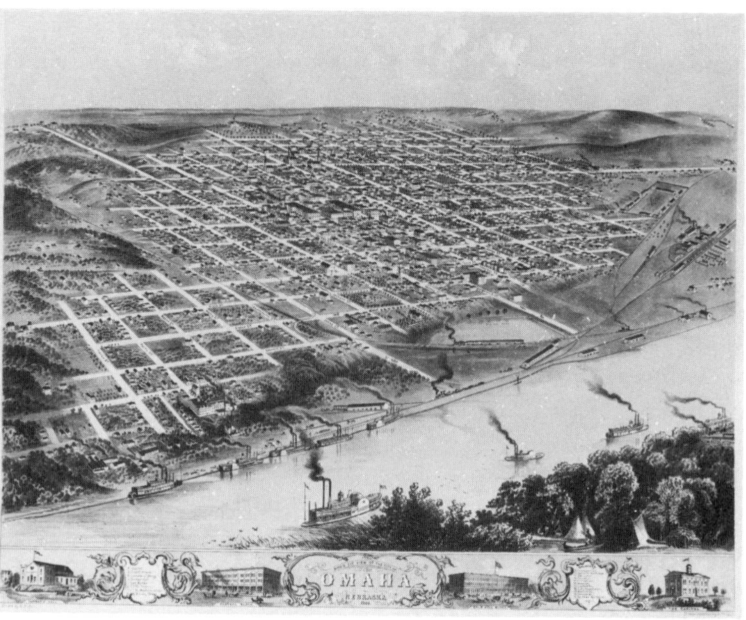

Nineteenth-century prints of western city views promoted civic pride and the economic possibilities of town development. Their implicit focus on urban growth distinguishes them from more traditional western paintings, which emphasize the virtues of America's rustic frontier roots.

Albert Ruger (1828–99). Bird's Eye View of the City of Omaha, Nebraska. *Lithograph, 1868. Courtesy Amon Carter Museum, Fort Worth, Texas.*

homesteaders, female and male, the monumental stature usually reserved for mountain men, cowboys, and cavalrymen. But western art is attuned to high plains and desert romance, not prairie realism. The farmer may be close to nature, but the land he possesses possesses him, rooting him in the earth and denying him the mythic resonance that comes with moving on. Russell, speaking for the cowboy artist tradition, scoffed at those who thought Dunn a western artist. He could not paint a horse, Russell insisted, and that alone disqualified him. Besides, the farmer was the cowboy's dreary opposite: he represented the triumph of the mundane. "When the nester turned this country grass side down the west we loved died," Russell wrote in 1919. "She was a beautifull girl that had many lovers but to day thair are only a fiew left to morn her the farmer plowed her under . . . the man betwene the plow handls was never a romance maker and when he comes history is dead." Russell's bias is still enshrined in western painting and western fiction. The farmer stands accused of obliterating the picture-and-story part of history—and in western art that is the part that matters.

Beginning in 1898, artists found pictures and stories aplenty in New Mexico. The layering of cultures in the Southwest made for a unique depth in time and for a vibrant present where native, Hispanic, and Anglo intermingled in a setting whose beauty, the writer Mary Austin observed, "takes the breath like pain." Here, in the art colonies that flourished in Taos and Santa Fe, an alternative vision found expression. It showed peoples accommodating, changing, enduring. Western art is predicated on a violent rupture between past and present. It is about transience and loss, whereas southwestern art is about continuity and survival. Explaining his decision to go north and paint for a while on the Crow reservation in Montana, Joseph Henry Sharp (1859–1953), a founding member of the Taos Society of Artists (1915), observed, "I went north because I realized that Taos would last longer." Although the "Vanishing American" lingered in the works of Sharp, Bert G. Phillips (1868–1956), and E. Irving Couse (1866–1936),

Depicting farmers as heroic subjects, the painter Harvey Dunn calls attention both to an important aspect of western development and to the biases of a western art tradition more concerned with mountain men and cowboys than the more numerous homesteaders.

Harvey T. Dunn (1884–1952). The Prairie Is My Garden. *Oil on canvas, ca. 1940. South Dakota Art Museum Collection, Brookings.*

the defining southwestern motif was permanence. "Of the Southwest of romance and story the vast, inscrutable mountains, the desolate open spaces remain," a journalist wrote in 1919. "The cowboy still herds his cattle on the mesa, and rides gallantly into a settlement to pass a weekend; hard-eyed prospectors seek hidden gold and silver among the mountains; grizzled frontiersmen trap animals for their fur; in short, . . . it is still the storied Southwest, sans hostile Indians." Grand as nature was, it made room for people. Life was on a human scale; it had texture and an ambling pace. Painters like Ernest L. Blumenschein (1874–1960), Oscar E. Berninghaus (1874–1952), Victor Higgins (1884–1949), and Walter Ufer (1876–1936) created languorous set pieces that caught and held the public's imagination into the 1930s. Indians, horses, and wide-open spaces do not necessarily western art make, however. The pastoral vision associated with Taos and Santa Fe stands outside the western art mainstream, promising something still there, awaiting discovery, whereas western art is about a time and a place that have no existence outside the artist's imagination—"The West That Has Passed," as Russell called it.

Western art also excludes women both as practitioners and subjects. Exceptions can be found (Mary Hallock Foote [1847–1938], an author and illustrator with wide experience in the western mining country, is often mentioned), but the rule stands with one great qualifier: Georgia O'Keeffe (1887–1986), the only woman included in most representative collections of western art and the only western artist represented in most collections of modernist American art. Her story has been retold so often that her life and personality are part of common lore. She is today a legendary figure on a level with Russell himself. He was the cowboy whose untutored genius, according to tradition, shaped western art at the end of the nineteenth century; she was the visionary genius who almost single-handedly dragged western art into the twentieth. Friends with Will Rogers, William S. Hart, and a posse of silent film stars, Russell was a deep-dyed traditionalist. Soul mate and wife of the experimental New York photographer Alfred Stieglitz, with whom she shared a stormy relationship at once suffocating and liberating,

For Victor Higgins and his fellow members of the Taos Society of Artists (founded 1915), the ancient Indian traditions and Hispanic history of northern New Mexico offered a rich source of images for an alternative strain of western art that celebrated cultural exchange and historical continuity.

Victor Higgins (1884–1949). Pueblo of Taos. *Oil on canvas, before 1927. Photograph by James O. Milmoe. Courtesy of The Anschutz Collection, Denver, Colorado.*

O'Keeffe was an instinctive modernist. Born the year Russell painted his miniature storytelling masterpiece *Waiting for a Chinook*, she visited Taos in 1929, three years after his death. She had located her heartland, her Montana—"for me it is the only place"—and after a protracted tug-of-war between New York and New Mexico, she moved to the West for good in 1949. Legendary ghosts like Billy the Kid had enticed her when she first journeyed west in 1912 to teach in Amarillo, Texas, but it was the land—with its fantastic formations, brilliant hues, and compelling emptiness—that caught and finally held her. In her work the human presence would be left implicit in adobe structures and crosses seemingly as old and durable as the hills.

O'Keeffe has been called a "visionary realist," to differentiate her from the run-of-the-mill realists who populate western art. But Russell, as noted, was a romantic realist himself. Modernists like O'Keeffe are said to evoke rather than describe, to interpret rather than document. Their art, unlike that of the representationalists, is about personal response, not the thing being responded to. However, this generalization risks creating a false dichotomy between western modernists and western traditionalists by implying that modernist works are acts of the imagination whereas traditionalist works are mere factual transcriptions. We might better regard Russell and O'Keeffe not as artistic opposites but as artists who created separate realities true to their times and places, and to themselves. Russell made a buffalo skull his personal insignia because it was visual shorthand for the Old West; O'Keeffe painted skulls and bones as one end of the desert

A modernist who found her most compelling subject matter in the southwestern landscape, Georgia O'Keeffe stands apart as the only woman to achieve widespread recognition in the male-dominated field of western art.

Georgia O'Keeffe (1887–1986). Ranchos Church—Taos. *Oil on canvas, 1930. © 1994 The Georgia O'Keeffe Foundation/ARS, New York. Courtesy Amon Carter Museum, Fort Worth, Texas.*

spectrum, with wildflowers at the other. Fiercely independent in her views, she made her isolation part of her art. Many came to Taos, and most left, just as many streamed through Russell's log cabin studio in Great Falls and his summer cabin on Lake McDonald in Glacier Park, eager to see what he saw but not to stay. Physically and emotionally, he identified with the scenes of his youth, as O'Keeffe would with the scenes of her maturity, and each evolved a distinctive way of expressing this attachment. The world now sees New Mexico through O'Keeffe's eyes, as it long has Montana through Russell's. Calendar reproductions widely broadcast his work during his lifetime, turning his paintings into western icons; he still merits at least one calendar each year. Similarly, O'Keeffe calendars are a growth industry today. As a hero of modernism and women's art, she is in danger of becoming a pinup girl for the 1990s and her strikingly original images the latest western clichés. Scholder has already offered a variation on her magisterial *Ranchos Church, Taos* (1930), and another native artist, David Bradley (b. 1954), has portrayed her as a southwestern "Whistler's Mother." Nevertheless, it is fitting that the painter who ushered western art into the twentieth century should also see it out. Nothing finally makes O'Keeffe a better exemplar of the western tradition than the very visionary realism that supposedly sets her apart.

Given all that western art excludes, it seems fitting to conclude with its own exclusion from American art. Western art did not start out as a regional expression, but it has certainly ended up as one. In the nineteenth century, it expressed an expanding, bombastic nation; today, it is a vital and accepted tradition only in the West. American art surveys may give it a passing nod in a separate chapter, serving merely to emphasize its isolation. Critics routinely ignore it, though some, aroused by a controversial exhibition at the Smithsonian Institution's National Museum of American Art in 1991,

"The West as America," now deplore on ideological grounds what they previously dismissed on artistic ones. The indifference to the actual art—be it traditional or modern, white or native—has led to an enclave existence for western art. Only two museums east of the Mississippi River (and none on the East Coast) house representative collections, though individual works are scattered throughout public collections. The Smithsonian's museums own most of Catlin's works, and the Frederic Remington Art Museum in Ogdensburg, New York, the artist's boyhood home, holds a substantial collection of his work. Otherwise, western art must be seen in the West—in Montana and Wyoming, California and Nebraska, Colorado and Oklahoma, and Texas most of all.

Remington, with a foot planted on both sides of the Mississippi, represents the uneasy current status of western art. In his noisy approval of the "winning of the West," he championed a cause now disapproved with a stridency equal to his own. If revisionism holds, much western art will come to seem as outdated as the triumphalist values it espouses, and images that once seemed timeless will appear increasingly (and embarrassingly) time-bound. But western art is about more than progress and conquest and "manifest destiny." It is also about open spaces and yearning and opportunity and hope. Big Rock Candy Mountains exist in the heart, not the head, and the West is still what Catlin called it a century and a half ago, a "phantom, travelling on its tireless wing." Revisionist history, with its stern lessons about the western past, may yet run that phantom to the ground. Meanwhile, western art continues to appeal to rainbow-chasers everywhere.

Bibliographic Note

Among the surveys of western art notable for scholarship and interpretation are the following: Robert Taft, *Artists and Illustrators of the Old West, 1850–1900* (New York, 1953); John C. Ewers, *Artists of the Old West* (Garden City, N.Y., 1965), which incorporates material from the seminal articles he published on individual artists in the Smithsonian's *Miscellaneous Collections* and *Annual Reports*, extending a tradition that reaches back to the 1920s and the pioneering scholarship of David I. Bushnell, Jr.; Frank Getlein, *The Lure of the Great West* (Waukesha, Wis., 1973); Peter Hassrick, *The Way West: Art of Frontier America* (New York, 1977); Dawn Glanz, *How the West Was Drawn: American Art and the Settling of the Frontier* (Ann Arbor, 1982); William H. Goetzmann and William N. Goetzmann, *The West of the Imagination* (New York, 1986), which draws on ideas first advanced in William H. Goetzmann's *Exploration and Empire: The Explorer and the Scientist in the Winning of the American West* (New York, 1966) and widely broadcast in a companion television series "The West of the Imagination," available on videocassette from Films for the Humanities, Princeton, N.J.; and Jules Prown et al., *Discovered Lands, Invented Pasts: Transforming Visions of the American West* (New Haven, 1992). The latter was produced in conjunction with an exhibition organized by the Thomas Gilcrease Institute in Tulsa, Oklahoma, and the Yale University Art Gallery. Many of the most useful—and sometimes controversial—works on western art have appeared as catalogs accompanying exhibitions. Some provide interpretive approaches (Chris Bruce et al., *Myth of the West* [Seattle, 1990], and William H. Truettner, ed., *The West as America: Reinterpreting Images of the Frontier, 1820–1920* [Washington, D.C., 1991]). Others focus on a group of artists (Charles C. Eldredge, Julie Schimmel, and William H. Truettner, *Art in New Mexico, 1900–1945: Paths to Taos and Santa Fe* [New York, 1986], and Ron Tyler et al., *American Frontier Life: Early Western Painting and Prints* [New York, 1987]) or photographers (Weston J. Naef et al., *Era of Exploration: The Rise of Landscape Photography in the American West, 1860–1885* [Buffalo, N.Y., 1975], and

Martha A. Sandweiss, ed., *Photography in Nineteenth-Century America* [Fort Worth, 1991]). Most explore the work of an individual. Recent examples include Michael Edward Shapiro et al., *George Caleb Bingham* (St. Louis, 1990), William Wallo and John Pickard, *T. C. Cannon, Native American: A New View of the West* (Oklahoma City, 1990), Rick Stewart, Joseph D. Ketner II, and Angela L. Miller, *Carl Wimar: Chronicler of the Missouri River Frontier* (Fort Worth, 1991), and Nancy K. Anderson and Linda S. Ferber, *Albert Bierstadt: Art and Enterprise* (New York, 1990). All of the major and most of the minor western artists and photographers have been the subject of at least one illustrated biography, many of them sponsored by museums like the Amon Carter in Fort Worth and the Buffalo Bill Historical Center in Cody, Wyoming. Recently, as the book on Bierstadt indicates, interest in the issue of patronage and western art has been growing. See, for example, Richard H. Saunders, *Collecting the West: The C. R. Smith Collection of Western American Art* (Austin, 1988), and Brian W. Dippie, *Catlin and His Contemporaries: The Politics of Patronage* (Lincoln, 1990). There is also more interest in modern western art, fueled by the extraordinary popularity of Georgia O'Keeffe. See Patricia Janis Broder, *The American West: The Modern Vision* (Boston, 1984), and for O'Keeffe, besides the many elaborate volumes showcasing her art, see Roxana Robinson's biography *Georgia O'Keeffe: A Life* (New York, 1989). A helpful compendium of biographical information on the entire range of western artists is Peggy Samuels and Harold Samuels, *The Illustrated Biographical Encyclopedia of Artists of the American West* (Garden City, N.Y., 1976). Native art has a huge literature of its own; a recent sampler is Edwin L. Wade, ed., *The Arts of the North American Indian: Native Traditions in Evolution* (New York, 1986). *American Indian Art Magazine* is useful for keeping abreast with activity in its field, whereas *Persimmon Hill* and *Southwest Art* are well-illustrated journals devoted to western art, past and present.

Vol. I. $2.50 a Year. Published Weekly by Beadle and Adams, No. 98 William St., New York. Price, Five Cents. No. 6.

THE PRAIRIE PILOT; or, THE PHANTOM SPY.

BY BUFFALO BILL.

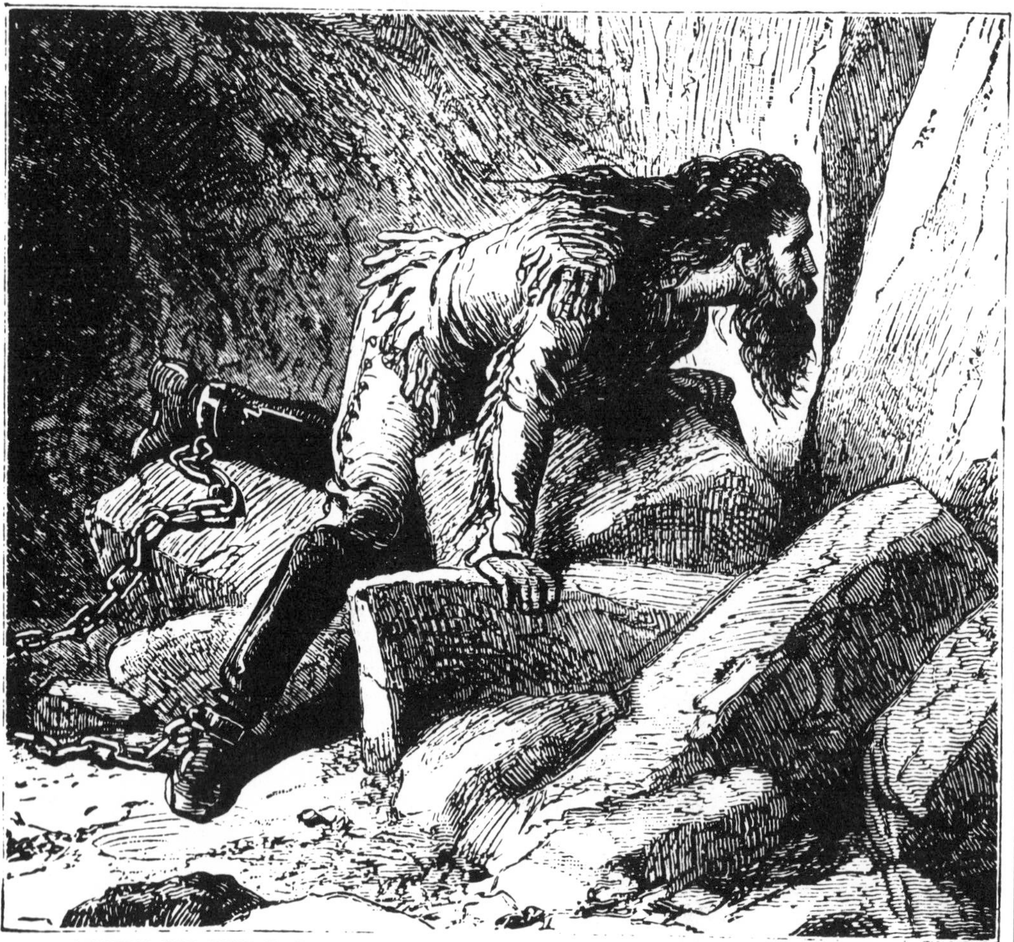

EXAMINING THE CREVICES IN THE WALL OF ROCK, PRAIRIE PILOT FOUND ONE THAT LOOKED INTO THE ADJOINING CAVERN.

Chapter Twenty

The Literary West

THOMAS J. LYON

Mapping the western literary range might seem to be a simple job: draw a line down the Mississippi River, and everything west of that is "western." But start talking with critics of western literature, and soon the good feeling of geographical neatness dwindles. Questions arise. Are the elegant detective novels of Ross Macdonald and Raymond Chandler, set mostly in Santa Barbara and Los Angeles, properly western? (Some critics rule out cities—perhaps on the idea that they are not open, western space.) When Henry Miller of Brooklyn and Paris, the author of *Tropic of Cancer*, moved to Big Sur in 1944, did he become a western writer? (Do we go by an author's geographical residence, or is the determinant something more complicated?)

Some students of the West, Walter Prescott Webb and Wallace Stegner among them, have refined the map by describing areas of low rainfall as diagnostically western. Aridity means space: the space between trees and shrubs, struggling to survive in marginal conditions, and also the space between human settlements, ground that is difficult to exploit or settle, thus remaining mostly wild and open. Relatively few people, living in small settlements within an extensive wilderness, comes close to defining the classic western situation.

But a literary map also needs to account for time and change. We need a theory elastic enough to acknowledge, for example, that although megalopolitan Los Angeles of the 1990s may not offer classic westernness to a writer, it nevertheless has been shaped by a definitively western process. It was a sleepy pueblo two centuries ago, became a cattle town in the nineteenth century, then began reaching out and appropriating water, and grew and grew, developing industry, and the end is not yet. The Los Angeles process is, in essence, the "westward movement," the expansion of European-American civilization at the expense of both traditional societies and natural ecology. Such a history is at heart dialectical, a history of conflict, and as Frederick Jackson Turner pointed out back in 1893, the point of interaction between the old (the wildland itself, or Indian culture, originally) and the new (explorers, mountain men, and pioneers, at first) is the frontier. In the traditional understanding, the frontier began at the East Coast and moved westward until closure was declared by the U.S. Bureau of the Census in 1890.

If frontier action defines writing as western, the western map expands considerably. Many critics, indeed, see James Fenimore Cooper's "Leatherstocking Tales," four of which take place in eighteenth-century New York State, as thematically western. Far enough back in time, even Virginia may be a kind of West—Leslie Fiedler once described John Smith's Pocahontas story as an early version of a prominent western

Late 19th-century dime novels depicted the West as a place of adventure and heroic masculine virtues. Although much regional writing is more self-critical, the success of the formulaic westerns is still reflected in an enduring strain of popular western literature.

William F. Cody (1846–1917). The Prairie Pilot; or, The Phantom Spy. *From* The House of Beadle and Adams and Its Dime and Nickel Novels; The Story of a Vanished Literature, Volume 1, *by Albert Johannsen. Copyright © 1950 by the University of Oklahoma Press, Norman.*

myth, called by Fiedler "Love in the Woods." But the problems with an exclusively dialectical approach that is not tied in some fashion to geography should be obvious. *Where* the action takes place also counts. It can be argued that the frontier is as old as post-Renaissance European expansion and is worldwide to boot; but on the ground of the American West in the nineteenth century, the drama of invasion achieved its best-known statement. The natives were more resistant than in some of Europe's other frontiers, the landscape was more sublime, and the advancing Anglo-Americans had had several generations of pioneering in which to build an identity as tough practitioners of expansion. What came out of this mix literarily was a durable pattern of romance and myth and a set of images that permeate the broader national culture.

The literature that uses the "frontier" set of myths and values unconsciously—that is, in effect, expansionist by faith, though perhaps tinged from time to time with a certain helpless regret, for example over the plight of the Indians or the loss of the wilderness—such literature has been enormously popular. It may answer some deep American needs, as John G. Cawelti has argued, creating formulas of resolution for the contradictions inherent in a history of violence and aggrandizement. But there is another western literature, one that should be called "postfrontier," whose stance toward the frontier and the frontier ethos is conscious, reflective, and analytic. In general, this writing is not characterized by expansionist sentiments or romance but by a regardful perspective on the environment, a sympathetic view toward Indians, and a realistic bent in historical and social descriptions. A complex self-consciousness stands behind this more mature regional literature. The western literary-critical divide, indeed, is between two literatures: popular and, if the term be allowed, serious. Unfortunately for the sake of neatness, the two kinds of western writing may blend, for example in Jack Schaefer's *Shane* (1949), where the popular myths are stated with such artistry and economy that the novel has been taken very seriously indeed. But in most cases the qualitative differences between frontier and postfrontier mentalities are easy enough to see and make for a reasonably reliable critical gauge.

Early Travel Accounts

The earliest significant writings produced in the West were the journals of the Lewis and Clark expedition, with those kept by Meriwether Lewis making the most substantial claim to literary standing. Lewis frequently went off by himself during the journey, looked over the land with what appears to have been a historical eye or at least a sense of the momentousness of the expedition, and often entered personal reflections in his record. Thus he seemed to have been writing, at times, more than just a log, and several of his more essayistic entries verge on literary territory. He responded to the utter wildness of the land with writing that, as clearly as any other American document, evokes an Edenic feeling. On 25 April 1805, for example, almost three weeks into what would prove to be the least-inhabited section of the continent that the expedition would see (the area of the upper Missouri River, in present-day North Dakota and Montana), Lewis wrote: "the whole face of the country was covered with herds of Buffaloe, Elk & Antelopes; deer are also abundant, but keep themselves more concealed in the woodland. the buffaloe Elk and Antelope are so gentle that we pass near them while feeding, without appearing to excite any alarm among them; and when we attract their attention, they frequently approach us more nearly to discover what we are, and in some instances pursue us a considerable distance apparenly with that view."

Lewis's account, beyond its considerable inherent interest, has provided a benchmark against which the subsequent history of the West may be measured literarily, as for example in Archibald MacLeish's bitterly satiric poem "Empire Builders," published in 1933. MacLeish denounced the "Makers Making America," who "fattened their bonds at her breasts till the thin blood ran from them," by using the literary technique of juxtaposing vignettes of capitalist exploitation with images of the purity and abundance that Lewis and Clark had seen. That original vision remains potent to Americans, who in the earliest moments of their consciousness as a people were imprinted, as it were, by newness and pristine nature.

Subsequent nineteenth-century travelers such as the naturalist Thomas Nuttall and the writers Washington Irving and Francis Parkman recorded some of the same awed perception of the wild that had vivified Lewis's journals, but inevitably, with the passage of time, a certain fading of the blossom becomes apparent. By the era of Mark Twain, as early as 1872 and *Roughing It*, it had even become possible to satirize romantic expectations of wilderness and savagery and to cast oneself, as author-persona, in the role of an easterner constantly being disillusioned by the true ordinariness of the West. Nevertheless, there had been genuine beauty and immense, intact wilderness in the Old West, and these realities had fitted remarkably well with the romantic temper of the early nineteenth century. The West was born, literarily, as romantic territory.

Nuttall, one of the most comprehensive of scientists in the "virtuoso" period of American natural history, made three trips into the West, in 1811, 1819, and 1834, but only his *Journal of Travels into the Arkansa Territory, during the Year 1819* (1821) survives. On this adventure, the British-born student recorded his delight in the flowers of the open prairie—a landscape he described as a "magnificent garden"—made satiric comments on many of the settlers he encountered at the fringes of civilization, and dramatized himself as a rapt, dedicated lover of nature. These ingredients signal a consciously put-together literary effort—rather a well-finished one in comparison with Meriwether Lewis's. The innocent note of discovery, the sense of an Adam-like figure walking out into a pristine and beautiful world, comes through strongly in Nuttall's descriptions.

Washington Irving, recently returned from Europe, traveled for some weeks in 1832 in what is now Oklahoma and thought the "glancing rays of the sun," shining through the leaves of a grove of ancient trees along the Arkansas River, were like "the effect of sunshine among the stained windows and clustering columns of a Gothic cathedral." He went on, "Indeed there is a grandeur and solemnity in our spacious forests of the West, that awaken in me the same feeling I have experienced in those vast and venerable piles, and the sound of the wind sweeping through them, supplies occasionally the deep breathings of the organ." In passages like this, not only in the record of his own tour but also in his subsequent western histories *Astoria* (1836) and *The Adventures of Captain Bonneville, U.S.A.* (1837), Irving helped solidify the romantic image of the West. When he described the mountain men as leading a "wild, Robin Hood kind of

Mocking the naïve dreams of Americans who sought quick and easy wealth in the West, Mark Twain challenged the region's romantic image in his travel classic Roughing It *(1872).*

Roswell Morse Shurtleff (1838–1915). The Miner's Dream. *Engraving, 1872. Illustration from Mark Twain,* Roughing It *(Hartford, Conn., 1872).*

life," or a group of Osage Indians as having "fine Roman countenances" and looking in general "like so many noble bronze figures," he was, somewhat ironically, helping to codify the untamed West under the rubric of a received aesthetic convention.

Francis Parkman too was an inheritor of an eastern, hence ultimately European, sensibility. One of his main reasons for traveling westward in 1846 was to see primitive Indians, and this tourist's motive fitted well with his predominantly visual or painter's-eye aesthetic. Some Kanzas Indians, encountered in western Missouri, "made a very striking and picturesque feature in the forested landscape," as Parkman wrote in *The Oregon Trail* (1849), and a little farther west, in Kansas, "the alternation of rich green prairies and groves that stood in clusters, or lined the banks of the numerous little streams, had all the softened and polished beauty of a region that has been for centuries under the hand of man." Similarly conventional vignettes dominate the account throughout, with Parkman's predilection for describing his subject in terms of the visual aesthetic reaching a strange kind of apotheosis in a buffalo-shooting incident near the end of the tour. Coming upon a large group of bison bulls lolling on a dusty patch of ground, he watched them for some time. He described their scruffy appearance and their seemingly pointless rolling in the deep dust at length and then suddenly concluded his account. "'You are too ugly to live,' thought I; and aiming at the ugliest, I shot three of them in succession. The rest were not at all discomposed at this; they kept on bellowing, butting, and rolling on the ground as before." Despite the distance from his subject demonstrated in this passage and perhaps as well in his overall tendency to framed views, Parkman as a historian was aware of the passing of the wild West and appeared to feel the loss personally. "Great changes are at hand in that region," he wrote. "Within a few years the traveller may pass in tolerable security through [the Indians'] country. Its danger and its charm will have disappeared altogether." In prefaces to later editions of *The Oregon Trail* in 1872 and 1892, in tones unmistakably tragic, Parkman treated the West as only a memory, a "withered" and "subdued" region.

It is easy to criticize the hyperbole inherent in the picturesque art of an Irving or a Parkman, but it is also undeniable that the West, at one time, truly was ecologically intact—that is, to European and eastern-seaboard eyes, it was wild. The journals of Warren Ferris and Osborne Russell, two mountain men of the 1820s and 1830s, describe enough solitude, rugged scenery, clear air, and high adventure to justify as romantic an attitude as might be wished. Ferris's rendering of Cache Valley, Utah, written in July 1830 and found in *Life in the Rocky Mountains* (1842), is apropos: "In this country, the nights are cold at any season, and the climate perhaps more healthy than that of any other part of the globe. The atmosphere is delightful, and so pure and clear, that a person of good sight has been known to distinguish an Indian from a white man, at a distance of more than a mile, and herds of buffalo may be recognized by the aid of a good glass, at even fifteen to eighteen miles."

Likewise, Osborne Russell, who kept a diary for nine years of Rocky Mountain adventure, recorded pristine vistas and, on many occasions, a paradise-like abundance of wildlife. His *Journal of a Trapper* (1914) also evinces his own appreciation for wilderness—he often climbed mountains just to see specific, beautiful views. He wrote of the Lamar River valley of Yellowstone, "I almost wished I could spend the remainder of my days in a place like this." The records of Ferris and Russell, along with those of other mountain-man diarists such as James Clyman and Jedediah Smith, amply attest to the West's base of authentic wildness.

But the glory did not last long. Signally, Parkman's first retrospective preface was published in 1872, the same year as *Roughing It*, Mark Twain's classic. Twain showed himself confronting real Goshute Indians (as opposed to Cooperian noble savages, perhaps), being duped into buying a "Mexican plug," and naïvely, ill-preparedly, rushing off to the latest mining strike, among other instructive episodes. *Roughing It* was an early sign that the romantic western mythos would not, henceforth, reign unchallenged.

The Romance of the Frontier

Over the last five decades of the nineteenth century, the overpowering realities of American industrial and population expansion removed most of the West's wildness. But so potent had been the original impression, and so needful of the myths of open land, freedom, individuality, and progress were the denizens of the new, urban-industrial America, that a resolutely frontier-minded body of popular literature began to flourish as early as 1860 and continued to hold sway in the mass imagination for many decades, in spite of—or perhaps because of—the actualities of history.

With the appearance of mass-market "dime novels" in the 1860s, the popular Western began to develop as a distinctive genre. These early paperbacks sold amazingly well almost from the start—one scholar has noted that after just four years in the publishing business, the House of Beadle and Adams had five million of its little books in circulation. Very soon, the dime-novel Western showed certain consistent, indeed programmatic, elements: a hero who represented a synthesis of civilization and wildness; an affirmative finding with regard to progress; an emphasis on action; and a setting of epical import—usually vast, wild, open spaces. Emphasis was laid on the utter self-reliance and individuality of the hero, whose natural nobility led him to do the right thing unerringly. By his actions in the plot, the typical hero supported civilization, dramatizing a faith in progress (this despite any and all contemporary evidence of corruption, uncertainty of economic opportunity, unfairness in distribution of wealth, or environmental degradation), thus lending overall unreality to the developing formula. As the scholar Daryl Jones has noted, "The dime novel operated at the level of fantasy, where conflicts irresolvable in the real world could find swift and clear-cut solutions." When the writer Edward Judson, known as "Ned Buntline," discovered William F. Cody in Nebraska in 1869 and later that year glamorized him as "Buffalo Bill, the King of the Border Men," a story published serially in the *New York Weekly*, the western dime novel had found its most theatrical and perhaps most influential icon. Prentiss Ingraham, Buffalo Bill's press agent, followed Buntline with no less than 121 "Buffalo Bill" novels, nine of them written in 1892 alone. The flood of titles, in synergy with the enormous popularity of Buffalo Bill's own "Wild West Show," helped to solidify the West as a pageant-like realm of adventure in the popular understanding and the Western itself as a formulaic or "automatized" text serving as a key to that never-never land. The characterization of the hero as a knight on horseback, the reliable, moralistic resolution of plot, the extraordinary emphasis on action, and the repetitive and sentimental description of landscape all combined by the end of the nineteenth century to create a dominant literary identity for the West. The rather astonishing proliferation of titles, all representing variations on a few central themes, helped to identify the Western as subliterary, with the result that modern criticism's interest in the genre has been mainly sociological or psycho-historical.

The cowboy, who had arrived on the dime-novel scene in the 1870s, achieved finished and potent description in Owen Wister's 1902 book *The Virginian*. This novel, which went through sixteen printings in its first year and remains in print today, drew all the elements of the mythic West together into an artistic whole, which in turn became definitive for the Westerns of the new century. Its author had had a profoundly rejuvenative experience in Wyoming in 1885, after suffering from nervous exhaustion in his home city of Philadelphia. He had seen beautiful country, had renewed himself through physical exercise, and had been awed by the casual, rugged cowboys on the ranch where he stayed. Over the following several summers, on western excursions, Wister apparently developed a moral geography in which the East represented a certain

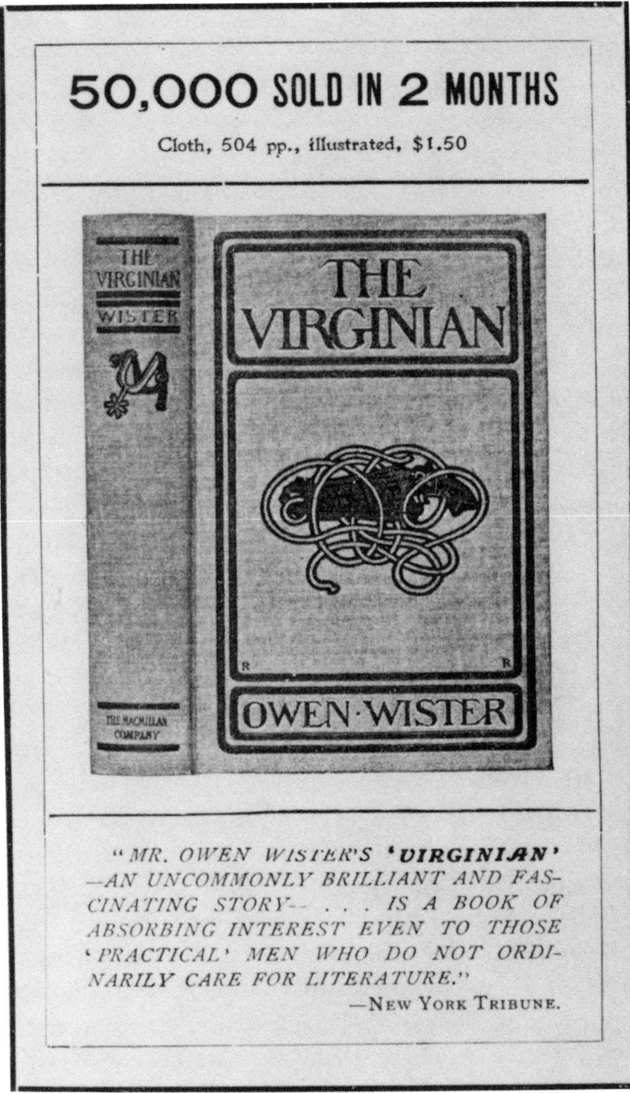

50,000 SOLD IN 2 MONTHS
Cloth, 504 pp., illustrated, $1.50
THE VIRGINIAN
WISTER
THE MACMILLAN COMPANY
THE VIRGINIAN
OWEN·WISTER
"MR. OWEN WISTER'S 'VIRGINIAN' —AN UNCOMMONLY BRILLIANT AND FASCINATING STORY— . . . IS A BOOK OF ABSORBING INTEREST EVEN TO THOSE 'PRACTICAL' MEN WHO DO NOT ORDINARILY CARE FOR LITERATURE."
—NEW YORK TRIBUNE.

An enormous success, appealing to both easterners and westerners, Owen Wister's The Virginian *sold 50,000 copies in just two months. Dedicated to Theodore Roosevelt, the book helped define the character of the romantic, manly cowboy hero who would come to dominate 20th-century Westerns.*

Publisher's Advertising Brochure for *The Virginian (New York: Macmillan Publishing Company, 1902). Owen Wister Papers, Manuscript Division, Library of Congress, Washington, D.C.*

effete and propriety-regarding mentality—and also intellectual refinement—whereas the West embodied health and decisive individualism. He himself seemed to yearn toward the perceived freedom of the West but at the same time enjoyed his eastern position as a member of the elite professional class, a lawyer who had been graduated from Harvard and who numbered among his many influential friends no less a personage than Theodore Roosevelt. This personal duality can be said to mirror the elements of the historical American dialectic of wilderness and civilization, and when Wister found an artistically satisfactory synthesis of it in the character of the Virginian, his formulation proved immediately and durably agreeable to the American public.

The Virginian, whose mythic status is well symbolized by his having no given name in the novel, is nominally a cowboy but is seldom pictured doing cowboy work. The main emphasis of the book is on his sheer attractiveness as "a horseman of the plains" (the novel's subtitle) and on the testing of his already-beautiful character. He is depicted from the narrator's eastern-tourist perspective—that is, from a suitably hero-making "camera angle"—in the book's opening paragraph: "Then for the first time I noticed a man who sat on the high gate of the corral, looking on. For he now climbed down with the undulations of a tiger, smooth and easy, as if his muscles flowed beneath his skin." The Virginian, for it is he, then proceeds to rope a horse that none of the other cowboys have been able to capture. "The others had all visibly whirled the rope, some of them even shoulder high. I did not see his arm lift or move. He appeared to hold the rope down low, by his leg. But like a sudden snake I saw the noose go out its length and fall true; and the thing was done." The narrator continues to regard the Virginian with wonder and awe and quite early in the novel records his impression that the hero is something more than an uncouth worker: "Here in flesh and blood was a truth which I had long believed in words, but never met before. The creature we call a *gentleman* lies deep in the hearts of thousands that are born without chance to master the outward graces of the type." This intimation of East-West synthesis is borne out in the subsequent plot, as the Virginian courts and wins a young, cultured schoolteacher who has come to Wyoming from Vermont; during the courtship he demonstrates a remarkable ability to read and understand European literature, always bringing to his criticism his innate and practical good sense. Tests of his character come when he has to deal with his former best friend's descent into cattle rustling and with the villainous opposition of yet another rustler, Trampas. The Virginian resolves both of these problems with acts of violence that are presented as absolutely righteous and indeed as marks of the hero's integrity. In the end, he marries the schoolteacher and is last seen as a holder of important coal-bearing lands, a man "with a strong grip on many various enterprises." The hero of the West is seen, in the end, to be a hero of America, bringing his vitality, beauty, and natural morality to bear in the universal movement of progress.

The western romance has had scores if not hundreds of practitioners after Wister, but in general these popular writers' attitudes toward their heroes have resembled Wister's with conspicuous consistency. Zane Grey, whose *Riders of the Purple Sage* (1912) rivals *The Virginian* for popularity among classic Westerns, apparently held social-Darwinist ideas that fitted well with a depiction of the hero as strong, manly, capable of violence when necessary, and above all successful. Like Wister, Grey had grown up in well-off circumstances in the East, had been frustrated and bored in his career there, and had been made an enthusiast of the West by a restorative trip, his to the Grand Canyon in 1907. In his fiction set in the West, the region itself—specifically, spacious and dramatic wilderness country—played a redemptive role. Gary Topping, one of Grey's most astute critics, wrote, "The basic Zane Grey plot is a drama in which a jaded, disillusioned, and perhaps physically frail or ill member of eastern society comes west to find a complete reorientation of values." In *Riders of the Purple Sage*, the person whom the wilderness makes new and strong is Bern Venters, who had journeyed originally from Ohio to southern Utah but had not yet developed independence. When he leaves his protectress's comfortable ranch and, by necessity, spends several weeks being tested by the rugged wilderness of the canyon country, he toughens. He "had gone away a boy—he had returned a man." Throughout *Riders*, Grey pays intense attention to the landscape, charging it with immense emotional and spiritual power. The wilderness is a kind of hovering presence behind all the novel's action. Whatever may be occurring, Grey does not fail to describe the lay of the land, the specific slant and color of the sunlight, and the look of the trees or sagebrush in the area. The descriptions in the aggregate build not a setting but a wild force, a wilderness that somehow encourages the finest qualities in heroic people. *Riders* presents a cowboy-gunman hero of unerring skills, a narrative larded with many chase scenes, and a casually dichotomous moral code by which violence is readily justified—all of these familiar elements of the western romance—but perhaps its most striking characteristic is its hymn to the land. In this, his second Western, Zane Grey struck one of the major western chords, one that also reverberates far back into American intellectual and literary history.

The western themes developed in the dime novels and then brought to fuller expression and more artistic shape by Wister and Grey continued with little fundamental change in both novels and short fiction well into the late twentieth century. Street and Smith's pulp magazine *Western Story* was perhaps the principal descendant of the dime novels when they died out in the 1920s, and in the words of a leading historian of the Western, Richard Etulain, *Western Story* "did much to stylize the new popular genre." Its editors "demanded stories with predictable plots and stereotyped characters." Other "pulp" magazines devoted to the genre arose and flourished; the popular "slicks" such as the *Saturday Evening Post* and *Collier's* also featured western fiction; and writers like Zane Grey and Max Brand continued to produce western novels in abundance. As the decades passed and the public's interest in Westerns did not abate, modifications to the Western were made by Ernest Haycox, some of whose heroes actually learned and developed (rather than simply appearing and remaining as perfected creatures), and by such writers as Eugene Manlove Rhodes and Luke Short, who along with Haycox introduced greater realism of historical and social detail into their western stories. In the post–World War II era, the most popular of western writers to highlight factual detail has, unquestionably, been Louis L'Amour. L'Amour was a self-taught scholar of western history who, in more than eighty novels beginning with *Westward the Tide* in 1950, inserted numberless mini-lectures on fine points of western events, lore, and equipment and who also, in the latter part of his career, transcended the antiquarian emphasis to concentrate on a broad-gauge sense of historical movements. His ecological valuation of nature, seen first in *Hondo* (1953), together with his sensitivity to Native Americans and Hispanic Americans, marks L'Amour as a maker of modern Westerns.

If the Western indeed sanctifies the outcome of American history by a ritual drama in which "civilization and savagery" are fruitfully blended (in the person of the hero and the issue of the plot) and in which violence is vindicated, one could hardly ask for a better epitome than the immensely popular *Shane* (1949), by Jack Schaefer. *Shane* begins, "He rode into our valley in the summer of '89"—a typically well-cut sentence that raises the hero to mythic, pronominal status, sets him in western, epic space (where else can a lone rider be seen "several miles away," as is revealed two sentences later?), and identifies the significant time of the action to come—the last year before the closing of the frontier. The mysterious horseman rides on into the valley, pauses at a fork in the road, then chooses the way that leads toward a row of homesteaders' new places rather than the road to a cattle baron's "big spread." He casts his lot with Joe Starrett, one of the farmers, and joins in the humble work of establishing a farm in a new land. He puts away his gun and his serge trousers, linen shirt, and silk neckerchief and appears to be starting a new life.

But there is trouble in the valley. Fletcher the cattleman, unwilling to lose what he regards as his own free range, has his cowboys harass the homesteaders. Eventually, when Fletcher goes so far as to hire a gunman, Shane faces a stirring moral choice. The valley is far from the machinery of established law and order (we are still in the frontier era, though just barely so), and a man in Shane's position—a man with the skills and tools of violence now desperately needed by the community—can count on no one and nothing but his own resources. Following his inward, natural goodness, he chooses to support the settlers and rides into confrontation with Fletcher and his hired gun. In a stylized scene that seems engineered to highlight Shane's essential morality, he kills those who would drive off the farmers and then rides, wounded, into exile, never to be seen again. He has made the valley safe for progress.

Shane presents the ritual western movement, and the mythically important role of the hero, in perhaps the clearest, most elegant version yet. Schaefer's choice of young Bob Starrett as narrator, "a kid then, barely topping the backboard of father's old chuck-wagon," makes the worshipful description of Shane only appropriate and allows a convincing, homey view of the pioneer family. The narrative as a whole unrolls with the simplicity and inevitability of myth. Despite its being such a definitive statement, however, *Shane* was by no means the final Western. By 1980, after three decades of output, Louis L'Amour's books alone had sold one hundred million copies. Furthermore, contemporary mutations to the form, including not only a minority version of the genre but also parodies, inversions of the standard plot, and even X-rated scenes, have since the time of *Shane* given a new vitality to the "horse opera." In many ways, the Western in the late twentieth century has become *un*predictable, but it is still doggedly alive. A century after the much-cited closing of the frontier, only a brave critic indeed would predict the demise of the Western anytime soon.

Closer Views of the Western Landscape

The frontier mind leans naturally toward romance and favors conventions that are, fundamentally, simplifications of experience. At the "frontier" level of literature, ecologically complex settings are enjoyed, to be sure, but are not investigated closely and are rendered predominantly as scenery; complex moral and psychological situations become straightforward dualities that can be resolved neatly, by characters who in turn are seen as unitary and functional to the all-important movement of the plot, rather than as humanly many-sided. The transition to a postfrontier outlook—and with it to a more mature and subtle literature—involves an opening of perception as a whole, so that a relation-seeing and complex view of existence replaces a simplistic or romantically abstract attention.

Slowing down and then looking closely at one's surroundings (as opposed to being on the move, with eyes on the horizon) are crucial stages in the transition. Simply to observe a place closely is to express the opposite of the tenor of "manifest destiny." In the West, the first writer to make a close study of the land and with that study develop a postfrontier philosophy was John Muir. This well-known figure, famous for environmental effectiveness, offers in his own intellectual history something like a textbook example of the opening and flowering of the relational-complex (that is, ecological) mode of consciousness. Born in Scotland in 1838 and brought to Wisconsin in 1849,

Muir as a youth experienced many of the stern and perhaps mind-narrowing realities of American pioneering. His father ruled the Muir household with rigid, Calvinistic ideas, and the whole family participated in the hard labor of converting wilderness into farmland. Muir grew up within the frontier-minded, American mainstream. But at the University of Wisconsin he encountered two liberating streams of thought—transcendentalism and science—and within a few more years, after working and botanizing in Canada, the Midwest, and the southeastern United States, he was recording in his journal a break with his father's creed. By the time of his first summer in the Sierra Nevada in 1869, Muir's journal shows that he had become a wilderness transcendentalist who took immense delight in nature and who believed that all the creation, including the grasses, rocks, and snowflakes, was a spiritually alive manifestation of a loving, universal God. That "baptismal" summer, to use Muir's term, was the foundation of his life as a postfrontier thinker. But it was in the succeeding years, the early 1870s, that Muir really *studied* the Sierra and melded scientifically precise knowledge with his own experience of an enlarged view, finally to synthesize a totalistic vision of human life and nature.

The impetus for his closer look was a desire to disprove the standard interpretation of the formation of Yosemite Valley. The theory was that in a massive and rather sudden subsidence, in "the wreck of matter and the crush of the worlds," as the California state geologist put it, the valley floor had sunk thousands of feet. Muir responded to this theory with two elements from his own background: first, his limited knowledge of the glaciation hypothesis, gained at the University of Wisconsin, and second, his deeply felt experience of the harmony of the natural world. The slow and stately flow of glaciers, as an agent of mountain sculpture, appealed much more to Muir's sense of things than did geologic cataclysm. But to prove his intuition, he had to go into the Sierra and find evidence—"learn the alphabet" of the range, as he put it. Now began the long, ascetic, solo expeditions into the mountains for which Muir became famous; and now too, many of his journal entries deepened past immediate ecstasy, becoming complex meditations on the nature of evolution, the paradoxical relationship between individual entities and the wholeness of nature, and the progress of the mind itself to ecological awareness. He became a philosopher of the wild. His ruling metaphor, drawn quite naturally from his observations of glaciers, was *flow*. This is a concept that gives precedence not to static entities but to process; it led Muir to an emphasis on cycles and ultimately to a holistic view of nature.

> There are no harsh, hard dividing lines in nature. Glaciers blend with the snow and the snow blends with the thin invisible breath of the sky.
>
> When I look on a glacier, I see the immeasurable sunbeams pouring faithfully on the outspread oceans, and the streaming, uprising vapors entering cool mountain basins and taking their places in the divinely beautiful six-rayed daisies of snow that go sifting, glinting to their appointed places on the sky-piercing mountains, joining ray to ray, forming glaciers amid the boom and thunder of avalanches, and at last flowing serenely back to the sea.

His study of the California mountains also impelled Muir to be a writer. His findings on the glacial history of the Sierra (which have been supported by modern research) were

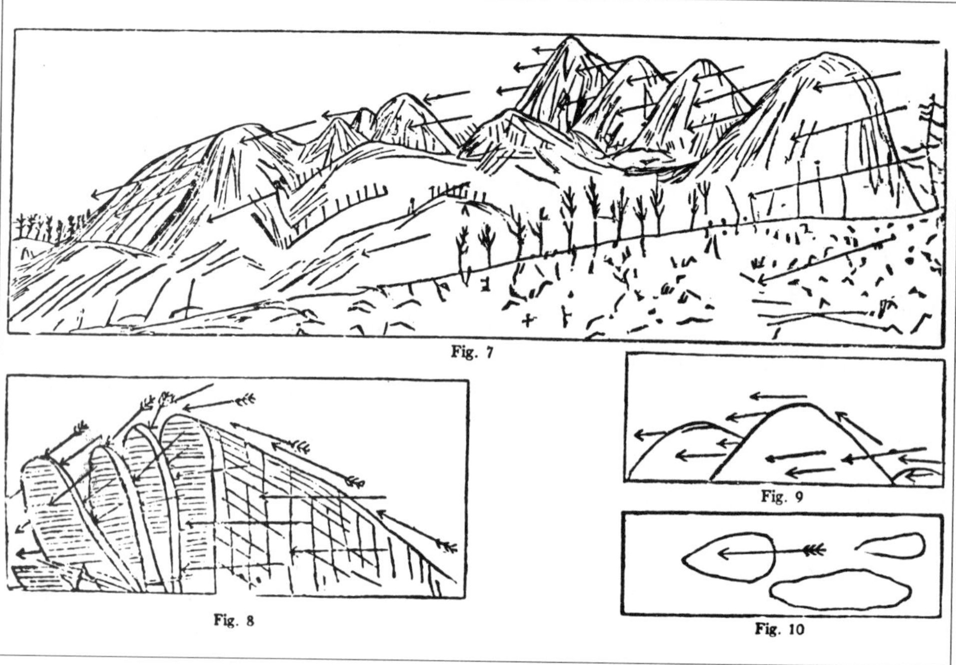

Combining two distinct approaches to the western landscape, John Muir drew like a scientist but wrote like a poet. His sketches of the Sierra Nevada documented the effects of glaciation, but his texts stressed the transcendental beauty of his subject. He wrote that Starr King, the rounded mountain shown here at the top right, "was one of the first to emerge from the glacial sea, and ere its newborn brightness was marred by storms, dispersed light like a crystal island over the snowy expanse in which it stood alone."

John Muir (1838–1914). Starr King Group of Domes, Sierras. *Pen and ink sketches, 1874. From William Colby,* John Muir's Studies in the Sierra *(1960).*

published in seven articles in the *Overland Monthly* in 1874–75 and helped to establish him as an authority on natural history. Over the following decades until his death in 1914, Muir elaborated on his studies and experiences of nature in a series of influential magazine articles and books, including *The Mountains of California* (1894), *Our National Parks* (1901), and *The Yosemite* (1912). By narrating his own revelatory experiences in the wild, exhorting his readers to recapture a similar range of experience, and thundering against short-sighted exploiters of nature, Muir helped to legitimatize a nonconsumptive, conservation-minded, postfrontier view of the West. His articles on Yosemite had almost immediate practical issue in the legislation establishing Yosemite National Park in 1890. Among other contributions to the "conservation" dimension of the Progressive Era, he influenced Theodore Roosevelt on forest and park policy and stood as a godfather to such important legislation as the Lacey Antiquities Act of 1906 and the enabling act for the National Park Service in 1916.

John Charles Van Dyke had nothing like Muir's impact on American history, but his minutely detailed aesthetic examinations of western landscapes, together with his forthright questioning of certain frontier-era assumptions, mark him as a writer of the mature phase in western literature. A librarian and art historian associated with Rutgers College in New Jersey, and already the author of influential texts on art appreciation, Van Dyke embarked on a remarkable California desert expedition in 1898, knowing little of the terrain he was about to enter and accompanied only by a horse and a small dog. Despite both chronic and acute health problems, he survived the several-month trek and in succeeding years made similar, impressive journeys in other western deserts and wild places. His precise itinerary, however, is always difficult to ascertain, because Van Dyke, of all wilderness travelers, tells the least about himself and his trips. In books like *The Desert* (1901), *The Mountain* (1916), and *The Grand Canyon of the Colorado*

(1920), he concentrates totally on the scene before him, on its line, form, color, and changing image as the sun proceeds through its great arc overhead. As the scholar Peter Wild has pointed out, there are no route descriptions, no extensively described ordeals, and rather surprisingly no other people mentioned in *The Desert*; one might add that there is vanishingly little of the author himself. What this traveler was seeking, apparently, was pure perception, for that is what his books deliver. "He saw his task as one of refining personal experience into an almost disembodied esthetic," Wild has written, and this aim has instructive, postfrontier literary dimensions. Van Dyke wanted nothing tangible from the desert—or the mountains, or the Grand Canyon later. These places were not "resources" to him. A second sign of his having transcended the frontier mentality is that in so casually effacing the intrepid-trekker aspect, Van Dyke demonstrated primary regard for the land itself. Muir apparently went through a similar annulment of the heroic self in his first summer in the Sierra, writing in his journal, "These blessed mountains are so compactly filled with God's beauty, no petty personal hope or experience has room to be." Past a certain crucial, inward divide, the focus turns outward, and both Muir and Van Dyke dramatize the liberation of energy inherent in this particular transcendence and the very close and precise seeing that seems to be concomitant. "In a few weeks we are studying bushes, bowlders, stones, sand-drifts," Van Dyke writes in *The Desert*, "—things we never thought of looking at in any other country." Years later, in *The Grand Canyon of the Colorado*, he restated the needed emancipation: "And we, if we would understand the Canyon, must largely eliminate the human element of it. It is insignificant."

The Desert was the first extended literary appreciation of the American arid lands. It described the desert as "the simplest in form and the finest in color" and "by all odds the most beautiful" of landscapes and addressed its finely detailed, indeed exhaustive, descriptions to a higher and finer sensibility than one that would simply *use* deserts. "The deserts should never be reclaimed. They are the breathing-spaces of the west and should be preserved forever," Van Dyke wrote, making here an attempt at the practical by promoting the benefits of dry and warm air. In the main, however, his moral point of view on the land appears to be founded in the experience of pure seeing. The unrivaled depth and precision of detail in Van Dyke's descriptions, and the seemingly tireless recall of fine points of light and shading—literary characteristics that might signal excess to the modern reader—may in truth be this writer's most revealing signposts to himself and where he had been. Without being overtly self-concerned, he was dramatizing what a freed mind might see.

Mary Austin's work, like that of Muir and Van Dyke, demonstrates the instructiveness of nature and landscape studies in any tracing of the transition to postfrontier western literature. She made the same fundamental move in consciousness whereby nature becomes the primary fact. "Not the law, but the land sets the limit" was the third sentence in her first book, and she reinforced this prescription later in that text with such ecologically minded declarations as "The manner of the country makes the usage of the life there, and the land will not be lived in except in its own fashion." Indeed, the whole of *The Land of Little Rain* (1903) is, as the title suggests, knit tightly around the accepted and basic reality of dryness. As an organizing principle for a book about southeastern California, aridity may seem obvious enough, but Austin's particular genius was to see, and record in process, the innumerable tiny adjustments of plants and animals that, taken together, make up arid biological communities. Her mind was alive to the relational aspect of existence, to detail, and to the patterns that connect. Thus her first book was an ecological text as much in its way of seeing as in its presentation of scientific fact.

Austin's responsiveness to the living and communal quality of nature may have had its origin in a powerful childhood experience. As a six-year-old in Illinois, she experienced a level of perception and consciousness that quite transcended the ordinary mind, and this, she wrote, became "the one abiding reality" of her life.

> It was a summer morning, and the child I was had walked down through the orchard and come out on the brow of a sloping hill where there were grass and a wind blowing and one tall tree reaching into infinite immensities of blueness. Quite suddenly after a moment of quietness there, earth and sky and windblown grass and the child in the midst of them came alive together with a pulsing light of consciousness. There was a wild foxglove at the child's feet and a bee dozing about it, and to this day [1931] I can recall the swift inclusive awareness of each for the whole.

Austin considered the level of being thus revealed to be her true or inner self, "I-Mary," whereas the social or historical identity—less intense, more guarded and conventional—she called "Mary-by-herself." It seems clear that over the years, her ability to recall the peculiar vividness of I-Mary's perception resulted in both the ecological sensitivity and emotional power of her best writing. Her challenge as an author was to create realistic and accessible literary vehicles for her transcendent sense, and she served a long and apparently difficult novitiate to this vocation. Newly married in 1891, and having moved to California only three years previously, Austin was taken by her husband to the remote Owens Valley, east of the Sierra Nevada. Here, while Stafford Austin examined the possibilities of water development, Mary Austin tried to "learn the secret of the mesa life" by taking the exact opposite of the frontiersman's path. She wrote down precise notes on what was around her: creosote bushes, dry lakes, deep echoing canyons, flocks of quail at a water hole. She apprenticed herself to the local Indians and claimed later that she "learned how to write" only after learning how the Indians managed to express a balance between their individual existences and the oneness of nature. The Indians, in their art and in the perceived maturity of their personalities, were a model to the young writer, showing her that a fruitful cooperation might be made between I-Mary and Mary-by-herself. Thus, where John Muir had taken small notice of Native Americans, and John Van Dyke perhaps even less, Mary Austin raised them to the level of tutelary spirits. "Psychologically the state called primitive," she wrote in 1923, "is one of deeply imbricated complexity."

After the publication of her first book, Austin was able to move out into the broader literary world, first to the writers' and artists' colony at Carmel, later to New York, and then to a good deal of travel in Europe. She left the West behind, but apparently, after several years of success among the literati, living in urban situations, she began to long for the missing dimension. "But this life of literary antics which I am leading isn't my real life, and this shell of hardness is only a shell—so far," she once wrote to a friend in

Arizona. In 1923, she made a long motor tour of New Mexico and Arizona, a journey that became (as perhaps she had planned) a return to sources. Once again, she noted the look of piñon-juniper woodland, the dry, exhilarating air, and the long, inspiring vistas of the sort of terrain in which she had learned her craft. In the book that came out of this trip, she spoke conclusively, as if final truths had come to her. She called *The Land of Journeys' Ending* (1924) "a book of prophecy;" its final prediction, among many somewhat oracular statements, is that the western environment itself, operating "subtly below all other types of adjustive experience," is working to create a new, land-harmonious culture. This will be "the *next* great and fructifying world culture." Rarely has the West been celebrated so decisively, but for Austin, a fructifying culture would not be based on exploiting resources or heightening the standard of living. She believed quite the opposite: that adjustment to ecological reality, and attention to the momentous potential in any experience of nature, constituted the true human hope.

Mature-phase Fiction of the West

Generally speaking, the maturation of western fiction is marked by increasing subtlety and complexity of characterization and a diminishment in the weight of plot and ideology. In short, to use the terms of Nathaniel Hawthorne and the critic Richard Chase, the romance gives way to the novel. In the modern western novel, ideas are presented, to be sure, but in less obvious fashion than in a romance. Ideas and themes emerge as the result of the interactions of more or less realistic characters, and thus the ideas have more sides to them, more dimension, and more loose ends. Where issues of justice, retribution, and violence, for example, are handled swiftly and rightfully through the simple goodness of a romance hero like the Virginian, in a mature-phase novel such as Walter Van Tilburg Clark's *The Ox-Bow Incident* (1940), these issues become intensely problematic. Human frailties, the arrogance of power, and the weight of peer pressure all help drive the action. In this novel, a terrible wrong is done, and there are no heroes of the old style. But it is not the lack of a happy ending that makes this book a novel; it is the writer's close attention to humanity.

In addition, postfrontier western fiction writers tend toward an analytic and critical view of American expansion into the West. They do not stand unconsciously inside the history of progress but ask what were, and are, the costs of its march. The West in modern hands, beginning approximately with Hamlin Garland's *Main-Travelled Roads* (1891), is for example not a promised land for upwardly mobile yeomen—the ungiving realities of capitalism need to be reckoned with. These realities, as in Garland's story "Under the Lion's Paw," can and do break good people down. Native Americans, for another example of the greater historical realism of postfrontier fiction, tend not to play stereotyped roles and are not simpleminded; cowboys are working men rather than knights; and the western landscape, though still very much able to arouse a transcendental response, tends to be treated in a less impressionistic and sentimental fashion. All of these improvements in realism are incorporated, in fiction, in the vehicle and test of characterization.

Some of the first moves away from the romantic image of the West were made in the decade immediately following the Civil War, when the growing city of San Francisco became established as a literary center. San Francisco's prominence came about largely through the writings of Bret Harte, Mark Twain, and others of a literary circle that would give birth to the "local color" movement in American fiction. Local color, in turn, became an important precursor to the development of realism as the dominant literary mode later in the century. In the middle 1860s, both Harte and Twain wrote numerous satirical pieces for San Francisco newspapers, emphasizing local peculiarities with great glee. Harte later published a number of rather sentimental, parable-like short stories of California life such as "The Luck of Roaring Camp" and "Tennessee's Partner," but he also edited the important San Francisco literary magazine *Overland Monthly* in its formative years from 1868 through 1871. *Overland* set its sights high, aiming to be the West's *Atlantic*, and published a number of quality essays and astute book reviews over the years, an accomplishment that moved California local color writing from sophomoric satire to more serious and interpretive levels. Twain, for his part, owed much to the sharp eye and ear he developed in his California (and Nevada) days as a journalist; his debunking style of realism and occasional taste for the grotesque, which in his later career were brought to bear on large philosophical issues, may be seen in embryonic form in the early story "The Celebrated Jumping Frog of Calaveras County," a reworking of a California tale he published in 1865.

Though not generally regarded as a western writer, Stephen Crane may be credited with an early and important contribution to the modernizing of western materials. "The Bride Comes to Yellow Sky" (1897), perhaps the most frequently anthologized of Crane's several western stories, exhibits a precisely realistic view of the West and draws on the old, romantic images strictly for comic and ironic effects. In the story, a drunken would-be gunfighter, wanting to "shoot up a town," instead is brought to a state of impotent irrelevance by the mere fact that the town marshal has become a married man. The chief figures, gunfighter and marshal, are of course stock characters of romance, but Crane penetrates the stereotypes to reveal two vulnerable human beings caught in historical transition. His awareness of western myths and the conventions of romance *as* myths and conventions allows him ironic insights, but the ironic view is not so pervasive as to hinder him from portraying authentic courage. The success of Crane's western stories may stem, in part, from the author's insight into his own mixed, ambiguous response to the West—attracted to the mythology and subvertingly conscious of the attraction.

Frank Norris and Jack London, with major publications spanning the period from 1899 to 1916, also made early advances in the scope, realism, and psychological depth of western fiction. Both flirted with high-definition romance at times in their careers—and London's success at swashbuckling adventure has had the ironic effect of obscuring his genuine penetration in several novels and stories—but both conceived of themselves as serious writers, and both attempted broad-gauge views of humanity and of important social issues. Norris, whose background was in the urban, upper middle class (Chicago, San Francisco), was the first western writer to use a city—in his case, San Francisco—as the setting in the action of a serious novel, *McTeague* (1899). This in itself was a major breakthrough. In *The Octopus* (1901), his best-known novel, Norris is concerned with what seems to be a morally dichotomous situation, in which honorable ranchers confront the "octopus" of railroad-based capitalism; but he does not portray the capitalist Cedarquist, or the frustrated mystic Vanamee, or the poet Presley, or the self-absorbed rancher Annixter (who is transformed by love) as one-dimensional characters. They are all quite convincing. Norris died tragically young, in the midst of ambitious novelistic plans; but in the work he did complete he helped move western fiction into the new area of urban life, and he demonstrated a sophisticated handling of powerful, sentiment-arousing materials.

Jack London's contribution to western fiction has probably been underrecognized. Books such as *The Call of the Wild* (1903) and *White Fang* (1906), together with his own celebrity as an adventurer, helped give him an image as a pop-litterateur, but in truth what drove London's fiction as much as animal zest and dramatic event was the author's lifelong intellectual curiosity. From his early days of borrowing armloads of books from the Oakland Free Public Library to his declaredly mind-opening reading of C. G. Jung in the last year of his life, London was a passionate student. Growing up poor, a working boy from the age of nine, he treated ideas as keys to a clarified, liberated realm. Thus developed his active interest in socialism, and his enlightened criticism of egoism and social-Darwinist success, seen for example in *The Sea-Wolf* (1904) and *The Iron Heel* (1907); thus his study of scientific farming, which bore fruit literally at his Beauty Ranch in California and literarily in *The Valley of the Moon* (1913); and thus, finally, his excited discovery and almost immediately successful literary use of Jungian psychology, seen for example in the late story "The Water Baby."

London himself exhibited many of the traditional, romantic western qualities, and he certainly gave expression to the romance of the frontier in several of his stories of the North. But he was not afraid to look within himself analytically. His outlook seems to have included elements of both rational skepticism and mystical confidence; in his most interesting fiction, these dimensions balance and inform each other. For example, two of the novels based on his Sonoma County farming experience—*The Valley of the Moon* (1913) and *The Little Lady of the Big House* (1916)—show London's intellectually critical understanding of the pastoral (read western, frontier-minded, romantic-escapist) urge and, just as firmly, his understanding of the danger of too much science, too much rational ordering of life. Saxon and Billy Roberts of *The Valley of the Moon*, who have escaped the urban distress of Oakland, find that they will need scientific knowledge and technique to make their newfound rural Eden a true success; and Dick Forrest, a hugely prosperous scientific farmer in *The Little Lady*, becomes so efficient that with ironic unconsciousness he destroys life and love all around him. Taken together, these novels demonstrate London's capacity for self-study—it should be noted that he himself began his Beauty Ranch in a mood of retreat from the modern, urban scene—and his artistically balanced handling in fiction of one of the strongest, most central of western themes, the escape to paradise.

A similar artistic equilibrium, giving equal voice both to the western-agrarian urge and to a sober recognition of its incompleteness as a philosophy for a whole life, distinguishes Willa Cather's breakthrough novel *O Pioneers!* (1913). In this novel, a strong woman of the second generation of Nebraska pioneers, Alexandra Bergson, sets her face toward success and achieves it, creating a productive farm on high ground that

Willa Cather's trip to Mesa Verde in 1915 inspired her book The Professor's House *(1925). Like many of her contemporaries, she found in the Southwest powerful reminders of America's rich historical legacy. The ruins of Mesa Verde, her hero proclaimed, "belonged to this country, to the state and to all the people."*

Unidentified photographer. Willa Cather at Cliff Palace. *Photograph, 1915. Private collection.*

had seemed to her father's generation only a mysterious, wild land with a mind of its own. In a profoundly transcendental scene early in the novel, as she and her younger brother drive a horse-drawn wagon into the country of her choice, Alexandra opens her heart to the land, wordlessly declaring complete allegiance.

> When the road began to climb the first long swells of the Divide, Alexandra hummed an old Swedish hymn, and Emil wondered why his sister looked so happy. Her face was so radiant that he felt shy about asking her. For the first time, perhaps, since that land emerged from the waters of geologic ages, a human face was set toward it with love and yearning. It seemed beautiful to her, rich and strong and glorious. Her eyes drank in the breadth of it, until her tears blinded her. Then the Genius of the Divide, the great, free spirit which breathes across it, must have bent lower than it ever bent to a human will before. The history of every country begins in the heart of a man or a woman.

Alexandra is depicted as a founder, and the secret of her success is precisely the pure, uncomplicated love she has for the country. In the great pastoral-romantic tradition, she has merged herself with nature. But as the novel develops, it becomes apparent that life in its entirety cannot be solved as neatly as some of the agricultural problems of early settlement. Alexandra raises her younger brother to enjoy advantages she herself never had, including education and travel, and she entertains a simplistic, accommodating regard for him. This is in keeping with her own directness and essential simplicity. "Her mind was slow, truthful, steadfast. She had not the least spark of cleverness." In a crucial chapter Cather described Alexandra's rather poignant repression of her own sexuality—a habit and characteristic that may have been necessary in her hard life as a builder but that keep her from an important recognition about Emil. Emil is in love with the young wife of a neighboring farmer, but Alexandra's chaste mind does not realize this. "That she [the young wife, Marie] was beautiful, impulsive, barely two years older than Emil, these facts had had no weight with Alexandra." In fact, she had innocently "omitted no opportunity of throwing Marie and Emil together." In a climactic scene, the jealous husband kills the lovers, shocking the pastoral community. What is left for the remaining characters is a rueful, tempered existence. The novel's greatness derives in large part from Cather's sympathetic but discerning dramatization of the minds and personalities of her chief characters. She did not write an essay on the limitations of the agrarian way, or the sacrifices involved in pioneering, but set some real-seeming people, products of a time and place, to interacting.

Cather herself took many years to come to terms with the Nebraska landscape of her later childhood and youth, so that with enough distance from it, but still remembering the power of the land, she could describe the place and its people with the complexity and insight of refined art. Born in Virginia in 1873, she had been brought to the prairie at age nine and apparently experienced great difficulty adjusting. After her college years in Lincoln at the University of Nebraska, she moved eastward, mainly working in journalism, and by 1909 had become managing editor of *McClure's*, one of the leading magazines of the time. In general outline, her career to this point shared in the "revolt from the village" of many nineteenth-century intellectuals; but a trip to the Southwest in 1912 appears to have renewed her sensitivity to wild country and to have reawakened her interest in the pioneer generations. Reflecting on the cliff dwellings of ancient Indians seems to have strengthened Cather's concern with the moral dimensions of living in any certain landscape—indeed with the moral and spiritual dimensions of establishing and maintaining civilizations in general. This moral approach to life and culture is the territory of her mature fiction, in large part, and certainly of her most broadly social western works: *O Pioneers!* (1913), *My Antonia* (1918), *The Professor's House* (1925), and *Death Comes for the Archbishop* (1927).

The Professor's House amply represents Cather's cultural and historical vision—in particular, her acute judgments on modern-day materialism—and also her poetic sense of the redemptive, creative contact with nature that may give an individual, and perhaps even a culture, a basis for right living. The house of the title is an older, undistinguished dwelling in which the professor, Godfrey St. Peter, has lived for many years, writing a prize-winning history of the Spanish explorations and conquests in North America and also helping to raise his and his wife's two daughters, now grown and married. In the first and longest of the novel's three sections, we learn that St. Peter does not want to leave the old house, though with his prize money a new one has been built; that his wife and daughters exhibit varying degrees of consumerist superficiality; that he once had a student, Tom Outland, who was killed in World War I and who has come to represent to St. Peter a fine, upright standard and mentality that his present circumstances seem distressingly to lack; and finally, that he himself, pulled between the remembered world of quality and earnestness on the one hand and the contemporary scene of greed, envy, and trivia on the other, is approaching a spiritual crisis.

When Mrs. St. Peter goes to Europe for a summer, the professor settles back into his study in the old house, planning to edit a diary that Outland had kept while exploring ancient cliff dwellings in New Mexico. Here the novel becomes, in the second section, "Tom Outland's Story"—and also becomes one of the most elemental and stirring evocations of place in all of western literature. Tom had been working as a cowboy and had discovered one day "a little city of stone, asleep," high up on Blue Mesa. During the course of his and his partner's amateur archaeology, and after a falling-out between the two, which begins a period of rich, renewing solitude on the mountain for Tom, it becomes apparent that Blue Mesa is a spiritually cogent place. It is, Cather wrote, "a world above the world." When Father Duchene, a country priest, tells Tom and his partner, Roddy, that "there is evidence on every hand that they [the ancient cliff dwellers] lived for something more than food and shelter," Cather's moral-geographical diagram becomes more sharply apparent. Blue Mesa is a benchmark by which the terrible fall into modern materialism can be decisively measured.

But the novel's treatment of this perhaps familiar, transcendental, and western theme is not programmatic. The piñon-scented air that Tom breathes is not abstractly presented, nor is the slanting sunlight at evening, nor the grasshoppers leaping against the door of the cowboys' line cabin. The landscape is felt and seen, never generalized. And as the "dig" progresses, we learn that the cliff city had its own human imperfections: the partners discover a probable murder victim and take to calling her "Mother Eve"—a good hint of Cather's knowing and complex view of humanity. Furthermore, at the modern end of the diagram, St. Peter himself, though profoundly affected by Tom and thus by Blue Mesa, is not above certain self-pleasuring foibles: he keeps an artificial little garden (quite different from the wild, infinity-suggesting Blue Mesa), enjoys dressing and eating well, and seems withal to be happy enough in his social position as an intellectual and a professor.

In the final section of *The Professor's House*, "The Professor," the novel's two worlds come to a climactic encounter within St. Peter. He is depicted, early in "The Professor," as regaining a certain primary, organic relationship with the earth. "He found he could lie on his sand-spit by the lake for hours and watch the seven motionless pines drink up the sun." "He was only interested in earth and woods and water." He has contacted the "Kansas boy," as the novel calls it, analogous to the western dimension of life that Outland and Blue Mesa represent; but he cannot see what this has to do with his present existence—after all, his wife will soon be returning, and there will be the new house to move into once and for all—and he sinks into a seemingly indifferent, stalled state. "He supposed he did his work; he heard no complaints from his assistants, and the students seemed interested." In the latter pages of the novel, thus, the modern relevance of Blue Mesa—that is, the West, the wild—comes into question. In St. Peter's new experience of nature, a great recovery is hinted at. Yet his spiritual recuperation, when it begins only tentatively to occur after a climactic, near-death incident, is rendered in strictly guarded prose. The novel's concluding sentence rises only to, "He thought he knew where he was, and that he could face with fortitude the *Berengaria* [his returning wife's ship] and the future."

Such restraint is perhaps the essence of Cather's modern and realistic outlook. Her style too, in its rigorous spareness, shows a highly conscious approach. Her refusal to sentimentalize or make too much of her wilderness and pioneering subjects, nevertheless granting them an undeniable, generative energy and moral reference, marks her work as mature western fiction. *The Professor's House* may indeed be, as John J. Murphy has written, "the keystone novel of her career." It is also, in both its sophisticated psychology and its carefully sculpted writing, a type-specimen of a regional literature that had come of age.

After Cather, the line of significant western fiction is continued by a remarkably productive—and in one important respect, closely related—group of writers born between 1890 and 1912: Harvey Fergusson (b. 1890); Vardis Fisher (b. 1895); A. B. Guthrie, Jr. (b. 1901); John Steinbeck (b. 1902); Frank Waters (b. 1902); Wallace Stegner (b. 1909); Walter Van Tilburg Clark (b. 1909); and Frederick Manfred (b. 1912). Among them they have covered nearly all of the thematic range of the West. But for all their immense variety when viewed in the aggregate, each has, in his most representative and consequential work, tried to cut through the peculiarly resistant western mythology and seek, in Stegner's words, "a usable continuity between past and present." These writers have attempted an illusionless point of view and a description of a historical West that plausibly connects with the mixed and real present.

To take just one example, Frank Waters presents a most interesting study on the theme of revision and realism. Although he has been at pains in several books to show western history from the Indians' point of view, and thus his work overturns many "majority" images and beliefs, and although he highly regards detail and fact, placing him again in the realists' camp, nevertheless in the larger dimensions of his worldview there are concepts most realists would regard as mystical, and perhaps mythical. He leans strongly toward an "Indian" sense that humans and nature make one nondual system and that the landscape, seen by the Lockeian eyes of Western civilization as so much matter, is in truth spiritually alive. His psychology is firmly Jungian. These elements of Waters's philosophy are brought to bear in his fiction, with the result that it tends to have prominent ideological content. One early critic, in fact, accused Waters of following a "regional imperative" of landscape-mysticism, to the detriment of novelistic quality. Critics of western fiction confront in Waters's work a radical and pointed example of the postfrontier outlook. The reader of a Frank Waters novel is asked to think beyond the dualistic world, beyond the "white man's" inheritance, psychologically. This requires, among other changes, a revised concept of self and personality—less heroic, less firmly defined and bordered—and in turn, perhaps, a different criterion for characterization. *Shane*'s opening words, "He rode into our valley," with their emphatic statement of individual action, are perhaps accepted without thought because they are part of our traditional frame of reference. Waters's attitude toward character is more relational and "ecological": of his character Martiniano in *The Man Who Killed the Deer* (1942), a person undergoing an emergence into a nonwhite philosophy of life, Waters wrote, "So little by little the richness and the wonder and the mystery of life stole in upon him." The entire process of life is seen differently by Waters: its large and beautiful impersonality is doing the acting, not Martiniano.

The strength of Waters's ideas and assumptions—clearly visible in just this one microcosmic example—may tend in some of his work to cast a fable- or parable-like aura. This is probably so in *People of the Valley* (1941) and *The Man Who Killed the Deer*. But he has also written novels distinguished by an exact and concrete particularization. *The Yogi of Cockroach Court* (1948), an unusual story in that it focuses on the process of meditation and the quest for enlightenment, grounds its subject in the gritty details of life among the underclass of a Mexican border town; in the end, Waters brings off a thoroughly realistic synthesis of these two realms. In one of his best novels, *The Woman at Otowi Crossing* (1966), the course of a woman's spiritual illumination is similarly, and even more successfully, set within an experienceable environment. Readers following Helen Chalmers's inward development understand and feel its progress because they are allowed to perceive her outward world in convincing itemization. The sound of the leaves of the cottonwood trees by her little house, the frosty air of a fall morning, the look of a heron standing on a sandbar of the Rio Grande—all come to the reader as intimate perceptions. In this novel, the northern New Mexico setting is more than setting; its details work in concert with a profound inner response, part and parcel of the character's growth toward a whole and enlightened view of the world. Waters's subtle depiction here gives a new dimension to the concept of realism.

It should be noted that several of the first generation of postfrontier interpreters of the West have written long autobiographical novels, suggesting again a commitment to analytic understanding. Manfred's trilogy *The Primitive* (1949), *The Brother* (1950), and *The Giant* (1951), Frank Waters's *The Wild Earth's Nobility* (1935), *Below Grass Roots* (1937), and *The Dust within the Rock* (1940), subsequently reshaped into *Pike's Peak* (1971), and Vardis Fisher's tetralogy *In Tragic Life* (1932), *Passions Spin the Plot* (1934), *We Are Betrayed* (1935), and *No Villain Need Be* (1936), which became *Orphans in Gethsemane* (1960), are the most extensive treatments, but Wallace Stegner's *The Big Rock Candy Mountain* (1943) and Walter Van Tilburg Clark's *The City of Trembling Leaves* (1945) also deserve recognition for their ardent quest for truth.

In the succeeding generations of western novelists, one does not see quite the intensity of preoccupation with overcoming myth and establishing a true West. It is as if the revolution has been secured, and now a writer is simply free—as any writer normally should be—to write about anything he or she pleases. The almost casual eclecticism of recent and contemporary western novelists may be taken as a sign of a freed literary territory. Among the many successors to the trailbreakers, the Texan Larry McMurtry (b. 1936) may be mentioned for his artistically successful uses of traditional western materials. With seeming ease, he has dealt with the pressure of the Texas mythological inheritance—which is something like the western legacy writ large. Prolific and immensely popular, he has proved difficult to assess or even categorize: his work includes novels honoring the land and the older-generation stewards of it (*Horseman, Pass By* [1961] and *Leaving Cheyenne* [1963]), novels satirizing small-town life and novels describing urban, existential displacement and ennui (*The Last Picture*

Show [1966], *Moving On* [1970], *Cadillac Jack* [1982], and *Texasville* [1987]), and recently, largely realistic novels making use of such hoary western subjects as the trail drive (*Lonesome Dove* [1985]) and Billy the Kid (*Anything for Billy* [1988]). He has left and come back to Texas, in life and also figuratively in critical essays on the state's literary heritage, and his own pronouncements on the thinness of traditional Texas subjects may have complicated the job of assessment. *Lonesome Dove* and *Anything for Billy* came after McMurtry had called for Texas writers "to turn from the antique myths of the rural past and to seek plots and characters and literary inspiration in modern Texas's urban, industrial present." It is clear that McMurtry will not be bound by either the frontier myth or a reactive, antifrontier myth; he appears to have broken through that particular controversy and to have assumed an absolute freedom of subject matter.

Poetry in the West

Western poetry, as a genre, has not been so strongly marked by the specifically regional reappraisal of history that characterizes the West's mature-phase fiction. Poetry in general, as is often noted, tends toward universal themes rather than regionally identifiable ones. Nevertheless, a distinctive western poetic temper does exist and can be seen in the critical retrospect of western poets—their broad-gauge critique of expansionist culture's way in the world—in their willingness to describe transcendental experience of nature, and finally, in their strong allegiance to place.

These defining elements may be seen in the work of the first major poets of the West, John G. Neihardt (1881–1973) and Robinson Jeffers (1887–1962). There had been a great deal of sentimental or genteel western poetry before Neihardt and Jeffers, with occasional pre-flashes of the regional temper: passages in the work of Joaquin Miller, Charles Warren Stoddard, and Edwin Markham still command attention and indeed almost sum up the western achievement before Neihardt and Jeffers. But Neihardt, closely associated with Nebraska and in fact that state's first poet laureate (1921), determined to write epic-scale verse on western themes and did so with critical success; and Jeffers created in his work a California coast that is both a seeable region and a philosophical standpoint from which humanity may be viewed to profound effect.

Neihardt's early verse was mostly lyrical and served as an apprenticeship to his greater ambition. "He believed," his biographer Lucile F. Aly has written, "that at thirty a poet should renounce subjective poetry for work that expressed a wider view of the world." The lyrics allow us to see, however, Neihardt's intensely mystical experience of life, his receptivity to the "Otherness" that fueled all of his poetry. *A Bundle of Myrrh* (1907), *The Stranger at the Gate* (1912), and *The Poet's Town* (1908–12), collected in *Lyric and Dramatic Poems of John G. Neihardt* (1965), express Neihardt's energetic and mystic sense of identity. In "April Theology," he declared,

> O, I know in my heart, in the sun-quickened, blossoming soul of me,
> This something called self is a part, but the world is the whole of me!

His epic-style retellings of western history—*The Song of Hugh Glass* (1915), *The Song of Three Friends* (1919), *The Song of the Indian Wars* (1925), *The Song of the Messiah*

(1935), and finally *The Song of Jed Smith* (1941), all collected in *A Cycle of the West* in 1949—are infused with a broad view of humanity in the cosmos, raising nineteenth-century western materials to classic and universal levels. Neihardt combined extensive historical research (including interviews with old-timers and Indians) with his own sense of nature and destiny, to form a grand-scale vision of the West. From his depiction of the Ashley-Henry expedition of 1822 to the terrible death of the Ghost Dance movement at Wounded Knee in 1890, Neihardt's characters rise to heroic stature by demonstrating such profoundly important, eminently human qualities as physical courage, forgiveness, endurance, and spiritual vision. They are Hugh Glass the mountain man and Wovoka the Paiute Ghost Dance dreamer, but they are also, in Neihardt's presentation, mankind on the earth, in splendid adventure and in deep travail. Neihardt's estimates of his people are evenhanded, compassionate, Shakespearean.

Robinson Jeffers self-published his first work in 1912 (*Flagons and Apples*), was critically praised for *Californians* in 1916, and broke through to his mature style with *Tamar* in 1924. It is interesting to note, in connection with Jeffers's emergence, that Frank Norris's character Presley, in *The Octopus*, had dreamed of becoming the poet of the West—"where the tumultuous life ran like fire from dawn to dark, and from dark to dawn again, primitive, brutal, honest, and without fear." Norris's implicit estimate, that there indeed had been no such West Coast poet yet, seems perfectly just. It fell to Jeffers, after World War I, to write in a way that was "primitive, brutal, honest, and without fear," although in reference to the term *primitive* it must be said that Jeffers was far beyond that state in both his knowledge of classical and modern history and his mastery of poetic form. His view of humanity's course on the earth was sharply critical. In "Original Sin," writing of the very early "man-brained and man-handed ground-ape," he declared his own detached position from our species, "As for me, I would rather / Be a worm in a wild apple than a son of man." A similar stance is seen in the famous "Hurt Hawks," where Jeffers stated he would "sooner, except the penalties, kill a man than a hawk." Several poems depend for their perspective, both visually and figuratively, on the poet's taking a position above and outside the usual concerns and the self-fascination, and indeed the whole history, of humankind. Jeffers referred to his personal philosophy as "Inhumanism," and despite the negative cast of the term itself, he regarded his view as positive, a "falling in love outward," a transpersonal recognition of the beauty and order of the world, a philosophy that revealed humanity in true perspective. Jeffers's may be the most thoroughly revisionist of historical views, for he attempts to consider mankind as a whole, on an evolutionary scale of time, as simply one naturalistic element among many.

The second great dimension of western poetry, transcendental experience of nature, seems to have been a bearing-point in Jeffers's own life. His wife reported that during the building of his stone house, when the poet daily handled rough, heavy, granite boulders, "there came to him a kind of awakening such as adolescents and religious converts are said to experience." The poem "Oh, Lovely Rock" records an instance when Jeffers felt as though he "were seeing through / the flame-lit surface into the real and bodily / And living rock." His character suggestively named "The Inhumanist," in "The Double Axe," cries out to the wilderness around him, "two or three times in my life / my walls have fallen—beyond love—no room for / love— / I have been you." But unlike

The poet Robinson Jeffers found his most compelling subject matter in the coastal landscape around his Carmel, California, home. From the rocky coast that inspired such other artists as the photographers Ansel Adams and Edward Weston, he drew rich material for his philosophical investigations of humanity and nature.

Ansel Adams (1902–84). Robinson Jeffers, Carmel, California. *Gelatin silver print, ca. 1935. Copyright © 1993 by the Trustees of the Ansel Adams Publishing Rights Trust. All Rights Reserved. The Ansel Adams Publishing Rights Trust, Carmel, California.*

Neihardt, Jeffers drew from such experience no positive interpretations about the human potential. His historical vision of mankind remained mordant. What was positive in him emerged in memorable, lyric descriptions of the California coast, either detached from humanity or seen in ironic juxtaposition to our species' foibles. The latter technique provided Jeffers some of his most dramatic effects, for example in "Apology for Bad Dreams," in which the poet stands on a high hill watching the sun set over the ocean, "the fountain / And furnace of incredible light flowing up from the sunk sun," and then sets against this natural glory what he sees below him: in a "little clearing a woman / Is punishing a horse." Jeffers goes on to describe the punishment in some detail and makes the moral implication of his scene-drawing quite clear at the end of the stanza: "What said the prophet? 'I create good: and I create evil: I am the Lord.'"

There are interesting general similarities between Jeffers's work and that of Gary Snyder (b. 1930), who is perhaps—with William Stafford—most significant among contemporary western poets. Both Jeffers and Snyder take human nature and all of human history for their province and view these things naturalistically, that is, from a point outside the usual human self-advertisement, and both write bell-clear descriptions of wild nature, deriving from profound contemplation. But Snyder, who has made a forty-year study and practice of Buddhism, has projected in some of his poetry a calmer detachment than has Jeffers. Jeffers *recommended* dispassion, but his poetic tone is as often tormented as serene. Working from a Buddhistic understanding of entity-hood—that "things" are not as definite and hard-bordered as the dualistic mind perceives them to be but are, in the Buddhist term, "empty"—Snyder at times seems free of the grip of history; he can for example write of Washington, D.C., that "the center, / The center of power is nothing!" In the same vein, he has argued in prose essays that since human nature is not an entity, it is not locked into a course or a destiny, as Jeffers had held. Humanity is free to draw on the past and chart a better future: "whatever is or ever was in any other culture can be reconstructed from the unconscious, through meditation."

But Snyder's work overall is not naïvely or blandly optimistic. His ability to criticize is shown clearly in such poems as "Mother Earth: Her Whales," a scorching indictment of the modern nation-state and its casual destruction of nature, or "Front Lines," which depicts his home territory in the Sierra Nevada as a war zone where logging companies, real estate brokers, and military jets overhead reveal the "cancer" of our time. Though his view of the human potential is Buddhistic and unfettered, allowing at times a near-transcendental confidence (see "Magpie's Song" in *Turtle Island* [1974], for example), his view of the human performance is unblinkingly realistic.

Snyder's strong attachment to his California home ground, and his proposal of "re-inhabitation," or living within ecological parameters and creating a sustainable society, are unmistakable signs of a postfrontier mentality. The poems and essays in *Turtle Island* and *Axe Handles* (1983) record a settled sense of place and stewardship that is the antithesis of the mobile, horizon-scanning frontier mind.

Neihardt, Jeffers, and Snyder show the western poetic temper quite clearly but do not of course cover the whole of the region's range of verse. Among other poets of the first rank, Thomas Hornsby Ferril (1896–1988) of Colorado, with dispassionate assessments of western history in *High Passage* (1926) and *Westering* (1934), made a significant contribution toward the demythifying of the West. Theodore Roethke (1908–63), who arrived in Washington State in 1947 and at the University of Washington became mentor to an entire generation of poets, wrote beautifully of the self in nature, giving to this transcendental theme a psychologically realistic inwardness. Richard Hugo (1923–82) performed an important, Roethke-like tutorship to dozens of poets at the University of Montana, from 1964 until his death. In its own right, his verse expresses an elegiac sense of the West that seems particularly appropriate to the tone and reality of the mid and later twentieth century. William Stafford (b. 1914), whose main poetic loci are Kansas and Oregon, has written some of the most penetrating meditations on humanity and nature, and on the inner life, that have been produced in the West. In *West of Your City* (1960), *Traveling through the Dark* (1962), *Allegiances* (1970), and *A Glass Face in the Rain* (1982), to name only a few of his many books, Stafford very quietly listens to the wilderness as if he could learn something from it, takes a wryly subversive view of all institutions, and proposes in between the lines an ethical revolution in which humankind might take a more modest position in nature. All of this is quintessentially western, in the postfrontier sense.

Native American Literature in the West

Perhaps no more thoroughly revisionist standpoint could be achieved, vis-à-vis the popular image of the West, than to credit the Native American literature of the region with real worth and standing. For the speakers and writers in the two to three hundred tribal communities in the West, before invasion, there was no "great American dialectic" of civilization and savagery at all, no "frontier"—their songs and stories were those of a people entirely at home. Their historical frame of reference was different from that of the European Americans, obviously—and after invasion it became something like the obverse of the whites'—but more important, their metaphysical reference was radically different. The oral literatures of all the different tribes, as various and highly elaborated as they became in widely differing living situations over thousands of years in the region, were marked at the deepest level by a shared worldview. This was a profoundly religious sense of existence as unified. The four major themes in Native American oral literature, as outlined by the modern scholars Larry Evers and Paul Pavich, are the sense of the sacred, the sense of the beautiful, the particular importance of place, and the centrality of community; all of these interweave in a coherent, holistic perception of the world. The "Great Spirit," variously and often inaccurately defined by European Americans, was the unifying reality behind Native American literature and has recently been described by two Native American scholars, Thomas E. Sanders and Walter W. Peek: "Wah'kon-tah is the sum total of all things, the collective totality that always was—without beginning, without end. Neither a force nor a spirit, it is the inexplicable sharing-togetherness that makes all things, animate and inanimate, of equal value, equal importance, and equal consequence because they are all Wah'kon-tah simultaneously, their forms collectively creating the form of Wah'kon-tah which is, obviously, incapable of being anthropomorphized."

Songs and narratives coming forth from such a view will project a different life, and a different concept of human personality, than the atomization and "individualism" that have been the condition and perhaps the pride of the European-American and "frontier" mind. In the native oral tradition, Evers and Pavich wrote, "The individual is constantly reminded that he is part of the whole, not any more important than any creature around him."

The formal structures of traditional songs and stories were highly organized and stylized, indicating long and cherished literary history. It is clear from this alone that literature was integral and important to native culture; the pervasiveness of songs specific to significant occasions, such as the following, a Papago "Death Song," is further evidence that native people were deeply literary, as they were religious:

> In the great night my heart will go out.
> Toward me the darkness comes rattling.
> In the great night my heart will go out.

After the white invasion of the West, from late in the nineteenth century until well into the twentieth, several Native Americans wrote or dictated autobiographies, acutely describing the tremendous and terrible changes they had seen, analyzing Indian-White relations, and in many cases proposing ways by which a fairer accommodation might be made between the cultures. Sarah Winnemucca (ca. 1844–91), an intelligent and forthright Paiute woman, was one of the first in the West to write a book in this genre; her *Life among the Paiutes: Their Wrongs and Claims* (1883) is, among its other qualities, distinguished by a perceptive awareness of her (white) readership. She engagingly

Combining a familiarity with traditional Native American stories and storytelling techniques with a literary style that borrows freely from other modern practices, Leslie Marmon Silko and other contemporary Indian writers have brought Native American fiction to a broad new audience of readers.

Robyn Stoutenburg (b. 1958). Leslie Marmon Silko. *Photograph, 1991. © 1993 Robyn Stoutenburg/Swanstock, Tucson, Arizona.*

explains her own life and role as an intermediary, dismisses various criticisms of her tribe with documentary logic, and narrates signal events in recent tribal history with a vivid dramatic sense. Her literary success is a testimony to a remarkable adaptability, all the more impressive when we consider, as H. David Brumble III has noted, that "as a young child she had lived with a stone-age, hunter-gatherer people," a people whose first contact with whites occurred in about 1848. Winnemucca's book, like those of the Omaha scholar Francis LaFlesche (*The Middle Five: Indian Boys at School* [1900]) and the Santee Sioux physician Charles Eastman (*Indian Boyhood* [1902] and *From the Deep Woods to Civilization* [1916]), readily and artistically makes use of one of the tools of the conqueror to demonstrate an earned standing in the modern world and to record continuity with the old. This autobiographical tradition has continued to the present and in its way is as varied and dramatic as the nominally more artistic genres of fiction and poetry. For just two further examples of the several hundred on record, *Black Elk Speaks* (1932) and *Lame Deer: Seeker of Visions* (1972) have been widely praised as artistically realized literature; both books have also played a major role in interpreting Native American spiritual thought to a modern, comparatively much more secular world.

The first novel published by a Native American was John Rollin Ridge's *The Life and Adventures of Joaquin Murieta, the Celebrated California Bandit* (1854), a book written in a heightened, romantic style yet unmistakably criticizing the race prejudice of the Anglo gold miners against whom the brave, sensitive Joaquin seeks revenge. But fiction did not, apparently, seem as congenial a mode as autobiography in the urgent and mostly calamitous situation of Native Americans, for it was not until well into the twentieth century that novels began to appear regularly. The first novel published by a native woman was *Co-ge-we-a* (1927), written by Mourning Dove (Cristal McLeod Galler, 1888–1936), an Okanogan. *Co-ge-we-a* takes as its territory the painful dilemmas of the mixed-blood, thereby personally and incisively commenting on the conflicting value systems of the two cultures and the unreasoning prejudice against both Indians and people of mixed lineage. Another important early working of the essentially tragic cross-cultural theme was *The Surrounded* (1936), by D'Arcy McNickle (1904–77), son of a Cree mother who had been adopted into the Salish Kootenai tribe and an Irish father. This novel is both naturalistic and autobiographical, depicting the profound cultural and psychological discontinuities within a mixed-blood protagonist. Like McNickle, young Archilde is sent away from the reservation and his mother to be educated at a "white" boarding school. The dichotomization of Archilde's life, expressed in several symbolic ways in the novel's settings, plot, and characters as well as directly in Archilde's own mind, allows McNickle to comment on the disastrous lack of communication between cultures and on the repressive power of the dominant society. Measured by its plot alone, *The Surrounded* is pessimistic, but it contains a resolute affirmation of the Native American perspective. In its naturalistic recognition of a conflict that will not diminish soon, and its persevering sense of the honor and rightness of the old ways, this novel foreshadows much of the succeeding Indian fiction.

The 1960s and 1970s, a time of turmoil and reassessment in many areas of American life, saw a rebirth of interest in things Indian (perhaps, as some have said, this was a recurrence of a cyclical phenomenon) and at the same time a remarkable creative outpouring—a true renaissance—in Native American writing. In the West, N. Scott Momaday (b. 1934), Leslie Marmon Silko (b. 1948), and James Welch (b. 1940) are widely regarded as important contributors to the new abundance and quality of Indian literature, with Momaday's novel *House Made of Dawn* (1968) usually regarded as bringer of the fire. That book, "the first non-linear, non-chronological, ritual novel written by an American Indian," as the scholar Paula Gunn Allen has said, alone demonstrates enough artistic sophistication, insight into culture and character, and psychologically evocative setting to call into question the stereotype of the "simple" Indian. In broad terms, the novel deals with the conflict of cultures that history itself has established as the "matter" of recent native narrative; but other layers of meaning, reaching back to connect with the traditional world, are also present and lend depth and universality to the story. For example, as Larry Evers has pointed out, the novel's movement from discord to harmony is common in old-time oral narratives, and furthermore this plot is often framed, as is *House Made of Dawn*, on the necessary process of reconnecting with the land. Abel, the protagonist, is alienated from place and culture simultaneously; it is the wholeness of the two with which he must align himself in the proper, ritual fashion.

Leslie Silko's highly regarded novel *Ceremony* (1977), like *House Made of Dawn*, centers on healing. Both novels open and close with ritually appropriate verbal frames, showing in this old way the tellers' awareness that story has a serious, communal, and moral function—a real standing; and both utilize narrative techniques drawn from Anglo and Native American traditions. Momaday and Silko seem to be clearly conscious of the dual literary legacies they work with and appear free to use at any time whatever methods or allusions may be appropriate. This technical free ranging itself suggests a comprehensive outlook; but the authors' ultimate philosophical standpoint, within the coherent, native way, gives their technical virtuosity a grounding and makes their work something more substantial than mere eclectic experimentation. Silko's protagonist Tayo, in *Ceremony*, has been deserted by his mother and traumatized in World War II and exhibits serious psychological problems; concurrently, the Laguna reservation suffers from severe drought. As the novel progresses, it becomes clear that health is not achieved in isolation: Silko's weaving of human personality and land, making a human ecology, reminds a reader that "health" and "whole" are indeed cognate terms. Speaking of comprehensiveness, it may be noteworthy that Betonie, the mixed-blood Navajo who as medicine man works for Tayo's healing, numbers among his tools old St. Louis, Seattle, New York, and Oakland phone books.

James Welch's fiction (*Winter in the Blood* [1974], *The Death of Jim Loney* [1979], and *Fools Crow* [1986]) has inspired a remarkable range of critical interpretation, with much attention given to the author's possible attitude toward his subject matter. The first two novels have seemed unremittingly bleak to many readers; the scholar Peter Wild has written, "*Winter in the Blood* appears to be a 'day-in-the-life-of' novel dogging the sadsack existence of yet another drunken, alienated Native American as he shuffles between his mother's little cow enterprise and the bar-studded towns surrounding the reservation." But as Wild usefully points out, such is only the surface of the book. Through remarkably sharp and memorable images (Welch is a poet as well as a novelist), odd bits of conversation, flashbacks whose significance only slowly begins to emerge, absurdist humor, and surreal dreams, the novel shows that the protagonist is slowly gaining some perspective on himself. The wisest character in the book, Yellow Calf, judges the world to be "cockeyed," and thus—assuming Welch agrees with Yellow Calf—one would expect that any changes for the better will be presented guardedly and will be difficult to see, just as they are in the real world. In the sophistication of his characterization, Welch demonstrates a quality of vision now widely seen in Native American writers: they work from a bicultural and conflicted base, true to history, but they go on to provide insights applicable to the human condition, anytime and probably anywhere.

Recent Indian poetry shows this same combination of a base in the ancient coherence plus a modern freedom of tone, form, and reference. The old world, it must be understood, is not always—or even often—explicitly described, but it is inescapably a generative reference point. The irony, humor, and the tragic sense too, often noted in modern Native American verse, all owe their existence to the brute fact of the two worlds. But there is often a more subtle knitting going on. In the traditional way, song and poetry were methods of naming and praising the wholeness of the world; perhaps in the modern world as well, the poet is attempting to make sense of things, to transcend fragmentation and offer a large enough view, and a view with heart, so that even this anguished, mechanical, and hurried world, so far from *all* human tradition, might be comprehended. Again, the distinctive native flavor derives in large part from the assumption of a communal and moral function for literature—modern Indian poetry shows comparatively little of the poet's obsessive concentration on the self that has trivialized so much American verse in recent years.

This long-term, characteristic depth of intention, and the modern freedom, may be exemplified by showing two Native American poems in juxtaposition. The first is a traditional chant of the Yokuts, the second a recent poem by the New Mexico writer Geary Hobson. Between them, these poems suggest much of what is distinctive and vital in Native American writing, whatever the age or genre.

My words are tied in one
with the great mountains,
with the great rocks,
with the great trees.
in one with my body
and my heart...
And you, day,
and you, night!
All of you see me
one with the world.

BUFFALO POEM # 1
(or)
ON HEARING THAT A SMALL HERD OF BUFFALO
HAS "BROKEN LOOSE" AND IS "RUNNING WILD"
AT THE ALBUQUERQUE AIRPORT—SEPTEMBER 26, 1975
—roam on, brothers . . .

Other Current Trends

Contemporary literary history is notoriously hard to sort out, but one generalization about recent and current western writing can quite safely be made: something like a creative explosion has been going on for the past twenty or so years, bringing the literary West surely into the modern time. Fiction and poetry by Native Americans, with enough good material for at least three major anthologies between 1975 and 1988 alone, new work by Japanese Americans, Chinese Americans, Filipino Americans, Mexican Americans, African Americans, and Armenian Americans, and a significant increase in the number of books by women writers in the West—all of this "minority" production indicates not a random flurry of activity but a general and positive release from old western stereotypes. Western writers work now as if on liberated terrain. A clear example of the new plasticity of response is furnished by the novels of Tony Hillerman of New Mexico (*The Blessing Way* [1970] and *A Thief of Time* [1988], among many others). In these remarkably original novels, close attention to Navajo thought is blended with clever detective-fiction plotting. Ivan Doig's play of an acute, present-day consciousness over historical materials (*Winter Brothers* [1980]) also demonstrates the contemporary western sophistication. Rediscoveries of ignored or forgotten writers such as Elinore Pruitt Stewart (*Letters of a Woman Homesteader* [1914]) and other western women (noted and honored in such anthologies as Joanna Stratton's *Pioneer Women: Voices from the Kansas Frontier* [1981] and Lillian Schlissel's *Women's Diaries of the Westward Journey* [1982]) are another important aspect of the new West. The proliferation of independent or "small" presses and the continued quality publishing by such established academic outlets as the University of Nebraska Press, the University of Oklahoma Press, the University of New Mexico Press, and the University of Arizona Press—all of whom have made it their business to publish western writers—are further evidence of ground firmly won for serious literature. Each year since the founding of *Western American Literature* in 1966, the number of review and bibliography pages in the journal has grown, reflecting both the increase in western titles and the acceptance of western writing as a field for scholarship. In the public and university-extension sector, conferences of scholars and writers, workshops, and short courses on western literature have all become common and regular features of the western literary landscape over the past twenty years, where as recently as the 1950s and early 1960s, the rare gathering of serious western writers had a trailbreaking, if not maverick, aura. The common note in all this activity is that the West has come into its own, literarily, and is no longer in thrall to the frontier mentality or the romantic western myths.

This release may be seen in certain formal experiments of recent years, such as the widely different minimalist and "magical-realist" schools of fiction, both of which have western adherents, and in jugglings and reversals of old western formulas such as Edward Abbey's *The Brave Cowboy* (1956) and E. L. Doctorow's *Welcome to Hard Times* (1975). But the new West most plainly shows its substance in what the writing says, thematically. Perhaps expectably, there is no clearer exposition of the modern reassessment than in nonfiction, and this is true in particular of writing dealing with nature. In the nature essay, the distinction between frontier mind and postfrontier mind is, as we have seen in the earlier time of John Muir and Mary Austin, fundamental, and the history of the West as an exploited region puts a sharp point on the inner, philosophical divide. Edward Abbey (*Desert Solitaire* [1968], *The Journey Home: Some Words in Defense of the American West* [1977], *Abbey's Road* [1979], and *Down the River* [1982]) is perhaps the most significant and influential of the writers who have engaged the philosophical issues of the nature essay and at the same time taken a new, hard look at the western environment. Although he denied the label of naturalist or nature writer, Abbey (1927–89) certainly examines the relationship of humanity and nature, and civilization and wilderness, with sharp insight. His attack on exploitation is as plain-spoken as one might wish, and such essays as "The Second Rape of the West" describe in precise detail, naming names, the current abusers of what has been called the "plundered province." In addition, his writing is made complex and often blackly humorous—very modern—by an ironic slant on history and humanity, which makes his essays quite different in tone from the usually genteel nature-essay tradition. In Abbey's complex and highly personal work, the nature essay becomes current indeed, though at root this author's allegiances are firmly in the established line.

Other important western essayists who have emphasized a postfrontier awareness include Wallace Stegner (*The Sound of Mountain Water* [1969]), Barry Lopez (b. 1945, *Of Wolves and Men* [1978]) and *Crossing Open Ground* [1988]), Ann Zwinger (b. 1925, *Run, River, Run* [1975]), and William Kittredge (b. 1932, *Owning It All* [1987] and *Hole in the Sky* [1992]). These writers describe—and demonstrate—the deep changes in consciousness and attitude that are required of inhabitants, as opposed to invaders, exploiters, or sentimentalists. In assessing history and speculating on what might be a proper, sustainable basis for human presence in the West, they look first to the details of particular places and the restraints of ecological setting. This shift in priority represents a major change of bearing, toward realism and accountability, and is fundamental to the new western literary outlook.

As a whole, western literature's distinctiveness is that it codifies this transition in full detail. It grew up on the frontier, where myths arise; it has, in its later development, indicated a new pattern: ecological adjustment, cultural refinement, and inward growth. When we consider that the presently dominant world culture—founded on mobility, power, and presumably infinite economic growth—is decidedly frontier-minded in the old sense, the maturation of western American literature assumes a significance far greater than heretofore granted.

Bibliographic Note

Obviously, the preceding offers only an outline. Anyone interested in fuller detail should consult Thomas J. Lyon et al., eds., *A Literary History of the American West* (Fort Worth, 1987), a volume of essays by seventy-one scholars. *LHAW* covers many of the writers mentioned above, but despite its size (1353 pages), the book is far from exhaustive, and an interested reader will find him- or herself piecing together bibliographies and likely scholarly trails from a number of sources. One should probably look first to "The West," in Clarence Gohdes and Sanford E. Marovitz, eds., *Bibliographical Guide to the Study of the Literature of the U.S.A.*, 5th ed. (Durham, N.C., 1984), and to Richard W. Etulain's *A Bibliographical Guide to the Study of Western American Literature* (Lincoln, 1982). In addition, the chapter bibliographies in Fred Erisman and Richard W. Etulain, eds., *Fifty Western Writers: A Bio-bibliographical Sourcebook* (Westport, Conn., 1982), offer good leads. The "Western Writers Series" of pamphlets, published at Boise State University under the editorship of Wayne Chatterton and James H. Maguire, is an excellent guide to a large number of western writers—at this writing, more than one hundred. The annual bibliography of critical books and articles in *Western American Literature*, published each February, and the listings under "Literature and the Arts," found in the "Recent Articles" section of each issue of the *Western Historical Quarterly*, will help keep a student of western literature up to date.

Provocative broad-gauge interpretations include Lucy Lockwood Hazard, *The Frontier in American Literature* (New York, 1927), Henry Nash Smith, *Virgin Land: The American West as Symbol and Myth* (Cambridge, Mass., 1950), Leslie Fielder, *The Return of the Vanishing American* (New York, 1968), and William Everson, *Archetype West* (Berkeley, 1976). For single essays characterizing the nature and history of western literature, it would be hard to surpass Wallace Stegner's "History, Myth, and the Western Writer," in *The Sound of Mountain Water* (1969; reprint, Lincoln, 1985), or Fred Erisman's amazingly comprehensive "The Changing Face of Western Literary Regionalism," in Gerald D. Nash and Richard W. Etulain, eds., *The Twentieth-Century West: Historical Interpretations* (Albuquerque, 1989).

Good anthologies—that is, collections put together with an eye to instructively typical material, and introduced knowledgeably—include J. Golden Taylor, ed., *Great Western Short Stories* (Palo Alto, Calif., 1967), J. Golden Taylor, ed., *The Literature of the American West* (Boston, 1969), Philip Durham and Everett L. Jones, eds., *The Western Story: Fact, Fiction, and Myth* (New York, 1975), Clinton F. Larson and William Stafford, eds., *Modern Poetry of Western America* (Provo, Utah, 1975), James D. Houston, ed., *West Coast Fiction* (New York, 1979), Max Apple, ed., *Southwest Fiction* (New York, 1981), Russell Martin and Marc Barasch, eds., *Writers of the Purple Sage: An Anthology of Recent Western Writing* (New York, 1984), Alexander Blackburn, Craig Lesley, and Jill Landem, eds., *The Interior Country: Stories of the Modern West* (Athens, Ohio, 1987), and William Kittredge and Annick Smith, eds., *The Last Best Place: A Montana Anthology* (Helena, 1988). These collections range across wide regions and times and do not limit themselves to either the popular or the serious-literary level. Perhaps the most comprehensive effort of all is James C. Work's *Prose and Poetry of the American West* (Lincoln,

1990), which with its more than seven hundred pages and its fifty-four authors (representing the time period 1540–1989), as well as the editor's astute historical introduction, is likely to set the textbook standard for years to come.

Notable collections of critical essays include Gerald W. Haslam, ed., *Western Writing* (Albuquerque, 1974), Merrill Lewis and L. L. Lee, eds., *The Westering Experience in American Literature* (Bellingham, Wash., 1977), Richard W. Etulain, ed., *The American Literary West* (Manhattan, Kans., 1980), the aforementioned Erisman and Etulain, *Fifty Western Writers*, and Judy Nolte Lensink, ed., *Old Southwest/New Southwest: Essays on a Region and Its Literature* (Tucson, 1987).

Turning now to more specialized genre and author studies, and framing the list on the preceding essay, I have found the following scholarly examinations to be useful. For ecological and historical definitions of the West applicable to literary criticism, Walter Prescott Webb's *The Great Plains* (1931; reprint, Lincoln, 1981) and *The Great Frontier* (1952; reprint, Austin, 1964), and Wallace Stegner's pithy lectures in *Where the Bluebird Sings to the Lemonade Springs: Living and Writing in the West* (New York, 1992) are invaluable. On the frontier and the "great American dialectic" of civilization and savagery, three landmark studies are Leo Marx, *The Machine in the Garden: Technology and the Pastoral Idea in America* (New York, 1964), Richard Slotkin, *Regeneration through Violence: The Mythology of the American Frontier, 1600–1860* (Middletown, Conn., 1973), and Frederick Turner, *Beyond Geography: The Western Spirit against the Wilderness* (New York, 1980). The latter two studies are strongly revisionist in character.

The early mass-market fiction of the West is covered by Daryl Jones in *The Dime Novel Western* (Bowling Green, Ohio, 1978), and the story of the popular formula is brought forward in two good collections of essays: Richard W. Etulain and Michael T. Marsden, eds., *The Popular Western* (Bowling Green, Ohio, 1974), and James K. Folsom, ed., *The Western: A Collection of Critical Essays* (Englewood Cliffs, N.J., 1979). Probably the most influential interpretation of the formula Western is John G. Cawelti's *The Six-Gun Mystique* (Bowling Green, Ohio, 1970), and students of the field will also want to consult Cawelti's *Adventure, Mystery, and Romance: Formula Studies as Art and Popular Culture* (Chicago, 1977). On Owen Wister, two good sources are G. Edward White, *The Eastern Establishment and the Western Experience: The West of Frederic Remington, Theodore Roosevelt, and Owen Wister* (New Haven, 1968), and Richard W. Etulain, *Owen Wister* (Boise, Idaho, 1973). Zane Grey has been taken seriously and studied to edifying effect by Ann Ronald, *Zane Grey* (Boise, Idaho, 1975), and Gary Topping, "Zane Grey," in Erisman and Etulain, *Fifty Western Writers*. Louis L'Amour is also covered well and fairly in *Fifty Western Writers*, in an essay by Michael T. Marsden. Readers interested in what might be made of Jack Schaefer's *Shane* should consult James C. Work, ed., *Shane: The Critical Edition* (Lincoln, 1984).

John Muir has recently come in for a good deal of instructive study, and useful sources in the wider vein include Michael P. Cohen, *The Pathless Way: John Muir and American Wilderness* (Madison, 1984), and Frederick Turner, *Rediscovering America: John Muir in His Time and Ours* (New York, 1985). Two recent biographies of Mary Austin—Augusta Fink's *I-Mary* (Tucson, 1983) and Esther L. Stineman's *Mary Austin: Song of a Maverick* (New Haven, 1989)—show that Austin, though few of her books remain in print, still has something to say to the modern age. John C. Van Dyke has dropped out of the common view and the literary canon even further than Austin, but Peter Wild's *John C. Van Dyke: The Desert* (Boise, Idaho, 1988) may help spark a rehabilitation.

On the novel in the West, it is useful to consult Richard Chase's distinction between romance and novel in *The American Novel and Its Tradition* (Garden City, N.Y., 1952). The foremost regionally specific studies of the genre have been James K. Folsom, *The American Western Novel* (New Haven, Conn., 1966), and John R. Milton, *The Novel of the American West* (Lincoln, 1980). Roy W. Meyer's *The Middle Western Farm Novel in the Twentieth Century* (Lincoln, 1965) is also valuable, and more widely applicable than its title might suggest. On San Francisco's early days, see Franklin Walker, *San Francisco's Literary Frontier* (New York, 1939). For a cogent analysis of Stephen Crane's West, see Frank Bergon, *Stephen Crane's Artistry* (New York, 1975). Warren French's *Frank Norris* (New York, 1962) will guide readers to further sources in the rich secondary material on this writer. Perhaps the most comprehensive scholar on Jack London is Earle Labor; his *Jack London* (New York, 1974) is a reliable "backgrounder," and his "Jack London's Agrarian Vision," *Western American Literature* 11 (Summer 1976): 83–101, is an excellent example of what might be termed the higher London criticism. Scholarship on Willa Cather represents something of an industry—as befits perhaps the most important novelist of the West—but a student who consults John J. Murphy's *Critical Essays on Willa Cather* (Boston, 1984), David Stouck's *Willa Cather's Imagination* (Lincoln, 1975), and Susan J. Rosowski's *The Voyage Perilous: Willa Cather's Romanticism* (Lincoln, 1986) would engage the most influential and recent Cather criticism. An in-depth study of one text, John J. Murphy's *My Antonia: The Road Home* (Boston, 1989), shows with remarkable critical insight the levels of complexity and layers of reference Cather wove into her work.

On Frank Waters, the series *Studies in Frank Waters*, edited by Charles L. Adams (University of Nevada, Las Vegas), is valuable and current. Terence Tanner's *Frank Waters: A Bibliography with Relevant Selections from his Correspondence* (Glenwood, Ill., 1983) is indispensable. On Larry McMurtry, the collection *Taking Stock: A Larry McMurtry Casebook*, edited by Clay Reynolds (Dallas, 1989), offers many interesting lines of thought. Lucile F. Aly's *John G. Neihardt: A Critical Biography* (Amsterdam, 1959), has established the main lines of interpretation on this important but often overlooked poet. Robinson Jeffers's life and work are surveyed perceptively by Frederic I. Carpenter, *Robinson Jeffers* (New York, 1962); interested scholars will want to delve into Robert J. Brophy's *Robinson Jeffers: Myth, Ritual, and Symbol in His Narrative Poems* (Cleveland, Ohio, 1973) and will keep up to date with the *Robinson Jeffers Newsletter*, edited by Brophy at California State University, Long Beach. Stanford University Press is currently bringing out *The Collected Poetry of Robinson Jeffers* under the editorship of Tim Hunt; two of the four projected volumes have been published (1987, 1989); further criticism and perhaps revaluation seem likely to follow this major western publication. Gary Snyder has been surveyed in Bob Steuding, *Gary Snyder* (Boston, 1976), spiritedly described by Kenneth White, *The Tribal Dharma* (Dyfed, Wales, 1975), and subjected to a variety of approaches in Patrick D. Murphy, ed., *Critical Essays on Gary Snyder* (Boston, 1990); students of this important poet will also want to read Scott McLean, ed., *The Real Work: Interviews and Talks, 1964–1979* (New York, 1980). William Stafford's *Writing the Australian Crawl: Views on the Writer's Vocation*, in the important "Poets on Poetry" series (Ann Arbor, 1978), should be referred to for explication of the poet's motives and values.

Among recent works on Native-American literature, Karl Kroeber, *Traditional Literatures of the American Indian: Texts and Interpretations* (Lincoln, 1981), Alan R. Velie, *Four American Indian Literary Masters: N. Scott Momaday, James Welch, Leslie Marmon Silko, and Gerald Vizenor* (Norman, 1982), Paula Gunn Allen, ed., *Studies in American Indian Literature* (New York, 1983), Brian Swann, ed., *Smoothing the Ground: Essays on Native American Oral Literature* (Berkeley, 1983), Andrew Wiget, *Native American Literature* (Boston, 1985), H. David Brumble, *American Indian Autobiography* (Berkeley, 1988), and Arnold Krupat, *The Voice in the Margin: Native American Literature and the Canon* (Berkeley, 1989), are regarded highly. Two anthologies with useful introductory notes are Thomas E. Sanders and Walter W. Peek, eds., *Literature of the American Indian*, abr. ed. (Beverly Hills, Calif., 1976), and Geary Hobson, ed., *The Remembered Earth: An Anthology of Contemporary Native American Literature* (Albuquerque, 1981). Duane Niatum, ed., *Harper's Anthology of 20th Century Native American Poetry* (San Francisco, 1988), and Simon Ortiz, ed., *Earth Power Coming: Short Fiction in Native American Literature* (Tsaile, Ariz., 1983), should also be consulted. Scholars interested in N. Scott Momaday will want to study his sometimes revealing remarks in Charles L. Woodard's *Ancestral Voice: Conversations with N. Scott Momaday* (Lincoln, 1989). Leslie Silko's work is introduced in Per Seyersted, *Leslie Marmon Silko* (Boise, Idaho, 1980), and James Welch's in Peter Wild, *James Welch* (Boise, Idaho, 1983).

A good critical source on Edward Abbey is James Hepworth and Gregory McNamee, eds., *Resist Much, Obey Little* (Salt Lake City, 1985); readers should also study Ann Ronald's thoughtful critique in *The New West of Edward Abbey* (Albuquerque, 1982).

Pageant of Lincoln
June·5·1915 · Lincoln·Neb.

Chapter Twenty-one

Speaking for the Past

CHARLES S. PETERSON

If historians speak for the past, who speaks for historians? Certainly historians seem endlessly willing to talk about themselves, but on occasion journalists and others too have paid attention. In 1989 and 1990, newspapers and magazines took note of disputes over what became known as the "new western history." National attention may have begun with an article in the *Washington Post* of 10 October 1989 under the wildly western headline: "Shootout in Academia over the History of U.S. West, New Generation Confronts Frontier Tradition." Not to be outdone by its East Coast rival, the *New York Times* published two pieces on this academic showdown in December 1989 and March 1990. In May, *U.S. News and World Report* had a cover announcing "The Old West: The New View of Frontier Life," and in October 1990, the cover of the *New Republic* sourly announced "Westward Ho Hum: What the New Historians Have Done to the Old West."

These reports from the academic front covered the most recent infighting but did not always explain how long the war of words had been under way. For decades an uneasy balance existed between the voices of popular impulse and those of academic professionalism. Without worrying about its subtleties, this constant shifting between general interest and scholarly discipline energized those who spoke for western history even as the very idea of the West changed. Now taken as the trans-Mississippi area, the West looks back beyond the hundred years of its immediate past to historiographic times far more distant and to geographies far wider than its present vastness. Not only has western history had to cope with successive Wests, but it has also had to respond to ongoing change about the meanings and uses of history.

All in all, the converging influences of personality, time, place, and thought make for a field of great appeal. With an ear turned to the seismic stirrings of popular taste, with an almost pathological need to be accepted professionally, and with subfields of growing complexity, western history is at once charming and productive. It has not yet lost its willingness to tell a good story nor abandoned an easy camaraderie, but western historians of widely varied backgrounds are putting together an impressive body of scholarly and popular literature and sustaining the institutional superstructure necessary to develop the story of one of the world's grand adventures.

As in so much else that deals with western history, getting a handle on all this is simplified by the centrality of the person and career of the frontier historian Frederick Jackson Turner. Born in Portage, Wisconsin, in 1861, Turner was a product of middle America and of a republic midway on its course to the present. Wilderness was still near. Old-timers remembered pioneering and Indian wars. Local pride and aspirations ran

The popular appeal of western history, celebrated in this poster as the progressive development from a simple Indian past to a modern urban present, has both inspired and troubled academic historians. If such popular history has created a broad interest in the West, it has also hampered investigation of alternative stories of the region's past.

Unidentified artist. Pageant of Lincoln. *Chromolithograph, 1915. Langdon Collection, John Hay Library, Brown University, Providence, Rhode Island.*

strong, and as the historian Ray A. Billington has recognized, the views of Thomas Jefferson and Alexis de Tocqueville were still "a common currency." Yet industrial cities were rising, and Populist protest and regional tensions stirred an uneasy land.

Reflecting the situation's complexity was Turner's own experience. As a student at the University of Wisconsin and at Johns Hopkins, he was introduced to history as a social tool and to the ascendant Teutonic germ theory, which rooted American (eastern) development in European influences. Turner returned to teach at Wisconsin, where he devoted himself to seeing the past of his own region in history's larger context. That process involved him with both the popular and the professional elements of history. By 1890 he was lecturing public-school teachers on "The Significance of History," which he described at two levels. On the one hand, history was common, fundamental, apparent. Wherever there remained "a chipped flint, . . . a pyramid, . . . a poem, . . . a coin," there was "history." On the other hand, history was an "ongoing encounter between past and present," written "anew" by each generation in response to the issues of its time and controlled and enlightened by a growing body of historical methodologies and interpretations.

In 1893 Turner delivered his celebrated paper "The Significance of the Frontier in American History." In it he declared, in striking and poetic terms, the centrality of western influences upon America's development. Written as "a programme" or interpretive tool, and "a protest against eastern neglect . . . of the West, and against" the West's own "antiquarian spirit," as he later informed a friend, this famous essay described the frontier both as a process and as the successive lines of advancing settlement. Sensing that the West exerted "a persistent pervasive influence in American life, which did not get its full attention from historians who thought in terms of" European influences or the slavery issue, nor yet from those who were fascinated with "the epic period of the West," Turner tried "to see it [the West] as a whole," to understand "its institutional, social, economic, and political side," and to apply to its study the analytical methods then developing in the social sciences.

At the deepest root of America's past had been "the existence of an area of free land, its continuous recession, and the advance of . . . settlement westward." Progressively distanced from European and even eastern influences, pioneers became a new race, an exceptional people. In Turner's overstated metaphor, democracy emerged fully formed from the American forest, strengthening liberty everywhere and making it the special attribute of Americans. With accessible land functioning as a safety valve, labor agitation, social discord, and violence were muted. Sectional distinctions were in no small part the product of geographical influences. A vital force since earliest settlement, the frontier had closed by 1890, thus ending, Turner warned, "a great historic movement."

The frontier thesis fell on ready ears. Before the end of 1893, Theodore Roosevelt and a few others praised it as a masterful summation of widely held views. In the years that followed, the public and the historical profession capitulated, suggesting Turner's ideas and spirit were firmly based in the general climate of opinion. Turner was widely sought as a lecturer and wrote often for popular journals. Professional papers, of which he delivered many, and monographic works, which were embarrassingly few, bore his

No historian did more to shape the debate about western history than Frederick Jackson Turner, shown here about 1893, the year he delivered his famous address on the significance of the frontier in American history.

Unidentified photographer. Frederick Jackson Turner. *Photograph, ca. 1893. The Huntington Library, San Marino, California.*

trademarks: sweeping generalizations, effective analysis, methodological soundness, and an optimistic and romantic affinity for America and especially the West.

Discovery and an Evolving Literature

To understand Turner's pivotal role it would be well to survey developments that contributed to western history in the centuries before 1900. Formative voices grew from discovery and conquest. Initiated by Renaissance generations, the first age of discovery extended for two centuries, circumnavigated the globe, and worked the Western Hemisphere into the formal record and popular consciousness of Europe. As a statement of history, it was preoccupied with national expansion and extension of the faith as well as filled with allusions to marvels and mythical expectations. Extraordinary only in its claims to priority was Christopher Columbus's first report, which included maritime data, a funding request, and information about the Indies and King Solomon's fabled treasure islands. Similar was Francis Drake's sixteenth-century account of his world-circling raid on the Spanish Empire. Suppressed in England by Queen Elizabeth, it was elsewhere "in every man's mouth," as a contemporary reported. Its distortions spread to the present as either a celebrated memory or a plague on scholarship depending on whom one asks.

The University of Texas historian William H. Goetzmann has explained that a second age of discovery began with the eighteenth century. Explorers no longer sought marvels but, guided by rationalism, set out to order geographical knowledge in terms of Newton's mechanistic model. The data they collected helped create a world culture of science while the "discovery of . . . peoples, who seemingly stepped out of Eden," fostered romantic history and national awareness. By the late 1800s recognition of evolution's workings had added "a linear, history-oriented time-line" to the literature of discovery.

Under these influences the record of exploration was dramatically enlarged in body and meaning. The product of Old World influences, it became increasingly relevant for the New. The United States was both the child of discovery and an active participant. Trade and seagoing industry gave America a recognizable place in the record of maritime discovery, but its unique domain lay in the historiography of land exploration and in a growing awareness of nature's importance.

The influence that discovering nature's purpose had upon the American record is nowhere more apparent than in the thought and policy of Thomas Jefferson. To Jefferson, man and nature were inseparably linked, integral parts of the benevolent design of nature's God. Closely related was the view that equality and independence were not only inalienable rights but, in the natural order of things, the special destiny of America. As the historian Daniel Boorstin has eloquently pointed out, Jefferson's ideas were future oriented rather than historical. In the natural circumstance of the American continent lay a sublime opportunity for individual development unhampered by the past and for the "fruition of a society built on largely naturalistic foundations." Like the agricultural development on which they depended, prosperity and republican virtues were the product of "the environment itself." It was a vision that not only partook of the Age of Reason's confidence in empirical observation but also was moved by the sentiments of romanticism, including nationalistic ardor and a penchant for optimistic expression, qualities still strong in Turner's teaching a century later.

Jefferson's 1803 purchase of Louisiana opened the last great West and vastly expanded America's stake in the historiography of exploration and conquest. A pattern for the record that evolved was Jefferson's *Notes on the State of Virginia* (1785), an "encyclopedic inventory" of old dominion Virginia (from seashore to the Mississippi). As the environmental historian John Opie has written, Jefferson used language that was both empirical and romantic to describe how "human activity" sprang "naturally from Virginia's topography, climate, plants, and animals" and pointed out its potential for future development. After failing to launch transcontinental exploration attempts before he became president, Jefferson improved on the purchase of Louisiana by dispatching the Lewis and Clark Expedition (1804–6) with instructions to record their discoveries in scientific detail.

The journals Lewis and Clark brought back were notable for their factual quality, yet in the fullest sense they were romantic and national history. They were published by Nicholas Biddle of Philadelphia in 1814 along with a "mountain" of supporting accounts, many of which were highly fictionalized. The report focused the attention of reading America on the West, an influence that continues as we approach the bicentennial of the famed expedition.

Something of its early effect may be seen in the experience of Parley Pratt, later a key figure in the Mormon movement westward. In a surge of romantic intensity, the youthful Pratt resolved during the fall of 1826, only months after Jefferson's death, to spend "the remainder of his days" in the "solitudes of the great West, among the natives of the forest." Taking "Lewis and Clark's tour up the Missouri and down the Columbia" to read, Pratt spent a solitary winter in a "holy retreat" west of Cleveland. The "storms of winter raged . . . the wind shook the forest, the wolf howled . . . and the owl chimed in harshly to complete" a "doleful music." To the end of his days Pratt looked West.

Others too looked West, creating a western record as they looked. Significant were Lewis and Clark's followers, official explorers who continued to write reports filled with useful description. One whose account rivaled theirs in popular appeal, and gave a southwestward tilt to American interest, was Zebulon Pike. Ordered to explore the southern reaches of the new purchase, Pike made an extended foray into northern New Spain in 1806–7. His report was filled with information about the Rio Grande and Chihuahua and sold widely in the United States and in numerous foreign editions. Perhaps its most lasting fame grew from Pike's description of the Great Plains as uninhabitable desert.

The national literature of discovery accumulated quickly. Exploration was given a formal place in the military tradition when the U.S. Topographical Bureau was organized during the War of 1812, followed by the Corps of Topographical Engineers in 1838. Later the Department of the Interior and universities incorporated scientific exploration into their functions. The West became a vast laboratory where scientific disciplines defined their perimeters and created a unique literature of western discovery.

Official discovery reports were more or less constant. Transcending journalism and technical data, many of them achieved status as history. Among the most widely read were John C. Frémont's reports of the 1840s and 1850s. Only slightly less acclaimed were exploration accounts from the Mexican War, the Pacific Railroad Surveys of the mid-1850s, and the Utah War of the late 1850s. Even more significant were reports generated by the Great Surveys in the decades after the Civil War. None of these became more deeply embedded in western historiography than John Wesley Powell's benchmark *Report on the Lands of the Arid Region of the United States* (1878). In it Powell called for the rationalization of natural resource utilization through classification and planned use. He also described the geographical regions of the United States and, improving on Pike, defined the West in terms of aridity.

In the meantime, an important western historiography grew out of economic enterprise and settlement in successive Wests. Among the earliest was John Smith's history of Jamestown, which Jefferson found to be "sensible, and well informed" but in style "barbarous and uncouth." More hopefully, as the historian John Higham noted of the Puritan William Bradford, early chroniclers "set down" straightforward narratives "of shared and remembered experience" that possessed "an effortless grasp of human motivation." Much more recently, local historians have written, in parlances humble, nostalgic, or mythic, of beginnings in homestead districts, mining camps, or railroad towns.

Closely related was what might be called the historical journalism of expeditionary enterprise. Existing from the earliest penetration of Europeans, expeditionary accounts took on added relevance with the growth of a market economy and advances in

Generously illustrated with images drawn by eyewitnesses, 19th-century western expeditionary reports such as John C. Frémont's narratives created visual and literary histories of the West's recent past.

After Charles Preuss (active 1845–53). Outlet of Subterranean River. *Lithograph, ca. 1845. From John Charles Frémont,* A Report of the Exploring Expedition to Oregon and North California in the Years 1843–'44 *(1845).*

publication. The first stirrings of this development may be seen soon after the War of 1812 as the fur trade and western enterprise proliferated. By the Mexican War in 1846, it was well established. Sensitive to sales, expeditionaries trafficked in the tastes of the reading public and became slaves to deadlines and profits, giving less attention to stylistic finish, moralistic overtones, and matters relating to public discourse than had earlier literary figures.

Among the best were Richard Henry Dana, Jr.'s *Two Years Before the Mast* (1840), a vivid account of California's coastal trade, and Josiah Gregg's *Commerce of the Prairies* (1844), a description of the Santa Fe trade. With their sights on the entertainment market, expeditionaries enlarged the use of stereotypes, controversy, and humor and increasingly turned to sensational and mythic topics that were proven sellers. For example, the British explorer Sir Richard Burton focused on Mormon polygamy in *The City of the Saints* (1862); and in Mark Twain's *Roughing It* (1872), stereotype and exaggerated humor—the "jackass rabbit" that ran so fast "we could . . . hear him whiz"—are transformed into social commentary of lasting value by a sensitive mix of realism and impressionism.

In the meantime, developments in New England were having an important bearing on western history. First was the evolution of the historical perspective, a subject later addressed by Turner. As Higham points out, the early Puritan commonwealth regarded history as a "vehicle of self analysis . . . and public discourse," only slightly less important than the sermon. This affinity for history persisted in secularized form throughout the eighteenth and nineteenth centuries among what have been called the patrician historians, typically men of private means and conservative social and political temper.

Toward the end of the patrician period, certain "men of letters" emerged whose romantic interests and artistic achievements threw a sophisticated light on western themes. Romantic historians sought to discover and re-create the spirit of the times about which they wrote to, in effect, bring an age to life on the written page. Heroic figures, elites, representative man, flesh-and-blood characters, creative imagination, drama, narrative force—all were tools of romantic historians' art. In their writing the master key was progress; the ill was evanescent, the good was enduring, the "march of destiny" spiraled upward.

In the strictest sense, romantic historians of the American West were few in number but long-lived and prolific. Among the earliest was Washington Irving, an expatriate New Yorker who returned from Europe in about 1830 to make western tours and the fur trade the object of his attention. Central also to romantic interest in western themes were the New Englanders William Prescott, George Bancroft, and Francis Parkman. Prescott's *History of the Conquest of Mexico* (1844) and *History of the Conquest of Peru* (1847) exploited the Hispanic turn in America's past, finding in it the heroic endeavor, the dramatic action, and the narrative power important in western tastes to this day. Romantic nationalism was epitomized in the works of George Bancroft, who brought poetry and prophecy to the heroic events and towering figures described in his twelve-volume *History of the United States* (1834–75).

No one wrote more fully in the "grand manner" to which romantic historians aspired than Francis Parkman. Tapping into the westward current at the time of the Mexican War, *The Oregon Trail* (1849) is an enduring classic. Back from his western tour, Parkman published *The History of the Conspiracy of Pontiac* in 1851 and the seven volumes of his monumental *France and England in the New World* between 1865 and 1892, immersing "himself most completely in" volumes dealing with the discovery of the West. In *Pioneers of France in the New World* (1865), *The Jesuits in North America in the Seventeenth Century* (1867), and *LaSalle and the Discovery of the Great West* (1869), he betrayed a fascination with beginnings and an appreciation for the influence of wilderness. As the historian William R. Taylor suggests, Parkman more than hinted at the environmental interpretation Turner made famous when he explained that "the hard practical wisdom of the forest" accounted for "the triumphant achievements of Champlain, La Salle, and a half-dozen others." Casting his work in terms of a titanic struggle between England and France on the one hand and against nature on the other, Parkman helped make the inevitability of progress an American article of faith, arguing that French culture was predestined to give way to English and, ultimately, to American influences.

In the mold of romantics adjusting to the industrial age was Hubert Howe Bancroft. Of Puritan stock, Bancroft grew up in Ohio, learned the book trade, and followed the gold rush to California, where he succeeded in the book business and by the mid-1870s had acquired a vast and superb collection of western Americana including Spanish materials. To this he applied the techniques of mass production and distribution and, assisted by scores of reductionists and writers, produced under his own byline the thirty-nine volumes of Bancroft's Works (1874–90), along with several lesser series. Bancroft's books were contested in authorship, florid in style, excessive in detail, and thin on integrating interpretation. But they also defined the West in terms of a "Western Shores" or Pacific regionalism and worked out thematic and proportional relationships that have persisted in western history. In addition they brought the Hispanic tradition squarely into the focus of western American history.

Bancroft was a vital link between romantic history and the eclecticism and exhaustive detail of western regionalism. His books were assembled from widely varying elements including an array of literary and historical conventions. Recognizable are biblical, classical, Puritan, Jeffersonian, Jacksonian, and Darwinian influences. But no tradition stands out more clearly than that of the romantics. This shows up in Bancroft's response to nature, his treatment of nationalism, his identification of all progress with developments on the California shore, his discussion of Hispanic themes, and perhaps most emphatically, in his penchant for uncritical, action-driven, male-dominated, fact-based narrative.

Whether embodied in Hernando Cortés, LaSalle, or the California forty-niner, the hero was the transcendent figure in romantic history. After 1840 the western hero also became the central figure in an immense popular literature as well as the symbol for a political era that could no longer claim direct revolutionary involvement but that still sought to distance itself from Europe. Beginning about 1825, politicians exploited the myth of the man of nature as democratic leader using the public media to popularize and give legitimacy to a long generation of western heroes. Eventually these included Andrew Jackson, Henry Clay, Davy Crockett, William Henry Harrison, Zachary Taylor, and Abraham Lincoln. In the process, these heroes gave official form to the country's western destiny, and for many, the westerner became the quintessential American.

Lincoln and Jackson were the most-admired political heroes, but with his willing contrivance Davy Crockett was the most extravagant and served to make the transition to the western hero as a literary figure. "Born in a cane brake, cradled in a sap trough, and clouted in coon skins," Crockett helped launch a myth in his own *Narrative of the Life of David Crockett of the State of Tennessee* (1834) and reveled in the antic legendry of *The Crockett Almanacks* (1835–38). Most significant, he died at the Alamo, following the Wild West into American memory through unnumbered fictional and film renditions.

Exploiting similar impulses was the Wild West of popular taste. In publication, as in politics, biography sold well. Writers turned again and again to well-known western figures including Kit Carson, George A. Custer, and Buffalo Bill Cody, around whom floods of myth-filled biography issued. Regional variations playing on major themes extended well into the twentieth century. Violent heroes like New Mexico's Billy the Kid and Mormon Orrin Porter Rockwell still asserted their hold, attracting biographers both serious and sensationalist. Even the political hero lived on in biographies like Carl Sandburg's superb *Abraham Lincoln: The Prairie Years* (1926) and Marquis James's Pulitzer prize–winning *The Raven: A Biography of Sam Houston* (1929), both written in the nineteenth century's grand tradition of history as literature.

From the early discovery accounts to the popular hero, an enormous body of history-related writing had come into existence. In a very real way this was the voice of western history in 1893 as Frederick Jackson Turner announced the closing of the frontier. It was, however, a voice so dispersed in space, so various in practical function, and so deeply rooted in nostalgia and popular culture as to be only partially realized as western history. It had almost no existence at all as an academic field of study. But developments were under way that soon made it possible for the voice of scholars to be heard on western history as well.

The University and the Professional Historian

In the late nineteenth century, universities multiplied throughout America. History was added piecemeal to the college curriculum, and eastern institutions appointed historians and established departments, giving history a new institutional focus. Initially, graduate work abroad was a common prerequisite, and the first generation of professors tended to reflect class circumstances that made foreign study possible. Soon, however, developing graduate programs enabled scholars to take their entire training in America. Of modest means, many of the new scholars broke with the conservatism of the patricians, tending instead to progressive or liberal social views. Tutorial relationships between student and mentor and debate between professors allowed for widespread exchange of ideas and made for continuity.

Turner's career may be taken as a metaphor for the changes that all this produced in western history. His youth in the Midwest and his education at Wisconsin and Johns Hopkins were typical. This was true also of the nostalgia that dominated his thought, as it was of his return to teach at Wisconsin. He was the creative thinker who linked West and nation to speak to the "ideals and problems of" his own time. He was also the consummate organization man who tended affairs on campus and made his influence felt in the world of history organizations. At both Wisconsin and Harvard, where he moved in 1910, he attracted hundreds of students, scores of whom became dedicated Turnerians.

Turner's threefold influence as sentimental representative of the West, analytical student of American processes, and consummate academic professional was mirrored in the careers of his students and followers. Many maintained regional loyalties, others broadened American history, and still others found ultimate meaning in the frontier interpretation. Perhaps no student reflected the sweep of Turner's thought, and western history's potential for both breadth and responsiveness, more than Carl L. Becker. The Iowa son of a westering German family, Becker did undergraduate work at Wisconsin and graduate study at Columbia under the "new historian" James Harvey Robinson but

Members of the Daughters of the Republic of Texas pose in front of the Alamo, which their organization restored and continues to maintain. The women exemplify the important role played by citizen historians in preserving western historical sites and archives. Often motivated by patriotism or local interests, such groups ensured the survival of major research archives for academic historians.

Hugo L. Summerville Studios (active 1910s–1920s). Officers of the Daughters of the Republic of Texas. *Photograph, 1925. Photography Collection, Harry Ransom Humanities Research Center, The University of Texas at Austin.*

returned to Wisconsin for his Ph.D. during Turner's day. Thereafter Becker pursued a remarkable teaching and publications career at Dartmouth, at Kansas, and at Cornell. Early on he wrote a rhapsodic essay on Kansas that was in the most atavistic Turnerian mode and did a dissertation on political parties in prerevolutionary New York that reflected Turner's interest in institutional studies. Becker's barometer-like responses to changes in public issues suggest that he endorsed without reservation Turner's conviction that each age rewrites the past in terms of its own moods and needs. But studies on the Revolution and the Age of Reason show that Becker was well prepared to pass from environmental determinism and frontier activism to the impact of institutions and ideas.

Quite different in personality was Edward Everett Dale. Once called the "rarest humorist" in the Mississippi Valley Historical Association, Dale was born in Texas and raised in western Oklahoma's cattle country, where he cowboyed and taught in rural schools for more than a decade before taking his Ph.D. at Harvard in 1922. Although Dale claimed Turner "opened a new Heaven and a new earth" when he introduced him to "American History," Oklahoma and its people remained the focus of his life, of his teaching career (at the University of Oklahoma), and of his prolific writing. Authentic, affectionate, and good-spirited, Dale was not by nature a critic. Like many western historians, he exhibited a professionalism gentled and humanized by native loyalty.

Also useful in understanding the sliding scale between regional enthusiasms and frontier history as scholarly profession was Frederic Logan Paxson. Not a Turner student but sometimes regarded as the truest disciple of all, Paxson was born and educated in Pennsylvania, where he also received his Ph.D. Following professional openings and personal inclinations, Paxson adopted the frontier process as almost his single interpretive tool. At the University of Colorado (1903–6) he taught the full history offering, gained a few professional contacts, and got into western history with several articles on Colorado. At Michigan (1906–10) he enlarged his professional circle and published *The Last American Frontier* (1910), a pleasing application of the frontier process. Riding its success, he took Turner's place at Wisconsin in 1910, continued to advance in his professional connections, worked with good students, and turned out a succession of well-received books including *History of the American Frontier, 1763–1893* (1924), a prize-winning work widely hailed as the masterful frontier synthesis Turner never wrote. Moving to the University of California in 1932, Paxson maintained his reputation as a Turnerian and headed the history department during its great western history era.

Another disciple who embraced Turner's frontier interpretation but went beyond him in connecting with the people was the University of Iowa political scientist Benjamin F. Shambaugh. An Iowa native, Shambaugh threw himself into the affairs of the Iowa Historical Society, which he initially set out to transform from an antiquarian club to a true research institution. However, after World War I, Shambaugh shifted sharply, promoting what the historian Alan M. Schroeder has rightly termed "a lighter, more impressionistic," from-the-ground-up approach to history. This change of emphasis was apparent in the *Palimpsest*, a "magazine of popular history," and in the annual Iowa Commonwealth Conference where, during the late 1920s, representatives of the general public, the press, and national figures mingled enthusiastically.

State Societies and Citizen Historians

Shambaugh's determination "to look at history from the bottom up" and his wholehearted turn to citizen involvement were major attributes of the era's spirit. Historical societies, archives, and individuals were busy throughout the West collecting the record of regional pasts and making it available through study and publication. Although citizen historians worked closely with the scholarly profession, in many ways they represented a second voice for western history that prolonged the influence of epic or classic western themes and emphasized regional treatment.

An early manifestation of this spirit at the personal level, as we have already seen, was the San Francisco collector-historian Hubert Howe Bancroft. Others, like Hiram Chittenden and Elliott Coues, found time during distinguished government careers to indulge personal interests in writing and editing. A Corps of Engineers officer, Chittenden produced the three-volume *American Fur Trade of the Far West* (1902), which is still regarded as the definitive work on the topic, along with studies of Pierre Jean De Smet and steam navigation on the Missouri. In a flurry of historiographical enterprise, surgeon-naturalist Coues edited widely, including thirteen important volumes on the fur trade and Spanish exploration.

Citizen history also took the form of publicly funded historical societies throughout the West. Many counties and municipalities joined in this distinctively western development, but attention here will be directed to the role of state societies. Launched in the 1850s, the Wisconsin Society developed quickly under the untiring work of Lyman C. Draper, a fabulously successful collector, who combined passions for retrieving the pioneer story with a knack for managing Wisconsin's political network. He also possessed the innovative leadership to guide Wisconsin in becoming the first state to make state history an official function and to fund its operations.

Also an important Draper legacy was his 1887 choice of Reuben Gold Thwaites as his successor; Thwaites proved to be his equal as a collector and promoter and his superior as an editor. In his twenty-six years as secretary Thwaites vastly enlarged the society's popular mission, located its library at the University of Wisconsin, and worked closely with Turner and other academics whose cooperation was symptomatic of the close relationship between citizens' history and the scholarly profession. Perhaps most significant, Thwaites published no fewer than 183 volumes of edited documents, including the 73-volume *Jesuit Relations* and the 31-volume *Early Western Travels*. Not surprisingly some of these were not edited to the exacting standards demanded by academic professionals, and some academics saw them as part of the fault line between professional discipline and the popular taste for quantity, beginnings, and narrative.

In 1914 Thwaites was succeeded at the Wisconsin Society by Milo Quaife. An Iowa native with a University of Chicago Ph.D., Quaife rivaled Thwaites's energy and improved on his editorial standards, but Quaife was uneasy in Madison's political community and with the membership of the Wisconsin Historical Society. After a frustrating decade he left for the Burton Historical Collection in the Detroit Public Library. There he continued to edit manuscripts and reprints, wrote numerous books and articles, and edited the *Mississippi Valley Historical Review*, demonstrating again the intimate connection between popular and professional history in the Midwest. Fostering history readable by the public and scholars alike, Quaife pushed the Midwest as a region but was also drawn to the newer West and stood firm against eastern biases. The promising appointment of Joseph Schafer as Quaife's successor at the Wisconsin Society also left something to be desired. A native son and Turner disciple, Schafer came with a strikingly original vision, the "Wisconsin Domesday" proposal, which called for a multivolume cooperative effort to evaluate the contributions of common people in Wisconsin's development. As the University of Oregon historian Richard Maxwell Brown has commented, the Domesday study was far ahead of its times, and in the 1920s and 1930s it found little support.

If they sometimes enjoyed lesser reputations, other state societies also reflected the varying lines of citizens' history. The Kansas State Historical Society, for example, was chartered as a private organization in 1875. In 1879 it became the state's official agent for history but for years functioned as a sister institution to Kansas's almost rabidly patriotic Grand Army of the Republic (GAR), which, like its counterparts elsewhere, was a pressure group composed of Civil War veterans. Challenged, like other post–Civil War state societies, to redefine the position of state loyalties, the Kansas Society was undeviatingly committed to the grandness of Kansas, frankly promotional, thoroughly interested in beginnings, successful in collecting Kansas newspapers, wild about heroes, artifacts, sites, trails, and Indian wars, and willing to bend the truth for good cause. Catching its spirit precisely was Carl Becker's early essay: "The Kansas spirit is the American spirit double distilled. It is a new grafted product of American individualism, American idealism, American intolerance. Kansas is America in microcosm. . . . Within

INTERPRETATION:
A Land of Myth and Memory

◆

The romance of the West has remained a central theme for artists in the twentieth century. For some, the romance survives in the vast landscapes of the contemporary West. For others, it exists only in myth and memory, a subject to be mourned with regret or mocked with gentle irony.

Once lauded as a realist, Frederic Remington turned to increasingly moody evocations of the West toward the end of his life. Using a looser, more impressionistic painting style, he suggested that his ever-popular cowboy subjects were actually "ghosts of the past," historical figures more compelling in myth than reality.

Frederic Remington (1861–1909). Ghosts of the Past. *Oil on canvas, ca. 1908. Gift of The Coe Foundation, Buffalo Bill Historical Center, Cody, Wyoming.*

With bleached bones floating over an arid landscape, Georgia O'Keeffe evoked the vast space and clear light of the southwestern desert. Though she worked in a modern style that distinguished her from many of her predecessors, she too celebrated the seductive romance of the West. "From the faraway nearby," it was an exotic but accessible place in which one could find physical beauty and psychological freedom.

Georgia O'Keeffe (1887–1986). From the Faraway Nearby. *Oil on canvas, 1937 © 1994 The Georgia O'Keeffe Foundation/Artists Rights Society (ARS), New York. Copyright © 1984/88 By The Metropolitan Museum of Art. The Alfred Stieglitz Collection, 1959, (59.204.2). The Metropolitan Museum of Art, New York.*

Directly addressing the romantic image of Native Americans preserved in earlier paintings and photographs, Fritz Scholder depicts Indian subjects who defiantly and humorously refuse to dress or behave in stereotypical ways.

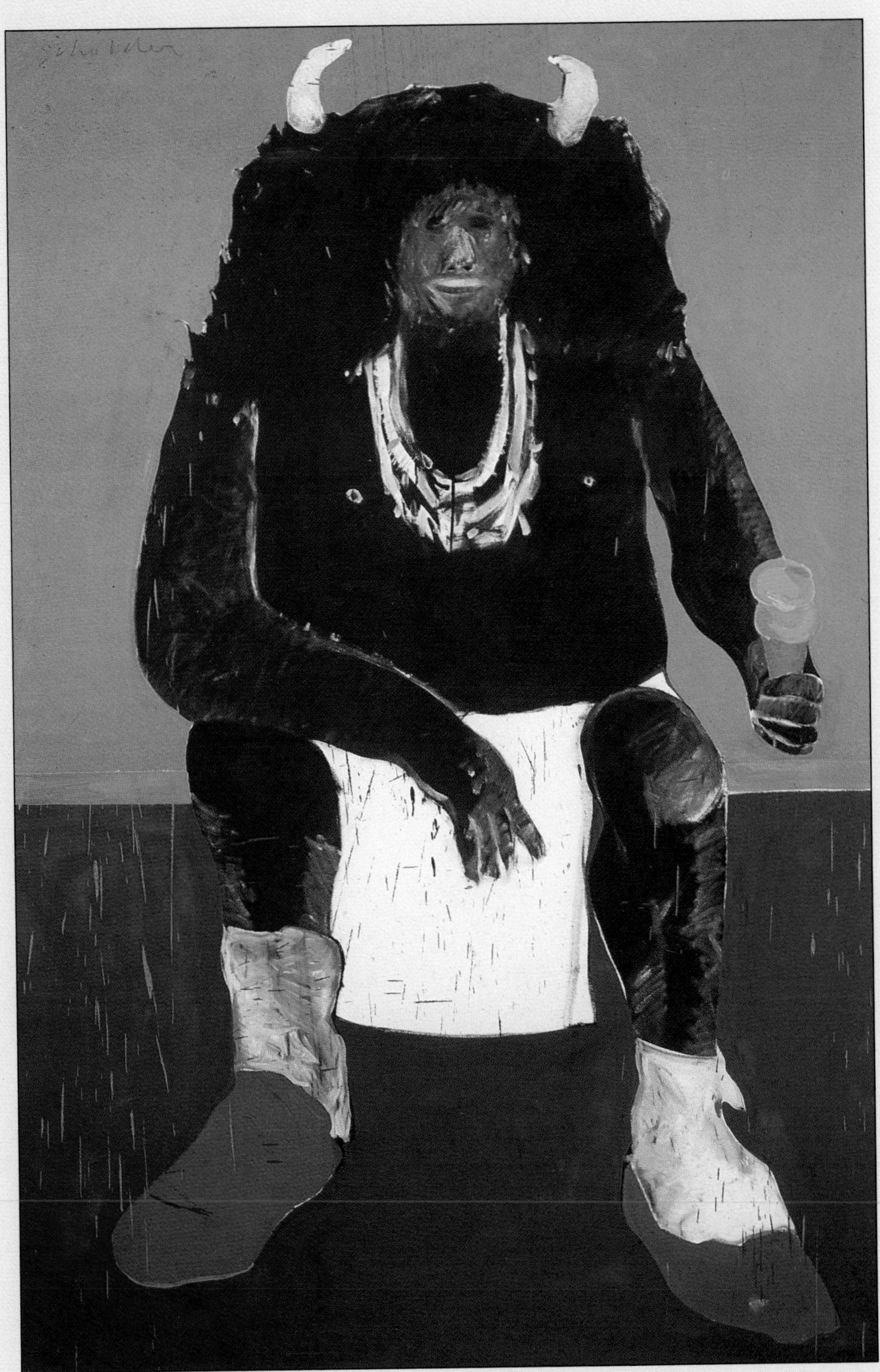

Fritz Scholder (b. 1937). Super Indian No. 2. *Oil on canvas, 1972. Courtesy of Nancy and Richard Bloch.*

Framing Monument Valley's distinctive rock formations through the windows of a viewing tower, the photographer Skeet McAuley ironically comments on a place made familiar through countless western films. If the Monument Valley of popular westerns is locked forever in a particular moment of the 19th century, this is a more dynamic site. Its history reaches into the past of "180 million years old" dinosaur track preserved at the lower left and extends into the present of the young Navajo worker who cleans the windows to clarify our view.

Skeet McAuley (b. 1951). Navajo Window Washer, Monument Valley Tribal Park, Arizona. *Type C print, 1984. © 1993 Skeet McAuley/Swanstock, Tucson, Arizona.*

its borders Americanism, pure and undefiled, has a new lease of life. It is the mission of this self-selected people to see to it that it does not perish from off the earth. The light on the altar, however neglected elsewhere must ever be replenished in Kansas."

First occupying a single showcase, the Kansas Society collections soon filled two rooms in the state capitol, then exploded to most of the basement as records and memorabilia accumulated, including a vile of Dead Sea water and a baby crocodile. In 1909 as the society threatened to take over an entire wing of the statehouse, the legislature authorized a magnificent five-story building. Dedicated in 1914 before twenty-five thousand cheering partisans, half of whom rolled into Topeka on chartered trains, the Memorial Building was shared with the GAR, which mercifully was a diminishing operation by this time. Several early directors were journalists noted for their Kansas propaganda but innocent of training in historical method. One, like Hubert Howe Bancroft, was said to have published works written largely by his subordinates. But the society flourished, making real contributions. Fulfilling its mandate to create a sense of community among a restless people, it proceeded from history dealing with classic, regionwide western themes to studies increasingly confined to the state itself in recent decades. With far-flung programs, a popular journal, respected directors, and a large membership, the modern society is mindful of the work of involved university professionals including James C. Malin's innovative locally based environmental studies, the history-from-the-bottom-up and for-the-sake-of-history-only of George L. Anderson, and the sound work of Homer E. Socolofsky and others. Well might Becker take comfort that the "light on the altar" is still "replenished."

In New Mexico, social and natural conditions made for a rich cultural texture that came to focus in the state historical society. Beginning with Native Americans and the conquistadores and extending through Zebulon Pike, the Mexican War, Spanish land grants, range wars, reclamation, and Los Alamos, a wide variety of writers and scholars worked what the western writer Wallace Stegner called the "broad borderlands of history." In their diversity they greatly enhanced state history, but they also modified its character, giving voice to points of view that went far beyond the conventions of academic professionals and hewed close to romantic western traditions.

Writers with national stature congregated at Taos in the years around 1900. More fully in the regionalist tradition were promoters such as the troubadour-poet Charles F. Lummis who was bent on bringing the entire Southwest to light. Beginning with Ralph Emerson Twitchell and moving through Harvey Fergusson, Paul Horgan, and, with shifting emphasis, Eleanor Adams, Myra Ellen Jenkins, Marc Simmons, and John Kessel, state historians described New Mexico's past, sometimes voluminously, often gracefully. Working from the record of American conquest and from rich Spanish archives, they came close to making the history of New Mexico and its related regions the quintessential western romance.

Meanwhile anthropologists and archaeologists called for "cooperative effort in all fields of history" and employed science as history's "lengthening arm." Known for their color and energy, they operated from bases in Washington, D.C., reveled in fieldwork at New Mexico's pueblos and prehistoric sites, and boasted worldwide connections. The live-in "Zuni" Frank Hamilton Cushing discomfited historians with his conviction that

Interest in local events and nostalgia for the imagined grandeur of the 19th-century past inspired a tradition of popular historical pageants in western communities that continues to the present.

Noel Photo Studio. West on the Lolo Trail, Lewiston, Idaho. *Photograph, 1936. National Museum of American History, Division of Community Life, Washington, D.C.*

the observable present could be used as historical evidence. Adolph Bandelier was considered an "anthropologist in the minds of historians and [a] historian in the minds of anthropologists," as the University of Arizona anthropologist Bernard Fontana has pointed out. And Edgar Lee Hewett was a globe-roaming, money-finagling, site-digging Museum of New Mexico director of almost heroic stature. Together they did much to establish the field of southwestern archaeology. They also complicated and enlivened the career of the New Mexico Historical Society and multiplied the voices of western history.

Regionalism and the Popular Impulse

Implicit in this essay and particularly relevant to the discussion of state history are questions of regionalism. It is a commonplace that the West itself is a region. No less obvious is the fact that it is composed of lesser regions that themselves divide into subregions. All of them vary with time and overlap according to the uses to which they are put. Turner's thesis takes, of course, a regional approach. As a means of understanding, regionalism is closely related to geography but is equally the product of how people identify with place and how one place affects another. Like nationalism, it rises from what Hans Kohn, a historian of Russian imperialism, called "the immense power of habitude" and, among other things, expresses itself in pride, in a sense of self-worth, and in group cohesiveness.

During much of the twentieth century, regionalism attracted wide attention. By extending and adapting a 1983 statement on "The New Regionalism in America" by Richard Maxwell Brown, we can divide western regionalism into three phases. First was the epic regionalism of the open or undifferentiated West that ended about 1890. Second were the six decades after 1890 during which what may be called sentimental and reform regionalism sought historiographic as well as political and economic redress for eastern domination. And third, after a post–World War II hiatus of perhaps three decades, during which regionalism's differentiations were obscured by an explosion of global, strategic, demographic, cultural, and environmental change, came a time of resurgent regionalism as historians sorted through the fallout.

To understand the early era of undifferentiated regionalism, we should recall that with the exception of California and Texas, all western states were territories before they were states. Before that, they had been Indian country, contested empire, unorganized territory, or simply part of the unknown. Across the entirety of this undifferentiated region moved the epic forces of conquest that inspired Prescott and Parkman and thrilled the denizens of the early historical society in Kansas. Wildness, discovery, westward expansion, the fur trade, transportation, and violent conquest—all predated the organization of states. Treated as history, they reached beyond state bounds to become epic regional themes. Vastly broadening the appeal of place, they brought common interests to all state historians, legitimated local studies for native-bred professional scholars, and made regionalists of them all.

Nostalgia for the epic adventures of the nineteenth century continued to dominate western regionalism during the first decades of the twentieth. Historians perpetuated H. H. Bancroft's romantic narrative style, emulated Thwaites and Quaife in publishing documents and new editions of rare books, and waffled between local promotion and a Turner-like mix of sentiment, radicalism, and analysis. In treating the past, they were also joined by people with a wide range of other interests, as we have seen in New Mexico.

Strong since the 1890s, regionalism's reformist mood intensified after 1918. The shock of war, industrial complexity, and depression transformed the disposition of many from the inward-looking and self-congratulatory doctrines of Turner to biting protest against things local, producing an indictment of rural isolationist America. The attack was especially virulent against the West, against the frontier, and against ideas and values associated with them. The western response sometimes added defensive elements to reformism as support formed behind insurgents and defenders. Illustrative were Van Wyck Brooks's despairing critique of the West in *The Ordeal of Mark Twain* (1920) and the spirited defense of Bernard DeVoto in *Mark Twain's America* (1932).

From as early as 1902, reform regionalism had taken an administrative turn in the rise of conservation and reclamation. Accelerating through the enlarged homestead and national park legislation of the Progressive Era, administrative regionalism came to its full reform expression in the New Deal. Relevant to the West's evolving historiography were the Public Records Survey and the Federal Writers' Project (FWP). Conceived as relief measures as well as a means of bolstering sagging pride, the Records and Writers' projects gave on-the-job historical training to thousands of writers and would-be writers in programs organized in state projects.

The effect on western historiography was dramatic. As Jerre Mangione explained in *The Dream and the Deal* (1972), the radical disposition of some prominent FWP figures brought heavy political fire. But influenced by gentler New Deal reformism, rank and filers of the western relief projects often shared popular views and exerted what was in the main a moderate if indeed not a conservative influence, giving the earlier romantic optimism of western regionalism a new lease on life as the New Deal sought to bolster self-confidence. Short-term outcomes for the West included the project's flagship undertaking, the *State Guide* series, as well as distinguished collecting and records survey programs, the promotion of citizen history, and a high interest among publishers and readers.

A longer-term outcome was the influence on a generation of regional historians whose careers were shaped by the project's populistic outlook and reform spirit. Western academics upon whose lives the FWP left a lasting imprint included George P. Hammond, state supervisor for the New Mexico writers' project, UNM professor of history, and longtime director of the Bancroft Library at the University of California. Supervisor of the Massachusetts state project was Ray A. Billington, later history professor at Northwestern University, author of the prestigious textbook *Westward Expansion* (1949), and research director at the privately endowed Huntington Library of San Marino, California. A professional in the broadest sense, Billington restructured and updated Turner in a brilliant and extended interpretive effort, helped transform the regionally oriented Mississippi Valley Historical Association into the more professional Organization of American Historians, held out for mixing buffs and professionals in the newly organized Western History Association, and maintained close contact with a wide range of citizen historians and publishers.

Suggestive of the FWP's influence beyond history was Harold Merriman, the Montana supervisor and English Department chairman at the University of Montana, who drew Pacific Northwest regionalists around him. In Arizona, the artist-illustrator, yarn-spinning free-lancer Ross Santee brought a contagious cowboy enthusiasm to the project, an enthusiasm that he continued to indulge throughout his long life. Idaho's state supervisor, the curmudgeon novelist Vardis Fisher, infuriated the project's Washington office but delighted westerners when he ignored directives and threats, and evaded efforts at physical restraint, to publish the *Idaho Guide* (January 1937) ahead of the guide for Washington, D.C., as well as those slated for powerful eastern states.

A Utah-related case may be used to illustrate the strengths and limitations of the FWP's influence on regional writing, highlighting the extending ripples cast by the project on the one hand and on the other pointing out that it provided no lasting center of gravity. The key figure was the Utah supervisor Dale Morgan, whose youthful energies surged beyond the immediate objectives of the project to a distinguished personal career in fur trade, western trails and maps, and Utah-Mormon history. Perhaps his best-known book was *Jedediah Smith and the Opening of the West* (1953). Without formal training in history, Morgan eschewed historical theories including Turner's. Facts, which he pursued with dogged determination, were the energizing thrust of his historical form. Yet, as Billington noted, Morgan "could write magnificently," combining "word sense with exactness of expression" and an intuitive grasp of history with impeccable scholarship. He also took the New Deal mandate to promote regional history as a continuing personal mission, placing himself at the center of a group of what may be described as Utah-Mormon reform regionalists.

Squarely in the reform tradition was the Utah-born pundit, editor, and historian Bernard DeVoto. He was known nationally for his shrill editorials in *Harper's*, for his "plundered province" approach to western exploitation, and for the neoromanticism of his great trilogy on the West: *The Year of Decision*, 1846 (1943), *Across the Wide Missouri* (1947), and *The Course of Empire* (1952). Back home in Utah, he indulged his penchant for pulling "shirt-tails out and setting them on fire," issued emancipation proclamations

Artists as well as writers received funds from the WPA programs that supported the documentation of regional history. Presented in post offices and other public buildings throughout the West, their paintings and murals generally depicted western history as the story of the triumph of the common man.

Everett C. Thorpe (1907–83). Early and Modern Provo. *Oil on canvas (study for mural in the Federal Building, Provo, Utah), 1940. Courtesy of Doris B. Thorpe.*

against economic and religious suppression, debated issues of Mormon history with Morgan and others, and helped aspiring Utah writers gain access to publishers. Equally in the reform tradition were several remarkable women, including Joseph Smith's biographer Fawn Brodie, Mountain Meadows Massacre historian Juanita Brooks, and novelist of polygamy Maureen Whipple, who were sustained by the tireless Morgan as they explored previously suppressed topics from personal situations largely beyond the reach of academic professionals. With scholarly credentials in the social sciences were Nels Anderson, Lowry Nelson, and Thomas O'Dea, who addressed Mormon themes from sociological perspectives. In a similar form was the historical geographer Donald Meinig who, in a benchmark article published while he was at the University of Utah, defined the Mormon cultural area as a significant western subregion. With creative writers, Morgan, who had majored in literature, maintained even closer touch, making vital contributions to the careers of William Mulder, Ray B. West, and Wallace Stegner, all of whom did significant histories in the reformist mood including Stegner's *Mormon Country*, a vastly underappreciated 1940s portrait of Utah, and his better-known *Beyond the Hundredth Meridian: John Wesley Powell and the Second Opening of the West* (1954). For Utah free-lancers, Morgan was a font of inspiration, a lexicon of information, and occasionally a ghostwriter. Although it has been little appreciated, this group and others who shared the FWP's regional reformism came close to being the established voice of Utah-Mormon history against which the so-called New Mormon History took its measure after 1965.

Although the Utah reform regionalists did impressive work, the ties that bound them slowly relaxed. The Writers' Project passed from the scene as World War II began. The exceptionalist spirit of regionalism, which they embraced wholeheartedly, seemed increasingly anachronistic in the post–World War II time of global crisis. As free-lance historians, they had no students to applaud them and perpetuate their work. Few enjoyed biographers, although Stegner's *The Uneasy Chair: A Biography of Bernard*

DeVoto (1974) is perhaps the best biography done on a Utahn. Death and other interests took them. DeVoto died in 1955. Little recognized at home, Morgan turned, about the same time, to more broadly regional themes at the Bancroft Library. He died in 1971. Social scientists and literature professors gravitated to other states and to professionally focused careers. Stegner's work assumed an increasingly literary turn, although, with a more regionally redefined western historiography taking form, he received the coveted Western History Prize in 1990. He continued to exert widespread influence on the environmental movement until his death in 1993. As the group aged, the thrust and nature of its work kept pace with the West's diminishing sense of its own exceptionalism. It was a process reenacted with variation throughout the region.

A New Regionalism: Paradoxes of Change, Success, and Identity

With these regional influences in mind we may return to the development of western history as a scholarly profession. In important ways scholarly historians were more subject to changing moods and interests than either the public or the more publicly attentive regionalists. Change among them has been explained by the University of California historian John D. Hicks and elaborated by the University of New Mexico professor Gerald Nash as resting on generational rhythms. Attention to successive generations helps explain the nature of the tensions through which western history was working.

From 1890 to 1920, as we have seen, was the era of the Turnerian synthesis. Encouraged by a responsive public, a small and homogeneous cohort of professional historians (two or three hundred strong at the most) hailed the frontier as "an ineradicable influence" for good. Defining it as the engine that powered the rise of an exceptional people, they lamented the frontier's closing, rejected European antecedents, excluded minorities, and ignored environmental costs. Bred to mounting complexity and disillusioned by the rise of industrial urbanism, world wars (hot and cold), the Communist revolution, the depression, and environmental woes, the generations that followed became increasingly pessimistic, questioned the role of the frontier, or, indeed, saw it as a destructive force. According to Richard Hofstadter's *The Progressive Historians* (1968), the very vagueness, impressionism, and overstatement that had given Turnerian doctrines "their plasticity and hence their broad acceptance" came under heavy attack in the decades after Turner's death in 1932. Also increasingly outmoded were his "assumptions about cultural transmission," his view of the West as "safety valve," and the "crippling isolationist implications" his theory had for foreign policy.

Temporarily checked by the affluence and professional opportunity of the years after 1945, pessimism among historians resurged sharply during and after the Vietnam conflict. Western history professionals, who by then numbered perhaps two thousand, hailed from a wide variety of backgrounds. Western history shifted its center of gravity to the Far West, setting up the Western History Association (1961) as the Mississippi Valley Historical Association became the Organization of American Historians (1965), and traditionalists refined nineteenth-century themes and worked to restructure Turner. In time many came to see the West as defunct, its history as passé, and the people attracted to its standard as second rank.

Even in the era of Turner's greatest ascendance there had been alternate voices.

Some, like Charles Beard and Arthur M. Schlesinger, Sr., respectively economic and urban scholars, foreshadowed a move away from Turner and the West. Among the founders of what the historiographers Rodman Paul and Michael Malone have termed the "classic tradition in western historiography," a few "operated outside the Turnerian nexus."

Among the earliest of these was Herbert Eugene Bolton, who "almost singlehandedly opened up and defined" the Spanish Borderlands as a "separate field of historical study." Born not many miles from Turner in Wisconsin in 1870, Bolton worked his way through college, studied under Turner, took a Ph.D. at the University of Pennsylvania, and by 1901 developed an interest in the northern frontiers of New Spain as a junior professor at the University of Texas. He moved first to Stanford and then to the University of California, his career taking on characteristics that became the hallmarks of his own work and that of his numerous followers. Not only did his interest in the Borderlands break with significant elements of the frontier thesis, but in his personal publications he went far toward doing for the Spanish what Parkman had done for the French (he once called it "Parkmanizing" the Spanish). Bolton also became the recognized father of research in Mexico's Spanish archives. Like Thwaites at the Wisconsin Historical Society, Bolton integrated a great collection of western Americana into scholarly research at the Bancroft Library, which he directed for many years after 1916. Also like Thwaites, he was one of western history's great editors, producing among other works *Anza's California Expeditions*, 5 vols. (1930), *Historical Memoirs of California, by Francisco Palou, O.F.M. . . .*, 4 vols. (1926), and *Kino's Historical Memoir of Pimeria Alta . . .*, 2 vols. (1919). He also inspired many of his students as editors. In his determination to put the record in context, he helped set two generations of western historians at work along the trails and through the historic districts as they undertook to make a coherent picture from fieldwork and the surviving records. His students were legion. He taught undergraduates by the thousands and directed the graduate work of some 300 M.A. students and 105 Ph.D.'s.

Nevertheless, Bolton was not universally acclaimed. A few saw him as little more than an antiquarian. He collected data and told its story, but he often stopped short of analysis and generalization. Responding to his work overall and especially to his 1932 address "The Epic of Greater America," other commentators found an unsound Pan-Americanism. Among others, John Francis Bannon and David J. Langrum have pointed out that defenders argued that he neither "ignored nor deprecated analysis or synthesis" but that like H. H. Bancroft before him, and many of his contemporaries, he "felt it was the job of" his generation "to furnish the facts." In sum, Bolton offered an alternative to Turner, yet he maintained much of the same spirit. More than Turner himself, he prolonged the West's taste for fact-loaded narrative history.

Radically different in style and in how he interacted with the scholarly profession and the public was Walter Prescott Webb. Always something of an outsider and a determined provincial, Webb was raised in a hand-to-mouth frontier existence at the southwestern edge of the Great Plains. After a delayed start he proceeded into history through a succession of windfalls and rebuffs. Throughout his career he remained at the University of Texas, where he indulged cross-disciplinary tastes and an instinct for advocacy and reform. Exposed early to "elements of late nineteenth-century social and cultural theory," Webb saw things in terms of a pronounced geographic determinism. However, he was influenced not by Turner but by Lindley Miller Keasbey, a mentor at the University of Texas, who had drawn from some of the same social science theoreticians. Profoundly aware of his own identity and less grounded in the historical method than most academic historians, Webb intuitively worked toward striking insights. These he supported with data gathered from a range of sources and personal observation. At its best, Webb's writing achieved a "feel of authenticity, a confidence and conviction" that, the historiographer Elliott West has said, reached beyond "thought to experience." He took criticism personally and, rather than bow to it, walked out on his Ph.D. program at the University of Chicago. Perhaps because of this experience he developed a "jaundiced view" of professional historians.

It is not surprising that Webb also broke from the crowd in his writing. Nowhere was this more apparent than in his first and perhaps greatest work, *The Great Plains* (1931). A compelling regional portrait that went beyond Turner in the forthrightness of its environmental determinism, it nevertheless served as a western fortification for Turner, in effect, applying exceptionalist doctrines to the very heartland of the West at a time when their influence was under attack elsewhere. Sharply criticized for *The Great Plains*'s tendency to proceed from intuition to evidence, Webb took refuge in the personal aspects of his approach and angrily refused to revise his book.

Nevertheless, the burden of Webb's message changed sharply over the years. Pausing along the way to celebrate the Texas Ranger and work up a handbook of Texas history, and always maintaining an old-boy touch with Texas, Webb shifted in mood from celebration to protest and finally to strident warning. The patent reform regionalism of *Divided We Stand: The Crisis of a Frontierless Democracy* (1935) lamented the ironies and injustices that made the West and South tributary to the Northeast. *The Great Frontier* (1952) detected the early warning signals of global disaster in the collapse of the "400 year boom" that had shaped the liberal tradition. *Divided We Stand* and *The Great Frontier* assailed distant enemies, but in "The American West: Perpetual Mirage," *Harper's Magazine* (1957), Webb indicted his own kind. Making a master key of aridity, he cried out against the thoughtless exploitation that was sweeping the West, bringing an avalanche of protest upon him.

Each of Webb's non-Texas works ran in important respects against the grain. Each was a statement in reform regionalism, and each was an act of scholarly independence. His later career might be said to reflect a spirit something like the despair of Mark Twain, and there can be no doubt that Webb suffered a loss of innocence. Change, however, brought more of sobriety and responsibility, perhaps even more of prophecy, than of despair. Indeed, it may be said that Webb personified both the strengths and the weaknesses of western history's play-off between the provincial influences of popular taste and the sophistication of professional method. He bored or infuriated many professional historians, but he continued to be heard. Important is the fact that in recognizing a vital interplay between region and globe and in shifting from the environmentalism of exceptionalism to the environmentalism of responsibility, Webb pointed the way for western history to make the transition from the buoyant and confident exceptionalisms of youth to maturity's hard responsibilities.

Serving to perpetuate and to broaden western history at a moment of otherwise narrowing interest was the innovative work of Henry Nash Smith in *Virgin Land: The American West as Symbol and Myth* (1950). Texas-born, and trained in cultural and literary criticism, Smith was overtly influenced by the Harvard Turnerian Frederick Merk's "richly factual" treatment of the "West's importance to any understanding of America" and was covertly disposed to many of Turner's ideas, as well as possessing a Turner-like penchant for bold analysis and overstated rhetoric. Concerned with the "ways that regional and cultural identity are shaped" and put off by what Princeton English Professor Lee Clark Mitchell describes as "the intellectual Balkanization that fractured mainstream scholarship," Smith "melded genres"—history and literature—to force an awareness that history is made as well as experienced. "As much an account of imagery and desire as of actual topographies or real events," history rested "upon what people thought they were doing as much as on what they actually had done." Drawing from cultural sources "'high' and 'low,'" Smith portrayed the West not simply as "a separate region of unique events and idiosyncratic assumptions, but as an essential component of . . . [the] national consciousness." Out of his work came a clearer understanding of how "popular beliefs, common assumptions, and government policies" coalesced in the form of symbols and images to become both self-definition and self-fulfilling prophecy for the western experience.

Virgin Land undertook to illuminate the national experience, yet it focused on the West and pertained primarily to the epic themes of conquest, heroic action, and agrarian development with which the westward movement had been identified. On the one hand, it clearly located the West in the mainstream of American history. On the other, it enlivened western history and offered a method by which it could be incorporated in the intellectual studies that were replacing the earlier emphasis on narrative and event-oriented history. *Virgin Land*, like *The Great Plains* and the Borderland studies, was a clear plus for western history, moving in the direction of relevance and broadened spectrums but perpetuating its earlier impetus.

The decades that followed 1950 were a paradoxical period for western history. In many respects it was a productive time, rich with creative energy and sound in the work accomplished. Yet, it was beset with ambiguity and pessimism, and for years its practitioners drifted without new direction in a sea of growing complexity. By the mid-1980s an awakening in regional interests was afoot among scholars and writers who had more than a little in common with the regionalists of the 1940s. This was apparent across a broad disciplinary spectrum, including novelists, journalists, and social scientists who were sensitive to classic western themes and to continuing Turnerian influences, especially as modified by the myth-and-symbol approach of Henry Nash Smith and his followers.

Nearer the traditional heart of western history was the work of a large number of mid- to late-twentieth-century historians who subjected the "individual topics" that made up the epic story of the nineteenth-century West to detailed and often brilliant examination. Considering this work, the historiographers Rodman Paul and Michael Malone called attention to solid progress in such classic fields as grazing, mining, agriculture, territorial history, Mormon history, comparative frontiers, and urban studies. Going further, others scholars broke with Anglo-American traditionalism to inquire into the role of Native Americans, women, and people of other national

The increasing academic attention paid to the historical experiences of ethnic groups in the West is paralleled in the emergence of new art forms depicting western history from the perspective of different cultural groups.

Barbara Carrasco (b. 1955). L.A. History, A Mexican Perspective. *Acrylic on masonite, 1981. Courtesy of the artist.*

backgrounds. Those concerned with women amended classic fields and worked new ground but initially did so in the nineteenth-century context. With newly developed quantitative techniques, ethnic groups were also considered, frequently in nineteenth-century settings but, in the case of southeastern Europeans and Hispanics, increasingly as manifestations of the twentieth century. Scattered pockets "of recent western history" were also worked vigorously, "among them Populism and the Progressives, conservation, reclamation," and preservation.

For decades the vitality and diversity of this work did surprisingly little to enhance western history as a field. Indeed to some it seemed only to cast further doubt on Turner's assumptions and to point up the inaccuracy and irrelevancies of the romantic tradition and the frontier model. Looking beyond the use of history as celebration, which had moved so many earlier historians, this detailed examination turned to a variety of new methodological and thematic alliances that modified and in many cases weakened their connections with the West.

Both the extent of this scholarship and the centrifugal forces acting upon it were related to the growing numbers of professional historians in the post–World War II years. With college enrollment stimulated by the GI Bill, baby boomers, and a series of Pacific wars, western universities fairly exploded in size. In the late 1940s and 1950s, history departments grew dramatically as Ph.D.'s pouring from distinguished universities were hired. Senior scholars, many of whom had local ties, launched new graduate programs in western history, and libraries vigorously competed for western materials as university after university sought to produce and market Ph.D.'s.

Just as this production line was put in place, markets collapsed. In the post-Vietnam climate, history enrollments diminished, and student interest shifted abruptly. The ensuing job crunch was catastrophic for overstocked history Ph.D.'s generally but was especially critical for graduates coming from institutions with recently founded western history programs, for whom head-to-head competition with candidates from established institutions added to the dilemma of finding employment even in public history and other fields outside the academy. As curricula broadened and faculties became increasingly national, western history's influence on departmental policy slipped sharply, leading old-timers to lament "The Insignificance of the Frontier" with John Caughey, longtime editor of the *Pacific Historical Review*.

While post–World War II momentum was still high, the Western History Association (WHA) had been organized in 1961 with a membership of some three thousand professors, writers, publishers, book collectors, public historians, archivists, and friends of the West. Under the editorship of Russell Mortensen, formerly director of the Utah Historical Society and director of the University of Utah Press, and Gregory Crampton, a Bolton Ph.D. at Utah, the WHA launched the *American West* in 1964. It was hoped that the lavishly illustrated, hybridized journal would help gather the WHA's variegated membership within a single loop and brand it as a popular movement. In a painful exhibition of centrifugal forces at work, WHA membership diminished, and surviving members lived with growing tension, yet the magazine quickly built a subscription list that increased from thirty thousand in the mid-1960s to more than one hundred thousand in the early 1980s. The *American West* first shook off the University of Utah and ultimately the WHA to become by the mid-1980s a glossy, environmentally sensitive, wide-circulation periodical of western life and tours—with only the thinnest, most commercialized veneer of western history. In 1970 the WHA responded to proposals coming from George Ellsworth and Leonard Arrington of Utah State University to set up the *Western Historical Quarterly*, which soon became a respected scholarly journal but made only an occasional bow in the direction of popular interests and did little to reverse tension and attrition in WHA membership.

Professional loyalists called for various fixes. Two proposals merit attention here. At first hearing, Caughey's "Insignificance of the Frontier" statement had a pessimistic ring indeed, but his "larger proposition" was positive and hopeful. Western history would always be "bobtailed and insufficient," he warned, unable to address the problems of the present, much less the future, unless it broke from its "self-imposed imprisonment in the early and antique West" into the twentieth century. A call for twentieth-century awareness made by the University of California professor John D. Hicks as early as the 1930s and 1940s was picked up and amplified by many, including Hicks's student Gerald Nash of the University of New Mexico. Publishing a useful twentieth-century text in 1974, Nash greatly facilitated course work in the twentieth-century history of the West. Simultaneously, graduate studies focusing on the recent past multiplied as the body of twentieth-century records grew.

Another thoughtful response to western history's malaise came from the University of Oregon historian Earl Pomeroy. California-born Pomeroy, like Caughey and Nash, was a UC Berkeley student during the high tide of the great triumvirate of Paxson, Bolton, and Hicks. Like Caughey and Nash, Pomeroy entertained a keen interest in the twentieth-century West, but Pomeroy went well beyond them, offering a larger critique and proposing shifts that would enable western historians to break out of the "self-imposed" constraints Caughey had described. Teaching at Wisconsin, the University of

North Carolina, and Ohio State during the 1940s, Pomeroy published books of lasting value on territorial government and America's dependencies in the Pacific. Later he moved to the University of Oregon, where in the 1950s and 1960s he issued a succession of important historiographic articles.

With the softer touch of one working from within, Pomeroy directed heavy criticism against western history yet paid homage to its early greats. Turner and H. H. Bancroft "sold," he declared, each after his own fashion, "the one as idea, the other as mass, as shelf space—to a public" ready to buy. As time demonstrated, the frontier idea—Turner's "environmental radicalism," Pomeroy called it—was oversold to generations of western historians. Failing to follow Turner in his "larger qualities: his concern for both analysis and synthesis, his effective English style, and the keenness of his mind," many "apostles" became more "orthodox" in their commitment to the frontier thesis than "the prophet" himself, their outlook becoming increasingly restricted as his orthodoxies became outdated. "By stressing differences rather than similarities, by offering escape rather than solution, by celebrating those aspects of the past that seemed least like the present, or least like other parts of the country," Pomeroy told his colleagues, they had been entrapped by the "dead past" of which Turner himself had repeatedly warned. They had failed, in effect, to let "conditions uppermost in their own time" lead them to write "history anew." Working from the large body of Turner criticism that had accumulated by the 1950s but tying his indictment to a call for constructive change, Pomeroy found particular fault with western history's rejection of what was continuous, similar, and comparable in favor of the discontinuities of space and time implicit in the "closed frontier" idea and in the West's strong regional identification. Also at fault were western history's preference for the romantic and heroic over the analytical, its infatuation with Turner's pastoral succession, its short-shrift treatment of cities, and its tendency to exalt the role of the individual at the expense of the territorial system, the federal government, and institutions generally.

Pomeroy's essays posed a significant challenge, not merely to western historians but to himself as well. Fortunately his subsequent books, especially *The Pacific Slope: A History of California, Oregon, Washington, Idaho, Utah, and Nevada* (1965), validated his right to speak. This work was a model of western history, the kind Pomeroy had been calling for. Stressing the interaction of the old and the new, the importance of social, political, and economic ties, the power of the metropolis, limits and growth, and environmental change, it was a masterwork of analysis and synthesis. Rich with insight and invitation to further work, it fully merited Yale westernist Howard Lamar's 1977 characterization as "easily the most complex and sophisticated history of the American West."

Among other things, *The Pacific Slope* was an expression of historical professionalism. A response to the issues of the twentieth century, it broke sharply with the kind of history that Pomeroy had inveighed against at various times as "atavistic," "antiquarian," or "accumulative rather than interpretive." Clearly he sought "solutions" to the issues of the era, rather than "escape." In many respects it was a rejection of the "multiple interest" formula that lay behind the organization of the WHA—a prime example of western history as a formal and sophisticated part of American history. Yet in dealing with the Pacific Slope, it exhibited an exemplary kind of regional connection.

The New Western History and the Quest for Relevance

In short Pomeroy was beginning to define a new and more interpretive point of view for western history. By the mid-1980s his call for analysis and relevance asserted itself in what was increasingly called the "new western history." "Thundering out of New Haven," as the cowboy fictionalist Larry McMurtry described it, the new western history was indeed deeply influenced by broader forces, including Yale's Howard Lamar and numerous other eastern historians and movements, but it also drew from influences from within western history. In one sense the movement experienced a "jump start" effect, as the large body of sound work in individual subfields of the Old West finally assumed the critical mass and momentum necessary to create a dynamic of its own. But beyond this accumulating body of detailed studies, the new western history was also the work of a number of young professionals, many of whom were moved by a passionate concern for late-twentieth-century issues, to which they applied widely varying interests and methodological approaches. To them, relevance and currency were necessary attributes. Most rejected the vagueness of moving frontiers for the physical West that lay beyond the hundredth meridian yet insisted that the western experience be studied in relation to larger national and global issues, an approach that accommodated discussion of the economic forces, twentieth-century developments, and social and natural influences in which they were interested. As expounded in the work of some, the new western history was sharply revisionist, punctuated with anger, confident in its presentist viewpoints, and quite ready to put down the old. Others saw their break with the past in less radical terms, or as the product of a rather clearly defined evolutionary process, and happily acknowledged the continuing worth of earlier viewpoints.

Not surprisingly it was the involved revisionists who gave the movement its most recognizable form. None played a more important role than Patricia Nelson Limerick, a California native, Yale Ph.D., and University of Colorado professor. Carried to the forefront by her *Legacy of Conquest: The Unbroken Past of the American West* (1987), and a taste for the limelight, the irrepressible Limerick based her argument on a neat turn of the shopworn proposition that the romantic spirit and exceptionalist doctrines of the frontier interpretation virtually precluded any understanding of the West's role in the contemporary world. But with much greater ingenuity she moved on, representing the western experience in terms of conflict's legacy, thus bringing to the surface of western history long-suppressed themes including the roles of gender, race, class, industrialization, urbanism, the environment, global influences, and the federal government. Although she could at times be blunt, or even offhanded, in her denunciation of earlier historians, as in her rejection of Ray Billington's career, Limerick's sense of her personal past, and her recognition that relevance required professionals to search for common grounds with the public, revealed an appreciation for popular tastes in history and a conviction that dialogue between professionals and the public was a necessary aspect of history's development.

Also at the forefront of the new western history was the University of Kansas environmental historian Donald Worster. Born in California of Dust Bowl refugee parents, Worster got his Ph.D. at Yale, taught at Hawaii and Brandeis, partook heavily of the developing spirit of the environmental movement, and wrote persuasively about the international development of the science of ecology, about the Dust Bowl, and about the natural and social problems inherent in harnessing the waters of the West. But more to the point of the new western history, Worster condemned western historians for failing "to see themselves as critical intellectuals" and called for a more "penetrating view" of the past, a view freed from the "unthinking acceptance" of myths and explanations, both official and private, a view that would allow them "to discover a new regional identity and a set of loyalties more inclusive and open to diversity than we have known and more compatible with a planetwide sense of ecological responsibility." Worster's was an enlightened statement and an impassioned, almost desperate, appeal. In it were strong elements of the environmentalist mentality, including certain elitist presumptions, the zeal of a crusader, and finally, what strikes one as a conviction that progress is the product of confrontation and partisan activism rather than of scientific objectivity.

Behind the eloquent leadership of Worster and Limerick, the new western history enlarged its impact. In addition to the passion of Worster's environmentalism and the excitement of Limerick's unbroken past, others made continuing and far-reaching contributions. Multiculturalists brought the West as a field of conflict to the fore. Native American history was addressed with renewed sensitivity and breadth. In biography, writers like the gifted and versatile Robert Utley continued to deal with mythic figures, but biography dealing with Hispanics, women, and twentieth-century figures also fared well in prize competitions and commercial markets. The twentieth century came into its own as a subject of historical curiosity. Understanding of the role of women made great strides, passing quickly from stereotype and celebration to penetrating gender-related analysis. Conflict was recognized as a major western legacy. Problems of power and hierarchy were frankly addressed. Frontier and empire were related as global manifestations. Historiography boomed as western historians sought their intellectual roots. And following the examples of Turner, Webb, Smith, and people like DeVoto, Stegner, and Meinig from related fields, as well as Limerick and Worster, western historians struggled to move away from what the latter termed the "camouflage" of "dull gray prose" into the light of stimulating words.

As a field of writing, western history has taken a more relevant stance. In moving toward this position, it has sometimes lashed out against many of the values and attitudes that underlie popular interests. Yet it has attracted a large if undefined public interest, suggesting that substantial elements of the public impulse have also shifted away from a western history of congratulation, escape, and development to one of cultural and ecological responsibility. After a century in which the voices of popular interest and professional effort have sustained and reworked each other, western history again finds itself the product of widely varying voices in which popular impulse and professional responsibility continue to take new forms. Together they promise well for the future.

Bibliographic Note

Of general works let me mention only two. Ray Allen Billington and Martin Ridge's magisterial *Westward Expansion: A History of the American Frontier*, 5th ed. (New York, 1982), first published in 1949, still eloquently anchors western history in America's frontier experience. Less comprehensive but updated in spirit and pointedly concerned with the trans-Mississippi West is Richard White's readable and thought-provoking *"It's Your Misfortune and None of My Own": A New History of the American West* (Norman, 1991).

Historiographical literature about the West is abundant. Couched primarily in terms of the frontier as a broad force in American history are general bibliographies of long-standing usefulness. Included among them are Frederick Jackson Turner and Frederick Merk, *References on the History of the West* (n.p., 1922), and R.W.G. Vail, ed., *The Voice of the Old Frontier* (Philadelphia, 1949). Working from the point of view of successive frontiers but more current in character is the "selective" bibliography that has been part of each of the five editions of Billington and Ridge's *Westward Expansion*. Impressive individually, Billington's bibliographies are even more useful cumulatively, since each new edition pares down the older listing of monographic literature to bring the current work up-to-date. Also serving to maintain the broader role of the frontier in American history are biographical works about frontier historians. For example, see Marcus Cunliffe and Robin W. Winks, ed., *Pastmasters: Some Essays on American Historians* (New York, 1969), which includes essays on Francis Parkman, Frederick Jackson Turner, and Vernon Louis Parrington, names closely associated with frontier history. Ray A. Billington, comp., *Allan Nevins on History* (New York, 1975), offers Nevins's views on Parkman, Turner, and George Bancroft. Clifford Lord, ed., *Keepers of the Past* (Chapel Hill, 1967), presents excellent biographical sketches dealing with the contributions made to state historical agencies by the western historians Lyman C. Draper, Reuben Gold Thwaites, and Edgar Lee Hewett.

Although the frontier idea remains strong and earlier Wests still attract much interest, recent historiographies tend more to the Far West. A useful beginning tool is Rodman W. Paul and Richard W. Etulain, comps., *The Frontier and the American West* (Arlington Heights, Ill., 1977), whose 2,973 entries, although true to the dualism of its title, give earlier frontiers short shrift in comparison to the West as trans-Mississippi region. In Michael P. Malone, ed., *Historians and the American West* (Lincoln, 1983), the editor's introduction and seventeen chapters authored by specialists tie strongly to today's West, yet in chapters dealing with Native Americans, "manifest destiny," and exploration, it too extends to earlier Wests. With a strong twentieth-century focus, Gerald D. Nash, *Creating the West: Historical Interpretations, 1890–1990* (Albuquerque, 1991), follows chronology to emphasize the West of recent times. Without targeting questions of West as region or as frontier, Roger Nichols, ed., *American Frontier and Western Issues: A Historiographical Review* (Westport, Conn., 1986), tends more to the latter than do other works with which it is contemporary. For a wide-ranging and still useful guide to periodical literature, see Oscar O. Winther, *The Trans-Mississippi West: A Guide to Its Periodical Literature* (Bloomington, 1942), for which supplements were issued in 1961 and 1970. The "Recent Articles" and "A Dissertation List" features of the *Western Historical Quarterly* (Logan, Utah, 1970–) have made running lists available in these two categories.

Biography also provides useful historiographic reading. For those with a taste for brief encounters, the "Dedications" in each issue of *Arizona and the West* (Tucson, 1961–85) offered biographical sketches describing and often celebrating the work of deceased western historians. During its first decade (1970–79) the *Western Historical Quarterly* gave the same approach a little different twist in autobiographical essays in which historians explained their interest in western American history and described their own work. Adding to the biographical approach to western historiography are jointly authored books sketching the lives and contributions of historians. Among the most useful are John R. Wunder, ed., *Historians of the American Frontier: A Bio-Bibliographical Sourcebook* (New York, 1988), in which fifty-seven early-twentieth-century figures are treated with what might be termed sympathetic objectivity. Similar in tone but written in greater depth are the eleven essays in Richard W. Etulain, ed., *Writing Western History: Essays on Major Western Historians* (Albuquerque, 1991). Different in the fact that they make a case for the new western history are the jointly produced essays in Patricia Nelson Limerick, Clyde A. Milner II, and Charles E. Rankin, eds., *Trails: Toward a New Western History* (Lawrence, 1991).

Taken together, these historiographies provide evidence that western history continues to have many different voices, some responding to popular and professional impulses and others gravitating around the question of whether the West is frontier or region.

BLACK HILLS
JEWELRY
BOOTS

Chapter Twenty-two

Selling the Popular Myth

ANNE M. BUTLER

Wall Drugstore, once a shabby soda fountain shop tucked deep into the South Dakota Badlands, exudes the rugged ambience of a popular culture that embraces the American West. An out-of-the-way pharmacy on the verge of collapse in 1936, the store rocketed into economic vitality when its shopkeepers, Dorothy and Ted Hustead, tied their advertising to the environmental factor that threatened to destroy them—western aridity.

During the viciously dry summer of 1936, a promotional gimmick—offering travelers a free drink of ice water—struck Dorothy Hustead. In a burst of old-time frontier ingenuity, the Husteads used their surrounding regional drabness to capture the attention and money of passersby. A few quaintly lettered road signs, patterned after the famous Burma Shave advertisements, helped the moribund drugstore change into a merchandising bonanza. Hundreds of people, drawn by the refreshing thought of cold water, stopped for a thirst-quenching drink and stayed to purchase western wares from an ever-growing selection.

The Husteads' operation expanded with each passing year. They enlarged the facilities for their famous ice-water well and added dining rooms, home-baked goods, jewelry counters, outdoor displays, a bookstore, a western clothing shop, a museum, and a chapel. Road-weary motorists stopped for water, or breakfast, or gas, but lingered to wander from room to room, shopping for cheap western knickknacks, a ten-gallon hat, a plaster of paris Indian statue, or cowboy art painted on velvet.

Wall Drugstore is both a geographic and a commercial entry point to the American West. Those who stop now do so less out of thirst and more out of curiosity. The aggressive merchandising of a tourist-trap set in neighborly association with the nearby Dakota Badlands surprises the visitor. Over fifty years, the focus of the store shifted from the critically necessary to the completely superfluous—the mass marketing of the commercial baubles of western tourism. The customers browse through rooms of western gewgaws, souvenirs, they believe, of the "real" West. The success of Wall Drug—based on regional commercialism, national imagination, and mass marketing—mirrors much of the twentieth-century selling of the West.

The Husteads, of course, are not the sole cultivators of this boulevard view of the American West. America's western vision grew from older, deeper roots. From the very earliest expansion, the nation built its epic lore. By the nineteenth century, almost three hundred years of exploration, settlement, and human exchanges had molded a popular

For more than 60 years, Wall Drug has successfully marketed the West as a commercial commodity. On a hot summer day, more than 20,000 visitors stop at the store in Wall, South Dakota, to shop in an indoor mall populated with life-size statues of trappers, cowboys, and dance-hall women and displays of Indian jewelry, boots, and mounted big-game trophies.

Unidentified photographer. Wall Drug Mall. *Gelatin silver print. Courtesy Wall Drug, Wall, South Dakota.*

culture that drew energy from song and story, event and myth, heroes and villains. Fact and fancy swirled and merged as a national population of increasing ethnic and cultural diversity struggled for identity. One nineteenth-century character, Davy Crockett, nicely illustrates this wedding of reality and fable, truth and legend. More than 150 years after his death at the Texas Alamo, Crockett maintains his status as an American folk hero. Crockett does so because his life—in both its real and its imagined events—captured important aspects of "Americanness." His gun-and-ax beginnings as a hunter and pioneer in Tennessee marked him as a tough, self-reliant outdoorsman. Although Crockett sprang from humble stock, his elevation to the U.S. Congress in 1827 underscored his place in the era of the Jacksonian self-made man. His adventuresome, impulsive migration to Texas thrust him across frontier worlds and linked him permanently with the Far West. His 1836 death at the Alamo during the Texas revolution defined him as a champion of Anglo-style democracy and a foe of its oppressors.

In all his exploits, some embroidered by his own words, Crockett seemed always "to stand tall" and to face challenge, hardship, and even death with an admirable stoicism, his basic beliefs unshaken. Crockett, by himself and with his compatriots at the Alamo, personified American notions of what it meant to cherish freedom and oppose tyranny. He carried the heritage of the American Revolution into new territories for white America and established a model of independence and feistiness in manhood. Thus, in his own day and for generations that followed, he reminded Americans that the West, a tangible place where one stood for cherished principles, against all odds, was for and of men.

Whether these virtues accurately captured Crockett's opinions and thoughts remains unimportant. His outward actions supposedly demonstrated his inner convictions. His character conformed to a mythical code by which American men can judge themselves. The historian Paul Andrew Hutton summed up the weight that is placed on Crockett's power as a figure of the American West. "We should focus on the positive forces he symbolized. . . . It is through his spirit of unbridled democracy and bold egalitarianism that this magnificent American is best remembered." Critics of Anglo-Texan chauvinism might not agree with Hutton, but his language outlines Crockett's embodiment of the western myth, especially for men and women from an Anglo-European background.

Crockett continues his hold on the national imagination, perhaps because he represents what some Americans want to believe about the West and about themselves. No historical revelation or scholarly analysis will shake Crockett from his pedestal, for his life provides the irrefutable evidence that the West of the popular imagination embraced an epic that belonged to white American males. His feats are tightly woven into the fabric of values that many Americans think they exhibit and cherish: independence, honesty, fair play, self-reliance, loyalty, courage, justice, love of freedom. These boastful Americans assert that if the stories of the frontier, the legends of a Davy Crockett, did not happen, they should have, for these myths give foundation to a modern sense of identity.

Nonetheless, Crockett, born of an eastern forested world, is but a John the Baptist in the face of modern western popular culture, squarely positioned in the trans-Mississippi West. The Far West, with its soaring mountains, shimmering deserts, and

A Tennessee congressman turned land speculator who met his end at the Alamo in 1836, Davy Crockett helped create the image that made him an enduring symbol of popular western culture. Though he often wore formal dress to fit into polite Washington society, he adopted buckskin leggings, moccasins, and a long rifle when he posed for John Gadsby Chapman in 1834. Walt Disney likewise portrayed him as an independent and resourceful frontiersman in a popular television series of 1954–55, but unlike his historical counterpart, the TV Crockett wore a coonskin cap. The Disney show catapulted Crockett to instant celebrity status and initiated the biggest run on raccoon pelts since the 1920s.

John Gadsby Chapman (1808–89). David Crockett. *Oil on canvas, 1834. Harry Ransom Humanities Research Center Art Collection, The University of Texas at Austin.*

Unidentified photographer. Fess Parker as Davy Crockett. *Gelatin silver print, ca. 1955. © The Walt Disney Company. The Disney Publishing Group, Burbank, California.*

windswept prairies, with its raging storms, blizzards, and droughts, with its spaciousness and starkness, reconfigured the national conception of wilderness. This dramatically beautiful region—difficult for newcomers to inhabit—demanded mental adjustments. So different from the lush East—really only a codicil of Europe—the arid West came to "explain" the philosophical and psychological development of America, seemingly the most singular of nations. In the search for standards by which to define national character and temperament, Americans often attributed moral and intellectual dimensions to western physical topography and the hardships it imposed. This magical transformation of wilderness into democratic political conventions and national moral principles enabled the movement into the Far West.

In this national mythology, westerners became larger-than-life figures, and the details of their humanity blurred. Gone were the fine lines of visage, the elements of personality, the shadings of culture. Western characters, shapes and voices with no hint of distinction, assumed all the variation of a string of paper doll cutouts. In a short time, America had simplistically defined its westerners and showed no inclination to relinquish the ensuing stereotypical perceptions of western people. The West meant cowboys and Indians, trappers and traders, miners and pioneers. One by one these characters assumed traits like those assigned to Davy Crockett, until they all marched through western history untarnished.

No Place for a Woman

White males controlled the central roles in America's popular West. They explored and settled a tough environment and battled its equally tough peoples. They handed over to the nation more than vast lands. On these champions, Americans hung their sense of winning and, perhaps even more important, their sense of adventure. Cowboys, miners, trappers, soldiers, outlaws, and even farmers seemed to prove to an adoring public that once life had held no tedium, no sense of entrapment. Rather, each day brought risks and challenges that enlivened the spirit and promoted American democratic principles. Whether a dashing cowboy or a raucous outlaw, these heroes—by their style, verve, and independence—"won" the West. Since historical reality never competed very forcefully against the fanciful images, Americans happily crowned their princely models of Wyatt Earp and Wild Bill Hickok. As the frontier era faded in the face of technology and change, twentieth-century Americans made no secret of their nostalgic longing for this grand era when gold was for the finding, land for the taking, and living for the bold.

Their imagined scenario depended on a regular cast of western characters against whom the hero figures regularly triumphed. An advancing white society drew Native American people into enough wars, economic clashes, and social destruction that a popular wisdom easily dismissed all tribes as inferior. Nonetheless, Native Americans, as supporting players in the western saga, were critically important props to document notions of the cultural superiority of white society. Without human adversaries for defeat, the conquering of the West remained incomplete. Although the barren, waterless land held its own terrors, these could not equal the fears generated by mounted foes from unknown cultural traditions. Only with the suppression of those who truly called the West home could white society rest easy in its new surroundings. Few heroic tales of the

West failed to include an obligatory cast of Indian people, either "noble" or "savage," foils who repeatedly reminded white America about the technological strength and power of its own culture.

Other indigenous populations fared badly too. Hispanic groups played a less central but no more dignified part in the tale of Anglo chauvinism. Cast as "colorful" standbys, Spanish-speaking people assumed rigidly defined roles in the popular vision of western history. Painted in the hues of docile peasants, ferocious bandits, or sensual fandango dancers, Mexican Americans provided the "humorous" proof of Anglo superiority. The internal contours of their world and their cultural responses as a community meant nothing in the construction of national myth.

African Americans also found their experiences distorted in the images of the American West. Indeed, the popular memory virtually expunged African-American life from the written and oral records of the western movement, which emerged solely as a tale of white society. Although many black families looked to the West for opportunity after the Civil War, their homesteading and ranching lives never assumed a place in the popular discourse. The "buffalo soldiers" rode with the U.S. Cavalry, five thousand or more black cowboys herded cattle, and nineteenth-century African-American communities sprang up in Colorado, Kansas, Texas, and Wyoming. Despite a well-documented involvement in the Far West, African Americans rarely saw public recognition given to their ancestors' presence in both the rural and the urban areas of the region.

If African Americans suffered from historical neglect, Asian Americans remained invisible in the popular perception of the West. White America vaguely connected Asian people to the 1860s effort to build railroads, but interest in the dynamics of their lives as a western minority group never materialized. Indeed, Asian workers, vital to the construction of the Central Pacific Railroad, were not visible in the famous photographs, taken by Andrew J. Russell on 10 May 1869, that celebrated the joining of the transcontinental lines at Promontory, Utah. Reduced to degrading caricatures, Asians hardly penetrated historical accounts. Their scantily noted experiences barely yielded enough data for a separate focus in the popular culture. Their lives, culture, and language all remained so rigidly peripheral to the interests of mainstream America that Japanese internment, which totally removed one group of citizens from public view during World War II, demanded no parallel excision of Issei elements from western popular culture.

Distortion and neglect characterized the presentation not only of these ethnic groups but also of western women. In the national imagination, nineteenth-century gender definitions intensified. Basic ideas about pioneer women sprang from a national conviction that frontier wives and mothers relentlessly followed a path dictated by courage, patience, and strength. In the American mind, these women, unquestioning of husbands' decisions, sustained families through any trauma of the pioneer West but contributed little of economic or political import.

This tale gathered moral authority from those who remembered pioneer family members. Kathleen Chapman, looking back on her Cripple Creek childhood, spoke for thousands of pioneer youngsters when she said, "I never knew Mama to be idle." The image of the passive, overworked wife and mother has proven a comfortable, unchallenging formula by which to assess the experiences of western women. However, under the scrutiny of historians, a more subtle picture is emerging. Katherine Harris has argued that among homesteading families in nineteenth-century northern Colorado, women as helpmates enjoyed an increase in family responsibility and economic decision making. Elizabeth Jameson stressed the flexibility and interdependence of work roles between men and women, a circumstance that ultimately transcended the family and affected women's roles in the public sector. Carolyn Stefanco expanded this idea with her assessment of the suffrage movement in Colorado, where women capitalized on their experience in church and social organizations to secure the franchise.

Furthermore, stereotypical versions of women's lives provided the basis for an unyielding attachment to racism, further enmeshing ethnic women in social bias and obscuring their historical past. Native-, Mexican-, African-, and Asian-American women struggled under the weight of gender constraints that have never evaporated. Within the popular language of the West, ethnic women rarely assumed any activist character but remained as shadow figures. Linguistic usage—generated from within the white community in academic texts, dime novels, and film—designated minority women as "dusky," "sensual," "earthy," "promiscuous," "exotic," "criminal," and "filthy." Within popular culture, ethnic women simply represented decorations that helped to reinforce the regional uniqueness of the West.

To some degree, this representation applied to all western women—they filled the background as decor. Simplistic definitions of womanly roles gave credibility to the romantic visions of the West, as understood by men. Although the "good" women of the West—presumably white, married, middle-class pioneers—received a sort of obligatory nod in the western scenario, their status could be truly clarified only if they appeared in contrast to the "bad" women of the West. These, of course, included the prostitutes, dance-hall girls, and female outlaws. Within that framework, the mule skinner Calamity Jane, the bandit Belle Starr, and the sharpshooter Annie Oakley fit perfectly, for each could be dismissed as the quirky exception that affirmed the maleness of western society. The secondary players—women and minorities—got their parts only so that they could add greater glamour to the leading men of the West. By the era of mass audiences for radio and television, central casting needed only a sidekick like Tonto to gallop behind the Lone Ranger, or a dance-hall madam like Miss Kitty at the Long Branch Saloon to adore Marshal Matt Dillon of *Gunsmoke.*

New Arenas

The American West has served as a vast national playground for collective and individual dreams. Davy Crockett, along with other historical and fictional figures, paraded through the national imagination. However, these frontier characters were only one aspect of the modern popular attachment to the American West. The place itself took on great significance. Thus, many tourists, from the United States and throughout the world, felt impelled to travel to this hallowed place, to drink in its meaning, to be restored by its atmosphere. In addition, the purchase of western symbols assured tourists they could return home with regional relics that validated their connection to the "real" America. Through keepsakes, the nationalistic values and virtues of the West could be exported to other parts of the country or of the world. Surrounded by their reminders of the West, these mostly middle-class tourists maintained the belief that purple sage and mounted riders meant Americanism.

But consumers did not need to leave home to see the West. Mass marketing brought images of the region to American households. Transformations in industrial production, especially improvements in the technology of printing, and expansion in communication media catapulted cultural images of an earlier age into the twentieth century. For example, after 1900, the production of western art, already a booming business of the nineteenth century, reached industrial proportions. A host of American illustrators took up pen and brush to depict the life and culture believed to be part of the western experience. These artists did so with a new dedication, for the advent of the twentieth century convinced Americans, once more, that rather than "capture" the West, they must "preserve" it. After all, only seven years earlier, the historian Frederick Jackson Turner, in his famous essay "The Significance of the Frontier in American History," had warned that the pioneer West and its democratic virtues had all but disappeared. The very essence of American social and political democracy appeared to be slipping through the nation's fingers.

Certain illustrators, even as they sought personal income, wanted to preserve within the American spirit this meaning of frontier life. Charles Russell and Frederic Remington, each with a highly successful art career, promoted both the frontier concepts and the artistic style that served as guides for others. Turn-of-the-century changes within the publishing industry provided these illustrators with a regular forum for selling works that resonated with their personal visions of national meaning. No fewer than ten thousand magazines appeared on the American market between 1890 and 1940. Publications such as *Cosmopolitan*, *Ladies' Home Journal*, *Collier's*, *Red Book*, and the *Saturday Evening Post* depended on a dozen or more western illustrators for the art that accompanied the western fiction pieces. Among the most successful, Newell Convers Wyeth, William H. D. Koerner, and Frank Schoonover sketched hundreds of western images for these often romantic short stories.

One of the most notable, William H. D. Koerner, exemplifies the impact of these artists on the American mind. During his career, Koerner turned out more than twenty-four hundred illustrations, approximately six hundred of which depicted western themes. He collaborated with a long list of western authors, illustrating the works of Oliver La Farge, Bernard DeVoto, and Zane Grey. His western images graced more than forty stories that appeared in the *Saturday Evening Post* between 1930 and 1935. Koerner's *Madonna of the Prairie* (1922), an earlier portrait of a pioneer woman, around whose head the arch of her Conestoga wagon seems to form a halo, remains one of the most widely recognized of western renderings. He cared deeply about the values that these authors of western fiction tried to portray, and their written tributes to him over his long life demonstrate the success with which authors felt he captured their philosophy.

Like Koerner, all the illustrators turned out appealing images of frontier people and heightened Americans' positive attitude toward western settlement. These artists presented westerners as individuals of character, refinement, and integrity. Perhaps more important, as the nation moved into a world that turned on mechanization and

Molly Wingate, the heroine of Emerson Hough's novel, The Covered Wagon, *traveled west with a wagon train and found true love when she arrived in Oregon. For this book jacket illustration (which subsequently influenced the film version of the story), William Koerner portrayed her as a beatific Madonna of the prairie and gave her a more contemporary look by depicting her with a fashionable 1920s hairstyle.*

William Henry David Koerner (1878–1938). Madonna of the Prairie. *Oil on canvas, 1922. Buffalo Bill Historical Center, Cody, Wyoming.*

industrial growth, these illustrated western stories kept national frontier concepts at center stage for American readers. The work of the illustrators could be found every week in any American home with magazine subscriptions. In the decades spanning World War I, the 1920s, the Great Depression, and World War II, western images accompanied uplifting frontier tales offered for the magazine-reading public.

Not all illustrators took pleasure in this easy market, and some, chafing under the commercial constraints of deadlines and style, determined to break away from the stereotypical magazine portrayals. One of these, Maynard Dixon, in the early twentieth century disappeared on extensive field trips deep into the Southwest in search of remote desert scenes. Dixon wanted to disassociate himself from "traditional," by which he meant commercial, western art and use his brush to capture the harsh grip of weather and environment. Desert aridity fascinated him, and his portrayals of it contradicted the rosy, watered horizons of a group of his colleagues who had, since the late nineteenth century and under the patronage of various railroads, turned out highly promotional western landscapes.

Inadvertently, Dixon convinced Americans that they had yet to understand the "authentic" West. So, his efforts to retreat from the popularized aspects of western art were not entirely successful. The 1939 decision of the U.S. Department of the Interior to place his murals in the Bureau of Indian Affairs pleased Dixon, but the placement of his work in this generic public building pointed to the hopeless task of separating art, popular image, and commercialization in the national visions of the American West.

While illustrators advertised the West and its inhabitants through art forms of every style, other popular entertainments, some of which originated in the nineteenth century, also added to the West's specialized image. For the American public, nothing brought the West more directly into their lives than the Wild West shows. From among fifty or more traveling shows, the one to popularize the American West most successfully was Buffalo Bill's Wild West and Congress of Rough Riders. More than any other figure, Buffalo Bill Cody, himself a mixture of one part reality and two parts fiction, bridged the nineteenth- and twentieth-century evolution of western popular culture. Almost single-handedly he pushed frontier western notions into the modern scenario and made them accessible to a general audience.

By the time Buffalo Bill got around to opening his "Wild West, Rocky Mountain, and Prairie Exhibition" in 1883, he already enjoyed wide fame as a western personality. A well-known scout for the frontier army, Cody assumed celebrity stature during the 1870s due to the literary vision of Edward Zane Carroll Judson, who used the pen name Ned Buntline. Buntline rode the crest of a new wave of western fiction that gave rise to the wildly popular dime novels.

Buntline's decision to develop a fictional frontier hero from the real-life Buffalo Bill was one of the most fortuitous moments in American marketing of the West. The public responded with such enthusiasm to Buntline's books that it proved to be a short hop for Cody from the published exploits in *Buffalo Bill: The King of the Border Men* to the New York stage as an actor in *The Scouts of the Plains; or, Red Deviltry As It Is.* However, it was the traveling Wild West show that made Buffalo Bill the king of international entertainment and cast him as the living definition of the West.

Cody's posters, plastered around the nation, advertised that each of the two daily shows could accommodate twenty thousand people and promised to present "actual scenes, genuine characters" from the West. These vast crowds loved his part-circus, part-pageant, all-West extravaganza. Audiences not only clamored for so-called historical depictions of frontier events—the attack on the Deadwood stage, the defeat of General George A. Custer at the Little Bighorn—but particularly cheered at those outdoor skills so singularly western. Dazzling rope tricks, dangerous bulldogging, and superior marksmanship belonged to the West, and Buffalo Bill filled his show with performers who excelled at these feats.

Americans flocked to Cody's shows convinced he brought them the true West, for, as he promised, he gathered into his performing ranks "genuine characters"—Native Americans of several tribes. As these men, most in flowing headdress and carrying elegantly feathered coup sticks, rode bareback into the arena on stunning painted horses, who among the enthralled viewers questioned how well these appearances replicated tribal life on the Great Plains? In addition, Cody's ability, in 1885, to produce the most famous Indian of all, Sitting Bull, further underscored the reliability of his West.

The Indians themselves, still reeling from recent political and social events that had uprooted traditional economies, found needed employment with Buffalo Bill. Closed out of political decision making within white America, denied access to industrial

Buffalo Bill Cody's touring show helped fix the American cowboy as a daring and dashing figure in the American imagination.

Unidentified artist. Buffalo Bill's Wild West and Congress of Rough Riders of the World. *Color lithograph, ca. 1896, Courier Lithography Co., Buffalo, New York. Buffalo Bill Historical Center, Cody, Wyoming.*

training, and segregated from educational centers, Native American people accepted one of the few jobs they could easily secure—that of entertainment figures. This depressed economic track continued to haunt Indian endeavors throughout the twentieth century, leaving them with few options. Economic opportunities within the context of white America rarely broadened and typically centered on Native Americans' willingness to "play" at being Indians. America's entrepreneurs, such as Buffalo Bill, happily hired Indians who "stayed" native and, thus, furthered their own stereotyping. The Indians who galloped into Buffalo Bill's arena did much more than add a dash of color; they embodied the reality that the twentieth-century forces of commercial capitalism marginalized native peoples who found themselves in a fixed and ungenerous economic structure.

In addition to presenting Indian peoples as glamorous showstoppers completely divorced from their own environment and outside their economic milieu, Cody brought Americans the perfect frontierswoman in the person of Annie Oakley. Her remarkable skills at marksmanship set her apart from effete eastern women, but her genteel appearance and stylish costumes revealed her to be a "true" woman. In one more western paradox, Annie Oakley, born in the East, a child of Ohio, defined the nature of western women—capable and refined—for the country.

These images of Native Americans and white women slid easily into the national consciousness. Indeed, why should any one have challenged the tone and content of the

Cody show? After all, Cody, a product of the West, knew the internal dynamics of the region, and he went directly to the source to secure his talent. His popularity knew no bounds. It was all right with his audiences if Buffalo Bill, transformed by fiction and fame into an elegant, patrician cowboy, became the national caretaker for western authenticity.

Cody not only commanded the admiration of millions of Americans but also saw his international reputation soar. One bronc rider who stayed with the show from 1900 until 1907 traveled around the world more than three and one-half times on tours that included performances in Outer Mongolia. The Cody cast appeared twice before Queen Victoria of England. In 1913, Prince Albert I of Monaco, who knew the American showman from these stunningly well received European tours, journeyed from his tiny principality to Cody, Wyoming, for the chance to "kill a grizzly or two" under the guidance of Buffalo Bill, the Prince of the Frontier.

The magnetism of the Buffalo Bill show could not draw audiences indefinitely. When the program diluted its strictly western themes, the demand for the Cody Wild West show faded. But by the time that happened, Americans had discovered the rodeo as a western entertainment replacement. Predicated on work skills and frontier amusement formed in the earliest days of cattle raising, the rodeo proved a natural for promoters. The thundering action, the magnificent animals, the death-defying cowboys—all swirling around in a sweaty, dust-filled arena—added up to a certain crowd pleaser. For the price of admission, Americans saw, close up, examples of modern men who reflected western toughness, raw courage, and natural strength, living proof that the pioneer spirit of the West remained intact. Once an informal ranch sport that branched into local and regional competitions for seasonally unemployed cowboys, rodeos hit the national big time in the twentieth century.

By 1910, with the sport spreading rapidly and the participants more full-time contestants than part-time cowboys, the competitors at a Denver fair organized the Bronco Busters' Union and demanded five dollars a day for the wild-horse riders. This early unionizing attempt exemplified the tone of rodeo competitions for the next several decades as gate revenues mounted around the country, but low wages and dangerous conditions remained unchanged for contestants. By 1936, both the Boston Garden and New York's Madison Square Garden hosted rodeos that ran for almost three weeks. In New York, audiences totaled nearly a quarter of a million, far surpassing the usual crowds of one hundred thousand associated with rodeos in Cheyenne, Wyoming. Once again, the "Wild West" had hit the eastern pocketbook.

Unlike several areas of western popularizing, the early rodeos made a place for women, who often competed in the same events as the men performers. As early as 1935, Tad Barnes Lucas, a female trick and bronc rider, commanded an impressive annual salary of twelve thousand dollars. Women bronc riders and trick ropers usually appeared under special individual contracts. However, the enigmatically named Cowboys Turtle Association, a union formed in 1936 to promote fairness for rodeo personnel, opened its membership in 1940 to all cowgirl performers.

Rodeo waned during World War II as many of its youthful stars went off to military service and as western communities restricted large public gatherings because of wartime exigencies. In those locations that permitted rodeos, women took up the slack, as they

Although she never ventured west of her home state of Ohio, except when on tour, Annie Oakley came to personify the spirited frontier woman of popular myth. Performing in Buffalo Bill's Wild West show for 17 years, she delighted audiences abroad and at home with her showmanship and skill as a sharpshooter.

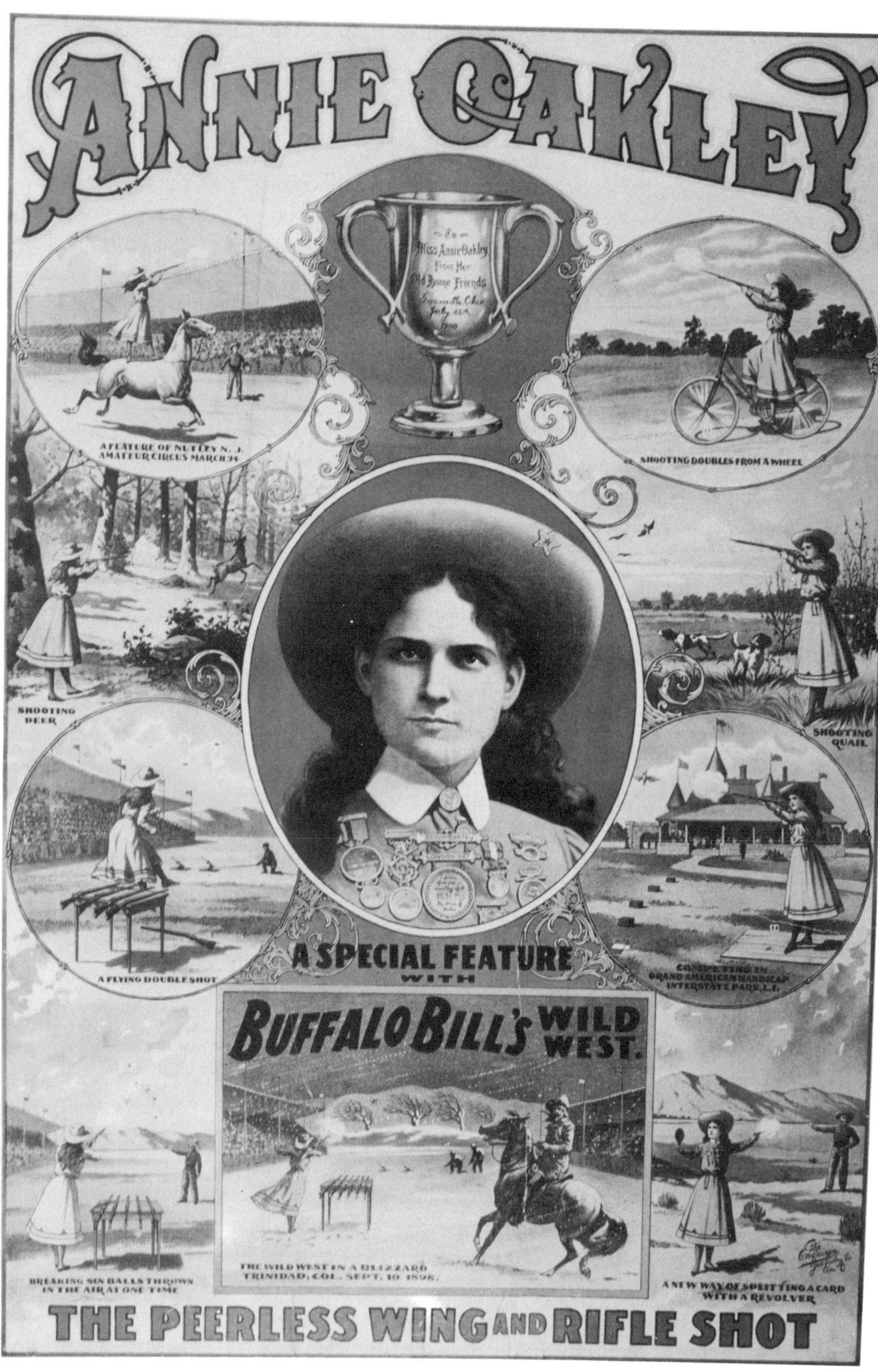

Unidentified artist. Annie Oakley: The Peerless Wing and Rifle Shot. *Lithograph, 1901, Enquirer Job Printing Co., Cincinnati, Ohio. Circus World Museum, Baraboo, Wisconsin.*

did in other war circumstances. In 1944, on the cover of *Hoofs and Horns*, the premier rodeo publication, mounted cowgirls smiled down on three rather aged cowboys welcoming a wounded veteran home from the war. In the post–World War II years, however, rodeos generally barred women competitors, until only barrel-racing events remained for distaff performance.

As public interest continued to accelerate during the postwar 1940s and 1950s, rodeo officials considered ways to further exploit the growing popularity. In 1959, a National Finals Rodeo Commission, which employed the promotional services of a well-known theatrical agency, brought the first rodeo competition to Dallas, Texas, in an event shown on national television.

Major American corporations savored the commercial opportunities in this decidedly nineteenth-century western event broadcast by a decidedly twentieth-century technology. R. J. Reynolds Tobacco, Levi Strauss Clothing, Wrangler Jeans, Justin Boots, Frontier Airlines, Ford Motor, Coors Beer, Schlitz Brewery, Heublein Whiskey, Nestea Foods, and Hesston Farm Machinery all invested in underwriting various aspects of rodeo competition during the 1970s, 1980s, and 1990s. Although market indicators pointed to climbing profits for sponsors, companies insisted they endorsed rodeos because these events reinforced American family values.

In the midst of cow wrangling and bull goring, that rationale remained blurry, yet huge trophies and giant purses drew an ever-increasing number of contestants, many of whom had never worked as cowhands but had trained at rodeo schools to become professional riders. When the national finals, held each December, took up residence in Las Vegas, Nevada, the transformation of rodeo from local pastime to major sport industry was complete. Set in the tinsel mecca of the United States, the national finals now meant a massive influx of fans, costly seats, high-stakes gambling, rodeo stars, ersatz cowboys, and media hype.

For those ersatz cowboys, the rodeo became one of the latest means by which young Americans found a "way out" from a less attractive life. In 1982, Charles Sampson, an African-American cowboy, ranked as the champion rodeo bull rider in the country, with winnings that totaled over ninety-one thousand dollars. Sampson, however, hardly acquired his expertise on the open range. The eleventh of thirteen children, Sampson grew up in Los Angeles, where he credited the YMCA and the Cub Scouts with fostering his interest in the rodeo. He made a clear decision to become a cowboy and studied to do just that. Although Sampson advanced as a national contender, hundreds of his peers found their competitive outlet in the small all-black rodeos sprinkled throughout the Southwest. In 1983, Sampson sustained a serious injury at a rodeo attended by President Ronald Reagan, another self-constructed media cowboy, and subsequently returned to bull riding in rodeos of the Southwest.

Local rodeos proved a popular sport for Native Americans and Mexican Americans as well as blacks. Within ethnic communities the rodeo served to strengthen the bonds of people who often lived great distances from each other. For example, the rodeo amplified earlier Indian social traditions. Building on the custom of tribal powwows, Indians added rodeos to their established social structure. However, within the context of mass marketing, the Indian rodeos, as well as the Mexican-American ones, have remained outside the commercial mainstream. Although ethnic groups have adopted some aspects of the "western" mode—clothing, jewelry, crafts, rodeos—they lack both internal capital and endorsement from big businesses and thus can point to only modest economic results.

Mexican Americans, however, were more successful in using their touring circuses as a source of ethnic support. Well into the 1950s, Mexican circuses brought regular ethnic entertainment to such southwestern cities as San Antonio and Tucson. The Escalante Circus, the Ortiz Brothers, and the Rivas Brothers—from the late nineteenth century—all enjoyed great popularity among Hispanic people. In the twentieth century, national entertainment agencies hired several of the performers from local Mexican circuses to tour on the vaudeville circuit throughout the United States.

The Mexican circuses continued to perform regularly along the Mexican-American border. Known as the *carpa*, these were intimate family-centered circuses, which brought entertainment to small towns in remote western regions. Although Anglos often enjoyed these performances, they overlooked the content of the Mexican-American comic routines that protested against language and social discrimination. The *carpa*, outside the context of the national imagination, points to the importance of localized entertainment, rich with cultural textures and popular with residents within the West.

Putting on the Hat

As American fans propelled rodeos—local and national—into a new status as western sport, they did so to a changing beat of music. New trends in American music began to waft out of southern communities, both white and black, in the 1920s. In the 1930s, as rodeos began to draw wider national attention, American music shifted again, this time with a regional tilt to the West, especially Texas and Oklahoma. Although music in many forms—folk songs, ballads, hymns—had always been central to the pioneer tradition, the transformation in the 1930s from southern-country to country-western represented just one more of the cultural impacts of the Great Depression. Until the 1940s, however, American taste in music continued to divide along urban-rural lines, and the new country-western style stayed popular with a limited audience. Within the West itself, the truly indigenous music of Native Americans and Mexican Americans never acquired the label of "western" music. Like Indian rodeos and Mexican circuses, each ethnic style of music retained an identity limited to the parameters of its own community.

Putting on a cowboy hat not only added a western element to what had been southern-country music but also increased the national audience. Tex Ritter's career furnishes a solid example of how commercial interests shaped music production. As a personality, this well-educated and savvy Texan solidified the singing cowboy image for easterners. Ritter cultivated his western persona in a folksy manner that belied his sharp business sense. He seemed to personify the down-home plain manner that increasingly appealed to an American population beleaguered by the Great Depression. Buoyed by Ritter's success, more and more musicians, often only as genuinely cowboy as the hats and boots they donned, invaded the national music scene—the Prairie Ramblers, the Monroe Brothers, the Blue Sky Boys, and Bob Wills and his western-swing band, the

Texas Playboys. All benefited, at least peripherally, from the jingling spurs and simple melodies of the wildly popular and true westerner Gene Autry. Of equal importance to the western music scene was the easy sound of that unauthentic cowboy Roy Rogers, who began life as Leonard Slye from Duck Run, Ohio. Rogers embodied all the personal virtues Americans imagined of cowboys. Clever handlers put those qualities on display for the nation, and Rogers's career took off, eventually landing him on the cover of *Life* magazine in 1943.

Regardless of where these musicians lived or traveled, when the New York songwriters picked up on the distinctive beat and sound, it was not long until all Americans knew tunes for more than "Tumbling Tumble Weeds." Few noticed that easterners increasingly wrote the popular lyrics, managed the publicity, controlled the record distribution, and owned the radio stations that broadcast the new performers. That commercial day soon came when the public cared not who produced or sang western music. In fact, country music no longer required that the performers even offer the pretense of being western cowboys. In the 1940s, both the Andrews Sisters and Bing Crosby cut highly successful country music hits, such as "Biting My Fingernails" and

Through radio shows, films, and later television, Gene Autry became the quintessential cowboy singer. In 82 films made between 1934 and 1959, in which he usually appeared as a fictionalized version of himself, he established the western musical as a film genre, paving the way for other cowboy stars such as Roy Rogers and Tex Ritter.

"Springtime in the Rockies." *Offset, © 1937, Allied Printing, Essex County, USA. Poster Collection, Prints and Photographs Division, Library of Congress, Washington, D.C.*

"Pistol Packin' Mama." Country music, born with rural southern roots, had used a few cowboy hats to switch from a "hillbilly" to a western identity, which quickly moved beyond region to a national sound.

From its heyday in the 1940s, commercial cowboy music took a substantial nosedive in the 1950s as competing styles within the genre struggled for commercial dominance. Then came the invasion of rock and roll and the eruption ten years later of a brash young group from England, the Beatles. Interestingly, it did not take rock and roll very long to turn to western themes, but these would veer away from the older cowboy lyrics of the nineteenth and early twentieth centuries. Instead, California as a sun-and-surf paradise emerged from the early music of the Beach Boys, only to be replaced in the mid-1970s by the more disillusioned Golden State vision of the Eagles. As groups proliferated and music escalated as a form of social comment, California and the West remained central to the themes, demonstrating once again the mythic power of western symbolism. No longer the Great Depression melancholy of Jimmie Rodgers's "California Blues," the musical message of the Pacific Coast now agonized over drugs, Haight-Ashbury, and America's failed revolution of the 1960s.

In this immense musical renovation, the guitar twangs of country-western music died away as listeners equated the sound with low-class, southern-white origins. American music, in songs like "If I Had a Hammer" and "Blowing in the Wind," took on more and more of the hues of a social protest—a protest that often placed the blame for America's ills in the heart of the South and in the homes of rural people. Country music, however, had not died. It merely waited for a changed climate of social impulses and a new generation of performers, all of whom would be handled by high-powered promoters. By the mid-1960s, as the nation absorbed the shock waves of political assassination, civil disorders, counterculture protests, and debilitating foreign entanglements, country-western music found its moment to go national, as never before. Although television altered production styles and personal costuming for performers, country music's original format—the radio—remained a powerful vehicle for selling the renovated sounds. No longer the music of the South and the West's rural whites, country music attracted the nation's yuppie commuters and blue-collar workers, who tuned their car and truck radios to those lyrics of heartbreak and loneliness, self-reliance and courage. An economic notch or two above the original country music crowd, these fans heaped their approval on a new generation of recording stars, such as Merle Haggard.

Haggard proved to be the perfect new champion of country music. Born to poor Oklahoma farmers, Haggard made the famous depression-style "Okie trek" to California, where he grew up in Bakersfield. After a troubled life that would one day be a press agent's dream, Haggard began to make a steady living with music. Haggard's 1969 hit "Okie from Muskogee," which celebrated people who did not smoke marijuana or burn a draft card, appealed to an entirely new national constituency of listeners: middle-class white Americans, who distrusted notions of social protest and increasingly endorsed conservative values and national conformity. That trend among Americans gained momentum until in the late 1980s and early 1990s two presidents, Ronald Reagan and George Bush, each attended star-studded performances at the Grand Ole Opry in Nashville, Tennessee, the eastern-based Vatican of country music. The importance of these visits—from presidents who pitched their political rhetoric in varying degrees of reactionary language targeted toward the white electorate of the South and the West—was not lost on the American public. Musical statement and political philosophy came together as never before.

Off to the West

Is it any surprise that many Americans longed to tour the West, to watch real cowboys leap into the saddle, to visit the haunting bluffs at the Little Bighorn, to stand at the north rim of the Grand Canyon, to canoe down the magnificent Snake River, to squint out over barren Death Valley? Is it any less surprising that businesses, in addition to the Santa Fe Railway and Frontier Airlines, eagerly hoped to make such adventures possible? Although western tours had always attracted eastern and European visitors, the tourist industry, born in the late nineteenth century, gathered unprecedented steam in the first third of the twentieth century.

The explosion of tourist interest owed much to that self-proclaimed westerner Theodore Roosevelt. In 1888, Roosevelt, who a few years earlier had retreated to the Dakota Territory after the deaths of his wife and his mother, wrote a series of articles about the West for *Century* magazine. A true celebrity, Roosevelt increased the enthusiasm of the American public for firsthand encounters in the wilderness.

Conveniently, the Roosevelt publicity corresponded to the economic misfortunes of many of his Dakota rancher friends, who had lost almost everything in the disastrous winter of 1886–87. Although their cattle profits plummeted, ranchers still owned enormous spreads of land and delighted at the chance to entertain paying guests, drawn largely from an eastern elite. Out of this series of coincidences evolved a new western feature, the dude ranch.

Here, urban visitors, outfitted in appropriate attire, had a chance to experience all the nuances of western life. They rode the range, hiked through national parks, camped in remote areas, and ate by smoky campfires. None of these activities required much adjustment from the cattlemen and their hands, whose daily tasks often followed this same pattern. The happy difference for ranchers concerned the economic benefits accrued from wealthy easterners who frequently stayed for an entire summer season.

The first influx of tourists guaranteed national publicity for the new concept in western vacationing. Among those who heaped praise on and brought attention to dude ranches were the author Mary Roberts Rinehart, the artist C. M. Russell, and the stage humorist Will Rogers. Through the 1930s, thousands of Americans visited the western dude ranches, adding to their own sense of the West and filling the coffers of western ranchers. At the peak of popularity, over 350 dude ranches sprawled from Arizona to Montana. A far cry from the luxury resorts of the modern West, complete with swimming pools and gourmet dinners, the first dude ranches offered no softening amenities and totally immersed visitors in the physical and spiritual meaning of western living, as defined by the ranchers and their hands. The owners of the dude ranches often required that guests plan for a minimum visit of two weeks, since the ranchers argued that to savor the pace and style of western life required an extended stay. Indeed, the ranchers shifted from innkeepers to interpreters of the West with remarkable ease. Once

By the early 20th century, the West's cowboy-and-Indian heritage had become a focal point for tourism.

In the Southwest, Fred Harvey built on the success of his regionally furnished railroad hotels with "Indian Detour" trips that brought visitors directly to the pueblos.

T. Harmon Parkhurst (1884–1952). Detour Buses at Santa Clara Pueblo. *Gelatin silver print, ca. 1935. Courtesy Museum of New Mexico (#3813), Santa Fe.*

Founded in 1884 as the West's first dude ranch, the Eaton Ranch in Wolf, Wyoming, introduced its guests to the frontier past with pack trips that combined a spirit of "roughing it" with the amenities of civilization. A camp brochure noted, "The tents of the ladies and the gentlemen are carefully segregated for mutual comfort . . . and a special row of tepees is arranged for married couples."

A. J. Baker. Ladies Row, Dude Ranch Camp in Glacier National Park. *Gelatin silver print, 1917. Montana Historical Society, Helena.*

merely the routines of a day-to-day existence, ranch schedules and daily chores transmuted into secular rituals for understanding the West.

The popularity of the dude ranches coincided with a growing attention to the spectacular scenery of the national parks. Owners of the dude ranches fanned this interest with highly publicized pack trips, which often led visitors into natural areas that were largely inaccessible except by horseback. As the general public expressed more interest in these remote regions, so did the corporate world. Inside Glacier National Park, the Great Northern Railway, like its southwestern counterpart the Atchison, Topeka, and Santa Fe at the Grand Canyon, built several hotels and oversaw the construction of trails to the interior. As early as 1903, Harry Child, who controlled the concessions in the Upper Geyser Basin of Yellowstone National Park, hired an architect to design a visitors' inn that looked out over Old Faithful. Rugged and rustic, the gigantic inn opened in 1904 to national acclaim that has never diminished. By 1920, almost one million tourists a year visited the national parks.

The national parks proved such a draw that in 1936 the businessman and politician-to-be Averell Harriman selected a wilderness site and used the Union Pacific Railroad to carve a private winter resort from the then-isolated Sun Valley, Idaho. The Union Pacific ran a skiers' special to Sun Valley until well into the 1960s. Everything about the trip and the resort catered to the rich and famous until other modes of transportation opened Sun Valley to greater numbers of less aristocratic visitors.

The commercial development of the interior reaches packaged a new western commodity for the American consumer—the wilderness. During the first four decades of the twentieth century, that wilderness remained largely the domain of the moneyed American. However, after World War II, accessibility to wilderness areas changed significantly. The expansion of the interstate highway system, along with a leap in the number of motel chains and campsites, gave American tourism a solid place in the West. A booming airline industry, the fad for oversized recreational vehicles, and the road appeal offered by snappy motorbikes and more gas-efficient cars put domestic travel within easy reach of most Americans. No longer did tourists need dude ranches as outposts from which to organize western sightseeing. Armed with a few guidebooks and a *Rand McNally Road Atlas*, Americans orchestrated their own getaways to the wilderness.

The nineteenth-century national parks, many of which encompassed magnificent western vistas, now threatened to become tourist merchandise. The goals of the conservationist John Muir, who founded the Sierra Club in 1892, changed in ways he could hardly have imagined. Whereas Muir had hoped to jolt Americans into a responsible wilderness awareness, he never anticipated the high-tech by-products of that heightened concern. With travel options maximized, the number of Americans pouring into the West increased at astonishing rates. In 1973, three million more people visited the national parks than in 1972. In the same year, almost four million people went to the Colorado Rockies for skiing. The economic complexities of environmentalism and wilderness preservation crystallized with stunning rapidity.

The national parks, always more closely wedded to economic interests than Americans wanted to realize, faced an avalanche of challenges. In the first place, local economic considerations had often outweighed aesthetic concerns when lands were set aside for the parks. In addition, nineteenth-century conservationists who wanted to preserve natural beauty understood little about ecosystems and biological integrity. As a result, the government often drew boundaries for national parks haphazardly, even before local logging and grazing companies complained about the borders and demanded further gerrymandering.

When millions of nature seekers added themselves to the equation in the post–World War II era, the full tangle of public land management, political pressure, and environmental issues hit the American West. Each year more and more Americans fled from urban pressures, to stack up at the entrances to Utah's Zion National Park or to inch bumper to bumper along the road in Wyoming's Yellowstone. This massive human intrusion placed at risk the very wilderness the visitors believed they cherished. Into the 1980s and 1990s, problems of litter, sewage disposal, wildlife protection, and public safety multiplied at unprecedented speeds. The embattled National Park Service, an agency of the Department of the Interior since 1916, struggled to cope with the flow of visitors even as it warded off conflicting complaints about its overwatchfulness or inattentiveness. As the twentieth century came to a close, less and less of the West fit the description of true wilderness, but more and more Americans believed they needed a pilgrimage to that special environment. Perhaps twentieth-century America revolved around the industry of urban areas, but those drawn to the scenic West appeared to think that the parks made it possible not only to step back into a wilderness paradise but also to capture a personal vision of America, such as the nation's frontier forebears were thought to have possessed.

Always the Cowboy

Vast numbers of people have experienced the West, through literature, art, entertainment, or tourism. Millions more may not have taken as active an interest in things western, but they, nonetheless, would know the singular icon whose image is emblematic for the region—the cowboy. Instantly and internationally recognizable, the cowboy is now a national symbol for America and not just for its western states. His glamour has drooped from time to time as foreign observers or Americans themselves followed other fads, but the cowboy as a beloved figure has demonstrated extraordinary staying power.

The cowboy figure started his heroic career on foot as that noble buckskin-clad pioneer in the eastern woodlands. However, once in the saddle—through story, song, and reality—he evolved into the cowboy of shirt, chaps, and neckerchief on the open western plains. His stature increased, and he quickly passed any other contenders in national prominence. Mountain men, pioneers, scouts, and gold miners cut their own swath in the evolution of western popular history, but they never won the public adoration heaped on the cowboy.

The American cowboy, most often a youthful bachelor, rode into the hearts of Americans with a laconic but honorable manner. No great thinker, he was guided in life by a few simple principles: he was always willing to right a wrong, to save a damsel in distress, and to defend the underdog. With an ever-constant good disposition, the cowboy did not seek violence, but when confronted, he knew how to respond quickly and thoroughly. He was white America's mounted warrior, the defender of a national code of honor, the champion of the open range. He carried a gun and rode a horse, but he did not always herd cattle.

The cowboy became a multimedia figure. He graced the pages of Owen Wister's *The Virginian* (1902) and the later stories of Louis L'Amour's Sackett family. The cowboy galloped across Charlie Russell's turn-of-the-century paintings and inspired the twentieth-century pieces of the European artist Americo Makk. As the Lone Ranger, he rode with youngsters through the pages of their comic books. As Hopalong Cassidy, he loaned his logo to the school lunch-box and the Hoppy thermos, making the sandwiches more appetizing and the milk more cooling. As the singing cowboy, Gene Autry, he crooned "Back in the Saddle Again" to listeners out in radio-land. With Tom Mix and Roy Rogers, he thundered across the silver screen during cliff-hanging Saturday-morning serials. He kept the watch with Marshal Matt Dillon over Dodge City and the Long Branch Saloon.

Actually, Matt Dillon of *Gunsmoke* fame was just one television westerner from a string of shows that proliferated between 1947 and 1960. The western motif proved a successful formula for the burgeoning new world of television. Writers and directors, working under production time constraints, quickly grasped the plot essentials. The underlying theme, as the film historian John Lenihan has pointed out, always concerned society's preoccupation with the conflicts between "individual freedom and social constraint." In the Western, the hero, in the face of nearly overwhelming odds, acted to secure the betterment of the community. This equation required simply a hero, a person to be saved, and an evil force to be defeated—vicious Indians, corrupt cattle ranchers, or heartless bandits. Audiences, well versed in the stock characters of the Western, comfortable with the cowboy as the protector of social values, and familiar with these plot lines, easily accepted the reduction of large movie sets onto the small screen. By 1959, Westerns captured the ratings as seven of the top-ten favorite shows on national television. Americans liked Matt Dillon and his cowboy friends so much that their televisions stayed tuned for *The Rifleman, Rawhide, Maverick, Wanted—Dead or Alive, Wagon Train, Have Gun, Will Travel, Lawman, Bat Masterson,* and *Bonanza.* Indeed, the early years of television saw an almost unending introduction of new Westerns with each season.

In the 1960s and 1970s, changing viewer demands, as well as a shift in the demographics of audiences, sounded the death knell for the Westerns' exaggerated popularity. Family viewing patterns altered, and competition forced network executives to survey carefully the program interests of the nation. Television concentrated on greater news coverage, as well as musical entertainment and sporting events, in part to satisfy an ever-growing population of teenage viewers. The Western faltered as television entertainment.

Yet, the Western survived to make another resurgence. The introduction of cable television in the early 1980s brought the shoot-outs and cattle stampedes back into American living rooms. Reruns of television shows and movie classics offered an available, cheap commodity for network markets that ballooned almost overnight. The Westerns added to their rejuvenated popularity through an appeal to viewers' nostalgia.

The introduction of the home videotape-player allowed television owners to accumulate a private film library, bringing the Westerns and their aging stars back to the attention of the American public. In this new age, the old "B" Westerns of the 1940s and 1950s lacked the sophistication of programming in the 1990s. However, they returned, much like long-lost friends, reminders of an earlier, simpler time. At least for the expanding population of elderly Americans, they provided an easy, inexpensive way to recapture the entertainment of youth, to delight in old stars made whole once more. There again was a dark-haired Jimmy Stewart, an athletic Burt Lancaster, a boyish Steve McQueen. Actor and viewer—everyone was young once more. In a make-believe world that never entirely died, the cowboy heroes rode across the screen, chasing Indians, saving young women, and winning the West.

Certainly this roller-coaster popularity of the film and television Western accounted for the cowboy careers of dozens of actors. Some fell into their cowboy star status by the accident of casting. Others fervently believed that cowboys personified moral integrity and made major contributions to democracy and American life. From William S. Hart to Kirk Douglas, from Joel McCrea to Michael Landon, from Randolph Scott to Jack Palance, from Hoot Gibson to Gary Cooper to Clint Eastwood—all had the western film to thank for major developments in their cinema success.

Typically, women have fared poorly in Westerns. Their screen roles tended to copy the prevailing social attitudes toward womanhood. Accordingly, film heroines of the 1920s and 1930s smiled and simpered as the "gentle tamers," those who brought "civilization" and order to the West. By the 1940s, with American women generally cast into more active and productive economic endeavors to meet the war needs, Hollywood followed suit with films about strong-minded women, such as the 1941 *Gangs of Sonora*, in which a female newspaper editor in Wyoming took on and defeated a corrupt politician. As quickly as the gender trend of self-reliant, independent women lost vogue within American society, Hollywood also dropped scripts that showed females surviving on their own. Within the western genre, women became an environmental adornment, as typified by Joanne Dru in the 1949 *She Wore a Yellow Ribbon*, which pitted John Wayne against the Apache nation.

Jane Russell's sensuous portrayal in Howard Hughes's 1943 film *The Outlaw* presaged a changed tone and texture in women's roles in the Western. By the 1960s, western films assumed greater and greater sexual explicitness, with rape and torture as common themes. Although one might argue that this trend introduced fundamental realities to the Western, often criticized in an earlier era as too sanitary, a counterpoint suggests that filmmakers simply used women characters as the vehicle through which to exploit western violence and gender aggression in more horrifying terms.

This strategy dominated the portrayal of ethnic women. Native-American women seldom emerged as personalities with identities and perspectives. Historically, Native-American women of film appeared only as background "pack animals" to further document the "savagery" of the men or as victims in scenes of village slaughter by marauding army troops, as for example in the 1970 film *Soldier Blue.* Thus Native-American people continued to play their main role in the Western—that of subjects for cultural or political extermination.

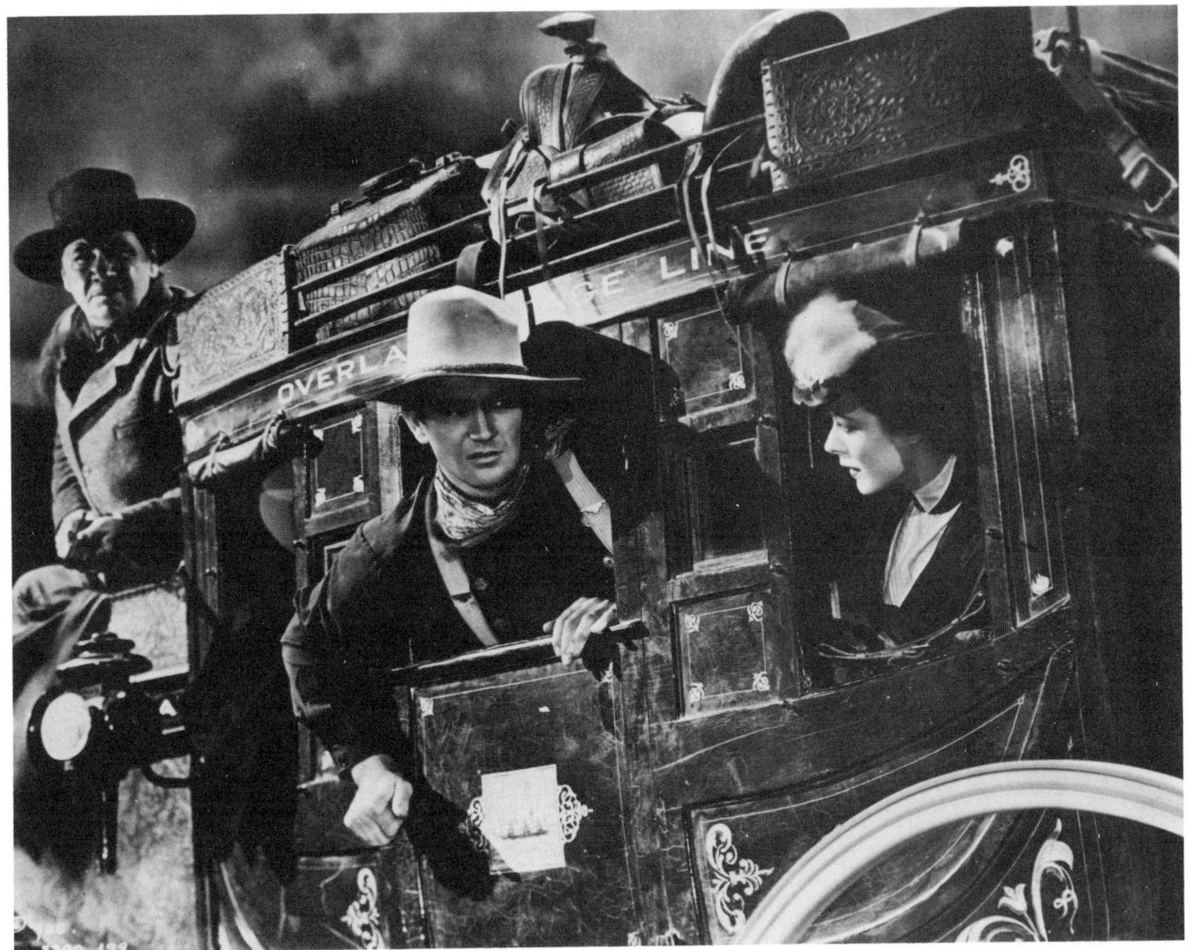

More than any other medium, film is responsible for the image of the West as a place locked in the 19th century and defined by stark encounters between whites and Indians, law and disorder. Although social and political trends have altered the content of western films, the strong, silent man of action—epitomized by John Wayne—remains the central figure.

Film still from Stagecoach *(1939), directed by John Ford. Courtesy of the Academy of Motion Pictures Arts and Sciences Library, Beverly Hills, California.*

In a further distortion, in those rare scripts that called for an actual woman character, directors typically cast Anglo actresses as Native Americans. Thus, Debra Pagent, Jennifer Jones, and Virginia Mayo, in their Indian and mixed-blood roles, helped implant an inaccurate physical image of Native-American women in the minds of moviegoers. In addition, the consistent refusal of the American film industry to hire appropriate ethnic females further codified the economic structures that unrelentingly kept Indian people from reaping personal or collective profits from western popular culture.

It appears Mexican-American women also had little to applaud from the film industry. Along with Native Americans, Mexican Americans found only stereotyping and cavalier treatment of their culture. Scorned as docile and lazy, their speech patterns and language ridiculed, Mexican Americans forcefully held their place in the West despite abuse that dated to the earliest contacts with Anglo culture. The strength and diversity of the Mexican-American culture rarely gained acknowledgment in an entertainment medium that appeared content to create its own image of an entire community. The film characterizations of Mexican women ranged from the elegant

Katy Jurado prostitute in the 1952 *High Noon* to the earthy Mexican prostitutes in the 1969 *The Wild Bunch* to the revolutionary Hispanic prostitute in the 1990 *Old Gringo*. In almost forty years of filmmaking, no appreciable refinement in the presentation of Mexican-American or Native-American women graced the American screen.

Despite these glaring weaknesses and offensive biases, the cowboy film continued to draw large audiences, and the intellectual analysis of the genre appealed to film critics, university professors, and cinema buffs. Social critics and scholars like John Cawelti and John Lenihan found something of the national forces and patterns in each era of cowboy films. In the deceptively simple stories, they saw the conflicts and dilemmas of American society—it contradictions, its goals, its fears, and its hopes for itself. Regardless of these sophisticated, often highly perceptive critiques, ordinary folk idolized screen cowboys for their bedrock devotion to American values and codes of behavior seemingly lost to the modern world. Scrubbed and polished in one era, rougher and dirtier in a more recent time, cowboys of film continued to suggest that within the confines of the rocky, barren West could be found valor, freedom, and, above all, justice.

Within that formula, American movie fans avidly followed the roles of their favorite stars and became conversant with Gary Cooper's angst in *High Noon* (1952), Lee Marvin's humor in *Cat Ballou* (1965), Marlon Brando's viciousness in *Missouri Breaks* (1976), or Kevin Costner's sensitivity in *Dances with Wolves* (1990). Cinema buffs can talk for hours about *A Fistful of Dollars* (1964), *Little Big Man* (1970), or *Shane* (1952). But no one did more to perfect the cowboy figure than John Wayne. Indeed, he so absorbed the flint and fiber of the character that Americans forgot where the actor stopped and the role began. Wayne's screen credentials constituted a catalogue of the film history of the American cowboy. In each—among them *Stagecoach* (1939), *Tall in the Saddle* (1944), *Red River* (1948), *Fort Apache* (1948), *Rio Bravo* (1959), *The Man Who Shot Liberty Valance* (1962), *The Sons of Katie Elder* (1965), *El Dorado* (1967), and *Rio Lobo* (1970)—Wayne offered American viewers a straightforward, easily comprehended performance. Audiences knew what to expect in a John Wayne cowboy film: obvious heroes and villains whose final encounters affirmed that in the West, tough problems had clear-cut, honorable solutions. The Duke, as he was known, helped to cement that philosophy and its cowboy values of decency and honesty, valor and integrity, in living technicolor.

As his list of films grew, Wayne, clearly attuned to the standard he represented, accepted scripts that were apparently more complicated in characterization but that did not undercut cowboy-Wayne virtues. *True Grit* (1969) and *The Cowboys* (1972) made his formula a bit more nuanced and definitely more compelling. Despite his artistic advances, his basic film plots remained uniform and his philosophical solutions unchanged. His cowboy characters typically projected a solid wisdom, convinced that truth and justice would carry the day. As his physical body aged, Wayne's personal values—always very American, very masculine, and painfully simplistic—held constant. Once said to be the most recognized film star in the world, John Wayne personified the cinematic definition of cowboy life.

By the time he made his final appearance in *The Shootist*, a 1976 film about a dying gunfighter in a dying West, Wayne had won over even his critics. A brilliant opening sequence that used clips from a dozen of his earlier films reminded audiences that John Wayne had ridden hard on America's cinematic range for more than forty years; almost half of his nearly two hundred movies had been Westerns. At the end of *The Shootist*, as his character died on the barroom floor, Wayne, himself clearly losing a battle with cancer, seemed to say that the West *had* changed, but not the values he endorsed in those four decades of filmmaking. The American public worshiped this film and its dying cowboy hero, fictional and real. In its aftermath, many wept with the actress Maureen O'Hara, a longtime Wayne friend and costar, when she appeared before a congressional committee to plead that the dying John Wayne be honored with the highly coveted U.S. Medal of Freedom. It seemed fitting, for Wayne himself vigorously held to the conservative, masculine, chauvinistic values that so clearly shaped western films and came to dominate American political thinking in the 1980s. That this amazing career of film achievement had really explained very little about the historical realities of the nineteenth-century West mattered not at all to Wayne's fans, Maureen O'Hara, or the U.S. Congress.

Rather, the ultimate expression of Americans' confidence in John Wayne and the cowboy mentality occurred one year after the Duke's 1979 death, when an elderly ex–film star whose personal identity also seemed carefully woven into the many cowboy roles he played was elected president of the United States. Ronald Reagan stepped out of his job as host for television's *Death Valley Days* and into the White House without any visible transformation, even given an earlier tenure in the California governor's mansion. Though more often seen in a tuxedo than his ranch clothes, Reagan, in many aspects of his presidency, made clear his attachment to the basic western script formula that the "good guys wear white hats."

The Reagan conviction that the West defined the country and democracy revealed itself clearly in the administration's rapid-fire acquisition of numerous works of western art. By 1982, the White House permanent collection boasted forty new pieces—paintings, prints, and sculptures—with western themes. Additionally, almost 130 pieces of art on loan from around the country decorated the halls and offices of the White House. In the official West Wing and Oval Office, Frederic Remington bronzes, Thomas Moran landscapes, and George Catlin Indians surrounded those privy to the inner circle. Upstairs, in their private quarters, Ronald and Nancy Reagan gazed on western landscapes painted by Thomas Moran and Thomas Hill.

The Reagans' choice for artistic decor well matched the general interests of the American public. During this same era, the craze for western design and culture reached new heights throughout the United States. Cowboys themselves, or many who claimed to be, led the charge. For example, the Cowboy Artists Association of America devoted itself to "preserving western tradition." The group, organized in 1966 because members wanted better exposure for their art, held an annual gathering featuring both a trail ride and an auction. By 1981, almost two thousand people flocked to the latter event, which collected approximately $1.46 million for eighty pieces of work.

After cowboy artists established their association, cowboy poets followed suit. By 1985, their organization set Elko, Nevada, as the site of its annual gathering. Within less than a decade of its founding, the cowboy poetry craze had swept the nation. Each year

A shared national myth about the West contributes to the ongoing popularity and success of western motifs in contemporary advertising.

The West.
It's not just stagecoaches and sagebrush.
It's an image of men who are real and proud.
Of the freedom and independence we all would like to feel.
Now, Ralph Lauren has expressed these feelings, in Chaps, his new men's cologne.
Chaps is a cologne a man can put on as naturally as a worn leather jacket or a pair of jeans.
Chaps. It's the West. The West you would like to feel inside of yourself.
Chaps. The new men's cologne by Ralph Lauren.
Chaps
RALPH LAUREN
Cologne
Chaps
RALPH LAUREN

Chaps: The New Men's Cologne by Ralph Lauren. *Advertisement, 1979. Courtesy of L'Oréal, Cosmetic and Fragrance Division of Cosmair, Inc.*

attendance at the national meeting expanded, and "cowboys," determined to "preserve our way of life," arrived from all over the country. In an event that quickly assumed commercial overtones, no one seemed ready with a definition of cowboy poetry or of the way of life to be preserved. Did poets or poems make cowboy poetry? Did one have to be a cowboy to write cowboy poetry? Was it necessary only to wear the appropriate attire to become a cowboy poet? Did cowboy poetry have any connection to nineteenth-century values, or did it focus on ranch humor? The answers to these questions remained blurry into the 1990s, but the popularity of and profits for the cowboy poets ballooned. Conferences, symposia, rodeos, and fairs began to invite the better known of these performers, whose subjects and rhymes often remained remarkably mundane and predictable.

Larger marketing schemes continued to sell western culture to the American consumer. In the mid-1970s, first-class eastern department stores sponsored annual trunk shows of silver and turquoise jewelry, complete with a Native American demonstrator. Leather boutiques that sold soft pouches, fringed jackets, and tooled vests dotted the nation's malls. On the clothing scene, the designer Ralph Lauren quickly dominated, setting a national style with his western line of stone-washed denim items, new clothes intended to look old, like the West itself.

By 1983, Lauren altered his western pitch and joined forces with J. P. Stevens, once only a textile company, in the manufacture of a wide assortment of designer products. With the emphasis on decor, advertisements displayed a western background of clay-red ceramic pots, wagon wheels, and giant cactus plants. Customers decorated entire bedrooms—drapes, rugs, bed linens—in color-coordinated matching Southwest patterns. Kitchen decor moved from an eastern country-cupboard style to western rustic ranch.

Lauren, the quintessential self-made man, often spoke about how he "believed in America." He acquired a ranch that encompassed thousands of Colorado acres in 1982, and he appeared, casually attired in plaid shirts and old boots, in his own magazine spreads. Consumers, who had never heard of Frederick Jackson Turner, agreed that Lauren, a child of New York who once considered becoming a history teacher, had rediscovered the meaning of the American West. However, the historian Ann Fabian noted, "Lauren found profits in a reinvented past, and he has been busy retelling in fashions and home decorations a story of white male migration, imperial conquest, and American development."

Between 1978 and 1988, Lauren's advertisements even more blatantly appealed to a distinctly white middle-class clientele that apparently wanted two things—stability and classiness. Sales exploded. Fashion-magazine advertisements sponsored by Lauren and other companies such as Dan Post Boots, Jacques Carcanques Jewelry, and Butterick Patterns framed lean, long-haired young women resting against sun-dappled rocks, old-time saloon fronts, and horse stalls. In various wordings, the advertisements mentioned the "pure America," where clothes in "pretty pioneer styles" had a "sense of adventure" and a "respect for tradition." Among these businesses, the West not only represented the best of the nation's history but also offered solid values, made a fashion statement, and defined style. These photographs captured rustic western images, but by the mid-1980s, slightly Indian-looking models in *Vogue* wore ultra-soft chamois dresses inset with gold panels and edged with gold fringe, frontier-basic in style but upscale in cost with price tags of almost six hundred dollars.

The West had galloped onto Madison Avenue. The country rushed into an era that, perhaps more than any earlier time of western adoration, cared little for the artistic renderings of the individual and only for the image that could be purchased. As a 1982 *Mademoiselle* spread entitled "Colors of the Earth" recommended, fashionable products appeared in "earth shades reminiscent of desert horizons, sun-drenched canyons, and mesas." As the trend mounted, designers who in 1978 had first noted a national demand for authentic "horn"—a nineteenth-century furniture crafted from Texas longhorn steers—profited greatly when prices for individual pieces rose to between five and twenty thousand dollars. Ubiquitous cowboys in Marlboro cigarette advertisements peppered urban billboards and every kind of national publication. Automobile companies set the sleekest new car model, along with an alluring woman, atop a high red-rock butte. Florists recommended a masculine cactus dish-garden for that recently promoted company executive. Architects advocated underground structures for residential and commercial needs, drawing on the frontier sod house for inspiration and design. Appealing to the public for deregulation, a western-based telephone company displayed a rodeo action photo and noted, "Take the bull by the horns or the bull takes you." Amid a clamor of western music, old and new, in 1978, the rusty voice of Willie Nelson rose above all others, urging, "Mommas, Don't Let Your Babies Grow Up to Be Cowboys."

In 1991, Billy Crystal, a stand-up comic from the Catskills, ignored Nelson's advice and tried his hand at cow-punching stardom. In his film *City Slickers*, Crystal teamed with old-time western star Jack Palance to warn young American males that unless they literally plunged back into a nineteenth-century cattle drive and learned "to be men," modern society threatened to emasculate them. In technology and sophistication, the nation had surpassed that nineteenth-century West, but American men needed to cherish the skills, the simplicity, the honesty, and the bravery—the maleness of its day. The grandeur of the scenery, the swelling music, the cowboy plot complete with self-reliance, heroism, and honesty, and the gentle, unchallenging women convinced viewers that two weeks in the West rearranged the values of these overpampered New York executives. *City Slickers* broke records at the box office, and Jack Palance won an Oscar for his role as the tough cowboy who snarled, "We are a dying breed."

In all these ventures, from the nineteenth through the twentieth century, the West meant something special to American consumers. From wilderness tales to splashy television productions, a substantial group of Americans derived national meaning from western expansion, and the experience touched the country's collective spirit. Although the fadlike aspects of western commercialism ebbed from time to time, the affection these Americans harbored for the West never totally died. In general, many Americans displayed remarkable consistency in their willingness to buy into the western myth in almost any form. Clearly, through mass marketing and commercial exploitation, powerful elements in western popular culture escaped the confines of region and embraced the nation. Yet, nagging problems linger about this bonding of national myth with commercial enterprise. Elements—some of them not especially edifying—lurk in the background of this apparent success story.

Essentially, across the span of the nation, many Americans—young and old, men and women—indulged their western fantasies so completely that they tolerated and encouraged misrepresentations of history. As a result, America saddled itself with a popular culture organized around problematic expressions of race, class, and gender. Yet, western popular culture persists in its suggestion that Americans shared a commonality of experience.

Profound contradictions mark the popular culture of the West, but there is no reason to assume that these cultural markers are permanently defined. Smug assumptions that the West is an easy place to understand have been proven wrong before. Revisionist thinking will continue, since the changing configurations of western history beckon to each new generation. Not just white Americans but also the sons and daughters of many cultures will have the opportunity, should they choose, to recast the symbols of the legacy of the West. After all, the images of the West belong to everyone, despite efforts, in both the past and the present, to kidnap them for only a select audience. If anything, the West gave all people a historical arena in which to imagine the opportunity and justice that define America's most positive national values. Although the formula for the popular story of the West often fell short of accuracy, the essentials empower those who demand a reinterpretation of the people and events. Those people, and there are many around the world, feel a sense of kinship with the place and purpose of the West. They can still reclaim its justice, its opportunity, its identity, and its spirit and rework western popular culture into larger, more attractive, more generous patterns. The American West deserves the effort, and who is to say that such a transformation will not happen?

Bibliographic Note

The literature of the popular culture of the American West is as vast and far-reaching as the subject itself. Not all subjects are included in this brief essay.

A number of general studies are useful for acquiring an overview of the broad subject of the West and popular culture: Anne Farrar Hyde, *An American Vision: Far Western Landscape and National Culture, 1820–1920* (New York, 1990), offers a gracefully written assessment of the construction of tourist centers for the American public; Lawrence W. Levine, *Highbrow/ Lowbrow: The Emergence of Cultural Hierarchy in America* (Cambridge, Mass., 1988), although not regional in its focus, is valuable for the background it lends; Barbara Novak, *Nature and Culture: American Landscape and Painting, 1825–1875* (New York, 1980), includes fascinating descriptions of the bonding of Americans to the wilderness through developments in art; William Cronon, George Miles, and Jay Gitlin, eds., *Under An Open Sky: Rethinking America's Western Past* (New York, 1992), is a collection of essays that touches on many useful topics, including popular culture and the modern West; Susan Armitage and Elizabeth Jameson, eds., *The Women's West* (Norman, 1987), is a compilation of essays that focuses on many cultural areas for western women; and Richard White's concluding chapter, "The Imagined West," in *"It's Your Misfortune and None of My Own": A New History of the American West* (Norman, 1991), discusses commercialism, the American mind, and the West. For a discussion of heroes in early America, see Michael A. Lofaro and Joe Cummings, *Crockett at Two Hundred: New Perspectives on the Man and the Myth* (Knoxville, 1989), John Mack Faragher, *Daniel Boone: The Life and Legend of an American Pioneer* (New York, 1992), Daryl Jones, *The Dime Novel Western* (Bowling Green, Ohio, 1978), and David M. Emmons, "Social Myth and Social Reality," *Montana: The Magazine of Western History* 39 (Autumn 1989): 2–9.

Much of the interesting and important work on popular culture in the West has appeared in journal articles in publications such as *American West* (*AW*), which began in 1963 and is now

American West: Its Land and Its People (*AWLP*), and *Montana: The Magazine of Western History* (*MTMWH*). Both have examined popular culture while still offering the reader solidly researched articles about the American West. Any student of western popular culture would do well to survey these and other journals. Examples of significant articles follow.

For art and the West, see these articles: Keith L. Bryant, "The Atchison, Topeka, and Santa Fe Railway and the Development of the Taos and Santa Fe Art Colonies," *Western Historical Quarterly* 9 (October 1978): 437–53; W. H. Hutchinson, "The Western Legacy of W.H.D. Koerner," *AW* 16 (September/October 1979): 32–44; Peter E. Palmquist, "The Life and Photography of Carleton E. Watkins," *AW* 17 (July/August 1980): 14–29; Clement E. Conger with William G. Allman, "Western Art in the White House," *AWLP* 19 (March/April 1982): 34–41; James K. Ballinger and Susan P. Gordon, "The Popular West," *AWLP* 19 (July/August 1982): 36–45; Ted Schwarz, "The Santa Fe Railway and Early Southwest Artists," *AWLP* 19 (September/October 1982): 32–41; Edith Hamlin, "Maynard Dixon: Painter of the West," *AWLP* 19 (November/December 1982): 50–59; and Richard W. Etulain, "Art and Architecture in the West," *MTMWH* 40 (Autumn 1990): 2–11.

For entertainment and films, see John G. Cawelti, *The Six Gun Mystique* (Bowling Green, Ohio, 1971). John H. Lenihan's *Showdown: Confronting Modern America in the Western Film* (Urbana, 1980), provides the reader with a highly readable, insightful analysis of the Western. Kristine Fredriksson's *American Rodeo: From Buffalo Bill to Big Business* (College Station, Tex., 1985), although not a scholarly analysis, contains a great deal of factual material about American rodeo. See also Richard D. McGhee, *John Wayne: Actor, Artist, Hero* (Jefferson, N.C., 1990), Lonn Taylor and Ingrid Maar, comps., *The American Cowboy* (Washington, D.C., 1983), Richard W. Etulain, comp., *Western Films: A Brief History* (Manhattan, Kans., 1983), and the following journal articles: Paul Andrew Hutton, "Celluloid Lawmen," *AWLP* 21 (May/June 1984): 58–65; Bodie Thoene and Rona Stuck, "Navajo Nation Meets Hollywood," *AWLP* 20 (September/October 1983): 38–44; Stan Steiner, "Real Horses and Mythic Riders," *AWLP* 18 (September/October 1981): 54–59; Paul Andrew Hutton, "Correct in Every Detail: General Custer in Hollywood," *MTMWH* 41 (Winter 1991): 29–57; "The Cowboy's West: A Special Issue," *AW* 14 (November/December 1977); "Darling Beautiful Western Girls: Sweethearts of the Wild West Shows," *AWLP* 22 (July/August 1985): 44–48; and Helene R. Day, "The Prince Who Would Be Mayor," *AWLP* 22 (January/February 1985): 54–58.

For tourism, see Alfred Runte's works: "The National Parks in Idealism and Reality," *MTMWH* 38 (Summer 1988): 75–76; *National Parks: The American Experience* (Lincoln, 1979); and *Yosemite: The Embattled Wilderness* (Lincoln, 1990). See also the following: David Lavender, "The Accessible Wilderness," *AW* 11 (January 1974): 18–27; the entire issue of *AW* 16 (July/August 1979), particularly Richard A. Bartlett, "The Presence of the Past," pp. 16–17, 59, and Lee Silliman, "'As Kind and Generous a Host as Ever Lived': Howard Eaton and the Birth of Western Dude Ranching," pp. 18–31; Peter Skafte, "Rubber Rafting Western Rivers: Yesterday and Today," *AWLP* 21 (March/April 1984): 26–34; and Lawrence R. Borne, "Dude Ranching in the Rockies," *MTMWH* 38 (Summer 1988): 14–27. Borne is also the author of *Welcome to My West: I. H. Larom, Dude Rancher* (Cody, Wyo., 1982) and *Dude Ranching: A Complete History* (Albuquerque, 1983).

For the subject of music, see Bill C. Malone, *Country Music U.S.A.: A Fifty-Year History* (Austin, 1968), the standard work in the field. See also Bill C. Malone and Judith McCulloh, eds., *Stars of Country Music: Uncle Dave Macon to Johnny Rodriguez* (Urbana, 1975), James N. Gregory, *American Exodus: The Dust Bowl Migration and Okie Culture in California* (New York, 1989), and Richard Aquila, "Images of the American West in Rock Music," *Western Historical Quarterly* 11 (October 1980): 415–32.

Other articles include the following: Robert M. Utley, "The Presence of the Past: Promontory Summit," *AW* 17 (March/April 1980): 34–39; Laurie K. Mercier, "Women's Economic Role in Montana Agriculture: 'You Had to Make Every Minute Count,'" *MTMWH* 38 (Autumn 1988): 50–61; Manya Winsted, "On the Trail of the Cowboys Fame and Fortune for CAA," *AW* 18 (March/April 1981): 56–59; William T. Anderson, "Wall Drug: South Dakota's Tourist Emporium," *AWLP* 22 (March/April 1985): 72–76; Thomas W. Pew, Jr., "Summit Powwow at Canyon de Chelly," *AWLP* 21 (November/December 1984): 38–44; Michael Wallis, "New Trails for Old Time Boots, *AW* 18 (January/February 1981): 38–47, 64; James Maurer, "Prairie Dugouts to Underground Dream Houses," *AWLP* 18 (November/December 1981): 34–41; Alma M. Garcia, "A Mexican American Community's Struggle for Educational Equality," *Journal of Ethnic Studies* 17 (Fall 1989): 133–39; LaVerne Harrell Clark, "The Girls' Puberty Ceremony of the San Carlos Apaches," *Journal of Popular Culture* 10 (Fall 1976): 431–48; Nicolas Kanellos, "A Brief Overview of the Mexican-American Circus in the Southwest," *Journal of Popular Culture* 18 (Fall 1984): 77–84; and Alessandra Stanley, "Presidency by Ralph Lauren," *New Republic*, 12 December 1988.

Chapter Twenty-three

Comparing Wests and Frontiers

WALTER NUGENT

The following pages will employ a wide-angle lens to compare, if only in snapshots, America's Wests with other frontiers, Wests, and regions. There is no doubt that the American West, however defined, was unique. But then, so were the Wests and frontiers of Canada, Australia, South Africa, Brazil, Argentina, and other places. Did they have anything in common as Wests and frontiers? Did similar processes take place in all or most of them? Do the processes of change in these other places tell us anything new about the American West? Is it worth looking at them, not only to satisfy our curiosity but also to get other perspectives on ourselves?

The American West itself has not always been where it is now. To the U.S. Census Bureau, the twentieth-century West has meant the thirteen westernmost states, with Montana, Wyoming, Colorado, and New Mexico forming the eastern tier. Unofficial definitions would usually agree, perhaps leaving out Hawai`i or Alaska, perhaps including the western halves of the Great Plains states. As recently as 1910 or 1920, however, the "West" still included Chicago and the Mississippi Valley, as it had for much of the nineteenth century. In the eighteenth century the term referred to any place west of the Appalachians, and in the seventeenth century it included the area west of the Tidewater in the Chesapeake or a few dozen miles from the Atlantic farther north. The West as a region, as it is now known, will stay in the same place because there is no place farther west to go. Historically, however, the western region has been a movable feast.

The places that are no longer Wests are now regarded simply as regions or sections—sizable areas with some sort of distinctiveness. These regions include (among others) New England, the Piedmont, Appalachia, and the Midwest. As transitory Wests, they were once called frontiers—the New England frontier, the southern frontier—and those frontiers-become-regions were themselves complex. A recent history of southern frontiers by John S. Otto, for example, identifies a frontier of occupance (very sparse white settlement), a grazing frontier, an extensive farming frontier, and an intensive farming frontier, as well as "huge frontier enclaves with unique agricultural adaptations." The common element in these southern frontiers is that they were all continuing to progress toward a stable level of rural settlement. Similar distinctions may be made in the history of the Midwest and Great Plains and isolated parts of the present West. Much

Frontier settlement in New Zealand, as in other areas around the world, has been marked by a European incursion into areas inhabited by indigenous peoples, with the subsequent transformation of the natural landscape and the disruption of indigenous cultural patterns.

Unidentified photographer. Land Clearance at Mangaonoho, Rangitikei. *Photograph, early 1900s. Child Collection (#G77631/1), Alexander Turnbull Library, Wellington, New Zealand.*

of today's West, however, never became settled in that way. Instead, its history has been marked by frontiers of exploitation and entrepreneurship (mining, stock raising, and lumbering most importantly), and by that definition, it becomes virtually impossible to point to exactly where the frontier in the present West ended, if indeed it has. But whether or not it is still a "frontier," the West is certainly one among several American regions.

In U.S. history, then, what are now regions or sections were once frontiers, and the present West is both. Each place's shift over time from "frontier" to "former frontier" or "postfrontier" has usually been called by American historians the "frontier process"; when this shift has occurred in Latin America it has been called "Third World development." They are different processes but are almost certainly related. Regions presently exist in all of the countries that had, or have, frontiers. Canada, for example, includes the Maritimes, Quebec, Ontario (a nineteenth-century farming frontier that soon became urban and industrial), the prairie provinces, British Columbia, and the North. Argentina has long opposed the city (Buenos Aires) and the pampas. Brazil is a mélange of regions, all once frontiers, among them the Nordeste, Minas Gerais, Rio and São Paulo states, the southernmost "gaucho" state of Rio Grande do Sul, and the still-developing frontiers of the Mato Grosso, Rondônia, and Amazonia in general. Australia divides between coast and "outback"—or, more precisely, between sheep, wheat, and mining frontiers on the one hand and vibrant cities on the other. In all of these countries—all developed by European-stock people since 1492—regions change, beginning as frontiers of settlement or exploitation and becoming settled-rural or urban-industrial, often both.

How then should we define "West"? In American terms it means the present West but also, when speaking historically, those regions that are frontiers no longer. In Canada and Australia the term "West" locates frontiers lying in that direction, but in Argentina, Brazil, and South Africa, "West" is inappropriate. In all of those countries, however, "frontier" seems to cover the place as well as the process by which they evolve into a nonfrontier region. That being so, let us look at several frontier histories, remembering that some of them were also "Wests" at one time or another and that one such region, the western half of the United States, roughly, will remain "the West" for foreseeable time.

"Frontier" was recently defined, waggishly, as a place where white people are scarce—not where they have always been scarce, as in Japan or China, but where they were (or are) scarce but tried not to be. That covers most places outside of Europe where Europeans and their descendants have tried to go. In this sense, Europeans began their expansionism close to the beginning of recorded history. The expansionist efforts of Alexander the Great and the Romans are early cases in point. The Roman Empire was studded with "frontier" army garrisons from Spain to Syria, northern Africa to Britain; and in some cases these were not just outposts but real settlements, such as the "Colonia" on the Rhine that has long since become Köln. Rome increasingly fell back on the device of acculturating the so-called barbarian tribes into Roman ways, rather than populating the empire with Romans themselves. There were simply not enough Romans to go around, and as population pressure from the barbarian tribes intensified, the western empire shrank and finally disappeared in the year 476. A similar process of expansion,

Columbus's voyages helped initiate a new era of European expansionism that led to the development of frontier cultures around the world. This first European depiction of Native Americans, published in a 1494 illustrated edition of Columbus's report of his 1492–93 voyage, suggests Europeans' fascination with the inhabitants of the New World but, with the elaborate ship, reasserts the superiority of European culture.

Unidentified artist. Insula hyspana. *Woodcut, 1493. From Carolus Verardus,* In laudem serenissimi Ferdinandi *(Basel, 1494). Beinecke Rare Book and Manuscript Library, Yale University, New Haven, Connecticut.*

stabilizing, and retreat took place in the eastern empire, but it continued another thousand years until the Ottoman Turks snuffed it out in 1453.

By then other Europeans showed signs of frontier activity that not only were expansionist but involved colonial settlements. The Norse voyages of the ninth to the eleventh centuries led to colonies in Iceland, Greenland, and present Newfoundland.

The Newfoundland settlement at L'Anse aux Meadows was probably the first colony created in North America. Iceland has had a continuous European-stock history since those voyages. The Norse colony at Greenland lasted over five hundred years until it ended shortly after 1400, probably because of a prolonged cycle of cold weather. The Norse people undoubtedly developed a frontier; they went "where white people were scarce," they were expansionist, and they created colonies of settlement rather than trading posts or army garrisons. Presumably they had contact with the native peoples and carried on economic activity involving the extractable resources of the frontier area (in this case, fishing). These elements—an increasing ethnic presence, expansionism, settlement, and economic extraction—are major parts of a definition of "frontier."

Beginning in the 1490s, European overseas activity expanded dramatically and on a geographic scale far greater than that of Alexander, the Romans, or the Norse people. With the voyages of Christopher Columbus, beginning in 1492, and of Vasco da Gama in 1497 around the Cape of Good Hope to India, followed less than thirty years later by Ferdinand Magellan's expedition around the entire world, the true period of European expansionism—and a whole new set of definitions of "frontier"—took shape. Those voyages introduced the four or five centuries of world history dominated by "the expansion of Europe," the "rise of the West," or in the words of the Texas historian Walter Prescott Webb, "the Great Frontier."

Portuguese and Spanish, then French and English, explorers and colonizers, under governmental blessing and often financial support, headed in several directions away from the traditional centers of European civilization. Even in the early twentieth century, Europeans and their descendants still expanded in economic, social, and cultural ways, but by 1950 they had generally stopped annexing and controlling territory. European empires that developed after 1500 reached their maximum extent by 1914 and have virtually all been dismantled since then, rapidly so after 1945.

In the Americas, the Antipodes, and South Africa, however, settlement accompanied empire building to such an extent that those areas became Neo-Europes. They are what Webb and others have regarded as the frontier lands of modern world history. As capitalism came into being, so did they. As nonanimal power sources—steam, electric, and nuclear—were created, so were they. The characteristics already noted of pre-1500 European frontiers continued into the modern period, but modern technology and economic organization, as well as the global geography involved, so changed the nature of frontier activity as to make it qualitatively different from the ancient cases. The great variety of post-1500 frontiers—colonial or imperial, permanent or temporary, agricultural or extractive, Latin or Anglo-Saxon, centralized or individualistic (and many other sets of opposites)—requires that we push somewhat further in the search for a precise definition of comparable frontiers.

Were Europeans, since Alexander the Great, the only expansionists? By no means entirely. But world history furnishes relatively few cases of expansionism outside of efforts by Europeans. Some of these cases, however, were spectacular. They include the expansion of Islam throughout the Middle East, North Africa, Iberia, and eastward to Indonesia beginning in the seventh century; Muslim expansion into Europe was checked in Spain only in 1492 and was not decisively thwarted until 1683, when the Pole Stefan Batory stopped the Ottomans at the gates of Vienna. From the eighth to the thirteenth centuries, Polynesians peopled the Pacific vastness, including previously uninhabited Hawai`i and New Zealand. The Mongol and Tatar incursions into present Russia and eastern Europe in the thirteenth and fourteenth centuries are another example. The consolidation of central power in China and, in quite recent times, the expansion of Japan into northern Honshu and Hokkaido are others. Nevertheless, Europeans created the lasting settler societies as well as the colonies and outposts that ringed the coasts of Africa and of eastern and southern Asia, at least for some time. In the case of the British Raj in India, Europeans ruled, for nearly two hundred years, one of the largest and most venerable world civilizations.

A distinction may be drawn, however, between Europeans' frontiers and their empires. Frontiers lasted; empires disappeared. The basis of the distinction is the demographic strength of the European incursion in relation to the indigenous people, especially if those people already possessed an advanced culture and technology for their time—and a biological resistance to European diseases. European "expansion" needs to be qualified: was it expansion involving settlement in fairly large numbers relative to the natives, or was it simply expansion in the form of economic outposts or military-naval garrisons? The Raj, despite its size and longevity, was essentially an outpost and garrison.

On smaller scales the British, French, German, Spanish, Portuguese, Belgian, Dutch, and Italian colonies of the nineteenth century in Africa, Asia, and Oceania were also outposts and garrisons. Despite a barrage of propaganda from advocates of colonies, no great masses of Europeans chose to emigrate to them. Instead, during the years of mass emigration from Europe (roughly 1870 to 1914), Italians emigrated to Argentina and Brazil, Spaniards went to Cuba, and Britons populated Canada and (in much smaller numbers) Australia, New Zealand, and South Africa, while just about every European emigrant group entered the United States. In time those thirty-some million European migrants became integrated into the Neo-Europes of North America and the southern temperate zone.

In certain colonial areas, such as Britain's in East Africa, France's in West Africa, and the Netherlands' in Indonesia, plantation settlements arose where a minority European planter elite developed a local resource like coffee or rubber using the local population as a labor force. These efforts were relatively transitory. The elite aside, they did not form frontiers of settlement in the New World or Antipodean sense. On some occasions a frontier of resource exploitation did become a frontier of settlement. This shift occurred in certain areas where an initially thin European presence, interested in exploiting a local resource, expanded demographically or intermarried with the local people so as to become a stable Neo-Europe, if a rather modified one. Cases in point include the fur-trade region of the American and Canadian West, Brazil's Minas Gerais (which began with diamond mining), and various gold rush locales. The distinction between frontiers and empires, as well as that between plantation and small-settler agriculture, was not always thorough or permanent, but it nonetheless helps sort out which places were frontiers and which were not.

"Frontier" thus means frontiers of European settlement in previously non-European areas, with some allowance also for frontiers of resource exploitation. In virtually every instance, the Europeans and their descendants encountered indigenous peoples. After the Europeans established themselves in some numbers (and it did not take many, once European contagious diseases took hold), the native peoples declined in numbers and retreated in space. As is now well known, the victories of Hernando Cortés over the Aztecs and Francisco Pizarro over the Incas in the early sixteenth century were greatly aided by horses and gunpowder, but their real secret weapons were pathogens against which the natives were defenseless.

The Varieties of American Frontier Experience

Within the area that is now the United States, a number of frontiers have existed since the early 1600s when Spanish, French, and British colonization began there. From these frontiers, distinguishable regions have developed such as New England, the Ohio Valley, and the Gulf Plains. The varieties of American frontier experience have been broad, and most of the main types have appeared elsewhere in the world: grain growing, lumbering, livestock raising, gold mining, coal mining, oil drilling, even town building. Each of these involved its own environmental problems, contacts with "Indians," and economic and social organization. All together, they compose the interior history of North America. This history, with its environmental adjustments and exploitation of land and other resources, is still continuing, although some important types of frontiers, most notably the frontier of new agricultural settlement, ended some time ago in North America.

For much of U.S. history the frontier meant a moving zone of contact, a series of places that shifted over time, generally in a western direction but sometimes north or even east, especially after the middle of the nineteenth century. This moving zone of contact, which thereafter became settled-rural and urban areas, expanded very slowly at first and then, almost continuously, accelerated. From the first English settlements shortly after 1600, the western edge of the frontier zone penetrated inland only a few dozen miles by 1700 and involved only about 250,000 people of European and African stock. By 1800 the edge of settlement had crossed the Appalachian chain at all points, had moved westward three to five hundred miles, and had enclosed to the east of it 5.3 million nonindigenous people. By 1870 the western edge of the zone had breached the ninety-eighth meridian at most points, putting the eastern half of the now territorially complete United States behind it, while significant oases of non-Indian populations lived west of the ninety-eighth, in northern California, Mormon Utah, the Denver area, and many smaller places mostly created and sustained by mining, lumbering, stock raising, military forts, and commerce.

By 1920 the area under cultivation had more than doubled since 1870, and the national population surpassed one hundred million. By then more than half of those millions lived in towns and cities. The settled area increased only marginally thereafter. In some places people had already begun retreating from their overly optimistic homesteading of a few years earlier. But urban and suburban development, in the West as elsewhere, continued rapidly. In three hundred years the tentative settlements on the Atlantic seaboard had expanded to encompass the entire continental land mass, and in the process almost everything the European-stock people encountered, including themselves, had changed, for better or worse.

The acceleration was all the more remarkable if one starts the time line with

Columbus. From the 1490s to the early 1600s, hardly any frontiers of any kind existed in the future United States. In that "long sixteenth century," Spain managed to subdue or exterminate the local people and establish firm control over the Caribbean islands, central Mexico, Peru, and related areas, and Portugal did the same along the Brazilian coast. But in the area that later became the United States, sixteenth-century European efforts were unimpressive: the unproductive or downright disastrous expeditions of Hernando de Soto, Francisco Vásquez de Coronado, and a few others (ca. 1539–42), the toy-like Spanish post at St. Augustine (1565), and the abortive English attempt to establish the Roanoke Colony in 1585.

Despite these inglorious episodes, Europeans almost without exception gained the impression that they had a right not only to invade such parts of North America as they wished but also to occupy and make use of these regions as they wished. They considered themselves limited only by their own power, and they showed little regard for any rights, or even for the lives, of the people who had been there since the Ice Age. This idea, seldom modified in the history of Anglo-American expansion except by the conscience-pleasing policy that Indian title could not be extinguished without some kind of treaty, guided action throughout the future of the expansionist process.

As the sixteenth century drew to an end and the seventeenth century began, permanent European settlement in the future area of the United States and Canada started in earnest. Juan de Oñate founded El Paso del Norte in 1598 and planted a line of settlements up the Rio Grande valley to Santa Fe (1609) and beyond; an English stock company put ashore settlers on the northern side of the James River in Virginia in 1607; and France founded Quebec in 1608. These three efforts involved three different modes of settlement, sets of institutions, relations with the home country, and relations with the Indians and the environment; that is, these were three different frontiers. But they had in common the intent to settle at least some Europeans, and they succeeded in that.

The first hundred years of English settlement brought some results: successful footholds in the Chesapeake and New England, and later in the seventeenth century in the Carolinas and along the Delaware River, as well as absorption of short-lived colonization efforts by the Swedes and Dutch. Contact with the Indians was often bloody, as evidenced by the Powhatans' near-extinction of the Jamestown colony in 1622 and, on the other hand, the Puritans' liquidation of the Pequots and other neighbors a few years later. Yet the impact of these rather few Europeans on the vast space of the Atlantic seaboard appeared minimal, even if the English had few doubts about the command in Genesis to use the land as they saw fit. Life was indeed a struggle, and often a short one, but the sense persisted that the struggle could be won and they had a right to win it. Before 1700 or 1720, however, the social and economic organization of the settlements was either peculiar, as in New England, or precarious, as in the Chesapeake. These first hundred years were in a true sense precolonial, because the kind of rural society that so typified American life for the next two hundred years had not yet taken shape. It did so in the early eighteenth century, as a frontier-rural mode that lasted into the twentieth.

Patterns of life in seventeenth-century New England and the Chesapeake have been described elsewhere in far more detail than is possible or necessary here. However, a brief

This image of a model American farm along the St. Lawrence River represents a particularly prosperous example of the kind of rural agricultural settlement that marked the American frontier of the late 18th century. An illustration for the travel account of a Scottish gentleman, the image suggests the kind of cultural borrowing that marked this rural life, with a "Virginia rail fence," "Dutch barn," supply of "Indian corn," and native canoes.

McIntyre, engraver. Plan of an American New Cleared Farm. *Engraving, 1793. From Patrick Campbell,* Travels in the Interior Inhabited Parts of North America *(Edinburgh, 1793). Beinecke Rare Book and Manuscript Library, Yale University, New Haven, Connecticut.*

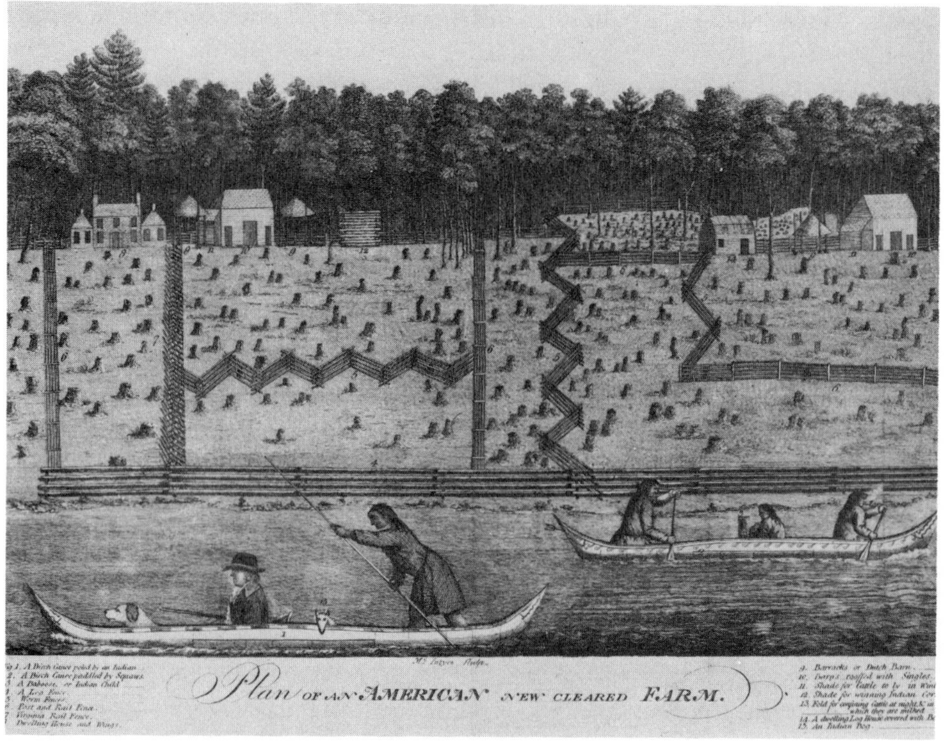

sketch will show how the farm-frontier future developed from these earliest patterns yet differed from them. In the Chesapeake, settlers arrived from England under an indenture of servitude that covered their passage and, after several years of labor, entitled them to land of their own. Individual ownership of land, by smallholders with secure title, became the common practice. Larger plantations also developed and from 1619 began using Africans, whose indenture was permanent; well before the century was over, slave labor was a fixture. Settlement of the Tidewater, much of it by planters raising tobacco for sale in England and shipping directly on oceangoing vessels at their own docks, came early.

But such activity was limited by the line of waterfalls that marked the eastern edge of the Piedmont. Into that backcountry, smallholding settlers rapidly moved. By 1676 they were strong enough to stage Bacon's Rebellion, their uprising to protest taxation and lack of frontier defense. Sometime very late in the seventeenth century, securely by 1720, Chesapeake settlers had finally begun to sustain and enlarge their own population, reproducing themselves more rapidly than they died off, and thus reducing the need for English and African immigrants to keep the settlement going. By the early eighteenth century they had moved well into the Piedmont and down the Great Valley of the Appalachians, subsisting on their own efforts, effective masters of their own small plots of wilderness. By the time of the American Revolution, Virginia was the largest colony in both size and population, home to one of every five Americans.

In New England, meanwhile, a quite different settlement pattern developed out of the Massachusetts Bay Colony, founded at Boston in 1630. This was not the first coastal settlement in New England, but it quickly became dominant. Full membership in the church congregation was identical with full membership in town governance; new settlements combined town and congregation and were created only as the theocratic central authority authorized them. Individuals and families held title to specific pieces of land, but only as granted and permitted by the town-congregation leadership. Private ownership, within the communal structure, typified New England settlement for the next three generations. By the early 1700s, however, the power of the theocracy had seriously eroded. Original family properties could be subdivided no further if the next generation was to earn much of a living from them. Population pressure, the demand for new land, and waning restraints from the theocracy produced a burst of settlements along the rivers leading inland, most notably the Connecticut. The backcountry of New England began to look more and more like the Chesapeake.

From 1720 through the American Revolution, the white English-stock population, reinforced by Rhinelanders and Scotch-Irish in Pennsylvania and western Virginia, pressed into areas new to them and "where whites were scarce." In the North, they cleared the forest and pushed back the Indians in New Hampshire and Vermont. In New York they proceeded up the Hudson and a short distance westward along the Mohawk River until Iroquois resistance stopped them. In Pennsylvania they followed military roads over the mountains to the head of the Ohio River. And in the South they swarmed across the rest of the Piedmont and southwestward to the tip of the Great Valley at Cumberland Gap. By 1776, from New England to South Carolina, they were poised to enter the middle of the continent. When the revolutionary war ended, they immediately did so. Slashing and burning the forest as they went, they produced enough to nourish themselves, and perhaps a little more to sell on a market if they could find one. They lived in large nuclear families and produced children at the biologically maximum rate. In law and politics they neither saw nor respected much human authority, British or colonial, but supported those who upheld their claim to land. They assumed, as always, that they had a right to take and use that land as they saw fit.

In 1783 the United States found itself in possession of not only the thirteen rebellious British colonies but also most of the land east of the Mississippi and south of the Great Lakes that had been French until 1763. Between 1785 and 1790 the new U.S. government established three basic policies regarding its transappalachian areas. First, Indian rights to land could be extinguished only by treaties—a policy never quite broken but greatly bent in governmental and popular practice. Second, western regions from the Appalachians to the Mississippi River would ultimately become states on an equal footing with the original thirteen—thus solving the problem of control that Britain failed to solve between 1763 and 1776 and encouraging western settlement while preserving political unity. Third, the eight states with colonial-era claims to western lands gave them up, allowing these lands to become the public domain of the whole United States; the Land Ordinance of 1785 prescribed how those lands would be surveyed and sold to smallholders. Despite initial major sell-offs to speculator-developers, the government maintained the principle of conveying the land to smallholders, with firm title, as rapidly as it could. By 1848 the area north of the Ohio River became five states, populated mostly by smallholders. The area to the south included a substantial number of large holdings, including new cotton plantations with many slaves in the Old Southwest in the 1820s and 1830s. Even there, however, smallholders (some with a few slaves, many with none) predominated numerically. The acquisition of the Louisiana Territory in 1803, effectively doubling the size of the country, promised to continue these patterns almost indefinitely. As it happened, the deforestation and cultivation of virtually the entire continent was completed by 1920, and the impressment of the native peoples into small reservations was an accomplished fact by 1890.

The uncontested heyday of the frontier-rural mode of life and the agricultural frontier of settlement lasted from the 1780s to the 1840s. In those decades population expanded 35 percent each decade, nearly unaided by migrants from Europe. Almost all of the increase took place on or near farms. The population of the United States doubled itself roughly every twenty-two years, from under three million in 1780 to twenty-three million in 1850. If ever there was a "frontier process," it worked, then, in the interaction of high fertility and abundant available land, with no Malthusian penalties for large families. The young simply left for cultivable areas farther west, depending on the military and the government to force and persuade the Indians to retreat. Crop and livestock technology remained premechanical and appropriate to near-subsistence levels, but forests and wildlife continued to disappear, slowly and inexorably.

The emphasis so far has been on the sameness of frontiers rather than on regional or other differences. This is, however, only a brief sketch, conflating many differences of time and place. One could easily list frontiers as early as the eighteenth century based on different crops or livestock, markets, climate, ownership-tenancy patterns, and the presence or absence of slavery. Here, however, stress has been on the broad similarities affecting people's lives, whether they were northern, southern, or transappalachian, once the frontier-rural mode of life became the American norm early in the eighteenth century.

From a broader perspective, American frontierspeople shared many characteristics, as already indicated. How their frontier-rural mode of life changed over the long run and how it shaded gradually into something quite different need to be pointed out. Their economic activity was initially subsistence agriculture, developing into some involvement with markets in the early nineteenth century. But by about the 1880s, when the American railway network was put in place, they had become participants in—or captives of—regional and even worldwide capitalist markets, placing Polish, Italian, Argentine, and American midwestern wheat farmers in competition with each other.

By that point, although thousands of homesteads were yet to be created on the Great Plains, one may question whether these settlers should be called frontierspeople at all or rather just farmers. Their legal relation to the land had been laid out in the Ordinance of 1785. It established the then-embryonic survey-and-sale system of transferring the public domain from the central government to smallholders, and the system was nurtured into maturity by successive land acts through the Homestead Act in 1862 to the Newlands irrigation law of 1902 and beyond. The survey-and-sale system, to be sure, was diluted by many policies inconsistent with smallholding such as grants to railroads, by inefficient or corrupt administration, and by a failure to adjust to aridity in the western half of the country. Later policies, such as subsidizing irrigation, building massive dams, and leasing rangeland cheaply to stockmen, hastened the West into corporate capitalism.

Nevertheless, the belief that a small farmer or an aspiring one could acquire secure title to a specific tract, and the law's encouragement of that belief, made Americans enviable compared with people on most other settlement frontiers. American demographic behavior was marked for generations by very high fertility, large nuclear families, youthfulness, and a strong propensity to migrate—all connected to the availability of accessible land. The rate of failure among would-be homesteading owner-operators is hard to define and is not definitely known but was probably high. Yet the homesteading urge and ideal remained enormously powerful well into the early twentieth century. By then, rural fertility, though higher than urban, had dropped substantially, family size had decreased, median age had risen (markedly in some settled-rural areas, which had once been frontiers), and migrants were heading for towns and cities rather than unplowed land.

Relations with the Indians continued to be wary and combative. But the U.S. government's successive policies regarding Native Americans—signing treaties (1790s to 1880s) to extinguish Indian title, building military forts in the Mississippi Valley and westward (1820s and later), physically removing eastern Indians to present-day Oklahoma and Kansas (1830s–40s), and finally subduing the Plains and southwestern Indians and confining them to reservations (1840s–90)—made life much easier for land-seeking frontierspeople than it had been in colonial times. Despite the 1787 Northwest Ordinance's pledge to deal with the Indians with "the utmost good faith," results were often brutal. As Alexis de Tocqueville noted in the early 1830s, "It is impossible to destroy men with more respect to the laws of humanity." Frontierspeople's relations with the physical environment continued to include deforestation and unlimited animal kill-offs (of which the destruction of the bison was the most visible but not the sole example; New England whalers did much the same to the great beasts of the oceans). On entering the arid region early in the twentieth century, settlers convinced themselves that "rain follows the plow." Thus they proceeded to disturb the fragile high plains until lack of rainfall forced many of them to retreat, leaving behind dashed hopes and a grievously damaged landscape. The reality of limits began buffeting the myth of limitlessness. In these several ways, the frontier-rural mode gradually changed and finally withered away by 1920, leaving behind a potent mythology that still shapes Americans' perceptions of reality.

In the brief years between the late 1840s and 1865, four events decisively changed American life. Each involved hundreds of thousands, even millions, of people. The first, in the Northeast, was urbanization—the sharp rise in the proportion of town and city dwellers in the population. Then the first massive influx of non-British immigrants began changing the populations of the Northeast, Midwest, and Far West—Irish, Germans, Scandinavians, and on the West Coast, a visible number of Chinese. The third event was the California gold rush of 1849–52, attracting a quarter million fortune-seekers, mostly young men, by land and sea; it was the first of many mining booms. The fourth was the end of slavery and the other outcomes of the Civil War. The frontier-rural way of life continued for several decades, as a moving zone of new settlement shifting ever westward to the Rockies. But because of these changes, in the context of the acquisition of Texas, Oregon, and much of northern Mexico in 1845–48, which brought the United States to the Pacific and provided it with most of its future West, the frontier-rural mode began to have competition from cities and from more blatantly exploitative frontiers.

In 1850 only 15 percent of the U.S. population lived in towns and cities of twenty-five hundred or more people. The proportion had, however, risen almost 5 percent since 1840 and would continue to rise on an average of 5 percent per decade (except for the 1930s) until peaking at about 75 percent urban in 1970. The American people were overwhelmingly (90 to 95 percent) rural until 1840, but thereafter became steadily less so. Moreover, the national figure masked great regional differences; Massachusetts reached an urban majority by 1860, and other northeastern industrial states quickly followed suit. If the end of the settlement frontier may be marked by the date at which a state or section begins exporting more people than it attracts, then New England's frontier days were over no later than the 1820s and New York's and Pennsylvania's by the 1840s. The same may be said for the seaboard South. The South's distinctiveness, however, lay not in urbanization, which made few significant inroads until the early twentieth century, or even in slavery. The region was distinguished by its climate and the fact that one-third of its people were black. The South held over 90 percent of America's black minority as late as 1910. The South would eventually be urbanized, but long after the other regions, especially the West. It is hard to imagine that the expansion of Atlanta, Miami, or any other large southern city after 1945 could have happened without air conditioning. Millions of southern black people, rural dwellers especially, moved to cities in the North, Midwest, and West from 1915 until their net migration flow stabilized in the late 1960s.

Urbanization and the end of slavery thus had enormous consequences. But more needs to be said about the gold rush because it was the first large-scale example of a different kind of frontier, nontraditional in 1850 but visibly outlasting the settlement frontier and in various ways still continuing in the West of the United States. That West—the Far West to some—has consisted of the Rocky Mountain, Great Basin, and Pacific Coast subregions. Within it, after the gold rush, social patterns developed that hardly resemble the 1720–1920 frontier-rural mode, and yet they are frontier patterns, in reality as well as in mythology.

The new type of frontier introduced by gold rush California differed most obviously from the traditional frontier-rural mode in that it was nonagricultural, attempting to exploit what was under the land rather than the surface. Instead of the farm frontier's moving line or zone of settlement, this frontier of resource exploitation became manifest unpredictably almost anywhere and often disappeared just as quickly. In certain cases of rich but technologically challenging mineral deposits, surface mining gave way to hard-rock, deep-shaft, hydraulic, or strip mining. With that, the mining camps became mining towns or cities, thereby losing their frontier character and becoming simply outposts of industrial capitalism.

In its fundamental demographic structure, the frontier of exploitation also differed from the settlement frontier. The 1850 census reported that of the ninety-three thousand people in California, 92 percent were male and 91 percent were in the fifteen-to-forty-four age range. Absent were children, older people, and almost as completely, women. Unattached youths often migrated from opportunity to opportunity. But the young men on mining frontiers were especially footloose and probably as a result were more violent and less careful about resources than people on settlement frontiers, wasteful though they were as well. In any event, mining frontiers and demographically similar ones such as cattle drives, cattle towns, lumber camps, and, in the twentieth century, construction sites, strip mines, oil fields, and other exploitative settlements comprised a kind of frontier essentially different from the traditional frontier-rural mode. If they were that different, do they deserve to be called frontiers at all? Certainly, for two reasons. First, settlement frontiers had their opportunistic and exploitative side too but exploited more slowly. Second, common usage defines cowboys, prospectors, and the like as frontierspeople. And so they were, and are; but they were not part of the frontier-rural tradition.

The frontiers of exploitation, in most cases, lost their bizarre demographic features in the twentieth century but otherwise lasted well beyond the demise of the settlement frontiers. Exploitative frontiers have been especially prominent in the Rocky Mountains, Great Basin, and Pacific subregions, but the type has not been absent from Texas oil fields or the northern Great Plains center-pivot irrigation that has mined irreplaceable water from the Ogallala Aquifer. Because of them, parts of the West lacked settlement frontiers, and frontiers of exploitation existed outside of the West at times. In other words, one can speak of a West without a frontier, a frontier without a West.

The settlement frontier itself met a new challenge after 1860 as it advanced onto the Great Plains. The vast area from the Atlantic Coast to the ninety-eighth meridian had been relatively well watered and was initially timbered or grassy. But settlers entering the arid plains in the late nineteenth and early twentieth centuries encountered decreasing rainfall and groundwater, river systems that were non-navigable and seasonally dry, and land that could not be farmed because it was mountainous or desert. Irrigation began to be touted as a solution from the late 1880s onward, but it never could, and never did, render productive more than a small fraction of the Great Plains and Rocky Mountains.

An early and important exception was Utah. After arriving in this semiarid region in the late 1840s, the Mormons based the survival of their community on irrigation. The Mormons extended irrigation, and themselves, to Idaho's Snake River valley and other areas in the next two generations by means of settlements planned and authorized by the church's central authority. In so doing, they created a town-congregation frontier of settlement paralleled in the American experience only by similarly theocratic New England over two hundred years earlier and at various times by much smaller and shorter-lived communitarian experiments. A complete typology of frontiers should include communitarian theocracies that flourished in inhospitable circumstances because of their ideological cohesion and commitment.

By the late twentieth century, the United States comprised many one-time settlement frontiers that were often losing population although producing more crops and livestock than ever. Farm population declined steadily for fifty years starting in the late 1930s, to about what it was in 1810—except that in 1810 the farm population was 93 percent of the whole and in 1990 only just over 2 percent. Agriculture had become agribusiness, nowhere more aggressively than in California, where massive irrigation networks, government-subsidized, enabled large corporate firms employing migrant workers of a succession of ethnicities to produce for market more crops and livestock than any other state. Ironically, California, the most productive agricultural state in the late twentieth century, scarcely experienced a settlement frontier of the eastern, southern, and midwestern type but included large landholdings since Spanish and Mexican times. Not coincidentally it was also the home of the first major gold rush and two of the most spectacular examples of wealthy urbanization in the world, the Bay Area and the Los Angeles Basin. Urban spread around well-watered Puget Sound and in the irrigated Arizona desert after 1940 was not quite as grand as in California. Together, however, they formed a latter-day frontier of exploitation—capitalist, federally influenced in many ways, multiethnic, and multiracial—in economic and social terms the leading edge of American culture.

Although farm settlement engaged far more Americans over the years than ranching or mining frontiers, the latter produced the legends and much of the cultural definition derived from frontier experience. No farm ballad ever competed with "Home on the Range" or "O My Darling Clementine"; no man with a pitchfork compares as a romantic figure to the forty-niner or even the Marlboro cowboy. American professional football teams call themselves Forty-Niners, Oilers, Broncos, and Cowboys but never Sowers or Reapers or even the more promising Threshers. Macho pursuits, rather than husbandry, seize the national imagination.

Canada's Several Frontiers and Wests

Culturally as well as geographically, Canada is closer to the United States than is any other society. Like the United States, Canada has had several frontiers, some of them closely paralleling the American experience. Canada had an urban frontier, if outposts such as Quebec (1608) and Montreal (1642), dominated by state, church, and commerce, can be called that. It very early had a frontier of exploitation in the fur trade, where *coureurs de bois* roamed from the St. Lawrence Valley westward beyond the upper Great Lakes before there were such places as Canada or the United States. And it had three distinct frontiers of settlement: that of the French *habitants* in New Brunswick and Quebec, beginning in the late seventeenth century; that of Anglophone Ontario in the early and mid-nineteenth century; and that of the prairies and Great Plains from the mid-1880s to 1930.

Canadian historians have long argued whether Canada had a frontier experience anything like that of the United States. It did and did not. The *habitant* and fur-trade frontiers had no counterparts at any time, but Ontario's mid-nineteenth-century farm frontier was very like nearby Ohio's, and Saskatchewan and Alberta were almost of a piece with adjacent North Dakota and Montana around 1910. Canada also, like the United States, had its exploitative frontiers, beginning with the Fraser River gold rush of 1858–67 and proceeding through its cattlemen's frontier of the late nineteenth century, the Yukon gold rush of the late 1890s, and the Alberta oil boom that began in

The generous land policy of the Canadian government drew more than four million settlers to western Canada during the first decade of the 20th century. Many homesteaders were ultimately defeated by the aridity of the lands. Others, like these Alberta ranch hands, became part of a successful cattleman's frontier that strongly resembled older American models.

Unidentified photographer. Canadian Cowboys, Near Thelma, Alberta. *Gelatin silver print, ca. 1910. Glenbow Archives, Calgary, Canada.*

1909, to the massive mid-century hydroelectric projects in Labrador, Quebec, Ontario, and the Northwest Territories—more a northern than a western frontier.

Canada has always differed from the United States in possessing two "core" cultures, French and English, and this duality affected its frontier and regional history. The European presence in Canada before about 1745 was exclusively French, except for a few English fur traders. French fur traders, missionaries, and soldiers dotted the vast wilderness from Quebec to the Great Lakes and southward to New Orleans and the border of New Spain, scarcely "filling" the region in any real sense. Some French mingled with the Indians in ways that were more benign than English or Spanish contacts. The majority of French, who probably totaled only seventy thousand at the time of the English victory at Quebec in 1759, dwelt in smallholdings hugging the banks of the St. Lawrence, seldom pushing far into a rugged interior. A scattering of towns served the needs of commerce, public administration, and the church. Partly because of their small numbers, but also by design and policy, the French got on well with the Indians, whom they welcomed to *réserves* amid their own riverine settlements, and ventured as guests and traders into the *pays d'en haut*, the backcountry that was the Indians' traditional land.

English fur trading into northern Canada followed the creation of the Hudson's Bay Company in 1670. But English Canada as a settlement frontier did not begin until the middle of the eighteenth century, when several important events occurred in close succession: the founding of Halifax, Nova Scotia, in 1749; the removal of the Acadians to Louisiana in the 1750s; the capture of the French naval base at Louisbourg in 1755; and James Wolfe's defeat of Louis Montcalm at the Quebec citadel in 1759. French authority ceased shortly after that, although the French cultural and economic presence in the fur trade and the Quebec settlements continued to flourish. French Canada, however, never produced an aggressively expanding line of settlement, and after 1759 it remained a self-contained entity. Its population grew not from immigration but from high birthrates in the nineteenth century. But unlike the young Anglo Canadians and Anglo Americans of that time, few young *Québecois* went west to establish frontier farms. Most gravitated instead to Montreal or migrated to New England in search of work. Like many labor-seeking migrants, they often stayed much longer than they had planned.

The American Revolution helped materially to create British Canada. Life became so difficult for American colonials loyal to the Crown that twenty-five thousand of them emigrated to New Brunswick and another six thousand to the upper St. Lawrence and the north shore of Lake Ontario. The latter group in particular built a true frontier of settlement, hacking homesteads out of Crown lands and producing children with amazing frequency. After the War of 1812, migrants from Britain and Ireland swelled the population. By 1850, much of the good farmland in Ontario had been taken up, and the frontier there had virtually closed by 1860.

Certain crucial differences between the Canadian and the American frontier experiences had begun to appear. First, the earliest English-speaking settlers in Ontario explicitly rejected the newly independent United States. They were Loyalists, and for their loyalty the Crown rewarded them with land grants and other compensation. The abortive attempt by Americans to capture Ontario in the War of 1812 reinforced in the Loyalists' children and newly arrived immigrants the idea that Britain was their protector and that they were part of the Empire. Although Canadian nationalism never became as aggressively convinced of its "mission" as the American brand, Canadians continued to believe that their land should be occupied and its borders made secure against further American incursions.

But an intractable problem prevented the extension of Ontario's frontier after 1860. Almost from the beginning, in Upper Canada, settlers had encountered the massive physiographic barrier of the Canadian Shield. With granitic rock lying only a few inches below the topsoil or actually cropping out at the surface and mixing with muskeg and scrub forest, much of Ontario's huge land mass could never be plowed and turned into farms. The Shield extended for hundreds of miles north of an east-west line across southern Ontario, even crossing the St. Lawrence into northern New York at one point; geologically it is eons older than the St. Lawrence and the Great Lakes. The Shield left open to farm settlement only extreme eastern Ontario, the land a few dozen miles north of Lake Ontario, and the southwestern peninsula above Lake Erie and east of Lake Huron and the Detroit and St. Clair rivers.

The Ontario frontier's natural extension was into Michigan and, from there, southwest and west. If the peace treaty of 1783 between Britain and the United States had been based on the actual military situation at the close of the revolutionary war, rather than on politics or diplomacy, then Michigan, Wisconsin, and adjacent areas would have become Canadian. The Shield prevented young Ontarians from continuing the frontier of settlement northwestward, whereas the international boundary meant that they moved out of Canada into Michigan, and beyond to the American Midwest.

The Canadian frontier of settlement was thus interrupted in space—and, as Canadian historians have pointed out, it was also interrupted in time for an entire generation or more, weakening any hold the frontier idea might have had in Canada. Whereas the American frontier experience was continuous from earliest times into the twentieth century, Canada's was fragmented into several nearly discontinuous pieces. The interruptions helped prevent the development of a national mythology of frontier-based exceptionalism, which was so characteristic of the United States.

The southwestward diversion of the settlement flow after 1860, together with the strength of industrial-urban development in New York State and New England, gave Canadian population history a peculiar sieve-like quality in the late nineteenth century. Despite high birthrates and substantial immigration, especially from Britain and Ireland, Canada suffered a net outflow of people for most of the 1870–1900 period. During much of that time, however, the Ottawa government under Prime Minister John A. Macdonald (1867–73, 1878–91) put into effect what he called the National Policy. It secured the area west of the Shield and north of the forty-ninth parallel for Canada, and it promoted there a second frontier of settlement that lasted until 1930. As the historian Carl Solberg noted, Canadians expected a lot from their government, and they got it; the National Policy included protective tariffs, a transcontinental railroad (the Canadian Pacific, or CPR) that bound the West to Ontario and Quebec, and a public land policy aimed directly at peopling the western prairies.

Macdonald's Liberal successor, Wilfrid Laurier (1896–1911), agreed with the basic aims of the National Policy. He and his aggressive interior minister, Clifford Sifton, employed an extensive network of recruiting agents in Europe, a cascade of booster propaganda, and a newfound willingness to admit non-British immigrants such as Ukrainians who were experienced in farming cold, steppe-like regions. As a result, the frontier of settlement in western Canada became a boom country, bringing in over four hundred thousand people *per year* between 1901 and 1911 and transferring sixty-eight million acres from the Dominion lands to settlers and another fifty-two million acres to railroads and other developers.

The law by which the Dominion lands were distributed resembled the American Homestead Act of 1862 but improved on it by requiring only three years' actual residence instead of five and by making available adjoining tracts of CPR land on attractive terms. It unambiguously identified each tract and clearly spelled out how to gain title. Canadian policy, like the American, in the long run worked all too well, enticing thousands of settlers after 1900 into arid and cold regions from which they later had to retreat. These areas included the shortgrass prairie of southern Alberta and Saskatchewan known as Palliser's Triangle, long regarded as too dry to cultivate, and also some settlements well north of Edmonton. Just after the Dominion lands policy ended in 1930, the Canadian geographer W.L.G. Joerg wrote: "The settlement of the Canadian West has exceeded all but the most optimistic estimates. Settlers have invaded Palliser's Triangle; country long designated as grazing country has been homesteaded; the northern forest has been attacked in some areas. Settlers with tractors are as far north as the 58th parallel. . . . All this has not been done, however, without grave human wastage. To the hardships of those pioneers who have succeeded we must add the losses

The caption on this image ironically comments on the "white only" immigration policy of the Australian government by calling attention to the mixed racial background of this rural Australian family. The Immigration Restriction Act, passed in 1901 and aimed at Asian settlers, was similar to the U.S. Chinese Exclusion Act of 1882.

Unidentified photographer. "White" Australia. *["Father. Born in Manila. Parents Cingalese and Filipino Brazilian. Mother. Dutch Hottentot Kanaka Chinese Aboriginal."] Photograph, ca. 1900. From the Davis-Royle Scrapbook, Mitchell Library, State Library of New South Wales, Sydney, Australia.*

of those who have failed." In some areas more than half of the attempted homesteads had to be abandoned, even before the economic disaster of the 1930s. As in the United States, a limited acreage east of the Rockies proved amenable to irrigation, sponsored in Canada by railroads and private developers rather than by government. Mormons also irrigated; their successful settlement around Cardston, Alberta (1906), replicated the earlier Mormon frontier in Utah and southern Idaho on a smaller scale.

In the meantime Canada also developed frontiers of exploitation. Canadian historians have often claimed that mining and cattle-raising frontiers were less violent north of the forty-ninth parallel than south of it and that Canada escaped most of the violence associated with American mining camps and cattle towns. They explain this by the presence of the Royal Canadian Mounted Police from 1874 onward, before the arrival of cattlemen, miners, or settlers. The Mounted Police negotiated with the Blackfeet for land and kept order thereafter, fortified by an effective judicial system. Central authority, evidenced also in the National Policy that included the CPR and kept the public lands under Dominion rather than provincial control until 1930, was more constant and continuing than in the United States. Otherwise the Canadian cattlemen's frontier resembled the American. Ranchers resisted farmer encroachment, maintained their presettler alliance with the Mounted Police, and made clearly known their Tory and pro-Empire views. They held the mineral rights on their land, and when oil was discovered early in the twentieth century they became oilmen too. In Canada, sheep ranchers prized individualism but also law and order; the stereotypical hero of the

Canadian West was no Wild Bill Hickok, or Billy the Kid, but was the red-coated man on horseback of the Mounted Police.

The depression of the 1930s forced retrenchment on Canada's several frontiers as it did in the United States. The Dominion Land Policy concluded, according to plan, in 1930, and Ottawa turned over the unoccupied remnants to the provinces. After 1940, agriculture more and more became agribusiness, another latter-day form of resource exploitation.

The Frontiers of Australia, New Zealand, and South Africa

In the nineteenth century, Australia became another frontier of settlement, as well as a series of frontiers of exploitation. Its development differed from that of America and Canada in that the two frontier types were somewhat mixed, since settlement very often involved raising sheep on large grazing tracts rather than crops on small homesteads. Moreover, Australia faced its own physiographic barrier: hills and mountains west of which was no verdant Mississippi Valley but, rather, a prairie that became increasingly arid, soon turning into an uninhabitable desert. Because its arid region is so vast, Australia had no "second chance" at frontier-making, as Canada had after Ontario became filled and then, decades later, when the CPR and settlement on the contiguous American northern Great Plains opened Manitoba, Saskatchewan, and Alberta. Australian settlement remained tied to its eastern and southern coasts. Several writers have attributed the strength of the Labor party in Australian politics to the absence of a frontier of small-farm settlement, which in the United States, at least in theory, siphoned potential workers away from strong labor unions toward farming. Australian egalitarianism, they say, stemmed from Chartism and the importing of undiluted British working-class culture, not from the interaction of Europeans, native peoples, and the environment, as in North America.

Physiography clamped strict limits on Australian frontiers. Although Australia's land mass is almost identical in size to the lower forty-eight American states, its population in the late 1980s was about sixteen million, less than two-thirds of Canada's, and was 85 percent urban. Only 9 percent of the land was considered arable; the lines of twenty-inch and twelve-inch average annual rainfall, about the minimum for raising wheat and livestock, are distressingly close to each other and to the seacoast, limiting farms and stock lands to the coastal fringe from southern Queensland through New South Wales and Victoria into South Australia, together with Tasmania and the extreme southwestern region around Perth. Although some hundreds of thousands of acres west of the mountains proved suitable for the commercial raising of sheep, cattle, and wheat, the great mass of the interior was and is too arid to permit settlement despite optimistic assessments as late as the 1920s. The ancestors of the 50,000 pure-blooded and 150,000 mixed-blood aborigines living in the late 1980s were swept back into the arid interior, displaced by farms and ranchers. They did not resist as the North American Indians did. Australia, as one of its historians noted, has a hinterland but not a heartland. It is a coastal, city-oriented society and has been so throughout its history.

That history began in 1788 when British prisoners were first transported to Botany Bay. The prison-colony period lasted until 1840 in New South Wales and a few years longer in Tasmania. It was followed by several decades of a squatters' frontier, overlaid almost immediately by a gold rush, principally in Victoria, beginning in 1851. "Squatter" applied to anyone, rich or poor, who moved onto Crown land to raise sheep, and by 1850 such people occupied much of the best land. Then came the gold rush, bringing in thousands of young men to a frontier of exploitation much like California's at the same historic moment. After they exhausted the easiest pickings, they formed a large cadre suddenly eager to turn to homesteading. They were blocked, however, by the squatters who had gotten to the land just a few years ahead of them.

Under popular pressure, New South Wales and Victoria passed homestead-type laws in 1861 and with later amendments began a period of "free selection" intended to provide small farmers with land. But squatters bent the laws to their purposes even more effectively than ranchers did in the United States. By the 1880s they owned, freehold, most of whatever land they wanted. The struggle between former squatters with large holdings and the smallholding farmer went on for some years, but from the 1870s through 1900, sheep raising meant large ranches, which were export oriented (especially after refrigerated ships began carrying Australian mutton to England in 1882) and capitalistic. These ranches composed the typically Australian "big man's frontier" so much in contrast to the smallholding frontiers of settlement in Canada and the United States. In Australia, as in the United States and Canada (and for that matter in Scotland and Ireland), something about sheep made people nationalistic. Yet a yearning for unfettered—frontier?—individualism raised the violent highwayman Ned Kelly to the status of a national hero, rather like Jesse James or, less lethally, Black Bart in the American West. Some similar taste raised the jolly swagman of the ballad "Waltzing Matilda" to national legend.

The "small men" gained a foothold only on the fringe, west of the big men in New South Wales and Victoria and well inland in South Australia. Wheat became their export staple. A survey-and-sale system, not unlike that in the United States and Canada though the grid pattern was less rigid, ensured title. The law also allowed a person more than one "selection," permitting him to grow wheat for a few years, move on just before final payment, and select again—thus creating a "hollow frontier," a crop-raising frontier of exploitation. But aridity took a greater toll. Overeager sheepmen and wheat growers alike had to retreat early in the twentieth century from millions of acres that proved to be climatically treacherous, as happened also on the American and Canadian Great Plains. Sheep and wheat have remained Australia's major exports, but the regions of cultivation stabilized not long after 1900, and Australia continued to become more urban than ever.

Although at the national level Australia and the United States had rather different frontier experiences, comparisons at a regional level, such as between northern California and Victoria, have been neatly drawn. Quite unlike the frontier-rural mode of the Midwest, California's frontiers consisted of a gold rush followed by wheat production in large landholdings and eventually production of many staples aided by massive irrigation works. In Victoria too, exploitative gold-seeking preceded rural settlement. Wool production followed, then wheat in the 1880s and 1890s and also irrigation. Though none of these enterprises reached the California scale, the sequence was similar. And each region centered on booming metropolises, San Francisco and Melbourne.

In New Zealand the small settler occupied a much more prominent place than in Australia. The New Zealand smallholder also concentrated on sheep and, like the Australian "big man," was willingly pulled into international markets by refrigerated ships after 1882. A royal governor arrived a week after the first English landing in January 1840, and Crown lands thereafter were distributed in relatively small parcels under strict government control. The many freeholders used near-universal suffrage and an abundance of local-government units to create and preserve, in the words of the historian Donald Denoon, a dictatorship not of the proletariat but of the smallholding sheepman. Large sheep-holdings began to develop on the earlier-settled South Island, but there and later on the North Island the smallholders managed to prevent land monopoly from developing as it did in Australia and, as we will shortly see, in Argentina.

New Zealand's physiography differs greatly from Australia's in being more temperate and mountainous, although the amount of potentially cultivable land is only a fraction of the total. Enough became available to allow settlement to continue through the 1920s. New Zealand's entire "frontier" history thus lasted only eighty or ninety years. It resembled North America's more than Australia's in one major respect: the bloodiness of the conflict between Europeans and the indigenous Maoris, who had reached New Zealand several centuries earlier from Polynesia. The first Maori War, 1843–48, opened the South Island and the second, 1860–70, the North. Treaties relieved the Maoris of most of their best land by 1900. Unprotected by immunities to European diseases, the Maoris declined steeply in number. After 1900 they made something of a comeback, as did Indians in North America after reaching their population nadir in the 1890s. Resources were used ruthlessly, according to a local historian: forests quickly cut for lumber or sheep pasture, gold and coal mined hydraulically, and farming done so extensively as to cause serious erosion—all practices well known in the United States.

South Africa, even more than Australia and New Zealand, is a special case. In the very broadest terms, Australia and New Zealand resembled the United States and Canada in becoming settler societies, with the settlers completely dominating the indigenous population and with settlement ultimately limited by geography more than by other humans. In South Africa the native peoples have continued greatly to outnumber the Europeans. Moreover, South Africa has included two distinct European core groups, which were at times at war with each other, a situation once present in Canada. But the Dutch and English in South Africa together have formed a small minority amid a large African majority. South Africa, therefore, has veered toward the plantation type of European colonization that flourished and then disappeared in Kenya, Zimbabwe, Indonesia, and elsewhere. It would be interesting to measure more precisely why the minority remained in power for so long in South Africa but not in those other places, that is, whether there is a "magic number" or ratio of just enough Europeans to retain control.

The European presence began in 1652 when the Dutch founded their way station to the East Indies at Cape Town; in 1806 the English captured the place to provide a pause on their trips to India. Despite that long history, Europeans were scarce during much of this time, numbering only twenty thousand in 1800 and one million in 1900. Whether Dutch or English, European South Africa for most of its first two hundred years

Aridity has been a major factor shaping settlement patterns in frontier cultures around the world. In Argentina, settlement became most dense in the damp northeastern quadrant of the nation, whereas in Australia, large communities developed only a thin, well-watered, coastal band. In Canada, as in the United States, arid climates east of the Rockies limited frontier settlement and growth.

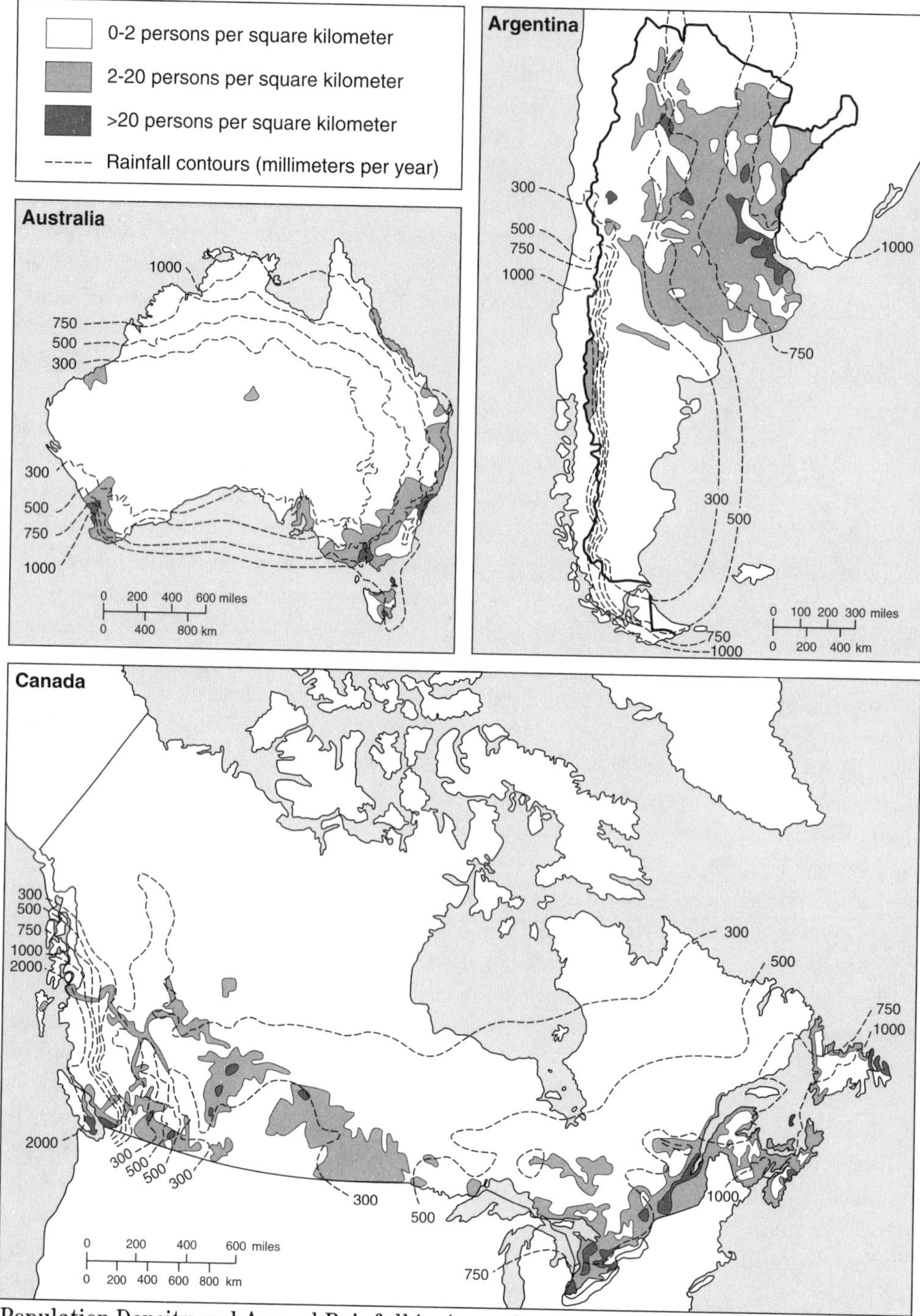

Population Density and Annual Rainfall in Argentina, Australia, and Canada

consisted of "company towns" and a few sheepherders in a riverless, mostly arid land, where they confronted much more numerous native peoples, many of whom in the interior possessed a roughly similar technology.

In their first decades near the Cape, the Dutch pushed back the native Khois and established commercial pastoralism nearby. By the close of the eighteenth century they had occupied further land and had encountered indigenous settled farmers and cattle-herders who more effectively resisted the whites. The Great Trek of 1836–46, involving about fourteen thousand Afrikaners, was only the best-known episode in a continuing push into the interior. (It has been compared, incidentally, to the Mormon migration of the late 1840s. Both were peoples who were united by culture and religion and who sought to create enclaves for themselves well out of reach of hostile neighbors.) The Voortrekkers became an Afrikaner symbol of independence and self-reliance, "frontier" virtues adapted by the Dutch-speaking minority, complete with wagon trains thrusting into the wilderness as happened on the American Overland Trail in the 1840s and 1850s.

Before 1869, however, the "frontier of settlement" by Europeans was pastoral, even to an extent nomadic, marked by continuous searching for new land to replace what they had overgrazed. The Nguni peoples whom the Europeans then encountered raised cattle in much the same way, but their contact with the whites involved them in market relations and increasingly stratified them into wealthy and poor—the latter becoming an increasingly victimized peasantry. The first gold strikes and the beginning of mining for gold and diamonds in 1869 greatly enriched the whites who controlled the mines and access to outside markets. Many Ngunis drifted into low-paid wage labor. European diseases never destroyed the people of southern Africa as they did the Native Americans; Africans possessed many Old World immunities. European force did not immediately triumph either; the Zulus put up a fierce, and for some time successful, resistance to both Dutch and English. But European capitalism succeeded in creating a native economic underclass, who by the twentieth century became a political and legal underclass as well. In its broadest, sketchiest outlines—as just given—South Africa's history involved a thin, moving frontier of pastoralism. Trekking, rather than true settlement, described its pre-1870 history, which was then followed by a frontier of exploitation of gold and diamonds tied to world markets and relying on semimigrant native wageworkers. A land act of 1913 deprived Africans of seven-eighths of the land area, legalizing white farming. Exploitation of grazing land, minerals, and the local people thus characterized the European presence in South Africa, as it did elsewhere, but peculiarly in that the Europeans were so few and the native people remained so numerous.

Other New World Frontiers: Argentina and Brazil

The European frontiers of Latin America began in the first third of the sixteenth century, when Spaniards conquered the Caribbean islands, Mexico, and Peru and Portuguese colonized the coast of Brazil. Much of this activity composed a frontier of exploitation of local resources such as gold and silver, using both native people and imported Africans as a labor force. Missionary activity complemented and at times ameliorated such enslavement. European diseases ravaged the mostly nonimmune indigenous peoples and provided the Spanish and Portuguese with their most effective weapon. But before achieving independence during and after the Napoleonic Wars, Latin America resembled New France in *not* being a major frontier of settlement. Beginning in the late nineteenth century, however, Argentina and Brazil exhibited significant frontier activity comparable to what was then taking place in North America.

Argentina and Australia have been compared by economic historians from those two countries. They "became strikingly alike," in the words of two Australians. "They developed at roughly the same rate, at about the same time, with a similar chemistry of foreign and domestic factors of production, to become competing societies of similar size, wealth and structure." Both battened on British capital investment, developed railroad networks in the late nineteenth century, and became major players in world markets—Australia with sheep and wheat, Argentina with cattle and wheat. Argentina enjoyed the stronger economy from 1890 to 1930, Australia thereafter; but until recently the two economies were seldom far apart. Both countries attracted European migrants—Australia from Britain, Argentina (in larger numbers) from Italy and Spain. Both opened up at much the same time all the frontiers of settlement they would ever have. Buenos Aires as a commercial and administrative center dominated Argentine life just as Sydney and Melbourne did Australian. Argentina, however, included an even larger "big man's frontier" than did Australia.

In other respects, the Argentine frontier paralleled the American and Canadian more closely, such as when the native people were subjugated. Just before the second Riel Rebellion in western Canada (1885), when Americans were conquering the Lakotas in the Montana and Dakota territories (1876–79), General Julio Roca in 1879 destroyed and dispersed the Araucanians and opened the Argentine pampas to white exploitation and settlement. But here the parallels break down, and a key contrast emerges between the Argentine and the North American, and indeed the Australian, frontiers. Because Roca and the government parceled out the pampas to fellow soldiers and friends, ownership became immediately and permanently concentrated in a few hands. Homestead-type laws did pass in 1876 and 1884 but produced only minimal and sporadic results. Land-distribution policy was left to the provinces, in effect meaning the large proprietors, while the central government maintained a laissez-faire stance on land as on other matters.

The Italians who flooded into Argentina mostly became tenant farmers rather than freeholders. Few became citizens. Unlike in North America, no law required homesteaders to be citizens. Thus Argentina's immigrants did not become a political counterforce to the *latifundistas*, demanding roads, schools, agricultural education, or cooperatives. "Argentine agriculture," wrote Carl Solberg, comparing Canada and Argentina, "was built on the systematic exploitation of the nation's tenant farmers rather than on modern production and marketing systems." After World War I the country fell behind Australia, Canada, and the United States in world markets. Climbing the agricultural ladder from migrant worker to sharecropper to tenant to freeholder was not unheard-of in Argentina, but immigrants more often made it to the top through the building trades, the military, the bureaucracy, the professions, and the urban-located occupations that provided services rather than goods. In Santa Fe province, group settlements of family farms, known as "colonies," succeeded well from the 1880s on, yet colonies did not become the national pattern; the *latifundistas* kept control of their ever more valuable real estate. As a Latin American economic historian observed, "The mediocre housing, poor social services, and lamentable infrastructural facilities in most of those melancholy little towns scattered across the pampean zone, were eloquent testimony to the rootlessness of Argentine farming and the weakness of the rural middle class."

Gauchos used their bolas to hunt ostriches—much prized for their feathers—across the pampas of Argentina in the 19th century. The common practice of burning the range to drive the birds out into the open created tension between the gauchos and the big ranchers who lost cattle and grazing lands to the fires.

Emeric Essex Vidal (1791–1861). Balling Ostriches. *Lithograph, 1820. From Vidal,* Picturesque Illustrations of Buenos Ayres and Montevideo *(London, 1820). Courtesy The Edward E. Ayer Collection, The Newberry Library, Chicago, Illinois.*

Argentina lacked a survey-and-sale system of land conveyance, a mortgage and credit market for small farmers, or much rural upward mobility. In fact it lacked a true frontier of settlement aside from the cattle barons and the tenant farmers. It also lacked a gold rush; there was no mining frontier as in North America, Australia, South Africa, or Brazil. National life, despite rapid development of wheat and cattle for world export markets from 1880 to World War I, was dominated rather by its great city, the "Paris of the South," Buenos Aires. In 1914 Argentina compared well in national and personal wealth to almost any country in the world. But after 1945, for lack of true, broadly based development, that was no longer the case.

The gaucho, the mounted cowboy bringing down his animals with his bola, became a stock figure of national definition, although in reality the "conquest of the desert" in 1879 removed the less tameable elements of the gaucho population and changed the rest into ranch hands as dependent for wages on the landowners as Texas cowhands were on ranchers there. The gaucho may have been the hero of ballads, but the *latifundistas* called the shots, both on the pampas and in the governments. It was much the same in Brazil, where the elite of Rio Grande do Sul (the southernmost, most temperate state) like to say they are gauchos, that is, just country boys, whereas that state has disproportionately provided Brazil with its presidents and other leaders.

Brazil, with half the land mass of South America, presents even greater contrasts. Larger in size and theoretically with more available land than the continental United States, its frontier development has been different in almost all important respects from

North America's. In fact, Brazil at the close of the twentieth century was the only large country in the world where the post-1492 frontier process—both of exploitation and of settlement—was still taking place. Brazil's colonial history began with Pedro Alvares Cabral's visit in 1500, a century earlier than the colonial periods of Canada and the United States. But for two centuries or more after Cabral, colonization seldom penetrated deeply inland and was oriented around the larger coastal cities and toward Europe. African slaves began arriving early, providing the labor force for the northeastern sugar economy, a plantation-style frontier of settlement not unlike the cotton plantations of antebellum Alabama and Mississippi. In the early eighteenth century an exploitation frontier emerged some two hundred miles inland from Rio de Janeiro, in the present state of Minas Gerais ("General Mines"), with the discovery of gold, diamonds, and other gems. The Amazon Basin's rubber boom of the early twentieth century formed another plantation-type exploitation frontier.

A real frontier of settlement, however, took very long to appear. The Portuguese Crown made enormous grants of land called *sesmarias* to favored people, and although the grants were seldom developed productively, they precluded mass settlement. The folk figure of the colonial period became the *bandeirante*, the swashbuckling paramilitary leader who roamed, raided, and then returned home. In his classic book comparing Brazilian retardation to North American progress, Vianna Moog claimed that the explanation lay in Brazil's adulation of the romantic but rootless *bandeirante*, whereas in the United States the Jeffersonian yeoman farmer and his family became the ideal and often the reality. Perhaps so. But Moog also pointed out the significance of Brazil's lack of a usable interior waterway system on anything like the scale of the Ohio-Missouri-Mississippi system, as well as its lack of coal, iron, and oil at a time when industrialization required them. From the 1820s through the 1870s, the Imperial Government tried to promote settlement on the land by supporting the establishment of German and Italian colonies in Rio Grande do Sul, Santa Catarina, and Paraná in the south. These colonies included several hundred thousand people by the late nineteenth century, and the Germans in particular continued in unabsorbed enclaves through the twentieth century. But except in the three southern states, the colonists hardly made a dent in Brazil's enormous space.

Change came when Brazil took an early and permanent lead in the world coffee market. Coffee plantations in the state of Rio de Janeiro prospered in the mid-nineteenth century, and then, as soil became increasingly "tired," planters moved west into São Paulo state. Their move was accelerated by the creation of a railroad network from Santos seaport up to São Paulo city in 1867 and then hundreds of miles inland in the next few years. The abolition of slavery in 1888 and the recruitment of tens of thousands of Italians and their families as a replacement labor force caused much of São Paulo and adjoining territory finally to become a settlement frontier.

That frontier, not surprisingly, resembled the Argentine more than the North American ones, except that the immigrants' prospects for improvement were even dimmer. A few thousand did settle in "nuclear colonies," but those lucky immigrants usually brought some capital that enabled them to do so. Much more commonly, the Brazilian settler lived in one or another kind of tenancy arrangement on the land of a *fazendeiro*, a large-holding coffee planter. The government recruited entire Italian families and helped them obtain labor contracts and train tickets out to the plantations, where their living conditions were usually so unenviable that in 1902 the Italian government forbade any further Brazilian recruitment.

Some small freeholding did emerge in São Paulo and the three southernmost states, but the frontier spread only gradually in the early twentieth century, entering the state of Goiás, just north of São Paulo, in the 1930s. The leap into the interior when Brasília was planned in the 1950s and opened in the early 1960s was an event more metaphorical than demographic, more a matter of national pride than of real regional development. Eventually, however, Brasília, together with the later construction of roads into Amazonas, western Mato Grosso, and other remote states and territories, produced a Brazilian frontier of exploitation and of settlement simultaneously. Slash-and-burn agriculture, and the large-scale forest burn-offs that the world began noticing in the 1980s, accompanied the movement of over one million people into Mato Grosso, Pará, Rondônia, and nearby areas in the 1970s. Some of these people obtained plots of land from the government; many others took up unoccupied land as *posseiros*, or squatters (American-style, not Australian). Neither planned colonization nor spontaneous migration led to a contiguous and self-sustaining frontier.

The Indians, who may have numbered five million in 1500, diminished to no more than 150,000 in 1950. They had disappeared from Brazil's coastal regions by the end of the eighteenth century, and with Amazonian development in the late twentieth century their remnant was threatened again. European disease, enslavement, and military raids nearly erased them. Though Indian reserves in remote Amazonia existed, the government agencies in charge—even when sympathetically run, which was not always the case under military rule in the 1970s—were under constant pressure from gold-seekers or other developers, a situation similar to the gold rush in the Dakota Black Hills in 1874–75 that touched off the Custer episode and the ensuing Sioux War. But the Brazilian Indians lacked the ability of the Lakotas to resist.

Despite the massive expansion of São Paulo, Rio de Janeiro, Belo Horizonte, and other cities, the West and Amazonia increased their share of Brazil's population from less than 10 percent in 1940 to about 15 percent in 1980. Very few new settlers lived in official colonies but rather as peasants, squatters, smallholders, and sharecroppers together with the rubber tappers, *mestiços*, and detribalized Indians who were already there. As a whole, however, they were not producing marketable surpluses as on other settlement frontiers. They were precapitalist semisubsistence settlers, more like American frontierspeople of the pre-1870 frontier-rural mode than later market-oriented ones.

An important difference between the continuing Brazilian frontier and the 1720–1920 settlement frontier in the United States was the lack of a survey-and-sale system that securely conveyed land to smallholders. The colonial large holdings were never revoked, despite brave words in the 1850 land law and other laws down through the 1970s. The Program of National Integration of the Medici and Geisel military administrations of the 1970s promised one hundred hectares, six months' start-up pay, cheap credit, a house and implements, and other encouragements to families who would become small farmers in Amazonia. Only seven thousand of the hoped-for one hundred thousand families appeared, and a third of them did not stay long. As an American anthropologist noted, "Laws that protect squatters' rights encourage land invasion and virtually assure access to land for those with the capacity to hire gunmen." In Rondônia in the far west, would-be settlers in the 1970s continued to seek clear title without much success.

In the meantime, in the late 1970s, the government began selling large tracts to private development companies—a practice the U.S. government resorted to in the 1780s when the fledgling survey-and-sale system of the 1785 Ordinance failed to earn revenue and settle the Ohio Valley quickly enough. The differences were that Brazil lacked a land survey and that the quality of land for western Brazilian settlers in the 1980s was much more crucial than for Americans in the 1780s because the Rondônia settlers, to succeed, had to produce surpluses for market very quickly. Capitalist market relations were much more complex in the 1980s, demanding an efficiency in Rondônia and Amazonas that the subsistence American frontierspeople of the late eighteenth century were spared. The Brazilian settlers of the 1970s and 1980s also lacked alternatives; there was almost no further "West" to go to, and city life meant dire poverty.

Treatment of the Indians fluctuated from benevolent paternalism, involving removal and reservations, to harsh repression. Wholesale deforestation topped a list of environmental problems. In a number of ways, the Brazilian frontier story of the twentieth century was a replay, at fast-forward, of most of the mistakes and regrettable features that marked the frontiers of settlement and exploitation of the United States since the eighteenth century.

Frontier, Region, and West Again

These pages have only begun to touch on some of the possibilities of internal and international comparisons in the histories of frontiers, regions, and Wests. Specialists in the history of any of the countries mentioned may rightly object to the sketchiness of the patterns and generalizations given here. However, one may at least see a simple classification scheme distinguishing frontiers of settlement and of exploitation (though concrete cases were often both), plus subtypes such as plantations run by Europeans using slave or wage labor and frontiers directed by central theocratic authority. Certain continuing kinds of interaction, particularly between European-stock people with native people and with the natural environment, seem to appear in all frontier histories, for good or for ill.

The regional uniqueness—and lack of it—in the present West of the United States still deserve clearer definition. The character of that West may help define others and also the ongoing Brazilian situation. Some scholars have looked at the history of frontiers as simply a special case of migration history, sharing the characteristics typical of migrations. Although that is too much of a reduction, it does point up the structural similarity among frontiers in all times and places and invites frontier and western historians to enrich their thinking with the scholarship on international migration. In a very few but very arresting works, theories devised to understand and guide social and economic development in the Third World have been applied to American frontier and western history; modernization, conquest, dependency theory, and center-periphery theory, as well as Antonio Gramsci's idea of hegemony of ruling elites, are cases in point. Insights from cultural anthropology and cultural geography have permitted historians to look with more empathy on native people and with less assurance that the advance of frontiers was beneficial to all concerned.

The term "frontier" will no doubt remain in use and at times be applied in bizarre ways, such as a frontier in space or on Mars—a total misapplication of the term because although exploitation may take place, no true settlement ever can. (If aridity defeated settlement of Australia and the upper Great Plains, it certainly will on Mars.) These would be extreme cases, if they ever happen, of imperial outposts as distinct from real frontiers. Comparisons provide instead the opportunity for interior-directed reflection. They blunt the force either of condemnation or of adulation of frontierspeople, who have had so many varied experiences since 1492; contact among Europeans and natives and the natural environment, inevitable in itself, often turned into exploitation and, it should be said, often into improvement and mutual benefit. These interactions have demonstrated no more and no less than the human condition.

Do comparisons with the histories of frontiers elsewhere help one understand what happened in the United States? No doubt they do, at a minimum to dilute the frontier-derived myth of moral superiority and self-righteous missionizing. Once one starts comparing, one has accepted that American history is not incomparable or unique. On the other hand, does the American story have relevance elsewhere? Surely it would be instructive in Brazil, whose ongoing frontier history has eerily repeated events that have happened elsewhere. Like all histories, comparative frontier histories tell of the good that people have achieved and the ill that they have committed. Thereby these histories indicate what we might expect of ourselves in the future.

Bibliographic Note

Introductions to comparative history in general include the essays by Raymond Grew and others in *Journal of Interdisciplinary History* 16 (Summer 1985) and the essays in C. Vann Woodward, ed., *The Comparative Approach to American History* (New York, 1968). The comparative study of frontiers and regions began with Paul Leroy Beaulieu, *De la Colonisation chez les Peuples Modernes* (Paris, 1874 and later eds.), a work intended to facilitate French imperialism but containing some useful distinctions. Walter Prescott Webb's *The Great Frontier* (Austin, 1952) conceived of Europe as metropolis and the rest of the world as frontier, a conception employed less sympathetically since the 1970s by Immanuel Wallerstein's books on "the modern world-system." Excellent essays on a dozen historical frontiers appear in Walker D. Wyman and Clifton B. Kroeber, eds., *The Frontier in Perspective* (Madison, 1957).

Comparative frontiers in recent times have been discussed from various angles by the following: W. Turrentine Jackson, "A Brief Message for the Young and/or Ambitious: Comparative Frontiers as a Field for Investigation," *Western Historical Quarterly* (hereafter cited as *WHQ*) 9 (January 1978): 5–18; Philip Wayne Powell et al., *Essays on Frontiers in World History*, edited by George Wolfskill and Stanley Palmer (College Station, Tex., 1983); David H. Miller and Jerome O. Steffen, eds., *The Frontier: Comparative Studies* (Norman, 1977); Herbert Heaton, "Other Wests Than Ours," *Journal of Economic History, Supplement VI* (1946): 50–62; Dietrich Gerhard, "The Frontier in Comparative View," *Comparative Studies in Society and History* 1 (1959): 205–29; Walter Nugent, "Frontiers and Empires in the Late Nineteenth Century," *WHQ* 20 (November 1989): 393–408; Donald W. Treadgold, "Russian Expansion in the Light of Turner's Study of the American Frontier," *Agricultural History* 26 (October 1952): 147–52; and William H. McNeill, *The Great Frontier: Freedom and Hierarchy in Modern Times* (Princeton, 1983).

Most of the foregoing range globally, but others have compared two or three Anglophone frontiers: Paul F. Sharp, "Three Frontiers: Some Comparative Studies of Canadian, American, and Australian Settlement," *Pacific Historical Review* 24 (November 1955): 369–77; H. C. Allen, *Bush and Backwoods: A Comparison of the Frontier in Australia and the United States* (East Lansing, Mich., 1959); Robin Winks, *The Myth of the American Frontier: Its Relevance to America, Canada, and Australia* (Leicester, Eng., 1971); and Morris W. Wills, "Sequential Frontiers: The Californian and Victorian Experience, 1850–1900," *WHQ* 9 (October 1978): 483–94. Comparisons that include Latin America are D.C.M. Platt and Guido di Tella, eds., *Argentina, Australia, and Canada: Studies in Comparative Development, 1870–1965* (New York, 1985), Daniel J. Elazar, *Jewish Communities in Frontier Societies: Argentina, Australia, and South Africa* (New York, 1983), and Carl E. Solberg's exemplary *The Prairies and the Pampas: Agrarian Policy in Canada and Argentina, 1880–1930* (Stanford, 1987). An important area of national self-definition and legend is treated in Richard W. Slatta, *Cowboys of the Americas* (New Haven, 1990); urbanism is discussed in David Hamer, *New Towns in the New World: Images and Perceptions of the Nineteenth-Century Urban Frontier* (New York, 1990). For comparisons of Southern Hemisphere societies by their own scholars, see John Fogarty, Ezequiel Gallo, and Hector Dieguez, *Argentina y Australia* (Buenos Aires, Arg., 1979), John Fogarty and Tim Duncan, *Australia and Argentina: On Parallel Paths* (Carlton, Victoria, Austral., 1984), and Donald Denoon, *Settler Capitalism: The Dynamics of Dependent Development in the Southern Hemisphere* (Oxford, Eng., 1983).

Regionalism and frontiers in U.S. history have been discussed copiously. A few of the best treatments include the following: Earl Pomeroy, *The Pacific Slope: A History of California, Oregon, Washington, Idaho, Utah, and Nevada* (New York, 1965); Carlos A. Schwantes, *The Pacific Northwest: An Interpretive History* (Lincoln, 1989); William G. Robbins, Robert J. Frank, and Richard E. Ross, eds., *Regionalism and the Pacific Northwest* (Corvallis, Oreg., 1983), especially Richard M. Brown's essay. See also several articles in the *WHQ*: Donald Worster, "New West, True West: Interpreting the Region's History," 18 (April 1987): 141–56; William Cronon, "Revisiting the Vanishing Frontier: The Legacy of Frederick Jackson Turner," 18 (April 1987): 157–76; Michael P. Malone, "Beyond the Last Frontier: Toward a New Approach to Western American History," 20 (November 1989): 409–27; and William G. Robbins, "Western History: A Dialectic on the Modern Condition," 20 (November 1989): 429–449. Also instructive is John S. Otto, *The Southern Frontiers, 1607–1860: The Agricultural Evolution of the Colonial and Antebellum South* (New York, 1989).

For Europeans' notion that they could do whatever they pleased to new lands and indigenous peoples, see David B. Quinn et al., *Essays on the History of North American Discovery and Exploration*, edited by Stanley H. Palmer and Dennis Reinhartz (College Station, Tex., 1988), and Antonello Gerbi, *Nature in the New World: From Christopher Columbus to Gonzalo Fernandez de Oviedo*, trans. Jeremy Moyle (Pittsburgh, 1985).

Among the many comparative discussions of Canada's frontiers are Michael S. Cross, ed., *The Frontier Thesis and the Canadas: The Debate on the Impact of the Canadian Environment* (Toronto, 1970), A. L. Burt, "If Turner Had Looked at Canada . . . ," in Wyman and Kroeber, *Frontier in Perspective*, and Marcus L. Hansen and John B. Brebner, *The Mingling of the Canadian and American Peoples* (New Haven, 1940). Australia and New Zealand as frontiers are discussed in the following: W.L.G. Joerg, *Pioneer Settlement* (1932; reprint, Freeport, N.Y., 1969); Ronald Lawson, "Toward Demythologizing the 'Australian Legend,'" *Journal of Social History* 13 (Summer 1980): 577–87; Peter J. Coleman, "The New Zealand Frontier and the Turner Thesis," *Pacific Historical Review* 27 (August 1958): 221–37; Brian Fitzpatrick, "The Big Man's Frontier and Australian Farming," *Agricultural History* 21 (January 1947): 8–12; J. W. McCarty, "Australia as a Region of Recent Settlement in the Nineteenth Century," *Australian Economic*

History Review 13 (September 1973): 148–67; Donald W. Meinig, "Colonisation of Wheatlands: Some Australian and American Comparisons," *Australian Geographer* 7 (August 1959): 205ff; and Geoffrey Blainey, *The Tyranny of Distance: How Distance Shaped Australia's History* (Melbourne, 1968).

The leading comparative treatment of South Africa is Howard Lamar and Leonard Thompson, eds., *The Frontier in History: North America and Southern Africa Compared* (New Haven, 1981). See also James Gump, "The Subjugation of the Zulus and Sioux: A Comparative Study," *WHQ* 19 (January 1988): 21–36, and W. K. Hancock, "Trek," *Economic History Review* 10, no. 3 (1958): 331–39.

Studded with comparative insights is Alistair Hennessy, *The Frontier in Latin American History* (Albuquerque, 1978). Other important comparative studies include the following: Matt S. Meier, ed., *Latin American Frontiers* (San Diego, 1981); David J. Weber, "Turner, the Boltonians, and the Borderlands," *American Historical Review* 91 (February 1986): 66–81; Carter Goodrich, "Argentina as a New Country," *Comparative Studies in Society and History* 7 (October 1964): 70–88; Sergio Villalobos R., *Relaciones Fronterizas en la Araucania* (Santiago, Chile, 1982); Mark Jefferson, *Peopling the Argentine Pampa* (New York, 1926); Mary Lombardi, "The Frontier in Brazilian History," *Pacific Historical Review* 44 (November 1975): 437–57; C. Vianna Moog, *Bandeirantes and Pioneers* (New York, 1964); Martin T. Katzman, *Cities and Frontiers in Brazil: Regional Dimensions of Economic Development* (Cambridge, Mass., 1977); Frederick C. Luebke, *Germans in the New World: Essays in the History of Immigration* (Urbana, 1990); and Marianne Schmink and Charles H. Wood., eds., *Frontier Expansion in Amazonia* (Gainesville, 1984). The essays by Thomas J. McCormick and Louis A. Pérez, Jr., in *Journal of American History*, June 1990, are a guide to world-systems theory.

Contributors

THE EDITORS

Clyde A. Milner II is Editor of the *Western Historical Quarterly* and Professor of History at Utah State University. He has written on a range of subjects including the work of eastern Quakers among the Plains Indians and the role of memory in creating a western identity. He is the editor of *Major Problems in the History of the American West* and the coeditor with Patricia Nelson Limerick and Charles E. Rankin of *Trails: Toward a New Western History.*

Carol A. O'Connor is Professor of History at Utah State University, where she has taught since 1977. A native of New York, she has authored *A Sort of Utopia: Scarsdale, 1891–1981,* as well as several articles on suburbs and suburbia. She is currently working on a study of middle-sized cities in the West's low-population states.

Martha A. Sandweiss is Director of the Mead Art Museum and Associate Professor of American Studies at Amherst College. A former curator of photographs at the Amon Carter Museum, she has written widely on western photography and art. She is the author of *Laura Gilpin: An Enduring Grace* and the editor of *Photography in Nineteenth-Century America.*

THE AUTHORS

Carl Abbott is Professor of Urban Studies and Planning at Portland State University. He has written extensively on the West in the twentieth century, with particular attention to issues of urban growth, regional development, and land-use planning. His books include *The New Urban America: Growth and Politics in Sunbelt Cities, The Metropolitan Frontier: Cities in the Modern American West,* and *Colorado: A History of the Centennial State.*

Allan G. Bogue is Frederick Jackson Turner Professor of History emeritus at the University of Wisconsin-Madison. He has been president of the Organization of American Historians, the Agricultural History Society, the Economic History Association, and the Social Science History Association. He is the author of many books and articles on American western and political history and is currently engaged in a biographical study of Frederick Jackson Turner.

Richard Maxwell Brown is Beekman Professor of Northwest and Pacific History emeritus at the University of Oregon. He was 1991–92 president of the Western History Association and a consultant to the National Commission on the Causes and Prevention of Violence in 1968–69. Among his publications are *Strain of Violence: Historical Studies of American Violence and Vigilantism* and *No Duty to Retreat: Violence and Values in American History and Society.*

Keith L. Bryant, Jr., is Professor of History and Head of the History Department at the University of Akron. His publications related to the economy of the West include *History of the Atchison, Topeka and Santa Fe Railway*, *Arthur E. Stilwell, Promoter with a Hunch*, and, with Henry C. Dethloff, *A History of American Business*.

Anne M. Butler is Coeditor of the *Western Historical Quarterly* and Professor of History at Utah State University. A specialist in western women's history, she is the author of *Daughters of Joy, Sisters of Misery: Prostitutes in the American West, 1865–1890*. She has also written on subjects as diverse as women in prisons and nuns in the West.

Kathleen Neils Conzen is Professor of History at the University of Chicago. A past president of the Immigration History Society, she is the author of *Immigrant Milwaukee, 1836–1860: Accommodation and Community in a Frontier City* and of numerous articles on definitions of ethnicity and issues of historical methodology. She is completing a study of German immigrant settlement in the Midwest.

William Cronon is Frederick Jackson Turner Professor of History, Geography, and Environmental Studies at the University of Wisconsin-Madison. A MacArthur Fellow, he is the author of *Changes in the Land: Indians, Colonists, and the Ecology of New England* and *Nature's Metropolis: Chicago and the Great West*.

Sarah Deutsch is Associate Professor of History at Clark University, where she teaches courses on social history and gender. She is the author of *No Separate Refuge: Culture, Class, and Gender on an Anglo-Hispanic Frontier in the American Southwest, 1880–1940* and *From Ballots to Breadlines: American Women, 1920–1940*. She is currently working on a study of women in Boston from 1870 to 1950.

Brian W. Dippie is Professor of History at the University of Victoria, British Columbia. A Canadian, he received his Ph.D. in American civilization from the University of Texas at Austin and specializes in the cultural history of the American West. His recent books include *Catlin and His Contemporaries: The Politics of Patronage* and *Charles M. Russell, Word Painter: Letters, 1887–1926*.

Jay Gitlin is a lecturer in American history at Yale University. He is completing a study of the French in the Mississippi Valley in the late eighteenth and early nineteenth centuries. He also performs regularly with the Bales-Gitlin band and is writing a book on the social history of American popular music.

Peter Iverson is Professor of History at Arizona State University. He has written *The Navajo Nation* and *Carlos Montezuma and the Changing World of American Indians* and has edited *The Plains Indians of the Twentieth Century* and with Albert Hurtado, *Major Problems in American Indian History*.

Thomas J. Lyon has taught at Utah State University since 1964, where he edits the journal *Western American Literature*. He edited *This Incomperable Lande: A Book of American Nature Writing* and *On Nature's Terms* and is coeditor, with Terry Tempest Williams, of the Utah Centennial Anthology, forthcoming.

Michael P. Malone is the President of Montana State University and Professor of American History. He is the author of several books, including *C. Ben Ross and the New Deal in Idaho, Battle for Butte: Mining and Politics on the Northern Frontier, 1864-1906*, and, with Richard Etulain, *The American West: A Twentieth-Century History*.

Walter Nugent is Andrew V. Tackes Professor of History at the University of Notre Dame. He has written several books, including *The Tolerant Populists, Money and American Society, 1865–1880, Structures of American Social History*, and *Crossings: The Great Transatlantic Migrations, 1870–1914*, as well as a number of essays on western history, comparative frontiers, and other topics. He is now working on migration and population change in the American West since 1890.

Gary Y. Okihiro is the Director of Cornell University's Asian American Studies Program and Associate Professor of History. He is the author of *Margins and Mainstreams: Asians in American History and Culture* and *Cane Fires: The Anti-Japanese Movement in Hawaii, 1865–1945*.

Charles S. Peterson is Professor of History emeritus at Utah State University and a member of the faculty at Southern Utah University. From 1971 to 1989 he was associated with the *Western Historical Quarterly*, initially as Associate Editor, then as Coeditor, and beginning in 1979 as Editor. He was Director of the Utah Historical Society and Editor of the *Utah Historical Quarterly*. He has written a number of books and more than a score of essays and articles on the subjects of Mormon history, environmental history, and agriculture.

F. Ross Peterson is Director of the Mountain West Center for Regional Studies and Professor of History at Utah State University. The author of *Prophet Without Honor: Glen H. Taylor and the Fight for American Liberalism* and *Idaho: A Bicentennial History*, he is currently working on a biography of Idaho Senator Frank Church.

George J. Sánchez is Associate Professor of History and American Culture at the University of Michigan, Ann Arbor. He has written *Becoming Mexican American: Ethnicity, Culture, and Identity in Chicano Los Angeles, 1900–1945*.

Carlos A. Schwantes is Director of the Institute for Pacific Northwest Studies and Professor of History at the University of Idaho, where he has taught since 1984. He is the author or editor of eleven books, including *The Pacific Northwest: An Interpretive History, In Mountain Shadows: A History of Idaho*, and *Railroad Signatures across the Pacific Northwest*.

Ferenc M. Szasz is Professor of History at the University of New Mexico in Albuquerque. He is the author or editor of over fifty articles and five books, including *The Divided Mind of Protestant America, 1880–1930* and *The Protestant Clergy in the Great Plains and Mountain West, 1865–1915*.

Margaret Connell Szasz is a member of the History Department at the University of New Mexico in Albuquerque. She has written widely in the field of American Indian history, including two books on American Indian education. Her most recent publication is *Between Indian and White Worlds: The Cultural Broker.*

David J. Weber, Dedman Professor of History at Southern Methodist University, is past president of the Western History Association and has been elected to membership in the Mexican Academy of History and the Society of American Historians. His books, many of them prizewinners, include *The Mexican Frontier, 1821–1846* and *The Spanish Frontier in North America.*

Elliott West is Professor of History at the University of Arkansas, Fayetteville. A specialist in the social history of the West and the frontier, he is the author of *Growing Up with the Country: Childhood on the Far-Western Frontier* and *The Saloon on the Rocky Mountain Mining Frontier.*

Richard White is the McClelland Professor of History at the University of Washington. His most recent books include two prizewinners: *"It's Your Misfortune and None of My Own": A New History of the American West* and *The Middle Ground: Indians, Empires, and Republics in the Great Lakes Region, 1650–1815.*

Victoria Wyatt is Associate Professor in the Department of History in Art at the University of Victoria, British Columbia. Her research focuses on the history and arts of native peoples of Alaska and the Pacific Northwest. She is the author of several articles and two books, *Images from the Inside Passage: An Alaskan Portrait by Winter and Pond* and *Shapes of Their Thoughts: Reflections of Culture Contact in Northwest Coast Indian Art.*

Index

Italicized page numbers refer to pictures and picture captions.

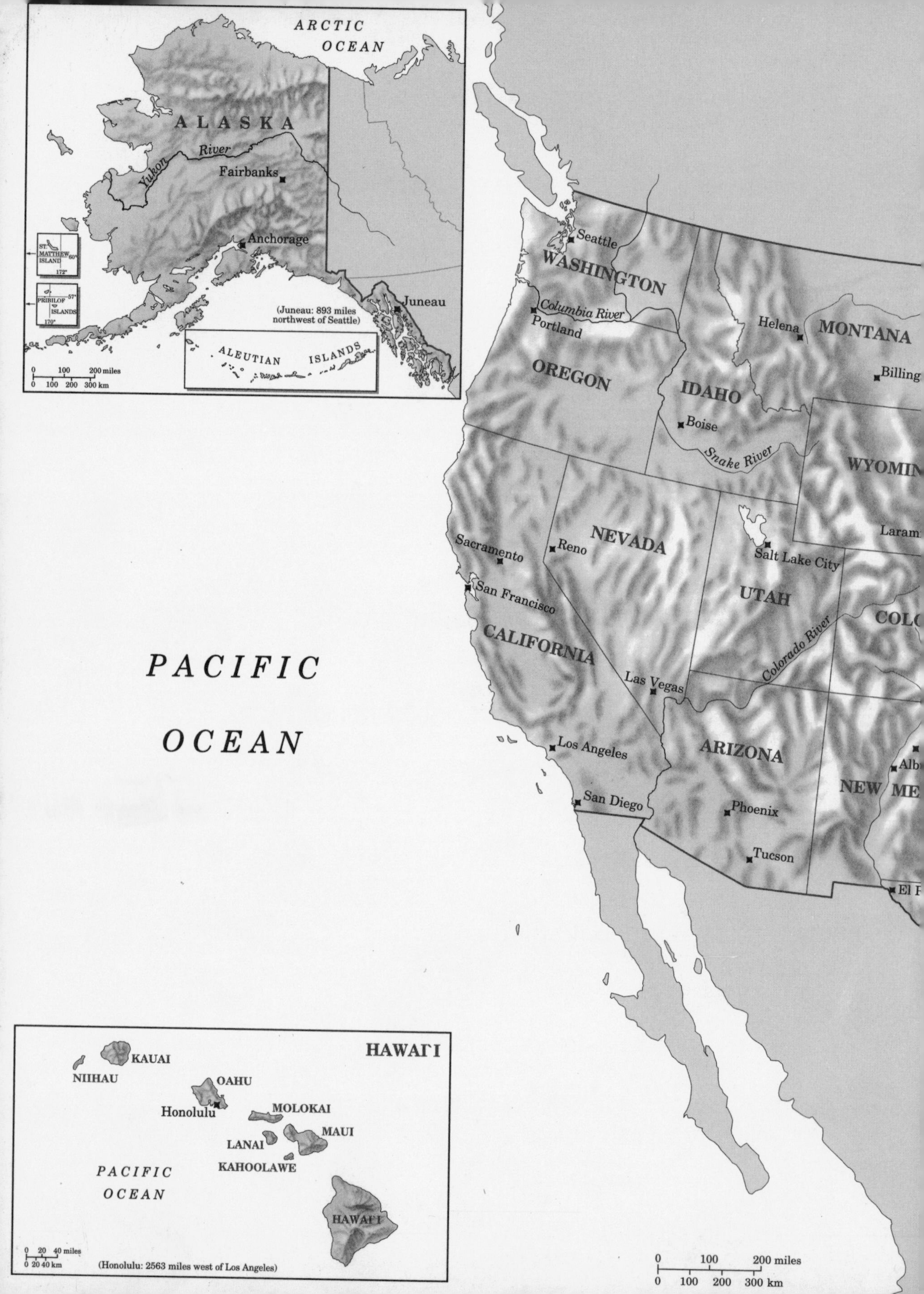

ARCTIC OCEAN
ALASKA
Yukon River
Fairbanks
Anchorage
ST. MATTHEW ISLAND
PRIBILOF ISLANDS
Juneau
(Juneau: 893 miles northwest of Seattle)
ALEUTIAN ISLANDS
0 100 200 miles
0 100 200 300 km
Seattle
WASHINGTON
Columbia River
Portland
Helena
MONTANA
Billing
OREGON
IDAHO
Boise
Snake River
WYOMIN
NEVADA
Reno
Sacramento
Salt Lake City
Laram
San Francisco
UTAH
COLO
CALIFORNIA
Colorado River
Las Vegas
PACIFIC OCEAN
Los Angeles
ARIZONA
Alb
NEW ME
San Diego
Phoenix
Tucson
El P
HAWAI'I
KAUAI
NIIHAU
OAHU
Honolulu
MOLOKAI
MAUI
LANAI
KAHOOLAWE
PACIFIC OCEAN
HAWAI'I
0 20 40 miles
0 20 40 km
(Honolulu: 2563 miles west of Los Angeles)
0 100 200 miles
0 100 200 300 km